McGraw Hill Education

DISCOVER MCGRAW-HILL NETWORKS™

AN AWARD-WINNING SOCIAL STUDIES PROGRAM DESIGNED TO FULLY SUPPORT YOUR SUCCESS.

» Aligned to the National Geography Standards

» Aligned to the National Council for the Social Studies Standards

» Engages you with interactive resources and compelling stories

» Provides resources and tools for every learning style

» Empowers targeted learning to help you be successful

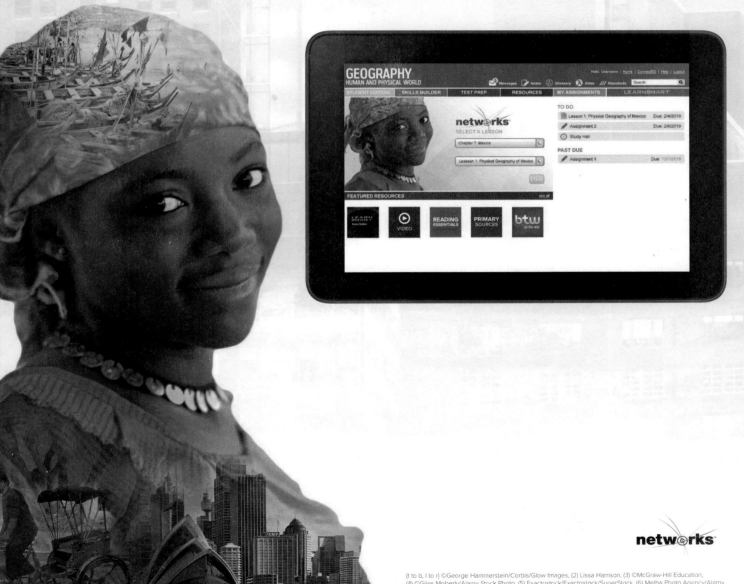

networks

UNDERSTANDING IS THE FOUNDATION OF ACHIEVEMENT

Clear writing, real-life examples, photos, interactive maps, videos, and more will capture your attention and keep you engaged so that you can succeed.

You will find tools and resources to help you read more effectively.

networks

FOCUS YOUR TIME AND YOUR EFFORT

LEARNSMART®

No two students are alike! We built LearnSmart® so that all students can work through the key material they need to learn at their own pace.

YOUR TIME MATTERS

LearnSmart with SmartBook™ adapts to you as you work, guiding you through your reading so you can make every minute count.

DISCOVER A PERSONALIZED READING EXPERIENCE

Every student experiences LearnSmart® differently. The interactive challenge format highlights content and helps you identify content you know and don't know.

RETAIN MORE INFORMATION

LearnSmart® detects content you are most likely to forget and will highlight what you need to review.

networks™

BE THE STUDENT YOU WANT TO BE

STUDENTS WHO UNDERSTAND THE WORLD WILL BE THE ADULTS WHO CAN CHANGE IT.

DISCOVER IT ALL ONLINE!

1. Go to connected.mcgraw-hill.com

2. Enter your username and password from your teacher.

3. Click on your book.

4. Select your chapter and lesson, or explore the Resource Library.

GO ONLINE AND START EXPLORING!

GEOGRAPHY
THE HUMAN AND PHYSICAL WORLD

netw⊕rks™
There's More Online!

Richard G. Boehm, Ph.D.

Mc Graw Hill Education

AUTHORS

Senior Author

Richard G. Boehm, Ph.D., was one of the original authors for *Geography for Life: National Geography Standards,* which outlined what students should know and be able to do in geography. He was also one of the authors of the *Guidelines for Geographic Education,* in which the Five Themes of Geography were first articulated. Dr. Boehm has received many honors, including "Distinguished Geography Educator" by the National Geographic Society (1990), the "George J. Miller Award" from the National Council for Geographic Education (NCGE) for distinguished service to geographic education (1991), "Gilbert Grosvenor Honors" in geographic education from the Association of American Geographers (2002), and the NCGE's "Distinguished Mentor Award" (2010). He served as president of NCGE, has twice won the *Journal of Geography* award for best article, and also received the NCGE's "Distinguished Teaching Achievement." Presently, Dr. Boehm holds the Jesse H. Jones Distinguished Chair in Geographic Education at Texas State University in San Marcos, Texas, where he serves as director of The Gilbert M. Grosvenor Center for Geographic Education. His most current project includes the production of the video-based professional development series, *Geography: Teaching With the Stars.* Available programs may be viewed at www.geoteach.org.

Contributing Author

Jay McTighe has published articles in a number of leading educational journals and has coauthored 10 books, including the best-selling *Understanding By Design* series with Grant Wiggins. McTighe also has an extensive background in professional development and is a featured speaker at national, state, and district conferences and workshops. He received his undergraduate degree from The College of William and Mary, earned a Masters degree from The University of Maryland and completed post-graduate studies at The Johns Hopkins University.

About the Cover: This woman is a member of one of the more than 250 ethnic groups in Nigeria, a country in West Africa. Nigeria is Africa's most populous country with over 180 million people and an estimated 521 different languages. Although the official language of Nigeria is English, the major native languages spoken in Nigeria represent three major families; Niger-Congo, the Hausa, and Kanuri.

Cover Photo Credits: (bkgd)©Giles Moberly/Alamy Stock Photo; (l to r, t to b)Lissa Harrison; (2)Exactostock/Exactostock/SuperStock; (3)Melba Photo Agency/Alamy; (4)Marco Simoni/Cultura RF/Getty Images

---

Discover McGraw-Hill Networks™, an award-winning Social Studies program designed to fully support your success.

- Aligned to the National Geography Standards
- Aligned to the National Council for the Social Studies Standards
- Engages you with interactive resources and compelling stories
- Provides resources and tools for every learning style
- Empowers targeted learning to help you be successful

Understanding is the foundation of achievement

Clear writing, real-life examples, photos, interactive maps, videos, and more will capture your attention and keep you engaged so that you can succeed. You will find tools and resources to help you read more effectively.

---

mheducation.com/prek-12

Send all inquiries to:
McGraw-Hill Education
8787 Orion Place
Columbus, OH 43240

ISBN: 978-0-07-668046-7
MHID: 0-07-668046-0

Printed in the United States of America.

6 7 8 9 10 LWI 22 21 20 19 18

ACADEMIC CONSULTANTS

Frederick L. Bein, Ph.D.
Professor of Geography
Indiana University-Purdue University
Indianapolis Adjunct Professor of Geography
Moi University
Indianapolis, Indiana

David Berger, Ph.D.
Ruth and I. Lewis Gordon Professor of Jewish
  History
Dean, Bernard Revel Graduate School
Yeshiva University
New York, New York

Randy Bertolas, Ph.D.
Professor of Geography
Wayne State College
Wayne, Nebraska

Lara Bryant, Ph.D.
Assistant Professor of Geography
Keene State College
Keene, New Hampshire

Tom Daccord
Educational Technology Specialist
Co-Director EdTech Teacher
Boston, Massachusetts

Charles Gritzner, Ph.D.
Professor of Geography, Retired
South Dakota State University
Brookings, South Dakota

P.P. Karan, Ph.D.
Professor of Geography
University of Kentucky
Lexington, Kentucky

Jeff Lash, Ph.D.
Associate Professor of Geography
University of Houston-Clear Lake
Houston, Texas

Elizabeth Leppman, Ph.D.
Writer
Map and Book Editor
Lexington, Kentucky

Joseph Manzo, Ph.D.
Professor of Geography
Concord University
Athens, West Virginia

Kent McGregor, Ph.D.
Associate Professor of Geography
University of North Texas
Denton, Texas

Olga Medvedkov, Ph.D.
Professor of Geography
Wittenberg University
Springfield, Ohio

Jerry Mitchell, Ph.D.
Research Associate Professor of Geography
University of South Carolina
Columbia, South Carolina

Shannon O'Lear, Ph.D.
Associate Professor of Geography
University of Kansas
Lawrence, Kansas

Justin Reich
Educational Technology Specialist
Co-Director EdTech Teacher
Boston, Massachusetts

David Rutherford, Ph.D.
Associate Professor of Geography & Public Policy
University of Mississippi
Oxford, Mississippi

Richard Sambrook, Ph.D.
Professor of Geography
Department Head
Eastern Michigan University
Ypsilanti, Michigan

Ginger Schmid, Ph.D.
Assistant Professor of Geography
Minnesota State University
Mankato, Minnesota

Joseph Stoltman, Ph.D.
Professor of Geography
Western Michigan University
Kalamazoo, Michigan

William Strong, Ph.D.
Professor Emeritus of Geography
University of North Alabama
Florence, Alabama

TEACHER REVIEWERS

Kimberly Coffelt
Social Studies Teacher
Southeast High School
Wichita, Kansas

Sara Heck
Social Studies Teacher
Teays Valley Local School District
Ashville, Ohio

Kenny Lee
Social Studies Curriculum
  Coordinator
Westerville City Schools
Westerville, Ohio

Barry Leonard
Social Studies Teacher
Graves County High School
Mayfield, Kentucky

Lorena McMenomy
Social Studies Educator
Hillcrest High School
Dallas, Texas

Jason Orendi
Social Studies Teacher
Boonsboro High School
Martinsburg, West Virginia

Jamie Reese
Social Studies Teacher
Buckhorn High School
New Market, Alabama

Adam Schwartz
Social Studies Teacher
Montgomery County Public
  Schools
Rockville, Maryland

Judi Shortt
Social Studies Teacher
Edmond Public Schools
Edmond, Oklahoma

Natalie Wojinski
Geography Teacher
Hercules High School
San Francisco, California

CONTENTS

HOW DO I STUDY GEOGRAPHY? xxx

SCAVENGER HUNT xxxii

REFERENCE ATLAS TABLE OF CONTENTS RA1

Geographic Dictionary RA2
World Time Zones RA4
World GDP Cartogram RA6
World Population Cartogram RA8
United States Physical RA10
United States Political RA12
North America Physical RA14
North America Political RA15
South America Physical RA16

South America Political RA17
Europe Physical RA18
Europe Political RA20
Africa Physical RA22
Africa Political RA23
Asia Physical RA24
Asia Political RA26
Oceania Physical/Political RA28
Polar Regions RA30

UNIT ONE   The World 1

How Geographers Look at the World 11

CHAPTER 1

ESSENTIAL QUESTION

How does geography help us interpret the past, understand the present, and plan for the future?

Why Geography Matters Distribution of
Political Power 12

LESSON 1 The Geographer's Tools 14

LESSON 2 The Geographer's Craft 26

The Physical World 37

CHAPTER 2

ESSENTIAL QUESTION

How do physical processes shape Earth's surface?

Why Geography Matters Economics and Resources:
Water Scarcity 38

LESSON 1 Planet Earth 40

LESSON 2 Forces of Change 44

LESSON 3 Earth's Water 51

Sue Flood/The Image Bank/Getty Images

CHAPTER 3

Climates of the Earth

Climates of the Earth................................**57**

ESSENTIAL QUESTION

Why is climate important to life on Earth?

Why Geography Matters Climate Change:
The Impacts on Humans.......................**58**

LESSON 1 Earth-Sun Relationships...............**60**

LESSON 2 Factors Affecting Climate.............**64**

LESSON 3 World Climate Patterns................**69**

©Tuul/Robert Harding World Imagery/Corbis

CHAPTER 4

The Human World

The Human World................................**75**

ESSENTIAL QUESTION

How do the characteristics and distribution of human populations affect human and physical systems?

Why Geography Matters The Human
Development Index...........................**76**

LESSON 1 Global Cultures.........................**78**

LESSON 2 Population Geography.................**82**

LESSON 3 Political Geography....................**87**

Case Study How Has Globalization Changed
Modern Culture?............................**92**

LESSON 4 Economic Geography.................**94**

Global Connections: Patterns of Resource
Distribution..................................**100**

LESSON 5 Urban Geography.....................**102**

CONTENTS

UNIT TWO The United States and Canada **109**

The United States **117**

ESSENTIAL QUESTION

How do physical systems and human systems shape a place?

Why Geography Matters Patterns of Immigration **118**

LESSON 1 Physical Geography of the United States **120**

LESSON 2 Human Geography of the United States **125**

Case Study The Environment: How Can Drought Lead to Conflict in the United States? **132**

LESSON 3 People and Their Environment: The United States **134**

Canada **141**

ESSENTIAL QUESTION

How do physical systems and human systems shape a place?

Why Geography Matters Energy Resources and Indigenous Rights **142**

LESSON 1 Physical Geography of Canada **144**

LESSON 2 Human Geography of Canada **149**

Global Connections: Two Decades of NAFTA **156**

LESSON 3 People and Their Environment: Canada **158**

UNIT THREE   Latin America 165

©Alison Wright/Corbis

CHAPTER **7**

Mexico **173**

ESSENTIAL QUESTION
How do physical systems and human systems shape a place?

Why Geography Matters Challenges of
  Urbanization............................... **174**

LESSON 1 Physical Geography of Mexico 176

LESSON 2 Human Geography of Mexico 180

LESSON 3 People and Their Environment:
  Mexico 187

Christian Aslund/Lonely Planet Images/Getty Images

CHAPTER **8**

Central America and the Caribbean **193**

ESSENTIAL QUESTION
How do physical systems and human systems shape a place?

Why Geography Matters Spatial Diffusion:
  The Columbian Exchange..................... **194**

LESSON 1 Physical Geography of Central America
  and the Caribbean 196

LESSON 2 Human Geography of Central America
  and the Caribbean 200

Case Study What Kind of Development Is
  Best for Haiti? 206

LESSON 3 People and Their Environment: Central
  America and the Caribbean 208

CONTENTS

South America **215**

ESSENTIAL QUESTION

How do physical systems and human systems shape a place?

Why Geography Matters Economic Geography:
Uneven Development 216

LESSON 1 **Physical Geography of
South America** 218

LESSON 2 **Human Geography of
South America** 223

Global Connections: Amazon in the Balance 230

LESSON 3 **People and Their Environment:
South America** 232

Europe 239

Northern Europe **247**

ESSENTIAL QUESTION

How do physical systems and human systems shape a place?

Why Geography Matters Volcanic Eruption
in Iceland 248

LESSON 1 **Physical Geography of
Northern Europe** 250

LESSON 2 **Human Geography of
Northern Europe** 255

LESSON 3 **People and Their Environment:
Northern Europe** 261

Vidler Steve/age fotostock

CHAPTER 11

Northwestern Europe **267**

ESSENTIAL QUESTION
How do physical systems and human systems shape a place?

Why Geography Matters Suburban Growth and Transportation **268**

LESSON 1 Physical Geography of Northwestern Europe **270**

LESSON 2 Human Geography of Northwestern Europe **275**

Case Study How Beneficial Is the European Union? ... **282**

LESSON 3 People and Their Environment: Northwestern Europe **284**

Steve Outram/Photographer's Choice/Getty Images

CHAPTER 12

Southern Europe **291**

ESSENTIAL QUESTION
How do physical systems and human systems shape a place?

Why Geography Matters Labor Migration from North Africa **292**

LESSON 1 Physical Geography of Southern Europe **294**

LESSON 2 Human Geography of Southern Europe **299**

LESSON 3 People and Their Environment: Southern Europe **305**

Matt Cardy/Alamy

CHAPTER 13

Eastern Europe **311**

ESSENTIAL QUESTION
How do physical systems and human systems shape a place?

Why Geography Matters The Breakup of Yugoslavia .. **312**

LESSON 1 Physical Geography of Eastern Europe **314**

LESSON 2 Human Geography of Eastern Europe **319**

LESSON 3 People and Their Environment: Eastern Europe **326**

CONTENTS

Russ Images/Alamy

CHAPTER 14

The Russian Core **333**

ESSENTIAL QUESTION

How do physical systems and human systems shape a place?

Why Geography Matters Russia's Shrinking
Population **334**

LESSON 1 Physical Geography of the
Russian Core **336**

LESSON 2 Human Geography of the
Russian Core **341**

Global Connections: Arctic Oil Frontiers **348**

LESSON 3 People and Their Environment:
The Russian Core **350**

UNIT FIVE North Africa, Southwest Asia, and Central Asia **357**

Kimberley Coole/Lonely Planet Images/Getty Images

CHAPTER 15

North Africa **365**

ESSENTIAL QUESTION

How do physical systems and human systems shape a place?

Why Geography Matters Choke Point: Suez Canal **366**

LESSON 1 Physical Geography of
North Africa **368**

LESSON 2 Human Geography of
North Africa **372**

Global Connections: The Arab Spring **378**

LESSON 3 People and Their Environment:
North Africa **380**

Cultura/Image Source

CHAPTER 16

The Eastern Mediterranean **387**

ESENTIAL QUESTION
How do physical systems and human systems shape a place?

Why Geography Matters The Israeli-Palestinian Conflict **388**

LESSON 1 Physical Geography of the Eastern Mediterranean **390**

LESSON 2 Human Geography of the Eastern Mediterranean **395**

LESSON 3 People and Their Environment: The Eastern Mediterranean **402**

Rebecca Erol/Alamy

CHAPTER 17

The Northeast **409**

ESSENTIAL QUESTION
How do physical systems and human systems shape a place?

Why Geography Matters A Stateless Nation: The Kurds **410**

LESSON 1 Physical Geography of the Northeast **412**

LESSON 2 Human Geography of the Northeast **416**

Case Study How Have Sunni and Shia Beliefs Led to Conflict? **422**

LESSON 3 People and Their Environment: The Northeast **424**

Morales/age fotostock

CHAPTER 18

The Arabian Peninsula **431**

ESSENTIAL QUESTION
How do physical systems and human systems shape a place?

Why Geography Matters Migrant Workers in the Arabian Peninsula **432**

LESSON 1 Physical Geography of the Arabian Peninsula **434**

LESSON 2 Human Geography of the Arabian Peninsula **438**

LESSON 3 People and Their Environment: The Arabian Peninsula **444**

CONTENTS

Central Asia **451**

ESSENTIAL QUESTION

How do physical systems and human systems shape a place?

Why Geography Matters Afghanistan's Troubled
History 452

LESSON 1 Physical Geography of
Central Asia 454

LESSON 2 Human Geography of
Central Asia 459

LESSON 3 People and Their Environment:
Central Asia 466

UNIT SIX   **Africa South of the Sahara** 473

The Transition Zone **481**

ESSENTIAL QUESTION

How do physical systems and human systems shape a place?

Why Geography Matters Diffusion: Muslim and
non-Muslim Cultures 482

LESSON 1 Physical Geography of the
Transition Zone 484

LESSON 2 Human Geography of the
Transition Zone 489

LESSON 3 People and Their Environment:
The Transition Zone 496

East Africa **503**

ESSENTIAL QUESTION

How do physical systems and human systems shape a place?

Why Geography Matters Export Crops and East Africa. 504

LESSON 1 Physical Geography of
East Africa 506

LESSON 2 Human Geography of East Africa 511

Case Study Should the Grand Ethiopian Renaissance
Dam Be Built? 518

LESSON 3 People and Their Environment:
East Africa 520

Pius Utomi Ekpei/AFP/Getty Images

CHAPTER 22

West Africa 527

ESSENTIAL QUESTION

How do physical systems and human systems shape a place?

Why Geography Matters Empowering Women in
West Africa 528

LESSON 1 **Physical Geography of
West Africa** 530

LESSON 2 **Human Geography of West Africa** 535

Global Connections: Conflict Diamonds 542

LESSON 3 **People and Their Environment:
West Africa** 544

Thomas Imo/Photothek/Getty Images

CHAPTER 23

Equatorial Africa 551

ESSENTIAL QUESTION

How do physical systems and human systems shape a place?

Why Geography Matters South Sudan:
Independence and Conflict 552

LESSON 1 **Physical Geography of
Equatorial Africa** 554

LESSON 2 **Human Geography of
Equatorial Africa** 559

LESSON 3 **People and Their Environment:
Equatorial Africa** 566

Kelly Cestari/ASP/Getty Images

CHAPTER 24

Southern Africa 573

ESSENTIAL QUESTION

How do physical systems and human systems shape a place?

Why Geography Matters Southern Africa and
HIV/AIDS 574

LESSON 1 **Physical Geography of
Southern Africa** 576

LESSON 2 **Human Geography of
Southern Africa** 581

LESSON 3 **People and Their Environment:
Southern Africa** 588

CONTENTS

UNIT SEVEN **South Asia** 595

India 603

ESSENTIAL QUESTION

How do physical systems and human systems shape a place?

Why Geography Matters India's Population Structure 604

LESSON 1 **Physical Geography of India** 606

LESSON 2 **Human Geography of India** 611

Case Study What Is the Future of Kashmir? 618

LESSON 3 **People and Their Environment: India** 620

Pakistan and Bangladesh 627

ESSENTIAL QUESTION

How do physical systems and human systems shape a place?

Why Geography Matters Flood-Prone Pakistan and Bangladesh 628

LESSON 1 **Physical Geography of Pakistan and Bangladesh** 630

LESSON 2 **Human Geography of Pakistan and Bangladesh** 634

Global Connections: South Asia on the Brink 640

LESSON 3 **People and Their Environment: Pakistan and Bangladesh** 642

Bhutan, Maldives, Nepal & Sri Lanka **649**

ESSENTIAL QUESTION

How do physical systems and human systems shape a place?

Why Geography Matters Nepal's Role as a
Buffer State **650**

LESSON 1 **Physical Geography of Bhutan, Maldives, Nepal & Sri Lanka** **652**

LESSON 2 **Human Geography of Bhutan, Maldives, Nepal & Sri Lanka** **657**

LESSON 3 **People and Their Environment: Bhutan, Maldives, Nepal & Sri Lanka** **663**

UNIT EIGHT   **East Asia** **669**

China and Mongolia **677**

ESSENTIAL QUESTION

How do physical systems and human systems shape a place?

Why Geography Matters China's Growing
Energy Demands **678**

LESSON 1 **Physical Geography of China and Mongolia** **680**

LESSON 2 **Human Geography of China and Mongolia** **685**

Case Study Are Limits on Growth Best for China? ... **692**

LESSON 3 **People and Their Environment: China and Mongolia** **694**

Japan **701**

ESSENTIAL QUESTION

How do physical systems and human systems shape a place?

Why Geography Matters Japan's Aging Population.... **702**

LESSON 1 **Physical Geography of Japan** **704**

Global Connections: The Tohoku Earthquake and
Tsunami **708**

LESSON 2 **Human Geography of Japan** **710**

LESSON 3 **People and Their Environment: Japan** **717**

CONTENTS

North Korea and South Korea **723**

ESSENTIAL QUESTION
How do physical systems and human systems shape a place?

Why Geography Matters Complementarity:
  Two Koreas **724**

LESSON 1 Physical Geography of North
    Korea and South Korea **726**

LESSON 2 Human Geography of North
    Korea and South Korea **730**

LESSON 3 People and Their Environment:
    North Korea and South Korea **737**

UNIT NINE **Southeast Asia and the Pacific World** **743**

Southeast Asia **751**

ESSENTIAL QUESTION
How do physical systems and human systems shape a place?

Why Geography Matters Emerging Markets in
  Southeast Asia **752**

LESSON 1 Physical Geography of
    Southeast Asia **754**

LESSON 2 Human Geography of
    Southeast Asia **759**

LESSON 3 People and Their Environment:
    Southeast Asia **766**

Australia and New Zealand 773

ESSENTIAL QUESTION

How do physical systems and human systems shape a place?

Why Geography Matters Australia and New Zealand:
Indigenous Peoples 774

LESSON 1 Physical Geography of Australia
and New Zealand 776

LESSON 2 Human Geography of Australia
and New Zealand 781

Global Connections: Non-Native Species:
Rabbits in Australia 788

LESSON 3 People and Their Environment:
Australia and New Zealand 790

Oceania 797

ESSENTIAL QUESTION

How do physical systems and human systems shape a place?

Why Geography Matters Samoa Hops the
International Date Line 798

LESSON 1 Physical Geography of Oceania 800

LESSON 2 Human Geography of Oceania 804

Case Study Who Owns the High Seas? 810

LESSON 3 People and Their Environment:
Oceania 812

Special Feature: Antarctica 819

World Religions Handbook 826

Gazetteer 846

English/Spanish Glossary 856

Index 883

FEATURES

Why Geography Matters

Distribution of Political Power **12**

Economics and Resources: Water Scarcity **38**

Climate Change: The Impacts on Humans **58**

The Human Development Index **76**

Patterns of Immigration **118**

Energy Resources and Indigenous Rights **142**

Challenges of Urbanization **174**

Spatial Diffusion: The Columbian Exchange **194**

Economic Geography: Uneven Development **216**

Volcanic Eruption in Iceland **248**

Suburban Growth and Transportation **268**

Labor Migration from North Africa **292**

The Breakup of Yugoslavia **312**

Russia's Shrinking Population **334**

Choke Point: Suez Canal **366**

Israeli-Palestinian Conflict **388**

A Stateless Nation: The Kurds **410**

Migrant Workers in the Arabian Peninsula **432**

Afghanistan's Troubled History **452**

Diffusion: Muslim and non-Muslim Cultures **482**

Export Crops and East Africa **504**

Empowering Women in West Africa **528**

South Sudan: Independence and Conflict **552**

Southern Africa and HIV/AIDS **574**

India's Population Structure **604**

Flood-Prone Pakistan and Bangladesh **628**

Nepal's Role as a Buffer State **650**

China's Growing Energy Demands **678**

Japan's Aging Population **702**

Complementarity: Two Koreas **724**

Emerging Markets in Southeast Asia **752**

Australia and New Zealand: Indigenous Peoples **774**

Samoa Hops the International Date Line **798**

EXPLORE the REGION

The United States and Canada **110**

Latin America **166**

Europe **240**

North Africa, Southwest Asia, and Central Asia **358**

Africa South of the Sahara **474**

South Asia **596**

East Asia **670**

Southeast Asia and the Pacific World **744**

Case Study

How Has Globalization Changed Modern Culture? **92**

How Can Drought Lead to Conflict in the United States? **132**

What Kind of Development Is Best for Haiti? **206**

How Beneficial Is the European Union? **282**

How Have Sunni and Shia Beliefs Led to Conflict? **422**

Should the Grand Ethiopian Renaissance Dam Be Built? **518**

What Is the Future of Kashmir? **618**

Are Limits on Growth Best for China? **692**

Who Owns the High Seas? **810**

FEATURES

Global Connections

Patterns of Resource Distribution 100

Two Decades of NAFTA 156

Amazon in the Balance 230

Arctic Oil Frontiers.................................. 348

The Arab Spring 378

Conflict Diamonds..................................... 542

South Asia on the Brink............................... 640

The Tohoku Earthquake and Tsunami 708

Non-Native Species: Rabbits in Australia 788

Connecting Geography to ....

Math: Mapmaking...................................... 18

Science: Astronomy.................................... 41

Science: Climatology................................... 66

Government: Geopolitics 91

Science: Soil Science.................................. 124

Science: Biology 148

Science: Environmental Science.......................... 189

Science: Climatology................................... 199

History: Before Columbus.............................. 202

Science: Biology 221

Science: Geology..................................... 254

Science: Cap-and-Trade Systems 288

Science: Demography.................................. 301

Government: Environmental Politics 330

Economics: Privatization 347

Government: Geopolitics 384

Economics: Development............................... 400

Science: Soil Erosion in Iraq 427

Science: Air-Conditioning 436

Government: Water Management 448

Government: Saving the Aral Sea....................... 470

Math: Geometry 487

History: Naming Places................................ 510

Sociology: Social Norms 540

History: Slave Forts................................... 561

Math: Output of the Kariba Hydroelectric Dam............ 579

Math: India's GDP 617

History: Natural Disaster 635

Science: Genetics 660

Sociology: *Hanami:* Viewing of Flowers 706

Science: The DMZ 729

Science: Tsunamis.................................... 756

Sociology: The Outback 780

Science: Carbon Dating................................ 805

Analyzing Primary Sources

Yi-fu Tuan, *Topophilia: a Study of Environmental Perception, Attitudes, and Values*, 1974 . . . . **31**

A Penguin's View of the World (political cartoon) **35**

Gina Christie, *BBC News*, May 17, 2010 . . . . **48**

Joel Achenbach, "The Next Big One," *National Geographic*, April 2006 . . . . **56**

Charles Q. Choi, *National Geographic News*, December 29, 2011 . . . . **72**

Pao K. Wang, "Chinese Historical Documents and Climate Change," www.accessscience.com **74**

Hillary Rodham Clinton, press statement, June 20, 2012 . . . . . **86**

Sam Quinones, PBS Frontline/World, "Guatemala/Mexico-Coffee Country," May 2003 . . . . **98**

Don Belt, "Europe's Big Gamble," *National Geographic*, May 2004 . . . . **108**

Brad Plumer, "Can Natural Gas Help Tackle Global Warming? A Primer," *Washington Post*, August 20, 2012 . . . . **138**

Michael Parfit, "Powering the Future," *National Geographic*, August 2005 . . . . **140**

Tavia Grant, *The Globe and Mail*, October 24, 2012 . . . . **153**

"Acid Rain Could Kill Maples Near Great Lakes," Associated Press, December 16, 2011 . . . . **160**

"Salmon Overfishing Warning Issued," CBC News, September 1, 2010 . . . . **164**

Tim Johnson, "Mexico's 'Maquiladora' Labor System Keeps Workers in Poverty," *The Miami Herald*, June 18, 2012 . . . . . . **186**

Jon Waterhouse, "How the Maya of Today Are Marking December 21," *National Geographic Explorers Journal*, December 19, 2012 . . . . **192**

Paolo Fortis, *Kuna Art and Shamanism: An Ethnographic Approach*, 2012 . . . . **204**

Silvia Lambiase, "Indigenous Communities in Central America Use Traditions to Protect Biodiversity," Platform for Agrobiodiversity Research, May 2011 . . . . **212**

Jonathan Watts, "Costa Rica Recognized for Biodiversity Protection," *The Guardian*, October 2010 . . . . **214**

Sara Shahriari, "Urban Population Boom Threatens Lake Titicaca," *The Guardian*, January 12, 2012 . . . . **234**

Pablo Neruda, "We are Many," *The Yellow Heart*, 1974 . . . . . . **238**

Katrin Bennhold, "Working Women Are the Key to Norway's Prosperity," *New York Times*, June 28, 2011 . . . . **260**

Marta Madina, "Oceana Concludes Expedition to Document Biodiversity and Fisheries in the Baltic," Oceana.org, June 7, 2011 . . . . **262**

Hans Chresten Jeppesen, *National Geographic News*, November 4, 2009 . . . . **266**

James Salter, "The Alps," *National Geographic Traveler*, October 1999 . . . . **273**

Charter of Fundamental Rights of the European Union, 2000 . . . . **290**

G. Fontanazza, "Importance of Olive-Oil Production in Italy," *Integrated Soil and Water Management for Orchard Development*, 2005 . . . . **297**

Federico García Lorca, *The Selected Poems of Federico García Lorca*, 2006 . . . . **310**

Anna Yukhananov, "Poland Makes Greatest Strides in Business Reforms—World Bank," Reuters, October 22, 2012 . **325**

"Deadly Heat Wave Grips Europe," CNN.com: Reuters (Budapest), July 25, 2007 . . . . **332**

Dwight Garner, "When Poets Rocked Russia's Stadiums," *New York Times*, June 3, 2010 . . . . **345**

Richard Stone, "The Long Shadow of Chernobyl," *National Geographic*, April 2006 . . . . **353**

David Braun, *National Geographic Daily News*, November 25, 2012 . . . . **356**

Rebecca Shaw, Nature Conservancy, *Saving Mediterranean Habitats Worldwide*, February 25, 2011 . . . . **371**

Hillary Rodham Clinton, U.S. Secretary of State, CNN News interview, January 30, 2011 . . . . **386**

World Wildlife Fund, "Lebanon's Forests: Facing New Threats" . . . **394**

United Nations, "Syria: Act Now to Stop Desertification, Says FAO," *Humanitarian News and Analysis*, June 15, 2010 . . **404**

Arthur James Balfour, The Balfour Declaration, November 2, 1917 . . . . **408**

Thomas K. Grose, "Iraq Poised to Lead World Oil Supply Growth, But Obstacles Loom," *National Geographic Daily News*, October 9, 2012 . . . . **415**

International Energy Agency, "Executive Summary," *Iraq Energy Outlook*, 2012 . . . . **430**

Dr. Shawki Barghouti, Director-General of the International Centre for Biosaline Agriculture, from "Desalination Threat to the Growing Gulf," *The National*, August 31, 2009 . . . . **446**

Joshua Hammer, "Days of Reckoning," *National Geographic*, September 2012 . . . . **450**

Strabo from *Geography*, Book XI, Chapter 8 (c. A.D. 21) **455**

Rupert Colville, spokesperson for the UN Office of the High Commissioner for Human Rights (OHCHR), June 10, 2011, UN News Centre **472**

A 36-year-old mother of nine at the Jamam refugee camp in South Sudan, from Médecins Sans Frontières, January 18, 2013 . . **491**

Analyzing Primary Sources

Robert Draper, "Shattered Somalia," *National Geographic Magazine,* September 2009 .... **502**

"Concerted Efforts Needed to Improve Post-Primary Education in Eastern and Southern Africa," United Nations Girls' Education Initiative, February 27, 2007 .... **515**

Stewart Edward White, American hunter, describing the Serengeti in Tanzania, 1913 .... **523**

IPP Media, describing Tanzania, September 28, 2012 .... **523**

President Clinton, Associated Press, March 1998 .... **526**

Muhammad Sanusi, a fisherman in Dogon Fili, "Lake Chad Fishermen Pack Up Their Nets," BBC News, January 15, 2007 .... **546**

Elizabeth Stevens, "Sahel Food Crisis: The Cost of Climate Change," October 2012 .... **550**

U.S. Department of State, International Travel for U.S. Citizens **563**

"Food Crisis Plagues War-Torn CAR," United Press International, February 18, 2013 .... **572**

Amanda Kendle, "Poverty Tourism: Exploring the Slums of India, Brazil and South Africa," www.vagabondish.com ... **586**

Nelson Mandela, speech delivered in London, February 3, 2005 .... **594**

Anika Gupta, "The Holy City of Varanasi," Smithsonian.com, August 20, 2009 .... **607**

Ishaan Tharoor, "Top 10 Most Influential Protests," *Time,* June 28, 2011 .... **612**

C. Rajagopalachar, Introduction to *Freedom's Battle: Being a Comprehensive Collection of Writings and Speeches on the Present Situation,* 1922 .... **626**

Jill McGivering, "'Elation' and 'Unease' at Helping Pakistan Flood Child," *BBC News,* September 4, 2010 .... **633**

Don Belt, "The Coming Storm," *National Geographic,* May 2011 .... **648**

Jamling Tenzing Norgay, *Touching My Father's Soul,* 2002 .... **654**

Tom Whittaker, *Higher Purpose: The Heroic Story of the First Disabled Man to Conquer Everest,* 2001 .... **668**

Carla Amurao, "From West to Far East: Rappin' and Rockin' the House," *Tavis Smiley: China Week,* PBS California, July 2011 .... **689**

Ian Johnson, "In China, the Forgotten Manchu Seek to Rekindle Their Glory," *The Wall Street Journal,* October 3, 2009 .... **695**

Made in China (political cartoon) **699**

Bill McKibben, "Can China Go Green?" *National Geographic,* June 2011 .... **700**

"Between the Folds," PBS Independent Lens, November 30, 2009 .... **714**

Excerpt from United States Bombing Survey, July 1, 1946, President's Secretary's File, Truman Papers **722**

Winston Churchill, "Why Should We Fear for Our Future?" House of Commons, August 16, 1945 .... **722**

Philip Jaisohn, letter dated September 14, 1950, to Lieutenant General John R. Hodge **732**

Human Rights Watch, "North Korea: Economic System Built on Forced Labor," June 13, 2012 .... **742**

Association of Southeast Asian Nations (ASEAN), Bangkok Declaration, 1967 .... **765**

"Mekong Giant Catfish," nationalgeographic.com **768**

President Lyndon B. Johnson, commenting on Vietnam to Senator Eugene McCarthy, February 1966 .... **772**

John Kerry, announcement speech, Patriots Point, South Carolina, September 2, 2003 .... **772**

Glenys Ward, stolengenerationstestimonies.com **796**

Kiribati president Anote Tong, statement to the Second Session of Global Platform for Disaster Risk-Reduction Conference, June 2009 .... **815**

Dr. Elizabeth Lindsey, www.elizabethlindsey.com **818**

MAPS

REFERENCE ATLAS MAPS **RA1**

World Time Zones **RA4**

World GDP Cartogram............................. **RA6**

World Population Cartogram **RA8**

United States Physical............................**RA10**

United States Political............................**RA12**

North America Physical...........................**RA14**

North America Political...........................**RA15**

South America Physical...........................**RA16**

South America Political...........................**RA17**

Europe Physical..................................**RA18**

Europe Political **RA20**

Africa Physical **RA22**

Africa Political **RA23**

Asia Physical **RA24**

Asia Political **RA26**

Oceania Physical/Political.........................**RA28**

Polar Regions **RA30**

UNIT 1: THE WORLD

Physical **2**

Political **4**

Population Density **6**

Economic Activity **8**

Climate..**10**

Redistricting in Ohio **12**

Great Circle Routes **15**

Common Map Projections...........................**16**

Latitude and Longitude.............................**17**

Small-Scale and Large-Scale Maps **19**

Mental Maps **20**

Perceptual Regions of the United States **29**

Continental Drift **46**

Plates and Plate Movement **47**

Global Desalination.................................**53**

World Zones of Latitude and Wind Patterns **65**

World Ocean Currents...............................**67**

World Biomes......................................**70**

Human Development **76**

World Language Families **79**

World Culture Hearths..............................**80**

UNIT 2: THE UNITED STATES AND CANADA

Physical **112**

Political **113**

Climate...**114**

Vegetation **114**

Economic Activity **115**

Population Density **116**

Largest Ancestry Reported by County 2010 ....**118**

The Continental Divide and the Fall Line **122**

Ethnic Populations in the United States.............**128**

Homes in Negative Equity in 2010 ....**130**

U.S. Water Withdrawals **135**

Acid Rain in the United States......................**136**

Glaciers of the Last Ice Age **145**

St. Lawrence Seaway System........................**146**

UNIT 3: LATIN AMERICA

Physical **168**

Political **169**

Climate...**170**

Vegetation **170**

Economic Activity **171**

Population Density **172**

Mexican Migration to the United States **185**

Environmental Deterioration in Mexico **188**

The Columbian Exchange **194**

Physical Geography: Central America and the Caribbean **197**

Dominant Ethnic Groups of South America **226**

Agricultural Land Use in South America.............**228**

UNIT 4: EUROPE

Physical **242**

Political **243**

Climate...**244**

Vegetation **244**

Economic Activity **245**

Population Density **246**

Acid Rain in Northern Europe **263**

Population Density of Northwestern Europe..........**278**

The European Union **280**

Overfishing in Northwestern Europe **285**

Oxygen Depletion in Coastal Marine Ecosystems..........**286**

Migration: North Africa to Southern Europe 292
Soil Erosion in Southern Europe 306
The Former Yugoslavia 312
Alpine Europe 315
Main-Danube Canal. 316
Ethnic Groups in Eastern Europe 322
Environmental Issues in the Russian Core. 351
Reach of Chernobyl Disaster 352

UNIT 5: NORTH AFRICA, SOUTHWEST ASIA, AND CENTRAL ASIA

Physical 360
Political 361
Climate. 362
Vegetation 362
Economic Activity 363
Population Density 364
North Africa: Invasions and Migrations 373
Eastern Mediterranean Precipitation. 393
The Distribution of Kurdish People 410
Physical Map: Turkey, Iran, and Iraq .413
Civilizations and Empires of the Northeast 417
Tigris-Euphrates River Basin. 425
Physical Geography: The Arabian Peninsula 435
Ethnic Groups of Afghanistan 452
Ethnic Groups in Central Asia 462

UNIT 6: AFRICA SOUTH OF THE SAHARA

Physical 476
Political 477
Climate. 478
Vegetation 478
Economic Activity 479
Population Density 480
The African Transition Zone 485
Kingdoms and Empires of the Transition Zone 490
The Sahel's Vulnerable Zone. 497
Sudan and South Sudan Conflicts 552
Physical Geography: The Congo Basin. 555
The Atlantic Slave Trade 560
World Distribution of HIV 2014 574
European Colonization of Africa 582

UNIT 7: SOUTH ASIA

Physical 598
Political 599
Climate. 600
Vegetation 600
Economic Activity 601
Population Density 602
South Asia: Monsoons. 608
Mountain Passes of Pakistan 631
Buffer States and Border Disputes in South Asia 650

UNIT 8: EAST ASIA

Physical 672
Political 673
Climate 674
Vegetation 674
Economic Activity 675
Population Density 676
Physical Geography of China and Mongolia 681
East Asian Monsoons 683
Physical Map of Japan 705
Japan's Nuclear Power Plants 718
Waterways of the Korean Peninsula .727

UNIT 9: SOUTHEAST ASIA AND THE PACIFIC WORLD

Physical 746
Political 747
Climate. 748
Vegetation 748
Economic Activity 749
Population Density 750
Volcanoes in Southeast Asia. 755
Patterns of European Settlement. 783
Samoa and the International Date Line. 798
Global Sea Level Trends 813

PRIMARY SOURCES

UNIT 1: THE WORLD

George J. Demko, *Why in the World: Adventures in Geography*, 1992 . . . . **36**

Richard Stone, "The Last Great Impact on Earth," *Discover*, September 1996 . . . . **41**

Joel Achenbach, "The Next Big One," *National Geographic*, April 2006 . . . . **46**

Corrado Sommariva, invitation to the International Desalination Association 2013 World Congress **53**

Richard A. Muller, from the Columbia Forum, "Physics for Future Presidents," January/February 2009 . . . . **63**

Curt Suplee, "El Niño/La Niña," *National Geographic Magazine*, March 1999 . . . . **67**

Doug Saunders, "The Great Shift From Farm to City," *Los Angeles Times*, June 19, 2011. . . . . **86**

Shogo Arai, Parliamentary Secretary for Foreign Affairs, "Japan and the Preservation of Intangible Cultural Heritage," 2004 . . . . **93**

Koïchiro Matsuura, Director-General of UNESCO, "Globalization, Intangible Cultural Heritage and the Role of UNESCO," 2004 . . . . **93**

Dusko Doder, "The Bolshevik Revolution," *National Geographic*, October 1992 . . . . **96**

UNIT 2: THE UNITED STATES AND CANADA

Greg Bluestein, Bill Rankin, and Scott Trubey, "High Court Grants Georgia Water-Wars Victory," *Atlanta Journal Constitution*, June 25, 2012 . . . . **133**

Florida governor Charlie Crist, letter to the United States Department of the Interior, May 28, 2009 . . . . **133**

UNIT 3: LATIN AMERICA

Paul Collier, *Haiti: From Natural Catastrophe to Economic Security*, A Report for the Secretary General of the United Nations, January 2009 . . . . **207**

Laurent Dubois and Deborah Jenson, "Haiti Can Be Rich Again," *New York Times*, January 8, 2012. . . . . **207**

Ian Sample, "Amazon's Doomed Species Set to Pay Deforestation's 'Extinction Debt,'" *The Guardian (UK)*, July 12, 2012 . . . . **231**

UNIT 4: EUROPE

Thorbjorn Jagland, the chairman of the Norwegian Nobel Committee, October 2012 . . . . **283**

Nigel Farage, UK Independence Party leader, October 2012. . **283**

Julian Scola, "Europe's Rivers Ready for Revival," World Wide Fund for Nature, April 20, 2001 . . . . **288**

Fen Montaigne, "The Great Northern Forest," *National Geographic*, June 2002 . . . . **353**

UNIT 5: NORTH AFRICA, SOUTHWEST ASIA, AND CENTRAL ASIA

Mona Moussavi, *ISS Voices*, October 10, 2012 . . . . **423**

The United States Institute of Peace, *U.S. Terrorism Report: MEK and Jundallah*, August 23, 2011. **423**

UNIT 6: AFRICA SOUTH OF THE SAHARA

Meles Zenawi, the former prime minister of Ethiopia, at the official commencement of the Millennium Dam, 2011 . . . . **519**

"Grand Ethiopian Renaissance Dam," International Rivers organization **519**

Nzinga Mbemba, the king of Kongo, quoted in *East Along the Equator*, 1987. . . . . **560**

UNIT 7: SOUTH ASIA

Shabir Choudhry, Director of the Institute of Kashmir Affairs . . **619**

Ministry of External Affairs, India **619**

UNIT 8: EAST ASIA

Liu Shaojie, Vice Director of the Population Commission in Henan province, China, *The Guardian*, October 25, 2011. . . . **693**

Liang Zhongtang, a former committee member of the National Family Planning Commission, *The Telegraph*, September 25, 2010. . . . . **693**

Li Jiamin, a specialist in population studies at Nankai University, *The Telegraph*, October 31, 2012 . . . . **693**

UNIT 9: SOUTHEAST ASIA AND THE PACIFIC WORLD

Colin Hanna, president of Let Freedom Ring, quoted in "Kill the Law of the Sea Treaty," *U.S. News*, May 10, 2012 . . . . **811**

Gustavo Fonseca, head of natural resources, quoted in "Oceans Special: GEF Rolls Out Investment Project to Address Issues for Areas in the High Seas," June 16, 2012 . . . . **811**

GEO@WORK

Explore specific examples of the principles and skills of geography applied to real-world challenges that impact people's lives. From agriculture, to urban planning, to wiping out disease, and managing changes in society—geography plays a key role in understanding relationships and generating solutions that make sense.

Videos

Every lesson has a video to help you learn more about your world!

Self-Check Quizzes

Every lesson has a self-check quiz to help you test your knowledge!

Games

Every lesson has games to help you review the material!

Vocabulary Flash Cards

Every chapter has vocabulary flash cards to help you remember important terms!

Infographics

Chapter 1
Lesson 1 GIS Layers

Chapter 2
Chapter Feature Salt vs. Freshwater
                Desalination Through Distillation
Lesson 1 The Solar System
          Water, Land, and Air
          Inside the Earth
          Underwater Landforms
Lesson 2 Forces of Change
Lesson 3 Water Cycle

Chapter 3
Lesson 1 The Earth's Seasons
          The Greenhouse Effect
          Earth's Tilt
Lesson 3 Latitude, Climate, and Vegetation

Chapter 5
Lesson 1 Western Topography
          Hurricanes in the United States
Lesson 3 How Acid Rain is Created

Chapter 6
Chapter Feature A Closer Look: Alberta
Lesson 3 Canada's Boreal Forest

Chapter 7
Lesson 1 Plate Tectonics in Mexico
          Vertical Climate Zones of Mexico

Chapter 8
Lesson 1 Biodiversity in Central America
Lesson 2 The Panama Canal

Chapter 9
Lesson 1 Impacts of El Niño in Latin America

Chapter 10
Chapter Feature Volcanic Eruption in Iceland
Lesson 1 Gulf Stream Effects in Northern Europe

Chapter 11
Lesson 1 Building Dikes and Polders

Chapter 12
Lesson 1 Rivers of Europe

Chapter 14
Chapter Feature Russia's Shrinking Population
Lesson 1 The Volga River Basin—Heart of Russia

Chapter 15
Lesson 1 North African Plates
Lesson 3 Egypt's Aswān High Dam

Chapter 16
Chapter Feature Israeli-Palestinian Conflict
Lesson 3 Managing Israel's Water Resources

Chapter 17
Lesson 3 Caspian Energy

Chapter 19
Lesson 3 Aral: A Sea Sacrificed

Chapter 20
Lesson 3 Great Green Wall of Africa

Chapter 21
Lesson 1 The Great Rift Valley

Chapter 22
Lesson 2 Today's Nigerian Family

networks ONLINE RESOURCES

Chapter 23
Lesson 1   Anatomy of the Rain Forest

Chapter 25
Chapter Feature   India's Population Structure
Lesson 2   All Aboard in India

Chapter 27
Lesson 1   Ecotourism in Nepal

Chapter 28
Chapter Feature   China's Growing Energy Demands
Lesson 3   Three Gorges Dam: Harnessing the Yangtze

Chapter 29
Chapter Feature   Japan's Aging Population

Chapter 30
Lesson 3   Facing Empty Oceans

Chapter 31
Lesson 3   An Oil That's Everywhere

Chapter 32
Lesson 1   Life in the Great Barrier Reef

Chapter 33
Lesson 1   From Volcano to Atoll

Interactive Charts, Graphs, and Tables

Chapter 2
Lesson 3   Water Resources

Chapter 4
Chapter Feature   Change in HDI Rank for Selected Countries,
                  2006–2011
                  Population Indicators in 2011
Lesson 1   Cultural Universals
Lesson 2   Projected Populations
Lesson 3   Population Pyramids
           WTO Members
Lesson 5   Ten Largest Cities
           Urbanization in India

Chapter 5
Chapter Feature   Ethnic Diversity in the United States, 2010
Lesson 1   Hurricane Information
Lesson 2   Minority Population Growth

Chapter 7
Lesson 2   Population Growth of Mexico City, 1960–2020
           Population Percentage of Mexico City

Chapter 9
Chapter Feature   Brazil's Urban Population Growth
                  The Core-Periphery Theory

Chapter 10
Lesson 2   Electricity Production 2011

Chapter 12
Chapter Feature   Money Sent by Workers in Southern Europe to
                  Their Home Countries
Lesson 2   The Effects of Geography on the Rise of Rome

Chapter 14
Lesson 1   Russia's Oil Reserves: A World Comparison
Lesson 3   The Nuclear Arms Race

Chapter 15
Lesson 3   Population: North Africa

Chapter 21
Lesson 2   Influences on African Religions

Chapter 22
Lesson 2   Ethnic Groups in West Africa

Chapter 24
Lesson 3   Wildlife Reserves in Southern Africa

Chapter 25
Chapter Feature   Compare Urban Population in India and U.S.
Lesson 2   India's Political Parties
           Religion in India

Chapter 26
Chapter Feature   Major Floods In Pakistan and Bangladesh

Chapter 27
Lesson 2   Major Exports of Nepal, Bhutan, Maldives, and Sri Lanka
Lesson 3   Forest Loss/Gain

Chapter 28
Lesson 2   Grain Production in China, 1950–1970
Lesson 3   Energy Use in China per Capita
           Daily Oil Production and Consumption in China,
           1993–2010

Chapter 29
Chapter Feature   Population Distribution in Japan

Chapter 30
Chapter Feature   North and South Korean Populations 1960–2013
Lesson 1   Average Monthly Temperatures, Seoul and P'yongyang,
           1971–2000
Lesson 2   North Korean Conflicts
Lesson 3   South Korean Exports

Chapter 31
Chapter Feature   Southeast Asian GDP by Country, 1980–2010
                  Major Exports of Indonesia, Malaysia, and
                  Vietnam

Chapter 33
Chapter Feature   Samoa's Trading Partners
                 Samoa: Before and After the Move
Lesson 3 Oceania: Water and Energy

Interactive Images and Slide Shows

Chapter 1
Lesson 2 Satellite Imagery

Chapter 4
Lesson 2 Times Square

Chapter 6
Lesson 1 Rock Formations
          St. Lawrence Seaway
Lesson 3 Hydroelectric Power

Chapter 7
Lesson 1 Sonoran Desert
Lesson 3 Challenges of Urbanization

Chapter 8
Chapter Feature   Columbian Exchange Trade

Chapter 9
Lesson 1 Landforms of Brazil
Lesson 3 South American Urbanization

Chapter 10
Lesson 1 Norwegian Landscape

Chapter 11
Chapter Feature   Paris the Megacity
Lesson 1 Windmill and Polders
Lesson 3 Air Pollution

Chapter 12
Lesson 1 Mediterranean Climate
Lesson 3 Costa Concordia

Chapter 13
Chapter Feature   Tito
Lesson 1 Karst Terrain
Lesson 3 Conservation in Eastern Europe

Chapter 14
Lesson 2 St. Basil's, the Kremlin, and Red Square

Chapter 15
Lesson 1 Mediterranean Climate and Agriculture

Chapter 16
Chapter Feature   Rabin, Arafat, and Clinton
Lesson 1 360° View: A Beach in Tel Aviv, Israel
Lesson 3 Oases

Chapter 17
Chapter Feature   The Kurds
Lesson 1 The Northeast's Physical Geography
Lesson 2 Foods of Ramadan
          Arabic Fast Food
Lesson 3 Life Along the Tigris and Euphrates

Chapter 19
Chapter Feature   Soviet Invasion of Afghanistan
Lesson 1 Ships Rest in what was the Aral Sea
Lesson 2 360° View: Samarkand, Uzbekistan
Lesson 3 The Aral Sea

Chapter 20
Lesson 1 Lake Chad

Chapter 21
Lesson 3 Animal Poaching in East Africa

Chapter 22
Lesson 1 West Africa's Variety of Land

Chapter 23
Lesson 1 The Congo River
Lesson 3 Harvest in Central Africa

Chapter 24
Lesson 1 360° View: Johannesburg, South Africa

Chapter 25
Lesson 3 Busy Street in Calcutta, India

Chapter 26
Lesson 3 Microloans in Bangladesh

Chapter 27
Lesson 1 Mount Everest

Chapter 28
Lesson 3 Floods in China

Chapter 29
Lesson 1 Mt. Fuji in Art
Lesson 2 Japan and the West

Chapter 31
Lesson 3 Teakwood

Chapter 32
Chapter Feature   Aboriginal and Maori Protests
Lesson 1 360° View: Australian Outback

Chapter 33
Lesson 1 360° View: Bora Bora in the South Pacific
Lesson 2 Fiji Hotel

netw⚙rks ONLINE RESOURCES

⌄ Interactive Maps

Chapter 2
Lesson 2 Tectonic Plate Boundaries

Chapter 4
Chapter Feature Expected Years of Schooling
                Gross National Income per Capita
                Life Expectancy at Birth
Lesson 2 Egypt Population Density
Lesson 3 Map of Terrorist Attacks
Lesson 4 World GDP
Lesson 5 Periphery Theory: Growth of the Suburbs

Chapter 5
Lesson 2 Foreign-Born Population as Percent of State
          Population, 2010
          U.S. Expansion

Chapter 6
Chapter Feature Canada's Natural Resources
Lesson 2 Canadian Explorations

Chapter 7
Chapter Feature Megacities of the World, 2010
Lesson 2 Civilizations of Mesoamerica

Chapter 8
Lesson 1 Sea Routes Before and After the Panama Canal
Lesson 2 Panama Canal Zone
          Voyages of Columbus, 1492–1502
Lesson 3 The 2010 Earthquake in Haiti

Chapter 9
Lesson 2 Cultures of South America, A.D. 700–1530
          European Colonies in Latin America, 1800

Chapter 11
Chapter Feature Population Density of Paris
Lesson 2 Division of Germany
          Europe After World War I 1919

Chapter 12
Lesson 2 Greek City-States, c. 500 B.C.

Chapter 13
Lesson 2 Major Nazi Death Camps

Chapter 14
Lesson 2 Russian Revolution and Civil War, 1917–1922
          Breakup of the Soviet Union 1991

Chapter 15
Chapter Feature Suez Canal
                Sea Routes Before and After the Suez Canal
Lesson 2 Egypt Population Density

Chapter 16
Chapter Feature Israel and Palestine

Chapter 18
Chapter Feature Migrant Workers in Saudi Arabia

Chapter 20
Chapter Feature Animist, Muslim, and Christian Believers in
                West Africa
Lesson 2 Crisis in Darfur
Lesson 3 Desertification of the Sahel

Chapter 21
Chapter Feature Resources in East Africa
Lesson 2 Independent Africa

Chapter 22
Chapter Feature Women in West Africa

Chapter 24
Chapter Feature HIV/AIDS in Africa

Chapter 25
Lesson 1 Monsoons in India
Lesson 2 Partition of India, 1947

Chapter 26
Chapter Feature Flooding in Bangladesh

Chapter 27
Chapter Feature China, India, Nepal: Boundary Disputes

Chapter 28
Lesson 1 China's River: Huang He

Chapter 29
Lesson 3 Fukushima's Nuclear Fallout

Chapter 30
Chapter Feature The Korean War

Chapter 31
Lesson 1 Volcanoes in Southeast Asia

Chapter 32
Lesson 2 Map of Australia and New Zealand
Lesson 3 How Invasive Species Get to Australia

Chapter 33
Lesson 2 Oceania in WWII

⌄ Primary Sources

Chapter 3
Lesson 3 Controversy Over Climate Change

Chapter 8
Chapter Feature Effects of the Columbian Exchange

Chapter 10
Lesson 3 Invasive Species in the Baltic

Chapter 11
Chapter Feature   Challenges in Paris's Suburbs

Chapter 14
Chapter Feature   Russia's Shrinking Population

Chapter 15
Chapter Feature   The Suez Crisis

Chapter 17
Chapter Feature   Saddam Hussein's Treatment of Kurds

Chapter 18
Chapter Feature   Problems Faced by Migrant Workers in
                  Saudi Arabia

Chapter 19
Chapter Feature   Afghanistan as Cultural Crossroads

Chapter 20
Chapter Feature   Religious Conflict and Cooperation in West Africa

Chapter 21
Chapter Feature   Tanzania's Cash Crops

Chapter 22
Chapter Feature   Ellen Johnson Sirleaf

Chapter 24
Chapter Feature   HIV/AIDS Prevention in Botswana

Chapter 29
Chapter Feature   Japan's Demographic Crisis

Time Lines

Chapter 5
Lesson 2 Terrorism and the United States

Chapter 6
Lesson 2 Canada: Expansion and Diversity

Chapter 7
Lesson 2 Mexican Independence and Change

Chapter 8
Lesson 2 Central America and the Caribbean: Paths to
          Independence

Chapter 9
Lesson 2 South America: Movements of Change

Chapter 10
Lesson 2 Northern Europe: Democracy and Independence

Chapter 11
Lesson 2 The Rise of Northwestern Europe

Chapter 12
Lesson 2 Southern Europe: Foundations of Western Civilization

Chapter 13
Chapter Feature   The Breakup of Yugoslavia
Lesson 2 Eastern Europe: The Road to a New Era

Chapter 14
Lesson 2 Soviet Era of the Russian Core

Chapter 15
Lesson 2 North Africa: Path to Independence

Chapter 16
Lesson 2 The Eastern Mediterranean:
          Israeli-Palestinian Conflict: An Elusive Peace Process

Chapter 17
Lesson 2 The Northeast: Independence and Turmoil

Chapter 18
Lesson 2 The Arabian Peninsula: Islam and Saudi Arabia

Chapter 19
Lesson 2 Central Asia: A Significant Crossroads

Chapter 20
Lesson 2 The Transition Zone: Droughts and Hunger

Chapter 21
Lesson 2 East Africa: A Cultural Crossroads

Chapter 22
Lesson 2 West Africa: Struggle for Power

Chapter 23
Lesson 2 Equatorial Africa: Conflict in the Congo

Chapter 24
Lesson 2 Southern Africa: Apartheid and Its Legacy

Chapter 25
Lesson 2 Women's Rights in India

Chapter 26
Lesson 2 Pakistan and Bangladesh: Unity and Division

Chapter 27
Chapter Feature   Nepal as a Buffer State
Lesson 2 Bhutan, Maldives, Nepal & Sri Lanka: Ethnic Strife

Chapter 28
Lesson 2 Modern China

Chapter 29
Lesson 2 Japan: Shifting Power

Chapter 30
Lesson 2 North Korea and South Korea: Focus of Rival Interests

Chapter 31
Lesson 2 Southeast Asia: Colonization and Independence

Chapter 32
Lesson 2 Australia and New Zealand: Migration and Settlement

Chapter 33
Lesson 2 Oceania: Colonization and Independence

How do I study Geography?

Geographers have tried to understand the best way to teach and learn about geography. In order to do this, geographers created the Five Themes of Geography. The themes acted as a guide for teaching the basic ideas about geography to students like yourself.

People who teach and study geography, though, thought that the Five Themes were too broad. In 1994, geographers created 18 national geography standards, *Geography for Life*. These standards were more detailed about what should be taught and learned. The Six Essential Elements act as a bridge connecting the Five Themes with the standards.

These pages show you how the Five Themes are related to the Six Essential Elements and the 18 standards.

5 Themes of Geography

1 Location

Location describes where something is. Absolute location describes a place's exact position on the Earth's surface. Relative location expresses where a place is in relation to another place.

2 Place

Place describes the physical and human characteristics that make a location unique.

3 Regions

Regions are areas that share common characteristics.

4 Movement

Movement explains how and why people and things move and are connected.

5 Human-Environment Interaction

Human-Environment Interaction describes the relationship between people and their environment.

(t to b) Thorsten Henn/Getty Images, David S. Boyer/National Geographic/Getty Images, John Lamb/The Image Bank/Getty Images, Glow Images

6 Essential Elements

18 Geography Standards

I. The World in Spatial Terms

Geographers look to see where a place is located. Location acts as a starting point to answer "Where Is It?" The location of a place helps you orient yourself as to where you are.

1 How to use maps and other tools

2 How to use mental maps to organize information

3 How to analyze the spatial organization of people, places, and environments

II. Places and Regions

Place describes physical characteristics such as landforms, climate, and plant or animal life. It might also describe human characteristics, including language and way of life. Places can also be organized into regions. Regions are places united by one or more characteristics.

4 The physical and human characteristics of places

5 How people create regions to interpret Earth's complexity

6 How culture and experience influence people's perceptions of places and regions

III. Physical Systems

Geographers study how physical systems, such as hurricanes, volcanoes, and glaciers, shape the surface of the Earth. They also look at how plants and animals depend upon one another and their surroundings for their survival.

7 The physical processes that shape Earth's surface

8 The distribution of ecosystems on Earth's surface

9 The characteristics, distribution, and migration of human populations

10 The complexity of Earth's cultural mosaics

IV. Human Systems

People shape the world in which they live. They settle in certain places but not in others. An ongoing theme in geography is the movement of people, ideas, and goods.

11 The patterns and networks of economic interdependence

12 The patterns of human settlement

13 The forces of cooperation and conflict

14 How human actions modify the physical environment

V. Environment and Society

How does the relationship between people and their natural surroundings influence the way people live? Geographers study how people use the environment and how their actions affect the environment.

15 How physical systems affect human systems

16 The meaning, use, and distribution of resources

VI. The Uses of Geography

Knowledge of geography helps us understand the relationships among people, places, and environments over time. Applying geographic skills helps you understand the past and prepare for the future.

17 How to apply geography to interpret the past

18 How to apply geography to interpret the present and plan for the future

HOW TO USE THE ONLINE STUDENT EDITION

TO THE STUDENT
Welcome to McGraw-Hill Education's **Networks** online student learning center. Here you will access your Online Student Edition as well as many other learning resources.

1 LOGGING ON TO THE STUDENT LEARNING CENTER

Using your Internet browser, go to connected.mcgraw-hill.com.

Enter your username and password or

Create a new account using the redemption code your teacher gave you.

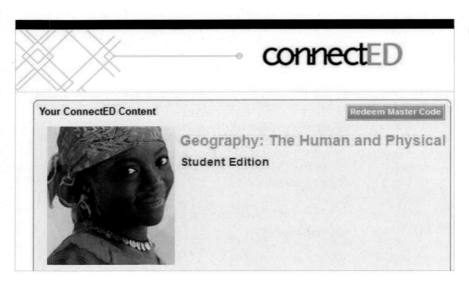

2 SELECT YOUR PROGRAM

Click your program to launch the home page of your Online Student Learning Center.

Using Your Home Page

1 HOME PAGE

To return to your home page at any time, click the Networks logo in the top left corner of the page.

2 QUICK LINKS MENU

Use this menu to access:

- My Notes (your personal notepad)
- Messages
- The online Glossary
- The online Atlas
- BTW (current events website for social studies)
- College and Career Readiness (CCR) materials

3 HELP

For videos and assistance with the various features of the Networks system, click Help.

4 MAIN MENU

Use the menu bar to access:

- The Online Student Edition
- Skills Builder (for activities to improve your skills)
- Assignments and Projects
- Resource Library
- Test Prep
- Collaborate with Classmates

5 ONLINE STUDENT EDITION

Go to your Online Student Edition by selecting the chapter and lesson and then click Go.

6 ASSIGNMENTS

Recent assignments from your teacher will appear here. Click the assignment or click See All to see the details.

7 RESOURCE LIBRARY

Use the carousel to browse the Resource Library.

8 MESSAGES

Recent messages from your teacher will appear here. To view the full message, click the message or click See All.

HOW TO USE THE ONLINE STUDENT EDITION

Using Your Online Student Edition

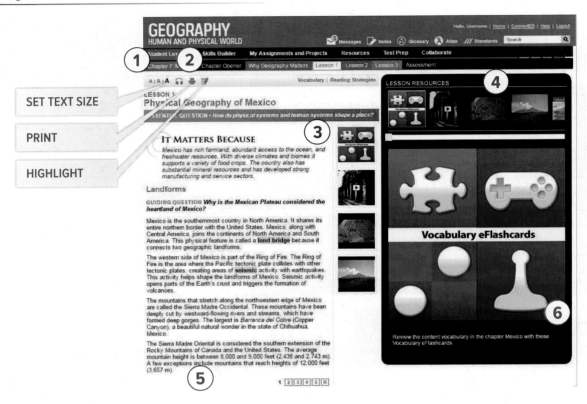

SET TEXT SIZE

PRINT

HIGHLIGHT

---

(1) LESSON MENU

• Use the tabs to open the different lessons and special features in a chapter or unit.

• Clicking on the unit or chapter title will open the table of contents.

(2) AUDIO EDITION

Click on the headphones symbol to have the page read to you. MP3 files for downloading each lesson are available in the Resource Library.

(3) RESOURCES FOR THIS PAGE

Resources appear in the middle column to show that they go with the text on this page. Click the images to open them in the viewer.

(4) LESSON RESOURCES

Use the carousel to browse the interactive resources available in this lesson. Click on a resource to open it in the viewer below.

(5) CHANGE PAGES

Click here to move to the next page in the lesson.

(6) RESOURCE VIEWER

Click on the image that appears in the viewer to launch an interactive resource, including:

• Lesson Videos
• Interactive Photos and Slide Shows
• Interactive Maps
• Interactive Charts and Graphs
• Games
• Self-Check Quizzes for each lesson

Reading Support in the Online Student Edition

Your Online Student Edition contains several features to help improve your reading skills and understanding of the content.

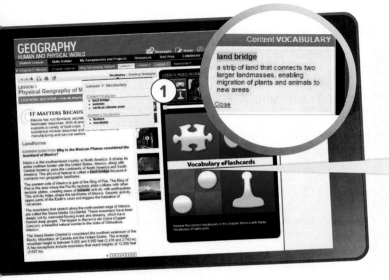

1 LESSON VOCABULARY

Click Vocabulary to bring up a list of terms introduced in this lesson.

VOCABULARY POP-UP

Click on any term highlighted in yellow to open a window with the term's definition.

2 MY NOTES

Click My Notes to open the note-taking tool. You can write and save any notes you want in the Lesson Notes tab.

Click on the Guided Notes tab to view the Guided Reading Questions. Answering these questions will help you build a set of notes about the lesson.

3 GRAPHIC ORGANIZER

Click Reading Strategies to open a note-taking activity using a graphic organizer.

Click the image of the graphic organizer to make it interactive. You can type directly into the graphic organizer and save or print your notes.

HOW TO USE THE ONLINE STUDENT EDITION

Using Interactive Resources in the Online Student Edition

Each lesson of your Online Student Edition contains many resources to help you learn the content and skills you need to know for this subject.

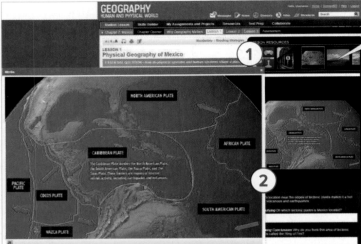

Networks provides many kinds of resources. This is an infographic.

1 LAUNCHING RESOURCES

Clicking a resource in the viewer launches an interactive resource.

2 QUESTIONS AND ACTIVITIES

When a resource appears in the viewer, there are usually 1 or 2 questions or activities beneath it. You can type and save your answers in the answer boxes and submit them to your teacher.

3 INTERACTIVE MAPS

When a map appears in the viewer, click on it to launch the interactive map. You can use the drawing tool to mark up the map. You can also zoom in and turn layers on and off to display different information. Drag the scale onto the map to measure distances. Many maps have animations and audio as well.

4 CHAPTER FEATURE

Each chapter begins with a feature called *Why Geography Matters.* Each feature highlights a topic in human geography relevant to the country or region of study. Each feature contains interactive digital assets such as maps, photos, slide shows, charts, graphs, and primary sources.

The map and other assets are interactive. You can click on each asset for an interactive version.

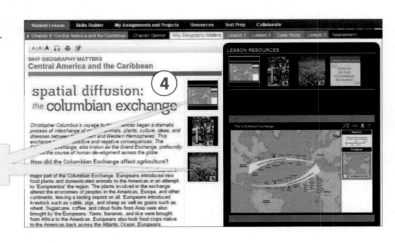

HOW TO USE THE ONLINE STUDENT EDITION

Assessment

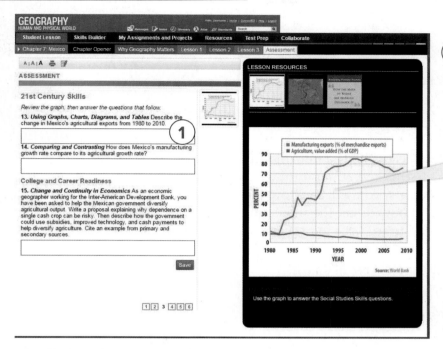

① CHAPTER ASSESSMENT

At the end of each chapter is the assessment tab. Here you can test your understanding of what you have learned. You can type and save answers in the answer boxes and submit them to your teacher.

When a question uses an image or graph or map, it will appear in the viewer.

Finding Other Resources

There are hundreds of additional resources available in the Resource Library. Click the Resources tab to enter the library.

② RESOURCE LIBRARY

Click the links to find collections of Primary Sources, Biographies, Skills Activities and the Reading Essentials and Study Guide.

You can search the Resource Library by keyword.

Click the star to mark a resource as a favorite.

SCAVENGER HUNT

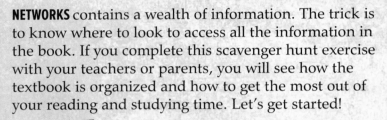

NETWORKS contains a wealth of information. The trick is to know where to look to access all the information in the book. If you complete this scavenger hunt exercise with your teachers or parents, you will see how the textbook is organized and how to get the most out of your reading and studying time. Let's get started!

1 How many chapters are in Unit 1?

2 What does Unit 3 cover?

3 Where can you find the Essential Question for each chapter?

4 Where can you find primary sources in your textbook?

5 How can you identify content vocabulary and academic vocabulary in the narrative?

6 Where do you find graphic organizers in your textbook?

7 You want to quickly find a map in the book about climate in Europe. Where do you look?

8 Where would you find the latitude and longitude for Dublin, Ireland?

9 If you needed to know the Spanish term for *primate city*, where would you look?

10 Where can you find a list of all the features in the book?

REFERENCE ATLAS

Geographic Dictionary RA2
World Time Zones RA4
World GDP: Cartogram RA6
World Population: Cartogram RA8
United States: Physical RA10
United States: Political RA12
North America: Physical RA14
North America: Political RA15
South America: Physical RA16
South America: Political RA17

Europe: Physical RA18
Europe: Political RA20
Africa: Physical RA22
Africa: Political RA23
Asia: Physical RA24
Asia: Political RA26
Oceania: Physical/Political RA28
Polar Regions RA30

ATLAS KEY

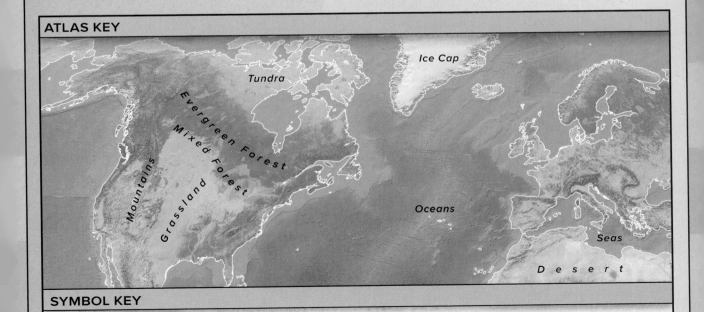

Ice Cap
Tundra
Evergreen Forest
Mixed Forest
Mountains
Grassland
Oceans
Seas
Desert

SYMBOL KEY

••••••• Claimed boundary	✪ National capital	Dry salt lake
—— International boundary (political map)	○ State/Provincial capital	Lake
—— International boundary (physical map)	• Towns	Rivers
	▼ Depression	Canal
	▲ Elevation	

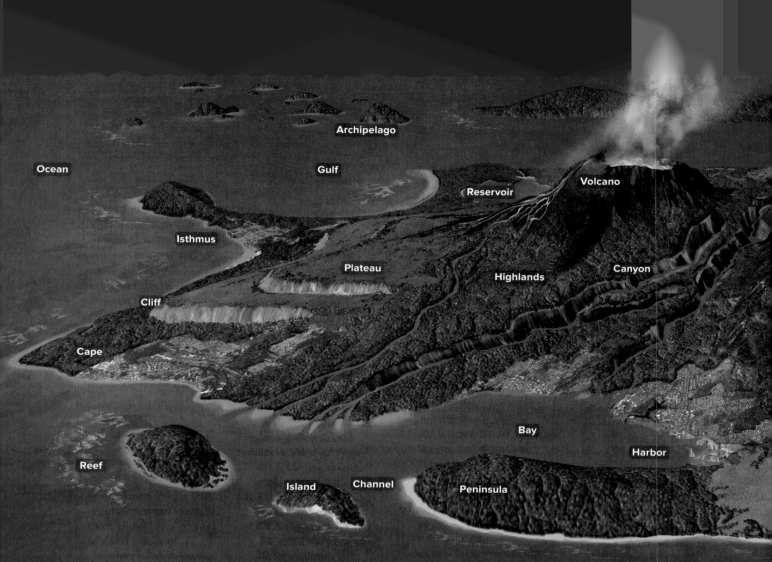

Archipelago
Ocean
Gulf
Reservoir
Volcano
Isthmus
Plateau
Highlands
Canyon
Cliff
Cape
Bay
Harbor
Reef
Island
Channel
Peninsula

archipelago a group of islands

basin area of land drained by a given river and its branches; area of land surrounded by lands of higher elevations

bay part of a large body of water that extends into a shoreline, generally smaller than a gulf

canyon deep and narrow valley with steep walls

cape point of land that extends into a river, lake, or ocean

channel wide strait or waterway between two land-masses that lie close to each other; deep part of a river or other waterway

cliff steep, high wall of rock, earth, or ice

continent one of the seven large landmasses on the Earth

delta flat, low-lying land built up from soil carried downstream by a river and deposited at its mouth

divide stretch of high land that separates river systems

downstream direction in which a river or stream flows from its source to its mouth

escarpment steep cliff or slope between a higher and lower land surface

glacier large, thick body of slowly moving ice

gulf part of a large body of water that extends into a shoreline, generally larger and more deeply indent-ed than a bay

harbor a sheltered place along a shoreline where ships can anchor safely

highland elevated land area such as a hill, mountain, or plateau

hill elevated land with sloping sides and rounded sum-mit; generally smaller than a mountain

island land area, smaller than a continent, completely surrounded by water

isthmus narrow stretch of land connecting two larger land areas

lake a sizable inland body of water

lowland land, usually level, at a low elevation

mesa broad, flat-topped landform with steep sides; smaller than a plateau

mountain land with steep sides that rises sharply (1,000 feet or more) from surrounding land; general-ly larger and more rugged than a hill

Mountain Peak

Desert

Oasis

Sound

Basin

Mountain Range

Source of River

Glacier

Tributary

Valley

Hills

Strait

Upstream

Lake

Downstream

River

Mouth of River

Escarpment

Lowland

Delta

Plain

Seacoast

mountain peak pointed top of a mountain

mountain range a series of connected mountains

mouth (of a river) place where a stream or river flows into a larger body of water

oasis small area in a desert where water and vegetation are found

ocean one of the four major bodies of salt water that surround the continents

ocean current stream of either cold or warm water that moves in a definite direction through an ocean

peninsula body of land jutting into a lake or ocean, surrounded on three sides by water

physical feature characteristic of a place occurring naturally, such as a landform, body of water, climate pattern, or resource

plain area of level land, usually at low elevation and often covered with grasses

plateau area of flat or rolling land at a high elevation, about 300 to 3,000 feet (90 to 900 m) high

reef a chain of rocks, coral or sand at or near the surface of the water

river large natural stream of water that runs through the land

sea large body of water completely or partly surrounded by land

seacoast land lying next to a sea or an ocean

sound broad inland body of water, often between a coastline and one or more islands off the coast

source (of a river) place where a river or stream begins, often in highlands

strait narrow stretch of water joining two larger bodies of water

tributary small river or stream that flows into a large river or stream; a branch of the river

upstream direction opposite the flow of a river; toward the source of a river or stream

valley area of low land usually between hills or mountains

volcano mountain or hill created as liquid rock and ash erupt from inside the Earth

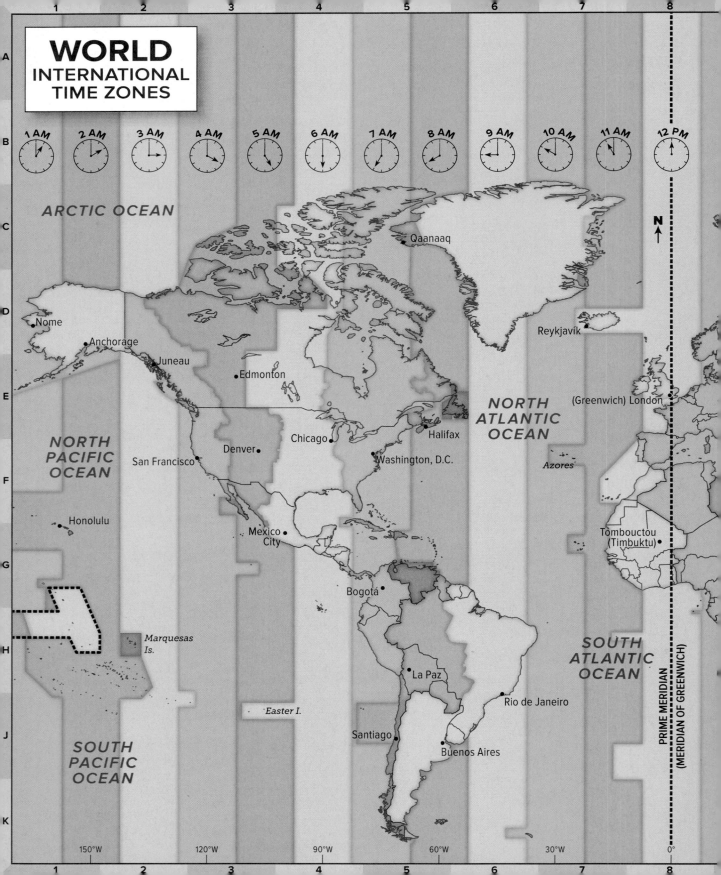

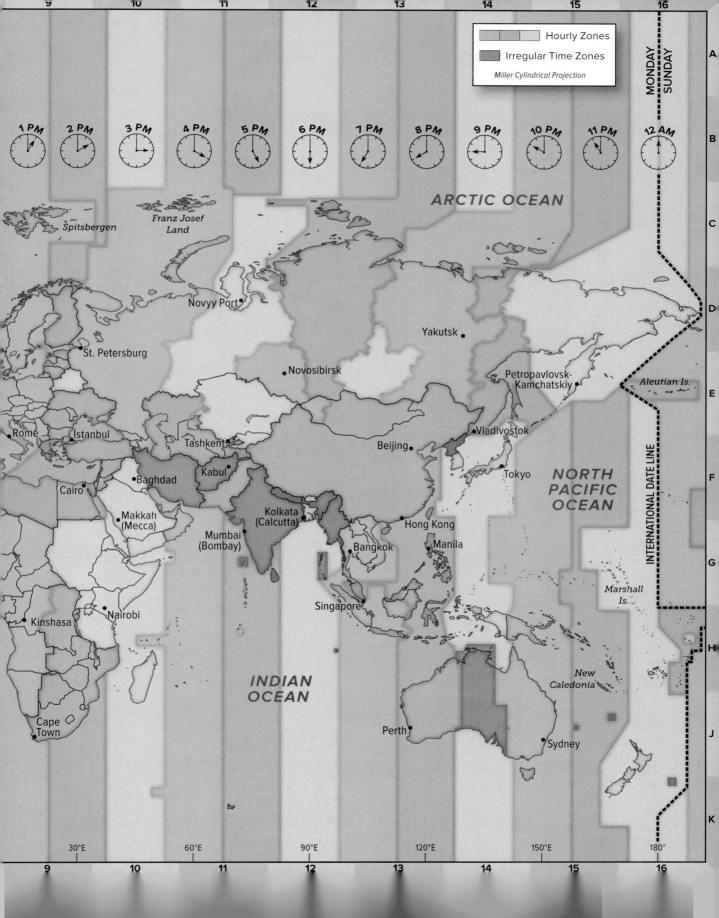

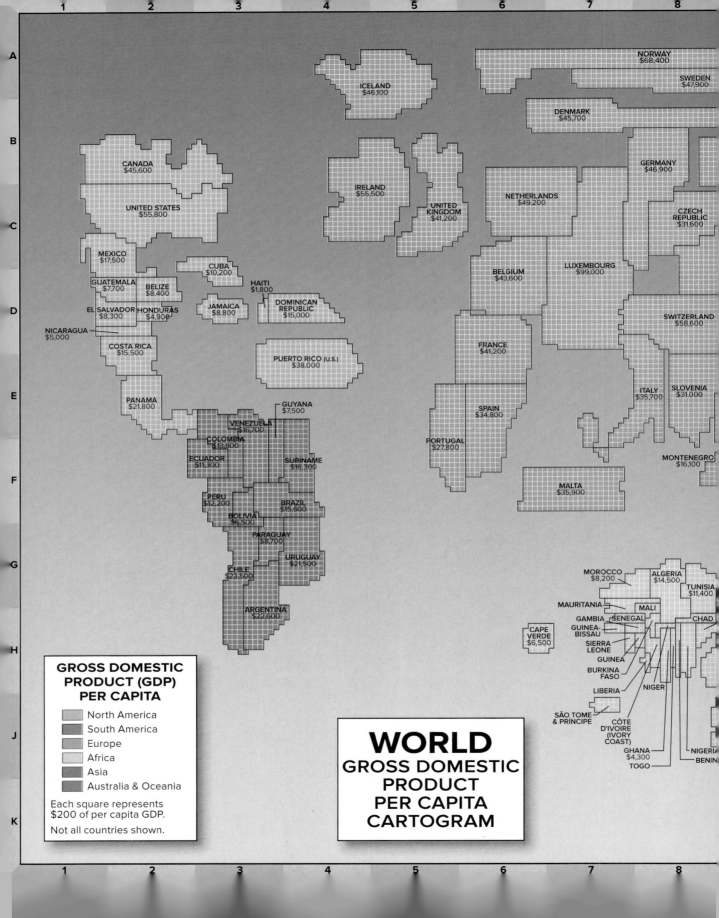

WORLD
GROSS DOMESTIC PRODUCT PER CAPITA CARTOGRAM

GROSS DOMESTIC PRODUCT (GDP) PER CAPITA

- North America
- South America
- Europe
- Africa
- Asia
- Australia & Oceania

Each square represents $200 of per capita GDP.

Not all countries shown.

NORWAY $68,400
SWEDEN $47,900
DENMARK $45,700
ICELAND $46,100
GERMANY $46,900
IRELAND $55,500
UNITED KINGDOM $41,200
NETHERLANDS $49,200
CZECH REPUBLIC $31,600
CANADA $45,600
UNITED STATES $55,800
BELGIUM $43,600
LUXEMBOURG $99,000
SWITZERLAND $58,600
MEXICO $17,500
CUBA $10,200
HAITI $1,800
GUATEMALA $7,700
BELIZE $8,400
JAMAICA $8,800
DOMINICAN REPUBLIC $15,000
FRANCE $41,200
ITALY $35,700
SLOVENIA $31,000
EL SALVADOR $8,300
HONDURAS $4,900
NICARAGUA $5,000
COSTA RICA $15,500
PUERTO RICO (U.S.) $38,000
SPAIN $34,800
MONTENEGRO $16,100
PANAMA $21,800
GUYANA $7,500
PORTUGAL $27,800
VENEZUELA $16,700
COLOMBIA $13,800
SURINAME $16,300
MALTA $35,900
ECUADOR $11,300
PERU $12,200
BRAZIL $15,600
BOLIVIA $6,500
PARAGUAY $8,700
URUGUAY $21,500
CHILE $23,500
ARGENTINA $22,600
MOROCCO $8,200
ALGERIA $14,500
TUNISIA $11,400
MAURITANIA
MALI
CAPE VERDE $6,500
GAMBIA
SENEGAL
GUINEA-BISSAU
CHAD
SIERRA LEONE
GUINEA
BURKINA FASO
NIGER
LIBERIA
SÃO TOME & PRÍNCIPE
CÔTE D'IVOIRE (IVORY COAST)
GHANA $4,300
NIGERIA
TOGO
BENIN

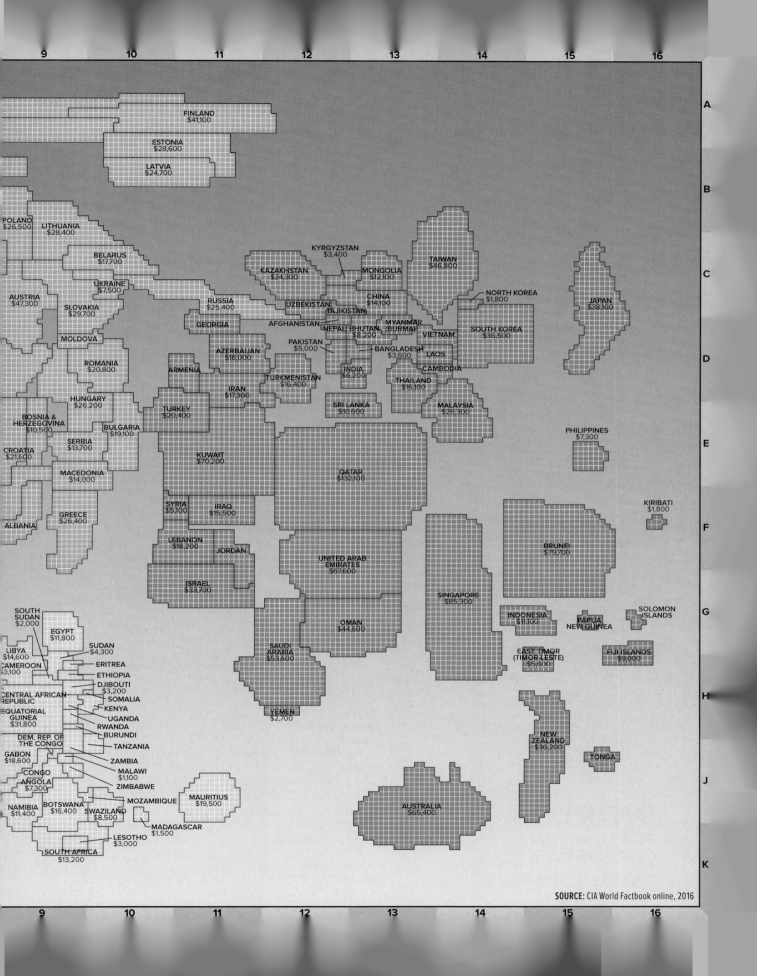

SOURCE: CIA World Factbook online, 2016

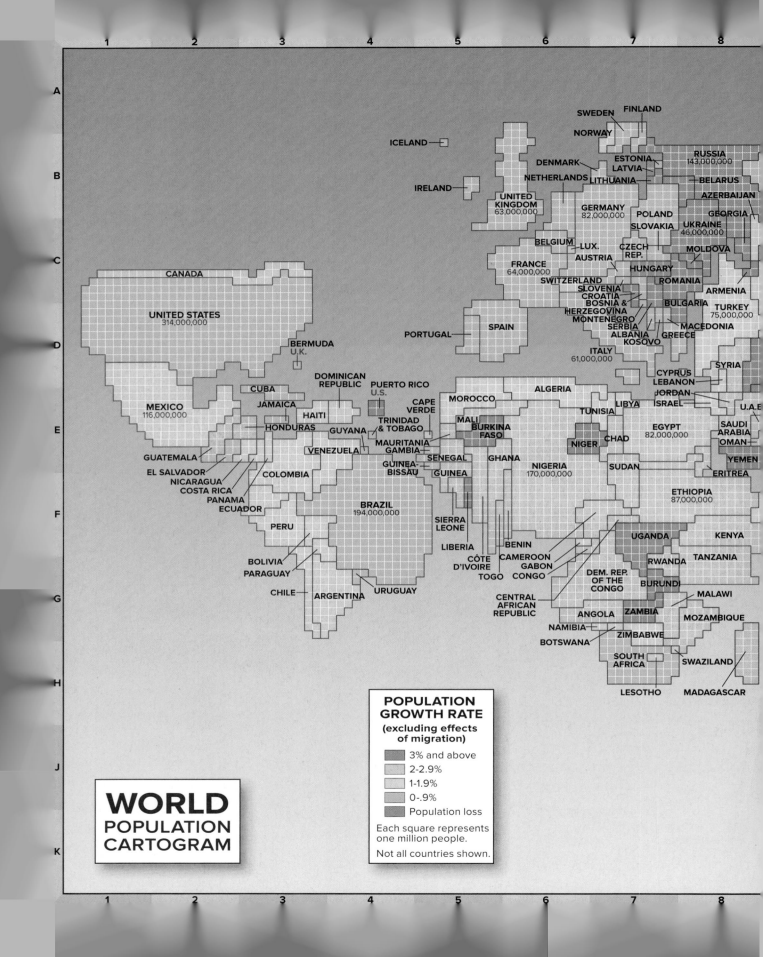

WORLD POPULATION CARTOGRAM

POPULATION GROWTH RATE
(excluding effects of migration)

- 3% and above
- 2-2.9%
- 1-1.9%
- 0-.9%
- Population loss

Each square represents one million people.

Not all countries shown.

Coordinates

1 2 3 4 5 6 7 8

A B C D E F G H J K

Country Labels

ICELAND

SWEDEN FINLAND NORWAY

DENMARK ESTONIA LATVIA RUSSIA 143,000,000 BELARUS AZERBAIJAN

IRELAND NETHERLANDS LITHUANIA GEORGIA

UNITED KINGDOM 63,000,000 GERMANY 82,000,000 POLAND SLOVAKIA UKRAINE 46,000,000 MOLDOVA

BELGIUM LUX. CZECH REP. AUSTRIA HUNGARY ARMENIA

FRANCE 64,000,000 SWITZERLAND SLOVENIA CROATIA BOSNIA & HERZEGOVINA MONTENEGRO SERBIA ALBANIA KOSOVO ROMANIA BULGARIA MACEDONIA GREECE TURKEY 75,000,000

CANADA

UNITED STATES 314,000,000

BERMUDA U.K.

PORTUGAL SPAIN ITALY 61,000,000 CYPRUS LEBANON JORDAN ISRAEL SYRIA U.A.E

DOMINICAN REPUBLIC PUERTO RICO U.S. ALGERIA MOROCCO

CUBA JAMAICA MEXICO 116,000,000 HAITI HONDURAS GUYANA CAPE VERDE TRINIDAD & TOBAGO VENEZUELA MALI BURKINA FASO TUNISIA LIBYA EGYPT 82,000,000 SAUDI ARABIA OMAN YEMEN

GUATEMALA EL SALVADOR NICARAGUA COSTA RICA PANAMA ECUADOR COLOMBIA MAURITANIA GAMBIA SENEGAL GUINEA-BISSAU GUINEA GHANA NIGER CHAD NIGERIA 170,000,000 SUDAN ERITREA

PERU BRAZIL 194,000,000 SIERRA LEONE LIBERIA ETHIOPIA 87,000,000

BOLIVIA PARAGUAY CÔTE D'IVOIRE BENIN CAMEROON GABON CONGO TOGO DEM. REP. OF THE CONGO UGANDA RWANDA KENYA TANZANIA

CHILE ARGENTINA URUGUAY CENTRAL AFRICAN REPUBLIC BURUNDI MALAWI

NAMIBIA ANGOLA ZAMBIA MOZAMBIQUE BOTSWANA ZIMBABWE

SOUTH AFRICA SWAZILAND LESOTHO MADAGASCAR

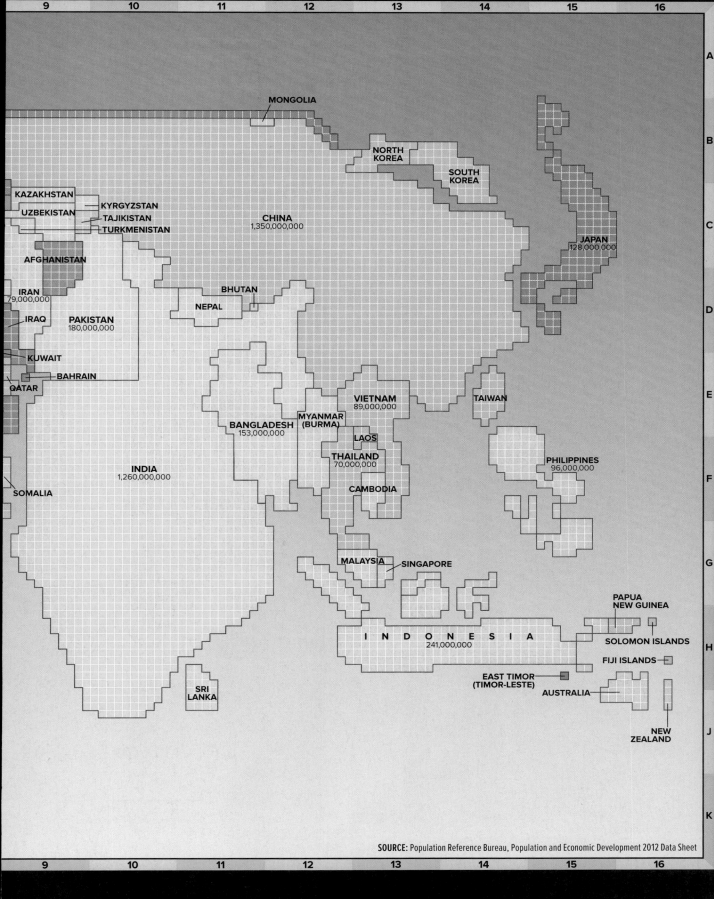

| 9 | 10 | 11 | 12 | 13 | 14 | 15 | 16 |

MONGOLIA

NORTH
KOREA

SOUTH
KOREA

JAPAN
128,000,000

KAZAKHSTAN

KYRGYZSTAN

UZBEKISTAN

TAJIKISTAN

TURKMENISTAN

CHINA
1,350,000,000

AFGHANISTAN

IRAN
79,000,000

BHUTAN

NEPAL

IRAQ

PAKISTAN
180,000,000

KUWAIT

BAHRAIN

QATAR

VIETNAM
89,000,000

TAIWAN

MYANMAR
(BURMA)

BANGLADESH
153,000,000

LAOS

THAILAND
70,000,000

PHILIPPINES
96,000,000

INDIA
1,260,000,000

CAMBODIA

SOMALIA

MALAYSIA

SINGAPORE

PAPUA
NEW GUINEA

INDONESIA
241,000,000

SOLOMON ISLANDS

FIJI ISLANDS

EAST TIMOR
(TIMOR-LESTE)

SRI
LANKA

AUSTRALIA

NEW
ZEALAND

SOURCE: Population Reference Bureau, Population and Economic Development 2012 Data Sheet

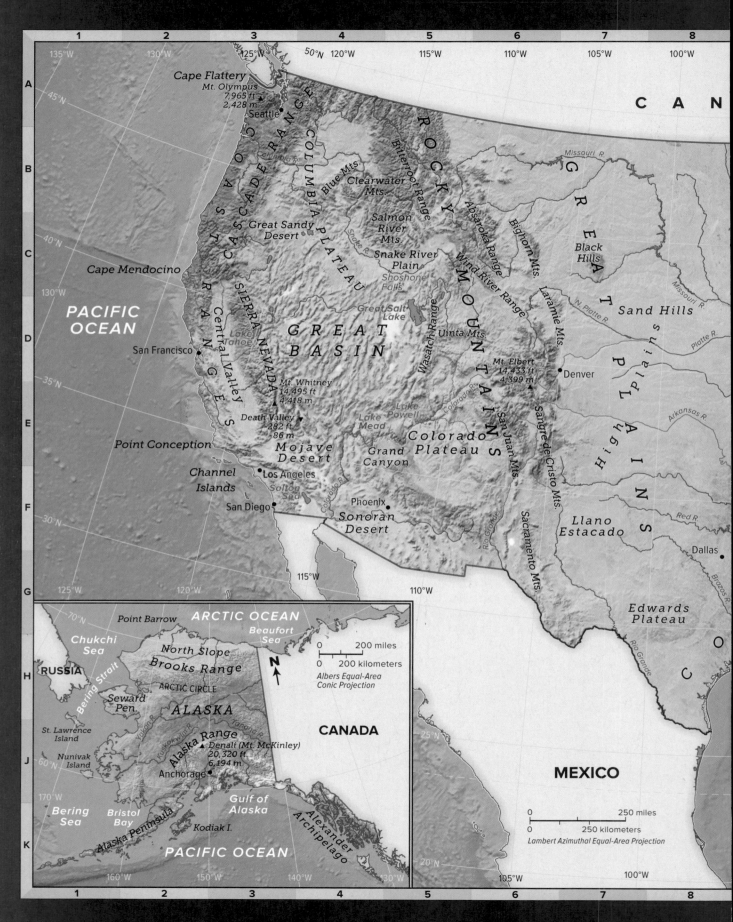

PACIFIC OCEAN

CANADA

CASCADE RANGE
COAST RANGES
SIERRA NEVADA
COLUMBIA PLATEAU
ROCKY MOUNTAINS
GREAT PLAINS

Cape Flattery
Mt. Olympus
7,965 ft.
2,428 m
Seattle

Cape Mendocino

San Francisco

Point Conception

Channel
Islands

San Diego

Los Angeles

Great Sandy
Desert

Blue Mts.

Clearwater
Mts.

Salmon
River
Mts.

Snake River
Plain

Shoshone
Falls

Great Salt
Lake

GREAT
BASIN

Central Valley

Lake
Tahoe

Mt. Whitney
14,495 ft
4,418 m

Death Valley
-282 ft.
-86 m

Mojave
Desert

Salton
Sea

Phoenix

Sonoran
Desert

Colorado R.

Lake
Mead

Lake
Powell

Grand
Canyon

Colorado
Plateau

Wasatch Range

Uinta Mts.

Colorado R.

Absaroka Range

Wind River Range

Bighorn Mts.

Laramie Mts.

San Juan Mts.

Sangre de Cristo Mts.

Sacramento Mts.

Mt. Elbert
14,433 ft
4,399 m

Denver

Black
Hills

Sand Hills

High Plains

Llano
Estacado

Edwards
Plateau

Dallas

Missouri R.

Missouri R.

N. Platte R.

Platte R.

Arkansas R.

Red R.

Rio Grande

Brazos R.

Rio Grande

MEXICO

135°W   130°W   125°W   50°N   120°W   115°W   110°W   105°W   100°W

45°N
40°N
35°N
30°N

130°W
125°W

115°W   110°W

25°N

20°N

105°W   100°W

Inset map (Alaska):

ARCTIC OCEAN

Point Barrow

Beaufort
Sea

Chukchi
Sea

RUSSIA

Bering Strait

St. Lawrence
Island

Nunivak
Island

Seward
Pen.

North Slope

Brooks Range

ARCTIC CIRCLE

ALASKA

Alaska Range
Denali (Mt. McKinley)
20,320 ft.
6,194 m

Anchorage

Kuskokwim R.

Yukon R.

Tanana R.

CANADA

Bering
Sea

Bristol
Bay

Alaska Peninsula

Kodiak I.

Gulf of
Alaska

Alexander
Archipelago

PACIFIC OCEAN

70°N
60°N

170°W   160°W   150°W   140°W   130°W

0   200 miles
0   200 kilometers
Albers Equal-Area
Conic Projection

N

0   250 miles
0   250 kilometers
Lambert Azimuthal Equal-Area Projection

UNITED STATES
PHYSICAL

PRINCIPAL HAWAIIAN ISLANDS

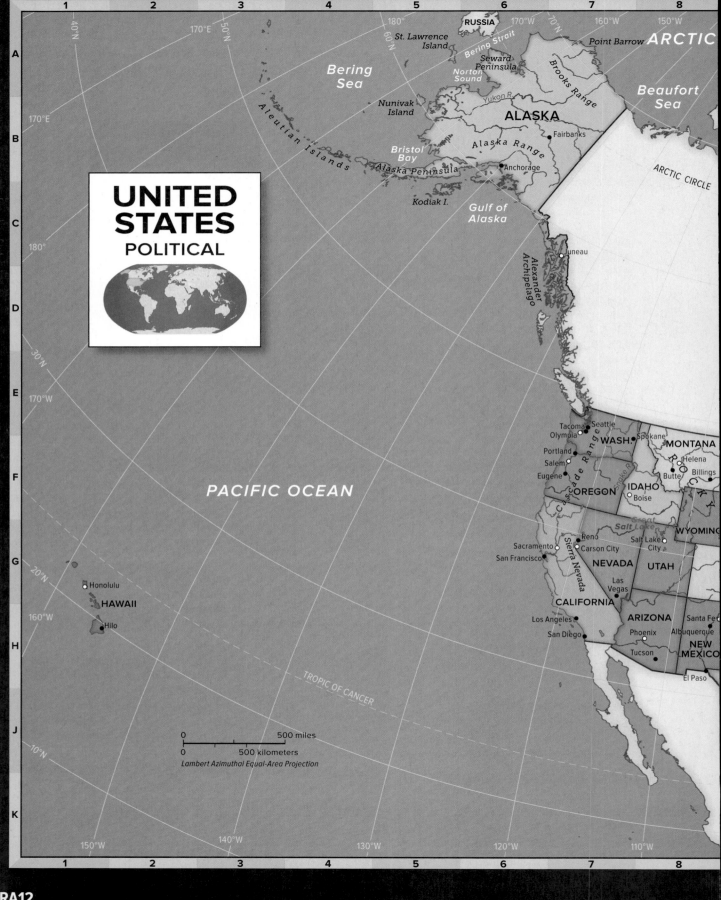

UNITED STATES
POLITICAL

RUSSIA

ARCTIC

Bering Strait Point Barrow

Bering Sea

St. Lawrence Island

Seward Peninsula

Norton Sound

Beaufort Sea

Nunivak Island

Brooks Range

Yukon R.

ALASKA

Aleutian Islands

Fairbanks

Bristol Bay

Alaska Range

ARCTIC CIRCLE

Alaska Peninsula

Anchorage

Kodiak I.

Gulf of Alaska

Alexander Archipelago

Juneau

PACIFIC OCEAN

Tacoma Seattle
Olympia Spokane
WASH. **MONTANA**
Portland Helena
Salem Billings
Eugene Butte
OREGON **IDAHO**
Boise **WYOMING**
Salt Lake
Reno Salt Lake City
Sacramento Carson City
San Francisco **NEVADA** **UTAH**
Sierra Nevada
Las Vegas
CALIFORNIA
Los Angeles **ARIZONA** Santa Fe
San Diego Phoenix Albuquerque
NEW MEXICO
Tucson
El Paso

Honolulu

HAWAII

Hilo

TROPIC OF CANCER

170°E
180°
170°E
180°
30°N
170°W
20°N
160°W
10°N
40°N
50°N
60°N
180°
170°W
70°N
160°W
150°W
150°W
140°W
130°W
120°W
110°W

0 — 500 miles
0 — 500 kilometers
Lambert Azimuthal Equal-Area Projection

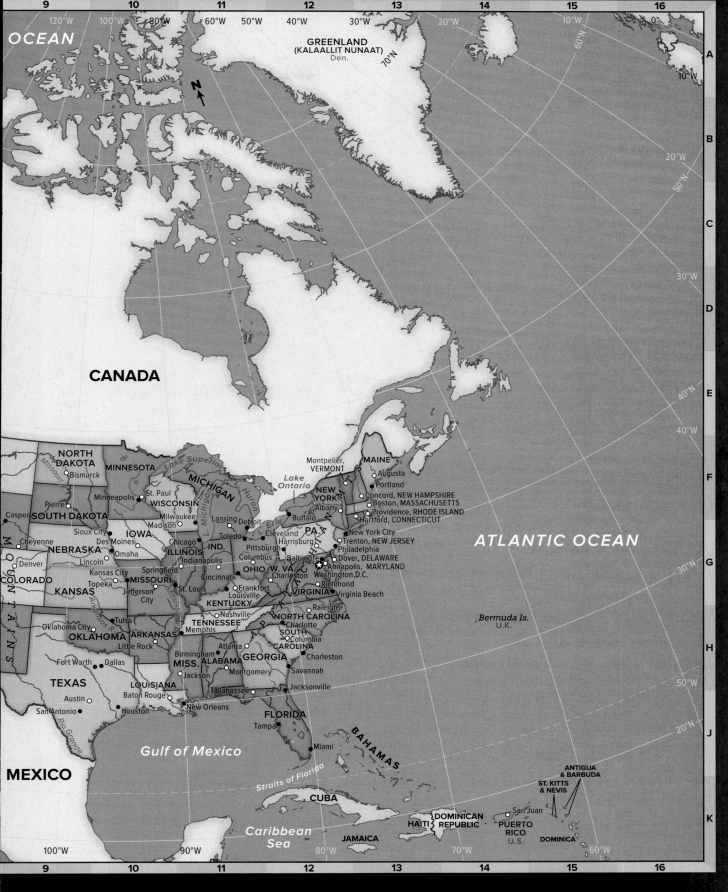

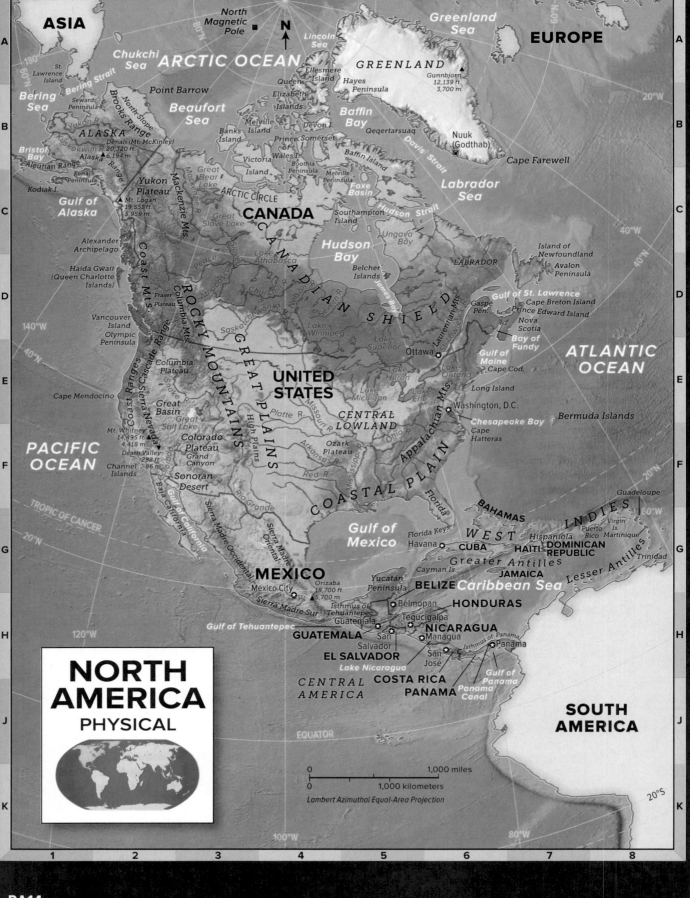

NORTH AMERICA
PHYSICAL

ASIA

EUROPE

ARCTIC OCEAN

North Magnetic Pole

N

Chukchi Sea

Lincoln Sea

GREENLAND

Greenland Sea

Gunnbjorn 12,139 ft. 3,700 m

Point Barrow

Beaufort Sea

Ellesmere Island

Queen Elizabeth Islands

Hayes Peninsula

Bering Strait

Seward Peninsula

Brooks Range

North Slope

ALASKA

Denali (Mt. McKinley) 20,320 ft. 6,194 m

Melville Island

Banks Island

Devon I.

Prince of Wales I.

Somerset I.

Baffin Bay

Qeqertarsuaq

Nuuk (Godthab)

Bering Sea

St. Lawrence Island

Bristol Bay

Aleutian Range

Kenai Peninsula

Kodiak I.

Gulf of Alaska

Mt. Logan 19,551 ft. 5,959 m

Yukon Plateau

Alaska Range

Mackenzie Mts.

Great Bear Lake

Victoria Island

Boothia Peninsula

Melville Peninsula

Foxe Basin

ARCTIC CIRCLE

Baffin Island

Hudson Strait

Davis Strait

Cape Farewell

Labrador Sea

Island of Newfoundland

Avalon Peninsula

Alexander Archipelago

Haida Gwaii (Queen Charlotte Islands)

Coast Mts.

ROCKY MOUNTAINS

Great Slave Lake

CANADA

CANADIAN SHIELD

Southampton Island

Hudson Bay

Ungava Bay

LABRADOR

Belcher Islands

James Bay

Gulf of St. Lawrence

Cape Breton Island

Prince Edward Island

Nova Scotia

Gaspé Pen.

Vancouver Island

Olympic Peninsula

Fraser Plateau

Columbia Plateau

Peace R.

Athabasca R.

Lake Athabasca

Churchill R.

Nelson R.

Saskatchewan R.

Severn R.

Lake Winnipeg

Lake Superior

Laurentian Mts.

Sudbury

Ottawa

Bay of Fundy

Gulf of Maine

ATLANTIC OCEAN

Cape Mendocino

Coast Ranges

Cascade Range

Sierra Nevada

Great Basin

Great Salt Lake

Columbia Plateau

Snake R.

UNITED STATES

CENTRAL LOWLAND

Lake Michigan

Lake Huron

Lake Erie

Lake Ontario

Appalachian Mts.

Washington, D.C.

Long Island

Cape Cod

Chesapeake Bay

Cape Hatteras

Bermuda Islands

PACIFIC OCEAN

Mt. Whitney 14,495 ft. 4,418 m

Death Valley -282 ft. -86 m

Channel Islands

Colorado Plateau

Grand Canyon

Sonoran Desert

High Plains

Platte R.

Missouri R.

Arkansas R.

Red R.

Ozark Plateau

Mississippi R.

Ohio R.

COASTAL PLAIN

Florida

GREAT PLAINS

TROPIC OF CANCER

Baja California

Gulf of California

Sierra Madre Occidental

Rio Grande

Sierra Madre Oriental

Gulf of Mexico

Florida Keys

Havana

CUBA

BAHAMAS

WEST INDIES

Hispaniola

HAITI

DOMINICAN REPUBLIC

Greater Antilles

JAMAICA

Cayman Is.

Guadeloupe

Puerto Rico

Virgin Is.

Martinique

Lesser Antilles

Caribbean Sea

Trinidad

MEXICO

Mexico City

Orizaba 18,700 ft. 5,700 m

Yucatan Peninsula

BELIZE

Belmopan

HONDURAS

Tegucigalpa

Sierra Madre Sur

Isthmus of Tehuantepec

Guatemala

Gulf of Tehuantepec

GUATEMALA

San Salvador

EL SALVADOR

NICARAGUA

Managua

Lake Nicaragua

COSTA RICA

San José

PANAMA

Isthmus of Panama

Panama

Panama Canal

Gulf of Panama

CENTRAL AMERICA

EQUATOR

SOUTH AMERICA

| 0 | 1,000 miles |
| 0 | 1,000 kilometers |

Lambert Azimuthal Equal-Area Projection

RA14

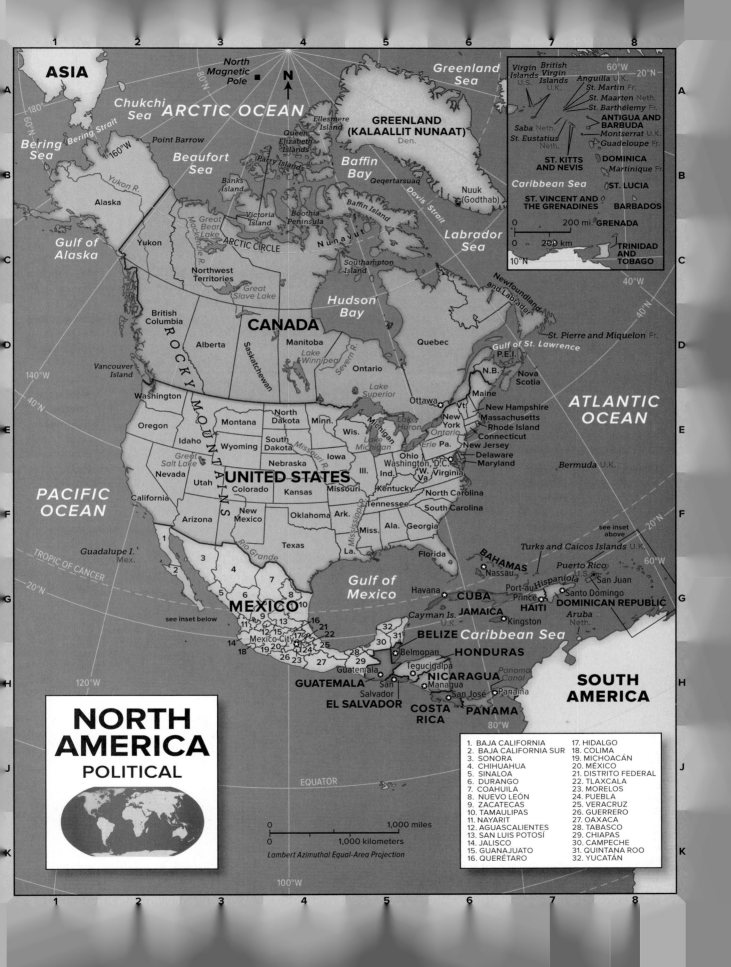

SOUTH AMERICA
PHYSICAL

Caribbean Sea

N

0   1,000 miles
0   1,000 kilometers
Lambert Azimuthal Equal-Area Projection

Caracas

VENEZUELA

Lake Maracaibo

Orinoco R.

GUYANA

SURINAME

Georgetown   Paramaribo

Bogotá

Angel Falls
Total drop
3,212 ft.   979 m

Cayenne

FRENCH GUIANA

GUIANA HIGHLANDS

Malpelo I.

COLOMBIA

Río Negro

Boundary claimed
by Suriname

Marajó
Island

Quito

ECUADOR

EQUATOR

0°

A M A Z O N

Amazon R.

Marañon R.

Amazon R.

B A S I N

S e l v a s

Purus R.

Madeira R.

Tapajos R.

Xingu R.

PERU

Ucayali R.

BRAZIL

Araguaia R.

Tocantins R.

São Francisco R.

*MATO GROSSO
PLATEAU*

B R A Z I L I A N

Lima

Machu
Picchu

Lake
Titicaca

La Paz

BOLIVIA

Brasília

H I G H L A N D S

Altiplano

Sucre

G R A N

Paraguay R.

Salar
de Uyuni

C H A C O

Paraguay R.

Paraná R.

20°S

TROPIC OF CAPRICORN

PARAGUAY

*Iguazú
Falls*

**ATLANTIC
OCEAN**

San Ambrosio I.

Asunción

San Félix I.

Paraná R.

P A M P A S

Uruguay R.

CHILE

Aconcagua
22,834 ft.
6,960 m

A N D E S

Uruguay R.

Juan Fernández Is.

Santiago

Buenos Aires

URUGUAY

Montevideo

ARGENTINA

Río de la Plata

Colorado R.

P A T A G O N I A

Negro R.

40°S

Chiloé Island

Valdés Peninsula
-131 ft.
-40 m

Gulf of
San Jorge

**PACIFIC
OCEAN**

Taitao
Peninsula

Wellington I.

Laguna
del Carbón
-344 ft.
-105 m

*Falkland Islands
(Islas Malvinas)*

Stanley

Tierra del Fuego

Strait of
Magellan

Cape Horn

South Georgia Island

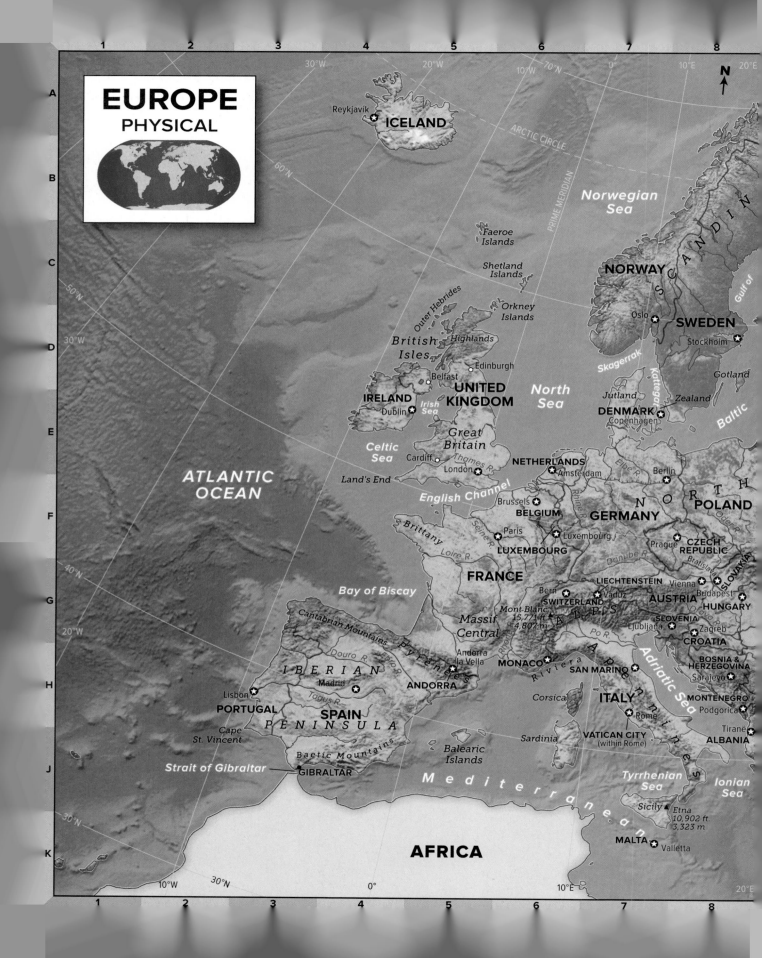

EUROPE
PHYSICAL

N

ARCTIC CIRCLE

Reykjavík
ICELAND

Norwegian Sea

Faeroe Islands

Shetland Islands

NORWAY

Outer Hebrides

Orkney Islands

British Isles

Highlands

Edinburgh

Oslo

SWEDEN

Stockholm

Gotland

Belfast

UNITED KINGDOM

North Sea

Skagerrak

Kattegat

Jutland

Zealand

IRELAND

Irish Sea

Dublin

DENMARK

Copenhagen

Baltic

Celtic Sea

Great Britain

Cardiff

Thames R.

London

NETHERLANDS

Amsterdam

Elbe R.

Berlin

N O R T H

POLAND

ATLANTIC OCEAN

Land's End

English Channel

Brussels

BELGIUM

Rhine R.

GERMANY

Brittany

Seine R.

Paris

Luxembourg.

LUXEMBOURG

Prague

CZECH REPUBLIC

Oder R.

Bratislava

SLOVAKIA

Loire R.

FRANCE

Danube R.

Bay of Biscay

Bern

SWITZERLAND

Vaduz

LIECHTENSTEIN

Vienna

AUSTRIA

Budapest

HUNGARY

Cantabrian Mountains

Mont-Blanc 15,771 ft. 4,807 m

A L P S

Massif Central

Ljubljana

SLOVENIA

Zagreb

Douro R.

Pyrenees

Andorra la Vella

Po R.

CROATIA

MONACO

Riviera

SAN MARINO

BOSNIA & HERZEGOVINA

Sarajevo

I B E R I A N

Madrid

ANDORRA

A p e n n i n e s

Adriatic Sea

MONTENEGRO

Podgorica

Lisbon

Corsica

ITALY

Rome

Tiranë

PORTUGAL

Tagus R.

SPAIN

P E N I N S U L A

Cape St. Vincent

VATICAN CITY (within Rome)

ALBANIA

Baetic Mountains

Sardinia

Balearic Islands

M e d i t e r r a n e a n

Tyrrhenian Sea

Ionian Sea

Strait of Gibraltar

GIBRALTAR

Sicily

Etna 10,902 ft. 3,323 m

MALTA

Valletta

AFRICA

30°W

20°W

70°N

10°W

0°

PRIME MERIDIAN

10°E

20°E

60°N

50°N

40°N

30°W

20°W

30°N

10°W

30°N

0°

10°E

20°E

S C A N D I N

Gulf of

Map labels

North Cape

Barents Sea

LAPLAND

Kola Peninsula

White Sea

Pechora R.

URAL MOUNTAINS

Europe/Asia boundary

ASIA

FINLAND

Lake Region

Northern Dvina R.

EUROPEAN PLAIN

RUSSIA

Bothnia

Lake Onega

Lake Ladoga

Helsinki

Gulf of Finland

Tallinn

ESTONIA

Sea

LATVIA

Riga

Moscow

Ural R.

LITHUANIA

Vilnius

Minsk

CENTRAL

RUSSIA

KAZAKHSTAN

RUSSIA

BELARUS

Volga R.

RUSSIAN

Caspian Depression

Warsaw

Don R.

UPLAND

Vistula R.

(Kyiv) Kiev

Dnieper R.

Dniester R.

UKRAINE

Carpathian Mts.

MOLDOVA

Sea of Azov

Caspian Sea

Tisza R.

Chișinău

Crimea

Mt. Elbrus 18,510 ft. 5,642 m

Caucasus Mountains

ROMANIA

Belgrade

Bucharest

Danube R.

AZERBAIJAN

SERBIA

GEORGIA

Baku

BALKAN

KOSOVO

Balkan Mts.

Black Sea

Pristina

Sofia

BULGARIA

Skopje

MACEDONIA

PENINSULA

TURKEY

Bosporus

GREECE

Dardanelles

Sea of Marmara

Athens

Aegean Sea

Peloponnese

Rhodes

Nicosia

ASIA

Crete

Sea

CYPRUS

400 miles

400 kilometers

Lambert Azimuthal Equal-Area Projection

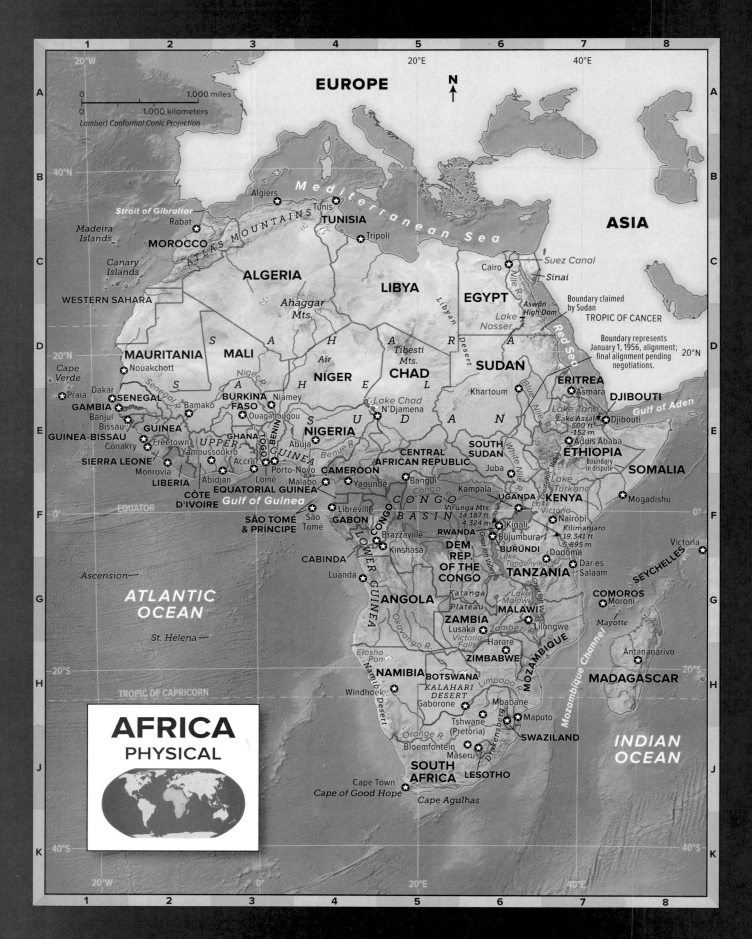

AFRICA
PHYSICAL

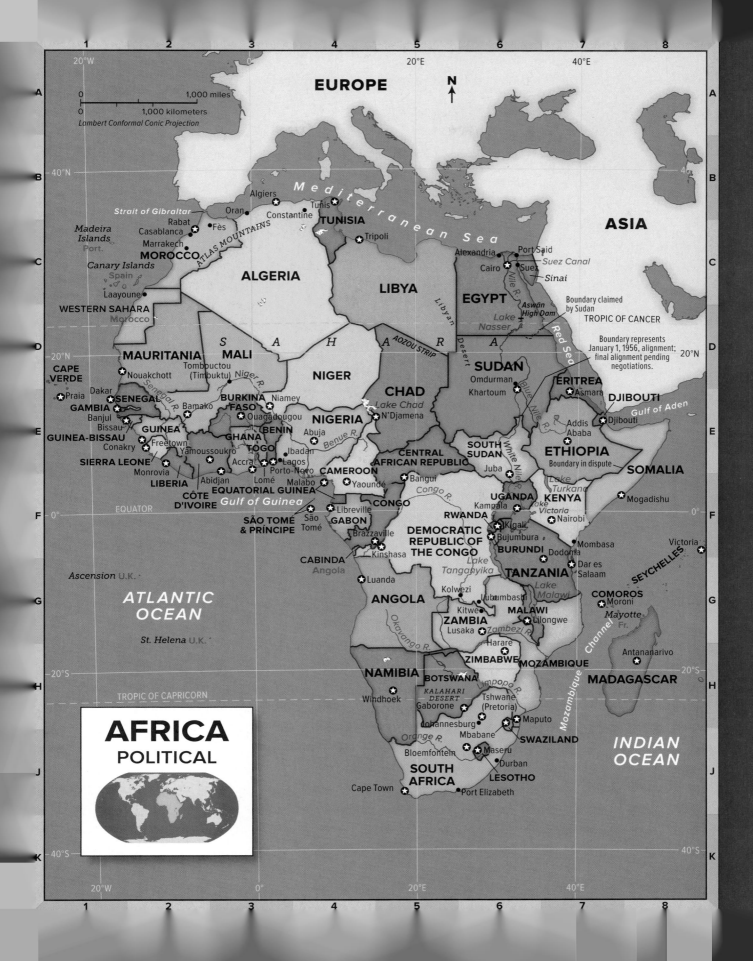

AFRICA
POLITICAL

Lambert Conformal Conic Projection

0   1,000 miles
0   1,000 kilometers

EUROPE

N

ASIA

Mediterranean Sea

Strait of Gibraltar

Algiers
Oran   Tunis
Constantine
TUNISIA
Tripoli
Rabat
Fès
Casablanca
Marrakech
MOROCCO

Madeira Islands Port.

Canary Islands Spain

Laayoune

WESTERN SAHARA
Morocco

ALGERIA

LIBYA

EGYPT

Alexandria
Port Said
Cairo   Suez
Suez Canal
Sinai

Aswān High Dam
Lake Nasser

Nile R.

Boundary claimed by Sudan
TROPIC OF CANCER

Boundary represents January 1, 1956, alignment; final alignment pending negotiations.

CAPE VERDE
Praia

MAURITANIA
Nouakchott

MALI

S   A   H   A   R   A

Tombouctou (Timbuktu)

AOZOU STRIP

Libyan Desert

NIGER

CHAD

SUDAN
Omdurman
Khartoum

ERITREA
Asmara
DJIBOUTI
Gulf of Aden
Djibouti

Red Sea
Blue Nile R.

20°N

Dakar
Senegal R.
SENEGAL
GAMBIA
Banjul
Bissau
GUINEA-BISSAU

Niger R.
Bamako
BURKINA FASO
Niamey
Ouagadougou

Lake Chad
N'Djamena
NIGERIA
Abuja

Benue R.

CENTRAL AFRICAN REPUBLIC

SOUTH SUDAN
Juba

White Nile

Addis Ababa
ETHIOPIA
Boundary in dispute

SOMALIA
Mogadishu

GUINEA
Conakry
Freetown
SIERRA LEONE
Monrovia
LIBERIA

GHANA
Yamoussoukro
TOGO
Accra
Abidjan
CÔTE D'IVOIRE

BENIN
Ibadan
Lagos
Porto-Novo
Lomé
Malabo
EQUATORIAL GUINEA

CAMEROON
Yaoundé

Bangui
Congo R.

UGANDA
Kampala
Lake Victoria
KENYA
Nairobi

Lake Turkana

SÃO TOMÉ & PRÍNCIPE
São Tomé

GABON
Libreville

CONGO

RWANDA
Kigali
BURUNDI
Bujumbura

Dodoma
Mombasa

SEYCHELLES
Victoria

CABINDA
Angola

Brazzaville
Kinshasa

DEMOCRATIC REPUBLIC OF THE CONGO

Lake Tanganyika

TANZANIA
Dar es Salaam

Lake Malawi

COMOROS
Moroni

Mayotte Fr.

Ascension U.K.

ATLANTIC OCEAN

St. Helena U.K.

EQUATOR
0°

Luanda

ANGOLA
Kolwezi
Lubumbashi
Kitwe

ZAMBIA
Lusaka

Okavango R.

Zambezi R.
Lilongwe
MALAWI

Mozambique Channel

MADAGASCAR
Antananarivo

NAMIBIA
Windhoek

BOTSWANA
Gaborone

KALAHARI DESERT

Harare
ZIMBABWE
MOZAMBIQUE

TROPIC OF CAPRICORN

Tshwane (Pretoria)
Maputo
Johannesburg
SWAZILAND
Mbabane

Limpopo R.

Orange R.

Bloemfontein
LESOTHO
Maseru
Durban

SOUTH AFRICA

Cape Town
Port Elizabeth

INDIAN OCEAN

20°S

40°S

20°W   0°   20°E   40°E

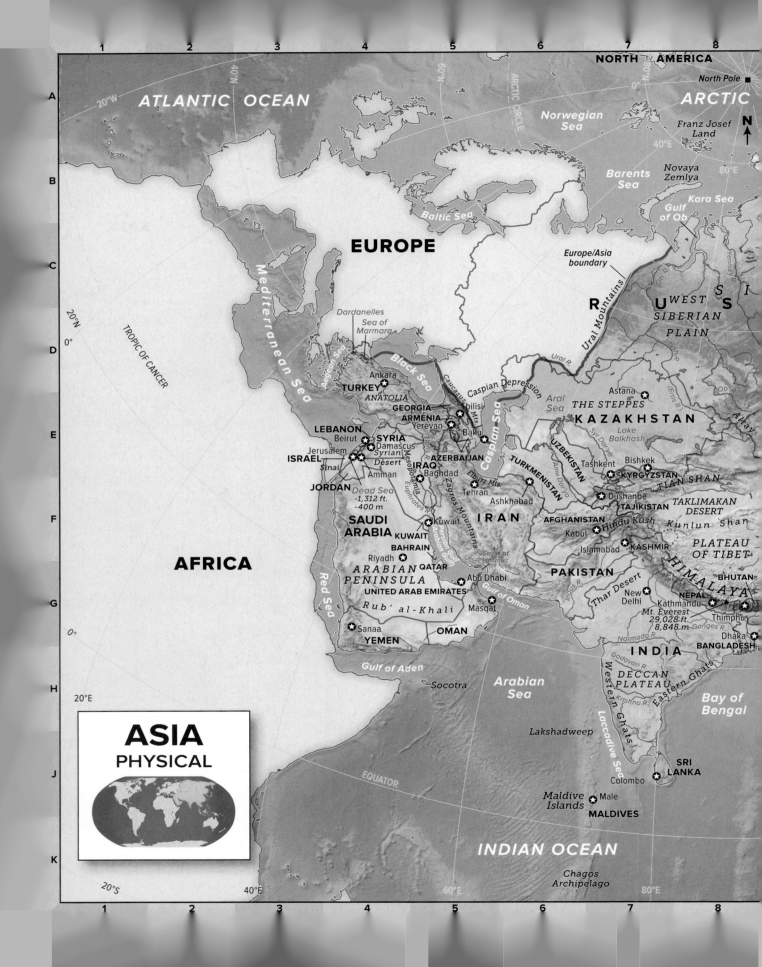

ASIA
PHYSICAL

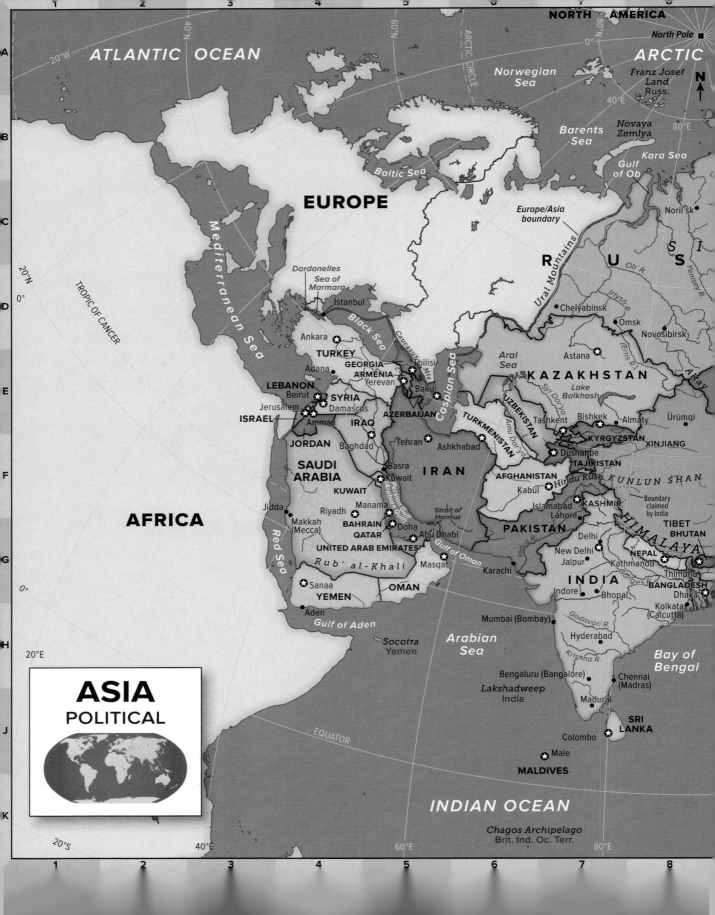

ASIA
POLITICAL

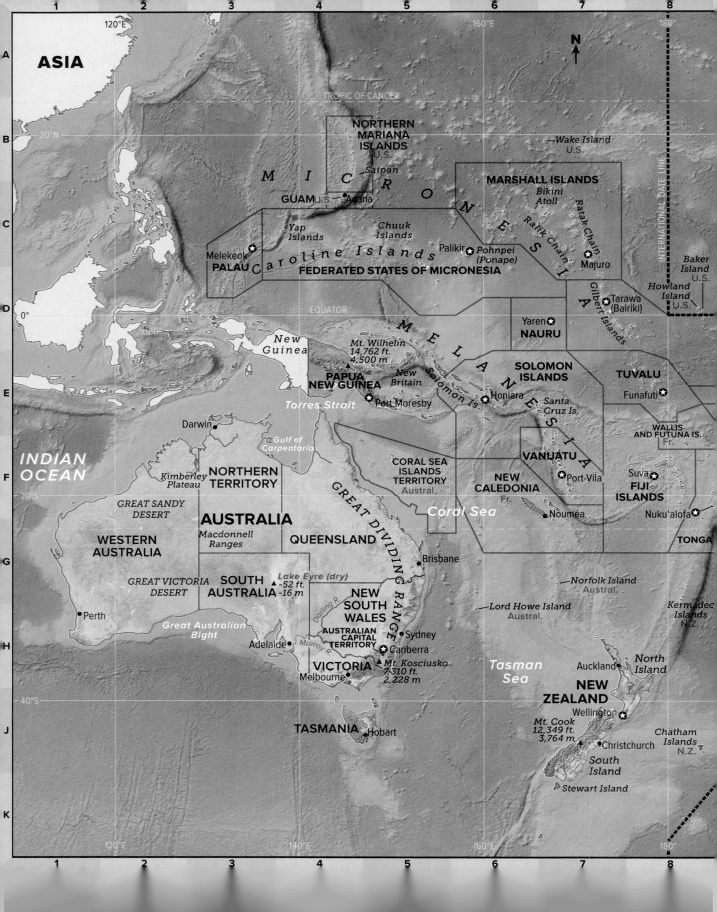

ASIA

120°E

140°E

160°E

180°

N

TROPIC OF CANCER

20°N

**NORTHERN
MARIANA
ISLANDS**
U.S.

Saipan

Wake Island
U.S.

INTERNATIONAL DATE LINE

MARSHALL ISLANDS

*Bikini
Atoll*

M I C R O N E S I A

GUAM U.S.
Agana

Ratak Chain

Ratik Chain

*Yap
Islands*

*Chuuk
Islands*

Caroline Islands

Palikir
Pohnpei
(Ponape)

Majuro

Melekeok

PALAU

FEDERATED STATES OF MICRONESIA

*Baker
Island*
U.S.

*Howland
Island*
U.S.

EQUATOR

0°

Gilbert Islands

Tarawa
(Bairiki)

Yaren
NAURU

M E L A N E S I A

*New
Guinea*

Mt. Wilhelm
14,762 ft.
4,500 m

**SOLOMON
ISLANDS**

TUVALU

Funafuti

**PAPUA
NEW GUINEA**

*New
Britain*

Solomon Is.

Honiara

*Santa
Cruz Is.*

Torres Strait

Port Moresby

**WALLIS
AND FUTUNA IS.**
Fr.

Darwin

Gulf of
Carpentaria

**CORAL SEA
ISLANDS
TERRITORY**
Austral.

VANUATU

Port-Vila

Suva

**FIJI
ISLANDS**

**INDIAN
OCEAN**

*Kimberley
Plateau*

**NORTHERN
TERRITORY**

**NEW
CALEDONIA**
Fr.

Nouméa

Coral Sea

Nuku'alofa

20°S

*GREAT SANDY
DESERT*

*Macdonnell
Ranges*

AUSTRALIA

QUEENSLAND

TONGA

**WESTERN
AUSTRALIA**

*GREAT VICTORIA
DESERT*

**SOUTH
AUSTRALIA**

Lake Eyre (dry)
-52 ft.
-16 m

**NEW
SOUTH
WALES**

Brisbane

Norfolk Island
Austral.

*Kermadec
Islands
N.Z.*

Perth

GREAT DIVIDING RANGE

Lord Howe Island
Austral.

Darling R.

**AUSTRALIAN
CAPITAL
TERRITORY**

Sydney

40°S

Adelaide

Murray R.

Canberra

*Tasman
Sea*

Auckland

*North
Island*

VICTORIA

Melbourne

Mt. Kosciusko
7,310 ft.
2,228 m

**NEW
ZEALAND**

Wellington

TASMANIA

Hobart

Mt. Cook
12,349 ft.
3,764 m

Christchurch

*Chatham
Islands
N.Z.*

*South
Island*

Stewart Island

Great Australian
Bight

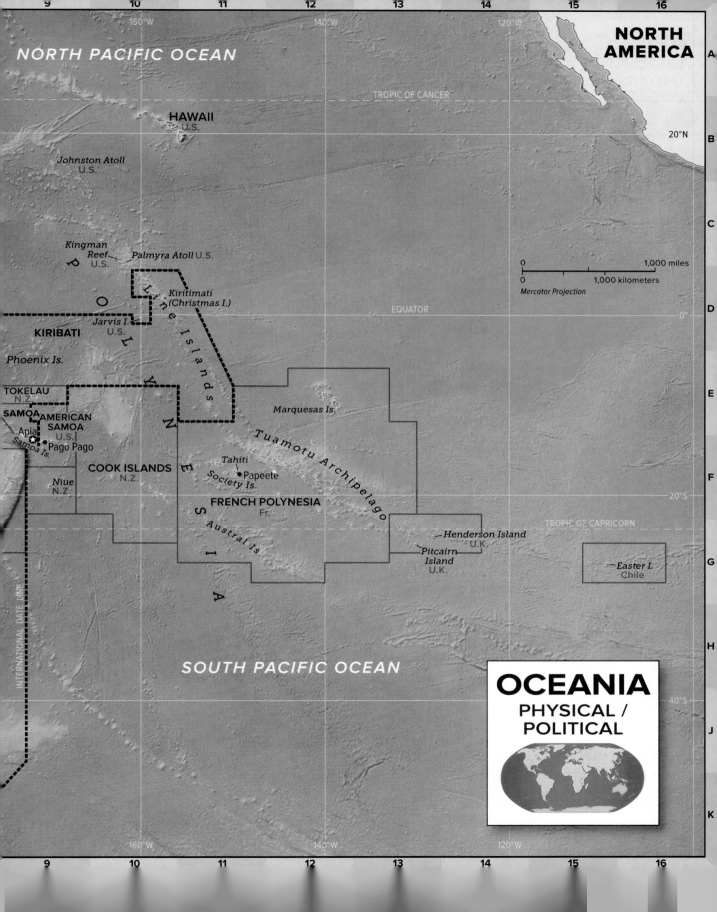

NORTH PACIFIC OCEAN

NORTH AMERICA

TROPIC OF CANCER

HAWAII
U.S.

20°N

Johnston Atoll
U.S.

P O L Y N E S I A

Kingman
Reef
U.S.

Palmyra Atoll U.S.

Kiritimati
(Christmas I.)

EQUATOR   0°

Jarvis I.
U.S.

KIRIBATI

Line Islands

Phoenix Is.

TOKELAU
N.Z.

Marquesas Is.

SAMOA

AMERICAN
SAMOA
U.S.

Apia
Pago Pago

Samoa Is.

Tuamotu Archipelago

COOK ISLANDS
N.Z.

Tahiti
Papeete

Niue
N.Z.

Society Is.

FRENCH POLYNESIA
Fr.

20°S

Austral Is.

TROPIC OF CAPRICORN

Henderson Island
U.K.

Pitcairn
Island
U.K.

Easter I.
Chile

INTERNATIONAL DATE LINE

SOUTH PACIFIC OCEAN

40°S

1,000 miles
0
0
1,000 kilometers
Mercator Projection

OCEANIA
PHYSICAL /
POLITICAL

The World

Chapter 1
How Geographers
Look at the World

Chapter 2
The Physical
World

Chapter 3
Climates of
the Earth

Chapter 4
The Human
World

World
Physical

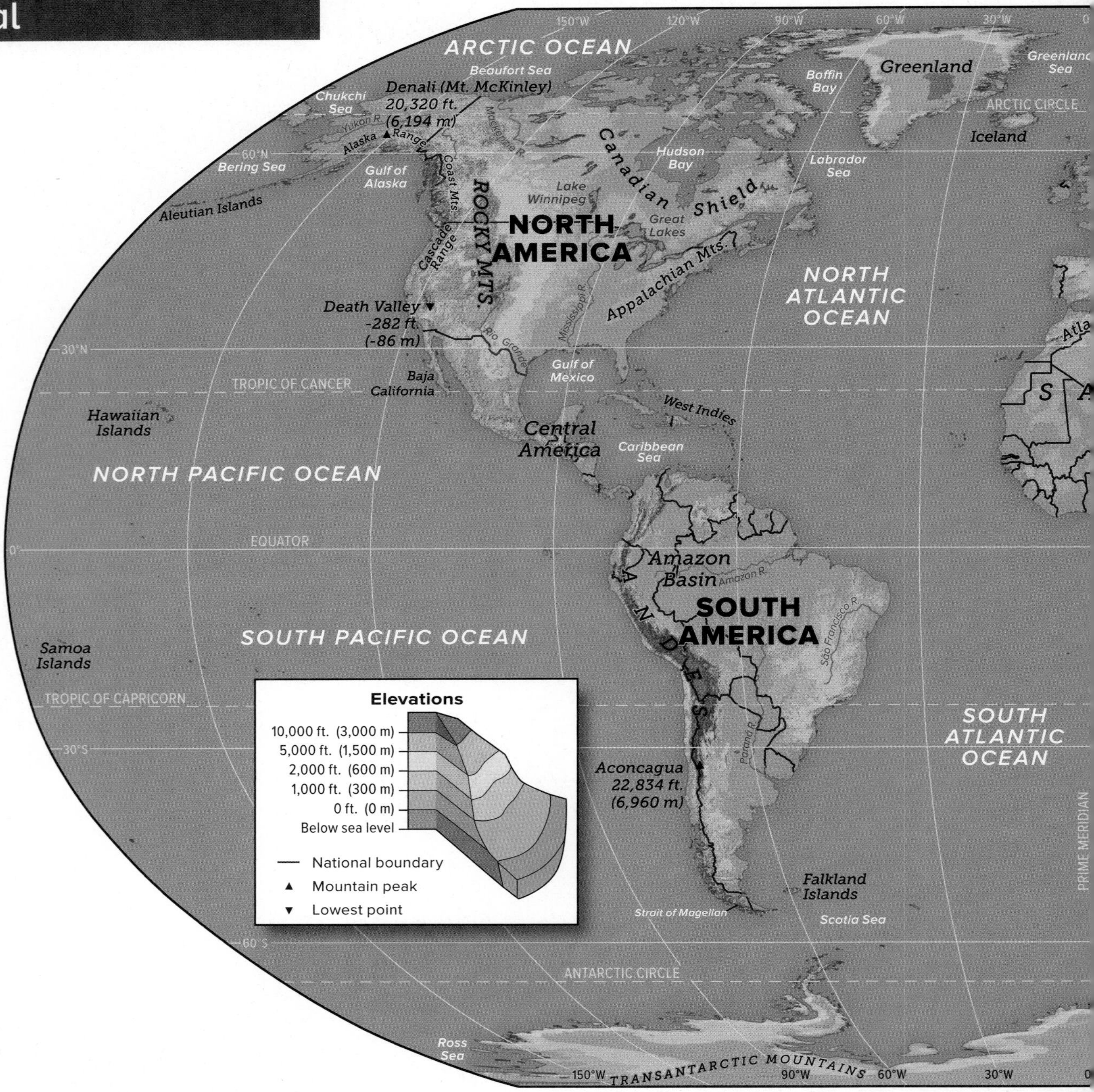

ARCTIC OCEAN

Beaufort Sea

Greenland

Greenland Sea

Denali (Mt. McKinley)
20,320 ft.
(6,194 m)

Chukchi Sea

Baffin Bay

ARCTIC CIRCLE

Iceland

Alaska ▲ Range

Bering Sea

Gulf of Alaska

Coast Mts.

Mackenzie R.

Hudson Bay

Labrador Sea

Aleutian Islands

NORTH AMERICA

ROCKY MTS.

Lake Winnipeg

Canadian Shield

Great Lakes

Cascade Range

Appalachian Mts.

NORTH ATLANTIC OCEAN

Death Valley ▼
-282 ft.
(-86 m)

Mississippi R.

Rio Grande

30°N

TROPIC OF CANCER

Gulf of Mexico

Baja California

West Indies

Hawaiian Islands

Central America

Caribbean Sea

S A

NORTH PACIFIC OCEAN

EQUATOR

Amazon Basin

Amazon R.

SOUTH AMERICA

São Francisco R.

Samoa Islands

SOUTH PACIFIC OCEAN

A N D E S

TROPIC OF CAPRICORN

Elevations

10,000 ft. (3,000 m)
5,000 ft. (1,500 m)
2,000 ft. (600 m)
1,000 ft. (300 m)
0 ft. (0 m)
Below sea level

— National boundary
▲ Mountain peak
▼ Lowest point

30°S

Aconcagua
22,834 ft.
(6,960 m)

Paraná R.

SOUTH ATLANTIC OCEAN

PRIME MERIDIAN

Falkland Islands

Strait of Magellan

Scotia Sea

60°S

ANTARCTIC CIRCLE

Ross Sea

150°W TRANSANTARCTIC MOUNTAINS 90°W 60°W 30°W

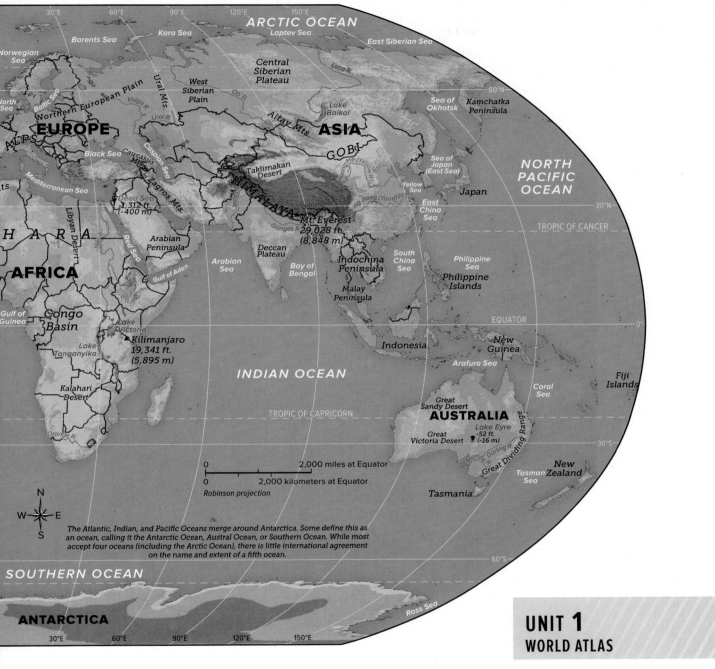

ARCTIC OCEAN

Norwegian Sea

Barents Sea

Kara Sea

Laptev Sea

East Siberian Sea

North Sea

Baltic Sea

Northern European Plain

Ural Mts.

West Siberian Plain

Central Siberian Plateau

Lena R.

Sea of Okhotsk

Kamchatka Peninsula

Volga R.

Ural R.

Ob' R.

EUROPE

ALPS

Black Sea

Caucasus Mts.

Caspian Sea

Altay Mts.

Lake Baikal

ASIA

60°N

Mediterranean Sea

Dead Sea -1,312 ft. (-400 m)

Zagros Mts.

GOBI

Huang He (Yellow R.)

Taklimakan Desert

Sea of Japan (East Sea)

NORTH PACIFIC OCEAN

.ts.

Red Sea

HIMALAYA

Yellow Sea

Japan

30°N

SAHARA

Libyan Desert

Nile R.

Arabian Peninsula

Mt. Everest 29,028 ft. (8,848 m)

Chang Jiang (Yangtze R.)

East China Sea

TROPIC OF CANCER

AFRICA

Gulf of Aden

Arabian Sea

Ganges R.

Deccan Plateau

Bay of Bengal

Indochina Peninsula

South China Sea

Philippine Sea

Philippine Islands

Gulf of Guinea

Congo Basin

Lake Victoria

Malay Peninsula

EQUATOR

0°

Kilimanjaro 19,341 ft. (5,895 m)

Lake Tanganyika

Indonesia

New Guinea

Fiji Islands

INDIAN OCEAN

Arafura Sea

Kalahari Desert

Great Sandy Desert

Coral Sea

TROPIC OF CAPRICORN

AUSTRALIA

Orange R.

Great Victoria Desert

Lake Eyre -52 ft. (-16 m)

Great Dividing Range

New Zealand

0   2,000 miles at Equator

0   2,000 kilometers at Equator

Robinson projection

Murray R.

Darling R.

Tasman Sea

Tasmania

N W E S

The Atlantic, Indian, and Pacific Oceans merge around Antarctica. Some define this as an ocean, calling it the Antarctic Ocean, Austral Ocean, or Southern Ocean. While most accept four oceans (including the Arctic Ocean), there is little international agreement on the name and extent of a fifth ocean.

30°S

60°S

SOUTHERN OCEAN

Ross Sea

ANTARCTICA

30°E   60°E   90°E   120°E   150°E

UNIT 1
WORLD ATLAS

MAP STUDY

1. **Physical Systems** Which part of North America has high elevations? Are they higher than the Andes of South America?

2. **Environment and Society** What types of physical features make Asia a crossroads for trade?

World
Political

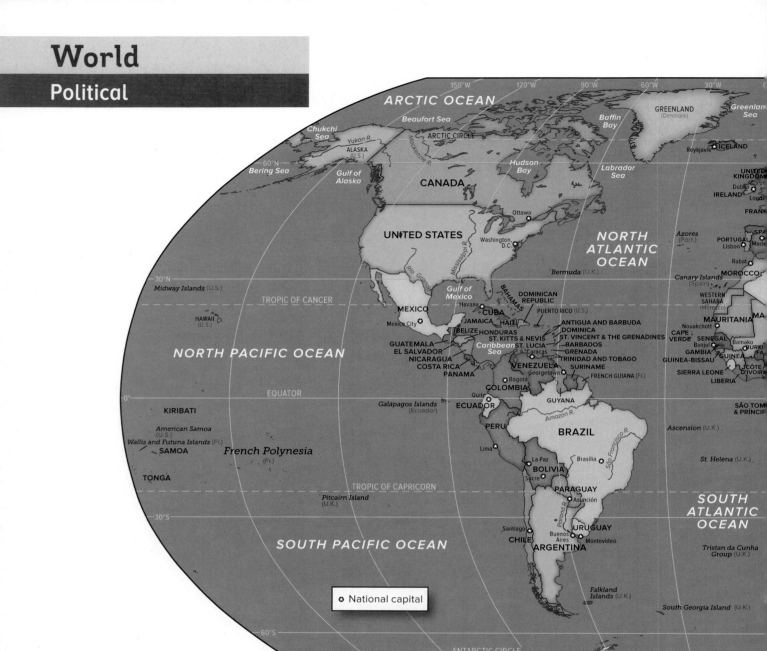

ARCTIC OCEAN

Beaufort Sea

Chukchi Sea

GREENLAND
(Denmark)

Greenlan Sea

Yukon R.

ARCTIC CIRCLE

Baffin Bay

Reykjavík ICELAND

ALASKA
(U.S.)

60°N
Bering Sea

Gulf of Alaska

Mackenzie R.

CANADA

Hudson Bay

Labrador Sea

UNITED KINGDOM

Dublin
IRELAND

Loud

Ottawa

FRAN

UNITED STATES

Washington D.C.

NORTH ATLANTIC OCEAN

Azores (Port.)

PORTUGAL
Lisbon

SP
Mad

30°N

Midway Islands (U.S.)

Bermuda (U.K.)

Rabat

MOROCCO

Canary Islands (Spain)

TROPIC OF CANCER

Mississippi R.

HAWAII
(U.S.)

Rio Grande

Gulf of Mexico

MEXICO

Havana

BAHAMAS

CUBA

DOMINICAN REPUBLIC

PUERTO RICO (U.S.)

WESTERN SAHARA (Morocco)

MA

MAURITANIA

Nouakchott

Mexico City

JAMAICA

HAITI

ANTIGUA AND BARBUDA

DOMINICA

CAPE VERDE

SENEGAL

Banjul

Bamako

BURKI
FA

BELIZE

HONDURAS

ST. KITTS & NEVIS

ST. VINCENT & THE GRENADINES

GUATEMALA

EL SALVADOR

NICARAGUA

COSTA RICA

PANAMA

Caribbean Sea

ST. LUCIA

BARBADOS

GRENADA

Caracas

TRINIDAD AND TOBAGO

VENEZUELA

Georgetown

SURINAME

FRENCH GUIANA (Fr.)

GAMBIA

GUINEA-BISSAU

GUINEA

SIERRA LEONE

LIBERIA

CÔTE D'IVOIRE

NORTH PACIFIC OCEAN

Bogotá

COLOMBIA

GUYANA

SÃO TOM & PRÍNCIP

EQUATOR

Galápagos Islands (Ecuador)

Quito

ECUADOR

KIRIBATI

PERU

Amazon R.

BRAZIL

Ascension (U.K.)

American Samoa (U.S.)

Lima

Wallis and Futuna Islands (Fr.)

SAMOA

French Polynesia (Fr.)

La Paz

BOLIVIA

Sucre

Brasília

São Francisco R.

St. Helena (U.K.)

TONGA

TROPIC OF CAPRICORN

Pitcairn Island (U.K.)

PARAGUAY

Asunción

SOUTH ATLANTIC OCEAN

30°S

Paraná R.

URUGUAY

Santiago

Buenos Aires

Montevideo

Tristan da Cunha Group (U.K.)

CHILE

ARGENTINA

SOUTH PACIFIC OCEAN

⊙ National capital

Falkland Islands (U.K.)

South Georgia Island (U.K.)

60°S

ANTARCTIC CIRCLE

150°W 120°W 90°W 60°W 30°W 0

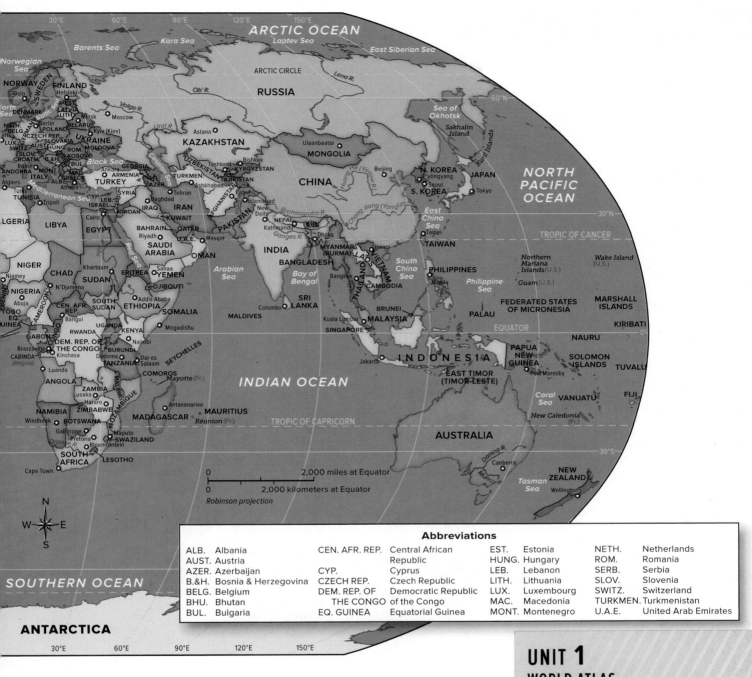

Abbreviations

ALB.	Albania	CEN. AFR. REP.	Central African Republic	EST.	Estonia	NETH.	Netherlands
AUST.	Austria			HUNG.	Hungary	ROM.	Romania
AZER.	Azerbaijan	CYP.	Cyprus	LEB.	Lebanon	SERB.	Serbia
B.&H.	Bosnia & Herzegovina	CZECH REP.	Czech Republic	LITH.	Lithuania	SLOV.	Slovenia
BELG.	Belgium	DEM. REP. OF	Democratic Republic	LUX.	Luxembourg	SWITZ.	Switzerland
BHU.	Bhutan	THE CONGO	of the Congo	MAC.	Macedonia	TURKMEN.	Turkmenistan
BUL.	Bulgaria	EQ. GUINEA	Equatorial Guinea	MONT.	Montenegro	U.A.E.	United Arab Emirates

UNIT 1
WORLD ATLAS

MAP STUDY

1. *Environment and Society* How do Earth's physical features appear to have shaped the borders of countries in Southeast Asia?

2. *Human Systems* List three countries whose capital cities are located on the coast.

World
Population Density

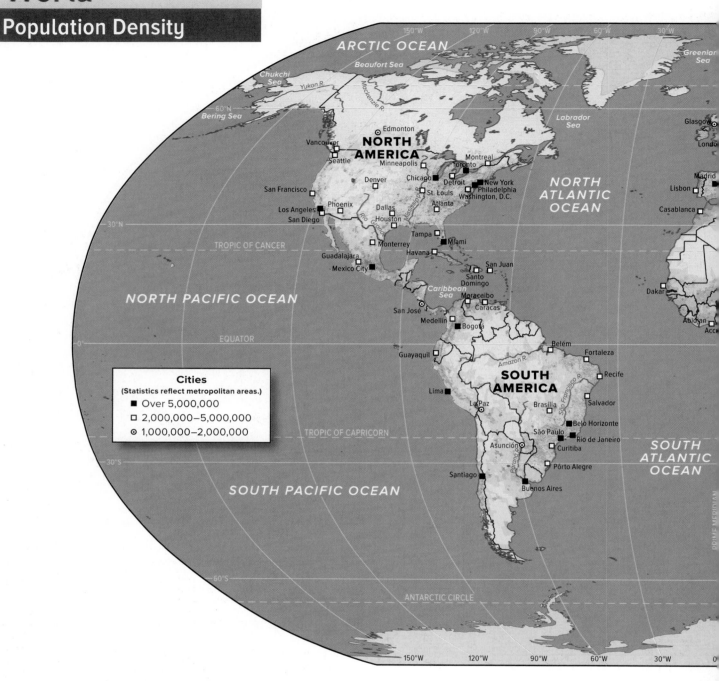

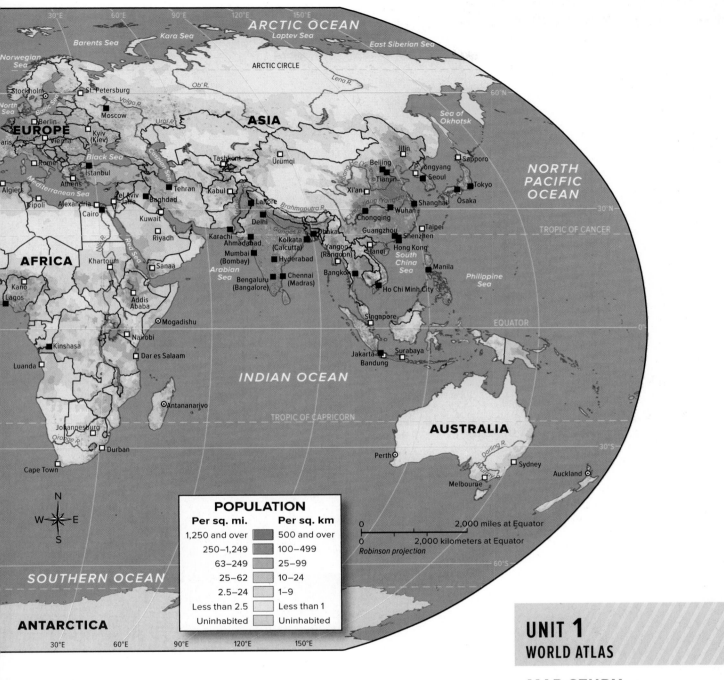

POPULATION

Per sq. mi.		Per sq. km
1,250 and over		500 and over
250–1,249		100–499
63–249		25–99
25–62		10–24
2.5–24		1–9
Less than 2.5		Less than 1
Uninhabited		Uninhabited

0   2,000 miles at Equator

0   2,000 kilometers at Equator
Robinson projection

MAP STUDY

1. *Human Systems* What cities in the United States have populations over five million people?

2. *Environment and Society* What generalizations can you make about the parts of the world that are least populated?

World
Economic Activity

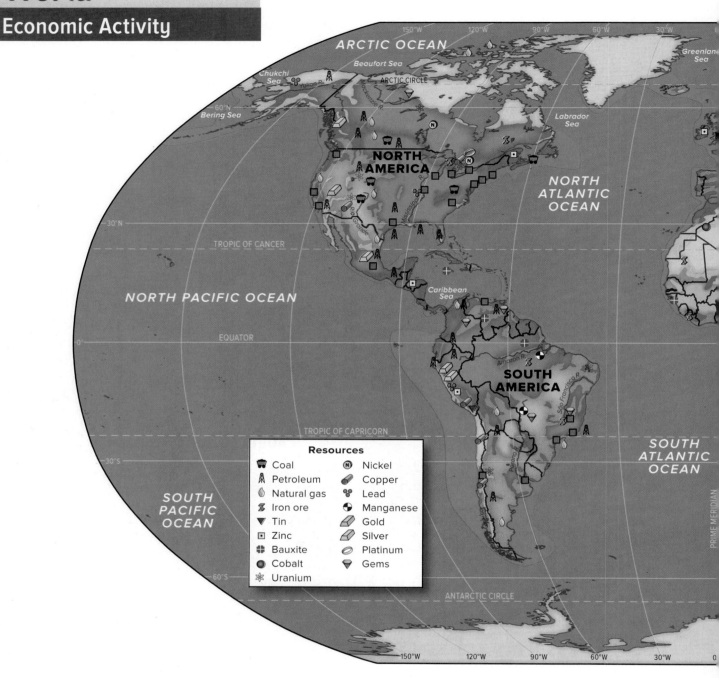

Resources

- Coal
- Petroleum
- Natural gas
- Iron ore
- Tin
- Zinc
- Bauxite
- Cobalt
- Uranium
- Nickel
- Copper
- Lead
- Manganese
- Gold
- Silver
- Platinum
- Gems

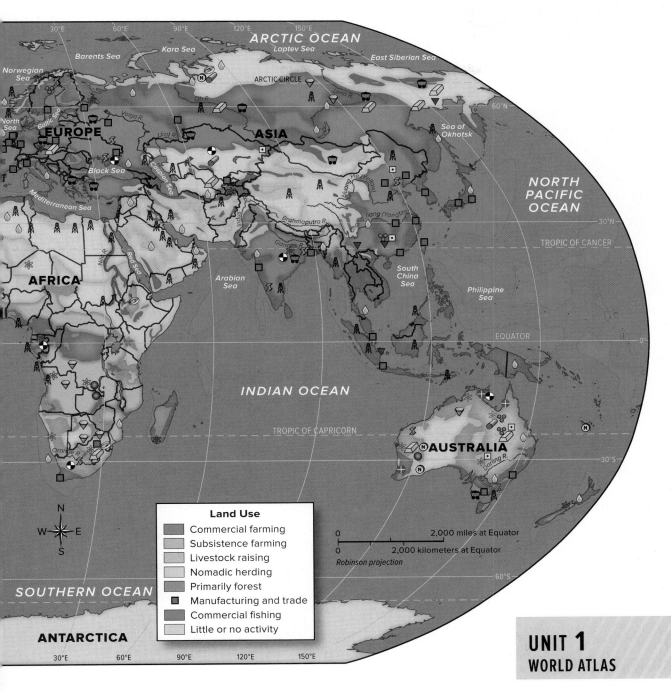

Land Use

- Commercial farming
- Subsistence farming
- Livestock raising
- Nomadic herding
- Primarily forest
- ■ Manufacturing and trade
- Commercial fishing
- Little or no activity

0   2,000 miles at Equator
0   2,000 kilometers at Equator
Robinson projection

UNIT 1
WORLD ATLAS

MAP STUDY

1. ***Environment and Society*** What are the primary land use activities in high latitude climate regions?

2. ***Human Systems*** In which areas of the world is commercial fishing a dominant economic activity?

World
Climate

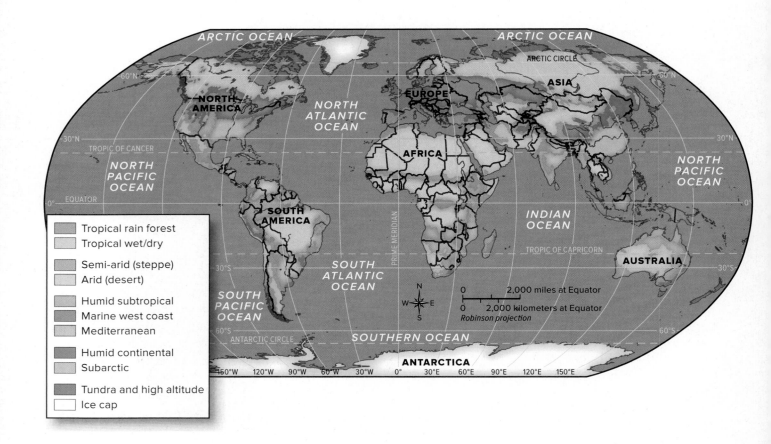

Legend:
- Tropical rain forest
- Tropical wet/dry
- Semi-arid (steppe)
- Arid (desert)
- Humid subtropical
- Marine west coast
- Mediterranean
- Humid continental
- Subarctic
- Tundra and high altitude
- Ice cap

MAP STUDY

1. *Physical Systems* Which two climate regions occur on and around the Equator?

2. *Physical Systems* What types of climate occur in the high latitudes?

How Geographers Look at the World

netw⊙rks

There's More Online about how geographers look at the world.

CHAPTER 1

Why Geography Matters
Distribution of Political Power

Lesson 1
The Geographer's Tools

Lesson 2
The Geographer's Craft

ESSENTIAL QUESTION • *How does geography help us interpret the past, understand the present, and plan for the future?*

Geography Matters...

Geographers are not the only ones who find value in geography skills. In fact, you use geography skills every day. The online maps you use to find the nearest grocery store, the GPS technology in your smartphone that tells your friends where you are, the weather reports that help you decide what to wear each day, the transportation systems that get you from place to place, and even your understanding of the people around you—their beliefs, values, and ways of life— are all examples of how geography is integrated into your daily life. Understanding the elements of geography in greater detail can help you become a better decision maker, planner, and citizen of the United States and the world.

◄ Philippe Cousteau, Jr., grandson of the famous marine explorer Jacques Cousteau, paddles in the waters of Blue Spring State Park in Florida while working on a documentary about Blue Spring and its manatees.

Daniel Wall/Orlando Sentinel/McClatchy-Tribune/Getty Images

11

distribution *of* political power

Political geography deals with the ways in which political processes and spatial environments interact and affect one another. Political boundaries such as the borders of countries, states, cities, and electoral districts are all part of political geography. Even election outcomes can be affected by the interaction of politics and geography.

Redistricting in Ohio

Ohio Districts, 2010

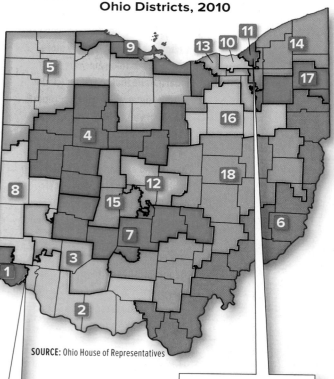

SOURCE: Ohio House of Representatives

Some counties can be divided between districts as a result of gerrymandering. For example, eastern Hamilton County is part of the Ohio 2nd district, while the rest of the county is part of the 1st district.

Gerrymandering can be used in ways that some perceive as positive. For example, districts are created in which a majority of the voters are a racial or ethnic minority, as in Ohio's 11th district.

Proposed Districts in Ohio

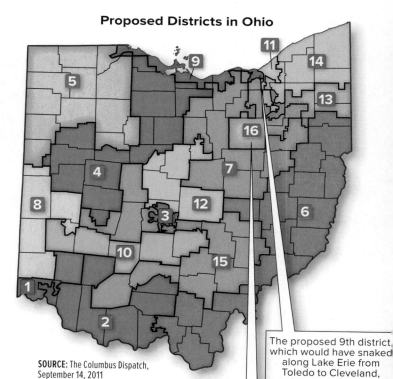

SOURCE: The Columbus Dispatch, September 14, 2011

During the process of reapportionment based on the 2010 Census, Ohio Republicans proposed a new congressional map that left them a good chance to hold many seats for the next decade.

The approved map of districts reflects some, but not all, of the Republicans' proposals.

The proposed 9th district, which would have snaked along Lake Erie from Toledo to Cleveland, packs concentrations of Democratic voters into one district.

The proposed 16th district moved a Democratic representative into a district that favored a Republican representative.

What is redistricting?

In the 1960s, after decades of inequalities, the U.S. Supreme Court ruled that each person's vote should be worth as much as any other person's vote in both federal and state elections. The government divided the states into districts, with the number of districts in each state determined by the state's population. In this way, states with larger populations get more votes than states with smaller populations, establishing what is called the "one person, one vote" requirement. In order to draw the boundary lines of congressional districts, the U.S. government uses the census, or official count of all the people living in the country. When new census data is collected, sometimes district boundaries must be redrawn to reflect changes in population size and distribution. This process is known as *redistricting*, and it ensures that each person's vote receives equal weight in government.

1. Human Systems What issues might have prompted the Supreme Court's ruling in the 1960s to establish the "one person, one vote" requirement?

How are electoral districts drawn?

Apportionment is the process used to decide how many representatives each state will have to represent it in the U.S. Congress. Census data is used to determine how many representatives each state will have. After the number of representatives for each state is determined, the state congressional districts are drawn so that there is one district for each representative. The government uses U.S. Census data along with geographic information systems (GIS) to create maps that draw district lines in such a way that each district in a state has a roughly equal population. For example, a district with people living fairly spread apart will cover a larger geographic area than a district that is densely populated, such as in urban areas. However, they both can be defined as districts with equal representation. In addition to population size, redistricting takes into account factors such as maintaining city boundaries, preserving boundaries from previous districts, and even avoiding contests between existing representatives.

2. Human Systems What term is used to describe the process the government uses in deciding how many representatives each state will have? What tools are used by the government to help make these decisions?

How does redistricting affect the distribution of political power?

Unfortunately, the process of redistricting is not perfect. This creates an opportunity for misuse by those who seek to manipulate the outcome of elections to favor particular candidates. *Gerrymandering* is a term used to describe the act of drawing political district boundary lines in such a way that one party or candidate has an advantage over another. *Packing* and *cracking* are methods of gerrymandering used to minimize the power of a particular group of voters. Packing concentrates members of a group in a single district, which helps the opposing party win other districts that are nearby. Cracking splits the voters among multiple districts, which dilutes their impact and prevents them from being a majority when they vote in the elections. Packing and cracking are commonly used together in such a way that they influence voting.

3. Places and Regions According to the maps, which districts would be significantly changed by the Ohio Republicans' plan? Which would remain the same?

THERE'S MORE ONLINE

READ about the effect of the census on redistricting • **WATCH** a video about the redistricting process

netw⊙rks

There's More Online!

☑ **IMAGE** Mental Map

☑ **IMAGE** Satellite Image of Earth

☑ **MAP** Common Map Projections

☑ **MAP** GIS Layers

☑ **MAP** Great Circle Routes

☑ **MAP** Latitude and Longitude

☑ **INTERACTIVE SELF-CHECK QUIZ**

☑ **VIDEO** The Geographer's Tools

Reading **HELP**DESK

Academic Vocabulary

- internal
- transmit

Content Vocabulary

- **great circle route**
- **map projection**
- **planar projection**
- **cylindrical projection**
- **conic projection**
- **absolute location**
- **relative location**
- **elevation**
- **relief**
- **thematic map**
- **global positioning system (GPS)**
- **geographic information systems (GIS)**
- **remote sensing**

TAKING NOTES: *Key Ideas and Details*

IDENTIFYING As you read the lesson, use a graphic organizer like the one below to identify the tools geographers use to look at the world.

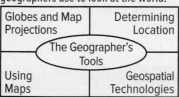

Globes and Map Projections	Determining Location
The Geographer's Tools	
Using Maps	Geospatial Technologies

14

LESSON 1
The Geographer's Tools

ESSENTIAL QUESTION • *How does geography help us interpret the past, understand the present, and plan for the future?*

IT MATTERS BECAUSE

The study of geography involves looking at every aspect of the Earth's systems. Aspects such as human economies, societies, and cultures, and plants, animals, climate, and the physical environment affect each other in many ways. Geography analyzes these diverse interactions to learn more about how Earth's systems are interconnected. Geographers gather information from various sources using a variety of tools to study these complex and interrelated Earth systems.

Globes and Map Projections

GUIDING QUESTION *How are globes and map projections related?*

A globe is a scale model of the Earth that depicts properties such as area, distance, and direction. Globes accurately display all these properties because they are round just like the Earth. A map is a flat representation of all or part of the planet. But, unlike globes, maps cannot show all the properties accurately.

Mapmakers, called cartographers, use mathematical formulas to transfer information from the three-dimensional globe to the two-dimensional map. However, when the curves of a globe become straight or only slightly curved lines on a map, distortion occurs in shape, distance, area, or direction.

A straight line of true direction on a map is not always the shortest distance between two points on Earth. The measured distance between any two points on a flat map will not have the same distance when measured on a round globe. To find the actual shortest distance between any two places, stretch a piece of string around a globe from one point to the other. The string will form an arc that is part of a great circle, an imaginary line that follows the Earth's curvature. **Great circle routes** therefore mark the shortest distance that an object can travel between two points. They are useful because they indicate actual distances between two locations. Determining a great circle route is important for travel and transportation. Ship captains and airplane pilots use great circle routes to reduce travel time and conserve fuel.

While globes are useful for portraying the entire Earth, their ability to display detailed features of a particular region are limited. Maps, however, are useful for showing more in-depth information. Cartographers convert the three-dimensional globe image onto a flat map by creating a **map projection**. But because map projections can distort one or more of the properties of size, shape, distance, area, or direction, the cartographer must choose the projection to use based on the purpose of the map. It is important to know which properties are distorted, and how much they are distorted, so you can use and interpret the map accurately.

There are many kinds of map projections, some with general names and some named after the cartographer who developed them. Three major categories of map projections are planar, cylindrical, and conic.

A **planar projection**, also known as an azimuthal projection, shows the Earth centered in such a way that a straight line coming from the center to any other point represents the shortest distance. Because a planar projection is most accurately represented from its center, it is often used for maps of the Poles.

A **cylindrical projection** is based on how a map would look if the globe was projected onto a cylinder. This type of projection is most accurate at the Equator because shapes and distances are increasingly distorted when moving away from the Equator and toward the Poles. A Mercator projection is a common example of a cylindrical projection. Because it displays true direction, a Mercator projection is useful for sea navigation.

A **conic projection** is the Earth's surface projected onto a map formed into a cone. Shape is relatively accurate on such projections, and straight lines drawn on them approximate great circle routes if distances are not great.

World maps used for general reference use the Winkel Tripel projection. This map projection cannot be used to determine precise distances, sizes, or shapes of specific global features. It does, however, provide a good balance between the overall size and shape of land areas shown.

A Robinson projection looks similar to a Winkel Tripel projection, although its east-west projections run in a straight line. The Robinson projection produces minor distortions, particularly in the polar areas that appear flattened on the map. The sizes and shapes near the eastern and western edges of the map are accurate, and outlines of the continents appear much as they do on the three-dimensional globe.

great circle route an imaginary line that follows the curve of the Earth and represents the shortest distance between two points

map projection a mathematical formula used to represent the curved surface of the Earth on the flat surface of a map

planar projection a map created by projecting an image of the Earth onto a geometric plane

cylindrical projection a map created by projecting Earth's image onto a cylinder

conic projection a map created by projecting an image of Earth onto a cone placed over part of an Earth model

Great Circle Routes

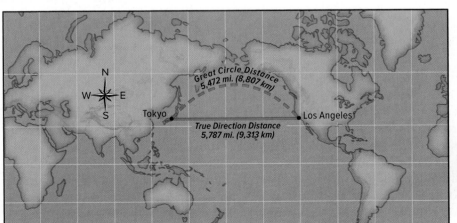

GEOGRAPHY CONNECTION

Great circle routes show the true distance between two places on Earth.

1. *THE USES OF GEOGRAPHY* Why do ship captains and airline pilots use great circle routes?

2. *THE WORLD IN SPATIAL TERMS* Why do distances appear longer on maps than on globes?

Winkel Tripel Projection

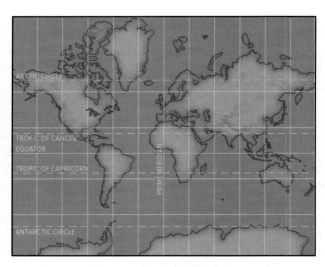

Mercator Projection

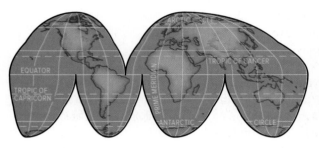

Goode's Interrupted Equal-Area Projection

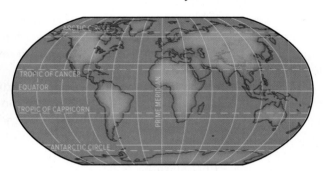

Robinson Projection

GEOGRAPHY CONNECTION

Each type of map projection has its advantages as well as some degree of inaccuracy. In addition to the major categories of map projections—planar, cylindrical, and conic—many other map projections can be used depending on the information the mapmaker wishes to show.

1. ***THE WORLD IN SPATIAL TERMS*** Which projection appears to have the least amount of distortion of distances and size of landmasses?

2. ***THE USES OF GEOGRAPHY*** Which projection is preferred for sea navigation? Why?

Goode's Interrupted Equal-Area projection resembles a globe that has been cut apart and laid flat. The process of creating this *interrupted projection* can be compared to slicing an orange peel in order to lay it flat on a page. Although this projection shows the true size and shape of Earth's landmasses, distances between land features are generally distorted.

☑ **READING PROGRESS CHECK**

Explaining Why is a trip from Tokyo to Los Angeles a longer distance than it appears to be on a map?

Determining Location

GUIDING QUESTION *How is location determined?*

Geography addresses the question of *where*. To answer this question, a geographer identifies a location. Both globes and maps use a grid system in order to form a pattern of lines that cross one another. These patterns are used to help find the location of places on the Earth's surface.

Lines of latitude, or parallels, circle the Earth parallel to the Equator. Although they run in an east-to-west direction, they measure distance to the north and south of the Equator. The measurements are in degrees. Parallels north of the Equator are called north latitude. Parallels south of the Equator are called south latitude. The Equator is defined as 0° latitude, the North Pole as 90° N, and the South Pole as 90° S.

Longitude lines, also called meridians, are lines that connect the North and South Poles. They run in a north-to-south direction, but they measure distance east and west of the Prime Meridian, which is identified as 0° longitude. Meridians run perpendicular to the lines of latitude, and they also use the measurement of degrees. Meridian lines located east of the Prime Meridian are identified as east longitude, and lines located west are known as west longitude. The longitude line located 180° from the Prime Meridian, on the opposite side of the Earth, is called the International Date Line.

The Equator divides the Earth in half, creating Northern and Southern Hemispheres. The Northern Hemisphere includes any location north of the Equator up to 90° N, while the Southern Hemisphere includes any location south of the Equator up to 90° S. Just as the Equator splits the Earth into Northern and Southern Hemispheres, the Prime Meridian and International Date Line split the globe into east and west halves. Locations east of the Prime Meridian are identified as part of the Eastern Hemisphere, and locations west of the Prime Meridian as part of the Western Hemisphere. All points on Earth are located in two of the four hemispheres: north or south and east or west.

An **absolute location** is an exact global address derived from the latitude and longitude lines that intersect at that place. For example, Tokyo, Japan, is located at approximately 36° N latitude and 140° E longitude. For a more precise reading of a location, a degree is divided into 60 minutes ('). Each minute is then divided into 60 seconds (") just as hours and minutes on a clock are divided to provide a more exact time. For example, the absolute location of the famous Tokyo Tower is 35°39' 30.96" N latitude and 139°44' 43.59" E longitude.

While absolute location identifies exact points using latitude and longitude, **relative location** uses a reference point to identify one place in relation to another. To find relative location, find a reference point—a location you already know—on a map. Then look in the appropriate direction for the new location. For example, locate the city of Paris on the map of France and use this as your reference point. The relative location of the city of Lyon can be described as southeast of Paris.

☑ READING PROGRESS CHECK

Listing List the four hemispheres of the Earth.

absolute location the exact position of a place on the Earth's surface

relative location location in relation to other places

Latitude and Longitude

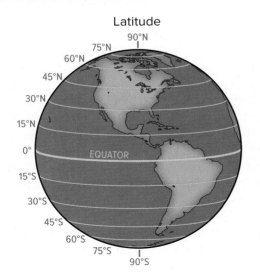

Latitude

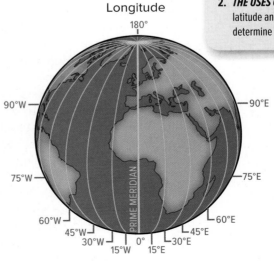

Longitude

Mathematics plays an integral role in the study of geography, especially the process of projecting Earth's spherical features onto a flat map. Early cartographers hand-calculated their map projections using geometry, trigonometry, and even calculus. Today, computer programs take much of the mathematical legwork out of the process of mapmaking. However, geographers still use mathematics for many tasks, such as calculating the area of a landmass, determining the volume of a body of water, and computing distances between locations.

MAKING CONNECTIONS
How does mathematics allow cartographers to create flat depictions of a spherical Earth?

elevation the height of a land surface above the level of the sea

Using Maps

GUIDING QUESTION *How do maps work?*

In addition to lines of latitude and longitude, maps include other important tools to help you understand the information they provide. Learning to use map tools will help you interpret the language of maps more easily.

Parts of a Map

The purpose of a map is identified by the map's title. For example, a map titled "Housing Developments in Washington, D.C." would show different details than a map titled "Topography of the Washington, D.C., area." The time period of a map is another important clue to understanding what the map shows. For example, a map titled "Europe Before World War I" would show country borders and national capitals of Europe that are quite different from Europe's current political borders. The map title is the first thing you should look at when reading a map because it provides context for the map's content.

An effective map will provide a legend, or key, to explain the meaning of various symbols used on the map. Geographic features represented on the map are identified by symbols, also called icons. Icons vary by map, depending on the details that are the focus of the map. Roads, highways, railroads, landmarks, parks, and buildings are all human-made features shown by icons. Dots are often used to represent cities. Sometimes the relative sizes of cities are shown using dots of different sizes. Capital cities can be identified by a star within a circle.

The compass rose indicates direction or orientation of a map. North, east, south, and west are the four cardinal directions. The intermediate directions—northeast, northwest, southeast, and southwest—may also be shown. The compass rose looks like intersecting arrows or points of a star.

Line symbols on a map emphasize various features of human activity, such as boundary lines, roads, streets, or routes of trade and transportation. On political maps, boundary lines highlight the borders between different countries and states. Line symbols can also represent physical features such as rivers, earthquake faults, and ocean shorelines.

Colors can be used to distinguish elements on a map. For example, a political map might make each country a different color. On a physical map, colors may indicate the various ranges of **elevation**, or the height above sea level. Colors are used for a variety of other purposes, including identifying water features such as oceans, lakes, or rivers; land features such as deserts, valleys, plains, or mountain ranges; and human-made features such as roads, parks, or streets. The meaning of a color can be identified by the map title or is provided in the legend.

All maps are drawn to a certain scale. Scale represents the consistent, proportional relationship between the measurements shown on a map and the actual measurements of the Earth's surface. Maps use scale to shrink what would be large distances and features of a region to a manageable size. When a map is scaled to fit on paper, every feature of the map is scaled by the same amount so that each feature will have the same proportion to every other feature on the map. However, not all parts of a map will be perfectly to scale because flat maps are subject to some distortion. A map's scale is identified by a scale bar, which compares distances shown by a map to actual distances on the Earth. For example, a scale bar might indicate that one inch (2.5 cm) on the map represents 100 miles (160.9 km).

The amount of scale portrayed on a map depends on its depiction as a *small-scale* or *large-scale* map. A small-scale map shows a larger area with fewer details. For instance, a small-scale map can focus on a specific country and its

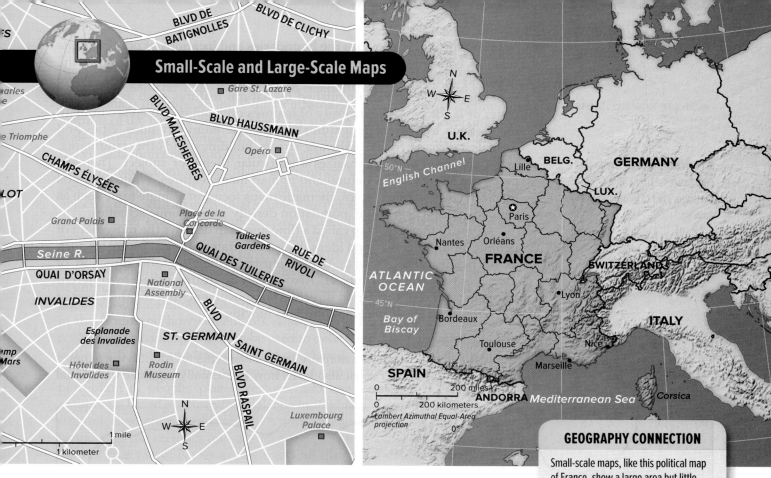

Small-Scale and Large-Scale Maps

GEOGRAPHY CONNECTION

Small-scale maps, like this political map of France, show a large area but little detail. Large-scale maps, like the map of the city of Paris, can show a small area with a great amount of detail.

1. **PLACES AND REGIONS** Using the scale bar on the map of France, what is the distance from Paris to Nice in miles?

2. **HUMAN SYSTEMS** What types of human-made features does the map of Paris include that the map of France does not?

neighboring countries to show boundary lines, major cities and capitals, important land or water features, or regional topography. For example, the scale bar on a map of France and its bordering countries could show a relationship of one inch (2.5 cm) as equal to 200 miles (321.9 km) in actual distance. On the other hand, a large-scale map can show a small area with a great amount of detail. It narrows in on an identified region to show more specific details. The map measurements of a large-scale map use much smaller distances than on a map of France. For example, a large-scale map of the city of Paris shows the layout of streets, major roads, bridges, parks, and important landmarks such as museums, hotels, and churches. The scale bar of a map of Paris could show a relationship of one inch (2.5 cm) as equal to one mile (1.6 km) in actual distance. This measurement is much more specific than for a small-scale map of France.

Types of Maps

A cartographer can choose from several types of maps in order to convey geographic information. Physical maps, political maps, and thematic maps each serve a unique purpose and are suited to showing different types of information. A physical map shows location and topography, or shape, of the Earth's land features. A study of a country's land and other physical features can help to explain the historical development of the country. For example, mountains may be barriers to transportation, and rivers and streams can provide access to a country's interior. Physical maps show water features such as rivers, streams, and lakes. They also show landforms such as mountains, plains, plateaus, and valleys.

Physical maps highlight general **relief** through shading and texture. Relief shows the differences in elevation between the various landforms of an area. An elevation key can use colors to indicate specific, measured differences in elevation above sea level.

relief the variation in elevation across an area of Earth's land

Biologists use GPS technology to track the movement and behavior of animals in the wild.

▲ CRITICAL THINKING

1. Drawing Conclusions How might wildlife biologists use GPS technology to protect endangered species such as the African elephant?

2. Classifying What other fields use GPS technology to gather information?

longitude, and even altitude. GPS technology in the United States relies on a system of 24 satellites that make 6 full orbits around the Earth every 12 hours. The European Union, as well as some individual countries such as Russia and China, have their own satellites that support GPS systems.

The satellites in all these systems send out radio signals that are picked up by GPS receivers on Earth. In a process called triangulation, a GPS receiver measures the precise time taken for radio signals from four or more satellites to travel to the receiver. The receiver then multiplies the time by the speed of a radio wave to calculate the distance between it and the satellite. When signals from the four or more satellites are processed in the same manner, the receiver's built-in computer determines the point at which at least four satellite signals intersect on the Earth. This intersection then identifies the receiver's latitude, longitude, and altitude. The more satellites that are used, the more accurate the location that is pinpointed.

GPS technology serves a commercial purpose for military machinery, space shuttles, aircraft, ships, submarines, trucks, trains, and ambulance fleets. Yet it can also aid in multiple forms of personal navigation. The GPS receiver in a car, for example, tracks the car's changing location on an electronic map to provide constantly updated directions based on where the car is located and where it is headed. Many current GPS receivers are battery-powered and are no larger than the palm of your hand, while GPS computer chips are smaller than your fingernail.

Many fields of science employ GPS technology. For example, seismologists, the scientists who study earthquakes, can use GPS to determine the size of earthquakes. Scientists first plant GPS receivers in the ground in regions vulnerable to tectonic, or earthquake-prone, activity. Once an earthquake hits, seismologists can quickly measure the strength of an earthquake by calculating how far the planted GPS receivers move. This measurement allows scientists to predict how likely the earthquake is to produce large ocean waves called tsunamis. Because tsunamis can cause devastating destruction to coastal communities, early warning would diminish loss of human life by advising people to flee as soon as possible.

Another function of GPS technology in the field of science is to track the migration of animals to determine any changing patterns within the animals' ecosystems. Biologists tag animals with GPS receivers so they can track their movement due to seasonal changes, changes of food or shelter, or threats to their habitat by human activity or by other animals.

Geographic Information Systems

Advances in technology have changed the way maps are made. An important tool in mapmaking today involves computers with software programs called **geographic information systems (GIS)**. But more than simply making maps, GIS can be used to perform advanced geographical analysis.

Many types of data can be entered into a GIS. These data come from a wide variety of sources such as maps, satellite images, printed text, and statistical databases. The primary and most important function of a GIS, however, is to link the location of a place with the characteristics, or attributes, found at that location. That function helps us not only identify and list the characteristics of places, but also analyze how places compare to one another and interact with one another. These patterns of interactions are known as spatial organization, and the study of them is called spatial analysis.

The locational data of places is stored in a GIS as latitude and longitude coordinates. These coordinates can be obtained from existing maps, GPS

geographic information systems (GIS) computer programs that process and organize details about places on Earth and integrate those details with satellite images and other pieces of information

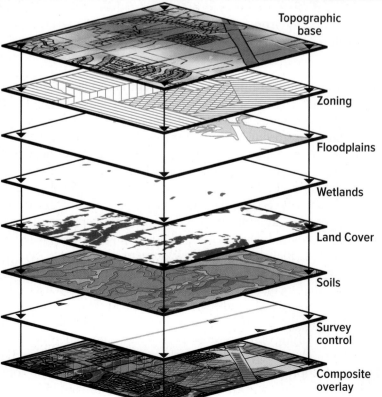

GIS Layers

Topographic base

Zoning

Floodplains

Wetlands

Land Cover

Soils

Survey control

Composite overlay

∨ DIAGRAM SKILLS

Geographic information systems allow different kinds of information to be saved on separate layers.

◀ **CRITICAL THINKING**

1. **Formulating Questions** What questions could a geographer answer using the layers of information in this sample GIS file?

2. **Evaluating** How does GIS allow cartographers to create maps and make changes quickly and easily?

The stark differences between North America and Africa at night are clearly seen with satellite images.

▲ **CRITICAL THINKING**

1. Drawing Conclusions What do these images tell you about North America? About Africa?

2. Analyzing How might the information from these images be used by geographers?

transmit to send from one place to another

receivers, and satellite images. The attribute data come from a wide range of sources. The GIS stores all these data in a digital database. Cartographers then select an appropriate map projection and program the GIS to produce thematic maps of the data. Each of the various types of attribute data in the database can be displayed on the map as a theme. The different themes can be saved as separate electronic layers that can be turned on or off. The GIS can show just one layer of information or multiple layers at the same time.

In this way, maps can be made—and changed—quickly and easily to display various complex types of information with a single map source. GIS is used by a diverse range of professional fields with work that relies on maps, including environmental and urban planners, marketing researchers, retail developers, environmentalists, and other professionals.

Satellites

A satellite is a natural or human-made object that orbits a planet or other large astronomical body. The first human-made satellite was launched in 1957 by the Soviet Union, and today hundreds of human-made satellites orbit the Earth. These satellites are useful for many purposes, such as navigation, collecting atmospheric data to make weather predictions, and communications for cellular phones and the Internet. Some satellites collect visual information of the Earth's surface as they orbit the Earth. Others carry instruments to observe the presence of and interaction between the land, ocean, air, and living things. Because scientists can gather very specific information regarding atmospheric phenomena, they can compare satellite images with ground research and knowledge of Earth's natural history to analyze how the environment has changed over time.

Different types of satellites are used to collect different types of information about the Earth. Once the data are collected, computers on a satellite store and then **transmit** those data by radio signals to receiving stations on Earth. Scientists on Earth receive the transmitted data and use their specialized knowledge to interpret the meaning of these data. The received data sometimes serve as inputs to computer models using mathematical formulas. Just as cartographers produce maps from data processed by GIS, so do scientists who use the converted data from satellite imagery to study Earth's natural and human-made processes.

Remote sensing is any technique used to measure, observe, or monitor a subject or process without physically touching the object under observation. For example, scientists use remote sensing when they analyze images from satellites, telescopes, and cameras in airplanes and spacecraft. Often, remote sensing collects images of things that could not be seen with the unaided human eye. It is also a useful process for obtaining information from locations that would otherwise be dangerous or difficult to reach, such as estimating precipitation rates in a desert region. The immediate and frequent flow of images from remote sensing allows cartographers to create detailed and relevant maps to estimate constant and changing environmental conditions, such as sediment buildup, air pollution, ocean surface roughness, surface temperatures, biomass volumes, mineral resources, and changes created by storms and floods.

remote sensing the science of obtaining information about an object or an area from a distance, typically from instruments in aircraft or satellites

Quality and Limitations of Geospatial Technologies

Geospatial technologies are excellent sources of information because they provide actual images and data related to a location and can provide a great amount of detail. While scientists can use observational and historical data to gather information about a place, geospatial technology acts as a primary source for compiling raw data. Because they are a relatively new innovation in comparison to traditional forms of mapmaking, the current uses of geospatial technologies can be limited. These informational technologies are constantly changing, and they will improve with the advancement of computer, aerospace, and Internet-based technology. Accelerated development in geospatial technology offers a number of possibilities for its use with government, private industry, scientists, and the general public. Because the economic, cultural, and political activities of the world's regions have become increasingly interconnected, information related to the world's physical and human systems needs to be readily available, consistent, and up-to-date. The combination of mental mapping with GPS, GIS, and aerial imagery can create a very detailed picture of places and regions.

Geospatial technologies allow access to a wealth of information about what sorts of features and objects are in the world and where those features and objects are located. This "geospatial information" can be very helpful for identifying and navigating, but by itself, does not help much in answering the "why" or the "why care" questions that lie at the heart of understanding and making decisions about the world in which we live. It is important to go beyond geospatial information to geographical understanding of peoples, places, and environments—and the connections among them—that are interesting as well as useful.

☑ **READING PROGRESS CHECK**

Identifying What type of data do geospatial technologies provide?

LESSON 1 REVIEW

Reviewing Vocabulary

1. *Identifying* What is the difference between absolute location and relative location?

Using Your Notes

2. *Describing* Using your graphic organizer, list and describe the four common map projections.

Answering the Guiding Questions

3. *Comparing* How are globes and map projections related?

4. *Explaining* How is location determined?

5. *Summarizing* How do maps work?

6. *Analyzing* How are geospatial technologies used to learn about the world?

Writing Activity

7. *Informative/Explanatory* If you were planning to open a sporting goods store, in what ways could GIS technologies help you choose a good location? Discuss the types of layers that might be helpful to your decision.

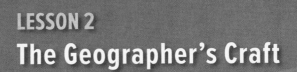

networks

There's More Online!

☑ **CHART** Skills for Thinking Like a Geographer

☑ **IMAGE** Hiker Reading a Map

☑ **IMAGE** Rice Paddy in Southern China

☑ **IMAGE** Scientist Studying a Glacier

☑ **MAP** Perceptual Regions of the United States

☑ **INTERACTIVE SELF-CHECK QUIZ**

☑ **VIDEO** The Geographer's Craft

Reading HELPDESK

Academic Vocabulary

- **primary**
- **obtain**
- **fluctuation**

Content Vocabulary

- **spatial perspective**
- **site**
- **situation**
- **formal region**
- **functional region**
- **perceptual region**

TAKING NOTES: *Key Ideas and Details*

IDENTIFYING As you read the lesson, use a graphic organizer like the one below to identify the basic elements of geography.

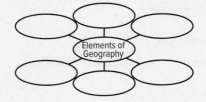

Elements of Geography

LESSON 2
The Geographer's Craft

ESSENTIAL QUESTION • *How does geography help us interpret the past, understand the present, and plan for the future?*

IT MATTERS BECAUSE

The root of the word geography *is an ancient Greek word meaning "earth description." Geographers study the location and relationships of Earth's physical and living features. They look for links between places and people and identify patterns in order to learn why those patterns exist or occur.*

A Geographic Perspective

GUIDING QUESTION *What is the spatial perspective?*

Geographers focus on understanding the world and answering questions about it. An important part of the geographical perspective is the spatial perspective. A **spatial perspective** focuses on how individual places, people, or objects are related to one another across the surface of the Earth. Using a spatial perspective, geographers examine why things are located where they are. Geographers also use the spatial perspective to examine what makes regions distinct based on changes and movements of people over time and changes in the physical environment. This spatial perspective can also be viewed from a local perspective, such as where to select a new home, or the global perspective, such as the economic activities of countries in the form of trade.

Other aspects of the geographic perspective include the ecological perspective and the perspective of experience. The ecological perspective focuses on understanding Earth as a complex set of interacting living and nonliving components. It involves thinking about the connections and interactions that operate among ecosystems and human societies. The perspective of experience considers how people make meaning from the world in which we live. As people live in locations on Earth, they have experiences and build memories that give places on Earth unique characteristics. Awareness and sensitivity to these uniquenesses are an important part of the perspective of experience.

Not all people who study or incorporate geography into their profession are labeled geographers. Geography skills can be applied to a variety of fields, including government, business, and education.

One broad cluster of career opportunities in geography is teaching and education. Teaching opportunities exist at all levels, from elementary to high school to university levels of education. Teachers with a background or training in geography topics are in demand in the United States for elementary and high schools. Students with formal geographic training from a university can find work in many diverse businesses, industries, and professional fields.

Because geography itself has many specialized fields, there are many ways that people use geography in their work. Those with a knowledge of physical geography can work as meteorologists who study the atmosphere and weather patterns; as emergency management officials dealing with natural hazards, such as earthquakes or hurricanes; as ecologists who study the interrelationships of organisms and their environment; as soil scientists; and as environmental managers. Work in the environmental field includes assessing the environmental impact of proposed development projects regarding air and water quality and wildlife.

spatial perspective a way of looking at the human and physical patterns on Earth and their relationships to one another

Skills for Thinking Like a Geographer

Skill	Examples	Tools and Technologies
Asking Geographic Questions helps you pose questions about your surroundings	• Why has traffic increased along this road? • What should be considered when building a new community sports facility?	• Maps • Globes • Internet • Remote sensing • News media
Acquiring Geographic Information helps you answer geographic questions	• Compare aerial photographs of a region over time. • Design a survey to determine who might use a community facility.	• Direct observation • Interviews • Reference books • Satellite images • Historical records
Organizing Geographic Information helps you analyze and interpret information you have collected	• Compile a map showing the spread of housing development over time. • Summarize information obtained from interviews.	• Field maps • Databases • Statistical tables • Graphs • Diagrams • Summaries
Analyzing Geographic Information helps you look for patterns, relationships, and connections	• Draw conclusions about the effects of road construction on traffic patterns. • Compare information from different maps that show available land and zoning districts.	• Maps • Charts • Graphs • GIS • Spreadsheets
Answering Geographic Questions helps you apply information to real-life situations and problem solving	• Present a report showing the results of a case study. • Suggest locations for a new facility based on geographic data gathered.	• Sketch maps • Reports • Research papers • Oral or multimedia presentations

One of the most important geographic tools is the ability to think geographically. The five skills identified above are key to geographic understanding.

▲ CRITICAL THINKING

1. *Analyzing* What types of patterns might you recognize by comparing aerial photographs of a region over a specific time period?

2. *Identifying the Central Issue* Provide a real-life example of how you might apply three of the skills described in the chart. Explain the issue, including which skills you chose, and how they would be used.

Those with knowledge of human geography find work in many areas such as health care, transportation, population studies, economic development, public policy, and international economics. Human geographers with a background in urban planning are hired by local and state government agencies to focus on projects such as housing and community development and parks and recreation planning. An economic geographer examines human economic activities. He or she may work at such tasks as market analysis and site selection for stores, factories, and restaurants. A regional geographer studies the features of a particular region and may assist government and businesses in making decisions about land use. Geographers can also find employment as writers and editors for publishers of textbooks, maps, atlases, news and travel magazines, and Web sites.

One of the most important tools of a geographic perspective is the ability to think geographically about the world. Geographers use five basic skills that are key to geographic understanding: asking, acquiring, organizing, analyzing, and answering geographic questions. Asking geographic questions provides information that can be used to better understand one's surroundings. As knowledge is acquired, it can be applied to recognize patterns and relationships that will help in real-life situations.

✔ **READING PROGRESS CHECK**

Listing List at least three fields outside of geography that use geography skills.

The Elements of Geography

GUIDING QUESTION *What are the elements of geography?*

Geography uses the geographic perspective to study the peoples, places, environments, histories, and cultures of the world's regions. Geographers study the interactions between peoples, places, and environments to explain why and how patterns of interaction occur. There are six overall elements geographers consider in their work: the world in spatial terms, places and regions, physical systems, human systems, environment and society, and the uses of geography.

Maps can be used to determine site and situation.

▼ **CRITICAL THINKING**

1. Speculating What could this hiker's map tell her about the site and situation of the landscape she is viewing?

2. Describing Explain how a map, such as the one shown, could be used to identify spatial relationships.

Thorsten Henn/Getty Images

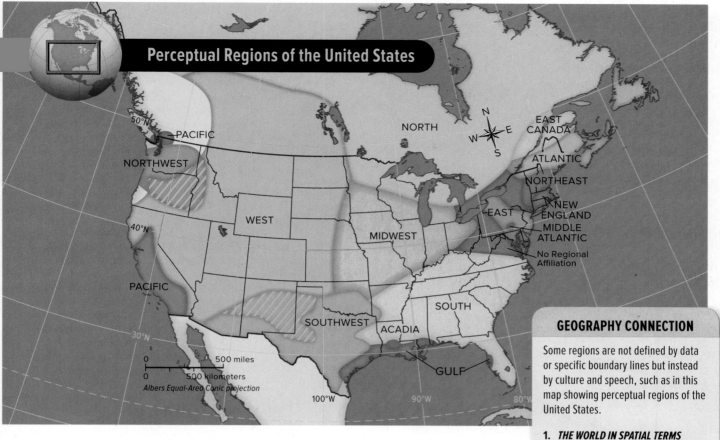

Perceptual Regions of the United States

PACIFIC
NORTHWEST
WEST
PACIFIC
SOUTHWEST
ACADIA
MIDWEST
SOUTH
GULF
NORTH
EAST CANADA
ATLANTIC
NORTHEAST
EAST
NEW ENGLAND
MIDDLE ATLANTIC
No Regional Affiliation

50°N
40°N
30°N
100°W
90°W
80°W

0   500 miles
0   500 kilometers
Albers Equal-Area Conic projection

GEOGRAPHY CONNECTION

Some regions are not defined by data or specific boundary lines but instead by culture and speech, such as in this map showing perceptual regions of the United States.

1. *THE WORLD IN SPATIAL TERMS* According to the map, in which region do you live?

2. *PLACES AND REGIONS* What aspects of culture do you think define a perceptual region in the United States?

The World in Spatial Terms

Spatial relationships link people and places based on their locations and relationships to each other. Location is a reference point for geographers in the same way that dates serve as reference points for historians.

One way of locating a place is by describing its absolute location—the exact spot at which the place is found on the Earth. To determine absolute location, geographers use the system of latitude and longitude. On a day-to-day basis, humans tend to identify a place based on relative location—a place's location in relation to another place. For example, New Orleans is located near the mouth of the Mississippi River. Knowing the relative location of a place helps you to think spatially. By creating a mental map based on relative location, you can orient yourself in space and develop an awareness of the world around you.

The broad or specific definition of a geographic location based on relative or absolute location also takes into account a place's site and situation. **Site** is the specific location of a place, including its physical setting. For example, the site of San Francisco is its location at the end of a peninsula in northern California. **Situation** refers to a more general location, defined by a place's geographic position in relation to other places and its connections to other regions. San Francisco's situation is a port city on the Pacific coast, close to California's agricultural lands.

site the specific location of a place, including its physical setting

situation the geographic position of a place in relation to other places or features of a larger region

Places and Regions

A place has physical and human significance. It has distinguishing characteristics defined by its features and surroundings. Geographers study and assess the similarities and differences between places to express what features are unique to each place. To interpret the Earth's complexity, geographers group places with similar characteristics into regions. A region can be defined by physical traits

Rice paddy fields in southern China reflect both human systems and physical systems at work.

▲ CRITICAL THINKING

1. Categorizing What type of region would you categorize the rice paddy fields as being? Explain your choice.

2. Formulating Questions Create one question you would ask the people living in the area that could help you determine whether the rice paddy is a formal or perceptual region.

formal region a region defined by a common characteristic, such as production of a product

primary of first rank, importance, or value

functional region a central place and the surrounding territory linked to it

such as climate, landforms, soils, vegetation, animal life, and natural resources. A region can also have human significance, as defined by characteristics such as language, religion, political or economic systems, and population distribution. Geographers identify three types of regions: formal, functional, and perceptual.

A **formal region** features a unifying characteristic, such as a product produced in that region. For example, the Corn Belt is a band of farmland stretching from Ohio to Nebraska in the United States. It is a formal region because corn is its **primary** crop. A **functional region** incorporates a central node and a surrounding area that is connected to the node by some defined function. For example, a cell tower provides the central node for a surrounding area in which cell phone users can obtain phone reception. A **perceptual region** uses a looser standard for characterization, defined more by commonly accepted tradition or value than by objective data. For example, the term "heartland" refers to a central area in the United States in which traditional values of family and hospitality are believed to predominate. A perceptual region could also be labeled a vernacular region. This refers to patterns native to a particular region in spite of boundary lines. The Creole dialect that is spoken in southern Louisiana is an example of a vernacular region. It is defined more by the culture and speech of the region than by a designation of state and city boundaries.

Physical Systems and Human Systems

Because geography can cover a broad range of themes, geographers divide their focus into major branches: physical geography and human geography. Physical geography—climate, land, water, plants, and animal life—looks at these processes and their significance to humans. Human geography, or cultural geography, analyzes human activities and their relationship to the cultural and physical environments. Political, economic, social, and cultural factors can include themes such as urban development, economic production and consumption, and population change. Because physical and human geography are still very broad in their focus, they can be further divided into subject areas. For example, climatology is the study of climate and long-term atmospheric conditions and their impact on ecology and society. Historical geography is the study of places and human activities over time based on the geographic factors that have shaped them.

Geographers study how physical features and processes of land, water, and climate interact with plants and animals to create, support, or change ecosystems. An ecosystem is a community of plants and animals that depend upon one another and their surroundings for survival. Geographers also study the processes by which people operate across Earth's surface—how they settle the Earth, form societies, and create permanent features. A recurring theme in geography is the ongoing movement of people, goods, and ideas. Human migration and settlement, as well as the exchange of ideas and practices among cultures, can over time transform societies, traditions, and the landscape in which humans live. In studying human systems, geographers look at how people compete or cooperate to change or control aspects of the Earth to meet their needs.

Environment and Society

The relationship between people and their physical environment is a theme embodied by human-environment interaction. Ways in which people use their surroundings, ways in which they change it voluntarily and involuntarily, and the consequences that result from such human-environment interaction are very important themes for geographers. Pollution, construction, human population growth, conservation of parks, and reintroduction of species into the wild are just a few of the ways humans change the physical environment. Yet the physical environment can also have an effect on humans. For example, physical barriers such as rivers, mountains, and deserts limit human movement and growth. Natural phenomena such as hurricanes, earthquakes, and heavy storms or droughts force humans to adapt their activities and lifestyles to the changing environment. By understanding how the Earth's physical features and processes shape and are shaped by human activity, geographers help societies make informed decisions about their relationship with their surrounding physical environment.

The Uses of Geography

Geography provides insight into how physical features and living things developed in the past. It also takes into account current trends regarding the physical and human environment in order to plan for future needs. Planning and policy making must account for interactions between humans and the natural environment. Data regarding physical features and processes can highlight suitable sites for resource extraction or for human habitation. Urban planners analyze trends in human growth within a specified region to determine where and what systems, such as schools, roads, public services, and businesses, are necessary for supporting a growing population. Geographers analyze past data in order to determine effective future actions to sustain and support both the natural environment and human development. Although people trained in geography are in great demand in the workforce, many of them do not have geographer as a job title. Geography skills are useful in so many different situations that geographers have more than a hundred different job titles.

perceptual region a region defined by popular feelings and images rather than by objective data

Scientists study Russell Glacier in Greenland and the causes of its rapid melting.

▲ CRITICAL THINKING

1. *Formulating Questions* List three questions the scientist might be asking about the melting glacier.

2. *Evaluating* What words would you use to describe the site where the scientist is located?

Geographers work in a variety of jobs in government, business, and education. They often combine the study of geography with other areas of study. For example, an ecologist must know the geographic characteristics of a place or region in which he or she studies living organisms. Similarly, a travel agent must have knowledge of the physical and human geography of a place in order to plan trips for clients.

✔ READING PROGRESS CHECK

Identifying What are the three types of regions?

Research Methods

GUIDING QUESTION *What methods do geographers use to conduct their work?*

To do their work, geographers use several research methods. Direct observation and measurement, mapping, interviewing, production and use of statistics, and the use of technology are all specialized research methods used by geographers.

Direct observation and measurement involves analysis of patterns of human activity that take place on the Earth's surface. Geographers using this method will visit a place to gather information about it from what they observe of the place and its geographic features. Geographers also employ remote sensing from satellite images and aerial photographs to locate specific information without having to visit the site in person. For example, aerial photographs or satellite images can be used to locate mineral deposits, to determine the size of freshwater sources, or to see the extent of urban sprawl.

Mapping is essential to geographers. Many findings from geography research can be shown visually and spatially on maps better than they can be explained through statistical methods or written documents. Complex information can be collected and shown in more easily understood terms through using maps that highlight features, patterns, and relationships of people, places, and things. Maps also are useful for making comparisons. For example, a geographer might compare population density maps or transportation networks maps of two counties in order to determine where to build new schools.

Interviewing requires a geographer to ask questions rather than just collect data, images, and on-site observations. Specifically, for human geographic studies, geographers may want to find out how people think or feel about certain places. They may also want to examine the ways in which people's beliefs and attitudes have affected the physical environment. To **obtain** such information, geographers interview their subjects. They can do this by selecting a particular group of people for study. Rather than contacting every individual in the group, however, geographers use a carefully selected sample of people whose answers represent the larger group.

Geographers also analyze *statistics*. Numerical data, such as temperature and snowfall, can provide insight into a region's climate trends. Geographers use computers to organize and present this information in an understandable manner, as well as to look in detail at the data for patterns and trends. For example, studies that identify age, ethnicity, and gender of specified regions can emphasize possible trends within a human population. After identifying such patterns and trends, geographers use statistical tests to see whether their ideas are valid.

obtain to gain or acquire, usually by planning or effort

☑ READING PROGRESS CHECK
Defining What do geographers do?

Geography and Other Subjects

GUIDING QUESTION *How is geography related to other subjects?*

Geography has important relationships to other subjects. Geographers use geographic tools and methods to understand historical patterns, economies, politics and political patterns, and the impact of societies and cultures on the landscape.

To visualize what places could have looked like in the past, geographers take into account historical perspectives of the place. For example, to gather information about how a city has changed over time, geographers can collect information from historical sources regarding census data, economic output, birth and death rates, natural disasters, disease, and major **fluctuations** in population size. Such data can address questions concerning how human activities have changed the natural vegetation, or how waterways are different today than in the past. Such historical perspectives provide insight as to which institutions or development should be constructed to avoid repeating past complications between human growth and the physical environment.

fluctuation a shift from a previous condition

Additionally, analysis of historical and current political patterns emphasizes changing boundary lines and government systems. Geographers are also interested in how the natural environment has influenced political decisions and how governments change natural environments. For example, in the 1960s the Egyptian government built the massive Aswān High Dam on the Nile River to help irrigate the land. The dam altered the Nile River valley and significantly impacted the region's people.

Human geographers, also called cultural geographers, use the ideas of sociology and anthropology to study human tendencies and past cultures and their influence on current traditions and social norms. Because people come from diverse cultural backgrounds, their interpretations of information and experiences differ depending on their frame of reference. For example, residents of a particular neighborhood may define boundaries based on location of activities, such as stores they frequent and people with whom they have contact.

The natural shelter and deep waters of Victoria Harbor make Hong Kong one of the world's largest shipping ports.

▲ **CRITICAL THINKING**

1. *Analyzing* What factors might a geographer study to learn about the economy of a particular country?

2. *Making Connections* How is interdependence important to economic activity?

However, the local governments create neighborhood boundaries to facilitate services and maintenance. Furthermore, other neighborhood boundaries may be created by police departments and school districts that have different needs to meet and different reasons for creating spatial divisions. Human geography can be used to study the relationships between people, places, and environments by mapping information about them into a spatial context using GIS and other geospatial technologies. Human geographers study the way people are rooted in particular places and how they have constructed various types of regions. Some geographers specialize in studying the feelings one has about a place, which is very closely connected with theories in the fields of psychology and behavioral science.

Geographers study economies to understand how the locations of resources affect the ways people make, transport, and use goods, and how and where services are provided. Geographers are interested in how locations are chosen for various economic activities. Where and what human groups choose to produce and consume depend on a variety of factors: location of natural resources for mining and extracting, fertile soil for farming, suitable climates for living and producing, and proximity to good transport routes and other cultures to establish trade relations.

Economic activity relies on not only a society's production and use of goods, but also on the transport of goods between cultures in the form of trade. Such interdependence between global economies is part of what defines relationships and communication between various cultures. The growth of technological and communication systems in today's world also affects these relationships. The ability to call a client halfway across the world, reduce production time by mechanical production instead of human labor, use the Internet to communicate ideas instantly, and send goods overnight via air delivery are examples of how innovations in technology have increased the speed and efficiency of the movement of information and goods.

✓ **READING PROGRESS CHECK**

Naming What is another name used for human geographers?

LESSON 2 REVIEW

Reviewing Vocabulary

1. *Understanding Relationships* How are site and situation different?

Using Your Notes

2. *Listing* Using your graphic organizer, write a definition of *geography* in your own words.

Answering the Guiding Questions

3. *Identifying* What is the spatial perspective?

4. *Categorizing* What are the elements of geography?

5. *Organizing* What methods do geographers use to conduct their work?

6. *Evaluating* How is geography related to other subjects?

Writing Activity

7. *Informative/Explanatory* Write a paragraph explaining how geography helps us interpret the past, understand the present, and plan for the future.

Directions: On a separate sheet of paper, answer the questions below. Make sure you read carefully and answer all parts of the questions.

Lesson Review

Lesson 1

1 **Describing** Describe the problems that arise when the curves of a globe become straight lines on a map.

2 **Comparing and Contrasting** Explain the similarities and differences between the Winkel Tripel projection and the Mercator projection.

3 **Describing** What is the importance of scale in reading maps?

4 **Listing** List three examples of things a map can show.

Lesson 2

5 **Explaining** Why is the U.S. Corn Belt considered a formal region?

6 **Discussing** What are two research methods used by geographers?

7 **Differentiating** What is the difference between physical geography and human geography?

8 **Summarizing** Why is human-environment interaction an important theme for geographers?

Applying Map Skills

Refer to the Unit 1 Atlas to answer the following questions.

9 **Places and Regions** Which continent has the most countries with the highest population densities?

10 **Environment and Society** Look at the map of world economic activities. Name three forms of land use for the continent of Africa.

11 **Physical Systems** What physical features in Africa might explain areas with little or no economic activity?

12 **The World in Spatial Terms** Look at the political map. What is the absolute location of Houston, Texas? What is the relative location of Houston?

DBQ Analyzing Primary Sources

Use the cartoon to answer the following questions.

PRIMARY SOURCE

13 **Analyzing Visuals** Why do we see the globe as being upside down? How is the penguin's point of view different? Why?

14 **Finding the Main Idea** What does this cartoon say about the broad definition of geography?

Exploring the Essential Question

15 **Understanding Historical Interpretation** Imagine you are studying an archaeological site to determine why a civilization came to end. What kind of geographic information might you want to gather in order to determine what happened to the civilization? Consider the skills for thinking geographically. Write a paragraph containing your response.

Need Extra Help?

If You've Missed Question	1	2	3	4	5	6	7	8	9	10	11	12	13	14	15
Go to page	14	15	18	18	30	32	30	30	2	2	2	2	35	35	32

Directions: On a separate sheet of paper, answer the questions below. Make sure you read carefully and answer all parts of the questions.

Critical Thinking

16 *Assessing* What are the different types of geospatial technologies? How has this advanced technology improved the way maps are created?

17 *Making Connections* What type of information would help you determine why traffic has increased on a certain road?

18 *Evaluating* How could geography be interpreted differently based on changing human perspectives?

19 *Drawing Conclusions* How can we use geography to make decisions for the future?

20 *Compare and Contrast* What differentiates mental mapping from other forms of visual mapping?

College and Career Readiness

21 *Examining Information* Describe three ways geographic knowledge or tools are used in professions other than geography.

Research and Presentation

22 *Research Skills* Using the Internet, conduct research to learn about how geography is used in the film industry. Write one paragraph describing a specific example of this use.

Writing About Geography

23 *Informative/Explanatory* Imagine you are a geographer working on a plan for a new community center. Using the different types of research methods as your guide, what factors would you use to develop the plan? Explain your choices in a paragraph.

21st Century Skills

Use the excerpt below to answer the questions that follow.

PRIMARY SOURCE

"Geography—real-world geography—is the art and science of location, or place. It is about spatial patterns and spatial processes. It is about which way the wind blows from Chernobyl, the Pacific 'ring of fire,' AIDS, terrorists, and refugees. It is about acid rain, El Niño, ocean dumping, cultural censorship, droughts and famines. . . .

Real-world geography also explores things in locations: why something is where it is and what processes change its distribution. Geography is the why of where of an ever-changing universe. Its surpassing objective is to discover the processes that move over space and connect places and continually transform the location and character of everything."

—George J. Demko, *Why in the World: Adventures in Geography,* 1992

24 *Geography Skills* What are some of the world issues that George Demko lists as concerns of geography?

25 *Primary and Secondary Sources* According to the excerpt, what is the main objective of geography?

26 *Identifying Cause & Effect* Identify how the physical environment affects human activity, and likewise how human activity affects the physical environment.

27 *Economics* How might the economic activities of a region affect its physical and human geography?

Need Extra Help?

If You've Missed Question	**16**	**17**	**18**	**19**	**20**	**21**	**22**	**23**	**24**	**25**	**26**	**27**
Go to page	21	22	28	31	20	33	36	32	36	36	36	36

The Physical World

ESSENTIAL QUESTION • *How do physical processes shape Earth's surface?*

Why Geography Matters
Economics and Resources: Water Scarcity

Lesson 1
Planet Earth

Lesson 2
Forces of Change

Lesson 3
Earth's Water

Geography Matters...

Because much of our world had not been explored before 1800, geography prior to that time focused on discovery, basic data collection, and observation. Geographers were busy determining the elevation of land surfaces, describing landforms, identifying weather and climate patterns, and classifying soils and ecosystems. The geographers of today use sophisticated tools and research methods to monitor the processes that shape Earth's surface, the natural resources essential to our survival, and the ways in which humans affect and are affected by the physical environment. Understanding the planet we call home helps us make better decisions about how to utilize, preserve, and protect the Earth and adapt to its changing environments.

◀ A climber rests alone on Mount Everest.

Harry Kikstra/Moment/Getty Images

economics *and* resources:
water scarcity

Water is a basic need for all humans. Providing freshwater to support human activity is a significant challenge in drought-prone areas around the world. Agriculture, industry, and thriving population centers all depend on a reliable supply of safe, fresh water to support human endeavors.

THERE'S MORE ONLINE

SEE a diagram of the desalination process • *READ* an infographic about freshwater resources

How does water scarcity impact economic activity?

North Africa and Southwest Asia are two areas of the world where there is a scarcity—lack of an adequate supply—of freshwater for agriculture, industry, drinking, bathing, and other human uses of water due to limited resources. A scarcity of freshwater limits the ability of a city, region, or country to engage in many economic pursuits that allow people to live comfortably and thrive. Without water, agricultural fields cannot be irrigated, industrial production cannot occur, and hospitals cannot provide the necessary quality of care.

1. **Environment and Society** Why is water scarcity a problem? Explain what the term *water scarcity* means and discuss how it affects human activities.

How is technology being used to solve the problem of water scarcity?

The town of Ashqelon is located on Israel's Mediterranean coast with an abundant supply of salt water lapping at its shore. Desalination (salt removal) technology is being used to turn the sea water into freshwater for Israeli citizens and industries that have faced recent droughts and population growth. Ashqelon, 40 miles (64.4 km) south of Tel Aviv, has the largest desalination plant of its kind in the region and one of the largest in the world. Seawater is pumped through a series of filters and reverse-osmosis membranes which allow the smaller water molecules to pass through while excluding the larger molecules of salt and other impurities. The plant provides approximately five percent of Israel's freshwater and is predicted to soon make Israel an exporter of water.

2. **Human Systems** How has desalination technology affected Israel's economy? What countries may import freshwater from Israel?

What are the environmental effects of desalination?

The freshwater provided by the Ashqelon desalination plant supports Israel's thriving agricultural and high-tech industries, both heavily dependent on a steady supply of freshwater. However, this water comes with several costs to the environment. Fossil fuels are used to power the pumps that circulate the water through the filter plant, contributing to pollution. The brine, or leftover material that the filters remove, is pumped back into the sea, resulting in a salty desert on the sea floor around the outflow pipes. The plant also occupies a coastal area that might otherwise harbor wildlife and offer recreation options.

3. **Environment and Society** Explain the costs and benefits of providing freshwater using reverse-osmosis desalination.

networks

There's More Online!

☑ **DIAGRAM** The Solar System

☑ **DIAGRAM** Underwater Landforms

☑ **DIAGRAM** Water, Land, and Air

☑ **INTERACTIVE SELF-CHECK QUIZ**

☑ **VIDEO** Planet Earth

SUN MERCURY BIOSPHERE

LESSON 1
Planet Earth

ESSENTIAL QUESTION • *How do physical processes shape Earth's surface?*

Reading HELPDESK

Academic Vocabulary

• **sphere**
• **theory**

Content Vocabulary

• **hydrosphere**
• **lithosphere**
• **atmosphere**
• **biosphere**
• **continental shelf**

TAKING NOTES: *Key Ideas and Details*

DESCRIBING Use a graphic organizer like the one below to list descriptions for the components that make life on Earth possible: the hydrosphere, lithosphere, atmosphere, and biosphere.

Component	Description
hydrosphere	
lithosphere	
atmosphere	
biosphere	

IT MATTERS BECAUSE

Physical processes shape Earth's surface. Understanding that Earth is part of a larger physical system called the solar system helps us see how it is possible for life on our planet to survive and thrive. Earth's physical systems are affected by natural forces such as earthquakes and volcanoes that can influence human activity on the planet.

Our Solar System

GUIDING QUESTION *In what physical system does Earth exist?*

Earth is part of our solar system, which is made up of the sun and all of the countless objects that revolve around it. At our solar system's center is the sun—a star, or ball of burning gases. About 109 times wider than Earth, the sun's enormous mass—the amount of matter it contains—creates a strong pull of gravity. This basic physical force keeps the Earth and the other objects revolving in orbit around the sun.

Except for the sun, **spheres** called planets are the largest objects in the solar system. At least eight planets are known to exist, and each is in its own orbit around the sun. Mercury, Venus, Earth, and Mars are the inner planets, or those nearest the sun. Earth, the third planet from the sun, is about 93 million miles (150 million km) away from the sun. Farthest from the sun are the outer planets—Jupiter, Saturn, Uranus, and Neptune.

The planets vary in size with Jupiter being the largest. Earth ranks fifth in size, and Mercury is the smallest. All of the planets except Mercury and Venus have moons—smaller spheres or satellites that orbit them. Earth has 1 moon, and Saturn has at least 18 moons. At least five other objects are dwarf planets. Pluto was known as the smallest planet in the solar system until 2003 when astronomers changed its status. Today Pluto is called a *dwarf planet*. Dwarf planets are small round bodies that orbit the sun, but do not have enough gravity to have cleared the area around their orbits of other orbiting bodies, thus making them too small to be considered planets.

The four inner planets are called *terrestrial planets* because they have solid, rocky crusts. Mercury and Venus are scalding hot, and Mars is a cold, barren desert. Only Earth has temperatures that are moderate enough to allow liquid water at the surface and to support a variety of life.

The four outer planets are called the *gas giant planets*. They are more gaseous and less dense than the terrestrial planets, although they are larger in diameter. Each gas giant is like a miniature solar system, with orbiting moons and encircling rings. Only Saturn's rings, however, are easily seen from Earth by telescope.

Thousands of smaller objects—including asteroids, comets, and meteoroids— revolve around the sun. Asteroids are small, irregularly shaped, planet-like objects. They are found mainly between Mars and Jupiter in the asteroid belt. A few asteroids follow paths that cross Earth's orbit. Others, like the recently discovered asteroid 2015 PDC, are held in a balance between the gravitational pull from the sun and an equal force from the Earth.

Comets, made of icy dust particles and frozen gases, look like bright balls with long, feathery tails. Their orbits are inclined at every possible angle to Earth's orbit. They may approach from any direction.

Meteoroids are pieces of space debris—chunks of rock and iron. When they occasionally enter Earth's atmosphere, friction usually burns them up before they reach the Earth's surface. Those that collide with Earth are called meteorites. Meteorite strikes, although rare, can significantly affect the landscape, leaving craters and causing other devastation. In 1908 a huge area of forest in the remote Russian region of Siberia was flattened and burned by a mysterious fireball. Scientific **theory**—a plausible general principle offered to explain observed facts—speculates that it was a meteorite or comet. A writer describes the effects:

PRIMARY SOURCE

❝The heat incinerated herds of reindeer and charred tens of thousands of evergreens across hundreds of square miles. For days, and for thousands of miles around, the sky remained bright with an eerie orange glow—as far away as western Europe people were able to read newspapers at night without a lamp.❞

—Richard Stone, "The Last Great Impact on Earth," *Discover*, September 1996

✔ **READING PROGRESS CHECK**

Drawing Conclusions What prevents most asteroids, comets, and meteoroids from colliding with Earth?

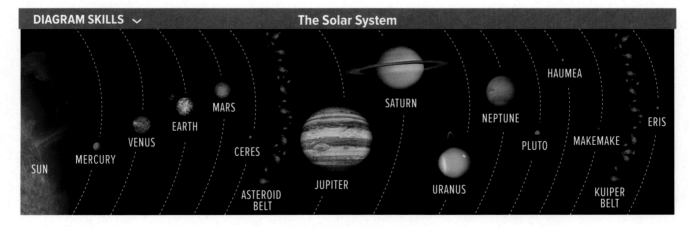

DIAGRAM SKILLS ⌄   **The Solar System**

HAUMEA

SATURN

NEPTUNE

ERIS

MARS

EARTH

MAKEMAKE

VENUS

CERES

PLUTO

MERCURY

SUN

JUPITER

URANUS

KUIPER BELT

ASTEROID BELT

Planet Earth is part of a solar system centered on the sun. Earth is one of at least eight planets orbiting the sun.

▲ **CRITICAL THINKING**

1. ***Classifying*** Which four planets are closest to the sun?

2. ***Drawing Conclusions*** Why might it be impossible for life to exist on Neptune? Think about where it is located.

(sun)NASA/GSFC/SDO, (Mercury, Earth, Mars, Uranus)Caltech/JPL/USGS/NASA, (Venus, Jupiter, Saturn, Neptune, Pluto)JPL/Caltech/NASA, (star maps)Caltech/JPL/NASA

sphere a globe-shaped body

theory a plausible general principle offered to explain observed facts

hydrosphere the water areas of the Earth, including oceans, lakes, rivers, and other bodies of water

lithosphere uppermost layer of the Earth that includes the crust, continents, and ocean basins

atmosphere a thin layer of gases that surrounds the Earth

biosphere the part of the Earth where life exists

Getting to Know Earth

GUIDING QUESTION *How does the biosphere support life on Earth?*

The Earth is a rounded object that is slightly wider around the center than from top to bottom. Earth has a larger diameter at the Equator—about 7,930 miles (12,760 km)—than from Pole to Pole, but the difference is less than 1 percent. With a circumference of about 24,900 miles (40,060 km), Earth is the largest of the inner planets in the solar system.

The surface of the Earth is made up of water and land. About 70 percent of our planet's surface is water. Oceans, lakes, rivers, underground water, and other bodies of water make up a part of the Earth called the **hydrosphere**.

About 30 percent of the Earth's surface is land, including continents and islands. Land makes up a part of the Earth called the **lithosphere**, the Earth's crust. The lithosphere also includes the ocean basins, or the land beneath the oceans.

The air we breathe is part of Earth's **atmosphere**, a thin layer of gases extending above the planet's surface. The atmosphere is composed of 78 percent nitrogen, 21 percent oxygen, and small amounts of argon and other gases.

All people, animals, and plants live on or close to the Earth's surface or in the atmosphere. The part of the Earth that supports life is the **biosphere**. Life outside the biosphere, such as on a space station orbiting Earth, exists only with the assistance of mechanical life-support systems.

DIAGRAM SKILLS ⌄   **Water, Land, and Air**

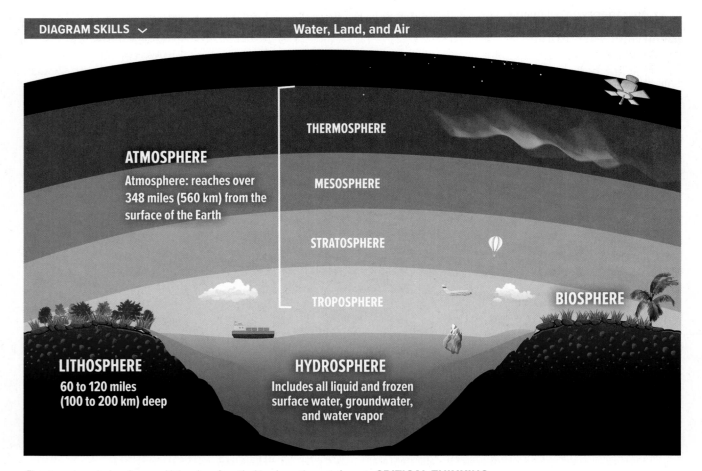

THERMOSPHERE

MESOSPHERE

STRATOSPHERE

TROPOSPHERE

ATMOSPHERE
Atmosphere: reaches over 348 miles (560 km) from the surface of the Earth

BIOSPHERE

LITHOSPHERE
60 to 120 miles (100 to 200 km) deep

HYDROSPHERE
Includes all liquid and frozen surface water, groundwater, and water vapor

The atmosphere, hydrosphere, and lithosphere form the biosphere, the part of Earth where life exists.

▲ **CRITICAL THINKING**

1. Classifying What are Earth's water systems called?

2. Drawing Conclusions How does human activity impact the biosphere?

Landforms are natural features at the surface of the Earth's lithosphere. Many of the Earth's landforms have a particular shape or elevation. Landforms often contain rivers, lakes, and streams.

Underwater landforms are as diverse as those found on dry land. In some places the ocean floor is a flat plain. Other parts feature mountain ranges, cliffs, valleys, and deep trenches. Seen from space, Earth's most visible landforms are the seven large landmasses called continents. Australia and Antarctica stand alone, while the others are joined in some way. Europe and Asia are parts of one landmass called Eurasia. A narrow strip of land called the Isthmus of Panama links North America and South America. At the Sinai Peninsula, the human-made Suez Canal separates Africa and Asia.

The **continental shelf** is an underwater extension of the coastal plain. Continental shelves slope out from land for as far as 800 miles (1,287 km). They descend gradually to a depth of about 660 feet (200 m), where a sharp drop marks the beginning of the continental slope. This area drops sharply to the ocean floor.

Great contrasts exist in the heights and depths of the Earth's surface. The highest point on Earth is in South Asia at the top of Mount Everest, which is 29,028 feet (8,848 m) above sea level. The lowest dry land point at 1,312 feet (400 m) below sea level is the shore of the Dead Sea in Southwest Asia. Earth's deepest known depression lies under the Pacific Ocean, southwest of Guam in the Mariana Trench, a narrow, underwater canyon about 36,198 feet (11,033 m) deep.

☑ **READING PROGRESS CHECK**

Categorizing What organisms might live in the hydrosphere?

Mt. Everest has such low oxygen levels and atmospheric pressure that hikers must prepare their bodies well in advance to adjust to altitude changes in order to avoid illness or death.

▲ **CRITICAL THINKING**

1. *Categorizing* In which part of the Earth's biosphere is Mt. Everest located?

2. *Predicting* Mt. Everest has such high altitudes that it is almost outside of which part of the biosphere?

continental shelf part of a continent that extends out underneath the ocean

---

LESSON 1 REVIEW

Reviewing Vocabulary

1. *Explaining* Define *continental shelf* and explain where it is located.

Using Your Notes

2. *Describing* Use your graphic organizer to describe the three parts of the Earth's biosphere.

Answering the Guiding Questions

3. *Identifying* In what physical system does Earth exist?

4. *Discussing* How does the biosphere support life on Earth?

Writing Activity

5. *Narrative* Consider the ratio of water and land on Earth. How might life on Earth be different if the proportions were reversed?

networks

There's More Online!

☑ **DIAGRAM** Forces of Change

☑ **DIAGRAM** Inside the Earth

☑ **MAP** Continental Drift

☑ **MAP** Tectonic Plate Boundaries

☑ **INTERACTIVE**
SELF-CHECK QUIZ

☑ **VIDEO** Forces of Change

Reading **HELP**DESK

Academic Vocabulary

- **create**
- **external**

Content Vocabulary

- **core**
- **mantle**
- **crust**
- **continental drift**
- **plate tectonics**
- **magma**
- **subduction**
- **accretion**
- **spreading**
- **fold**
- **fault**
- **faulting**
- **weathering**
- **erosion**
- **glacier**
- **moraine**

TAKING NOTES: *Key Ideas and Details*

IDENTIFYING Use a graphic organizer like the one below to decribe the processes of plate tectonics.

Force of Change	How it Works	Example

LESSON 2
Forces of Change

ESSENTIAL QUESTION • *How do physical processes shape Earth's surface?*

IT MATTERS BECAUSE
Plate tectonics acts upon the Earth's internal and external structures to help create the continents, ocean basins, and mountain ranges. Plate tectonics operates by folding, lifting, bending, and breaking parts of the Earth's surface. Other forces such as weathering and erosion also help shape the Earth's surface.

Earth's Structure

GUIDING QUESTION *How is Earth's structure related to the creation of continents, oceans, and mountain ranges?*

For hundreds of millions of years, the surface of the Earth has been in motion. Pressures generally build up slowly inside the Earth and are then released in sudden events such as volcanic eruptions and earthquakes. Other forces that change the Earth, such as wind and water, occur on the surface.

The Earth is composed of three main layers—the core, the mantle, and the crust. At the very center of the planet is a super-hot but solid inner **core**. Scientists believe that the inner core is made up of iron and nickel that is under enormous pressure. Surrounding the inner core is another band also composed of iron and nickel called the liquid outer core. Even though the liquid outer core is composed of the same elements as the inner core, it is liquid because the pressure is not as great as it is in the inner core.

Next to the outer core is a thick layer of hot, dense rock called the **mantle**. The mantle consists of silicon, aluminum, iron, magnesium, oxygen, and other elements. This dense mixture is soft enough to slowly but continually rise, cool, sink, warm up, and rise again, releasing 80 percent of the heat generated from the Earth's interior.

The outer layer is the **crust**, a hard rocky shell forming the Earth's surface. This relatively thin layer of rock ranges from about 2 miles (3.2 km) thick under oceans to about 75 miles (120.7 km) thick under mountains. The crust is broken into more than a dozen great slabs of

Jim Kruger/Getty Images

rock called plates that rest—or more accurately, float—on a partially melted layer in the upper portion of the mantle. The plates carry the Earth's oceans and continents.

If you had seen the Earth from space 500 million years ago, the planet probably would not have looked at all like it does today. Many scientists believe that most of the landmasses forming our present-day continents were once part of one gigantic supercontinent called Pangaea (pan•JEE•uh). The maps on the next page show that over millions of years, this supercontinent has broken apart into smaller continents. These continents in turn have drifted and, in some places, recombined. The theory that the continents were once joined and then slowly drifted apart is called **continental drift**.

The term **plate tectonics** refers to all of the physical processes that create many of the Earth's physical features. Many scientists theorize that plates moving around the globe have produced Earth's largest features—not only continents, but also oceans and mountain ranges. Most of the time, plate movement is so gradual—only about 1 inch (2 to 3 cm) a year—that it cannot be felt unless there is an earthquake strong enough to detect the movement. As they move, the plates may crash into each other, pull apart, or grind and slide past each other. Whatever their actions, plates are constantly changing the face of the planet. They push up mountains, **create** volcanoes, and produce earthquakes. Plates spread apart because **magma**, or molten rock, is pushed up from the mantle and ridges are formed. When plates bump together, one may slide under another, forming a trench.

Many scientists estimate that plate tectonics has been shaping the Earth's surface for 2.5 to 4 billion years. According to some scientists, plate tectonics will have sculpted a whole new look for the planet millions of years from now that could make it difficult for us to recognize.

Scientists, however, have not yet determined exactly what causes plate tectonics. They theorize that heat rising from the Earth's core may create slow-moving currents within the mantle. Over millions of years, these currents of molten rock may shift the plates around, but the movements in the mantle are extremely slow and difficult to detect.

core innermost layer of the Earth made up of a super-hot but solid inner core and a super-hot liquid outer core

mantle thick middle layer of the Earth's interior structure consisting of hot rock that is dense but flexible

crust outer layer of the Earth, a hard rocky shell forming Earth's surface

continental drift the theory that the continents were once joined and then slowly drifted apart

plate tectonics the term scientists use to describe the activities of continental drift and magma flow, which create many of Earth's physical features

create to bring into being or cause to exist

magma molten rock that is located below Earth's surface

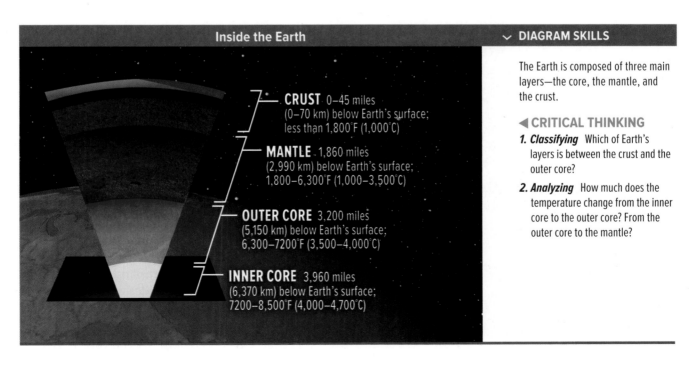

Inside the Earth

CRUST 0–45 miles (0–70 km) below Earth's surface; less than 1,800°F (1,000°C)

MANTLE 1,860 miles (2,990 km) below Earth's surface; 1,800–6,300°F (1,000–3,500°C)

OUTER CORE 3,200 miles (5,150 km) below Earth's surface; 6,300–7200°F (3,500–4,000°C)

INNER CORE 3,960 miles (6,370 km) below Earth's surface; 7200–8,500°F (4,000–4,700°C)

∨ DIAGRAM SKILLS

The Earth is composed of three main layers—the core, the mantle, and the crust.

◀ CRITICAL THINKING

1. *Classifying* Which of Earth's layers is between the crust and the outer core?

2. *Analyzing* How much does the temperature change from the inner core to the outer core? From the outer core to the mantle?

Eyewitness: Icelandic Volcano

"I woke up on Friday with a weird feeling that something just wasn't right. It wasn't light as it normally is—we don't really have night-time at this time of year.

I looked outside and there was a thick, black cloud of ash directly above us. It was exactly like the middle of winter. What is even more surreal was the absolute bright daylight on either side of our village."

—Gina Christie, BBC News, May 17, 2010

DBQ *DRAWING CONCLUSIONS*
Why do you think it could have been dark at Gina Christie's house, yet bright across town?

Many of these events occur as a series of small jumps, felt as minor tremors on the Earth's surface. A few, however, occur as sudden and violent movements of Earth's surface.

Earthquakes and Volcanoes

Sudden, violent movements of the lithosphere along fault lines are known as earthquakes. These shaking activities dramatically change the surface of the land and the floor of the ocean. During a severe earthquake in Alaska in 1964, a portion of the ground lurched upward 38 feet (11.6 m).

Earthquakes often occur where plates meet. Tension builds up along fault lines as the plates stick. The strain eventually becomes so intense that the rocks suddenly snap and shift. This movement releases stored-up energy along the fault. The ground then trembles and shakes as shock waves surge through it, moving away from the area where the rocks first snapped apart.

Disastrous earthquakes have occurred in Kōbe, Japan; in the U.S. cities of Los Angeles and San Francisco; near the Indonesian island of Sumatra; and in Oaxaca, Mexico. These places are located along the *Ring of Fire*, one of the most earthquake-prone areas on the planet. It is a zone of earthquake and volcanic activity around the perimeter of the Pacific Ocean. Here the plates that cradle the Pacific meet the plates that hold the continents surrounding the Pacific. North America, South America, Asia, and Australia are affected by their location on the Ring of Fire.

Volcanoes are mountains formed by lava or by magma that breaks through the Earth's crust. Volcanoes often rise along plate boundaries where one plate plunges beneath another, as along the Ring of Fire. In such a process, the rocky plate melts as it dives downward into the hot mantle. If the molten rock is too thick, its flow is blocked and pressure builds. A cloud of ash and gas may then spew forth, creating a funnel through which the red-hot magma rushes to the surface. There the lava flow may eventually form a large volcanic cone topped by a crater—a bowl-shaped depression at a volcano's mouth.

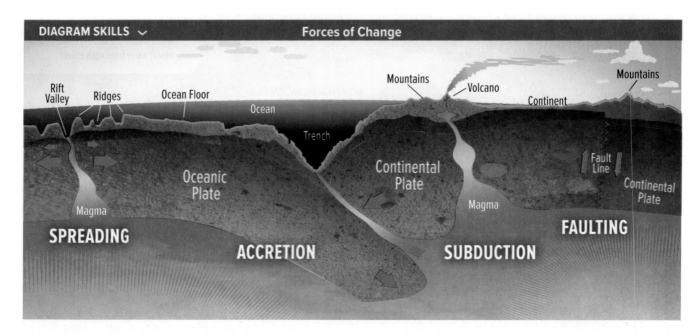

DIAGRAM SKILLS ⌄   **Forces of Change**

Rift Valley   Ridges   Ocean Floor   Ocean   Mountains   Volcano   Continent   Mountains

Trench   Fault Line

Magma   Oceanic Plate   Continental Plate   Magma   Continental Plate

SPREADING   **ACCRETION**   **SUBDUCTION**   **FAULTING**

The forces of subduction, accretion, spreading, and faulting shape our planet, creating the landforms we see today.

▲ **CRITICAL THINKING**

1. *Analyzing* How does the process of accretion create deep trenches on the Earth's surface?

2. *Speculating* What observable evidence might you see after movement along a fault line?

Volcanoes also arise in areas away from plate boundaries. Some areas deep in the Earth are hotter than others, and magma often blasts through the crust and creates volcanoes at the surface. As a moving plate passes over these hot spots, molten rock flowing out of the Earth may create volcanic island chains, such as the Hawaiian Islands. At some hot spots, molten rock may also heat underground water, resulting in hot springs or geysers like Old Faithful in Yellowstone National Park.

☑ **READING PROGRESS CHECK**

Explaining How are volcanoes formed and where are they typically located?

External Forces of Change

GUIDING QUESTION *What external forces shape Earth's surface?*

External forces, such as wind and water, also change the Earth's surface. Wind and water movements involve two processes. **Weathering** breaks down rocks, and **erosion** wears away the Earth's surface by wind, glaciers, and moving water.

Weathering and Erosion

The Earth is changed by two basic kinds of weathering. Physical weathering occurs when large masses of rock are physically broken down into smaller pieces. For example, water seeps into the cracks in a rock and freezes, expanding and causing the rock to split. Chemical weathering changes the chemical makeup of rocks. For example, rainwater that contains carbon dioxide from the air easily dissolves certain rocks such as limestone. Many of the world's caves have been and continue to be formed by this process.

Wind erosion carries small particles of dust, sand, and soil from one place to another. Plants help protect the land from wind erosion. However, in dry places where people have cut down trees and plants, winds pick up large amounts of soil and blow it away. Wind erosion can provide some benefits; the dust carried by wind often forms large deposits of mineral-rich soil. Another cause of erosion is **glaciers**, large bodies of ice that move across the Earth's surface. Glaciers form

The Colorado River has been shaping the main gorge of the Grand Canyon for thousands of years.

▲ **CRITICAL THINKING**

1. Comparing and Contrasting How does weathering differ from erosion?

2. Classifying What are the three different types of erosion?

external arising outside of

weathering chemical or physical processes that break down rocks into smaller pieces

erosion the movement of weathered rock and material by wind, glaciers, and moving water

glacier a large body of ice that moves across the surface of the Earth

Jim Kruger/Getty Images

over time as layers of snow press together and turn to ice. Their great weight causes them to move slowly downhill and spread outward. As they move, glaciers pick up rocks and soil in their paths, changing the landscape. They can remove forests, carve out valleys, alter the courses of rivers, and wear down mountaintops.

When glaciers melt and recede in some places, they leave behind large piles of rocks and debris called **moraines**. Some moraines form long ridges of land, while others form dams that hold water back and create glacial lakes.

There are two types of glaciers. Ice sheets are flat, broad sheets of ice. Today, ice sheet glaciers cover most of Greenland and all of Antarctica. They advance a few feet each winter and recede in the summer. Large blocks of ice often break off from the coastal edges of sheet glaciers to become icebergs floating in the ocean. Mountain glaciers, which are a more common type of glacier today, are located in high mountain valleys where the climate is cold. They scar the Earth's surface, gouging out rounded, U-shaped valleys as they move downhill. As they melt, rocks and soil are deposited in new locations.

Water erosion begins when springwater and rainwater flow downhill in streams, cutting into the land and wearing away the soil and rock. The resulting sediment grinds away the surface of rocks along the stream's path. Over time, the eroding action of water forms first a gully and then a V-shaped valley. Sometimes, valleys are eroded even further to form canyons. The Grand Canyon is an example of the eroding power of water. Oceans also play an important role in water erosion. Pounding waves continually erode coastal cliffs, wear rocks into sandy beaches, and move sand away to other coastal areas.

moraine piles of rocky debris left by melting glaciers

Soil Building

Soil is the product of thousands of years of weathering, erosion, and biological activity. Soil development begins when weathering breaks down solid rock into smaller pieces. Worms and other organisms help break down organic matter—dead plant and animal material—that comes to rest on these particles. Living organisms also add nutrients to the soil and create passages for air and water.

Five factors influence soil formation, with *climate* being the most significant. Wind, temperature, and rainfall determine the type of soil that can develop. *Topography*—the shape and position of Earth's physical features—affects surface runoff of water, drainage, and the rate of erosion. *Geology* determines the parent material (original rock), which influences depth, texture, drainage, and nutrient content of soil. *Biology*, living and dead plants and animals, adds organic matter to the soil. The length of *time* the other four factors have been interacting also affects soil formation. These factors combine to produce different types of soils.

☑ **READING PROGRESS CHECK**

Listing List the five factors that influence soil formation.

LESSON 2 REVIEW

Reviewing Vocabulary
1. *Explaining* Explain the difference between subduction and accretion and the significance of each.

Using Your Notes
2. *Describing* Use your graphic organizer to explain how plate tectonics folds, lifts, bends, and breaks parts of the Earth's surface.

Answering the Guiding Questions
3. *Discussing* How is Earth's structure related to the creation of continents, oceans, and mountain ranges?

4. *Defining* How does plate tectonics affect Earth's surface?

5. *Identifying* What external forces shape Earth's surface?

Writing Activity
6. *Informative/Explanatory* Describe three kinds of erosion that shape Earth's surface. How does erosion help create soil?

networks

There's More Online!

☑ **INFOGRAPHIC**
The Water Cycle

☑ **MAP** Global Desalination

☑ **INTERACTIVE**
SELF-CHECK QUIZ

☑ **VIDEO** Earth's Water

Reading **HELP**DESK

Academic Vocabulary

- **constant**
- **enormous**

Content Vocabulary

- **water cycle**
- **evaporation**
- **condensation**
- **precipitation**
- **desalination**
- **groundwater**
- **aquifer**

TAKING NOTES: *Key Ideas and Details*

IDENTIFYING As you read the lesson, use a concept map like the one below to write descriptions of the freshwater sources on Earth, including how they are used by humans.

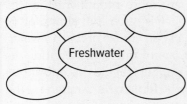

LESSON 3
Earth's Water

ESSENTIAL QUESTION • *How do physical processes shape Earth's surface?*

IT MATTERS BECAUSE
The amount of water on Earth remains fairly constant and moves in the water cycle. Salt water covers much of the Earth's surface. Although there is only a small amount of freshwater on Earth, it is necessary to sustain life.

The Water Cycle

GUIDING QUESTION *What drives the Earth's water cycle?*

As you recall, oceans, lakes, rivers, and other bodies of water make up the Earth's hydrosphere. Almost all of the hydrosphere is salt water found in the oceans, seas, and a few large saltwater lakes. The remainder is freshwater found in lakes, rivers, glaciers, and groundwater.

The total amount of water on Earth does not change, but it is constantly moving—from the oceans to the air to the land and finally back to the oceans. This regular movement of water is called the **water cycle**. The sun drives the water cycle by evaporating water from the surfaces of bodies of water. **Evaporation** is the changing of liquid water into vapor, or gas. The sun's energy causes evaporation. Water vapor rising from bodies of water and plants is gathered in the air. The amount of water vapor the air holds depends on its temperature. Warm air holds more water vapor than cool air.

When warm air cools, it cannot retain all of its water vapor, so the excess water vapor changes into liquid water—a process called **condensation**. Tiny droplets of water come together to form clouds. When clouds gather more water than they can hold, they release moisture, which falls to the Earth as **precipitation**—rain, snow, or sleet, depending on the air temperature and wind conditions. This precipitation sinks into the ground and collects in streams and lakes to return to the oceans. Soon most of it evaporates, and the cycle begins again.

The amount of water that evaporates is approximately the same amount that falls back to Earth. This amount varies little from year to year. Thus, the total volume of water in the water cycle is fairly **constant**.

☑ **READING PROGRESS CHECK**
Explaining What causes evaporation?

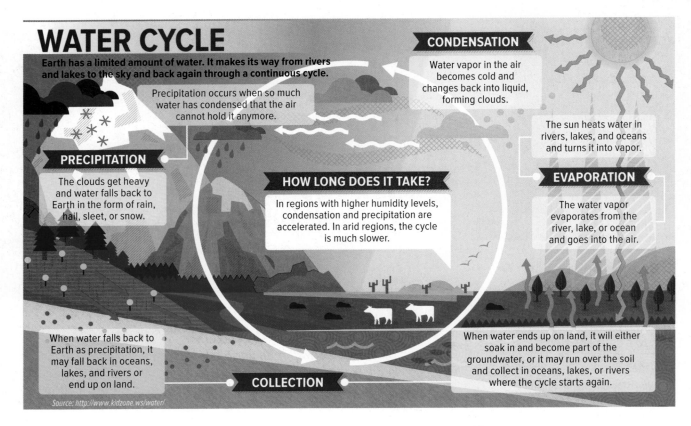

WATER CYCLE

Earth has a limited amount of water. It makes its way from rivers and lakes to the sky and back again through a continuous cycle.

CONDENSATION
Water vapor in the air becomes cold and changes back into liquid, forming clouds.

Precipitation occurs when so much water has condensed that the air cannot hold it anymore.

The sun heats water in rivers, lakes, and oceans and turns it into vapor.

PRECIPITATION
The clouds get heavy and water falls back to Earth in the form of rain, hail, sleet, or snow.

HOW LONG DOES IT TAKE?
In regions with higher humidity levels, condensation and precipitation are accelerated. In arid regions, the cycle is much slower.

EVAPORATION
The water vapor evaporates from the river, lake, or ocean and goes into the air.

When water falls back to Earth as precipitation, it may fall back in oceans, lakes, and rivers or end up on land.

COLLECTION

When water ends up on land, it will either soak in and become part of the groundwater, or it may run over the soil and collect in oceans, lakes, or rivers where the cycle starts again.

Source: http://www.kidzone.ws/water/

The water cycle depicts the movement of water from ocean to air to ground and back to the ocean.

▲ **CRITICAL THINKING**

1. *Categorizing* What are three types of precipitation? Where might this precipitation end up?

2. *Speculating* How might contaminated water end up affecting people even if they live far away from the source?

water cycle regular movement of Earth's water from ocean to air to ground and back to the ocean

evaporation the process of converting liquid into vapor, or gas

condensation the process of excess water vapor changing into liquid water when warm air cools

precipitation moisture that falls to the Earth as rain, sleet, hail, or snow

constant unchanging

enormous gigantic; exceedingly large

Bodies of Salt Water

GUIDING QUESTION *What is salt water?*

Seen from space, the Earth's oceans and seas are more prominent than its landmasses. About 70 percent of the Earth's surface is water, but almost all of this is salt water. Freshwater makes up only a small percentage of Earth's water.

About 97 percent of the Earth's water consists of one huge, continuous body of water that circles all the continents. Geographers divide this **enormous** expanse into five oceans: the Pacific, the Atlantic, the Indian, the Arctic, and the Southern. The first four lie in large basins between the continents, while the Southern Ocean extends from the coast of Antarctica north to 60° S latitude. The Pacific, the largest of the oceans, covers more area than all the Earth's land combined. The Pacific Ocean is also deep enough in some places to cover Mount Everest, the world's highest mountain, with more than 1 mile (1.6 km) to spare.

Seas, gulfs, and bays are bodies of salt water smaller than oceans. These bodies of water are often partially enclosed by land. As one of the world's largest seas, the Mediterranean Sea is almost entirely encircled by southern Europe, northern Africa, and southwestern Asia. The Gulf of Mexico is nearly encircled by the coasts of the United States and Mexico.

Although 97 percent of the world's water is found in oceans, the water is too salty for drinking, farming, or manufacturing. The world's growing population and increasing urbanization require freshwater for such activities. Governments, planners, and scientists continue to look for ways to meet the world's growing need for freshwater. Today, some of these efforts focus on ways to remove the salt from ocean water or groundwater in a process known as **desalination**.

Desalination is a controversial topic, however. Supporters point out that it is one of the most promising solutions to the problem of freshwater shortages.

desalination the removal of salt from seawater to make it usable for drinking and farming

❝Desalination is a promise fulfilled. Today, hundreds of million people around the world have access to clean water thanks to desalination. Just as importantly, desalination is also a promise for the future, with its unique ability to deliver a reliable, sustainable and new source of water to our thirsty planet.❞

—Corrado Sommariva, invitation to the International Desalination Association 2013 World Congress

Critics of the process argue that it does not come without economic and environmental costs. Desalinated ocean water is one of the most expensive forms of freshwater available because of the costs associated with collecting the ocean water, removing the salt, and distributing the new freshwater. For example, desalinated water in the United States can cost up to five times more than other sources of freshwater. Such high costs make desalination an impossible option for many less developed countries, where limited funds are already stretched too thin.

Some countries, such as Saudi Arabia and the United Arab Emirates, use desalination because other freshwater sources are scarce and because they have the financial and energy resources to support such ventures. In fact, about three-fourths of the world's desalinated water is produced in North Africa and Southwest Asia. Desalination plants in the United States—mostly in California, Florida, and Texas—also produce a small amount of freshwater.

The environmental concerns surrounding desalination are related to ocean and marine biodiversity. When ocean water is collected, marine life is drawn up in the intake pipe and eventually destroyed during the desalination process.

GEOGRAPHY CONNECTION

Desalination is an option for some countries to meet freshwater needs.

1. ***PLACES AND REGIONS*** In which general region of the world are many of the desalination plants located?

2. ***ENVIRONMENT AND SOCIETY*** Which energy resource allows North Africa and Southwest Asia to be the world leader in freshwater production?

Global Desalination

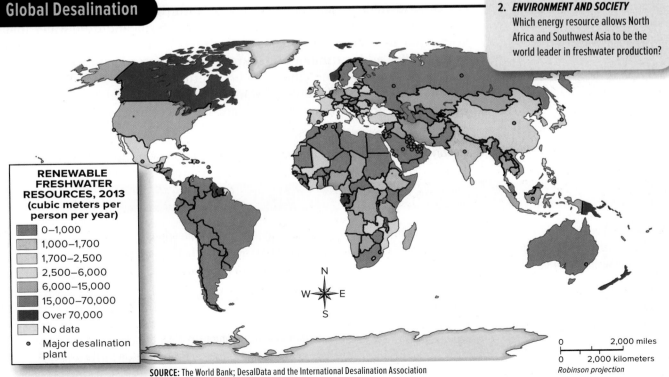

RENEWABLE FRESHWATER RESOURCES, 2013 (cubic meters per person per year)
- 0–1,000
- 1,000–1,700
- 1,700–2,500
- 2,500–6,000
- 6,000–15,000
- 15,000–70,000
- Over 70,000
- No data
- • Major desalination plant

0   2,000 miles
0   2,000 kilometers
Robinson projection

SOURCE: The World Bank; DesalData and the International Desalination Association

Glacial lakes, like this one in Argentina, are most abundant in high-altitude areas that were once occupied by glaciers.

▲ CRITICAL THINKING

1. Identifying Other than Argentina, where else could you find glacial lakes?

2. Classifying What are the different sources of freshwater?

groundwater water located underground within the Earth that supplies wells and springs

aquifer underground water-bearing layers of porous rock, sand, or gravel

In addition, wastewater from the process affects coastal water quality. Besides being very salty, this brine may be warmer in temperature and contain chemicals, which have detrimental effects on water and marine organisms.

☑ READING PROGRESS CHECK

Identifying What is the name of the process that removes salt from ocean water?

Bodies of Freshwater

GUIDING QUESTION *Why is freshwater important to life on Earth?*

Only about 3 percent of Earth's total water supply is freshwater, and most is not available for human consumption. More than two-thirds of Earth's freshwater is frozen as glaciers and ice caps. Another sixth is found beneath the surface. Lakes, streams, and rivers contain less than one-third of 1 percent of Earth's freshwater.

A lake is a body of water completely surrounded by land. Most lakes contain freshwater, although some, such as Southwest Asia's Dead Sea, are saltwater remnants of ancient seas. Many lakes are found where glacial movement has cut deep valleys and built up dams of soil and rock that held back melting ice water. North America has thousands of glacial lakes.

Flowing water forms streams and rivers. Meltwater, an overflowing lake, or a spring may be the source, or the beginning, of a stream. Streams may combine to form a river, a larger stream of higher volume that follows a channel along a particular course. When rivers join, the major river systems that result may flow for thousands of miles. The smaller streams or rivers that flow into larger rivers are called *tributaries*. Rain, runoff, and water from these tributaries swell rivers as they flow toward a lake, gulf, sea, or ocean. The place where the river empties into another body of water is its mouth.

Groundwater, freshwater that lies beneath the Earth's surface, comes from rain and melted snow that filter through the soil and from water that seeps into the ground from lakes and rivers. Wells and springs tap into groundwater and are important sources of freshwater for people in many rural areas and in some cities. An underground porous rock layer often saturated by very slow flows of water is called an **aquifer** (A•kwuh•fuhr). Aquifers and groundwater are important sources of freshwater.

☑ READING PROGRESS CHECK

Identifying What portion of Earth's freshwater is found below the surface?

LESSON 3 REVIEW

Reviewing Vocabulary

1. *Summarizing* Summarize the water cycle using the following terms: evaporation, condensation, and precipitation.

Using Your Notes

2. *Identifying* Use your web diagram to identify the bodies of freshwater that are necessary to sustain life on Earth.

Answering the Guiding Questions

3. *Explaining* What drives Earth's water cycle?

4. *Stating* What is salt water?

5. *Analyzing* Why is freshwater important to life on Earth?

Writing Activity

6. *Informative/Explanatory* Many cities and towns develop near sources of water. Write a paragraph describing the sources of water used by your community.

54

Directions: On a separate sheet of paper, answer the questions below. Make sure you read carefully and answer all parts of the questions.

Lesson Review

Lesson 1

1 ***Describing*** Describe the larger physical system of which Earth is a part.

2 ***Explaining*** How do the hydrosphere, lithosphere, and atmosphere work together to form the biosphere?

3 ***Hypothesizing*** What conditions would have to exist in order for a space station to support life?

Lesson 2

4 ***Evaluating*** How have earthquakes and volcanoes influenced the size and location of Earth's physical features?

5 ***Analyzing*** How has erosion been beneficial and harmful to agricultural communities?

6 ***Summarizing*** How do physical processes such as tectonic forces, erosion, and soil building affect different regions? Give examples.

Lesson 3

7 ***Identifying Central Issues*** Explain how the statement "The total amount of water on Earth does not change" relates to the water cycle.

8 ***Explaining*** How have technological innovations like desalination allowed people living in areas without adequate freshwater sources to adapt to their environment?

9 ***Identifying Cause and Effect*** How can pollution in streams affect the oceans?

21st Century Skills

Review the diagram. Then answer the questions that follow.

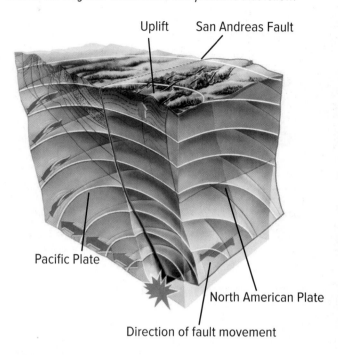

Uplift   San Andreas Fault

Pacific Plate

North American Plate

Direction of fault movement

10 ***Using Graphs, Charts, Diagrams, and Tables*** Describe the movement of the San Andreas Fault.

11 ***Compare and Contrast*** How does accretion differ from subduction? What are the effects of each?

College and Career Readiness

12 ***Global Analysis*** Discuss the ways in which the United States could help less developed countries prepare for natural disasters, such as earthquakes and volcanic eruptions.

Need Extra Help?

If You've Missed Question	❶	❷	❸	❹	❺	❻	❼	❽	❾	❿	⓫	⓬
Go to page	40	42	42	48	49	45	51	53	54	55	47	48

Richard Leech/National Geographic Stock

Directions: On a separate sheet of paper, answer the questions below. Make sure you read carefully and answer all parts of the questions.

Critical Thinking

13 *Making Generalizations* Explain how weathering and erosion help create soil.

14 *Identifying Cause and Effect* How would life on Earth be different if there were no precipitation?

15 *Comparing and Contrasting* What is the difference between groundwater and aquifers, and why is the distinction important?

16 *Defending* How can wind erosion actually be helpful in some areas?

Applying Map Skills

Refer to the Unit 1 Atlas to answer the following questions.

17 *Physical Systems* Which coast of South America appears to have the most signs of plate tectonic movement due to its landforms?

18 *The World in Spatial Terms* Which continents are located along the Ring of Fire?

19 *Places and Regions* Use what you know of the world's oceans to draw a mental map showing the continents and oceans. Be sure to label the five oceans.

Exploring the Essential Question

20 *Making Connections* Write a one-page essay explaining how knowledge of plate tectonics can help governments prepare for natural disasters. Provide at least two specific examples.

Writing About Geography

21 *Informative/Explanatory* Describe at least three landforms found in the region of the world where you live. Include the name of each landform and the processes that created it.

DBQ Analyzing Primary Sources

Use the document to answer the following questions.

The center of the Earth is filled with intense heat and pressure. These natural forces drive numerous changes such as volcanoes and earthquakes that renew and enrich Earth's surface. The physical processes can also disrupt, and often destroy, human life. As a result, scientists are working to learn how to predict them.

PRIMARY SOURCE

❝*Scientists are doing everything they can to solve the mysteries of earthquakes. They break rocks in laboratories, studying how stone behaves under stress. They hike through ghost forests where dead trees tell of long-ago tsunamis. They make maps of precarious, balanced rocks to see where the ground has shaken in the past, and how hard. They dig trenches across faults, searching for the active trace. They have wired up fault zones with so many sensors it's as though the Earth is a patient in intensive care.*❞

—Joel Achenbach, "The Next Big One," *National Geographic,* April 2006

22 *Analyzing* How does Achenbach describe the work done to solve earthquake mysteries?

23 *Interpreting* What did Achenbach mean by "wired up fault zones...as though the Earth is a patient in intensive care"?

Research and Presentation

24 *Research Skills* Using the Internet, research the U.S. Geological Survey and the specific role it plays in contributing to our knowledge of earthquakes. Create a multimedia presentation to share your findings.

Need Extra Help?

If You've Missed Question	**13**	**14**	**15**	**16**	**17**	**18**	**19**	**20**	**21**	**22**	**23**	**24**
Go to page	50	51	54	49	4	4	4	45	43	56	56	48

Climates of the Earth

ESSENTIAL QUESTION • *Why is climate important to life on Earth?*

A young Inuit boy on Canada's Baffin Island wears warm clothing to protect against the cold Arctic conditions.

Sue Flood/The Image Bank/Getty Images

networks

There's More Online about climates of the Earth.

CHAPTER 3

Why Geography Matters
Climate Change: The Impacts on Humans

Lesson 1
Earth-Sun Relationships

Lesson 2
Factors Affecting Climate

Lesson 3
World Climate Patterns

Geography Matters...

The relationships between the Earth and the sun influence climate and weather patterns all over the Earth. The various climates on Earth support different communities of plants and animals that are called biomes. Human activity is also affected by Earth's climates, such as knowing that only certain vegetables can grow in our backyard gardens. Fortunately, our planet is just the right distance from our sun to make life on Earth possible.

climate change:
the impacts *on* humans

Climate scientists have amassed data on weather patterns and climate since the 1980s. Most agree that the average surface air temperature across the globe has increased about 1.4°F (0.8°C) since then. Predictions show that warming of the planet, or global warming, will accelerate through the end of the twenty-first century.

THERE'S MORE ONLINE

What is climate change and global warming?

Climate change is a significant change in temperature, precipitation, and wind patterns that lasts for at least several decades. These changes may be difficult to notice, but there are several signs of climate change that are observable, such as heat waves, flooding, droughts, melting glaciers, and extreme weather events. Global warming refers to warming of Earth's average temperature that has occurred since the mid-1800s. It represents only one aspect of climate change. Global warming does not simply mean that the climate everywhere on Earth will become warmer. An important part of what it means is that some weather patterns will become more extreme.

1. **Physical Systems** How is global warming different from climate change?

What are the causes of climate change?

Although average global surface temperature does fluctuate due to natural causes, most climate scientists agree that human activity is a major contributing factor to recent warming of the planet. As global population and industrialization increase, a variety of human activities increase the presence of so-called "greenhouse gases" in Earth's atmosphere. In contrast, the greenhouse effect is a naturally occurring process in which these gases act as a blanket, trapping the sun's heat within the atmosphere. The human-produced greenhouse gases amplify the effect and increase the warming of Earth. These gases—primarily carbon dioxide and methane—are released into the atmosphere through the burning of fossil fuels and wood, along with some industrial processes. Agricultural practices and the decay of organic waste in landfills also contribute to greenhouse gases.

2. **Physical Systems** Describe two factors contributing to climate change.

What are the potential impacts of climate change on people?

As Earth's surface temperature warms and its climate becomes less stable, people will be affected in several ways. Melting glaciers will cause sea levels to rise, likely rendering some coastal regions and islands uninhabitable. Increased droughts, heat waves, and wildfires will affect some regions, while other areas will be impacted by severe flooding. Storms and other extreme weather events may increase in their frequency and intensity as well. As a result, people may face increased health and safety risks along with threats to their homes, cities, and critical infrastructure systems.

3. **Environment and Society** How will climate change affect humans? Consider water supplies, agriculture, power and transportation systems, the environment, and human health.

networks

There's More Online!

☑ **DIAGRAM** The Earth's Seasons

☑ **INFOGRAPHIC**
The Greenhouse Effect

☑ **INTERACTIVE**
SELF-CHECK QUIZ

☑ **VIDEO** Earth-Sun Relationships

Sun

LESSON 1
Earth-Sun Relationships

ESSENTIAL QUESTION · *Why is climate important to life on Earth?*

Reading HELPDESK

Academic Vocabulary

- predictable
- eventually

Content Vocabulary

- **weather**
- **climate**
- **axis**
- **revolution**
- **equinox**
- **solstice**
- **midnight sun**
- **greenhouse effect**

TAKING NOTES: *Key Ideas and Details*

DESCRIBING As you read the lesson, use a chart like the one below to list the characteristics of Earth-sun relationships and describe their effects on climate.

Earth-Sun Relationships	Effects on Climate

It Matters Because

Daily life on Earth is influenced by the dynamic relationship between the Earth and the sun. The amount of direct sunlight reaching Earth's surface plays an important role in affecting the temperature of different places. The Earth's rotation determines when we receive sunlight, giving us day or night. The Earth's tilt and its revolution around the sun result in the four seasons we experience.

Climate and Weather

GUIDING QUESTION *How do the relationships between the Earth and the sun affect climate?*

There is an important difference between climate and weather. **Weather** is the condition of the atmosphere in one place over a short period of time, such as hours or days. For example, when you look outside the window in the morning to decide what to wear that day, you are checking the weather. **Climate**, on the other hand, refers to the average weather conditions as measured over many years. Climate is the reason why you decide to buy certain types of clothing to wear based on where you live.

Earth's Tilt and Rotation

The relationship between the Earth and the sun directly affects climate. An important aspect of the Earth-sun relationship is that the Earth's **axis** is tilted. The axis runs from the North Pole to the South Pole through the center of the planet. Currently, the Earth is tilted at about 23½°.

Because of the Earth's tilted axis, not all places on Earth receive the same amount of direct sunlight at the same time. For this reason, Earth's tilt affects the temperature of a particular place. Temperature is the measure of how hot or cold a place is. Temperature is measured in degrees on a set scale. The most common scales for measuring temperature are Fahrenheit (°F) and Celsius (°C).

Why is it usually warmer during the day than it is during the night? This depends on which side of the planet is facing the sun. The Earth rotates on its axis, making one complete rotation every 24 hours, or one day. The Earth's rotation from west to east ensures that every part of the world receives sunlight in a **predictable** pattern during those 24 hours. The side of the planet not facing the sun is colder, and the side of the planet facing the sun is warmer.

Earth's Revolution

While the Earth rotates on its axis, it also revolves around the closest star to us, the sun. It takes the Earth one year, approximately 365 days, to complete one **revolution** around the sun. The Earth's revolution, combined with its tilted axis, affects the amount of sunlight that reaches different locations on the Earth at different times of the year. People who live in the Northern Hemisphere experience summer when the Northern Hemisphere is tilted toward the sun and is receiving the most direct sunlight. The seasons are reversed north and south of the Equator. When it is summer in the Northern Hemisphere, it is winter in the Southern Hemisphere because the Southern Hemisphere is tilted away from the sun and receives less direct sunlight. Likewise, when it is fall in the Northern Hemisphere, it is spring in the Southern Hemisphere.

Twice a year (around March 21 and September 23), the direct sunlight falls on the Equator. This day is called an **equinox**, meaning "equal night," because daytime and nighttime hours are equal. On the equinox, equal amounts of light reach the Northern and Southern Hemispheres. The two equinoxes mark the shift in seasons between winter and spring and between summer and fall.

In addition to the Equator, there are two other lines of latitude that run parallel to the Equator and mark important changes in the Earth's seasons. As the Earth proceeds in its revolution around the sun, the direct rays of the sun

weather condition of the atmosphere in one place during a short period of time

climate weather patterns typical for an area over a long period of time

axis an imaginary line that runs through the center of the Earth between the North and South Poles

predictable expected or able to be foreseen

revolution in astronomy, the Earth's yearly trip around the sun, taking 365¼ days

equinox one of two days (about March 21 and September 23) on which the sun is directly above the Equator, making day and night equal in length

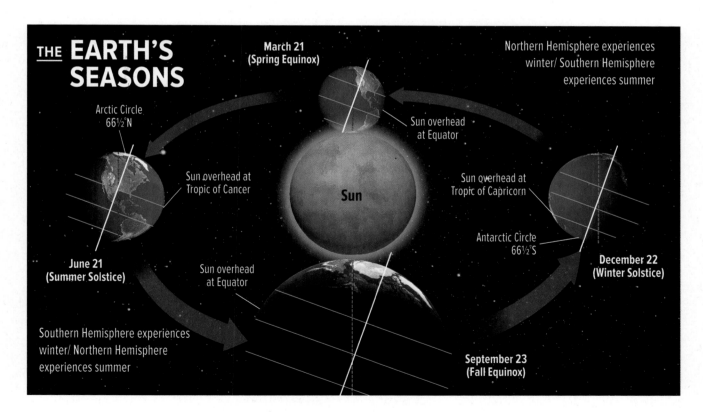

THE EARTH'S SEASONS

March 21
(Spring Equinox)

Northern Hemisphere experiences winter/ Southern Hemisphere experiences summer

Arctic Circle
66½°N

Sun overhead at Equator

Sun overhead at Tropic of Cancer

Sun

Sun overhead at Tropic of Capricorn

Antarctic Circle
66½°S

June 21
(Summer Solstice)

Sun overhead at Equator

December 22
(Winter Solstice)

Southern Hemisphere experiences winter/ Northern Hemisphere experiences summer

September 23
(Fall Equinox)

The angle of the sun's rays as they strike the Earth affects the Northern and Southern Hemispheres differently.

▲ **CRITICAL THINKING**

1. Contrasting Explain the difference between solstices and equinoxes.

2. Drawing Conclusions In what months do the sun's rays directly strike the Equator? The Tropics of Cancer and Capricorn?

solstice one of two days (about June 21 and December 22) on which the sun's rays strike directly on the Tropic of Cancer or Tropic of Capricorn, marking the beginning of summer or winter

midnight sun continuous daylight, a time when the sun is visible at midnight during the summer in either the Arctic or Antarctic Circle

eventually strike the Tropic of Cancer, the latitude line at 23½° N that passes through Mexico, North Africa, and India. The sun usually hits the Tropic of Cancer around June 21, bringing the longest day of sunlight to the Northern Hemisphere. This date is known as the summer **solstice** and marks the beginning of the summer season in the Northern Hemisphere. By about September 23, the Earth has revolved so that the direct rays of the sun hit the Equator again. This equinox marks the end of summer and the beginning of the fall season in the Northern Hemisphere.

As the Earth continues in its revolution, the direct rays of the sun eventually strike the Tropic of Capricorn—the latitude line at 23½° S running through South America, the southern tip of Africa, and Australia—around December 22. This marks the winter solstice, bringing the shortest day of sunlight to the Northern Hemisphere and signaling the beginning of the winter season.

The most dramatic variation in the amount of sunlight occurs near the Poles. For six months of the year, one Pole gets continuous sunlight while the other Pole receives none. From about March 20 to about September 23, the polar area north of the Arctic Circle (66½° N) experiences continuous daylight or twilight. The polar area south of the Antarctic Circle (66½° S) experiences continuous daylight or twilight for the other six months of the year. Continuous daylight, a phenomenon also known as **midnight sun**, is caused by the tilt of the Earth's axis as it revolves around the sun. The Poles are very sparsely populated, so many people remain unaffected by midnight sun. However, parts of northern North America and northern Europe—such as Alaska, Sweden, Denmark, Norway, Finland, and others—have become popular tourist destinations particularly because of midnight sun.

☑ **READING PROGRESS CHECK**

Contrasting What factor distinguishes weather from climate?

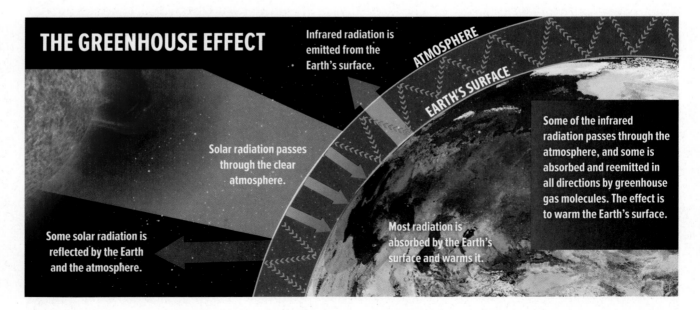

THE GREENHOUSE EFFECT

Infrared radiation is emitted from the Earth's surface.

ATMOSPHERE

EARTH'S SURFACE

Solar radiation passes through the clear atmosphere.

Some of the infrared radiation passes through the atmosphere, and some is absorbed and reemitted in all directions by greenhouse gas molecules. The effect is to warm the Earth's surface.

Some solar radiation is reflected by the Earth and the atmosphere.

Most radiation is absorbed by the Earth's surface and warms it.

Solar radiation, when combined with greenhouse gases like carbon dioxide and methane, results in the warming of the Earth's surface. Without this greenhouse effect, Earth would be too cold for most living things.

▲ **CRITICAL THINKING**

1. ***Interpreting*** What processes of the greenhouse effect contribute to the warming of the Earth's surface?

2. ***Sequencing*** What happens to the solar radiation that is not absorbed by the Earth?

The Greenhouse Effect

GUIDING QUESTION *What is the greenhouse effect?*

Even on the warmest days, only some of the sun's rays pass through the Earth's atmosphere. The atmosphere reflects some of the radiation back into space. Enough radiation reaches Earth's surface, however, to warm the air, land, and water. Once Earth's surface absorbs the radiation from the sun and is warmed, it radiates this heat energy back again into the atmosphere.

Normally, the atmosphere provides just the right amount of insulation to promote life on the planet. The 50 percent of the sun's radiation that reaches the Earth is converted into infrared radiation, or heat. As shown in the infographic, clouds and greenhouse gases—atmospheric gases such as water vapor, methane, and carbon dioxide (CO_2)—absorb the reradiated heat energy and trap it so that most of it cannot escape back into space. The atmosphere is therefore like a greenhouse. It traps enough radiation to warm the land, water, and air and help plants grow while reflecting some radiation to ensure that the Earth does not overheat.

This **greenhouse effect** is the warming of the Earth that occurs when the sun's radiation passes through the atmosphere, is absorbed by the Earth, and is radiated as heat energy back into the atmosphere where it cannot escape into space. Without the greenhouse effect, Earth's average temperature would be below 0°F (-17°C) and life as we know it could not exist.

To understand the planet's natural greenhouse effect, consider that according to the laws of physics the radiation Earth receives from the sun must be equally balanced by the heat Earth radiates back out to space. If Earth gave back less energy than it received, the planet would **eventually** become too warm to support life. Likewise, if Earth gave back more energy than it received, the planet would be too cold for life.

greenhouse effect the capacity of certain gases in the atmosphere to trap heat, thereby warming the Earth

eventually taking place later; in the end

PRIMARY SOURCE

❝The greenhouse effect is one of the fundamental facts of atmospheric science. It is real; that fact is beyond dispute. Without it, the entire surface of the ocean would be frozen solid. Life—at least the kind that depends on liquid water and warmth—could not survive. We owe our existence to the greenhouse effect. So why are we worried about it? . . . The answer is that some of the heat radiation leaks out through the atmosphere, because there is not enough water vapor, carbon dioxide, and other gases to absorb all of the IR [infrared heat radiation]. Think of the atmosphere as a leaky blanket.❞

—Richard A. Muller, from the Columbia Forum, "Physics for Future Presidents," January/February 2009

Richard A. Muller, Professor of Physics, University of California Berkeley

☑ READING PROGRESS CHECK

Assessing How does the greenhouse effect influence Earth's surface temperature?

LESSON 1 REVIEW

Reviewing Vocabulary
1. ***Making Connections*** What is the relationship between weather, climate, axis, temperature, revolution, equinox, and solstice?

Using Your Notes
2. ***Discussing*** Using your graphic organizer, discuss how temperature is affected by the tilt of the Earth.

Answering the Guiding Questions
3. ***Identifying Cause and Effect*** How do the relationships between the Earth and the sun affect climate?

4. ***Explaining*** What is the greenhouse effect?

Writing Activity
5. ***Informative/Explanatory*** Write a paragraph explaining the differences in weather you would expect in Alaska and Florida.

networks

There's More Online!

☑ **DIAGRAM** The Rain Shadow Effect

☑ **MAP** World Zones of Latitude and Wind Patterns

☑ **MAP** World Ocean Currents

☑ **INTERACTIVE SELF-CHECK QUIZ**

☑ **VIDEO** Factors Affecting Climate

LESSON 2
Factors Affecting Climate

ESSENTIAL QUESTION · *Why is climate important to life on Earth?*

Reading HELPDESK

Academic Vocabulary

- crucial
- derive

Content Vocabulary

- current
- prevailing wind
- Coriolis effect
- doldrums
- El Niño
- windward
- leeward
- rain shadow

TAKING NOTES: *Key Ideas and Details*

DESCRIBING As you read the lesson, use a web diagram like the one below to list the factors that cause both wind and ocean currents.

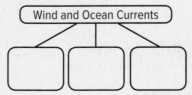

IT MATTERS BECAUSE

The climate of a particular place may have extreme weather and temperature ranges that are caused by several geographic features. Both the latitude and the elevation of a place, along with wind and ocean currents, influence its climate. Two sides of a mountain range may also have two different climates. While one side receives more precipitation as air rises, the opposite side has drier, warmer air as air descends.

Latitude, Climate, and Elevation

GUIDING QUESTION *How are climate patterns related to each zone of latitude?*

The Earth's annual revolution around the sun creates predictable climate patterns. These patterns correspond with bands, or zones, of latitude.

The low latitude zone is between 30° S and 30° N. This zone includes the Tropic of Capricorn, the Equator, and the Tropic of Cancer. Portions of the low latitude zone receive direct rays from the sun year-round and therefore have warm to hot climates.

The high latitude zone includes the Earth's polar areas, which stretch from 60° N to 90° N and from 60° S to 90° S. When either the Northern or the Southern Hemisphere is tilted toward the sun, its polar area receives nearly continuous, but indirect, sunlight.

The midlatitude zone, between 30° N and 60° N in the Northern Hemisphere and between 30° S and 60° S in the Southern Hemisphere, is home to climates that have the most variable weather on Earth. The midlatitudes generally have a temperate climate, or one that varies from fairly hot to fairly cold, with dramatic seasonal weather changes. Warm/hot and cool/cold air masses move across the midlatitudes. These movements and the interactions between the different air masses affect weather in this latitude zone throughout the year.

At all latitudes, elevation influences climate because of the relationship between the elevation of a place and its temperature. The Earth's atmosphere thins as altitude increases. Less dense air retains less heat. As elevation increases, temperatures decrease by about 3.5°F (1.9°C) for each 1,000 feet (305 m). This effect occurs at all latitudes.

For example, in Ecuador, the city of Quito (KEE•toh) is nearly on the Equator. However, Quito is located in the Andes at an elevation of more than 9,000 feet (2,743 m), so average temperatures there are more than 20 degrees cooler than in the coastal lowlands. Sunlight is bright in places with high elevation because the thinner atmosphere filters fewer rays of the sun. But even in bright sunlight, the world's highest mountains, such as the Andes, are cold and snowy.

✓ READING PROGRESS CHECK

Explaining What happens to temperature as elevation increases?

Winds and Ocean Currents

GUIDING QUESTION *How do winds and ocean currents affect climate?*

How do wind and water work together to affect weather? Air moving across the surface of the Earth is called wind. Winds occur because sunlight heats the Earth's atmosphere and surface unevenly. Warm temperatures cause air to rise and create areas of low pressure. Cool temperatures cause air to sink, which creates areas of high pressure. Air moves along the pressure gradient from areas of high pressure to low pressure, so the cool air then flows in to replace the warm rising air. These movements cause winds to distribute the sun's energy around the planet. For this reason, wind patterns are **crucial** to a region's climate.

Ocean **currents** also help distribute energy around the planet. As they circulate, cold water from the polar areas moves slowly toward the Equator. These are cold ocean currents because they consist of cooler water flowing into warmer water. The opposite is also true: warm water moves away from the Equator and these are warm currents because they consist of warmer water flowing into cooler water.

crucial vitally important

current cold or warm stream of seawater that flows in the oceans, generally in a circular pattern

GEOGRAPHY CONNECTION

Global wind patterns are affected by latitude.

1. *PHYSICAL SYSTEMS* In what direction are winds deflected in the Northern Hemisphere?

2. *PLACES AND REGIONS* In what latitude zone are the warm winds deflected toward the west?

World Zones of Latitude and Wind Patterns

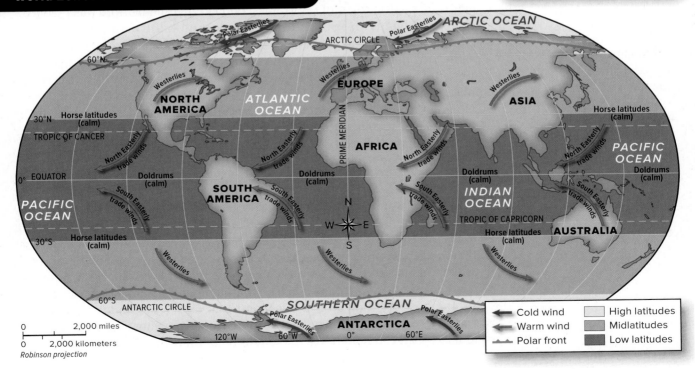

← Cold wind	High latitudes
← Warm wind	Midlatitudes
Polar front	Low latitudes

0 — 2,000 miles
0 — 2,000 kilometers
Robinson projection

We know that climate is influenced by location or geography. Climatology, the study of atmospheric changes that define average climates and their change over time, is closely connected to the work of the geographer. Changes in climate can be due to natural processes and human activities. Questions climatologists ask are framed in the same way a geographer explains phenomena using cause-and-effect relationships. Both the geographer and the climatologist interpret information and predict future patterns.

COMPARING What similarities does the work of the geographer have with that of the climatologist?

prevailing wind wind in a region that blows in a fairly constant directional pattern

Coriolis effect the resulting deflection of prevailing winds caused by the Earth's rotation

doldrums a frequently windless area near the Equator

derive to acquire

Patterns of Wind and Ocean Currents

As winds blow because of pressure differences on Earth's surface, warm tropical air moves toward the Poles and cool polar air moves toward the Equator. This movement of air creates the global winds that blow in fairly constant patterns called **prevailing winds**. The direction of prevailing winds is determined by latitude and is also affected by the Earth's movement. As the Earth rotates from west to east, the paths of the global winds are deflected to the right in the Northern Hemisphere and deflected to the left in the Southern Hemisphere. This phenomenon is called the **Coriolis effect** and causes prevailing winds to blow diagonally rather than along strict north, south, east, or west directions. The strength of the Coriolis effect is proportional to the speed of Earth's rotation at different latitudes.

Winds are generally named for the direction from which they blow, but they sometimes were given names from the early days of sailing. Named for their ability to move trading ships through the region, the prevailing winds of the low latitudes are called trade winds. They blow from the northeast toward the Equator from about latitude 30° N and from the southeast toward the Equator from about latitude 30° S. Westerlies are the prevailing winds in the midlatitudes, blowing diagonally from west to east between about 30° N and 60° N and between about 30° S and 60° S. In the high latitudes, the polar easterlies blow diagonally east to west, pushing cold air toward the midlatitudes.

Near the Equator, the horizontal movement of the trade winds subsides as the warm air rises. This rising air leaves a narrow, generally windless band called the **doldrums**. Two other narrow bands of calm air encircle the globe just north of the Tropic of Cancer and just south of the Tropic of Capricorn. These bands result from descending, high pressure air. In the days of wind-powered sailing ships, crews feared being stranded in these windless areas. With no moving air to lift the sails, ships were stranded for weeks in the hot, still weather. Food supplies dwindled, and perishable cargoes spoiled as the ships sat, helpless and windless. To lighten the load so the ships could take advantage of the slightest breeze, sailors would toss excess cargo and supplies overboard, including livestock being carried to colonial settlements. The *horse latitudes,* the calm areas at the edges of the Tropics, **derived** their name from this practice.

Just as winds move in patterns, the cold and warm ocean currents move through the ocean in patterns. Ocean currents are caused by many of the same factors that cause winds, including the Earth's rotation, changes in air pressure, and differences in water temperature. The Coriolis effect is also observed in ocean currents. Ocean currents affect climate in the coastal lands along which they flow. Cold ocean currents cool the lands they pass, while warm ocean currents bring warmer temperatures. For example, the warmer North Atlantic Current flows near western Europe. This current gives western Europe a relatively mild climate in spite of its northern latitude.

Influences on Weather

Wind and water work together to affect weather in an important way. Driven by temperature, condensation creates precipitation, or water falling to the Earth in the form of rain, sleet, hail, or snow. The sudden cloudburst that cools a steamy summer day is an example of how precipitation both affects and is affected by temperature. Water vapor forms in the atmosphere from evaporated surface water. The high temperature causes the air to rise. As the air rises, however, it cools and this results in condensation of the water vapor into liquid droplets, forming clouds. Further cooling causes rain to fall, which can help lower the temperature on warm days.

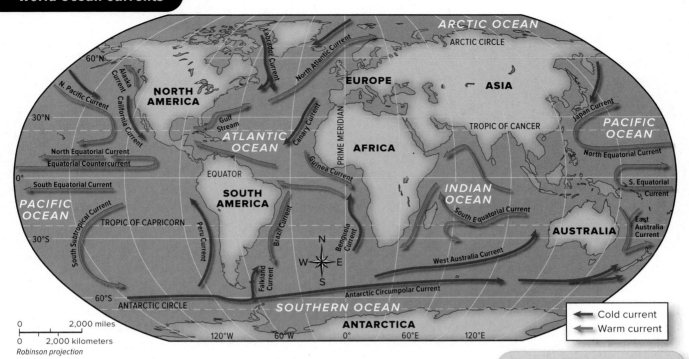

One example of the way climate is affected by recurring events that alter weather patterns is the **El Niño** (ehl NEE•nyoh) phenomenon. El Niño is a periodic change in the pattern of ocean currents, water temperatures, and weather in the mid-Pacific region. El Niño does not occur every year, but its frequency has increased since the 1970s. In an El Niño year, the normally low atmospheric pressure over the western Pacific is replaced by higher pressure, and the normally high pressure over the eastern Pacific drops. This reversal causes the trade winds to diminish or even to reverse direction. The change in wind pattern reverses the equatorial ocean currents, drawing warm water from near Indonesia east to Ecuador, where it spreads along the coasts of Peru and Chile.

These changes in the Pacific influence climates around the world. Precipitation increases along the coasts of North and South America, making the winters warmer and increasing the risk of floods. In Southeast Asia and Australia, drought and occasional massive forest fires occur. Climates in the midlatitudes are affected as well; for example, winter rains are heavier along the west coast of the United States during El Niño years. The costs in human and economic terms, such as damaged homes and flooded crops, make learning about and preparing for an El Niño year vitally important.

PRIMARY SOURCE

❝ It rose out of the tropical Pacific in late 1997, bearing more energy than a million Hiroshima bombs. By the time it had run its course eight months later, the giant El Niño of 1997–98 had . . . killed an estimated 2,100 people, and caused at least 33 billion [U.S.] dollars in property damage. ❞

—Curt Suplee, "El Niño/La Niña," *National Geographic Magazine,* March 1999

✔ **READING PROGRESS CHECK**

Describing What happens to global winds at the Equator?

GEOGRAPHY CONNECTION

Ocean currents, when combined with wind patterns and the effects of the sun, influence Earth's weather and climate.

1. ***PHYSICAL SYSTEMS*** Which type of current is located near the Equator?

2. ***PLACES AND REGIONS*** Why would the ocean temperatures along the western coast of southern Africa be colder than the ocean temperatures along the eastern coast?

El Niño a periodic reversal of the pattern of ocean currents and water temperatures in the mid-Pacific region

The Rain Shadow Effect

The rain shadow effect is influenced by landforms and climate.

▶ **CRITICAL THINKING**

1. *Interpreting* Why does air lose moisture as it rises over mountains?

2. *Speculating* Winds are blowing from the west toward the east of a mountain range running from north to south. Which side of the mountain range is greener and has more trees, the western side or the eastern side?

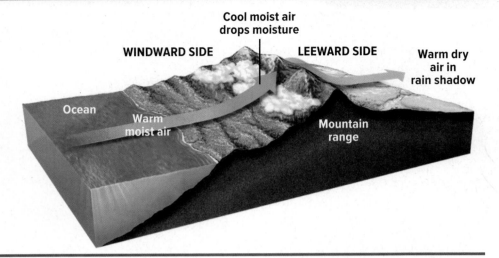

Cool moist air drops moisture

WINDWARD SIDE　　**LEEWARD SIDE**　　**Warm dry air in rain shadow**

Ocean

Warm moist air

Mountain range

Landforms and Climate

GUIDING QUESTION *How can landforms and bodies of water affect climate?*

While one can generalize about the climates of places located in the same latitude zones, they vary based upon the presence or absence of certain physical features. Large bodies of water, for example, are slower to heat and cool, so they tend to keep temperatures in surrounding lands moderate. Coastal lands receive the benefit of this influence and experience less changeable weather. Conversely, the interiors of the continents tend to experience extremes in seasonal temperatures.

Yet another physical feature that affects the climates in the latitude zones are the mountain ranges. As the diagram shows, mountain ranges push wind upward and, as a result, the rising air cools and releases moisture in the form of precipitation. Most of the precipitation falls on the **windward** side of the mountain, or the side of the mountain range facing the wind. After the precipitation is released, winds become warmer and drier as they descend on the opposite, or **leeward**, side of the mountains. The hot, dry air produces little precipitation in an effect known as a **rain shadow**. The rain shadow effect often causes dry areas—and even deserts—to develop on the leeward sides of mountain ranges.

windward being in or facing the direction from which the wind is blowing

leeward being in or facing the direction toward which the wind is blowing

rain shadow result of a process by which dry areas develop on the leeward sides of mountain ranges

✓ **READING PROGRESS CHECK**

Discussing What happens to winds after they release precipitation?

LESSON 2 REVIEW

Reviewing Vocabulary

1. *Describing* How are prevailing winds influenced by the Coriolis effect?

Using Your Notes

2. *Making Connections* Which of the factors from your graphic organizer do you think have the strongest effect on the climate where you live?

Answering the Guiding Questions

3. *Explaining* How are climate patterns related to each zone of latitude?

4. *Identifying Cause and Effect* How do wind currents and ocean currents affect climate?

5. *Evaluating* How can landforms and bodies of water affect climate?

Writing Activity

6. *Informative/Explanatory* Suppose you are on a ship sailing in the low latitudes. Write a paragraph explaining what might happen as you drift near the Equator.

LESSON 3
World Climate Patterns

IT MATTERS BECAUSE
Climate patterns vary from region to region. However, factors such as wind and air pressure can create zones where a climate becomes quite dramatic. For example, most of Australia has a dry climate, but when trade winds meet during the summer months, Australia's northern coast sees intense thunderstorms.

Climate Regions and Biomes

GUIDING QUESTION *How are world climates and biomes organized?*

Climates in the world are organized into four climate zones: tropical, dry, midlatitude, and high-latitude climates. These climates support different kinds of biomes. A **biome** is a major type of ecological community defined primarily by distinctive **natural vegetation** and animal groups. The characteristics of biomes may **overlap** with one another.

Tropical Climates

Tropical climates are found in or near low latitudes in areas otherwise referred to as the Tropics. The two most widespread kinds of tropical climate regions are wet climates and dry climates.

Tropical rain forest climates have an **average daily temperature** of 80°F (27°C), and since the warm air is humid, or saturated with moisture, it rains almost daily. Annual rainfall averages from 50 to 260 inches (125 to 660 cm). This continual rain tends to strip the soil of nutrients. The biome in these climates are the tropical rain forests, characterized by thick vegetation that grows in layers. Tall trees form a canopy over shorter trees and bushes, and shade-loving plants grow on the completely shaded forest floor. The world's largest tropical rain forest is in the Amazon River basin. Similar climate and vegetation exist in other parts of South America, the Caribbean, Asia, and Africa. Due to the vast amount of plant food, wildlife is also abundant, and scientists estimate that more than half of all of the plant and animal species exist in the tropical rain forests. As with all biomes, plants and animals vary within a given rain forest, with differentiation often occurring. For example, trees that grow in the mountains of the Amazon rain forest do not grow in the lowlands of the same rain forest.

Reading **HELP**DESK

Academic Vocabulary

- overlap
- distinct

Content Vocabulary

- **biome**
- **natural vegetation**
- **average daily temperature**
- **oasis**
- **prairie**
- **coniferous**
- **deciduous**
- **mixed forest**
- **permafrost**

TAKING NOTES: *Key Ideas and Details*

IDENTIFYING Use a web diagram like the one below to take notes about the Earth's four climate zones.

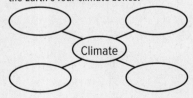

biome major type of ecological community defined primarily by distinctive plant and animal groups

natural vegetation plant life that grows in a certain area if people have not changed the natural environment

overlap to partly cover

average daily temperature the average of the daily high temperature and the overnight low; often used for comparison across climate regions

distinct recognizably different

oasis small area in a desert where water and vegetation are found

Tropical wet/dry climates have pronounced dry and wet seasons, with high year-round temperatures. These regions, also called savannas, have fewer plants and animals than the tropical rain forest climates. One distinguishing characteristic of a tropical wet/dry climate is that sunlight is not blocked by trees and is able to reach much of the ground surface. This makes for more highly specialized plant and animal species. Tropical savannas are found in Africa, Central and South America, Asia, and Australia. Each maintains **distinct** types of plant and animal life. A key factor determining the types of plants and animals is the length and severity of the dry season.

Dry Climates

The two main types of dry climates are semi-arid (or steppe) and arid (or desert), both of which occur in low latitudes and midlatitudes. Geographers distinguish these dry climates based on the amount of rainfall and the vegetation in each.

Steppes are usually located away from oceans or large bodies of water and therefore are less humid. However, they do receive an average of 10 to 30 inches (25 to 76 cm) of rainfall per year. Steppes experience warm summers and harshly cold winters. Some steppes have heavy snowfall, while others are susceptible to droughts and violent winds. Steppes are found on almost every continent and are home to a diverse variety of grasses.

Deserts are extremely dry areas that receive about 10 inches (25 cm) of rainfall or less per year and support a very small amount of plant and animal life. Only plants that can live without much water and tolerate unreliable precipitation and extreme temperatures live in the desert. Deserts are usually hot and dry, although some deserts experience snowfall in the winter. In some desert areas, underground springs support an **oasis**, an area of lush vegetation. Temperatures tend to vary widely from day to night, as well as from season to season.

World Biomes

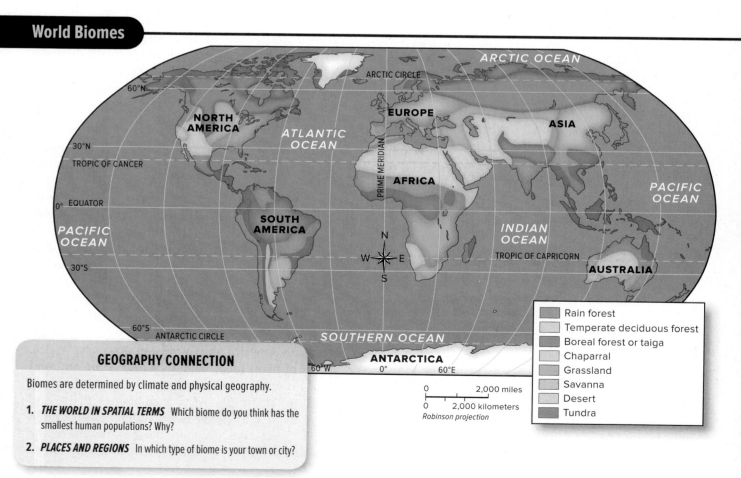

Rain forest
Temperate deciduous forest
Boreal forest or taiga
Chaparral
Grassland
Savanna
Desert
Tundra

0   2,000 miles
0   2,000 kilometers
Robinson projection

GEOGRAPHY CONNECTION

Biomes are determined by climate and physical geography.

1. **THE WORLD IN SPATIAL TERMS** Which biome do you think has the smallest human populations? Why?

2. **PLACES AND REGIONS** In which type of biome is your town or city?

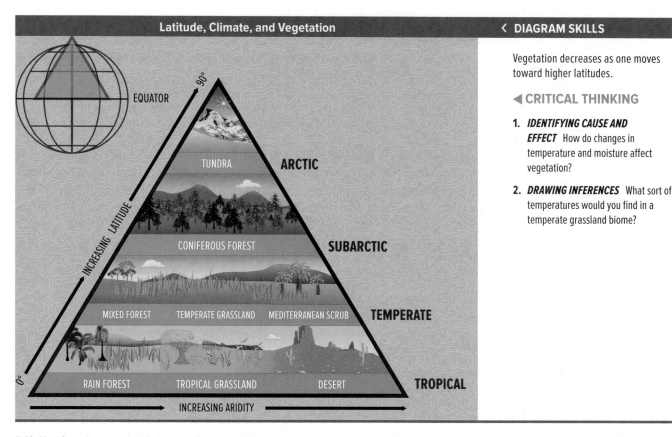

EQUATOR

90°

INCREASING LATITUDE

0°

TUNDRA **ARCTIC**

CONIFEROUS FOREST **SUBARCTIC**

MIXED FOREST   TEMPERATE GRASSLAND   MEDITERRANEAN SCRUB **TEMPERATE**

RAIN FOREST   TROPICAL GRASSLAND   DESERT **TROPICAL**

INCREASING ARIDITY

Vegetation decreases as one moves toward higher latitudes.

◄ **CRITICAL THINKING**

1. *IDENTIFYING CAUSE AND EFFECT* How do changes in temperature and moisture affect vegetation?

2. *DRAWING INFERENCES* What sort of temperatures would you find in a temperate grassland biome?

Midlatitude and High-Latitude Climates

The midlatitude climates include four temperate climate regions: humid subtropical, marine west coast, Mediterranean, and humid continental. Midlatitude climates experience variable weather patterns due to two conflicting air masses. Tropical air masses move from the Equator to the Poles, and polar air masses move in the opposite direction from the Poles to the Equator.

Humid subtropical climates, which include the southeastern United States as well as parts of Brazil, China, Japan, Australia, and India, are characterized by short, mild winters and nearly year-round rain. The wind patterns and high pressure from nearby oceans keep humidity levels high. Vegetation consists of **prairies** and evergreen and deciduous forests. **Coniferous** trees, most of which are evergreens, have cones. **Deciduous** trees, most of which have broad leaves, change color and drop their leaves in autumn.

Marine west coast climates, including the southern coast of Chile, parts of Australia, the British Isles, and the Pacific coast of North America, are mainly between the latitudes of about 30° N and 60° N and about 30° S and 60° S. Ocean winds bring cool summers and cool, damp winters. Abundant rainfall supports both coniferous and deciduous trees, often resulting in **mixed forests**.

Lands surrounding the Mediterranean Sea, in addition to the southwestern coast of Australia and central California, have mild, rainy winters and hot, dry summers. The natural vegetation includes thickets of woody bushes and short trees known as Mediterranean shrubs. Geographers classify as *Mediterranean* any coastal midlatitude area with similar climate and vegetation.

In some midlatitude regions of the Northern Hemisphere, landforms influence climate more than winds, precipitation, or ocean temperatures do. *Humid continental* climate regions do not experience the moderating effect of ocean winds because of their northerly continental, or inland, locations. The farther north one travels, the longer and more severe are the snowy winters, and the shorter and

prairie an inland grassland area

coniferous referring to vegetation having cones and needle-shaped leaves, including many evergreens, that keep their foliage throughout the winter

deciduous falling off or shed seasonally or periodically; trees such as oak and maple, which lose their leaves in autumn

mixed forest forest with both coniferous and deciduous trees

permafrost permanently frozen layer of soil beneath the surface of the ground

"Scientists analyzed summertime storm activity in the eastern U.S. from 1995 to 2009.... They discovered that tornadoes and hailstorms occurred at a rate of about 20 percent above average during the middle of the week. In contrast, the phenomena occurred at a rate of roughly 20 percent below average on the weekend....The team then investigated Environmental Protection Agency air-quality monitoring data and noted that human-made, summertime air pollution over the eastern U.S. peaks midweek. The cycle is linked to more human-made pollution created during the five-day workweek, such as commuters driving to and from work."

—Charles Q. Choi,
National Geographic News,
December 29, 2011

DBQ **ANALYZING** Do you think there is a cause-and-effect relationship between increased air pollution and increased tornado activity? Explain.

cooler are the summers. Vegetation is similar to that found in marine west coast areas, with evergreens outnumbering deciduous trees in the northernmost areas.

In high-latitude climates, freezing temperatures are common all year because of the lack of direct sunlight. As a result, the amount and variety of vegetation is limited here. Just south of the Arctic Circle are the *subarctic* climate regions. Winters here are bitterly cold, and summers are short and cool. Subarctic regions have the world's widest temperature ranges. In parts of the subarctic, only a thin layer of surface soil thaws each summer. Below is permanently frozen subsoil, or **permafrost**. Brief summer growing seasons may support needled evergreens.

Closer to the Poles are *tundra* climate regions. Winter darkness and bitter cold last for months, while summer has only limited warming. The layer of thawed soil is even thinner than in the subarctic. Trees cannot establish roots, so vegetation is limited to low bushes, very short grasses, mosses, and lichens (LY• kuhns).

Snow and ice, often more than 2 miles (3 km) thick, constantly cover the surfaces of *ice cap* regions. Lichens are the only form of vegetation that can survive in these areas where monthly temperatures average below freezing.

✔ **READING PROGRESS CHECK**

Making Connections Why do high-latitude climates have limited vegetation?

Climate Change

GUIDING QUESTION *What causes climates to change over time?*

Climate change refers to major changes in the factors used to measure climate over an extended period of time. For example, scientists have concluded that the average global temperature has increased by 1.4°F (0.8°C) over the last century. Some indicators of climate change include rising global temperatures, severe weather changes such as intense heat waves and changes in precipitation, an increase in severe weather events, and rising sea levels. Scientists search for answers by studying the interrelationships among ocean temperatures, greenhouse gases, wind patterns, and cloud cover. While scientists continue to disagree about the causes of climate change, we do know that the Earth undergoes natural and predictable cycles of cooling and warming caused by factors such as solar flares and volcanic activity.

However, increased global temperatures can also be attributed to greenhouse gas emissions. Burning fossil fuels releases gases that mix with water in the air, forming acids that fall in rain and snow. Acid rain can destroy forests. Fewer forests may result in climate change. The exhaust released from burning fossil fuels in automobile engines and factories is heated in the atmosphere by the sun's ultraviolet rays, forming smog, a visible chemical haze in the atmosphere.

✔ **READING PROGRESS CHECK**

Explaining How does the burning of fossil fuels create smog?

LESSON 3 REVIEW

Reviewing Vocabulary
1. ***Identifying*** Use the following terms in one sentence that describes humid subtropical climate regions: prairie, coniferous, deciduous.

Using Your Notes
2. ***Describing*** Use your graphic organizer to describe the Earth's four climate zones.

Answering the Guiding Questions
3. ***Categorizing*** How are world climates organized?

4. ***Drawing Conclusions*** What causes climates to change over time?

Writing Activity
5. ***Informative/Explanatory*** Write a paragraph detailing how the four major climate regions are related to the three zones of latitude.

Directions: On a separate sheet of paper, answer the questions below. Make sure you read carefully and answer all parts of the questions.

Lesson Review

Lesson 1

1 *Listing* What are the two types of solstices?

2 *Explaining* Why are the seasons different in the Northern and Southern Hemispheres?

3 *Describing* Describe what is meant by "midnight sun."

Lesson 2

4 *Defining* How are winds named?

5 *Describing* Describe the Coriolis effect. How does it differ based on latitude?

6 *Discussing* What effects do mountain ranges have on climates in a region?

Lesson 3

7 *Identifying* What are the four temperate climate regions into which midlatitude climates are classified?

8 *Explaining* How do geographers distinguish between the types of dry climate?

9 *Listing* What are three indicators of climate change?

Critical Thinking

10 *Making Generalizations* Has human activity had a positive or negative effect on the global climate? Explain.

11 *Comparing and Contrasting* How are tropical rain forest and tropical wet/dry climates alike? How are they different?

12 *Drawing Conclusions* How might human activities negatively affect plants and animals?

21st Century Skills

13 *Creating and Using Graphs, Charts, Diagrams, and Tables* All of the cities in the table are located in the Tropics except for Telluride. Why does Quito have a lower average temperature than the other tropical cities?

14 *Compare and Contrast* Use the information in the table to compare and contrast how elevation and latitude affect temperatures on Earth. Provide two examples.

The Influence of Elevation on Temperature			
	Elevation	**Latitude & Longitude**	**Average Temperature**
Quito, Ecuador	9,233 ft. (2,811 m)	0°09' S 78°29' W	58°F (14°C)
Nairobi, Kenya	5,327 ft. (1,623 m)	1°19' S 36°55' E	67°F (19°C)
Bjumbura, Burundi	2,568 ft. (782 m)	3°19' S 29°19' E	77°F (25°C)
Manaus, Brazil	276 ft. (84 m)	3°09' S 59°59' W	81°F (27°C)
Telluride, Colorado	8,760 ft. (2,670 m)	37°57' N 107°49' W	39°F (3°C)

Source: www.weatherbase.com

College and Career Readiness

15 *Clear Communication* Use the Internet to research careers in climatology and meteorology. Then write a job posting for one of the careers you learned about. Be sure that the job posting describes the required skills and experience necessary for the position.

Need Extra Help?

If You've Missed Question	1	2	3	4	5	6	7	8	9	10	11	12	13	14	15
Go to page	62	61	62	65	66	68	71	70	72	72	69	72	73	73	66

Directions: On a separate sheet of paper, answer the questions below. Make sure you read carefully and answer all parts of the questions.

Applying Map Skills

Refer to the Unit 1 Atlas to answer the following questions.

16 *Environment and Society* How is human activity affected by specific climate regions? Using the maps in the unit atlas, choose a climate region and provide an example of how it influences human activity.

17 *Human Systems* Using the maps in the unit atlas, what connections can you make between densely populated areas and climate regions?

Exploring the Essential Question

18 *Making Connections* Choose one of the biomes discussed in this chapter: tropical rain forest, tropical wet/dry, semi-arid (steppe), arid (desert), humid subtropical, marine west coast, Mediterranean, humid continental, subarctic, tundra, and ice cap. Use what you have learned about the biome to create a poster illustrating the plants and animals in the biome. Conduct additional research to include statistics on average rainfall and amount of direct sunlight. Posters should also illustrate the interactions of human systems within this biome. Posters should be highly visual and can include photos of typical vegetation, graphs/charts, and maps.

Research and Presentation

19 *Research Skills* Use Internet and library resources to gather information about the leatherback sea turtle. Create a multimedia presentation outlining the leatherback sea turtle's migration and mating patterns in map and graph/chart forms. Compare and contrast historical and current data regarding how many leatherback sea turtles existed in the past and exist today. Be sure to (1) explain why the turtle follows these migratory and mating patterns and (2) how climate change and human activity have impacted these patterns.

DBQ Analyzing Primary Sources

Use the document to answer the following questions.

PRIMARY SOURCE

"It has been demonstrated that climate changes at millennial, centennial, and even decadal scales, as many studies in the twentieth century have revealed. Such studies were largely done by examining environmental data such as tree rings, pollen assemblages, lake sediment, and ice cores. These data are objective and usually continuous, but they are often difficult to interpret. For example, a narrow tree ring could mean either a dry spell or cold spring, or both. Consequently, conclusions obtained this way are often associated with considerable uncertainties or ambiguities."

—Pao K. Wang, "Chinese historical documents and climate change", www.accessscience.com

20 *Analyzing* According to the excerpt, how did scientists learn that climates have changed in the past?

21 *Interpreting* What example is provided to explain why climate change data is often difficult to interpret?

Writing About Geography

22 *Informative/Explanatory* Use standard grammar, spelling, sentence structure, and punctuation to write a one-page essay explaining how the Earth's rotation on its axis and its revolution around the sun creates certain types of climate conditions in low, mid-, and high latitude regions. Be sure to cite specific examples in your essay.

Need Extra Help?

If You've Missed Question	16	17	18	19	20	21	22
Go to page	4	4	69	72	74	74	60

The Human World

ESSENTIAL QUESTION • *How do the characteristics and distribution of human populations affect human and physical systems?*

networks

There's More Online about the human world.

CHAPTER 4

Why Geography Matters
The Human Development Index

Lesson 1
Global Cultures

Lesson 2
Population Geography

Lesson 3
Political Geography

Lesson 4
Economic Geography

Lesson 5
Urban Geography

Geography Matters...

Have you ever traveled far from home and had new experiences of sight, sound, and taste? Examining these varieties in the human experience is the work of the human geographer. Human geographers study relationships between humans and their natural environment and analyze geographic patterns. They look at present-day events with an eye for trends, knowing that human actions and decisions can have major impacts on people near and far.

◄ A woman works on a tea plantation in Kerala, India.

the human development index

The relative social and economic status of countries is often described in the media and among academics as either "less developed," "newly industrialized," or "more developed." What precisely is meant by the term development? And how is development measured?

Human Development

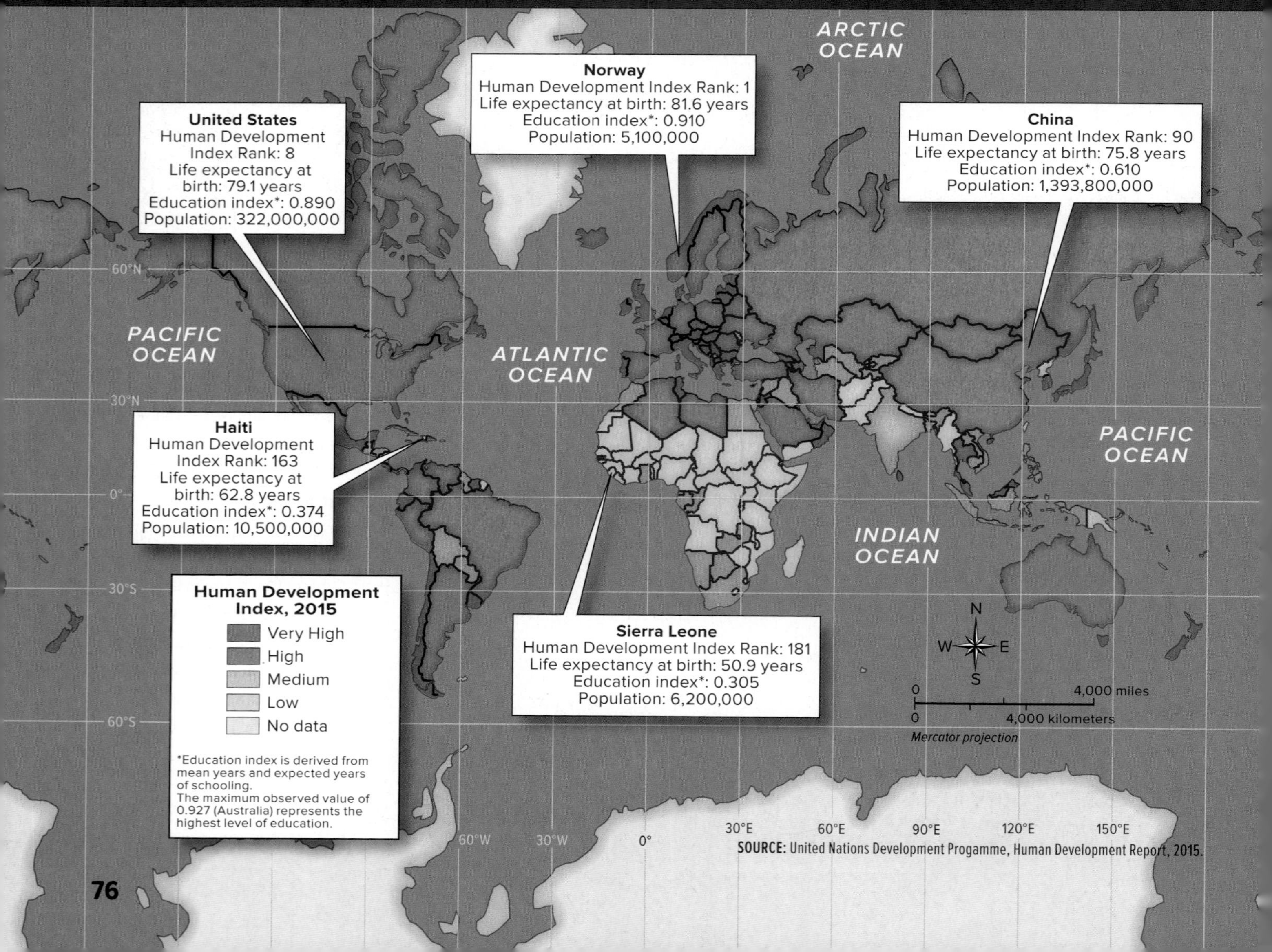
ARCTIC OCEAN
Norway
Human Development Index Rank: 1
Life expectancy at birth: 81.6 years
Education index*: 0.910
Population: 5,100,000
United States
Human Development Index Rank: 8
Life expectancy at birth: 79.1 years
Education index*: 0.890
Population: 322,000,000
China
Human Development Index Rank: 90
Life expectancy at birth: 75.8 years
Education index*: 0.610
Population: 1,393,800,000
PACIFIC OCEAN
ATLANTIC OCEAN
PACIFIC OCEAN
Haiti
Human Development Index Rank: 163
Life expectancy at birth: 62.8 years
Education index*: 0.374
Population: 10,500,000
INDIAN OCEAN
Sierra Leone
Human Development Index Rank: 181
Life expectancy at birth: 50.9 years
Education index*: 0.305
Population: 6,200,000
Human Development Index, 2015
Very High
High
Medium
Low
No data
*Education index is derived from mean years and expected years of schooling.
The maximum observed value of 0.927 (Australia) represents the highest level of education.
N
W
E
S
0
4,000 miles
0
4,000 kilometers
Mercator projection
60°N
30°N
0°
30°S
60°S
60°W
30°W
0°
30°E
60°E
90°E
120°E
150°E

SOURCE: United Nations Development Progamme, Human Development Report, 2015.

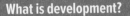

What is development?

Although different societies and cultures have their own perspectives and values that shape the types of human development that occur, most agree that development includes technological progress that leads to a better standard of living for people. The modern notion of development as a means of improving people's lives gained favor during the Industrial Revolution. Advances in science and technology led to increased agricultural and industrial production, which in turn provided a greater food supply and more manufactured products that people needed. Such new inventions and products led to easing some people's workloads, providing comfort and convenience, and allowing many to begin acquiring wealth. Broadly speaking, development is improvement in people's living situations and economic prospects.

1. **Human Systems** Why is human development an important topic? Define the term *human development* and describe the role of technology as a component of it.

Why do we measure development?

Governments rely on measures of development when making important policy decisions and addressing the specific needs of both their own citizens and those of other countries. For example, local governments use measures of development as a way to bring the needs of people to the attention of the central government in the hopes of having funds and resources allocated appropriately. Geographers and other social scientists are also interested in measures of development because they study the spatial patterns of human need and well-being, looking for cause-and-effect relationships. Identifying these types of relationships helps governments, aid agencies, and others make changes to improve the economic, social, or environmental conditions in which people live. Although there are many ways to measure development, many governments today use the Human Development Index (HDI), a statistic used to rank countries by their level of human development. The index, devised by Pakistani economist Mahbub ul Haq in 1990, was designed to include measurements of improvement in people's well-being, rather than focusing solely on national economic indicators.

2. **Places and Regions** How is measuring development useful? Explain how measures of development are used.

How does the Human Development Index (HDI) measure development?

The HDI ranks countries' level of human development by measuring three dimensions of development: health, education, and living standards. The measure of health is determined by life expectancy at birth. Access to education is measured by examining mean, or average, years of schooling and expected years of schooling. Living standards are measured by gross national income per capita. The Human Development Index uses the data from these three dimensions of development to calculate one composite statistic for each country. The HDI is published annually by the United Nations Development Programme, and it is used to measure how economic policy decisions affect people's quality of life.

3. **The Uses of Geography** Why do you think the HDI incorporates more than one dimension of development? What do these dimensions indicate about the level of development in a country? Provide examples to support your answer.

THERE'S MORE ONLINE

VIEW a graph of changes in HDI ranking over time • **COMPARE** population indicators of several countries

networks

There's More Online!

☑ **CHART** Cultural Universals

☑ **MAP** World Language Families

☑ **MAP** World Culture Hearths

☑ **INTERACTIVE SELF-CHECK QUIZ**

☑ **VIDEO** Global Cultures

Reading **HELP**DESK

Academic Vocabulary

- similar
- major

Content Vocabulary

- culture
- language family
- ethnic group
- culture region
- cultural diffusion
- culture hearth

TAKING NOTES: *Key Ideas and Details*

IDENTIFYING Use a graphic organizer like the one below to take notes as you read about global cultures in this lesson.

Elements of Culture	Cultural Change

LESSON 1
Global Cultures

ESSENTIAL QUESTION • *How do the characteristics and distribution of human populations affect human and physical systems?*

IT MATTERS BECAUSE

The world's people organize communities, develop ways of life, and adjust to the differences and similarities they experience. As the world becomes increasingly interconnected, cultures spread and are shared. To help understand this cultural diversity, geographers divide the Earth into culture regions, which are defined by the presence of common cultural elements such as language and religion.

Elements of Culture

GUIDING QUESTION *What are the elements of culture?*

Geographers study **culture**, the way of life of a group of people who share **similar** ways of thinking, believing, and living, expressed in common elements or features. For example, a particular culture can be understood by looking at language, religion, daily life, history, art, government, technology, and economy.

Language is a key element in a culture's development. Through language, people communicate information and experiences and pass on cultural values and traditions. Even within a culture, however, there are language differences. Some people may speak a dialect, or a local form of a language, that differs from the main language. These differences may include variations in the pronunciation and meaning of words.

Linguists, scientists who study languages, organize the world's languages into **language families**—large groups of languages having similar roots. Seemingly diverse languages may belong to the same language family. For example, English, Spanish, and Russian are all members of the Indo-European language family.

Religion is another important element of a culture. Religious beliefs vary significantly around the world. For many people, religion provides an important sense of identity. It also influences many aspects of daily life, from the practice of moral values to the celebration of holidays and festivals. Throughout history, religious symbols and stories have shaped cultural expressions such as literature, painting and sculpture, architecture, and music.

Purepix/Alamy Stock Photo

A social system develops to help the members of a culture work together to meet basic needs. In all cultures, the family forms an important group. Most cultures are also made up of social classes, groups of people ranked according to ancestry, wealth, education, or other criteria. Moreover, cultures may include people who belong to different ethnic groups. An **ethnic group** is made up of people who share a common language, history, or place of origin.

Geographers also analyze governments to help understand a culture. Governments of the world share certain features, such as maintaining order within the country, providing protection from outside dangers, and supplying other services to the people. Governments can be categorized by levels of power—national, regional, and local—and by type of authority—a single ruler, a small group of leaders, or a body of citizens and their representatives.

Economic activities also influence and shape a culture. People must make a living, whether in farming, industry, or by providing services. Geographers study how a culture utilizes its natural resources to meet such needs as food and shelter. They also analyze the ways in which people produce, obtain, use, and sell goods and services.

To organize their understanding of cultural development, geographers divide the Earth into culture regions. Each **culture region** includes areas that have certain traits in common. They may share similar economic systems, forms of government, or social groups. Their histories, religions, and art forms may share similar influences.

✅ **READING PROGRESS CHECK**

Explaining Why are social groups important to the development of a culture?

culture way of life of a group of people who share similar culture traits, including beliefs, customs, technology, and material items

similar comparable

language family group of related languages that have all developed from one earlier language

ethnic group group of people who share common ancestry, language, religion, customs, or place of origin

World Language Families

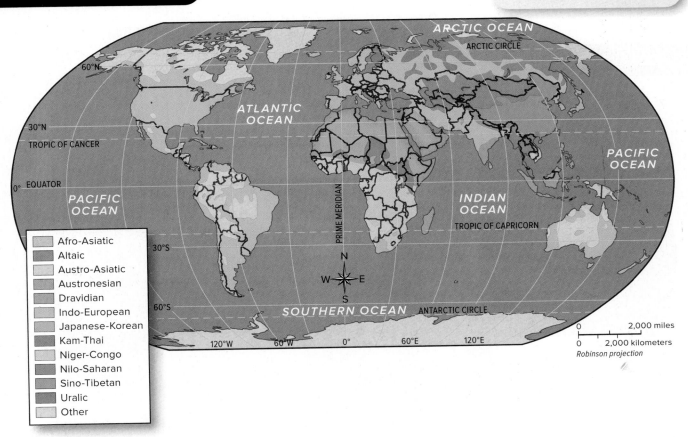

Afro-Asiatic
Altaic
Austro-Asiatic
Austronesian
Dravidian
Indo-European
Japanese-Korean
Kam-Thai
Niger-Congo
Nilo-Saharan
Sino-Tibetan
Uralic
Other

Cultural Change

GUIDING QUESTION *What are two major ways in which cultures change over time?*

culture region division of the
Earth in which people share a similar
way of life, including language,
religion, economic systems, and
values

cultural diffusion the spread
of culture traits, material and
non-material, from one culture to
another

major greater in importance or
interest

culture hearth a center
where cultures developed and from
which ideas and traditions spread
outward

Cultures are dynamic and continually changing. Internal factors—new ideas, lifestyles, and inventions—create change within cultures. Change can also come through spatial interaction such as trade, migration, and war. The spread of new knowledge from one culture to another is called cultural diffusion. Cultural diffusion has been a major factor in cultural development since the dawn of human history, and the pace of cultural change has accelerated in contemporary times. The earliest humans were small groups of hunters and gatherers, who moved from place to place in search of animals to hunt, plants to gather, water, and useful materials. As they migrated, they helped spread culture traits from one group and place to another.

Cultural Change in History

The world's first civilizations arose in culture hearths—early centers of civilization whose ideas and practices spread to surrounding areas. The map shows that some of the most influential culture hearths developed in areas that make up the modern countries of Egypt, Iraq, Pakistan, China, and Mexico.

These five culture hearths had certain geographic features in common. They all emerged from farming settlements in areas with a mild climate and fertile land. In addition, all five culture hearths were located near a major river or source of water. Making use of favorable environments, the people dug canals and ditches to irrigate the land. All of these factors contributed to what is known as the agricultural revolution, a major shift from food gathering to food production that enabled people to grow surplus crops.

Surplus food set the stage for the rise of cities and civilizations and the development of long-distance trade. The increased wealth from trade led to the rise of cities and complex social systems. These new cities needed a well-organized

---

World Culture Hearths

GEOGRAPHY CONNECTION

Culture hearths are centers where the world's first civilizations arose.

1. *PLACES AND REGIONS* Where in Asia were the first major settlements located?

2. *HUMAN SYSTEMS* What water feature do most of the culture hearths have in common?

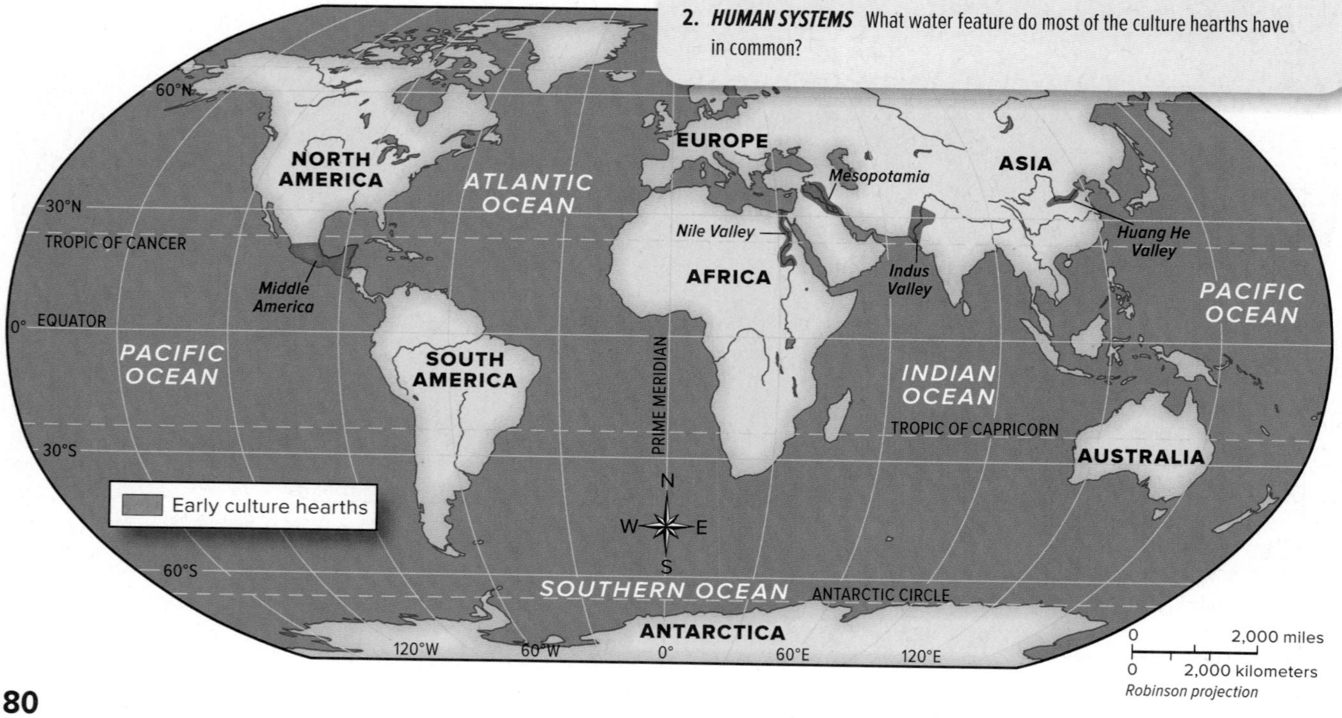

Early culture hearths

government to coordinate harvests, plan building projects, and manage an army for defense. Officials and merchants created writing systems to record and transmit government and trade information.

Cultural diffusion has increased rapidly during the last 250 years. In the 1700s and 1800s, some countries began to industrialize, using power-driven machines and factories to mass-produce goods. This period is known as the Industrial Revolution. With new production methods, these countries produced goods quickly and cheaply, and their economies changed dramatically. These developments also led to social changes. As people left farms for jobs in factories and mills, cities grew larger.

Cultural Change in the Contemporary World

At the end of the twentieth century, the world experienced a new turning point—the information revolution. Computers now make it possible to store huge amounts of information and instantly send it all over the world, thus allowing more rapid spread of ideas and traditions among the cultures of the world. The Internet has been responsible for communication and socialization around the world via social networking sites and other sites that allow users to share many types of information and stay connected with others. Consequently, the world feels much smaller than it might have previously.

Cultural contact among different peoples promotes cultural change as ideas and practices spread. Computer technology certainly accelerates the spread of cultural change, but other connections among people do as well. Trade and travel are important avenues by which cultural change occurs. Migration has also fostered cultural diffusion. People migrate for many reasons. Positive factors—better social and economic conditions and religious or political freedoms—may draw people from one place to another. Most people move from one place to another in search of better economic opportunity. Negative factors— wars, persecution, and famines—also motivate people to migrate. In some instances, as in the case of enslaved Africans brought to the Americas, mass migrations have been forced. Regardless of the reasons, migrants carry their cultures with them, and their ideas and practices often blend with those of the people already living in the migrants' adopted countries.

Contact between different cultures usually leads to change in both systems.

▲ CRITICAL THINKING

1. *Describing* In what way is this picture an example of cultural contact?

2. *Speculating* What types of ideas and practices might be exchanged due to this example of cultural contact?

☑ READING PROGRESS CHECK

Identifying Where were the five earliest culture hearths located?

LESSON 1 REVIEW

Reviewing Vocabulary

1. *Explaining* Explain the relationships between culture, ethnic group, culture region, cultural diffusion, and culture hearth.

Using Your Notes

2. *Describing* Use your graphic organizer to describe the external factors that change cultures.

Answering the Guiding Questions

3. *Identifying* What are the elements of culture?

4. *Drawing Conclusions* What are two major ways in which cultures change over time?

Writing Activity

5. *Explanatory* Research to find three different definitions of culture. Write a paragraph comparing the definitions.

networks

There's More Online!

- ☑ **DIAGRAM** Demographic Transition Model
- ☑ **GRAPH** Projected Populations
- ☑ **INTERACTIVE SELF-CHECK QUIZ**
- ☑ **VIDEO** Population Geography

Reading **HELP**DESK

Academic Vocabulary

- **community**
- **trend**

Content Vocabulary

- **birthrate**
- **death rate**
- **natural increase**
- **migration**
- **demographic transition**
- **doubling time**
- **population pyramid**
- **population distribution**
- **population density**

TAKING NOTES: *Key Ideas and Details*

DESCRIBING As you read, use a graphic organizer like the one below to take notes on the demographic transition model and challenges of growth.

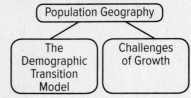

LESSON 2

Population Geography

ESSENTIAL QUESTION · *How do the characteristics and distribution of human populations affect human and physical systems?*

IT MATTERS BECAUSE

Earth's human population increased dramatically during the nineteenth century and much of the twentieth century. Although that growth has begun to slow in recent decades, there are far more people on Earth today than ever before. The result is that issues of population growth or decline are crucial to all countries. Geographers play an important role in examining ways to plan for the future and solve problems of tomorrow.

Population Growth

GUIDING QUESTION *What factors influence population growth?*

More than 7 billion people now live on Earth, and most of the population inhabits about 30 percent of the planet's land area. Global population continues to grow and is expected to level off at nearly 10 billion by the year 2050. Such rapid growth was not always the case. From the year 1000 until 1800, the world's population increased slowly. Then the number of people on Earth more than doubled between 1800 and 1950. It doubled again between 1950 and about 2000.

The Demographic Transition Model

Scientists in the field of *demography*, the study of populations, use statistics to learn about population growth. The **birthrate** is the number of births per year for every 1,000 people. The **death rate** is the number of deaths per year for every 1,000 people. **Natural increase**, or the growth rate of a population, is the difference between an area's birthrate and its death rate. **Migration**, or the movement of people from place to place, must also be considered when examining population changes.

The **demographic transition** model uses birthrates and death rates to show how populations in countries or regions can change over time. The model was first used to show the relationship of declining birthrates and death rates to industrialization in Western Europe. Death rates can fall quickly as a result of more abundant and reliable food supplies, improved health care, access to medicine and technology, and better living conditions. Birthrates decline more slowly because the declines result from changes in cultural traditions that can often take longer.

Today, most of the world's industrialized and technologically developed countries have experienced the transition from high birthrates and death rates to low birthrates and death rates. These countries have reached what is known as *zero population growth,* in which the birthrate and death rate are equal. When this balance occurs, a country's population does not grow as a result of natural increase, although it can still grow as a result of migration.

Although birthrates have fallen significantly in many countries in Asia, Africa, and Latin America over the past 40 years, they are still higher than in the industrialized world. Families in these regions traditionally are large because of cultural beliefs about marriage, family, and the value of children. For example, a husband and wife in a rural agricultural area may choose to have several children who will help farm the land. The high number of births often continues after death rates decrease as a result of improved living conditions. This causes the population to greatly increase. As a result, the **doubling time**, or the number of years it takes a population to double in size, has been reduced to below 50 years in some parts of Asia, Africa, and Latin America. In contrast, the average doubling time of a more developed country can be more than 300 years.

Challenges of Growth

Rapid population growth presents many challenges that affect individual countries and the global **community**. As the number of people increases, so does the difficulty of producing enough food to feed them. In Africa, for example, food shortages in

birthrate number of births per year for every 1,000 people

death rate number of deaths per year for every 1,000 people

natural increase the growth rate of a population; the difference between birthrate and death rate

migration the movement of people from place to place

demographic transition the model that uses birthrates and death rates to show how populations in countries or regions change over time

doubling time the number of years it takes for a population to double in size

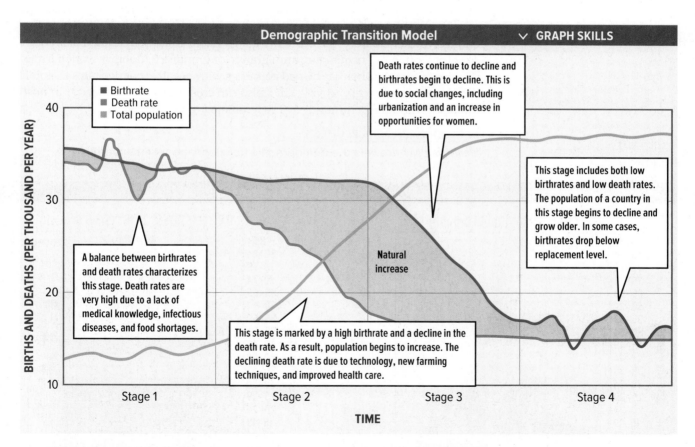

Demographic Transition Model ⌄ GRAPH SKILLS

Death rates continue to decline and birthrates begin to decline. This is due to social changes, including urbanization and an increase in opportunities for women.

This stage includes both low birthrates and low death rates. The population of a country in this stage begins to decline and grow older. In some cases, birthrates drop below replacement level.

A balance between birthrates and death rates characterizes this stage. Death rates are very high due to a lack of medical knowledge, infectious diseases, and food shortages.

Natural increase

This stage is marked by a high birthrate and a decline in the death rate. As a result, population begins to increase. The declining death rate is due to technology, new farming techniques, and improved health care.

- ■ Birthrate
- ■ Death rate
- ■ Total population

BIRTHS AND DEATHS (PER THOUSAND PER YEAR)

40
30
20
10

Stage 1   Stage 2   Stage 3   Stage 4

TIME

Changes in population trends can be identified by examining the relationships between birthrates and death rates.

▲ **CRITICAL THINKING**

1. Analyzing What characteristics of Stage 2 create an increase in population?

2. Identifying What effect does an increase in opportunities for women have on populations? Of what stage is it a characteristic?

community people with common interests living in a particular area

population pyramid a diagram that shows the distribution of a population by age and gender

trend a general movement

production and availability continue to be an issue. In better circumstances, Africa's agricultural sector would respond to rising prices by increasing food supply. In Africa, however, lack of government investment and other support of agriculture—along with warfare, poor access to rural areas, and weather and pests that can ruin crops—have combined to bring hunger to the region.

In addition, populations that grow rapidly use resources more quickly. Some countries face shortages of water, housing, and clothing, for instance. Rapid population growth strains these limited resources. Another concern is that the world's population is unevenly distributed by age, with the majority of some countries' populations being infants and young children who cannot contribute to food production. This population structure can be seen with a **population pyramid**.

While some experts are pessimistic about the long-term effects of rapid population growth, others are optimistic that, as the number of humans increases, the levels of technology and creativity will also rise. For example, scientists continue to study and develop ways to boost agricultural productivity. Fertilizers can improve crop yields. Irrigation systems can help increase the amount of land available for farming. New varieties of crops have been created to withstand severe conditions and yield more food.

In the late 1900s some countries in Europe began to experience a **trend** called *negative population growth,* in which the annual death rate exceeds the annual birthrate. Hungary, for example, shows a change rate of about -0.3. This situation has economic consequences different from, but just as serious as, those caused by high growth rates. In countries with negative population growth, it is difficult to find enough workers to keep the economy going. Labor must be recruited from other countries, often by encouraging immigration or granting temporary work permits. The use of foreign labor has helped countries with negative change rates maintain their levels of economic activity. But it also can create tensions between the "host" population and the communities of newcomers.

✓ READING PROGRESS CHECK

Identifying Where was the demographic transition model first used?

Stages Of Growth: Population Pyramids

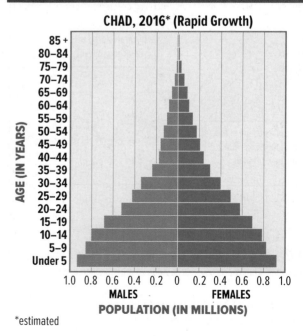

CHAD, 2016* (Rapid Growth)

AGE (IN YEARS): 85+, 80–84, 75–79, 70–74, 65–69, 60–64, 55–59, 50–54, 45–49, 40–44, 35–39, 30–34, 25–29, 20–24, 15–19, 10–14, 5–9, Under 5

1.0 0.8 0.6 0.4 0.2 0 0.2 0.4 0.6 0.8 1.0
MALES — FEMALES
POPULATION (IN MILLIONS)

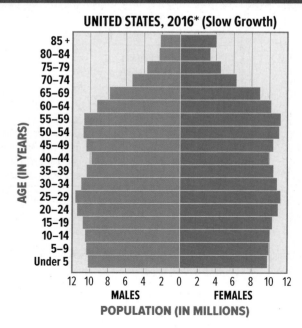

UNITED STATES, 2016* (Slow Growth)

AGE (IN YEARS): 85+, 80–84, 75–79, 70–74, 65–69, 60–64, 55–59, 50–54, 45–49, 40–44, 35–39, 30–34, 25–29, 20–24, 15–19, 10–14, 5–9, Under 5

12 10 8 6 4 2 0 2 4 6 8 10 12
MALES — FEMALES
POPULATION (IN MILLIONS)

*estimated

Population Distribution

GUIDING QUESTION *What influences population distribution?*

Not only do population growth rates vary among Earth's regions, but the pattern of human settlement, or **population distribution**, is uneven as well. Population distribution is related to the Earth's physical geography. Only about 30 percent of Earth's surface is made up of land, and much of that land is inhospitable. High mountain peaks, barren deserts, and frozen tundra make human activity difficult in many places. Almost everyone on Earth lives on a relatively small portion of the planet's land—a little less than one-third. Most people live where fertile soil, available water, and a climate without harsh extremes make human life possible.

Of all the continents, Europe and Asia are the most densely populated. Asia alone contains about 60 percent of the world's people. Many people throughout the world live in metropolitan areas—cities and their surrounding urbanized areas—where populations are highly concentrated. Today, most people in Europe, North America, South America, and Australia live in or around urban areas.

Geographers determine how crowded a country or region is by measuring **population density**—the number of people living on a square mile or square kilometer of land. To determine population density in a country, geographers divide the total population of the country by its total land area.

Population density varies widely from country to country. Canada, with a low population density of about 10 people per square mile (4 people per sq. km), offers wide-open spaces and the choice of living in thriving cities or quiet rural areas. In contrast, Bangladesh has one of the highest population densities in the world—about 3,362 people per square mile (1,298 people per sq. km).

Countries with populations of about the same size do not necessarily have similar population densities. For example, both Cuba and Somalia have about 11.1 million people. Somalia, with a larger land area, has only 46 people per square mile (18 people per sq. km). However, Cuba has 261 people per square mile (101 people per sq. km).

population distribution
the variations in population that occur across a country, a continent, or the world

population density the average number of people living on a square mile or square kilometer of land

---

GRAPH SKILLS

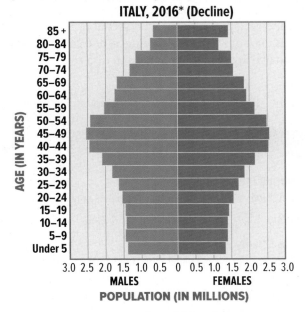

ITALY, 2016* (Decline)

AGE (IN YEARS)

85 +
80–84
75–79
70–74
65–69
60–64
55–59
50–54
45–49
40–44
35–39
30–34
25–29
20–24
15–19
10–14
5–9
Under 5

3.0 2.5 2.0 1.5 1.0 0.5 0 0.5 1.0 1.5 2.0 2.5 3.0
MALES · FEMALES
POPULATION (IN MILLIONS)

Source: U.S. Census Bureau, International Data Base.

Population pyramids include information about the age and gender distribution of a country's population. The population structure of a country provides insight into population growth trends and the country's stage in the demographic transition.

◄ CRITICAL THINKING

1. *Comparing and Contrasting* How does the shape of the population pyramid for a country undergoing rapid growth differ from that of a country experiencing population decline?

2. *Analyzing* What does the shape of the U.S. population pyramid tell you about the country's population structure?

World Refugee Day 2012

"The United States is strongly committed to protecting and assisting refugees and we offer resettlement to more refugees each year than all other countries in the world combined. Since 1975, more than three million refugees have made new homes in the United States, and nearly half of them have become U.S. citizens.

Refugees are contributing in ways large and small to business, academia, the arts, science and technology. Today we celebrate the success of refugees who have built new lives here and in other resettlement countries, but we also recognize the millions of refugees who remain displaced in camps, cities, and rural settlements around the world."

—Hillary Rodham Clinton, press statement, June 20, 2012

DBQ *SPECULATING* Why might refugees choose to migrate to the United States? What difficulties might refugees face?

Because the measure of population density includes all the land area of a country, it does not account for uneven population distribution within a country. In Egypt, for example, overall population density is 232 people per square mile (90 people per sq. km). In reality, over 90 percent of Egypt's people live along the Nile River. The rest of Egypt is desert. Thus, some geographers describe a country's population density in terms of land that can be used to support the population rather than total land area. When Egypt's population density is measured this way, it is about 7,501 people per square mile (3,196 people per sq. km)!

The Earth's population is moving in great numbers. People are moving from city to city or from rural villages to cities. The resulting growth of city populations brought about by such migration and the changes that come with this increase are called *urbanization.* The primary cause of urbanization is the desire of rural people to find jobs and a better life in more prosperous urban areas. Rural populations have certainly grown in some countries, but the amount of farmland has not increased to meet the growing number of people. As a result, many rural migrants find urban jobs in manufacturing and service industries.

About half of the world's people live in cities. Between 1960 and 2016, the population of metropolitan Mexico City rose from about 5 million to more than 22 million. Other cities in Latin America, as well as in Asia and Africa, have seen similar growth. Some cities contain a large part of their country's entire population. For example, about one-third of Argentina's population lives in Buenos Aires.

PRIMARY SOURCE

*❝*Never in human history have so many people changed their locations and lifestyles so quickly. Each month, there are 5 million new city dwellers created through migration or birth in Africa, Asia, and the Middle East. . . .*❞*

—Doug Saunders, "The great shift from farm to city," *Los Angeles Times*, June 19, 2011

Population movement also occurs between countries. Some people emigrate from the country of their birth and move to another. They are known as emigrants in their homeland and immigrants in their new country. In the past 40 years, millions of people have left Africa and Asia to find jobs in the wealthier countries of Europe. Economic factors that attract people to a place are called *pull factors.* Religious and political freedoms are also examples of pull factors that attract people to another country. Environmental pull factors are things that make a place a desirable place to live, such as mountains, warm climates, and sea coasts. Other factors of migration, called *push factors,* trigger out-migration; these also include lack of economic opportunities and religious or political persecution. Some people are forced to flee their country because of wars, food shortages, or other problems. They are refugees, or people who flee to escape persecution or disaster.

✓ **READING PROGRESS CHECK**

Explaining What factors influence the migration of people from one country to another?

LESSON 2 REVIEW

Reviewing Vocabulary

1. *Explaining* Define demographic transition model and doubling time. How are they related to each other?

2. *Calculating* Define population density and how it is calculated.

Using Your Notes

3. *Describing* Use your graphic organizer to explain why population growth varies from place to place.

Answering the Guiding Questions

4. *Discussing* What factors influence population growth?

5. *Stating* What influences population distribution?

Writing Activity

6. *Narrative* What human-made structures might be present in countries that have large numbers of people concentrated in relatively small areas? Write a paragraph with supporting details to explain your answer.

networks

There's More Online!

- ☑ **CHART** WTO Members
- ☑ **IMAGE** Rhine River Valley
- ☑ **IMAGE** Four Corners
- ☑ **IMAGE** Pakistan-India Border
- ☑ **IMAGE** Terrorist Attack on Mumbai
- ☑ **INTERACTIVE SELF-CHECK QUIZ**
- ☑ **VIDEO** Political Geography

LESSON 3
Political Geography

ESSENTIAL QUESTION • *How do the characteristics and distribution of human populations affect human and physical systems?*

Reading **HELP**DESK

Academic Vocabulary

- unique
- authority

Content Vocabulary

- **unitary system**
- **federal system**
- **autocracy**
- **monarchy**
- **oligarchy**
- **theocracy**
- **democracy**
- **natural boundary**
- **cultural boundary**
- **geometric boundary**

TAKING NOTES: *Key Ideas and Details*

DESCRIBING As you read, use a graphic organizer like the one below to take notes on the features of government.

Features of Government	
Levels of Government	Types of Government
Geography and Government	Conflict and Cooperation

IT MATTERS BECAUSE

Governments and economies of countries around the world are becoming increasingly interconnected. Some countries or groups of countries, such as the European Union, have strong economies that allow them to help improve standards of living in other countries.

Features of Government

GUIDING QUESTION *What influences the level and type of a country's government?*

Today the world includes nearly 200 independent countries that vary in size, military might, natural resources, and world influence. Each country is defined by characteristics such as territory, population, and sovereignty, or freedom from outside control. These elements are brought together under a government. A government must make and enforce policies and laws that are binding upon all people living within its territory.

Levels of Government

The government of each country has **unique** characteristics that relate to that country's historical development. To carry out their functions, governments are organized in a variety of ways. Most large countries have several different levels of government. These usually include a national or central government, as well as the governments of smaller internal divisions such as provinces, states, counties, cities, towns, and villages.

A **unitary system** of government gives all key powers to the national or central government. This structure does not mean that only one level of government exists. Rather, it means that the central government creates state, provincial, or other local governments and gives them limited sovereignty. The United Kingdom and France both developed unitary governments as they emerged from smaller territories during the late Middle Ages and early modern times.

A **federal system** of government divides the powers of government between the national government and state or provincial governments. Each level of government has sovereignty in some areas. The United States developed a federal system after the thirteen colonies became independent from Great Britain.

Prime minister Justin Trudeau is the leader in the Canadian federal system.

▲ CRITICAL THINKING

1. Identifying Over what type of government does the prime minister of Canada preside?

2. Contrasting How is this form of government different from an oligarchy?

unique being the only one; without a like or an equal

unitary system form of government in which all key powers are given to the national or central government

federal system form of government in which powers are divided between the national government and state or provincial governments

autocracy system of government in which one person rules with unlimited power and authority

authority power to influence or command thought, opinion, or behavior

monarchy a form of autocracy with a hereditary king or queen exercising supreme power

oligarchy system of government in which a small group holds power

Another similar government structure is a confederation, or a loose union of independent territories. The United States at first formed a confederation, but this type of political arrangement failed to provide an effective national government for the new nation. As a result, the U.S. Constitution established a strong national government while preserving some state government powers. Today, other countries with federal or confederate systems include Canada, Switzerland, Mexico, Brazil, Australia, and India.

Types of Governments

Governments can be classified by asking the question: Who governs the state? Under this classification system, all governments belong to one of the three major groups: (1) autocracy, or rule by one person; (2) oligarchy, or rule by a few people; or (3) democracy, or rule by many people.

Any system of government in which the power and authority to rule belongs to a single individual is an **autocracy** (aw•TAH•kruh•see). Autocracies are the oldest and one of the most common forms of government. Most autocrats achieve and maintain their position of **authority** through inheritance or by the ruthless use of military or police power.

Several forms of autocracy exist. One is an absolute or totalitarian dictatorship in which the decisions of a single leader determine government policies. The government under such a system can come to power through a revolution or an election. The totalitarian dictator seeks to control all aspects of social and economic life. Examples of totalitarian dictatorships include Adolf Hitler in Nazi Germany, Saddam Hussein in Iraq, Raul Castro of Cuba, and Kim Jong Un of North Korea.

Monarchy (MAH•nuhr•kee) is another form of autocratic government. In a monarchy, a king or queen exercises the supreme powers of government. Monarchs usually inherit their positions. Absolute monarchs have complete and unlimited power to rule. The king of Saudi Arabia, for example, is an absolute monarch. Absolute monarchs are rare today, but from the 1400s to the 1700s, kings or queens with absolute power ruled most of Western Europe.

Today, some countries, such as the United Kingdom, Canada, Japan, Jordan, and Thailand, have constitutional monarchies. Their monarchs share governmental powers with elected legislatures or serve as ceremonial leaders.

An **oligarchy** (AH•luh•GAHR•kee) is any system of government in which a small group holds power. The group derives its power from wealth, military power, social position, or a combination of these elements. Today the governments of communist countries, such as China, are mostly oligarchies. Leaders in the Communist Party and the military control the government.

Sometimes religion is the source of power in an oligarchy. A **theocracy**, for example, is a government of officials believed to be divinely inspired. In a theocracy, a divine power is thought to be the head of the government. Government officials receive their inspiration, guidance, and authority to rule from this divine power. For example, Islamic sharia law is imposed in parts of North Africa and Southwest Asia today.

Both dictatorships and oligarchies sometimes claim they rule for the people. Such governments may try to give the appearance of control by the people. For example, they might hold elections but offer only one candidate. Such governments may also have some type of legislature or national assembly elected by or representing the people. These legislatures, however, only approve policies and decisions already made by the leaders. As in a dictatorship, oligarchies usually suppress all political opposition.

A **democracy** is any system of government in which leaders rule with the consent of the citizens. The term *democracy* comes from the Greek *demos* (meaning "the people") and *kratia* (meaning "rule"). The ancient Greeks used the word *democracy* to mean government by the many in contrast to government by the few. The key idea of democracy is that people hold sovereign power.

Direct democracy, in which citizens themselves decide on issues, exists in some places at local levels of government. No country today has a national government based on direct democracy. Instead, democratic countries have representative democracies, in which the people elect representatives with the responsibility and power to make laws and conduct government. An assembly of the people's representatives may be called a council, a legislature, a congress, or a parliament.

Many democratic countries, such as the United States and France, are republics. In a republic, voters elect all major officials, who are responsible to the people. The head of state—or head of government—is usually a president elected for a specific term. Not every democracy is a republic. The United Kingdom, for example, is a democracy with a monarch as head of state. This monarch's role is ceremonial, however, and elected officials hold the actual power to rule.

✔ READING PROGRESS CHECK

Contrasting How does an autocracy differ from an oligarchy?

theocracy system of government in which those who rule are regarded as divinely inspired

democracy system of government in which leaders rule with consent of the citizens

Geography and Government

GUIDING QUESTION *How does geography influence a country's government?*

Governments can be greatly influenced by geography. Geographic areas can actually determine how political and administrative units are drawn up and how they will be governed. Democratic countries have entities based on location, which are divided into local bodies that might have different laws.

A government must consider the cultural and religious beliefs of its citizens in order to govern effectively. In autocracies, governments frequently suppress

Types of Boundaries

Natural Boundary: The Rhine River Valley near Ruggell forms the border between Switzerland (left bank) and Liechtenstein (right bank).

Geometric Boundary: The Four Corners site in the United States occurs where the states of Utah, New Mexico, Arizona, and Colorado all meet.

Cultural Boundary The Wagah border post that separates northern India from eastern Pakistan is an example of a cultural boundary.

▲ **CRITICAL THINKING**

1. Identifying Which of the boundary types might change slowly over time without the consent of the governments involved? Explain your choice.

2. Analyzing Describe the Four Corners site and explain why it is considered a geometric boundary.

Firefighters extinguish the flames coming out from the Taj Palace Hotel in Mumbai, India, during a terrorist attack in the city in November 2008.

▲ CRITICAL THINKING

1. Defining What elements of the definition of terrorism are shown in the photo?

2. Speculating What are some of the long-term effects that people might have who survived but witnessed the bombing?

natural boundary

a fixed limit or extent defined along physical geographic features such as mountains and rivers

cultural boundary

a geographical boundary between two different cultures

geometric boundary

a boundary that follows a geometric pattern

cultural and religious groups in order to maintain order and power. In democracies, governments usually take account of cultural and religious beliefs in order to protect their people's freedoms and ensure their well-being.

Geography influences governments as they develop policy to provide people with goods and services. Governments must also know where their citizens are moving, why they are moving there, and how that affects their relationship with the environment. Infrastructures, such as roads, bridges, and power plants, must be built based on the geographic distribution of people using both current demographic data and future projections.

Several geographic factors influence the development of political boundaries. A **natural boundary** follows physical geographic features such as mountains and rivers. For example, the Mississippi River forms the borders between several U.S. states. Natural boundaries are often more defensible and easy to identify.

Other boundaries develop to separate areas with cultural differences, such as places with different religions or languages. These **cultural boundaries** geographically divide two identifiable cultures. For example, when Britain partitioned India and created Pakistan, it created a religious cultural boundary. Muslims were reorganized into Pakistan and Hindus into India.

At other times, cultural and natural landforms are not considered when boundaries are drawn. In these cases, treaties might create **geometric boundaries** to separate countries or nations. Geometric boundaries—which often follow straight lines and do not account for natural and cultural features—exist between Libya, Egypt, and Algeria.

Political boundaries, referred to as borders, are not always permanent. Many areas of the world have seen changing borders as the result of wars and territorial disputes. Border disputes arise from unsettled territorial claims or as a result of one state desiring the resources of a neighboring state. In February 1848, Mexico and the United States signed a treaty which ended the war between them and gave large portions of the Southwest, including present-day California, to the United States. Several days earlier, gold had been discovered near the present-day capital of Sacramento. This started the gold rush and sped up California's statehood.

☑ READING PROGRESS CHECK

Identifying What is the cause of many boundary disputes?

Conflict and Cooperation

GUIDING QUESTION *How do cooperation and conflict shape the division of Earth's surface?*

Cooperation and conflict have contributed to and resulted from the political geographic divisions of the world. Global cooperation is frustrated by many factors, including border disputes, tensions over larger territories, multiple ethnic groups within one state, competition for fewer resources, and control of strategic sites.

Nationalism often contributes to political conflicts. Nationalism is a belief that the individual's loyalty and devotion to the nation or state surpasses other individual or group interests. After passing through the new countries of Latin America, nationalism spread in the early nineteenth century to central Europe and from there, toward the middle of the century, to eastern and southeastern Europe. This period is considered the age of nationalism in Europe. Asia and Africa saw a rise in nationalism at the beginning of the twentieth century as powerful movements took place. Nationalism can breed conflict if it reaches fanatical levels. This can and often does lead to war.

Terrorism is also a type of political conflict. Terrorism inspires fear and is any violent and destructive act committed to intimidate a people or a government.

Terrorist attacks are usually carried out in such a way as to maximize the severity and length of the psychological impact. Not usually government supported, each act of terrorism is devised to have an impact on many large audiences. Terrorists also attack national symbols to show power and to attempt to shake the foundation of the country or society they are opposed to. For example, there was a series of terrorist attacks on September 11, 2001, at the World Trade Center in New York City, the Pentagon near Washington, D.C., and in the sky over western Pennsylvania. In 2012 there was an attack on the U.S. embassy in Libya. Terrorist acts frequently have a political purpose. They desire change so badly that failure to achieve change is seen as a worse outcome than the deaths of civilians.

Terrorism can be influenced by geographic factors, as in the Arab-Israeli conflict in which many innocent lives were lost. In 1947 the Palestine Mandate was divided and the State of Israel was established. These and subsequent events have polarized Arabs and Israelis for over 60 years, resulting in ongoing violent conflicts in the region.

Alliances and cooperation can also be explored from a geographic perspective. Treaties and international organizations are examples of how countries work together to resolve conflicts and establish ways to share resources. For example, much of the acid rain in Canada comes from pollution in the United States. As a result, in the early 1990s the two countries signed a cooperative agreement that would reduce acid rain.

The United Nations (UN) is an international organization whose stated aims are facilitating cooperation in international law, international security, economic development, social progress, human rights, and aspirations to achieve world peace. The UN was founded in 1945 after World War II to stop wars between countries and to provide a platform for international dialogue. The North Atlantic Treaty Organization (NATO) is an alliance of 16 sovereign Euro-Atlantic countries dedicated to maintaining democratic freedom by means of collective defense. The World Trade Organization (WTO) is an international body that oversees trade agreements and settles trade disputes among countries. A country's membership in the WTO is an important step in its development, and less developed countries strive to become members. In 2000, hoping that trade might open China to democratic change, the United States granted full trading privileges to China and supported its entrance into the WTO. The following year China was admitted to the WTO.

☑ READING PROGRESS CHECK

Discussing What is the function of the United Nations?

Connecting Geography to GOVERNMENT

Geopolitics

The *how* and *why* of the creation of political divisions—such as zones, countries, states within countries, and territories— are examples of the connection between geography and politics. *Geopolitics* examines how political units are influenced by geographic factors, such as a country's size, location, and resources. Geopolitical issues influence government and foreign policies, and guide political and economic decisions. It is crucial in a world with terrorism, globalization, and technological advances that governments understand how people and places are interconnected.

EXPLORING ISSUES Discuss a specific example of how geography and politics are related.

LESSON 3 REVIEW

Reviewing Vocabulary
1. *Contrasting* Describe the difference between a unitary system and a federal system.

Using Your Notes
2. *Displaying* Use your graphic organizer to summarize features of government. Include both the levels of government and types of governments.

Answering the Guiding Questions
3. *Identifying* What influences the level and type of a country's government?

4. *Discussing* How does geography influence a country's government?

5. *Expressing* How do cooperation and conflict shape the division of Earth's surface?

Writing Activity
6. *Informative/Explanatory* Write a one-page essay explaining the human and physical geographic characteristics that can influence a country's foreign policy.

HOW HAS GLOBALIZATION **CHANGED** MODERN CULTURE?

Globalization—the widening exchange of culture traits such as trade, technology, and ideas—is a process that has been taking place for as long as trade has existed. As the world becomes increasingly interconnected economically and socially, largely thanks to advances in communication and transportation, the process of globalization has reached every corner of the Earth.

There are many benefits to globalization. For example, technology developed in one country and shared with others helps to increase economic efficiency and standards of living. International trade has also allowed countries to specialize in the goods they produce well and trade for the goods they do not, enabling economies to grow more rapidly and giving people access to products and resources they may not otherwise have. Globalization has helped us become more aware of the cultures and lifestyles of others around the world, giving us new and unique perspectives on the diversity of life.

While globalization has many benefits, it also has drawbacks. As ideas, products, and even lifestyles are shared between cultures, traditional cultural heritage can become diluted by outside influences. Language, artistic traditions, clothing styles, and even behaviors can all be altered through interactions with other cultures. Language, for example, is one way in which globalization can lead to the permanent loss of some of the world's rare languages, such as Xyzyl, a language spoken in northwest Mongolia. Experts predict that there may be more than 500 languages that are spoken by fewer than 10 people. Many groups, including the United Nations Educational, Scientific and Cultural Organization (UNESCO), urge the protection of traditional culture in the face of increasing globalization.

Protect Cultural Heritage

PRIMARY SOURCE

" Today, with the rapid advance of globalization, the loss of intangible cultural heritage . . . can now be observed throughout the world. The threat of extinction to intangible cultural heritage is particularly noticeable in developing countries in Asia, Africa and the Middle East, today. Therefore, while modernization and industrialization remain urgent issues, it is at the same time essential to preserve and transmit these traditional cultures.

Taking into consideration the fact that every culture has been more or less influenced by others, and has forged a cultural identity within history, it goes without saying that the openness of one culture to others is very significant. However, the rapid flow of people, products and information—or rapid cultural interpenetration caused by globalization—menaces minority cultures, especially their intangible cultural heritage, which should be handed down from generation to generation. It is therefore most necessary that measures be taken to prevent this loss. "

—Shogo Arai, Parliamentary Secretary for Foreign Affairs, "Japan and the Preservation of Intangible Cultural Heritage," 2004

Increase Cultural Awareness

PRIMARY SOURCE

" Paradoxically, it is precisely in the context of increasing globalization that more and more peoples and communities of the world have begun to recognize the importance of their cultural heritage—whether tangible or intangible—as a contribution to the world's cultural diversity. Communities in every land have come to realize that their cultural heritage, which is by nature fragile, plays a crucial role in their identity and that their engagement in safeguarding activities contributes to a sense of continuity. As a result, while globalization has undeniably contributed to the dissemination of cultures, its effects on cultural diversity can, if we are not careful, be negative. . . .

But more is needed in order to respond to peoples' growing awareness of the importance of their culture . . . . "

—Koïchiro Matsuura, Director-General of UNESCO, "Globalization, Intangible Cultural Heritage and the Role of UNESCO," 2004

What do you think?

1. **Drawing Conclusions** Arai states that the loss of cultural heritage is most noticeable in developing countries. Why might the culture of more developed countries have such a strong impact on less developed countries?

2. **Identifying Central Issues** According to Matsuura, how has globalization increased cultural awareness?

3. **Hypothesizing** What steps do you think can be taken to ensure that cultural heritage is not diminished as a result of globalization? How might international organizations like UNESCO play a role?

netw⊙rks

There's More Online!

☑ **IMAGE** New York Stock Exchange

☑ **IMAGE** Alaskan Oil Pipeline

☑ **MAP** World GDP

☑ **TABLE** Economic Activities and Economic Development

☑ **INTERACTIVE SELF-CHECK QUIZ**

☑ **VIDEO** Economic Geography

Reading **HELP**DESK

Academic Vocabulary

- **regulate**
- **incentive**

Content Vocabulary

- **traditional economy**
- **market economy**
- **free enterprise**
- **capitalism**
- **mixed economy**
- **command economy**
- **more developed country**
- **newly industrialized country**
- **less developed country**

TAKING NOTES: *Key Ideas and Details*

ORGANIZING As you read about economic geography, complete a web diagram like the one below to list the major concepts of economic systems, development, and world trade.

Economic Geography
- Economic Systems
- Economies and World Trade
- Economic Development

LESSON 4

Economic Geography

ESSENTIAL QUESTION • *How do the characteristics and distribution of human populations affect human and physical systems?*

IT MATTERS BECAUSE

The growth of the global economy continues to make the world's peoples increasingly interdependent, or reliant on each other. Natural resources are extracted and traded around the world. Other trade items could be goods, services, and even labor. Countries with varying levels of economic development have become increasingly interdependent through this world trade.

Economic Systems

GUIDING QUESTION *What are the three main types of economic systems?*

All economic systems must make three basic economic decisions: (1) what and how many goods and services should be produced, (2) how they should be produced, and (3) who gets the goods and services that are produced. These decisions are made differently in the three major economic systems—traditional, market, and command.

In a **traditional economy**, habit and custom determine the rules for all economic activity. Individuals are not free to make decisions based on what they would like to have. Instead, their behavior is defined by the customs of their elders and ancestors. For example, it was a tradition in the Inuit society of northern Canada that a hunter would share the food from the hunt with the other families in the village. Today, traditional economies exist in very limited parts of the world. One of the few advantages in a traditional economy is that the roles of individuals are clearly defined. There are also many disadvantages to this type of society. These societies are often very slow to change. When new technologies are introduced, these ideas and techniques are discouraged.

In a **market economy**, individuals and private groups make decisions about what things to produce. People, as shoppers, choose what products they will or will not buy, and businesses produce more of what they believe consumers want. A market economy is based on the concept of **free enterprise**, the idea that private individuals or groups have the right to own property or businesses and make a profit with only limited government interference. In a free enterprise system,

All economic systems must make decisions about the types and quantity of goods and services, including how they are going to be produced and sold.

◄ CRITICAL THINKING
1. *Identifying* Which type of system does the image on the left represent?
2. *Describing* What features in the image on the right represent characteristics of a traditional economy?

people are free to choose what jobs they will have and for whom they will work. People have the ability to make as much money as they can and do what is in their best interest. Another positive aspect of market economies is that the government tries to stay out of the way of businesses and there is a great variety of goods and services for consumers. An economic system organized in this way is referred to as **capitalism**.

One major problem with this type of economy is that it does not always provide the basic needs to everyone in the society. The weak, sick, disabled, and old sometimes have trouble providing for themselves and often slip into poverty.

No country in the world, however, has a pure market economy system. Today, the U.S. economy and others like it are described as mixed economies. A **mixed economy** is one in which the government supports and **regulates** free enterprise through decisions that affect the marketplace. In this arrangement, the government's main economic task is to preserve the free market by keeping competition free and fair and by supporting public interests. Governments in modern mixed economies also influence their economies by spending tax revenues to support social services.

In a **command economy**, the government owns or directs the means of production—land, labor, capital (machinery, factories), and business managers—and controls the distribution of goods. Believing that such economic decision making benefits all of society and not just a few people, countries with command economies try to distribute goods and services equally among all citizens. Public taxes, for example, are used to support social services, such as housing and health care, for all citizens. However, citizens have no voice in how this tax money is spent.

Socialism and communism are examples of command economies because they involve heavy government control. However, in practice these two economic systems are mixed economies. In a socialist economy, the government owns some, but not all, of the basic productive resources. The government also provides for some of the basic needs of the people, such as education and health care. Communism is an extreme form of socialism in which all property is collectively, not privately, owned. Under communism, the government decides how much to produce, what to produce, and how to distribute the goods and services produced. One political party—the Communist Party—makes decisions and may even use various forms of coercion to ensure that the decisions are carried out at lower political and economic levels.

traditional economy
a system in which tradition and custom control all economic activity; exists in only a few parts of the world today

market economy
an economic system based on free enterprise, in which businesses are privately owned and production and prices are determined by supply and demand

free enterprise a system in which private individuals or groups have the right to own property or businesses and make a profit with limited government interference

capitalism a system in which factors of production are privately owned

mixed economy a system of resource management in which the government supports and regulates enterprise through decisions that affect the marketplace

regulate to govern or direct according to rule

command economy a system of resource management in which decisions about production and distribution of goods and services are made by a central authority

Recent history has demonstrated, however, that communist economies lack the free decision making and **incentives** that foster business innovation and generation of products that people need and want. Customers can be limited in their choices, and economies can stagnate. As a result of these problems, command economies often decline. The Soviet Union, as described below by a Russian observer, provided an example of this situation:

PRIMARY SOURCE

❝In 1961, the [Communist] party predicted . . . that the Soviet Union would have the world's highest living standard by 1980. . . . But when that year came and went, the Soviet Union still limped along, burdened by . . . a stagnant economy.❞

—Dusko Doder, "The Bolshevik Revolution," *National Geographic*, October 1992

By 2000, Russia and the other countries that were once part of the Soviet Union were developing market economies. Communist China and Vietnam have also allowed some free enterprise to promote economic growth, although their governments tightly control political affairs.

Socialism allows a wider range of free enterprise. It has three main goals: (1) an equitable distribution of wealth and economic opportunity; (2) society's control, through its government, of decisions about public goods; and (3) public ownership of services and factories that are essential. Some socialist countries, like those in Western Europe, are democracies. Under democratic socialism people have basic human rights and elect their political leaders.

☑ **READING PROGRESS CHECK**

Assessing On what idea is a market economy based?

Economic Development

GUIDING QUESTION *What influences economic development?*

Most natural resources are not evenly distributed throughout the Earth. This uneven distribution affects the global economy. As a result, countries specialize in the economic activities best suited to their resources.

Geographers and economists classify all of the world's economic activities into four types. Primary economic activities—such as farming, grazing, fishing, forestry, and mining—involve taking or using natural resources directly from the Earth. Such activities take place near the natural resources that are being gathered or used. For example, coal mining occurs at the site of a coal deposit.

Secondary economic activities use raw materials to make a tangible product that is new and more valuable than the original raw material. Such activities include manufacturing automobiles, assembling electronic goods, producing electric power, or making pottery. These activities occur close to the resource or to the market for the finished good.

Tertiary economic activities do not involve directly acquiring and remaking natural resources. Instead, these activities provide services to people and businesses. Doctors, teachers, lawyers, bankers, truck drivers, and store clerks all provide professional, wholesale, or retail services.

Quaternary economic activities are concerned with the processing, management, and distribution of information. They are vitally important to modern economies that have been transformed in recent years by the information revolution. Just as with tertiary economic activities, people performing these activities include "white collar" professionals working in education, government, business, information processing, and research.

Economic Activities and Economic Development			⌄ CHART SKILLS
Country	Level of Economic Development	Major Economic Activities	GDP per Capita (purchasing power parity)
United States	More Developed	service industries, commercial agriculture, industrial supplies	$56,300
Sweden	More Developed	service industries, iron and steel, precision machinery	$48,000
South Korea	More Developed	electronics, telecommunications, automobiles	$36,700
Mexico	Newly Industrialized	service industries, food and beverages, consumer goods manufacturing	$18,500
China	Newly Industrialized	mining and ore processing, textiles, petroleum	$14,300
South Africa	Newly Industrialized	service industries, mining, automobile assembly	$13,400
Belize	Less Developed	tourism, oil, agriculture	$8,600
Pakistan	Less Developed	agriculture, textiles, crude oil production	$4,900
Zimbabwe	Less Developed	mining, steel production, wood products	$2,100

Source: *CIA World Factbook,* 2015

The major economic activities of a country have a direct relationship to the country's level of development. The level of development also has a relationship to standard of living, as shown in this table with GDP per capita (gross domestic product per person).

▲ **CRITICAL THINKING**

1. *Classifying* Which countries in the table have a major economic activity that is considered a primary economic activity?

2. *Evaluating* What information in the table illustrates the standard of living in each country? What is the relationship between economic development and standard of living?

Economic activities, including *industrialization*, or the spread of industry, help influence a country's level of economic development. Those countries having more technology and manufacturing, such as the United States and Canada, are called **more developed countries**. Most people work in service or information industries and enjoy a high standard of living. Because of modern techniques, only a small percentage of workers in more developed countries is needed to grow enough food to feed entire populations. For similar reasons, relatively small percentages of the people are employed in manufacturing industries in more developed countries.

Newly industrialized countries have moved from primarily agricultural activities to primarily manufacturing and industrial activities. This transition to manufacturing and industry often brings improvements in socioeconomic development. Examples of newly industrialized countries are Mexico, Malaysia, and Turkey.

Those countries that, according to the United Nations, exhibit the lowest indicators of socioeconomic development are **less developed countries**. In many less developed countries, which are primarily in Africa, Asia, and Latin America, agriculture remains dominant. Even though some commercial farming occurs, most farmers in these countries engage in subsistence farming, growing only enough food for family needs. Some countries' involvement in light industry grows out of a history of cottage industries, businesses that employ workers in their homes. As a result, most people in less developed countries remain poor, as economic development typically reduces poverty.

more developed country
a country that has a highly developed economy and advanced technological infrastructure relative to other less developed nations

newly industrialized country a country that has begun transitioning from primarily agricultural to primarily manufacturing and industrial activity

less developed country
a country that, according to the United Nations, exhibits the lowest indicators of socioeconomic development

✓ **READING PROGRESS CHECK**

Listing List the four types of economic activities and explain how these economic activities relate to a country's level of development.

Economies and World Trade

GUIDING QUESTION *What stimulates and supports world trade?*

World trade is the exchange of capital, labor, goods, and services across international borders or territories, involving the import and export of goods. In most countries, such trade represents a significant share of gross domestic product (GDP). Trade among countries has been present throughout history, but its economic, social, and political importance has increased in recent centuries.

The unequal distribution of natural resources is one factor that promotes a complex network of trade among countries. Countries export their specialized products, trading them to other countries that cannot produce those goods. When countries cannot produce as much as they need of a certain good, they import it, or buy it from another country. That country, in turn, may buy the first country's products, making the two countries trading partners.

Other factors affecting world trade are differences in labor costs and education. Multinationals often base their business decisions on these factors. They locate their headquarters in a more developed country and locate their manufacturing or assembly operations in less developed or newly industrialized countries with low labor costs. In recent decades, many less developed countries have allowed multinationals to build factories or form partnerships with local companies.

ANALYZING PRIMARY SOURCES

Coffee Country

Coffee has a long tradition as an important export crop for both Guatemala and Mexico, employing thousands of workers and bringing income to the region. However, a coffee crisis is causing families that have grown coffee for generations to leave their fields and head for the city or for the border, leaving entire coffee estates abandoned.

"The problem is what is known as the international coffee crisis. Simply put, there's too much cheap coffee flooding the market these days. It comes from countries such as Brazil, and more recently Vietnam, which have been using massive agribusiness techniques.

However, some small family farmers have found a way to prosper by following the environmental guidelines for what is known as fair trade. They sell their coffee directly to buyers, thereby cutting out the middleman. The difficulty is that the success depends upon the taste of the coffee.

And so the tasting process becomes a critical make-or-break step for many farmers, and failure of the taste test can mean taking a loss on an entire season.

'In the current crisis,' [reporter Sam] Quinones observes, 'peasant coffee growers have to learn the Starbucks lesson and focus on quality. Consumers, meanwhile, have to be willing to pay extra for the best coffee, searching out regional coffees the way they do with wine.'

Even those consumers who like good coffee don't know where it comes from. And many haven't even heard of the fair trade concept. Until all that changes, the international coffee crisis may not be going away anytime soon."

—PBS Frontline/World, "Guatemala/Mexico- Coffee Country," May 2003

A woman working on a large coffee estate separates the green coffee seeds from the ripe ones.

DBQ ▲ CRITICAL THINKING

1. *Identifying Cause and Effect* In what ways could coffee purchased in countries far from Guatemala or Mexico affect people living there?

2. *Problem Solving* How does fair trade help the coffee farmers in this region?

Oil is a major commodity of trade in the world markets.

◄ **CRITICAL THINKING**

1. Contrasting How is an import different from an export?

2. Evaluating Explain why a country would need to import oil.

International trade is, in principle, not different from domestic trade. The main difference is that international trade is typically more costly due to additional costs such as tariffs, time costs due to border delays, and costs associated with country differences such as language, the legal system, or other cultural barriers.

Another difference between domestic and international trade is that factors of production such as capital and labor are typically more mobile within a country than across countries. Thus, international trade is mostly restricted to trade in goods and services, and only to a lesser extent to trade in capital, labor, or other factors of production. Trade in goods and services can serve as a substitute for trade in factors of production.

Sometimes, a country can import goods that make extensive use of that factor of production and thus embody it. An example is the import of labor-intensive goods by the United States from China. Instead of importing Chinese labor, the United States imports goods that were produced with Chinese labor. Emerging markets are nations with social or business activity in the process of rapid growth and industrialization. The economies of China and India are considered to be the largest. The seven largest emerging and developing economies by either nominal GDP or GDP (PPP) are China, Brazil, Russia, India, Mexico, Indonesia, and Turkey. The ASEAN–China Free Trade Area, launched on January 1, 2010, is the largest regional emerging market in the world.

✓ **READING PROGRESS CHECK**

Contrasting Explain the differences between international trade and domestic trade.

LESSON 4 REVIEW

Reviewing Vocabulary
1. Identifying What are the advantages and disadvantages to a command economy?

Using Your Notes
2. Describing Use your graphic organizer to describe the three basic economic decisions that are made by economic systems.

Answering the Guiding Questions
3. Listing What are the three main types of economic systems?

4. Describing What influences economic development?

5. Discussing What stimulates and supports world trade?

Writing Activity
6. Informative/Explanatory Write an essay explaining the advantages and disadvantages a less developed country might experience by joining a free trade agreement.

PATTERNS OF RESOURCE
DISTRIBUTION

Earth's natural resources are unevenly distributed across the globe. As a result, a country might have abundant access to one resource but may have a limited supply of another resource. Saudi Arabia, for example, is a country rich in petroleum, but it has very limited access to freshwater. This uneven distribution of resources is a major reason countries trade with one another, exchanging the resources they have for the ones they need. Access to natural resources can also be a cause of conflict in some parts of the world.

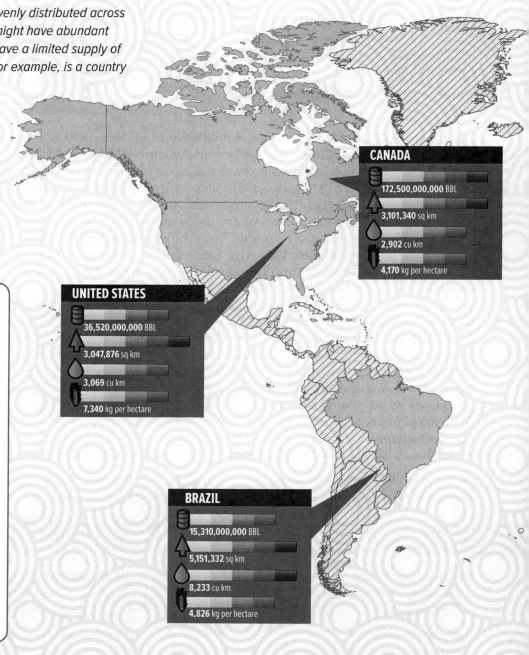

CANADA
- 172,500,000,000 BBL
- 3,101,340 sq km
- 2,902 cu km
- 4,170 kg per hectare

UNITED STATES
- 36,520,000,000 BBL
- 3,047,876 sq km
- 3,069 cu km
- 7,340 kg per hectare

BRAZIL
- 15,310,000,000 BBL
- 5,151,332 sq km
- 8,233 cu km
- 4,826 kg per hectare

LEGEND

Proven Oil Reserves

Renewable Water Resources

Forest Area (timber)

Cereal Yield (grains)

GLOBAL SCALE
Relative amount compared to total world resources.

VERY LOW   LOW   MED   HIGH   VERY HIGH

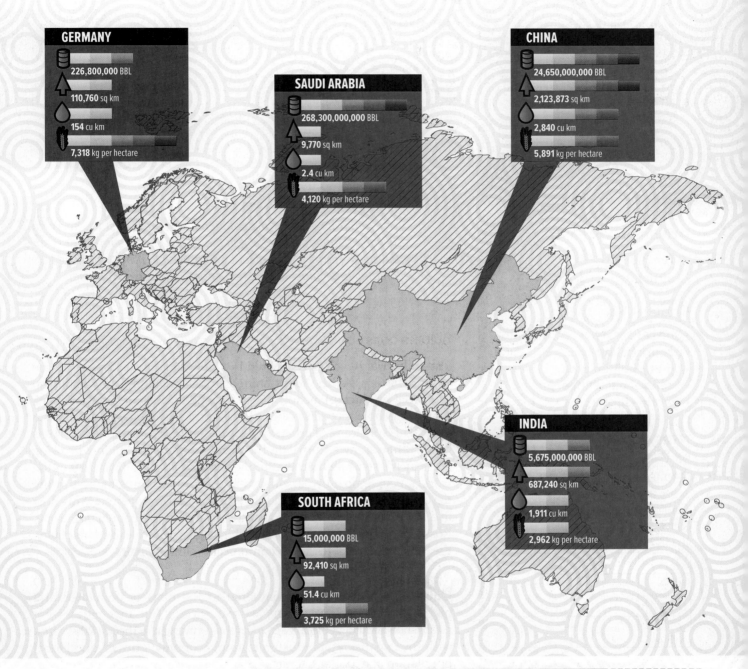

GERMANY
226,800,000 BBL
110,760 sq km
154 cu km
7,318 kg per hectare

SAUDI ARABIA
268,300,000,000 BBL
9,770 sq km
2.4 cu km
4,120 kg per hectare

CHINA
24,650,000,000 BBL
2,123,873 sq km
2,840 cu km
5,891 kg per hectare

INDIA
5,675,000,000 BBL
687,240 sq km
1,911 cu km
2,962 kg per hectare

SOUTH AFRICA
15,000,000 BBL
92,410 sq km
51.4 cu km
3,725 kg per hectare

Making Connections

1. *Analyzing* Which of the four resources on this map do you think Canada supplements with imports? From which country might Canada import this resource?

2. *Drawing Conclusions* Using one of the countries featured on this map as an example, explain how a country's physical geography, including climate, location, and landforms, influences its access to natural resources.

- ⬤ City
- ⬤ Town
- ⬤ Market
- • Village

Reading HELPDESK

Academic Vocabulary

- function
- structure

Content Vocabulary

- urban sprawl
- connectivity
- metropolitan area
- central place theory
- world cities

TAKING NOTES: *Key Ideas and Details*

DESCRIBING Complete a graphic organizer similar to the one below describing the nature of cities, patterns of urbanization, and challenges of urban growth.

Urban Geography		
The Nature of Cities	Patterns of Urbanization	Challenges of Urban Growth

LESSON 5
Urban Geography

ESSENTIAL QUESTION • *How do the characteristics and distribution of human populations affect human and physical systems?*

IT MATTERS BECAUSE

Urban geography is a branch of human geography concerned with cities and the people who live in them. An urban geographer analyzes patterns of settlement and growth in urban areas and evaluates the impact of cities on people and the environment.

The Nature of Cities

GUIDING QUESTION *How does a city's function influence its structure?*

The Industrial Revolution ushered in a new age of urbanization in the world's history. As the focus shifted from agricultural production to industrial production, people began moving to cities in large numbers. As more people moved to cities, the physical size of cities also began to grow. This spreading of urban areas onto undeveloped land near cities is called **urban sprawl**. Currently, the world's urban population is growing at a much faster rate than that of the rural population. Over half of the world's people now live in cities, and this proportion is highest in the developed regions of the world. Eighty-two percent of Americans now live in urban areas, and more than two-thirds of the people of Europe, Russia, Japan, and Australia do as well. This growth is due to **connectivity**, the directness of routes and communication linking pairs of places.

The Function of Cities

Only recently have people gathered in the densely populated and highly structured settlements we call cities. The first cities were established about 5,000 years ago, but it has only been in the last 200 years—with the expansion of industrialization, economic growth, and global population at exponential rates—that cities have grown significantly in size and number. At the start of the twentieth century, only about one person in ten lived in a city. Today, the proportion of urban and rural dwellers is approximately equal. By 2025, it is expected that nearly two-thirds of the world's population will live in urban areas.

All cities serve a variety of **functions**. For example, manufacturing, retail, and service centers are often located in urban areas. These functions are the economic base of a city, generating employment and wealth. The larger a city is, the more numerous and highly specialized its functions are likely to be. Smaller cities and towns have fewer functions, which tend to be of a more general nature. In the field of health care, for example, clinics are found in a wide range of places, but specialized teaching hospitals tend to be located only in larger cities.

Cities also tend to be centers of culture and creativity. Artists, musicians, architects, philosophers, scientists, and writers gravitate toward cities where there are patrons, communities of other artists, universities, clients, and a skilled workforce. Today's urban centers of culture have changed over time, mostly based on their economic or political strength with the outside world.

There are several reasons cities can support a variety of functions. The large population of a city means there are plenty of workers available to support a variety of industries. The large population also means there is a large market of consumers to sustain the demand for specialized functions. From a functional perspective, infrastructure facilitates the production of goods and services, and also the distribution of finished products to markets. In addition, it provides basic social services such as schools and hospitals. Roads provide adequate transportation, and safe buildings provide secure housing.

Urban areas have both advantages and challenges. The diversity of peoples and activities encourages innovation and creativity, but overcrowding, crime, poverty, social conflict, and pollution can become challenges. An urban area differs from country to country, but each is considered a **metropolitan area**—a region that includes a central city and its surrounding suburbs.

urban sprawl spreading of urban developments on land near a city

connectivity the directness of routes linking pairs of places

function a special purpose

metropolitan area region that includes a central city and its surrounding suburbs

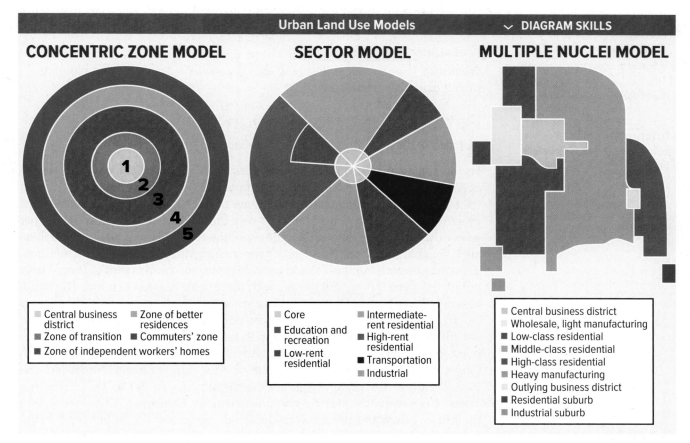

Urban Land Use Models ⌄ **DIAGRAM SKILLS**

CONCENTRIC ZONE MODEL

1
2
3
4
5

- Central business district
- Zone of transition
- Zone of independent workers' homes
- Zone of better residences
- Commuters' zone

SECTOR MODEL

- Core
- Education and recreation
- Low-rent residential
- Intermediate-rent residential
- High-rent residential
- Transportation
- Industrial

MULTIPLE NUCLEI MODEL

- Central business district
- Wholesale, light manufacturing
- Low-class residential
- Middle-class residential
- High-class residential
- Heavy manufacturing
- Outlying business district
- Residential suburb
- Industrial suburb

Urban geographers use urban land use models like these to describe the internal structure of cities. They examine location patterns of people and businesses within the urban setting.

▲ **CRITICAL THINKING**

1. Identifying What district is typically largest in the multiple nuclei model?

2. Describing What zone is located between the commuters' zone and the zone of independent workers' homes in the concentric zone model?

Landfills are examples of the type of infrastructures a city must have in order to support large populations.

▲ CRITICAL THINKING

1. Defining How does a landfill fit the definition of an infrastructure?

2. Identifying What are a few other infrastructures a city must maintain for its citizens?

structure something constructed or arranged in a definite pattern of organization

central place theory geographical theory that seeks to explain the number, size, and location of human settlements in an urban system

The Structure of Cities

Urban **structure** is the arrangement of land use in urban areas. Sociologists, economists, and geographers have developed several models explaining where different types of people and businesses tend to exist within the urban setting. Urban structure can also refer to the urban spatial structure, which concerns the arrangement of public and private space in cities and the degree of connectivity and accessibility.

The *concentric zone model* was the first to explain distribution of social groups within urban areas. Based on a single city, Chicago, it was created by sociologist Ernest Burgess in 1924. According to this model, a city grows outward from a central point in a series of rings. A second theory of urban structure was proposed in 1939 by economist Homer Hoyt. The *sector model* proposed that a city develops in sectors instead of rings. Certain areas of a city are more attractive for various activities, whether by chance or geographic and environmental reasons. As the city grows and these activities flourish and expand outward, they do so in a wedge shape and become a sector of the city.

Geographers C. D. Harris and E. L. Ullman developed the *multiple nuclei model* in 1945. According to this model, a city contains more than one center around which activities revolve. Some activities are attracted to particular nodes while others try to avoid them. For example, a university node may attract well-educated residents, pizzerias, and bookstores, whereas an airport may attract hotels and warehouses. Other businesses may also form clusters for automobile repair, tire stores, or arts districts. Incompatible activities will avoid clustering in the same area, thus explaining why heavy industry and high-income housing rarely exist together in the same neighborhood.

✓ READING PROGRESS CHECK

Contrasting In economic terms, how are the functions of larger cities different from those of smaller cities?

Patterns of Urbanization

GUIDING QUESTION *What influences the location and growth of cities?*

Factors that led to the early growth of cities are still influential today. The growth of American cities, for example, began in the late 1700s. This growth was directly related to certain influential factors. Some of the same factors that led to such growth hundreds of years ago are the very same factors influencing the further growth, development, and urbanization of cities today. People will go where there are navigable water sources, such as river crossings and fertile deltas. In addition, if the area is mountainous it could provide protection from enemies. People have populated areas throughout history with these same reasons in mind. The basics of survival are food and water sources and security from enemies.

In addition to factors that promote population, there are factors that can shrink population in a region. For example, if an industry is no longer needed, the city's population will move on in order to seek a livelihood. Ghost towns are a prime example of this. The railroads provided the transportation for prospectors and entrepreneurs to get to these thriving areas. Consequently, during the gold rush in California, the towns were booming. When the mines were depleted, there was no longer any reason for these towns to exist and the people moved on.

The **central place theory** is a spatial theory in urban geography that attempts to explain the reasons behind the distribution patterns, size, and number of cities and towns around the world. It attempts to illustrate how settlements locate in relation to one another, the amount of market area a central

Kevin Leigh/Photolibrary/Getty Images

place can control, and why some central places function as hamlets, villages, towns, or cities. It also attempts to provide a framework by which those areas can be studied both for historical reasons and for the locational patterns of areas today.

A **world city** is a city generally considered to play an important role in the global economic system. World cities possess such features as having international diverse cultures, an active influence on and interaction in world affairs, a large population, a major international airport, and an advanced transportation system. The world city concept comes from geography and urban studies. It can be seen as a type of "point of entry" for studying the changes that come from globalization.

One important world city that has resulted from the geography of its location is İstanbul. It straddles a strait, thus placing it on vital land and sea trade routes. In addition, it can easily defend itself against enemy factions because of its water location. Farming is good because of its fertile soil, and consequently İstanbul has been under attack and conquered many times. Its desirable location is the very thing that has made it vulnerable.

New trends in cities have emerged, including the development of suburban business districts and major diversified centers, to name just two. Typically, suburbia refers to an outlying community around a city. These communities are business districts in their own right. The name that is now most commonly used to describe these terms is "edge cities."

These new suburban cities have sprung up all over and are home to glistening office towers and huge retail complexes. They are always located close to major highways. "Boomers" are the most common type of edge cities, having developed around a shopping mall or highway interchange such as Pasadena, California. On the suburban fringe of Phoenix, Arizona, Sun City is a "greenfield"—a new, master-planned city built on undeveloped land. In contrast, "uptown" edge cities are historic activity centers built over an older city or town, such as the Rosslyn-Ballston Corridor in Virginia.

world cities cities generally considered to play an important role in the global economic system

☑ READING PROGRESS CHECK

Summarizing What does the central place theory attempt to explain?

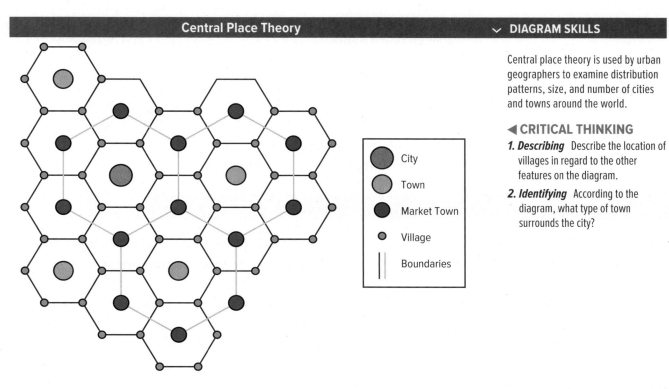

Central Place Theory

⌄ DIAGRAM SKILLS

City
Town
Market Town
Village
Boundaries

Central place theory is used by urban geographers to examine distribution patterns, size, and number of cities and towns around the world.

◀ CRITICAL THINKING

1. *Describing* Describe the location of villages in regard to the other features on the diagram.

2. *Identifying* According to the diagram, what type of town surrounds the city?

Challenges of Urban Growth

GUIDING QUESTION *What problems do urban areas face?*

The urbanization process refers to much more than simple population growth. It involves changes in the economic, social, and political structures of a region. Rapid urban growth is responsible for many environmental and social changes, and its effects are strongly related to issues of pollution and economics. These changes are not always positive. The rapid growth of cities strains their capacity to provide services such as energy, education, health care, transportation, sanitation, and physical security.

The more developed countries experienced urbanization during the nineteenth and twentieth centuries along with the Industrial Revolution. During this time, urbanization resulted from and contributed to industrialization. New job opportunities in the cities motivated people to migrate from rural areas to cities. At the same time, migrants provided cheap, plentiful labor for the emerging factories. Today, the circumstances are rather different in less developed countries. People are forced out of rural areas because of insufficient land on which to grow subsistence crops. Meanwhile, there are not enough jobs to accommodate the many migrants looking for employment, creating a large surplus labor force. This influx keeps wages low and can lead to poverty in many urban areas. Foreign investment companies from more developed countries see such situations as attractive. By employing these workers they can produce goods for far less.

Modern cities all over the world face many of the same problems: poor housing, homelessness, pollution, and social problems such as addiction, crime, and gang violence. People often live in old houses or other structures without electricity or sanitation. Some live on the streets with little access to adequate food or shelter. In addition, cars and industries pollute city air and water. Unemployment continues to grow. Larger multiethnic cities continue to face conflicts between different cultural groups.

One of the major effects of rapid urban growth is urban sprawl—scattered development that increases traffic, saps local resources, and destroys open space. Urban sprawl is responsible for changes in the physical environment and can also diminish the local character of the community. Small local businesses find it difficult to compete with larger stores and restaurants. Some cities are trying to be proactive and establish measures aimed at fighting urban sprawl by limiting construction and using innovative land-use planning techniques or community cooperation. One new form of land use is called "smart growth" or "New Urbanism," in which cities plan the communities' growth in a strategic way for livable and walkable neighborhoods.

✓ READING PROGRESS CHECK

Defining In what ways does rapid growth strain cities?

LESSON 5 REVIEW

Reviewing Vocabulary
1. *Defining* What is urban sprawl, and what is it responsible for?

Using Your Notes
2. *Listing* Use your graphic organizer to list the three models of urban structures.

Answering the Guiding Questions
3. *Explaining* How does a city's function influence its structure?

4. *Describing* What influences the location and growth of cities?

5. *Discussing* What problems do urban areas face?

Writing Activity
6. *Argument* Consider the negative aspects of urban sprawl. What are some possible solutions the government could provide to bolster infrastructure and services? Develop an argument to support your ideas.

Directions: On a separate sheet of paper, answer the questions below. Make sure you read carefully and answer all parts of the questions.

Lesson Review

Lesson 1

1 *Describing* Describe how culture affects the daily lives of people.

2 *Explaining* Explain the elements of culture geographers use to organize the world into culture regions.

3 *Drawing Conclusions* How did the agricultural revolution influence cultural diffusion?

Lesson 2

4 *Evaluating* What factors contribute to the uneven distribution of the world's population?

5 *Analyzing* How is the demographic transition model used to explain a country's population growth?

6 *Summarizing* How do the effects of zero population growth and negative population growth differ? How are they similar?

Lesson 3

7 *Comparing and Contrasting* What different roles might local citizens have in government decision making under a unitary system, a federal system, and a confederation?

8 *Explaining* Explain the different ways in which an autocracy, an oligarchy, and a democracy exercise authority.

9 *Defining* List and describe the three types of political boundaries.

Lesson 4

10 *Listing* What are the characteristics of a mixed economy?

11 *Contrasting* Contrast a more developed country with a less developed country. What are the primary differences?

12 *Analyzing* What are the advantages and disadvantages of a capitalist economy?

Lesson 5

13 *Explaining* Explain two functions of urban areas.

14 *Inferring* What are some issues associated with urban sprawl? Discuss some solutions to address these issues.

15 *Identifying Central Issues* What factors have influenced the site and growth of cities? Give examples.

21st Century Skills

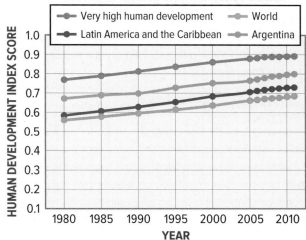

RECENT TRENDS IN DEVELOPMENT

Source: United Nations Development Programme

16 *Using Graphs, Charts, Diagrams, and Tables* According to the graph, how does the Human Development Index score of Argentina compare to that of the world?

College and Career Readiness

17 *Economics* Imagine that you are interviewing people for a position studying economic geography. Write a series of interview questions (with sample answers) that assesses the interviewees' understanding of economic activities and economic development.

Need Extra Help?

If You've Missed Question	1	2	3	4	5	6	7	8	9	10	11	12	13	14	15	16	17
Go to page	78	78	80	85	82	83	87	88	90	95	97	95	102	106	104	107	94

Directions: On a separate sheet of paper, answer the questions below. Make sure you read carefully and answer all parts of the questions.

Writing About Geography

18 *Informative/Explanatory* Use standard grammar, spelling, sentence structure, and punctuation to write a one-page essay discussing problems in today's urban areas. Include examples suggesting solutions and their feasibility.

DBQ Analyzing Primary Sources

Use the document to answer the following questions.

Governments and economies of countries around the world are becoming increasingly interconnected. Some countries, such as members of the European Union, have had strong economies that allowed them to help improve standards of living in other countries.

PRIMARY SOURCE

" I drove east on the highway that connects the capital of Talinn with Narva, on the Russian border. Nearly everywhere I looked I saw the handiwork of the European Union, starting with the road itself. The EU has already invested millions of euros to improve the highway, which serves as the main link to St. Petersburg, Russia. This highway passes the town of Sillamäe, once a 'closed' city run by the Soviet military, which enriched uranium for weapons programs in a huge factory overlooking the sea. The EU is here, too, kicking in more than a million dollars to help prevent the radioactive waste from leaching into the Baltic Sea."

—Don Belt, "Europe's Big Gamble," *National Geographic, May 2004*

19 *Speculating* Why would the EU want to invest in the improvement of other countries?

20 *Interpreting* What did Belt mean by "I saw the handiwork...starting with the road itself"?

Applying Map Skills

Use the Unit 1 Atlas to answer the following questions.

21 *Places and Regions* The Ural Mountains are a natural border that divides Asia from what other continent?

22 *Human Systems* How would you describe the population density pattern of Australia?

23 *The Environment and Society* Which of the countries in South America have petroleum resources?

Exploring the Essential Question

24 *Making Connections* Recall what you have learned about the demographic transition model. In which types of economic activities would a country in Stage 2 most likely be involved?

25 *Researching* Research and compare two countries in the ways they depend on the environment for the products they export.

Critical Thinking

26 *Making Generalizations* Explain the factors that influence a country's ability to control territory.

27 *Identifying Cause and Effect* What cultural changes have resulted from the information revolution?

28 *Comparing and Contrasting* What is the difference between a culture region and a culture hearth?

Research and Presentation

29 *Research Skills* Use Internet and library resources to gather information about a city plagued by urban sprawl. Your research should focus on the past, present, and future of this city. Create a multimedia presentation outlining that city's experiences along the road to urbanization. Be sure to (1) describe the human systems that affected the city's growth and (2) explain how other cities in the region might learn from its experience.

Need Extra Help?

If You've Missed Question	18	19	20	21	22	23	24	25	26	27	28	29
Go to page	106	108	108	2	6	8	82	97	89	81	79	102

The United States and Canada

Chapter 5
The United States

Chapter 6
Canada

UNIT **2**

① Culture Immigration to the United States and Canada from other areas of the world has had a dramatic effect on the cultures of the two countries. Parades celebrating these cultural roots are common throughout the region.

EXPLORE the REGION

The region of the **UNITED STATES** and **CANADA** stretches from the Pacific Ocean in the west to the Atlantic Ocean in the east. The two countries share many physical features—mountains frame their eastern and western edges, cradling a central area of vast plains. The region is a land of immigrants. Many made this land their home by choice. Others were forced to come as exiles or enslaved workers. Along with native peoples, these groups have shaped the cultures of the region.

THERE'S MORE ONLINE

③ Lakes and Rivers Long rivers, such as the Mississippi River, have played an important role in trade and industry in both the United States and Canada.

② Mountains The Rocky Mountains are the longest mountain range in North America, stretching from British Columbia in Canada to New Mexico in the United States.

④ Economy Today the service industry employs most of the workers in the United States and Canada. Many of these jobs are located in urban centers such as Toronto, Canada's largest city.

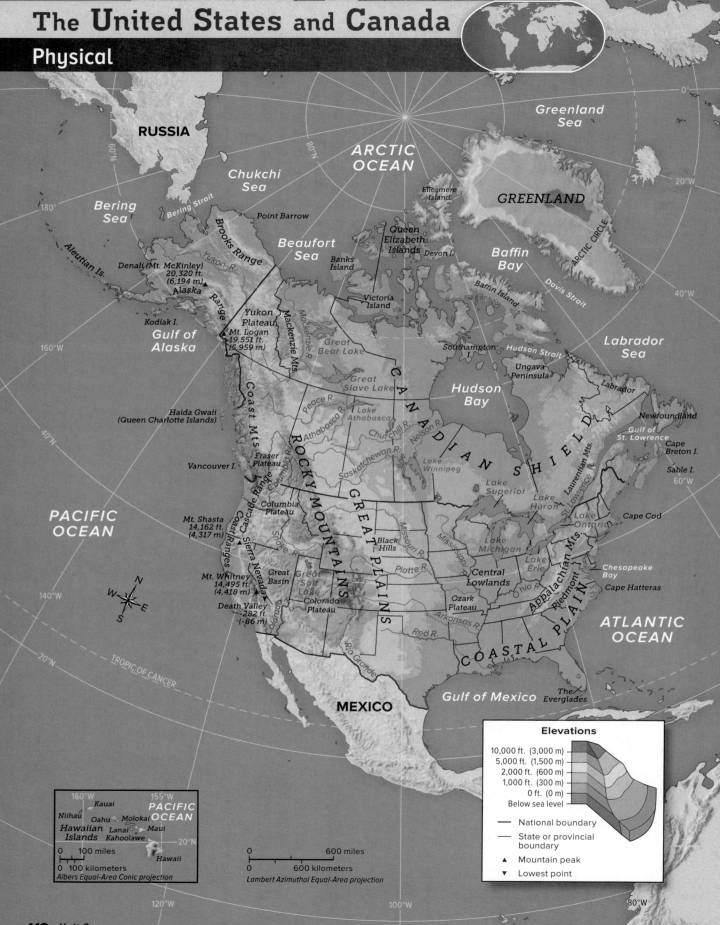

The **United States** and **Canada**
Physical

RUSSIA

ARCTIC OCEAN

Greenland Sea

Chukchi Sea

Bering Sea

Ellesmere Island

GREENLAND

Bering Strait

Point Barrow

Brooks Range

Beaufort Sea

Queen Elizabeth Islands

Devon I.

Baffin Bay

Aleutian Is.

Denali (Mt. McKinley) 20,320 ft. (6,194 m)

Alaska Range

Yukon R.

Banks Island

Victoria Island

Southampton I.

Baffin Island

Davis Strait

ARCTIC CIRCLE

Kodiak I.

Yukon Plateau

Mackenzie Mts.

Great Bear Lake

Hudson Strait

Labrador Sea

Gulf of Alaska

Mt. Logan 19,551 ft. (5,959 m)

Great Slave Lake

Hudson Bay

Ungava Peninsula

Labrador

Haida Gwaii (Queen Charlotte Islands)

Coast Mts.

Peace R.

Athabasca R.

Lake Athabasca

Churchill R.

Nelson R.

C A N A D I A N   S H I E L D

Newfoundland

Gulf of St. Lawrence

Cape Breton I.

Vancouver I.

Fraser Plateau

Columbia R.

Saskatchewan R.

Lake Winnipeg

Lake Superior

Laurentian Mts.

St. Lawrence R.

Sable I.

PACIFIC OCEAN

Columbia Plateau

R O C K Y   M O U N T A I N S

G R E A T   P L A I N S

Lake Huron

Lake Ontario

Cape Cod

Mt. Shasta 14,162 ft. (4,317 m)

Cascade Range

Snake R.

Missouri R.

Black Hills

Mississippi R.

Lake Michigan

Lake Erie

Appalachian Mts.

Chesapeake Bay

Coast Ranges

Sierra Nevada

Platte R.

Central Lowlands

Piedmont

Cape Hatteras

Mt. Whitney 14,495 ft. (4,418 m)

Great Basin

Great Salt Lake

Ozark Plateau

Ohio R.

Death Valley -282 ft. (-86 m)

Colorado R.

Colorado Plateau

Arkansas R.

C O A S T A L   P L A I N

ATLANTIC OCEAN

TROPIC OF CANCER

Red R.

Rio Grande

N
W E
S

MEXICO

Gulf of Mexico

The Everglades

Elevations

10,000 ft. (3,000 m)	
5,000 ft. (1,500 m)	
2,000 ft. (600 m)	
1,000 ft. (300 m)	
0 ft. (0 m)	
Below sea level	

— National boundary
— State or provincial boundary
▲ Mountain peak
▼ Lowest point

160°W   155°W

Kauai

Niihau   Oahu   Molokai

Hawaiian Islands   Lanai   Maui

Kahoolawe

PACIFIC OCEAN

20°N

Hawaii

0   100 miles
0   100 kilometers
Albers Equal-Area Conic projection

0   600 miles
0   600 kilometers
Lambert Azimuthal Equal-Area projection

The United States and Canada
Political

RUSSIA

ARCTIC OCEAN

GREENLAND (DENMARK)

Bering Sea

Bering Strait

Beaufort Sea

Baffin Bay

ARCTIC CIRCLE

Gulf of Alaska

Alaska

Yukon R.

Yukon

Northwest Territories

Mackenzie R.

Nunavut

Labrador Sea

Newfoundland and Labrador

Hudson Bay

British Columbia

Alberta

Manitoba

CANADA

Sask.

Ontario

Quebec

Gulf of St. Lawrence

P.E.I.

St. Lawrence R.

N.B.

Nova Scotia

Maine

Wash.

N. Dak.

Minn.

Ottawa

Vt.

N.H.

Mass.

Oregon

Montana

Wis.

Michigan

N.Y.

R.I.
Conn.

Idaho

S. Dak.

Missouri R.

Iowa

Mississippi R.

Ohio

Pa.

N.J.

Wyoming

Ind.

Washington, D.C.

Del.
Md.

Nevada

Utah

Colorado

Nebraska

Ill.

W. Va.

Va.

UNITED STATES

Mo.

Ky.

California

Kansas

Tenn.

N.C.

ATLANTIC OCEAN

PACIFIC OCEAN

Arizona

New Mexico

Okla.

Ark.

S.C.

Miss.

Ala.

Ga.

Rio Grande

Texas

La.

Fla.

TROPIC OF CANCER

Gulf of Mexico

MEXICO

⊙ National capital

Hawaii

Kauai

Niihau

Oahu

Molokai

Lanai

Maui

Kahoolawe

Hawaii

PACIFIC OCEAN

0   100 miles

0   100 kilometers

Albers Equal-Area Conic projection

0   600 miles

0   600 kilometers

Lambert Azimuthal Equal-Area projection

EQUATOR

UNIT 2
REGIONAL ATLAS

MAP STUDY

1. **Physical Systems** Compare the overall elevation of the western and eastern parts of the region.

2. **Human Systems** Which U.S. state is on the farthest northern latitude?

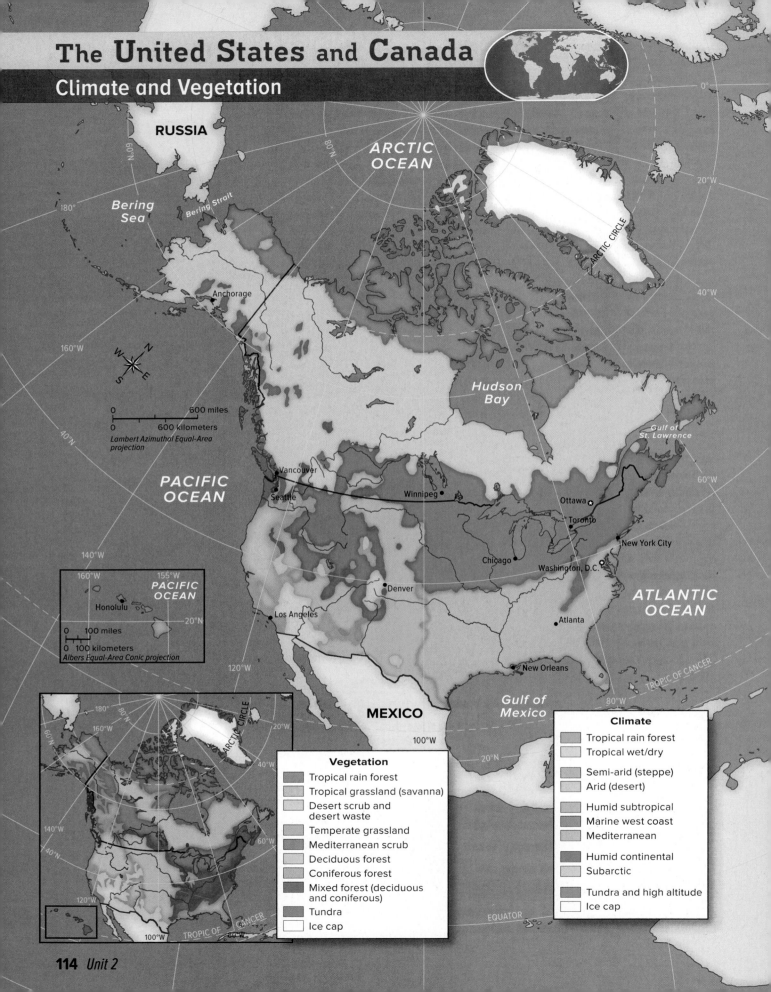

The United States and Canada
Climate and Vegetation

RUSSIA

ARCTIC OCEAN

Bering Sea

Bering Strait

Anchorage

ARCTIC CIRCLE

Hudson Bay

Gulf of St. Lawrence

Vancouver

Seattle

Winnipeg

Ottawa

Toronto

New York City

Chicago

Washington, D.C.

PACIFIC OCEAN

ATLANTIC OCEAN

Denver

Atlanta

0   600 miles
0   600 kilometers
Lambert Azimuthal Equal-Area projection

PACIFIC OCEAN

Honolulu

0   100 miles
0   100 kilometers
Albers Equal-Area Conic projection

MEXICO

Los Angeles

New Orleans

Gulf of Mexico

100°W

TROPIC OF CANCER

EQUATOR

Vegetation
- Tropical rain forest
- Tropical grassland (savanna)
- Desert scrub and desert waste
- Temperate grassland
- Mediterranean scrub
- Deciduous forest
- Coniferous forest
- Mixed forest (deciduous and coniferous)
- Tundra
- Ice cap

Climate
- Tropical rain forest
- Tropical wet/dry
- Semi-arid (steppe)
- Arid (desert)
- Humid subtropical
- Marine west coast
- Mediterranean
- Humid continental
- Subarctic
- Tundra and high altitude
- Ice cap

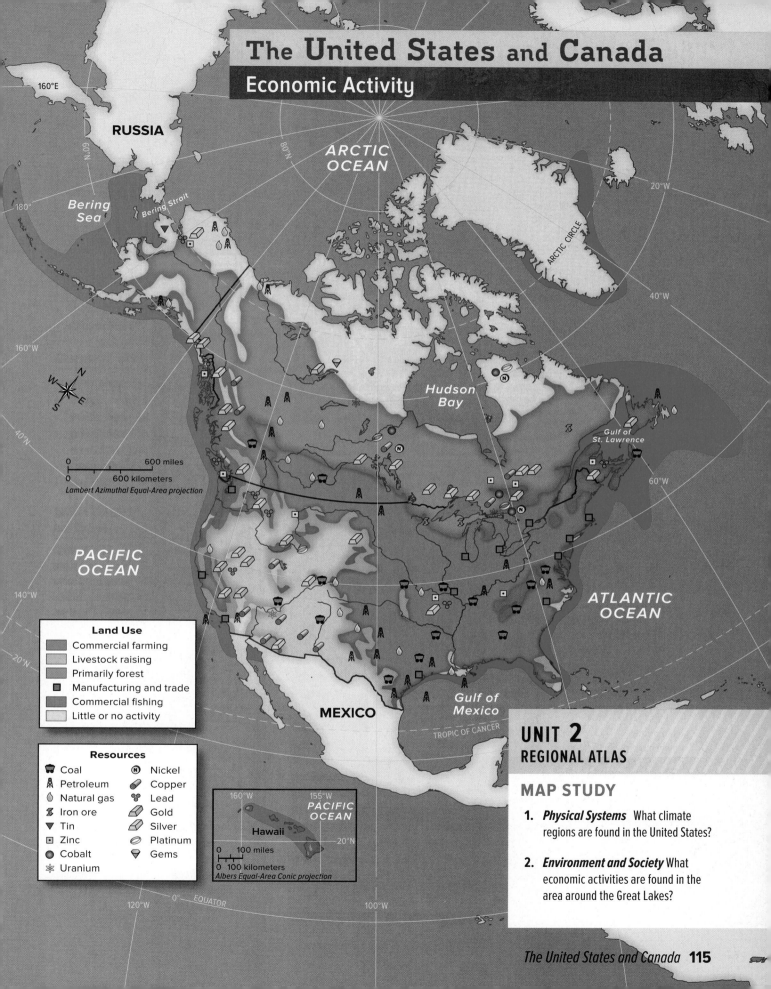

The United States and Canada
Economic Activity

RUSSIA

ARCTIC OCEAN

Bering Sea

Bering Strait

ARCTIC CIRCLE

Hudson Bay

Gulf of St. Lawrence

PACIFIC OCEAN

600 miles
600 kilometers
Lambert Azimuthal Equal-Area projection

N
W   E
S

ATLANTIC OCEAN

Land Use
- Commercial farming
- Livestock raising
- Primarily forest
- Manufacturing and trade
- Commercial fishing
- Little or no activity

MEXICO

Gulf of Mexico

TROPIC OF CANCER

Resources

Coal		Nickel	
Petroleum		Copper	
Natural gas		Lead	
Iron ore		Gold	
Tin		Silver	
Zinc		Platinum	
Cobalt		Gems	
Uranium			

PACIFIC OCEAN

Hawaii

0   100 miles
0   100 kilometers
Albers Equal-Area Conic projection

EQUATOR

UNIT 2
REGIONAL ATLAS

MAP STUDY

1. *Physical Systems* What climate regions are found in the United States?

2. *Environment and Society* What economic activities are found in the area around the Great Lakes?

160°E

180°

160°W

140°W

120°W

100°W

80°N

20°W

40°W

60°W

40°N

20°N

The United States and Canada **115**

The United States and Canada
Population Density

RUSSIA

ARCTIC OCEAN

GREENLAND (DENMARK)

Bering Sea

Bering Strait

ARCTIC CIRCLE

Hudson Bay

Edmonton
Calgary
Vancouver
Seattle
Portland

PACIFIC OCEAN

Montreal
Ottawa
Boston
Toronto
Buffalo
New York City
Minneapolis-St. Paul
Detroit
Pittsburgh
Philadelphia
Milwaukee
Baltimore
Chicago
Indianapolis
Washington, D.C.
San Francisco
Sacramento
St. Louis
Cincinnati
San Jose
Denver
Kansas City

ATLANTIC OCEAN

Las Vegas
Los Angeles
San Bernardino
Memphis
Atlanta
San Diego
Phoenix
Dallas-Ft. Worth
Austin
Orlando
San Antonio
Tampa
Miami
Houston

MEXICO

Gulf of St. Lawrence

Gulf of Mexico

TROPIC OF CANCER

CUBA

POPULATION

Per sq. mi.	Per sq. km
1,250 and over	500 and over
250–1,249	100–499
63–249	25–99
25–62	10–24
2.5–24	1–9
Less than 2.5	Less than 1
Uninhabited	Uninhabited

Cities
(Statistics reflect metropolitan areas.)

- ■ Over 5,000,000
- □ 2,000,000–5,000,000
- ⊙ 1,000,000–2,000,000

PACIFIC OCEAN

Hawaii

0 100 miles
0 100 kilometers
Albers Equal-Area Conic projection

0 600 miles
0 600 kilometers
Lambert Azimuthal Equal-Area projection

EQUATOR

UNIT 2
REGIONAL ATLAS

MAP STUDY

1. *Environment and Society* In what area is Canada's greatest population density located?

2. *Places and Regions* What generalizations can you make about the region's major population centers?

The United States

ESSENTIAL QUESTION • *How do physical systems and human systems shape a place?*

◄ Native Americans such as this Navajo girl keep their traditions alive through dance, dress, and spiritual practices.

National Geographic Image Collection/Alamy

networks

There's More Online about the United States.

CHAPTER 5

Why Geography Matters
Patterns of Immigration

Lesson 1
Physical Geography of the United States

Lesson 2
Human Geography of the United States

Lesson 3
People and Their Environment: The United States

Geography Matters...

What do the words *the United States* bring to mind? The images are endless for a land that offers so many opportunities. Some people hope to travel across its four million miles of highways, while others seek the "American Dream" of economic opportunity or religious and cultural freedom. It is not easy to characterize the American experience because the United States is a land of many cultures and peoples. The country is also known for its varied landscape of big cities, rugged mountains, wide plains, and forests that cover vast stretches of land.

patterns of immigration

One of the defining attributes of the United States is that it is largely a country of immigrants and their descendants. About 13 percent of people in the United States are foreign born, while Native Americans, Alaska Natives, and Native Hawaiians make up about 2 percent of the population. The remaining population is descended from immigrants.

Largest Ancestry Reported by County 2010

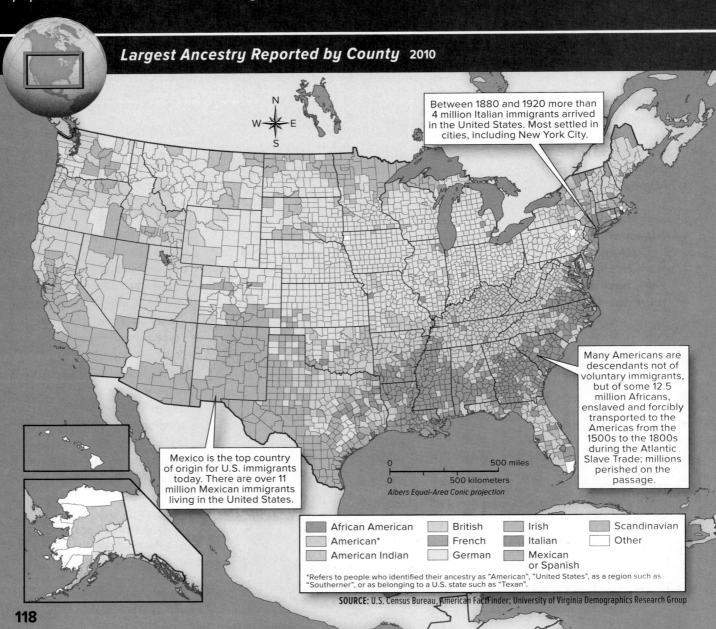

Between 1880 and 1920 more than 4 million Italian immigrants arrived in the United States. Most settled in cities, including New York City.

Many Americans are descendants not of voluntary immigrants, but of some 12.5 million Africans, enslaved and forcibly transported to the Americas from the 1500s to the 1800s during the Atlantic Slave Trade; millions perished on the passage.

Mexico is the top country of origin for U.S. immigrants today. There are over 11 million Mexican immigrants living in the United States.

0 — 500 miles
0 — 500 kilometers
Albers Equal-Area Conic projection

African American	British	Irish	Scandinavian
American*	French	Italian	Other
American Indian	German	Mexican or Spanish	

*Refers to people who identified their ancestry as "American", "United States", as a region such as "Southerner", or as belonging to a U.S. state such as "Texan".

SOURCE: U.S. Census Bureau, American FactFinder; University of Virginia Demographics Research Group

Where have immigrants to the United States come from?

The history of immigration to the United States can be divided into four distinct periods. The earliest period, the colonial period, was characterized by an influx of English and other European peoples, many of whom came as indentured servants to work for little or no pay. During the mid-nineteenth century, the majority of immigrants originated from the countries of northern and western Europe. The third major period of immigration was primarily from southern and eastern Europe. The start of the fourth wave of immigrants is marked by the abolishment in 1965 of the quotas that limited immigration from the Eastern Hemisphere based on country of origin. This opened the doors to a wave of immigrants from Asia and Africa. At the same time, worsening economic conditions in many Latin American countries led to a rise in the number of Latino immigrants to the United States. Today, Mexicans are the largest single immigrant group in the United States.

1. Places and Regions Describe the four periods into which immigration to the United States can be divided.

Why did they immigrate to the United States?

Two sets of factors influence immigration patterns: push factors and pull factors. Push factors motivate people to leave their countries of origin. They can be quite varied, including religious and ethnic persecution, lack of economic prospects or personal security, civil strife, and war. Environmental conditions such as drought, famine, and flooding and events such as earthquakes and volcanic eruptions can also push people to migrate to a new country. In the nineteenth century, for example, many Irish immigrants came to the United States to escape the Irish potato famine of 1845 to 1852. Similarly, many Chinese began migrating to the United States during this same time because of the population explosion and food shortages in China. Pull factors are those which draw people to a new country, such as employment opportunities, mild climates, sea coasts, mountains, political and religious freedom, and better living conditions, including access to education and health care. Pull factors drew people to the United States during all four major periods of immigration, as people came seeking work, freedom, and a better way of life.

2. Human Systems What are the reasons for immigration to the United States? Define push factors and pull factors and list at least three examples of each.

How has immigration shaped human geography in the United States?

Successive waves of immigrants have absorbed, challenged, and ultimately reshaped American culture. As people move, their cultural traits and ideas move with them, creating and modifying the cultural landscape in their new home. These imprints on the United States include spoken languages, religious beliefs and institutions, family and ethnic traditions, architectural styles, art, music, and food. These cultural modifications can be seen in the ethnic neighborhoods of both small and large cities where grocery stores, churches and other places of worship, schools, restaurants, and other businesses reflect the immigrant culture of each community. Examples of ethnic neighborhoods include Little Tokyo in Los Angeles, California; Little Havana in Miami, Florida; Polish Hill in Pittsburgh, Pennsylvania; and Arabian Village in Detroit, Michigan.

3. Human Systems Write a paragraph describing three examples of how immigrants have left their imprint on the cultural landscape of your community or a place you have visited.

THERE'S MORE ONLINE

VIEW a graph of the ethnic makeup of the United States • **WATCH** a video about U.S. ethnic diversity

networks

There's More Online!

- ☑ **DIAGRAM** Western Topography
- ☑ **INFOGRAPHIC** Hurricanes in the United States
- ☑ **MAP** The Continental Divide and the Fall Line
- ☑ **INTERACTIVE SELF-CHECK QUIZ**
- ☑ **VIDEO** Physical Geography of the United States

LESSON 1

Physical Geography of the United States

ESSENTIAL QUESTION • *How do physical systems and human systems shape a place?*

Reading **HELP**DESK

Academic Vocabulary

- shift
- alter

Content Vocabulary

- tributary
- headwaters
- divide
- fall line
- hurricane
- fossil fuel

TAKING NOTES: *Key Ideas and Details*

IDENTIFYING Use a graphic organizer like the one below to take notes on the physical geography of the United States as you read.

Physical Geography of the United States		
Landforms	Water Systems	Climate, Biomes, and Resources

It Matters Because

With over 3.5 million square miles (9 million sq. km) of land, the United States is the third-largest country in the world. Its natural environment is diverse and makes the country one of the world's most productive regions.

Landforms

GUIDING QUESTION *How has tectonic activity helped create so many of the landforms in the United States?*

Many of the landforms of the United States can be traced back to glacial activity and the tectonic plate movement of the Earth's crust. The Pacific and Rocky Mountain ranges in the west and the Appalachian Mountains in the east are the result of powerful tectonic plate activity. The tectonic forces **shifted** giant rock slabs upward.

Considered young in geologic terms, the Pacific Ranges of the United States consist of the Sierra Nevada, the Cascade Range, the Coast Range, and the Alaska Range. Denali (Mt. McKinley) in the Alaska Range, at 20,320 feet (6,194 m), is the highest point in the United States. The Rocky Mountains begin in New Mexico and stretch northward over 3,000 miles (4,828 km). Between the Pacific Ranges and the Rocky Mountains is an area of plateaus and dry basins that was formed by volcanic lava seeping upward through cracks in the Earth's crust. These lava flows **altered**, or changed, the land forming the Columbia Plateau. Farther south are the flat-topped mesas of the Colorado Plateau and the spectacular gorge of the Grand Canyon, which plunges more than a mile into the Earth at its deepest points.

Extending eastward from the Rockies, the landscape flattens considerably to form the Great Plains, which stretch from 300 miles (483 km) to over 700 miles (1,126 km) wide. The land of the plains is higher in the west and slopes downward until it reaches the Central Lowlands. The plains continue eastward to the base of the Appalachian Mountains, the oldest mountain range on the North American continent. This range extends 1,500 miles (2,414 km), from Canada into the state of

Alabama. As tectonic plates within the Earth's crust collided and pushed upward, they formed the Appalachians. The resulting peaks were shaped further by ice and water. Between the Appalachian Mountains and the Atlantic Coastal Plain lies the Piedmont, a low-rolling, fertile plateau cut by many rivers.

The Hawaiian Islands, located about 2,400 miles (3,862 km) off the western coast of the mainland United States, were formed when magma erupted from a spot on the seafloor, called a hot spot. This hot spot created a string of 8 major and 124 smaller islands that make up the Hawaiian Island chain.

shift to change the place, position, or direction of

alter to change partly

☑ **READING PROGRESS CHECK**

Explaining How were the Pacific Ranges formed?

Water Systems

GUIDING QUESTION *How have rivers and lakes been important to the economic development of the United States?*

Many lakes, rivers, and **tributaries** play a crucial role in many aspects of life. The Mississippi River is one of the longest rivers in North America. It flows 2,350 miles (3,782 km) from its **headwaters**, or source, in Minnesota and reaches a width of 1.5 miles (2.4 km) at its mouth, where it empties into the Gulf of Mexico. The Colorado River and the Rio Grande both have their headwaters in the Rocky Mountains where many tributaries merge to form these two major waterways.

A physical feature called a **divide** determines the direction of river flow. The Continental Divide is a high ridge in the Rocky Mountains. Waterways to the west of the divide flow into the Pacific Ocean. Waterways to the east of the divide flow toward the Arctic Ocean, Hudson Bay, Atlantic Ocean, and Mississippi River system.

In the eastern United States, the **fall line** marks the place where the higher land of the Piedmont drops to the lower Atlantic Coastal Plain. Along the fall line, eastern rivers break into rapids and waterfalls, preventing ships from the Atlantic Ocean from traveling farther inland. Many cities, such as Philadelphia, Baltimore, and Washington, D.C., were established along the fall line.

tributary a smaller river or stream that feeds into a larger river

headwaters the source of a stream or river

divide a high point or ridge that determines the direction rivers flow

fall line a boundary in the eastern United States where the higher land of the Piedmont drops to the lower Atlantic Coastal Plain

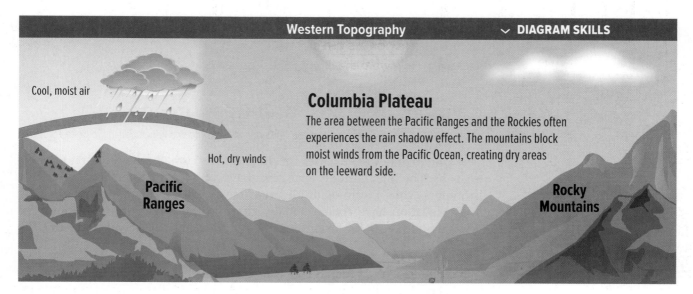

Western Topography ∨ **DIAGRAM SKILLS**

Cool, moist air

Hot, dry winds

Pacific Ranges

Columbia Plateau
The area between the Pacific Ranges and the Rockies often experiences the rain shadow effect. The mountains block moist winds from the Pacific Ocean, creating dry areas on the leeward side.

Rocky Mountains

The shape and location of the Cascade Range affect conditions on the Columbia Plateau.

▲ **CRITICAL THINKING**

1. Drawing Conclusions What type of vegetation would you expect to find in the flat plateau areas between mountain ranges?

2. Analyzing How does the rain shadow effect impact the Columbia Plateau?

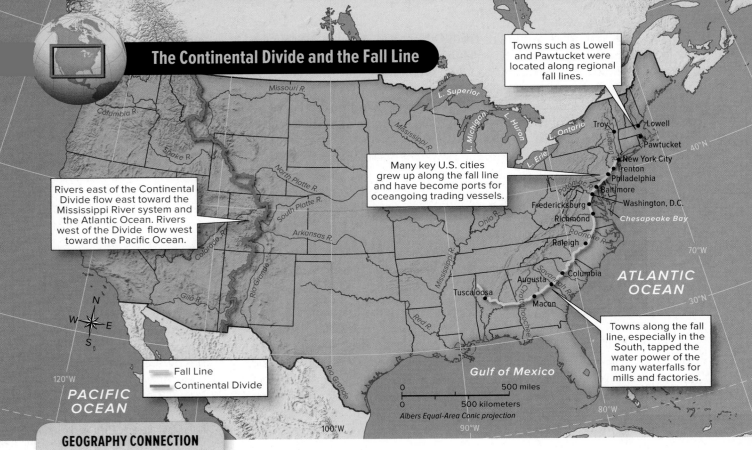

The Continental Divide and the Fall Line

Towns such as Lowell and Pawtucket were located along regional fall lines.

Many key U.S. cities grew up along the fall line and have become ports for oceangoing trading vessels.

Rivers east of the Continental Divide flow east toward the Mississippi River system and the Atlantic Ocean. Rivers west of the Divide flow west toward the Pacific Ocean.

Towns along the fall line, especially in the South, tapped the water power of the many waterfalls for mills and factories.

Fall Line
Continental Divide

ATLANTIC OCEAN

PACIFIC OCEAN

Gulf of Mexico

0   500 miles
0   500 kilometers
Albers Equal-Area Conic projection

GEOGRAPHY CONNECTION

The Continental Divide and the fall line are important to the flow of rivers and river traffic in the United States.

1. **PHYSICAL SYSTEMS** How does the Continental Divide affect the flow of rivers in the western United States?

2. **PLACES AND REGIONS** Describe how the fall line influenced economic development in the eastern United States.

Formed when glacier basins filled with water, Lake Superior, Lake Huron, Lake Erie, Lake Ontario, and Lake Michigan make up the Great Lakes. The glaciers uncovered major deposits of natural resources, including iron ore and coal, that later spurred explosive economic growth. The Great Lakes serve many economic and recreational purposes, but none more valuable than the Great Lakes–St. Lawrence Seaway System, a series of canals, rivers, and waterways linking the Great Lakes with the Atlantic Ocean. The seaway helped make cities along the Great Lakes, such as Chicago, powerful trade and industrial centers.

☑ **READING PROGRESS CHECK**

Explaining How were the Great Lakes formed?

Climate, Biomes, and Resources

GUIDING QUESTION *What factors cause variations in climate and vegetation in the United States?*

The United States has a variety of climates. The climates differ for a number of reasons. The high latitudes of Alaska have long, cold winters and brief, mild summers while the midlatitude areas have temperate climates. Places with high elevation have cooler climates than do those with low elevation. The United States even has tropical climates in the states of Hawaii and Florida.

Climate Regions and Biomes

The Southeast has a humid subtropical climate that is rainy with long, muggy summers and mild winters. Since it borders large bodies of water—the Atlantic Ocean and Gulf of Mexico—there is no dry season. Maple, oak, and pine trees are plentiful, and many types of mammals, reptiles, and amphibians are common.

Wetlands and swamps such as Florida's Everglades shelter a great variety of vegetation and wildlife. In late summer and early autumn, **hurricanes**—ocean

hurricane a large, powerful windstorm that forms over warm ocean waters

122

storms hundreds of miles wide with sustained winds of about 74 miles per hour (119 km per hour) or more—can pound the region's coastlines.

The climate of the Great Plains reflects its location in the center of the continent. Because this area is far from the moderating influences of the ocean waters, it experiences very cold winters and hot summers. This is known as a continental climate. Moreover, parts of this area have a humid continental climate because they receive significant precipitation. This interior climate also extends into the hills and plateaus between the Mississippi River and the Appalachian Mountains. In the Great Plains and the eastern United States, violent spring and summer thunderstorms called supercells often spawn tornadoes with winds that can reach 200 miles (339 km) per hour.

Some areas west of the Great Plains have a semiarid climate with a mixture of vegetation, depending on latitude and elevation. These are transitional climates that occur between the humid continental climates and the arid climates of the Colorado Plateau. Animals in semiarid regions include deer, bison, coyotes, and wolves.

To the west of the semiarid regions, dry air moves down the leeward side of the mountains, creating an arid climate. This is called the rain shadow effect. Plants in arid climates, such as scrub bushes and cacti, have developed long root systems and other adaptations that allow them to survive with little water.

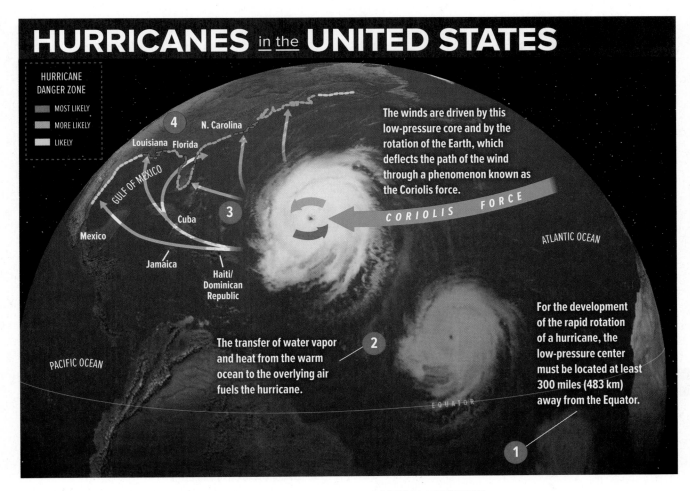

HURRICANES in the UNITED STATES

HURRICANE DANGER ZONE
- MOST LIKELY
- MORE LIKELY
- LIKELY

The winds are driven by this low-pressure core and by the rotation of the Earth, which deflects the path of the wind through a phenomenon known as the Coriolis force.

CORIOLIS FORCE

The transfer of water vapor and heat from the warm ocean to the overlying air fuels the hurricane.

For the development of the rapid rotation of a hurricane, the low-pressure center must be located at least 300 miles (483 km) away from the Equator.

Tropical storms that are known as hurricanes in the United States are called typhoons in the western North Pacific Ocean and cyclones in the South Pacific and Indian Oceans.

▲ CRITICAL THINKING

1. Interpreting Describe the air pressure inside the core of a hurricane and how proximity to the Equator affects the development of this type of storm.

2. Analyzing Visuals Which parts of the United States are most vulnerable to damage from hurricanes?

Soil Science

Soil, the thin outer layer of Earth's crust, is necessary for human survival. We get most of our food directly or indirectly from plants growing in soil. Our water supply is filtered by the soil, and soil products supply many engineering materials needed for buildings and infrastructure. Pedologists (soil scientists) study Earth's soil types, their physical and chemical characteristics, and how they can be used to meet human needs without degrading or depleting this resource. Many pedologists work for government agencies, universities, and private industry.

DRAWING INFERENCES Which industries are likely to benefit from the advice of soil scientists? Why?

fossil fuel a resource formed in the Earth by plant and animal remains

---

A Mediterranean climate is found in central and southern California. Such a climate is confined to coastal areas and is characterized by mild, wet winters and summers that are warm to hot and dry. The vegetation consists of twisted, drought-resistant broad-leafed trees, known as chaparral (SHA•puh•RAL).

The Rockies and the Pacific Ranges have a high altitude climate characterized by cold, snowy winters and warm, dry summers. Coniferous forests cover the middle elevations, and lichens and mosses grow in higher elevations. In early spring, a warm, dry wind called the chinook (shuh•NUK) blows down the eastern slopes of the Rockies. Mountain goats and mountain lions are common. The interplay of ocean currents and westerly winds with the Pacific Ranges gives the Pacific coast from northern California to southern Alaska a marine west coast climate. Parts of this region receive more than 100 inches (254 cm) of rain each year. Ferns, mosses, grasses, and coniferous forests grow here.

Large parts of Alaska have a subarctic climate with frigid winter temperatures of −70°F (−57°C) in some places. Conifers such as pine and spruce are able to survive the cold. Many animals thrive in the harsh climate of the subarctic, including grizzly bears, bald eagles, wolves, and bobcats.

Natural Resources

The United States is rich with natural resources including water, **fossil fuels**, timber, fish, and more. Fossil fuels were formed over hundreds of millions of years from the fossilized remains of plants and animals. This makes them nonrenewable. Also, because they must be retrieved from the ground, there can be damage to the environment when they are extracted from the Earth. Fossil fuels include coal, petroleum, and natural gas. They can be found in great supply in Texas and Alaska, which rank first and second respectively in U.S. petroleum reserves. The United States has the largest known coal reserves in the world and soon may be the largest producer of oil.

The United States has plentiful mineral resources as well. The Rocky Mountains yield gold and silver. Other minerals include copper, lead, phosphates, uranium, bauxite, iron, mercury, nickel, silver, tungsten, and zinc. Free and abundant access to this natural wealth helped to speed the industrialization of the United States and has helped to create one of the most prosperous countries in the world.

Fish are also an important natural resource. Commercial fishing in the Atlantic and Pacific Oceans and the Gulf of Mexico is important to the U.S. economy. The large commercial fishing companies and small family businesses provide employment for many people, as well as food for domestic consumption and for export.

☑ READING PROGRESS CHECK

Identifying What factors contribute to the marine west coast climate?

LESSON 1 REVIEW

Reviewing Vocabulary
1. ***Explaining*** Explain how fossil fuels are formed.

Using Your Notes
2. ***Describing*** Use your graphic organizer to list three major water systems in the United States.

Answering the Guiding Questions
3. ***Discussing*** How has tectonic activity helped create so many of the landforms in the United States?

4. ***Expressing*** How have rivers and lakes been important to the economic development of the United States?

5. ***Identifying*** What factors cause variations in climate and vegetation in the United States?

Writing Activity
6. ***Narrative*** Write a paragraph that describes the path of the Mississippi River as if you were writing for a travel magazine. Describe the climate, landforms, and biomes along its path.

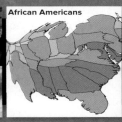
African Americans

networks

There's More Online!

☑ **MAP** Ethnic Populations in the United States

☑ **MAP** Homes in Negative Equity, 2010

☑ **MAP** U.S. Expansion

☑ **TIME LINE** Terrorism and the United States

☑ **INTERACTIVE SELF-CHECK QUIZ**

☑ **VIDEO** Human Geography of the United States

LESSON 2
Human Geography of the United States

ESSENTIAL QUESTION • *How do physical systems and human systems shape a place?*

Reading HELPDESK

Academic Vocabulary

- **conflict**
- **immigrate**

Content Vocabulary

- **Underground Railroad**
- **dry farming**
- **Manufacturing Belt**
- **Sunbelt**
- **megalopolis**
- **jazz**
- **postindustrial**
- **foreclosure**

TAKING NOTES: *Key Ideas and Details*

IDENTIFYING Use a graphic organizer like the one below to take notes about the major economic activities of the United States.

United States: Economic Activities

IT MATTERS BECAUSE

Urban lifestyles predominate in the United States, but traditional and rural values are still respected. The country has also been enriched by the tens of millions of immigrants who have come to America hoping to improve their lives.

History and Geography

GUIDING QUESTION *How did physical geography and a spirit of independence influence the development of the United States?*

The physical environment has played a significant role in the patterns of settlement in the United States. The largest city, New York City, is located on one of the world's finest harbors as a result of physical geography. Similarly, many people today choose to live in California for its favorable climate and beautiful landscapes.

Growth, Division, and Unity

Scientific studies suggest that there were at least three migrations of people from Asia to Alaska. They began about 15,000 years ago and occurred by land and by boat. The lives of Native Americans, the descendants of these early peoples, were shaped by location and climate.

Native Americans occupied North America undisturbed until the mid-1500s when European immigration began. The Spanish explored the southern region, setting up farms, ranches, military posts, and missions. The French settled in the northeast and were involved in the fur trade.

After 1670, Britain controlled much of the land along the Atlantic coast, divided into three colonial regions. The New England Colonies had rocky soil and a short growing season, but the area's harbors and an abundant supply of timber and fish made shipbuilding and fishing important industries. The Middle Colonies had the fertile soil, mild winters, and warm summers needed for growing cash crops for export. The mild climate, rich soils, and open land of the coastal plain of the Southern Colonies were a favorable environment for plantation agriculture.

conflict a competition or struggle

In 1763 France was forced to give up much of its North American empire to Great Britain. **Conflicts** soon arose between Native Americans and colonial settlers. Settlers arriving in the British colonies took the land of Native Americans. Loss of hunting and farming lands, combined with European diseases, reduced Native American populations and severely disrupted their cultures.

In the 1760s, the British government angered the colonists by imposing new taxes and limiting their freedoms. The thirteen colonies eventually fought for independence from Britain in the American Revolution (1775–1783). The outcome was an independent federal republic called the United States of America.

During the 1800s, the United States more than doubled its territory. The country gained valuable land and natural resources. For Native Americans, however, expansion led to the steady loss of lands and restrictions on traditional ways of life.

Industrialization transformed the United States during this time. The first factories harnessed the power of waterfalls along the fall line in the Northeast. Later, large supplies of coal in the Midwest were used to fuel cheap steam power, thus making manufacturing profitable. As a result, the Midwest became a leading center of industry, using the Great Lakes and rivers for transportation.

In the South, cotton became a major cash crop as the textile industry grew in the Northeast. Land was cleared for more plantations, and the labor of enslaved African Americans became critical to the Southern economy. By the 1800s, however, some people were working to end slavery, and many African Americans made their way north to freedom along the **Underground Railroad**—a network of safe houses.

Tensions between the industrialized North and the agricultural South mounted steadily until they erupted in the American Civil War in 1861. After four bloody years, the North triumphed. Slavery was abolished after the war and the country began rebuilding.

Underground Railroad
a network of safe houses in the United States that helped thousands of enslaved people escape to freedom

(tr)Spencer Platt/Getty Images News/Getty Images, (b)Roberto Schmidt/AFP/Getty Images

---

TIME LINE ⌄

TERRORISM
and the United States ➡

President George W. Bush was the first U.S. president to use the term "War on Terror."

▶ **CRITICAL THINKING**

1. Sequencing What events led to the passage of the USA PATRIOT Act?

2. Assessing Why do you think Congress renewed the USA PATRIOT Act in March 2006?

1990 ➡

1991 U.S. forces play dominant role in war against Iraq after invasion of Kuwait

1995 Domestic terrorist bombs federal building in Oklahoma City, killing 168 people.

Coordinated suicide attacks by al-Qaeda on multiple high-profile U.S. targets

2001

2001 U.S. leads military campaign against Afghanistan; USA PATRIOT Act enacted

Changes and Challenges

In the late 1800s, the government encouraged the movement of people to the Great Plains to speed up the settlement of the United States. New immigrants wanted land, and there was an increasing need for food in the growing cities. Due to the dry conditions on the Great Plains, settlers developed **dry farming**. Steel plows and steam tractors made farming easier, and fewer people were needed for farm work. At the same time, the Industrial Revolution brought people to cities in the Northeast and Great Lakes regions, or the **Manufacturing Belt.**

Europeans, Chinese, Mexicans, and others **immigrated** to the United States. Many helped build the railroads. Joining the Central Pacific and Union Pacific railroads created a transcontinental railroad. A network of railways moved manufactured goods from east to west and food products from west to east.

Two world wars spurred economic growth. Assembly lines increased efficiency and improved the standard of living. The population became more mobile and urbanized. By the 1990s, many manufacturing activities were less important than the rising high-tech industries.

Social changes also took place. Immigration from Latin America and Asia increased. Minority groups began to participate in business and politics. Native Americans negotiated with the government over land claims.

Terrorism became a major concern of many Americans after September 11, 2001, when Islamist terrorists hijacked four passenger planes, crashing them into the World Trade Center, the Pentagon, and a Pennsylvania field. After such devastation, the United States launched a war on terrorism focused on Afghanistan and Iraq.

✔ **READING PROGRESS CHECK**

Explaining Why did the Midwest become a center of industry?

dry farming a farming method used in dry regions in which crops are grown that rely only on the natural precipitation

Manufacturing Belt
a concentrated region of manufacturing industries in the northeastern and midwestern United States

immigrate to change residence from a country to begin living permanently in another country

---

Missile attacks on Baghdad mark the start of a U.S.-led campaign to topple Iraqi leader Saddam Hussein. U.S. forces advance into Baghdad in early April.

Senate report says U.S. and its allies went to war in Iraq on flawed information. Report on 9/11 attacks highlights deep institutional failings in U.S. intelligence services.

April—Two bombs explode near the finish line at the Boston Marathon, killing three people and injuring more than 260. Two brothers with radical Islamic views are responsible for the attack.

2003

2004

2011 May—U.S. forces kill terrorist leader Osama bin Laden in Pakistan

2013

➡2002

➡2012

2002 President Bush signs into law a bill creating a Department of Homeland Security, aimed at protecting the country against terrorist attacks.

2006 U.S. Congress renews USA PATRIOT Act. The government agrees to curbs on information gathering.

2012 September—U.S. ambassador to Libya and three others are killed in attack on U.S. consulate in Benghazi

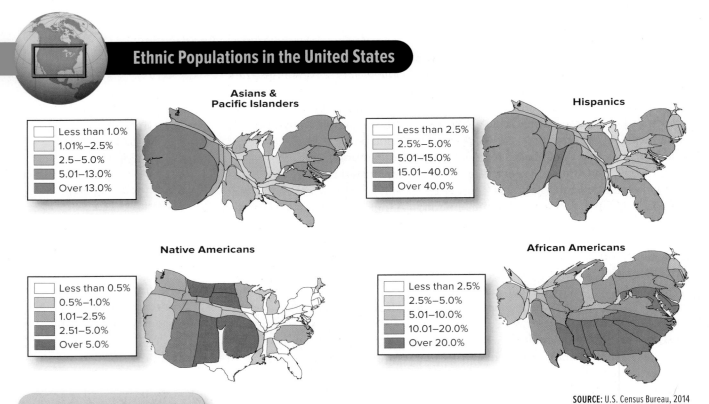

Ethnic Populations in the United States

Asians & Pacific Islanders

- Less than 1.0%
- 1.01–2.5%
- 2.5–5.0%
- 5.01–13.0%
- Over 13.0%

Hispanics

- Less than 2.5%
- 2.5–5.0%
- 5.01–15.0%
- 15.01–40.0%
- Over 40.0%

Native Americans

- Less than 0.5%
- 0.5%–1.0%
- 1.01–2.5%
- 2.51–5.0%
- Over 5.0%

African Americans

- Less than 2.5%
- 2.5–5.0%
- 5.01–10.0%
- 10.01–20.0%
- Over 20.0%

SOURCE: U.S. Census Bureau, 2014

GEOGRAPHY CONNECTION

The United States Census captures the country's ethnic diversity.

1. **PLACES AND REGIONS** Which region of the United States has the greatest percentage of Native Americans, the Central Plains or the East?

2. **PLACES AND REGIONS** How would you characterize the distribution of the Asian American population in the United States?

Sunbelt a mild climate region in the southern and southwestern portions of the United States

Population Patterns

GUIDING QUESTION *What factors influence population patterns in the United States?*

More than 320 million people live in the United States today. While about 2.5 million are Native Americans, a majority are immigrants or the descendants of immigrants from Europe, Asia, Africa, and Latin America. Some arrived only recently, while others belong to families whose ancestors came to the region centuries ago.

The average population density of the United States is about 91 people per square mile (35 people per sq. km). Outside of large urban areas, however, the population is widely distributed. The Northeast and Great Lakes regions are densely populated because they are the historic centers of industry and commerce. The Pacific coast attracts people looking for a mild climate and economic opportunities, resulting in a population cluster there. The least densely populated areas of the country include the subarctic region of Alaska, the dry Great Basin, and parts of the arid and semiarid Great Plains.

The population structure of the United States is changing. The U.S. Census Bureau projects that the population aged 65 and older will likely grow from some 48 million in 2010 to about 88 million by the year 2050. This increase presents challenges to the federal government as costs for Social Security and Medicare rise. The health care sector, the business sector, and families will also be affected.

Since the 1960s, the Manufacturing Belt has suffered a decline in population and economic strength as manufacturers relocated. Many businesses moved to **Sunbelt** states in the South and Southwest. Over the years, as mechanized agriculture has required fewer workers, the United States also has experienced urbanization, the movement of people from rural areas to cities. Today, most people in the United States live in the country's 381 metropolitan areas. A metropolitan area is a city with a population of at least 50,000 people including outlying communities, called suburbs.

Many U.S. population clusters lie in coastal areas with strong economies linked to world trade. Pacific coast cities provide important links to the rest of the world, especially to the growing Asian economies. The **megalopolis** that stretches from Santa Barbara, California, to Mexico is also an important corridor for world trade. Along the Atlantic coast, a chain of closely linked metropolitan areas from Boston, Massachusetts, to Washington, D.C., form the Boswash megalopolis. The Great Lakes region has three megalopolises: one centered in Chicago, Illinois, another in Buffalo, New York, and a third in Detroit, Michigan.

☑ READING PROGRESS CHECK

Speculating Provide a specific example of how the aging of the population might affect Americans.

Society and Culture Today

GUIDING QUESTION *How has immigration influenced the culture of the United States?*

Today, the United States has one of the most diverse populations in the world. Some immigrants have come to the United States to seek political and religious freedom or to find economic opportunities. Others are fleeing wars or natural disasters. Rich natural resources, industry, and economic wealth make the United States an attractive destination.

Throughout history, immigrants have often faced discrimination, but they have invariably enriched their new country through their hard work and talents and by bringing greater cultural diversity to the country. In 2014 the Census Bureau reported that 13.3 percent of the total U.S. population was foreign born and that more than half of the foreign-born population came from Latin America.

Immigrants have contributed to the country's diverse religious beliefs. Since the country's founding, religious freedom has been a core value in the United States. Today, most Americans who are members of an organized religion are Christian, with the majority being Protestant. Judaism, Islam, and Buddhism are among the other religions practiced in the United States. About 22 percent of the U.S. population today is not affiliated with any organized religion.

Family and Status of Women

Although population patterns in the United States continue to change, the family remains a vital institution. About half of adults are married, but some are single people living alone, or single mothers or fathers living with their children. More and more women work outside the home. Women have also continued to make gains in college completion rates, exceeding the graduation rate of men for the people between the ages of 25 and 34.

The Arts

The history of music in the United States can be traced back to Native American traditions. Europeans later brought their own folk and religious music. At the start of the 1900s, a distinct form of music known as **jazz** developed in African American communities throughout the United States. Jazz blended African rhythms with European harmonies. By the second half of the century, country music and rock-and-roll had become popular, not only in North America but around the world. Blues, punk rock, and hip-hop all have their origins in the United States, though hip-hop also traces its roots to the dance hall musicians of Jamaica.

Many styles of art can also be found in the United States. In the early 1900s, a group of American artists known as the Ashcan School painted the grim realities of urban America. In the mid-1900s, many artists adopted European abstract styles, which express artists' emotions and attitudes without depicting recognizable images.

megalopolis a large population concentration made up of several large and many smaller cities, such as the area between Boston and Washington, D.C.

jazz musical form that developed in the United States in the early 1900s, blending African rhythms and European harmonies

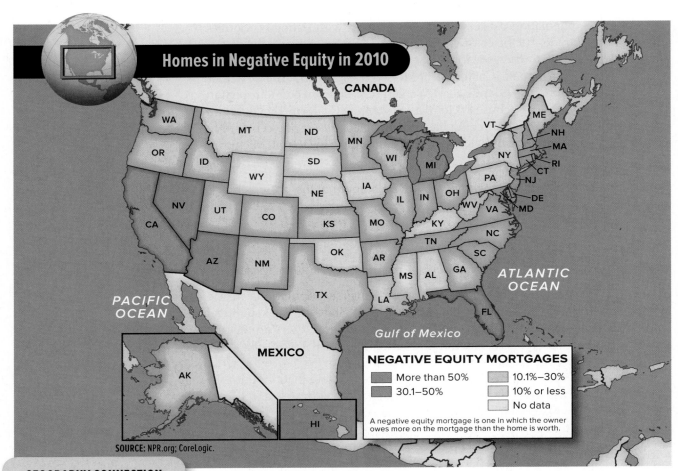

Homes in Negative Equity in 2010

CANADA

WA

MT

ND

MN

OR

ID

WY

SD

WI

MI

VT

ME

NH

MA

NY

NV

UT

CO

NE

IA

IL

IN

OH

PA

NJ

CT

RI

CA

KS

MO

KY

WV

VA

MD

DE

AZ

NM

OK

AR

TN

NC

SC

MS

AL

GA

TX

LA

FL

AK

MEXICO

HI

PACIFIC OCEAN

ATLANTIC OCEAN

Gulf of Mexico

NEGATIVE EQUITY MORTGAGES

- More than 50%
- 30.1–50%
- 10.1%–30%
- 10% or less
- No data

A negative equity mortgage is one in which the owner owes more on the mortgage than the home is worth.

SOURCE: NPR.org; CoreLogic.

GEOGRAPHY CONNECTION

During the foreclosure crisis, millions of Americans were unable to pay their home mortgages.

1. ***PLACES AND REGIONS*** Which states were least affected by negative equity mortgages?

2. ***HUMAN SYSTEMS*** Which states are shown with a more than 50% negative equity statistic?

The American-born graffiti art movement, begun by disenfranchised urban youth, has grown into a worldwide phenomenon. Graffiti is now created by artists who are commissioned by governments and private citizens to create enormous and intricate works of art.

✔ READING PROGRESS CHECK

Identifying What is the most common religion practiced in America today?

Economic Activities

GUIDING QUESTION *How is the U.S. economy an important part of the global economy?*

The United States has always been based on a free market economy and experiences ups and downs. Between the two world wars in the twentieth century, a long and devastating depression affected tens of millions of Americans. As the 1950s began, the manufacturing sector became a driving force of the economy. Manufacturers helped the economy by turning from wartime production to the manufacturing of cars, televisions, and appliances.

Today, the U.S. economy is a free market economy that allows people to profit from owning their own businesses. This freedom—coupled with laws that protect private property rights, employment opportunities, and the health and safety of workers—has created a great economic power. The country's wealth, measured in terms of gross national product (GNP), is due to universal education, technology and innovation, abundant natural resources, high agricultural output, and highly developed industries. The country has important reserves of natural gas and petroleum and is also a world leader in coal exports.

Resources, Power, and Industry

Agriculture in the United States has undergone many changes since the 1950s. The average size of farms has grown and continues to grow. Manufacturing has also evolved. The number of manufacturing jobs has declined while time efficiency and productivity have increased for most U.S. factories.

While agriculture and manufacturing are still important, the **postindustrial** economy is dominated by high-tech, biotechnology, and service industries. The service sector has grown more than any other part of the economy in recent decades. In the high-tech industry, California's Silicon Valley and cities such as Seattle, Washington, and Austin and Dallas, both in Texas, are leaders in software development. The North Carolina cities of Raleigh, Durham, and Chapel Hill form the Research Triangle region, known for attracting biotechnology companies.

In recent years, businesses in the United States have turned to offshoring, the practice of setting up plants abroad to produce parts or products for domestic use and international sale. While offshoring decreases the costs of goods, some people argue that it takes jobs away from American workers.

Good transportation and reliable communications are crucial to the economy of the United States. The automobile is still the most commonly used personal transportation method in the country. Its use has resulted in large investments in the building and maintenance of highways, roads, and bridges. The country also relies on air travel as a major method of transportation. A large percentage of the freight in the United States is transported by truck. The country's long-distance communications are carried via wireless, microwave, and satellite relays. Cellular and digital services have made mobile communication the norm, with fewer and fewer households using traditional landline telephones.

postindustrial economy that emphasizes services and technology rather than industry and manufacturing

The Economic Downturn

In 2008 the United States entered a serious economic downturn caused by an excessive number of ill-advised home mortgage loans. This resulted in a record number of homes going into **foreclosure**. The downturn was called the subprime mortgage crisis. At the same time, the stock market became unstable and unemployment rose. The bad loans led to the failure of some large banks and required large government investments to save many other banks. Because the U.S. economy and financial system are so important to the global economy, many other countries also spiraled into an economic downturn. The U.S. and global economies are improving, gaining the strength that is needed for robust health.

foreclosure legal proceeding in which a borrower's rights to a property are relinquished due to his or her inability to make payments on the loan

✔ **READING PROGRESS CHECK**

Describing What are the characteristics of the U.S. free market economy?

LESSON 2 REVIEW

Reviewing Vocabulary
1. *Defining* Define Sunbelt and megalopolis, and describe the locations of each.

Using Your Notes
2. *Identifying* Use the notes from your graphic organizer to write a paragraph describing the economic activities of the United States.

Answering the Guiding Questions
3. *Describing* How did physical geography and a spirit of independence influence the development of the United States?

4. *Explaining* What factors influence population patterns in the United States?

5. *Discussing* How has immigration influenced the culture of the United States?

6. *Expressing* How is the U.S. economy an important part of the global economy?

Writing Activity
7. *Informative/Explanatory* Write an essay describing the postindustrial economy of the United States.

HOW CAN DROUGHT LEAD TO CONFLICT IN THE UNITED STATES?

The southeastern United States has a humid subtropical climate characterized by high year-round precipitation. This part of the country is considered *water rich*—having large amounts of water available as a natural resource. Surface streams account for the majority of water resources. Despite the abundant water resources in the region, however, many large cities are located far from surface streams. As a result, they experience periodic drought.

Much of this region also has a population growth rate calculated to exceed the U.S. average. Consequently, the demand for water is rapidly increasing. Water quality in this region is a concern because water resources are affected by industrial discharges, surface mining, and contamination by salt water. With a growing urban population, an increasing demand for water, and region-specific problems that affect water quality, cities and states often disagree over where and how water should be diverted and distributed. Moreover, after an almost three-year rainfall deficit from 2005 to 2008, Georgia, South Carolina, and Tennessee experienced drought conditions resulting in damages to crops, potentially dangerous drops in reservoir levels, and increased risk of fire. This sudden and unexpected lack of water threw the region into drought-related conflict.

To address the drought, the state of Georgia requested that the U.S. Army Corps of Engineers limit the flow of water from the Lake Lanier reservoir to the Chattahoochee River and the Apalachicola River. These rivers provide water for Georgia, Florida, and Alabama. While this step would fill Lake Lanier and provide much-needed drinking water for the city of Atlanta, it would also limit the water supply to Alabama and Florida.

Chris Rank/Bloomberg/Getty Images

Foster Economic Development

PRIMARY SOURCE

" The U.S. Supreme Court on Monday secured metro Atlanta's claim to water from Lake Lanier, handing Georgia an enormous legal victory in the tri-state water dispute. . .

'We can legally drink the water of Lake Lanier,' Williams said to booming applause throughout the banquet hall.

The much-anticipated decision could have monumental ramifications for economic development across the state and growth of the metro region.

Some companies have been hesitant to move to or expand in Atlanta, given the uncertainty of water supply, Williams said.

'That danger is gone now,' Williams said. 'It's time to sit down with our friends in Alabama and Florida. . . . We can go sit down and resolve this. "

—Greg Bluestein, Bill Rankin, and Scott Trubey,
"High court grants Georgia water-wars victory,"
Atlanta Journal Constitution, June 25, 2012

Preserve Local Economies

PRIMARY SOURCE

" This drought and possible further flow reductions have threatened the very existence for some 1,300 families of 3rd and 4th generation oystermen and a way of life that is an integral part of the community. Florida's Apalachicola River and Bay is the most productive contained commercial fishery in Florida. The local economy depends on the entire ecosystem, which has annual seafood landings reaching millions of dollars dockside. "

—Florida governor Charlie Crist,
letter to the United States Department of
the Interior, May 28, 2009

What do you think? DBQ

1. **Drawing Conclusions** According to the Atlanta Journal Constitution article, how might Atlanta's economy benefit from an increased water supply?

2. **Identifying Central Issues** Provide an example of the importance of water to the local economy in Florida.

3. **Evaluating** Who do you think provides the stronger argument, the writers of the Atlanta Journal Constitution article or Governor Crist? Explain your answer.

netw✺rks

There's More Online!

☑ **INFOGRAPHIC** Formation of Acid Rain

☑ **IMAGE** Wind as Renewable Energy

☑ **MAP** U.S. Water Withdrawals

☑ **MAP** Acid Rain in the United States

☑ **INTERACTIVE SELF-CHECK QUIZ**

☑ **VIDEO** People and Their Environment: The United States

LESSON 3
People and Their Environment: The United States

ESSENTIAL QUESTION • *How do physical systems and human systems shape a place?*

Reading HELPDESK

Academic Vocabulary

- **contribute**
- **abandon**

Content Vocabulary

- **clear-cutting**
- **acid rain**
- **smog**
- **eutrophication**
- **aqueduct**

TAKING NOTES: *Key Ideas and Details*

SUMMARIZING Use a web diagram like the one below to take notes as you read about people and their environment in the United States.

IT MATTERS BECAUSE

Although the United States is a land of unparalleled opportunity, its natural resources are not limitless, and its environment is not immune to potential harm. In many ways, modern life—with its ravenous use of natural resources, destruction of habitats, and contamination of the environment—poses a threat to American society. Efforts have occurred among concerned citizens and their federal, state, and local governments to ensure that U.S. resources will continue to exist in the future.

Managing Resources

GUIDING QUESTION *Why are water and timber resources in the United States in need of responsible management?*

Forests are one of the United States's major natural resources. However, **clear-cutting**, or the removal of whole forests when harvesting timber, occurs in many areas today. Clear-cutting has destroyed much of the country's old-growth forests. As a result, forest ecosystems are less diverse. In addition, wildlife is endangered and the land is subject to erosion and flooding.

In addition to threats posed by the destruction of the forests, people in some areas of the country also face water shortages and groundwater depletion. This is due partly to the fact that people in the United States consume much more freshwater than people in any other country. People use water in all aspects of daily life—at home and in manufacturing, energy production, and agriculture. The Environmental Protection Agency (EPA) calculates that each U.S. home uses an average of 400 gallons (1515 l) of water daily. The EPA also estimates that industry accounts for 46 percent of water usage overall.

Pollution threatens many wetland areas, which include marshes, ponds, and swamps. Wetlands also disappear when they are converted to agricultural or urban land uses. Wetlands are important because they hold valuable water supplies and fisheries and in many cases buffer

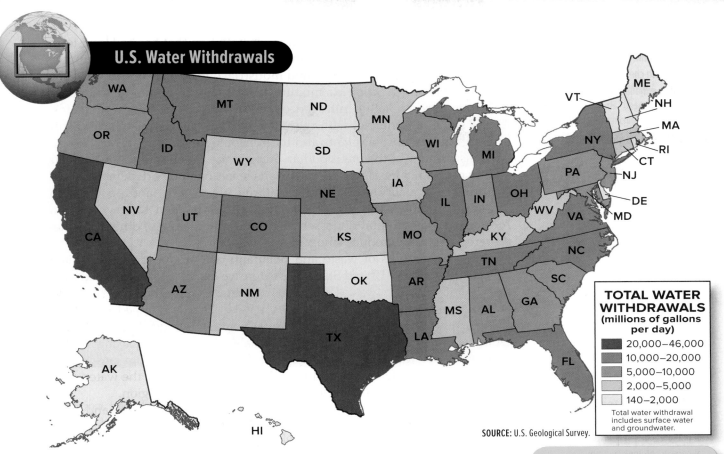

U.S. Water Withdrawals

TOTAL WATER WITHDRAWALS (millions of gallons per day)

- 20,000–46,000
- 10,000–20,000
- 5,000–10,000
- 2,000–5,000
- 140–2,000

Total water withdrawal includes surface water and groundwater.

SOURCE: U.S. Geological Survey.

---

coastal areas from storms and floods. In New Orleans, Louisiana, for example, the building of levees, or raised embankments, around the city has destroyed wetlands that once protected the area from flooding.

Successful resource management must include understanding and respecting the balances that exist in natural ecosystems. Overfishing, which occurs when the number of fish that are caught exceeds the number that can be resupplied by natural reproduction, has depleted many of the region's fisheries. The accidental or deliberate introduction of non-native plant and animal species, on the rise because of increased global travel and trade, also causes environmental problems. These include blocked waterways, crop destruction, and displacement of crucial native species. Efforts to reverse the damage to the environment have begun, but the country has a long way to go toward achieving the sustainable use of its natural resources.

✓ **READING PROGRESS CHECK**

Explaining What are two causes of wetland habitat destruction in the United States?

Human Impact

GUIDING QUESTION *How can human activity lead to air and water pollution?*

While economic growth and industrial development have dramatically improved the standard of living in the United States, an unfortunate consequence has been the polluting of the air and water. **Acid rain**, precipitation carrying high amounts of acidic material, affects a large area of the eastern United States. Acid rain corrodes stone and metal buildings, damages crops, and pollutes the soil. It is especially damaging to the region's waters, as plant life and fish cannot survive in highly acidic waters. Over time, lakes can become biologically dead, or unable to support most organisms.

clear-cutting the removal of all trees in a stand of timber

acid rain precipitation carrying large amounts of dissolved acids, which kills wildlife and damages buildings, forests, and crops

Directions: On a separate sheet of paper, answer the questions below. Make sure you read carefully and answer all parts of the questions.

Applying Map Skills

Use the Unit 2 Atlas to answer the following questions.

16 *Physical Systems* Describe the general differences in elevation you encounter as you travel from California east to Oklahoma.

17 *Places and Regions* Using your mental map of the United States, list the bodies of water that surround Florida.

18 *Human Systems* What vegetation type is found in the most central part of the United States?

Exploring the Essential Question

19 *Making Connections* Imagine you are writing a science fiction novel about a person who has the power to change history by going back in time to sources that contributed substantially to two types of major pollution in the United States. Write a summary of the plot in your novel to explain the changes you would make to humans' actions—and the impact of these changes.

Research and Presentation

20 *Research Skills* Use the Internet and other resources to gather information about the effects of NAFTA on the physical environment of member countries. What cooperation issues other than trade does NAFTA cover? Create a multimedia presentation to share your findings.

Writing About Geography

21 *Informative* Use standard grammar, spelling, sentence structure, and punctuation to write a one-page essay that explains the effects of a specific type of pollution on your town or state. Discuss steps your town or state is taking to alleviate the problems.

DBQ Analyzing Primary Sources

Use the document to answer the following questions.

In the United States today, years of industrial emissions, automobile exhaust gases, and the extraction of natural resources have taken their toll on the environment. Curbing dependency on nonrenewable energy sources is an important issue. More and more people are exploring alternative energy sources that are both clean and renewable.

PRIMARY SOURCE

"Freedom! I stand in a cluttered room surrounded by the debris of electrical enthusiasm: wire peelings, snippets of copper, yellow connectors, insulated pliers. For me these are the tools of freedom. I have just installed a dozen solar panels on my roof, and they work. A meter shows that 1,285 watts of power are blasting straight from the sun into my system, charging my batteries, cooling my refrigerator, humming through my computer, liberating my life.

The euphoria of energy freedom is addictive. . . . Maybe that's because for me, as for most Americans, one energy crisis or another has shadowed most of the past three decades."
—Michael Parfit, "Powering the Future," *National Geographic,* August 2005

22 *Analyzing* What is the author installing and what will they be used for?

23 *Interpreting* What did Parfit mean by " the euphoria of energy freedom"?

Need Extra Help?

If You've Missed Question	**16**	**17**	**18**	**19**	**20**	**21**	**22**	**23**
Go to page	112	112	114	136	138	135	140	140

Canada

ESSENTIAL QUESTION • *How do physical systems and human systems shape a place?*

CHAPTER 6

Why Geography Matters
Energy Resources and Indigenous Rights

Lesson 1
Physical Geography of Canada

Lesson 2
Human Geography of Canada

Lesson 3
People and Their Environment: Canada

Geography Matters...

Although Canada borders the United States, many Americans do not know much about our neighbor. The world's second-largest country in land area, Canada stretches across the northern half of North America. The vast land is rich in resources, especially energy resources. It has many climate regions, each with different landscapes and peoples. Most Canadians, however, live in the southern part of the country, near the border with the United States.

◀ A Canadian woman plays ice hockey on a pond. Hockey is a popular sport in Canada.

Alexander Nicholson/Taxi/Getty Images

141

energy resources
and indigenous rights

*As an energy-rich country, **Canada** has massive coal reserves and enough petroleum and natural gas to supply its population. Extracting these resources, however, affects local populations. In Canada's northern frontier, this often pits the rights of the indigenous peoples who live there against those involved in the extraction.*

What are the economic benefits of extracting energy resources?

What are the human and environmental impacts of extracting energy resources?

How have groups worked together to address this issue?

Canada has the third-largest proven oil reserves in the world after Saudi Arabia and Venezuela. Canada's economy depends on the United States as an export market. Pipelines connect Canadian oil production regions with refining and export centers in the United States. Canada also has a surplus of natural gas and electricity available to export.

Energy reserves fuel economic growth. Besides profiting from exports, the Canadian economy benefits from the jobs created by the extraction of fossil fuels. Canadian households and businesses also enjoy access to an abundant supply of affordable energy. All of these factors boost the country's economic growth.

1. **Environment and Society** Write a paragraph making the case for the extraction of an energy resource.

The three main sources of oil production in Canada are the oil sands of Alberta, the conventional resources in western Canada, and the offshore oil fields in the Atlantic. The oil sands currently produce the majority of Canadian oil. These "tar sands" require traditional pit mining for surface deposits, but extracting deeper reserves involves the injection of steam underground. Oil sands extraction is very energy intensive and poses environmental risks, such as oil spills from pipelines.

The cost of extraction to indigenous communities has generally exceeded any benefit they receive. Land usually cannot be used for agriculture after extraction. The wildlife upon which many indigenous peoples rely for food may also be harmed by extraction. Indigenous peoples have even been forced to leave their land so companies could use the land for resource extraction.

2. **Human Systems** Describe the causes of conflict between resource extraction companies and indigenous peoples.

The Canadian government has addressed land rights of some indigenous communities for years. Currently, the government has ownership of their reserves. Indian Oil and Gas Canada is a government organization that manages and regulates oil and gas resources on indigenous peoples' lands.

The government has also cooperated with indigenous peoples to address conflicts and concerns over land rights. In 2012 it held hearings over a challenge by the Athabasca Chipewyan to a tar sands project proposed by Shell Oil Canada. The hearings provided an opportunity for the Athabasca Chipewyan to voice their concerns about the environmental damage the project might cause to their lands.

3. **Human Systems** How can groups work together to ease the negative impacts of resource extraction?

A CLOSER LOOK:
ALBERTA

The Canadian province of Alberta is the source of most of Canada's mineral wealth. In addition to oil sands, it leads the nation in production of coal, natural gas, and crude oil. Alongside its resources, Alberta also has one of the largest indigenous populations in Canada. Nearly 200,000 indigenous people call Alberta home.

	CRUDE OIL & TAR SANDS
	COAL
	NATURAL GAS
▲	INDIGENOUS SETTLEMENTS

THE INDIGENOUS POPULATION

BLOOD TRIBE	SADDLE LAKE FIRST NATION	SAMSON CREE NATION	SIKSIKA NATION	LITTLE RED RIVER CREE NATION
7,555	5,883	5,550	3,634	3,341

Energy and Indigenous Rights in Alberta

There are over 5,000 square miles of indigenous reserves and settlements in Alberta. This, along with Alberta's vast mineral reserves, has led to some disputes over land rights and resource ownership. To attempt to resolve these disputes fairly, Alberta developed Canada's first mineral rights consultation policy in 2005.

CANADA'S ENERGY PRODUCTION

VERY LOW   LOW   MED   HIGH   VERY HIGH

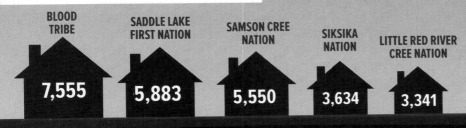

TAR SANDS	**1.5** MILLION BARRELS PER DAY	**#1** RANK IN THE WORLD
OIL	**4.3** MILLION BARRELS PER DAY	**#5** RANK IN THE WORLD
COAL	**68.9** MILLION METRIC TONNES PER YEAR	**#12** RANK IN THE WORLD
NATURAL GAS	**15.68** BILLION CUBIC FEET PER DAY	**#5** RANK IN THE WORLD

Reading **HELP**DESK

Academic Vocabulary

- series
- controversy

Content Vocabulary

- **timberline**
- **chinook**
- **tar sands**
- **fishery**
- **overfishing**
- **aquaculture**

TAKING NOTES: *Key Ideas and Details*

CATEGORIZING As you read the lesson, use a graphic organizer like the one below to take notes on the physical geography of Canada.

Physical Geography of Canada		
Landforms	Water Systems	Climate, Biomes, and Resources

LESSON 1
Physical Geography of Canada

ESSENTIAL QUESTION • *How do physical systems and human systems shape a place?*

IT MATTERS BECAUSE

Canada is a country of great physical variety and natural wealth. This wealth includes breathtaking landforms shaped by water, wind, ice, and tectonic activity over millions of years. It also includes abundant water and energy resources as well as a variety of wildlife.

Landforms

GUIDING QUESTION *How do landforms link the geography of Canada and the United States?*

Canada covers approximately the northern third of North America, stretching from the Pacific to the Atlantic. It is made up of 3 territories and 10 provinces. A province is a political unit similar to a state.

Mountains on Canada's eastern and western edges cradle a central region of plains. When people first arrived on the plains, they found a sea of grass and dark, fertile soil that later became some of the world's most productive farmland. To the east of the plains stand the rocks of the Laurentian Highlands, the Canadian Shield, and the ancient, rounded Appalachian Mountains. To the west are the younger Rocky Mountains. A variety of climates are found in Canada, from the frozen tundra and cold, subarctic climates in the north to the steppe and humid continental climates along the border with the United States.

The western third of Canada is mountainous. Collisions between tectonic plates millions of years ago thrust up a **series** of sharp-peaked mountains called the Pacific Ranges. These ranges include the Cascade Range and the Coast Range. Mount Fairweather, on the border of Alaska and western British Columbia, at 15,300 feet (4,663 m) is one of the tallest coastal mountains in the world.

Like the Pacific Ranges, the Rocky Mountains farther east grew as geologic forces heaved slabs of rock upward. The Rocky Mountains link the United States and Canada, stretching more than 3,000 miles (4,828 km) from New Mexico to Alaska. The Canadian Rockies extend northward from the Rockies in the United States.

East of the Rockies, the land falls in elevation and flattens into the Great Plains and Interior Lowlands, which extend across the center of the region and south into the United States. These plains and lowlands are generally flat, but diverse in terms of landscape types. They can include hills, escarpments, low mountains, forests, and river valleys. Water from these lowlands drains into the Atlantic Ocean through the St. Lawrence River. The region was shaped by glacial activity during the last ice age. It therefore has deep arable soil that makes it important for agriculture.

At the eastern edge of the Interior Lowlands lies the Canadian Shield. The Canadian Shield is a giant core of rock anchoring the continent and centered on Hudson Bay and James Bay. This stony land makes up the eastern half of Canada and parts of the northeastern United States. Erosion and glacial ice from the last ice age have smoothed the surface and created lakes, rivers, and streams.

The heavily eroded Appalachian Mountains are North America's oldest mountains. They extend from Canada's Maritime Provinces south through the eastern United States to the state of Alabama. The Appalachians were formed by tectonic plate movements. Over time they were shaped by ice and running water. Coastal lowlands lie to the east of the Appalachians.

☑ **READING PROGRESS CHECK**

Making Connections Why are landforms an important part of Canada's geography?

Water Systems

GUIDING QUESTION *How are water systems important to the Canadian economy?*

Freshwater lakes and rivers have helped make Canada prosperous. Abundant water satisfies the needs of cities and rural areas. It also provides power for homes and industries and moves resources across the country. The Mackenzie River, which flows from the Great Slave Lake to the Arctic Ocean, drains much of Canada's northern interior. The Fraser River flows southwest and drains into the Pacific Ocean just south of Vancouver.

series a number of things of the same type following one after the other in space or time

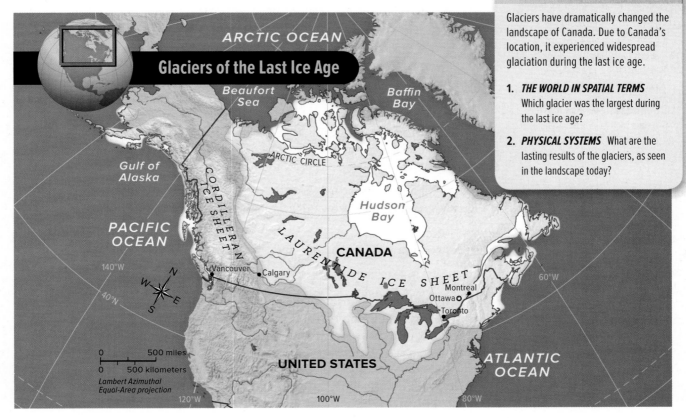

GEOGRAPHY CONNECTION

Glaciers have dramatically changed the landscape of Canada. Due to Canada's location, it experienced widespread glaciation during the last ice age.

1. ***THE WORLD IN SPATIAL TERMS*** Which glacier was the largest during the last ice age?

2. ***PHYSICAL SYSTEMS*** What are the lasting results of the glaciers, as seen in the landscape today?

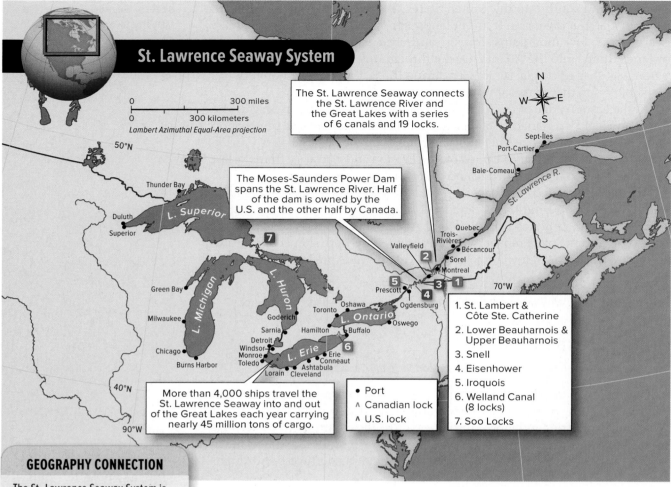

St. Lawrence Seaway System

The St. Lawrence Seaway connects the St. Lawrence River and the Great Lakes with a series of 6 canals and 19 locks.

The Moses-Saunders Power Dam spans the St. Lawrence River. Half of the dam is owned by the U.S. and the other half by Canada.

More than 4,000 ships travel the St. Lawrence Seaway into and out of the Great Lakes each year carrying nearly 45 million tons of cargo.

1. St. Lambert & Côte Ste. Catherine
2. Lower Beauharnois & Upper Beauharnois
3. Snell
4. Eisenhower
5. Iroquois
6. Welland Canal (8 locks)
7. Soo Locks

- Port
- ∧ Canadian lock
- ∧ U.S. lock

0 — 300 miles
0 — 300 kilometers
Lambert Azimuthal Equal-Area projection

GEOGRAPHY CONNECTION

The St. Lawrence Seaway System is an important link in Canada's transportation network.

1. **PLACES AND REGIONS** List the bodies of water that are connected by the St. Lawrence Seaway System.

2. **HUMAN SYSTEMS** How does the system contribute to the economy of the region?

The St. Lawrence River flows about 760 miles (1,223 km) from Lake Ontario to the Gulf of St. Lawrence in the Atlantic. It forms part of Canada's border with the United States. The cities of Quebec, Montreal, and Ottawa grew up along the St. Lawrence and its tributaries. Niagara Falls, on the Niagara River, forms another part of the Canada-U.S. border. The falls are a key source of hydroelectric power.

In northern Canada, naturally occurring dams created by glacial ice formed Great Bear Lake and Great Slave Lake. Elsewhere, moving glaciers tore at the earth, leaving behind basins that became the Great Lakes. Deposits of coal, iron ore, and other minerals nearby favored industrial development and urban growth here.

☑ **READING PROGRESS CHECK**

Identifying Central Issues How does water help the Canadian economy?

Climate, Biomes, and Resources

GUIDING QUESTION *What factors cause variations in climate and vegetation in Canada?*

Due to its great expanse of latitude, Canada experiences a large variation in climate and vegetation types. It varies from the bitter cold of the high-latitude areas to the radically changing seasons of the interior regions. Canada is characterized by biomes common to the midlatitude and high latitude.

Canada is rarely affected by natural weather hazards. The waters off the Pacific and Atlantic coasts are too cold to support hurricanes. Also, the extreme temperatures needed for thunderstorms, tornadoes, and hail are less common.

Climate Regions and Biomes

Ocean currents play a key role in Canada's climates. The Gulf Stream, for example, is a warm, northward-flowing ocean current off the southeastern coast of Canada. It moderates coastal temperatures and carries nutrients as far north as Newfoundland. Merging with the colder air and water of the southward-flowing Labrador Current off eastern Canada, the potential for fog-bound coastal conditions is always present. Off the Pacific coast, the cold Alaska Current flows southward and parallel to British Columbia. It churns up nutrients from the ocean floor, inviting whales to follow a moving feast during their migrations.

Climate and vegetation are variable in the southern third of Canada—from about 40° N to 50° N latitude. The area's humid continental climate ranges from hot and humid to cool and wet. The farther north one travels, the more severe and snowy the winters become, with shorter and cooler summers. Coniferous evergreen trees tend to outnumber deciduous trees in the northernmost areas.

Canada's Pacific coast has a marine west coast climate. The Pacific Ranges force moist ocean air upward, where it cools and releases moisture, causing heavy rainfall. Winters are overcast and rainy. Summers are cloudless and cool, appealing to many varieties of warblers and other small birds that go north to nest. Ferns, mosses, and coniferous forests grow here. The soils in Canada's forests are quite acidic. This is due to the leaching (removal) of minerals out of the topsoil during rainfall. With fewer minerals, the soils are unfit for agriculture. The exceptions are the soils in the mixed and deciduous forests.

Areas between the Pacific Ranges and Rocky Mountains have a drier climate due to the rain shadow effect. This area has a semi-arid (steppe) climate, with grasslands and coniferous forests. Spruce and fir trees cover the middle elevations of the ranges. Beyond the **timberline**, the elevation above which trees cannot grow, lichens (LY• kuhns) and mosses are found. These higher elevations of the Canadian Rockies and Pacific Ranges are high-latitude climate regions. Rocky Mountain sheep, mountain goats, elk, mule deer, and black bears are common in the southern latitudes of these mountains. The semi-arid (steppe) climate predominates east of the Rocky Mountains, again because of the rain shadow effect. In addition, in spring, a warm, dry wind called the **chinook** (shuh•NUK) blows down the eastern slopes of the Rockies and melts the snow.

In the north, there are subarctic, tundra, and ice cap climates. In these high-latitude climates, freezing temperatures are common all year because of a lack of direct sunlight. The amount and variety of vegetation is limited.

Boreal forest, also known as taiga, occupies the bulk of Canada's northern areas. It is the world's second-largest area of uninterrupted forest. The boreal forest has long winters and moderate to high precipitation each year. Covered mostly with coniferous evergreen trees, the region is an important source of pulpwood and lumber. The forest is also home to moose, black bears, beavers, Canada lynx, wolves, snowshoe hares, Canada jays, blue jays, crows, and ravens.

Just south of the Arctic Circle lie the subarctic climate regions. These regions have Earth's widest temperature ranges, varying by as much as 120°F (49°C) from their bitterly cold winters to cool, short summers. In parts, only a thin layer of surface soil thaws each summer. Below is permanently frozen subsoil, or permafrost.

timberline elevation above which it is too cold for trees to grow

chinook a seasonal warm wind that blows down the Rockies in late winter and early spring

tar sands sand or sandstone naturally impregnated with petroleum

fishery an area in which fish or sea animals are caught

A bull moose grazes in Bowron Lake Park in British Columbia.

▼ CRITICAL THINKING

1. *Analyzing Visuals* What features of the moose make it suited for its biome?

2. *Speculating* How might large populations of moose be a threat to the forests in some parts of Canada?

Using aquaculture, fish can be raised in concrete or earthen ponds, coastal pools, or floating cages in the open ocean, where they are protected from predators and environmental hazards. Aquaculture is an important industry in Canada. People raise Atlantic salmon, cod, tilapia, rainbow trout, and more. The industry provides income and jobs and can help meet the increasing global demand for seafood and restore threatened wild populations of species.

HYPOTHESIZING How might aquaculture provide economic benefits for countries lacking land-based resources for international trade?

overfishing harvesting fish to the point that species are depleted and the value of the fishery reduced

aquaculture the cultivation of seafood

controversy the presence of opposing views

A few needled evergreens survive here, along with herds of woodland caribou that are endangered in some areas.

Closer to the North Pole is the tundra climate region. Winter darkness and bitter cold last for months, and the brief summer period brings only limited warming effects. The layer of thawed soil is even thinner here than in the subarctic, with vegetation limited to low bushes, short grasses, mosses, and lichens. Winter temperatures can fall to –70°F (–57°C).

Natural Resources

Abundant natural resources such as energy, minerals, timber, and fish have made Canada wealthy. Their extraction, however, has led to depletion and environmental problems. Energy resources include coal, petroleum, and natural gas. Canada's petroleum and natural gas reserves lie largely in or near Alberta. The Athabasca **Tar Sands** contain large deposits of extremely heavy crude oil, much of it in semisolid form. Large inputs of energy and water are required to transform these substances into synthetic crude oil for use by humans.

Mineral resources are also plentiful in Canada. The Rocky Mountains yield gold, silver, and copper. Parts of the Canadian Shield are rich in iron ore and nickel. Canada's mineral inventory includes more than one-third of the world's production of potash (a mineral salt used in fertilizers), 4 percent of its copper and gold, and 5 percent of its silver. Because mining involves heavy equipment, uses large quantities of water, and moves a great deal of rock and other natural materials, it can damage land, water, and air systems.

Timber is a vital resource in Canada. Trees once covered much of Canada. Today, forests cover only about 34 percent of the country. Trees are a renewable resource, but only if people take steps to protect forests and their ecosystems. Cutting down trees can have a negative effect on the other plants and animals. Many species have been lost. Positive efforts to preserve forests include replanting trees, cooperating to protect the many species of native forest animals facing extinction, and preserving old-growth forests.

The coastal Atlantic and Pacific waters have long been important **fisheries**, or places for catching fish and other sea animals. The Grand Banks, once one of the world's richest fishing grounds, covers about 139,000 square miles (360,000 sq. km) of Canada's southeast coast. In recent years, **overfishing** has caused fish stocks to drop dangerously. Canada is now working to protect key species. **Aquaculture**, or fish farming, is a growing economic activity. This has caused some **controversy**, as crowded conditions can lead to disease among farmed species. The industry is working to minimize risks to consumers.

✅ **READING PROGRESS CHECK**

Specifying What are Canada's natural energy resources?

LESSON 1 REVIEW

Reviewing Vocabulary
1. *Summarizing* Explain the significance of timberline, chinook, fishery, overfishing, aquaculture, and tar sands.

Using Your Notes
2. *Listing* Using your notes from the graphic organizer, describe the major Canadian landforms.

Answering the Guiding Questions
3. *Applying* How do landforms link the geography of Canada and the United States?

4. *Analyzing* How are water systems important to the Canadian economy?

5. *Identifying* What factors cause variations in climate and vegetation in Canada?

Writing Activity
6. *Informative/Explanatory* Write a short paragraph explaining how tectonic forces have influenced landscapes in Canada.

Reading **HELP**DESK

Academic Vocabulary

- **trace**
- **advocate**

Content Vocabulary

- **Inuit**
- **First Nations**
- **dominion**
- **Quebecois**
- **separatism**
- **Loyalist**
- **emigrate**

TAKING NOTES: *Key Ideas and Details*

IDENTIFYING As you read the lesson, use a graphic organizer like the one below to take notes on the human geography of Canada.

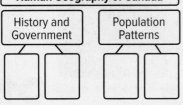

Human Geography of Canada

History and Government — Population Patterns

LESSON 2
Human Geography of Canada

ESSENTIAL QUESTION • *How do physical systems and human systems shape a place?*

IT MATTERS BECAUSE

Diversity has been important in shaping Canada. The indigenous peoples who have lived on the land for centuries, the French and British colonists who settled the area later, and the waves of immigrants from Europe, Asia, Africa, and Latin America have all influenced the human geography of this country.

History and Government

GUIDING QUESTION *How did British and French cultures influence Canada's history?*

Many Canadians can **trace** their roots back to countries from around the world. Some are descendants of the country's native peoples. Their earliest ancestors arrived in North America thousands of years ago.

Unity, Expansion, and Diversity

About 200,000 native peoples were living in what is now Canada when Europeans arrived off the coast of Newfoundland in 1497. In the next 200 years the native populations declined, as Europeans claimed their lands and diseases from Europe spread. After 1950 high birthrates and access to improved medical care contributed to population growth.

The main indigenous groups that exist in Canada today are the **Inuit**, the Métis, and **First Nations** peoples. The Inuit are the indigenous people of the Canadian Arctic. Métis are persons who have both indigenous and French Canadian ancestry. *First Nations* is a term that refers to indigenous peoples of Canada who are neither Inuit nor Métis. The current population of First Nations peoples is about 700,000 of the 1.25 million indigenous people in Canada today.

French explorers helped establish claims to land in the region in the early 1600s. This land was named New France, and part of it became the province of Quebec. Territorial rivalry between Great Britain and France began in 1670, when the British chartered the Hudson's Bay Company to seek a northwest passage to the Pacific Ocean. A clash of French and British interests along the Atlantic coast led to wars.

Britain and France were first drawn to the North American continent to gain riches from precious metals and beaver pelts that came mainly from the Atlantic coast. Exploration inland soon followed, however.

trace to follow or study in detail or step by step

Inuit a member of the Arctic native peoples of North America; once known as Eskimo

First Nation one of the indigenous peoples of Canada who are neither Inuit nor Métis

dominion a partially self-governing country with close ties to another country

Quebecois Quebec's French-speaking inhabitants

Both European powers sought to claim more resources and more colonies. The British eventually drove the French from the Hudson Bay area, capturing Quebec in 1759 and winning control of New France in 1763. The Quebec Act, passed by the British in 1774, gave French settlers the right to keep their language, religion, and laws. The act also extended Canadian territory south to the Ohio River. This angered American colonists and brought them closer to war with the British.

During the early 1800s, English- and French-speaking communities feuded over colonial government policies. Fears of a takeover by the United States, however, forced both to work together. In 1867 the colonies of Quebec, Ontario, Nova Scotia, and New Brunswick united as provinces of the **Dominion** of Canada, a new country within the British Empire. Manitoba, British Columbia, Alberta, Saskatchewan, Prince Edward Island, and Newfoundland became provinces over the next 100 years.

Canada was created as a dominion, a partially self-governing country with close ties to Great Britain. It gained full independence in 1931, but the British government kept the right to approve changes to Canada's constitution. This legislative link to Great Britain finally ended in 1982 with passage of the Constitution Act. Today, Canada is a constitutional monarchy.

The executive part of Canada's federal government includes the governor-general, the prime minister, and the cabinet. The British monarch serves as the head of state and appoints a governor-general to act in his or her place. The national legislature, or Parliament, includes the Senate and the House of Commons. Canada's prime minister is the actual head of government. Nine judges sit on the Supreme Court of Canada, the country's highest court.

In the 1800s, Canada acquired lands stretching from the Atlantic to the Pacific and from the Arctic to the U.S. border. The British government encouraged immigration to Canada. Between 1815 and 1855, a million people arrived from Great Britain. This made the French-speaking citizens a minority and fueled French nationalism among **Quebecois** (kay•beh•KWAH), Quebec's French-speaking inhabitants. This nationalism has been present throughout Canada's history.

Widespread immigration from other parts of the world began in Canada in the late 1800s. Some came for the Klondike Gold Rush, but many more were attracted by the fertile soil of the prairies of Alberta, Saskatchewan, and Manitoba.

TIME LINE

CANADA
Expansion
and Diversity ➔

Canada's history features steady expansion and multiculturalism.

▶ **CRITICAL THINKING**

1. *Identifying* What are two events that show how or why Canada's settlement expanded?

2. *Explaining* How has Canada attempted to address past mistreatment of native peoples?

1776 ➔

1776 Loyalist refugees from the American War of Independence settle in Canada.

1800s Thousands of immigrants from England, Scotland, and Ireland arrive each year.

1885 Canadian Pacific Railroad is completed.

1898 Gold rush along upper Yukon River; Yukon Territory is given separate status.

➔**1900**

Immigrants from Germany, Scandinavia, Ukraine, Japan, and China arrived to settle the land. In the 1800s Canada also sheltered African Americans who had escaped slavery in the United States. Canada never practiced slavery. These refugees, many of whom escaped via the Underground Railroad, were safely beyond the reach of American laws once they arrived in Canada.

Westward expansion in Canada came at a price, however, as immigrants pushed First Nations peoples off their lands. The injustice was formally recognized in 1998, when the Canadian government apologized to native peoples for their mistreatment. The government established a "healing fund" to make reparations.

In the 1900s Canada became an industrialized, urban country. Mineral resources were utilized, and hydroelectric projects and transportation systems were developed. World War II stimulated the economy, making it a crucial military and industrial power. After the war, Canada sought to improve federal assistance to its citizens through pensions, unemployment insurance, and medical care.

Modern Challenges

The United States has more trade with Canada than any other country. Being neighbors and sharing a border has permitted international trade that is beneficial to the economies of both countries. In 1994 the North American Free Trade Agreement (NAFTA) eliminated tariffs and other trade barriers between Canada, Mexico, and the United States. Although the open border and a history of cooperation have benefited both countries, some Canadians continue to dislike the effect free trade has had on their culture. Canadians struggle to maintain a separate identity while being bombarded by U.S. popular culture.

Conflicts continue as French-speaking Canadians seek greater protection for their language and culture. Many desire Quebec's independence and strongly support **separatism**—the breaking away of one part of a country to create a separate, independent country. In 2012 members from the Parti Québécois, a political party in Quebec that **advocates**—or publicly supports—separatism, were elected to political office in Quebec.

separatism the breaking away of one part of a country to create a separate, independent country

advocate to publicly recommend or support

✔️ **READING PROGRESS CHECK**

Identifying What factors led to the decline of the native populations of Canada?

Great Britain transfers final legal powers to Canada. A new constitution is adopted.

1982

1998

A referendum rejects Quebec's independence from Canada by 1 percent.

Truth and Reconciliation Commission begins hearings about past policies harmful to indigenous peoples and their cultural identities.

2010

1992

North American Free Trade Agreement (NAFTA) enacted

1999

Nunavut is formed, becoming the first Canadian territory with a majority indigenous population.

➡️**2012**

2012

Prime Minister Harper calls for fresh start in government relations with First Nations peoples.

Population Patterns

GUIDING QUESTION *What explains Canada's diverse mix of people and settlement patterns?*

Canada is a highly developed country with bustling cities. It also has sparsely populated areas of beautiful, pristine wilderness. These rugged natural areas sometimes have difficulty supporting communities. But they are as much a part of the country's cultural identity as are the busy metropolitan areas.

Nunavut is the cultural and economic homeland for the Inuit people. Today, about 60,000 Inuit from the Arctic live mostly in settlements of 25 to 500 people, with just a few larger towns. In recent decades, mining, oil exploration, and pipeline construction, along with a decline in the demand for fur, have increased their reliance on government services.

Immigrants to Canada came in search of political and religious freedom, economic and educational opportunities, and refuge from wars. For example, **Loyalists**, or American colonists who remained loyal to the British government, fled to Canada after the American Revolution. They settled in the Maritime Provinces of Nova Scotia, New Brunswick, and Prince Edward Island. Some immigrant groups settled in areas that let them keep their familiar ways of life.

The ethnic origins of Canadians vary from province to province. Today, more than one-fourth of Canadians identify themselves as being of mixed ethnic origins. In addition, more than 1 million or 4 percent of all Canadians identify themselves as of North American Indian, Inuit, or Métis ancestry.

Rugged terrain and a cold arctic climate with a short growing season make much of Canada difficult for human settlement. About 90 percent of the population lives within 100 miles (160 km) of the U.S.-Canada border. Average population density is about 10 people per square mile (4 people per sq. km). More densely populated areas are clustered near the coasts, the Great Lakes, and in places that support agriculture, fishing, and trade. Over the past 100 years, most internal migration has been westward to the Prairie Provinces of Manitoba, Saskatchewan, and Alberta. This move was due in part to the discovery of oil and natural gas since the 1960s.

According to 2015 estimates, nearly 70 percent of the country's population is between the ages of 15 and 64. Since Canadians enjoy a long life expectancy and low infant mortality rates, Canada has an aging population. This presents economic challenges, as a larger share of the population reaches retirement and requires more government-funded health care.

Approximately 81 percent of Canada's 35 million inhabitants live in urban areas. These urban areas developed in places with intensive commercial agriculture in the Great Lakes and St. Lawrence lowlands areas. Smaller settlements were merged into larger metropolitan areas. Today, these cities are important centers for commerce, education, and transportation. As Canada's largest city, Toronto is an industrial and financial center. Montreal is an industrial and shipping center. On the Pacific coast, Vancouver handles nearly all of the trade between Canada and Asia. Edmonton grew with the development of the petroleum industry.

Loyalist an American colonist who remained loyal to the British government

Nunavut, home to many Inuit people, was established as a separate territory in 1999.

▼ **CRITICAL THINKING**

1. Describing What kinds of settlements do many Inuit people live in?

2. Drawing Conclusions Why do you think it was important for the Inuit people to have Nunavut declared a separate territory?

✓ **READING PROGRESS CHECK**

Paraphrasing How has Canada's physical geography influenced its settlement patterns?

Yvette Cardozo/Getty Images

Society and Culture Today

GUIDING QUESTION *Why is Canada often called a multicultural society?*

Canada has a truly multicultural society, largely due to immigration. There are many reasons people decide to **emigrate**, or leave their countries to settle in another. Canada's high standard of living and government policies protecting multiculturalism are appealing to people of diverse ethnic origins. Historically these immigrants have arrived primarily from Europe. Recent migration patterns, however, feature large increases in immigration from Asia and Central and South America, and an overall decrease in immigration from Europe.

Language and religion reflect Canada's diverse immigrant population. There are two official languages—English and French. The British introduced English to most of Canada. In the province of Quebec, however, French was established as the dominant language. Languages spoken also include German, Italian, and Chinese. Native languages include Cree and Inuktitut, the language of the Inuit. Christians make up the largest religious group in Canada. Other religions practiced in Canada by small minorities include Islam, Buddhism, Hinduism, Judaism, and Sikhism.

Education and health care are supported by the government. The literacy rate in Canada is 99 percent, and there is an extensive network of public and private schools. Attending school is required for children ages 6 to 16. Each province is responsible for organizing and administering public education. The federal government sets standards, with each province responsible for financing and managing its own system. In some cases, rising costs resulting in part from an aging population have created a need to limit benefits or raise taxes.

Family and Status of Women

Today, the average Canadian family contains three members. Two parents live in 39 percent of Canadian families, but the percentage is decreasing just as in the United States. The overall composition of the Canadian family is changing and families have fewer children. Since women are increasingly joining the Canadian workforce, there are a growing number of dual-income households.

The government of Canada has passed several laws protecting the rights of women. The Status of Women in Canada (SWC) was created in 1971 to increase women's participation and equality in all aspects of life. In 1982 the Canadian Charter of Rights and Freedom ensured gender equality in employment, public life, and education. Canadian women have the same literacy rate as men and outnumber men in getting secondary and university-level educations. They still, however, face barriers to equality in the labor market.

The Arts

The arts in Canada have been influenced primarily by native and European cultures. For example, First Nations artists combine old and modern techniques with traditional color schemes and motifs that often relate to the natural world. In the twentieth century, museums and scholars began to focus on art created by native peoples. Immigration has also added distinctive features to Canadian art.

French explorers, missionaries, and settlers wrote the earliest Canadian literature. Their writings had strong historical and religious themes. Popular modern Canadian fiction writers include Margaret Atwood, author of *The Handmaid's Tale,* and Yann Martel, author of *Life of Pi.*

Toronto is highly regarded in the realms of theater and music. It is the third-largest production center in the English-speaking world, after London and New York City. The world-renowned Toronto Symphony Orchestra and the top-ranked National Ballet of Canada call Toronto home.

Analyzing PRIMARY SOURCES

Canada Slips in Gender Equality Rankings

"Canada is losing ground in a key global measure of gender equality, sliding out of the world's top 20 chiefly due to a lack of female representation in politics.

'In the future, talent will be more important than capital or anything else,' said Klaus Schwab, the forum's [World Economic Forum] founder and executive chairman. 'To develop the gender dimension is not just a question of equality; it is the entry card to succeed and prosper in an ever more competitive world.' "

—Tavia Grant, *The Globe and Mail*, October 24, 2012

DBQ *IDENTIFYING CENTRAL ISSUES* What is the chief cause of Canada's slide from the world's top 20 in gender equality?

emigrate to leave one's own country to settle permanently in another

Canadian industries generally faced some difficult times as a result of the 2008 global recession, but have since shown signs of bouncing back.

▶ **CRITICAL THINKING**

1. *Contrasting* Which industry's performance is most unlike the others?

2. *Summarizing* What has been the general trend in Canadian industry in recent years?

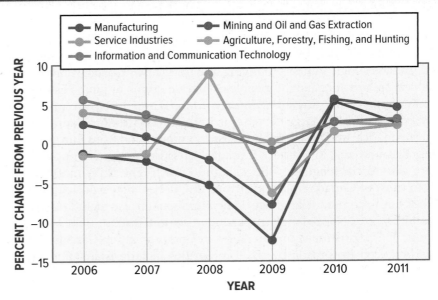

Source: Statistics Canada, 2012

Canada is also at the forefront of contemporary art. The "Vancouver School" artists focus on photoconceptualism, and the Nova Scotia College of Art and Design is known for innovation in photography and graphic design.

☑ READING PROGRESS CHECK

Explaining How has the government worked to improve the status of women?

Economic Activities

GUIDING QUESTION *What factors define Canada as a postindustrial market economy?*

Like the United States, Canada maintains a market economy. Today more than 50 percent of Canada's exports were made up of agricultural, energy, forestry, and mining products. Thus, natural resources are essential to the Canadian economy. Canada also has the third-largest oil reserves in the world and makes a substantial amount of money from exporting oil and natural gas.

Resources, Power, and Industry

Canada is a leader in the production of uranium, iron ore, coal, petroleum, copper, and silver. Canada is the world's fourth-largest producer of hydroelectricity, relies heavily on coal-fueled power, and can meet its own petroleum needs while still having a surplus of oil and natural gas. These surpluses are exported worldwide. The tar sands in Alberta are a major source of crude oil.

Natural resources and agriculture make up over 50 percent of Canada's exports. Despite this, those economic activities provide Canada with only about 4 percent of its GDP. Services such as transportation and communication, retail, health care, and others provide about 70 percent of GDP. The Canadian economy produces goods and services at home as well as being linked to the global economy through imports, exports, and trade agreements such as NAFTA.

Canada continues to be one of the world's important suppliers of agricultural products, specifically wheat, corn, and other grain crops. Canada's Prairie Provinces are major grain producers. They are also home to many cattle ranches.

The fishing industry used to heavily support the economy of the Atlantic coast. In the past few decades, however, overfishing has reduced this industry's impact. Forestry is important for certain provinces, such as British Columbia.

The last few decades have seen a rapid growth of high-tech and electronics industries. Canada has 32 million Internet users—89 percent of its population. Another high-tech industry in Canada is its space and aircraft industry.

The development of reliable transportation systems has been essential to the economic growth of Canada due to its large land area. The majority of Canada's transportation systems are located in the southern part of the country. The Trans-Canada Highway runs 4,860 miles (7,821 km) from Victoria, British Columbia, to St. John's, Newfoundland. Communication networks have also promoted development of the economy. Cellular and digital services using satellites have made telephone communication more accessible in distant places. Business transactions and personal communications can be completed instantaneously using e-mail and the Internet.

The Economy Today

Since Canada's economy is linked to the global market, it suffered some losses in the 2008 global recession. However, it was not as seriously affected as the United States. This was largely because Canadian banks were more conservative in extending credit. The housing market stayed healthy as a result. Also, Canada's federal deficit is one of the smallest in the Western world. Unemployment rates have been generally lower than those in the United States and much of Europe.

Trade is still central to the Canadian economy. In addition to forestry products, leading exports are automobiles, auto parts, crude petroleum, natural gas, electricity, aluminum, machinery, and equipment. Canada tends to import manufactured goods, such as chemical products, textiles, and foods. Exports to the United States amount to over half of all Canadian trade. Canada exports more to the United States than it imports from it, resulting in a trade surplus.

While natural resources and trade are important to Canada's economy, the service sector employs more people than all other sectors combined, with 76 percent of the over 19 million Canadians in the labor force. For this reason, Canada is considered a postindustrial society. The service industry includes retail jobs in the increasingly common large chain stores such as Walmart and Best Buy. Tourism is also a fast-growing service industry because Canada offers natural and cultural diversity in addition to national parks and historic sites.

☑ READING PROGRESS CHECK

Analyzing Why was Canada less affected than the United States by the 2008 global recession?

LESSON 2 REVIEW

Reviewing Vocabulary

1. *Making Connections* Use the following vocabulary terms in a paragraph about the early history of Canada: First Nation, Inuit, Loyalist, and dominion.

Using Your Notes

2. *Describing* Use your graphic organizer to describe the ethnic makeup of Canada today.

Answering the Guiding Questions

3. *Sequencing Information* How did British and French cultures influence Canada's history?

4. *Outlining* What explains Canada's diverse mix of people and settlement patterns?

5. *Inferring* Why is Canada often called a multicultural society?

6. *Determining Importance* What factors define Canada as a postindustrial market economy?

Writing Activity

7. *Narrative* Suppose you are an immigrant writing a letter to relatives about your new home in Canada. Explain your reasons for settling where you live.

Two Decades of NAFTA

In 1994 Canada, the United States, and Mexico signed the North American Free Trade Agreement (NAFTA) to reduce or eliminate tariffs on many goods traded among the three countries. Since its implementation, NAFTA has had mixed results. While it has helped to increase trade throughout North America, the gains for the three countries have not been equal. Supporters of NAFTA point to lower product prices and increased industrial integration among the countries. Opponents of the agreement argue it has led to job displacement and has done little to improve the quality of life in Mexico.

CANADA–U.S.
$315.3 Billion

U.S.–CANADA
$280.9 Billion

MEXICO–CANADA
$24.6 Billion

U.S.–MEXICO
$198.4 Billion

MEXICO–U.S.
$262.9 Billion

CANADA–MEXICO
$5.5 Billion

CANADA

"Few Canadians speak up for enhancing ties with Mexico. But before leaving Ottawa this month, Emilio Goicoechea, Mexico's ambassador, wrote a rebuttal urging Canada to stay the trilateral course. Trade between the two has grown fivefold since 1994. . . . Some Canadian companies have invested in Mexico: Bombardier has factories making aircraft parts and trains, while Scotiabank is Mexico's seventh-biggest bank."

—The Economist
"No mariachis, please," February 12, 2009

UNITED STATES

"[A] worker . . . whose 16.6% tariff was eliminated by Nafta, saw wage growth rate about 11 percentage points less than a worker in a business that didn't depend on high tariffs. So . . . Nafta did hurt workers at the lower end of education spectrum. Blue-collar workers in vulnerable industries suffered large absolute declines in real wages as a result of the agreement. . . ."

—Bob Davis
"The Battle Over NAFTA Continues,"
Wall Street Journal, November 22, 2010

MEXICO

"...NAFTA has made it easier to buy American products in Mexico. I can shop at H-E-B and other stores in Mexico and buy Procter & Gamble and other American-made products that we could not buy in Mexico before NAFTA. When I lived in Mexico and wanted American products before NAFTA, the taxes imposed on American products made them unaffordable."

—Eduardo Bravo
"NAFTA Has Fueled Job Growth,"
SanAntonio Express-News, November 19, 2012

Making Connections

1. **Places and Regions** Which NAFTA member country exports the largest volume of goods to the other two member countries?

2. **Human Systems** How might a person's role in the economy, such as an industrial worker, a business owner, or a policy maker, impact his or her views on NAFTA?

3. **The Uses of Geography** Using what you know about NAFTA, write a one-page letter to the editor of a local newspaper explaining why you think the overall effect of NAFTA has been positive or negative.

*Interact with **Global Connections** Online*

networks

There's More Online!

- ☑ **IMAGE** Human Impact on Wildlife
- ☑ **INFOGRAPHIC** Canada's Boreal Forest
- ☑ **INTERACTIVE SELF-CHECK QUIZ**
- ☑ **VIDEO** People and Their Environment: Canada

CANADA

LESSON 3
People and Their Environment: Canada

ESSENTIAL QUESTION • *How do physical systems and human systems shape a place?*

Reading**HELP**DESK

Academic Vocabulary

- **extract**
- **diminish**

Content Vocabulary

- **old-growth forest**

TAKING NOTES: *Key Ideas and Details*

IDENTIFYING As you read, use a web diagram like the one below to outline issues that arise as a result of interactions between the Canadian people and their environment.

IT MATTERS BECAUSE
While Canada once had an abundance of wildlife and natural resources, human activity has caused many of these resources to decline and some to even disappear. Human activities also cause pollution and contribute to climate change.

Managing Resources

GUIDING QUESTION *How do economic activities in Canada put the country's natural resources at risk?*

Canada engages in numerous economic activities that involve **extracting**, or removing, natural resources. Logging, which involves cutting down trees for human use, has resulted in more trees being cut down in a given amount of time than can grow. For example, a logger can cut down 100 trees in one day. However, it will take decades for the same number of trees to grow back. Logging, when left unchecked, can lead to the complete destruction of forests and the animals that live in those forests.

Canada's boreal forest is one of the largest forest and wetland ecosystems remaining on Earth. It is home to some of the biggest populations of wolves, grizzly bears, and caribou. It also has lakes that support many types of fish and trees that shelter billions of birds.

The boreal forest has been shrinking because of logging, mining, and oil and gas extraction. As of 2013, only 12 percent of it was federally protected. Many nature advocates and conservation groups are calling on the Canadian government to protect more of the boreal forest from these devastating activities, which can deplete the animal and plant populations. Parts of the boreal at high risk are old-growth forests.

Old-growth forests are complex forests that have developed over a long period of time and are relatively untouched by human activity. They are increasingly rare and also need to be protected.

The same is true for wetlands. Home to large numbers of plant and animal species, wetlands are considered one of the most productive ecosystems in the world. Wetlands cover about 14 percent of Canada, mainly in Ontario, Manitoba, and the Northwest Territories. Just like old-growth forests, Canadian wetlands are at risk. Wetlands have been drained so the land can be used for industrial, commercial, and agricultural purposes. If wetlands are not protected, even more will be destroyed.

Overfishing is caused by catching more fish than an ecosystem can replenish naturally. As with logging, if the fish cannot multiply at the same rate as they are caught, then the fish supply will seriously **diminish**, or become less. Overfishing can lead to the extinction of fish species if it is not stopped or regulated. This could have happened with Pacific salmon, which migrate in a route that takes them through waters in both Canada and the United States. Overfishing by both countries seriously depleted the salmon stock, and fishing industries in both countries suffered. In 1999 Canada and the United States signed an agreement to promote salmon conservation and harvest-sharing principles. While conservation efforts and fishery regulation have been implemented, some species of Pacific salmon are still endangered.

extract to remove

old-growth forest complex forest that has developed over a long period of time and is relatively untouched by human activity

diminish to make less or cause to appear less

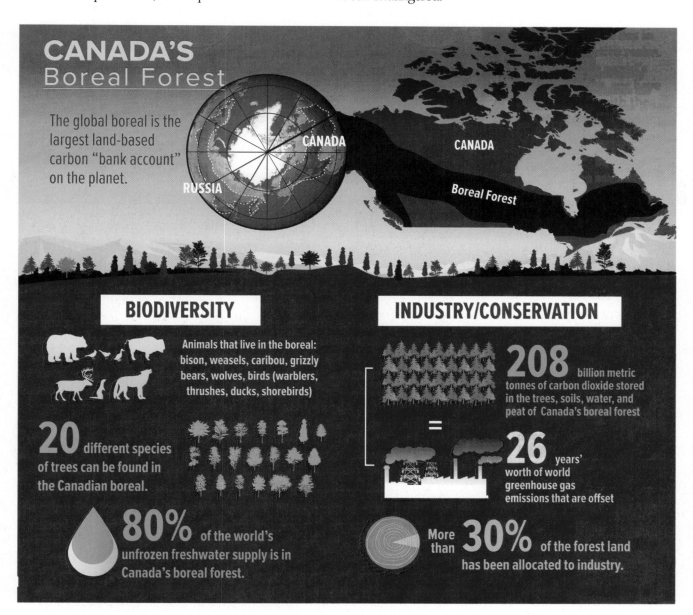

CANADA'S Boreal Forest

The global boreal is the largest land-based carbon "bank account" on the planet.

RUSSIA CANADA CANADA Boreal Forest

BIODIVERSITY

Animals that live in the boreal: bison, weasels, caribou, grizzly bears, wolves, birds (warblers, thrushes, ducks, shorebirds)

20 different species of trees can be found in the Canadian boreal.

80% of the world's unfrozen freshwater supply is in Canada's boreal forest.

INDUSTRY/CONSERVATION

208 billion metric tonnes of carbon dioxide stored in the trees, soils, water, and peat of Canada's boreal forest

=

26 years' worth of world greenhouse gas emissions that are offset

More than **30%** of the forest land has been allocated to industry.

Canada's boreal forest in the far north latitudes is situated just south of the treeless tundra of the polar region.

▲ **CRITICAL THINKING**

1. Analyzing Explain how Canada's boreal forest affects all life on Earth.

2. Assessing What are some threats facing Canada's boreal forest today?

Canada is the world's fourth-largest producer of hydroelectricity—electricity generated from flowing or falling water. While the cost of hydroelectricity is low, making it a great source of renewable energy, it often requires damming. Many environmental groups and indigenous communities protest dam building because damming interrupts the flow of rivers and impacts local ecosystems. In some cases, these protests have succeeded in persuading companies to change their plans and work toward projects that do less harm to local ecosystems.

The continued extraction of natural resources—particularly coal, oil, and natural gas from the tar sands in Alberta—fuels a growing concern among many people over climate change. It requires a large amount of energy to convert the heavy crude oil found in the tar sands into the more useful synthetic crude oil. Furthermore, the process of converting synthetic crude oil into usable substances, such as gasoline and kerosene, is also energy intensive and releases large amounts of carbon dioxide into the air. Additionally, exporting these substances via airplanes and ships releases still more carbon dioxide. Carbon dioxide is believed to contribute directly to climate change and the rising of the Earth's temperatures. Like people around the world, many Canadians are looking for ways to develop renewable sources of energy that do not contribute to climate change.

☑ **READING PROGRESS CHECK**

Assessing What natural resources are most at risk in Canada?

ANALYZING PRIMARY SOURCES

Acid Rain: Sugar Maples at Risk

As acid rain and other effects of air and water pollution escalate, the risk to plant life is increasing. In addition to damaging existing plants, acid rain can slow the growth of new plant life or, as scientists are learning by studying sugar maple trees in Canada, stop new growth altogether. Over time, entire forests can be obliterated as a result of acid rain destruction.

❝Sugar maple abundance already has dropped in parts of the northeastern U.S. and southeastern Canada over the past 40 years, primarily because of high acid levels in soils.

The upper Great Lakes region has mostly escaped the damage because its soils are rich in calcium, which provides a buffer against acid. But in an article published this month in the *Journal of Applied Ecology,* scientists say they've discovered another way that acid rain harms sugar maple seedlings in upper Great Lakes forests.

Donald Zak of the University of Michigan says acid rain prevents dead maple leaves from decaying on the forest floor, creating a barrier that hampers growth of new trees.❞

—"Acid Rain Could Kill Maples Near Great Lakes," Associated Press, December 16, 2011

Industrial smokestacks emit pollutants, including sulfur and nitrogen oxides, that rise into the upper atmosphere, leading to acid rain.

DBQ ▲ **CRITICAL THINKING**

1. ***Identifying Central Issues*** According to Donald Zak, how does acid rain interfere with seedling development?

2. ***Speculating*** Maple syrup from sugar maple trees is an important export product in parts of Canada. How might acid rain affect the economy in areas where maple syrup is harvested?

Human Impact

GUIDING QUESTION *How do human activities impact the environment in Canada?*

Human activities, such as overfishing, logging, and mining, can seriously impact the environment in Canada. Agricultural, industrial, mining, and forestry activities have also contributed to pollution of the ocean waters. Acid rain is another serious threat that can even contaminate lakes, streams, and rivers far from the source of the pollution. These acids can come from natural sources, such as volcanoes, but human activities such as burning fossil fuels also release such acids into the air. Acid rain causes lakes, streams, and rivers to be contaminated. Since Canada is connected to the United States through a number of wind and water systems, pollution in the United States is negatively affecting the environment in Canada. For example, emissions in the United States can result in acid rain in Canada, threatening timber and water resources.

Acid rain contributes to water pollution, but it is not the only threat to water supplies. Many companies and factories put their waste, such as heavy metals and pesticides, into water sources. Water pollution affects plants, animals, and small organisms living in the water. Contaminated water can kill trees, plants, and animals. It can also be harmful to humans. The United States and Canadian governments worked together to produce the Great Lakes Water Quality Agreement and the Clean Water Act in the 1970s to address the serious issue of water pollution that threatens the Great Lakes. The result has been greatly improved water quality and the resurgence of some fish populations.

✔ **READING PROGRESS CHECK**

Describing How have human activities affected water systems in Canada?

Addressing the Issues

GUIDING QUESTION *How is the Canadian government working to address environmental issues?*

Climate change is an issue the provinces, not the federal government, address. In 2011 Canada ranked ninth in greenhouse gas emissions, behind China, the United States, Russia, India, Japan, Germany, Iran, and South Korea.

Canadian Wildlife Service scientists evaluate the condition of a tranquilized mother polar bear and her young triplet cubs near western Hudson Bay. This is part of an effort to help protect animals living in Canada's boreal forest.

▲ **CRITICAL THINKING**

1. *Speculating* What data do you think the scientists are checking to assess the health of the polar bear and her cubs?

2. *Hypothesizing* How might the bears' health be an indicator of the health of their habitat?

The main sources of greenhouse gas emissions are transportation, electricity generation, and producing and refining fossil fuels and petroleum. In 2011 Canada pulled out of the Kyoto Protocol. The Kyoto Protocol is an international agreement signed by 37 industrialized countries that agreed to reduce their greenhouse gas emissions. The Canadian government failed to meet its goals and chose to remove itself from the agreement.

Individual provinces have taken actions to address the climate change issue. In 2009 Ontario passed sweeping legislation to support renewable energy and promote conservation. The new law—the Green Energy and Green Economy Act (GEGEA)—promotes the development of renewable "green" energy projects within the province. The act also advocates for efficient energy and energy conservation efforts in homes, schools, and offices. GEGEA offers all of Ontario's residents—including home owners and large companies—financial incentives to develop small- and large-scale renewable energy sources. As an added benefit, it was projected to produce some 50,000 jobs, both directly and indirectly, in its first three years. Ontario is also the leader in Canadian wind and solar energy projects. So far, Nova Scotia is the only province besides Ontario to set formal goals for wind power projects. Such projects can attract billions of dollars in international investments and also help create jobs.

Hydroelectricity is the most efficient source of renewable energy in Canada and accounts for more than half of all the electricity produced. It is far from a perfect solution, however. The places suitable for the installation of hydroelectric facilities are limited. In addition, the damming of rivers can seriously disrupt the surrounding ecosystems.

The Canadian government has begun to explore wind and solar power options for energy production in the southern provinces. Organizations outside the government are also working hard to address environmental issues in Canada—and to put pressure on the Canadian government to address them more effectively.

There are positive efforts to report. Canada is taking steps to protect its renewable resources. Protection of Pacific salmon is one example, although it remains to be seen whether those efforts will enjoy ultimate success. Still, the effort illustrates how the government and others can take aggressive action to protect and preserve precious natural resources.

☑ **READING PROGRESS CHECK**

Describing What kinds of renewable energy sources is the Canadian government exploring?

LESSON 3 REVIEW

Reviewing Vocabulary

1. *Defining* Provide a definition and example for the following term: old-growth forest.

Using Your Notes

2. *Listing* Using your web diagram, identify a way in which Canadians are impacting their environment and one way in which they are trying to manage that impact.

Answering the Guiding Questions

3. *Identifying* How do economic activities in Canada put natural resources at risk?

4. *Evaluating* How do human activities impact the environment in Canada?

5. *Identifying Central Issues* How is the Canadian government working to address environmental issues?

Writing Activity

6. *Informative/Explanatory* With a partner, select one natural resource from the following list: water, oil, natural gas, and coal. Write a paragraph describing why this resource is valuable. How has human activity affected this resource, and what are the ways the Canadian government can do more to address the conservation of this resource?

Directions: On a separate sheet of paper, answer the questions below. Make sure you read carefully and answer all parts of the questions.

Lesson Review

Lesson 1

1 **Describing** Describe how tectonic forces have played a role in shaping Canada's physical geography.

2 **Discussing** Provide two examples of Canadian lakes that were formed by glacier activity and describe the process.

3 **Explaining** How does Canada's climate change as one travels farther north?

Lesson 2

4 **Assessing** Describe the ways in which physical geography and natural resources have influenced Canadian culture and economy.

5 **Finding the Main Idea** How have the cultures of various indigenous groups influenced Canada's cultural diversity?

6 **Describing** Describe some ways that the Canadian family has changed in recent years.

Lesson 3

7 **Summarizing** What efforts have been made to promote renewable "green" energy by the Canadian government?

8 **Making Inferences** Why is it in the best interest of industries to use natural resources responsibly?

9 **Evaluating** Why do many environmental groups and indigenous communities feel that hydroelectricity does more harm than good?

Exploring the Essential Question

10 **Diagramming** Create a cause-and-effect chart that shows how physical features and climate have influenced the human settlement of Canada. How have these relationships shaped Canada today?

21st Century Skills

Review the circle graph. Then answer the questions that follow.

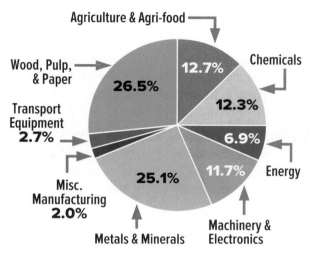

CHINESE IMPORTS FROM CANADA BY SECTOR, 2011 (as a share of total imports from Canada)*

- Agriculture & Agri-food
- Chemicals — 12.3%
- 12.7%
- Energy — 6.9%
- Machinery & Electronics — 11.7%
- Metals & Minerals — 25.1%
- Misc. Manufacturing — 2.0%
- Transport Equipment — 2.7%
- Wood, Pulp, & Paper — 26.5%

Source: Global Trade Atlas (Chinese statistics)
*May not add up to 100% due to rounding.

11 **Using Graphs, Charts, Diagrams, and Tables** China imports the most products from Canada in which two categories?

12 **Geography Skills** What does the high percentage of exports of wood imply about Canada's physical geography?

Applying Map Skills

Use the Unit 2 Atlas maps to answer the following questions.

13 **Physical Systems** List all the climate regions found in Canada and describe their location in relation to latitude.

14 **The World in Spatial Terms** Which parts of Canada are the most densely populated and which are the least densely populated? Why do you think this is the case?

15 **Human Systems** Use your mental map of Canada to describe the relative location of Ottawa.

Need Extra Help?

If You've Missed Question	1	2	3	4	5	6	7	8	9	10	11	12	13	14	15
Go to page	144	146	146	149	153	153	162	158	162	147	163	163	116	112	147

Directions: On a separate sheet of paper, answer the questions below. Make sure you read carefully and answer all parts of the questions.

Critical Thinking

16 *Making Connections* What is the connection between the semi-arid (steppe) climate region and the Pacific Ranges?

17 *Drawing Inferences* What can you infer from the fact that the Canadian government has recently begun working with indigenous populations to ensure that their culture and ways of life are preserved?

18 *Exploring Issues* What are different methods for preventing overfishing and the destruction of old-growth forests? Be sure to include examples of what a government can do and what a citizen can do.

19 *Evaluating Counter Arguments* Given the threat of environmental damage from reliance on fossil fuels, why do you think Canada continues to rely on fossil fuels for energy and as a commodity for trade?

College and Career Readiness

20 *Decision Making* You are a recent college graduate seeking a teaching job in Canada. Based on your knowledge of the landforms, climates, and cultural features of Canada, in which area of the country would you choose to live and why? Explain your answer using details from the chapter.

Writing About Geography

21 *Argument* Environmentalists working on resource management in Canada face many challenges, such as overfishing, loss of old-growth forests, acid rain, and climate change. Which challenge do you feel is the most significant? Write a one-page essay explaining your position. Be sure to describe the problem and to explain why you feel it is the most significant.

DBQ Analyzing Primary Sources

Use the document to answer the following questions.

PRIMARY SOURCE

"*A conservation group is warning against allowing too much fishing of sockeye salmon on the Fraser River despite expectations that this year's run will be one of the biggest in 100 years.*

Society director Craig Orr said in a release headlined 'Don't succumb to sockeye fever' that most of this year's sockeye are from one place—the Adams River—while stocks from many other sources are severely depleted.

'We should all rejoice in this year's bounty, but remember that returns of Fraser River sockeye in this decade have been extremely low for reasons not yet understood,' said Orr.

Environmentalist Vicky Husband, an advisor to Watershed Watch, said the future of the whole fishery ecosystem has to be taken into account.

'We've endured a century of over-fishing and collapse of smaller sockeye populations,' said Husband."

—"Salmon Overfishing Warning Issued," CBC News, September 1, 2010

22 *Identifying Cause and Effect* Why does Husband feel that overfishing threatens the entire fishery ecosystem?

23 *Making Inferences* Why does Orr use the phrase "sockeye fever" to warn against overfishing of salmon on the Fraser River?

Research and Presentation

24 *Research Skills* Use the Internet to research the major immigrant groups who came to Canada during different periods of its history. As you read, take notes. Use your notes to share what you learned with a classmate.

Need Extra Help?

If You've Missed Question	**16**	**17**	**18**	**19**	**20**	**21**	**22**	**23**	**24**
Go to page	147	151	158	142	146	149	164	164	152

Latin America

Chapter 7
Mexico

Chapter 8
Central America and
the Caribbean

Chapter 9
South America

UNIT **3**

① Culture Latin America is a region where cultures have collided. Native cultures are blended with those of Europeans. Today, the faces, clothing, and customs of many Latin Americans reveal their mixed heritage.

EXPLORE the REGION

Spanning more than 85 degrees of latitude, **LATIN AMERICA** encompasses Mexico, Central America, the Caribbean, and South America. It is a region of startling physical contrasts, from the high peaks of the Andes to the lush rain forests of the Amazon. Latin America's human geography reflects a shared colonial legacy, but does present some contrasts between urban and rural, rich and poor, more developed and less developed.

THERE'S MORE ONLINE

HUGHES HervA©/hemis.fr/Getty Images

3 Rain Forests Like a snake slithering through the grass, the Nanay River meanders through the Peruvian rain forest.

4 Mountains The jagged peaks of Chile's Torres del Paine are part of the Andes, the world's longest mountain chain.

2 Cities Many people migrate to urban centers such as Mexico City where work offers the possibility of economic advancement. This rapid growth forces cities to look for ways to provide their growing populations with necessary resources.

(br)© Ken Hornbrook 2010/Getty Images

Latin America
Physical

ATLANTIC OCEAN

40°N

UNITED STATES

Bermuda Islands

SIERRA MADRE OCCIDENTAL

Gulf of California

MEXICAN PLATEAU

Rio Bravo

SIERRA MADRE ORIENTAL

Baja California

Gulf of Mexico

TROPIC OF CANCER

Bahamas

WEST INDIES

20°N

Yucatán Peninsula

Greater

Cuba

Hispaniola

Jamaica

Antilles

Puerto Rico

Guadeloupe

SIERRA MADRE DEL SUR

Caribbean Sea

Martinique

Lesser Antilles

Mosquito Coast

Lake Maracaibo

Trinidad

Lake Nicaragua

Orinoco R.

Guiana Highlands

Angel Falls

Isthmus of Panama

ANDES

Llanos

0   1,000 miles

0   1,000 kilometers

Lambert Azimuthal Equal-Area projection

Galápagos Islands

EQUATOR 0°

Rio Negro

Amazon R.

Marajó Island

Cape São Roque

AMAZON BASIN

SELVAS

Madeira R.

Tapajós R.

Xingu R.

Araguaia R.

Tocantins R.

São Francisco R.

Caatinga

Elevations

10,000 ft. (3,000 m)
5,000 ft. (1,500 m)
2,000 ft. (600 m)
1,000 ft. (300 m)
0 ft. (0 m)
Below sea level

—— National boundary
▲ Mountain peak
▼ Lowest point

La Montaña

Marañón R.

PACIFIC OCEAN

Lake Titicaca

Altiplano

MATO GROSSO PLATEAU

BRAZILIAN HIGHLANDS

Campos

20°S

TROPIC OF CAPRICORN

Atacama Desert

ANDES

Gran Chaco

Paraguay R.

Paraná R.

Cape São Tomé
Cape Frio

Aconcagua 22,834 ft. (6,960 m)

Juan Fernández Islands

PAMPAS

Uruguay R.

Río de la Plata

Colorado R.

MAP STUDY

1. *Environment and Society* What physical features could present barriers to the development of Latin America?

2. *Human Systems* What European countries still control territory in Latin America?

Chiloé Island

PATAGONIA

Laguna del Carbón -344 ft. -105 m

ATLANTIC OCEAN

Falkland Islands (Islas Malvinas)

Strait of Magellan

Tierra del Fuego

South Georgia Island

Cape Horn

140°W   120°W   100°W   80°W   60°W   40°W   20°W

60°S

Latin America
Political

UNITED STATES

Tijuana
Ciudad Juárez
Chihuahua
Monterrey
MEXICO
Guadalajara
Mexico City
Veracruz
Puebla
Orizaba
Gulf of California
Rio Bravo

Gulf of Mexico

BAHAMAS
Nassau
Havana
CUBA
Cayman Is. (U.K.)
Port-au-Prince
DOMINICAN REPUBLIC
Santo Domingo
BELIZE
Belmopan
JAMAICA
Kingston
HAITI
GUATEMALA
Guatemala
San Salvador
EL SALVADOR
HONDURAS
Tegucigalpa
NICARAGUA
Managua
San José
COSTA RICA
Panama
PANAMA

TROPIC OF CANCER

Caribbean Sea
Aruba (Neth.)
Neth. Antilles (Neth.)
Caracas
Orinoco R.
Port-of-Spain
TRINIDAD & TOBAGO
VENEZUELA
GUYANA
Georgetown
Paramaribo
Cayenne
FRENCH GUIANA (Fr.)
SURINAME
Medellín
Bogotá
Cali
COLOMBIA
Quito
ECUADOR
Galápagos Islands (Ecuador)

EQUATOR

Rio Negro
Manaus
Amazon R.
Marajó Island
Belém
Madeira R.
BRAZIL
Fortaleza
Recife
PERU
Lima
Brasília
Salvador
Arequipa
BOLIVIA
Lake Titicaca
La Paz
Santa Cruz
Sucre
São Francisco R.
Tocantins R.
Belo Horizonte
PACIFIC OCEAN
PARAGUAY
Asunción
Paraguay R.
Paraná R.
São Paulo
Rio de Janeiro
Curitiba
Uruguay R.
Pôrto Alegre

TROPIC OF CAPRICORN

Valparaíso
Santiago
ARGENTINA
Rosario
Buenos Aires
URUGUAY
Montevideo
Río de la Plata
Juan Fernández Islands (Chile)
CHILE
Bahía Blanca

N W E S

1,000 miles
1,000 kilometers
Lambert Azimuthal Equal-Area projection

- ⊙ National capital
- ○ Department capital
- ● Major city

Strait of Magellan
Falkland Islands (Islas Malvinas)
Administered by United Kingdom
(Claimed by Arg.)
South Georgia Island (U.K.)

ATLANTIC OCEAN

40°N
20°N
0°
20°S
40°S

140°W 120°W 100°W 80°W 60°W 40°W 20°W

Inset (Caribbean)

60°W
20°N
Virgin Islands (U.S.)
British Virgin Islands (U.K.)
Anguilla (U.K.)
St. Martin (Fr.)
St. Maarten (Neth.)
St. Barthélemy (Fr.)
Puerto Rico (U.S.)
Saba (Neth.)
St. Eustatius (Neth.)
ANTIGUA AND BARBUDA
Montserrat (U.K.)
Guadeloupe (Fr.)
ST. KITTS AND NEVIS
DOMINICA
Martinique (Fr.)
Caribbean Sea
ST. LUCIA
ST. VINCENT AND THE GRENADINES
BARBADOS
GRENADA
TRINIDAD & TOBAGO
0 200 mi.
0 200 km
10°N

Bermuda (U.K.)

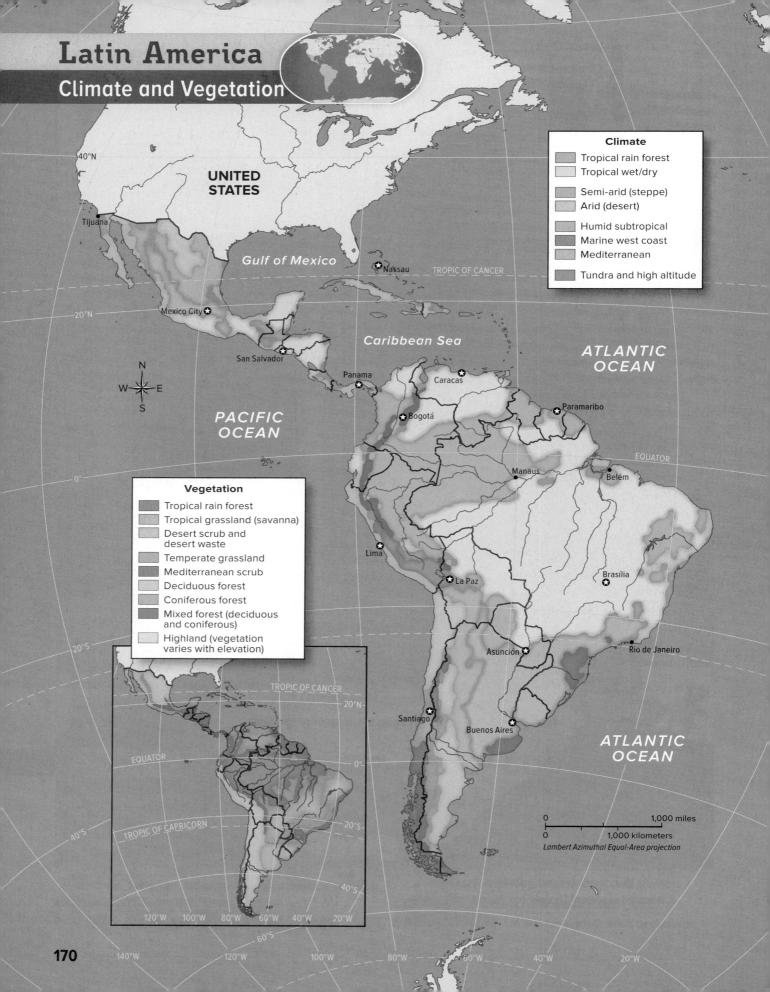

Latin America
Climate and Vegetation

UNITED STATES

Tijuana

Gulf of Mexico

Mexico City

Nassau

TROPIC OF CANCER

San Salvador

Panama

Caribbean Sea

Caracas

Bogotá

Paramaribo

ATLANTIC OCEAN

PACIFIC OCEAN

EQUATOR

Manaus

Belém

Lima

Brasília

La Paz

Asunción

Rio de Janeiro

Santiago

Buenos Aires

ATLANTIC OCEAN

Climate
- Tropical rain forest
- Tropical wet/dry
- Semi-arid (steppe)
- Arid (desert)
- Humid subtropical
- Marine west coast
- Mediterranean
- Tundra and high altitude

Vegetation
- Tropical rain forest
- Tropical grassland (savanna)
- Desert scrub and desert waste
- Temperate grassland
- Mediterranean scrub
- Deciduous forest
- Coniferous forest
- Mixed forest (deciduous and coniferous)
- Highland (vegetation varies with elevation)

N W E S

TROPIC OF CANCER

EQUATOR

TROPIC OF CAPRICORN

0   1,000 miles
0   1,000 kilometers
Lambert Azimuthal Equal-Area projection

40°N

20°N

0°

20°S

140°W   120°W   100°W   80°W   60°W

20°N

0°

20°S

40°S

60°S

Latin America
Economic Activity

Land Use

- Commercial farming
- Subsistence farming
- Livestock raising
- Primarily forest
- ◘ Manufacturing and trade
- Commercial fishing
- Little or no activity

UNITED STATES

40°N

Gulf of Mexico

TROPIC OF CANCER

20°N

Caribbean Sea

ATLANTIC OCEAN

N
W E
S

0   1,000 miles
0   1,000 kilometers
Lambert Azimuthal Equal-Area projection

Resources

Coal		Cobalt	
Petroleum		Nickel	
Natural gas		Copper	
Iron ore		Lead	
Zinc		Manganese	
Bauxite		Gold	
Uranium		Silver	
		Gems	

0°

PACIFIC OCEAN

20°S

TROPIC OF CAPRICORN

40°S

140°W   120°W   100°W   80°W   60°W

60°S

UNIT 3
REGIONAL ATLAS

MAP STUDY

1. *Physical Systems* What are the predominant types of natural vegetation in the Tropics?

2. *Places and Regions* What generalizations can you make about the locations of the region's manufacturing areas?

Latin America
Population Density

ATLANTIC OCEAN

PACIFIC OCEAN

Caribbean Sea

Gulf of Mexico

TROPIC OF CANCER

EQUATOR

TROPIC OF CAPRICORN

40°N

20°N

0°

20°S

40°S

60°S

120°W

100°W

80°W

60°W

40°W

N
W · E
S

0   1,000 miles
0   1,000 kilometers
Lambert Azimuthal Equal-Area projection

Cities and places:
Mexicali, Tijuana, Hermosillo, Torreón, Culiacán, Monterrey, Matamoros, Havana, Mazatlán, Guadalajara, León, Tampico, Mérida, Mexico City, Veracruz, Puebla, Acapulco, Guatemala, San Salvador, Tegucigalpa, Managua, San José, Panama, San Juan, Port-au-Prince, Kingston, Santo Domingo, Fort-de-France, Barranquilla, Maracaibo, Caracas, Cartagena, Valencia, Mérida, Port-of-Spain, Medellín, Ciudad Guayana, Georgetown, Paramaribo, Bogotá, Cali, Quito, Guayaquil, Iquitos, Manaus, Belém, São Luís, Piura, Fortaleza, Teresina, Trujillo, Pucallpa, Rio Branco, Pôrto Velho, Natal, Recife, Lima, Cuzco, Maceió, Ica, Cuiabá, Salvador, Arequipa, La Paz, Santa Cruz, Goiânia, Brasília, Uberlândia, Belo Horizonte, Antofagasta, Salta, Ribeirão Prêto, Campinas, Vila Velha, Asunción, São Paulo, Rio de Janeiro, San Miguel de Tucumán, Curitiba, Valparaíso, Mendoza, Florianópolis, Santiago, Rosario, Córdoba, Pôrto Alegre, Buenos Aires, Concepción, Montevideo, Bahía Blanca, Mar del Plata, Punta Arenas

UNIT 3
REGIONAL ATLAS

MAP STUDY

1. *Human Systems* Which parts of Latin America are the most densely populated? What might account for this?

2. *Places and Regions* What generalizations can you make about the locations of South America's cities?

Mexico

ESSENTIAL QUESTION · *How do physical systems and human systems shape a place?*

Why Geography Matters
Challenges of Urbanization

Lesson 1
Physical Geography of Mexico

Lesson 2
Human Geography of Mexico

Lesson 3
People and Their Environment: Mexico

Geography Matters...

Places reflect their relationship between humans and their environment. Mexico today is a result of history, geography, and increased globalization. Cultures have collided in Mexico for centuries. Indigenous civilizations flourished, followed by Europeans who brought new laws, languages, and religions. Today, the faces and customs of many Mexican people reflect their mixed heritage. Economic forces have pulled Mexico into the global economy with a promise of prosperity dependent on natural resources and industrial growth.

◀ Traditional Mexican dress is colorful and reflects the style of the region.

©Alison Wright/Corbis

173

challenges
of urbanization

*Rapid urban growth brings challenges to city governments around the world as they struggle to provide housing, services, infrastructure, and jobs, as well as curb pollution. Governments have limited funds to spend on basic upkeep and services. As a result, cities like **Mexico City** experience challenges such as environmental problems and poverty.*

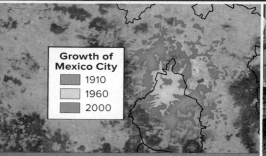

Growth of Mexico City
- 1910
- 1960
- 2000

Why has Mexico City grown so fast?

Today's Mexico City is the result of years of rural-to-urban migration by people looking for better economic opportunities. The first influx of these economic migrants coincided with rapid industrialization in the late nineteenth century. The pull forces of industrial jobs and the push forces of rural land policies drew people to Mexico City. The rural poor moved to the city as land was purchased around them by wealthy landowners. Similar factors continue to bring economic migrants to Mexico City seeking a better life for themselves and their families.

1. **Human Systems** What are the pull factors influencing migration to Mexico City? How are these different from the push factors that bring people to the city?

What is the social impact of rapid growth?

Economic migrants move to the city expecting to find jobs. Unfortunately, unemployment is common. Some migrants find temporary jobs or work in the informal sector—"underground economies" that are not taxed or regulated by the government. People often do not have access to health care and education. Lack of infrastructure—housing, electrical grids, sewer facilities, and roads—to support the growing population leads to the development of shantytowns. The influx of people to Mexico City puts enormous pressures on the natural environment. Underground water aquifers are being depleted, causing the city to sink. Inadequate sewer facilities lead to polluted land and water. Full of rubbish, landfills have been closed. Unregulated by the government, shantytowns are built in environmentally sensitive areas such as hill slopes.

2. **Environment and Society** What challenges has rural-to-urban migration created for the government of Mexico City?

What can be done?

Government agencies and other groups continue to establish initiatives and special projects to address these challenges. The government of Mexico and public-private partnerships are investing in sustainable and environmentally friendly housing development. Plan Verde (Green Plan) includes a range of programs to promote environmental sustainability by easing traffic congestion, reducing greenhouse gas emissions, and encouraging public transportation, cycling, and walking options. The Mexico City Climate Action Program provides funding for sustainable housing as well as renewable energy programs.

3. **Human Systems** Write a paragraph explaining how environmentally friendly policies could improve life in Mexico City.

THERE'S MORE ONLINE

VIEW a map of the world's megacities • *WATCH* a video of Mexico's urban sprawl

Reading **HELP**DESK

Academic Vocabulary

- **feature**
- **inevitable**

Content Vocabulary

- **land bridge**
- **seismic**
- **vertical climate zone**

TAKING NOTES: *Key Ideas and Details*

SUMMARIZING As you read the lesson, use a graphic organizer like the one below to take notes on the physical geography of Mexico.

Physical Geography of Mexico		
Landforms	Water Systems	Climates, Biomes, and Resources

LESSON 1
Physical Geography of Mexico

ESSENTIAL QUESTION • *How do physical systems and human systems shape a place?*

IT MATTERS BECAUSE

Mexico has rich farmland, abundant access to the ocean, and freshwater resources. With diverse climates and biomes it supports a variety of food crops. The country also has substantial mineral resources and has developed strong manufacturing and service sectors.

Landforms

GUIDING QUESTION *Why is the Mexican Plateau considered the heartland of Mexico?*

Mexico is the southernmost country in North America. It shares its entire northern border with the United States. Mexico, along with Central America, joins the continents of North America and South America. This physical feature is called a **land bridge** because it connects two geographic landforms.

The western side of Mexico is part of the Ring of Fire. The Ring of Fire is the area where the Pacific tectonic plate collides with other tectonic plates, creating areas of **seismic** activity with earthquakes. This activity helps shape the landforms of Mexico. Seismic activity opens parts of the Earth's crust and triggers the formation of volcanoes.

The mountains that stretch along the northwestern edge of Mexico are called the Sierra Madre Occidental. These mountains have been deeply cut by westward-flowing rivers and streams, which have formed deep gorges. The largest is *Barranca del Cobre* (Copper Canyon), a beautiful natural wonder in the state of Chihuahua, Mexico.

The Sierra Madre Oriental is considered the southern extension of the Rocky Mountains of Canada and the United States. The average mountain height is between 8,000 and 9,000 feet (2,438 and 2,743 m). A few exceptions include mountains that reach heights of 12,000 feet (3,657 m).

Between these two mountain ranges is the inland Mexican Plateau. Moderate, consistent temperatures make this area an attractive place to live. It is the largest and most densely populated region of

Mexico. The Mexican Plateau is broken into two parts, the huge *Mesa del Norte* (Northern Plateau) and the smaller but heavily populated *Mesa Central* (Central Plateau).The dry Northern Plateau is home to several large cities. The Central Plateau is considered the breadbasket—or major grain-producing region—of Mexico. It is less arid than the Northern Plateau and **features** several smaller valleys. Most of the food grown in Mexico comes from this area.

The Gulf Coastal Plain is a wide stretch of land east of the Sierra Madre Oriental. These mountains extend from the Texas-Mexico border along the Gulf of Mexico to the Yucatán Peninsula. In the south, a series of mountain ranges and plateaus called the Southern Highlands reach from just south of Mexico City to the southwest edge of Mexico's border with Guatemala.

The variety of landforms in Mexico—from large plateaus and valleys to long mountain ranges and highlands—has made it possible to support large communities of people. There are ample grazing and farmlands on the North and Central Plateaus, in the Southern Highlands, and along the coastlines. The population density in Mexico is greater near parts that have the most agriculture, especially on the Central Plateau and in the Southern Highlands.

☑ READING PROGRESS CHECK

Interpreting How has the geography of Mexico affected the way people use the land?

land bridge a strip of land that connects two larger landmasses, enabling migration of plants and animals to new areas

seismic relating to or caused by an earthquake

feature to have as a characteristic or as a prominent attribute

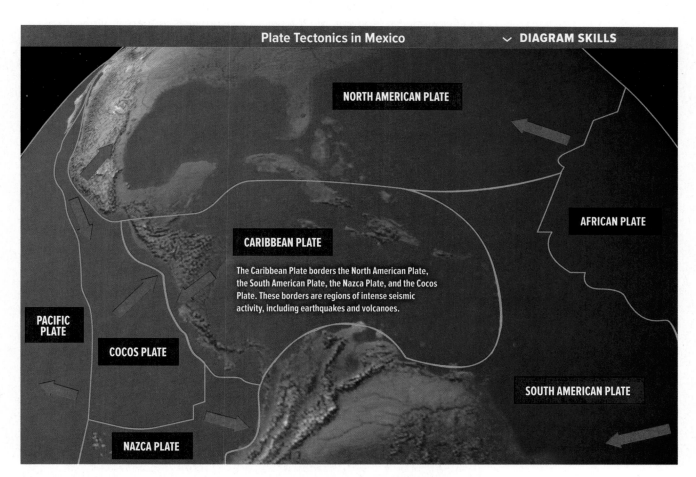

Plate Tectonics in Mexico ⌄ **DIAGRAM SKILLS**

NORTH AMERICAN PLATE

AFRICAN PLATE

CARIBBEAN PLATE

The Caribbean Plate borders the North American Plate, the South American Plate, the Nazca Plate, and the Cocos Plate. These borders are regions of intense seismic activity, including earthquakes and volcanoes.

PACIFIC PLATE

COCOS PLATE

SOUTH AMERICAN PLATE

NAZCA PLATE

Mexico's location near the edges of tectonic plates makes it a hot spot for volcanoes and earthquakes.

▲ CRITICAL THINKING

1. Classifying On which tectonic plates is Mexico located?

2. Drawing Conclusions Why do you think this area of tectonic activity is called the Ring of Fire?

Water Systems

GUIDING QUESTION *Why does Mexico have few major rivers and natural lakes?*

Northern Mexico is generally characterized by a dry climate. This makes permanent waterways rare. The high mountain ranges and plateaus create temperate **vertical climate zones** that do not collect the volume of water that is more common in tropical regions. The few rivers and natural lakes that exist are found in the central part of the country and are generally small. One important exception is the Rio Grande. Known as the Río Bravo del Norte in Mexico, it forms part of the border between Mexico and the United States.

The Lerma River is one of Mexico's most important rivers. It begins in the Toluca Basin, on the Central Plateau west of Mexico City. The Lerma River feeds into Lake Chapala, the largest natural lake in Mexico.

The Gulf of Mexico is the large body of water that forms Mexico's east coast. It supports diverse sea life including an ancient sea creature known as the manatee. The Gulf of Mexico is famous for shrimp and supplies the fishing industry in both the United States and Mexico. The waters in the Gulf of Mexico are relatively sheltered from ocean currents, so the beaches are calm and the waters are warm.

On the western side of Mexico, the Gulf of California divides the Baja Peninsula from the northern coast of Mexico. This body of water supports a remarkable diversity of aquatic animals. These include several types of whales, the giant Pacific manta ray, endangered leatherback sea turtles, and great white sharks.

☑ **READING PROGRESS CHECK**

Describing What is the importance of the Río Bravo del Norte to Mexico?

Climate, Biomes, and Resources

GUIDING QUESTION *How does climate affect human activities in Mexico?*

The climate of a particular region **inevitably** affects the way of life that people have in each place. For example, people who graze cattle on the Northern Plateau anxiously await rain each year. The farmers in the valleys of the Central Plateau

vertical climate zone

a climate zone that occurs as elevation increases, with its own natural vegetation and crops

inevitable incapable of being avoided or evaded

Differences in elevation create distinct climate zones in Mexico and other high-altitude areas in Latin America.

CRITICAL THINKING ▶

1. *Analyzing Visuals* Which climate zones are found above 6,000 feet (1,829 m)?

2. *Synthesizing* How might increasing elevation affect the type of resources found in each vertical climate zone?

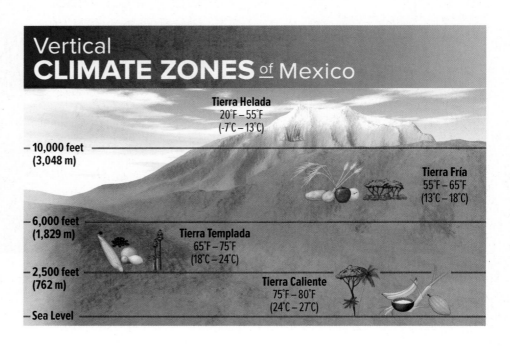

Vertical
CLIMATE ZONES of Mexico

Tierra Helada
20°F – 55°F
(-7°C – 13°C)

— 10,000 feet
(3,048 m)

Tierra Fría
55°F – 65°F
(13°C – 18°C)

— 6,000 feet
(1,829 m)

Tierra Templada
65°F – 75°F
(18°C – 24°C)

— 2,500 feet
(762 m)

Tierra Caliente
75°F – 80°F
(24°C – 27°C)

— Sea Level

depend on their climate to help them get enough water for their crops. The three factors that influence Mexico's climates are the regional high-pressure systems, the northeast trade winds, and the vertical climate zones.

Regional high-pressure systems keep the north and central parts of Mexico dry and cause occasional droughts. The northeast trade winds are responsible for the pattern of tropical storms. Vertical climate zones create the temperate or mild climates found throughout the southern part of Mexico. The elevation of the highlands keeps them at a comfortable temperature that is also helpful for growing certain crops.

Northern Mexico is defined as a chaparral biome. It has mild, rainy winters and hot, dry summers. The plant life consists of cacti, shrubs, and shrub oak. High winds and low-growing plants make the soil in this region good for grasslands. The soil is also thin and rocky, however, so it is not good for crops.

Southern Mexico has a variety of biomes. In the lower altitudes along the coasts, the climate is hot. On the east coast, daily rainfall and high humidity occur. This biome supports very diverse plant and animal life and is essentially a rain forest.

Minerals are an important part of Mexico's economy, especially silver. Mexico is the world's leading producer of silver. In the area called the "Silver Belt" on the Mexican Plateau, both industrial and precious minerals are mined. Zinc, bauxite (the ore of aluminum), lead, gold, mercury, cadmium, and such trace minerals as antimony, manganese, and copper are also important. Timber, fish, and agricultural products are also a significant part of Mexico's economy.

Mexico is a leading petroleum-producing country. Petroleum exports account for a large share of foreign-exchange earnings. Mexico ranks thirteenth in the world for crude oil exports. About three-fourths of Mexico's electricity is generated by thermal power plants that are fired mainly by oil and natural gas. Another one-tenth of Mexico's electric power is created by nuclear power and renewable resources of wind, solar energy, and biomass (plant materials and animal waste used as a source of fuel).

✓ READING PROGRESS CHECK

Assessing How do vertical climate zones affect the economic activity of the Southern Highlands?

Copper mining takes place in the Mexican copper belt in the west part of the country.

▲ CRITICAL THINKING

1. *Analyzing Visuals* Describe the negative impacts of strip mining as depicted in this photo.

2. *Comparing* Where is Mexico's copper belt located in comparison to the "Silver Belt"?

LESSON 1 REVIEW

Reviewing Vocabulary
1. *Discussing* Write a paragraph that discusses the geography of vertical climate zones.

Using Your Notes
2. *Describing* Use your graphic organizer from the lesson to describe three of Mexico's water systems.

Answering the Guiding Questions
3. *Drawing Conclusions* Why is the Mexican Plateau considered the heartland of Mexico?

4. *Interpreting* Why does Mexico have few major rivers and natural lakes?

5. *Making Connections* How does climate affect human activities in Mexico?

Writing Activity
6. *Informative/Explanatory* Write a paragraph describing Mexico's location along the Ring of Fire and how that creates natural hazards.

networks

There's More Online!

☑ **GRAPH** Female Labor in Mexico

☑ **IMAGE** Maquiladora Along the U.S.-Mexico Border

☑ **IMAGE** Women in Mexican Labor Force

☑ **MAP** Mexican Migration to the United States

☑ **INTERACTIVE SELF-CHECK QUIZ**

☑ **TIME LINE** Mexican Independence and Change

☑ **VIDEO** Human Geography of Mexico

Reading **HELP**DESK

Academic Vocabulary

- culture
- diverse

Content Vocabulary

- mestizo
- conquistador
- cash crop
- syncretism
- megacity
- primate city
- extended family
- gross domestic product
- maquiladora
- free trade zone

TAKING NOTES: *Key Ideas and Details*

PARAPHRASING Use a graphic organizer like the one below to describe the human geography of Mexico.

Human Geography of Mexico	
History and Government	Population Patterns

LESSON 2

Human Geography of Mexico

ESSENTIAL QUESTION · *How do physical systems and human systems shape a place?*

IT MATTERS BECAUSE

Mexico's human geography reflects influences from the Maya and Aztec civilizations, the introduction of Spanish culture during the colonial era, and cultural and social elements shared from recent interaction with the United States and other countries.

History and Government

GUIDING QUESTION *What influenced Mexico's political and social structures?*

Variations in the physical geography of Mexico led to the development of diverse **cultures,** languages, and civilizations among the indigenous peoples of Mexico. These differing peoples developed cultures to suit the environments in which they lived. These cultures can be seen in the regional distinctions of Mexico today.

The northern half of Mexico, on the inland plateau and in the mountains, originally had a small population of mostly independent groups of nomadic people. Agriculture was used to supplement hunting, herding, and gathering of food. Some of these seminomadic groups still live in their traditional homelands, separated from most outside influences. The Tarahumara people in the Sierra Madre Occidental are one example of an indigenous group who still live in northern Mexico.

The southern half of Mexico was geographically more **diverse.** It could support large-scale agriculture and produce the variety and abundance of foods necessary to maintain empires and cities. Centered in the Yucatán Peninsula, the Maya civilization was one of the earliest and largest civilizations in Mexico. The Maya built huge stone cities, which were abandoned a few hundred years before the arrival of the first Spanish explorers. The Maya ruled a vast territory and engaged in long-distance trade with other cultures, including Teotihuacán and the Zapotec. Their descendants still live in and around the areas of their former empire. Many of these Mayan people maintain their culture, speak their ancestral languages, and practice the same cultural traditions.

The Aztec Empire arose in central Mexico. The Aztec ruled from their capital, Tenochtitlán (tay•NAWCH•teet•LAHN), the site of present-day Mexico City. They had conquered other peoples in the area when the

Spanish arrived in 1519. Although **mestizos**, people of mixed Spanish and indigenous heritage, now densely populate the region, there are groups that trace their ancestry to the Aztec.

After the conquest of the Aztec by Spanish **conquistador** Hernán Cortés and his men, the Spanish took the wealth of Mexico's gold and silver resources. They also found value in the variety of food available to the local people, quickly taking corn, tomatoes, chocolate, and other native crops on the return trips to Spain. Large tracts of land in Mexico were given to the Spanish settlers of Mexico. These landowners began growing **cash crops** such as cacao (chocolate) and maize (corn) in large quantities, which they exported to Spain. This further enriched the Spanish. Mexico remained a part of the Spanish Empire for nearly three centuries. It was governed by Spain under a highly structured political system ruled by officials called viceroys who were appointed by the Spanish monarch.

In the late 1700s, throughout Mexico and the rest of Latin America, people started to protest European rule. In 1821 Mexico became the first Spanish territory to win its independence. Mexico was free from Spain, but the political system was ruled by a small group of wealthy landowners, army officers, and Catholic clergy who remained in power. Power struggles, public dissatisfaction, and civic revolts made the new republic fragile and chaotic. During this time a new type of leader emerged, the caudillo (kow•DEE•yoh), or military dictator. For brief periods in the 1800s, the government moved toward democratic principles. However, the caudillos found ways to return to power.

The long and bloody Mexican Revolution overthrew the caudillos and established a new constitution in 1917. This brought reforms and established the current Mexican government as a federal republic. Power was divided into three branches of government—legislative, executive, and judicial—and a president could only be elected for one six-year term. However, the rule of law did not last for long. In 1929 one political party, the *Partido Revolucionario Institucional* (PRI), was elected and established a corrupt monopoly on the political system of Mexico. The PRI went on to control the political establishment for nearly 70 years. Not until 2000 was the opposition party, *Partido Acción Nacional* (PAN), able to win the presidency. In 2012, however, the PRI was reelected.

Over the past few decades, drug cartels have come to control different regions of Mexico. New cartels have been forming or breaking away from older and larger cartels. These new cartels compete with old cartels for power and control of drug-producing territories. The result is internal warfare in Mexico. The cartels have increasingly incited street gun battles, massacres in the mountains, and other acts of violence and terror.

Struggles for additional reforms in the government continue. Indigenous communities, small farmers, and groups of underpaid laborers are continuing to pressure the government for greater inclusion in the political system. Corruption remains a common reality in the government. A small group of very wealthy landowners still controls most of Mexico's wealth.

☑ **READING PROGRESS CHECK**

Exploring the Issues Why do drug cartels have such a powerful influence in Mexico?

Population Patterns

GUIDING QUESTION *What factors have shaped Mexico's population patterns?*

For unknown reasons, the Maya had abandoned their cities in the Yucatán and southern Mexico by the time the Spanish conquistadors arrived. They were mainly living as subsistence farmers in small communities, where many still live today.

culture the customary beliefs, social forms, and material traits of a racial, religious, or social group

diverse differing from one another

mestizo refers to people of mixed indigenous and European descent

conquistador Spanish for "conqueror"; Spanish soldier who participated in conquest of indigenous peoples of Latin America

cash crop farm product grown to be sold or traded rather than used by the farm family

maquiladora in Mexico, a manufacturing plant owned by a foreign company

free trade zone an area of a country in which trade restrictions do not apply

Communications are also essential to the Mexican economy. The use of cell phones has increased rapidly since the mid-1990s. The infrastructure and availability of high-speed Internet exists, although not in all areas. People in Mexico City, in some areas along the border with Texas, and in business centers like Monterrey and Guadalajara have the most access. The vast majority of Mexicans, however, are left out of the digital age. About 58 million people—45 percent of Mexicans—had no access to the Internet in 2016.

NAFTA, Trade, and Maquiladoras

In 1992 Mexico, the United States, and Canada signed the North American Free Trade Agreement (NAFTA). NAFTA is a comprehensive agreement that eliminated most trade restrictions. As a result, trade among the three countries grew by 10 to 15 percent annually. Mexico's economy has been transformed by these increases in trade and the flow of investment.

NAFTA has also been a source of controversy and concern. Mexico is more dependent on the economy of its northern neighbor than the United States is on the Mexican economy. Mexico has protested the harmful effects of subsidized agricultural exports from the United States that may be forcing Mexican small landholders off their farms and into service-based or industrial jobs. Meanwhile, many U.S. workers are concerned about the loss of their jobs to workers in Mexico.

During the past 50 years, American and Japanese firms have built manufacturing plants in Mexico. Many of these factories, known as **maquiladoras**, are located close to the U.S.-Mexico border. Maquiladoras are located in **free trade zones**. Such areas benefit foreign corporations by allowing them to hire low-cost labor and produce duty-free exports. They also offer the host country employment opportunities and investment income. Critics of maquiladoras charge that the system often ignores labor laws, thus encouraging low-paying or dangerous jobs.

The illegal drug trade is both an influential and dangerous part of the Mexican economy. Drug cartels often reinvest the money they make into their communities—both through private loans to small businesses and in the form of bribes to police and politicians. This makes them a powerful social and economic force. Many rural mountain communities have relied for generations on the poppy and marijuana fields, controlled by cartels, that support their families. The government has not been able to discourage the growing of these crops because no legal crop can match their cash value for these isolated farmers.

✓ **READING PROGRESS CHECK**

Assessing What factors contribute to Mexico's higher standard of living compared to other Latin American countries?

LESSON 2 REVIEW

Reviewing Vocabulary

1. ***Classifying*** In what ways is Mexico City both a megacity and a primate city?

Using Your Notes

2. ***Summarizing*** Use your graphic organizer on the human geography of Mexico to write a paragraph summarizing society and culture in Mexico today.

Answering the Guiding Questions

3. ***Drawing Conclusions*** What influenced Mexico's political and social structures?

4. ***Hypothesizing*** What factors have shaped Mexico's population patterns?

5. ***Evaluating*** How does Mexican society and culture reflect the country's colonial past?

6. ***Explaining*** How has Mexico's place in the global economy changed over time?

Writing Activity

7. ***Informative/Explanatory*** Write a paragraph discussing how maquiladoras involve Mexico in world trade.

Reading HELPDESK

Academic Vocabulary

- **corporate**
- **ignorance**

Content Vocabulary

- **deforestation**
- **sustainable development**
- **land subsidence**

TAKING NOTES: *Key Ideas and Details*

IDENTIFYING Use a web diagram similar to the one below to take notes as you read about the issues that relate to people and their environment in Mexico.

People and Their Environment: Mexico → Managing Resources, Human Impact, Addressing the Issues

(l)Yuri Cortez/AFP/Getty Images, (tcl)Jorge Uzon/AFP/Getty Images, (tcr)Eco Images/Universal Images Group/Getty Images, (tr)Stephen Alvarez/National Geographic/Getty Images

LESSON 3
People and Their Environment: Mexico

ESSENTIAL QUESTION • *How do physical systems and human systems shape a place?*

IT MATTERS BECAUSE

The ways that people extract and use resources today can have substantial impacts on their well-being in the future. Resource management and sustainable development are important so that future generations can continue to benefit from an area's natural resources. Mexico has been so focused on increasing economic development that conservation of resources has not been as high a priority. However, concerned Mexican citizens are working hard to find ways to preserve the land and all that it provides.

Managing Resources

GUIDING QUESTION *Why are Mexico's resources in jeopardy?*

Mexico has many natural resources. These include petroleum, silver, copper, gold, lead, zinc, natural gas, and timber. Industrial access to these resources enriches the economy and creates jobs and new investment opportunities. Obtaining and using these resources, however, often results in significant problems that threaten Mexico's environmental health.

Mexico's many ecosystems are experiencing the effects of global climate change and environmental destruction. For example, semi-arid regions are seeing longer droughts and more desertification, or the development of desertlike conditions. Forested areas throughout Mexico are experiencing new patterns in rainfall. Many changes in the environment are a result of increasing migration to urban centers. This puts pressure on the surrounding environment. As cities grow, the surrounding land is cleared and developed.

At the same time, environmental degradation itself is increasing urbanization. The destruction of rural resources forces migration to urban areas. People move to the cities to seek employment away from the hardship of living in regions destroyed by poor environmental management.

About one-third of Mexico is covered in large forests ranging from deciduous and coniferous forests to tropical rain forests. Forest destruction and the loss of biodiversity, however, is occurring at an alarming rate. As Mexico's economy grows, so does the demand for timber

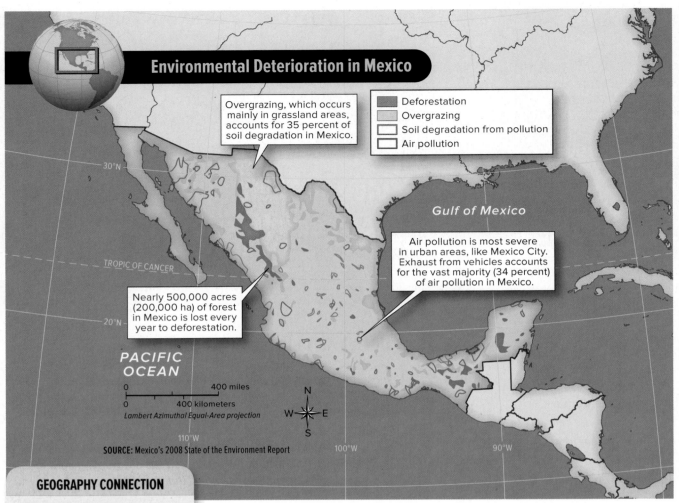

Environmental Deterioration in Mexico

Overgrazing, which occurs mainly in grassland areas, accounts for 35 percent of soil degradation in Mexico.

Legend:
- Deforestation
- Overgrazing
- Soil degradation from pollution
- Air pollution

Air pollution is most severe in urban areas, like Mexico City. Exhaust from vehicles accounts for the vast majority (34 percent) of air pollution in Mexico.

Nearly 500,000 acres (200,000 ha) of forest in Mexico is lost every year to deforestation.

Gulf of Mexico

TROPIC OF CANCER

PACIFIC OCEAN

0 — 400 miles
0 — 400 kilometers
Lambert Azimuthal Equal-Area projection

SOURCE: Mexico's 2008 State of the Environment Report

GEOGRAPHY CONNECTION

Mexico has beautiful landscapes with abundant resources, but environmental deterioration threatens its lands.

1. **THE WORLD IN SPATIAL TERMS** Which coast of Mexico experiences the most industrial contamination?

2. **ENVIRONMENT AND SOCIETY** Which threat affects the largest area of land in Mexico?

corporate formed into an association and endowed by law with the rights and liabilities of an individual

deforestation the loss or destruction of forests, mainly for logging or farming

ignorance lack of knowledge, education, or awareness

resources, which are an important part of Mexico's export economy. The **corporate** logging industry is one source of the problem, but not the only cause of **deforestation**. The growing population demands more food resources. In response, ranchers and farmers are clearing the forests and creating new areas for growing food and grazing cattle.

In 2000 Mexico developed the National Biodiversity Strategy and Action Plan. The plan has four major objectives: to conserve and protect the biodiversity components, to value the different components of biodiversity, to promote knowledge of biodiversity, and to encourage sustainable and diversified use of biodiversity components. Reducing public **ignorance** of the consequences of environmental mismanagement should reduce the loss of Mexican forest resources. **Sustainable development** projects that utilize natural resources responsibly are the only solution to the demands of a growing population.

Mexico has numerous mountain ranges, dry northern plains, vast southern jungles, and many large cities. Consequently, only 12 percent of Mexico's land is arable, or suitable for farming. Producing enough agricultural products is difficult in Mexico. With only a few major rivers and lakes, water resources are precious.

The demand for water resources in the northern part of the country is so high that desertification is a growing problem. Climate change has meant that recent years have seen an increase in drought throughout northern Mexico. Both ranchers and farmers have suffered from the water shortages.

More than 50 percent of Mexico's population lives below the poverty line, many in substandard conditions in large urban centers. Both rural and urban areas struggle to provide basic resources like clean water, electricity, and garbage removal. Human needs are putting heavy demands on land, water, and timber resources. Pollution of water, air, and land is a growing concern for Mexico.

Mexico City in particular is facing serious problems with its water supply. Providing water to more than 22 million residents is a challenge for the city's struggling infrastructure. The natural underground reserves of water have been pumped dry. Once water is removed, the clay soil compacts in the empty space and the water cannot be replaced. Over the years, these empty water reservoirs in and around Mexico City have been collapsing. This creates sinkholes, or depressions in the land, and relevels the surface. Buildings in Mexico City are tilting because the land underneath them has been emptied of water. This process is called **land subsidence**.

The poor are the greatest victims of the urban water crisis. Municipal water supplies often do not reach their settlements on the outskirts of the city. Citizen groups have been working to improve the water supply through advocacy and education, and by encouraging the government to privatize water management.

☑ **READING PROGRESS CHECK**

Exploring the Issues Describe how large-scale urbanization has affected Mexico City.

Human Impact

GUIDING QUESTION *How do human activities impact Mexico's environment?*

Rapid urban growth in the last century and high rates of poverty have made social development a constant challenge in Mexico. As portions of the economy grow, access to consumer goods and the number of consumers increase. As a result, waste accumulation is a growing challenge. Without infrastructure to support proper waste disposal, pollution is an enormous problem in urban areas.

Mexico's economy is still growing. When people achieve a new economic status, they invest in material goods such as electronics and cars. Cars are a major contributing factor to air pollution problems. Mexico City is located in a valley. Carbon emissions from cars are often trapped in the valley. This creates a toxic haze over the city. As a result, the sky around Mexico City is often a dull gray or brown. The government has been making emission regulations a priority to try to reduce the level of pollution in the air. Until these regulations take effect, citizens will continue to suffer from health problems related to air pollution, such as an increased risk of asthma and chronic lung infections.

☑ **READING PROGRESS CHECK**

Analyzing What are the causes and consequences of air pollution in an urban environment?

Addressing the Issues

GUIDING QUESTION *How are governments in Mexico addressing environmental issues?*

The last 20 years have seen a rise in political action and activity by many Mexican citizens. As a result of government investment in health and education, a better educated and more literate population has begun to demand more from their government and from society. These citizens want better living conditions and have also shown a concern for protecting the environment.

Mexico has enacted new regulations to try to curb the destruction of natural resources. The government is also working to support farms and businesses that contribute to the economy and are interested in protecting those resources.

sustainable development technological and economic growth that does not deplete the human and natural resources of a given area

land subsidence the sinking or settling of land to a lower level in response to various natural and human-caused factors

Private citizens of Mexico and nongovernmental organizations are making efforts to fill the needs of protecting the environment.

▶ CRITICAL THINKING

1. **Analyzing Visuals** What cause do you think the surfers may be promoting?

2. **Identifying Cause and Effect** Explain how efforts by grassroots organizations can lead to positive change in environmental issues.

For example, the Border 2020 Program is an environmental program that emphasizes regional and local approaches for decision making, priority setting, and project implementation. It also addresses the environmental and public health problems in the U.S.-Mexico border region. The program empowers citizens by encouraging meaningful relationships and participation among scientists, communities, and local business owners.

The Reducing Emissions from Deforestation and Forest Degradation (REDD+) program is a program designed to use market and financial incentives to reduce the emission of greenhouse gases. Yet some groups worry that for indigenous peoples and other forest communities, REDD+ poses significant risks. That is because it enables companies to buy carbon credits rather than reduce pollution at home. This could lead to indigenous lands being taken in exchange for permits that allow industries to continue to pollute.

Another program aimed at protecting the environment and reducing pollution is *Muévete en Bici*. Launched in 2007 by Mexico City mayor Marcelo Ebrard, the program closes major thruways to auto traffic on Sundays and gives the right of way to tens of thousands of cyclists in a 14-mile (22.5-km) loop. The mayor followed the Sunday rides with the city's *Ecobici* program in 2010. This gives subscribers unlimited access to bicycles at stations for $25 a year. At the end of 2014, the program had 6,200 bicycles at 444 stations for 20 million bicyclists. Encouraging the use of bicycles contributes to the reduction of air pollution in Mexico City by reducing the number of cars on the roads.

✓ READING PROGRESS CHECK

Describing Describe two specific steps taken by the Mexican government to address pollution and resource management.

Yuri Cortez/AFP/Getty Images

LESSON 3 REVIEW

Reviewing Vocabulary

1. *Describing* Define sustainable development and provide at least one example of how it pertains to Mexico.

Using Your Notes

2. *Making Connections* Using your graphic organizer, write a paragraph discussing how Mexico manages its resources.

Answering the Guiding Questions

3. *Making Generalizations* Why are Mexico's resources in jeopardy?

4. *Speculating* How do human activities impact Mexico's environment?

5. *Identifying* How are governments in Mexico addressing environmental issues?

Writing Activity

6. *Argument* Write a letter designed to persuade the government of Mexico to address an environmental problem discussed in this lesson.

Directions: On a separate sheet of paper, answer the questions below. Make sure you read carefully and answer all parts of the questions.

Lesson Review

Lesson 1

1 *Explaining* Describe how Mexico's location on the "Ring of Fire" has helped to shape its landscape.

2 *Describing* Describe the Mexican Plateau and its importance to Mexico in terms of agriculture.

3 *Drawing Conclusions* What industries would be most affected when an oil spill occurs in the Gulf of Mexico?

Lesson 2

4 *Evaluating* How has colonialism and indigenous culture shaped the human geography of Mexico?

5 *Analyzing* How has the Catholic Church influenced Mexican culture?

6 *Summarizing* How does family shape Mexican society?

Lesson 3

7 *Analyzing* Describe how the growth of Mexico's urban middle class has affected the economy of Mexico.

8 *Explaining* What are two results of rural-to-urban migration in Mexico?

9 *Identifying Cause and Effect* What are the causes and effects of deforestation in Mexico?

Critical Thinking

10 *Making Generalizations* Has the maquiladora system had a positive or negative effect on Mexico's people? Explain.

11 *Identifying Cause and Effect* Explain how the growth of the middle class in Mexico has contributed to an increase in awareness and political action in regard to environmental issues.

12 *Drawing Conclusions* How might wars between drug cartels affect the economy of Mexico?

21st Century Skills

Review the graph, then answer the questions that follow.

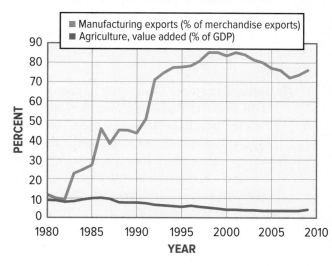

MEXICAN MANUFACTURING AND AGRICULTURE

- ■ Manufacturing exports (% of merchandise exports)
- ■ Agriculture, value added (% of GDP)

Source: World Bank

13 *Using Graphs, Charts, Diagrams, and Tables* Describe the change in Mexico's agricultural exports from 1980 to 2010.

14 *Comparing and Contrasting* How does Mexico's manufacturing growth rate compare to its agricultural growth rate?

College and Career Readiness

15 *Change and Continuity in Economics* As an economic geographer working for the Inter-American Development Bank, you have been asked to help the Mexican government diversify agricultural output. Write a proposal explaining why dependence on a single cash crop can be risky. Then describe how the government could use subsidies, improved technology, and cash payments to help diversify agriculture. Cite an example from primary and secondary sources.

Need Extra Help?

If You've Missed Question	1	2	3	4	5	6	7	8	9	10	11	12	13	14	15
Go to page	176	177	178	181	182	183	189	187	187	186	189	186	191	191	179

Directions: On a separate sheet of paper, answer the questions below. Make sure you read carefully and answer all parts of the questions.

Research and Presentation

16 **Research Skills** Use Internet and library resources to gather information about a particular art form popular in Mexico. Specifically, your research should focus on a description of the art form, the type of materials used in the art form, and the cultural significance of the art form.

Exploring the Essential Question

17 **Making Connections** Choose one of the places discussed in this chapter: Northern Plateau, Mexico City, Yucatán Peninsula, Sierra Madre Occidental or Oriental, or Southern Highlands. Use what you have learned about human systems—history, politics, population, society, culture, and economics—to create a poster illustrating how human systems have impacted your chosen place. Remember to consider the interactions of human systems. Posters should be visual and can include photos, graphs, charts, and maps.

Applying Map Skills

Refer to the Unit 3 Atlas to answer the following questions.

18 **Environment and Society** What generalizations can you make about the location of Mexico's mining areas?

19 **Physical Systems** What is the predominant type of vegetation along the Tropic of Cancer in Mexico?

20 **Human Systems** Using your mental map, imagine you are traveling with your family from the southern tip of Texas to the Mexican Plateau. What major river would you see during your travels? Explain how you visualize this body of water as you read a map, and then explain how this body of water would appear if you were to fly over it during your trip.

DBQ Analyzing Primary Sources

Use the document to answer the following questions.

PRIMARY SOURCE

"In the most of simple terms, this time is solstice. December 21, 2012, marks the end of the 13th Baktun [each Baktun is 144,000 days—or nearly 400 years on the Maya calendar], and it marks the beginning of the 14th Baktun. The significance of 21 December, 2012, this calendar's end, and this particular 13/14 Baktun transition, is that it marks the end of a 26,000 year galactic cycle, and begins the calendar of the next 26,000 years galactic cycle. By the very detailed prophecies of the Mayas, this means leaving the calendar of Night and beginning the calendar of Day."

—Jon Waterhouse, "How the Maya of Today Are Marking December 21,"
National Geographic Explorers Journal, December 19, 2012

21 **Determining Importance** What was the significance of December 21, 2012, on the Mayan calendar?

22 **Identifying** How many days and years does each Baktun represent on the Mayan calendar?

23 **Making Connections** What events in Mexico's history might be symbolic of the transition of "leaving the calendar of Night and beginning the calendar of Day"?

Writing About Geography

24 **Argument** Use standard grammar, spelling, sentence structure, and punctuation to write a one-page essay suggesting suitable locations for constructing new cities to relieve the population pressures that exist in Mexico City. Be sure to describe the types of resources required to sustain large populations.

Need Extra Help?

If You've Missed Question	16	17	18	19	20	21	22	23	24
Go to page	184	176	171	170	168	192	192	192	176

Central America and the Caribbean

ESSENTIAL QUESTION • *How do physical systems and human systems shape a place?*

Christian Aslund/Lonely Planet Images/Getty Images

networks

There's More Online about Central America and the Caribbean.

CHAPTER 8

Why Geography Matters
Spatial Diffusion: The Columbian Exchange

Lesson 1
Physical Geography of Central America and the Caribbean

Lesson 2
Human Geography of Central America and the Caribbean

Lesson 3
People and Their Environment: Central America and the Caribbean

Geography Matters...

Central America and the Caribbean is one of the most biologically diverse regions in the world. Many types of landforms and climates support abundant plant and animal species. Cultures here have been influenced by ancient civilizations, colonial pasts, political upheavals, and by a modern mix of peoples migrating in and out of the region. No single influence, culture, or history can characterize this region of diversity and biological and cultural wealth.

◀ Caribbean music is a diverse blend of African, European, and indigenous influences.

spatial diffusion:
the columbian exchange

Christopher Columbus's voyage to the Americas began a dramatic process of interchange of peoples, animals, plants, cultures, ideas, and diseases between the Eastern and Western Hemispheres. This exchange had both positive and negative consequences. The Columbian Exchange, also known as the Grand Exchange, profoundly altered the course of human development across the globe.

The Columbian Exchange

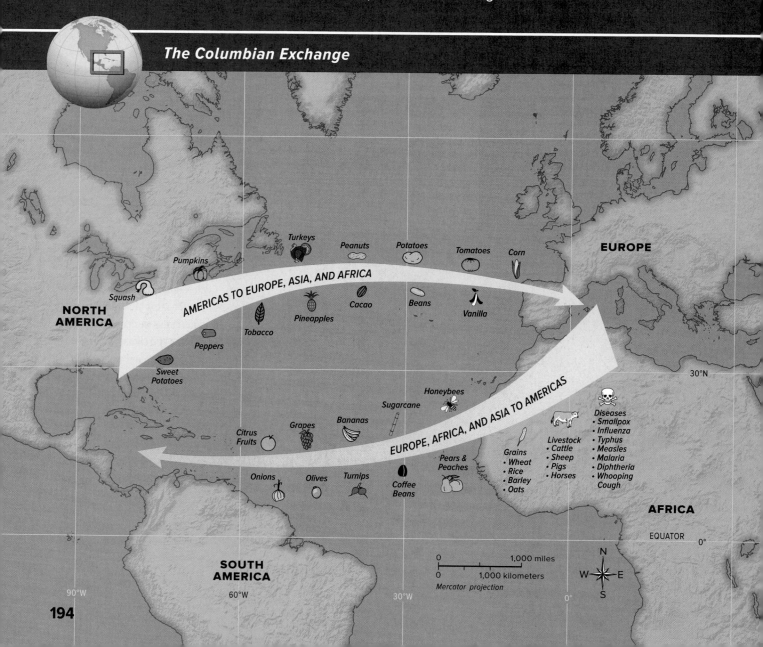

AMERICAS TO EUROPE, ASIA, AND AFRICA

EUROPE, AFRICA, AND ASIA TO AMERICAS

EUROPE

NORTH AMERICA

AFRICA

SOUTH AMERICA

Turkeys
Peanuts
Potatoes
Tomatoes
Corn
Pumpkins
Squash
Cacao
Beans
Vanilla
Pineapples
Tobacco
Peppers
Sweet Potatoes

Honeybees
Sugarcane
Bananas
Citrus Fruits
Grapes
Onions
Olives
Turnips
Coffee Beans
Pears & Peaches

Grains
• Wheat
• Rice
• Barley
• Oats

Livestock
• Cattle
• Sheep
• Pigs
• Horses

Diseases
• Smallpox
• Influenza
• Typhus
• Measles
• Malaria
• Diphtheria
• Whooping Cough

30°N

EQUATOR   0°

90°W   60°W   30°W   0°

0   1,000 miles
0   1,000 kilometers
Mercator projection

N W E S

194

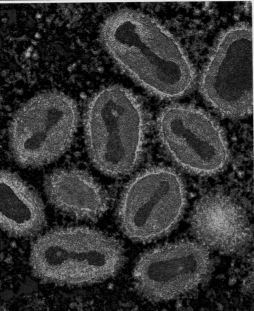

How did the Columbian Exchange affect agriculture?

New foods were exchanged between Europe and the Americas as a major part of the Columbian Exchange. Europeans introduced new food plants and domesticated animals to the Americas in an attempt to "Europeanize" the region. The plants involved in the exchange altered the economies of peoples in the Americas, Europe, and other continents, leaving a lasting imprint on all. Europeans introduced livestock such as cattle, pigs, and sheep as well as grains such as wheat. Sugarcane, coffee, and citrus fruits from Asia were also brought by the Europeans. Yams, bananas, and rice were brought from Africa to the Americas. Europeans also took food crops native to the Americas back across the Atlantic Ocean. Europeans discovered a variety of valuable Native American crops, including beans, squash, chili peppers, sunflowers, tomatoes, sweet potatoes, avocados, and cacao; but potatoes and corn were most important.

1. **Human Systems** Describe how the exchange of food left a lasting imprint in the Americas and Europe.

What was an unintended consequence of the Columbian Exchange?

The unintentional transmission of infectious diseases had serious and long-lasting effects on the peoples and cultures on both sides of the Atlantic Ocean. Human populations that develop in isolation from one another are often highly vulnerable to new infectious diseases introduced by outsiders. Lack of antibodies to the new diseases and lack of access to medicine can result in widespread illness and death. Smallpox, malaria, typhus, cholera, and a host of other diseases were introduced to the Americas by the Europeans. The native populations had no experience with these diseases and therefore had no resistance to them. As a result, native populations were decimated. Although it is difficult to determine the full extent of the population loss, estimates suggest that 80–95 percent of the indigenous peoples may have died in the first 100 to 150 years following the arrival of the Europeans in 1492. Other unintentional consequences included transmission of diseases from the Americas to Europe, although much fewer in number.

2. **Environment and Society** Write a summary of how the culture and history of Central America and the Caribbean might have been different if diseases had not been transmitted during the Columbian Exchange.

How were the lifestyles of indigenous peoples altered?

Contact with Europeans not only wiped out most of the native population of the Americas through disease, but also permanently altered the culture of those who survived. For example, because the Americas lacked large animal species, indigenous peoples had no beasts of burden. Asians and Europeans had long since domesticated several large animals for use in agriculture, trade, and warfare. Horses, long extinct from the Americas, were reintroduced by the Spanish, whose horse-mounted soldiers quickly overran and dominated indigenous forces. The horse transformed the Americas through warfare, hunting, and culture. Of even greater significance was that people themselves crossed the Atlantic. The presence of new peoples and their impact on the indigenous peoples of the Americas, due to the Columbian Exchange, initiated one of the largest cultural transformations in human history.

3. **Human Systems** What advantages did horses give Spanish invaders in the Americas?

THERE'S MORE ONLINE

VIEW a slide show of goods that were exchanged • **READ** a quote about the effects of the Columbian Exchange

netw⊙rks

There's More Online!

☑ **MAP** Physical Geography: Central America and the Caribbean

☑ **INFOGRAPHIC** Biodiversity in Central America

☑ **INTERACTIVE SELF-CHECK QUIZ**

☑ **VIDEO** Physical Geography of Central America and the Caribbean

LESSON 1

Physical Geography of Central America and the Caribbean

Reading HELPDESK

Academic Vocabulary

- **energy**
- **exhibit**

Content Vocabulary

- **isthmus**
- **archipelago**
- **biodiversity**

TAKING NOTES: *Key Ideas and Details*

SUMMARIZING Use a graphic organizer like the one below to take notes on the natural resources of Central America and the Caribbean.

Natural Resources: Central America and the Caribbean

ESSENTIAL QUESTION • *How do physical systems and human systems shape a place?*

IT MATTERS BECAUSE

Central America acts as a land bridge connecting North and South America. The region boasts dense rain forests, coastal plains, and high mountains, and is one of the world's great biodiversity hotspots. The more than 7,000 islands of the Caribbean exhibit their own unique landforms and living things. Shaped in part by the Ring of Fire, this region is a volatile zone of earthquakes and volcanic eruptions.

Landforms

GUIDING QUESTION *Why are the majority of Central America's people concentrated in the Central Highlands?*

Much of Central America is hilly or mountainous, although swamps and lowlands extend along both coasts. These landforms create three distinct belts: the Pacific Lowlands, the Caribbean Lowlands, and the Central Highlands. The narrow plains of the Pacific Lowlands extend from Guatemala to Panama. The Caribbean Lowlands are also narrow, except in Nicaragua and Honduras. Central America's most distinctive landforms—mountains—form the Central Highlands. The region climbs steadily higher west of the Caribbean Lowlands. It rises up to the western plateau highlands where mountains and some 40 volcanic cones attain elevations of more than 12,000 feet (3,700 m). These are the volcanic highlands, or the Volcanic Axis. These mountains are an extension of the Sierra Madre of Mexico. Volcanic eruptions and earthquakes are not uncommon. The weathered lava produces fertile soil, making these highlands rich agricultural zones and areas of dense population.

Its location on the Ring of Fire brings potential hazards to living there. Yet humans have thrived in the mountains, valleys, and plateaus of the Central Highlands for thousands of years. The region's cooler climates, adequate rainfall, and rich natural resources—water, volcanic soil, timber, and minerals—attracted the area's earliest peoples and is where the majority of the people live today.

Another distinctive feature is the **Isthmus** of Panama. It extends west to east, connecting North and South America and separating the Caribbean Sea from the Gulf of Panama. The mountains, swampy coastal lands, and dense rain forests make contact between people difficult in this area.

In the Caribbean, many of the more than 7,000 islands are the tops of mountains that are part of the mainland's Central Highlands. The islands of the Greater and Lesser Antilles, however, are part of an **archipelago.** This archipelago is composed of the crests and peaks of a mountain range formed from collisions between the Caribbean plate and other tectonic plates. Tectonic activity continues to change the landscape. In 2010, for example, an earthquake struck Haiti's capital, Port-au-Prince, collapsing buildings and killing large numbers of people.

isthmus a narrow strip of land connecting two larger land areas

archipelago a group or chain of islands

✓ **READING PROGRESS CHECK**

Identifying What factors attracted people to settle in the highlands?

Water Systems

GUIDING QUESTION *How are Central American rivers and lakes important to the human systems of the area?*

Inland lakes and waterways play a vital role in Central America, aiding growth and development. The water systems also provide transportation, drinking water, drainage, irrigation, and a source of hydroelectric power. Lake Nicaragua is Central America's largest freshwater lake. It is the only one in the world to contain oceanic animal life such as sharks, swordfish, and tarpon. Nearby Lake Managua has commercially viable fish and alligators. It is drained by a river that flows into Lake Nicaragua and fed by streams from the highlands. Nicaragua's capital city, Managua, is located on the southern shore of Lake Managua.

> **GEOGRAPHY CONNECTION**
>
> A wide array of landforms and varied elevations give Central America and the Caribbean a diverse physical landscape.
>
> 1. *PHYSICAL SYSTEMS* What is the name of the most dominant mountain range in Central America that runs from north to south?
>
> 2. *PLACES AND REGIONS* Name the bodies of water that border the islands of the Caribbean.

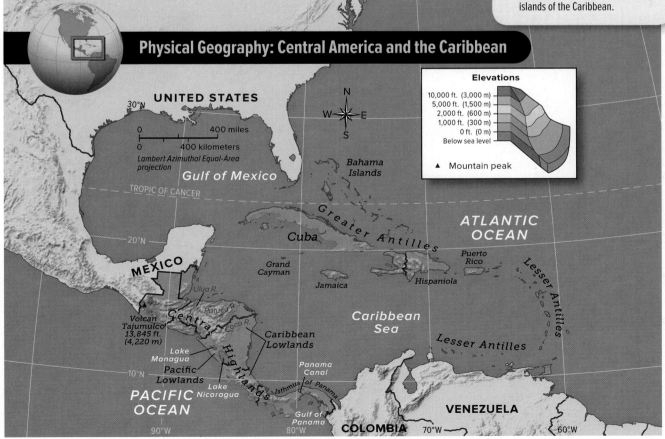

Physical Geography: Central America and the Caribbean

The Panama Canal bisects the Isthmus of Panama. It is one of the world's most important human-made waterways. The canal allows ships to travel between the Atlantic and Pacific Oceans without making the long trip around South America's Cape Horn. International traffic through the Canal Zone is dominated by ships traveling between East Asia and the eastern United States. Approximately two-thirds of trading ships are sailing to or from U.S. ports.

Many of Central America's rivers provide commercial water routes because they are short and steep. The San Juan River, an outlet for Lake Nicaragua, drains into the Caribbean Sea. El Salvador's only navigable river, the Lempa River, generates hydroelectricity from the **energy** of moving water.

The warm, clear waters of the Antillean regions of the Caribbean Sea contain coral reefs. These reefs **exhibit** a wide variety of reef fish. Other common types of characteristic marine life include manatees, manta rays, spiny lobsters, dolphins, and numerous species of sea turtles. Commercial fishing of sardines and tuna and the utilization of other marine resources have increased international trade.

energy usable power

exhibit to demonstrate or show openly

✅ READING PROGRESS CHECK

Describing Describe the importance of the Panama Canal to international trade.

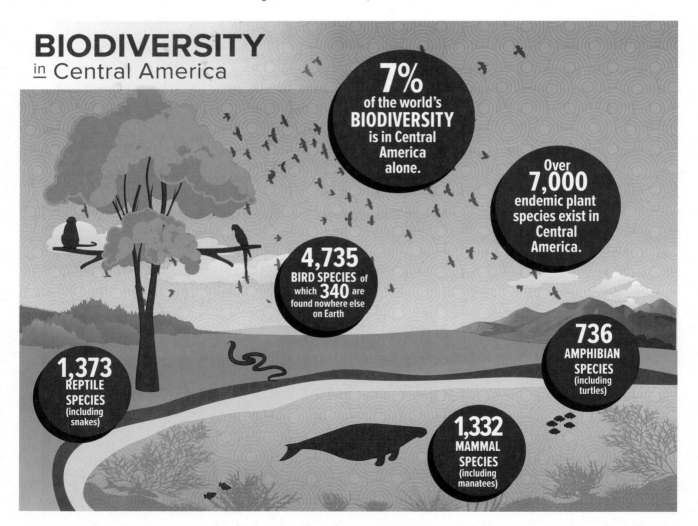

BIODIVERSITY
in Central America

7%
of the world's
BIODIVERSITY
is in Central
America
alone.

Over 7,000 endemic plant species exist in Central America.

4,735
BIRD SPECIES of
which **340** are
found nowhere else
on Earth

736
AMPHIBIAN
SPECIES
(including
turtles)

1,373
REPTILE
SPECIES
(including
snakes)

1,332
MAMMAL
SPECIES
(including
manatees)

The natural richness of Central America is evident in its diverse animal and plant life.

▲ CRITICAL THINKING

1. *Analyzing* In what way might Central America's position as a land bridge linking two continents affect its biodiversity?

2. *Drawing Conclusions* What might be one effect of the loss of biodiversity in Central America?

Climate, Biomes, and Resources

GUIDING QUESTION *How does its location in the Tropics affect the climate of Central America and the Caribbean?*

Climates in the region are dependent on many factors, such as proximity to the sea, elevation, latitude, and local topography. For much of the region, its location in the Tropics and the prevailing winds that carry warm, moist air from the Atlantic Ocean result in high temperatures and abundant rainfall year-round. A tropical rain forest climate and rain forests dominate much of Central America. The tropical forest biome has a continuous canopy of trees and diversity of species. These species include tall trees, ferns, and mosses, as well as birds, bats, small mammals, and insects. Costa Rica and Panama are global **biodiversity** hotspots, rich in natural resources that may one day provide medicines and other products.

A tropical wet/dry climate is typical of Caribbean islands. Moist winds sweep in from the east. This leaves leeward island areas—those not exposed to wind—dry. Thus some islands may have high temperatures and rainfall but also an extended dry season. Grasslands flourish, although soils may not be very fertile or suitable for large-scale agriculture. On mountainous islands, windward slopes facing moist winds help cool the warm, humid air and increase precipitation. Intense hurricanes strike each year from June to November in the northern Caribbean and the Gulf of Mexico.

Elevation affects the climate and ecosystems of some parts of Central America more than distance from the Equator. Such areas have vertical climate zones in which animals, plants, and climates change as altitude increases. The *tierra caliente*, or "hot land," lies at elevations below 2,500 feet (762 m). Bananas and sugarcane are grown here. The *tierra templada*, or "temperate land," lies between 2,500 to 6,000 feet (762 to 1,829 m) and is the most densely populated. Ecosystems with broadleaf evergreens at lower altitudes give way to evergreens at upper elevations. Land at 6,000 to 10,000 feet (1,829 to 3,048 m) is known as the *tierra fria,* or "cold land." Winter frosts are common here, and potatoes and barley are grown. Above the tree line is the *tierra helada,* or "frozen land." This is an area of permanent ice and snow.

Natural resources extracted or used throughout Central America include nickel, iron ore, fish, timber, and petroleum. Guatemala and certain islands refine petroleum, while Belize extracts crude oil. Much commercial fishing also takes place around the Caribbean islands.

☑ **READING PROGRESS CHECK**

Categorizing How is the climate of Central America different from that of the Caribbean islands?

Connecting Geography to **SCIENCE**

Climatology

Predicting when hurricanes will form and where they will land is often difficult. In recent years, climatologists have used El Niño to help predict hurricanes. El Niño is a climatic change that affects the equatorial Pacific. It includes uncommonly warm waters off the western coast of Peru and Ecuador. Often the increased wind levels during El Niño prevent hurricanes from forming, so there tend to be fewer Atlantic hurricanes during El Niño. Knowing when El Niño winds will develop can help predict when hurricanes are a threat and when they are not.

IDENTIFYING How do climatologists use El Niño to help them predict the behavior of hurricanes?

biodiversity biological diversity in an environment as indicated by numbers of different species of plants and animals

---

LESSON 1 REVIEW

Reviewing Vocabulary

1. ***Making Connections*** Write a sentence to describe the difference between an isthmus and an archipelago.

Using Your Notes

2. ***Listing*** Use your graphic organizer to list the natural resources of Central America and the Caribbean and their importance to the region.

Answering the Guiding Questions

3. ***Describing*** Why are the majority of Central America's people concentrated in the Central Highlands?

4. ***Examining*** How are Central American rivers and lakes important to the human systems of the area?

5. ***Interpreting*** How does its location in the Tropics affect the climate in Central America and the Caribbean?

Writing Activity

6. ***Informative/Explanatory*** Write a paragraph describing how altitude affects the climates and biodiversity of Central America and the Caribbean.

networks

There's More Online!

- ☑ **IMAGE** Kuna Molas
- ☑ **IMAGE** The Panama Canal
- ☑ **INTERACTIVE SELF-CHECK QUIZ**
- ☑ **TIME LINE** Central America and the Caribbean: Paths to Independence
- ☑ **VIDEO** Human Geography of Central America and the Caribbean

LESSON 2
Human Geography of Central America and the Caribbean

ESSENTIAL QUESTION • *How do physical systems and human systems shape a place?*

Reading **HELP**DESK

Academic Vocabulary

- **attribute**
- **obvious**

Content Vocabulary

- **population pressure**
- **dialect**
- **patois**
- **matriarchal**
- *latifundia*
- *minifundia*
- **cottage industry**
- **ecotourism**

TAKING NOTES: *Key Ideas and Details*

IDENTIFYING Use a graphic organizer like the one below to take notes as you read the lesson.

IT MATTERS BECAUSE

A study of the human geography of Central America and the Caribbean highlights how history, geography, and the blending of native and outside cultures have shaped the region. The region's mixing of languages, religions, economic practices, and social customs combine traditional and modern. Such diversity provides both benefits and challenges as the region works toward increased economic and social development.

History and Government

GUIDING QUESTION *How did colonialism influence the history and government of Central America and the Caribbean?*

The voyages of Christopher Columbus to the Americas from 1492 to 1504 triggered a period of conquest and colonization in Central America and the Caribbean. Spain founded the region's first permanent European settlement on the island of Hispaniola in 1493. Rodrigo de Bastidas made Spain's first claim to Central America in March 1501. Other Spanish conquistadors followed. They attempted to subdue the local people and establish permanent colonies. Vasco Nuñez de Balboa explored extensively, crossing the isthmus, claiming land for Spain, and finding enough gold and pearls to establish the first profitable colony in the Americas, called Castilla del Oro.

The king of Spain replaced Balboa with Pedro Arias Dávila, known as Pedrarias. Pedrarias expanded the colony, but was notorious for enslaving and murdering indigenous people. In 1519, as governor of Panama, he established Panama City on the Pacific coast and later moved the capital there. In 1524 Pedrarias sent Francisco Hernández de Córdoba to Nicaragua to conquer the region. Córdoba established Granada on Lake Nicaragua. He also founded León not far from Lake Managua and attempted to rule Nicaragua himself. Pedrarias had Córdoba executed and named himself governor of Nicaragua. While Pedrarias and Córdoba fought to conquer southern Central America, Cristóbal de Olid set sail to Honduras. Pedro de Alvarado, who had served with Cortés in the conquest of Mexico, set out overland to conquer Guatemala and El Salvador. When gold was discovered in Honduras, various Spanish forces fought each other for control of the region. In Costa Rica, the native population strenuously resisted Spanish efforts at conquest. Spain did not establish a permanent colony there until 1561.

The physical geography influenced colonization of the region. Remote areas of Central America remained outside of Spanish control. This allowed Great Britain to colonize Belize and the Mosquito Coast of Nicaragua. British Honduras (Belize) was the only non-Spanish colony in Central America. In time, however, France, the Netherlands, and Portugal also established colonies in the Caribbean and other parts of the Americas.

By the mid-1600s, forced labor, starvation, and European diseases had nearly killed the entire indigenous population. For the Europeans, the drastically reduced numbers of indigenous people meant a labor shortage. They began bringing Africans, who had been forcibly captured, enslaved, and transported by ship, to the Caribbean to meet the demand for workers.

In the late 1700s, Africans and indigenous people began to take organized action to free themselves from slavery and European control. François Toussaint-Louverture, a soldier born to enslaved parents, led a revolt of enslaved Africans in Haiti. By 1804, Haiti had won its independence from France. Haitian independence inspired downtrodden people throughout Latin America, but frightened many of the elite who had much to lose. Most Caribbean colonies did not gain independence until the 1900s. Cuba gained self-rule in 1898 as a result of the Spanish-American War, but remained under the protection of the United States until 1902. Some islands remain under foreign control or continue to have foreign ties today.

During the 1800s, several Central American colonies struggled for independence from Spain. In 1823 the federation of the United Provinces of Central America was formed. Eventually it divided into five separate countries: Costa Rica, El Salvador, Guatemala, Honduras, and Nicaragua. In 1903 Panama declared its independence from Colombia and signed a treaty with the United States creating the Panama Canal Zone. The Panama Canal opened in 1914 and was controlled by the United States. Today, under Panamanian control, the canal is being upgraded to reduce traffic congestion and to allow larger ships to pass.

In the 1900s, many Central American and Caribbean countries faced political, social, and economic upheaval. In Panama, the Canal Zone and related industries brought new wealth to the upper classes. Most Panamanians benefited little from the canal, however. In Cuba, a revolution in 1959 produced a communist state. Fidel Castro ruled until 2008 when he handed power to his brother Raúl, who initiated some economic reforms. Armed conflict, civil wars, unstable economies, and poor social conditions were all problems in the region. By the end of the twentieth century, however, Central American governments sought ways to increase international trade and diversify their economies. In the twenty-first century, many Central Americans have exercised their right to vote, demanding positive change. For example, Guatemala ended its history of rule by military regime and continues its political and economic recovery, including regular elections since 1996.

✔ READING PROGRESS CHECK

Explaining What role did Haiti play in the movement for independence from European control in the region?

The Panama Canal provides a passage through the isthmus by means of a system of canal locks.

▼ CRITICAL THINKING

1. Speculating What major challenges would builders of the Panama Canal have faced?

2. Identifying Which bodies of water does the Panama Canal link?

WHAT KIND OF DEVELOPMENT IS BEST FOR HAITI?

The Human Development Index ranks Haiti as the poorest country in the Western Hemisphere. Most of the population lives in absolute poverty, while a large percentage is unemployed or underemployed. Health care resources, including basic sanitation systems, are lacking. Although education is required for children between the ages of six and twelve, only a small percentage of children actually attend school because of a lack of facilities and staff.

In geography, the term *development* refers to improvements in the social and economic welfare of people as well as improvements in production and technology. The ways in which a country experiences development vary greatly—from NGO (nongovernmental organization) programs to government initiatives to foreign business arrangements. Funding in the form of foreign aid, investments, and loans makes development possible, but such support often includes special requirements or limits.

A devastating magnitude 7.0 earthquake on January 12, 2010, reduced much of Port-au-Prince, Haiti, to rubble. Southern areas of the country were also affected. Buildings of all kinds—from shantytowns to hospitals to national landmarks—were damaged or completely collapsed. Telephone service and electricity were knocked out. According to reports, more than 300,000 people lost their lives and over 600,000 were displaced. Estimates show the total cost of the disaster to be between $8 billion and $14 billion. In October 2010, after the earthquake, one of history's worst cholera outbreaks hit Haiti. Cholera is a waterborne disease that is spread through contaminated water. The outbreak was made worse by the earthquake's destruction and lack of public sewage systems. Such loss has only served to magnify Haiti's need for economic and social development.

Create Market Opportunities

PRIMARY SOURCE

❝ Haiti has duty-free, quota-free access to the American market guaranteed for the next nine years, with generous rules of origin well-suited to the garment industry. . . .

Of course, market access is not enough: costs of production must be globally competitive. . . . In garments the largest single component of costs is labour. Due to its poverty and relatively unregulated labour market, Haiti has labour costs that are fully competitive with China, which is the global benchmark.

Market access and costs of production are not the only factors of importance: transport to market is also a fundamental consideration. . . . Haiti is on the doorstep of its market. Since it is the only low-wage economy in the region, it has a transport advantage over competing low-wage economies of several thousand miles. . . . ❞

—Paul Collier, *Haiti: From Natural Catastrophe to Economic Security,* A Report for the Secretary General of the United Nations, January 2009

Develop Local Agriculture

PRIMARY SOURCE

❝ Municipal governments should construct properly equipped marketplaces for the women who sell rural produce. The Haitian state should develop trade policies aimed at protecting the agricultural sector, and take the lead in fixing roads and ports, confronting deforestation and improving systems of water management. . . ❞

The return on the investment in the rural economy would be self-reliance, the alleviation of dangerous overcrowding in cities and, most important, a path toward ending Haiti's now chronic problems of malnutrition and food insecurity. . . . ❞

—Laurent Dubois and Deborah Jenson, "Haiti can be rich again," *New York Times,* January 8, 2012

What do you think? DBQ

1. ***Drawing Conclusions*** According to Collier, why is the garment industry an ideal market opportunity for Haiti?

2. ***Identifying Central Issues*** Why do Dubois and Jenson argue that protecting the agricultural sector is best for Haitian development?

3. ***Evaluating*** Who do you think makes the stronger argument, Collier or Dubois and Jenson? Explain.

Reading HELPDESK

Academic Vocabulary

- **cooperation**
- **sustainable**

Content Vocabulary

- **sedimentation**
- **reforestation**

TAKING NOTES: *Key Ideas and Details*

IDENTIFYING Use a graphic organizer like the one below to take notes as you read about the issues relating to people and their environment in Central America and the Caribbean.

LESSON 3

People and Their Environment: Central America and the Caribbean

ESSENTIAL QUESTION • *How do physical systems and human systems shape a place?*

IT MATTERS BECAUSE

People and governments in Central America and the islands of the Caribbean struggle to develop modern economies without destroying the natural environment and the invaluable resources and biodiversity it provides. Balancing the needs of humans with sustainable environmental practices is both an important goal and a major challenge for countries of the subregion.

Managing Resources

GUIDING QUESTION *How do growing human needs affect resources and the environment in Central America and the Caribbean?*

Countries in Central America and the Caribbean face a daunting challenge: How can they preserve and manage their resources while developing their economies and meeting the increasing needs of a growing population? Water shortages, access to freshwater, and legal issues related to waterways are a big concern for several countries in the subregion. Urban populations in Central America and the Caribbean increase at a rate that strains their cities' ability to provide freshwater for personal and industrial use.

Water has been at the center of border disputes throughout Central America's history. The latest dispute is part of a conflict that is two centuries old. It involves a wetlands area that is part of a nature reserve owned by Costa Rica. The conflict erupted in October 2010 when Nicaragua began a controversial dredging project to redirect the San Juan River—a Nicaraguan-controlled waterway that forms part of the border—and posted Nicaraguan troops at the site. Costa Rica and other neighboring countries asked to have the troops withdrawn. The Nicaraguan government refused, saying it was reclaiming a natural resource. Costa Rica took the dispute to the United Nations. The United Nations ordered all troops to withdraw. This allowed Costa Rica to monitor potential damage to the wetlands environment, but did not stop Nicaragua from clearing sand from the river's channel.

To increase the power supply necessary for industry, some countries in Central America and the Caribbean turn to the construction of hydroelectric power plants. Such plants are located within human-made dams that raise the level of water so that when intake channels are opened, water flows rapidly through the dams, turning water turbines. The water turbines capture the energy of fast-falling or flowing water and produce electricity from generators.

Costa Rica benefits from the power generated by hydroelectric plants. The plants supply energy to Costa Rica's industries, allowing them to increase production capabilities and become more competitive in international trade. Today about 65 percent of Costa Rica's electricity is produced from hydroelectric plants. Hydroelectric dams promote clean energy use by harnessing the power of water, a renewable resource.

However, hydroelectric systems threaten the natural environment of the areas from which they extract energy. The interaction of plant and animal species within an ecosystem is very sensitive and complex. Fish habitats, for example, are affected by factors such as water level, water velocity, and the availability of food and shelter from predators. The flooding created by hydroelectric dams dramatically alters the habitats and devastates native fish populations. Hydroelectric projects on El Salvador's Lempa River provide most of the country's power needs. These dams harm the natural environment, however, because of changing water levels and the construction of roads through otherwise unaffected landscapes. Similarly, the hydroelectric project on the Patuca River in Honduras blocks fish migrations, alters habitats, and threatens the survival of the region's indigenous population who depend on the environment.

Central America boasts a biodiversity hotspot with naturally fertile soil. Yet much of the subregion's timberland has been cleared by slash-and-burn farmers. In this process of cultivation, all plants are cut down and any trees are stripped of bark. After the plants and trees have dried out, they are set on fire. The ash from the fire adds nutrients to the soil, making agriculture more productive. Unfortunately, frequent rains leach away the beneficial nutrients and within a year or two the soil loses its fertility. Crop yields decline due to poor soils, and farmers move on to clear new parts of the forest. The spent land supports little growth of vegetation. Within just a few years, huge swaths of centuries-old rain forests have disappeared.

Deforestation is a major threat to biodiversity as shown here in Panama and throughout Central America and the Caribbean.

◀ **CRITICAL THINKING**

1. *Analyzing Visuals* What features of the land shown in the forest signify that deforestation has taken place?

2. *Predicting* How does the destruction of rain forests affect biodiversity?

Slash-and-burn cultivation is not the only activity that contributes to deforestation, or the clearing or destruction of forests. Commercial logging operations are key components of the region's economies. The logging companies harvest trees for timber and other products, which are sold as exports.

As Central American forests are depleted, habitats are lost, resources are threatened, and Earth's biodiversity dwindles. Deforestation severely alters biologically rich ecosystems of tropical rain forests and threatens various species of plants and organisms from which key medicines are derived. Scientists are trying to save species from extinction by creating corridors of vegetation that connect remaining areas of the forest to provide animals with access to habitats. Deforestation may also mean that less carbon dioxide is captured in plants. This would cause higher levels of carbon dioxide to remain in the atmosphere.

☑ READING PROGRESS CHECK

Gathering Information What are the effects of slash-and-burn farming?

Human Impact

GUIDING QUESTION *Why is soil erosion and soil decline such an issue in Central America and the Caribbean?*

As cities in Central America and the Caribbean experience rapid urbanization, population growth exceeds available resources like housing, running water, sewage systems, and electricity. Rural workers migrating to cities often cannot find jobs or adequate housing. Thousands are forced to live in slums or shantytowns on the edges of cities. Such communities often rest on dangerous slopes and near delicate wetlands. Mud slides, floods, and other natural disasters can wipe out entire communities. Because they lack running water and underground sewage systems, these areas are unsanitary. As a result, disease can spread rapidly. Cities that expand rapidly may also encroach on natural landscapes that have previously been untouched by human impact.

The clearing of land has led to soil erosion in some parts of Central America and the Caribbean.

CRITICAL THINKING ▼

1. *Identifying Cause and Effect* How is a growing population and increased urbanization related to soil erosion?

2. *Describing* What other factors have contributed to soil erosion in Central America and the Caribbean?

The rapid growth of cities creates high pollution rates from overloaded sewage, electric, and water systems. Air pollution affects people in cities without adequate clean-air laws. Vehicles clog city streets and release massive amounts of exhaust (greenhouse gases) into the air. Industrial pollutants from factory smokestacks built by multinational firms under free-trade agreements also foul the air. Similarly, runoff from chemical fertilizers and pesticides used on commercial farms damages oceans and freshwater areas, including the Caribbean's coral reefs.

City sewage is a major contributor to water pollution in Central America.

As cities in Central America and the Caribbean grow, the demand for food also increases. The effect of expanding agriculture on the natural environment can be devastating. To increase food supplies, Central American farmers and ranchers clear forested regions, which not only destroys forests but also causes the soil to erode and lose fertility over time. In the Caribbean, small islands colonized for intensive plantation agriculture became prone to soil erosion because of the elimination of natural vegetation. Furthermore, after more than 300 years of commercial agriculture, soil fertility has decreased to the point where large applications of fertilizer are necessary for plant production.

Another factor affecting vegetation loss and soil erosion is the construction of hotels and other structures to support the tourism industry. As rain falls, it washes soil eroded by construction and farming down hills and into the sea, resulting in the **sedimentation** of the reef system. Ultimately, sedimentation may kill coral beds in the water. When fertilizers also become part of the runoff into the sea, they damage coral further and threaten the hundreds of fish species, marine turtles, and sharks that live in the reefs. Communities that depend on the reefs for their livelihoods and food security are also affected.

☑ READING PROGRESS CHECK

Assessing How has human activity influenced soil erosion and its threat to coral reefs?

Addressing the Issues

GUIDING QUESTION *Why is biodiversity protection so important in Central America?*

Central America and the Caribbean face many international challenges, including conflicts over natural resources and the need to prepare for wide-scale natural disasters. Regional **cooperation** in addressing issues that reach beyond national borders will help the subregion move forward. Government, private industries and grassroots efforts all contribute to protecting the environment.

If the issue of deforestation is not addressed, the rain forests of this subregion will be greatly reduced within 40 years. Although the threats to the world's rain forests are well known, the proposed strategies for preserving them are hotly debated. For instance, Costa Rica and other countries with rain forests listen to the advice of scientists and environmentalists, but they still face pressing social and economic realities. If Costa Rica were to ban the use of rain forest lands, for example, how would it provide for the people who would no longer have a way to support themselves? How would the country handle

▲ CRITICAL THINKING

1. *Analyzing Visuals* What items in the photograph could pose a threat to sea animals?

2. *Describing* In what way could the pollution shown be an effect of rapid urbanization? Explain.

sedimentation the action or process of forming or depositing sediment

cooperation a common effort

sustainable a method of harvesting or using a resource so that the resource is not depleted or permanently damaged

reforestation planting young trees or seeds on lands where trees have been cut or destroyed

Orlando Sierra/AFP/Getty Images

Indigenous Communities Protect Biodiversity

"José López Hernández, a member of the Oxlajuj No'j tribe, would chop down trees in Santa Maria de Jesus, Guatemala, before IDB (Inter-American Development Bank) started a project to restore local traditions and culture as a way to prevent further land degradation and to preserve the region's biodiversity.

Now Hernández is a leader of his indigenous community and is working, together with another 600 families, to plant 60,000 new trees by the end of the year using traditional organic methods."

—Silvia Lambiase, "Indigenous communities in Central America use traditions to protect biodiversity," Platform for Agrobiodiversity Research, May 2011

DBQ *HYPOTHESIZING*
Based on the passage above, what can be inferred about the traditional organic methods in regard to biodiversity?

population growth in coastal areas if vast parts of the interior were not open to settlement? Some of these answers lie in **sustainable** development—technological and economic growth that does not deplete the human and natural resources of a given area. Given time, of course, rain forests would regenerate on their own, but with a considerable loss of biodiversity. Laws requiring **reforestation**—the planting of young trees or the seeds of trees on the land that has been stripped—can help. Developing new methods of farming, mining, and logging and combining conservation with responsible tourism protect the forests and boost local economies.

Initiatives are also underway across Central America to develop "green businesses." The businesses include environmentally-friendly organic food production, renewable energy, and sustainable tourism. Large industries are looking at ways to increase energy efficiency and improve production practices in an effort to be more environmentally friendly.

Governments, international agencies, and grassroots groups are beginning to address the needs of the subregion's urban areas. Work is being done to create programs that limit rural-to-urban migration and to improve the infrastructure of cities. Programs that encourage cooperation with other countries to establish incentives for industries to reduce pollution have also been implemented. A United Nation's initiative in Central America called Reducing Emissions from Deforestation and Forest Degradation, or REDD+, is a program of environmental credits. PRODESEC (Economic Development Program for the Dry Region of Nicaragua) is a Nicaraguan plan aimed at providing new opportunities for rural families. Environmental laws and programs, however, have not reduced the risks of increased pollution associated with industrial growth.

Human activity is not the only source of potential problems for Central America and the Caribbean. The physical geography of the region leaves the area vulnerable to extreme weather-related disasters, such as hurricanes and floods, as well as devastating earthquakes. In order to increase the region's emergency preparedness, governments are cooperating in the use of sophisticated technology, such as satellite imaging and computer modeling, to forecast the direction and severity of natural disasters such as hurricanes.

Scientists are also gathering detailed information about volcanic eruptions in the Caribbean. Since a 1997 major volcanic eruption in Montserrat, scientists are closely monitoring the volcano, which continues to be active. Montserrat's volcano is similar to those on other continents, so the information and lessons learned will help produce more detailed forecasts and predictions around the world.

✅ **READING PROGRESS CHECK**

Exploring Issues Why is reforestation so important?

LESSON 3 REVIEW

Reviewing Vocabulary
1. *Determining Importance* What is the significance of sedimentation and reforestation?

Using Your Notes
2. *Describing* Use your graphic organizer to describe the threats to coral reefs from development and pollution.

Answering the Guiding Questions
3. *Assessing* How do growing human needs affect resources and the environment in Central America and the Caribbean?

4. *Explaining* Why is soil erosion and decline such an issue in Central America and the Caribbean?

5. *Applying* Why is biodiversity protection so important in Central America?

Writing Activity
6. *Informative/Explanatory* Think about the physical environment of the state in which you live. In two paragraphs, compare the ways urban populations in Central America and those in your state have modified their physical environments.

Directions: On a separate sheet of paper, answer the questions below. Make sure you read carefully and answer all parts of the questions.

Lesson Review

Lesson 1

1 *Evaluating* Which Central American waterways provide an important link among the region's countries? Give examples.

2 *Identifying* List some natural resources of the Central America and Caribbean subregion that are important to the global economy.

Lesson 2

3 *Summarizing* Describe the ways in which each of the following factors have influenced history and government in Central America and the Caribbean: indigenous cultures, colonialism, slavery, and struggles for freedom.

4 *Identifying Cause and Effect* What urban challenges have been caused by the migration of many people in Central America and the Caribbean to capital cities and major ports?

Lesson 3

5 *Interpreting* What does the phrase "sustainable development" mean? What is one example of sustainable development?

6 *Identifying Central Issues* Why is deforestation a major problem in Central America? Include reasons why many people cut down forests and reasons many others want to restore them.

Exploring the Essential Question

7 *Speculating* How do physical systems and human systems shape a place? Give specific examples showing why the highland regions have attracted human settlement and migration throughout Central America.

21st Century Skills

8 *Analyzing* Why is regional cooperation particularly important to address international challenges in Central America?

9 *Compare and Contrast* Create a chart of the countries in Central America, ranking them in order of population. Use the Internet to research the latest statistics.

10 *Making Connections* What circumstances might make environmental protection a low priority for some Central American people? List one government program aimed at meeting the needs of the people and protecting the environment.

College and Career Readiness

ETHNIC GROUPS IN THE CARIBBEAN

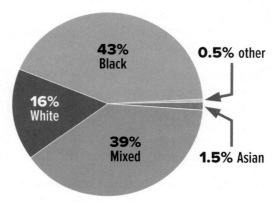

43% Black

0.5% other

16% White

39% Mixed

1.5% Asian

Source: Encyclopedia Britannica Almanac 2009

Use the circle graph to answer the questions below.

11 *Draw Conclusions* Based on what the circle graph shows about ethnic groups in the Caribbean, how do you think these ethnic groups are reflected in the culture of the subregion?

12 *Compare and Contrast* How does the percentage of White compare to the percentage of Mixed in the Caribbean?

Need Extra Help?

If You've Missed Question	**1**	**2**	**3**	**4**	**5**	**6**	**7**	**8**	**9**	**10**	**11**	**12**
Go to page	197	199	200	202	212	210	199	211	202	211	213	213

Directions: On a separate sheet of paper, answer the questions below. Make sure you read carefully and answer all parts of the questions.

Critical Thinking

13 *Exploring* Describe how physical features have kept some of Central America's people isolated.

14 *Drawing Conclusions* Write a paragraph describing the series of events that led to the exploration, conquest, and colonization of Central America and the Caribbean.

15 *Explaining* Describe how elevation can affect the climate and ecosystems more than distance from the Equator in some parts of tropical Central America. Provide specific examples.

Applying Map Skills

Use your Unit 3 Atlas to answer the following questions.

16 *Physical Systems* What is the predominant type of natural vegetation found in Central America?

17 *Places and Regions* What generalizations can you make about the population density in Honduras compared to the population density of El Salvador?

18 *Human Systems* Use your mental map of Central America to describe how the Panama Canal saves travel distance and time for ocean travel from the west coast of North America to the Caribbean.

Research and Presentation

19 *Explaining* Choose a country in Central America or the Caribbean. Write a paragraph describing the landforms within that country. Explain how those landforms affect trade both within the country and with other countries.

Writing About Geography

20 *Narrative* Use the Internet to research and write a three-paragraph account of a cultural event, such as a festival, holiday, or music or dance concert, important to people living in Central America or the Caribbean.

DBQ Analyzing Primary Sources

Read the excerpt and use it to answer the following questions.

At a summit on biodiversity in Japan, the World Future Council announced that Costa Rica had won the 2010 Future Policy award.

PRIMARY SOURCE

"*Costa Rica channels funds from a fuel tax, car stamp duty and energy fees to pay for nature reserve management and environmental services like clean air, fresh water and biodiversity protection.*

Landowners are paid to preserve old-growth forests and to plant new trees. As a result, forest cover has risen from 24% in 1985 to close to 46% today.

It has also established a national commission on biodiversity, comprising scientists, civil servants and indigenous representatives, which proposes policies to the government and promotes green education among the public.

'We are declaring peace with nature,' said Mario Fernández Silva, the ambassador of Costa Rica, referring also to his country's abolition of its army in 1958. 'We feel a strong sense of responsibility about looking after our wealth of biodiversity. Our attitude is not progressive; it is conservative. Our view is that until we know what we have, it is our duty to protect it.'"

—Jonathan Watts, "Costa Rica Recognized for Biodiversity Protection," The Guardian, October 2010

21 *Assessing* Why do you think Costa Rica decided to tax fuel, cars, and energy rather than other goods and services?

22 *Making Decisions* What choices do you think individuals can make to help prevent the loss of rain forests and what incentives can governments provide?

Need Extra Help?

If You've Missed Question	13	14	15	16	17	18	19	20	21	22
Go to page	196	200	199	170	172	168	199	204	214	214

South America

ESSENTIAL QUESTION • *How do physical systems and human systems shape a place?*

Why Geography Matters
*Economic Geography:
Uneven Development*

Lesson 1
*Physical Geography of
South America*

Lesson 2
*Human Geography of
South America*

Lesson 3
*People and Their
Environment: South America*

Geography Matters...

South America is a land of beautiful natural wonders, including the vast Amazon rain forest. However, not all of South America is a lush jungle paradise. The region also includes snow-covered mountains and boasts the world's driest place—the Atacama Desert.

South America is diverse in culture as well as in landscape. The people of this subregion reflect a unique blend of indigenous and colonial heritage, the result of centuries of migration, trade, and conquest.

◄ Carnival in Brazil is held over five days preceding Lent, with parades featuring elaborate floats, drummers, and dancers.

Viviane Ponti/Lonely Planet Images/Getty Images

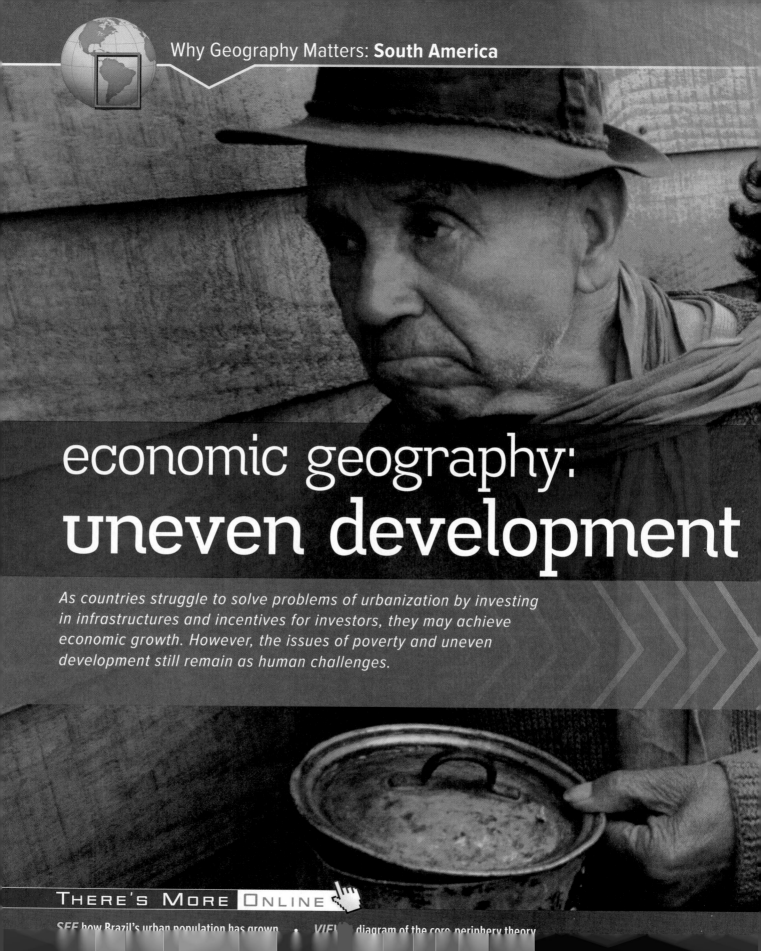

economic geography:
uneven development

As countries struggle to solve problems of urbanization by investing in infrastructures and incentives for investors, they may achieve economic growth. However, the issues of poverty and uneven development still remain as human challenges.

THERE'S MORE ONLINE

SEE how Brazil's urban population has grown • VIEW diagram of the core-periphery theory

How do core-periphery relations fuel uneven development?

Rural workers in search of better job opportunities tend to seek work in the core—cities, such as São Paulo, Brazil. Within a country, rural areas act as a periphery, contributing to burgeoning urban populations. The cities, or core, receiving the migrants often experience explosive growth. As the core grows, it begins to engulf surrounding smaller towns and cities. While the core offers job opportunities, it also presents challenges to governments, which struggle to maintain the infrastructure and services needed to support such a large, quickly growing urban population.

1. The World in Spatial Terms

Describe the migration of rural workers to urban areas in terms of both push factors and pull factors.

What are the effects of uneven patterns of development?

Many government programs focus on the metropolitan core areas and provide insufficient support to the periphery areas. This can leave the periphery with inadequate transportation, sanitation, education, and other essential services. Another result of uneven development is brain drain, or the process of educated and professional individuals leaving the periphery to go to the core for job opportunities. This deprives areas in the periphery of their most educated residents. For example, in the state of Ceará, home to some of the poorest Brazilians, there is a lack of incentives for establishing non-farming activities. This is a push factor for many residents to seek better opportunities in nearby core areas. Although the metropolitan areas have shown economic growth, the smaller urban areas and rural towns are still areas of widespread poverty.

2. Human Systems

How might the migration of workers from rural areas to large cities negatively impact rural areas?

How can the issue of uneven development be addressed?

Uneven patterns of development within countries are a result of unequal distribution of resources, capital, and infrastructure between the core and the periphery. Some governments work to create programs promoting non-farming activities in the periphery so there is more incentive for skilled workers to stay in these areas to develop business and industry.

3. Environment and Society

Imagine you are a candidate for mayor of a large city plagued by problems of overcrowding, poor sanitation, crime, and unregulated land use (squatting) as a result of rural workers flocking to your city for better jobs. Write a campaign speech detailing how you would address these problems if elected.

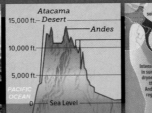
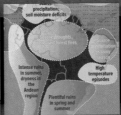

LESSON 1
Physical Geography of South America

ESSENTIAL QUESTION • *How do physical systems and human systems shape a place?*

IT MATTERS BECAUSE

The diverse landscapes of South America are very different from what we know in the United States. They have led to the people in South America developing very different lifestyles that are adapted to their physical environment.

Landforms

GUIDING QUESTION *How has South America's rugged landscape both attracted and isolated people?*

The Isthmus of Panama connects North America to South America. The subregion of South America spans 4,700 miles (7,564 km) from north to south, passing through the Equator near its widest point of 3,300 miles (5,311 km). Like Central America and Mexico, the most characteristic of South America's many landforms are its mountains. The Andes are the world's longest mountain chain. Some peaks in the Andes rise more than 20,000 feet (6,096 m) above sea level. The Andes consist of **cordilleras,** groups of several mountain ranges that run parallel to one another. Although known by different names, they are an extension of the Rocky Mountains that run from Canada south through the western United States and into Mexico and Central America. Because the cordilleras have established natural barriers between surrounding areas, many indigenous communities developed as isolated groups. As a result, some mountain villages exhibit centuries-old social customs.

The Andes encircle the **altiplano**, which means "high plain." The altiplano is an area that includes southeastern Peru and western Bolivia. It is the second-largest mountain plateau in the world.

In southern Argentina, hills and flatlands form the plateau of Patagonia. The presence of the Andes to the west produces a rain shadow that causes Patagonia to be dry, barren, and windy. The Patagonian region also extends across the Andes to southern Chile. Patagonia boasts dramatic valleys, glaciers, and fjords. The rugged Andes and Patagonia's landscape are a result of its location along the Ring of Fire.

Heavy tectonic activity in the subregion changes and reshapes the landscape. But despite the threats of natural disasters, people have chosen to settle in the Andean highlands for thousands of years. The climates are cooler, the volcanic soil is good for agriculture, and natural resources are concentrated here.

In contrast to the high peaks of the western Andes, eastern South America is defined by broad plateaus and valleys. The Amazon Basin, located along the eastern base of the Andes, is the lowlands area drained by the Amazon River. Just south is the Mato Grosso Plateau, a sparsely populated plateau of forests and grasslands extending across Brazil, Bolivia, and Peru. Farther east are the Brazilian Highlands, a vast area spanning several climate and vegetation zones. Warm climates and open spaces make the Brazilian Highlands good for raising livestock. The Eastern Highlands plunge to the Atlantic Ocean, forming a steep slope called an **escarpment**. This escarpment presents obstacles for inland development. As a result, most of Brazil's population lives along the coast.

Narrow coastal lowlands hem the Atlantic and Pacific coasts of South America. South America's inland grasslands—the **llanos** (LAH•nohs) of Colombia and Venezuela and the **pampas** of Argentina and Uruguay—provide grazing for cattle. Ranchers on large estates employ cowhands, called *llaneros* or gauchos, to drive herds across the rolling plains. Known for its fertile soil, the pampas are one of the world's breadbaskets, producing wheat and corn.

cordillera parallel chains or ranges of mountains

altiplano Spanish for "high plain," a region in Peru and Bolivia encircled by the Andes

escarpment a steep cliff or slope between a higher and lower land surface

llanos fertile grasslands found in inland areas of Colombia and Venezuela

pampas grassy, treeless plains of southern South America

☑ READING PROGRESS CHECK

Explaining What features of South America's landscape have hindered or encouraged development?

The Andes extend along the western part of South America. They are the world's longest mountain chain and one of the highest.

▲ CRITICAL THINKING

1. *Explaining* How do physical features affect human development?

2. *Describing* In what ways have the populations of the Andes region become dependent on their environment?

Water Systems

GUIDING QUESTION *How are South America's rivers important for economic development?*

Waterways are important for the subregion's economic development because they provide ways to transport goods and people within and between the countries of South America. As the Western Hemisphere's longest river and the world's second longest, the Amazon River flows about 4,000 miles (6,400 km) through the heart of South America. It begins in the headwaters of the Peruvian Andes, flows across the lowlands of the Amazon Basin in the interior of Brazil, and drains into the Atlantic Ocean. Hundreds of smaller rivers join the Amazon as it flows from the Andes to the Atlantic Ocean. Together these rivers form the Amazon Basin. The basin drains an area of more than 2 million square miles (5.2 million sq. km).

The Paraná, Paraguay, and Uruguay Rivers form the second-largest river system in Latin America, draining the rainy eastern half of South America. These rivers flow through the Pantanal, one of the world's largest tropical wetlands. After coursing through inland areas, the rivers flow into a broad estuary where the ocean tide meets a river current. This estuary, the Río de la Plata, or "River of Silver," flows into the Atlantic Ocean.

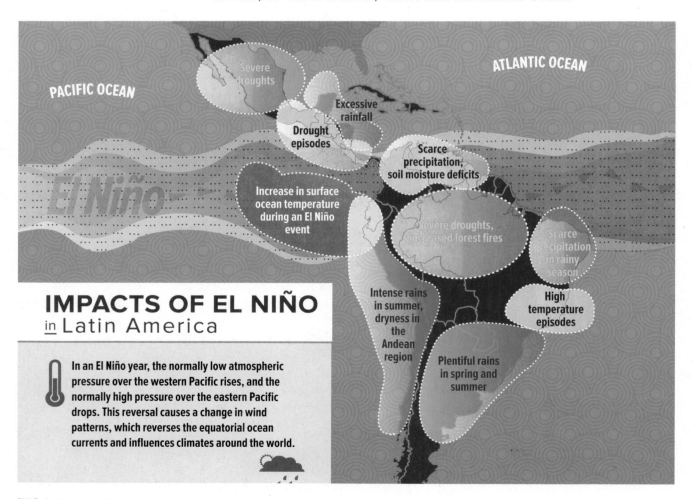

IMPACTS OF EL NIÑO in Latin America

In an El Niño year, the normally low atmospheric pressure over the western Pacific rises, and the normally high pressure over the eastern Pacific drops. This reversal causes a change in wind patterns, which reverses the equatorial ocean currents and influences climates around the world.

El Niño 's disruption of the ocean-atmosphere system in South America has a great impact on climate and economies in the region.

▲ CRITICAL THINKING

1. Sequencing Describe the order of events beginning with changes in the atmospheric pressure systems that lead to the impacts caused by El Niño.

2. Speculating Explain how the environmental impacts of El Niño could have negative economic effects on countries.

Though Latin America has few large lakes, some of its largest lakes are located in South America. Lake Maracaibo (MAH•rah•KY•boh) in Venezuela and Lake Titicaca (TEE•tee•KAH•kah), which run through Bolivia and Peru, are South America's largest lakes. Lake Titicaca is also the world's highest large lake.

☑ **READING PROGRESS CHECK**

Identifying Which rivers drain the eastern part of South America?

Climate, Biomes, and Resources

GUIDING QUESTION *How does climate affect human activities in South America?*

Diverse climates make South America a region of astonishing contrasts. Steamy rain forests, arid deserts, grassy plains, and sandy beaches can all be found in the subregion. The dense, nearly impenetrable vegetation of South America's tropical rain forests represents a tremendous resource and supports many communities.

Climate Regions and Biomes

The vertical climate zones found in the highland areas of Central America and Mexico also exist in the highlands of South America. The Andes are distinct not only because of their dramatic height, but also because they have such cold climates despite their proximity to the otherwise tropical equatorial zone. The range in elevation has produced a wide variety of climate and ecological zones.

The temperate climate of the *tierra templada* is found in areas of Peru, Brazil, and Colombia. Many Andean communities are located in the *tierra fría*. People in the highlands subsist on potato, barley, and quinoa crops that grow well in this colder climate. South America's colonial cities in Peru, Bolivia, and Colombia were developed over historical indigenous cities at high altitudes to extract valuable mineral resources found in the Andes. Consequently, several South American capitals are located in the *tierra fría* zone. The *tierra helada* and the *puna,* the highest vertical climate zones located above the tree line, are zones of permanent snow and ice on the peaks of the Andes.

The El Niño **phenomenon** also affects climate in South America. El Niño creates unusually warm ocean conditions on the west coast that extend as far north as Ecuador and as far south as Chile. As in Central America and the Caribbean, El Niño can have negative effects on coastal weather, fishing, and agriculture.

Tropical wet (rain forest) and tropical wet/dry (savanna) are the **predominant** climates of eastern South America, which is home to the Amazon rain forest, the world's largest rain forest. It is located primarily in Brazil but also extends into Peru, Colombia, Venezuela, Ecuador, Bolivia, Guyana, Suriname, and French Guiana. The Amazon shelters more species of plants and animals per square mile than anywhere else on Earth. It covers one-third of South America and is the world's wettest tropical plain. Heavy rains drench the densely forested lowlands.

A tropical wet/dry climate is typical of north-central South America. These areas have high temperatures and abundant rainfall, but also experience an extended dry season. In many tropical wet/dry areas, grasslands flourish. Some of these grasslands, such as the llanos of Colombia and Venezuela, are covered with scattered trees and are considered transition zones between grasslands and forests. A humid subtropical climate exists in much of southeastern South America. Winters here are short with cool to mild temperatures. Summers are long, hot, and humid. Rainfall is generally uniform throughout the year, but it can be heavier during the summer.

Much of the inland parts of Peru, Bolivia, and Chile experience an arid climate. In these areas, cold air and high elevations result in very little precipitation. Shifting winds and the rain shadow effect of the Andes produce

Connecting Geography to **SCIENCE**

Biology

The Galápagos Islands are located about 600 miles (965 km) off Ecuador's Pacific coast. They are home to the northernmost species of penguin, the Galapagos penguin. These small penguins feed on fish and marine crustaceans, but their population is vulnerable to disruptions in their food supply. Episodes of El Niño can trigger such shortages. El Niño causes the ocean surface to warm, which reduces the number of plankton. Plankton form the base of the food chain on which the penguins rely. Fish and crustacean populations that feed on plankton dwindle as a result. This, in turn, limits the food supply of the penguins.

PROBLEM SOLVING What can humans do to ensure the survival of Galapagos penguins when El Niño events occur? Is this a desirable goal? Explain your answer.

phenomenon a fact or event of scientific interest that can be scientifically explained or described

predominant main or most common

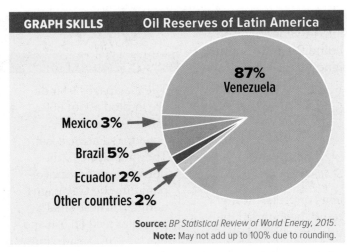

87%
Venezuela

Mexico **3%** →

Brazil **5%** →

Ecuador **2%** →

Other countries **2%** →

Source: *BP Statistical Review of World Energy, 2015.*
Note: May not add up to 100% due to rounding.

Oil reserves in Latin America are not evenly distributed, as shown in the circle graph.

▲ **CRITICAL THINKING**

1. *Contrasting* How do the oil reserves in Venezuela compare to those of Ecuador?

2. *Making Connections* How might the uneven balance of oil reserves be reflected in the region's economic development?

aridity in the southeastern part of Argentina. The low vegetation here is adapted to the low moisture conditions. The cold, oceanic Peru Current creates dry coastal deserts along the Pacific coast of Chile and Peru. These desert areas are called the Atacama. It is so arid that in some places no rainfall has ever been recorded. A dense fog known as *camanchaca* is the only appreciable source of precipitation.

Natural Resources

South American countries are among the world's leading producers of energy resources. Energy resources have supported major growth in economies such as that of Venezuela, which holds most of the subregion's oil reserves. Because of South America's substantial natural resources and active tectonic plates that allow for oil extraction, countries continue to search for additional, yet untapped oil reserves.

South America also has an abundance of mineral resources. For example, the foothills along Venezuela's Orinoco River contain large amounts of gold, and Peru is known for silver. Mines in Colombia have been producing the world's finest emeralds for more than 1,000 years. South America's non-precious minerals also have significant economic value. Chile is the world's largest exporter of copper. Peru and Chile together hold about 40 percent of the world's known copper reserves.

Countries in South America do not have equal access to the continent's natural resources. The size of Venezuela's oil reserves in comparison to the rest of the subregion's countries is a clear example of this unbalanced distribution of energy resources. The physical geography within and surrounding each country largely dictates its access to natural resources. Alongside political borders dictated by physical geography, distribution of natural resources within South America was a decisive factor in defining countries' political borders. Extraction of resources also relates to a country's infrastructure, level of economic development, and relationships between people and their government. Countries with low capital, social and political divisions, and lack of advanced technology for extracting resources have been at a disadvantage in comparison to countries that can cope with such factors.

☑ READING PROGRESS CHECK

Identifying Name three predominant climates in South America and describe their locations.

LESSON 1 REVIEW

Reviewing Vocabulary

1. *Identifying* Name and describe the two inland grasslands areas of South America.

Using Your Notes

2. *Listing* Using your graphic organizer, describe the llanos and the Atacama. Include the climates of the regions and any adaptations that plants and animals must have to survive.

Answering the Guiding Questions

3. *Assessing* How has South America's rugged landscape both attracted and isolated people?

4. *Describing* How are South America's rivers important for economic development?

5. *Drawing Conclusions* How does climate affect human activities in South America?

Writing Activity

6. *Informative/Explanatory* How has varied access to natural resources in South America promoted different rates of development among countries? Identify specific examples of natural resources in South American countries.

networks

There's More Online!

☑ **IMAGE** Moche Funeral Mask

☑ **MAP** Dominant Ethnic Groups of South America

☑ **MAP** Agricultural Land Use in South America

☑ **TIME LINE** South America: Movements of Change

☑ **INTERACTIVE SELF-CHECK QUIZ**

☑ **VIDEO** Human Geography of South America

LESSON 2

Human Geography of South America

ESSENTIAL QUESTION • *How do physical systems and human systems shape a place?*

Reading HELPDESK

Academic Vocabulary

- **widespread**
- **implement**

Content Vocabulary

- **quipu**
- **brain drain**
- **uneven development**

TAKING NOTES: *Key Ideas and Details*

SUMMARIZING Use a graphic organizer like the one below to take notes on the human geography of South America as you read.

Human Geography of South America

History and Government	→	
Population Patterns	→	
Society and Culture Today	→	

IT MATTERS BECAUSE

The society and governments of South America have been shaped by the subregion's rich history of indigenous peoples and their interactions with Europeans, Africans, and Asians. The modern economic geography—based on agricultural practices, the continent's natural resources, and industrial development—has resulted in an imbalance in overall wealth and development.

History and Government

GUIDING QUESTION *How have indigenous peoples and Europeans contributed to the creation of modern governments in South America?*

South America's diverse population is the result of centuries of blending among hundreds of indigenous groups, Europeans, Africans, and Asians. Some areas in South America are a microcosm of these diverse cultures. In other areas—many of them remote and isolated—indigenous peoples live much as their ancestors did hundreds of years ago, virtually untouched by the influence of other cultures or modern technology.

Early Cultures and European Conquest

Before the Inca established their empire in the Andes, other early indigenous groups—such as the Moche, Mapuche, and Aymara—developed societies that were based primarily on agriculture. The Inca later established a highly developed civilization in the area. At its height, the Inca Empire stretched from present-day Ecuador to central Chile.

The Inca were skilled engineers. They built temples and fortresses and laid out a network of roads that crossed mountain passes and penetrated forests. Inca farmers cut terraces into the slopes of the Andes and built irrigation systems. Machu Picchu, Peru's most well-known archaeological site, is a grand display of Inca engineering that is remarkably preserved. With no Inca written language, knowledge was passed on to each generation through storytelling. The Inca used **quipus** (KEE•poos) to account for financial and historical records.

quipu knotted cords of various lengths and colors used by the Inca to keep financial records

Silver and gold were important resources in the Inca culture. The precious metals and the wealth of farmers of the Inca Empire attracted Spanish conquistadors to Peru. After defeating the Inca army and its rulers, they looted the empire's capital and network of cities. The Inca connected their vast empire with a network of roads that extended throughout the empire. This allowed the Spanish conquerors to move quickly through the region. Spanish conquistadors expanded into Colombia, Argentina, and Chile. The Portuguese settled on the coast of Brazil, and the British, French, and Dutch later settled in parts of northern South America. The effects of epidemics caused by diseases introduced by the Europeans and the hardships of intensive labor on colonial plantations drastically reduced indigenous populations. To meet the resulting labor shortage, European colonists imported enslaved Africans.

Independence and Movements for Change

In the 1800s, independence movements arose in South America. These were inspired by the French and American Revolutions, as well as by the struggles for independence in Mexico and the Caribbean. By the mid-1800s, led by revolutionaries such as Simón Bolívar of Venezuela and José de San Martín of Argentina, most South American countries had won independence.

The postcolonial period was politically and economically unstable for most of the newly independent countries. They lacked a tradition of self-government. Power remained in the hands of the wealthy and elite classes of residents, despite written constitutions. With military backing, caudillos, or dictators, throughout South America seized power in the nineteenth century. Caudillos often gained power illegally and with much bloodshed among civilians.

Dictatorships have given way to democratically elected governments across South America. Today, however, these countries are struggling with many issues. These include political corruption and violence, wide gaps between the rich and poor, unemployment, and protecting the rights of indigenous groups.

☑ **READING PROGRESS CHECK**

Identifying What early indigenous civilization dominated much of western South America?

TIME LINE ⌄

SOUTH AMERICA
Movements of Change ➜

The history of South America traces the rise of ancient empires that were nearly destroyed by colonialism, followed by wars of independence. Today, contemporary protests are fueled by social inequalities.

▶ **CRITICAL THINKING**

1. Analyzing What did Bolivia lose as a result of the War of the Pacific?

2. Comparing How were the policies of Perón similar to those of Chávez?

1850 ➜

1879–1883
Chile fights Bolivia and Peru in the War of the Pacific. Dispute arises over Bolivian-held Atacama Desert along the Pacific coast. Peru and Bolivia lose the war, ceding land to Chile. Bolivia is left landlocked.

1932–1935
Bolivia tries to gain access to the Atlantic coast by taking land from Paraguay, resulting in the Chaco War. A peace treaty grants Paraguay most of the disputed region.

Population Patterns

GUIDING QUESTION *How has South America's physical geography influenced its population patterns?*

South America is the world's fourth-largest continent. Its 12 countries are home to about 415 million people. Like much of the rest of the developing world, population growth is steady, and so is the migration of people into large, urban areas.

South America's once high rate of population growth is beginning to slow. Urban populations now include fewer children as well as women and men with increased levels of education. Most people live on or near the coasts and along major rivers of the continent. These coastal regions offer favorable climates, fertile land, and access to transportation. The rain forests, deserts, and mountainous areas of South America's interior have discouraged human settlement.

South American countries tend to have low population densities. Ecuador, the most densely populated country in South America, has an average of 152 people per square mile (59 people per sq. km). Brazil has a population of 205 million. However, because Brazil has about 3.2 million square miles (8.4 million sq. km), its average population density is about 64 people per square mile (24 people per sq. km). Despite population densities that are lower overall than other world regions, much of the economic and structural development is concentrated in major cities. Today about 84 percent of the subregion's population lives in urban areas.

In highly populated urban areas such as São Paulo, Buenos Aires, and Bogotá, finding employment and suitable living conditions is difficult for migrants arriving in the city. Rural-to-urban migrants seek higher wages, better living conditions, and sometimes an escape from the violence of drug cartels or criminal groups. Countries across the region are experiencing **brain drain** to North America and Europe as people search for a better life.

brain drain the loss of highly educated and skilled workers to other countries

✓ READING PROGRESS CHECK

Explaining Why do most South Americans live along the continent's coasts?

(l)©Bettmann/Corbis, (c)©John Van Hasselt/Sygma/Corbis, (r)Paul Zimmerman/WireImage/Getty Images

---

1946
Juan Perón is elected president of Argentina. He is also founder and leader of the Peronist movement, which combines populist and nationalistic policies.

➔ **1950**

1973
Perón is reelected president after 18 years in exile.

1998
Hugo Chávez is elected president of Venezuela. Key factors of his ideology *(chavismo)* are nationalism, a centralized economy, and a strong military.

➔ **2000**

2006
Michelle Bachelet is elected Chile's first female president; Cristina Kirchner is elected president of Argentina the following year.

2010
Dilma Rousseff is elected Brazil's first female president.

Dominant Ethnic Groups of South America

Caribbean Sea

VENEZUELA
GUYANA
SURINAME
FRENCH GUIANA (Fr.)
COLOMBIA
ECUADOR
Galápagos Islands (Ecuador)
PERU
BRAZIL
PACIFIC OCEAN
BOLIVIA
PARAGUAY
CHILE
URUGUAY
ATLANTIC OCEAN
ARGENTINA
EQUATOR
TROPIC OF CAPRICORN

0   1,000 miles
0   1,000 kilometers
Lambert Azimuthal Equal-Area projection

African
Mestizo
European
Indigenous

Falkland Islands (U.K.)
South Georgia Island (U.K.)

GEOGRAPHY CONNECTION

South America's population has been shaped by ethnic diversity, physical geography, migration, and urban growth.

1. *PLACES AND REGIONS* What landforms are found in the areas where the majority of indigenous people live?

2. *HUMAN SYSTEMS* What countries in South America have large African populations?

Society and Culture Today

GUIDING QUESTION *Why is South America one of the world's most culturally diverse areas?*

South America is home to an ethnically diverse population. Today many indigenous cultural groups inhabit the subregion, especially in rural or less populated areas. Most indigenous groups—of which there are more than 350—live in the Andes region of Ecuador, Peru, and Bolivia.

The Spanish and Portuguese were the first Europeans in South America. Enslaved Africans were later brought as laborers. After South American countries gained their independence, other European groups—French, Dutch, Italians, and Germans—moved to South America. In fact, Argentina's population is 97 percent European, as the majority of Argentines are descendants of Spanish and Italian immigrants. Towns in the lakes region of southern Chile exhibit architecture, cuisine, and traditions influenced by its German population.

Immigrants from Asia also arrived in South America. In Guyana, almost half of the population is of South Asian descent. People of Chinese descent have immigrated to Peru. Many people of Japanese descent live in Brazil, Argentina, and Peru. Spanish, Portuguese, Dutch, French, and English are each spoken in different parts of South America. In countries with South Asian populations, such as in Guyana and Suriname, people also speak Urdu, Javanese, and Caribbean Hindustani, a dialect of Hindi. As a result, people in South America are often bilingual. During the colonial period, some European languages blended with indigenous languages to form completely new languages.

The majority of South Americans are Roman Catholic. Carnival is celebrated in the week before the Roman Catholic observance of Lent, a 40-day period of fasting and prayer before Easter. People from around the world come to Rio de Janeiro to participate in Carnival celebrations. In addition, tens of millions of people practice syncretism, a combination of mixed religions, such as Macumba and Candomblé, which combine West African religions with Roman Catholicism. Other minority religions include Protestant Christianity, Hinduism, Buddhism, Shinto, Islam, Judaism, and Eastern Orthodox Christianity.

Education varies greatly throughout South America. Many countries support public education through high school, and literacy rates have risen steadily. Public universities provide higher education at little or no cost to students in many countries. Many children leave school before completion, however, to help support the family by selling goods in markets or engaging in household or farming duties.

In countries with stable economies and high standards of living, people have access to better health care and live longer, healthier lives. This results in a situation of **uneven development** among countries. The health of a country's people is linked to poverty, lack of sanitation, infectious diseases, and malnutrition. These conditions persist in rural areas and especially in the slums on the outskirts of cities where millions of people live in overcrowded conditions.

uneven development
condition in which some places do not benefit as much as others from social and economic advancement

Family and the Status of Women

In urban upper and middle classes, the family unit is likely to consist of a nuclear household—father, mother, and dependent children—rather than an extended family. Loyalty and responsibility toward the extended family, however, remain very strong. The *compadre* relationship, in which parents and godparents share in the upbringing of a child, is valued in parts of Latin America. However, changes brought about by urban society have diminished its overall importance.

The elevation of women's rights has grown as countries have established more stable governments and economies in the past few decades. The result is an increasing proportion of women entering the workforce. Work codes in Chile and Colombia provide benefits for pregnant employees, and women's earnings are increasing in comparison to men's wages. Although women can still face discrimination and mistreatment, there are signs of change. Some countries now provide shelters for abused women and enforce stricter penalties for offenders.

The Arts

Indigenous arts survive in many different forms. The massive buildings of the ancient Inca at Cuzco and Machu Picchu reveal a mastery of stone and engineering that are still studied today for their ingenuity. Traditional arts and crafts dating from before the arrival of the Europeans—such as weaving, ceramics, and metalworking—have been passed from generation to generation.

Music also has ancient ties. Panpipes are one of the most common pre-Columbian musical instruments from the Andean region. Musical traditions later mixed Native American, African, and European influences to create unique styles. The Brazilian samba, Chilean cumbia, and the Argentine tango complement the Cuban salsa and Dominican merengue to exemplify the diversity of music developed from a mixture of cultural and geographical roots.

This funeral mask from the Moche culture of South America is pre-Columbian, meaning it dates from before the arrival of Europeans.

▼ CRITICAL THINKING

1. *Theorizing* Why might geographers use the term *pre-Columbian* when studying cultures? Think about the significance of the arrival of Europeans.

2. *Making Connections* What features do you recognize in the mask as being similar to a popular fad in youth cultures today?

✓ READING PROGRESS CHECK

Specifying How has South America's role as a cultural melting pot contributed to its unique cultural elements?

Economic Activities

GUIDING QUESTION *How have South America's abundant natural resources contributed to its economic development?*

Several countries in the region have combined their abundance of natural resources with current changes in government and improved economic conditions. Argentina is not only rich in natural resources, but also has a highly literate population and a diversified economic base that have helped to fuel a strong expansion in its economy. Brazil, the largest country in South America, has been undergoing continuous growth and development since 1970. Taking advantage of its natural resources, Brazil has become a powerful country in economic terms. Chile has established

©Bowers Museum/Corbis

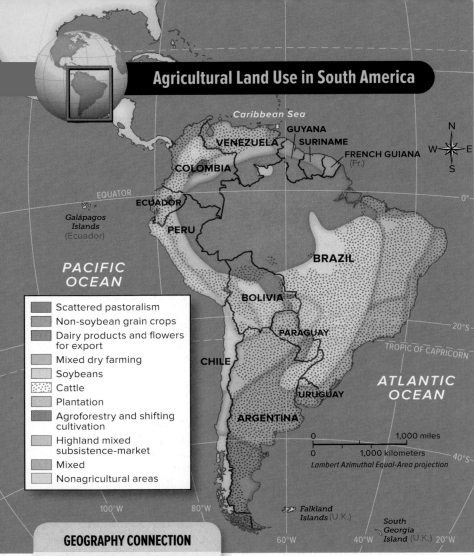

Agricultural Land Use in South America

Caribbean Sea

GUYANA
VENEZUELA
SURINAME
FRENCH GUIANA
(Fr.)
COLOMBIA

EQUATOR

ECUADOR

Galápagos
Islands
(Ecuador)

PERU

BRAZIL

PACIFIC
OCEAN

BOLIVIA

PARAGUAY

CHILE

20°S

TROPIC OF CAPRICORN

ATLANTIC
OCEAN

URUGUAY

ARGENTINA

0   1,000 miles
0   1,000 kilometers
Lambert Azimuthal Equal-Area projection

40°S

Falkland
Islands (U.K.)
South
Georgia
Island (U.K.) 20°W

100°W   80°W   60°W   40°W

Legend:
- Scattered pastoralism
- Non-soybean grain crops
- Dairy products and flowers for export
- Mixed dry farming
- Soybeans
- Cattle
- Plantation
- Agroforestry and shifting cultivation
- Highland mixed subsistence-market
- Mixed
- Nonagricultural areas

GEOGRAPHY CONNECTION

South America's rich agricultural tradition presents some modern-day challenges to development.

1. **ENVIRONMENT AND SOCIETY** What physical feature in Argentina and Uruguay creates good areas for raising cattle?

2. **PLACES AND REGIONS** What generalization can you make about the location of agroforestry and shifting cultivation in South America?

widespread covering a wide area; prevalent

free-trade agreements with the United States, Turkey, Australia, and other countries to enhance its economic activities.

Resources and Industry

Land and water use in South America follow physical geography. Forestry prevails in the Amazon Basin. Ranching is **widespread** in the grasslands of the south. Herding llamas and alpacas occurs in the high Andean regions. Fishing occurs in major lakes, rivers, and along coastlines. Agriculture remains highly important in South America. More than 20 percent of the subregion's workforce is employed in the primary sector that includes farming, ranching, and fishing. As in other subregions of Latin America (Mexico and Central America and the Caribbean), the legacy of the hacienda system still exists in South America. Larger commercial and smaller subsistence agriculture exist side by side.

Agriculture dominates much of east-central Brazil and the nearby areas of Paraguay, Uruguay, and Argentina. South America's contribution to agricultural global trade includes grains, soybeans, coffee, cocoa, citrus, cattle, sugarcane, tobacco, and cotton. In fact, Brazil is the world's largest exporter of coffee. Brazil and Paraguay also cultivate today's fastest-growing crop in the global economy: soybeans. Paraguay is the fourth-largest producer of soybeans in the world.

Additionally, the coca plant thrives in the northwestern parts of South America. Coca use is popular among Bolivia and Peru's working class for its effects as a legal stimulant and appetite suppressant. Yet coca's derivative can also be used to make the illegal drug cocaine. Peru, Bolivia, and Colombia have nonetheless legalized coca farming. They did so because, sold in its legal form, it is a large source of profit for these countries.

Natural resources include timber, gold, silver, copper, iron ore, and tin. South America contains about one-fifth of the world's iron ore, which is used for steel making and machine building. Many countries are heavily dependent on exporting their natural resources. Energy resources include petroleum and natural gas. Venezuela, Ecuador, and Argentina are leading exporters.

Manufacturing is growing rapidly, but the region's geography varies greatly. Most manufacturing is concentrated in urban areas, especially the primate cities. Because the largest cities lie mainly along the coasts where transportation is the best, the vast interior of South America has few manufacturing plants.

The major road systems in South America include the Pan-American Highway that stretches through Chile as it links many cities north to south, and the Trans-Andean Highway that links cities in Chile and Argentina east and west. The Trans-Amazonian Highway was built by Brazil to access the Amazon rain forest for developing timber and mineral resources. The Transoceanic Highway was designed to link the Amazon River ports with Peru's ports on the Pacific to transport agricultural products to the global markets in Asia and Europe. Argentina and Brazil have well-developed rail systems, which are important modes of transportation along with the inland waterways. All South American capital cities and major cities have domestic and international airports.

Economic Integration

The increased global demand for raw natural resources and manufacturing has had an impact on the overall economic growth in the region's countries. The economic growth has affected Brazil, Chile, and Argentina more than other countries. Ecuador, Peru, Venezuela, and Bolivia have struggled to modernize their economies and improve standards of living. For example, Bolivia is one of South America's most impoverished and least developed countries. Political reforms in the 1990s stimulated economic growth, but Bolivia continues to struggle to improve conditions for its people.

The separatism that characterized South American countries in the past is giving way to new trade partnerships and cooperation on infrastructure that are mutually beneficial. In South America's current economy, investments flow more freely from one country to another. For example, Colombia has taken advantage of its more stable economy by promoting free-trade agreements with other countries. Colombia **implemented** the U.S.-Colombia Free Trade Agreement with the United States in 2012. Colombia is negotiating free-trade agreements with other Latin American countries, such as Chile and Mexico, as well as with countries outside South America. Because Colombia is an open, free-market economy, it has signed many trade agreements.

Economic growth has been steady and strong across the region and has permitted countries to pay their foreign debt. If the economy does not grow, then the debt payments reduce the amount of money available to the government for essential services such as roads, water, flood control, and health care. Consistent, stable economic policies adopted by some South American countries in recent decades have contributed to steady economic growth and improved standards of living for people.

implement to carry out or accomplish by concrete measures

☑ **READING PROGRESS CHECK**

Inferring What is the benefit of cooperation between countries in South America?

LESSON 2 REVIEW

Reviewing Vocabulary

1. *Summarizing* Describe the concept of brain drain and how it relates to the countries of South America.

Using Your Notes

2. *Describing* Use your notes from the graphic organizer to describe South America's contributions to the arts. Include specific examples from the lesson.

Answering the Guiding Questions

3. *Evaluating* How have indigenous peoples and Europeans contributed to the creation of modern governments in South America?

4. *Making Connections* How has South America's physical geography influenced its population patterns?

5. *Explaining* Why is South America one of the world's most culturally diverse areas?

6. *Examining* How have South America's abundant natural resources contributed to its economic development?

Writing Activity

7. *Informative/Explanatory* Write a paragraph describing how countries in South America have established transnational and international relationships. Include examples of relations among the countries within South America and countries around the world.

AMAZON

IN THE BALANCE

The Amazon rain forest is the largest and most diverse rain forest on Earth, holding an abundance of plant and animal life. This rain forest and the life in it is under constant threat. Deforestation caused by logging, farming, and ranching threatens its depletion and the extinction of the plants and animals that call it home. In some ways the loss of the Amazon rain forest threatens us all.

FULL OF LIFE

In every 4 square miles of the Amazon rain forest, there are **1,500** flowering plant species, **750** tree species, and **900** tons of living plants. The Amazon is also home to **30 million** insect species, **175** lizard species, and **500** mammalian species. **One-third** of the world's birds also live in the Amazon.

CARBON CONTROL

The Amazon rain forest holds nearly **half** of the 247 billion tons of carbon contained in the world's tropical rain forests. Carbon is stored in the woods and roots, but when trees are **cut, burned, or decompose**, carbon is released into the atmosphere. This has **consequences** for the Earth's climate. According to NASA "...the average global temperature on Earth has **increased** by about 0.8^0Celsius (1.4^0Fahrenheit) since 1880. Two-thirds of the warming has occurred since 1975, at a rate of roughly $0.15-0.20^0$C per decade."

LONG-TERM EFFECTS OF DEFORESTATION

"Forest clearing in Brazil has already claimed **casualties**, but the animals lost to date in the rainforest region are just **one-fifth** of those that will slowly die out as the full impact of the loss of habitat takes its toll. In parts of the eastern and southern Amazon, 30 years of **concerted deforestation** have shrunk viable living and breeding territories enough to condemn 38 species to regional **extinction** in coming years, including 10 mammal, 20 bird and 8 amphibian species, scientists found."

—Ian Sample, *The Guardian (UK)*, July 12, 2012

ON THE RANGE

Cattle ranching is the **leading** cause of deforestation in the Brazilian Amazon. It accounts for **60%** of forest clearing. Subsistence and commercial agriculture, development, and **logging** also drive deforestation.

Making Connections

1. **Making Predictions** What may happen to the rain forest if land clearing continues unchecked?

2. **Drawing Conclusions** How might the continued deforestation and clearing of the Amazon rain forest have a direct effect on your life?

3. **Environment and Society** What circumstances might make environmental protection of the Amazon rain forest a low priority for some Latin American countries?

*Interact with **Global Connections** Online*

networks

There's More Online!

- ☑ **GRAPH** Per Capita CO₂ Emissions
- ☑ **IMAGE** EcoTaxi
- ☑ **IMAGE** Large-Scale Monoculture
- ☑ **INTERACTIVE SELF-CHECK QUIZ**
- ☑ **VIDEO** People and Their Environment: South America

LESSON 3
People and Their Environment: South America

(t)|©Paulo Fridman/Corbis, (tc)Nilton Ricardo/Brazil Photos/Alamy, (tc)Mauricio Lima/AFP/Getty Images, (tc)Alexandre Cappi/LatinContent/Getty Images, (tr)Steve Allen/Getty Images

Reading HELPDESK

Academic Vocabulary

- voluntary
- alternative

Content Vocabulary

- oxisol
- monoculture

TAKING NOTES: *Key Ideas and Details*

IDENTIFYING Use a graphic organizer like the one below to take notes as you read about the issues that relate to people and their environment in South America.

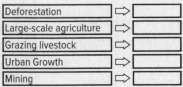

	Impact on Environment
Deforestation	⇨
Large-scale agriculture	⇨
Grazing livestock	⇨
Urban Growth	⇨
Mining	⇨

ESSENTIAL QUESTION • *How do physical systems and human systems shape a place?*

IT MATTERS BECAUSE
The use of natural resources for economic advancement has benefited the livelihood and stability of many South American countries. It has also caused damage to the continent's natural environment and dramatic changes to its biodiversity.

Managing Resources

GUIDING QUESTION *How has the management of forest and agricultural resources impacted the environment in South America?*

South America is home to some of the largest reserves of forest and agricultural resources in the world. Countries are taking advantage of such resources to improve their economies and overall wealth. Extensive exploitation of resources comes at a cost, however. Issues such as deforestation, soil erosion, desertification, and pollution have been significant problems for all countries in South America.

Like the rain forests of Costa Rica and Panama, the rain forests of South America are threatened by intensive human activity. Rain forests harbor at least half of all animal and plant species on Earth. Deforestation is occurring at a rapid rate in the Amazon rain forest. This has reduced the diversity of plants and animals there. Brazil has the world's largest remaining expanses of tropical rain forest, but almost 20 percent of the Amazon rain forest has already been destroyed. The loss of biodiversity is also occurring in Brazil's lesser-known Atlantic Forest, one of Earth's richest and most threatened habitats. The Atlantic Forest now covers less than 10 percent of its original area.

Soil erosion in South America has diminished the ability of soils to produce food and vegetation. Intensive farming, construction, logging, fires, and overgrazing all increase the rate of soil erosion. Certain soil types in South America—such as the volcanic soils and

oxisols found in the humid tropical lowlands—are especially vulnerable to erosion. The oxisols, sometimes known as laterites, can degrade into a baked clay-like form when too much of the natural vegetation cover is removed. The removal of topsoil occurs as a result of intensive agriculture, especially on landscapes that have been cultivated for long periods of time.

Large-scale agriculture also worsens the progression of soil erosion. **Monoculture**, the growth of a single type of crop on agricultural or forested land, depletes the soil of its nutrients. It disrupts the natural cycle of growth and breakdown of plants, animals, and bacteria. Without these natural processes, soil cannot rebuild its nutrients. Vast monoculture soybean crops in Brazil, for example, are quickly depleting soil fertility.

The general process of soil erosion summarizes the progression and effects of desertification. Because ecosystems are dynamic and respond to changes in environmental conditions, processes such as soil erosion and desertification are drastically changing the landscapes and ecology of South America's extensive croplands and grazing lands. The primary cause of desertification is not drought, but rather mismanagement of land by human activities such as overgrazing of livestock and deforestation. Rain-fed crops in drylands, such as wheat and corn, can lead to desertification. After wheat and corn are harvested, the lands left uncovered between planting seasons become vulnerable to erosion by climatic forces such as wind and rain. Wind can create heavy dust storms by sweeping up uncovered topsoil. This deprives extensive land areas of the nutrients from organic matter contained by topsoil. Though rainfall in drylands is uncommon, heavy downpours do occur. This results in otherwise fertile and nutrient-rich topsoil being washed away. For large-scale irrigation on drylands, salinization is also a significant issue. Once fields are irrigated, water dries quickly and leaves behind salt. These salts collect and reduce the ability of plant roots to absorb water and grow.

Desertification also occurs in rangelands, which support a large population of grazing animals such as cattle and sheep. Grazing livestock consume plants almost to ground level. This weakens plants' ability to grow.

oxisol a thick, weathered soil of the humid tropics that is largely depleted of fertility and nutrients

monoculture the cultivation or growth of a single crop over a wide area for a consecutive number of years

Large-scale monoculture, like this soybean farm in Brazil, depletes soil fertility.

▼ **CRITICAL THINKING**

1. Making Connections What factors contribute to the decline of soil fertility with the practice of monoculture?

2. Speculating Why is monoculture still practiced even though its negative impacts are known?

"Because of its size and history, El Alto [Bolivian city connected to Lake Titicaca via the Pallina River] is a political powerhouse, yet the chronic poverty and lack of access to services widely faced by Bolivia's indigenous peoples persist there, and tackling pollution is a struggle. Changing the waste disposal habits of the sparsely populated countryside is one obstacle. But at the heart of the matter is weak enforcement of environmental laws and inadequate infrastructure."

—Sara Shahriari, "Urban Population Boom Threatens Lake Titicaca," *The Guardian*, January 12, 2012

DBQ *IDENTIFYING CENTRAL ISSUES* What two issues are central to the inadequate treatment of wastewater?

The movement of herds of livestock also destroys plant roots that bind the soil. When rain comes, water often washes away unprotected topsoil. This process of desertification in rangelands is a serious issue in places with a strong livestock industry: Argentina, Uruguay, Brazil, and Paraguay.

☑ **READING PROGRESS CHECK**

Summarizing How does overgrazing worsen processes of desertification?

Human Impact

GUIDING QUESTION *Why does urban growth and industrialization create environmental problems in South America?*

Large-scale economic production and urban growth have created multiple forms of environmental pollution. São Paulo, Brazil, is an example of the significant amount and effect of multiple sources of urban pollution. With a population of more than 20 million, São Paulo is the largest city in the Southern Hemisphere. São Paulo has attracted millions of poor migrants. Many have settled in favelas, or slums on the outskirts of the city consisting of crudely built shacks. These favelas are disconnected from the services of the established city. They are thus particularly problematic as sources of sewage and unrestricted residential growth.

Rapid urban growth also requires cities to find methods for disposing of human waste and sewage. Many urban regions, particularly those in less developed countries, lack the funding and organization to build extensive networks of piped water, drains, and sewage treatment plants. One example is the Bolivian city of El Alto, which is located along the Pallina River that flows into Lake Titicaca. El Alto has seen rapid growth resulting in increased pollution from human waste and sewage. The polluted Pallina River had once been a source of clean water for the people who lived on its banks, but now the waters that flow into Lake Titicaca are contaminated.

Illegal mining has further damaged the natural land and water features of South America. Since 2007, the price of gold has doubled in value. For gold-producing countries such as Peru, the sixth-largest producer in the world, this has presented a potential for immense wealth and economic development. Yet the rise in the value of gold has also encouraged illegal mining activity, especially in countries with an impoverished majority population and ineffective regulatory procedures in government. In Peru, for example, tens of thousands of people have set up camp in the Amazon rain forest in search of vast gold reserves. Alongside individual prospecting, large-scale mining by use of bulldozers and barges has also increased. Rapid deforestation has resulted from rapid migration, makeshift housing, and industrial-scale mining operations. Also, because miners use mercury and other toxic compounds to separate gold from ore, high levels of mercury and cyanide pollution in rivers have been reported.

Though clashes between security forces and miners have created hostility and further chaos, countries affected by illegal mining have made efforts to regulate the vulnerable regions. Colombia is cracking down on illegal mining because anti-government guerrilla groups have been using profits from illegal mining to finance their efforts. Meanwhile, Brazil has employed military personnel to fight illegal gold mining along its northern borders.

☑ **READING PROGRESS CHECK**

Specifying What are three sources of pollution in South America?

Addressing the Issues

GUIDING QUESTION *How are people and governments addressing environmental issues in South America?*

Addressing issues related to human impact on the natural environment is important not only for protecting regional biodiversity, but also for preserving the livelihood of human populations. Many South American countries, such as Uruguay, Paraguay, and Guyana, depend on crops and agriculture as their economic base. These countries are very vulnerable to changing weather patterns and infertile lands.

South American countries that recognize the impact of deforestation are passing laws to protect their lands. For example, in response to the high rate of deforestation in Paraguay, the country's government passed the Zero Deforestation Law in 2004. This law prohibits forested areas from being converted to landscapes for other uses in the eastern region of Paraguay. The law's enforcement has dramatically reduced Paraguay's deforestation rate.

At a local level, farmers can implement management strategies to slow the process of soil erosion. Specifically, soil erosion due to the formation of oxisols can be prevented with careful application and management of lime and fertilizers. Cover crops, which are plants that cover topsoil after crops have been harvested, prevent potential soil erosion from wind and water. At large scales, plant cover can contribute to more regular rainfall patterns, reducing the occurrence of drought and further soil erosion.

Soil conservation efforts will slow land degradation, but restoring soil fertility is much more difficult. On average, a millimeter of soil is generated in about 100 years. Therefore, soil erosion is a more rapid process than soil generation. Once soil has begun to erode, the amount by which it can be restored to fertility depends on how much it has degraded. Lightly degraded soils can be improved by using farm practices. Severely eroded land is generally abandoned because required resources to restore the soil are often too costly. Furthermore, without coordinated prevention efforts among farmers and government, areas of infertile soil will continue to reduce the region's biodiversity. Land available for food production will also be reduced.

Countries across South America are taking steps to reduce air pollution at the local level by establishing regulations. The results of the regulations can be seen in the reduction of greenhouse gas emissions in the continent's

Per Capita CO$_2$ Emissions

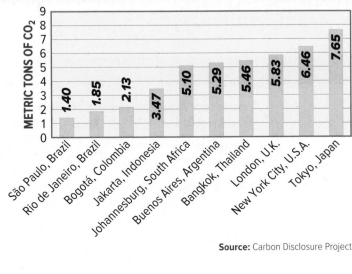

Source: Carbon Disclosure Project

GRAPH SKILLS ⌄

Many South American countries have established regulations in an effort to reduce greenhouse gas emissions.

◀ **CRITICAL THINKING**

1. **Comparing** How do the carbon dioxide emissions from the South American cities shown on the graph compare to other cities on the graph?

2. **Hypothesizing** What factors other than population do you think contribute to the emission levels of greenhouse gases such as carbon dioxide?

Countries throughout South America are developing programs for air quality management. They tend to be focused on primary metropolitan areas such as São Paulo. EcoTaxis, like the one shown here, are part of World Car-Free Day and help to reduce emissions in the city.

▲ **CRITICAL THINKING**

1. Analyzing Why do governments focus air-quality management efforts on large cities? Do you think that is an effective approach?

2. Drawing Conclusions How would an EcoTaxi help combat air pollution?

voluntary of one's own choice or consent

alternative different from the usual or regular

largest cities. For example, although São Paulo is one of the world's largest cities, it has much lower levels of carbon dioxide emissions than other large cities. Other large cities in South America also have lower greenhouse gas emissions than might be expected given their high populations.

In order to fix current problems and reduce future ones, South American countries must address the specific factors related to urbanization. The problems include urban sprawl, longer distances that residents travel within cities, increased use of cars, and inefficient public transport systems. Argentina is a global role model for setting **voluntary** greenhouse gas emissions targets. Furthermore, in 2010 Argentina and Uruguay formed a joint effort to monitor pollution along the Uruguay River, which defines the Argentina-Uruguay border.

Another issue for establishing effective policies is that regulations established by governments or international agreements could restrict countries' access to natural resources for export production. South American countries that lack the money to invest in **alternative** export resources are reluctant to establish environmental restrictions. Therefore, to promote the growth of sustainable economies, more developed countries must work closely with less developed countries in South America. Assistance in the area of enforcing and creating policies is of great importance.

Human modifications of the physical environment can have significant global impacts. Therefore, the international community is motivated to work together to find solutions to increasing environmental changes. Despite measures taken to produce effective policies, however, there is a weak foundation of air and water quality management among South American countries. Efforts are focused more on primary metropolitan areas rather than smaller urban regions, which are now among the urban areas of South America with the most rapid growth rates. Environmental awareness and policy making is relatively recent and therefore will require time, resources, and cooperation among the world's institutions in order to ensure an improved quality of life for South Americans.

☑ **READING PROGRESS CHECK**

Drawing Conclusions Why is it important for countries to work together to create regulations for future environmental use?

LESSON 3 REVIEW

Reviewing Vocabulary

1. Describing How are oxisols formed in soil?

Using Your Notes

2. Describing Use your graphic organizer to describe how mining has affected the environments of South America.

Answering the Guiding Questions

3. Exploring Issues How has the management of forest and agricultural resources impacted the environment in South America?

4. Evaluating Why does urban growth and industrialization create environmental problems in South America?

5. Discussing How are people and governments addressing environmental issues in South America?

Writing Activity

6. Informative/Explanatory Describe the process of physical change due to desertification. Include two specific factors that cause this change.

Directions: On a separate sheet of paper, answer the questions below. Make sure you read carefully and answer all parts of the questions.

Lesson Review

Lesson 1

① *Describing* Which landform specific to South America is identified by the term *llanos*?

② *Explaining* Describe the general features of three of the vertical climate zones found in South America.

③ *Summarizing* What are some important natural resources South American countries rely on for international trade?

Lesson 2

④ *Assessing* Explain how an empire as large and developed as the Inca Empire was so quickly defeated by Spanish conquistadors.

⑤ *Analyzing* Why do South American countries have such a diverse mix of ethnic groups? Compare the diversity of population between two South American countries.

⑥ *Identifying* What are two examples of projects or agreements between South American countries that have been developed to boost the countries' economies?

Lesson 3

⑦ *Comparing* Describe how the processes of desertification and soil erosion are related. Provide specific examples of how they are negatively affecting the land in South America.

⑧ *Identifying Cause and Effect* How do large South American cities such as São Paulo contribute to increased pollution?

⑨ *Identifying Central Issues* How has illegal mining negatively affected the physical environment in South America?

21st Century Skills

Use the following chart to answer the questions below.

Foreign-Born U.S. Population from South America by Country of Birth, 2010	
Country of Origin	Number of U.S. Residents
Brazil	340,000
Colombia	637,000
Ecuador	443,000
Peru	429,000
Other South American countries	882,000

Source: U.S. Census Bureau 2010

⑩ *Using Graphs, Charts, Diagrams, and Tables* From which country do most South American-born people in the United States originate?

⑪ *Identifying Cause and Effect* Describe a few of the push factors that lead to migration from South America to the United States.

⑫ *Drawing Conclusions* How might the figures in this table compare to the number of foreign-born residents from Mexico and Central America? Why?

Exploring the Essential Question

⑬ *Sequencing* Create a chart that depicts three ways in which South America's rugged landscape both attracts and isolates communities.

College and Career Readiness

⑭ *Reaching Conclusions* Imagine you are an ornithologist (a person who studies birds) working in Ecuador. Using the Internet, research why Charles Darwin found this part of South America—specifically the Galápagos Islands—to be an ideal spot for his research on evolution and species diversity. Use his research on finches as a specific example. Why would this research be important?

Need Extra Help?

If You've Missed Question	❶	❷	❸	❹	❺	❻	❼	❽	❾	❿	⓫	⓬	⓭	⓮
Go to page	219	221	222	224	226	229	232	234	234	237	237	237	218	221

Directions: On a separate sheet of paper, answer the questions below. Make sure you read carefully and answer all parts of the questions.

Critical Thinking

15 *Making Generalizations* Describe specific ways that South American countries have overcome physical barriers to make transportation possible.

16 *Identifying Cause and Effect* What issues did the newly independent countries of South America face in the postcolonial period? What have been the lasting effects?

17 *Exploring Issues* What are some possible solutions to setbacks that South American countries have faced in becoming important traders in the international economy?

18 *Drawing Conclusions* What are some regulatory measures farmers and governments can take to feed growing populations without continuing to deplete the land of fertile soil?

DBQ Analyzing Primary Sources

Use the excerpt from one of South America's most beloved poets to answer the questions below.

> **PRIMARY SOURCE**
>
> "...While I'm writing, I'm far away;
> and when I come back, I've gone.
> I would like to know if others
> go through the same things that I do,
> have as many selves as I have,
> and see themselves similarly;
> and when I've exhausted this problem,
> I'm going to study so hard
> that when I explain myself,
> I'll be talking geography."
>
> —Pablo Neruda, "We are Many,"
> *The Yellow Heart,* 1974

19 *Speculating* What do you think Neruda meant by "I'll be talking geography" as a method of explaining himself, in regard to his life?

20 *Drawing Conclusions* Based on the topic of the poem, what aspects of Pablo Neruda's life made him a famous poet throughout the world?

Applying Map Skills

Use the Unit 3 Atlas to answer the following questions.

21 *Human Systems* Which countries form the northern edge of South America?

22 *Physical Systems* Use the scale on the physical map of South America to estimate the length in miles across the widest part of Brazil.

23 *Places and Regions* Use your mental map of South America to make generalizations about the location of major population centers in relation to physical features and landforms.

Research and Presentation

24 *Research Skills* Use Internet and library resources to gather information about the factors that have shaped a South American country's current political situation. Create a time line outlining the country's historical and current political status. Cover relevant events and figures, from the country's colonization, to any rebellions it faced, to its declaration of independence, to its changes in the government systems in the twentieth and twenty-first centuries. Share the time line you create with your class.

Writing About Geography

25 *Informative/Explanatory* Use standard grammar, spelling, sentence structure, and punctuation to write a one-page essay suggesting possible solutions among South American countries to address the various types of environmental issues within the region. What kinds of regulations and agreements would control increased environmental degradation from human activities such as farming, construction, urban growth, and illegal mining?

Need Extra Help?

If You've Missed Question	**15**	**16**	**17**	**18**	**19**	**20**	**21**	**22**	**23**	**24**	**25**
Go to page	228	224	229	235	238	238	169	168	172	224	235

Europe

Chapter 10
Northern
Europe

Chapter 11
Northwestern
Europe

Chapter 12
Southern
Europe

Chapter 13
Eastern
Europe

Chapter 14
The Russian
Core

UNIT **4**

1 Culture Europe's cultural heritage is as varied as its landscape. Centuries-old traditions, such as the running of the bulls in Pamplona, Spain, illustrate how the region's history is intertwined with its modern-day culture.

EXPLORE the REGION

EUROPE is a peninsula of peninsulas, with many pieces of land extending into the Atlantic Ocean and the Mediterranean Sea. Over the centuries, Europeans have taken advantage of their location, using the seas as a source of food and an avenue for trade and exploration. Europe is home to many different languages and culture groups. Throughout history, conflicts between competing nations have caused destruction in the region. In recent years, however, most of Europe has joined together in an economic and political union that has fostered peace and prosperity.

THERE'S MORE ONLINE

3 **Rivers** The Danube River flows from southern Germany to the Black Sea. Like many European rivers, it is an important commercial route as well as a scenic highlight.

4 **Mountains** The Alps stretch 750 miles (1,200 km) through eight countries, separating Southern Europe from the north. The mountain range's imposing peaks are the source of many of the region's rivers.

2 **Economy** Tourism is an important sector of the European economy. It comprises a wide variety of destinations and employs a large percentage of the European workforce. Venice, with its beautiful canals, is a top choice among travelers.

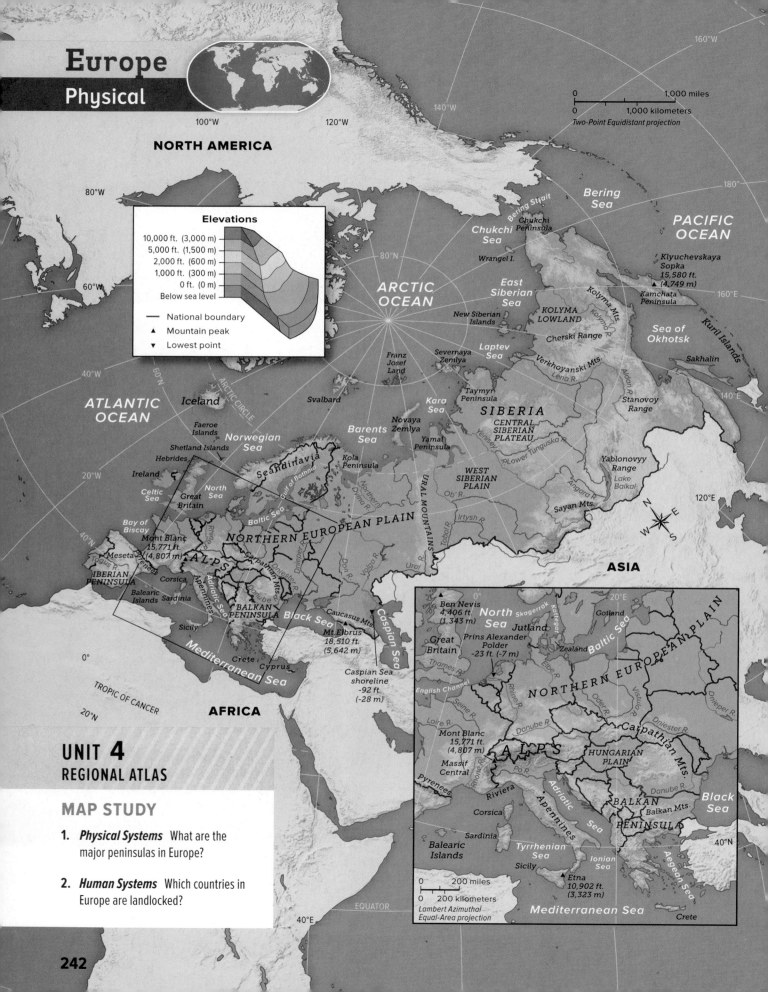

Europe
Physical

NORTH AMERICA

Elevations

10,000 ft. (3,000 m)
5,000 ft. (1,500 m)
2,000 ft. (600 m)
1,000 ft. (300 m)
0 ft. (0 m)
Below sea level

— National boundary
▲ Mountain peak
▼ Lowest point

0   1,000 miles
0   1,000 kilometers
Two-Point Equidistant projection

PACIFIC OCEAN

Bering Strait
Bering Sea
Chukchi Peninsula
Chukchi Sea
Wrangel I.
Klyuchevskaya Sopka 15,580 ft. ▲ (4,749 m)
Kamchata Peninsula
East Siberian Sea
New Siberian Islands
KOLYMA LOWLAND
Kolyma Mts.
Kolyma
Sea of Okhotsk
Kuril Islands
ARCTIC OCEAN
Franz Josef Land
Severnaya Zemlya
Laptev Sea
Taymyr Peninsula
Cherski Range
Verkhoyanski Mts.
Lena R.
Sakhalin
Aldan R.

ATLANTIC OCEAN
Iceland
Svalbard
Kara Sea
Novaya Zemlya
Barents Sea
Yamal Peninsula
SIBERIA
CENTRAL SIBERIAN PLATEAU
Stanovoy Range
Faeroe Islands
Norwegian Sea
Shetland Islands
Hebrides
Ireland
Celtic Sea
Great Britain
North Sea
Scandinavia
Kola Peninsula
Gulf of Bothnia
Baltic Sea
Northern Dvina R.
WEST SIBERIAN PLAIN
Lower Tunguska R.
Yenisey
Yablonovyy Range
Lake Baikal
Bay of Biscay
Mont Blanc 15,771 ft. ▲ (4,807 m)
Meseta
Pyrenees
NORTHERN EUROPEAN PLAIN
Rhine
Carpathian Mts.
Dnieper R.
Dniester R.
Don R.
Volga R.
Ural R.
URAL MOUNTAINS
Ob' R.
Tobol R.
Irtysh R.
Zingara R.
Sayan Mts.
ASIA
IBERIAN PENINSULA
Tagus R.
ALPS
Corsica
Apennines
Adriatic Sea
Danube
BALKAN PENINSULA
Black Sea
Caucasus Mts. ▲
Mt Elbrus 18,510 ft. ▲ (5,642 m)
Caspian Sea
Balearic Islands
Sardinia
Sicily
Crete
Cyprus
Mediterranean Sea
Caspian Sea shoreline -92 ft. (-28 m)
TROPIC OF CANCER
AFRICA

EQUATOR

Inset map:

Ben Nevis 4,406 ft. ▲ (1,343 m)
North Sea
Skagerrak
Kattegat
Gotland
20°E
Jutland
Great Britain
Prins Alexander Polder -23 ft. (-7 m) ▼
Zealand
Baltic Sea
NORTHERN EUROPEAN PLAIN
Thames R.
English Channel
Elbe R.
Oder R.
Vistula R.
Dnieper R.
Dniester R.
Seine R.
Rhine R.
Loire R.
Danube R.
Carpathian Mts.
Mont Blanc 15,771 ft. ▲ (4,807 m)
ALPS
HUNGARIAN PLAIN
Massif Central
Rhône R.
Po R.
Danube R.
Pyrenees
Riviera
Adriatic Sea
BALKAN PENINSULA
Balkan Mts.
Black Sea
Corsica
Apennines
Sardinia
Tyrrhenian Sea
Balearic Islands
Sicily
Etna 10,902 ft. ▲ (3,323 m)
Ionian Sea
Aegean Sea
Mediterranean Sea
Crete

0   200 miles
0   200 kilometers
Lambert Azimuthal Equal-Area projection

UNIT 4
REGIONAL ATLAS

MAP STUDY

1. *Physical Systems* What are the major peninsulas in Europe?

2. *Human Systems* Which countries in Europe are landlocked?

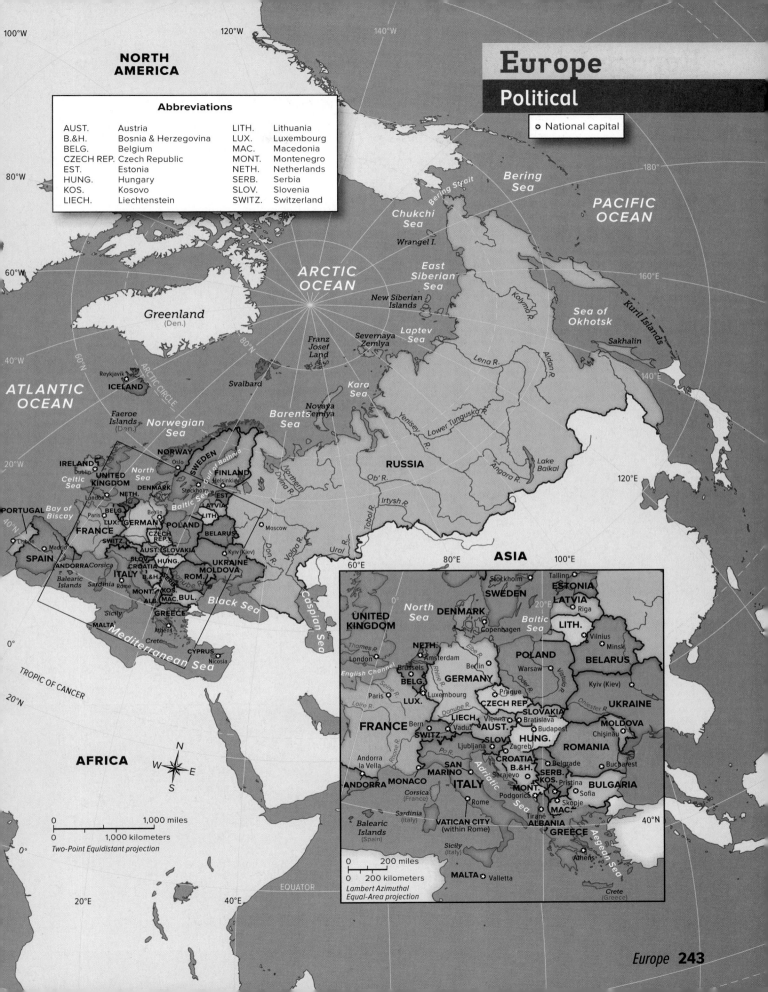

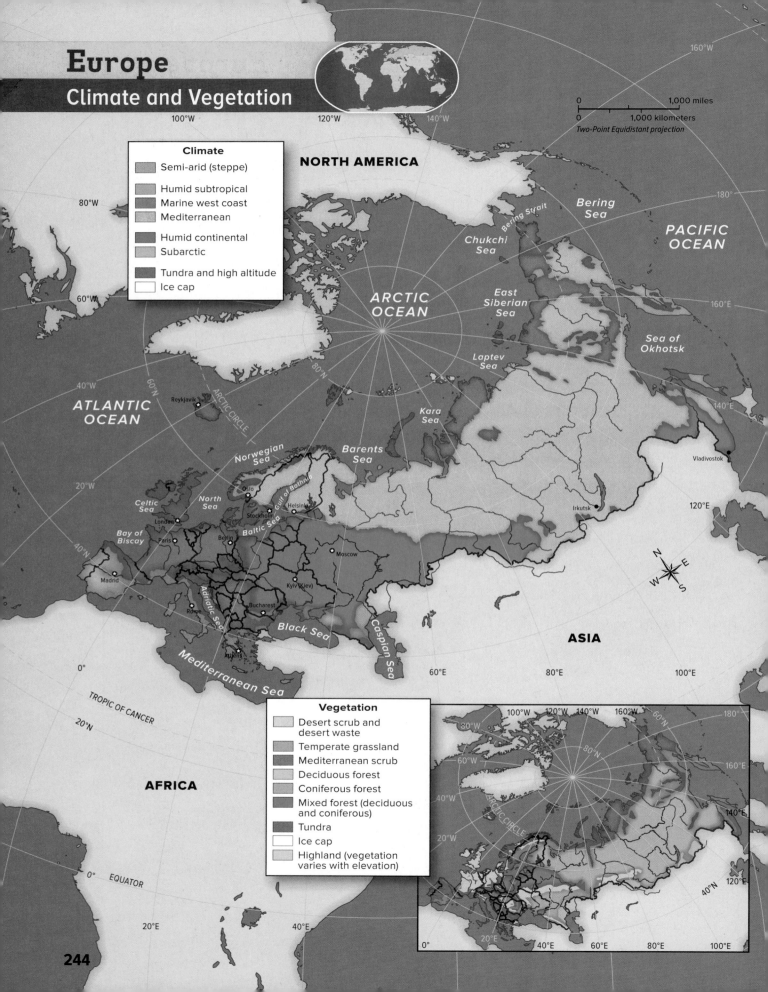

Europe
Climate and Vegetation

Climate
- Semi-arid (steppe)
- Humid subtropical
- Marine west coast
- Mediterranean
- Humid continental
- Subarctic
- Tundra and high altitude
- Ice cap

Vegetation
- Desert scrub and desert waste
- Temperate grassland
- Mediterranean scrub
- Deciduous forest
- Coniferous forest
- Mixed forest (deciduous and coniferous)
- Tundra
- Ice cap
- Highland (vegetation varies with elevation)

0 — 1,000 miles
0 — 1,000 kilometers
Two-Point Equidistant projection

NORTH AMERICA
ASIA
AFRICA

ARCTIC OCEAN
ATLANTIC OCEAN
PACIFIC OCEAN

Bering Sea
Bering Strait
Chukchi Sea
East Siberian Sea
Laptev Sea
Sea of Okhotsk
Kara Sea
Barents Sea
Norwegian Sea
North Sea
Celtic Sea
Bay of Biscay
Baltic Sea
Gulf of Bothnia
Adriatic Sea
Black Sea
Caspian Sea
Mediterranean Sea

Reykjavik
Oslo
Stockholm
Helsinki
London
Paris
Berlin
Madrid
Rome
Athens
Bucharest
Kyiv (Kiev)
Moscow
Irkutsk
Vladivostok

ARCTIC CIRCLE
TROPIC OF CANCER
EQUATOR

160°W
140°W
120°W
100°W
80°W
60°W
40°W
20°W
0°
20°N
40°N
60°N
80°N
180°
160°E
140°E
120°E
100°E
80°E
60°E

N
S
E
W

244

NORTH AMERICA

100°W

120°W

140°W

160°W

180°

Land Use

- Commercial farming
- Livestock raising
- Nomadic herding
- Primarily forest
- Manufacturing and trade
- Commercial fishing
- Little or no activity

Resources

- Coal
- Petroleum
- Natural gas
- Iron ore
- Tin
- Zinc
- Bauxite
- Cobalt
- Uranium
- Nickel
- Copper
- Lead
- Manganese
- Gold
- Silver
- Platinum
- Gems

80°W

60°W

40°W

20°W

80°N

60°N

40°N

0°

ARCTIC CIRCLE

ARCTIC OCEAN

Bering Strait

Chukchi Sea

Bering Sea

PACIFIC OCEAN

East Siberian Sea

160°E

180°

Laptev Sea

Kolyma R.

Sea of Okhotsk

Kara Sea

Barents Sea

Lena R.

Aldan R.

140°E

Norwegian Sea

Yenisey R.

Lower Tunguska R.

Lake Baikal

120°E

ATLANTIC OCEAN

Celtic Sea

North Sea

Gulf of Bothnia

Baltic Sea

N. Dvina R.

Pechora R.

Ob' R.

Angara R.

Bay of Biscay

Volga R.

Don R.

Tobol R.

Irtysh R.

Ural R.

Caspian Sea

ASIA

Black Sea

Danube R.

Mediterranean Sea

0°

TROPIC OF CANCER

20°N

AFRICA

20°E

40°E

60°E

80°E

EQUATOR

INDIAN OCEAN

0
1,000 miles

0
1,000 kilometers

Two-Point Equidistant projection

UNIT 4
REGIONAL ATLAS

MAP STUDY

1. *Environment and Society* Describe the relationship between land use and climate.

2. *Human Systems* In which parts of Europe is nomadic herding practiced?

Europe **245**

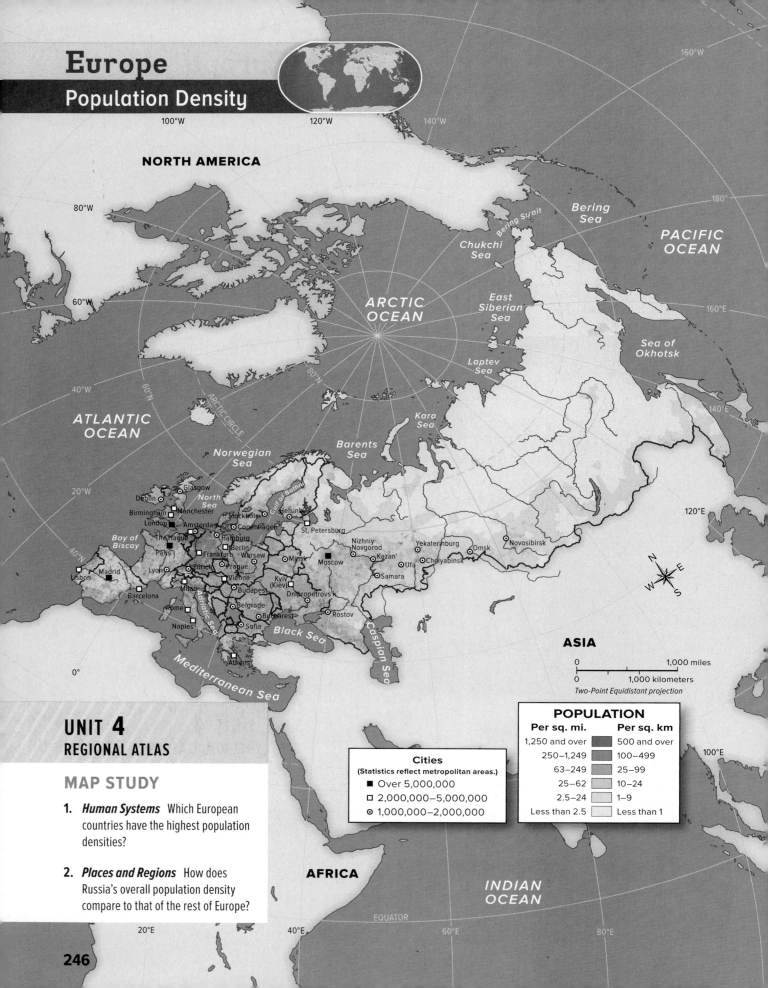

Europe
Population Density

NORTH AMERICA

100°W   120°W   140°W   160°W

80°W   180°

Bering Strait   Bering Sea   PACIFIC OCEAN

Chukchi Sea

60°W   160°E

ARCTIC OCEAN

East Siberian Sea

Laptev Sea

Sea of Okhotsk

40°W   140°E

ATLANTIC OCEAN

60°N

Kara Sea

80°N

20°W   Norwegian Sea   Barents Sea

Glasgow
Dublin   North Sea   Gulf of Bothnia
Birmingham   Manchester   Stockholm   Helsinki
London   Amsterdam   Copenhagen   St. Petersburg   120°E
The Hague   Hamburg   Nizhniy Novgorod   Yekaterinburg   Omsk   Novosibirsk
Bay of Biscay   Berlin   Moscow   Kazan'   Chelyabinsk
Paris   Frankfurt   Warsaw   Minsk   Ufa
Lyon   Zürich   Prague   Samara
Madrid   Vienna   Kyiv (Kiev)
Lisbon   Milan   Budapest   Dnipropetrovsk
Barcelona   Belgrade   Rostov
Rome   Bucharest   Caspian Sea
Naples   Sofia
Athens   Black Sea

Mediterranean Sea

ASIA

0   1,000 miles
0   1,000 kilometers
Two-Point Equidistant projection

100°E

AFRICA   INDIAN OCEAN

EQUATOR

20°E   40°E   60°E   80°E

UNIT 4
REGIONAL ATLAS

MAP STUDY

1. *Human Systems* Which European countries have the highest population densities?

2. *Places and Regions* How does Russia's overall population density compare to that of the rest of Europe?

Northern Europe

ESSENTIAL QUESTION • *How do physical systems and human systems shape a place?*

netw⊙rks

There's More Online about Northern Europe.

CHAPTER 10

Why Geography Matters
Volcanic Eruption in Iceland

Lesson 1
Physical Geography of Northern Europe

Lesson 2
Human Geography of Northern Europe

Lesson 3
People and Their Environment: Northern Europe

Geography Matters...

The countries of Northern Europe rank among the highest in the world in happiness surveys. Maybe it is the beautiful landscapes, high standard of living, cutting-edge cities, and eco-friendly planning that lead their citizens to this contentment. The region is also a destination for visitors to view the amazing phenomenon of the northern lights—a spectacular array of colorful lights appearing in the sky that can be seen in the northern reaches of Europe, attracting people from all over the world.

◄ This hockey player proudly wears Sweden's flag painted on his face.

Tomasz Tomaszewski/National Geographic Stock

volcanic eruption
in Iceland

The island of Iceland is a geologically young country located within the Mid-Atlantic Ridge, a 10,000-mile- (16,093-km-) long undersea mountain range. This ridge sits at the juncture of two tectonic plates, resulting in frequent seismic activity, including earthquakes, volcanic eruptions, and geothermal venting. Iceland experiences all of these as a result of its location along the Mid-Atlantic Ridge.

What was the physical process that occurred?

How did this affect transportation systems in Europe?

What were the resulting economic effects?

In 2010 the long-dormant Eyjafjallajökull, one of Iceland's largest volcanoes, became increasingly active. The world watched as several small earthquakes were followed by intense and frequent quakes. On March 21, lava erupted from a 0.3-mile- (500-m-) long vent in the Fimmvörduháls Pass. The volcano, often called Eyja (EYE•YAH) for short, spewed fountains of lava up to 328 feet (100 m) high and released lava flows up to 66 feet (20 m) thick. The glacier above the lava melted, sending mud and ice into the rivers. This caused flooding that damaged roads and farmland. Huge plumes of smoke and ash were sent nearly seven miles (11 km) into the atmosphere, where they were carried by wind toward continental Europe.

1. **Physical Systems** How did the eruption of the volcano lead to massive flooding?

The airborne ash posed a dangerous threat to jet engines. Volcanic ash consists of tiny pieces of glass. When those pieces get sucked into a jet engine, they melt and cause the engine to seize up. Within two days of the 2010 eruption, air travel authorities across northern and central Europe were forced to cancel thousands of commercial flights in the largest disruption of peacetime air travel in history. Millions of passengers were stranded or delayed as the ripple effects of the eruption spread across air travel systems worldwide. Many Europeans were stranded abroad. Other transportation systems, such as trains and ferry boat operations, benefited as passengers who could not travel by air made other arrangements.

2. **Human Systems** Imagine you were a traveler stranded in Northern Europe by the eruption of Eyja. How would you cope with the situation? Write a paragraph detailing your experience.

Airlines lost hundreds of millions of dollars because of canceled flights, and Europe's ability to respond effectively to a crisis was called into question. Tourism in Europe was also affected. Many tourists ended up spending their money elsewhere. Countries that relied heavily on air travelers to sustain their tourist sectors were negatively affected by the sudden disruption. Exports and imports were impeded as well. Countries with perishable exports that depend on air travel suffered extra losses. Iceland eventually experienced an increase in tourism, however, as visitors sought out views of the now-famous volcano.

3. **Environment and Society** How could individual governments across a region cooperate to minimize the economic disruption caused by events like the eruption of Eyja?

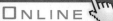

THERE'S MORE ONLINE

WATCH a video of Eyjafjallajökull's eruption • *HEAR* a news story about the volcano's effect on tourism

EYJAFJALLAJÖKULL

ENGINE TROUBLES

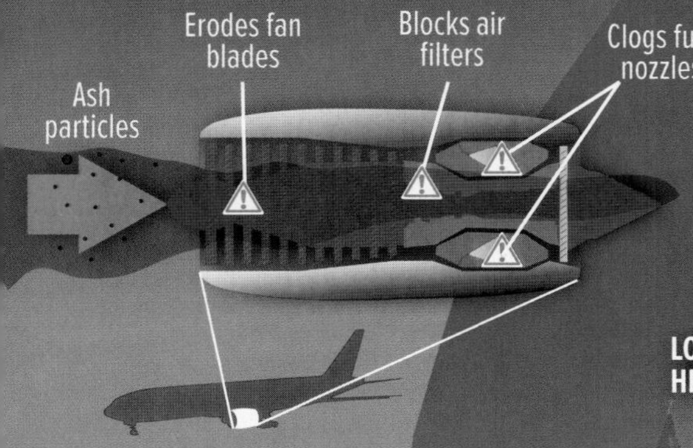

Ash particles
Erodes fan blades
Blocks air filters
Clogs fuel nozzles

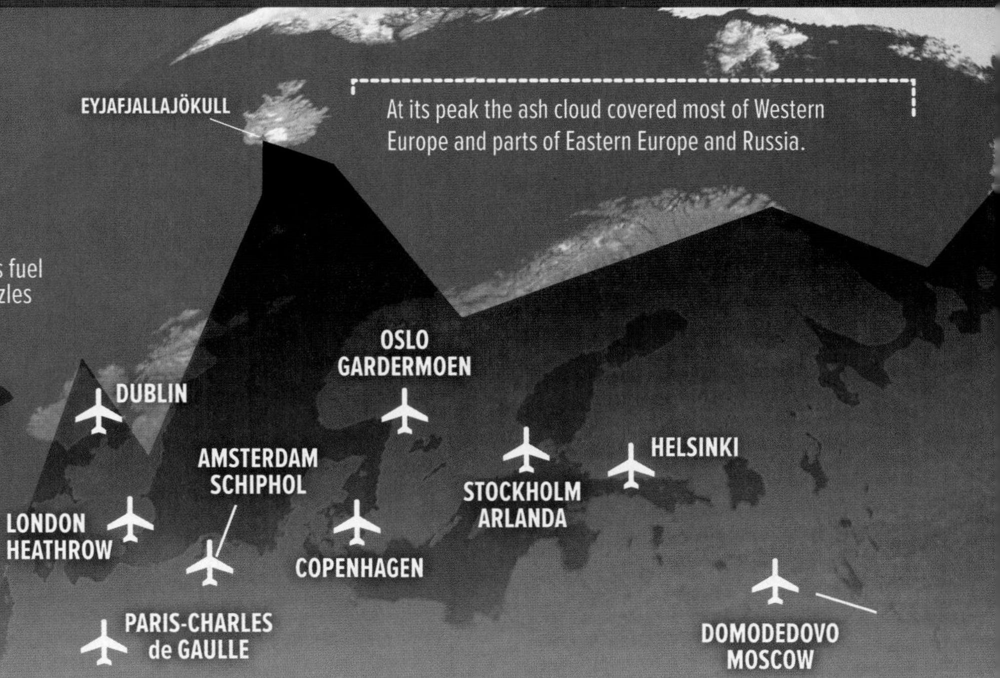

EYJAFJALLAJÖKULL
At its peak the ash cloud covered most of Western Europe and parts of Eastern Europe and Russia.
DUBLIN
OSLO GARDERMOEN
AMSTERDAM SCHIPHOL
HELSINKI
STOCKHOLM ARLANDA
LONDON HEATHROW
COPENHAGEN
PARIS-CHARLES de GAULLE
DOMODEDOVO MOSCOW

THE TIME LINE

DAY 1 — APRIL 14, 2010

THE ERUPTION

Eyjafjallajökull erupted on April 14, 2010, releasing a 4-mile high- (6.4 km-) plume of smoke into the sky.

DAY 2 — APRIL 15, 2010

SPREADING ASH

The drifting ash cloud forced northern European countries to close their airspace; 6,000 flights were cancelled.

DAY 3 — APRIL 16, 2010

GROWING CONCERN

France, Lithuania, and Hungary closed their airspace. Iceland's airlines were unaffected.

DAY 4 — APRIL 17, 2010

AT A STANDSTILL

All of Europe's major airports were closed; 17,000 flights were cancelled.

DAY 5 — APRIL 18, 2010

STILL STRANDED

80 percent of Europe's airline traffic was still grounded. Millions of passengers worldwide were stranded.

DAY 6 — APRIL 19, 2010

TRAVEL RESUMES

Selected zones of European airspace reopened to flight traffic.

GLOBAL GDP LOSS

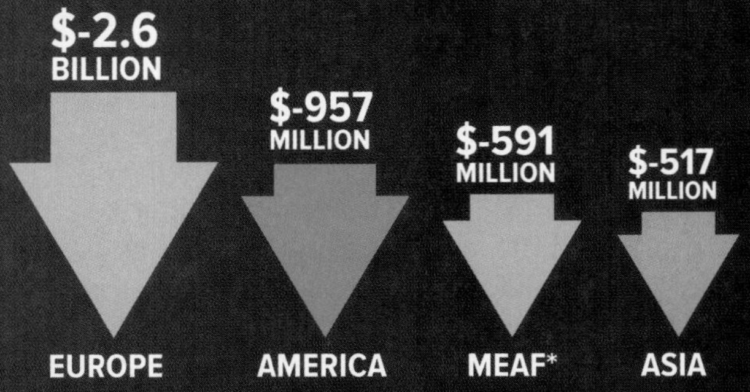

$-2.6 BILLION
EUROPE
$-957 MILLION
AMERICA
$-591 MILLION
MEAF*
$-517 MILLION
ASIA

Airlines lost an estimated **$200 million** per day.

6.8 million passengers were stranded worldwide.

*MIDDLE EAST AND AFRICA

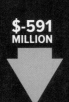

networks

There's More Online!

☑ **DIAGRAM** Gulf Stream Effects on Northern Europe

☑ **IMAGE** Fjords of Northern Europe

☑ **IMAGE** Geyser in Northern Europe

☑ **INTERACTIVE SELF-CHECK QUIZ**

☑ **VIDEO** Physical Geography of Northern Europe

LESSON 1
Physical Geography of Northern Europe

ESSENTIAL QUESTION • *How do physical systems and human systems shape a place?*

Reading HELPDESK

Academic Vocabulary

• **emerge**
• **migrate**

Content Vocabulary

• **glaciation**
• **fjord**
• **geothermal energy**
• **hot spring**
• **geyser**

TAKING NOTES: *Key Ideas and Details*

LISTING Use a graphic organizer like the one below to list the landforms, water systems, climate regions, and resources of Northern Europe.

Landforms
Water Systems
Climate Regions
Resources

IT MATTERS BECAUSE

Europe is a large peninsula made up of numerous peninsulas, such as the Scandinavian Peninsula and the Jutland Peninsula found in Northern Europe. The unique physical geography, shaped by glaciers and plate tectonics, and the cold northern climate have influenced the lives of people in this subregion.

Landforms

GUIDING QUESTION *How did the last ice age impact the landforms of Northern Europe?*

Glaciation has been the primary process by which the landforms of Northern Europe came to be as they are today. During the last ice age, the process of glaciation scoured the land and shaped the landforms. Ice filled the valleys and carved out long, narrow, steep-sided **fjords** (fee•AWRDS) that are now filled with seawater. Plains were scraped flat by the glaciers that covered the land, while the mountains in the region were made steeper and more rugged.

The ice that covered Northern Europe during the last ice age was over one mile (1.6 km) thick. It was so heavy that it pressed the land down into the Earth's mantle. Over time, as the ice melted and lessened the weight on the land beneath, the land began to rise in a process called continental rebound. The entire land surface continues to rise today. When the ice sheet melted about 10,000 years ago, it also gouged the surface of the land and left in its wake innumerable islands, rivers, and streams as well as countless lakes.

Northern Europe is made up of five countries. Norway and Sweden are found on the scenic Scandinavian Peninsula. The Jutland Peninsula forms the mainland part of Denmark and extends into the North Sea. Although not situated on the Scandinavian Peninsula, Denmark is considered part of the cultural region called Scandinavia. Finland lies in the eastern part of the region, and the island country of Iceland is located in the North Atlantic Ocean.

Most of Norway and northern Sweden are mountainous, but in southern Sweden lowlands slope gently to the Baltic Sea. Glaciers from the last ice age left behind thousands of sparkling lakes in these two countries as well as in Finland. Many deep fjords lie on the Atlantic coastline of the Scandinavian Peninsula.

Svalbard is an archipelago in the Arctic Ocean that constitutes the northernmost part of Norway. This group of islands is located about 400 miles (644 km) north of the mainland, midway between mainland Norway and the North Pole. Today glaciers and snowfields cover more than 50 percent of the islands' land. Glaciers cut the former plateau into fjords and valleys, and the geological processes of folding and faulting cause the mountainous landscape to **emerge**. The landforms of Svalbard were created through repeated ice ages and the folding and faulting associated with continental drift and plate tectonics. Norway's strongest earthquake, measuring 6.5 on the Richter scale, occurred in Svalbard on March 6, 2009.

Finland is mostly flat with a few hills and mountains. Over 10 percent of its area is covered with inland waters such as lakes and rivers. Its rugged coastline is deeply indented with bays and inlets, and the offshore region is dotted with thousands of islands.

Formerly a possession of Denmark, Iceland is an island country in the North Atlantic. Iceland is located 186 miles (300 km) east of Greenland and 621 miles (1,000 km) west of Norway. It is situated on a geological hot spot along the Mid-Atlantic Ridge. The island is very geologically active. With about 200 volcanoes, volcanic activity is frequent. Earthquakes are also frequent but rarely result in serious damage.

Although considered to be a European country, Iceland sits partly on ocean crust shared with the North American continent, as it straddles the Mid-Atlantic Ridge that marks the boundary between the Eurasian and North American tectonic plates. The tectonic activity caused by these plates' separating is the source of the abundant **geothermal energy** in the region. Iceland's many rivers and waterfalls are also harnessed to produce hydropower. These two natural resources provide Iceland with sustainable and inexpensive sources of energy.

Additional physical features include Iceland's numerous mountains, countless **hot springs**, rivers, small lakes, waterfalls, glaciers, and **geysers**. Glaciers cover roughly 11 percent of the island. The largest, Vatnajökull, is nearly 1,300 feet (400 m) thick and covers about 8 percent of the island. It is by far the largest glacier in Europe. The word *geyser* is derived from a geyser in Iceland named Geysir.

During the last ice age, glaciers deposited sand and gravel on the Jutland Peninsula's flat western side and carved fjords on the coastline of the east. Flat plains make up most of the Jutland Peninsula's interior in Denmark. The Kingdom of Denmark also includes the Faeroe Islands and Greenland in the North Atlantic.

glaciation a process by which glaciers form and spread

fjord a long, steep-sided glacial valley now filled by seawater

emerge to rise from an obscure position or condition; to become visible

geothermal energy a form of energy conversion that captures heat energy from within Earth

hot spring a spring whose water issues at a temperature higher than that of its surroundings

geyser a spring that throws forth intermittent jets of heated water and steam

Fjords and geysers are both signature features of the Northern European landscape.

▼ **CRITICAL THINKING**

1. *Comparing* How are geysers similar to volcanoes?

2. *Hypothesizing* How might fjords have influenced the history of Northern Europe?

During the last ice age, glaciers scraped across the land, making the mountains steeper and more rugged.

▲ **CRITICAL THINKING**

1. *Analyzing Visuals* Describe the landforms shown in the picture.

2. *Hypothesizing* How might the same landscape have looked during the last ice age?

At 839,399 square miles (2.1 million sq. km), Greenland is the world's largest island. The Faeroe Islands, an island group and archipelago, are located about halfway between Iceland and Norway. They are made up of volcanic rocks with high and rugged cliffs.

☑ **READING PROGRESS CHECK**

Identifying What landforms were created by glaciation during the last ice age in Northern Europe?

Water Systems

GUIDING QUESTION *Why is the landscape of Northern Europe dotted with so many lakes?*

Continental glaciers covered much of Northern Europe during the last ice age. The scouring action of these glaciers created a landscape dotted with hundreds of thousands of lakes. After the glaciers melted, the debris left behind on a flat landscape blocked rivers and trapped water like dams.

The landscape of Iceland is geologically young. It is characterized by impressive waterfalls and an abundance of small lakes and numerous, swift-moving rivers that are filled by glacier meltwater and heavy rainfall. The majority of rivers in Iceland consist of meltwater from glaciers. Thus, they contain large amounts of glacial debris that makes the water cloudy. The longest river is called Thjórsá. Located in the southern region of Iceland, it extends for 143 miles (230 km).

Many of the rivers of the Scandinavian Peninsula are short and do not provide easy connections between cities. Norway's chief rivers stem from the mountains of Norrland. These rivers mostly flow toward the southeast with many falls and rapids, eventually emptying into the Gulf of Bothnia or the Baltic Sea. The country's longest river is the Klar-Göta. It flows 447 miles (719 km) until it reaches Lake Vänern. The Glåma River drains an area of 16,236 square miles (42,051 sq. km), running almost the entire length of Norway from north to south.

The Kemi River in Finland is harnessed for hydroelectric power. It rises near the Russian border and flows generally southwest for about 300 miles (483 km) to the Gulf of Bothnia in the town of Kemi. The Muonio and Torne Rivers flow along the border of Finland and Sweden. Sweden has many small hydroelectric power plants that harness the power of the country's rivers.

✔ READING PROGRESS CHECK

Explaining From what landforms do most rivers originate?

Climate, Biomes, and Resources

GUIDING QUESTION *How does Northern Europe's location affect its climate and vegetation?*

Latitude, mountain barriers, wind patterns, and distance from large bodies of water influence Northern Europe's climate patterns. Climate regions include marine west coast, humid continental, subarctic, and tundra. Location influences vegetation patterns. Natural vegetation varies from forests and grasslands to tundra plants. In Iceland, the Gulf Stream creates a mild climate even though the country is located in higher latitudes.

Climate Regions and Biomes

Strong interrelationships exist between climate and plant and animal life in Northern Europe. The arctic tundra regions lie in the extreme northern parts of Scandinavia and Iceland. Due to dry conditions, poor soil quality, extremely cold temperatures, and frozen ground, vegetation in this climate is limited. Arctic tundra plants must adapt to the short, cold growing seasons. The frozen ground prevents plants with deep roots, like trees, from growing. Animals in the alpine zone **migrate** to lower elevations in the winter to escape the cold and find food. South of the tundra biome is the subarctic climate region. This covers most of the northern half of Scandinavia. It has long, very cold winters and short, cold to mild summers. The vegetation is limited to only the few species that can tolerate the cold conditions. Along the Atlantic coast and in southern Sweden, the climate is the marine west coast type that has milder

migrate to move from one place to another

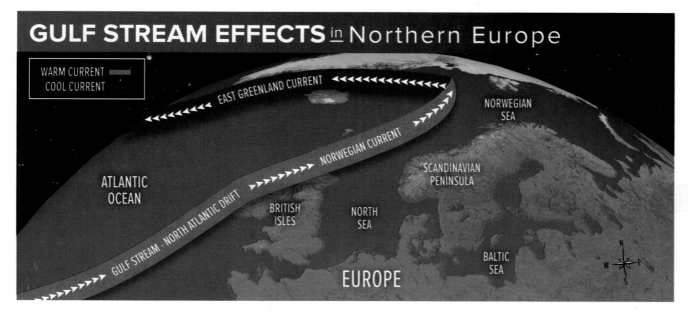

GULF STREAM EFFECTS in Northern Europe

WARM CURRENT
COOL CURRENT

EAST GREENLAND CURRENT

NORWEGIAN SEA

NORWEGIAN CURRENT

ATLANTIC OCEAN

SCANDINAVIAN PENINSULA

GULF STREAM - NORTH ATLANTIC DRIFT

BRITISH ISLES

NORTH SEA

BALTIC SEA

EUROPE

The Gulf Stream is a warm ocean current that originates in the Gulf of Mexico. As it crosses the Atlantic Ocean, the Gulf Stream splits into two currents, the North Atlantic Drift and the Canary Current. The North Atlantic Drift warms the climates of Northern Europe.

▲ CRITICAL THINKING

1. *Analyzing Visuals* In what direction is the Gulf Stream's North Atlantic Drift moving as it approaches Europe?

2. *Identifying* Over what other part of Europe does the Gulf Stream flow before it reaches Scandinavia?

CHAPTER 10 Assessment

Directions: On a separate sheet of paper, answer the questions below. Make sure you read carefully and answer all parts of the questions.

Critical Thinking

14 Making Generalizations How does tectonic activity affect the islands of Northern Europe?

15 Identifying Central Issues What is the relationship between glaciation and hydroelectric power in Northern Europe?

16 Evaluating Write a paragraph discussing the economic characteristics of the Nordic model and why it has been considered successful.

17 Drawing Conclusions How has Northern Europe's culture been affected by its physical environment?

Applying Map Skills

Use the Unit 4 Atlas to answer the following questions.

18 Place and Regions Using the political map of Europe, which Northern European countries lie partially within the Arctic Circle?

19 Physical Systems Describe the elevation of Finland compared to that of Norway as shown on the physical map of Europe.

20 Environment and Society Use your mental map to list the seas that border the countries of Northern Europe.

Exploring the Essential Question

21 Identifying Cause and Effect Create a flow chart that shows the geologic events that led to the eruption of the Eyjafjallajökull volcano in Iceland. How did it affect the economy and transportation? How is this event an example of how physical systems and human systems shape a place?

DBQ Analyzing Primary Sources

Use the document to answer the following questions.

In Denmark, some families have decided to generate their own electricity as a way of protecting the environment.

PRIMARY SOURCE

"We like using computers, and watching the television. And we like taking hot showers like everybody else. By having a windmill we can do that with a good conscience, because we know that the power we consume doesn't expose the environment. Now we produce electricity without producing carbon dioxide. It feels right to take our little part of the responsibility of getting a better environment."

—Hans Chresten Jeppesen, *National Geographic News*, November 4, 2009

22 Analyzing Describe what the Jeppesen family is doing to help limit their impact on global warming.

23 Speculating How could the governments of countries in Northern Europe and elsewhere encourage more people to take action like the Jeppesen family?

Research and Presentation

24 Research Skills Use the Internet and other resources to gather information about religions in Northern Europe and how they have changed throughout the centuries. Create a multimedia presentation to share your findings.

Writing About Geography

25 Informative/Explanatory Use standard grammar, spelling, sentence structure, and punctuation to write a one-page essay discussing current population trends in Northern Europe, including birthrates, life expectancy, and international migration.

Need Extra Help?

If You've Missed Question	14	15	16	17	18	19	20	21	22	23	24	25
Go to page	251	251	260	255	243	242	242	248	266	266	258	259

Northwestern Europe

ESSENTIAL QUESTION • *How do physical systems and human systems shape a place?*

Why Geography Matters
Suburban Growth and Transportation

Lesson 1
Physical Geography of Northwestern Europe

Lesson 2
Human Geography of Northwestern Europe

Lesson 3
People and Their Environment: Northwestern Europe

Geography Matters...

The countries of Northwestern Europe have long been at the crossroads of many cultures. There you will see a colorful mix of the old and the new. People in the subregion enjoy a higher standard of living than many other subregions. This is reflected in high levels of education, long life expectancies, and industrialized economies. The people of Northwestern Europe include an ethnic mix. Some countries are made up of two or more ethnicities that have blended over the centuries. Switzerland, for example, has three official languages.

◀ A young girl from Ireland is dressed for traditional folk dancing.

Vidler Steve/age fotostock

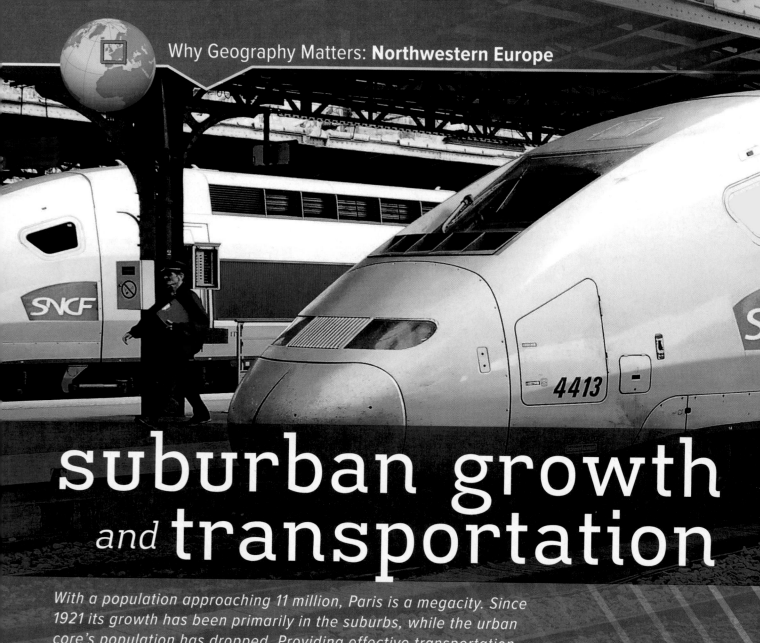

suburban growth
and transportation

With a population approaching 11 million, Paris is a megacity. Since 1921 its growth has been primarily in the suburbs, while the urban core's population has dropped. Providing effective transportation for surburban populations is an ongoing challenge. Important to meeting this challenge is understanding the city's geography.

THERE'S MORE ONLINE

READ a news article about suburbanization in Paris • **SEE** a map of Paris's population density

What is fueling suburban growth in Paris?

Various factors contribute to suburban growth in Paris. Housing prices have risen dramatically in the urban core. This is largely a result of gentrification, a process in which older areas of a city are revived by wealthier residents. This has caused many people to look for more affordable housing in locations outside the city. Deindustrialization has also occurred in the urban core, a process in which older manufacturing industries shut down. Also, many housing units have been converted into office space. These two processes have led residents to migrate to suburbs to find new jobs and places to live.

1. **Human Systems** What trends have contributed to the population growth in the Paris suburbs while the urban core population has declined?

How does this affect transportation needs in Paris?

The Paris Metro subway system has served the city since 1900 and is considered one of the world's finest. The Metro trains are both fast and frequent. However, as population growth has shifted from the urban core to the suburbs, the geographic pattern of the Metro has presented problems. Rail lines have been built to connect the city to the suburbs in a "hub and spoke" arrangement. This makes travel between suburbs and city easier, but rail travel across the region typically requires a trip into the urban core to change trains. This takes extra time, and residents increasingly use automobiles to commute, causing significant congestion. Official campaigns to discourage daily automobile use have not had much success because, even with reduced traffic congestion, it can still be more efficient to drive than to take the train.

2. **Places and Regions** How has this population shift affected the transportation needs of Paris's residents?

How is Paris responding to these needs?

The Paris Metro system is slated for a significant upgrade aimed at serving the growing suburban population and alleviating traffic problems mainly caused by commuters. A highlight of the planned system expansion (the Metro Grand Paris Plan) is a 96-mile (154.5-km) automated rail line linking suburban areas and connecting to existing Metro lines. Transportation planners think that the suburb-to-suburb expansion of the Metro system will eliminate needless trips to the city center. They also predict that the fast travel times on the new lines will encourage travelers to use the Metro system rather than automobiles.

3. **Physical Systems** How are transportation planners responding to regional travelers' needs?

networks

There's More Online!

- ☑ **IMAGE** The Alps of Northwestern Europe
- ☑ **INFOGRAPHIC** Netherlands' Reclaimed Lands: Polders
- ☑ **MAP** Gulf Stream Effects on Climates in High-Latitude Northern Europe
- ☑ **INTERACTIVE SELF-CHECK QUIZ**
- ☑ **VIDEO** Physical Geography of Northwestern Europe

LESSON 1

Physical Geography of Northwestern Europe

ESSENTIAL QUESTION • *How do physical systems and human systems shape a place?*

Reading **HELP**DESK

Academic Vocabulary

- **consist**
- **significant**

Content Vocabulary

- **loess**
- **dike**
- **polder**
- **mistral**
- **foehn**
- **avalanche**

TAKING NOTES: *Key Ideas and Details*

PARAPHRASING Use a graphic organizer like the one below to take notes on the climate regions of Northwestern Europe.

Climate	Influences on Climate

It Matters Because

Together, the United Kingdom, Ireland, France, the Netherlands, Belgium, Switzerland, Germany, Austria, Liechtenstein, Monaco, and Luxembourg make up the subregion known as Northwestern Europe. Within these countries lie some of Europe's most iconic landscapes, from the banks of the Seine River that runs through Paris to the jagged, awe-inspiring Alps.

Landforms

GUIDING QUESTION *How did the Northern European Plain affect the development of Europe?*

Northwestern Europe's landscape **consists** of plains interrupted by mountains. Scoured by Ice-Age glaciers, the Northern European Plain, or Great European Plain, is an area of relatively flat and low-lying land. It stretches from southeastern England and western France to central France and across Germany. The plain's fertile soil and wealth of rivers originally drew farmers to the area. The southern edge is especially fertile because it is covered by deposits of **loess**, a fine, rich, wind-borne sediment left by glaciers.

In contrast, the Alps are a high and jagged mountain range that lies to the south of the Northern European Plain. Created by the folding of the Earth's crust and shaped by glaciation, the Alps mountain system forms a crescent that runs from southern France through Switzerland and Austria to the Balkan Peninsula. Mont Blanc, the highest peak in the Alps, stands in France on the border with Italy at a height of 15,771 feet (4,807 m).

The Central Uplands lie between the Alps and the Northern European Plain—in parts of eastern France, southern Belgium, and southern Germany. This landform is made up of low rounded mountains, hills, and high plateaus with scattered forests. The Central Uplands are rich in natural resources.

The British Isles lie northwest of the mainland. They consist of the two large islands—Great Britain and Ireland—and thousands of smaller islands. The rugged coastline of the British Isles features rocky cliffs that drop to deep bays. Mountains, plateaus, and valleys make up most of northern and western Great Britain. Low hills and rolling plains dominate in the south and in Ireland.

✅ **READING PROGRESS CHECK**

Explaining What makes the Northern European Plain good land for agriculture?

Water Systems

GUIDING QUESTION *How have the rivers in Europe's heartland contributed to the region's development?*

Water plays a crucial role in the lives and economic activities of many people who live in Northwestern Europe. Most of Northwestern Europe lies within 300 miles (483 km) of a sea or ocean coast, so ocean transportation is important. In addition, people here depend on the many rivers that flow across the subregion for transportation, trade, and recreational activities.

The Alps are the location of major water sources. Eleven **significant** European lakes surround the Alps. Most are in valleys that were formed during the geological uplift of the Alps. They are long, narrow, deep lakes and have provided good places for people to settle because of water power for industry and convenient water routes for transportation. The spectacular scenery also makes the lake areas popular as tourist attractions.

In the Netherlands, water can be friend or foe. Approximately 25 percent of the country lies below sea level. There are extensive coastal dunes, but they have not always been helpful in keeping out the North Sea waters.

consist to be composed of or made up of

loess fine, yellowish, brownish topsoil made up of particles of silt and clay, carried and deposited by the wind

significant important

Mark Harris/Getty Images

The Alps are considered relatively young mountains in geologic terms.

◄ **CRITICAL THINKING**

1. ***Identifying Central Issues*** How do the Alps benefit the people of Northwestern Europe?

2. ***Speculating*** What features show that the Alps are relatively young?

dike large bank of earth and stone that holds back water

polder low-lying area from which seawater has been drained to create new land

Since the Middle Ages, the Dutch have built **dikes**, or large banks of earth and stone, to hold back the water. With the dikes as protection, they have reclaimed land from the sea. These reclaimed lands, called **polders**, once were drained and kept dry by the use of windmills. Today, other power sources run pumps to remove water. Polders provide hundreds of thousands of acres of land for farming and settlement. Still, stormy seas have breached the dikes, creating devastating floods.

The rivers of Northwestern Europe have differing characteristics. Although relatively short, England's Thames River allows oceangoing ships to reach the port of London. On the European mainland, however, the relatively long rivers provide links between inland areas as well as to the sea. The Rhine River, the most important river in Northwestern Europe, flows from the Swiss Alps through France and Germany and into the Netherlands. It connects many industrial cities of the interior to the port of Rotterdam on the North Sea.

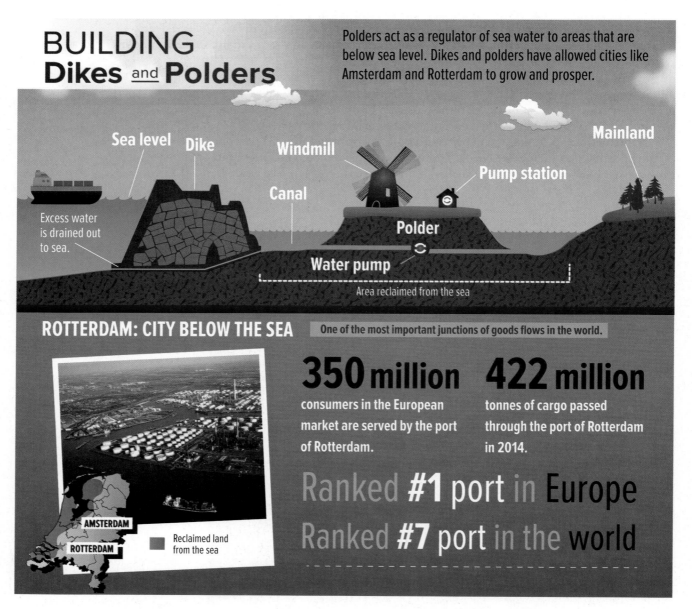

BUILDING
Dikes and Polders

Polders act as a regulator of sea water to areas that are below sea level. Dikes and polders have allowed cities like Amsterdam and Rotterdam to grow and prosper.

Sea level   Dike   Windmill   Mainland   Pump station   Canal

Excess water is drained out to sea.

Polder

Water pump

Area reclaimed from the sea

ROTTERDAM: CITY BELOW THE SEA   One of the most important junctions of goods flows in the world.

AMSTERDAM
ROTTERDAM

Reclaimed land from the sea

350 million consumers in the European market are served by the port of Rotterdam.

422 million tonnes of cargo passed through the port of Rotterdam in 2014.

Ranked **#1** port in Europe
Ranked **#7** port in the world

Polders provide hundreds of thousands of acres for farming and settlement in the Netherlands, but the dikes built to protect them are no guarantee against flooding.

▲ CRITICAL THINKING

1. *Synthesizing* What natural resources were needed to create polders?
2. *Drawing Conclusions* What geographic factors pushed the Dutch—and enabled them—to reclaim so much land from the sea?

The Seine River, whose source is in central France, flows northwest through Paris and empties into the English Channel. It carries most of France's inland waterway traffic. The Loire River is the longest river in France. From its source in southern France, it flows north and then west before emptying into the Atlantic Ocean. Silting, shallowness, and fluctuations in volume throughout its course limit the use of the Loire for navigation. Canals connect the river with both the Seine and Rhône river systems. Important to conservationists and the environment, these systems form unique habitats for migratory birds.

The Rhône River begins in the Swiss Alps and flows through Switzerland and France. It is the only major river in the subregion that flows directly to the Mediterranean Sea. Important for hydroelectric power, the Rhône is 505 miles (813 km) long and has a drainage basin of 37,750 square miles (97,775 sq. km).

Another important European river that runs through the subregion is the Danube River. Both historically and economically important, the Danube River runs from southern Germany's Black Forest through Austria and into Eastern Europe. After flowing some 1,770 miles (2,850 km), it empties into the Black Sea.

☑ READING PROGRESS CHECK

Listing For what activities are Europe's waterways used?

Climate, Biomes, and Resources

GUIDING QUESTION *Why does most of Northwestern Europe have a generally mild climate?*

Several factors affect climate in Northwestern Europe. These factors include the presence of the Alps and the location of the subregion near or along large bodies of water. The winter storms that originate over the North Atlantic Ocean also affect Northwestern Europe's climate.

Climate Regions and Biomes

Northwestern Europe generally has a mild climate compared with other regions located at the same latitude. This results from the North Atlantic Current, a powerful warm-ocean current. It is warm because it is a continuation of the Gulf Stream that emerges from the tropical waters of the Caribbean. This warm water flows along the coast of Northwestern Europe and carries warm, maritime air that bathes the coasts and also blows far inland across the Northern European Plain.

When the moist Atlantic winds reach the Alps, the winds rise up the slopes and the temperature of the air cools. This cooling air produces snow that covers the mountains in winter. Local winds in a region sometimes cause changes in the normal weather pattern. For example, the **mistral**, a strong north wind from the Alps, can send gusts of bitterly cold air into southern France. At other times, dry winter winds called **foehns** (FUHRNS) blow down from the mountains into valleys and plains. Foehns can trigger **avalanches**, which are destructive masses of ice, snow, and rock sliding down mountainsides. Avalanches represent a serious natural hazard in the Alps. They destroy everything in their paths, threatening skiers, hikers, and villages.

As is the case anywhere, climate influences the distribution of biomes in Northwestern Europe. Most of the subregion has a marine west coast climate with mild winters, cool summers, and abundant rainfall. This type of climate produces soils that are often rich in humus—a material formed from decaying leaves and other organic matter that makes soil extremely fertile. A mild Mediterranean climate is found in Monaco, which enjoys an average annual temperature of 61°F (16°C) and receives about 60 days a year of rain.

Natural vegetation in Northwestern Europe includes varieties of deciduous and coniferous trees. Deciduous trees, or those that lose their leaves seasonally,

mistral a strong northerly wind from the Alps that can bring cold air to southern France

foehn a dry wind that blows from the leeward sides of mountains, sometimes melting snow and causing avalanches; term used mainly in Europe

avalanche a large mass of ice, snow, and rock that slides down a mountainside

Although coal creates pollution, it still has the advantages of being abundant and easy to ship and burn.

▲ CRITICAL THINKING

1. *Drawing Inferences* Why do you think Europeans are using less coal, even if it is relatively plentiful?

2. *Analyzing Visuals* Based on this photograph and prior knowledge, what effects might coal mining have on the environment?

such as ash, beech, and oak, thrive in the subregion's marine west coast climate. Coniferous trees—cone-bearing fir, pine, and spruce—are found in cooler, alpine mountain areas up to the timberline, the elevation above which trees cannot grow. The wildlife in this region includes deer, brown bears, badgers, squirrels, and numerous songbirds.

Natural Resources

Northwestern Europe's abundant supply of coal and iron ore fueled the development of modern industry in the 1700s. Today, people in the subregion still rely on coal, but it is being replaced by oil, natural gas, nuclear, and hydroelectric energy sources. Vast oil and natural gas deposits under the North Sea contribute greatly to Europe's energy needs. France, which lacks large oil and gas reserves, has invested heavily in nuclear power. People in the Netherlands rely on natural gas for a majority of their energy needs, but continue to use some wind power. Mountainous Switzerland and Austria get most of their electricity from renewable sources, such as hydroelectric plants. They also have substantial timber resources. The peat bogs of Ireland serve as a source of fuel, especially in the rural countryside where it is used to heat homes. In contrast to much of the subregion, Germany has relatively few natural resources and imports more than half of its energy needs.

☑ READING PROGRESS CHECK

Drawing Conclusions Why does Northwestern Europe have a generally mild climate compared with other places at the same latitude?

LESSON 1 REVIEW

Reviewing Vocabulary

1. *Explaining* Explain the relationship between foehns and avalanches.

Using Your Notes

2. *Expressing* Use your notes to describe the factors that influence the climate regions of Northwestern Europe.

Answering the Guiding Questions

3. *Identifying* How did the Northern European Plain affect the development of Europe?

4. *Describing* How have the rivers in Europe's heartland contributed to the region's development?

5. *Explaining* Why does most of Northwestern Europe have a generally mild climate?

Writing Activity

6. *Informative/Explanatory* Write an essay explaining which geographic factors contribute to climate differences between the highlands area of the Alps and the Northern European Plain.

networks

There's More Online!

- ☑ **MAP** Population Density of Northwestern Europe
- ☑ **IMAGE** Paris World of Fashion
- ☑ **MAP** The European Union
- ☑ **INTERACTIVE SELF-CHECK QUIZ**
- ☑ **TIME LINE** The Rise of Northwestern Europe
- ☑ **VIDEO** Human Geography of Northwestern Europe

LESSON 2
Human Geography of Northwestern Europe

ESSENTIAL QUESTION • *How do physical systems and human systems shape a place?*

(tl)Universal History Archive/Getty Images, (cc)SZ Photo/Scherl/Alamy, (frl)Library of Congress Prints and Photographs Division [LC-USZ62-77142]

Reading **HELP**DESK

Academic Vocabulary

- **comprehensive**
- **focus**

Content Vocabulary

- **Industrial Revolution**
- **industrial capitalism**
- **communism**
- **Holocaust**
- **Cold War**
- **devolution**
- **guest worker**
- **agribusiness**

TAKING NOTES: *Key Ideas and Details*

IDENTIFYING Use a graphic organizer like the one below to take notes on the population, society, and culture of Northwestern Europe.

```
        Human Geography of
        Northwestern Europe
         /              \
   Population       Society and
   Patterns        Culture Today
```

IT MATTERS BECAUSE

The countries of Northwestern Europe have a long and complex history shaped by migration, ethnic differences, wars and revolutions, and efforts at peaceful integration. From the Industrial Revolution to the establishment of the European Union, this subregion has had an enormous impact on world events past and present.

History and Government

GUIDING QUESTION *How have new ideas influenced the development of governments and economies in Northwestern Europe?*

Northwestern Europe was shaped by thousands of years of migrations and invasions. Over the centuries, a variety of ethnic groups came into contact in this subregion. Additionally, Northwestern Europe was profoundly influenced by Christianity, beginning with the arrival of the Romans and its inclusion in their empire.

The Rise of Northwestern Europe

Most of Northwestern Europe was once part of the Roman Empire, one of the largest empires in history. The Romans built towns, roads, and cities throughout Europe and brought stability and general prosperity to the subregion. However, the collapse of the Roman Empire during the A.D. 400s left the subregion vulnerable to invading Germanic groups for the next several hundred years.

During Roman times, Christianity was established as the official religion of the empire, and this had long-lasting effects on the peoples and cultures of this subregion. Beginning in the A.D. 1000s, armies that consisted primarily of Northwestern Europeans fought the Crusades. The Crusades were a series of religious wars against Islamic states of the eastern Mediterranean. The goal of the Crusades was to regain the Holy Land, the birthplace of Christianity, from Muslim rule. European forces did not win permanent control of the region.

How beneficial is THE EUROPEAN UNION?

In November 2012, the European Union (EU) was awarded the Nobel Peace Prize. The world's reaction was decidedly mixed. The prize was awarded, not to an individual or an organization, but to this entire group of 27 countries that, at the time, was in the midst of an economic crisis. A nagging question was raised once again: Just how beneficial is the European Union?

The EU was officially created in November 1993, with the goal of establishing a more unified and economically and politically stable Europe. EU supporters argue that its single market has provided companies a strong business arena. Businesses do not have to worry about currency exchanges and complex international tax laws. Competition, innovation, and productivity have increased, leading to lower costs and a wider variety of consumer products. Moreover, students can study abroad in any of the EU member countries. Most importantly, after two major wars in the last century, Europe has enjoyed a long era of peace.

However, EU detractors point out that the single market has always been inherently unstable. As evidence, they point to the financial crisis Europe experienced beginning in 2010. The EU became entangled in a multibillion euro crisis, and EU members hotly debated the merits and ramifications of bailouts. Since 2008, the EU has scrambled to support failing economies in Greece, Italy, and Spain. In 2010 and 2011, for example, the EU made loans to Greece to rescue its faltering economy. The loans were conditional on the adoption of strict austerity measures by the government of Greece, which led to a worsening recession and increased social unrest. Such social unrest in countries suffering from failing economies has pitted the poorer southern European countries against the generally more affluent northern European countries. EU bureaucracy also slowed down solutions. Some worry that the EU has gone beyond the goal of economic unity, and that larger countries dominate at the expense of smaller ones. Others worry about the stability of the EU since member countries can even decide to withdraw from it, as the UK voted to do in a referendum in June 2016.

The EU Is a Stabilizing Force

PRIMARY SOURCE

" The union and its forerunners have for over six decades contributed to the advancement of peace and reconciliation, democracy and human rights in Europe. . . . The dreadful suffering in World War II demonstrated the need for a new Europe. Over a seventy-year period, Germany and France had fought three wars. Today war between Germany and France is unthinkable. This shows how, through well-aimed efforts and by building up mutual confidence, historical enemies can become close partners. . . . The stabilizing part played by the EU has helped to transform most of Europe from a continent of war to a continent of peace. "

—Thorbjorn Jagland, the chairman of the Norwegian Nobel Committee, October 2012

The EU Is Driving Europe into Recession

PRIMARY SOURCE

" You only have to open your eyes to see the increasing violence and division within the EU which is caused by the Euro project. Spain is on the verge of a bail-out, with senior military figures warning that the Army may have to intervene in Catalonia. In Greece people are starving and abandoning their children through desperate poverty and never a week goes by that we don't see riots and protests in capital cities against the troika [a committee composed of the European Commission, the European Central Bank, and the International Monetary Fund that organized loans to rescue the economies of faltering EU countries] and the economic prison they have imposed. . . . The last attempt in Europe to impose a new flag, currency and nationality on separate states was called Yugoslavia. The EU is repeating the same tragic mistake. . . . Rather than bring peace and harmony, the EU will cause insurgency and violence. "

—Nigel Farage, UK Independence Party leader, October 2012

What do you think? DBQ

1. **Finding the Main Idea** According to Jagland, how has the EU benefited Europe?

2. **Identifying Central Issues** What arguments does Farage give to support his opinion that the EU has failed?

3. **Evaluating** Who do you think makes the stronger argument, Jagland or Farage? Explain.

netw☺rks

There's More Online!

☑ **IMAGE** Fishing Vessel in Northwestern Europe

☑ **IMAGE** Environmental Activism

☑ **GRAPH** Overfishing in Northwestern Europe

☑ **MAP** Oxygen Depletion in Coastal Marine Ecosystems

☑ **INTERACTIVE SELF-CHECK QUIZ**

☑ **VIDEO** People and Their Environment: Northwestern Europe

Reading **HELP**DESK

Academic Vocabulary

- **isolate**
- **ensure**

Content Vocabulary

- **acid deposition**
- **Kyoto Protocol**
- **cap-and-trade**

TAKING NOTES: *Key Ideas and Details*

SUMMARIZING Use a graphic organizer like the one below to take notes about the human impact on the environment in Northwestern Europe and how people are addressing the issues.

Humans and Their Environment

Human Impact	Addressing the Issues

LESSON 3
People and Their Environment: Northwestern Europe

ESSENTIAL QUESTION • *How do physical systems and human systems shape a place?*

IT MATTERS BECAUSE
Humans have been changing the environment of Northwestern Europe for thousands of years. Today, very little of the region has escaped substantial alteration from its natural condition. Nevertheless, a healthy and beautiful environment is important to the region's people. They have been successful in finding ways to reverse the environmental damage. But water quality and pollution are still major issues facing Northwestern Europe.

Managing Resources

GUIDING QUESTION *How has modern development resulted in challenges to the management of resources in Northwestern Europe?*

With its highly developed and industrialized economy, Northwestern Europe consumes large amounts of natural resources and generates considerable waste products. While many natural resources are imported from other parts of the world, some of them are extracted from mines, quarries, forests, and other operations within the subregion. Such activities cause pollution and land-use changes that require careful management.

Pollution contaminates marine and animal life and creates health hazards for humans. France and other countries that border the Mediterranean Sea use the sea for waste disposal. In the past, bacteria broke down most of the waste. However, growing populations and tourism along the coasts have increased the environmental problems. Small tides and weak currents tend to keep pollution where it is discharged.

Overfishing has been a problem in the subregion as well. Global fishing levels are estimated to be four times greater than the amount of fish left to catch. Adding to the problem, native species of seaweed and shellfish often compete with foreign species carried into local waterways by ships. Unsustainable fishing remains a major issue for the European Union (EU), where 75 percent of stocks are overfished and catches are only a fraction of what they were 15 to 20 years ago. However, in many fishing grounds that affect Great Britain and Germany, progress has been made. Reports show that overfishing in the northeast Atlantic, the North Sea, and the Baltic Sea declined from 72 percent in 2010 to 41 percent in 2014.

As new roads and railways intersect Northwestern Europe, the fragmentation of the landscape limits the migration of wildlife and causes some animal populations to become **isolated**. The highest levels of fragmentation in the subregion are found in the Benelux countries (Belgium, the Netherlands, and Luxembourg), followed by Germany and France. Fragmentation also reduces the development of healthy ecosystems upon which human and animal life rely.

isolate to place or keep by itself

☑ **READING PROGRESS CHECK**

Evaluating How has industry affected Northwestern Europe's environment?

Human Impact

GUIDING QUESTION *In what ways have human activities impacted the environment in Northwestern Europe?*

The degradation of marine and coastal ecosystems is occurring in Northwestern Europe. This trend has intensified due largely to overfishing, agriculture, pollution, tourism, industrial chemicals, and coastal development.

Air pollution is a problem throughout Europe due to manufacturing industries and the heavy use of vehicles. Industrial fumes and vehicle exhaust cause eye irritation, asthma, and respiratory infections in people who live in industrial areas of Northwestern Europe. Factories built during the Communist era in East Germany, for example, still give off soot, sulfur, and carbon dioxide into the air. Some former Communist countries are closing polluting factories. Yet they are putting more cars on the road, emitting sulfur, nitrogen oxides, and carbon.

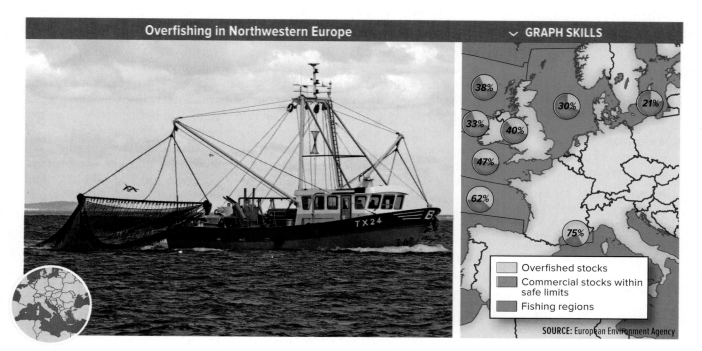

Overfishing in Northwestern Europe

⌄ **GRAPH SKILLS**

38%
30%
21%
33%
40%
47%
62%
75%

☐ Overfished stocks
☐ Commercial stocks within safe limits
☐ Fishing regions

SOURCE: European Environment Agency

Overfishing in the coastal waters of Northwestern Europe has led to reduced catches for fishers of the subregion.

▲ **CRITICAL THINKING**

1. *Analyzing Visuals* What body of water suffers the most from overfishing?

2. *Making Predictions* What might be the consequences if overfishing in this subregion continues?

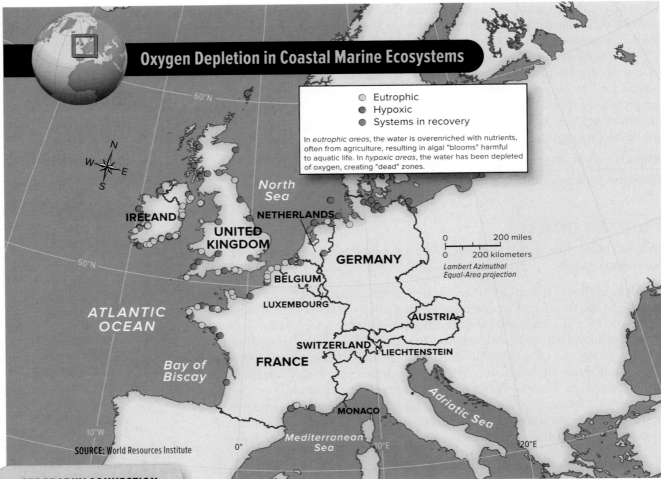

Oxygen Depletion in Coastal Marine Ecosystems

○ Eutrophic
● Hypoxic
● Systems in recovery

In *eutrophic areas*, the water is overenriched with nutrients, often from agriculture, resulting in algal "blooms" harmful to aquatic life. In *hypoxic areas*, the water has been depleted of oxygen, creating "dead" zones.

0   200 miles
0   200 kilometers
Lambert Azimuthal Equal-Area projection

SOURCE: World Resources Institute

GEOGRAPHY CONNECTION

Marine coastal ecosystems in Northwestern Europe have been polluted, resulting in areas of over- enriched nutrients that are dangerous to aquatic life and areas where aquatic life cannot survive.

1. **HUMAN SYSTEM** Which type of oxygen depletion is the result of the agricultural pollutants?

2. **PLACES AND REGIONS** How does France compare to the United Kingdom in terms of areas in recovery?

acid deposition wet or dry airborne acids that fall to the ground

In the 1970s and 1980s, industries in Northwestern Europe built taller smokestacks to carry pollution away from their communities. This worked locally, but pollution drifted across national borders. This pollution, containing acidic chemicals, combines with moisture in the air and falls as acid rain. Polluted clouds drift from the industrial belt of the subregion, and **acid deposition**—wet or dry acid pollution that falls to the ground—withers forests and damages rivers in other areas. Acid pollution even damages buildings, especially those made of limestone. During winter, snow carries the industrial pollution to the ground. During spring, meltwater—the result of melting snow and ice—carries the acid into lakes and rivers. As acid concentrations build, fish and other aquatic life die. Many lakes in Northwestern Europe have a declining fish population or even no fish at all.

✓ **READING PROGRESS CHECK**

Identifying Central Issues What human activities have led to withered forests and damaged rivers and buildings?

Addressing the Issues

GUIDING QUESTION *What actions have been taken to address environmental issues in Northwestern Europe?*

In recent decades, Europeans have made concerted efforts to clean up the environment. Countries in the EU can face legal action if they do not respect environmental protection laws. Individual countries are also addressing the consequences of pollution. For example, cities in Northwestern Europe now protect buildings and statues with acid-resistant coatings. In 2014 the mayor of Paris called for diesel cars to be banned from France's capital city by 2020 as part of plans to reduce pollution.

The EU and others continue to develop ways to protect the environment. Many power plants now burn natural gas instead of lignite coal. Natural gas burns cleaner and results in less pollution. Some countries are also developing alternative fuel sources, such as solar and wind.

Global climate change is another process that has implications for Northwestern Europe. Scientists have not identified the exact cause of climate change, but it appears clear that emissions of greenhouse gases from human activities are partly responsible for it.

In response, all the countries in the Northwestern Europe have ratified the **Kyoto Protocol**—an amendment to the international treaty on climate change designed to reduce the amount of greenhouse gases emitted by specific countries. The Kyoto Protocol sets emissions targets for participating countries and establishes a system of **cap-and-trade**.

Under a cap-and-trade system, a limit is set on the amount of air pollution that can be emitted. Businesses that produce amounts lower than the cap receive credits for the difference between their actual emissions and the cap.

Kyoto Protocol an amendment to the international treaty on climate change designed to reduce the amount of greenhouse gases emitted by specific countries

cap-and-trade a method for managing pollution in which a limit is placed on emissions and businesses or countries can buy and sell emissions allowances

Many Northwestern Europeans, such as these protesters, are deeply concerned about damage to the environment in their subregion.

▲ **CRITICAL THINKING**

1. *Speculating* What are some environmental issues that concern citizens in Northwestern Europe that may be the cause of this protest?

2. *Identifying Cause and Effect* Describe how protests like the one pictured can lead to change in government policies.

Cap-and-Trade Systems

In 2005 the European Union began a cap-and-trade system aimed at reducing the impact of pollutants that are linked to the burning of fossil fuels. The program sets a maximum overall limit (cap) on greenhouse gas emissions. Sources of the emissions receive permits to emit (emissions allowances), which can be traded. Greenhouse gas producers in the program are responsible for monitoring emissions and using permits to pay for emission privileges. The program is designed to provide both flexibility and an incentive to reduce emissions. Between 2005 and 2010, greenhouse-gas emissions dropped 13 percent.

CONSIDERING ADVANTAGES AND DISADVANTAGES What are some potential pros and cons of a cap-and-trade system? Discuss which industries may have objections to a strict cap-and-trade program and provide effective counter-arguments.

ensure to make sure or certain

Then the businesses can "trade" their pollution credits, so that a business that produces pollution at levels higher than the cap may purchase credits from businesses who earned credits. This creates an economic incentive to reduce the emissions, and the cap can be lowered over time, so that emissions continue to decline.

Protecting water resources in the subregion has also become a high priority. By the end of 2015, as required by the EU Water Framework Directive, several rivers in Europe will be required to have major restoration to meet the new regulations. Progress has been made, but the rivers in the region still suffer from floodplain drainage, industrial discharge, and the heavy use of fertilizers.

PRIMARY SOURCE

❝It will require a substantial investment but the costs of reviving Europe's rivers will be more than repaid by long-term savings in flood damage, water treatment and public health. WWF will be keeping a close eye on EU countries to make sure they live up to these commitments. ❞
—Julian Scola, "Europe's Rivers Ready for Revival," World Wide Fund for Nature, April 20, 2001.

In 1992 the Convention on Biological Diversity (CBC) was signed by 159 governments at the United Nation's Earth Summit in Rio de Janeiro, Brazil, creating the first treaty to provide a legal framework for biodiversity conservation. Following the convention, the United Kingdom created the UK BAP (Biodiversity Action Plan) a national strategy to conserve, protect and enhance biological diversity. It describes the United Kingdom's biological resources, including species and habitats, with a detailed plan for protecting their resources.

To address air pollution and improve air quality, the government of the United Kingdom allows community governments to forward fines imposed by the European Union to the local offenders. Local authorities can remove the most polluting vehicles from the roads. Governments of the countries in Northwestern Europe are not only working on solutions to environmental problems together, but with nongovernmental organizations (NGOs) as well. For example, the European Union's common fisheries policy is designed to **ensure** fisheries are more sustainably managed to prevent overfishing.

Three NGOs that are working to protect, restore, and improve the marine and coastal ecosystems are Greenpeace, Oceana, and Seas At Risk. The mission of all three of these international organizations is to limit further degradation of marine areas and loss of species, and to provide awareness, education, and concrete steps to arrest further damage. Seas At Risk, in particular, focuses its efforts on the North Atlantic and the Irish and North Seas.

✔ **READING PROGRESS CHECK**

Explaining How is the EU encouraging countries to develop new ways to protect the environment?

LESSON 3 REVIEW

Reviewing Vocabulary
1. *Explaining* Explain the Kyoto Protocol.

Using Your Notes
2. *Describing* Use your graphic organizer to describe how Northwestern Europe is addressing the issues of pollution in the region.

Answering the Guiding Questions
3. *Identifying* How has modern development resulted in challenges to the management of resources in Northwestern Europe?

4. *Describing* In what ways have human activities impacted the environment in Northwestern Europe?

5. *Explaining* What actions have been taken to address environmental issues in Northwestern Europe?

Writing Activity
6. *Argument* Imagine that you live in a polluted area of Northwestern Europe. Write a letter to the editor of a local newspaper advocating steps to halt environmental damage.

Directions: On a separate sheet of paper, answer the questions below. Make sure you read carefully and answer all parts of the questions.

Lesson Review

Lesson 1

1 *Describing* Describe the geography of the Northern European Plain and its appeal to early farmers.

2 *Explaining* Explain how water plays an important role in the lives of Northwestern Europeans.

3 *Drawing Conclusions* What are some factors affecting climate in Northwestern Europe? Why are they significant?

Lesson 2

4 *Evaluating* How did the Industrial Revolution change the human geography of Northwestern Europe?

5 *Discussing* Describe the impact of the aging population on population patterns in Northwestern Europe.

6 *Summarizing* What are the main service industries that fuel Northwestern Europe's economy?

Lesson 3

7 *Identifying Central Issues* Describe the problem of overfishing in Northwestern Europe and provide some viable solutions to reduce its impact.

8 *Explaining* What is the main reason for poor air quality and air pollution in parts of Northwestern Europe?

9 *Comparing and Contrasting* Explain cap-and-trade and the Kyoto Protocol. How are they related?

Exploring the Essential Question

10 *Making Connections* Write a paragraph that explains the effect of the Seine River on the location of cities in France.

21st Century Skills

Use the following graph to answer the questions below.

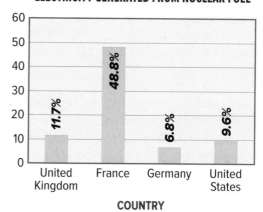

ELECTRICITY GENERATED FROM NUCLEAR FUEL

United Kingdom: 11.7%
France: 48.8%
Germany: 6.8%
United States: 9.6%

COUNTRY

Source: CIA World Factbook (2012 estimates)

11 *Compare and Contrast* How does France compare to Germany and United Kingdom in regard to nuclear fuels generated?

12 *Identifying Cause and Effect* What factors may contribute to the higher percentage of nuclear fuels generated in Northwestern Europe, as compared to the amount generated in the United States?

College and Career Readiness

13 *Economics* As a consultant for a global conglomerate, you have been hired to research the importance of transportation and communications systems to the economy of Northwestern Europe. What are the strengths and weaknesses of the current systems, and how can they be improved?

Need Extra Help?

If You've Missed Question	**1**	**2**	**3**	**4**	**5**	**6**	**7**	**8**	**9**	**10**	**11**	**12**	**13**
Go to page	270	271	273	276	278	280	284	286	287	273	289	289	280

Directions: On a separate sheet of paper, answer the questions below. Make sure you read carefully and answer all parts of the questions.

DBQ Analyzing Primary Sources

Use the document to answer the questions that follow.

PRIMARY SOURCE

"The peoples of Europe, in creating an ever closer union among them, are resolved to share a peaceful future based on common values.

Conscious of its spiritual and moral heritage, the Union is founded on the indivisible, universal values of human dignity, freedom, equality and solidarity; it is based on the principles of democracy and the rule of law. It places the individual at the heart of its activities, by establishing the citizenship of the Union and by creating an area of freedom, security and justice."

—Charter of Fundamental Rights of the European Union, 2000

14 *Identifying* According to the excerpt, what core values are held by the European Union?

15 *Paraphrasing* In your own words, summarize the meaning of the Charter of Fundamental Rights of the European Union.

Research and Presentation

16 *Research Skills* Use the Internet and other resources to gather information about one of Northwestern Europe's most famous cities, such as Paris, London, Brussels, or Amsterdam. What is the city most known for? How is it alike and different from other cities in the subregion? Create a multimedia presentation to share your findings.

Applying Map Skills

Use your Unit 4 Atlas to answer the following questions.

17 *Places and Regions* Describe the location and extent of the Alps in relation to the Northern European Plain.

18 *Physical Systems* Use the physical map of Europe from the unit atlas to list the bodies of water that border Great Britain.

19 *The World in Spatial Terms* Use your mental map of Europe to describe the location of Ireland and Great Britain in relation to continental Europe.

Critical Thinking

20 *Identifying Central Issues* How have urbanization and immigration shaped the population patterns of Northwestern Europe?

21 *Making Generalizations* What are the characteristics of internal migration and suburbanization, and how did they affect population patterns?

22 *Drawing Conclusions* Discuss the ways in which human activities have affected marine and coastal ecosystems in Northwestern Europe.

Writing About Geography

23 *Informative/Explanatory* Use standard grammar, spelling, sentence structure, and punctuation to write a one-page essay about Paris's rise to megacity status. Use library and Internet resources to gather information to discuss the city's population and how it has changed over the decades. Describe the factors regarding the decline in out-migration and a higher natural population rate increase.

Need Extra Help?

If You've Missed Question	14	15	16	17	18	19	20	21	22	23
Go to page	290	290	278	242	242	242	278	278	286	269

Southern Europe

ESSENTIAL QUESTION · *How do physical systems and human systems shape a place?*

◀ Greece's many islands and long coastlines have always supported a strong tradition of fishing.

Steve Outram/Photographer's Choice/Getty Images

Why Geography Matters
Labor Migration from North Africa

Lesson 1
Physical Geography of Southern Europe

Lesson 2
Human Geography of Southern Europe

Lesson 3
People and Their Environment: Southern Europe

Geography Matters...

Have you ever heard of the Seven Wonders of the World? Europe is the continent where that idea was born. There have been many lists, so no one knows for certain what all of the Seven Wonders were. The cultures of Southern Europe have created architectural wonders. Among them are the Roman Colosseum, the Parthenon of ancient Greece, and the elaborate Spanish palace called the Alhambra. The natural wonders of Southern Europe are also abundant, from the Mediterranean Sea to Basque Country—home of the largest cavern in Europe.

labor migration
from North Africa

The migration of people from place to place is an important theme in the study of geography. Throughout history people have migrated in search of better opportunities. One modern example is the migration of North African workers to the countries of Southern Europe. This has had important consequences both for the North African migrants and the countries of Southern Europe.

Migration: North Africa to Southern Europe

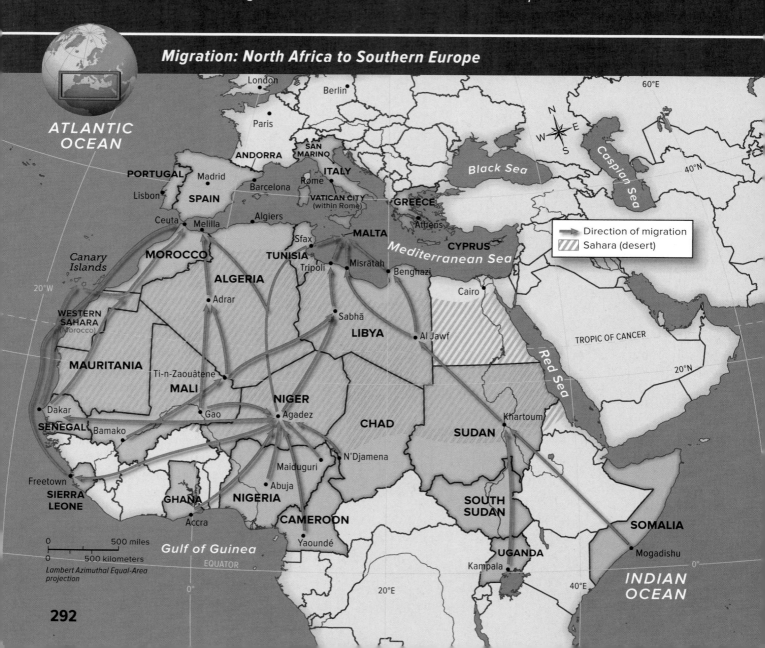

Why do people from North Africa migrate to Southern Europe?

Several factors cause North African workers to migrate to Southern Europe in search of employment. The economic opportunities in Southern Europe represent major pull factors. Migrants are drawn to the Southern European countries in search of employment. Demographic changes have made this trend greater. The population of Southern Europe has aged. Young workers are needed to fill the spaces left as older people retire and stop working, thus increasing the demand for migrants. Other factors push people out of North Africa. As countries in North Africa face increasingly high unemployment and wages decline for workers, many choose to leave their home countries—a push factor. In addition, political instability and conflict arising from the "Arab Spring" uprisings and the Libyan civil war have forced many to flee to Southern Europe.

1. **Human Systems** Describe the causes of North African migration to Southern Europe. Include both push and pull factors.

How is Southern Europe affected by this migration?

An economic boom has resulted in increased demand for low-skilled labor in several Southern European countries. Italy and Spain, with their long Mediterranean coastlines, have become popular entry points for undocumented migrants as they fill the need for service industry jobs. During the prosperous economic times, the labor of these migrants helped fuel the engine of economic development by providing sources of cheap labor. However, besides providing labor, North African migrants (whether documented or not) have influenced the cultural development of the receiving countries of Southern Europe. The migrants have enriched the cultural landscape of their new countries. They have brought their religious beliefs, social customs, arts, cuisine, and dress to Southern Europe. In some places in Southern Europe the change has come at the expense of the migrants. The arrival of migrants from North Africa has led to racism and to a backlash against them as they try to establish their new lives in Southern Europe.

2. **Human Systems** How does North African migration to Southern European countries benefit the receiving countries?

How is North Africa affected by this migration?

North Africa is affected both negatively and positively by the migration to Southern Europe. Many North African migrants send some of their pay home to help support their families. The money sent by a migrant to his or her home country is called a remittance. These remittances have reduced poverty in their countries of origin. The remittances also help the country even when the money goes to families that are relatively wealthy since it contributes to the economy of the North African home country. Other positive effects include more funds being available for North Africans to invest in children's health and education and a reduction in child labor. Not all of the effects are positive, as the migrations can cause labor shortages in the home country.

3. **Environment and Society** What are some of the potential long-term effects of the remittances by North African migrants? Write a paragraph discussing the possible impacts on North African society and on the economies of the home countries.

(l)©David Turnley/Corbis, (c)Directphoto/age fotostock/SuperStock, (r)Andrew Woodley/Alamy

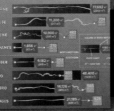

networks

There's More Online!

☑ **IMAGE** Mining in Spain

☑ **INFOGRAPHIC** Rivers of Europe

☑ **IMAGE** Massifs in Southern Europe

☑ **INTERACTIVE SELF-CHECK QUIZ**

☑ **VIDEO** Physical Geography of Southern Europe

LESSON 1
Physical Geography of Southern Europe

ESSENTIAL QUESTION • *How do physical systems and human systems shape a place?*

Reading HELPDESK

Academic Vocabulary

- **access**
- **resource**

Content Vocabulary

- **massif**
- **tungsten**

TAKING NOTES: *Key Ideas and Details*

LISTING Use a fishbone graphic organizer like the one below to list the landforms and rivers of Southern Europe.

IT MATTERS BECAUSE

The seaports of the Iberian Peninsula, the mountains and rivers of the Italian Peninsula, and the rocky islands of Greece have played important roles in the history of Southern Europe. Its geographic location has made this subregion—which includes the countries of Italy, Spain, Andorra, Greece, Portugal, Vatican City, Malta, Cyprus, and San Marino—important for trade and agriculture stretching back more than 3,000 years.

Landforms

GUIDING QUESTION What two types of physical features dominate Southern Europe's physical geography?

Geographically, Europe is a continent made up of peninsulas. The southern part of the region includes three major peninsulas: the Iberian Peninsula, the Italian Peninsula, and the Balkan Peninsula.

Extending off southwestern Europe, the Iberian Peninsula is the location of Spain and Portugal. This landmass separates the Atlantic Ocean from the Mediterranean Sea, leaving only the 20-mile-(32-km-) wide Strait of Gibraltar to connect them. Coastal plains give way to the Meseta, a large plateau that makes up most of the interior of the peninsula. To the north, the Iberian Peninsula is separated from the rest of Europe by the Pyrenees, mountains that have isolated the peninsula's residents for centuries. The independent principality of Andorra is located high in the Pyrenees between modern-day Spain and France. Andorra owes its political autonomy to its isolated location in the mountains.

The most southwestern of Europe's mountain ranges, the Pyrenees stretch from the Bay of Biscay on the Atlantic side of Spain to the Mediterranean Sea. At its widest the Pyrenees range is a daunting 80 miles (128 km) across. The Pyrenees are characterized by flat-topped **massifs**—a body of mountain ranges formed by fault-line activity. The forces of plate tectonics are responsible for the rise of these massifs, and earthquakes occur as the mountains are built up.

Italy occupies the Italian Peninsula, which extends from the south of Europe into the heart of the Mediterranean Sea. Plains cover only about one-third of the Italian Peninsula. The largest is the plain of Lombardy along the Po River in the north. The coastline of Italy varies from high, rocky cliffs to long, sandy beaches, and has several well-sheltered ports to support trade. The Apennine Mountains run down the spine of the peninsula all the way through the center of the large island of Sicily off the southwestern tip of Italy. The range is about 1,245 miles (2,000 km) long.

To the north of the Italian Peninsula lie the majestic Alps, the most recognizable range of mountains on the European continent. They loom over Southern Europe and form a natural barrier between the Italian Peninsula and Northern Europe. Because the Alps are the highest mountain range in Europe, they are also the source of Europe's largest and most important rivers. These rivers flow north into France and Germany or south where they empty into the Mediterranean, Adriatic, and Black Seas.

In southeastern Europe, the Balkan Peninsula is bounded by the Adriatic and Ionian Seas to the west and the Aegean and Black Seas to the east. Greece is the southernmost country on the Balkan Peninsula. The numerous mountains on this peninsula have limited the area's potential for communication and development. However, this has been offset by the region's easy **access** to the sea. Greece is known for the large numbers of islands—nearly 2,000—that spread out from its coastline in the Aegean Sea.

The islands that lie south of the mainland of the Iberian, Italian, and Balkan Peninsulas are geographically and politically important to Spain, Italy, and Greece. They serve as trading posts in the Mediterranean. Rugged mountains form the larger islands of Sicily, Sardinia, Corsica, Crete, and Cyprus. Tectonic activity is characteristic of this region. Sicily, the largest of these islands, is dominated by Mount Etna. At 10,700 feet (3,261 m), Mount Etna is Europe's tallest active volcano. Smaller island groups include Spain's Balearic Islands in the Mediterranean, Italy's Lipari Islands in the Tyrrhenian Sea, and the three islands of Malta in the Mediterranean.

✔ READING PROGRESS CHECK

Identifying What seas surround the Balkan Peninsula?

massif a body of mountain ranges formed by fault-line activity

access a way to approach or enter

Massifs, such as these in the Dolomite range of the Alps in northeast Italy, developed as a result of tectonic activity.

▼ CRITICAL THINKING

1. Identifying Cause and Effect Why do massifs form?

2. Drawing Inferences Other than the Alps, where else can massifs be found in Southern Europe?

Water Systems

GUIDING QUESTION *How do Southern Europe's rivers compare to those of Northwestern Europe?*

The two major rivers on the Iberian Peninsula are the Tagus and the Ebro. Both rivers play crucial roles in the economy and ecology of the region. However, their roles are limited because they, like all rivers on the Iberian Peninsula, are generally too shallow for large ships.

The Tagus River begins near the eastern edge of Spain and travels westward for 626 miles (1,007 km) through Portugal to the Atlantic Ocean. In northern Spain, nearly 200 tributaries, mostly from the rainy Pyrenees, feed the Ebro River, Spain's longest river. The steep gorges and rocky terrain that this river flows through make it inaccessible to boats. However, the Ebro has been dammed to provide a significant portion of Spain's hydroelectric power. In addition, the

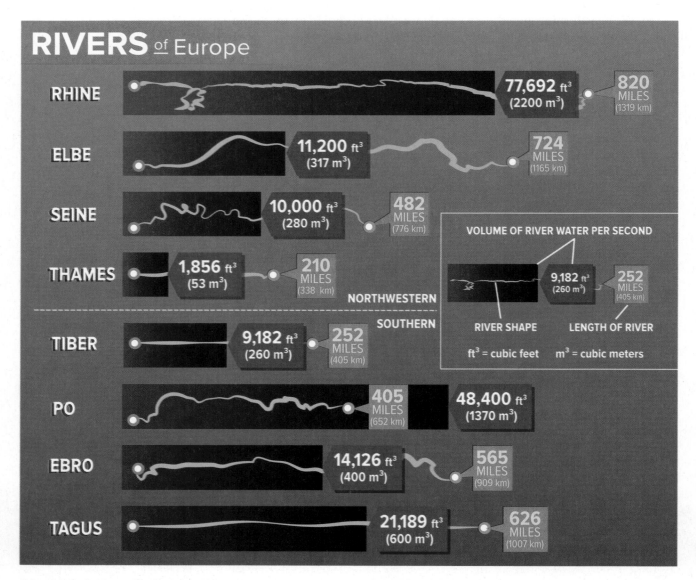

RIVERS of Europe

RHINE — 77,692 ft³ (2200 m³) — 820 MILES (1319 km)

ELBE — 11,200 ft³ (317 m³) — 724 MILES (1165 km)

SEINE — 10,000 ft³ (280 m³) — 482 MILES (776 km)

THAMES — 1,856 ft³ (53 m³) — 210 MILES (338 km)

NORTHWESTERN

VOLUME OF RIVER WATER PER SECOND
9,182 ft³ (260 m³) — 252 MILES (405 km)
RIVER SHAPE   LENGTH OF RIVER
ft³ = cubic feet   m³ = cubic meters

SOUTHERN

TIBER — 9,182 ft³ (260 m³) — 252 MILES (405 km)

PO — 405 MILES (652 km) — 48,400 ft³ (1370 m³)

EBRO — 14,126 ft³ (400 m³) — 565 MILES (909 km)

TAGUS — 21,189 ft³ (600 m³) — 626 MILES (1007 km)

Although critically important to the ecology of the region, the rivers of Southern Europe have less importance as transportation and trade routes than the rivers of Northwestern Europe.

▲ CRITICAL THINKING

1. **Contrasting** How are the rivers of Southern Europe different from the rivers of Northwestern Europe in regard to length?

2. **Evaluating** Is there a correlation between length of rivers in Europe and the volume of water flow? Provide an example to support your answer.

resulting reservoirs provide water to an impressive network of irrigation canals that support the agriculture of Spain.

The Apennines form a mountain range that runs down the center of the long and narrow Italian Peninsula. This has created rivers that are steep, short, and relatively narrow and shallow, and not suitable for transportation by boat. In the north, however, the Po River runs through the plain of Lombardy. Although it is Italy's longest and most significant river, it is still only 405 miles (652 km) long.

Venice is located at the mouth of the Po on the Adriatic Sea. The city has built a complicated system of dikes and canals to help control the river's outflow. Efforts to claim marshy areas of the Po Delta for small farms failed due to floods, especially in the 1950s and 1960s. Nevertheless, the drainage basin of the Po River forms Italy's largest and most fertile agricultural plain, covering 27,062 square miles (70,091 sq. km).

The Tiber River is Italy's second-longest river and has great historical significance. A mere 252 miles (405 km) long, this short river is nevertheless very important to Italy's economic history. It is the primary water source for the capital, Rome. Civitavecchia, on the lower part of the river, is a significant port and naval harbor for Rome. The Tiber River empties into the Tyrrhenian Sea.

Greece, on the southern tip of the Balkan Peninsula, has a mountainous terrain. Its rivers are short, unsuitable for navigation, and unusable for irrigation. In their upper courses, near their sources, the rivers flow in broad, gently sloping valleys. However, in their middle courses they plunge through basins into narrow gorges. In their lower courses, as they near their mouths, they meander across coastal plains into marshy deltas. Northeastern Greece is home to the Maritsa River, located in a low valley full of marshes. The Maritsa marks Greece's border with Turkey. In northeastern Greece, the two main rivers are the Vardar and the Aliákmon.

Glacial movement in the last ice age did not reach Southern Europe's peninsulas. As a result, the landforms of these countries lack the natural lakes or reservoirs found in Northern Europe and Northwestern Europe. The climate of Southern Europe is also much drier than farther north, another reason why there are fewer rivers and lakes in this subregion.

☑ **READING PROGRESS CHECK**

Describing Describe the importance of the Po River.

Climate, Biomes, and Resources

GUIDING QUESTION *Why does Southern Europe's climate make it popular with tourists and ideal for agricultural activities?*

Southern Europe's location on the Mediterranean Sea influences the climate and biomes of the subregion. The climate also makes the subregion a popular vacation destination. The subregion is particularly suited for growing grapes, olives, and shrub herbs and raising goats and other livestock.

Climate Regions and Biomes

The Alps separate two major climate zones: the marine west coast climate to the north and the warm Mediterranean climate of Italy and the Balkans to the south. The Alps block most Atlantic winds from the north, causing less precipitation to fall in Southern Europe. Generally, Southern Europe experiences the warm, dry summers and the mild, rainy winters characteristic of the Mediterranean climate.

The Mediterranean climate of Southern Europe results, in part, from the warm waters of this sea. Average yearly rainfall across Southern Europe is less than 30 inches (76 cm), and most of the yearly rainfall occurs in the winter months.

Spain has more mineral resources than Italy due to their different geological histories.

▲ CRITICAL THINKING

1. *Making Generalizations* Why does Italy have fewer mineral resources than Spain?

2. *Analyzing Visuals* Based on the photo, what might be a drawback to plentiful mineral resources?

tungsten an extremely rare heavy-metal element essential in high-tech industry

resource a usable stock or supply

The plants and animals native to the subregion are well-suited for less water and the long summer dry period. The coastal areas are covered in chaparral, or shrubs and shrub trees that are drought resistant. Most agriculture in Southern Europe takes place on coastal plains that receive more rainfall and develop a thicker topsoil as a result of runoff and sediment left by rivers.

The ecosystems in the Mediterranean are diverse and ecologically sensitive to climate change. Gorges channel water away from the land, leaving much of the region warm, dry, and covered in scrub plants. The coastal plains have rich sedimentary soil and a high diversity of plant life, and support most of the regional agriculture. However, they are prone to flooding.

Natural Resources

Italy has few mineral resources. Portugal, however, has large deposits of copper. Northern Spain, along the Pyrenees and the Atlantic Ocean, is rich in coal, tin, and **tungsten**. Tungsten is an extremely rare heavy-metal element which is essential in high-tech industries. Spain's mining operations in search of this valuable natural element are unlikely to be reduced.

Both Italy and Spain have benefited from the production of hydroelectricity. Greece has many rivers suitable for producing hydroelectricity. However, Greece has not fully developed these **resources**.

✓ READING PROGRESS CHECK

Describing Describe the ecosystem of the Mediterranean climate.

LESSON 1 REVIEW

Reviewing Vocabulary

1. *Explaining* Explain what massifs are and how they are formed.

Using Your Notes

2. *Listing* Use your graphic organizer to list the major landforms in Southern Europe.

Answering the Guiding Questions

3. *Identifying* What two types of physical features dominate Southern Europe's physical geography?

4. *Comparing* How do Southern Europe's rivers compare to those of Northwestern Europe?

5. *Expressing* Why does Southern Europe's climate make it popular with tourists and ideal for agricultural activities?

Writing Activity

6. *Informative/Explanatory* Write a paragraph describing how the rivers of Southern Europe can be an asset for the countries in the subregion.

networks

There's More Online!

- ☑ **IMAGE** Classic Greek Architecture
- ☑ **IMAGE** Classic Roman Architecture
- ☑ **TIME LINE** Foundations of Western Civilization
- ☑ **INTERACTIVE SELF-CHECK QUIZ**
- ☑ **VIDEO** Human Geography of Southern Europe

Reading **HELP**DESK

Academic Vocabulary

- **issue**
- **accurate**

Content Vocabulary

- **city-state**
- **Renaissance**
- **complementarity**

TAKING NOTES: *Key Ideas and Details*

IDENTIFYING Use a graphic organizer like the one below to take notes about the human geography of Southern Europe.

Southern Europe

History and Government →

Population Patterns →

Society and Culture Today →

LESSON 2
Human Geography of Southern Europe

ESSENTIAL QUESTION · *How do physical systems and human systems shape a place?*

IT MATTERS BECAUSE

The European countries in the region of the Mediterranean Sea have a rich cultural heritage. Greece was the birthplace of classical civilization, the Roman Empire was born in Italy, and Spain and Portugal became leaders in the Age of Exploration. Today the region is highly developed and plays an important role in geopolitics and the world economy and as a cultural center.

History and Government

GUIDING QUESTION *What characteristics of early civilizations are evident in Southern Europe today?*

Evidence of the cultural inheritance of Southern Europe is visible in the ruins of ancient civic architecture, such as the Parthenon. Those majestic ruins are a reminder of the lasting impact that past civilizations have had on this region and around the world. The civilizations of ancient Greece and Rome laid the foundations for European—and Western—civilization.

Early History

Ancient Greece was a collection of **city-states,** each independent with its own form of government and society. They were linked only by their common language and shared cultural identity. The classical period of Greek history reached its height in the 400s B.C. The city-state of Athens introduced the concept of democracy to the world. The city-state of Sparta was built on the glory of war. Sparta's generals created battle strategies that are still studied and used today. The mythology of the Greeks and Romans has influenced and inspired the art and culture of Western civilization for over 2,000 years.

The Roman Republic, founded in Italy, was established on the rule of law and the balance of power. It was the foundation of the largest empire of the ancient world. The empire reached the height of its power in 27 B.C. and experienced a resurgence around 200 years later, in A.D. 180.

Southern Europe **299**

(bl)©Bettmann/Corbis, (tcl)Fine Art Images/SuperStock/Getty Images, (tc)Photodisc, (tcr)©Bettmann/Corbis, (tr)Angel Navarrete/Bloomberg/Getty Images

Since the 1990s, Spain's economy has grown as a result of lower costs of operating factories and a large supply of workers. Membership in the EU provided a market for Spanish products. Although it was severely affected by the global recession, Spain has continued its important economic role in the region.

Italy went through an industrial reconstruction after World War II. Areas along the Po River valley became the leading industrial region of the country. The EU opened other countries for the import of Italian Fiat cars and trucks, clothing, and home furnishings. Today Italy continues to thrive on high-tech engineering and metallurgical manufacturing. Certain portions of Italy's agricultural market play an important part in the export market, namely olive oil and wine.

Greece's economy is one of the least developed in Southern Europe. Natural resources are limited and industrialization has been slow to develop. A large sector of the economy is tourism. EU membership has been beneficial to agriculture and industrialization, since EU investments and subsidies compensate for low productivity. Today Greece faces major challenges to reduce its public spending. Generous social programs, tax evasion, and persistently high unemployment continue to be issues facing Greece.

Southern Europe faces many challenges in meeting its energy needs. The EU in general is the world's largest energy importer and is focusing more on natural gas due to its plans to reduce carbon dioxide emissions. While regional resources are limited, just south across the Mediterranean are rich oil- and natural gas-producing countries in North Africa. Pipelines on the seafloor deliver important energy supplies.

Future Prospects

Prior to 2009, membership in the EU was good for countries' economies. Member countries were doing well, and trade among them was growing annually. Easier migration within the EU across the borders between countries was filling the labor shortage. Living standards increased steadily.

In the twenty-first century the EU faces its first great test of economic stability within the subregion. The Southern European members are dealing with economic issues such as too much national debt, too high unemployment, and too much governmental spending. Those problems have stretched resources and goodwill very thin across all of Europe. Tensions have grown as the EU has provided special assistance to the Southern European members through loans and subsidies.

☑ **READING PROGRESS CHECK**

Summarizing Why is the Greek economy one of the least developed in Southern Europe?

LESSON 2 REVIEW

Reviewing Vocabulary

1. *Drawing Conclusions* How did city-states influence the political history of Italy and Greece?

Using Your Notes

2. *Summarizing* Use your lesson graphic organizer to describe society and culture today in Southern Europe.

Answering the Guiding Questions

3. *Identifying* What characteristics of early civilizations are evident in Southern Europe today?

4. *Expressing* How have migration and aging populations affected Southern Europe's population patterns?

5. *Discussing* How have religion, the arts, and Southern Europe's rich intellectual traditions shaped society and culture today?

6. *Describing* What are the characteristics of Southern Europe's economy today?

Writing Activity

7. *Informative/Explanatory* Write a paragraph describing the contributions Southern Europe has made to the arts, both classical and modern art. Include specific time periods and forms of art and architecture.

networks

There's More Online!

- ☑ **IMAGE** Southern Europe Addressing the Issues
- ☑ **IMAGE** Pollution Around the Mediterranean Sea
- ☑ **MAP** Soil Erosion in Southern Europe
- ☑ **INTERACTIVE SELF-CHECK QUIZ**
- ☑ **VIDEO** People and Their Environment: Southern Europe

LESSON 3
People and Their Environment: Southern Europe

ESSENTIAL QUESTION • *How do physical systems and human systems shape a place?*

Reading **HELP**DESK

Academic Vocabulary

- promote
- factor

Content Vocabulary

- pollution hot spot

TAKING NOTES: *Key Ideas and Details*

PARAPHRASING Use a web diagram like the one below to take notes about the human impact on the environment of Southern Europe.

IT MATTERS BECAUSE

The Mediterranean Sea is essential to the culture and economy of Southern Europe. The human population growth of the region is slowing after 100 years of rapid expansion. Maintaining the balance between human needs and the environment is often challenging with such a large population.

Managing Resources

GUIDING QUESTION *What are the threats that require closer management of resources in Southern Europe?*

Southern Europe's landscape is full of natural wonders—beautiful rivers, seas, and forests. Across most of the region, the climate is suitable for outdoor activity year-round, encouraging tourists and residents to spend a good deal of time exploring the outdoors. Human settlement has greatly increased over the last few decades, which has resulted in a number of concerns for the area's resources and environment.

One major environmental concern in the region is the presence of large algae blooms in the Adriatic Sea. This sea forms a smaller part of the Mediterranean Sea between Italy and Greece. The algae blooms are evidence of the effects of human activity on delicate marine biomes. These visible clusters of organisms appear when there is an imbalance in the ecological structure. Warmer water, chemical fertilizer runoff from agriculture, and human settlements are all possible causes for an algae bloom. The bloom is potentially harmful because it can use up the dissolved oxygen in the water, killing fish and other marine life. Toxins produced by particularly harmful algae blooms can kill marine life and humans who consume affected marine life.

The Mediterranean climate is known for its long, dry summers and wet, mild winters. However, changes in the global climate have made summers unpredictable. Some summers turn into droughts and others produce unseasonable rain. Both scenarios can result in excessive soil erosion. Loss of topsoil and soil erosion can threaten the natural balance of an ecosystem by removing potential footholds for vegetation. When bare rock is exposed, the land becomes vulnerable to other environmental threats, such as fire or desertification. When there is no plant life to hold dry soil to steep hillsides, rain can wash away already thin topsoil on the rocky terrain of much of the subregion. Too much rain can also **promote** the growth of vegetation that can become fuel for a fire.

Dry season fires are a natural part of Mediterranean climate ecosystems. They are necessary to maintain the native vegetation. However, fire is a contributing **factor** in the deforestation of human-modified areas. Climate change is also a significant contributing factor to deforestation. The Mediterranean region is experiencing longer and hotter summers with less humidity and more wind. Both low humidity and more wind provide conditions that increase the incidence of forest fires.

Higher altitude areas are also affected by climate change. Parts of the Pyrenees and the Alps that normally experience sufficient snowfall and cold temperatures to maintain glaciers and forests of coniferous trees are changing. The glaciers are shrinking, while the forests are being attacked by beetles and other insects that flourish in the warmer temperatures. Awareness of these effects of climate change is needed to protect the environments that support life in the Mediterranean areas.

☑ **READING PROGRESS CHECK**

Making Connections How is climate change related to soil erosion in Southern Europe?

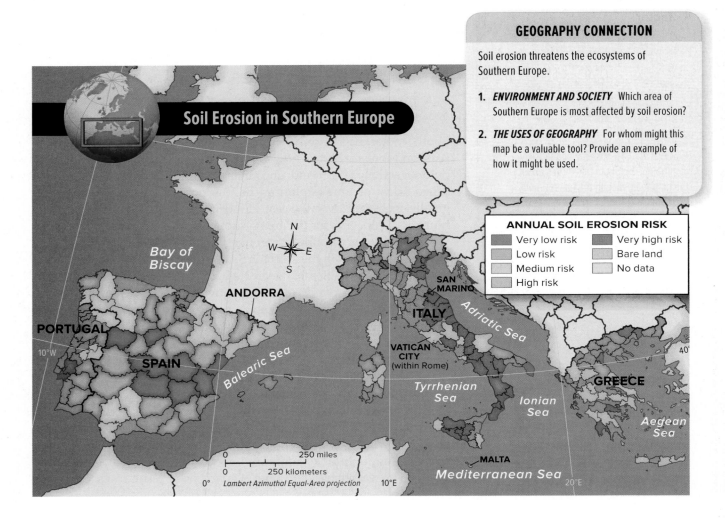

Soil Erosion in Southern Europe

GEOGRAPHY CONNECTION

Soil erosion threatens the ecosystems of Southern Europe.

1. *ENVIRONMENT AND SOCIETY* Which area of Southern Europe is most affected by soil erosion?

2. *THE USES OF GEOGRAPHY* For whom might this map be a valuable tool? Provide an example of how it might be used.

ANNUAL SOIL EROSION RISK

- Very low risk
- Low risk
- Medium risk
- High risk
- Very high risk
- Bare land
- No data

Human Impact

GUIDING QUESTION *What are the environmental concerns regarding tourism in Southern Europe?*

Southern Europe is well known for its food culture. In the coastal areas, this means fish and other seafood are common in the local diets. As a result, overfishing has occurred. Fish stocks are not being replenished as quickly as they are being depleted. As the fish stocks have declined, the overall health of the marine biome has declined as well.

Two sectors of the economy in Southern Europe that have seen tremendous growth in the last 50 years are industrial manufacturing and tourism. Industrial pollution is a major threat to the surrounding seas and to the quality of water and agricultural resources. Development associated with tourism causes major damage to coastal ecosystems, loss of natural habitat, overuse of freshwater resources, as well as pollution and waste.

Tourism inundates regions that are normally sparsely populated with huge numbers of people. More people make a greater impact on the local environment. Waste management issues are the primary concern. One popular form of vacationing is via cruise ship. Oil spills and oil pollutants from ships of all kinds are a growing concern in the subregion.

Pollutants and other human impacts in the water systems have given rise to **pollution hot spots** in the Mediterranean Sea. These hot spots are locations where such impacts have led to the degradation, and even the death, of the local ecosystem. These spots do not support marine life. The water is warmer and more vulnerable to the development of algae and other microorganisms, some of which are harmful to humans.

☑ **READING PROGRESS CHECK**

Summarizing How are manufacturing and tourism affecting the environment?

Industrialization and development along coastal Southern Europe has led to numerous problems in the marine biomes of the region.

▲ **CRITICAL THINKING**

1. Analyzing Visuals How might the human activity in this photo affect the nearby waterway?

2. Formulating Questions If you were an ecologist, what are two questions you would ask to evaluate the health of this waterway?

pollution hot spot a location where pollution and other human activities have led to the degradation, or even death, of an ecosystem

Addressing the Issues

GUIDING QUESTION *How are groups, governments, and others addressing environmental issues in Southern Europe?*

As early as 1975, European countries in the Mediterranean region recognized that rapid population expansion, industrialization, and a general increase in human activity were threatening the region's natural beauty and resources. The Mediterranean Action Plan was created at that time to help plan a way to curb damage to the environment. Each participating country set goals for reducing its environmental impact on the region, and governments in the region have been diligent in addressing the issues. Legislation has been enacted in all participating countries to try to regulate the effects of industrialization and population growth.

In 1990 the European Union (EU) created the European Environmental Agency. Its purpose is to help evaluate threats to the region and create plans to effectively deal with environmental issues. This agency makes independent reports to the European Union, which can then work with member countries to enact action plans to address environmental concerns.

In Venice, Italy, demonstrators protest against the docking of huge cruise ships—*grandi navi*—in the city's bay, or lagoon.

▲ CRITICAL THINKING

1. *Identifying Central Issues* What types of damage might very large cruise ships cause?

2. *Speculating* Why might some people in Venice want cruise ships to keep docking in the lagoon, even if they cause problems?

Many nongovernmental organizations (NGOs) are dedicated to addressing issues that affect the region and that may have a lasting impact on the welfare of the global environment. These include the World Wildlife Fund (WWF), Earthwatch, and the Nature Conservancy. These organizations work to solve the problems, including industrial pollution, algae blooms, and soil erosion, which are prominent in Southern Europe. There are many ways for individuals to get involved with organizations that protect the regions of the world that are essential to a region's cultural and natural heritage.

Deforestation is also a growing concern in Europe. The Forest Stewardship Council (FSC), an international NGO, is leading a grassroots campaign in Europe to reforest or protect forest biomes. Specifically, they are working to reduce soil erosion and the risk of forest fires and fighting causes of global warming. Certification is one way to ensure that forest resources are being used responsibly. When resource extractors obtain certification, consumers can be assured that they support products and services that avoid exploiting forest resources. A public awareness campaign has helped to some extent, although it is hard to measure by how much. Teaching the public about these issues helps people make decisions that support responsible stewardship of the environment.

☑ READING PROGRESS CHECK

Identifying What is the role of the European Environmental Agency?

LESSON 3 REVIEW

Reviewing Vocabulary
1. *Making Predictions* How might pollution hot spots in the waters of Southern Europe affect future economic activity in the region?

Using Your Notes
2. *Explaining* Use your web diagram to describe the major impacts humans are having on the environment in Southern Europe.

Answering the Guiding Questions
3. *Evaluating* What are the threats that require closer management of resources in Southern Europe?

4. *Explaining* What are the environmental concerns regarding tourism in Southern Europe?

5. *Discussing* How are groups, governments, and others addressing environmental issues in Southern Europe?

Writing Activity
6. *Informative/Explanatory* In a paragraph, explain how soil erosion poses a threat to the ecosystems of the subregion. What can be done to lessen the negative effects of climate change and the problem of soil erosion in Southern Europe?

Directions: On a separate sheet of paper, answer the questions below. Make sure you read carefully and answer all parts of the questions.

Lesson Review

Lesson 1

1 *Analyzing* Several of the rivers of Southern Europe are not navigable because they are not wide or deep enough for large ships to travel on. Why are rivers still important to the region?

2 *Explaining* What factors of the Mediterranean climate of Southern Europe make it ideal for growing grapes and olives?

3 *Describing* Describe the geographic locations of the three major Southern European peninsulas.

Lesson 2

4 *Evaluating* How has Southern Europe contributed to the arts? Provide specific examples.

5 *Identifying Central Issues* How have changes in population growth affected demographics in Italy?

6 *Understanding Relationships* What effect does tourism have on the economies of Southern Europe?

Lesson 3

7 *Identifying Cause and Effect* What are the causes and consequences of deforestation in the region?

8 *Summarizing* What key factors have contributed to pollution in the Mediterranean Sea?

9 *Analyzing* In what ways have the governments of Southern Europe taken steps to reduce pollution?

Exploring the Essential Question

10 *Making Connections* Create a poster illustrating one of the three peninsulas of Southern Europe. Show how the peninsula has been influenced by human systems, including history, art, politics, and economics. Use photos, graphs, charts, and maps.

21st Century Skills

Use the chart to answer the questions below.

The Labor Force in Southern Europe			
	% of Labor Force in Agriculture	% of Labor Force in Industry	% of Labor Force in Services
Greece	12.6	15	72.4
Italy	3.9	28.3	67.8
Portugal	8.6	23.9	67.5
Spain	2.9	15	58.4

Source: CIA World Factbook

11 *Economics* In which two countries is agriculture still a relatively significant contributor to the economy?

12 *Compare & Contrast* What generalization can you make about these four countries and their level of economic development?

Critical Thinking

13 *Making Generalizations* What long-term effects will Italy's aging population have on the country?

14 *Comparing and Contrasting* Compare the levels of development and industrialization in northern Italy to those in Greece. What factors help explain the differences?

15 *Predicting* Why should Southern European countries be concerned about high rates of unemployment among educated youth?

Need Extra Help?

If You've Missed Question	**1**	**2**	**3**	**4**	**5**	**6**	**7**	**8**	**9**	**10**	**11**	**12**	**13**	**14**	**15**
Go to page	296	297	294	303	301	304	306	307	307	294	309	309	301	303	303

Directions: On a separate sheet of paper, answer the questions below. Make sure you read carefully and answer all parts of the questions.

DBQ Analyzing Primary Sources

Use the document to answer the following questions.

Federico García Lorca (1898–1936) was a Spanish poet and playwright who was part of a group of avant-garde Spanish artists that also included Salvador Dalí.

PRIMARY SOURCE

Rider's Song

*Córdoba
Far away and alone.*

*Black pony, big moon,
and olives in my saddle-bag.
Although I know the roads
I'll never reach Córdoba.*

*Through the plain, through the wind,
black pony, red moon.
Death is looking at me
from the towers of Córdoba.*

*Ay! How long the road!
Ay! My valiant pony!
Ay! That death should wait me
before I reach Córdoba.*

*Córdoba.
Far away and alone.*

—Federico García Lorca, *The Selected Poems of Federico García Lorca*, 2006

16 **Analyzing** The North African poets of the Middle Ages used literary forms of allusion and repetition similar to those used by the author of this poem. Why would there be similarities between Spanish and North African poetry?

17 **Interpreting** What aspects of Spain's physical geography can be identified in this poem?

Applying Map Skills

Use your Unit 4 Atlas to answer the following questions.

18 **Places and Regions** What climate and vegetation types are found in Greece?

19 **Human Systems** What generalizations can you make about the population density of Spain in regard to its physical geography? What city is the exception?

20 **Physical Systems** Use your mental map of Southern Europe to describe the location of the major mountain ranges in Italy.

College and Career Readiness

21 **Reaching Conclusions** Imagine you are an economic geographer working for a nongovernmental organization (NGO). Write a recommendation for the Greek government explaining the benefits of investing in locally produced power. What natural resources can the Greek government use?

Research and Presentation

22 **Research Skills** Use Internet and library resources to gather information about the country of Andorra. Focus specifically on the unique government structure of this small country. Create a presentation explaining the history of this country's government. How did it become a dual principality and how does it maintain the system today?

Writing About Geography

23 **Informative/Explanatory** Use standard grammar, spelling, sentence structure, and punctuation to write a one-page essay suggesting possible solutions to the problem of pollution in the Mediterranean Sea. Who will need to participate for the solution to be successfully implemented?

Need Extra Help?

If You've Missed Question	16	17	18	19	20	21	22	23
Go to page	310	310	244	246	242	298	294	307

Eastern Europe

ESSENTIAL QUESTION • *How do physical systems and human systems shape a place?*

Why Geography Matters
The Breakup of Yugoslavia

Lesson 1
Physical Geography of Eastern Europe

Lesson 2
Human Geography of Eastern Europe

Lesson 3
People and Their Environment: Eastern Europe

Geography Matters...

Borders in Eastern Europe have quite a history. Political changes have made the subregion's borders a source of division and conflict. Physical features also play a role. The Danube River divides countries physically, but it also helps connect Eastern and Western Europe economically. The Eastern European countries of today were once part of large powerful empires. Most modern countries of the subregion did not gain independence until after World War I. The subregion today is a unique land with a turbulent past, but with richness in literature, art, and culture.

◄ A young Roma woman in Slovakia holds her child.

Matt Cardy/Alamy

311

the breakup of Yugoslavia

The former Yugoslavia was historically a crossroads. Because the countries of Eastern Europe have a long history of belonging to different empires at different times, they have a complex mix of ethnic and religious influences. Conflict was common, and the region became known as the "powder keg" of Europe.

The Former Yugoslavia

15°E   20°E   25°E

AUSTRIA   HUNGARY

Danube R.

N W E S

SLOVENIA
Ljubljana ◉

◉ Zagreb

CROATIA

ROMANIA

Ethnic and religious diversity characterized the former Yugoslavia and contributed to its breakup by 1995. The map shows the countries today that were once part of Yugoslavia.

45°N

0   100 miles
0   100 kilometers
Lambert Azimuthal Equal-Area projection

BOSNIA AND HERZEGOVINA

Belgrade ◉

Sarajevo ◉

SERBIA

Adriatic Sea

BULGARIA

ITALY

MONTENEGRO   Priština ◉

Nationalism was also a strong force at work. In 2008 Kosovo was the last country to declare its independence.

Ethnic and religious tensions resulted in bloody conflict between Serbs and Bosnians in the 1990s.

Podgorica ◉

KOSOVO

◉ Skopje

MACEDONIA

—— Former boundary of Yugoslavia

40°N

ALBANIA

GREECE

Why was Yugoslavia considered an ethnic melting pot?

How did this contribute to the breakup of Yugoslavia?

How did the breakup change the political geography of Europe?

Yugoslavia (a combination of the Slavic words meaning "south" and "Slavs") was, for a time, a true ethnic mix. The country did not exist until 1918 with the dissolution of the Austro-Hungarian and Ottoman Empires. Originally created as the Kingdom of Serbs, Croats, and Slovenes, it united several distinct Slavic ethnic groups in its population: Bosnians, Bulgarians, Croats, Macedonians, Montenegrins, Serbs, and Slovenes. It also included non-Slavic peoples such as Albanians. Yugoslavia was religiously diverse as well. Eastern Orthodox Christians, Muslims, Catholics, Protestants, and Jews composed the population. Animosities between some of the ethnic groups stretched back hundreds of years. After World War II, under the firm Communist rule of Josip Broz Tito, the country enjoyed internal peace. Despite the country's wide ethnic and cultural differences, Tito managed to prevent several nationalist movements from instigating rebellions that threatened his rule, at times arresting or executing protest leaders.

1. **Human Systems** What were the conditions that helped set the stage for ethnic conflict in Yugoslavia?

The 1970s brought an economic crisis to Yugoslavia. The crisis was triggered by government borrowing to support export growth. A general recession, however, made Western economies unable to purchase those exports. In 1974 a new constitution decreased the federal government's power, giving more powers to the country's six republics and two autonomous provinces. Nonetheless, Tito's authority kept the country together until his death in 1980, after which ethnic and nationalistic tensions quickly escalated. The republics of Croatia and Slovenia demanded looser ties to central authority. In March and April of 1981, riots broke out in Kosovo, an autonomous province with a majority Albanian-speaking Muslim population that was located within the republic of Serbia. Serbia had a majority Serbo-Croatian-speaking Eastern Orthodox population. Kosovo demanded outright secession from Yugoslavia or republican status within the federation. In 1989 Serbian politician Slobodan Milošević became president of Serbia and responded to new demonstrations by taking away what remained of Kosovo's autonomy and incorporating it into Serbia.

2. **Human Systems** How did Yugoslavia's ethnic makeup contribute to the conflicts that caused the breakup of the country?

Political maneuvering failed to pacify the competing groups. During the 1990s, a series of wars culminated in Europe's bloodiest period since World War II. Genocide, mass murder, and ethnic cleansing—a policy aimed at removing an ethnic or religious rival from a region through violence and terror—characterized the conflicts. Many of the political and military leaders, including Slobodan Milošević, were subsequently charged with war crimes. By 1995, Yugoslavia ceased to exist. In its place were five Yugoslav successor countries: Bosnia-Herzegovina, Croatia, Macedonia, Slovenia, and the Federal Republic of Yugoslavia, which later became the countries of Serbia and Montenegro. In 2008 the republic of Kosovo declared its independence. International recognition of Kosovo is still disputed by some countries, including Russia.

3. **Places and Regions** Write a paragraph describing how the political map of Europe was transformed by the breakup of Yugoslavia.

THERE'S MORE ONLINE

SEE a photo of Josip Broz Tito • *EXPLORE* a time line of the breakup of Yugoslavia

networks

There's More Online!

- ☑ **IMAGE** European Bison
- ☑ **MAP** Alpine Europe
- ☑ **MAP** Main-Danube Canal
- ☑ **INTERACTIVE SELF-CHECK QUIZ**
- ☑ **VIDEO** Physical Geography of Eastern Europe

Reading **HELP**DESK

Academic Vocabulary

- **economy**
- **comprise**

Content Vocabulary

- **karst**

TAKING NOTES: *Key Ideas and Details*

CATEGORIZING Use a graphic organizer like the one below to take notes about the landforms and waterways of Eastern Europe.

Physical Geography of Eastern Europe

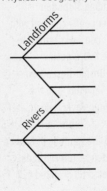

LESSON 1
Physical Geography of Eastern Europe

ESSENTIAL QUESTION • *How do physical systems and human systems shape a place?*

IT MATTERS BECAUSE

Eastern Europe is made up of a diverse group of countries. They range from the Baltic countries in the north to the Balkan countries in the south. The entire subregion includes Estonia, Latvia, Lithuania, the Czech Republic, Slovakia, Poland, Hungary, Bulgaria, Moldova, Albania, Romania, Serbia, Montenegro, Bosnia and Herzegovina, Croatia, Slovenia, Macedonia, and Kosovo. These many countries are united by a shared history as a crossroads where different cultures from Asia and Europe have encountered one another. Such encounters have been both facilitated and inhibited by Eastern Europe's diverse physical geography.

Landforms

GUIDING QUESTION *How do mountains and plains define Eastern Europe?*

The physical geography of Eastern Europe is characterized by mountains and plains, which influence the human geography of the subregion's countries. Mountains dominate in the south, part of Europe's Alpine system, with the curve of the Carpathian Mountains in Slovakia and northern Romania, along with the Balkan Mountains of Bulgaria and the coastal ranges of the Dinaric Alps. All of these are the eastern extension of the Swiss Alps. While not as high as the Swiss Alps, these ranges create mountainous and hilly landscapes.

Lowlands within these mountainous areas are restricted to relatively small plains and deltas formed by deposits of rock fragments and particles carried down by rivers that erode the mountains. The Dinaric Alps span the countries of Slovenia, Croatia, Bosnia and Herzegovina, Macedonia, Albania, and Montenegro. They run parallel to the Adriatic coast (sometimes referred to as the Dalmatian coast). This region exhibits **karst** topography, which refers to limestone bedrock sculpted into steep-sided cliffs and rocky columns. Karst terrain is characterized by caves, sinkholes, underground rivers, and the absence of surface rivers, streams, and lakes. These features are due to the soluble limestone rock that underlies the region. The limestone is dissolved by organically produced acids in groundwater, creating depressions and holes in the earth. Forestry and mining are the primary economic activities for people living in the Dinaric Alps.

The geologically young Carpathian Mountains run from Slovakia to Romania. These mountains are less compact than the Swiss Alps to the west and are distinguished by mountains separated by large basins. The water from the Carpathians generally flows into the Black Sea. This region is sparsely populated. Its inhabitants depend on agriculture and forestry, especially on the Transylvanian Plateau, for income.

The Balkan Peninsula is the easternmost of Europe's three great southern peninsulas. It is dominated by the Balkan Mountains, an extension of the Alpine-Carpathian ranges to the north. Due to its rugged landscape and deep snow during winter, traveling over land in the Balkan Peninsula is difficult. The mountains form a major land divide between two points of human transit, the Danube River to the north and the Maritsa River to the south. While the mountains can be crossed through several passes, people have historically tended to move by way of rivers and seas in order to avoid the mountains. The Balkan Mountains are also a climate barrier between the continental climate of the Danube River valley and the transitional climate south of the mountains.

North of the Carpathians, the landscape displays broad, extensive plains, such as the lowland areas that dominate Poland and the Baltic countries. These lowlands are part of the Northern European Plain. They are home to a number of navigable rivers in Eastern Europe, mainly the Elbe, Oder, and Vistula Rivers. Another prominent lowland, the Hungarian Plain, extends from southeastern Hungary into eastern Croatia, northern Serbia, and western Romania. In these lowlands, farmers cultivate grains, fruit, and vegetables. They also raise livestock along the banks of the Danube River.

karst terrain dominated by limestone bedrock and characterized by rocky ground, caves, sinkholes, underground rivers, and the absence of surface streams and lakes

☑ **READING PROGRESS CHECK**

Explaining How have the Balkan Mountains defined the region?

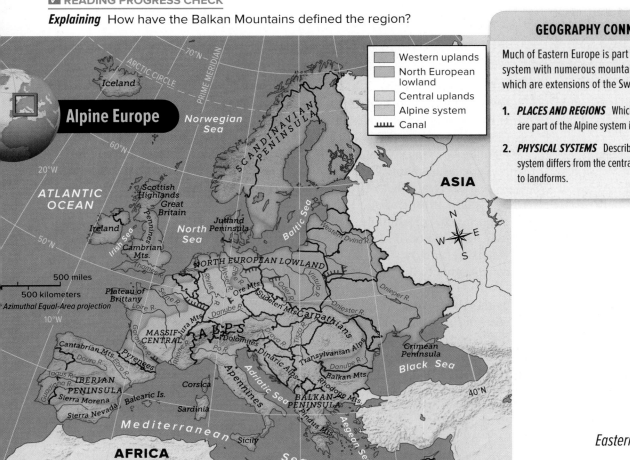

GEOGRAPHY CONNECTION

Much of Eastern Europe is part of Europe's Alpine system with numerous mountain ranges, most of which are extensions of the Swiss Alps.

1. ***PLACES AND REGIONS*** Which mountain ranges are part of the Alpine system in Eastern Europe?

2. ***PHYSICAL SYSTEMS*** Describe how the Alpine system differs from the central uplands in regard to landforms.

Eastern Europe **315**

Main-Danube Canal

Dozens of hydroelectric power plants along the Main River help Germany meet its power needs.

The 106-mile- (171-km-) long Main-Danube Canal, completed in 1992, connects the Main River to the Danube River.

The entire waterway connecting the North Sea with the Black Sea is 2,200 miles (3,500 km) long and runs through 11 countries.

200 miles
200 kilometers
Lambert Azimuthal Equal-Area projection

GEOGRAPHY CONNECTION

The Main-Danube Canal is an important waterway for Eastern Europe.

1. *HUMAN SYSTEMS* What bodies of water are linked by the canal?

2. *PLACES AND REGIONS* Which Eastern European countries are situated along the Main-Danube waterway?

Water Systems

GUIDING QUESTION *Why are the Danube and Vistula Rivers important to economic activity in Eastern Europe?*

A number of large rivers flow through the northern part of Eastern Europe and are extremely important to economic activities in the region. The Danube River is the second-longest river in Europe. It starts in the Black Forest of western Germany and empties into the Black Sea, passing through nine countries. The Danube has played a vital role in the settlement and development of Europe. Historically, its banks have formed the boundaries between great empires and were most famously the frontier of the Roman Empire.

Today, the Danube River maintains its geographical and political importance. For example, it divides Budapest—the capital of Hungary—into its two main parts: Buda and Pest. The Danube is officially an international waterway and has served as a commercial highway for many countries, contributing to their economic development.

The Danube is connected to the Main River, a tributary of the Rhine River, by the Main-Danube Canal. The waterway was completed in 1992 after over 30 years of construction. Many of the waterway's locks on the Main River also have hydroelectric power stations. The Main-Danube Canal provides an important connection that links the North Sea and Northwestern Europe with Eastern Europe and the Black Sea. The canal allows for goods—such as food and animal feed, ores, iron and other metals, and fertilizers—to be transferred in both directions. Tourism along the canal also contributes to its economic importance.

The Vistula River, the largest river in Poland, flows from south to north and empties into the Baltic Sea. A number of major cities and industrial centers—such as Warsaw, the capital of Poland—lie on the banks of the Vistula River.

The Vistula is connected by canal to the Oder River, which originates in the Czech Republic. The Oder also empties into the Baltic Sea by way of western Poland via a heavily industrialized canal. The river forms part of the border between Poland and Germany. In supplementing the overburdened railway and highway systems that link the industrialized areas of the south with ports on the Baltic Sea, the Oder is crucial to the Polish **economy**.

Several major seas surround Eastern Europe and serve as vital channels of trade and economic development. The Black Sea is an inland sea that hugs the eastern coast of the Balkan Peninsula and has served to link Europe to Asia for centuries. A narrow channel connects the Black Sea to the Mediterranean Sea that borders the region to the south. The Mediterranean connects to the Atlantic Ocean, creating an important link for trade between Eastern Europe and the rest of the world. The Baltic Sea in the north, between mainland Europe and the Scandinavian Peninsula, has historically been used as a trade route for importing oil and coal and exporting minerals, timber, and wood products. The Adriatic Sea between the Balkan Peninsula and the Italian Peninsula is a primary mode of importing and exporting goods and provides fish to the surrounding countries.

☑ **READING PROGRESS CHECK**

Summarizing Explain how waterways facilitate trade between Eastern Europe and other regions of the world.

Climate, Biomes, and Resources

GUIDING QUESTION *What are the general climate conditions in much of Eastern Europe?*

Eastern Europe is located within the midlatitudes. Much of the region has a humid continental climate characterized by cold, snowy winters and hot summers. Eastern Europe does not benefit from the warm ocean currents that moderate the climates in Western Europe. Summer and winter temperatures inland vary more widely than they do on the coasts.

Climate Regions and Biomes

The Baltic Sea region and the Northern European Plain are marked by long, cold winters with temperatures reaching an average low of 14°F (−10°C) in mid-winter. The summers are relatively short in this area of continental climate. In the forests of the Baltic Sea region, evergreens outnumber deciduous trees, while an equal mix of evergreens and deciduous trees are found across the Northern European Plain.

The region south of the Northern European Plain is sometimes referred to as the Danube region and **comprises** the Carpathian and Balkan Mountains together with the Hungarian Plain. The Danube region also has a continental climate. Winters and summers, however, are about equally long, and the region enjoys moderate average temperatures. Some of the coastal regions of the landlocked Black Sea have micro-biomes that are a combination of humid continental and humid subtropical climates. These regions are characterized by mild winters and a great deal of precipitation during the warm summers. The basin northeast of the Black Sea is typified by a steppe climate, with cold winters and hot, dry summers, and a grassland biome. Subtropical air flowing across the Mediterranean and Black Seas ensures warm, moist summers on the southwestern shores.

economy an ordered system for the production, distribution, and consumption of goods and services

comprise to contain; to consist of

Large land mammals such as these European bison can be found in the forests of Eastern Europe.

▲ **CRITICAL THINKING**

1. ***Explaining*** How would deforestation affect the European bison in Eastern Europe?

2. ***Speculating*** Why might the European bison be a target for hunters in Eastern Europe?

Areas around most of the Adriatic Sea have a Mediterranean climate characterized by mild, rainy winters and hot, dry, and sunny summers. The natural vegetation here is predominantly shrubland interspersed with woodlands and some forests. The Mediterranean climate supports crops such as olives and grapes.

The northern Adriatic has a humid subtropical climate, which means it has short, mild winters and year-round rain. The vegetation in this climate consists of prairies, inland grasslands, and evergreen and deciduous forests. The Adriatic Sea has many fish species that are endemic, or restricted to a particular region, especially in the northern part of the sea. Many are endangered due to overfishing. To the east of the Adriatic Sea, the Dinaric Alps experience cold and dry continental air, cooling Croatia's coastal cities such as Split and Dubrovnik.

Natural Resources

Eastern Europe has large quantities of natural resources. The Carpathian Mountains, for example, contain reserves of natural gas, oil, and coal. The Baltic Mountains do not have as many natural resources. However, some countries, such as Latvia, have begun to take advantage of water as a potential source of hydroelectric power. Hydroelectricity accounts for about 70 percent of Latvia's total electric capacity. Similarly, Romania, which has diminishing petroleum reserves, uses the Danube River to generate hydroelectric power. Romania obtains over one-third of its electricity from hydroelectric plants.

Poland has vast coal, natural gas, iron, zinc, lead, and copper reserves, as well as smaller amounts of silver. Poland is famous for its amber, often called "Baltic gold." The raw amber was transported for centuries along the ancient Amber Route from the Baltic Sea to the Adriatic coast.

Many Eastern European countries have bauxite reserves. Bauxite is the main ore used to make aluminum. Although the region is not the world's leading source of bauxite, the demand for aluminum is rising, making bauxite an important resource for Eastern European countries.

☑ **READING PROGRESS CHECK**

Listing What landforms are found in the Danube region?

©Raymond Gehman/Corbis

LESSON 1 REVIEW

Reviewing Vocabulary

1. ***Summarizing*** What is karst terrain and what causes it to form?

Using Your Notes

2. ***Listing*** Use your graphic organizer from the lesson to list the important waterways in Eastern Europe.

Answering the Guiding Questions

3. ***Applying*** How do mountains and plains define Eastern Europe?

4. ***Stating*** Why are the Danube and Vistula Rivers important to economic activity in Eastern Europe?

5. ***Identifying*** What are the general climate conditions in much of Eastern Europe?

Writing Activity

6. ***Informative/Explanatory*** Write a short paragraph explaining how the geography of the Baltic and Adriatic Seas is important for trade.

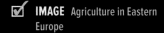

networks

There's More Online!

☑ **IMAGE** Agriculture in Eastern Europe

☑ **IMAGE** Religious Conflict in Eastern Europe

☑ **MAP** Eastern European Migration

☑ **INTERACTIVE SELF-CHECK QUIZ**

☑ **Time Line** The Road to a New Era

☑ **VIDEO** Human Geography of Eastern Europe

Reading HELPDESK

Academic Vocabulary

- **persistent**
- **ethnic**

Content Vocabulary

- **shatter belt**
- **Balkanization**
- **ethnic cleansing**

TAKING NOTES: *Key Ideas and Details*

PARAPHRASING Use a graphic organizer like the one below to take notes about the population patterns of Eastern Europe.

Population Patterns

- Historical: The Slavs
- Geographic Factors: Location
- Industrialization
- Migration/ Immigration

LESSON 2
Human Geography of Eastern Europe

ESSENTIAL QUESTION • *How do physical systems and human systems shape a place?*

IT MATTERS BECAUSE

Eastern Europe has undergone many changes in the past few decades. The political geography of the subregion has changed drastically as the result of the dissolution of the Soviet Union in 1991. Eastern European economies are being integrated into the European Union and world markets. They have made variable progress in the difficult transition from communist-based command economies to market economies.

History and Government

GUIDING QUESTION *How have political and ethnic struggles shaped the Eastern Europe of today?*

Eastern Europe is known as a **shatter belt** region because it has **persistently**, or constantly, experienced political and territorial splintering and fracturing along cultural and **ethnic** lines. Shatter belt regions are more likely to engage in interstate wars and undergo internal conflicts such as civil wars. During the Cold War period (1945–1991), Eastern Europe was dominated by the Soviet Union. After 1991 the countries had to determine their own futures. The change resulted in old animosities recurring between and within some countries. For others, it began a period of unprecedented political cooperation.

Early Peoples, Empires, and Conflict

The earliest Slavs migrated from Asia thousands of years ago and settled in Eastern Europe alongside Celtic and Germanic tribes. By the A.D. 400s and 500s, they had spread across the region. The Slavic peoples living on the Balkan Peninsula established independent states. The mountainous terrain allowed them to resist invading armies. The Slavic peoples to the east of the Balkan Peninsula, however, experienced invasions from peoples coming from Asia, particularly the Mongols in the 1200s. They ultimately settled in sparsely populated and widely separated communities in the forests and plains north of the Caucasus Mountains.

shatter belt a region where political alliances are constantly splintering and fracturing based on ethnicity

persistent continuing, existing, or acting for a long time

ethnic of or relating to large groups of people classed according to common traits and customs

Balkanization division of a region into smaller regions

Around A.D. 106 the Romans conquered the lands between the Carpathian Mountains and the Danube River and named the area Romania. As part of the eastern half of the Roman Empire, the area became part of the Byzantine Empire that emerged after the fall of Rome. The Byzantine Empire lasted for a thousand years before falling to the Ottoman Turks in 1453. The Ottoman Empire then ruled much of southern Eastern Europe until the end of World War I.

Conflict, Union, and Division

The Balkan Peninsula, like much of Eastern Europe, has long been a region of instability. The ethnic conflict in the region contributed to the start of World War I when a Serbian nationalist assassinated Archduke Francis Ferdinand, heir to the Austro-Hungarian throne, in 1914.

The breakup of the Ottoman and Austro-Hungarian Empires after World War I left regions of peoples without formally recognized countries. A new map of Europe was drawn based on ethnicity, population, politics, and economic strengths. One new country, Yugoslavia, combined many ethnic groups in lands that had been contested between the Austro-Hungarian and Ottoman Empires for hundreds of years.

As a shatter belt, the Balkan Peninsula had seen the division of the larger regions or countries into smaller regions or countries for many centuries, often as a result of war. Because of this long history, this process of division has become known as **Balkanization**. Yugoslavia attempted to reverse that pattern, since it took many smaller regions based on ethnicity and combined them into one country.

After World War II, Eastern Europe fell under the control of the communist Soviet Union. The control of communist Eastern Europe in contrast to democratic Western Europe brought about the Cold War, an ideological, political, and geographical war. Based on the devastation of Russia and other Soviet lands in both world wars, the Soviets used Eastern Europe as a "buffer zone." The eastern countries of Europe separated the Soviet Union from Western Europe's democracies. This buffer zone provided military protection and led to very different types of social, economic, and political developments between Eastern and Western Europe.

TIME LINE ⌄

THE ROAD
to a New Era ➡

Once dominated by powerful states outside the region, the many ethnic groups of Eastern Europe struggled for independence in the twentieth century. Struggles for control and autonomy within the region triggered repeated armed conflicts.

▶ **CRITICAL THINKING**

1. *Explaining* How did ethnic tensions lead to the outbreak of World War I in the early twentieth century?

2. *Explaining* How did ethnic tensions cause political boundaries in Eastern Europe to be redrawn in the late twentieth and early twenty-first centuries?

The first of two brief wars in the Balkans is started by an alliance between Bulgaria, Greece, Serbia, and Montenegro.

Archduke Francis Ferdinand is assassinated by a Serbian nationalist, triggering World War I.

1912

1914

➡ **1900**

1918

1939

World War I ends, signaling the end of the Ottoman and the Austro-Hungarian Empires.

World War II begins and Nazi Germany rapidly takes over Eastern Europe.

The Road to a New Era

From the 1950s to the 1980s, revolts against communist rule periodically swept Eastern Europe. In 1989 large-scale public demonstrations contributed to the fall of the communist governments in the east. By 1991 the nationalist protests and financial crisis led to new, ethnic-based countries within Yugoslavia. Slovenia, Croatia, Bosnia-Herzegovina, and Macedonia all seceded from Yugoslavia, leaving just Serbia and Montenegro. Power struggles by ethnic and political groups resulted in civil war. What followed was a horrific practice called **ethnic cleansing**. Ethnic cleansing was used mainly by military units, but civilians did participate. The "cleansing" resulted in certain ethnic groups being killed or expelled from the disputed territories where they lived. Ethnic cleansing affected all groups, but Bosnian Croats and Bosnian Muslims were affected the most. Military aircraft from NATO countries bombed Serbian military and civilian targets to stop Serbia's support of the ethnic cleansing. Eventually international peacekeeping efforts were able to dislodge Serb leaders and end the conflict. Balkanization continued, and Montenegro declared independence from Serbia in 2006, with Kosovo declaring independence in 2008. In less than 20 years, the country of Yugoslavia was divided into eight smaller independent countries.

ethnic cleansing the expelling from a country or genocide of an ethnic group

☑ **READING PROGRESS CHECK**

Expressing What did the public demonstrations of 1989 in Yugoslavia lead to?

Population Patterns

GUIDING QUESTION *How have wars, migrations, and changing political borders influenced the population patterns of Eastern Europe?*

Most Eastern Europeans are ethnically Slavic. Slavs are descended from Indo-European peoples who migrated from Asia and settled in the region. Eastern Europe contains west Slavs, including Poles, Czechs, and Slovaks, and south

©Yevgeny Khaldei/Corbis

1949 The United States and its allies in Western Europe form the North Atlantic Treaty Organization (NATO).

1945 World War II ends and the Cold War begins.

➜ 1950

1989 Public demonstrations lead to the fall of the communist governments in Eastern Europe. Yugoslavia descends into civil war.

1992 The UN recognizes independent Slovenia, Croatia, and Bosnia as member states.

1993 Macedonia is admitted to the UN. Czechoslovakia splits into the Czech Republic and Slovakia. The European Union is created.

➜ 2000

2004 The EU admits the Czech Republic, Hungary, Latvia, Lithuania, Poland, Slovakia, and Slovenia.

2006 Montenegro declares itself independent from Serbia.

2007 Bulgaria and Romania join the European Union.

2008 Kosovo declares independence from Serbia.

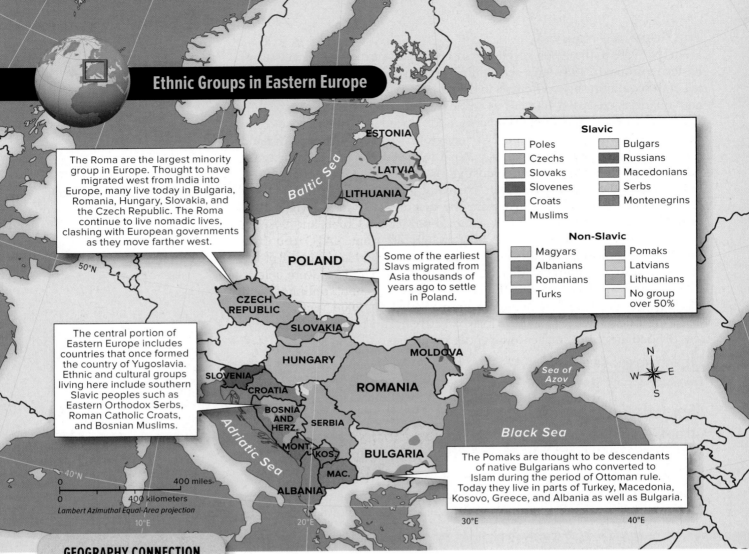

Ethnic Groups in Eastern Europe

The Roma are the largest minority group in Europe. Thought to have migrated west from India into Europe, many live today in Bulgaria, Romania, Hungary, Slovakia, and the Czech Republic. The Roma continue to live nomadic lives, clashing with European governments as they move farther west.

Some of the earliest Slavs migrated from Asia thousands of years ago to settle in Poland.

The central portion of Eastern Europe includes countries that once formed the country of Yugoslavia. Ethnic and cultural groups living here include southern Slavic peoples such as Eastern Orthodox Serbs, Roman Catholic Croats, and Bosnian Muslims.

The Pomaks are thought to be descendants of native Bulgarians who converted to Islam during the period of Ottoman rule. Today they live in parts of Turkey, Macedonia, Kosovo, Greece, and Albania as well as Bulgaria.

Slavic

Poles		Bulgars	
Czechs		Russians	
Slovaks		Macedonians	
Slovenes		Serbs	
Croats		Montenegrins	
Muslims			

Non-Slavic

Magyars		Pomaks	
Albanians		Latvians	
Romanians		Lithuanians	
Turks		No group over 50%	

400 miles
400 kilometers
Lambert Azimuthal Equal-Area projection

GEOGRAPHY CONNECTION

Historical migrations to Eastern Europe resulted in a subregion that is very ethnically diverse.

1. **PLACES AND REGIONS** Which Eastern European countries have the largest populations of Roma today?

2. **HUMAN SYSTEMS** What major religions do the southern Slavic people practice?

Slavs, including Serbs, Croats, Slovenes, and Macedonians. Slavs are an ethnic group defined mainly by their language. They speak languages today that have words and grammar that vary by country, but are traced to the same Indo-European root languages.

Another ethnic group, the Roma, traces its language to Indo-European origins. Roma are thought to have migrated from northern India to Europe centuries ago. Today there are several million Roma who live in Europe. They generally have less education, poorer health care, and shorter life expectancies than other groups.

Population density and distribution in Eastern Europe are influenced by geographic factors. For example, Poland has fertile soil and ample water resources that support large populations. Montenegro, by contrast, is the least populous country in Eastern Europe. It is located along the Dalmatian coast with a jagged coastline of narrow coastal plains backed by rugged limestone mountains and plateaus. The capital, Podgorica, has the main population concentration. Located along the Adriatic Sea, it provides access to the Mediterranean and therefore global transportation routes.

Industrialization throughout the 1900s led to urbanization in Eastern Europe. The majority of Eastern Europe's population today lives in and around large towns and cities. However, during much of the twentieth century, about half the population lived in rural areas. The cities that developed in the subregion were located near navigable water, such as the Danube and Vistula Rivers. Water transportation is not as expensive as other forms of transport and provided the cities with advantages as industrial and trade centers. During the Cold War, Eastern Europe was the major trading partner for the Soviet Union. New trade connections were necessary after 1991, and cities have new economic objectives based on tourism and services. Joining the European Union (EU) has been a major advantage to those Eastern European countries that met the requirements for membership.

Eastern European countries faced many difficult economic and political circumstances during and following World War II. Following the war there were large-scale internal and external migrations. Eastern European populations experienced heavy declines. Many Jewish people emigrated to Israel and to countries on other continents. People left in great numbers to escape communism. With the fall of communism, Eastern European countries underwent additional changes in economics, politics, and social conditions that continue to affect them today.

✅ **READING PROGRESS CHECK**

Describing How did industrialization affect population distribution in Eastern Europe?

Society and Culture Today

GUIDING QUESTION *How has conflict affected society and culture in Eastern Europe?*

Since almost all of Eastern Europe was a part of the Soviet bloc, the region has a recent history of free education for the population. As a result, literacy rates are improving throughout the region. However, in the transition to democratic governments, some countries have faced economic challenges. A lack of funds has affected the education systems in some countries. The health care system, known as "cradle to grave," was affected by the change to the market economy and by the lower birthrates and higher numbers of senior citizens. Despite this, most Eastern Europeans continue to have access to quality health care provided by national governments. The European outlook on health care makes it a basic human right.

Religious and ethnic differences were at the heart of conflict in the Balkan Peninsula in the 1990s. Tensions between Serbs and other ethnic groups such as Croats and Bosnians erupted into armed violence. Layered on top of such tensions were also religious divisions between people practicing Eastern Orthodox Christianity and Islam. Roman Catholicism and Judaism are also practiced in the region.

Family and Status of Women

In traditional Eastern European culture, the family is the basic social unit and serves to reinforce social values. However, twentieth-century industrialization, World War I, and new political movements introduced greater mobility and attracted rural populations to urban areas to find work. Since the beginning of industrialization and urbanization in Europe, families have tended to live in small houses or apartments. Eastern European families today have fewer children, and extended family members often live in other places. Women have made gains in education and, as a result, are employed in professional jobs.

The Arts

Many traditional forms of art and music exist throughout Eastern Europe. Folk and classical music are particularly important among Czech, Hungarian, Slovak, and Slovene peoples. In larger cities,

A crucifix stands against a bullet-riddled wall in the former Yugoslavia.

▼ **CRITICAL THINKING**

1. Describing In what ways has war shaped Eastern European culture and politics?

2. Explaining How has religion played a role in the conflicts of Eastern Europe?

contemporary music from Western Europe and the United States is also popular. Some areas have thriving nightclub scenes. Some of the large countries also have a long musical history and continue to perform the works of famous composers in historic concert halls and opera houses.

Literature is a valued art form in Eastern Europe. The region has produced world-famous writers, such as Czech-born Franz Kafka (1883–1924), known for his novel *The Metamorphosis*. The most common themes in Eastern European literature reflect the region's history. Many authors wrote about the despair and darkness that fell over the region due to war, ethnic conflict, and economic depressions.

✅ **READING PROGRESS CHECK**

Explaining What are the common themes in Eastern European literature?

Economic Activities

GUIDING QUESTION *How have Eastern European economies developed since the fall of communism?*

Eastern European countries have made the transition to market economic systems over the past 20 years. The change to a market economy has presented both problems of production and consumption—of providing items that others want to buy and managing the changes in ownership from government to private corporations. The countries located nearest to Western Europe benefited early from the market system and received investments from countries such as Germany, Sweden, and the United Kingdom. Slovenia, for example, prospered after the fall of communism because it developed economic relationships with Germany. Political instability in the 1990s resulted in countries such as Macedonia having less success attracting investment than countries neighboring on Western Europe.

Agriculture and Industry

While Eastern Europe was largely industrialized under communist rule, many parts maintained their agricultural roots. The countries with Mediterranean climates such as Bulgaria, Serbia, Croatia, Macedonia, Bosnia and Herzegovina,

Growing crops such as these potatoes in Romania reflects the agricultural roots of Eastern Europe.

▼ **CRITICAL THINKING**

1. *Analyzing Visuals* How would you describe the type of farming you see in the picture?

2. *Contrasting* How was Eastern European society changed in terms of industrialization during the era of communist rule?

Diana Mayfield/Lonely Planet Images/Getty Images

and Montenegro produce olives, citrus fruits, dates, and grapes. The types of agricultural products explain why the geography of the Balkans is a transition zone between Southern Europe and the northern regions of Eastern Europe. Farther north in Eastern Europe the climate is continental and farmers grow wheat, rye, and other grains. They also raise livestock.

Fishing is an industry in Eastern Europe that has depended largely on rivers and lakes. While large rivers like the Danube provide fish such as carp and perch, many rivers have been seriously overfished. The EU has introduced limits on fish catches and species protection rules in order to restore this important resource.

While Eastern Europe continues to produce small quantities of natural resources such as coal, it has become a global center for low-cost manufacturing of electronics products. International investors are attracted by high educational levels among the population and low labor costs. Automobile production has also been one of the region's main successes. Tourism is yet another success story, and many countries have been able to rebuild their economies and attract foreign investment through hotel and hospitality tourist services.

Communication systems throughout Northwestern and Eastern Europe have become increasingly linked. Railway, airway, and highway systems link major cities, which facilitates trade and tourism. Seas and rivers also connect Eastern Europe with international shipping by water at a low cost. Most countries have world-class Internet services in urban centers and in many rural regions.

The European Union

The Czech Republic, Hungary, Latvia, Lithuania, Estonia, Poland, Slovakia, Slovenia, Bulgaria, and Romania all were members of the European Union (EU) by 2012. The global economic crisis that began in 2008 had a major effect on the EU and the rest of Eastern Europe. The economic downturn was most serious in Greece, Portugal, Ireland, Italy, and Spain. While the consequences of a harsh economy were most serious in those countries, the EU functioned as a unit to help all member countries through the economic crisis. More prosperous countries, such as Germany, helped weaker countries after much debate and negotiations among the EU members. It was the first major crisis for the EU, and the leaders of member countries demonstrated the belief that the EU was strong enough to survive. Their success will be judged sometime in the future if the EU continues as a strong regional organization in Europe.

☑ **READING PROGRESS CHECK**

Describing How did the European Union help all member countries through the economic crisis that began in 2008?

LESSON 2 REVIEW

Reviewing Vocabulary

1. *Explaining* Explain the significance of Balkanization and ethnic cleansing.

Using Your Notes

2. *Summarizing* Use your graphic organizer to summarize the population patterns of Eastern Europe.

Answering the Guiding Questions

3. *Identifying* How have political and ethnic struggles shaped the Eastern Europe of today?

4. *Assessing* How have wars, migrations, and changing political borders influenced the population patterns of Eastern Europe?

5. *Evaluating* How has conflict affected society and culture in Eastern Europe?

6. *Discussing* How have Eastern European economies developed since the fall of communism?

Writing Activity

7. *Narrative* Suppose you are a farmer in Eastern Europe. Write one paragraph describing how your life has changed since the fall of communism.

networks

There's More Online!

☑ **GRAPH** Water Pollution in Europe

☑ **IMAGE** Białowieza Forest

☑ **IMAGE** Communist-Era Pollution

☑ **INTERACTIVE SELF-CHECK QUIZ**

☑ **VIDEO** People and Their Environment: Eastern Europe

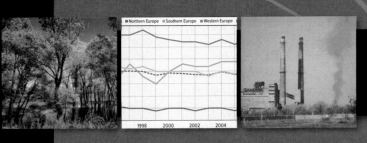

LESSON 3
People and Their Environment: Eastern Europe

Reading **HELP**DESK

Academic Vocabulary

- emphasis

Content Vocabulary

- **reforestation**
- **meltwater**

TAKING NOTES: *Key Ideas and Details*

SUMMARIZING Use a graphic organizer like the one below to identify ways humans are affecting the environment in Eastern Europe.

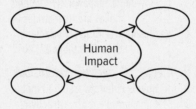

ESSENTIAL QUESTION • *How do physical systems and human systems shape a place?*

IT MATTERS BECAUSE

Natural resources, people, and culture combine to influence the nature of human impact on the environment in a given region. For example, some countries in Eastern Europe have followed plans of rapid and intense industrialization without regard to the effects on the environment. In addition, dams, dikes, and other types of development have damaged water quality, posing a threat to wetlands, fish, and bird populations.

Managing Resources

GUIDING QUESTION *Why do forest resources in Eastern Europe need to be managed effectively?*

It is believed that about 80 percent of Europe was once covered by forest, two-thirds of which has been removed over time. Historically, people cut down trees to create space for cities and farms. Today some countries in Eastern Europe, such as Bulgaria and Albania, have a problem with illegal logging, or the unlicensed cutting and selling of wood. Selling illegally logged wood can be a lucrative business, especially in countries with stagnating economies. Individuals and groups who engage in illegal logging seldom get caught. This is because many wood-processing companies, formerly in charge of responsible logging and forest maintenance, went out of business in the 1990s. One reason for the popularity of illegal logging is the high cost of electricity, which leads many people to burn wood for heat. This generates air pollution, especially in urban areas. Romania and Bulgaria rank at the top of the United Nations Development Programme's list of countries where citizens have died from urban air pollution. Illegal logging has also led to huge losses in the biodiversity of many areas.

As communist rule declined, Eastern European countries rushed to take part in the global economy. Individuals left rural farmlands to work in cities, and government interest in **reforestation**, or replanting trees, grew. People in Eastern Europe also began to focus on preserving what little forest remains. For example, the Białowieza Forest, located in eastern Poland and western Belarus, is the site of debate between the Polish government and environmental activists. Most of the forest is home to some of the tallest trees in Europe, which are over 100 years old. It is also home to wolves, lynx, and the European bison. The forest is nationally protected and has been for centuries. However, in nonprotected forestry-managed areas, logging is ruining the habitats of many birds and animals and destroying trees that have stood longer than Poland has been a country. While environmental activists are fighting to save the trees, the logging industry provides jobs to many otherwise unemployed Polish workers. The workers argue that the wood would likely be stolen if they did not legally log in the area. The debate continues over whether forests should be preserved in order to protect endangered wildlife and preserve biodiversity or be used to create jobs.

The Białowieza Forest is one of the last forests left preserved in its ancient state in Europe.

✅ **READING PROGRESS CHECK**

Expressing What is the debate over forest preservation in Eastern Europe?

Human Impact

GUIDING QUESTION *What human activities result in acid rain and water pollution in Eastern Europe?*

The communist governments in Eastern Europe placed an **emphasis** on rapid industrialization and heavy manufacturing. Eastern Europe's high concentration of industry has had a devastating impact on the environment. For example, the black triangle is a heavily industrialized area of Poland, eastern Germany, and the Czech Republic that relies primarily on burning coal as a power source. Soot covers the ground, and the air smells of sulfur from smokestacks. Although efforts are now under way to clean up the environment, the black triangle still bears the scars of poorly planned development from the Communist years.

Excessive reliance on coal burning in this small geographic region led to air, soil, and water pollution and caused people to develop respiratory diseases and cancer. In the 1970s and 1980s, Eastern European industries built smokestacks to carry pollution away from industrial sites. Air pollution then began to drift across national borders and made the problem more widespread. This coal pollution, which contains acid-producing chemicals, combines with moisture in the air and falls to the ground as acid rain. Polluted clouds drift away from the black triangle and other industrial areas and spread the pollution to forests in Northwestern Europe, degrading those environments as well.

Acid rain not only affects forests but lakes and rivers as well. In the cold Eastern European winters, snow carries industrial pollution to the ground. When this snow melts in the spring, **meltwater**, the result of melting snow and ice, carries the acid into lakes and rivers. As acid rain and meltwater pollution increased, fish and other aquatic life became endangered and even extinct. Some rivers and forests in the Czech Republic, for example, can no longer sustain life.

▲ **CRITICAL THINKING**

1. *Hypothesizing* What characteristics of the trees in the Białowieza Forest make them sought after for logging?

2. *Making Connections* Why is it important to preserve the Białowieza Forest?

reforestation the action of renewing forest cover (as by natural seeding or by the artificial planting of seeds or young trees)

emphasis importance

meltwater water formed by melting snow and ice

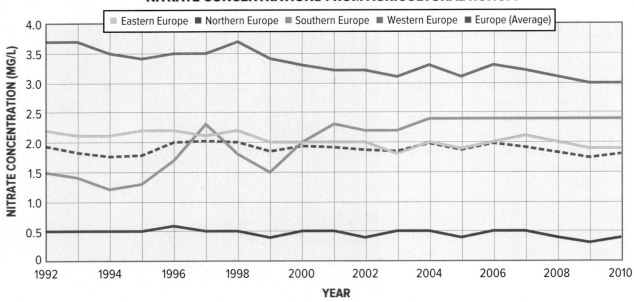

NITRATE CONCENTRATIONS FROM AGRICULTURAL RUNOFF

◻ Eastern Europe ◼ Northern Europe ◼ Southern Europe ◼ Western Europe ◼ Europe (Average)

Source: European Environment Agency (EEA)

Nitrates from agricultural fertilizer contribute to water pollution in Eastern Europe.

▲ **CRITICAL THINKING**

1. *Comparing and Contrasting* How does the level of nitrate concentration in Eastern Europe compare to the European average?

2. *Interpreting* Based on what you know of Northern Europe, why do you think it has the lowest nitrate concentration?

Agricultural pollution in Eastern Europe is also a huge problem. As water moves over and through the ground, it picks up chemical fertilizers and pesticides. This agricultural runoff deposits pollutants into water sources and has severely polluted both the Danube and Vistula Rivers. Outdated mining methods have also released different kinds of polluted mine drainage into rivers and lakes.

Before 1989, Eastern European countries had virtually no laws to protect the environment against pollution, which increased until it affected public health. Poor air quality in Eastern Europe also has global consequences, and climate change has been accelerated by the industrialization of the twentieth century. But as Eastern European countries joined the European Union (EU), they have come under the environmental protection standards and cleanup required of member countries. The EU has also set strict emissions regulations for industries and vehicles. This often involves equipping smokestacks and vehicle exhaust systems with devices that remove sulfur and nitrogen compounds from their emissions.

☑ **READING PROGRESS CHECK**

Identifying What is the black triangle?

Addressing the Issues

GUIDING QUESTION *How are Eastern Europeans working to address the environmental issues related to resource management and human activity?*

In the past few decades, Eastern Europeans have made serious efforts to clean up the environment. Potential EU member countries have to meet certain environmental standards to be admitted into the EU. Furthermore, member states can face legal action if they do not respect international environmental laws. The EU also created a system of low-emission zones, which are areas where vehicles whose pollutant emissions are too high cannot enter. This emission system is currently in place in the Czech Republic and in Hungary. Since cleanup costs amount to billions of dollars, Eastern European countries are now seeking advanced technology and investment from EU countries in Western Europe.

Pollution that crosses national borders presents a complex situation. For example, pollution in the Danube River threatens wildlife in its outlet, the Black Sea. Many Eastern European governments recognize that improving water quality is necessary. Directing and financing cleanup, however, is difficult when the process involves many countries. Furthermore, the involvement of such a wide area makes tracking down one main source of contamination extremely challenging for regulators.

Non-EU Eastern European countries are also contributing to global efforts to protect the environment. The Baltic countries signed the Helsinki Convention for the Protection of the Marine Environment of the Baltic Sea Area. The Convention marked the first time the international community had regulated land-based sources of pollution, such as agricultural runoff, in a shared marine environment. The International Maritime Organization (IMO) has listed the Baltic and Black Seas as areas that need a high level of regulation and protection in order to prevent sea pollution. Since fish are the most important resource in the Black Sea, conservation efforts by the surrounding countries include banning dolphin fishing and preventing industrial waste from entering the sea. In 1994 Bulgaria, Georgia, Romania, Russia, Turkey, and Ukraine signed the Convention for the Protection of the Black Sea Against Pollution. The Convention addressed programs to control pollution, sustain fisheries, and protect marine life.

The World Wildlife Fund (WWF), an international nongovernmental organization (NGO), has opened up offices in Bulgaria, Hungary, Latvia, Poland, and Romania. It supports conservation work such as cleaning up the Danube

Johann Brandstätter/Alamy

In the 1970s and 1980s, smokestacks were built to carry pollution away from industrial sites in Eastern Europe. Air pollution began to drift across the borders and made the problem more widespread.

◀ CRITICAL THINKING

1. Assessing Why is air pollution a major concern in Eastern Europe?

2. Exploring the Issues What steps are countries taking to resolve the growing environmental crisis?

Environmental Politics

Earth's natural resources are not confined within country borders, so it is important for governments to cooperate internationally. The Regional Environmental Center (REC) for Central and Eastern Europe was created in 1990 as a collaboration between the United States, Hungary, and the European Commission. The program provides financial assistance to help solve environmental problems in Central and Eastern Europe, with over $12 million in funding thus far given by the United States. Cooperation among the governments of all countries is promoted with a free exchange of information and the establishment of public participation in environmental decision-making.

SPECULATING How might the United States be able to provide help beyond financial assistance to the countries of Eastern Europe through the REC?

and Vistula Rivers, stopping illegal logging, preventing overfishing, and creating protected areas so plants and animals can no longer be harmed directly by human activity. It is particularly devoted to preserving the Danube-Carpathian region and protecting biodiversity in the region. These efforts have been supported by various governments and are largely successful. The Bulgarian government established the Srebarna Nature Reserve. It protects a lake that feeds into the Danube and is also a bird sanctuary. The reserve was made a UNESCO World Heritage Site in 1983. Another park in Bulgaria is the Rila National Park that protects abundant varieties of animals, particularly birds.

The Balkan countries lag behind larger Eastern European countries such as the Czech Republic and Hungary in terms of addressing environmental issues. This is largely due to political instability in the Balkans throughout the 1990s and early 2000s. However, Albania, Bosnia and Herzegovina, Macedonia, Montenegro, and Serbia are all looking to join the EU. The desire for EU membership provides an incentive to more wisely manage their resources and to minimize their environmental impacts. The major challenge that the Balkan countries face is working across political borders. For example, after the breakup of Yugoslavia produced six new Balkan states, the countries now share 13 river basins. Other ecologically important areas are located in the mountains and cross international borders. The remoteness of these mountainous border areas has helped to protect them in the past. The sharing of such important resources requires cooperation because an entire ecosystem needs to be protected in order for the efforts to be successful.

To ensure effective resource management and to meet EU standards, the Balkan countries will need to cooperate. They first need to regulate their outdated mining methods and to adopt modern technologies. The Balkan countries have a wealth of precious minerals at their disposal and need to ensure that their mining techniques are environmentally friendly.

In addition to technological improvement, education about environmental sustainability is also needed. Overall, environmental awareness among people in the Balkans is very low. Efforts to show how environmental protection can develop local economies can help. For example, the Durmitor National Park in Montenegro includes the Tara River canyon. The Tara River flows through Europe's deepest gorges and is one of the last wild rivers in Europe. It is an attractive destination for hikers and nature tourists and helps to support the local tourism industry.

☑ **READING PROGRESS CHECK**

Explaining Why do the Balkan countries lag behind other Eastern European countries in terms of addressing environmental issues?

LESSON 3 REVIEW

Reviewing Vocabulary

1. *Describing* Explain what meltwater is and describe how it can carry pollution to new locations.

Using Your Notes

2. *Explaining* What are some ways humans have affected their environment negatively in Eastern Europe?

Answering the Guiding Questions

3. *Evaluating* Why do forest resources in Eastern Europe need to be managed effectively?

4. *Stating* What human activities result in acid rain and water pollution in Eastern Europe?

5. *Explaining* How are Eastern Europeans working to address the environmental issues related to resource management and human activity?

Writing Activity

6. *Informative/Explanatory* Imagine that you live in a polluted area in Eastern Europe. Write a letter to the editor of a newspaper there and suggest ways to halt environmental damage.

Directions: On a separate sheet of paper, answer the questions below. Make sure you read carefully and answer all parts of the questions.

Lesson Review

Lesson 1

1 *Stating* Which rivers have historically served as a link for the people in Eastern Europe, and what has been created to supplement the passages?

2 *Contrasting* How do the mineral resources of the Carpathian Mountains differ from those of the Baltic Mountains?

Lesson 2

3 *Making Connections* Why are the Balkans considered a shatter belt region? Be sure to use the term *Balkanization* in your response.

4 *Finding the Main Idea* How have political activities affected settlement and population patterns in Eastern Europe?

Lesson 3

5 *Making Generalizations* What efforts have been made to preserve forests in Eastern Europe?

6 *Exploring Issues* How has industrialization resulted in acid rain and water pollution in Eastern Europe?

Critical Thinking

7 *Assessing* Describe the term *ethnic cleansing* and how it was used in Eastern Europe.

8 *Classifying* What were many of the new borders based upon after World War I?

9 *Analyzing* What impact has EU membership had on the environment in member countries? Give examples from the chapter to support your answer.

10 *Contrasting* Write a paragraph describing how the role of tourism in the economy of Eastern European countries today is different from how it was during the Soviet era. Include factors you think may draw tourists to the countries of Eastern Europe today.

11 *Sequencing* Prepare an itinerary for a 10-day trip to Eastern Europe, including the countries you plan to visit and the natural features you plan to see in each country. Make sure your itinerary reflects a logical sequence of travel given the location of the countries.

21st Century Skills

Use the chart to answer the following questions.

Poland's Top Trading Partners: Percent of Total Export Value, 2014	
Germany	25.0%
United Kingdom	6.3%
Czech Republic	5.9%
France	5.6%
Italy	4.6%
Russia	4.2%
Netherlands	4.1%
Belgium-Luxembourg	2.6%
Sweden	2.6%
Spain	2.5%

Source: http://atlas.media.mit.edu/en/visualize/tree_map/hs92/export/pol/show/all/2014/

12 *Economics* Where does Poland send the greatest percentage of its exports?

13 *Draw Conclusions* Refer to your response to question 12 above. Explain the significance of its location to the comparative size of the percentage. Based on this significance, why do you think the percentage is not higher for the Czech Republic?

Need Extra Help?

If You've Missed Question	**1**	**2**	**3**	**4**	**5**	**6**	**7**	**8**	**9**	**10**	**11**	**12**	**13**
Go to page	316	318	319	319	327	327	321	320	328	325	314	331	331

Directions: On a separate sheet of paper, answer the questions below. Make sure you read carefully and answer all parts of the questions.

DBQ Analyzing Primary Sources

Use the document to answer the following questions.

In 2007 Europe experienced exceptionally hot temperatures during the summer.

PRIMARY SOURCE

"Up to 500 people are estimated to have died across Hungary last week, partly due to a heat wave gripping central and southeast Europe. . . .

Record-breaking high temperatures also killed 12 Romanians, one man in Macedonia and another man on the island of Corfu . . . while firefighters, soldiers and volunteers battled wildfires across a tinderbox southeastern Europe. . . .

'Extreme events such as we have seen in recent weeks herald the specter of climate change and it would be irresponsible to imagine that they won't become more frequent,' Nick Reeves, executive director of The Chartered Institution of Water and Environmental Management, a scientific group, said."

—"Deadly Heat Wave Grips Europe,"
CNN.com: Reuters (Budapest), July 25, 2007

14 *Speculating* What do you think spurred the wildfires?

15 *Interpreting* How does the quote effectively communicate the impact of global warming on Eastern Europe?

Research and Presentation

16 *Problem Solving* Based on information you have gathered, create a multimedia presentation to inform others of the main issues covered by the Baltic Sea Action Plan and the specific challenges it faces. Within your presentation, provide at least one solution to help address at least one of the challenges. For the presentation, include maps and images of places, as well as charts, diagrams, or graphs.

Applying Map Skills

Use your Unit 4 Atlas to answer the following questions.

17 *Physical Systems* What type of vegetation is most dominant in Eastern Europe?

18 *The World in Spatial Terms* Use your mental map of the region to list the bodies of water that connect Eastern Europe to the Atlantic Ocean and the North Sea.

19 *Environment and Society* Which types of natural resources can be found in Poland? Which type is most abundant near Warsaw?

College and Career Readiness

20 *Comparing and Contrasting* You are working for the World Wildlife Fund and have been asked to create a two-page summary of its mission and its activity in Eastern Europe. The purpose of the summary is to inspire college students in Eastern Europe to support the fund's efforts. Use Internet sources to gather information and write this summary, making the summary engaging and exciting for the college students who make up its intended audience.

Exploring the Essential Question

21 *Drawing Conclusions* Draw a map of one country that has been discussed in this chapter. Using the map from the book, label the physical features that exist in the country, such as lakes, mountains, and rivers. Then, using the information you have drawn, write a paragraph describing how geographic features and human systems are related.

Writing About Geography

22 *Narrative* Based on details in the chapter, write a one-page narrative to describe the landscape and your experience as you take an imaginary hike across the Balkan Peninsula. Make certain the descriptive details within your narrative make sense within the context of the geography of the Balkan Peninsula.

Need Extra Help?

If You've Missed Question	**14**	**15**	**16**	**17**	**18**	**19**	**20**	**21**	**22**
Go to page	332	332	329	244	242	245	329	314	314

The Russian Core

ESSENTIAL QUESTION • *How do physical systems and human systems shape a place?*

Why Geography Matters
Russia's Shrinking Population

Lesson 1
Physical Geography of the Russian Core

Lesson 2
Human Geography of the Russian Core

Lesson 3
People and Their Environment: The Russian Core

Geography Matters...

People rely on natural resources to provide for their everyday needs. The harsh geography of the Russian Core has stimulated empire building. For centuries, the people of the subregion have sought out waterways that connect to other lands and resources. This subregion has a violent political history, as well as a rich cultural heritage. After the fall of the Communist regime, the Russian Core today is claiming a new identity in the global economy.

◀ Russian girl wearing a colorful head scarf during a traditional religious festival

Russ Images/Alamy

Russia's shrinking population

The breakup of the Union of Soviet Socialist Republics (USSR) in 1991 signaled the beginning of a decline in Russia's population. Government officials and demographers (scientists who study population data) expressed deep concern over reduced population size and its threat to the country's society and economy. Leaders hope that improvements in the economic circumstances will help reverse the decline.

What happened to Russia's population after 1991?

When Vladimir Putin became Russia's president and prime minister in 1999, he inherited a demographic crisis with potentially dire outcomes for the country. Population declined by five million between 1992 and 2002 and has continued to decline, dropping from 145.2 million in 2002 to 144.3 million in 2015. Life expectancy plummeted to its lowest peacetime level ever from a high of 70.1 years in 1986–1987 to only 59 years for men and 72 years for women in 1992. Demographers voiced concerns about the country's ability to maintain its army, industry, and agricultural production.

1. **Human Systems** What problems can a steep population decline cause for a country? Explain the effects on labor, government services, prosperity, and the environment.

Why did Russia's population decline?

An important factor contributing to Russia's negative population growth rate during the 1990s was a steep decline in the birth (fertility) rate. At the same time, there was an increase in the death (mortality) rate and a sharp drop in life expectancy. In 1991, for the first time in the postwar history of Russia, the number of deaths exceeded that of births. The total population decreased by 30,900 in 1992 and by 307,600 in 1993. The life expectancy decreased largely due to poor nutrition and health care. Poverty, stress, alcoholism, heavy smoking, suicide, diseases such as HIV/AIDS, strict immigration policies, and homicide all made the trend more severe. The trend was especially strong among working-age men.

2. **Human Systems** Why did Russia's population begin to decline after the collapse of the Soviet Union? Summarize the factors contributing to the decline.

What is being done to address the issue?

Russian president Putin directed his government to halt the drop in population, offering subsidies and financial incentives for women to have more children. He has also proposed new initiatives to battle alcoholism and to provide better education and housing opportunities for all Russians. Another proposal is an attempt to use the immigration policy to lure Russians living abroad to return and to attract skilled foreign workers. This last component of Putin's proposal is one that Russian leaders have historically avoided. They feared a loss of national identity if foreigners were allowed to enter Russian society.

3. **Places and Regions** Why have Russian immigration officials historically resisted allowing non-Russians to immigrate?

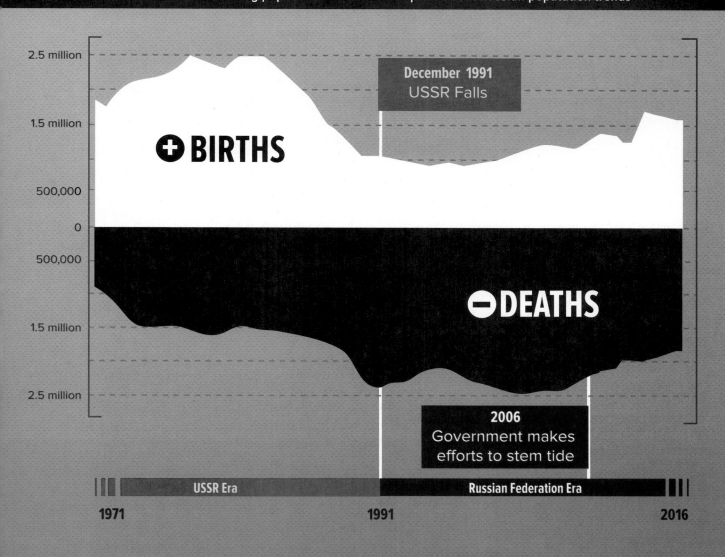

⊕ **BIRTHS**

December 1991
USSR Falls

⊖ **DEATHS**

2006
Government makes
efforts to stem tide

USSR Era | Russian Federation Era

1971 | 1991 | 2016

CAUSES OF THE DECLINE

 AGING POPULATION
Russians overall are getting older. In 2015, about 14% were age 65 or older.

 MOSCOW CALLING
Despite an increase in incentives, Russia has difficulty attracting immigrants.

 ALCOHOLISM
Since the fall of the USSR, alcoholism has increased. In 2015, 1 of every 6 alcohol-related deaths in the world occurred in Russia.

 HIV/AIDS
Uncommon in the USSR, HIV boomed after its fall and continues to spread. The number of Russians living with HIV more than doubled from 2006 to 2015.

 BRAIN DRAIN
Due to a lack of employment, many educated Russians leave the country. The vast majority who do are under 35 years old.

 FEWER BIRTHS
A low fertility rate of 1.8 births per woman and a high infant mortality rate of 9.3 deaths per 1,000 live births contribute to Russia's shrinking natural increase.

networks

There's More Online!

- ☑ **IMAGE** Russia's Shipping Industry
- ☑ **IMAGE** The Ural Mountains
- ☑ **INFOGRAPHIC** The Volga River Basin—Heart of Russia
- ☑ **INTERACTIVE SELF-CHECK QUIZ**
- ☑ **VIDEO** Physical Geography of the Russian Core

Reading **HELP**DESK

Academic Vocabulary

- **corresponding**
- **challenge**

Content Vocabulary

- **chernozem**
- **permafrost**
- **continentality**

TAKING NOTES: *Key Ideas and Details*

PARAPHRASING Use a chart like the one below to take notes about the landforms, waterways, climates, biomes, and resources of the Russian Core.

Physical Geography of the Russian Core		
Landforms	Water Systems	Climate, Biomes, and Resources

LESSON 1

Physical Geography of the Russian Core

ESSENTIAL QUESTION · *How do physical systems and human systems shape a place?*

IT MATTERS BECAUSE

The lowlands, plains, and plateaus of the Russian Core—Russia, Ukraine, and Belarus—helped shape human activity for centuries. In some cases, the harsh climate of the subregion made it a forbidding environment for people to settle and live in. Some were drawn to the few mountainous areas found in this region for their beauty and abundant natural resources despite difficult winters. Certainly the diverse topography of the Russian Core makes for some of the most beautiful geography in the world.

Landforms

GUIDING QUESTION *How do interconnected mountain ranges and plains shape human activities in the Russian Core?*

Russia has two notable mountain ranges, the Ural Mountains and the Caucasus Mountains. The other distinct landforms that dominate Russia are plains, hills, and plateaus. The Ural Mountains form a natural barrier between European Russia and Siberian Russia. Rich in iron ore and mineral fuels, they run in a north-to-south direction and span 1,230 miles (2,000 km) from the Arctic Ocean almost to the Caspian Sea. The Caucasus Mountains run east to west along the southwestern portion of the country, forming a natural barrier between Russia and countries to the south. The highest peak of the Caucasus Mountains is Mount Elbrus, which soars to 18,510 feet (5,642 m). It is the highest mountain in Russia and Europe. Mount Elbrus is also a popular tourist destination for skiing and mountain climbing.

To the west of the Ural Mountains is the Northern European Plain. The southern part of the plain has navigable waterways and rich, black soil called **chernozem** that supports agriculture. As a result, the majority of Russia's population lives here. This vast plain is an extension of the plain that begins in France and stretches across Northwestern Europe. This eastern part of the plain is sometimes called the Russian Plain

(t)Evgeny Prokofyev/Alamy, (tr)©Pashkov Andrey/Alamy

336

because as it reaches the Russian Core it broadens and becomes very expansive.

East of the Ural Mountains and extending to the Pacific Ocean are the vast stretches of plains and plateaus that make up much of Siberia. The West Siberian Plain covers about one-third of Siberia and is one of the largest low-lying flatlands in the world. It is known for its harsh continental climate and some of the world's largest swamps and wetlands.

Ukraine is the second-largest country in Europe, occupying the southwest portion of the Russian Plain. Its two main landforms are vast plains and plateaus, with mountains found only in small areas in the west and south, accounting for less than five percent of the country's landmass. In the south, the Isthmus of Perekop connects the Crimean Peninsula to the mainland.

Belarus is the smallest of the three Slavic republics that were once part of the Soviet Union. Belarus is a landlocked country and lies entirely on the Northern European Plain. Glacier scarring accounts for the flat terrain and some 11,000 lakes found in Belarus. There are also numerous swamps and rivers.

☑ READING PROGRESS CHECK

Naming Which mountains form a natural boundary between European Russia and Siberian Russia?

Water Systems

GUIDING QUESTION *What role do rivers of the Russian Core play in the economic activities of the region?*

The combined waterways of the Russian Core have played important roles in its social and economic development from early times. The rivers and lakes of Russia, Ukraine, and Belarus are key to the subregion's growth, expansion, and success. Included in the water system is the Volga River, one of the world's greatest rivers.

The Volga River and its many tributaries make up the Volga River system. Draining most of western Russia, the Volga travels 2,293 miles (3,690 km), making it the longest river in Europe. It starts in the Valdai Hills west of Moscow and travels across much of southern Russia before emptying into the Caspian Sea. The Volga River system is an important commercial, transportation, and hydroelectric resource for millions of Russians.

In Ukraine, the Dnieper River is the longest in the country, stretching 609 miles (908 km) beginning in the northwestern plains and traversing southeast to drain into the Black Sea. Other major waterways in Ukraine include the Southern Bug, Dniester, and even a small portion of the Danube River. Like the Volga in Russia, the Dnieper provides hydroelectric power and transportation as well as enabling commerce. But the most important function of these rivers collectively is to supply water to Ukrainians. This is done through an intricately constructed series of canals.

The Ural Mountains extend for well over a thousand miles, cutting through many varied landscapes.

▲ CRITICAL THINKING

1. *Classifying* Why are the Ural Mountains considered a natural boundary?

2. *Speculating* How might the Ural Mountains have influenced settlement patterns in the areas they span?

chernozem rich, black topsoil found in the Northern European Plain, especially in Russia and Ukraine

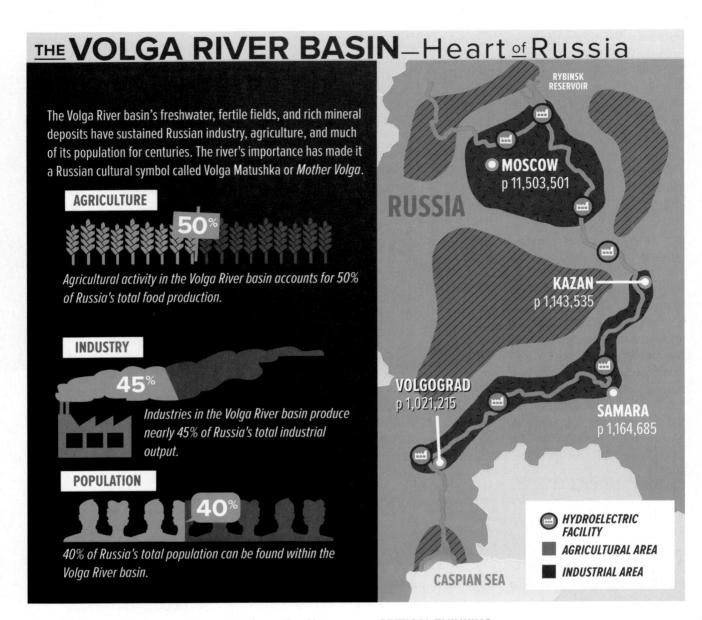

THE **VOLGA RIVER BASIN**—Heart of Russia

The Volga River basin's freshwater, fertile fields, and rich mineral deposits have sustained Russian industry, agriculture, and much of its population for centuries. The river's importance has made it a Russian cultural symbol called Volga Matushka or *Mother Volga*.

AGRICULTURE

50%

Agricultural activity in the Volga River basin accounts for 50% of Russia's total food production.

INDUSTRY

45%

Industries in the Volga River basin produce nearly 45% of Russia's total industrial output.

POPULATION

40%

40% of Russia's total population can be found within the Volga River basin.

RYBINSK RESERVOIR

MOSCOW p 11,503,501

RUSSIA

KAZAN p 1,143,535

VOLGOGRAD p 1,021,215

SAMARA p 1,164,685

CASPIAN SEA

HYDROELECTRIC FACILITY

AGRICULTURAL AREA

INDUSTRIAL AREA

Western Russia's Volga River connects Moscow to the Caspian Sea. Often referred to as "Mother Volga," the river is vital to Russia.

▲ **CRITICAL THINKING**

1. Assessing   Based on the infographic, describe the importance of the Volga River to Russia in regard to hydropower generation.

2. Identifying Central Issues   Why is the Volga River called *Mother Volga*? Provide at least two reasons.

There are few freshwater lakes across Ukraine. Small saltwater lakes can be found in the Black Sea Lowland and Crimea, while *limans*, or large saline lakes, are found along the coast. In recent years Ukraine has created some artificial lakes and reservoirs.

Belarus was once an important passage route for inland navigation between the Baltic Sea and the Black Sea. Each with its own tributaries, the largest rivers in the country include the Dnieper River, Berezina River, Pripyat' River, Neman River, Bug River, and Western Dvina. Narach is the country's largest lake, measuring 31 square miles (79.6 sq. km).

Waterways are also a prominent feature of Siberia. Rivers such as the Ob', Irtysh, Yenisey, and Lena begin in the south of Siberia and flow northward, emptying into the Arctic Ocean. Ice in the Arctic Ocean blocks the rivers from reaching the ocean for much of the year. This is problematic in the springtime when the ice blockage prevents the surging rivers from reaching the ocean. Instead, the water floods out across the low-lying plains, causing extensive areas of swamps and floodlands.

One of the most renowned Siberian lakes is Lake Baikal. Located in southeast Russia, Lake Baikal is the oldest lake in the world (25 million years). It is also the deepest at 5,715 feet (1,742 m). It holds one-fifth of all unfrozen freshwater found on the planet. Known as the "Galapagos of Russia" because of its age and isolation, Lake Baikal has many unusual freshwater marine species, which are of exceptional value to scientists who study how species evolve.

✔ READING PROGRESS CHECK

Identifying Which river provides western Russia with hydroelectric power?

Climate, Biomes, and Resources

GUIDING QUESTION *What are the general climate conditions in much of the Russian Core?*

Russia's vast expanse of land extends from east to west, and all of the country lies in the same high latitude range. The result is that the dominant characteristic of the subregion is cold, snowy winters. There is, however, variation in how cold and how long the winters last, and the warmth and length of the summers. These variations in climate have shaped settlement patterns from the earliest times.

Climate Regions and Biomes

The tundra occupies the parts of the subregion that are farthest north and covers about 10 percent of Russia. Here, the sky stays dark for many weeks before and after the winter solstice that occurs around December 22 each year. Then, for several weeks during the summer, there is continuous sunlight. Its short growing season and thin, acidic soil lying just above the **permafrost** limit the kinds of plants that can grow there. Only mosses, lichens, algae, and dwarf shrubs thrive in the tundra.

permafrost a permanently frozen layer of soil beneath the surface of the ground

South of the tundra lies the subarctic climate zone. While not as severe as the tundra climate, this zone only has four months of the year when the temperature rises above 50°F (10°C). The rest of the year is cold. The biome here is boreal forest, or taiga, which consists of broad expanses of coniferous evergreen trees.

Russia's midlatitude climates, found in the western region, are not as severe and have milder winters and warmer summers. Although still relatively cold, these climates are where most Russians live and where much of Russia's agricultural production takes place. The natural biome here is deciduous forest, although much of it has been cleared for agriculture and construction.

An area between the Black and Caspian Seas north of the Caucasus Mountains and along Russia's border with Kazakhstan make up Russia's steppe climate region. The steppe is a broad, open grassland in which seas of grass stretch to the horizon in every direction. The region's chernozem soil supports the production of wheat, barley, rye, oats, and other crops. Sunflowers, mint, and beans also flourish here.

The southern part of Ukraine lies in a humid continental climate zone where warmer, humid air from the Atlantic Ocean makes the climate milder than farther north. The greatest amount of precipitation falls in the warmer summer season, with the maximum precipitation occurring in late June and July.

Belarus has a humid continental climate moderated by maritime influences from the Atlantic Ocean. Average January temperatures range from the mid-20°sF (about −4°C) in the southwest to about 17°–19° F (about −8°C) in the

Russia's shipping industry includes both river- and sea-going vessels.

▲ CRITICAL THINKING
1. **Speculating** Why is Russia's coal, although abundant, difficult to produce?
2. **Describing** Where does Russia rank in the production of nickel, aluminum, and platinum-group metals?

corresponding showing a direct connection between two things

continentality effect of extreme variation in temperature and very little precipitation within the interior portions of a landmass

challenge to arouse or stimulate especially by presenting with difficulties

northeast, but thaw days are frequent. **Corresponding** to these temperature differences, the frost-free period decreases from more than 170 days in the southwest to 130 days in the northeast. Maximum temperatures in July are generally in the mid-60°sF (about 18°C). Rainfall is moderate, although higher than over most of the vast Russian Plain.

Warmer air from the Atlantic Ocean moderates temperatures in western Russia. Most of Russia, however, lies well within the Eurasian landmass, far away from any moderating ocean influences. As a result, much of the country's interior has more extreme variations in temperature and little precipitation. This climatic effect within the interior areas of a landmass is called **continentality**.

Natural Resources

Russia's physical geography benefits but also **challenges** its people. The country has an abundance of natural resources, including one-fifth of the world's forests. Much of this wealth, however, lies in remote and climatically unfavorable areas and is difficult to tap or utilize. For example, Russia holds large petroleum deposits and 16 percent of the world's coal reserves; however, the country's biggest coalfields lie in remote areas of eastern Siberia. Russia is also a leading producer of natural gas, but much of it is located in northern Siberia. It also leads the world in nickel production and ranks among the top three producers of aluminum, gemstones, and platinum-group metals. Russia's rivers make it a leading producer of hydroelectric power.

There are considerable resources of iron and other ores in parts of Ukraine. Other resources such as mountain wax, granite, and graphite are among the country's most abundant. Ukraine also produces various salts and has a rich base for metallurgical, porcelain, and chemical industries.

Natural resources in Belarus include fuel sources such as peat, oil, and natural gas deposits. The country also contains small quantities of a variety of rocks and minerals, such as chalk, sand, clay, gravel, granite, and limestone.

☑ READING PROGRESS CHECK
Classifying How does Russia's location in the high latitudes affect its climate?

Pashkov Andrey/Alamy

LESSON 1 REVIEW

Reviewing Vocabulary
1. *Explaining* Explain the significance of permafrost and continentality.

Using Your Notes
2. *Listing* Use your chart to list and describe the locations of the major landforms of the Russian Core.

Answering the Guiding Questions
3. *Identifying* How do interconnected mountain ranges and plains shape human activities in the Russian Core?

4. *Describing* What role do rivers of the Russian Core play in the economic activities of the region?

5. *Explaining* What are the general climate conditions in much of the Russian Core?

Writing Activity
6. *Informative/Explanatory* Think about the locations of Russia's seas. Write a paragraph describing how the locations of these seas affect Russia's economy.

   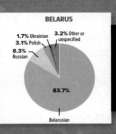

BELARUS

1.7% Ukrainian
3.1% Polish
3.2% Other or unspecified
8.3% Russian
83.7%
Belarusian

networks

There's More Online!

☑ **GRAPHS** Ethnic Composition of the Russian Core

☑ **GRAPH** GDP Per Capita in the Russian Core

☑ **TIME LINE** Soviet Era of the Russian Core

☑ **INTERACTIVE SELF-CHECK QUIZ**

☑ **VIDEO** Human Geography of the Russian Core

LESSON 2
Human Geography of the Russian Core

ESSENTIAL QUESTION · *How do physical systems and human systems shape a place?*

Reading **HELP**DESK

Academic Vocabulary

- **acquire**
- **trigger**
- **decline**

Content Vocabulary

- **czar**
- **Russification**
- **satellite**
- **perestroika**
- **glasnost**
- **black market**
- **privatization**

TAKING NOTES: *Key Ideas and Details*

IDENTIFYING Use a graphic organizer like the one below to take notes on society and culture today in the Russian Core.

Human Geography of the Russian Core: Society and Culture Today

Lasting effects of Soviet government policies → ☐

Family and Status of Women → ☐

The Arts → ☐

It Matters Because

The Russian Revolution, the Soviet era, and the collapse of the Soviet Union have had profound effects on the Russian Core's economies, politics, and people. This subregion's countries have been forced to redefine who they are and to restructure their economies based on new ideas about political and economic systems.

History and Government

GUIDING QUESTION *How have the Russian Core's historical roots and modern ideas influenced the history and government of the region?*

For most of the last century, Russia was part of the Soviet Union. Ruled by a Communist government, it challenged the United States and other democracies for global influence. Then the Soviet Union collapsed, and Russia, Belarus, and Ukraine emerged as independent countries.

Early History

Russia's historical roots go back to the A.D. 600s when Slav farmers, hunters, and fishers settled near the waterways of the Northern European Plain. The Slav communities were once organized into a loose union of city-states known as Kievan Rus. Ruled by princes, the leading city-state, Kiev, controlled a prosperous trading route, using Russia's western rivers to link seaports and trade centers on the Baltic and Black Seas.

In the early 1200s, Mongols from Central Asia invaded Kiev and many of the Slav territories. Although the Mongols allowed the Slavs self-rule, they continued to control the area militarily for more than 200 years.

In the late 1600s, **Czar** Peter I—known as Peter the Great—came to power, determined to modernize Russia. Under his rule, Russia enlarged its territory, built a strong military, and developed trade with Western Europe. To **acquire** seaports, Russia gained land along the Baltic Sea from Sweden. St. Petersburg became the new capital and a major port city. It was carved out of the wilderness along the Gulf of Finland, providing access to the Baltic Sea and giving Russia "a window on the West."

(t)©Swim Ink/Corbis, (tc)©The Dmitri Baltermants Collection/Corbis, (tr)©Peter Turnley/Corbis

The Russian Core **341**

czar **czar** ruler of Russia until the 1917 revolution; originally from Latin word *Caesar*, title of Roman emperors

acquire to gain possession

Russification in nineteenth and twentieth century Russia and the Soviet Union, a government program that required everyone in the empire to speak Russian and to become a Christian; assignment of some Russian-speaking people to non-Russian ethnic regions

trigger to set off

satellite a country controlled by another country, notably Eastern European countries controlled by the Soviet Union by the end of World War II

During the late 1700s, Empress Catherine the Great continued to expand Russia's empire and gained a long-desired warm-water port on the Black Sea. A later royal family of Russia, the Romanovs, continued expansion, resulting in acquisitions in both Eastern and Central Europe. These new territories and their people brought many non-Russians under Russian rule. A program known as **Russification** was begun to make those peoples more "Russian."

Czar Alexander II's limited reforms caused many former serfs to move to cities. There they faced the poor conditions and low wages of factory work. Non-Russian peoples from newly colonized regions of the Russian Empire faced prejudice. The government also insisted colonized people become like Russians. Russification became government policy, and people were required to speak Russian and follow Eastern Orthodox Christianity to receive jobs and benefits.

Beginning in 1891, under Czar Alexander III, Russia expanded into Siberia with the construction of the Trans-Siberian Railroad. Nearly 6,000 miles (9,700 km) long, it connected Moscow to Vladivostok. Once completed in 1916, the railroad opened Russia's Asian eastern region to settlement.

Revolution and Change

One of the biggest proponents for greater economic equality was the German philosopher Karl Marx, the founder of modern communism. He advocated two principles: the public ownership of all land and means of production, and a classless society with an equal sharing of wealth. During World War I, which began in 1914, numerous strikes and demonstrations were organized by Russian workers who suffered hardships because of the war. They protested, demanding "bread and freedom." This unrest **triggered,** or set off, the Russian Revolution of 1917. Czar Nicholas and his family were murdered, signaling the demise of Europe's last absolute monarchy. What emerged was the Communist-controlled Union of Soviet Socialist Republics (USSR), or the Soviet Union.

The Soviet Union played a pivotal role in the Allied victory over Germany during World War II. Following the war, the Soviet Union occupied much of Eastern Europe. Several countries in the region were controlled as **satellites** under Communist rule. Following World War II, the Communist Soviet Union was engaged in a political and ideological war with the West, particularly the United

©Swim Ink/Corbis

TIME LINE ⌄

SOVIET ERA
of the Russian Core ➔

Discontent with inequality in Russian society led to revolution and freedom from generations of czarist rule in the early 1900s. This was followed by new ideas about political and economic systems that led to changes in Russia in the 1990s.

▶ CRITICAL THINKING

1. *Explaining* What did the Soviet Union gain from the Nazi-Soviet Nonaggression Pact?

2. *Describing* How did post-World War II agreements change Russia's influence in Europe?

Lenin dies. Joseph Stalin emerges as new leader.

German-Soviet Nonaggression Pact gives Soviet Union a sphere of influence in Estonia, Latvia, Lithuania, and eastern Poland.

1924

1910 ➔

1917

1922

1939

Germany invades western Poland, starting World War II. The Soviets occupy eastern Poland.

Revolution forces Czar Nicholas II to abdicate the throne. Vladimir Lenin becomes leader of Russia.

The Union of Soviet Socialist Republics (USSR) is established.

States, in the Cold War. Tensions during this time brought the world to the brink of nuclear war and shaped modern economic and political policies.

The enormous costs of the Soviet Union's military and inefficient economic policies weakened its role in the world. In 1985 Mikhail Gorbachev, a reform-minded official, became the leader of the Soviet Union. He instituted a policy of economic restructuring called **perestroika** (PEHR•uh•STROY•kuh) and a policy of greater political openness called **glasnost** (GLAZ•nohst). Political reform was set in motion. Satellites controlled by the Soviet Union began to replace their Communist governments in 1989. In 1991 a failed coup led to the collapse of the Soviet government and regions of the country began declaring their independence. Today there are 15 independent countries, including Russia, Ukraine, and Belarus.

The newly independent countries emerged from a long period of both imperial Russian and Soviet domination. Belarus is a good example. Although Belarusians share an ethnic identity and language, they had never enjoyed unity and political sovereignty, except during a brief period in 1918. Belarusian history is a study of regional conflicts and territorial claims by neighbors to the east and west.

Since the beginning of Russian occupation, Chechnya has unsuccessfully sought independence. Several oil and gas pipelines vital to the Russian economy run through Chechen territory. In May 2000, Russian president Vladimir Putin established direct rule of Chechnya to try to stop the rebels. In 2003 a new constitution was passed that gave Chechnya a significant amount of autonomy. The new Chechen government is now struggling to recover from the violence waged by the Chechen rebels and to bring stability and peace to Chechnya.

In November 2013, Ukrainian president Viktor Yanukovych abandoned an agreement that would strengthen trade ties with the EU and would instead seek closer cooperation with Russia. This announcement led to violent protests as many people felt the trade agreement would bring Ukraine back under Russian influence.

These protests continued into 2014. As tensions and violence increased, Yanukovych fled Ukraine and an interim president was named. In the Crimean region of Ukraine, which has a majority Russian population, pro-Russians demonstrated to secede from Ukraine. Then the fears of many Ukrainians were realized as Russian troops moved into Crimea. They claimed to be protecting those of Russian descent from the interim Ukrainian government.

perestroika in Russian, "restructuring"; part of Gorbachev's plan for reforming the Soviet economy and government

glasnost Russian term for new openness in areas of politics, social issues, and media; part of Gorbachev's reform plans

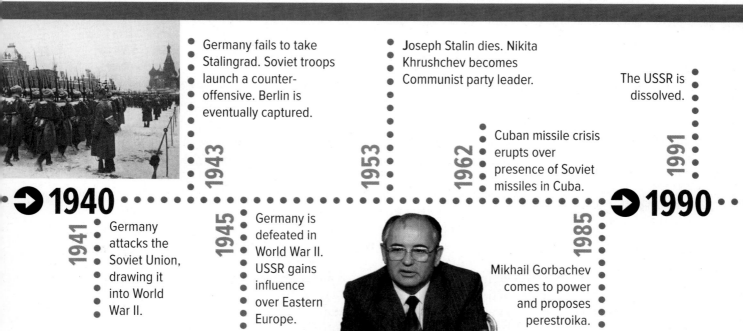

1940

1941 Germany attacks the Soviet Union, drawing it into World War II.

1943 Germany fails to take Stalingrad. Soviet troops launch a counter-offensive. Berlin is eventually captured.

1945 Germany is defeated in World War II. USSR gains influence over Eastern Europe.

1953 Joseph Stalin dies. Nikita Khrushchev becomes Communist party leader.

1962 Cuban missile crisis erupts over presence of Soviet missiles in Cuba.

1985 Mikhail Gorbachev comes to power and proposes perestroika.

1990

1991 The USSR is dissolved.

BELARUS

1.7% Ukrainian
3.1% Polish
8.3% Russian
3.2% Other or unspecified
83.7%
Belarusian

RUSSIA

1.0% Chechen
1.0% Chuvash
1.1% Bashkir
1.4% Ukrainian
3.7% Tatar
14.1% Other or unspecified
77.7%
Russian

UKRAINE

0.5% Crimean Tatar
0.5% Moldovan
0.6% Belarusian
Russian
17.3%
3.3% Other or unspecified
77.8%
Ukrainian

Source: CIA World Factbook

The diversity of people in Russia has led to many ethnic groups demanding greater self-rule or independence. In some places, like Chechnya, some groups have resorted to violent methods, such as terrorism.

▲ **CRITICAL THINKING**

1. *Analyzing* Which groups make up over 10 % of the Ukrainian population?

2. *Categorizing* Which country shown has the largest population of "Other" ethnic groups? What factors may account for the higher percentage?

decline to become less in amount or number

Russian troops took over government buildings and cut off Crimea from the rest of Ukraine. Elected officials were ordered out of office at gunpoint and replaced with pro-Russia politicians. These new leaders announced a referendum to secede, a vote backed by Russian president Vladimir Putin who was unhappy that Ukraine wanted to strengthen ties to Europe. Although many consider it to be illegitimate and unconstitutional, the referendum resulted in Crimea becoming part of Russia.

✅ **READING PROGRESS CHECK**

Explaining How was St. Petersburg important to the expansion of the Russian Empire?

Population Patterns

GUIDING QUESTION *What factors have shaped population patterns in the Russian Core?*

About 80 percent of all Russians live west of the Ural Mountains. This is due in part to the rich soils, waterways, and a milder climate than that of eastern Russia, also known as Siberia. Densely settled western Russia includes the country's industrialized cities. The major industrial city is Moscow, Russia's capital. Following the reestablishment of Russia as a country in 1991, the controls over where people could live were removed. People migrated where there were jobs and other opportunities. As a result, Russian cities experienced increases in population. By 2000 the population of ethnic Russians began to **decline** due to a greatly reduced birthrate. The decline has continued into the twenty-first century.

A shrinking population is a major problem for Russia. To offset falling birthrates, there are government programs that pay monthly allowances to parents with babies and children. The total fertility rate of 1.8 babies per woman in 2015 is below the 2.1 needed to naturally maintain the population. Reducing mortality rates is also necessary. Death rates, particularly among males between the ages of 25 and 45, have increased to levels that outpace new births. As a result of having fewer younger people, Russia's overall population has been simultaneously decreasing and growing older.

✅ **READING PROGRESS CHECK**

Making Connections Why do most Russians live west of the Ural Mountains?

Society and Culture Today

GUIDING QUESTION *How have past policies affected the Russian Core today?*

The policies of the past have had lasting effects on Russia's culture. The Soviet government severely discouraged religious practices and discriminated against different ethnic groups. It actively promoted atheism, or the belief that God or another supreme being does not exist. In the late 1980s, however, the government relaxed its restrictions on religion. Since then, millions of Russians are rediscovering their religious faiths and traditions. Similarly, more than 100 languages are spoken in the country today, but Russian is the official language.

During the Soviet era, education was free and mandatory. The emphasis was on math, science, and engineering rather than on language, history, and literature. The curriculum changed dramatically after Russian independence. Schools began to emphasize a more balanced approach to learning, including language, history, and literature. Today, students have a choice of different types of schools, but the country's economy has limited funding for schools.

Family and Status of Women

Living conditions in Russia affect family life. Due to a housing shortage, most people live in large apartment blocks. While space in the traditional apartments is tight, new housing developments since 1991 offer space and living conditions similar to those in Western Europe and the United States.

The status of women in Russian society is a combination of the past and the present. Women have always had a very important role working in industry. During World War II, women not only fought in every branch of the Soviet military but also took jobs in the wartime industrial workforce. That trend continued into the 1970s. In the 1990s, there were increasing financial pressures and shrinking government programs in Russia. This meant more women entered the workforce out of necessity. Full-time employment of women contributed to Russia's declining birthrate. Higher education, better-paying jobs, and feminist groups and social organizations have resulted in greater rights for women. As a consequence, many women marry later in life, many have careers, and they expect to be treated as equals in a society that has traditionally been dominated by men.

The Arts

Russian arts are characterized by a list of well-known artists. Painters such as Viktor Vasnetsov and composers such as Pyotr (Peter) Tchaikovsky contributed to the richness of Russian culture. The works of poets Aleksandr Pushkin, Boris Pasternak, and Anna Akhmatova and novelists Lev (Leo) Tolstoy and Fyodor Dostoyevsky have made Russian literature famous. Russian ballet and theater also have an international reputation.

The period of communism limited individual artistic expression and directed artists to glorify the government's achievements in their works, an approach known as socialist realism. Artists who did not follow these guidelines were punished. Accounts of imprisonment and punishment among artists during the period are shown in works such as Aleksandr Solzhenitsyn's *The Gulag Archipelago.*

Beginning in the mid-1980s, activity in the arts renewed as loosening government controls allowed the printing of previously unpublished works and new materials. During the height of Communist repression, books were smuggled from Russia and printed in other languages, becoming best sellers in other countries. Many were not available to Russian readers until after 1991.

✓ **READING PROGRESS CHECK**

Analyzing What contributed to a resurgence of the arts in the 1980s?

Economic Activities

GUIDING QUESTION *How has the economy of the Russian Core changed since 1991?*

Under Communist leaders, the Soviet Union operated as a command economy in which the government made key economic decisions. The government owned banks, factories, farms, mines, and transportation systems. The government decided what and how much to produce, where and how to produce it, and who would benefit from the profits. It also controlled the pricing of most goods and decided where they would be sold.

Unemployment was nearly nonexistent, but wages were low. Some people could not afford consumer goods, or goods needed for everyday life. Even when people had enough money, such goods were hard to find. Scarce products could be bought on the **black market**, illegal trade in which scarce or illegal goods are sold at high prices. Most workers, however, could not afford such high prices.

When Mikhail Gorbachev came to power in 1985, the Soviet command economy was under strain. He began to move the country toward a market

black market illegal trade of scarce or illegal goods, usually sold at high prices

economy, in which businesses became owned by private individuals and companies. Gorbachev reduced government controls, allowed people to start small businesses, and encouraged foreign investment. Boris Yeltsin, Gorbachev's successor, expanded this process.

Russia's economy continued to change after 1991. Yeltsin removed governmental price controls and encouraged **privatization**—a shift to private ownership—of state-owned companies. This favored wealthy people or businesses with money to purchase large companies. Those who made money sometimes invested their profits outside the country in Europe and the United States. Most Russians were not able to benefit from this new wealth at the time.

The Russian economy experienced many successes throughout the 1990s. More consumer goods were available to meet pent-up demands. Greater demands resulted in prices that soared, and many people could not afford to buy the goods. Between 1990 and 1995, Russia experienced an economic depression as its GDP fell by 50 percent. Following a 1998 financial crisis, the ruble—Russia's currency—lost 71 percent of its value. The international community issued loans and credit to help the country.

Russia has experienced steady economic growth since 1998, due in part to increases in productivity, wages, consumption, and a growing middle class. The anchor in the economic success has been its vast supplies of natural resources that are marketed to industrialized countries.

Resources, Power, and Industry

The fertile triangle is the most productive agricultural area of western Russia, extending into Ukraine. With its tip at St. Petersburg and its base stretching from Novosibirsk to Odessa, the triangle has rich soils and a favorable climate. Ukraine's crop production is highly developed. It is referred to as a breadbasket because of the grain crops it grows. Yields of grain and potatoes are among the highest in Europe.

The Russian manufacturing and service sectors are expanding, while the oil and gas sector has grown rapidly. Russia's most important industry is petroleum extraction and processing, and the country is one of the world's largest producers of crude oil. While energy resources dominate Russian exports, minerals also provide important export income. Russian forests produce one-fifth of the world's softwood, and Russian supertrawlers, or fish-factory ships, process catches from the Atlantic and Pacific Oceans for sale on the global market.

Belarus is the one country that has remained closely aligned with Russia since 1991. The two countries remain industrially integrated as a carryover from the past. After 1991 Belarus experienced less privatization of industry and land. Instead, Belarus looked to Russia to be a strong trading partner in exchanging its agricultural products for oil and natural gas.

Trade and Interdependence

Russia has focused on becoming a full partner in the global community by expanding trade and building international relationships. The country is a major source of oil and natural gas. These resources account for more than 50 percent of Russia's exports.

Relationships with neighboring countries are important to Russia. Ukraine imports petroleum, petroleum products, and natural gas, as well as

---

privatization a change to private ownership of state-owned companies and industries

The gross domestic products (GDPs) of the Russian Core countries have fluctuated since the breakup of the Soviet Union.

▼ **CRITICAL THINKING**

1. *Analyzing Visuals* Which Russian Core country has consistently had the highest GDP since 1990?

2. *Assessing* In what year did the GDP per capita of all three Russian Core countries decline?

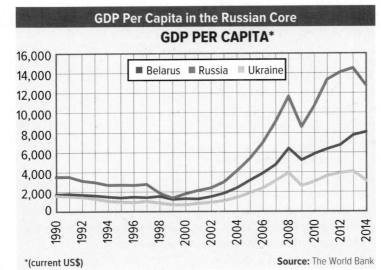

GDP Per Capita in the Russian Core

GDP PER CAPITA*

■ Belarus ■ Russia ■ Ukraine

*(current US$)   **Source:** The World Bank

many other products. Ukraine has the major seaport on the Black Sea and exports grain, sugar, iron ore, coal, and manganese. In the twenty-first century, Russia and Ukraine remain dependent upon one another for trade. However, Ukraine is wary of past Russian dominance of the Ukrainian people. Therefore, Ukraine's government has discussed making application to join the European Union (EU), which would switch its closest relationships from Russia to Europe. However, acceptance to the EU is dependent on economic, political, and social criteria on which Ukraine must make progress. In the meantime, Ukraine is developing its economy and its national identity within the region.

A major highway system links Moscow with other major cities in Russia, but winters are severe on roadways and repairs are both necessary and expensive. Because of its size and climate extremes, Russia depends on railroads and waterways for much of its transportation needs. Major cities are found where the Trans-Siberian Railroad and other railroads cross large rivers. Millions of tons of goods travel along thousands of miles of navigable inland waterways, which include seaports and inland cities. Pipelines are effective in transporting petroleum products, although constructing and maintaining them is challenging in large regions, such as Siberia, with a harsh climate.

Russia, Ukraine, and Belarus experience the effects of pressure for rapid modernization of their economies. Some of the world's best software companies are operated by people in Ukraine. Belarus is a crossroads for trade from Western Europe to Moscow and the rest of Russia. It has had to improve highways, communications, and repair services for long-haul trucks and railroads that transport goods both to and from Russia. All three countries are in the process of identifying their future roles in Europe and the world. All have traditions and cultures going back for centuries. However, they are each new countries, having become independent after 1991, and face political, economic, and social changes.

✅ **READING PROGRESS CHECK**

Identifying What economic transition has Russia been making since the mid-1980s?

Connecting Geography to ECONOMICS

Privatization

After 1991, private ownership of companies in Russia was allowed. The government established a stock exchange and gave vouchers to citizens enabling them to purchase shares in private companies, but these were often sold for cash instead. In 2001 the sale of land was legalized. These changes were a challenge for Russians who were accustomed to government control of nearly every aspect of their lives. It was complicated by a poor economy, inflation, and a drop in personal income. Corruption and organized crime have become issues in Russia, and they hinder investment by international businesses.

SPECULATING Why has the transition from a command to a market economy after 1991 affected Russia in both positive and negative ways?

LESSON 2 REVIEW

Reviewing Vocabulary

1. *Explaining* Explain the significance of black market and privatization.

Using Your Notes

2. *Describing* Use your graphic organizer to describe how the policies of the Soviet era affect society and culture in the Russian Core today.

Answering the Guiding Questions

3. *Identifying Cause and Effect* How have the Russian Core's historical roots and modern ideas influenced the history and government of the region?

4. *Identifying* What factors have shaped population patterns in the Russian Core?

5. *Describing* How have past policies affected the Russian Core today?

6. *Explaining* How has the economy of the Russian Core changed since 1991?

Writing Activity

7. *Informative/Explanatory* Write a paragraph describing the conditions in Russian society that led to the Revolution of 1917. Be sure to include the influence of German philosopher Karl Marx in your answer.

Arctic Oil
Frontiers

Some of Earth's largest oil and natural gas fields lie under the icy waters of the Arctic Ocean. As extraction technology has improved and as demand for these resources has increased, states that border the Arctic Ocean have staked their claims and energy companies have begun to plan for drilling operations. But are energy companies prepared for the hazards involved in Arctic drilling? Are the rewards worth the risk?

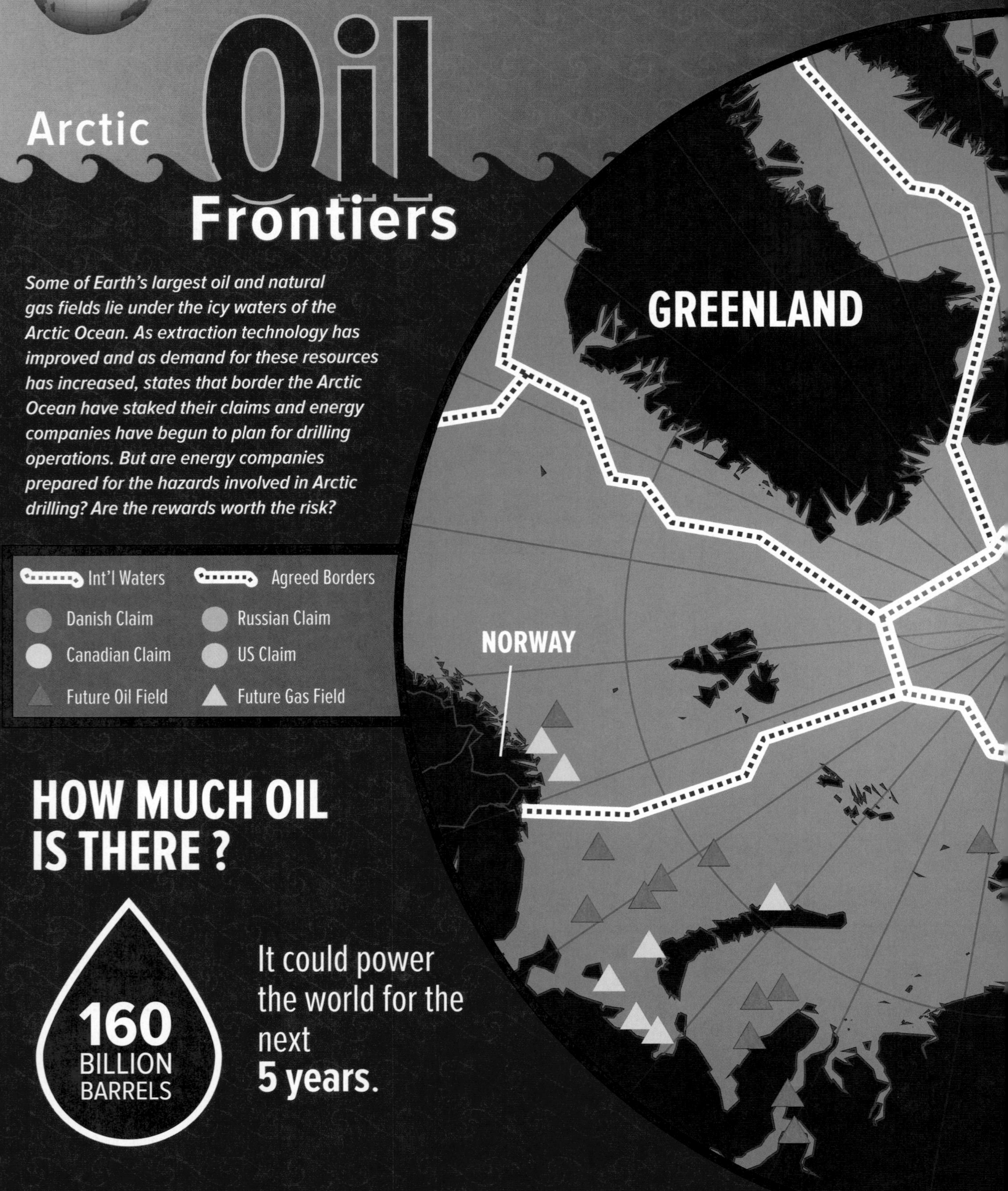

- Int'l Waters
- Agreed Borders
- Danish Claim
- Russian Claim
- Canadian Claim
- US Claim
- Future Oil Field
- Future Gas Field

HOW MUCH OIL IS THERE ?

160 BILLION BARRELS

It could power the world for the next **5 years.**

GREENLAND

NORWAY

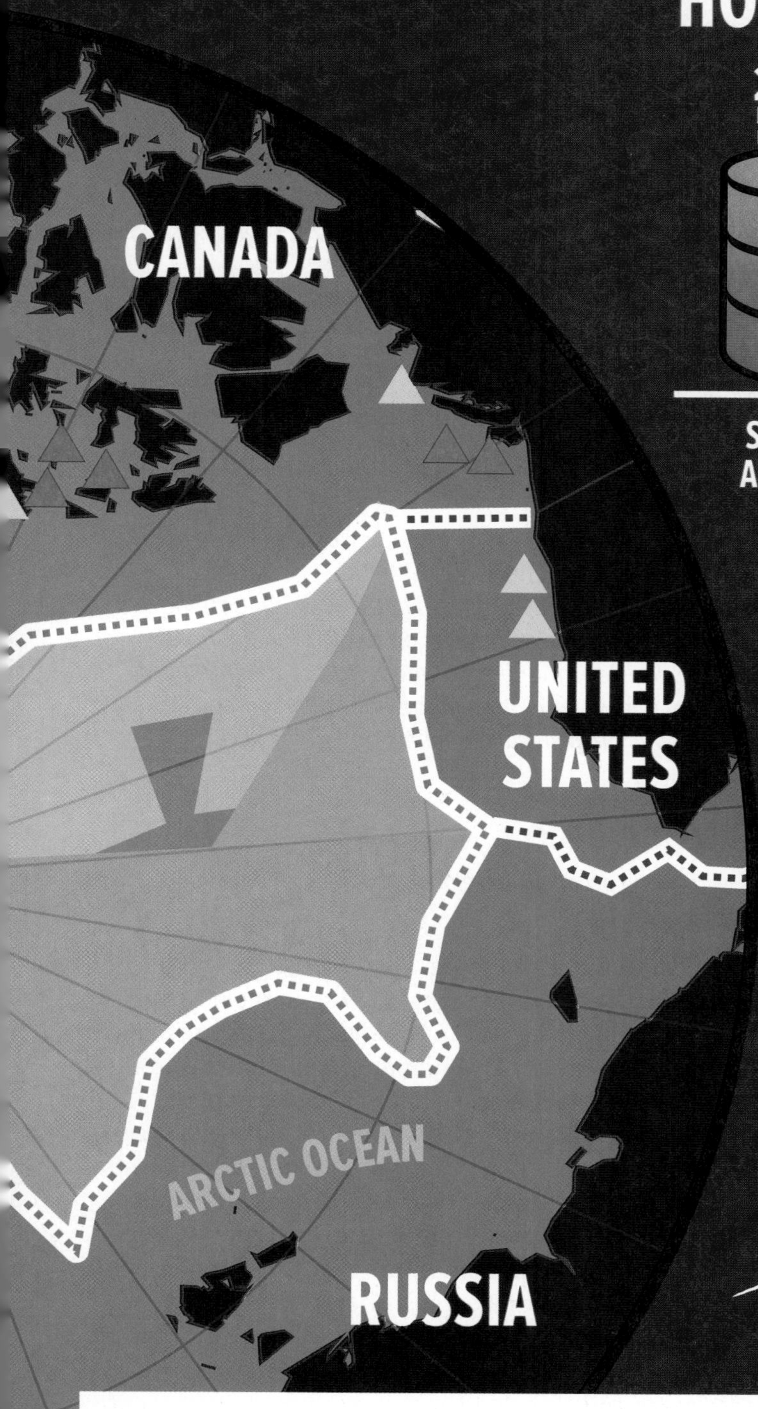

HOW DOES IT COMPARE?

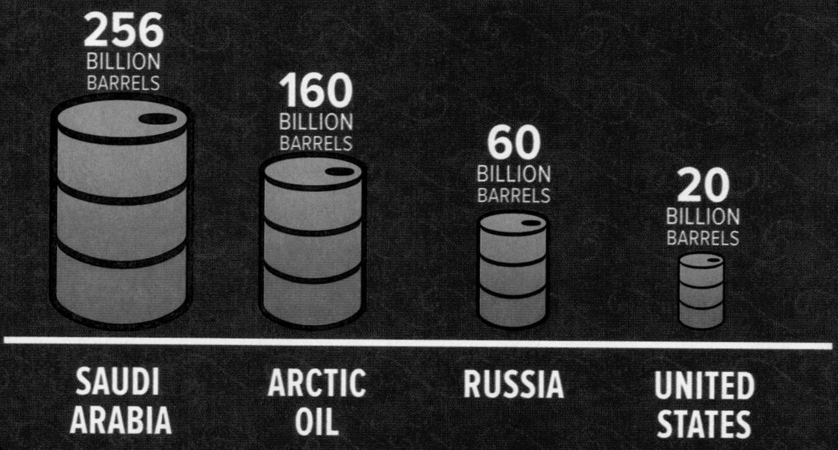

256 BILLION BARRELS

160 BILLION BARRELS

60 BILLION BARRELS

20 BILLION BARRELS

| SAUDI ARABIA | ARCTIC OIL | RUSSIA | UNITED STATES |

CANADA

UNITED STATES

ARCTIC OCEAN

RUSSIA

IS IT SAFE TO EXTRACT?

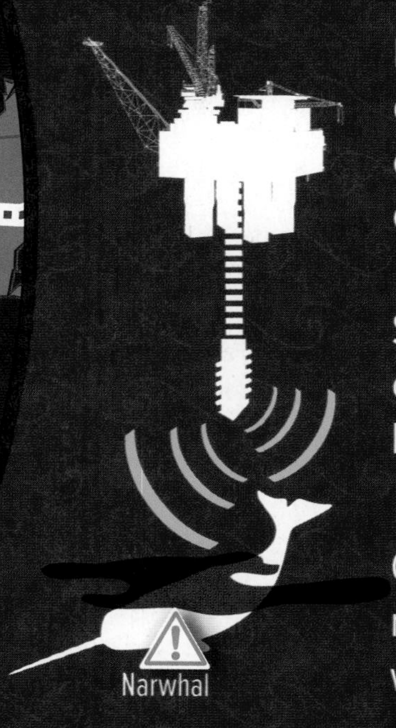

Extensive construction could alter the ecologically pristine environment.

Sound vibrations could disturb marine habitats.

Narwhal

Oil spills could harm marine and coastal wildlife.

Making Connections

1. **Speculating** Besides harm to the environment, what other risks with Arctic drilling can you foresee?

2. **Human Systems** What patterns do you notice about the claims made on the Arctic Ocean by bordering states? Explain why you think these claims are fair or unfair.

3. **Considering Advantages and Disadvantages** Using what you know about the risks and rewards of Arctic drilling, write a one-page summary statement explaining why you think plans for drilling in the Arctic Ocean are positive or negative.

*Interact with **Global Connections** Online*

There's More Online!

☑ **IMAGE** Grassroots Efforts

☑ **MAP** Environmental Issues in the Russian Core

☑ **MAP** Extent/Reach of Chernobyl Disaster

☑ **INTERACTIVE SELF-CHECK QUIZ**

☑ **VIDEO** People and Their Environment: The Russian Core

Reading HELPDESK

Academic Vocabulary

- **radical**
- **stable**

Content Vocabulary

- **nuclear wastes**
- **radioactive material**
- **pesticide**

TAKING NOTES: *Key Ideas and Details*

SUMMARIZING Use a chart like the one below to take notes about the human impact on the environment in the Russian Core.

The Human Impact on the Environment: The Russian Core	
Industrialization	Nuclear Wastes
Forest Destruction	Global Warming

LESSON 3
People and Their Environment: The Russian Core

ESSENTIAL QUESTION • *How do physical systems and human systems shape a place?*

IT MATTERS BECAUSE
Industrialization and economic progress in the Russian Core have had both positive and negative impacts on the subregion. Russia has to deal with considerable past damage to the environment. At the same time, it is working to manage natural resources and encourage economic growth while minimizing further harm.

Managing Resources

GUIDING QUESTION *How do economic development and environmental protection cause conflict in the Russian Core?*

In general, Russia's economy is expanding, with robust growth in the areas of manufacturing and services and rapid growth in the oil and gas sector. Over time, this expansion has led to conflicts with organizations seeking to protect the environment and human populations from the negative aspects of industrialization and economic development.

Russia's most important industry—petroleum extraction and processing—has made the country one of the world's largest producers of crude oil. It is no secret that Russia, by Arctic drilling, is trying to access the vast minerals, oil, and gas that lie beneath the Arctic Ocean. The area north of the Arctic Circle contains more than 90 billion barrels of oil and more than 1.5 trillion cubic feet of gas. There is much resistance, however, as environmentalist groups such as Greenpeace are protesting the drilling. Additionally, pipelines built to transport oil and gas pass through wilderness areas and threaten the surrounding environment. In 2006 Russia began constructing a highly controversial pipeline to carry oil from eastern Siberia to the Pacific Ocean. It will bring Russia billions of dollars from countries in the Asia-Pacific region. The pipeline will pass through a protected wilderness area near Lake Baikal (by•KAWL). President Vladimir Putin ordered that the proposed route be diverted farther away from the lake, but environmentalists still fear the irreversible damage that could be caused by an oil spill.

Another source of conflict is the subregion's fishing industry. Fish are a crucial component to the Russian diet and economy. Salmon from the Pacific Ocean and herring, cod, and halibut from the Arctic Ocean support a flourishing fishing industry. Russia's ocean fishing fleet contains what are known as supertrawlers that tow huge trawl nets—large enough to scoop up a whale. The ships can catch and process more than 40 tons (36 t) of fish a day. Because supertrawlers want only certain kinds of fish, everything else hauled up in the nets gets discarded. Millions of fish and other marine animals die unnecessarily every year as a result of supertrawling, and there are organizations working actively to stop this practice.

Another fish product from the region is the world-famous Russian caviar that comes from the eggs of sturgeon fish that inhabit freshwater rivers and lakes. The sturgeon population has declined, however. It is in danger because of dams built on the Volga River, which have interrupted the migration and disturbed the habitat of these fish.

☑ READING PROGRESS CHECK

Assessing Why is a new pipeline linking eastern Siberia to the Asia-Pacific region controversial?

Human Impact

GUIDING QUESTION *What impact did Soviet-era ideas and actions have on the environment of the Russian Core?*

The environmental damage caused by Soviet-era industrialization, including the region's affiliation with nuclear technologies, continues to pose risks to natural resources and human health. The Soviets' disregard for the environmental effects of industrialization has damaged Russia's water, air, soil, and forests.

Environmental Issues in the Russian Core

Legend:
- ☢ Nuclear contamination
- Industrial pollution
- ◊ Water pollution
- 🌲 Extensive deforestation

Two-Point Equidistant projection

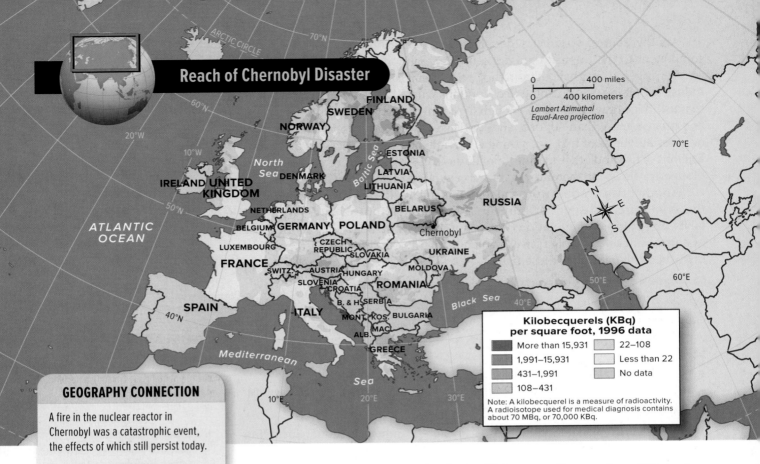

Reach of Chernobyl Disaster

0 — 400 miles
0 — 400 kilometers
Lambert Azimuthal Equal-Area projection

Kilobecquerels (KBq) per square foot, 1996 data

More than 15,931		22–108	
1,991–15,931		Less than 22	
431–1,991		No data	
108–431			

Note: A kilobecquerel is a measure of radioactivity. A radioisotope used for medical diagnosis contains about 70 MBq, or 70,000 KBq.

GEOGRAPHY CONNECTION

A fire in the nuclear reactor in Chernobyl was a catastrophic event, the effects of which still persist today.

1. **THE WORLD IN SPATIAL TERMS** Which countries outside the Russian Core had Kbq levels of 431–1,991 or greater?

2. **PHYSICAL SYSTEMS** Based on the data shown, which way were the winds blowing that carried the nuclear fallout?

nuclear wastes by-products of producing nuclear power and weapons

radioactive material material contaminated by residue from the generation of nuclear energy and weapons

radical fundamental or extreme; drastic

Between 1949 and 1987, the Soviet Union set off more than 600 nuclear explosions to test and improve its nuclear technologies and weapons. Soviets developed and then stockpiled nuclear weapons throughout the Cold War. Today, the condition and fate of those weapons concern Russia and the rest of the world. **Nuclear wastes** are the by-products of producing nuclear power and weapons. Some nuclear wastes can remain radioactive for thousands of years, posing danger to people and the environment. The Soviets placed most nuclear wastes in storage facilities, but some **radioactive materials**—materials contaminated by residue from the generation of nuclear energy and weapons—were dumped directly into the Barents, Baltic, and Bering Seas. **Radical** changes in these practices have been slow in coming.

The first significant environmental disaster in the Soviet Union related to nuclear technologies occurred in 1986. A fire in a nuclear reactor in the town of Chernobyl (chuhr•NOH•buhl), 60 miles (97 km) north of Kiev, Ukraine, released tons of radioactive particles into the local environment. The amount of radioactivity released was 400 times more than what was released when the nuclear bomb was dropped on Hiroshima, Japan, in 1945. Radiation covered tens of thousands of square miles of farmland and forests in the Soviet republics of Belarus, Ukraine, and Russia. Because of prevailing winds, other countries suffered as well. Millions of people were exposed to deadly levels of radiation because Soviet officials were slow to alert the public to the crisis and did not evacuate people soon enough. Thousands of people died as a direct result of radiation poisoning. Tens of thousands more continue to suffer from cancer, stomach diseases, cataracts, and immune system disorders. Approximately 350,000 people were displaced from their homes. Today, there is a tightly controlled exclusion zone surrounding the Chernobyl Nuclear Power Plant.

After the accident, international pressure prompted Soviet leaders to improve nuclear safety standards and to shut down dangerous plants. Despite concerns, 35 nuclear reactors continue to provide some of the country's electricity. Experts think that many remaining Soviet-era reactors are poorly designed and unsafe. Russia plans to expand its nuclear power industry by building more reactors and new power plants. In 2012 over 10 percent of Russia's total electricity output was generated through radioactive decay of nuclear fuel, and Ukraine ranked ninth in the world with about 24 percent of its electricity produced by nuclear fuels.

In addition to the impact of nuclear technologies on the Russian Core, the region's industrialization—which began during the Soviet era and continues today—has polluted much of Russia's lakes, rivers, and soils. Fertilizer runoff, sewage, and radioactive material all contribute to poor water quality. The waters of the Moskva and Volga Rivers pose health risks, and dams along the Volga River trap contaminated water. Pollution also threatens the Caspian Sea.

Lake Baikal, the world's deepest and oldest lake, called the Pearl of Siberia, is considered a natural wonder with more than 1,500 different native species of aquatic plants and animals. In 1957 the Soviet Union announced a plan to build a paper-pulp factory along Lake Baikal's shores. Although this plan was opposed by people in the area, their protests were ignored and the factory was built. This factory and others that followed dumped industrial waste into the lake.

For decades, toxic waste dumps and airborne pollution have also posed problems for Russia's soil. In the rush to industrialize, little effort had been made to store toxic wastes properly. Thus, aging storage containers cracked and toxic wastes leaked into the soil. Petroleum pipelines often broke, allowing petroleum to ruin the land. In addition, overuse of fertilizers and **pesticides**—chemicals used to kill crop-damaging insects, rodents, and other pests—pollute farmland and water.

Russia's forests are also at risk from harmful practices. About one-fifth of the world's forest lands lie in Russia—75 percent of them in Siberia. Second only to the Amazon rain forest in the amount of oxygen returned to the atmosphere, the Russian boreal, or northern, forest also supplies much of the world's timber. The boreal is the southern part of the taiga biome. As a result of commercial logging, illegal poaching of trees, and wildfires, however, Russian forests shrink by almost 40 million acres (16 million ha) each year—a rate of loss higher than that of the Amazon Basin.

PRIMARY SOURCE

❝ If the tropical forests, which contain half the planet's woodlands, are one lung of the Earth, then the boreal forest is the other. Both play a vital role in regulating climate as they—along with the ocean, Earth's largest carbon repository—filter out billions of tons of carbon dioxide and other greenhouse gases during photosynthesis, storing the carbon in trees, roots, and soils. ❞

—Fen Montaigne, "The Great Northern Forest," *National Geographic*, June 2002

The widespread effects of global warming are becoming visible in western Siberia because climate change more readily affects temperatures at higher northern latitudes. The warming is likely to produce an unprecedented thawing of the world's largest peat bog that would release into the atmosphere billions of metric tons of methane, a powerful greenhouse gas. Where permafrost once covered the subarctic regions of western Siberia, shallow lakes are now more prevalent. In addition, air pollution has contributed to increased temperatures in both the subregion and around the world. The human population is also affected in several ways. Increased health risks due to heat waves and changes in infectious diseases are present. In addition, activities such as hunting and travel over snow and ice are affected.

✔ READING PROGRESS CHECK

Evaluating What factors contribute to poor air quality in Russia?

Analyzing
PRIMARY SOURCES

The Long Shadow of Chernobyl

"Early estimates that tens or hundreds of thousands of people would die from Chernobyl have been discredited. But genetic damage done 20 years ago is slowly taking a toll. No one can be sure of the ultimate impact, but an authoritative report estimated last year that the cancer fuse lit by Chernobyl will claim 4,000 lives. Alexei Okeanov of the International Sakharov Environmental University in Minsk, Belarus, who studies the health effects of the accident, calls it 'a fire that can't be put out in our lifetimes.'"

—Richard Stone, "The Long Shadow of Chernobyl," *National Geographic*, April 2006

DBQ **INTERPRETING** How does the description of the Chernobyl disaster as a lit fuse and a "fire that can't be put out" help you understand the problem?

pesticide chemicals used to kill crop-damaging insects, rodents, and other pests

Driving vehicles fueled by electricity and natural gas provides a way for Russian citizens to help in reducing carbon emissions.

▲ CRITICAL THINKING

1. Identifying Cause and Effect How does driving an electric or natural gas-fueled vehicle help protect the peat bogs in western Siberia?

2. Assessing In what ways can governments help to increase the use of electric and natural gas-fueled vehicles?

stable not changing or fluctuating

Addressing the Issues

GUIDING QUESTION *How are people and the governments in the Russian Core working to change ideas about the environment?*

The balance between using natural resources and preserving the environment can be viewed from several different perspectives. Solutions that are available are environmental protection and wise management through monitoring pollution levels, conservation, and international regulation. For example, people have come together to oppose a mining operation in remote Kamchatka in eastern Russia. Environmental groups have demanded that the mining company meet strict environmental standards. The possible threat to the area's salmon spawning grounds prompted the local fishing industry to support the effort. The mine also caused concern among local residents because it was close to a protected wildlife area. Even with growing environmental awareness, economic pressure continues to open other sensitive regions to development.

In September 2012, a British parliamentary committee called for a halt to drilling in the Arctic Ocean until necessary steps are taken to protect the region from the potentially catastrophic consequences of an oil spill. Another successful outside effort was Greenpeace's organization of an independent forum on the protection of sites in the Russian Core. Other outside groups are working to protect wildlife and the region's ecosystems.

In the case of Lake Baikal, all was not lost. Efforts from concerned citizens and environmental organizations have met with moderate success. In response to ongoing protests, the most serious polluters have been closed, and others are working to reduce pollution. Pollution levels in the lake are now relatively low compared with many lakes in Europe.

Additionally, the World Bank's Sustainable Forestry Pilot project is helping Russia manage its forests. Using land more wisely, protecting forests, planting new trees, and increasing private investment all help Russia's environment and economy. Increased employment opportunities in the forest industry and more **stable** economies will be possible only if steps are taken to conserve the forests.

✓ READING PROGRESS CHECK

Analyzing How is Russia trying to reverse past damage to its natural resources as well as manage them responsibly today?

LESSON 3 REVIEW

Reviewing Vocabulary

1. *Explaining* Explain the difference between nuclear waste and radioactive material.

Using Your Notes

2. *Describing* Use your chart to describe the major threats to the forests of the Russian Core.

Answering the Guiding Questions

3. *Explaining* How do economic development and environmental protection cause conflict in the Russian Core?

4. *Discussing* What impact did Soviet-era ideas and actions have on the environment of the Russian Core?

5. *Explaining* How are people and governments in the Russian Core working to change ideas about the environment?

Writing Activity

6. *Informative/Explanatory* Think about the challenges Russia faces concerning water quality. Write a paragraph explaining why Russians do not use more water from Lake Baikal to supply their freshwater needs.

Andrey Rudakov/Bloomberg/Getty Images

Directions: On a separate sheet of paper, answer the questions below. Make sure you read carefully and answer all parts of the questions.

Lesson Review

Lesson 1

1 *Describing* Describe Russia's midlatitude climate regions. Why do such climate regions support most of the country's agricultural production?

2 *Explaining* How do Russia's climates and short growing season affect food production?

3 *Assessing* Discuss the importance of rivers in the Russian Core.

Lesson 2

4 *Evaluating* What area of Russia has the greatest population density? Why is the population concentrated there?

5 *Drawing Inferences* Why were satellite states eager to declare their independence from the Soviet Union in 1991? Provide an example.

6 *Summarizing* How did the fall of the Soviet Union affect religion in Russia?

Lesson 3

7 *Discussing* How are Russia's fishing practices affecting the fish populations in the oceans and rivers of the Russian Core? Provide specific examples.

8 *Exploring the Issues* Describe the Chernobyl disaster and how it affected the people of the region.

9 *Explaining* How has industrialization of the Russian Core damaged the environment? What steps have local and international governments taken to combat the problem?

Critical Thinking

10 *Identifying Central Issues* What actions has Russia taken to become part of the global economy?

11 *Identifying Cause and Effect* How did the transition from a command economy to a market economy affect the Russian people?

12 *Explaining* How did the migration of the Slavs and their interactions with other peoples influence the history of Russia?

13 *Exploring the Issues* Discuss one way in which Russia is working with international agencies to manage its forests.

Research and Presentation

14 *Research Skills* Use the Internet and other resources to investigate Russian and American cooperative involvement with the International Space Station. What have both countries contributed to this project? Who are some of the key individuals?

21st Century Skills

Telephone Use in Russia (2014)		
	Main Lines	**Mobile Cellular**
Number of Telephones	39.4 million	221 million
World Ranking	7	7

Source: CIA World Factbook

Use the chart to answer the following questions.

15 *Assessing* How does the number of main lines compare to the number of cellular users in Russia?

16 *Speculating* How might the disparity in cell phone usage compared to main line usage be a reflection of the population structure of Russia today?

Need Extra Help?

If You've Missed Question	**1**	**2**	**3**	**4**	**5**	**6**	**7**	**8**	**9**	**10**	**11**	**12**	**13**	**14**	**15**	**16**
Go to page	339	338	337	334	343	344	351	352	353	347	346	341	354	343	355	355

Directions: On a separate sheet of paper, answer the questions below. Make sure you read carefully and answer all parts of the questions.

DBQ Analyzing Primary Sources

Use the document to answer the following questions.

PRIMARY SOURCE

"As director of the Phoenix Fund, a small, environmental NGO that he has headed for 12 years, [Sergei] Bereznuk and his team of six people are carrying out an impressive range of activities to preserve the Amur [Siberian] tiger over a territory of 166,000 km². These include support of anti-poaching units, awareness-raising among local people, reversing habitat reduction due to fires and logging and resolution of human-animal conflicts, along with providing compensation for damage and monitoring invasive industrial projects in the region.

'Poaching remains the principal threat to the tigers' survival,' Rolex adds. The animals are killed in retaliation, mainly for loss of cattle and wild prey and as hunting trophies. There is also demand for their skin, bones and body parts, used primarily in Chinese traditional medicine. Despite international laws banning the sale of tiger parts there is a lucrative market that fuels poaching. In their campaign to reduce the slaughter, Bereznuk and the Vladivostok-based Phoenix Fund provide anti-poaching teams with software—the Management Information System (MIST)—developed specifically for this purpose by the Wildlife Conservation Society. Up-to-date, relevant and timely information is an integral part of effective protected area management."

—David Braun, *National Geographic Daily News*, November 25, 2012

17 *Exploring the Issues* What are the major threats to the Siberian tiger?

18 *Interpreting* In what ways does technology assist the protection efforts of the Phoenix Fund?

Exploring the Essential Question

19 *Making Connections* Write a paragraph describing how the landforms of the Russian Core have helped shape it as a place.

Applying Map Skills

Use the Unit 4 Atlas to answer the following questions.

20 *Human Systems* What rivers serve as links to the Arctic Ocean in the Russian Core?

21 *Environment and Society* What natural resources are most abundant in the Russian Core?

22 *Physical Systems* Sketch your mental map of Russia. Then label the tundra climate region and label the major lines of latitude.

College and Career Readiness

23 *Examining Information* Use the Internet and other resources to research education in Russia today and how it compares with education during the Soviet era. Write a few paragraphs comparing and contrasting education during the two time periods.

24 *Change and Continuity of Groups* As a social scientist conducting research in the Russian Core, you have been asked to write a brief report on the causes of demographic decline. Can this decline be reversed? What is being done to solve this critical problem?

Writing About Geography

25 *Informative/Explanatory* Use standard grammar, spelling, sentence structure, and punctuation to write a one-page essay considering how physical geography has influenced culture in Russia.

Need Extra Help?

If You've Missed Question	**17**	**18**	**19**	**20**	**21**	**22**	**23**	**24**	**25**
Go to page	356	356	336	242	245	244	345	344	341

North Africa, Southwest Asia, and Central Asia

UNIT 5

| **Chapter 15** North Africa | **Chapter 16** The Eastern Mediterranean | **Chapter 17** The Northeast | **Chapter 18** The Arabian Peninsula | **Chapter 19** Central Asia |

1 Culture Southwest Asia is the birthplace of three major world religions—Christianity, Islam, and Judaism. Islam is the dominant religion in the region, as Muslims form the majority in most countries.

EXPLORE the REGION

Stretching from the Atlantic Ocean in the west to the borders of China in the east, **NORTH AFRICA, SOUTHWEST ASIA,** and **CENTRAL ASIA** are three distinct but similar subregions. This is a vast area of contrasts, with plentiful oil and scarce water, bustling cities and large expanses of uninhabited land, soaring mountains and flat plains, ancient landmarks and modern technology, extreme wealth and struggling poverty.

3 Deserts   The Rub' al-Khali covers more than 250,000 square miles (650,000 sq. km) and is one of the major deserts found in the region.

4 History   Home to ancient kingdoms and empires that rose and fell thousands of years ago, the region has left a lasting mark on history as the source of such common practices as farming wheat and writing with an alphabet.

2 Oil   Much of the energy used around the world comes from this region, which includes six of the top ten countries with the most oil reserves and five of the top ten oil producers.

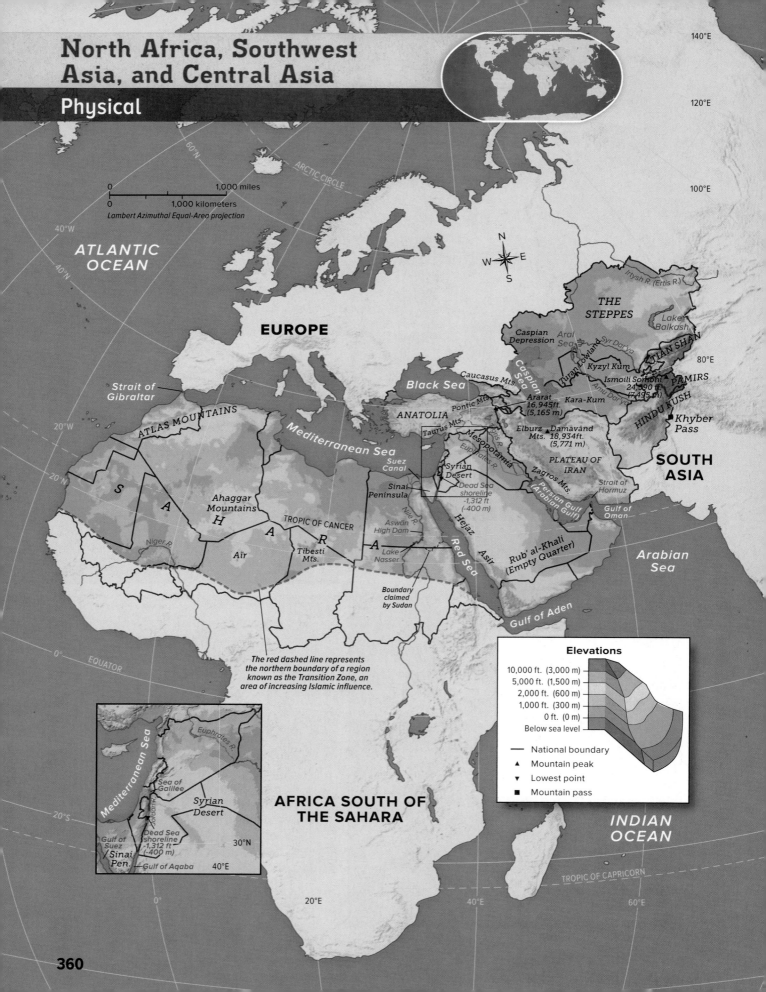

North Africa, Southwest Asia, and Central Asia
Physical

140°E
120°E
100°E

ARCTIC CIRCLE

0   1,000 miles
0   1,000 kilometers
Lambert Azimuthal Equal-Area projection

40°W

ATLANTIC
OCEAN

60°N

40°N

N
W   E
S

THE
STEPPES

Irtysh R. (Ertis R.)

Lake
Balkash

EUROPE

Caspian
Depression

*Aral
Sea*

Turan Lowland

Syr Darya

Kyzyl Kum

TIAN SHAN

80°E

Black Sea

Caucasus Mts.

Caspian
Sea

Ararat
16,945 ft.
(5,165 m)

Kara-Kum

Ismoili Somoni
24,590 ft.
(7,495 m)

PAMIRS

Amu Darya

HINDU KUSH

Strait of
Gibraltar

ATLAS MOUNTAINS

20°W

Mediterranean Sea

ANATOLIA

Pontic Mts.

Taurus Mts.

Tigris R.

Mesopotamia

Euphrates R.

Elburz
Mts.

Damāvānd
18,934 ft.
(5,771 m)

PLATEAU OF
IRAN

Zagros Mts.

SOUTH
ASIA

Khyber
Pass

20°N

S
A
H
A
R
A

Ahaggar
Mountains

TROPIC OF CANCER

Niger R.

Aïr

Tibesti
Mts.

Suez
Canal

Sinai
Peninsula

Syrian
Desert

Dead Sea
shoreline
-1,312 ft
(-400 m)

Nile R.

Aswān
High Dam

Lake
Nasser

Hejaz

Red Sea

Asir

Persian Gulf
(Arabian Gulf)

Strait of
Hormuz

Gulf of
Oman

Arabian
Sea

Rub' al-Khali
(Empty Quarter)

Boundary
claimed
by Sudan

Gulf of Aden

0°

EQUATOR

The red dashed line represents
the northern boundary of a region
known as the Transition Zone, an
area of increasing Islamic influence.

Elevations

10,000 ft. (3,000 m)
5,000 ft. (1,500 m)
2,000 ft. (600 m)
1,000 ft. (300 m)
0 ft. (0 m)
Below sea level

—— National boundary
▲ Mountain peak
▼ Lowest point
■ Mountain pass

20°S

Mediterranean Sea

Euphrates R.

Sea of
Galilee

Syrian
Desert

Jordan R.

Gulf of
Suez

Dead Sea
shoreline
-1,312 ft
(-400 m)

Sinai
Pen.

Gulf of Aqaba

30°N

40°E

AFRICA SOUTH OF
THE SAHARA

INDIAN
OCEAN

TROPIC OF CAPRICORN

20°E
40°E
60°E

North Africa, Southwest Asia, and Central Asia
Political

Capital city
City

0   1,000 miles
0   1,000 kilometers
Lambert Azimuthal Equal-Area projection

ATLANTIC OCEAN

ARCTIC CIRCLE

EUROPE

100°E

Astana ⊛
Lake Balkash
Irtysh R. (Ertis R.)
KAZAKHSTAN
Almaty •
Aral Sea
Syr Darya
Bishkek ⊛
KYRGYZSTAN
80°E
Tashkent ⊛
UZBEKISTAN
TAJIKISTAN
Dushanbe ⊛
Amu Darya
TURKMENISTAN
Ashkhabad ⊛
Kabul ⊛
AFGHANISTAN

Black Sea
GEORGIA
İstanbul •
Tbilisi ⊛
Yerevan ⊛ Baku ⊛
ARMENIA
Ankara ⊛
TURKEY
AZERBAIJAN
Caspian Sea
Tehran ⊛
Mashhad •
Baghdad ⊛
IRAQ
IRAN
Shīrāz •
Mashhad
SOUTH ASIA

Strait of Gibraltar
Rabat ⊛
Casablanca •
Oran •
Algiers ⊛
Tunis •
MOROCCO
TUNISIA
Mediterranean Sea
Tripoli •
Laayoune •
WESTERN SAHARA (Morocco)
20°W
ALGERIA
LIBYA
Benghazi •
Alexandria • Cairo ⊛
EGYPT
Nile R.
KUWAIT
Kuwait ⊛
UNITED ARAB EMIRATES
Persian Gulf (Arabian Gulf)
Manama ⊛
BAHRAIN
Doha ⊛
QATAR
Abu Dhabi ⊛
Masqat ⊛
Gulf of Oman

MAURITANIA
Nouakchott ⊛
MALI
Tombouctou (Timbuktu) •
TROPIC OF CANCER
Luxor •
Red Sea
Jidda •
Riyadh ⊛
SAUDI ARABIA
OMAN
Arabian Sea

BURKINA FASO
NIGER
CHAD
Boundary claimed by Sudan
ERITREA
Sanaa ⊛
YEMEN
Aden •
Gulf of Aden
60°E
INDIAN OCEAN

SUDAN

EQUATOR

The red dashed line represents the northern boundary of a region known as the Transition Zone, an area of increasing Islamic influence.

TURKEY
Aleppo •
SYRIA
Euphrates R.
Tripoli •
Homs •
Beirut ⊛
LEBANON
Damascus ⊛
IRAQ
Tel Aviv-Jaffa •
Jerusalem ⊛
Amman ⊛
GAZA STRIP
WEST BANK
ISRAEL
JORDAN
SAUDI ARABIA
EGYPT
30°N
40°E

AFRICA SOUTH OF THE SAHARA

20°S
30°N

TROPIC OF CAPRICORN

20°E
0°
40°E

UNIT 5
REGIONAL ATLAS

MAP STUDY

1. *Environment and Society* How would you describe the elevation of the region? What impact would it have on human settlement patterns?

2. *Human Systems* Which countries have areas that are part of the African Transition Zone?

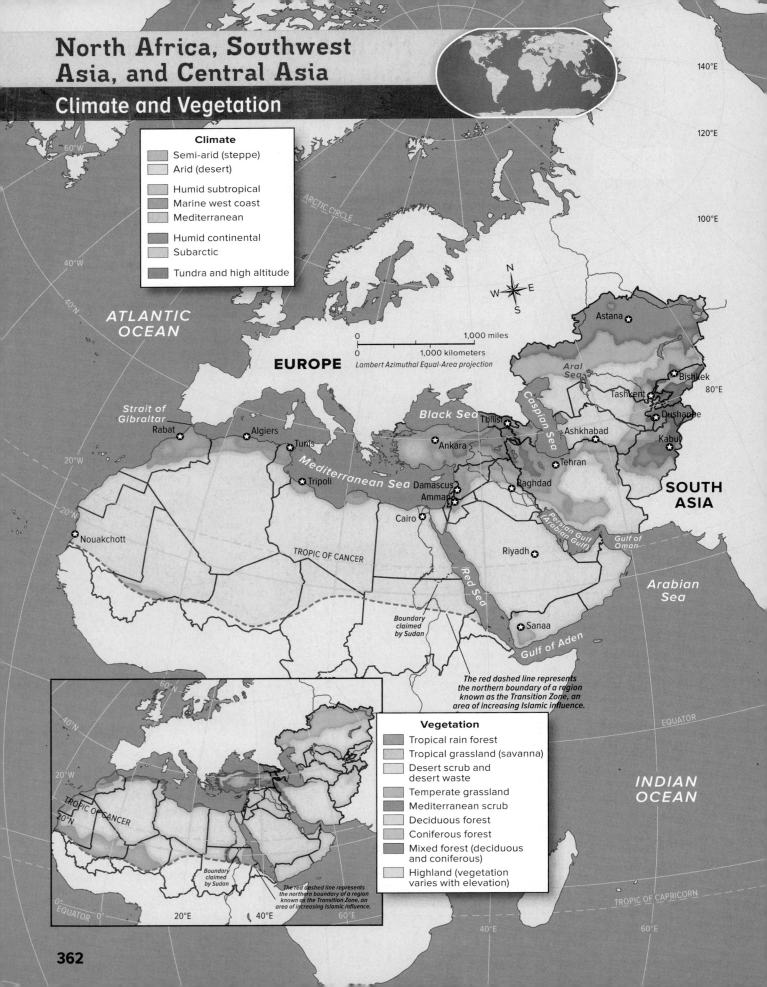

North Africa, Southwest Asia, and Central Asia

Climate and Vegetation

Climate

- Semi-arid (steppe)
- Arid (desert)
- Humid subtropical
- Marine west coast
- Mediterranean
- Humid continental
- Subarctic
- Tundra and high altitude

ATLANTIC OCEAN

ARCTIC CIRCLE

140°E

120°E

100°E

EUROPE

0   1,000 miles
0   1,000 kilometers
Lambert Azimuthal Equal-Area projection

60°W

40°W

20°W

60°N

40°N

20°N

Strait of Gibraltar

Rabat

Algiers

Tunis

Tripoli

Mediterranean Sea

Nouakchott

TROPIC OF CANCER

Cairo

Red Sea

Black Sea

Ankara

Damascus
Amman

Baghdad

Riyadh

Sanaa

Gulf of Aden

Tbilisi

Caspian Sea

Ashkhabad

Tehran

Aral Sea

Astana

Bishkek

Tashkent

Dushanbe

Kabul

80°E

SOUTH ASIA

Persian Gulf (Arabian Gulf)

Gulf of Oman

Arabian Sea

Boundary claimed by Sudan

The red dashed line represents the northern boundary of a region known as the Transition Zone, an area of increasing Islamic influence.

EQUATOR

INDIAN OCEAN

TROPIC OF CAPRICORN

40°E

60°E

Vegetation

- Tropical rain forest
- Tropical grassland (savanna)
- Desert scrub and desert waste
- Temperate grassland
- Mediterranean scrub
- Deciduous forest
- Coniferous forest
- Mixed forest (deciduous and coniferous)
- Highland (vegetation varies with elevation)

60°N

40°N

20°N

TROPIC OF CANCER

20°W

0°

20°E

40°E

60°E

0° EQUATOR

Boundary claimed by Sudan

The red dashed line represents the northern boundary of a region known as the Transition Zone, an area of increasing Islamic influence.

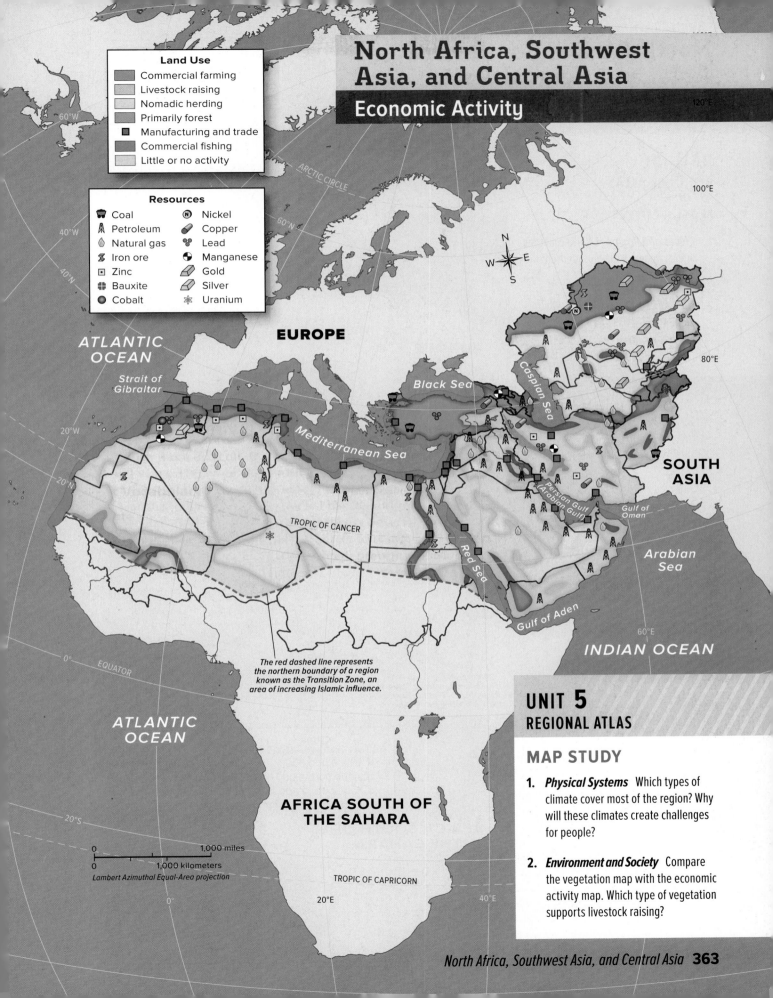

North Africa, Southwest Asia, and Central Asia

Economic Activity

120°E

100°E

80°E

Land Use
- Commercial farming
- Livestock raising
- Nomadic herding
- Primarily forest
- Manufacturing and trade
- Commercial fishing
- Little or no activity

Resources
- Coal
- Petroleum
- Natural gas
- Iron ore
- Zinc
- Bauxite
- Cobalt
- Nickel
- Copper
- Lead
- Manganese
- Gold
- Silver
- Uranium

ARCTIC CIRCLE

60°N

EUROPE

ATLANTIC OCEAN

Strait of Gibraltar

Black Sea

Caspian Sea

Mediterranean Sea

SOUTH ASIA

TROPIC OF CANCER

Persian Gulf (Arabian Gulf)

Gulf of Oman

Red Sea

Arabian Sea

Gulf of Aden

INDIAN OCEAN

EQUATOR

ATLANTIC OCEAN

The red dashed line represents the northern boundary of a region known as the Transition Zone, an area of increasing Islamic influence.

AFRICA SOUTH OF THE SAHARA

1,000 miles

1,000 kilometers

Lambert Azimuthal Equal-Area projection

TROPIC OF CAPRICORN

20°E

40°E

UNIT 5
REGIONAL ATLAS

MAP STUDY

1. **Physical Systems** Which types of climate cover most of the region? Why will these climates create challenges for people?

2. **Environment and Society** Compare the vegetation map with the economic activity map. Which type of vegetation supports livestock raising?

North Africa, Southwest Asia, and Central Asia **363**

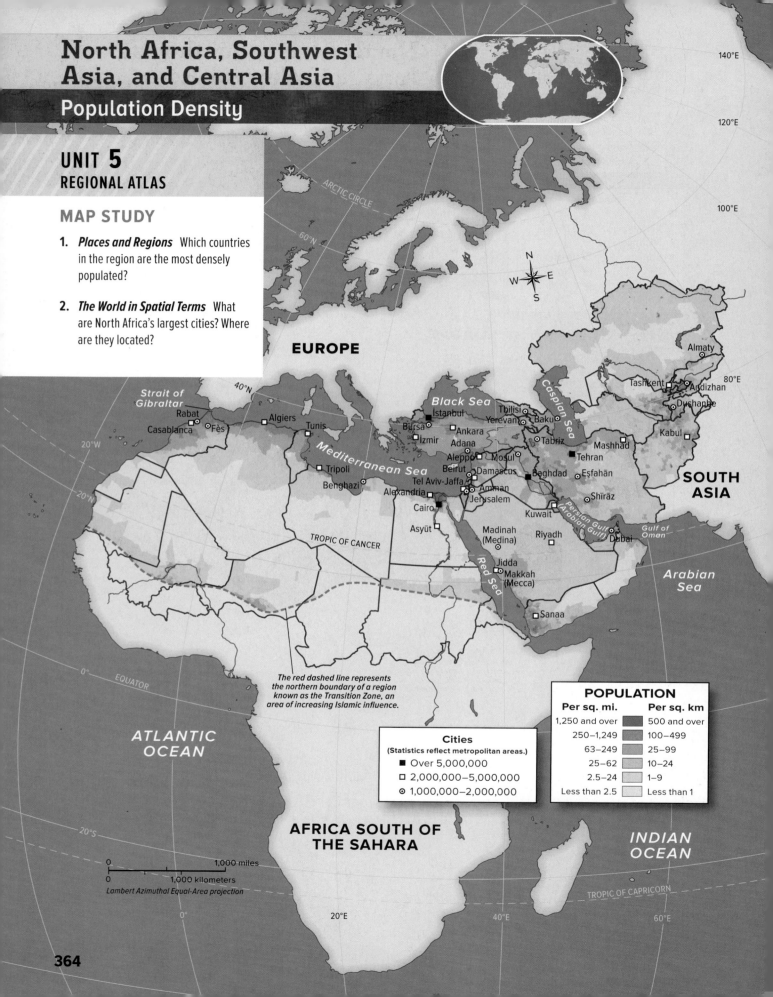

North Africa, Southwest Asia, and Central Asia
Population Density

UNIT 5
REGIONAL ATLAS

MAP STUDY

1. *Places and Regions* Which countries in the region are the most densely populated?

2. *The World in Spatial Terms* What are North Africa's largest cities? Where are they located?

EUROPE

ARCTIC CIRCLE

60°N

40°N

Strait of Gibraltar

Rabat
Casablanca
Fès
Algiers
Tunis

Black Sea

İstanbul
Bursa
Ankara
İzmir
Adana
Aleppo
Mosul
Beirut
Damascus
Tel Aviv-Jaffa
Jerusalem
Amman

Tbilisi
Yerevan
Baku
Tabriz

Caspian Sea

Tashkent
Andizhan
Dushanbe
Kabul

Almaty

80°E

Mashhad
Tehran
Eşfahān
Baghdad
Shīrāz

SOUTH ASIA

Mediterranean Sea

Tripoli
Benghazi
Alexandria
Cairo
Asyūt

20°W

20°N

TROPIC OF CANCER

Madinah (Medina)
Riyadh

Kuwait

Persian Gulf (Arabian Gulf)

Dubai

Gulf of Oman

Arabian Sea

Red Sea

Jidda
Makkah (Mecca)

Sanaa

The red dashed line represents the northern boundary of a region known as the Transition Zone, an area of increasing Islamic influence.

0°
EQUATOR

ATLANTIC OCEAN

AFRICA SOUTH OF THE SAHARA

20°S

INDIAN OCEAN

TROPIC OF CAPRICORN

Cities
(Statistics reflect metropolitan areas.)

- ■ Over 5,000,000
- □ 2,000,000–5,000,000
- ⊙ 1,000,000–2,000,000

POPULATION

Per sq. mi.		Per sq. km
1,250 and over		500 and over
250–1,249		100–499
63–249		25–99
25–62		10–24
2.5–24		1–9
Less than 2.5		Less than 1

0 — 1,000 miles
0 — 1,000 kilometers
Lambert Azimuthal Equal-Area projection

20°E
40°E
60°E
0°

North Africa

ESSENTIAL QUESTION • *How do physical systems and human systems shape a place?*

Why Geography Matters
Choke Point: Suez Canal

Lesson 1
Physical Geography of North Africa

Lesson 2
Human Geography of North Africa

Lesson 3
People and Their Environment: North Africa

Geography Matters...

North Africa is home to the Sahara, the world's largest hot desert, and the Nile Delta, the outflow of the world's longest river. North Africa also has access to thousands of miles of coastline and many favorable trade routes. There is a long history of trade across the subregion, with various ethnic groups moving in and out of North Africa. Today the culture of North Africa has developed as an eclectic mix of the indigenous Berbers and those of their trading partners throughout Europe and Asia.

◀ This young man is a Berber, one of the oldest ethnic groups in the region. Many Berbers live in the Atlas Mountains and in the Sahara.

Kimberley Coole/Lonely Planet Images/Getty Images

365

netw⊙rks

There's More Online!

- ☑ **IMAGE** Traditional Crafts of North Africa
- ☑ **MAP** North Africa: Invasions and Migrations
- ☑ **TIME LINE** North Africa: Path to Independence
- ☑ **INTERACTIVE SELF-CHECK QUIZ**
- ☑ **VIDEO** Human Geography of North Africa

LESSON 2
Human Geography of North Africa

ESSENTIAL QUESTION · *How do physical systems and human systems shape a place?*

Reading HELPDESK

Academic Vocabulary

- principal
- partner

Content Vocabulary

- domesticate
- hieroglyphics
- geometric boundary
- nationalism
- nomad
- bedouin

TAKING NOTES: *Key Ideas and Details*

IDENTIFYING Use a graphic organizer like the one below to take notes on the indigenous peoples and independence movements of North African countries.

Human Geography of North Africa

Early Peoples and Civilizations	Invasions and Independence

IT MATTERS BECAUSE

For centuries, North Africa has been the birthplace and meeting place of cultures, from the Maghreb area in the west to the Nile Valley in the east. Today the subregion exhibits a cultural vibrancy that rests on this deep historical foundation. Ethnic groups, religious changes, and political movements such as the Arab Spring make this a dynamic and important part of the world.

History and Government

GUIDING QUESTION *How have the Sahara, the Nile River valley, and multiple invasions influenced the history and government of North Africa?*

North Africa has a long history. Some of the oldest human civilizations began in the fertile Nile River valley. People from nearby places, including Greece, Rome, Arabia, and Europe, have all influenced North Africa.

Early Peoples and Civilizations

Hunters and gatherers settled throughout North Africa about 10,000 years ago. The subregion's farmers were among the first to **domesticate** plants and animals. Under this process, plants and animals once wild became raised by people for food, clothing, and transportation.

The Egyptian civilization developed in the fertile Nile River valley about 6,000 years ago. Annual floods from the Nile deposited rich soil on the floodplain. During dry seasons, Egyptians used sophisticated irrigation systems to water crops. This enabled farmers to grow two crops each year. The Egyptians also developed a calendar with a 365-day year and a system of writing known as **hieroglyphics** (HY•ruh•GLIH•fihks). The large pyramids the Egyptians built were used as tombs for their rulers and demonstrate their abilities in mathematics and engineering.

In the A.D. 600s, invasions of Arab armies moved westward from the Arabian Peninsula. These invasions heavily influenced the cultures of North Africa. As Arab rule spread across North Africa, so did the Muslim religion. Muslim and Jewish exiles fleeing persecution by the Spanish Inquisition infused Morocco with Spanish culture in the 1400s. The Arabs retained control over much of North Africa until the early 1500s, when the Ottoman Empire began expanding westward from what is now Turkey. The Ottomans were Muslims, or followers of the Islamic faith. With time, the Ottomans were expelled from North Africa as ethnic groups took up military efforts to end their rule. Resistance to Ottoman rule was finally ended with World War I in 1918.

Location enabled other cultural influences on the subregion. Europe was geographically nearby and established colonial control as the Ottomans left. Algeria was invaded by the French in the mid-1800s. French influence was imprinted on the country, but unrest simmered beneath the colonial surface. Italy colonized Libya, and France occupied additional territory south of the Atlas Mountains. There were disagreements over who would rule North Africa. As a result, the European colonial powers drew **geometric boundaries** that followed straight lines and did not account for natural and cultural features. The boundaries often created conflict between groups of people. Tensions occurred because the local methods of governance were not the same as European methods of governance.

Independence and Power Struggles

In the 1800s, an educated, urban middle class developed in North Africa. This new middle class came in contact with worldwide anticolonial thought and adopted European ideas of **nationalism**. This development stirred demands for self-rule that provided the basis for the modern countries that emerged.

Egypt gained independence from the United Kingdom in 1922. Algeria won independence from the France in 1962 when a nationalist movement led a violent revolt. Since independence, Algeria has developed its resources and raised its living standards. Libya won independence from Italy in 1951 and was then ruled by a Western-friendly monarchy until 1969 when a coup led by Colonel Muammar

domesticate to adapt plants and animals from the wild for human use

hieroglyphics an ancient writing system used in Egypt in which pictures and symbols represent words or sounds

geometric boundary a fixed limit or extent that typically follows straight lines

nationalism a belief in the right of a nation to be an independent state

---

GEOGRAPHY CONNECTION

North Africa's location near Europe and Southwest Asia has made it vulnerable to numerous migrations and invasions over the centuries.

1. *PLACES AND REGIONS* Which cities were affected by Islamic invasions until the 750s?

2. *HUMAN SYSTEMS* Which later migration route is similar to that of Islamic invasions?

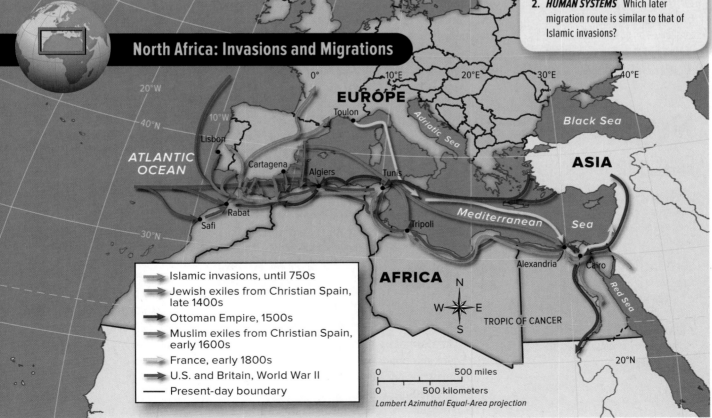

North Africa: Invasions and Migrations

- ➤ Islamic invasions, until 750s
- ➤ Jewish exiles from Christian Spain, late 1400s
- ➤ Ottoman Empire, 1500s
- ➤ Muslim exiles from Christian Spain, early 1600s
- ➤ France, early 1800s
- ➤ U.S. and Britain, World War II
- — Present-day boundary

0   500 miles
0   500 kilometers
Lambert Azimuthal Equal-Area projection

al-Qaddafi overthrew the monarchy. Tunisia, one of the major fronts during World War II, separated from France in 1956. Its history is similar to Libya, with autocratic regimes ruling until 2011. Morocco also won independence from France in 1956. Today Morocco is a constitutional monarchy. Though most countries adopted constitutions, their elected presidents often ruled as dictators once they were in office. Their rule often stifled human rights and free speech.

Political and ethnic tensions have risen often in North Africa. A protest by a Tunisian citizen that resulted in his death in 2010 led to riots. The act was to protest the lack of economic opportunities. News of the event resulted in revolts against oppressive, nondemocratic governments in Egypt, Libya, and Tunisia. The movement became known as the Arab Spring. Citizen protests in Morocco did not overthrow the government, but the king was forced to concede political and social reform. Beginning in 2011, a series of new governments were elected in Egypt, Libya, and Tunisia.

✓ READING PROGRESS CHECK

Examining What movement led the demand for self-rule in North African countries?

Population Patterns

GUIDING QUESTION *How have indigenous ethnic groups, migrations, and climate shaped population patterns in North Africa?*

While European connections continue in the coastal regions of North Africa, a widespread influence throughout the subregion is a mix of indigenous and Arab cultures. The Berbers were the indigenous people of North Africa before the Arab invasions. They maintain their traditional non-Arabic language. Most of the 15 million Berbers live today as farmers. They had previously been pastoral **nomads**, groups of people who move from place to place with herds of animals depending on the season and the availability of grass for grazing and water. The other **principal** ethnic group in North Africa is the Arab people. United by language, Arabs first migrated from the Arabian Peninsula to North Africa in the

nomad a member of a wandering pastoral people

principal most important, consequential, or influential

Transcendental Graphics/Hulton Archive/Getty Images

TIME LINE ⌄

NORTH AFRICA
Path to Independence ➜

The countries of North Africa share a common history of colonialism, independence, and journeys toward democracy.

CRITICAL THINKING ▶

1. Describing Discuss the effects of colonization on the recent history of North Africa.

2. Identifying Which countries changed governments as a result of the Arab Spring?

1800s
France and Great Britain take control over areas of North Africa.

1800 ➜

1922
Great Britain grants Egypt independence, but maintains control of the Suez Canal.

1950 ➜

1952
Military coup topples government of Egypt; military leader Gamal Abdel Nasser later becomes president of Egypt.

1954
Algerian nationalist group, the National Liberation Front, launches guerrilla war against France.

A.D. 600s. Some of the Arabic-speaking peoples are nomadic **bedouins** (BEH•duh•wuhns) who migrated to North Africa from deserts in Southwest Asia.

Geographic factors, especially the availability of water, have influenced settlement in the subregion. Because water is scarce, people have for centuries settled along seacoasts and rivers. The Mediterranean and Atlantic coasts and the Nile River valley hold most of the subregion's people. The Nile River valley is one of the world's most densely populated areas. Other population centers in North Africa include Casablanca, Tunis, Tripoli, and Cairo. Egypt's most populous city and its capital, Cairo, dominates the country's social and cultural life. Cities have grown rapidly as people migrate from rural areas in search of a better life. Cities face problems providing services because growth has occurred quickly.

Emigration to regions outside North Africa is high due to higher availability of employment opportunities in other regions. The region's close proximity and historical ties to European countries such as Spain and France pull migrants from Morocco, Algeria, and Tunisia. Migrants from Egypt, an ex-British colony, migrate to the United Kingdom and other English-speaking countries such as the United States, Canada, and Australia.

bedouin member of the nomadic desert peoples of North Africa and Southwest Asia

☑ **READING PROGRESS CHECK**

Naming Who are the people indigenous to North Africa?

Society and Culture Today

GUIDING QUESTION _How have Islam and the Arabic language helped define much of the culture and society of North Africa?_

When the Arabs invaded North Africa, they brought Islam. Calls to worship occur five times each day in countries with Muslim populations. A muezzin, or crier, delivers the call to prayer for each local mosque. Following the movements of the imam, or prayer leader, individuals bow and kneel, touching their foreheads to the ground in the direction of the holy city of Makkah (Mecca) in Saudi Arabia.

1956 March—France recognizes independence of Morocco and Tunisia.

1956 July—President Nasser of Egypt nationalizes Suez Canal, leading to Suez Crisis.

1962 Brutal Algerian war of independence with France ends; Algeria becomes an independent country.

1969 Libyan army overthrows king. Muammar al-Qaddafi becomes head of state.

2010 ➜

2010 Young Tunisians launch protests that ultimately forces president to flee.

2011 Arab Spring leads to regime change in Tunisia, Libya, and Egypt.

Islamic patterns are found in weaving and embroidery throughout the subregion.

▲ CRITICAL THINKING

1. *Analyzing Visuals* Describe the designs shown in the picture above.

2. *Making Connections* Why are there no animals or people in the designs of traditional art in North Africa?

As Islam spread, so did the Arabic language. Non-Arab Muslims learned Arabic in order to read the Quran, Islam's holy book. Moroccans speak an Arabic dialect that is a result of Berber influence. Egyptians speak Arabic with a different accent than do Moroccans, Tunisians, and Algerians. People in Algeria and Morocco also speak French as a result of colonization and influence on education.

Family and Status of Women

Since achieving independence in the 1950s and 1960s, North African countries have been redefining their political, social, and cultural systems. Class status influences family size. Upper-class families have fewer children than lower-class families. Historically, extended family members have lived nearby, so children usually grow up with many cousins, aunts, and uncles. Today, families are living more spread out and do not necessarily live close to one another.

Decades of urbanization have led nomadic peoples to settle in villages and cities. Very few groups still roam the desert as they did centuries ago. They have access to schools, transportation, and health care near where they live. Families have plans that educated children will earn enough money to support the older members of the family.

Although women participated in winning independence, women's issues were not priorities in the years immediately following independence. Government rule often stifled human rights. This affected women's issues as well, and many women are not permitted to accept jobs even when they have an education level higher than men.

The Arts

From the earliest times, the peoples of North Africa have expressed themselves through the arts. The ancient Egyptians built huge pyramids that still stand as marvels of construction. One of Algeria's most popular forms of music, *raï*, involves various instruments and poetic lyrics. Contemporary Egyptian music blends traditional Arabic and Western styles. Egypt has experienced a revival of folkloric dance and traditional crafts. The arts of weaving, embroidery, and metalworking are heavily influenced by Islamic patterns found in local architecture. Elaborate geometric designs are featured in these works because of the traditional Islamic prohibition against the representation of living beings.

✔ READING PROGRESS CHECK

Assessing What has been the primary influence of language in North Africa?

Economic Activities

GUIDING QUESTION *What factors have helped many countries in North Africa become rising middle-income countries?*

North Africa, unlike other regions of the continent, has historically engaged in a range of manufacturing processes. By the end of the 1800s, much of the region, however, was regarded by European countries as a source of raw materials. Although manufacturing increased during World War II, output accelerated quickly only after the decline of European control. Despite expansion since the 1950s, manufacturing output among the countries varies greatly.

Only a small part of the subregion's land is arable, or suitable for farming. Agriculture plays a smaller role in countries with desert climates. It is also less important in countries with economies based on oil, such as Libya and Algeria.

Areas of North Africa that have a Mediterranean climate are best suited for growing a variety of crops. Barley, wheat, citrus fruits, grapes, vegetables, olives, figs, dates, and almonds are grown. When rainfall is below normal, however, harvests of major crops such as wheat and barley seldom meet local needs, which increases dependence on imported food. Seafood is an important food source, and fishing vessels catch sardines and mackerel from the Atlantic Ocean.

Wealth from oil, natural gas, and mining has helped develop economies in the subregion. Petroleum and oil products are North Africa's main export commodities, or economic goods. Natural gas is also a major export. Coal and copper mining and cement production are important as well.

Tourism plays a significant role in the subregion's economies. Certain areas serve as popular travel destinations because of their historical importance, blend of religious history, or pleasant climates. The Valley of the Kings and the pyramids draw tourists to Egypt. Port el-Kantaoui near Sousse, Tunisia, is a vacation beach destination. Carthage, an ancient city, is rich in historical artifacts.

Advancements in transportation and communications have brought improvements to the subregion. However, the physical environment and government control have limited some development here. Water transportation is very important to North Africa. Ships load and unload cargo and passengers at ports on the Mediterranean Sea, Red Sea, and Atlantic Ocean, including oil, natural gas, minerals, and tourists. Because of its accessibility to Europe and low-cost labor, Morocco has become an appealing trading **partner**—especially for Spain, France, and Italy. Other North African countries also have significant trade relations with countries outside the region.

partner one associated with another, especially in an action

Despite economic progress within North Africa, its countries suffer from high rates of unemployment, poverty, wealth disparity, and political upheaval. Key economic challenges include reducing socioeconomic disparities in government spending, increasing available jobs for the young, and establishing more diverse industries. The Arab Spring has resulted in a number of stressful aftereffects in North Africa as well. These include high food and fuel prices that have strained many of the resources and budgets of the subregion's governments. Political instability, public unrest regarding dishonest elections, and the protection of human rights are concerns expressed in the new Arab Spring democracies.

✓ **READING PROGRESS CHECK**

Explaining How have North Africa's natural resources influenced economic growth?

LESSON 2 REVIEW

Reviewing Vocabulary

1. *Locating* Where did the bedouin people originate?

2. *Defining* What political concept introduced by North Africa's middle class is a response to the region's history of colonialism?

Using Your Notes

3. *Summarizing* Use your graphic organizer to describe the early peoples of North Africa.

Answering the Guiding Questions

4. *Gathering Information* How have the Sahara, the Nile River valley, and multiple invasions influenced the history and government of North Africa?

5. *Explaining* How have indigenous ethnic groups, migrations, and climate shaped population patterns in North Africa?

6. *Summarizing* How have Islam and the Arabic language helped define much of the culture and society of North Africa?

7. *Describing* What factors have helped many countries in North Africa become rising middle-income countries?

Writing Activity

8. *Informative/Explanatory* Identify two natural resources in North Africa crucial to economic activity and trade. In a paragraph, provide specific examples of how these resources increased trade and interdependence between North Africa and the world.

The Arab Spring

The Arab Spring was a series of protests and revolutions that began to spread across the Arab world in late 2010. All protesters shared the same general goals: less corruption in their country's leadership and more openly democratic forms of government. The outcomes of the Arab Spring have differed between countries. In some places, like Egypt, the protests have resulted in major changes in government. In others, like Syria, protests and government repression have broken down into full-scale civil wars. The lasting outcomes of the Arab Spring, positive or negative, are yet to be fully realized.

 MAJOR PROTESTS   **MINOR PROTESTS**   **PROTESTS SUPPRESSED**

 CHANGE IN GOVERNMENT   **LEADER KILLED**   **CIVIL WAR**

MOROCCO   **ALGERIA**

TUNISIA

LIBYA   **EGYPT**

TUNISIA December 2010–December 2014

As a protest against the government, Mohammed Bouazizi set himself on fire in Sidi Bouzid. Then protests against the authoritarian rule of President Zine El Adidine Ben Ali began in Tunis. Ben Ali fled the country. A national unity government was formed, followed by the ratification of a new constitution in January 2014. Benji Caid Essebsi was elected as the first president under the new constitution at the end of 2014.

LIBYA February 2011 – September 2014

Calls for protest on the Internet resulted in demonstrations in Benghazi, Tripoli, and across the country. Clashes between protesters and the government turned into a civil war which ended with a rebel victory. The Qaddafi regime was replaced with a transitional government. New parliament elections in 2012 and 2014 resulted in rival governments. The UN has been working to reconcile the governments.

EGYPT January 2011 – December 2015

Protests in Cairo, Alexandria, and across the country erupted against President Hosni Mubarak in January 2011. Mubarak stepped down shortly after, allowing Egypt to hold its first ever democratic elections in late 2011. The new president, Mohammad Morsi, was himself removed from power after violent protests. He was replaced with an interim president in July 2013. In January 2014 Abdel el Sisi was elected president and a new legislature was elected in December 2015.

SYRIA February 2011 – January 2016

Angry with corruption in their government, Syrian protesters organized a "Day of Rage" for mid-February 2011 on Facebook and Twitter. Sporadic protests in Damascus and across the country popped up. Government suppression turned the protests into a civil war. International pressure has failed to bring a resolution to the conflict. Millions of people have been displaced.

THE RISE OF ISIL

The Islamic State of Iraq and the Levant (ISIL) formed and gained strength after the withdrawal of U.S. troops from Iraq in 2011. ISIL has ignored international borders and taken advantage of political instability by rapidly seizing control of large areas of northern Iraq and eastern Syria. ISIL has brutally targeted its religious enemies and engaged in public executions. Refugees have fled from ISIL-controlled areas in large numbers.

YEMEN January 2011 – March 2016

Inspired by events in Tunisia and Egypt, Yemeni people protested against the corrupt government of President Ali Abdullah Saleh. Civil war erupted in Yemen and Saleh stepped down in late 2011. The Huthis, a Shia minority, expanded their influence and forced new leaders to flee. Saudi Arabia intervened but the Huthis still control parts of Yemen.

SYRIA

IRAQ

LEBANON

JORDAN

KUWAIT

SAUDI ARABIA

OMAN

YEMEN

Making Connections

1. **Places and Regions** What were the two most common outcomes of states experiencing major protests?

2. **Human Systems** Many protest groups used public sites like Facebook and Twitter to organize and communicate. How might this practice have inspired and encouraged protests elsewhere?

3. **The Uses of Geography** Create a time line of events for one of the states above, detailing some of the protests, changes in government, or conflicts. Then, write a short summary explaining the overall positive or negative effects of the movement on that state.
*Interact with **Global Connections** Online*

networks

There's More Online!

- ☑ **IMAGE** Managing the Great Man-Made River project
- ☑ **IMAGE** Oasis in Tinghir, Morocco
- ☑ **INFOGRAPHIC** Aswān High Dam
- ☑ **INTERACTIVE SELF-CHECK QUIZ**
- ☑ **VIDEO** People and Their Environment: North Africa

Reading HELPDESK

Academic Vocabulary

- **subsequent**
- **relevant**

Content Vocabulary

- **aquifer**

TAKING NOTES: *Key Ideas and Details*

Summarizing Use a graphic organizer like the one below to summarize ways that North Africa is managing its resources.

LESSON 3
People and Their Environment: North Africa

ESSENTIAL QUESTION • *How do physical systems and human systems shape a place?*

IT MATTERS BECAUSE

The relationship of humans with their environment is constantly changing. While humans have achieved impressive advancements in energy and power production, such innovation comes at a cost. For example, the need for freshwater and the process of oil extraction have often negatively affected the North African environment. The subregion's water scarcity and oil wealth make trade an essential activity for many North African countries.

Managing Resources

GUIDING QUESTION *Why are water resources in such demand in North Africa?*

As populations and economies grow, so does demand for water resources. Clean drinking water ensures the continued livelihood and development of human populations. Much water in North Africa is used for irrigation, which allows for continued food production. Places with fewer water resources, such as Western Sahara and Libya, are more dependent on trade to import food. This not only leaves populations dependent on outside food sources, but also creates a vulnerability to changes in market prices for food. Agricultural land loss due to urbanization and windblown sands is also an issue affecting food production.

Much of the freshwater in North Africa comes from rivers, oases, and **aquifers**, or underground sources of water. Aquifers can be problematic for countries with coastal access. As water is removed from the coastal aquifers, it is replaced by salty seawater. Water contaminated by increased salinity in the aquifers becomes unusable for agriculture. The contaminated water is also unfit to drink.

The source of water for consumption and agriculture in North Africa varies by country. Egypt, for example, gathers surface water from large rivers, particularly the Nile. Other countries, such as Algeria and Libya, have more limited water resources. As another form of water extraction, these countries use a process called desalination. Desalination is a process in which the salt is removed from salt water. Building and maintaining desalination plants, however, can be expensive for most countries.

Libya's Great Man-Made River is an ambitious effort to supply freshwater to this country that consists of 95 percent desert and relies on food imports to feed its population. Nearly 40,000 years ago, when North Africa had a more temperate climate, rainwater collected in underground reservoirs beneath Libya. It was not until the mid-1950s that oil exploration in the southern Libyan Desert exposed these vast quantities of fresh, clean groundwater. The Great Man-Made River is a pumping and pipeline system that attempts to utilize this extensive source of groundwater.

The Great Man-Made River provides 70 percent of the Libyan population with water for drinking and irrigation. It was funded by profits from Libya's nationalized oil sector. The project uses two pipelines to pump water from large aquifers beneath the Sahara to farms along Libya's coastal region. The first two phases of the project were completed in 1993 and 1996. The third phase of the project, completed in 2009, links the pipelines of the first two phases. Two **subsequent** phases will extend the distribution network to the northeast coast and unite the two transporting systems into a single network.

However, environmental challenges exist for Libya's Great Man-Made River. Scientists fear that the pipelines could drain aquifers in Libya and neighboring countries. They also fear that pumping aquifers near the Mediterranean could draw in salt water from the sea. In most areas, the rate of water extraction far exceeds the rate at which water is replaced. Such rates of extraction cannot be sustained in the coming decades. Although providing water for the current population is important, an equally **relevant** concern is having water for future generations.

☑ **READING PROGRESS CHECK**

Specifying What is a project that has provided water to the region's inhabitants?

aquifer underground water-bearing layers of porous rock, sand, or gravel

subsequent later; following after another event

relevant having to do with the matter at hand

This dense palm oasis in Tinghir, Morocco, is irrigated by a network of pipes and irrigation canals.

CRITICAL THINKING ▼

1. *Analyzing Visuals* What environmental challenges does irrigation pose in North Africa?

2. *Hypothesizing* What type of water source is likely to be used to irrigate the oasis shown?

Human Impact

GUIDING QUESTION *How have modern economic activities impacted North Africa's environment?*

North Africa faces many challenges related to economic activity and its effects on the environment. Oil and fishing are two very profitable economic sectors for the subregion. High-volume production for these economic sectors, however, can alter the physical environment. The plants and animals in the area also suffer when habitats are lost or damaged.

Mass tourism is another main cause behind ecological loss in the subregion. Areas along the coast that receive the most visits from tourists have suffered the greatest losses. Some locations that were once pristine are now severely damaged beyond repair. Increased use of water in hotels, swimming pools, and golf courses—especially during the summer—is also a concern since the subregion suffers from water shortages.

North Africa's location puts it at risk for oil spills since tankers pass the subregion's coasts. Oil spills can occur on land or on the seas. Following an oil spill at sea, winds can sweep the oil slicks toward coastlines, affecting the ecosystems of the land as well as the marine environments. Past oil spills have

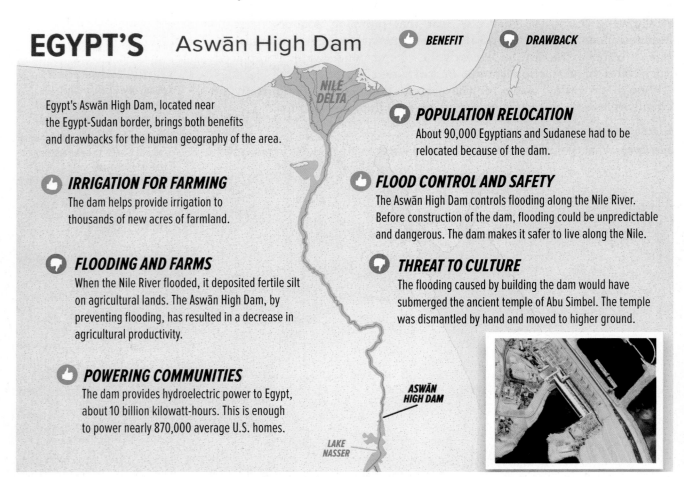

EGYPT'S Aswān High Dam

🔼 **BENEFIT**   🔽 **DRAWBACK**

Egypt's Aswān High Dam, located near the Egypt-Sudan border, brings both benefits and drawbacks for the human geography of the area.

🔼 **IRRIGATION FOR FARMING**
The dam helps provide irrigation to thousands of new acres of farmland.

🔽 **FLOODING AND FARMS**
When the Nile River flooded, it deposited fertile silt on agricultural lands. The Aswān High Dam, by preventing flooding, has resulted in a decrease in agricultural productivity.

🔼 **POWERING COMMUNITIES**
The dam provides hydroelectric power to Egypt, about 10 billion kilowatt-hours. This is enough to power nearly 870,000 average U.S. homes.

🔽 **POPULATION RELOCATION**
About 90,000 Egyptians and Sudanese had to be relocated because of the dam.

🔼 **FLOOD CONTROL AND SAFETY**
The Aswān High Dam controls flooding along the Nile River. Before construction of the dam, flooding could be unpredictable and dangerous. The dam makes it safer to live along the Nile.

🔽 **THREAT TO CULTURE**
The flooding caused by building the dam would have submerged the ancient temple of Abu Simbel. The temple was dismantled by hand and moved to higher ground.

NILE DELTA

ASWĀN HIGH DAM

LAKE NASSER

The Aswān High Dam and other modern dams constructed farther up the Nile control the river's flow, resulting in both positive and negative effects for the people of the subregion.

▲ **CRITICAL THINKING**

1. *Assessing* How does the Aswān High Dam benefit the people of North Africa?

2. *Posing Questions* What questions would you ask the government agency in charge of the dam about its effects on the environment of North Africa?

exposed the lack of safety precautions related to oil extraction and the necessity of establishing a faster response to oil spills. The need for greater cooperation between countries to prevent future spills is crucial to limit the disastrous effects of oil spills.

The combination of oil by-products, sewage, and industrial waste threatens the North African coast. Marine biodiversity is threatened by the pollutants. In Libya and Algeria, waste from petroleum refineries and raw sewage seeps into the rivers and coastal waters, damaging ecosystems.

The European Union's 2010 bilateral trade agreement with Morocco allows for greater freedom in trade with Europe. The drawback of increased trade is that fish populations and other marine life may be endangered. Many fish populations have been in rapid decline in the Mediterranean during the past decade. For example, bluefin tuna have been overfished for decades and are at serious risk of extinction if overfishing and unsustainable fishing practices in the region are not stopped. Marine turtles have also been negatively affected by the destruction and disturbance of nesting sites.

In 1970 Egypt completed the Aswān High Dam, located about 600 miles (966 km) south of Cairo, Egypt. The dam controls the Nile's floods. It also created Lake Nasser and provides irrigation for about 3 million acres (1.2 million ha) of land while also increasing the Egyptian fishing industry. Electricity for Egypt is generated by the dam. The completion of the dam marked the first time in history that the Nile's annual flood could be controlled by humans.

Despite the economic boosts of the Aswān High Dam and Lake Nasser, these construction projects have had negative consequences. These negative impacts affect the health and livelihood of Egypt's people and their livestock. For instance, when the dam was built, it changed the ecosystem. Some plants and animals could adapt, but others could not. Ultimately, the new ecosystem provided the perfect environment for an increase in waterborne diseases such as malaria and dysentery. In addition, fertile silt has accumulated behind the dam, soil that formerly went downstream and revitalized farmland.

There are three basic concerns in much of North Africa that relate to human impact on the environment: population growth, agricultural performance, and environmental degradation. Population growth—coupled with longer life spans and the ability to feed the population—is increasingly a challenge to the subregion. Making the matter worse is the low agricultural productivity faced by many farmers in North Africa.

Land degradation from desertification is an environmental issue in all of the North African countries and is human induced. Soil erosion as a result of farming, overgrazing, and destruction of vegetation contributes to desertification. Increasingly, evidence shows that land degradation is a driver of climate change. The other causes of land degradation include drought, population pressure, and local agricultural and land use policies.

✓ READING PROGRESS CHECK

Identifying Cause and Effect How has tourism in North Africa negatively affected the environment?

The Great Man-Made River project uses pipelines to pump water from large aquifers beneath the Sahara to farms along Libya's coastal region.

▲ CRITICAL THINKING

1. Assessing How was the Great Man-Made project funded?

2. Evaluating What is the risk caused by pumping water from the underground aquifers?

Regional cooperation and international aid is important during times of regional unrest and government changes. Fuel scarcities due to the Libyan conflict in 2011 affected the amount of energy supplied to the Great Man-Made River and desalination plants. During the conflict, water shortages in Tripoli resulted from power outages, attacks on the staff managing the water network, and a reservoir held hostage by pro–al-Qaddafi fighters. United Nations agencies, the International Committee of the Red Cross, and neighboring countries such as Greece and Malta delivered millions of liters of water to Tripoli. Government resistance to change and the dependence of large populations on projects such as the Great Man-Made River exemplify how vulnerable Libya and other North African countries are to political and environmental crises.

EXPLAINING How can water supplies be vulnerable during times of political unrest and conflict?

Addressing the Issues

GUIDING QUESTION *How are environmental issues in North Africa being addressed?*

In North Africa, the environmental impacts of economic activities and growing populations is of increasing concern. Natural resources are important to economic activity, but resource extraction and harvesting are also major contributors to pollution and degradation in the region. The Middle East and North African countries (MENA) hold approximately 61 percent of the world's oil reserves. About 45 percent of the world's natural gas reserves are also located in these countries. Because the world relies on access to oil, global markets are affected by changes in the government policies of these countries. Therefore, cooperation among MENA countries can settle issues of fluctuating oil supplies to better meet the demand. Cooperation among countries can also help establish preventative measures related to oil spills and cleanup. The Partnership for African Fisheries (PAF) is working to strengthen the fishing industry. One goal of the partnership is to enforce stricter regulations to prevent overharvesting of certain species.

Determining water rights between countries is also a concern in North Africa. For example, Libya's Great Man-Made River receives water supplies from the Nubian Sandstone Aquifer System and other aquifers in the region. The aquifers are located beneath multiple political borders in the region. As a result, a joint authority for the management of the aquifer system was established in 1992. The joint authority encourages cooperation between Egypt and Libya in managing the Nubian Sandstone Aquifer System.

At the international level, the World Bank is developing plans with several countries to invest in modern irrigation practices to address the needs of agriculture. The plans call for increasing water supply sources in North Africa. The World Bank is also working to develop water policies and institutions that will improve water quality in the region. Groundwater management and wastewater collection and treatment are central to these plans. Projects in Egypt and Morocco hope to increase irrigation efficiency and implement the reuse of treated wastewater. Improving the way farmers' associations are managed is another goal of projects in North Africa. In Algeria, the World Bank is promoting the creation of independent public companies that offer water supply delivery service to small cities and towns. The World Bank also supports organizations such as the Arab Water Council as a means to gather and share information related to water supply issues.

☑ **READING PROGRESS CHECK**

Classifying Name two industries in North Africa that have required stricter regulation and enforcement to preserve natural resources.

LESSON 3 REVIEW

Reviewing Vocabulary
1. ***Defining*** What is an aquifer, and why is it important to the region?

Using Your Notes
2. ***Summarizing*** Use your graphic organizer to describe two human-made structures built to manage North Africa's resources.

Answering the Guiding Questions
3. ***Evaluating*** Why are water resources in such demand in North Africa?

4. ***Identifying Cause and Effect*** How have modern economic activities impacted North Africa's environment?

5. ***Synthesizing*** How are environmental issues in North Africa being addressed?

Writing Activity
6. ***Argument*** Using examples of economic activities or human-made projects, identify three ways in which natural resources in North Africa have been managed to profit and sustain the subregion's economies.

Directions: On a separate sheet of paper, answer the questions below. Make sure that you read carefully and answer all parts of the questions.

Lesson Review

Lesson 1

1 *Identifying Cause and Effect* Briefly explain how tectonic activity has shaped the North African landscape. Provide an example of a land formation that illustrates its effects.

2 *Explaining* Why does such a great percentage of Egypt's population make a home on such a small percentage of the country's land? In your response, identify the area where this large population concentration lives.

3 *Analyzing* Explain whether this statement is accurate: "Sand covers most of the Sahara, and the Sahara and other deserts in North Africa are unable to support any substantial vegetation."

Lesson 2

4 *Exploring Issues* Provide a summary of why it has been difficult to maintain peace in North Africa.

5 *Making Connections* Discuss influences that have led to a change in lifestyle for nomadic peoples in North Africa.

6 *Identifying Cause and Effect* Identify at least three types of crops grown in North Africa, and name the type of climate that provides desirable conditions for crop growth. Explain why harvests of major crops sometimes fail to meet people's needs.

Lesson 3

7 *Explaining* Discuss how the location of North Africa makes the countries vulnerable to oil spills and what steps can be taken to prepare for such events.

8 *Considering Advantages and Disadvantages* Explain why petroleum refining and the Aswān High Dam can both be assessed as providing benefits and detriments.

9 *Diagramming* Create a graphic organizer to compare and contrast these organizations: Middle Eastern and North African countries (MENA) and Partnership for African Fisheries (PAF). Include the reasons for the formation of each organization as well as each organization's actions.

21st Century Skills

Use the graph to answer the questions that follow.

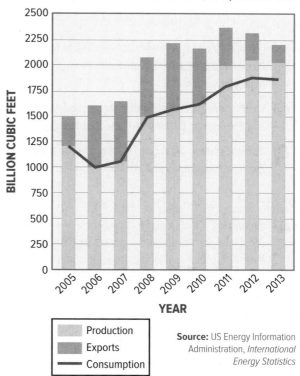

EGYPT: NATURAL GAS PRODUCTION, CONSUMPTION AND EXPORTS, 2005–2013

Source: US Energy Information Administration, *International Energy Statistics*

10 *Creating and Using Graphs, Charts, Diagrams, and Tables* Explain the relationship between Egypt's natural gas production and consumption during 2005.

11 *Compare and Contrast* Explain Egypt's natural gas consumption between 2005 and 2013.

12 *Identifying and Explaining Continuity and Change* Did Egypt's natural gas consumption rise, fall, or remain the same from 2009 to 2010? What impact, if any, did this have on exports of natural gas? What is a logical reason for this impact?

Need Extra Help?

If You've Missed Question	**1**	**2**	**3**	**4**	**5**	**6**	**7**	**8**	**9**	**10**	**11**	**12**
Go to page	368	369	370	372	374	377	382	382	384	385	385	385

Directions: On a separate sheet of paper, answer the questions below. Make sure that you read carefully and answer all parts of the questions.

Critical Thinking

13 *Interpreting Significance* Discuss the impact of groundwater availability on exports and the economy in general in North Africa.

Applying Map Skills

Refer to the Unit 5 Atlas to answer the following questions.

14 *Places and Regions* What are the three major climate regions in North Africa?

15 *The World in Spatial Terms* Use your mental map of North Africa to describe the location of the Atlas Mountains.

16 *Human Systems* What generalizations can be made about the population density of North Africa?

Exploring the Essential Question

17 *Organizing* Create a plan for a website to teach students about how physical and human geography have shaped North Africa. Your goal for this project is to provide general categories with hyperlinks. Create a document that describes each page category for your site and a design for the home page. Include at least three links for each page of the website. Include the following aspects of North Africa: landforms, water systems, climates, biomes, history and government, language and religion, economic activities and trade, and environmental issues.

College and Career Readiness

18 *Reaching Conclusions* Imagine you are an intern working for a company that is developing a report to analyze the pros and cons of desalination in Algeria and Libya. Write an overview for the company, briefly explaining the process of desalination and why it is important in Algeria and Libya. Your overview should include a chart listing the pros and cons of the process for these countries. Based on the overview, write a paragraph explaining whether you think desalination should be pursued in Algeria and Libya.

DBQ Analyzing Primary Sources

Use the quote to answer the following questions.

PRIMARY SOURCE

"Egypt is a large, complex, very important country . . ."

—Hillary Rodham Clinton, U.S. Secretary of State, CNN News interview, January 30, 2011

19 *Evaluating* Secretary of State Hillary Rodham Clinton made this statement in 2011. Explain whether this statement would have been equally accurate during earlier periods of history.

20 *Identifying Central Issues* What events in Egypt in 2011 would have given rise to the focus of Secretary of State Clinton and the world?

Research and Presentation

21 *Problem Solving* Utilize Internet and library sources to research information regarding the Great Man-Made River and create a multimedia presentation. Your presentation should explain the following: how and why the Great Man-Made River began, the accomplishments of the first three phases, the accomplishments of phases completed or in progress after the first three phases, and suggestions that could effectively be incorporated into a current or future phase. Within your presentation, provide an analysis as to whether the Great Man-Made River has provided effective solutions for the people of Libya. Include maps and images of places, as well as charts, diagrams, or graphs in your presentation.

Writing About Geography

22 *Argument* Use standard grammar, spelling, sentence structure, and punctuation to write a paragraph explaining why Algeria should diversify its sources of revenue.

Need Extra Help?

If You've Missed Question	13	14	15	16	17	18	19	20	21	22
Go to page	380	362	360	364	365	380	372	374	381	371

The Eastern Mediterranean

ESSENTIAL QUESTION • *How do physical systems and human systems shape a place?*

◄ The Eastern Mediterranean is home to many ethnicities and religions.

Cultura/Image Source

networks

There's More Online about the Eastern Mediterranean.

CHAPTER 16

Why Geography Matters
Israeli-Palestinian Conflict

Lesson 1
Physical Geography of the Eastern Mediterranean

Lesson 2
Human Geography of the Eastern Mediterranean

Lesson 3
People and Their Environment: The Eastern Mediterranean

Geography Matters...

Because its fertile land supported the development of agriculture, the Eastern Mediterranean became one of the birthplaces of civilization. Many of its cities have been continuously inhabited for thousands of years, and several of the world's most widespread religions have roots here. This subregion is marked by a deep, complex history of political and religious conflict. Modern tensions have developed many distinct ideologies that reside in close proximity.

Israeli-Palestinian conflict

Conflict between Israelis and Palestinians has not evolved from one single factor such as a dispute over religion or land. Rather, the conflict has resulted from years of differences over cultural, religious, economic, and geographic issues. At the heart of the conflict lie the West Bank, the Gaza Strip, and East Jerusalem.

Israel and the Palestinian Territories

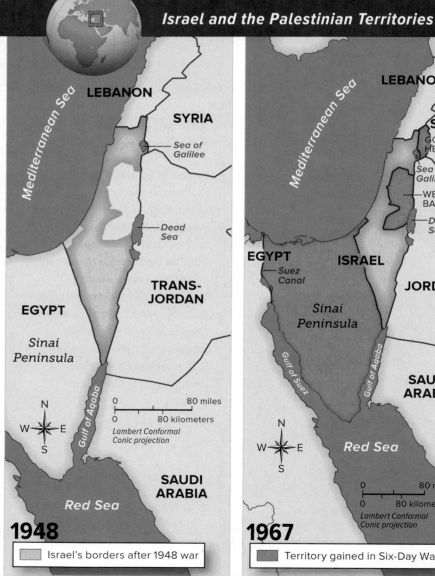

1948

Israel's borders after 1948 war

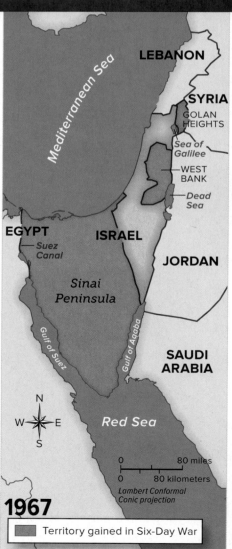

1967

Territory gained in Six-Day War

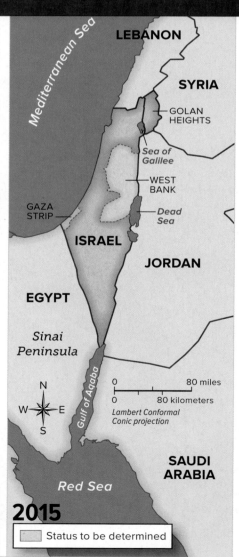

2015

Status to be determined

What is the conflict?

From the 1880s through the 1940s many European Jews immigrated to the Palestine Mandate. This immigration took place in the context of rising nationalisms in the late 19th century and amidst widespread anti-Semitism, as part of the Zionist movement. Zionism was the modern political movement that aimed at re-establishing a Jewish safe haven in the historic homeland, the Land of Israel. Prior to World War I the Ottoman Empire controlled Palestine, and the British received a Mandate from the United Nations to control it. Arab immigration to the region also increased in this period. Arabs were in the majority at this time. In 1947, the UN voted to partition the Palestine Mandate into two states, Arab and Jewish. The Arabs rejected the UN partition, whereas the Jews accepted it and the Zionists established the state of Israel. In 1948 the British withdrew and the State of Israel was created. Israel was then attacked by five Arab states in the first of several wars. At the end of the war, Egypt controlled Gaza, while Jordan controlled the West Bank and East Jerusalem, including the Old City, preventing Jews from accessing the Western Wall of the ancient Temple complex, the holiest site in Judaism. In 1967 Israel captured the West Bank, Gaza, and East Jerusalem, and these areas have become the focus of demands for the creation of a Palestinian state. Agreements like the Oslo Accords of the mid-1990s have attempted to resolve the conflict. The Oslo Accords of 1993 and 1995 provided for different levels of authority in the Gaza Strip and the West Bank. Israel granted partial self-rule to Palestinians in these two areas. Moderate Palestinians accepted Israel's right to exist. The issue of who was to control Jerusalem remained unsettled. Although the accords are still in effect, they are unpopular with both the Palestinian and Israeli public.

1. Human Systems   How did the Zionist movement lead to the creation of the State of Israel?

What is the Israeli perspective?

The Land of Israel is the historic home of the Jewish people from ancient times. Jews had sovereignty there for over 1,000 years until A.D. 70. Even after the destruction of Jerusalem by the Romans, and the dispersion of many Jews, some Jews continued to live there, and the yearning for return to the Land of Israel was never abandoned. In the late 1880s, Jews began returning to the land in greater numbers and began to develop the institutions necessary for statehood.

After centuries of persecution culminating in the Holocaust, this hope was finally realized. The Jews accepted a 1947 UN-sponsored plan to partition the British Palestine Mandate into two states. The Arabs did not accept this and instead attacked Israel. During a brief 1967 war, Israel captured areas known as the West Bank, East Jerusalem, the Sinai Peninsula, the Gaza Strip, and Golan Heights. Its control of these areas was opposed by Palestinian Arabs and neighboring Arab countries, which declared shortly after the war that they would not make peace with Israel.

In 1973 Egypt and Syria attacked Israel, launching the Yom Kippur War. Israel withdrew from the Sinai Peninsula in 1982. In the 1993 Oslo Accords, Israel agreed to limited self-rule in the Palestinian territories. Israel withdrew from the Gaza Strip in 2005. It continues to control the West Bank and East Jerusalem. Attacks on Israel by anti-Israel Islamist groups such as Hezbollah and Hamas resulted in Israel's construction of a security barrier. Numerous attempts have been made to find a peaceful solution to the Arab-Israeli conflict, but so far none have been successful. Three major unresolved issues are the control of Jerusalem, the future of the Israeli settlements in the West Bank, and the Palestinian refugees' right of return to the areas from which they were displaced. The fact that reasonable Israeli peace initiatives have been regularly rejected reinforces other evidence that many Palestinian and other Arab leaders continue to strive not for peace but for the end of Israel as a Jewish state.

2. Places and Regions   What issues make the conflict between the Israelis and the Palestinians so complex?

What is the Palestinian perspective?

Many Palestinian Arabs have lived in the region for centuries. Jerusalem is the third holiest city in Islam and is of religious importance to Christians and thus is significant to Palestinian Muslims and Christians. Britain had promised self-rule for Arabs in exchange for support during World War I, but betrayed its promise. While the UN partition plan provided for two states, Arabs viewed the plan as unfair and rejected it. Zionists established the state of Israel. While neighboring Arab states promised the Palestinian Arabs victory and the destruction of Israel, the war that followed resulted in Palestinian Arabs fleeing or being driven out from their homes and forced to live in refugee camps in Lebanon and Jordan. Those that remained in Israel are citizens, but many feel they are not treated fairly and have been denied the right of self-determination. The defeat of the Arab states in 1967, resulting in Israel gaining Gaza and Sinai from Egypt, and the West Bank from Jordan, further dashed Palestinian hopes. Palestinians believe that the building of Israeli settlements in the West Bank and Gaza violates international law and makes it more and more difficult to bring about the creation of a viable Palestinian state. Some Palestinians feel that attacks on Israeli civilians and soldiers are legitimate responses in pursuit of nationalist objectives, while others advocate a diplomatic and non-violent approach resulting in a two state solution. The leading Palestinian groups governing Palestinian society are the moderate Palestinian Authority (controlled by the Fatah party) in the West Bank, and the anti-Israeli militant Islamist group Hamas in the Gaza Strip. Palestinian society includes both secular and religious elements that differ in their views of how Palestinian society should be run and how to approach relations with Israel. Settlement building activity in the West Bank undermines Palestinians' confidence that Israel seeks peace and is willing to accept a Palestinian state.

3. Human Systems   How has conflict resulted in changing borders and demographics?

THERE'S MORE ONLINE

EXPLORE a map of Israel and the Palestinian territories   •   **VIEW** an image of the Oslo Accords

networks

There's More Online!

- ☑ **IMAGE** Fish Farm in Israel
- ☑ **IMAGE** Jordan River Valley
- ☑ **MAP** Eastern Mediterranean Precipitation
- ☑ **INTERACTIVE SELF-CHECK QUIZ**
- ☑ **VIDEO** Physical Geography of the Eastern Mediterranean

Reading **HELP**DESK

Academic Vocabulary

- **range**
- **transform**

Content Vocabulary

- **rift valley**
- **kibbutz**
- **moshav**

TAKING NOTES: *Key Ideas and Details*

SUMMARIZING Use a graphic organizer like the one below to write about the water systems of the Eastern Mediterranean.

Water Systems of the Eastern Mediterranean		
The Jordan River	The Sea of Galilee	The Gulf of Aqaba

LESSON 1
Physical Geography of the Eastern Mediterranean

ESSENTIAL QUESTION • *How do physical systems and human systems shape a place?*

It Matters Because

Landforms, waterways, and climate affect settlement and economic activities in any region. These factors interact in the Eastern Mediterranean, which has been home to people for centuries. It has a diverse and changing landscape. While the subregion is generally temperate, water scarcity greatly affects agricultural production and other aspects of life.

Landforms

GUIDING QUESTION *How have physical features affected the human geography of the Eastern Mediterranean?*

The Eastern Mediterranean subregion includes the countries of Syria, Jordan, Lebanon, and Israel and the Palestinian territories. This area is also known as the Levant, French for "rising," referring to the sun rising in the East. Some of the most prominent landforms in the area are the Anti-Lebanon Mountains, the Syrian Desert, the Jordan Rift Valley, and the Negev Desert.

Syria, the northernmost country in the Eastern Mediterranean, is bordered by Turkey on the north, Iraq on the east, Jordan on the south, and Lebanon and Israel on the southwest. To the southwest of Syria is a territory called the Golan Heights. This territory consists of a rocky plateau that is officially part of Syria, but most of it has been occupied by Israel since 1967.

The Anti-Lebanon mountain **range** runs along the border between Syria and Lebanon. The mountains have low population density since they have thin soil that is not useful for agriculture. They are frequented mainly by nomadic herders. The highest point in the Anti-Lebanon Mountains is Mount Hermon. At 9,232 feet (2,814 m), it is also the highest point on the land surrounding the Mediterranean Sea. The southern and western slopes of Mount Hermon, which are located in the Golan Heights, have been developed for tourist activities. A favorite tourist activity is skiing. The summit of the mountain is located in both Lebanon and Syria.

To the east of the Anti-Lebanon mountain range is the Syrian Desert. It covers parts of Saudi Arabia, Jordan, and Iraq. The desert is composed of gravel, not sand. Sparsely populated by nomadic tribes, the desert is used by the general public as a roadway. Southern Syria also includes high lava plains. Some of Syria's great cities, such as Aleppo, lie on the steppes, a semi-arid region east of the coastal country. Another feature is a narrow plain on the eastern edge of the Mediterranean Sea that stretches from the Turkish to the Lebanese borders.

The Syrian Desert takes up four-fifths of Jordan's total territory. The Jordan Rift Valley is located on the western border of the country. A **rift valley** is a valley formed by the separation of tectonic plates. The Jordan Rift Valley is deep, reaching 1,312 feet (400 m) below sea level on the coast of the Dead Sea. This is the lowest point on the Earth's land surface. Between the Syrian Desert and the Jordan Rift Valley is a large plateau where most cities in Jordan are located.

Lebanon is mountainous. The Anti-Lebanon Mountains, which run from north to south, cover the eastern part of the country. The Lebanon Mountains lie in the western part of the country. They also run north to south and cover almost the entire length of the country. Both the Anti-Lebanon and Lebanon Mountains run parallel to the eastern coast of the Mediterranean Sea.

The northern and central parts of Israel, which border Lebanon, are hilly regions. The beaches of western Lebanon that border the eastern Mediterranean Sea continue south into Israel. The Galilee Mountains in the north of Israel are low in comparison with the Anti-Lebanon Mountains. The Galilee area consists of rocky terrain that is unsuitable for agriculture. The Negev Desert occupies most of southern Israel. The western side of the desert connects Israel with the Sinai Peninsula. The Negev Desert lies on top of fault lines and is geographically unique because of the erosion craters that dot the landscape.

☑ **READING PROGRESS CHECK**

Identifying What factors have limited the settlement of the Anti-Lebanon Mountains?

Water Systems

GUIDING QUESTION *What role has water played in the human systems of the Eastern Mediterranean?*

There are several water bodies in the Eastern Mediterranean subregion. These include the Euphrates and Jordan Rivers, the Sea of Galilee, the Dead Sea, the Gulf of Aqaba, and the Mediterranean Sea itself. The Euphrates River, which originates in Turkey, is the most important river in Syria and provides the entire country with water. A complex irrigation network has watered the valley and supported farming there for some 7,000 years. The Euphrates helps irrigate Syria today. The river was dammed in the 1970s. Lake al-Assad was formed behind the dam. It was named after Syria's dictator.

The Jordan River, which has a tributary in Syria, flows through all of the countries in the Eastern Mediterranean. The river flows along

range a series of things in a row; a series of mountains

rift valley a valley formed by the separation of tectonic plates

Located in the Jordan River valley, the Dead Sea is fed by the Jordan River north of the sea as well as other smaller streams and springs.

▼ CRITICAL THINKING
1. ***Analyzing Visuals*** How can you tell that the Dead Sea is located in the Jordan Rift Valley?

2. ***Classifying*** What type of border does the Jordan River form between Jordan, Israel, and the West Bank?

Spacephotos/age fotostock

This fish farm in the Gulf of Aqaba in Israel once used cages to raise fish.

▲ CRITICAL THINKING

1. *Analyzing Visuals* What other activities can be seen in the photo?

2. *Assessing* How might the fish in the cages have been contaminated given their location?

the Syrian-Lebanese border, through the Anti-Lebanon Mountains and near Mount Hermon. It then flows south into the Sea of Galilee. There the river becomes a natural border between the West Bank and Jordan before it empties into the Dead Sea. The river is important for irrigation and agriculture. Farmers grow oranges, bananas, beets, and other vegetables on the river's eastern banks.

Many of the streams in the desert regions of the Eastern Mediterranean subregion, mainly in inland areas, flow only intermittently, appearing suddenly and disappearing just as quickly. In the subregion's deserts, runoff from infrequent rainstorms creates wadis (WAH•dees), or streambeds that remain dry until a heavy rain. Irregular rainstorms often produce flash flooding. During a flash flood, wadis fill with so much sediment that they can rapidly become mudflows, or moving masses of wet soil. The mudflows pose dangers to humans and animals and can destroy agriculture.

The Sea of Galilee in western Israel supports some marine life, even though it is relatively salty for a freshwater body. The sea has supported communities of people for millennia due to its low elevation and moderate climate. The Dead Sea, which lies more than 1,300 feet (400 m) below sea level, is the lowest body of water on Earth. The salty waters make objects so buoyant that people can remain afloat in the sea without swimming.

The Gulf of Aqaba, located south of Israel and Jordan, connects the Eastern Mediterranean countries with the Red Sea and the Indian Ocean. The Mediterranean Sea allows people in the Eastern Mediterranean countries to navigate to Turkey and North Africa, as well as a number of countries in Western and Eastern Europe. Syria has a relatively short coastline that stretches between Turkey in the north and Lebanon in the south, while Lebanon and Israel have longer Mediterranean coastlines. The coastal areas are more densely populated than the interior areas of Syria and Lebanon.

Lake al-Assad is a large source of fish. It is also used for irrigation and agriculture in the surrounding lands and provides drinking water for the city of Aleppo. The southern part of the lake is home to trees such as the Aleppo pine. The lake is also a resting point for migrating birds.

☑ READING PROGRESS CHECK

Identifying What bodies of water does the Gulf of Aqaba connect?

Climate, Biomes, and Resources

GUIDING QUESTION *What defines the climate of the Eastern Mediterranean?*

Rainfall is limited throughout the Eastern Mediterranean subregion. Much of the area contains semi-arid (steppe) and arid (desert) climates. Areas with slightly greater moisture include the Mediterranean, highland, and humid subtropical climates.

Climate Regions and Biomes

The coastal regions of the Eastern Mediterranean have a Mediterranean climate with hot, dry summers and mild, rainy winters. Mediterranean climates support the growth of citrus fruits, olives, apples, apricots, and grapes. Natural vegetation is varied and includes oaks, evergreen conifers, and many types of wildflowers. Thickets of woody bushes and short trees known as Mediterranean shrub cover many parts of the region. Wildcats, wild boars, gazelles, hyenas, hares, badgers, and tiger weasels are found in the Eastern Mediterranean. The coastal areas are also important stopping places for more than 400 species of migratory birds.

Interior areas of the subregion have much less rainfall. As one moves inland, the climate becomes humid subtropical and then **transforms**, or changes, into semi-arid steppe and arid desert. The semi-arid climate zones usually receive about 14 inches (36 cm) of rain each year. Areas with less than 10 inches (25 cm) of rain per year are classified as deserts. Animal life in the deserts of the subregion includes geckos, lizards, and vipers. Plants in the deserts have adapted to the dry conditions and include a few scattered tree species, scrubs, and herbs. Overall, the landscape in Israel and Lebanon is extremely varied and can change within short distances. Syria and Jordan are dominated by semi-arid and arid climates.

Soils vary across the subregion just as climates do. Eastern Mediterranean coastal soils are alluvial, making them richer than the inland soils. East of the coastal mountain ranges, in the more arid inland parts of Syria and Jordan, topsoil is often lost to strong winds. These dry areas need to be irrigated in order to produce agricultural products because of the lack of precipitation. The Syrian Desert, for example, receives less than 5 inches (13 cm) of rainfall each year. Similarly, the coast of the Dead Sea is desert and cannot support many types of plant life aside from halophytes, or plants that can grow in salty soil.

transform to change in form or appearance

kibbutz a communal farm or settlement in Israel

moshav a cooperative settlement of small individual farms in Israel

GEOGRAPHY CONNECTION

Differences in annual precipitation across the Eastern Mediterranean have affected and continue to affect vegetation and human activities.

1. **PLACES AND REGIONS** How does the annual precipitation in Israel compare to that of Jordan?

2. **THE WORLD IN SPATIAL TERMS** Describe how the physical geography of Israel and Lebanon influence precipitation in the two countries.

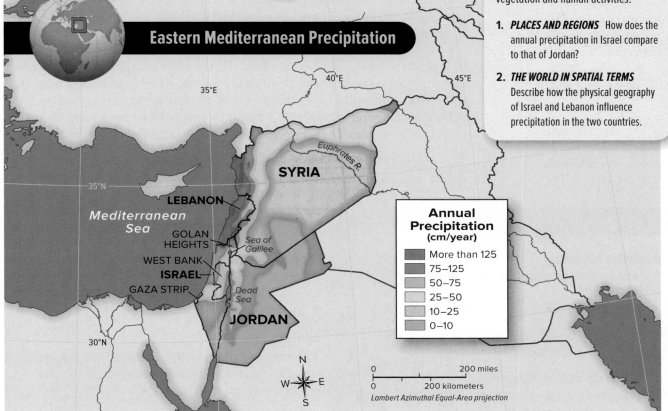

Eastern Mediterranean Precipitation

Annual Precipitation (cm/year)

- More than 125
- 75–125
- 50–75
- 25–50
- 10–25
- 0–10

200 miles
200 kilometers
Lambert Azimuthal Equal-Area projection

"In 2002 alone, there were over 15,000 fires that became out of control in this small country. The yearly fires are a huge threat to Lebanon's already vulnerable forests.

Most forest fires in Lebanon start when a fire set by a farmer to clear his orchards and fields of grasses and stubble gets out of control. The problem has become worse over the last decade because of changing land use.

In the past, villagers used scrubby bushes from the forests as fuel for cooking, which kept the dry undergrowth thin. Now there's electricity, so the undergrowth stays thick—perfect fuel for fires."

—World Wildlife Fund,
"Lebanon's Forests: Facing
New Threats"

DBQ *ANALYZING PRIMARY SOURCES* How has technological progress resulted in threats to the forests?

Natural Resources

Many countries in the Eastern Mediterranean were once heavily wooded in the past, but have since cut down most of their trees. Cedar trees have been used by people in the region dating back to around 2500 B.C. The cedar trees are prized natural resources due to their fragrance and commercial value as a source of lumber. Most are found in the western part of the Anti-Lebanon Mountains. The cedar appears on the flag of Lebanon and is esteemed as a symbol of happiness and prosperity.

Minerals are also important to the economies of the Eastern Mediterranean subregion. Minerals—such as bromine and magnesium, gypsum from the Negev, and marble in the Galilee area—are produced. In addition, Syria produces chrome and manganese ores, asphalt, iron ore, and rock salt. Jordan has fewer mineral resources.

Recently, large oil and natural gas reserves were discovered below the waters of the Mediterranean Sea in this region, including offshore parts of Israel. There have been other finds in the Palestinian territories. These finds are still being explored, and currently, production of oil and gas is of minor importance to these countries. While petroleum exports could enrich the region, they also would make the Eastern Mediterranean economies fluctuate, or rise and fall, in the global market. Some countries in the subregion seek to diversify their economies so that they are not reliant on single commodities. Many countries are also turning to tourism as a source of revenue.

All Eastern Mediterranean countries have relied on agriculture for centuries. However, the Israeli government has always made a specialized effort to create farming communities. They supply these communities with farming equipment and irrigation education. The two types of communities are the **kibbutz** and the **moshav**. The communities have supported the cultivation of crops such as peanuts, sugar beets, and cotton in addition to dairy farming. Extensive farm mechanization has recently contributed to the increase in the value of Israel's agricultural production. Through these efforts, Israel produces a major portion of its food supply and imports the remainder.

The biggest challenge for agriculture in the Eastern Mediterranean is the lack of water. Many farmers have begun to use drip irrigation as an alternative to more costly and wasteful flood-field or spraying irrigation methods. Thin soil makes cultivation of certain crops very difficult in certain parts of the subregion. Underground springs created by cracked limestone make irrigation of the lower slopes of Lebanon and Israel very successful, however.

☑ **READING PROGRESS CHECK**

Naming What is the dominant climate of the Eastern Mediterranean?

LESSON 1 REVIEW

Reviewing Vocabulary

1. *Specifying* What geographical characteristics define the Negev Desert?

Using Your Notes

2. *Summarizing* Use your graphic organizer to discuss the water systems of the Eastern Mediterranean subregion.

Answering the Guiding Questions

3. *Making Generalizations* How have physical features affected the human geography of the Eastern Mediterranean?

4. *Evaluating* What role has water played in the human systems of the Eastern Mediterranean?

5. *Synthesizing* What defines the climate of the Eastern Mediterranean?

Writing Activity

6. *Informative/Explanatory* Write a paragraph describing plants found in the Eastern Mediterranean subregion, including the climate zones and biomes where they are found.

networks

There's More Online!

☑ **IMAGE** Family and Religion in the Eastern Mediterranean

☑ **GRAPHS** Ethnic Groups of the Eastern Mediterranean

☑ **GRAPH** GDP per Capita

☑ **INTERACTIVE SELF-CHECK QUIZ**

☑ **TIME LINE** Israeli-Palestinian Conflict: An Elusive Peace Process

☑ **VIDEO** Human Geography of the Eastern Mediterranean

LESSON 2
Human Geography of the Eastern Mediterranean

ESSENTIAL QUESTION · *How do physical systems and human systems shape a place?*

Reading HELPDESK

Academic Vocabulary

- **rely**
- **output**

Content Vocabulary

- **monotheism**
- **prophet**
- **mosque**
- **stateless nation**

TAKING NOTES: *Key Ideas and Details*

SUMMARIZNG Use a graphic organizer like the one below to record details about how the settlement of Israel affected population patterns in the Eastern Mediterranean.

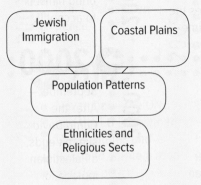

Jewish Immigration

Coastal Plains

Population Patterns

Ethnicities and Religious Sects

IT MATTERS BECAUSE

The Eastern Mediterranean is the birthplace of Judaism, Christianity, and Islam. Throughout history, including more recent times, the subregion has been characterized by political instability and conflict based on a combination of ethnic, religious, cultural, economic, and political factors.

History and Government

GUIDING QUESTION *How have Judaism, Christianity, and Islam shaped the politics and culture of the Eastern Mediterranean?*

As a bridge between Europe, Africa, and Asia, the Eastern Mediterranean has been influenced by cultural groups from each of these continents throughout history. The rise of influential religions has profoundly affected its history, geography, and culture.

Civilizations and Religion

The subregion has been under the control of powerful cultures and empires over the centuries. The capital city of Syria, Damascus, is one of the oldest continuously settled urban centers in the world. Judaism, Christianity, and Islam share the same territory, and they also share many beliefs, especially **monotheism**, or belief that there is only one God.

Judaism is the oldest of these monotheistic religions. Jews trace their origin to the ancient Israelites who created the kingdom of Israel along the Eastern Mediterranean coast. The kingdom, which included what is now Israel and the West Bank, had Jerusalem as its capital and religious center. Judaism has both ritual and ethical requirements. Ritual requirements include daily prayer, observing the Sabbath, holidays and dietary laws, and studying Jewish texts. Jewish ethical requirements include giving to charity, loving your neighbor, being kind to strangers, healing the world, pursuing justice, avoiding gossip, and seeing the dignity and worth of every individual. These laws are described in the Hebrew Bible, which contains the Torah, the books of the prophets, and the sacred writings. Although

CONSERVATION

🌧 **RAINFALL** is scarce, leaving little fresh drinkable water.

〰 **DRIP IRRIGATION** helps conserve water and is used in more than 90% of Israel's agriculture.

LESSON 3
People and Their Environment: The Eastern Mediterranean

ESSENTIAL QUESTION • *How do physical systems and human systems shape a place?*

Reading HELPDESK

Academic Vocabulary

- **preliminary**
- **adequate**

Content Vocabulary

- **fertilizer**
- **pesticide**
- **desertification**
- **overgrazing**

TAKING NOTES: *Key Ideas and Details*

SUMMARIZING Use a graphic organizer like the one below to take notes on the ways agricultural pollutants affect the Eastern Mediterranean.

Eastern Mediterranean: People and Their Environment	
Agricultural Pollutants	Process of Desertification

IT MATTERS BECAUSE

The Eastern Mediterranean subregion has a long and important history of human habitation and human impact on the environment. That impact has become an increasingly serious problem that some people and governments are currently addressing.

Managing Resources

GUIDING QUESTION *What resources are at risk in the Eastern Mediterranean?*

The Mediterranean Sea is at risk for a number of reasons. Overfishing, contamination, a rise in sea surface temperatures due to climate change, and the introduction of invasive species have all caused environmental damage. The Suez Canal connects the Red Sea to the Mediterranean Sea, allowing invasive species to enter. Invasive fish from the Red Sea have left Eastern Mediterranean reefs almost bare of native species and have greatly changed the ecology of the region. Also, the waters of the Eastern Mediterranean have been nearly depleted of fish due to overfishing. Another problem caused by human activity is water pollution. Sewage, oil spills, and chemical **fertilizers** and **pesticides** used for agriculture contaminate the Mediterranean Sea and have limited its use as a water source. The pollutants have also contributed to the vast depletion of marine life. Water pollution affects the limited number of freshwater supplies in the area and adds to the region's problem of water scarcity.

Several tectonic plates meet in the Eastern Mediterranean Sea. This contact zone is an interesting deep-sea terrain with unique landforms and aquatic life. Scientists have only recently begun to study the ecosystems of these deep waters. They still have much to learn. However, because the fish resources of the continental shelf (the areas between the shoreline and the deeper waters) are seriously depleted, Mediterranean commercial fishers are turning to these deep-sea habitats as a new source of fish. This is putting additional strain on the natural resources of the Eastern Mediterranean subregion.

(l)Paul Gapper/Alamy, (tc)©Tomasz Grzyb/Demotix/Corbis, (tr)©Pallava Bagla/Corbis

Air pollution is also a serious problem in the Eastern Mediterranean. High levels of urbanization and immigration to the area in the past few decades have increased air pollution. Unplanned urbanization not only strains city services but also amplifies the air pollution problem produced largely by using old buses and taxis for transportation in densely populated urban communities. The unregulated burning of fossil fuels and wood for heating also contributes to air pollution. Air pollution problems have been difficult to address, mainly due to long-term unrest in the subregion and a resulting lack of environmental policies and services.

Heat waves in the subregion compound the problems of air pollution in these cities. Air pollutants have also been linked to health hazards. In the past few years, an increasing number of patients have been admitted to hospitals for respiratory diseases. Many cases of respiratory diseases are a result of poor air quality.

The wetlands of the Eastern Mediterranean are another resource at risk. They serve as an important stopping place for over 250 species of migratory birds. In Lebanon, the Aammiq Wetland is the most significant remaining freshwater wetland in the country. The area covered by the Aammiq today is only a remnant of the marshes and lakes that once existed. Bird populations in the region also face serious threat from hunters. The government of Lebanon banned hunting in 1994, but environmental groups report that the hunting laws are ignored. Soaring aquatic birds and raptors are most at risk. This is a worldwide issue since these birds migrate to and from breeding grounds outside the subregion.

☑ READING PROGRESS CHECK

Identifying What risks to human health are posed by pollution in the Eastern Mediterranean?

fertilizer a chemical or natural substance added to soil or land to increase its fertility

pesticide a chemical used to kill insects, rodents, and other pests

Although desalination is expensive, more countries are using it as alternative water sources become scarce.

▼ CRITICAL THINKING

1. Assessing What is Israel's annual desalination output? How much of the country's drinking water is obtained from desalination?

2. Evaluating Describe the progress Israel has made in regard to water conservation.

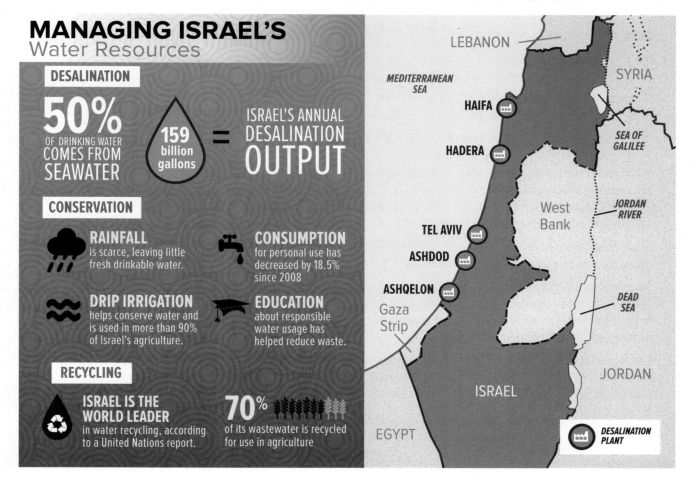

MANAGING ISRAEL'S
Water Resources

DESALINATION

50% OF DRINKING WATER COMES FROM SEAWATER

159 billion gallons = ISRAEL'S ANNUAL DESALINATION **OUTPUT**

CONSERVATION

RAINFALL is scarce, leaving little fresh drinkable water.

CONSUMPTION for personal use has decreased by 18.5% since 2008

DRIP IRRIGATION helps conserve water and is used in more than 90% of Israel's agriculture.

EDUCATION about responsible water usage has helped reduce waste.

RECYCLING

ISRAEL IS THE WORLD LEADER in water recycling, according to a United Nations report.

70% of its wastewater is recycled for use in agriculture

LEBANON
MEDITERRANEAN SEA
SYRIA
HAIFA
SEA OF GALILEE
HADERA
West Bank
JORDAN RIVER
TEL AVIV
ASHDOD
ASHQELON
Gaza Strip
DEAD SEA
JORDAN
ISRAEL
EGYPT

🏭 DESALINATION PLANT

Human Impact

GUIDING QUESTION *What human activities have affected the physical environment of the Eastern Mediterranean?*

desertification process in which arable land becomes desert

overgrazing grazing so heavily that the vegetation is damaged and the ground erodes

Long-term deforestation and desertification are other problems in the Eastern Mediterranean. **Desertification** refers to a process in which arable land, or land that is suitable for growing crops, turns into desert. Desertification is caused by a number of factors, such as **overgrazing** and deforestation. Overgrazing in certain areas of Syria and Jordan has led to the desertification of grasslands and deforestation of the mountain ranges. This change has contributed to climate change in the subregion. Overgrazing is also a significant cause of soil erosion.

Lebanon's legendary forests have a long history of exploitation. During the Middle Ages the forests were cleared for farmland, and trees were used for fuel and construction. In the early 1900s, the Ottomans controlled much of the Eastern Mediterranean. They pursued a policy of very aggressive deforestation in order to fuel their railways and keep up with industrialized Europe. The forests continue to face threats from overgrazing, unregulated tourism, and a high occurrence of forest fires. Lebanon had once been almost completely covered in forests, but by 2012 just 5 percent of the land was covered with forest. The Lebanese cedar—long a prized symbol of the country—survives today only in a small number of patches, although there have been efforts to conserve

ANALYZING PRIMARY SOURCES

Desertification in Syria

The UN has stated that 80 percent of Syria's land is at risk for desertification. The causes include both human-made and natural factors. Abdulla Tahir Bin Yehia, head of the Food and Agriculture Organization (FAO) in Syria, discusses the causes of desertification.

❛❛'Traditionally, communities had methods to avoid desertification, such as rotation or leaving an area unused. This allowed the vegetation to grow back,' said Bin Yehia. 'But modernization and centralization takes the decision out of their hands.'

He said rising demand for meat from a growing and increasingly affluent population was also contributing to land degradation.

'Syria's estimated livestock stands at 14–16 million. But it is only that low because many died during the drought. Prior to this the national herd stood at around 21 million. We need to study how much livestock the land can take,' said Bin Yehia.

Desertification can be irreversible, such as when an aquifer dries out and the land sinks in on itself, destroying the structure. Flora and fauna species that lose their natural habitat can become extinct.❜❜

—United Nations, "Syria: Act Now to Stop Desertification, Says FAO," *Humanitarian News and Analysis*, June 15, 2010

Modern farming practices in Syria do not allow fields to lie fallow. This puts them at risk for desertification when droughts occur.

DBQ ▲ **CRITICAL THINKING**

1. Assessing How did traditional methods protect the land from desertification?

2. Describing How can desertification become irreversible?

and protect these trees. This deforestation destroys animal habitats. As a result, large mammals such as wolves and wild boars—as well as migratory birds that stop in the forests on their way from Europe to Africa—are endangered.

Agriculture is one of Israel's most highly developed economic activities. It remains successful despite the lack of freshwater resources in the country. Israeli lands are not naturally suited to large-scale agricultural production. Only about 15 percent of Israel's land is arable. Israel is a pioneer in water management, including drip irrigation, water desalination, and water purification treatment. It is also a world leader in water reclamation, recycling 75 percent of all wastewater and using it for agricultural irrigation. Another major source of water is from the Sea of Galilee in northern Israel. It is an important freshwater source for the subregion. In Israel, water is carried through canals in the National Water Carrier project built in the 1960s. Desalination plants and water recycling programs also provide water for human use and for agricultural irrigation. Israel has the world's biggest seawater desalination plant with a seawater treatment capacity of 624,000 million cubic liters a day. In 2014, desalination was expected to provide 80 percent of Israel's drinkable water. Modern drip irrigation developed in Israel is now used worldwide in arid climates. New Israeli agricultural methods to increase crop yields will also become important as land is exhausted.

 READING PROGRESS CHECK

Explaining Why has Lebanon experienced so much deforestation?

This landfill in the coastal city of Saida, Lebanon, has become a major environmental hazard.

▲ CRITICAL THINKING
1. *Hypothesizing* Why might a city locate a landfill along the coast?
2. *Sequencing* How do landfills cause the pollution of waterways?

Addressing the Issues

GUIDING QUESTION *How are environmental issues being addressed in the Eastern Mediterranean?*

In recent years, Eastern Mediterranean governments have made real attempts to address serious environmental issues. In 2011 the Israeli Clean Air Law came into effect. This legislation provides a framework for reducing air pollution by imposing responsibilities on national and local government authorities and the country's industrial plants. The law aims to improve air quality and protect biodiversity. Also in 2011, the Ministry of the Environment in Lebanon, working with the United Nations Environmental Program, launched a project whose goal is sustainable management of marine and coastal biodiversity.

The Jordanian government has instituted an environmental education program in schools to address that country's severe water shortage. The importance of maintaining biodiversity in the country is also taught as part of the program. The Royal Society for the Conservation of Nature (RSCN) is an organization in Jordan with broad authority to manage the natural resources of the country. It has outlawed semiautomatic hunting weapons in Jordan. The government has determined how many and what types of animals can be hunted in any given season. The RSCN has also addressed biodiversity by

New agricultural methods, such as those in this high-tech greenhouse in Israel, are being developed to increase crop yields.

▲ CRITICAL THINKING

1. Assessing Why is it important for governments to collaborate with international organizations in developing new agricultural methods?

2. Speculating How could research help increase agricultural yields?

preliminary something done in preparation for something more important

adequate satisfactory or acceptable

planning a system of wildlife reserves. One success story is that of the Arabian oryx, a type of antelope with large horns. The Arabian oryx was extinct in the wild but was reintroduced to the wild in the Azraq wetlands in the deserts of eastern Jordan in 1978. The populations have since thrived under legal protection from hunters and habitat loss.

A number of international organizations have offices throughout the Eastern Mediterranean. These include the United Nations Development Program (UNDP), the World Wildlife Fund (WWF), and the International Union for the Conservation of Nature (IUCN). Their urgent projects address desertification, unsustainable water extraction and use, biodiversity and habitat loss, and threats to sensitive marine ecosystems. The **preliminary** plan for most of these organizations is to address environmental issues by introducing legislation. They have been unsuccessful in getting most governments to take environmental projects seriously because most projects that are put on legislative agendas are severely underfunded. Local nongovernmental organizations, such as the Society for the Protection of Nature in Lebanon, also address environmental issues. However, these organizations are generally smaller and less effective than international environmental organizations.

In 2003 the WWF began a campaign to protect the cedar trees of Lebanon from fires and destruction caused by overgrazing. Fires threaten about 28 percent of Lebanon's forests. The project, which has met with some success, utilizes information and communication technologies (ICTs) to achieve its goal of protecting the endangered forests.

All governments in the subregion could more **adequately** address air and water pollution. Investing in energy-efficient public transportation systems would help alleviate the air pollution caused by urbanization. Furthermore, marine life could gradually increase in the Mediterranean Sea if countries would employ sustainable fishing methods and legislate to protect the sea from agricultural runoff and other pollution.

☑ READING PROGRESS CHECK

Describing How has the Jordanian government addressed the issue of water scarcity in the country?

©Pallava Bagla/Corbis

LESSON 3 REVIEW

Reviewing Vocabulary
1. Naming Define overgrazing and desertification and write two sentences describing the connection between them.

Using Your Notes
2. Explaining Use your notes from your graphic organizer to describe the impact of agricultural pollutants in the Eastern Mediterranean.

Answering the Guiding Questions
3. Stating What resources are at risk in the Eastern Mediterranean?

4. Categorizing What human activities have affected the physical environment of the Eastern Mediterranean?

5. Differentiating How are environmental issues being addressed in the Eastern Mediterranean?

Writing Activity
6. Informative/Explanatory Pick one of the international organizations discussed in the lesson. Using the Internet to search its website, pick one of its Eastern Mediterranean projects. Write a paragraph identifying the main issue that the project seeks to address, why the issue is important, and what the organization plans to do about the issue.

Directions: On a separate sheet of paper, answer the questions below. Make sure you read carefully and answer all parts of the questions.

Lesson Review

Lesson 1

1 *Making Generalizations* How do water systems and water scarcity affect economic activities in the Eastern Mediterranean?

2 *Describing* How does the climate change as one travels inland, away from coastal regions, in the Eastern Mediterranean?

3 *Explaining* Why do some economies in the Eastern Mediterranean subregion often fluctuate, or rise and fall, in the global market?

Lesson 2

4 *Exploring Issues* How have political activities affected settlement and population patterns in the Eastern Mediterranean?

5 *Analyzing* What activities have encouraged economic growth in the economies of the Eastern Mediterranean?

6 *Making Connections* Describe the three major world religions practiced in the Eastern Mediterranean. Why is the Eastern Mediterranean an important connection for them?

Lesson 3

7 *Finding the Main Idea* How have human activities affected the Eastern Mediterranean environment?

8 *Making Connections* What efforts have been made to preserve natural resources in the Eastern Mediterranean?

9 *Evaluating* How successful have international organizations been at addressing environmental issues in the Eastern Mediterranean?

21st Century Skills

Use the graph to answer the following questions.

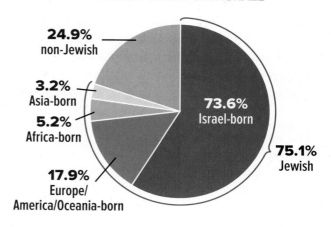

ETHNIC GROUPS IN ISRAEL

24.9% non-Jewish
3.2% Asia-born
5.2% Africa-born
73.6% Israel-born
75.1% Jewish
17.9% Europe/America/Oceania-born

Source: CIA World Factbook

10 *Creating and Using Graphs, Charts, Diagrams, and Tables* From what part of the world have most non-native Israeli Jews emigrated?

11 *Time, Chronology, and Sequencing* What factors might be responsible for attracting non-native Jews to Israel?

Exploring the Essential Question

12 *Drawing Conclusions* Using a map of the Eastern Mediterranean, describe the locations of the physical features that exist in those countries—such as lakes, mountains, and rivers—and the major capital cities. Based on this information, write a paragraph describing how geographic features and urbanization in those countries are related.

College and Career Readiness

13 *Problem Solving* Think about the issue of water scarcity in the Eastern Mediterranean. As an environmental geographer, where would you recommend desalination plants be built? Consider population centers, energy needs, and water sources. Explain your reasoning.

Need Extra Help?

If You've Missed Question	1	2	3	4	5	6	7	8	9	10	11	12	13
Go to page	391	393	394	398	400	395	402	405	406	407	407	390	391

Directions: On a separate sheet of paper, answer the questions below. Make sure you read carefully and answer all parts of the questions.

DBQ Analyzing Primary Sources

Use the document to answer the following questions.

In 1917, when Great Britain ruled parts of the Eastern Mediterranean subregion, the government issued the Balfour Declaration. This letter promised that the British government would support the Zionist effort to secure a homeland for the Jewish people. In time, the Declaration helped to lay the foundation for the establishment of the present-day state of Israel.

PRIMARY SOURCE

"Foreign Office
November 2nd, 1917
Dear Lord Rothschild,
I have much pleasure in conveying to you, on behalf of His Majesty's [the British] Government, the following declaration of sympathy with Jewish Zionist aspirations which has been submitted to, and approved by, the Cabinet:

'His Majesty's Government view with favour the establishment in Palestine of a national home for the Jewish people, and will use their best endeavours to facilitate the achievement of this object, it being clearly understood that nothing shall be done which may prejudice the civil and religious rights of existing non-Jewish communities in Palestine, or the rights and political status enjoyed by Jews in any other country.'

I should be grateful if you would bring this declaration to the knowledge of the Zionist Federation.

Yours sincerely
Arthur James Balfour"

—The Balfour Declaration, November 2, 1917

14 *Evaluating* How did the Balfour Declaration contribute to the formation of a Jewish state?

15 *Identifying Central Issues* What did the British say about the rights of the people who already lived in this area and about the rights of Jews elsewhere?

Critical Thinking

16 *Identifying Cause and Effect* What factors have created both Palestinian and Jewish refugees in the subregion?

17 *Evaluating* How have demographics in the subregion changed since 1947? Be sure to use examples from the chapter and include information on Jordan, Lebanon, and Syria.

18 *Drawing Inferences* What might result if countries in the subregion do not focus on immediate environmental issues?

Applying Map Skills

Refer to the Unit 5 Atlas to answer the following questions.

19 *Human Systems* Which capital cities of the Eastern Mediterranean are located near major waterways?

20 *Physical Systems* Name the major mountain ranges in the Eastern Mediterranean.

21 *Human Systems* Think about your mental map of Israel. Using this mental map, describe the location of major areas of conflict.

Research and Presentation

22 *Evaluating* Use Internet and library resources to gather information about the factors that have shaped Lebanon's current political system. Create a multimedia presentation outlining the country's current political structure, its interaction with the global economy, and the most pressing environmental issues the country is facing. The presentation should include facts, figures, and images regarding ethnic and religious diversity in the country.

Writing About Geography

23 *Informative/Explanatory* Use standard grammar, spelling, sentence structure, and punctuation to write a one-page essay describing the origins of the Israeli-Palestinian conflict. What role might international organizations play in resolving this conflict?

Need Extra Help?

If You've Missed Question	14	15	16	17	18	19	20	21	22	23
Go to page	408	408	399	399	405	361	360	361	397	396

The Northeast

networks

There's More Online about the Northeast.

CHAPTER 17

ESSENTIAL QUESTION • *How do physical systems and human systems shape a place?*

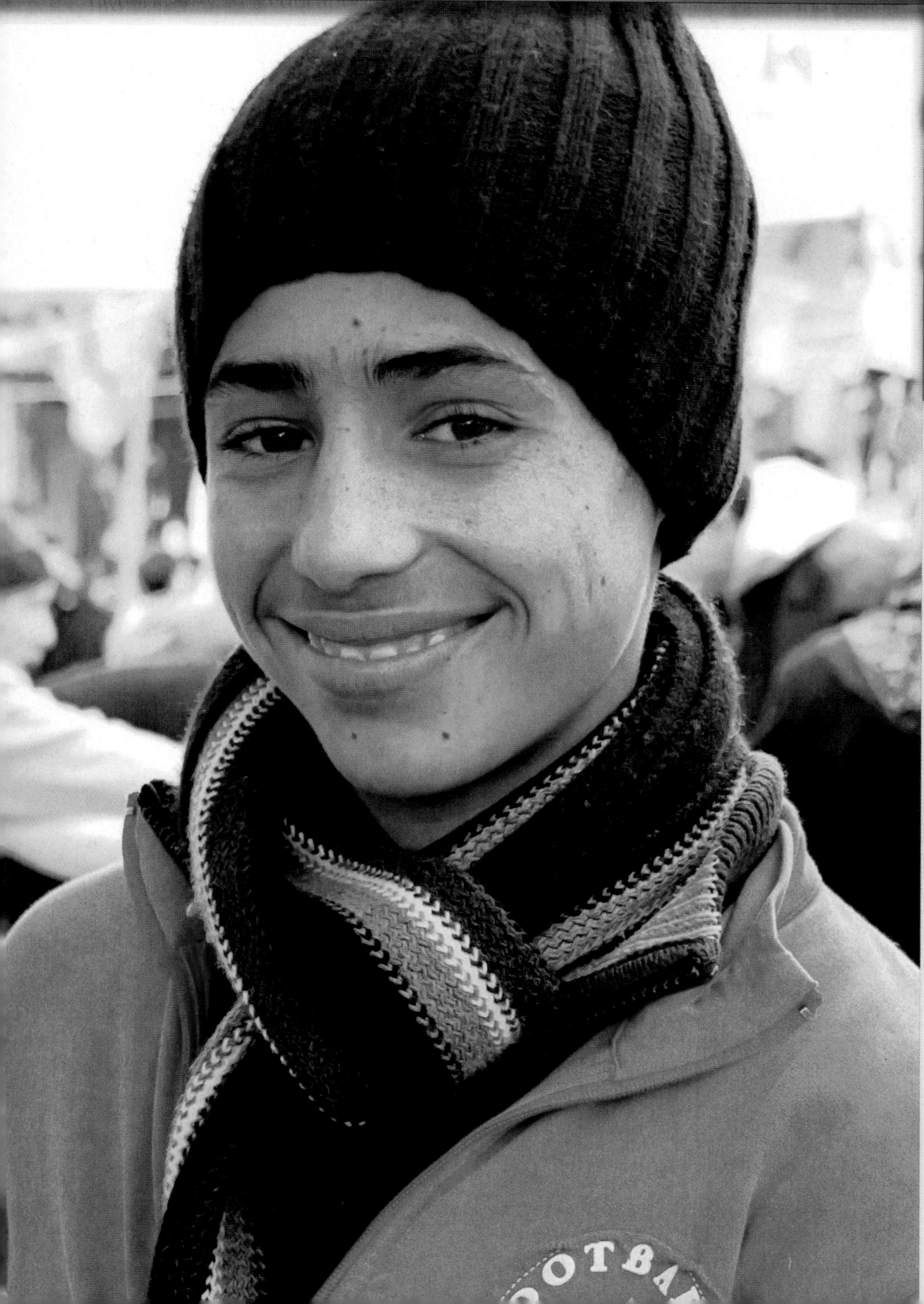

Why Geography Matters
A Stateless Nation: The Kurds

Lesson 1
Physical Geography of the Northeast

Lesson 2
Human Geography of the Northeast

Lesson 3
People and Their Environment: The Northeast

Geography Matters...

In the Northeast, cultures are fundamentally tied to religion. Islam is the most widely practiced of them all. For hundreds of years, it has defined people's lives in the subregion. The Northeast is not homogeneous, however. Many different ethnic groups live in the Northeast, including Arabs, Persians (Iranians), Turks, and Kurds.

◀ Turkish youth in an outdoor market in İstanbul, Turkey

Rebecca Erol/Alamy

a stateless nation:
the Kurds

A nation is a group of people who share a common language, religion, culture, and common institutions. When a national group has a state or country of its own, it is called a nation-state. Albania, Iceland, and Japan are examples of nation-states. Other nations, such as the Kurds of the Northeast, do not have their own country. They are considered "stateless nations."

The Distribution of Kurdish People

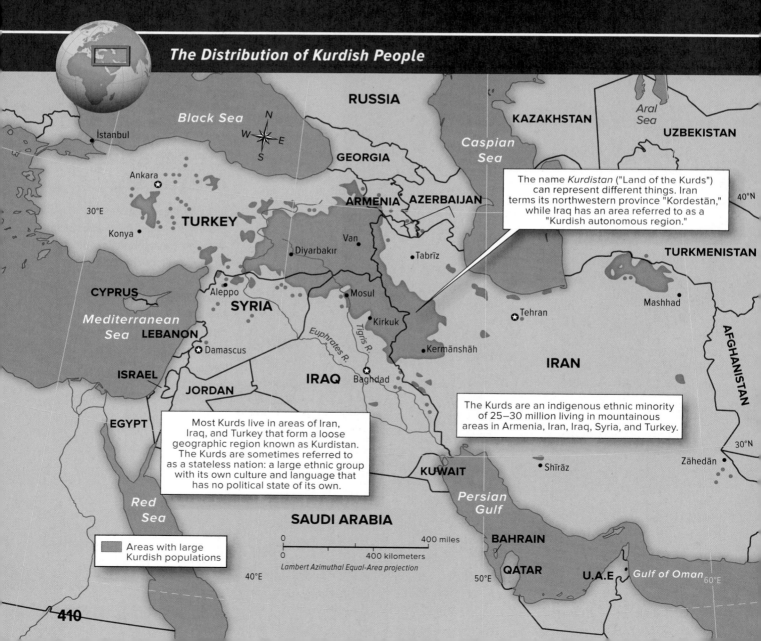

The name *Kurdistan* ("Land of the Kurds") can represent different things. Iran terms its northwestern province "Kordestān," while Iraq has an area referred to as a "Kurdish autonomous region."

The Kurds are an indigenous ethnic minority of 25–30 million living in mountainous areas in Armenia, Iran, Iraq, Syria, and Turkey.

Most Kurds live in areas of Iran, Iraq, and Turkey that form a loose geographic region known as Kurdistan. The Kurds are sometimes referred to as a stateless nation: a large ethnic group with its own culture and language that has no political state of its own.

Areas with large Kurdish populations

0   400 miles
0   400 kilometers
Lambert Azimuthal Equal-Area projection

What is a stateless nation?

Who are the Kurds?

What is the nation-state of Kurdistan?

The term *stateless nation* refers to a group that identifies itself as a nation based on common ethnic, linguistic, and religious identity, but that lacks majority status in any nation-state. Some stateless nations are native minority populations within a larger state. Examples of such minority populations include the Uyghur people in China and the Catalan people in Spain. Others are dispersed across several countries in numbers insufficient to form a majority in any one state. The Yoruba people of Nigeria, Benin, and Togo are an example of this latter category. Another such group is the Kurdish people. The Kurds are an Islamic pastoral people who have lived in parts of the Northeast for at least three thousand years, but have never achieved the status of a nation-state. Today they constitute a stateless nation of an estimated 25 to 30 million people. The Kurds are the largest ethnic group in the world that does not have its own country.

1. Human Systems Why are the Kurds considered a stateless nation?

The Kurds are the fourth-largest ethnic group in the Northeast, after Arabs, Persians, and Turks. The majority are Sunni Muslims, although some are Shia Muslims, Christians, or followers of other faiths. Kurds speak dialects of the Indo-European language family that are related to Persian. They live in an area that straddles mountain and plateau regions of eastern Turkey, northwestern Iran, northern Iraq, northeastern Syria, southern Armenia, and eastern Azerbaijan. There are also Kurdish communities in Georgia, Kazakhstan, Lebanon, and parts of Europe. For centuries, their lands were part of the vast multiethnic Ottoman and Persian Empires. After World War I, an independent Kurdish state was proposed, but that plan was abandoned. The Kurds found their lands divided among several new nation-states. Since then, Kurdish nationalist movements have led to conflicts with those countries. During the Iran-Iraq War (1980–1988), the Iraqi regime of Saddam Hussein embarked on a campaign to crush Kurdish resistance. This included the use of chemical weapons against Iraqi Kurdish civilians. Land mines planted during the war continue to take the lives of Kurds who live along the Iran-Iraq border.

2. Places and Regions Why do the Kurds want their own state?

Kurdistan literally means "Land of the Kurds." Calls for an independent Kurdish state date back to at least the 1800s. In 1880 a Kurdish sheikh in Turkey wrote to an American missionary, explaining the need for an independent Kurdish state: "We are . . . a nation apart. We want our affairs to be in our own hands." In the twenty-first century, however, there is still no Kurdish nation-state. Some Kurds (particularly educated urban residents) have formed a nationalist movement aimed at gaining autonomous (self-governing) rule. After years of brutal suppression and the fall of the regime of Saddam Hussein, Iraqi Kurds were able to gain autonomous civil authority over a section of northern Iraq, and in 2005 a Kurdish parliament was formed. Kurdish populations in Syria, Turkey, and Iran—encouraged by the success of the Iraqi Kurds—continue to push for more political autonomy.

3. Human Systems What progress have Kurds made toward achieving the goal of an autonomous nation-state?

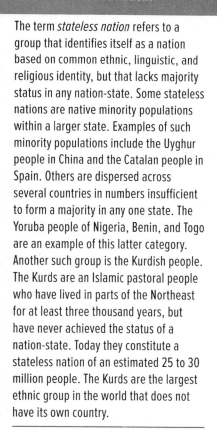

THERE'S MORE ONLINE

READ a primary source quote about Kurds in Iraq • *VIEW* an image of the Kurdish people

networks

There's More Online!

☑ **IMAGE** A River Flows Through Coastal Turkey

☑ **MAP** Physical Map: Turkey, Iran, and Iraq

☑ **INTERACTIVE SELF-CHECK QUIZ**

☑ **VIDEO** Physical Geography of the Northeast

LESSON 1
Physical Geography of the Northeast

Reading HELPDESK

Academic Vocabulary

• **exceed**
• **sustain**

Content Vocabulary

• **pastoralism**

TAKING NOTES: *Key Ideas and Details*

IDENTIFYING As you read about the physical geography of the Northeast, use a graphic organizer like the one below to record the major features of the subregion.

Landforms of the Northeast

ESSENTIAL QUESTION • *How do physical systems and human systems shape a place?*

IT MATTERS BECAUSE
The mountains that dominate the Northeast have shaped and continue to shape human history. Some of the earliest civilizations sprang from this ancient land. Today, the Northeast subregion is home to the countries of Iran, Iraq, and Turkey.

Landforms

GUIDING QUESTION *What features dominate the physical geography of the Northeast?*

The Northeast is a largely mountainous area, but it also features a range of other significant landforms and environments, particularly the broad plain of the Tigris and Euphrates Rivers in Iraq. The subregion occupies a part of the Earth where several tectonic plates converge. The movement of these plates has produced the many mountain ranges that dominate the Northeast. Movement along the fault lines produces earthquakes.

The North Anatolian Fault extends much of the distance across Turkey. At about 750 miles (1,200 km) long, it rivals the San Andreas Fault of California in length. As the plates slide past each other in an east-west direction, they can rupture and cause destructive earthquakes. Over the last seven decades, the North Anatolian Fault has been among the most active faults in the world, with nearly a dozen quakes **exceeding** 6.7 on the Richter scale of earthquake magnitude. Recently, the East Anatolian Fault in eastern Turkey has also become more active, causing several large quakes in the twenty-first century. Earthquakes are also frequent in Iran. In recent years, several devastating quakes have taken tens of thousands of lives.

Ancient volcanic activity also contributed to the mountainous terrain of the Northeast. Even today, a number of volcanoes dot the region. Turkey alone has over a dozen volcanoes within its borders.

The Pontic Mountains and the Taurus Mountains rise from the Turkish landscape. The highest peak in the Pontic range rises to 12,900 feet (3,932 m). The Taurus Mountains have many peaks which

exceed 10,000 feet (3,048 m). Between these ranges lies the Anatolian Plateau, or central massif. It stands 2,000 to 5,000 feet (610 to 1,524 m) above sea level. The plateau also has extensive sections of relatively flat terrain. East of the Pontic range, camel-backed Mount Ararat—at a height of 16,945 feet (5,165 m)—overlooks the Turkish-Iranian border.

To the east of Turkey, in Iran, lie the mountains of the Elburz and Zagros ranges. The Elburz Mountains cover the northern portion of Iran with a series of peaks that often exceed 9,000 feet (2,743 m). The tallest of these mountains is Iran's highest peak—the volcanic Mount Damāvand, which rises to 18,934 feet (5,771 m). The Zagros Mountains were formed by the collision of the Eurasian and Arabian plates. This mountain system stretches for 932 miles (1,500 km) from southwestern Iran and extending into Iraq at its northwestern end. The tallest peak in this range, Zard Kūh in Iran, is 14,921 feet (4,548 m) high.

In the western part of the subregion lies the Anatolian Peninsula, the location where the continents of Europe and Asia meet. This great peninsula is surrounded to the north by the Black Sea, to the east and south by the eastern Taurus Mountains and the Mediterranean Sea, and to the west by the Aegean Sea. It was historically referred to as Asia Minor. Today, it constitutes the Asian part of Turkey.

Over thousands of years, rivers created an alluvial plain that covers the central and southern third of Iraq. This alluvial plain is a low-lying area that is frequently marshy and covered with lakes and swamps. It is this land that produced and **sustained** the earliest civilizations in human history.

exceed to be greater than

sustain to give support to

✔️ READING PROGRESS CHECK

Describing How has tectonic activity shaped the Northeast?

GEOGRAPHY CONNECTION

The Northeast is a largely mountainous area that includes some of the most important waterways in the world.

1. **PHYSICAL SYSTEMS** Name the two major mountain ranges in Turkey and describe their locations.

2. **ENVIRONMENT AND SOCIETY** On the map, find the Dardanelles, Sea of Marmara, and Bosporus. How are these features important to trade?

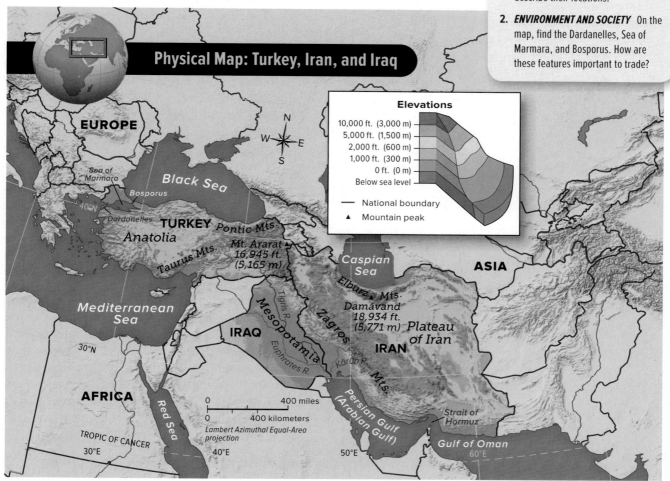

Physical Map: Turkey, Iran, and Iraq

Elevations
10,000 ft. (3,000 m)
5,000 ft. (1,500 m)
2,000 ft. (600 m)
1,000 ft. (300 m)
0 ft. (0 m)
Below sea level

— National boundary
▲ Mountain peak

EUROPE
Sea of Marmara
Bosporus
Black Sea
Dardanelles
TURKEY
Pontic Mts.
Anatolia
40°N
Mt. Ararat 16,945 ft. (5,165 m)
Taurus Mts.
Caspian Sea
ASIA
Mediterranean Sea
Tigris R.
Zagros
Elburz Mts.
Damāvand 18,934 ft. (5,771 m)
Plateau of Iran
30°N
Mesopotamia
IRAQ
Euphrates R.
Karūn R.
IRAN
Mts.
AFRICA
Red Sea
Persian Gulf (Arabian Gulf)
Strait of Hormuz
TROPIC OF CANCER
30°E
0   400 miles
0   400 kilometers
Lambert Azimuthal Equal-Area projection
40°E
50°E
Gulf of Oman
60°E

A river flows through the Turkish plains near the Mediterranean coast.

▲ CRITICAL THINKING

1. *Analyzing Visuals* Based on the photo, what human purposes might this river serve?

2. *Interpreting Significance* What is the significance of this river's close proximity to the sea?

Water Systems

GUIDING QUESTION *How is the Tigris-Euphrates river system important to the human geography of the Northeast?*

The waterways of the Northeast have supported the birth of great civilizations and played vital roles in the fates of nations and empires. Linking the Aegean Sea and the Black Sea is a key waterway known as the Turkish Straits. It marks the border between Asia and Europe and consists of three smaller waterways called, from west to east, the Dardanelles, the Sea of Marmara, and the Bosporus. Through history, control of this waterway has produced military and commercial advantages for those who possessed it. The ability to stop or profit from traffic between the Mediterranean Sea and the Black Sea has been fought over repeatedly.

The Dardanelles is a narrow strait. At one point, it is only three-fourths of a mile (1 km) in width and at no point along its 38 miles (61 km) is it greater than 4 miles (6 km) wide. It opens on its eastern end to the small but deep Sea of Marmara. To the east of that point begins the Bosporus, which is another long, narrow strait. The Bosporus runs for 19 miles (31 km) before emptying into the Black Sea. The Black Sea encompasses a vast area, covering about 180,000 square miles (466,200 sq. km).

In the mountains of far eastern Turkey, within 50 miles (80.5 km) of each other, two great rivers begin their descent to the sea. These are the Tigris and the Euphrates. The rivers bound a region known since ancient times as Mesopotamia, which means "the land between the rivers." It is in this region that some of the world's earliest advanced civilizations arose. From its source, the Euphrates flows through the Taurus Mountains and southwest toward the Mediterranean coast in southern Turkey. But before reaching the Mediterranean, it turns southeast and continues in this direction for most of its 2,235-mile (3,596 km) journey across Syria and Iraq.

The shorter Tigris, at 1,180 miles (1,900 km) in length, flows from a mountain lake. It moves steadily southeast into Iraq. There, the Tigris and the Euphrates gradually converge. They almost touch near the Iraqi capital, Baghdad. They then separate and converge again. The two rivers eventually meet to form the Shatt al Arab. This river then flows the final 120 miles (193 km) to the Persian Gulf.

The dominant water feature in northern Iran is the Caspian Sea, which is slowly shrinking because of evaporation and a reduction of water flowing into the sea from the Volga River. This great inland sea forms the northern border of the country, where its waters are salty. The Caspian Sea serves as an important transportation link between Iran and other Asian countries. The Caspian also yields a number of valuable resources, notably oil and natural gas. Iran also has an active fishery in the Caspian Sea. Southern Iran is bounded by the Persian Gulf (or Arabian Gulf) and the Gulf of Oman. Between these lies the narrow Strait of Hormuz, a vital and strategic outlet from the Persian Gulf to the Indian Ocean.

✓ READING PROGRESS CHECK

Explaining What is the significance of the Dardanelles, the Bosporus, and the Sea of Marmara?

Climate, Biomes, and Resources

GUIDING QUESTION *How do mountains influence climate in Turkey and Iran?*

The climate of the Northeast is heavily influenced by two major factors: mountain ranges and proximity to major bodies of water. Coastal and highland areas near mountain ranges usually receive the most rainfall. This is because moist, warm air is driven off the sea by the prevailing winds.

Climate Regions and Biomes

In Turkey, the areas along the coast enjoy a Mediterranean climate. Temperatures are subtropical, and typical Mediterranean scrub plants are native to the subregion. The summers are warm and dry, and the winters are mild and rainy. A similar climate prevails in the northern part of Iran along the Caspian Sea. The most valuable soils in Turkey are the alluvial soils found in the valleys, deltas, and basins of the lowland areas.

The interior of Turkey features a semi-arid steppe climate, which means it is generally dry with mild temperatures. Grasslands are the dominant vegetation type in the interior areas. This climate supports **pastoralism**—the raising of livestock—which is widely practiced in these areas. Western Iran also has an area of semi-arid steppe, although the temperatures are generally higher than in Turkey.

The high mountains of the subregion form barriers that block moisture from the major bodies of water from penetrating far inland. Thus, apart from the coasts, much of the Northeast experiences arid or semi-arid climates. Deserts cover western and southern Iraq and eastern and southern Iran.

Natural Resources

The most significant resources of the Northeast today are fossil fuels and the vital waterways that transport them. Iraq and Iran are among the world's leading producers of petroleum and natural gas. Iraq has many untapped mineral resources. Known reserves include sulfur and phosphates. Iran's major natural resources include chromium, copper, iron ore, lead, manganese, zinc, and sulfur. In addition to hydroelectric power, Turkey's natural resources include coal, iron ore, copper, mercury, and gold.

Turkey's geographic location is also an asset. In 2006 a pipeline that carries oil from an oil field in the Caspian Sea to the Mediterranean opened. Turkey benefits substantially from the transit fees. Other pipeline projects are planned to transport oil to Europe through Turkey from other countries.

✔ **READING PROGRESS CHECK**

Identifying What are the key natural resources of the Northeast subregion?

pastoralism the raising of animals for food and other products

LESSON 1 REVIEW

Reviewing Vocabulary

1. *Identifying* Write a sentence or two explaining pastoralism and how the geography of Turkey supports it.

Using Your Notes

2. *Listing* Use your graphic organizer to describe the significance of the North Anatolian Fault.

Answering the Guiding Questions

3. *Describing* What features dominate the physical geography of the Northeast?

4. *Evaluating* How is the Tigris-Euphrates river system important to the human geography of the Northeast?

5. *Making Connections* How do mountains influence climate in Turkey and Iran?

Writing Activity

6. *Informative/Explanatory* Write a paragraph explaining why the alluvial soils of the Northeast have been so central to the development of the human geography of the subregion.

networks

There's More Online!

- ☑ **IMAGE** Women Wearing Burkas
- ☑ **MAP** Civilizations and Empires of the Northeast
- ☑ **TIME LINE** Independence and Turmoil
- ☑ **INTERACTIVE SELF-CHECK QUIZ**
- ☑ **VIDEO** Human Geography of the Northeast

LESSON 2
Human Geography of the Northeast

ESSENTIAL QUESTION • *How do physical systems and human systems shape a place?*

Reading **HELP**DESK

Academic Vocabulary

- **assume**
- **participate**

Content Vocabulary

- **natural boundary**
- **culture hearth**
- **cuneiform**
- ***qanat***
- **ziggurat**
- **embargo**

TAKING NOTES: *Key Ideas and Details*

IDENTIFYING As you read about the human geography of the Northeast, use a graphic organizer like the one below to list the subregion's historical civilizations and empires and record their characteristics.

The Northeast

Ancient Civilizations	Characteristics

It Matters Because

In the Northeast, cultures are fundamentally tied to religion. Islam is the most widely practiced religion in the subregion. For hundreds of years, it has defined people's lives. Today, Islam is the fastest-growing religion in the world.

History and Government

GUIDING QUESTION *How have ancient civilizations and the discovery of oil impacted the Northeast?*

The Northeast saw the rise of several great civilizations. The earliest was Sumer, followed by the Babylonian, Persian, and Ottoman civilizations. More recently, the subregion's rich oil resources have helped make it the focus of the international quest to obtain and control energy supplies.

Civilizations and Empires

The Tigris and Euphrates Rivers formed a **natural boundary** around a historical region known as Mesopotamia. Mesopotamia was one of the world's first **culture hearths**. A culture hearth is a center from which cultures develop and then spread to other places.

The rich alluvial soils of Mesopotamia supported the development of agriculture. As humans learned to raise their own food, they were able to settle in permanent villages. Farmers could produce a surplus, which supported growing populations. In larger, settled communities, people now had time to develop new tools and organize and govern communities. In this way, civilizations emerged.

Mesopotamia was home to the Sumerian civilization that developed some 5,000 years ago. The Sumerians grew crops year-round and used canals to irrigate their fields. To keep records, they developed an early system of writing based on **cuneiform** which consists of wedge-shaped symbols pressed into clay tablets. In time the Sumerians created a system of mathematics and a code of law. About 1900 B.C., the Babylonian civilization emerged and dominated Mesopotamia.

416

(l)Thomas Hartwell/TIME & LIFE Images/Getty Images, (tcl)Tom Stoddart Archive/Getty Images, (tcr)Janet Wishnetsky/Alamy, (tr)Ali Al-Saadi/AFP/Getty Images

The Persian Empire arose to the east of Mesopotamia, in what is now Iran. By 500 B.C. the Persians had conquered nearly all of the present-day territory of Iraq, Jordan, Israel, and Turkey. One of their great engineering achievements was the building of **qanats**, or underground canals. *Qanats* reduced the evaporation of water as it flowed from the mountains to farmlands.

The Ottoman Empire has its roots on the Anatolian Plateau. At its peak, the empire extended into North Africa, western Asia, and southeastern Europe. The defeat of the Ottomans and Germans in World War I ended some 600 years of rule.

Oil and the Modern Era

From the remnants of the Ottoman Empire, the modern country of Turkey was established. Turkey is aligned militarily with Western Europe as a member of the North Atlantic Treaty Organization (NATO). It has applied for EU membership.

Once part of the Ottoman Empire, Iraq became an independent state in 1932. It has experienced considerable turmoil since. This includes a war with Iran (1980–1988) and two wars against the United States and coalition forces. The most recent occurred between 2003 and 2011. It resulted in the arrest, conviction, and execution of dictator Saddam Hussein and the establishment of a parliamentary democracy.

The last Persian Empire lost most of its territory to the Ottoman Empire, but the core of Persia—modern-day Iran—remained an independent country. In 1979 a revolution overthrew the secular government headed by the shah, and an Islamic Republic was formed. Islamic scholars, known as mullahs, came to power. They continue to dominate Iranian politics. Recently, international tensions have risen as Iran is suspected of developing nuclear weapons.

Fossil fuels are a major natural resource in the subregion. Oil was discovered near the Persian Gulf in 1908. By 1911 the first commercial oil well was operating in Iran. Over time, the wells produced large quantities of oil and came under the control of countries and national oil companies, benefiting their economies.

natural boundary a boundary created by a physical feature, such as a mountain, river, or strait

culture hearth a center in which cultures develop and from which they are spread

cuneiform a system of writing using wedge-shaped symbols that were pressed into clay tablets

qanat an underground canal first built by the ancient Persians

GEOGRAPHY CONNECTION

The Northeast produced several great civilizations and empires.

1. *PLACES AND REGIONS* Which present-day country in the Northeast was not part of the Ottoman Empire?

2. *PHYSICAL SYSTEMS* Which area benefited from its physical geography, which included the Tigris and Euphrates Rivers?

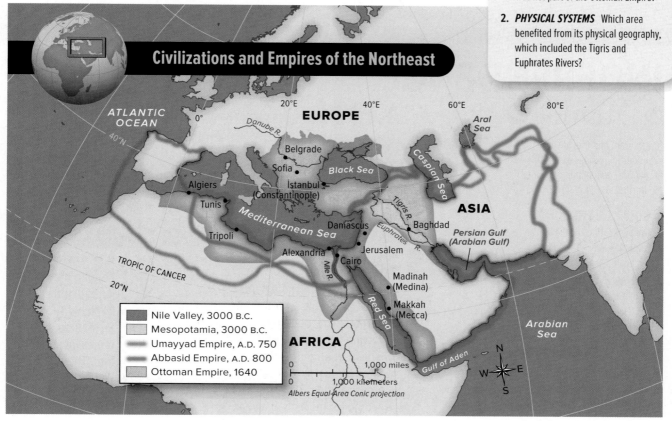

Civilizations and Empires of the Northeast

Nile Valley, 3000 B.C.
Mesopotamia, 3000 B.C.
Umayyad Empire, A.D. 750
Abbasid Empire, A.D. 800
Ottoman Empire, 1640

0 1,000 miles
0 1,000 kilometers
Albers Equal-Area Conic projection

assume to gain or acquire

In 1960 Iran and Iraq joined with several oil-producing neighbors to form the Organization of Petroleum Exporting Countries (OPEC). This group agreed to regulate oil production to keep oil prices high. As the global demand for oil grew, OPEC **assumed**, or gained, more power over global oil prices.

☑ READING PROGRESS CHECK

Explaining How did agriculture contribute to the formation of the Sumerian civilization?

Population Patterns

GUIDING QUESTION *How have ethnic diversity and Islam shaped the population patterns of the Northeast?*

The majority of people in the Northeast are Muslims. Their religion, Islam, has a significant role in their cultures. While Islam is the primary religion, the population is ethnically diverse. Major ethnic groups include Turks, Iranians, Arabs, and Kurds.

This ethnic diversity persists because over the past 8,000 years many peoples have occupied the subregion. Each group added its own customs and beliefs to the local culture. Turkic peoples migrated to the western Anatolian Plateau from Central Asia in the A.D. 1000s. The name *Iran* means "land of the Aryans" (AR•ee•uhnz), from the ancient Indo-European people who settled in Iran.

Today the majority of people living in Iraq are Arabs, with ethnic ties to the Arabian Peninsula. The Kurds are an ethnic group that has lived for thousands of years in the mountainous border areas of Turkey, Iraq, and Iran. Kurds have no country of their own, though they call the land they live in Kurdistan. Some Kurds **participate** in efforts to achieve independence and have their own country.

participate to take part in

Turkey and Iran each have populations of about 80 million people, while about 31 million people live in Iraq. Population density ranges from about 246 people per square mile (95 people per sq. km) in Turkey to about 117 people per square mile (45 people per sq. km) in Iran. All three countries are increasingly urban.

Thomas Hartwell/TIME & LIFE Images/Getty Images

TIME LINE ∨

INDEPENDENCE
and Turmoil ➔

Turkey, Iran, and Iraq have experienced decades of struggles for political and military power in the Northeast.

▶ CRITICAL THINKING

1. Analyzing How did the Iranian Revolution affect Iran and its neighbors?

2. Describing What role has the United States played in events in this region?

1979 ➔

July—Saddam Hussein becomes president of Iraq and begins work to secure his power as a dictator.

1979 · January—Shah of Iran forced to flee due to Iranian Revolution. Ayatollah Ruhollah Khomeini takes over as Iran's supreme leader.

1979 · November—Iranian militants seize U.S. embassy in Tehran to protest U.S. decision allowing the shah to seek medical treatment in the United States. Dozens of embassy workers are held hostage for more than a year.

1979

1980 · Fearing the spread of the Iranian Revolution, Iraqi forces invade Iran. The Iraq-Iran War lasts eight years.

The cities of Istanbul, Turkey; Tehran, Iran; and Baghdad, Iraq, dominate social and cultural life in their respective countries. These cities have become overcrowded due to the rapid influx of villagers seeking opportunities available in the urban centers. Iran has tried to address this problem by moving some of its government offices to towns away from the capital, Tehran.

Over 3 million refugees from Syria, Iraq, and other countries in North Africa and Southwest Asia fled to Turkey in 2014-2015. As the numbers continue to grow, Turkey is faced with hosting close to hundreds of thousands of refugees in camps, with ongoing costs for health, education, food, and social and other services. Refugees living outside the camps live in poor conditions, struggling to afford rent and food with their already depleted resources.

☑ READING PROGRESS CHECK

Identifying What are the major ethnic groups in the Northeast?

Society and Culture Today

GUIDING QUESTION *How do religion and language influence life in the Northeast today?*

The legacies of the ancient civilizations of the Northeast can be seen today. Languages and customs differ depending on ancestral homes. Arabic is the most commonly spoken language in Iraq. Most Iraqi Arabs are Shia Muslims, but about 35 percent are Sunni Muslims. A small percentage is Christian or another faith. Iranians speak Farsi, also called Persian, and the majority are Shia. Turks speak Turkish, and most practice Islam. Modern Turkish culture blends Turkish, Islamic, and Western elements. Most Kurds are Sunni Muslims and speak Kurdish, a language distinct from, but related to, Farsi.

Turkey, Iraq, and Iran all support systems of free public education. The amount of education a child receives in a lifetime is 13 years in Iran, 12 years in Turkey, and 10 years in Iraq. Iran has made great strides in educating its

1984
Kurdish nationalist organization, Kurdistan Workers' Party (PKK), launches armed campaign against Turkey. Actions lead to reprisals by the Turkish government and further PKK attacks.

➔1985
Iraq invades Kuwait. Persian Gulf War begins in early 1991 with air strikes by a U.S.-led coalition. The war ends six weeks later in an Iraqi defeat.

1990

1990s
International fears grow over Iran's nuclear program and development of nuclear, chemical, and biological weapons.

2003
U.S.-led coalition invades Iraq, overthrows Saddam Hussein, and turns over power to an interim Iraqi government.

2011
February—Street demonstrations erupt in Iran during the "Arab Spring."

➔2011

2011
December—United States completes its withdrawal of troops from Iraq.

population since the Islamic Revolution. Before the revolution, its literacy rate was under 50 percent. Today, the literacy rate is 87 percent in Iran, about 80 percent in Iraq, and 95 percent in Turkey.

Health care varies in the subregion. Iraq is still struggling to rebuild hospitals after years of war. In other countries, hospitals are government-owned and may suffer from doctor shortages, especially in rural areas. In recent years, private hospitals in Turkey have begun competing with state hospitals. The competition has led to improvements in state-owned hospitals.

Family and Status of Women

Family life usually includes the extended family. In Turkey, the average household consists of a nuclear family with 3.8 persons, but the extended family remains important. The average household size in Iran is similar, with 3.6 persons per household. In recent years, family size in Iran has decreased due to a government-supported family planning program and the economic effects of embargoes intended to stop Iran's nuclear program. The average household in Iraq is much larger than in Turkey and Iran, with an average of 6.7 persons.

The conservative Islamic culture of the Northeast defines the roles and opportunities available to women. Government policies in some countries have improved the status of women. In Turkey, a 2002 law gave women equal rights with men for marriage, divorce, and property. In Iran, however, women's rights were curtailed after the establishment of the Islamic Republic in 1979. Many public places were segregated by sex, and female government workers were required to observe an Islamic dress code. In the more secular state of Iraq, women and men have equal rights under the law. However, lack of women's rights and gender equity are issues that hinder development in much of the Northeast.

The Arts

ziggurat a large temple built by the Sumerians

The early civilizations of the Northeast created sculptures, fine metalwork, and large buildings. In Mesopotamia the Sumerians built large, mud-brick temples called **ziggurats**. These temples were shaped like pyramids and rose high above the flat landscape. Sumerian architecture was also characterized by the construction of arches and domes.

Literature in the Northeast is based on strong oral tradition, epics, and poetry. The *Rubáiyát* by the Persian poet Omar Khayyám, who lived some 900 years ago, is the most famous example. *Osman's Dream* is an Ottoman epic dating to the 1200s that foretells of a great future empire.

The subregion also has a long tradition of fine carpet weaving that dates back to ancient Persia. Iran is still a center for the production of Persian rugs. These spectacular rugs are works of art and considered the finest made in the world today.

☑ **READING PROGRESS CHECK**

Differentiating In what ways does Iran differ from Iraq and Turkey in terms of society and culture?

Economic Activities

GUIDING QUESTION *On what natural resources do the economies of the Northeast depend?*

Oil and natural gas are the major economic resources in the Northeast. There are other valuable resources and economic activities in the subregion, however. For example, the Caspian Sea supports Iran's fishing industry. Agriculture produces many fruits and nuts, such as figs and pistachios, of very high quality and value.

Resources, Power, and Industry

Both Iran and Iraq produce far more oil than they need for domestic consumption, and so are leading exporters of oil. Revenues from oil exports are enormously important to their economies. As members of OPEC, Iran and Iraq have considerable influence in global affairs and in the patterns of movement of money from their management of this key natural resource. For example, OPEC raised oil prices during the 1970s and placed an **embargo** on oil shipments to the United States and other industrialized countries. OPEC actions can also have positive effects. OPEC restored stability to the oil market and supported economic recovery during the global recession in 2009. Events unrelated to OPEC also show the influence that Iraq and Iran can have in global affairs. For example, in 2006 and 2008, oil prices rose sharply due to the ongoing conflict in Iraq and tensions over Iran's nuclear ambitions.

embargo a ban on trade

Turkey does not have the oil or natural gas resources that its regional neighbors do, but it does produce hydroelectric power. Other key industries in Turkey include textiles, food processing, and manufacturing autos and electronics. Turkey's service sector also forms a significant part of the economy.

Trade and Interdependence

The countries of the Northeast and the rest of the world depend on one another. Industrialized countries need oil from the region; the region needs industrial products for its markets. Iran operates oil-refining facilities, although Iraq's oil-refining facilities were closed or greatly restricted during the 2003–2011 U.S.-led war. Natural gas and oil pipelines are examples of the advanced transportation systems that crisscross this subregion. Much of this network is designed to move oil and gas from the fields to various ports. Since Turkey has access to the Mediterranean and the rest of Europe, many pipelines from oil fields end in Turkish ports. Other lines deliver oil to ports on the Persian Gulf and the Gulf of Oman. Much of the oil shipped from the Persian Gulf must pass through the Strait of Hormuz. The narrow strait is a choke point for shipping. Military action, a shipping accident, or political decisions could possibly close the strait. During most years, there are about 17 supertankers passing through the strait each day.

Television and radio broadcasting is expanding in the subregion, though government control of the media in many places limits programming. Advances in satellite technology are improving communication services. Wireless service and solar-powered radiophones are bringing telephone service to more people. Cell phones are now common, and many people have computer and Internet access.

☑ **READING PROGRESS CHECK**

Identifying What is the most significant natural resource in the Northeast?

LESSON 2 REVIEW

Reviewing Vocabulary
1. *Applying* Write a paragraph about the history of the Northeast that includes the words *culture hearth*, *cuneiform*, qanat, *natural boundary*, *embargo*, and *ziggurat*.

Using Your Notes
2. *Summarizing* Use your notes from the graphic organizer to write two paragraphs summarizing the history of the Northeast.

Answering the Guiding Questions
3. *Explaining* How have ancient civilizations and the discovery of oil impacted the Northeast?

4. *Describing* How have ethnic diversity and Islam shaped the population patterns of the Northeast?

5. *Making Connections* How do religion and language influence life in the Northeast today?

6. *Identifying* On what natural resources do the economies of the Northeast depend?

Writing Activity
7. *Informative/Explanatory* Write a paragraph comparing and contrasting modern Turkey, Iran, and Iraq. Include aspects of their economies, cultures and population patterns.

HOW HAVE SUNNI AND SHIA BELIEFS LED TO CONFLICT?

Within Islam, there are two main branches—Sunni and Shia. The divisions between these groups go back to the early years of the religion. The death of Muhammad in A.D. 632 led to a dispute over the succession of leadership in Islam. What we today call the Shia insisted that the Muslim leader, or caliph, be a descendant of Ali, the son-in-law and cousin of Muhammad, and his wife Fatima, Muhammad's daughter. Conversely, the Sunni believed that the leader should be selected by the Islamic community on the basis of personal qualities and abilities.

The split between the two groups has persisted over the centuries. Today Sunnis are the majority of Muslims. It is estimated that between 85 and 90 percent of all Muslims worldwide are Sunni. Only in Iran, Iraq, Bahrain, and Azerbaijan do Shias make up the majority of Muslims.

Iran is a country dominated by Shia Muslims. With this dominance has come some conflict—and not all of it having to do purely with matters of deeply held religious faith. In many places, Sunnis and Shias have vied for political power and sought to advance their people's interests at the expense of others.

After the Iranian Revolution, Islamic law was reintroduced. Although the new government was a republic with elements of a parliamentary democracy, it was also a theocracy. Shia clerics have final approval of legislation, with the intention of ensuring compliance with Islamic law. Iran does permit and have religious minorities. These include a small but significant Sunni Muslim population (many of whom are Kurds), Zoroastrians, Christians, and Baha'is. However, relations between the government and its Sunni minority have been complicated, and members of religious minorities are treated as second-class citizens.

Charges of Sunni Mistreatment in Iran

PRIMARY SOURCE

" Iran's Sunni Muslims . . . face widespread discrimination by Iranian authorities. The United States Commission on International Religious Freedom 2012 Report states that 'Sunni leaders have reported . . . abuses and restrictions on their religious practice, including detentions and abuse of Sunni clerics . . . and bans on Sunni teachings in public schools and Sunni religious literature, even in predominantly Sunni areas'. Sunnis are not allowed to build mosques in large cities and have been banned from conducting separate Eid prayers.

Attacks in Sunni-populated areas are not unusual. In May, one person was killed and two injured after police forces opened fire on protestors in Sistan-Baluchistan province. They were protesting against the recent arrest of local Sunni clerics. Last year, at least 12 people believed to be Sunni protestors were killed in Ahwaz, a southwestern city in the Khuzestan province, in clashes with security forces. Several Sunni mosques have been destroyed—the Abu Hanifa Mosque in Sistan-Baluchistan was bulldozed in 2008—or converted into parks. "

—Mona Moussavi, *IISS Voices*, October 10, 2012

Charges of Sunni Terrorism in Iran

PRIMARY SOURCE

" Jundallah [a pro-Sunni group] was designated as a Foreign Terrorist Organization on November 4, 2010. Since its inception in 2003, Jundallah, a violent extremist organization that operates primarily in the province of Sistan va Balochistan of Iran, has engaged in numerous attacks resulting in the death and maiming of scores of Iranian civilians and government officials. Jundallah's stated goals are to secure recognition of Balochi cultural, economic, and political rights from the government of Iran and to spread awareness of the plight of the Balochi situation through violent and nonviolent means. In October 2007, Amnesty International reported that Jundallah has by its own admission, carried out gross abuses such as hostage-taking, the killing of hostages, and attacks against non-military targets. "

—The United States Institute of Peace, *U.S. Terrorism Report: MEK and Jundallah*, August 23, 2011

What do you think? DBQ

1. *Drawing Conclusions* How might religious differences affect how countries interact?

2. *Identifying Cause and Effect* How might the actions of a group like Jundallah promote the denial of rights to Sunni groups in Iran?

3. *Identifying Cause and Effect* How might the actions of the Iranian government promote the activities of groups like Jundallah?

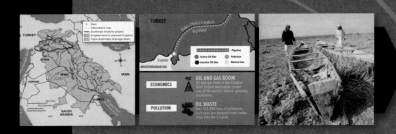

netw⊙rks

There's More Online!

☑ **IMAGE** Drained Marshland

☑ **INFOGRAPHIC** Water Pollution in the Northeast

☑ **MAP** Tigris-Euphrates River Basin

☑ **INTERACTIVE SELF-CHECK QUIZ**

☑ **VIDEO** People and Their Environment: The Northeast

Reading HELPDESK

Academic Vocabulary

- link
- monitor

Content Vocabulary

- **feeder stream**
- **marsh**

TAKING NOTES: *Key Ideas and Details*

IDENTIFYING As you read about people and their environment in the Northeast, use a graphic organizer like the one below to record the environmental dangers facing the subregion.

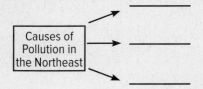

Causes of Pollution in the Northeast

LESSON 3

People and Their Environment: The Northeast

ESSENTIAL QUESTION • *How do physical systems and human systems shape a place?*

IT MATTERS BECAUSE

The Northeast contains valuable natural resources. Human utilization of those resources, however, can sometimes place other aspects of the environment at risk. Countries in the subregion attain variable levels of success in maximizing the benefits from their resources while minimizing the costs associated with obtaining them.

Managing Resources

GUIDING QUESTION *How and why are water resources at risk in the Northeast?*

Turkey has long utilized its many rivers for hydroelectric power and other purposes. In recent years, Turkey has been engaged in a regional development project called the Southeast Anatolia Project. The project is known as GAP in Turkey. GAP involves the construction of nearly two dozen dams along the Tigris and Euphrates Rivers and a number of their **feeder streams**, or tributaries. The plan also calls for a large number of hydroelectric stations capable of generating huge amounts of energy. The project will divert water to irrigate about 4.2 million acres (1.7 million ha) of land. Finally, GAP is also an economic development program that provides a number of benefits to a poor region of Turkey.

Such massive projects threaten existing water systems in a number of ways. The dams change the course and ecology of rivers. They also inundate lands that were once dry and divert water that would otherwise flow downstream. The effect on water quality and quantity is significant. Many ecosystems and habitats are disturbed or even destroyed.

Southern Iraq is characterized by **marshes**. The lower portions of the Tigris and Euphrates feature the Mesopotamian Marsh, which covers 15,000 square miles (38,850 sq. km). It is the largest wetlands area in the Northeast. In recent decades the Iraqi government has undertaken the draining of many of these marshlands. Marsh drainage was used to create farmland and divert water for agricultural purposes.

Some marshlands were also drained for political reasons. The people who historically lived on these lands—the Marsh Arabs—had developed a culture and economy based upon the marsh. The Marsh Arabs became involved in uprisings against the regime of Saddam Hussein in the early 1990s. During the uprisings they took refuge in the tall reeds. The Iraqi government retaliated and drained large portions of the marshlands. The Marsh Arabs, whose cultural **link** to the land went back thousands of years, were forced to relocate.

More recently, another threat to the marshes has become a concern: drought. The problem is made worse by the heavy diversion of water for irrigation projects for agriculture in places like Turkey. These factors have lowered water levels in many wetlands areas to dangerous levels. Drought has also led to wetlands destruction in Iran.

☑ **READING PROGRESS CHECK**

Explaining How did politics play a part in the destruction of the Mesopotamian Marsh in the 1990s?

Human Impact

GUIDING QUESTION *What human activities result in air, land, and water pollution in the Northeast?*

Human activity impacts the environment in many ways. A major threat is the pollution of air, land, and water. Human activity can also lead to the loss or degradation of natural resources, such as soils and forests. War has also had

feeder stream a tributary that feeds a larger river

marsh a wetland typically covered with grasses

link a connecting structure

GEOGRAPHY CONNECTION

The Tigris and Euphrates Rivers define the boundaries of the historical region known as Mesopotamia.

1. *PLACES AND REGIONS* In what country is the Southeast Anatolia Project located?

2. *THE WORLD IN SPATIAL TERMS* What region of the Tigris-Euphrates Basin has the most irrigated land?

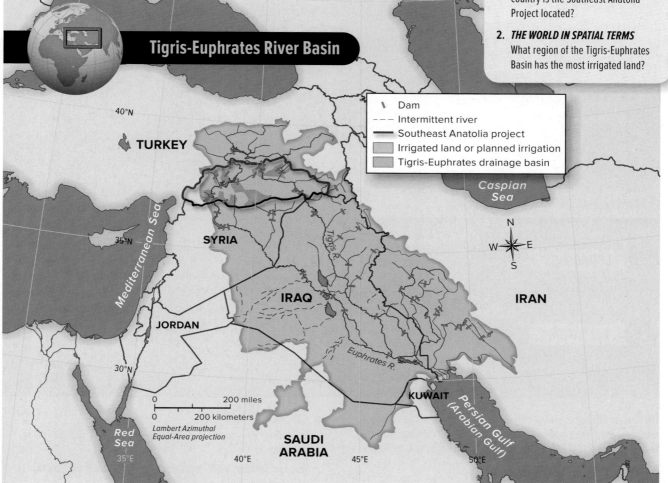

Tigris-Euphrates River Basin

- ↘ Dam
- --- Intermittent river
- ▬ Southeast Anatolia project
- ▢ Irrigated land or planned irrigation
- ▢ Tigris-Euphrates drainage basin

negative effects on the environment. During the Persian Gulf War in 1991, Iraqi troops retreating from Kuwait set fire to more than 700 oil wells. Huge black clouds of smoke polluted the area. Iraqi troops also dumped about 250 million gallons (946 million l) of oil into the Persian Gulf. Thousands of fish and other marine life died when the oil spill spread along the coastal areas of the Persian Gulf. Smoke from the oil well fires threatened millions of birds. Oil pollution from routine shipping also adversely affects the environment of the Persian Gulf.

Water pollution in the Caspian Sea is an issue. The Caspian Sea is the world's largest inland body of water, measuring 750 miles (1,200 km) from north to south with an average width of 200 miles (320 km). It has historically served as a key fishery, notable for its sturgeon and the valuable caviar. Today, however, pollution threatens these fish populations.

There are several sources of biological and chemical pollutants in the Caspian Sea. Untreated waste from the Volga River flows into the sea. The river also receives chemicals that run off from industries located along its shores, which then flow into the Caspian. In Iran it has been common practice to dump garbage from seaside cities and towns directly into the Caspian Sea. Runoff from farms—carrying pesticides, fertilizers, and detergents—also flows into the sea, adding to what is a major environmental problem.

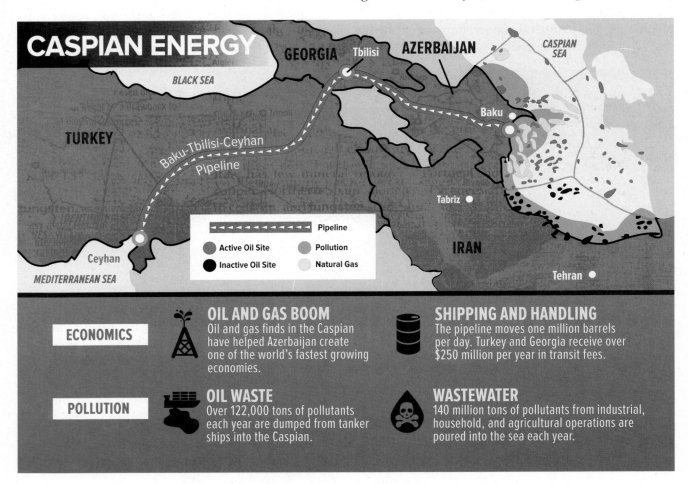

CASPIAN ENERGY

Pipeline
Active Oil Site
Inactive Oil Site
Pollution
Natural Gas

ECONOMICS

OIL AND GAS BOOM
Oil and gas finds in the Caspian have helped Azerbaijan create one of the world's fastest growing economies.

SHIPPING AND HANDLING
The pipeline moves one million barrels per day. Turkey and Georgia receive over $250 million per year in transit fees.

POLLUTION

OIL WASTE
Over 122,000 tons of pollutants each year are dumped from tanker ships into the Caspian.

WASTEWATER
140 million tons of pollutants from industrial, household, and agricultural operations are poured into the sea each year.

The Caspian Sea has experienced a boom in oil and natural gas finds in recent decades. The processes used to extract these resources, however, have contributed to the increasing levels of pollution in the Caspian Sea.

▲ **CRITICAL THINKING**

1. *Evaluating* How might the pollution of water sources limit their value to people?

2. *Problem Solving* What steps could governments or individuals take to reduce water pollution in the Northeast?

Perhaps the biggest threat to the Caspian's waters comes from the large oil-extraction operations that take place throughout the sea. Much of this work is carried on by countries other than Iran that rim the Caspian. The effects of pollution from oil and gas operations have been devastating. Each year, thousands upon thousands of tons of hydrocarbons and heavy metals are spilled into the Caspian. This pollution is likely to get worse as Iran and other countries increase their exploitation of the large reserves of oil and gas beneath the Caspian Sea.

The Black Sea, which forms Turkey's northern border, is also plagued by severe pollution. The contributors are agricultural runoff and industrial pollution related to the shipping of oil. Much of this pollution comes from countries other than Turkey. Of particular concern to Turkey is the danger of an oil spill in the Bosporus. This major shipping channel sees a great deal of traffic by oil tankers. With a population of more than 10 million, Turkey's largest city—İstanbul—straddles the Bosporus. Thus, the potential for environmental disaster is high. In addition, Turkey's ability to regulate traffic in the strait is limited by international agreements designed to ensure the free flow of shipping.

In Iraq, the Tigris and the Euphrates have also suffered a serious decline in water quality. Some of this comes from agricultural runoff, as water used for irrigation flows back into the rivers. Pollution from industrial sources and sewage also threatens the rivers.

Air pollution is a problem in all of the countries of the Northeast, especially in urban areas. The population of these areas has been growing. With that increase have come concerns about air quality. Streets and roadways are frequently clogged with cars, many of which are older or poorly maintained models that produce heavy exhaust. Iran's capital, Tehran, has especially suffered from severe air pollution. In recent years there have been many instances of school and public office closures due to high air pollution levels. The Iranian government has blamed air pollution for thousands of deaths in the city.

In Iraq problems with air pollution raised the concern of U.S. military officials, who worried about the effects on American soldiers stationed there. Iraq's frequent dust storms produce large amounts of airborne particulates, many contaminated with substances such as lead, a toxin still used in Iraqi gasoline. Iraq has traditionally done little to ensure good air quality.

The health of forests and soils is threatened in several parts of the Northeast. In Iraq, for example, soils in many areas have become increasingly saline—salty—to the point that many areas have been abandoned for agricultural use. High salinity results from heavy irrigation in a place where evaporation happens quickly. Iraq has also had major problems with overgrazing and deforestation, which have left the soil exposed to seasonal rains. Many tons of soil have been carried down the Tigris and the Euphrates and been deposited in the Persian Gulf.

Iran has sizable forests in the northern part of the country. These, however, have been heavily exploited, especially in recent decades. Loss of forest resources is a major concern in the country. In 1963 Iran nationalized its forests, but that failed to stop deforestation.

Turkey also has seen many of its historical forest resources destroyed. Areas once covered with forests are now open plains. The loss of forests has brought a loss of soil. Erosion in hillier areas is a serious problem for a country that depends heavily on agriculture.

☑ **READING PROGRESS CHECK**

Describing In what ways does oil extraction pose an environmental threat to the Northeast?

Connecting Geography to **SCIENCE**

Soil Erosion in Iraq

Iraq is an ancient land that has supported agriculture and pastoralism longer perhaps than any other land on Earth. The removal of trees and the cultivation of soil began thousands of years ago. This has exposed soils in the mountain and foothill regions of northern Iraq to erosion. Massive amounts of soil have flowed down these mountains and hills over the centuries. These soils have been deposited at the mouth of the Shatt al Arab, the river formed where the Tigris and Euphrates meet and an area long contested by Iraq and Iran. Over the years, the river's delta has extended many miles into the Persian Gulf. The Iranian city of Ābādān, for example, was located on the Persian Gulf 1,000 years ago. Today it is some 30 miles (50 km) inland, situated on an island in the Shatt al Arab.

IDENTIFYING CAUSE AND EFFECT
What led to the extension of the alluvial deposits at the mouth of the Shatt al Arab?

Iraqi Marsh Arabs examine their canoes in what had once been marshland.

▲ CRITICAL THINKING

1. *Analyzing Visuals* How would you describe the land shown in the photo that was once covered with marshes?

2. *Making Connections* How might the changed landscape affect the lives and livelihoods of the Marsh Arabs?

monitor to watch closely, evaluate

Addressing the Issues

GUIDING QUESTION *How are environmental issues being addressed in the Northeast?*

In Turkey there are many active groups working to address environmental issues. Opposition to GAP has been strong. In Turkish universities, scientists have questioned some of the claimed benefits of the project. Activists have also sought to prevent the destruction of archaeological treasures that would occur if land is flooded for GAP. Supporters of GAP point out that hydropower is a relatively clean way to address rising energy needs. Environmentalists have placed a heavy emphasis on reducing Turkey's historical dependence on coal, which is highly polluting. There is lively debate in Turkey about the best ways to balance competing interests.

Political forces from outside Turkey are also having an impact. Turkey continues to make its case for entry into the European Union (EU). In an effort to attain this goal, it has strengthened certain environmental laws to meet EU standards.

The loss of wetlands in Iraq has prompted action from Iraqis as well as from people around the world. Organizations such as Wetlands International and the International Wetlands Conference are working on solutions to the threats. Iran has also been acting to protect its many valuable wetlands. In 1971 Iran hosted the Ramsar Convention. This meeting produced an international treaty, the Ramsar Convention on Wetlands, to **monitor**—or closely watch—and preserve wetlands resources around the world. Today, this convention brings together governments and nongovernmental organizations (NGOs) from around the world to protect wetlands of all types.

The Iranian government has also been vocal in raising concerns about pollution in the Caspian Sea. Many of these complaints, however, have been aimed at other nations who contribute to the problem. In 2013, for example, Iran threatened to sue the government of Azerbaijan over pollution coming from its oil platforms in the Caspian.

☑ READING PROGRESS CHECK

Evaluating What factors must Turkey balance when evaluating the benefits of its GAP project?

LESSON 3 REVIEW

Reviewing Vocabulary

1. *Applying* Using the words *feeder stream* and *marsh*, explain the environmental challenges facing the Northeast.

Using Your Notes

2. *Describing* Use your graphic organizer to write a paragraph describing the major causes of pollution in the Northeast.

Answering the Guiding Questions

3. *Analyzing* How and why are water resources at risk in the Northeast?

4. *Describing* What human activities result in air, land, and water pollution in the Northeast?

5. *Problem Solving* How are environmental issues being addressed in the Northeast?

Writing Activity

6. *Argument* Write an argument either for or against the development of the Southeast Anatolia Project, or GAP, in Turkey.

Directions: On a separate sheet of paper, answer the questions below. Make sure that you read carefully and answer all parts of the questions.

Lesson Review

Lesson 1

1 *Analyzing* What is the meaning of the word *Mesopotamia*? Based on the location of Mesopotamia, explain whether this name is logical for the region.

2 *Interpreting Significance* Identify the dominant water feature in northern Iran. Explain its significance to the area.

3 *Making Decisions* Imagine you are considering the possibility of pursuing agricultural interests in the Northeast subregion. Would you select the steep slopes or the river basin? Explain.

Lesson 2

4 *Identifying Cause and Effect* In Mesopotamia, humans learned to raise their own food. How and why did this change population and settlement patterns?

5 *Describing* When did Turkic peoples migrate to the Anatolian Peninsula? Provide an overview of the cultural blend that constitutes modern Turkish culture and identify the language spoken and religion practiced by most Turks.

6 *Identifying Perspectives* Explain why other countries in the Northeast could benefit by following Turkey's example of an alternative energy source. Within your response, identify the specific alternative energy source used in Turkey and explain at least one likely reason that Turkey developed it.

Lesson 3

7 *Identifying Central Issues* Discuss one human factor and one factor of the physical environment that has posed threats to marshlands in Iraq.

8 *Summarizing* Summarize the threats to the Caspian Sea, explaining which is likely the greatest threat and why.

9 *Making Connections* Identify the causes and consequences of soil erosion in the Northeast.

21st Century Skills

Use the graph below to answer the following questions.

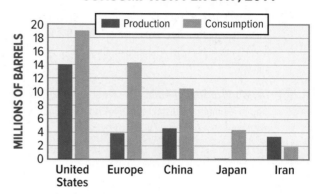

CRUDE OIL PRODUCTION AND CONSUMPTION PER DAY, 2014

Source: U.S. Energy Information Association
Note: Consumption data for China, Iran, and Europe is 2013 data.

10 *Using Graphs, Charts, Diagrams, and Tables* Which country shown on the graph produced the most barrels of crude oil per day in 2014? How did this relate to its consumption?

11 *Compare & Contrast* Which country or area produced the fewest barrels of crude oil per day in 2014? How did this relate to its consumption? Discuss the relationships among production, consumption, and likelihood of exports. Then tie this to the likely imports of the country that produced the most barrels of crude oil per day.

Critical Thinking

12 *Exploring Issues* Write a one-page essay explaining this statement: "The modern-day Northeast is a product of early historical events in the region, though more recent events threaten the human and physical geography of the area."

Need Extra Help?

If You've Missed Question	**1**	**2**	**3**	**4**	**5**	**6**	**7**	**8**	**9**	**10**	**11**	**12**
Go to page	414	414	415	416	419	421	424	426	427	429	429	419

Directions: On a separate sheet of paper, answer the questions below. Make sure that you read carefully and answer all parts of the questions.

Applying Map Skills

Refer to the Unit 5 Atlas to answer the following questions.

13 **The World in Spatial Terms** Use your mental map of Turkey and Iraq to describe the course of the Euphrates River.

14 **Environment and Society** Use the Unit Atlas to explain where the greatest threat of pollution of the Black Sea would most likely exist—İstanbul or Baghdad— and why.

15 **The World in Spatial Terms** Use the scale on the physical map to estimate the distance in miles from Tehran to Ankara.

Exploring the Essential Question

16 **Making Connections** You are a television broadcast journalist who has been assigned a broadcast segment focusing on the following statement: "The ongoing conflicts between Shia Muslims and Sunni Muslims can be viewed through the perspectives of the physical and human geography of the region." Your broadcast report must open with a broad overview of the interaction between physical and human systems of the subregion. Write your report, and deliver it orally to the class. Include visuals, such as maps and photos.

College and Career Readiness

17 **Examining Information** You are working for a firm that has interests in Turkey. Your boss is concerned about the possibility of oil spills in the Bosporus Strait, and you have been asked to write a memo to summarize the following: the likelihood of oil spills in the Bosporus; the reasons for this likelihood; issues with cleaning up after an oil spill in the Bosporus. Within your memo, you have been asked to reference at least two other oil spills that have occurred in the world—with an explanation as to the causes and the mitigation after the spills. Prepare the memo for your boss, clearly and logically supporting your statements.

DBQ Analyzing Primary Sources

Use the excerpt to answer the following questions.

PRIMARY SOURCE

"Iraq's energy sector holds the key to the country's future prosperity and can make a major contribution to the stability and security of global energy markets. Iraq is already the world's third-largest oil exporter and has the resources and plans to increase rapidly its oil and natural gas production as it recovers from three decades punctuated by conflict and instability. Success in developing Iraq's hydrocarbon potential and effective management of the resulting revenues can fuel Iraq's social and economic development. Failure will hinder Iraq's recovery and put global energy markets on course for troubled waters."

—International Energy Agency, "Executive Summary," *Iraq Energy Outlook*, 2012

18 **Interpreting Significance** Reread the following statement from the quote: "Failure will hinder Iraq's recovery and put global energy markets on course for troubled waters." What did the writer most likely mean by this statement?

19 **Drawing Conclusions** Why would oil exports alone fail to hold the key to Iraq's future prosperity?

Research and Presentation

20 **Gathering Information** Use Internet and library sources to research the Southeast Anatolia Project, known as GAP, in Turkey. Identify the purpose of GAP and note the reasons for opposition to GAP within Turkey.

Writing About Geography

21 **Narrative** Use standard grammar, spelling, sentence structure, and punctuation to write a paragraph that describes what you might see around you if you were to fly in an airplane from Ankara to an area near the summit of Mount Ararat.

Need Extra Help?

If You've Missed Question	13	14	15	16	17	18	19	20	21
Go to page	360	360	360	409	427	430	430	424	413

The Arabian Peninsula

ESSENTIAL QUESTION • *How do physical systems and human systems shape a place?*

networks

There's More Online about the Arabian Peninsula's geography.

CHAPTER 18

Why Geography Matters
Migrant Workers in the Arabian Peninsula

Lesson 1
Physical Geography of the Arabian Peninsula

Lesson 2
Human Geography of the Arabian Peninsula

Lesson 3
People and Their Environment: The Arabian Peninsula

Geography Matters...

Once a land of nomadic herders, the Arabian Peninsula entered the modern world during the twentieth century with a sudden burst of wealth and power created by the vast deposits of oil beneath its land. It is a region of contrasts. Its teeming cities perch on the edge of the sea or on inland oases, surrounded by seemingly endless deserts. The oil beneath those deserts has created almost boundless wealth. However, it is not the most precious resource on the peninsula. The most sought-after resource of all is water.

◄ Herding sheep and goats is an ancient tradition on the Arabian Peninsula.

Morales/age fotostock

431

migrant workers *in the* Arabian Peninsula

The Arabian Peninsula is flanked by the Red Sea and the Persian Gulf, far from the island countries of Indonesia and the Philippines. Yet this peninsula has drawn thousands of migrants from these and other countries who migrate here in search of work. This "guest" population, however, faces several other difficulties in addition to living thousands of miles from their homes and families.

Karim Sahib/AFP/Getty Images

432

Why are migrant workers coming to the Arabian Peninsula?

Migrant workers are pushed out of their homelands because of increasing poverty and unemployment. This was especially true after the global financial crisis that broke out in 2007–2008. Many migrant workers flock to the Arabian Peninsula because of the promise of wages that are much higher than they could ever earn in their native countries. Even Yemen, which has high levels of poverty, sends many workers to Saudi Arabia.

In 2010 Saudi Arabia was the world's second-largest source of jobs for migrant workers, followed by the United States. Another reason Muslim migrants in particular are attracted to the Arabian Peninsula is that they will be able to fulfill one of the pillars of Islam—the pilgrimage to Makkah (Mecca).

1. **Human Systems** What are the push and pull factors influencing migration to the Arabian Peninsula? How do similar religious beliefs affect the cultural geography of the subregion?

Who makes up the majority of migrant workers, and what challenges do they face?

Migrant workers make up a large percentage of the population of the Arabian Peninsula. For example, over 30 percent of the total population of Saudi Arabia are workers from other countries. The millions of migrant workers include women who travel each year in search of domestic work. They come from such Asian countries as the Philippines, Sri Lanka, and Indonesia, and from some African countries including Ethiopia, Egypt, and Madagascar.

Many are products of the *kafala* system, in which an employer sponsors a migrant worker by paying travel expenses and providing room and board. In some cases, however, the system is abused, and the sponsor exercises tight control over the migrant worker. Sometimes sponsors confiscate, or take control of, workers' passports. Some of the challenges migrant workers face include unsanitary living conditions, hunger, low wages, a lack of health care, and discrimination. All too often, they become the victims of human rights violations. Some critics say the system closely resembles slavery; female migrant workers, especially, are treated like property.

2. **Human Systems** What challenges do female migrant workers face, and what influence does the *kafala* system have on their rights?

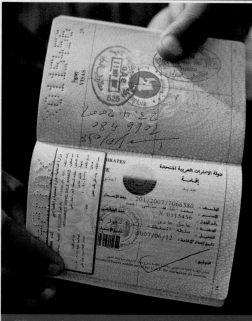

What actions can be taken to help migrant workers?

Migrant workers have few rights and are not protected under most labor laws. Many work nearly 100 hours per week with little or no compensation for overtime work; few are given leisure time. Under the *kafala* system, they are indebted to their sponsors, who prevent them from leaving to visit family in their home countries. The workers are also prevented from seeking other employment.

International groups such as Amnesty International and other organizations are pressuring governments and organizations in the region to take action to crack down on human rights abuses. The United Nations is also investigating court rulings that have favored and empowered employers in labor disputes that involve migrant workers.

3. **Human Systems** Write a paragraph explaining how government and community action could improve the lives of female migrant workers in the Arabian Peninsula.

THERE'S MORE ONLINE

VIEW a map of migrant worker home countries • *READ* about issues of migrant workers' rights in Saudi Arabia

networks

There's More Online!

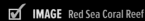

- ☑ **IMAGE** Red Sea Coral Reef
- ☑ **MAP** Physical Geography: The Arabian Peninsula
- ☑ **INTERACTIVE SELF-CHECK QUIZ**
- ☑ **VIDEO** Physical Geography of the Arabian Peninsula

Reading HELPDESK

Academic Vocabulary

- **nonetheless**
- **estimate**

Content Vocabulary

- **peninsula**
- **arid**
- **dune**
- **monsoon**
- **simooms**
- *shamal*

TAKING NOTES: *Key Ideas and Details*

IDENTIFYING As you read about the landforms of the Arabian Peninsula, use a graphic organizer like the one below to identify the characteristics of each major area.

Arabian Peninsula

Area	Characteristics

LESSON 1
Physical Geography of the Arabian Peninsula

ESSENTIAL QUESTION • *How do physical systems and human systems shape a place?*

IT MATTERS BECAUSE

The physical systems of the Arabian Peninsula affect the distribution of its two greatest resources, oil and water. The location and types of landforms contribute to extremes of climate. Human activities shape the peninsula through the movement of people, the creation of settlements, and the use of natural resources.

Landforms

GUIDING QUESTION *What are the major physical characteristics of the Arabian Peninsula?*

About 20 to 30 million years ago, Africa and Arabia formed a single tectonic plate. Over the millennia, tectonic forces created a rift valley which separated the plate into two landmasses. The valley filled with water, becoming the Red Sea. Positioned almost equally north and south of the Tropic of Cancer, this area east of the Red Sea is now the Arabian Peninsula.

The Red Sea became the western and southwestern boundaries of the newly formed **peninsula**. A peninsula is an area of land almost entirely surrounded by water. The waters of the Gulf of Aden, the Arabian Sea, the Gulf of Oman, and the Persian Gulf form natural boundaries on the southern and eastern parts of the peninsula. The southern borders of Israel, Jordan, and Iraq make up its northern boundary.

On the west, the Arabian Shield runs from north to south along the Red Sea coast. A shield is a landform composed of hard, ancient rocks. Here the Shield forms a tall mountain range. Along the western side of the Shield there is a steep escarpment, or cliff, that looms above the narrow strip of coastal plain that runs the length of the Red Sea. Extinct volcanoes cover the surface of the Shield. The lava beds produced by once-active volcanoes create the wide black bands typical of the western Arabian landscape.

To the east of the peaks of the Arabian Shield lies the Najd, the central plateau of the Arabian Peninsula. Unlike most plateaus, it does not end abruptly with a steep cliff. Instead, it slopes gently from west to east and ends at sea level on the Persian Gulf coast.

The Arabian Peninsula is a region dominated by the **arid**, or very dry, land of its interior. The 1.2 million square miles (3 million sq. km) of the peninsula are mostly unsuitable for human settlement. In much of the peninsula, silt and dust left by winds and sediment deposited by ancient seas have created soil suitable for agriculture. **Nonetheless**, a lack of water has meant that less than two percent of the peninsula is used for agriculture.

The Arabian desert has two areas with distinct characteristics. The northern desert, An Nafūd, consists of reddish, crescent-shaped **dunes**—ridges of sand formed by wind. This desert stretches across an area of 25,000 square miles (65,000 sq. km). Far to the south of An Nafūd is the Rub' al-Khali. Known as "the Empty Quarter," the Rub' al-Khali has ten times the area of the northern desert. It is the world's largest uninterrupted area of sand. It is also one of the driest regions on Earth. These four areas—the central plateau, the northern desert, the Rub' al-Khali, and the Arabian Shield—make up most of the Arabian Peninsula.

✔ READING PROGRESS CHECK

Describing How did tectonic activity form the Red Sea?

Water Systems

GUIDING QUESTION *How do the waters surrounding the Arabian Peninsula affect life there?*

Thousands of miles of mostly tropical seas surround three sides of the Arabian Peninsula. The Red Sea runs south along the western edge of the Arabian Shield and then to the southeast, where it meets the waters of the Gulf of Aden. The Arabian Sea lies to the southeast of the peninsula. On the northeast lie the Gulf of Oman and the Persian Gulf. These bodies of water isolate the peninsula from

peninsula a portion of land nearly surrounded by water

arid excessively dry

nonetheless in spite of; nevertheless

dune a mound or ridge of sand formed by the wind

GEOGRAPHY CONNECTION

The Arabian Plate broke free of Africa millions of years ago. It is still moving northeast at a rate of approximately .039 inches (4 mm) per year.

1. ***PHYSICAL SYSTEMS*** What physical characteristics might the Arabian Peninsula share with the African continent?

2. ***PHYSICAL SYSTEMS*** What will eventually happen to the Persian Gulf as the Arabian Plate moves northeast over time?

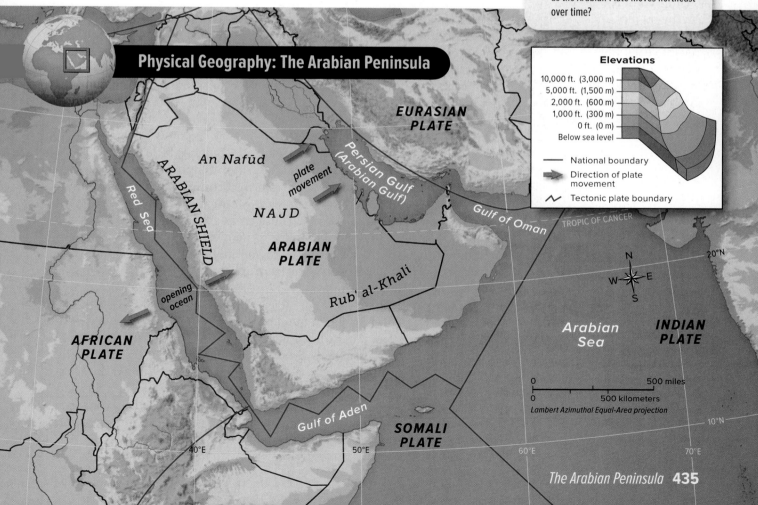

Physical Geography: The Arabian Peninsula

EURASIAN PLATE

An Nafūd

plate movement

Persian Gulf (Arabian Gulf)

Red Sea

ARABIAN SHIELD

NAJD

ARABIAN PLATE

Gulf of Oman

TROPIC OF CANCER

opening ocean

Rub' al-Khali

AFRICAN PLATE

Arabian Sea

INDIAN PLATE

Gulf of Aden

SOMALI PLATE

Elevations

10,000 ft. (3,000 m)
5,000 ft. (1,500 m)
2,000 ft. (600 m)
1,000 ft. (300 m)
0 ft. (0 m)
Below sea level

— National boundary

➡ Direction of plate movement

∿ Tectonic plate boundary

0 _____ 500 miles
0 _____ 500 kilometers
Lambert Azimuthal Equal-Area projection

N W E S

20°N

10°N

40°E   50°E   60°E   70°E

In 1902 an American named Willis Carrier invented air-conditioning. At first it was used commercially in the United States in public buildings like movie theaters and hotels. It was a boon to productivity in the summertime. Air-conditioning allowed cities in the southern United States to attract industries from other parts of the country. On the Arabian Peninsula, air-conditioning made it possible for people from temperate climates to work during the summers in the hot, steamy coastal cities or on the oil pipelines that cross the deserts. In recent years, the demand for air-conditioning on the Arabian Peninsula has increased as industries and tourism have expanded.

ANALYZING How does air-conditioning affect productivity in hot weather?

monsoon a seasonal wind that brings warm, moist air from the oceans in summer and cooler, dry air from inland in winter

simoom a hot, dry, suffocating wind that blows from time to time in the Arabian Peninsula

shamal a northwesterly wind in the Persian Gulf area

its neighbors to the east and west. However, they are vital to the life of the peninsula. The Red Sea and the Persian Gulf, in particular, are necessary to the economies of the region.

The Red Sea, on the western side of the Arabian Peninsula, is the northernmost tropical sea in the world. It is also an important shipping route. Ships entering from the Suez Canal in the north carry goods through the Red Sea to the Gulf of Aden. Most of the products transported between Europe and Asia pass through the long and narrow Red Sea.

On the east, the Persian Gulf serves the international shipping needs of all the oil-producing countries of the peninsula. The warm waters of the Persian Gulf are crowded with traffic. Huge tankers bring oil from Saudi Arabia, Kuwait, Qatar, Bahrain, and the United Arab Emirates to the rest of the world.

Although oil is plentiful on the Arabian Peninsula, freshwater is scarce. The subregion receives little rainfall. Rapidly growing populations are quickly using up groundwater, the water in the earth that supplies wells and springs. Dotted across the Arabian landscape are small oases, which are gradually going dry as groundwater is depleted. Small, temporary rivers form in wadis when there is rain. Wadis are streambeds that remain dry when no rain falls. They are not reliable sources of water because rainfall is unpredictable.

☑ **READING PROGRESS CHECK**

Locating Where are the principal bodies of water that border the Arabian Peninsula?

Climate, Biomes, and Resources

GUIDING QUESTION *How does the climate of the Arabian Peninsula affect its biomes?*

For the most part, the Arabian Peninsula is a desert with an extremely dry climate. In much of the peninsula, the average rainfall is less than 4 inches (10 cm) per year. The temperature in some places has been recorded as high as 129°F (54°C). The noontime sun is directly overhead at least once a year for land on or south of the Tropic of Cancer. The heat of the sun combined with the lack of moisture for cloud cover makes the peninsula one of the hottest places on Earth.

Climate Regions and Biomes

A hot and dry desert biome like the Rub' al-Khali has little rainfall. There is no surface water. The little rain that falls often evaporates before it reaches the ground. The Rub' al-Khali has few forms of animal or plant life. Some of the desert areas in the northern part of the peninsula have a greater abundance of plants and animals. These areas are similar to the Mojave and Sonoran Deserts in the United States.

The dry heat of the desert areas in summer contrasts with the extreme humidity of some coastal areas, which are affected by their closeness to water. Dew and fog add to the humidity. Seasonal **monsoon** winds bring heavy rains in the summer.

Wind is a significant factor in the climate of the Arabian Peninsula. The heat of the deserts in the summer sometimes generates hot, suffocating winds called **simooms**. The simooms last only about 20 minutes, but they destroy plant and animal life. These winds are so dangerous, in fact, that the word *simoom* actually means "poison" in Arabic. In areas near the Persian Gulf, sand- and dust-laden winds called **shamals** occur in midwinter and early summer. The word *shamal* means "north," and the winds blow from the north or northwest. These winds carry millions of tons of sand and silt into the Rub' al-Khali.

The Red Sea is an aquatic biome. The coral reefs shelter about 1,200 different species of fish and many species of plants. About 10 percent of the species found in the Red Sea are not found in any other marine environment. This aquatic biome is unique because its corals can endure extreme heat and salinity. The salinity is high because of high rates of evaporation and a lack of rain over the area. Violent, dust-laden winds churn the waters in ways that would prove fatal to other reefs. In spite of their hardiness, the reefs are subject to the same threats as other aquatic biomes: heavy pollution and overuse.

Natural Resources

At one time, the Earth was a huge aquatic biome. Animal and plant life teemed in ancient seas. Over time, the remains of these organisms were buried under the sedimentary rock. Millions of years of intense heat and pressure eventually transformed the remains of these organisms into crude oil. This fossil fuel is one of the world's most sought-after resources.

The Arabian deserts lie over a vast deposit of crude oil. This deposit is **estimated** to be at least 25 percent of the world's proven reserves, and oil is the region's principal export. Discovered in Saudi Arabia in 1938, rapid expansion of the oil industry did not occur until after World War II. Of the seven countries of the Arabian Peninsula, only Bahrain does not have significant petroleum reserves.

The Arabian Peninsula has few other significant resources. Some of these resources include fish, pearls, and salt and gypsum produced from saline flats. Mineral resources include iron ore, gold, copper, limestone, and some marble.

The resource that is most crucial to the peninsula, however, is freshwater. Drinkable water is rapidly being depleted because the population has grown dramatically. Aquifers beneath the desert sands cannot be replenished at the rate they are being used. The countries of the area are looking for ways to increase their supplies of this critical resource.

☑ **READING PROGRESS CHECK**

Classifying What characteristics of the Rub' al-Khali make it a desert biome?

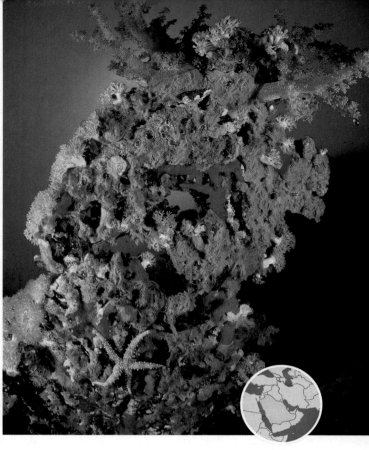

The Red Sea is known for its extensive coral reefs.

▲ **CRITICAL THINKING**

1. Hypothesizing Why would it be important to protect the Red Sea reefs?

2. Describing What resources does the Red Sea provide?

estimate to judge approximately; to determine roughly

LESSON 1 REVIEW

Reviewing Vocabulary

1. *Comparing* What do *shamals* and simooms have in common?

2. *Contrasting* How do monsoons differ from *shamals* and simooms?

Using Your Notes

3. *Analyzing* Use your graphic organizer on the Arabian Peninsula's landforms to write a paragraph analyzing the characteristics that make some areas of the peninsula more suitable for some activities than others.

Answering the Guiding Questions

4. *Differentiating* What are the major physical characteristics of the Arabian Peninsula?

5. *Evaluating* How do the waters surrounding the Arabian Peninsula affect life there?

6. *Making Connections* How does the climate of the Arabian Peninsula affect its biomes?

Writing Activity

7. *Explanatory* In a paragraph, discuss why the Red Sea and the Persian Gulf are vital to the economy of the Arabian Peninsula.

Reading HELPDESK

Academic Vocabulary

- **document**
- **strategy**

Content Vocabulary

- **sheikdoms**
- **Sunni**
- **Shia**
- **Ibadhism**
- **shari'ah**
- **hajj**
- **choke point**

TAKING NOTES: *Key Ideas and Details*

ORGANIZING As you read the lesson, take notes on how the people of the Arabian Peninsula have been governed in the past and how they are governed today.

Arabian Peninsula: Government

Past	Today

LESSON 2

Human Geography of the Arabian Peninsula

ESSENTIAL QUESTION • *How do physical systems and human systems shape a place?*

IT MATTERS BECAUSE

History and physical geography have had profound effects on the Arabian Peninsula. A similar history, religion, and culture have shaped the lives of its people. The discovery of vast oil and natural gas resources in the twentieth century produced widespread changes and major challenges to the human population, as well as effects on the physical environment.

History and Government

GUIDING QUESTION *How did history, culture, and geography help influence the types of governments on the Arabian Peninsula?*

The people of the Arabian Peninsula are overwhelmingly ethnic Arabs and religious Muslims. Religion, ethnicity, and a shared culture unite and define the subregion. From the earliest times, the people of the peninsula lived in tribes—social groups based on family relationships. Individual tribes controlled specific areas of the region, called **sheikdoms**. Some tribes were sedentary, and others were nomadic.

The bedouin nomads of Saudi Arabia had a tribal structure but were not confined to a single place. Bedouins occupied most of the Arabian Peninsula. Primarily herders of camels, goats, and sheep, the bedouin tribes moved throughout the Arabian Peninsula and beyond. Different tribes of nomads became powerful from time to time and dominated much of the area. Nevertheless, there was no formal government. Each tribal leader had absolute power, and there was no authority above him.

Similarly, powerful families controlled the settled areas along the eastern and southern coasts. These lands were more vulnerable to foreign control than the vast area of the interior. Throughout history, foreigners invaded these areas. The ancient Greeks established a trade center in the Persian Gulf near what is now Kuwait. The Ottoman Empire was a strong presence on the Arabian Peninsula for several hundred years until the Saud family finally drove them from the northern part of the peninsula in the early twentieth century. Despite the interference of foreigners, however, the power of the families remained.

sheikdom territory ruled by an Arab tribal leader

Today, the majority of governments in the Arabian Peninsula are monarchies. Saudi Arabia, Oman, and Qatar are absolute monarchies. Kuwait and Bahrain are constitutional monarchies. The United Arab Emirates is a federation of emirates, formed from seven smaller states that were headed by emirs, or Arab rulers. In 1990 North Yemen and South Yemen were united to form a republic. However, southern secessionists and northern insurgents have continued to oppose the union. By 2012, civil unrest and violence had destabilized the government. The activities of al-Qaeda militants increased in number in recent years, and ocean piracy along the coasts remains a problem.

☑ READING PROGRESS CHECK

Drawing Conclusions How has the role of nomadic tribes affected the way that most people are governed in the Arabian Peninsula?

Population Patterns

GUIDING QUESTION *How have climate and history helped determine the population patterns of the Arabian Peninsula?*

The harsh climate of the Arabian Peninsula has had a strong influence on human settlement. Most of the subregion's population lives along the coasts of the Persian Gulf and the Red Sea. Bedouin herders once moved from oasis to oasis in the arid interior of the peninsula. Today, most bedouins have settled in towns and cities.

The population of the Arabian Peninsula was once almost exclusively Arab. Today, as the table shows, some countries in the subregion are more heterogeneous. The makeup of the population began to change with the development of the petroleum industry. The oil fields and construction jobs attracted guest workers from South Asia, mainly from India and Bangladesh. Poverty and unemployment in South Asia pushed these workers toward the oil-wealthy Arab countries. The prospect of high wages continues to pull guest workers to the subregion despite the long work hours and crowded housing conditions.

Arabian Peninsula Country Statistics				
Country	**Population**	**Median Age**	**Principal Ethnic Groups**	**GDP (PPP) $**
Bahrain	1.4 million	31.8	Bahrainis, Asians	64.9 billion
Kuwait	3.8 million	29	Kuwaiti, other Arabs, South Asians, Iranians	288.8 billion
Oman	4.2 million	25.1	Arabs, Baluchis, South Asians, Africans	171.7 billion
Qatar	2.4 million	32.8	Arabs, South Asians, Iranians	324.2 billion
Saudi Arabia	31.6 million	26.8	Arabs, Asians, Africans	1.679 trillion
United Arab Emirates	9.6 million	30.3	South Asians, other Arabs and Iranians, Emiratis	641.9 billion
Yemen	26.7 million	18.9	predominantly Arabs	75.5 billion

Source: World Population Data Sheet, 2015; CIA World Factbook, 2015

⌄ CHART SKILLS

This chart shows some population statistics for countries of the Arabian Peninsula. It also gives each country's GDP, or gross domestic product. GDP is a measure of a country's wealth.

◄ CRITICAL THINKING

1. *Differentiating* In what ways does Yemen's population differ from that of other Arabian Peninsula countries?

2. *Hypothesizing* Why might few immigrants be attracted to Yemen?

Before the discovery of oil in 1932, the cities of the Arabian Peninsula were already important economic centers. Their populations were small, however, compared to the numbers of people living in rural areas and in desert towns and villages. In addition, there were smaller clans of nomads grazing sheep, camels, and goats where there was vegetation for the animals.

Today about 80 percent of the population in the Arabian Peninsula lives in cities. Most of these cities, like Jidda (Saudi Arabia), Kuwait (Kuwait), and Doha (Qatar), are located on either the Red Sea or the Persian Gulf. They thrive because they are important centers of their countries' oil industries.

One major city is located in the interior, away from the coast. Riyadh, a center of oil refining, is also the capital of Saudi Arabia and home to the royal palace of the king. Other important cities are near the coast. Makkah (Mecca) is the holiest city of Islam. Millions of religious pilgrims visit the city every year. Lying 50 miles (80 km) from the Red Sea, it houses the Grand Mosque. Madinah (Medina), which is farther inland, is several hundred miles north of Makkah. The tomb of Muhammad is in Madinah. The city is considered a holy city second in importance only to Makkah.

✔ READING PROGRESS CHECK

Drawing Conclusions Why does most of the population live along the coast?

Society and Culture Today

GUIDING QUESTION *How does Islam affect life on the Arabian Peninsula?*

Arabic is the language common to all Arab people of the Arabian Peninsula. Many people also speak a second language, often English. Immigration to the peninsula has added to the number of languages spoken, but most immigrants learn the Arabic necessary to function in their jobs and everyday life.

Hassan Ammar/AFP/Getty Images

TIME LINE ⌄

ISLAM and
Saudi Arabia ➔

In the seventh century, Islam originated in Makkah, a city on the Arabian Peninsula. Islam had a profound influence over the culture and governance of the peoples of the Arabian Peninsula.

▶ CRITICAL THINKING

1. Describing Why did Islam split into two branches?

2. Analyzing Why do millions of Muslims travel to Saudi Arabia each year?

According to Islamic teachings, an angel tells Muhammad that he is chosen to be Allah's prophet.

Muhammad leads his followers from Madinah to Makkah. Muslims dedicate the Kaaba to worship of Allah.

500 ➔

610

630

570

Muhammad is born in the city of Makkah.

622

To escape persecution, Muhammad and his followers leave Makkah and travel to Madinah.

632

Muhammad dies. Much of the Arabian Peninsula is united under Islam.

The majority of immigrants are Muslims. When Muhammad founded Islam, the new religion spread rapidly across the peninsula. Arab armies conquered Persia, much of the Byzantine Empire, North Africa, and Spain. Their military successes eventually ended, but Islam continued to diffuse across Asia and Africa.

The unifying fervor of the early years of Islam was broken by a growing conflict within the religion. After Muhammad's death, arguments arose over who would succeed him. As a result, Islam broke into two branches, the **Sunni** and the **Shia**. The conservative sect of **Ibadhism** broke off from the main branches of Islam in A.D. 657. Ibadhites became the major religious group in Oman.

Islamic law, or **shari'ah**, governs every aspect of a Muslim's life. Shari'ah specifies what a person owes to Allah, or God, and what he or she owes to other human beings. A Muslim's duty to Allah is satisfied by observing such practices as making a religious journey to Makkah, called the **hajj**, at least once in one's lifetime. Shari'ah also shapes criminal and civil laws that govern the responsibility of an individual to other people.

With the exception of Yemen, the countries of the Arabian Peninsula have universal health care. However, medical practice varies greatly. Qatar spends large amounts on health care, and its citizens are the most satisfied with their health care system. People in Yemen, which does not offer universal health care, receive little or no preventive care. Overall, in Yemen it is estimated that 48 percent of women and 37 percent of men have the necessary access to health care.

Education for males is compulsory in all of the countries of the subregion. Boys and girls attend separate schools, and higher education is not as available to women as it is to men. For example, in Saudi Arabia there were no schools for girls until the 1960s, a situation that changed by the twenty-first century.

Saudis must attend school from the ages of six to eleven. Saudi Arabia has a literacy rate of 95 percent. Qatar, with a literacy rate of 97 percent, requires children to attend school until they are seventeen. In Yemen, a much poorer

Sunni a branch of Islam that regards the first four successors of Muhammad as his rightful successors

Shia a branch of Islam that regards Muhammad's son-in-law Ali and the imams as his rightful successors

Ibadhism a conservative form of Islam distinct from Sunni and Shia sects

shari'ah Islamic law derived from the Quran and the teachings of Muhammad

hajj in Islam, the yearly pilgrimage to Makkah that Muslims must make at least once in a lifetime

661 Islam divides into two main branches—Sunni and Shia.

Abdul Aziz Al-Saud captures the Hejaz, unites warring tribes.

More than 3 million people travel to Makkah to perform the annual hajj.

1925

2012

1000 ➜

2000 ➜

1744 Muhammad bin Saud establishes the first Saudi state.

1932 Modern Kingdom of Saudi Arabia is established as an Islamic state with the Quran as its constitution.

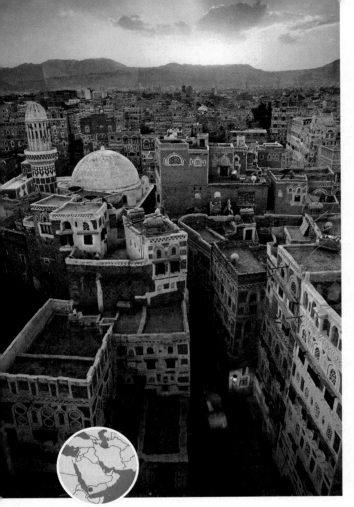

Sanaa, Yemen, is a UNESCO World Heritage City and famous for its ancient architecture.

▲ CRITICAL THINKING

1. Analyzing Visuals What clues reveal that this is an old city?

2. Analyzing Sanaa's population has increased dramatically and is growing almost twice as fast as the country as a whole. Why do you think this is so?

country without oil resources, students are supposed to stay in school until they are age fourteen. Only about 70 percent of Yemen's population is literate.

Family and the Status of Women

Most marriages in the subregion are arranged between families. Although not required, the arranged marriages of cousins have been a tradition. Once they are married, most women are expected to stay at home to take care of the children and provide a comfortable home for their husbands. In most countries, a married woman has domestic help, usually female Asian guest workers. The male head of household is the ultimate authority in the home.

Throughout the peninsula, men and women socialize separately. They are also segregated in public, particularly in Saudi Arabia. Usually women wear an *abaya*, a long black robe, over their clothing. In most of the countries, women cover their hair but are permitted to show their faces. In most of Saudi Arabia, however, women are considered immodest if their faces are not covered by a veil.

There has been political progress for women in several countries in the subregion. Most countries guarantee women equal rights, although equality is not enforced. Women in Saudi Arabia are denied voting rights, but King Abdullah recently granted women the right to vote in municipal elections. Kuwaiti women secured the right to vote in 2005. Elsewhere in the region, women's suffrage is permitted, but women have little political power.

The Arts

The Arabian Peninsula has produced many art forms. There is a strong oral tradition, and the culture is rich in poetry and folk art. The founding of Islam was accompanied by the flowering of religious art. In order to glorify the words of Allah, artists produced rich decoration called arabesques, often created in intricate geometric designs. Inlaid tile designs were used on buildings as far away as Spain and in other places in the Mediterranean.

Calligraphy is another characteristic Islamic art form. The word *calligraphy* means "beautiful writing." Arab calligraphers have used their script in the pages of the Quran, as well as to decorate textiles, walls, walkways, and building facades.

Islamic architecture is uniquely beautiful. Yemen's architecture features tall buildings, sometimes tinted in various colors, that have stood for over 2,000 years. The capital city, Sanaa, and several Yemeni towns have been designated as World Heritage sites because of their extraordinary architecture.

☑ READING PROGRESS CHECK

Identifying How does Islam affect the daily life of people on the Arabian Peninsula?

Economic Activities

GUIDING QUESTION *What impact does the oil industry have on life on the Arabian Peninsula?*

Oil was discovered on the Arabian Peninsula in 1932, but it was not until 1938 that the largest Saudi Arabian oil fields were discovered. U.S. and British oil companies became the major producers of Saudi oil. With time, Saudi Arabian engineers were trained, often in the United States and the United Kingdom, so they could fill management positions in the oil fields. By the end of the twentieth century, the Saudi Arabian government was in control of the production and export of its oil. In 1960 Saudi Arabia joined with Kuwait and other oil-producing countries to form the Organization of the Petroleum Exporting Countries (OPEC). Later Qatar and the United Arab Emirates and other countries outside the peninsula joined OPEC. The members of OPEC meet regularly to set production quotas. Every five years, OPEC **documents** its **strategy** for dealing with the long-term demands for oil.

The Arabian Peninsula is one of the most strategically important regions in the world for oil. Maintaining good relations with its neighbors is important. This is because two strategic waterways, the Strait of Hormuz on the Persian Gulf and the Bab el Mandeb, a strait on the Red Sea, are two of the world's leading oil-shipping **choke points**. Choke points are strategically located narrow passages between two bodies of water. Blocking either of these straits would prevent the export by shipping of the crude oil and petroleum products from the Arabian Peninsula.

Oil has produced great wealth for some countries on the peninsula. Some countries have used this wealth wisely. Saudi Arabia, with its enormous oil income, has made large investments in infrastructure and education. Bahrain, with relatively small reserves of oil, has successfully diversified its economy into banking and finance. Oman and the United Arab Emirates have invested their oil revenue. The investments help to soften the effects of fluctuations in oil prices in the international markets. Despite its oil wealth, Kuwait has not invested in its infrastructure. Yemen's oil production is small compared to other countries in the subregion. It has an unstable government and poorly developed economy. It has also been affected by internal division and conflict between Sunni and Shia groups, each wanting to control the country.

Although petroleum is the major product of the peninsula, there are other important industries. Cement manufacturing is a leading industry, as is ship repair. Bahrain and Oman have fishing industries, and commercial aquaculture is being developed. Tourism is important to both Bahrain and the United Arab Emirates. The Red Sea coastline of Saudi Arabia is a growing magnet for tourists because of its white sand beaches and its magnificent coral reefs. Yemen has a small textile industry.

Although arable land is scarce and precipitation is limited, a variety of fruits, vegetables, and some grains are grown. Agriculture consumes most of the water in the subregion. Food crop production is being developed that uses less water and has a high yield. Water is the subregion's most important natural resource issue.

document to put in writing in order to have as a record

strategy a plan or method

choke point a strategic, narrow waterway between two larger bodies of water

☑ **READING PROGRESS CHECK**

Explaining How has Saudi Arabia used its oil wealth to improve life in the country?

LESSON 2 REVIEW

Reviewing Vocabulary
1. *Determining Importance* What is the significance of shari'ah?

Using Your Notes
2. *Explaining* Use your graphic organizer on the governments of the Arabian Peninsula to explain how the governments of the subregion have or have not changed throughout history.

Answering the Guiding Questions
3. *Inferring* How did history, culture, and geography help influence the types of governments on the Arabian Peninsula?

4. *Evaluating* How have climate and history helped determine the population patterns of the Arabian Peninsula?

5. *Explaining* How does Islam affect life on the Arabian Peninsula?

6. *Drawing Conclusions* What impact does the oil industry have on life on the Arabian Peninsula?

Writing Activity
7. *Argument* In a paragraph, discuss the advantages and disadvantages of monarchy as a form of government for a country in the Arabian Peninsula. Conclude with your opinion supported by reasons.

networks

There's More Online!

☑ **IMAGE** Dubai's Air Pollution

☑ **IMAGE** Golf Course in the Desert

☑ **IMAGE** Drilling Equipment

☑ **INTERACTIVE SELF-CHECK QUIZ**

☑ **VIDEO** People and Their Environment: The Arabian Peninsula

Reading **HELP**DESK

Academic Vocabulary

- **furthermore**
- **coincide**

Content Vocabulary

- **geopolitics**
- **desalination**

TAKING NOTES: *Key Ideas and Details*

IDENTIFYING As you read about environmental concerns of the Arabian Peninsula, use a graphic organizer like the one below to identify the causes and effects of the environmental problems.

Environmental Concerns

Cause	Effect

LESSON 3
People and Their Environment: The Arabian Peninsula

ESSENTIAL QUESTION • *How do physical systems and human systems shape a place?*

IT MATTERS BECAUSE
The petroleum industry has had an enormous impact on the environment of the Arabian Peninsula. The growth of the population, due partly to the arrival of millions of foreign workers, is putting a serious strain on the subregion's water supplies. In addition to the scarcity of water, environmental concerns such as pollution and climate change are an ongoing challenge to the people and governments of the peninsula.

Managing Resources

GUIDING QUESTION *How has the extraction of petroleum resources impacted the environment of the Arabian Peninsula?*

Oil is the most widely used energy source in the modern world. For industrialized nations, oil is one of the most sought-after natural resources. Petroleum products are used for the generation of power and the operation of vehicles. Oil is a fossil fuel, created over millions of years, and thus is a nonrenewable resource. The relative scarcity of this resource makes it an important factor in the economies of the Arabian Peninsula. The peninsula's vast petroleum reserves also make it a focus of international interest.

Events in the oil-producing countries generate concern throughout the world. Unrest, such as that generated by the series of uprisings known as the Arab Spring, is magnified in importance by the effect it might have on oil production or transportation.

Physical, human, and economic geography can influence government policy. Such **geopolitics** determines how world powers respond to shifts in power on the peninsula. The 1991 Persian Gulf War is an example of such a response. In late 1990, Iraq invaded Kuwait, a neighbor of Saudi Arabia. The presence of Iraqi forces in Kuwait was not only a violation of international law, but it was also a threat to oil-rich Saudi Arabia. The United States, heading a large multinational force, pushed Iraq's forces back and restored the control of Kuwait to the Kuwaitis.

Oil extraction comes with a price. Oil pollution, whether from extraction, processing, or spills, releases hundreds of chemical gases into the air. These pollutants linger in the air until rain carries them to the ground, where they contaminate the groundwater and the soil.

One of the most dramatic examples of oil pollution occurred during the Persian Gulf War. In a successful attempt to sabotage Kuwait's oil industry, Iraqi forces set fire to oil wells and released 250 million gallons of oil into the Persian Gulf. Hundreds of miles of coastline were soaked with oil. The fishing industry, a vital part of the Persian Gulf economy, was dealt a serious blow.

The extraction and processing of petroleum has other consequences. Climate studies show that the average temperature of the peninsula has increased over the past three decades. All of the countries of the subregion have been affected by climate change.

Rainfall has decreased over much of Saudi Arabia. Persian Gulf countries like Oman are experiencing flash floods that erode the topsoil. Drought has affected the fertility of the soil and altered times for planting crops. Farmers report that increased temperatures have led to an increase of insect pests and plant diseases. Higher temperatures also increase the evaporation of water used for irrigation.

Kuwait has experienced both a sharp drop in rainfall and a dramatic rise in average temperature. The country has experienced tornado winds and an increase in sandstorms and flooding. Offshore, seawater temperatures have risen at three times the global rate.

Climate change is also causing a rise in sea level. The archipelago country of Bahrain is particularly vulnerable to rising sea levels. A significant rise would be a disaster for the small country, whose population lives mainly on the coasts, as well as for its marine ecology.

Most scientists think that these changes can be attributed, at least in part, to human activities. The burning of fossil fuels releases carbon dioxide into the air. These carbon emissions are called greenhouse gases because they contribute to

geopolitics government policy as it is influenced by physical, human, and economic geography

Air pollution creates a dense blanket over Dubai, United Arab Emirates.

▼ **CRITICAL THINKING**
1. **Drawing Inferences** How does air pollution affect the lives of people?
2. **Explaining** Why is air pollution such a problem on the Arabian Peninsula?

"You have thousands of tankers entering the Gulf and washing their tanks illegally. Between the tankers, pollution from urban centres and the brine disposed of from desalination plants, the Gulf is almost dead."

—Dr. Shawki Barghouti, director-general of the International Centre for Biosaline Agriculture, quoted in "Desalination Threat to the Growing Gulf," *The National*, August 31, 2009

DBQ *ANALYZING* What does the speaker mean by "the Gulf is almost dead"?

furthermore besides; in addition

A golf course has been carved out of the barren desert of the Arabian Peninsula.

▶ **CRITICAL THINKING**

1. Assessing Is maintaining a golf course in the middle of a desert a good use of water? Why or why not?

2. Analyzing Visuals How can you tell that this golf course is not suited to the environment?

the warming climate. Heavy use of fossil fuels for desalination, air-conditioning, and oil processing add greenhouse gases to the atmosphere. Carbon emissions were the topic of a special meeting of a United Nations convention on climate change in December 2012. The meeting was held in Qatar, which has the highest rate of carbon dioxide emissions per capita of any country in the world.

☑ **READING PROGRESS CHECK**

Identifying How have the advantages of the Arabian Peninsula's resources created a problem for its environment?

Human Impact

GUIDING QUESTION *How has a growing population created a set of related environmental challenges in the subregion?*

In 2015 the Arabian Peninsula was home to approximately 80 million people, and the population is expected to reach 120 million by 2050. Saudi Arabia's population alone is projected to grow by 333 percent over the next 34 years. The population of Kuwait is increasing at an annual rate of 1.62 percent, and Oman's is increasing at a rate of 2.07 percent. Yemen's population is growing by about 2.47 percent each year. As the number of people increases, so does the human impact on the environment.

The need for water is critical. Surface water created by rainfall exists in only a few places, and the evaporation rate is high because of the hot, arid climate. **Furthermore**, climate change is reducing the amount of rainfall in some areas, while increasing it in others. As weather patterns change, severe flooding can occur. These increases in rainfall overwhelm the existing drainage system, eroding the soil and endangering the human population.

Every country in the peninsula extracts increasingly higher annual percentages of water from its aquifers every year. This increase is not sustainable. The aquifers contain fossil water—water trapped in the underground rocks millions of years ago. Only minute amounts of surface water reach these

Barry Iverson/arabianEye/Getty Images

depths over time. Not only is the water virtually a nonrenewable resource, removing it from the aquifers draws in seawater. The seawater contaminates the aquifer and makes it unfit for drinking. The impact of the rising population on water resources has reached a crisis stage in many countries. For example, Sanaa, the capital city of Yemen, is estimated to have only a three-year supply of water left.

Yemen shows that an increasing population and the effects of climate change can contribute to political instability. Warring factions, coupled with widespread corruption and a weak government, prevent Yemen from enacting solutions to its problems. A 40 percent unemployment rate and widespread poverty are made worse by decreased rainfall and a resulting water crisis. Without governmental leadership, individuals and groups are extracting water as quickly as they can in order to make easy short-term profits. Meanwhile, the large rural population, which depends on agriculture, is starved for water. Yemen is at the point of collapse.

The Arabian Peninsula is facing growing desertification. There are many factors that **coincide** to turn arable land into desert. Reduced rainfall can gradually extend a desert into a once fertile area. Mismanagement of the land is another factor. As the population of farmers has increased, so have the herds of sheep and goats. Domestic animals take a heavy toll on vegetation that anchors topsoil. Increasing temperatures reduce the fertility of the soil, and unusually violent rainstorms wash it away. All the countries of the peninsula are affected to some degree by these conditions, but the four Persian Gulf States—Bahrain, Kuwait, Qatar, and the United Arab Emirates—are the most desertified countries on the Arabian Peninsula.

✓ READING PROGRESS CHECK

Identifying Which challenges to the environment are related to population?

Addressing the Issues

GUIDING QUESTION *How have people and governments on the Arabian Peninsula addressed the environmental challenges they face?*

Water use in most countries of the Arabian Peninsula is similar to that of more developed countries with far greater supplies of freshwater. Saudi Arabia uses the same amount of water per capita as other more developed countries. Qatar uses almost twice that amount.

One response to the need for water has been to build **desalination** plants. Desalination removes salt from seawater, as well as minerals from undrinkable groundwater. Current methods of desalination require enormous amounts of energy. For example, desalination accounts for 25 to 30 percent of the energy used in Saudi Arabia.

As groundwater has been depleted, the countries of the subregion have greatly increased their dependence on desalination. Qatar is an extreme example of this dependence. It relies on desalination for 99 percent of its water needs. Yemen, on the other hand, has just recently begun to use desalination.

Drilling equipment sits at a farm in Yemen. Yemen faces a water crisis due to overconsumption and mismanagement. Illegal drilling in aquifers is rampant.

▲ CRITICAL THINKING

1. Analyzing What factors have led to Yemen's water shortage?

2. Making Predictions What are some ways Yemen might address water shortages?

coincide to happen in the same place and at the same time

desalination the removal of salt from seawater or from brackish groundwater to make it usable for drinking and irrigation

©Bryan Denton/Corbis

Yemen, like the other countries of the Arabian Peninsula, has a limited water supply. Farmers use 90 percent of that water every year, irrigating their crops as they have done for centuries. Plentiful groundwater supplies exist in eastern Yemen, but the cost of transporting it to other parts of the country is too high.

Yemen's water problems stem in part from the lack of a stable government. In other parts of the peninsula, governments reach out to farmers, teaching them how to use water effectively. Yemen does not have an agricultural service, nor does the government have the ability to regulate consumption of water. Without a stable government, there can be no countrywide solutions to the water problem.

ASSESSING What should be government's role in managing a country's resources?

Desalination is not a perfect solution. Desalinated water must be blended with water from aquifers to make it drinkable. Even then, many people prefer the taste of imported bottled water to that of their local tap water.

Perhaps more important, desalination has environmental costs. Disposal of the brine that results from the desalination process presents problems. If it is collected in pools, it can seep into the groundwater and contaminate the aquifer. If it is piped into the sea, it increases the salinity of the water, interfering with the ecology of the coastal waters.

Historically, all of the governments of the Arabian Peninsula have subsidized the cost of water, and cheap water has led to overconsumption. To cater to wealthy citizens and attract tourists, the Persian Gulf countries have built resorts and spas that feature swimming pools and waterfalls, among other water-intensive attractions. Even the less wealthy are accustomed to using water freely to wash their cars and water their gardens.

By far the greatest overconsumption, however, has been in the practice of agriculture. During the 1970s, the demand for meat and dairy products increased as the urban population grew richer. An effort was made to supply meat from local ranches, but within 10 years most meat was imported. Many governments became alarmed by increasing prices on the world market and concerned about their countries' dependence on foreign food. In response, the governments of Kuwait, Bahrain, the United Arab Emirates, and Saudi Arabia joined in an effort to encourage agriculture. However, about 85 percent of the region's annual water supply is used for agriculture, and much of it is wasted due to inefficient irrigation. Yemen provides an example of impractical water use. Much of its groundwater goes toward khat cultivation, the leaves of which are chewed as a mild stimulant.

Today, many countries on the Arabian Peninsula, particularly Saudi Arabia, are promoting a variety of changes in how agriculture is practiced. Greenhouse agriculture, which uses much less water than other methods of growing food, is being encouraged. Scientists are developing salt-tolerant plants that can make use of water that is too salty for drinking. Several countries are also looking abroad to find new solutions to their food problems. They are investing in land overseas. They plan to use the natural resources and labor of the host countries to produce food for import to the peninsula. Qatar, for example, is purchasing farmland in such distant places as Sudan, Australia, Kenya, Brazil, Argentina, Turkey, and Ukraine.

✓ **READING PROGRESS CHECK**

Explaining What are the advantages and disadvantages of desalination as a solution for water scarcity issues in the subregion?

LESSON 3 REVIEW

Reviewing Vocabulary

1. *Explaining* Write a paragraph explaining how geopolitics can create conflict between countries.

Using Your Notes

2. *Summarizing* Use your graphic organizer on the environmental concerns in the Arabian Peninsula to write a paragraph summarizing the environmental situation in the subregion.

Answering the Guiding Questions

3. *Evaluating* How has the extraction of petroleum resources impacted the environment of the Arabian Peninsula?

4. *Making Connections* How has a growing population created a set of related environmental challenges in the subregion?

5. *Explaining* How have people and governments on the Arabian Peninsula addressed the environmental challenges they face?

Writing Activity

6. *Argument* In a paragraph, take the position of a water resource manager of a city in the subregion, and explain to citizens why it is necessary to conserve water.

Directions: On a separate sheet of paper, answer the questions below. Make sure you read carefully and answer all parts of the questions.

Lesson Review

Lesson 1

1 *Describing* Describe how the Red Sea interacted with a single tectonic plate to create the Arabian Peninsula.

2 *Explaining* Why is wind such a significant factor in the climate of the Arabian Peninsula? How do simooms and *shamals* affect plant and animal life there?

3 *Contrasting* The two greatest resources in the Arabian Peninsula are oil and water. Why is freshwater even more crucial than oil to countries in the subregion?

Lesson 2

4 *Identifying Cause and Effect* How did the development of the petroleum industry affect the makeup of the population on the Arabian Peninsula?

5 *Evaluating* Why is the Arabian Peninsula considered one of the most strategically important regions in the world?

6 *Identifying* What countries in the Arabian Peninsula belong to OPEC? What is the role of that organization?

Lesson 3

7 *Making Connections* How does the extraction of oil cause damage to air, land, and water throughout the Arabian Peninsula?

8 *Summarizing* What are some of the drawbacks associated with current methods of desalination?

9 *Explaining* Why is the 1991 Persian Gulf War an example of the importance of oil production to geopolitics?

Critical Thinking

10 *Hypothesizing* Oil is a nonrenewable resource. How might the economies of the Arabian Peninsula be different if oil were renewable?

11 *Drawing Conclusions* Why do you think millions of immigrants have been drawn to the Arabian Peninsula even though their employers may treat them unfairly?

21st Century Skills

Use the graph to answer the following questions.

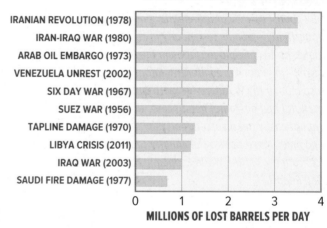

HUMAN EFFECTS ON OIL SUPPLY

- IRANIAN REVOLUTION (1978)
- IRAN-IRAQ WAR (1980)
- ARAB OIL EMBARGO (1973)
- VENEZUELA UNREST (2002)
- SIX DAY WAR (1967)
- SUEZ WAR (1956)
- TAPLINE DAMAGE (1970)
- LIBYA CRISIS (2011)
- IRAQ WAR (2003)
- SAUDI FIRE DAMAGE (1977)

0   1   2   3   4
MILLIONS OF LOST BARRELS PER DAY

Sources: Energy Information Administration (EIA), Bank of America Merrill Lynch Global Commodities Research, Money Morning staff research

12 *Understanding Relationships Among Events* According to the graph, how do war, revolution, and social unrest affect global oil production?

13 *Drawing Conclusions* What other factors may also have affected daily oil production during these events?

College and Career Readiness

14 *Multicultural Societies* Tourism has become an important way to diversify the economy in Bahrain and the United Arab Emirates. Choose one of these countries. Imagine that you have been hired by a travel agency to design a three-day tour of that country. Research the major tourist destinations, cultural highlights, and other activities. Write a travel itinerary for each day of the trip.

Need Extra Help?

If You've Missed Question	**1**	**2**	**3**	**4**	**5**	**6**	**7**	**8**	**9**	**10**	**11**	**12**	**13**	**14**
Go to page	434	436	437	439	443	443	445	447	444	444	439	449	444	443

Directions: On a separate sheet of paper, answer the questions below. Make sure you read carefully and answer all parts of the questions.

DBQ Analyzing Primary Sources

Use the document to answer the following questions.

Strict interpretations of Islamic law and tradition on the Arabian Peninsula have created many restrictions for women. This is particularly the case in Yemen, the country with the least developed infrastructure.

PRIMARY SOURCE

"*By almost all indicators—health, education, and economic opportunity—women fare poorly in Yemen. It has one of the highest rates of child marriage in the world, and 60 percent of Yemeni women are illiterate. Infant mortality rates are also among the world's worst, attributed to the lack of prenatal and postnatal health care. Unlike men, women cannot easily get divorced, and they have limited property and inheritance rights. The country ranks dead last among 135 countries in the World Economic Forum's Global Gender Gap Report.*

Al Abdeli is an assistant professor of accounting at a university in Taizz, and she enjoys more freedoms than most of her peers. She credits this to growing up in Taizz and having an open-minded father who 'did not go to university but is knowledgeable about the world.' She is also a poet who for years openly expressed her loathing for Ali Abdullah Saleh's regime. 'I put some of my dreams in my poetry and implanted them in the minds of my students,' she said."

—Joshua Hammer, "Days of Reckoning," *National Geographic,* September 2012

15 **Analyzing** How do traditions of marriage and divorce contribute to the difficulties women face in Yemen?

16 **Problem Solving** How might greater access to education improve the condition of women in Yemen?

17 **Interpreting** What dreams do you think Al Abdeli implanted in the minds of her students?

Applying Map Skills

Use the Unit 5 Atlas to answer the following questions:

18 **Physical Systems** Use your mental map of the Arabian Peninsula to describe the location of the Red Sea, the Arabian Sea, and the Persian Gulf.

19 **Places and Regions** Describe the overall population density pattern of the Arabian Peninsula.

20 **Environment and Society** What physical and human features help explain why the Bab el Mandeb and the Strait of Hormuz are called choke points?

Exploring the Essential Question

21 **Diagramming** Design two diagrams showing how vast deposits of crude oil under the sands of the Arabian deserts have affected the human and physical systems of the Arabian Peninsula. In each diagram, label the factors involved, and explain how the human and physical systems interact.

Research and Presentation

22 **Research Skills** Using the Internet and other resources, research the history and culture of nomadic bedouin tribes in the Arabian Peninsula. Describe how bedouin culture evolved to meet the demands of life in the enormous central deserts of the interior. How do bedouins live today? Create a multimedia presentation to share your findings.

Writing About Geography

23 **Argument** Use standard grammar, spelling, sentence structure, and punctuation to write an essay arguing the benefits of a diverse economy for countries in the Arabian Peninsula. Give examples of how countries in the subregion have successfully diversified their economies, and discuss the potential consequences of relying too heavily on oil.

Need Extra Help?

If You've Missed Question	15	16	17	18	19	20	21	22	23
Go to page	442	441	442	360	364	360	439	438	443

Central Asia

ESSENTIAL QUESTION • *How do physical systems and human systems shape a place?*

Why Geography Matters
*Afghanistan's
Troubled History*

Lesson 1
*Physical Geography of
Central Asia*

Lesson 2
*Human Geography of
Central Asia*

Lesson 3
*People and Their
Environment: Central Asia*

Geography Matters...

Central Asia can be a challenging area in which to live. Much of it is arid. Farmers rely on snowmelt from the mountains to provide water to farm land in the valleys. Rocky highlands prevent farming in many areas but provide pasture for sheep. Forbidding deserts stretch over large expanses.

However, much of Central Asia is quite beautiful. Snow-covered mountains tower over deserts of stark beauty. Dazzling tiles adorn buildings. In some cities, centuries-old mosques stand alongside gleaming glass skyscrapers.

◄ This man in Kabul, Afghanistan, faces traditional and modern ways.

©Alison Wright/Corbis

Afghanistan's troubled history

Located on trade routes that link areas from East Asia, South Asia, Southwest Asia, and Europe, Afghanistan has been "the crossroads of Central Asia" for thousands of years. Becoming an independent country in 1747, Afghanistan has had a troubled history. In recent decades, Afghanistan has faced the external forces of invasion, as well as the internal forces of ethnic and tribal diversity. Both threaten the country's stability.

Ethnic Groups of Afghanistan

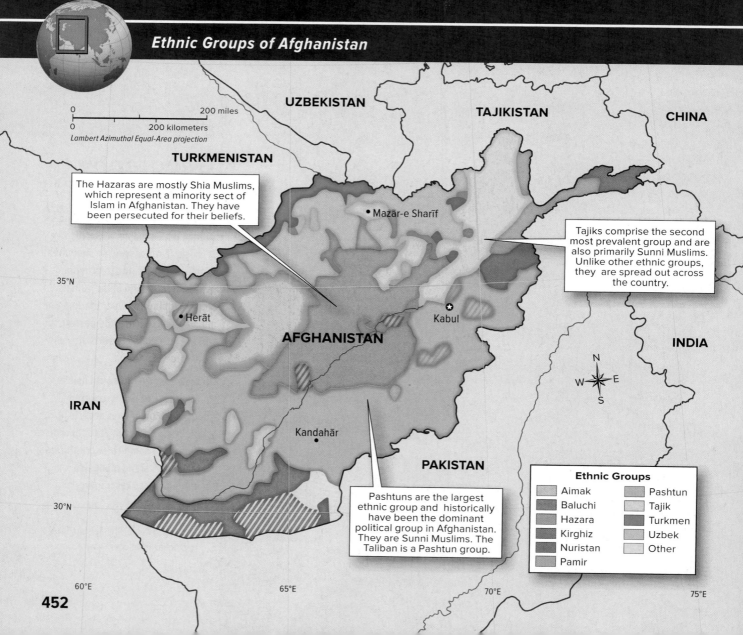

0 — 200 miles
0 — 200 kilometers
Lambert Azimuthal Equal-Area projection

UZBEKISTAN

TAJIKISTAN

CHINA

TURKMENISTAN

The Hazaras are mostly Shia Muslims, which represent a minority sect of Islam in Afghanistan. They have been persecuted for their beliefs.

• Mazār-e Sharīf

Tajiks comprise the second most prevalent group and are also primarily Sunni Muslims. Unlike other ethnic groups, they are spread out across the country.

35°N

• Herāt

AFGHANISTAN

Kabul ✪

INDIA

IRAN

Kandahār •

PAKISTAN

N
W E
S

Pashtuns are the largest ethnic group and historically have been the dominant political group in Afghanistan. They are Sunni Muslims. The Taliban is a Pashtun group.

Ethnic Groups

Aimak		Pashtun	
Baluchi		Tajik	
Hazara		Turkmen	
Kirghiz		Uzbek	
Nuristan		Other	
Pamir			

30°N

Who are the people of Afghanistan?

Afghanistan has a complex human geography. The country is home to dozens of different ethnic groups and subgroups. The map shows only the largest. While Pashtuns form the largest group, even they do not make up a majority of the country's people. Afghanistan's people speak about 50 languages. The two most widely spoken are Pashto and Dari. Pashto is the language of the Pashtun. Dari is a Farsi, or Persian, dialect. Dari often serves as a lingua franca (common language) in Afghanistan.

Ethnic differences are not necessarily strong divisions. People from different ethnic groups marry one another and live in the same areas. Perhaps the strongest tie Afghanistan's people feel is to their local tribes. For many people, their bond to the tribe is stronger than their feeling of national identity, which has made it difficult to organize them into a single modern country.

1. Human Systems How might tribal loyalties affect a leader's ability to govern Afghanistan?

Who has invaded and tried to rule Afghanistan?

For thousands of years, Afghanistan has been the target of foreign invaders. Persians ruled the area until Alexander the Great conquered it around 330 B.C. Alexander was followed by the Scythians, then the Huns, and then the Turks. Later, Arabs conquered the area. Persians ruled again, followed by the Turks once more. Then two different Mongol conquerors invaded Afghanistan.

The Soviet Union invaded Afghanistan in 1979. However, the Soviets withdrew after ten years of costly fighting. In the 1990s, a fundamentalist Islamic group took control of Afghanistan. They began to shelter Islamic terrorists. The terrorists established training camps for their fighters, some of whom attacked the United States on September 11, 2001. When the government refused to close the terrorist camps and even refused to hand over the terrorist leader, the United States and other countries sent troops to fight them. The government was quickly defeated and an elected government eventually took office. However, rebels have continued to fight against the new regime.

2. Places and Regions Why do you think Afghanistan has been invaded by outside forces so frequently?

Why does Afghanistan continue to suffer?

Afghanistan is among the poorest countries in the world. More than three-fourths of its workforce makes its living by farming using traditional methods. As a result, incomes are low. For example, more than a third of the people live in poverty. Just over a third of the workforce is unemployed.

Afghanistan's economy has begun to recover since the fundamentalists were ousted, and the economy has the potential for development. The country has reserves of natural gas and petroleum, precious metals such as gold and silver, and gems. However, conflict has hampered efforts to develop the country's natural resources. For example, continuing conflict has interfered with infrastructure development. Government corruption has also been a significant problem in trying to develop the economy and a stable society. As a result, Afghanistan still depends heavily on aid from the international community.

3. Environment and Society What opportunities and what challenges does Afghanistan face today?

THERE'S MORE ONLINE

READ about Afghanistan's place in the world • **VIEW** an image of the Soviet invasion of Afghanistan

networks

There's More Online!

- ☑ **IMAGE** Mountains in Tajikistan
- ☑ **IMAGE** Resort on Caspian Sea
- ☑ **IMAGE** Herding in Armenia
- ☑ **IMAGE** Farming in Afghanistan
- ☑ **INTERACTIVE SELF-CHECK QUIZ**
- ☑ **VIDEO** Physical Geography of Central Asia

Reading HELPDESK

Academic Vocabulary

- **unique**
- **reverse**

Content Vocabulary

- **steppe**
- **cereal**

TAKING NOTES: *Key Ideas and Details*

IDENTIFYING As you read about the physical geography of Central Asia, use a graphic organizer like the one below to identify major features of the subregion.

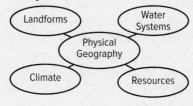

LESSON 1

Physical Geography of Central Asia

ESSENTIAL QUESTION · *How do physical systems and human systems shape a place?*

IT MATTERS BECAUSE

Central Asia has long been home to varied peoples. It has also served as the location of trade routes connecting Asia and Europe. For thousands of years, the people of this subregion have carved a life from a rugged land of mountains, deserts, and grasslands.

Landforms

GUIDING QUESTION *What are the major landforms of Central Asia?*

Central Asia stretches from the Black Sea to China. It is bounded on the north by Russia, on the east by China and Pakistan, and on the south by Iran and Turkey. It is a distinct subregion defined by its physical and human geography. The land is challenging, marked mainly by highlands, a dry climate, and vast expanses of desert. About half of the countries of the subregion are landlocked, meaning they have no access to the sea. The many cultures of the subregion share some **unique**, or distinctive, characteristics. They have been formed by the ethnic traditions and ways of life of the people and largely by Islam. All the countries in Central Asia, except Afghanistan, have something of a common history as well. They were part of the Soviet Union, gaining their independence in the 1990s.

Central Asia can be divided into two sections. The western section, sometimes called the Caucasus, lies between the Black Sea and the Caspian Sea. It consists of the three small countries of Georgia, Armenia, and Azerbaijan. The eastern section, reaching from the eastern shores of the Caspian Sea to China and Pakistan, includes the six "stans": Kazakhstan, Uzbekistan, Turkmenistan, Kyrgyzstan, Tajikistan, and Afghanistan. The suffix -*stan* is an ancient Persian word that means "land of" or "place of." In most of these countries, the first part of the name identifies the major ethnic group living there.

The western section gets its name from the Caucasus Mountains, which straddle the area between the Black Sea and Caspian Sea. The mountains actually consist of two ranges. The higher peaks, called the

Greater Caucasus, are to the north. They include Mount Shkhara, in Georgia. At 17,063 feet (5,201 m), it is the third-highest peak in the Caucasus. The lower mountain range, the Lesser Caucasus, is to the south.

The Central Asian countries have several high mountain ranges. Like the Caucasus, they were formed and continue to rise because of the collision of tectonic plates. The eastern ranges include the high Tian Shan of Kazakhstan and Kyrgyzstan; the Altay Shan, which run through those countries and Tajikistan; the Pamirs of Tajikistan; and the Hindu Kush of Afghanistan. Some of the peaks in these chains, like Nowshak in the Hindu Kush, soar over 24,000 feet (7,315 m) above sea level. These mountains dominate the eastern part of the subregion. They are separated by high plateaus and basins. Some of these basins, like the Fergana Valley of eastern Uzbekistan, have water and fertile soil. The Fergana, which covers an area about the size of Massachusetts, is thickly settled and marked by farms. Other basins are vast, dune-covered *kums* (KOOMZ), or deserts.

While the eastern part of the subregion has mountains that soar miles high, the western part slopes downward. The Turan Plain, around the Aral Sea, is only 200 to 300 feet (61 to 91 m) above sea level. The Caspian Depression, on the north shore of the Caspian Sea in southwestern Kazakhstan, is below sea level.

The Kara-Kum, or black-sand desert, covers most of Turkmenistan. The Kyzyl Kum, or red-sand desert, blankets the western half of Uzbekistan. These deserts are each about the size of Arizona. The deserts are sparsely settled. Afghanistan has a fertile plateau to the north of the Hindu Kush and a dry plateau to the south.

The Caucasus section is dominated by the Caucasus Mountains. Small patches of lowland plains lie along the shores of the Black and Caspian Seas. The mountains are flanked north and south by foothills and divided by valleys.

Tectonic activity built the mountains of this subregion. As is typical along plate boundaries, earthquakes frequently strike here. Afghanistan, surrounded by active plate boundaries, has an estimated 500 significant earthquakes a year. Tajikistan averages a major destructive quake every 10 or 15 years. A 1948 earthquake in Turkmenistan destroyed the capital city of Ashkhabad and killed about 176,000 people. Forty years later, a quake shook Armenia and caused such damage that its manufacturing capacity fell by 25 percent.

✅ **READING PROGRESS CHECK**

Explaining How do the landforms of Central Asia affect where people live?

Analyzing
PRIMARY SOURCES
Ancient Geography Text

"In proceeding from the [Caspian] Sea towards the east, on the right hand are the mountains . . . extending as far as India. They . . . stretch to this part from the west in a continuous line, bearing different names in different places."

—Strabo from *Geography*, Book XI, Chapter 8, c.A.D. 21

DBQ *MAKING CONNECTIONS*
Through what present-day countries would you pass if following the route through Central Asia as described by this ancient geographer?

unique distinctive

Fertile areas of Central Asia are dwarfed by the arid deserts and high, rocky mountains.

◀ **CRITICAL THINKING**

1. *Analyzing Visuals* Do you think this is the Kara-Kum or the Kyzyl Kum? Why do you think so?

2. *Hypothesizing* What challenges might this landscape pose to the people of the subregion?

Resorts, such as this one in Baku, Azerbaijan, dot some parts of the coast of the Caspian Sea, where people go to vacation.

▲ **CRITICAL THINKING**

1. Identifying Which countries in the subregion border on the Caspian Sea?

2. Summarizing What economic benefits does the Caspian Sea provide?

reverse the opposite

Water Systems

GUIDING QUESTION *What water features are important to the people of Central Asia?*

Central Asia is home to a large inland sea, the Caspian Sea, and a large lake, Lake Balkhash. The Caspian is the world's largest inland sea. Thousands of years ago it was connected to the Black Sea and, through it, to the Mediterranean. When the climate is warmer and precipitation is low, the Caspian Sea loses more water from evaporation than it gains from rivers emptying into it. In cooler, rainier times, the **reverse**, or opposite, is true.

Three countries from the subregion, plus Russia and Iran, border the Caspian Sea. Sturgeon fishing has long been an important industry. The sturgeon's eggs are eaten as caviar, an expensive delicacy. In recent years, oil and natural gas found under the Caspian Sea have become very important. Along some coastal areas of the sea are beaches and resort hotels.

East of the Caspian Sea is the Aral Sea. Until the 1960s, it was one of the world's largest inland seas. Today the sea is several small, shallow, disconnected bodies of water separated by dry land. This drying of the Aral Sea began in the 1960s when water from the main tributary was diverted to irrigate fields. This diminished the water supply to the Aral Sea. Also, the sea sits in a hot, dry region where high temperatures evaporate the water faster than rainfall replaces it.

Similar problems have hurt Lake Balkhash, in eastern Kazakhstan. The Ile River provided most of the lake's water. When the river was dammed, the lake's water level began to drop. The water also grew more saline, or salty.

The subregion's main rivers are the Amu Dar'ya and Syr Dar'ya. Both form in the subregion's center, fed by mountain rivers. Both rivers flow into the Aral Sea. Turkmenistan's Karakum Canal connects the Amu Dar'ya to the Caspian Sea.

☑ **READING PROGRESS CHECK**

Identifying Cause and Effect How have humans affected the water systems of Central Asia?

Michael Runkel/Alamy

Climate, Biomes, and Resources

GUIDING QUESTION *How does the climate affect human activity in Central Asia?*

Central Asia sits in the midst of continental Asia. It is far from any major oceans. As a result, the subregion has mainly dry climates, although the Caucasus area is generally wetter. The subregion's sparse scrub vegetation reflects this dryness.

While vegetation in the Kara-Kum and Kyzyl Kum deserts is sparse, these deserts are not completely desolate. Scrub grasses can grow in some areas. For centuries, the people of the region have lived there by practicing pastoralism, or the raising of livestock. The people drive herds of sheep, horses, camels, and cattle to graze on brush. Karakul sheep provide milk, meat, wool, and pelts. The wool of the Karakul sheep is used to weave carpets common to the region. The pelts of newborn sheep, often referred to as Persian lamb, are used for clothing.

Some farming is carried out on oases. Oases are areas where underground water naturally comes to the surface. Farming also takes place where irrigation has supplied water to the area, expanding the amount of arable land.

Steppes border desert regions across eastern Kazakhstan. **Steppes** form the wide, grassy plains of Central Asia. Precipitation in this semi-arid climate region usually averages from 8 to 12 inches (20 to 30 cm) annually. This amount is enough to support shrubs and short grasses, providing pasturage for sheep, horses, goats, and camels. Forests are rare in Central Asia. For example, only 3 percent of Kazakhstan's land is wooded. Small forests dot the islands of the Amu Dar'ya and on some of the foothills of the region's mountains. Temperatures are warm in summer and cold in winter. Temperatures are even hotter in Uzbekistan.

Because the soil is fertile, some parts of the steppe have been turned over to growing crops. Much of this farmed area is watered by irrigation. The major crops are **cereals**—grains like barley, oats, or wheat grown for food. Inhabitants also grow a great deal of cotton on the irrigated land of the steppes. Nomads use the steppes to graze their herds. In spring and summer, nomads lead their sheep and goats up into the mountains, where grasses grow in that season and where temperatures are cooler than in the lowlands.

Central Asia also has other climate zones. Areas on the shores of the southern part of the Caspian Sea have a Mediterranean climate. The Caucasus region has a semi-arid climate to the north of the mountains and a humid subtropical climate to the south. Although it is a small area, the Caucasus has great variation in climate. Western Georgia, on the shores of the Black Sea, receives from 40 to

steppe wide, grassy plains of Eurasia; also, similar semi-arid grassy areas elsewhere

cereal any grain like barley, oats, or wheat that is grown for food

People in Central Asia have adapted their agricultural practices to the climate and biomes where they live.

▼ **CRITICAL THINKING**

1. *Explaining* What seasonal pattern of movement do the subregion's pastoral nomads follow?

2. *Hypothesizing* Why can farmers of the Caucasus grow fruits and vegetables more easily than people in the rest of Central Asia?

100 inches of rain (102 to 254 cm) a year. Winds over the Black Sea provide much of this moisture. Eastern Georgia, on the other hand, receives only 16 to 28 inches (41 to 71 cm) of rain on average. Winters in the mountain foothills are mild. This climate allows the people of the Caucasus to grow different crops than people can on the steppes of the region. Many farmers here raise fruits and vegetables. Tea and citrus fruits are commercially important crops. The greater rainfall and milder weather in Georgia also supports forests of deciduous trees such as oak and beech on the mountainsides.

Higher areas, like the Caucasus Mountains and the mountains of the eastern subregion, have highland climates. It is generally wetter and colder than in other parts of the subregion. The highland climate varies, however, with elevation and exposure to wind and sun.

The major resources in the subregion are oil and natural gas. The largest deposits are found under the Caspian Sea, benefiting Kazakhstan, Turkmenistan, and Azerbaijan. Pipelines carry the oil to other parts of the world, allowing these landlocked nations to export these valuable resources.

Kazakhstan has the most abundant resources of the Central Asian countries. In addition to oil and gas, it has uranium, copper, zinc, iron ore, and other metals. Oil and oil products generate three-fifths of Kazakhstan's export earnings. Metals account for nearly another fifth. According to one estimate, Turkmenistan has the fourth-largest natural gas reserves in the world. While all those reserves have not yet been tapped, natural gas is a major export. In addition, Uzbekistan has the world's fourth-largest gold reserves; gold and natural gas are major exports.

Kyrgyzstan also has large supplies of gold, its number one export. Tajikistan also has gold as well as silver, tungsten, and uranium. These resources are not fully developed, however. None of them figures as a major export of Tajikistan. Afghanistan possesses natural resources, but few have been developed.

Among the Caucasus countries, Azerbaijan has the largest oil and gas resources. Its economy relies heavily on oil and oil products, which form 90 percent of its exports.

Georgia has no oil, but is taking advantage of its location. It has benefited by allowing pipelines to be built through it so that Caspian Sea oil can be transported to Europe. Georgia relies on hydroelectric power for electricity. The Georgians mine the metals manganese, copper, and gold. With its favorable climate, Georgia also grows fruits, vegetables, and other agricultural products. Tiny Armenia has few natural resources.

☑ READING PROGRESS CHECK

Synthesizing What is the main natural resource of Central Asia?

LESSON 1 REVIEW

Reviewing Vocabulary

1. *Explaining* Write a paragraph explaining the connection between the steppes and cereals in Central Asia.

Using Your Notes

2. *Summarizing* Use your graphic organizer on the physical systems of Central Asia to write a paragraph explaining how landforms and climates in the subregion are related.

Answering the Guiding Questions

3. *Identifying* What are the major landforms of Central Asia?

4. *Identifying* What water features are important to the people of Central Asia?

5. *Describing* How does the climate affect human activity in Central Asia?

Writing Activity

6. *Informative/Explanatory* In a paragraph, discuss how the landforms and climates of Central Asia shape people's lives in the subregion.

Reading HELPDESK

Academic Vocabulary

- **restrain**
- **virtually**

Content Vocabulary

- **mujahideen**
- **Taliban**
- **exclave**
- **enclave**

TAKING NOTES: *Key Ideas and Details*

ORGANIZING As you read about the human geography of Central Asia, use a graphic organizer like the one below to identify examples of the influence of native and foreign cultures on the subregion.

	Influences
History	
People/Culture	
Economy	

LESSON 2
Human Geography of Central Asia

ESSENTIAL QUESTION • *How do physical systems and human systems shape a place?*

IT MATTERS BECAUSE

Over the centuries Central Asia has been a crossroads of cultures and empires, as well as a home for people seeking independence. While other countries have ruled the subregion, the diverse peoples of Central Asia have created their own distinct cultures.

History and Government

GUIDING QUESTION *How have the countries of Central Asia been governed over the centuries?*

The peoples of Central Asia have a long and very old history. Some established their own countries in ancient times. Other peoples were herders and moved often in search of pasture. The subregion has been conquered many times by various empires. With some countries recently independent, the countries of Central Asia are working to find economic and political stability.

Kingdoms and Conquests

Both Georgia and Armenia formed kingdoms more than 2,000 years ago. Both places were Christianized under the Roman Empire in the A.D. fourth century. They were the first countries in the world to declare Christianity as the national religion. The Armenian and Georgian churches remain important today.

The peoples in the eastern part of the subregion lived largely as nomads in ancient times. Beginning about 100 B.C., the Silk Road was established for trade between Europe and China. The Silk Road crossed Central Asia. Locations along the road, such as Samarqand in present-day Uzbekistan, became rich trading centers.

The Silk Road made Central Asia a crossroads, and because Central Asia is surrounded by powerful neighbors, it has come under the control of various empires. In the seventh through the ninth centuries, Arabs conquered such areas as Azerbaijan, Turkmenistan, Uzbekistan, and Tajikistan. They brought Islam to the region. That religion later spread to other parts of the subregion. Armenia and Georgia remained Christian in a vast area dominated by Islam.

The Arts

Georgia and Armenia have literary histories that reach back more than two thousand years, and classic works from then and later are still treasured in both countries. The building styles of both these ancient kingdoms influenced the architecture of the Byzantine Empire. The six countries in the eastern subregion have a rich oral literature centered on epic poems that relate the adventures of cultural heroes. Like several countries in Southwest Asia, they also have a tradition of weaving brilliantly colored and intricately designed rugs. Bukhara and Samarqand in Uzbekistan have some magnificent mosques and tombs that are many centuries old.

☑ **READING PROGRESS CHECK**

Identifying Cause and Effect How are society and culture in Central Asia today different in urban and rural areas?

Economic Activities

GUIDING QUESTION *How do the people of Central Asia use resources to create economies?*

The countries of Central Asia are less developed. Most are not very industrialized and rely instead on agriculture and natural resources exports. Monetary output for these countries is a fraction of that of the United States, a more developed country.

Much of the land in this subregion has limited uses for agriculture. Yet for most of these countries, nearly half the workforce lives by farming. The high percentage of agricultural workers, the lack of good land, and the reliance on traditional farming methods help explain the countries' low economic output.

While many areas are not well suited for farming, the steppes of Central Asia do provide fertile soil for growing crops and supporting the grasslands that nomadic herders use in colder months for grazing their livestock. Uzbekistan is one of the world's largest cotton producers and one of the top exporters of cotton. Uzbekistanis also produce fine clothing from silk, a skill they learned as a result of their location along the Silk Road. Turkmenistan is another major cotton producer and is also a wheat producer. Uzbekistan and Turkmenistan rely on irrigation to grow crops in dry regions. Kazakhstan is among the world's top ten wheat producers. Its nomadic herders and farmers also produce dairy products, meat, and wool. Kyrgyzstan produces grains, wool, meat, and two export crops—cotton and tobacco. Tajikistan is mountainous and must import more than half of its food. Georgia's humid subtropical climate is good for growing citrus fruits, grapes, and hazelnuts. Farmers in Azerbaijan grow wheat, cotton, rice, fruits and

A traditional potato farmer in Afghanistan cultivates his field by hand.

▼ **CRITICAL THINKING**

1. Identifying Which countries in Central Asia have the highest per capita economic output?

2. Comparing and Contrasting How does the workforce of these countries compare to that of other countries in the subregion?

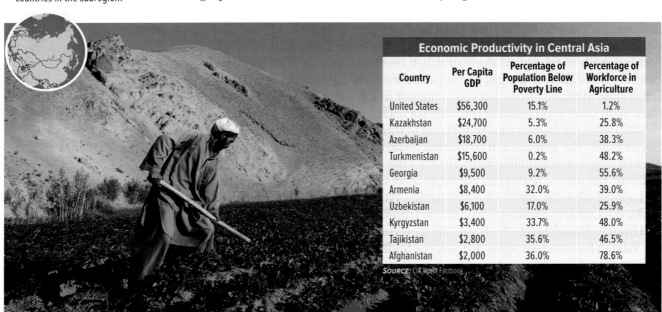

Economic Productivity in Central Asia			
Country	**Per Capita GDP**	**Percentage of Population Below Poverty Line**	**Percentage of Workforce in Agriculture**
United States	$56,300	15.1%	1.2%
Kazakhstan	$24,700	5.3%	25.8%
Azerbaijan	$18,700	6.0%	38.3%
Turkmenistan	$15,600	0.2%	48.2%
Georgia	$9,500	9.2%	55.6%
Armenia	$8,400	32.0%	39.0%
Uzbekistan	$6,100	17.0%	25.9%
Kyrgyzstan	$3,400	33.7%	48.0%
Tajikistan	$2,800	35.6%	46.5%
Afghanistan	$2,000	36.0%	78.6%

SOURCE: CIA World Factbook

vegetables, and tea, with cotton as a major export. Armenians grow fruits and vegetables and raise livestock.

Energy resources fuel Kazakhstan's economy. Oil accounts for some three-fourths of its export earnings. Oil and gas sales helped the country rebound from the worldwide economic crisis of 2008. Kazakhstan aims to substantially increase production.

Uzbekistan and Turkmenistan produce and export natural gas. Uzbekistan also exports gold and uranium. Kyrgyzstan is a major gold exporter. Azerbaijan, an important oil producer, is improving its exports of natural gas. Pipelines and train lines built through Georgia have helped Azerbaijan export its oil and gas to European markets. The lines have also benefited Georgia. Georgia has developed hydroelectric power and exports some of the electricity.

Manufacturing has a small role in Central Asia. Uzbekistan's factories make automobiles, most of which are exported to Russia. Kazakhstan produces tractors and agricultural equipment. Tajikistan produces aluminum and cement. Armenia's economy was geared more toward manufacturing when it was part of the Soviet Union. Much of that activity ended after independence. Armenia still has economic ties to Russia and depends on that country for natural gas. In addition, some businesses in Armenia are Russian owned. Tajikistan relies on Russia as a source of employment income, with nearly a million Tajikistani guest workers sending their earnings back home.

When the countries of Central Asia were part of the Soviet Union, they were also part of its command economy. Some of the countries in the subregion have not yet moved to a market economy. In Turkmenistan, for instance, the government owns the land, which it leases to farmers while also telling them what crops to grow. Along with government control of the economy, Turkmenistan is plagued by government corruption. These same problems have hurt Azerbaijan's economy, as has too much reliance on oil and gas revenues.

Uzbekistan has made more of a shift to a market economy. Still, the largest firms are government owned. Kazakhstan and Kyrgyzstan have moved more fully to market economies. Kyrgyzstan and Georgia have adopted reforms aimed at fixing government corruption, which has helped their economies develop.

✓ READING PROGRESS CHECK

Describing Explain how one country in Central Asia has used its resources to build its economy.

Central Asia is not as rich in energy resources as nearby Southwest Asia, although some countries such as Azerbaijan have sizable reserves of oil and natural gas.

▲ CRITICAL THINKING

1. Hypothesizing In which country of Central Asia do you think this photograph was taken? Why?

2. Analyzing Why have the energy-producing countries in the subregion focused on building pipelines in recent years?

LESSON 2 REVIEW

Reviewing Vocabulary

1. *Explaining* Write a paragraph explaining the difference between an enclave and an exclave.

Using Your Notes

2. *Summarizing* Use your graphic organizer on Central Asia's human geography to write a paragraph summarizing the influence of Soviet rule on the subregion's history, people, and economies.

Answering the Guiding Questions

3. *Summarizing* How have the countries of Central Asia been governed over the centuries?

4. *Explaining* How have population patterns changed over time in Central Asia?

5. *Describing* What are some aspects of the cultures of the people of Central Asia?

6. *Explaining* How do the people of Central Asia use resources to create economies?

Writing Activity

7. *Persuasive* In a paragraph, discuss whether you think traditional culture, Islam, or Soviet rule had the strongest influence on the subregion.

There's More Online!

☑ **IMAGE** Central Asian Steppe

☑ **INFOGRAPHIC** Aral: A Sea Sacrificed

☑ **IMAGE** Kazakh Factory

☑ **INTERACTIVE SELF-CHECK QUIZ**

☑ **VIDEO** People and Their Environment: Central Asia

Reading **HELP**DESK

Academic Vocabulary

- **concentrated**
- **expose**

Content Vocabulary

- **radioactive material**

TAKING NOTES: *Key Ideas and Details*

EXPLORING ISSUES As you read about the environment of Central Asia, use a graphic organizer like the one below to identify environmental issues and steps being taken to address them.

Issue	Steps

LESSON 3
People and Their Environment: Central Asia

ESSENTIAL QUESTION • *How do physical systems and human systems shape a place?*

IT MATTERS BECAUSE
Because of the generally dry climate of Central Asia, water and other resources have to be managed carefully. Mistakes in the past have posed several challenges to some aspects of the environment. Since some resources are shared by several countries, they need to work together to solve these problems and manage their resources.

Managing Resources

GUIDING QUESTION *How do the people and governments manage the natural resources of Central Asia?*

Several countries in Central Asia are well supplied with natural resources, including fossil fuels and minerals. Others have few of these resources, and must try to find alternative sources to meet their energy and other needs. The strong role of agriculture in all the economies places demands on soil and water, while the fairly dry climate of the eastern subregion puts limits on the growth of agricultural output.

Kazakhstan, Azerbaijan, Turkmenistan, and Uzbekistan all have substantial oil or natural gas reserves, or both. Though these countries are landlocked, pipelines connect them to other parts of the world. One carries oil from Azerbaijan to the Black Sea coast of Georgia. Turkmenistan recently opened a pipeline to transport its natural gas to China, an important market because of its growing economy and increasing demand for energy.

The mountainous terrain of Georgia and Kyrgyzstan gives them swiftly flowing rivers that can be tapped for hydroelectric power. With very little of their own oil and gas, these countries rely on hydroelectric power. In fact, they generate enough electricity to not only meet their own needs, but to export power to neighboring countries as well.

Tajikistan often faces energy shortages, leaving many of its people without electricity for long periods of time. The country hopes to expand its hydroelectric power by building a dam on a tributary of the Amu Dar'ya, but this project has created conflict. Uzbekistan fears that the dam would limit its own ability to use the river's water to irrigate its profitable cotton fields. In response, Uzbekistan cut off sale of natural gas to Tajikistan. Tajikistan's government complained that Uzbekistan also shut down rail service carrying needed food imports. Another concern is the impact of reduced water flow on the quality of the river's water. The World Bank has urged Tajikistan to halt construction until further study of the dam's environmental impact is completed.

Despite small amounts of arable land, farming remains an important economic activity in all countries of the subregion. Herding can take place in hilly areas that are not suitable for farming, but some livestock herds graze on flatter grasslands. Given the low levels of rainfall in much of the area, carefully managing water resources for farming, grazing, and other uses is vitally important. Underscoring this need is the estimate by World Bank experts that more than three-fourths of the water used for irrigation in the subregion is lost. Inefficient practices will have to change if governments in the subregion want to farm more productively and save water.

Deforestation and desertification are problems in Afghanistan. People needing fuel sources have been cutting down trees to burn the wood. The loss of trees along with overgrazing by livestock are contributing to desertification.

Heavy reliance on exporting just one or two resources—however vital they may be—can be risky for an economy. For example, because of their dependence on oil exports, both Azerbaijan and Kazakhstan suffered in recent years when oil prices fell. Aware of this problem, Azerbaijan has taken steps to develop a more varied economy. It has increased investment in tourism as well as information and communications industries to develop new sources of jobs and revenue.

Of course, the reverse is also true: lack of resources poses problems for other countries. Armenia depends on natural gas from Russia. When the price goes up or supply falters, its economy suffers. Armenia could try to buy natural gas from neighboring Azerbaijan, but relations between those two countries are poor because of the dispute over Nagorno-Karabakh. Armenia recently opened a new pipeline that allows it to import natural gas from Iran. However, overreliance on Iran could become a problem in the future.

☑ **READING PROGRESS CHECK**

Identifying What is one example of how countries in the subregion are taking steps to properly manage their resources?

Irrigation is essential to supplying the cotton fields growing on the Central Asian steppe in Uzbekistan.

◄ **CRITICAL THINKING**

1. *Explaining* Why is it necessary to irrigate these fields?

2. *Identifying* Where does the water come from?

Human Impact

GUIDING QUESTION *How have modern economic activities impacted Central Asia?*

The Soviet era left a legacy of environmental issues resulting from Soviet economic policies. The subregion's major bodies of water face severe environmental challenges. Pollution in the Caspian Sea is severe. Pollution and overfishing threaten fish, like sturgeon, which provide important exports as well as serve important roles in maintaining the ecology of the sea. Scientists from the area say that the Apsheron Peninsula of Azerbaijan is the most environmentally damaged place on Earth. The damage is the result of air, water, and soil pollution from the oil and gas industries and from the overuse of pesticides and other harmful chemicals for agriculture.

Perhaps nowhere has the environmental damage of the Soviet era been more profound than in the area around the Aral Sea. This body of water, once the fourth-largest inland sea in the world, is now a fraction of its former size—in fact, it is no longer one sea but several very small bodies of water. The Soviets took water from the Syr Dar'ya and Amu Dar'ya—the two rivers that feed the Aral Sea—to irrigate fields in their push to boost cotton production. The result was a sharp reduction in the supply of water reaching the sea. Evaporation during the hot summers increased the loss of water from the sea. Over decades, the sea's levels

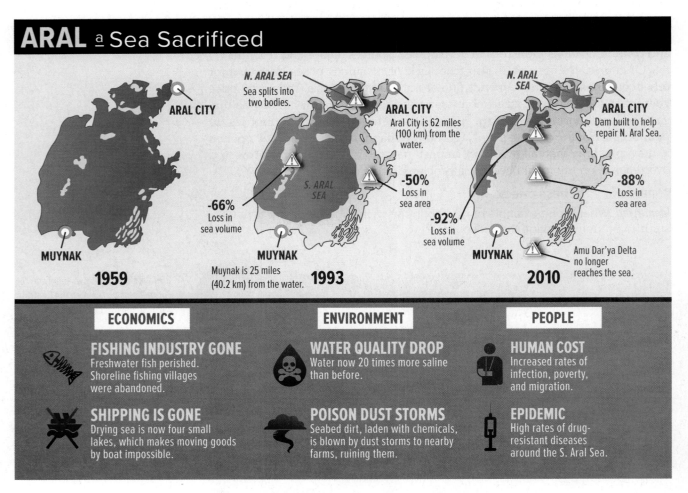

ARAL a Sea Sacrificed

1959

ARAL CITY

MUYNAK

1993

N. ARAL SEA
Sea splits into two bodies.

ARAL CITY
Aral City is 62 miles (100 km) from the water.

-66%
Loss in sea volume

S. ARAL SEA

-50%
Loss in sea area

MUYNAK
Muynak is 25 miles (40.2 km) from the water.

2010

N. ARAL SEA

ARAL CITY
Dam built to help repair N. Aral Sea.

-88%
Loss in sea area

-92%
Loss in sea volume

MUYNAK

Amu Dar'ya Delta no longer reaches the sea.

ECONOMICS

FISHING INDUSTRY GONE
Freshwater fish perished. Shoreline fishing villages were abandoned.

SHIPPING IS GONE
Drying sea is now four small lakes, which makes moving goods by boat impossible.

ENVIRONMENT

WATER QUALITY DROP
Water now 20 times more saline than before.

POISON DUST STORMS
Seabed dirt, laden with chemicals, is blown by dust storms to nearby farms, ruining them.

PEOPLE

HUMAN COST
Increased rates of infection, poverty, and migration.

EPIDEMIC
High rates of drug-resistant diseases around the S. Aral Sea.

The Aral Sea is in peril because too much water has been diverted from the rivers that feed into the sea. Most of the diverted water has been used to grow cotton.

▲ **CRITICAL THINKING**

1. Analyzing Visuals Which part of the sea is now in the most danger?

2. Making Connections How did the climate of the area around the Aral Sea contribute to the environmental problem there?

decreased dramatically, and water quality also fell as levels of salt and of harmful chemicals in the water became more **concentrated**, or stronger.

While those problems were bad enough, worse occurred. Winds blew salt and harmful chemicals from the dry seabed to nearby areas. These substances damaged the soil where they landed, ruining farmland. The salt and chemicals in the air also caused cancer and respiratory diseases in people.

Problems with the Caspian and Aral Seas are not the only environmental issues in Central Asia. Kazakhstan was home to nuclear bases during the Soviet era. During the Cold War, the Soviet government tested nuclear, chemical, and biological weapons there. As a result, northeastern Kazakhstan remains severely affected by radiation given off by the **radioactive material** after the Soviets tested nearly 500 nuclear weapons. In many instances, local populations were not warned or evacuated before the testing took place. In 1989 it was discovered that this weapons testing had caused radiation leaks. Scientists think many years will pass before all the resulting contamination disappears.

Soviet planners also chose Kazakhstan as a site for heavy industry, which polluted the air with toxic chemicals. Scientists have linked increased infant mortality in Kazakhstan directly to industrial pollution. The people of Kyrgyzstan, another site of heavy Soviet industry, have suffered similar effects. Soviet-era agriculture also caused serious environmental damage from heavy use of pesticides all across the region.

Armenia has been strongly criticized by environmental activists for continuing to run a nuclear power plant in an area prone to earthquakes. Activists worry that a quake could lead to a serious nuclear accident. The country, however, relies on that plant to generate 40 percent of its electricity, and the government has chosen to build a new plant rather than to abandon nuclear power completely. Armenia has also been criticized for deforestation of the Teghut Forest in northern Armenia as a result of mining activity there.

Land mines put into the ground during periods of war are a serious problem in Afghanistan. In 2001 and 2002, an average of 176 people were killed or injured as a result of the inadvertent explosion of land mines or other ammunition each month. More than three-fourths of those victims were children. While these rates of death and injury dropped dramatically in the following ten years, the problem remains and tragedies continue to take place.

Land mines and unexploded ammunition do not just **expose** humans to major life and health threats. They also hamper economic activity. They prevent farmers from working fields because of the fear of mines. They also interfere with the delivery of goods and services when explosive devices on roadways explode.

☑ **READING PROGRESS CHECK**

Identifying Cause and Effect What were the causes and effects of one of the environmental challenges facing the subregion?

While Central Asia is not highly industrialized, factories like these in Kazakhstan have created a serious pollution problem.

▲ **CRITICAL THINKING**

1. ***Making Connections*** Is this kind of pollution more likely to affect urban or rural areas? Why?

2. ***Hypothesizing*** Based on the information in this lesson, how concerned do you think Soviet leaders were about environmental issues? Explain your response.

concentrated less dilute, stronger

radioactive material material contaminated by residue from the generation of nuclear energy or the testing of nuclear weapons

expose to make vulnerable

Efforts to save the Aral Sea began in the 1990s when the mayor of a town on the northern shore had a dam built. The dam prevented the loss of water from the northern part of the sea by blocking it from flowing into the more seriously damaged southern part. Later, the government of Kazakhstan built a higher, sturdier dam. The World Bank has helped Kazakhstan with this program. Since then, sea levels in the north have risen, salt levels have declined, and fish have returned to the Northern Aral Sea from the Syr Dar'ya. The southern portion of the sea is still in serious danger, however, in part because Uzbekistan has not acted to solve the problem.

DEFENDING Do you think Kazakhstan was selfish in saving the northern part of the Aral Sea and doing nothing for the southern part? Why or why not?

Addressing the Issues

GUIDING QUESTION *How are environmental issues in Central Asia being addressed?*

The environmental problems of Central Asia must be addressed for the health and economic future of the countries in the subregion. Governments in the subregion, which often do not have enough resources, have sought outside help in developing and implementing solutions. As the dispute between Tajikistan and Uzbekistan over water resources highlights, some of the problems cross borders, which means regional governments must cooperate.

The governments of the United States and other countries and also nongovernmental organizations (NGOs) have taken part in efforts to solve environmental problems in Central Asia. For example, the United States Agency for International Development (USAID) has joined with the affected countries to work to protect the river basin of the Syr Dar'ya. A group formed by several universities in Europe and most of the governments of the eastern subregion is dedicated to managing water resources. One of the early steps in this program has been to found water-management centers in each of the countries involved. The program's goal is to train local people in water-management techniques. The group also hopes to build cooperation among the countries involved.

In 2006 the countries around the Caspian Sea agreed to the legally binding Convention for the Protection of the Marine Environment of the Caspian Sea. The agreement commits them to reducing agricultural and industrial pollution into that troubled sea. It also commits the countries to conserve fish and other marine life. In addition, another international agreement called the Convention on International Trade in Endangered Species has placed all sturgeon species from the Caspian Sea on its list of threatened animals. Strict quotas have been set for sturgeon fishing. Sturgeon eggs, or caviar, are very profitable, however, which has led to poaching. Poaching is the illegal harvesting of wild animals.

Afghanistan has joined with other countries and NGOs to try to solve the landmine problem. Many of the NGOs use an approach called "community-based demining." Experienced workers recruit and train local people in the skills needed to identify and to disable mines so they cannot explode. The NGOs work closely with local communities to find people to do this dangerous work and to support their activities. One possible future benefit of this approach is that trained Afghanis will be available to help other countries with the same problem.

☑ READING PROGRESS CHECK

Evaluating Why do the countries of the subregion need to cooperate with each other and with other countries to address their environmental issues?

LESSON 3 REVIEW

Reviewing Vocabulary
1. *Explaining* Write a paragraph explaining why radioactive materials are so dangerous.

Using Your Notes
2. *Explaining* Use your graphic organizer on Central Asia's environmental issues to write a paragraph explaining how each solution chosen aims to solve the targeted environmental challenge.

Answering the Guiding Questions
3. *Explaining* How do the people and governments manage the natural resources of Central Asia?

4. *Summarizing* How have modern economic activities impacted Central Asia?

5. *Explaining* How are environmental issues in Central Asia being addressed?

Writing Activity
6. *Argument* In a paragraph, discuss which environmental problem in the subregion you think is the worst, and defend your reasoning.

Directions: On a separate sheet of paper, answer the questions below. Make sure that you read carefully and answer all parts of the questions.

Lesson Review

Lesson 1

1 ***Identifying Cause and Effect*** Explain how and why the Aral Sea has changed since the 1960s.

2 ***Describing*** What makes up the western section and the eastern section of Central Asia?

3 ***Examining*** Despite a lack of oil resources, Georgia has profited from oil, and it has managed to create power to generate its electricity. How has it been successful in these endeavors?

Lesson 2

4 ***Considering Advantages and Disadvantages*** Why is it accurate to describe Central Asia as an ethnic mosaic? Explain why this "ethnic mosaic" is both a benefit and a detriment.

5 ***Making Connections*** Why was Afghanistan never a Soviet republic?

6 ***Analyzing*** Explain whether this statement is accurate: "The percentage of the workforce in Central Asia earning its living by farming is disproportionate to the amount of fertile farmland in the region."

Lesson 3

7 ***Identifying Cause and Effect*** What was the impact on Azerbaijan and Kazakhstan when oil prices recently fell dramatically? What has Azerbaijan done since then to prevent a similar outcome should oil prices drop again?

8 ***Summarizing*** Summarize the nuclear and chemical pollution in Kazakhstan. Explain the origins of this pollution.

9 ***Evaluating*** What are the impacts of land mines and unexploded ammunition in Afghanistan?

21st Century Skills

Use the cartoon below to answer the questions that follow.

"Wow! The Ukraine, Moldavia, Uzbekistan, Kazakhstan, Byelorussia, Tadzhikistan, Kirgizia, Turkmenistan . . ."

10 ***Using Primary Sources*** The teacher in the cartoon has just discussed information about these countries and the Soviet Union. What do all of these countries have in common, as related to the Soviet Union, that would likely have given rise to the comment in the caption?

11 ***Identifying Perspectives*** Create a new editorial cartoon to illustrate another point about one or all of the countries mentioned in the caption above. The cartoon should clearly depict a point about one or all of these countries. Write a short paragraph to explain your cartoon.

Need Extra Help?

If You've Missed Question	**1**	**2**	**3**	**4**	**5**	**6**	**7**	**8**	**9**	**10**	**11**
Go to page	456	454	458	462	460	464	467	469	469	454	454

Directions: On a separate sheet of paper, answer the questions below. Make sure that you read carefully and answer all parts of the questions.

Critical Thinking

12 *Exploring Issues* Write a one-page essay to explain this statement: "Early historical and geographical events in Central Asia are clearly evident in the region today, though events in recent history have given rise to perils to the human and physical geography, perils that would not have been foreseen during the early history of the region."

Applying Map Skills

Refer to the Unit 5 Atlas to answer the following questions.

13 *The World in Spatial Terms* Use your mental map of Central Asia to generally describe the subregion's two sections and their relationship to the Black Sea, Caspian Sea, China, Pakistan, and the Caucasus Mountains.

14 *Human Systems* Which subregion of North Africa, Southwest Asia, and Central Asia is the most densely populated?

15 *The World in Spatial Terms* Identify the sea that is located partially in Kazakhstan and partially in Uzbekistan.

Exploring the Essential Question

16 *Making Connections* Summarize the conflicts over resources between Tajikistan and Uzbekistan. Explain how this conflict is emblematic of the types of conflicts faced in the subregion. Your summary should reference the ways physical and human systems interact to shape a place.

College and Career Readiness

17 *Change and Continuity in Economics* You have been hired to conduct research for a company that provides data about the world's oil producers. Use your research to develop a memo and a graph to explain Kazakhstan's current level of oil production in relation to the world's top ten oil producers. Explain where Kazakhstan's oil production falls within the graph and whether it is likely to change during the next five years compared to other countries included in the graph.

DBQ Analyzing Primary Sources

Read the excerpt and use it to answer the following questions.

PRIMARY SOURCE

Despite Government efforts, deficiencies in the administration of justice continue to pose a major impediment to the attainment of justice . . . Trials monitored by OHCHR staff in Kyrgyzstan—from city courts to the Supreme Court—continue to reveal concerns about due process, the independence of the judiciary, security for defendants, their lawyers and court officials . . . Reports of continuing discriminatory practices towards . . . national and ethnic minorities are deeply troublesome.

—Rupert Colville, UN Office of the High Commissioner for Human Rights (OHCHR), June 10, 2011, UN News Centre

18 *Interpreting Significance* What do Colville's comments indicate about the pursuit of justice in Kyrgyzstan?

19 *Drawing Conclusions* Why would the UN and news agencies be interested in comments from Rupert Colville regarding 2010 events in Kyrgyzstan?

Research and Presentation

20 *Gathering Information* Research the work of the various organizations to protect the Syr Dar'ya river basin. With a partner, create a multimedia presentation to explain the problems affecting the river basin, current efforts to solve the problems, and additional solutions you would propose. Include map and images and charts, diagrams, or graphs in your presentation.

Writing About Geography

21 *Argument* Use standard grammar, spelling, sentence structure, and punctuation to write an editorial explaining why nuclear power plants in Armenia should be shut down. Be certain to anticipate and rebut the opposing viewpoint.

Need Extra Help?

If You've Missed Question	**12**	**13**	**14**	**15**	**16**	**17**	**18**	**19**	**20**	**21**
Go to page	454	360	360	360	466	465	472	463	470	469

Africa South of the Sahara

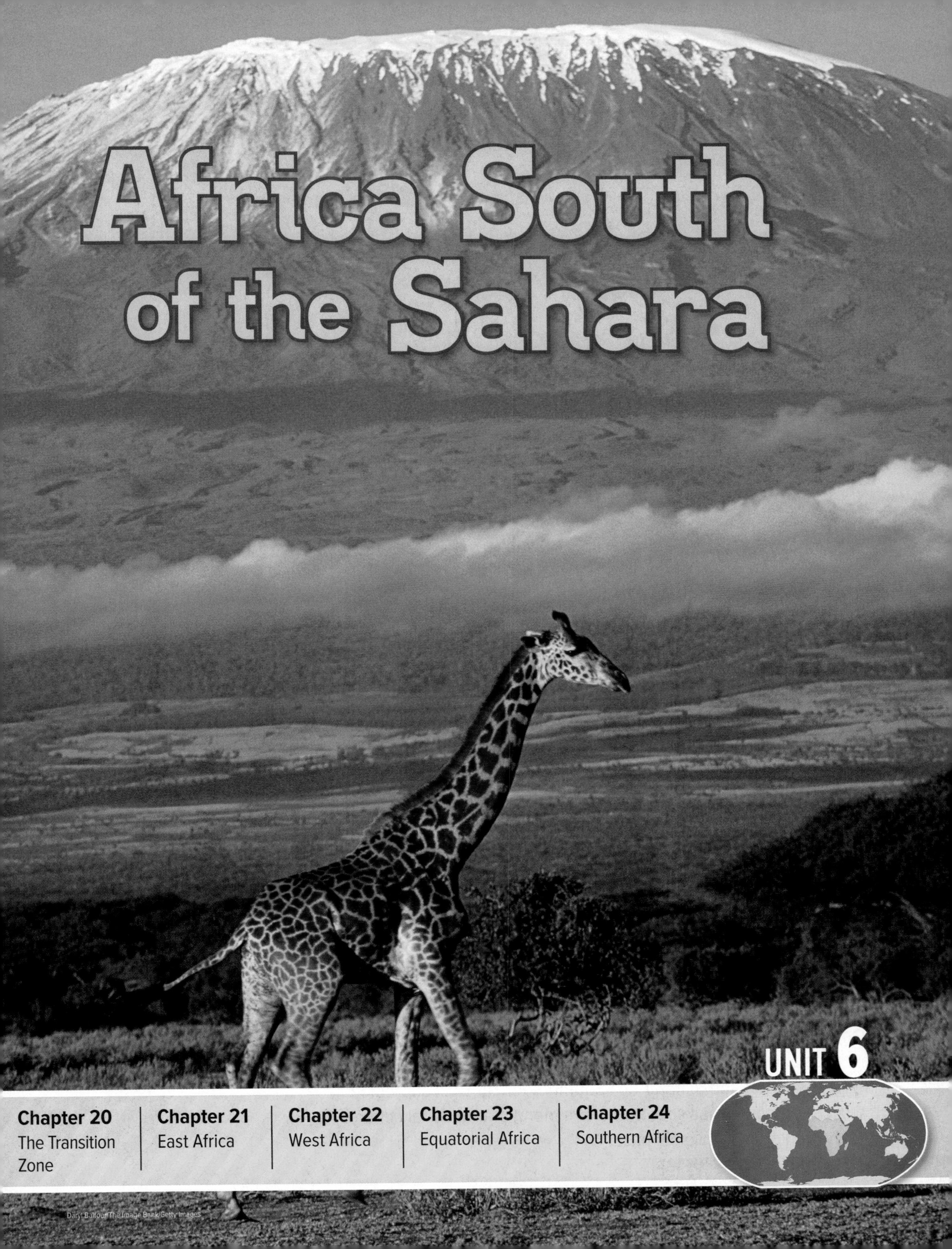

UNIT 6

Chapter 20
The Transition Zone

Chapter 21
East Africa

Chapter 22
West Africa

Chapter 23
Equatorial Africa

Chapter 24
Southern Africa

Daryl Balfour The Image Bank/Getty Images

1 Culture **Women of the Masai ethnic group sing to welcome tourists to their village in Kenya's Maasai Mara National Reserve.**

EXPLORE the REGION

AFRICA SOUTH OF THE SAHARA includes some of the driest and wettest spots on Earth. A region of physical and human diversity, Africa south of the Sahara includes the Transition Zone, with its mix of Islamic and African influences; the ethnic complexity of West Africa; the savannas and mountains of East Africa; the dense, lush rain forests of Equatorial Africa; and mineral-rich Southern Africa. Each subregion has a mix of large and small countries and a blend of traditional cultures deeply marked by a colonial past.

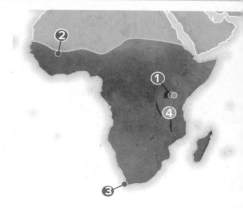

THERE'S MORE ONLINE

② **The People**   Africa south of the Sahara has hundreds of ethnic groups, each with rich cultural and artistic traditions and distinct ways of life.

③ **Colonial Legacy**   Beginning in the late 1800s, European powers took possession of most of Africa. The lasting impact of European colonization is still evident today in language, culture, politics, and economic and social issues.

④ **The Resources**   Africa south of the Sahara is famous for its captivating wildlife, for being rich in mineral resources, and for its lush rain forests. Countries in the region work to balance the use of these resources, seeking to spur economic development while still preserving the environment.

Africa South of the Sahara

Physical

Elevations

10,000 ft. (3,000 m)
5,000 ft. (1,500 m)
2,000 ft. (600 m)
1,000 ft. (300 m)
0 ft. (0 m)
Below sea level

— National boundary
▲ Mountain peak
▼ Lowest point

EUROPE

N
W E
S

CENTRAL ASIA

Mediterranean Sea

TROPIC OF CANCER

NORTH AFRICA

Boundary represents January 1, 1956, alignment; final alignment pending negotiations.

SOUTHWEST ASIA

20°N

S A H A R A

Nile R.

Red Sea

Gulf of Aden

Cape Verde Islands

Senegal R.

Niger R.

S A H E L

Lake Chad

Darfur

Blue Nile R.

Lake Tana

Great Rift Valley

Lake Assal -500 ft. (-152 m)

ETHIOPIAN HIGHLANDS

Boundary in dispute

Yobe R.

Benue R.

Chari R.

White Nile R.

Gulf of Guinea

Bioko
Príncipe
São Tomé
Pagalu

Lake Volta

EQUATOR 0°

The red dashed lines represent the northern and southern boundaries of a region known as the Transition Zone, an area of increasing Islamic influence.

Ubangi R.

CONGO BASIN

Congo R.

Lake Turkana

Ruwenzori

Mt. Kenya 17,058 ft. (5,199 m)

Serengeti Plain

Kilimanjaro 19,341 ft. (5,895 m)

Lake Victoria

Seychelles

Amirante Is.

Great Rift Valley

Lake Tanganyika

Pemba I.
Zanzibar I.

Farquhar Is.

ATLANTIC OCEAN

BIÉ PLATEAU

Katanga Plateau

Lake Malawi

Comoro Is.

INDIAN OCEAN

Okavango R.

Lake Kariba

Zambezi R.

Victoria Falls

Madagascar

Mauritius

Réunion

Mozambique Channel

20°S

Namib Desert

Okavango Delta

Kalahari Desert

Limpopo R.

TROPIC OF CAPRICORN

Orange R. Drakensberg

0   1,000 miles
0   1,000 kilometers
Lambert Azimuthal Equal-Area Conic projection

Cape of Good Hope

Cape Agulhas

60°N

40°N

20°N

0°

20°S

40°S

40°W   20°W   0°   20°E   40°E   60°E

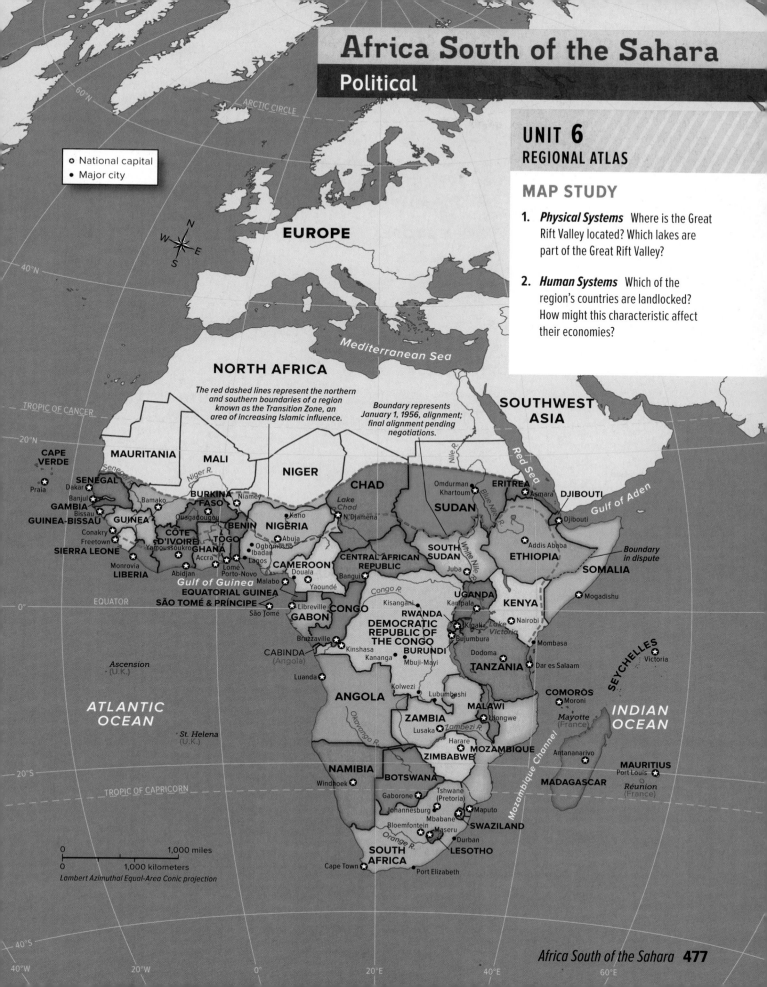

Africa South of the Sahara
Political

○ National capital
● Major city

EUROPE

N
W · E
S

Mediterranean Sea

NORTH AFRICA

The red dashed lines represent the northern and southern boundaries of a region known as the Transition Zone, an area of increasing Islamic influence.

Boundary represents January 1, 1956, alignment; final alignment pending negotiations.

SOUTHWEST ASIA

UNIT 6
REGIONAL ATLAS

MAP STUDY

1. **Physical Systems** Where is the Great Rift Valley located? Which lakes are part of the Great Rift Valley?

2. **Human Systems** Which of the region's countries are landlocked? How might this characteristic affect their economies?

ARCTIC CIRCLE

60°N

40°N

TROPIC OF CANCER

20°N

CAPE VERDE
Praia

MAURITANIA

MALI

NIGER

CHAD

Omdurman
Khartoum

ERITREA
Asmara

DJIBOUTI
Djibouti

Gulf of Aden

Senegal R.
Niger R.

Dakar
SENEGAL
Banjul
GAMBIA
Bissau
GUINEA-BISSAU
Conakry
Freetown
SIERRA LEONE
Monrovia
LIBERIA

Bamako
BURKINA FASO
Niamey
Ouagadougou
GUINEA
CÔTE D'IVOIRE
Yamoussoukro
GHANA
Abidjan
Accra
TOGO
BENIN
Lomé
Porto-Novo

Kano
NIGERIA
Abuja
Ogbomosho
Ibadan
Lagos

Lake Chad
N'Djamena

SUDAN

Addis Ababa

Boundary in dispute

SOUTH SUDAN
Juba

ETHIOPIA

SOMALIA

Mogadishu

EQUATOR
0°

CAMEROON
Douala
Yaoundé
Malabo
EQUATORIAL GUINEA
SÃO TOMÉ & PRÍNCIPE
São Tomé

Gulf of Guinea

CENTRAL AFRICAN REPUBLIC
Bangui

Congo R.
Kisangani

UGANDA
Kampala

KENYA
Nairobi

Mombasa

Nile R.
Blue Nile R.
White Nile R.

Red Sea

CONGO
Libreville
GABON
Brazzaville
Kinshasa
CABINDA (Angola)

RWANDA
Kigali
DEMOCRATIC REPUBLIC OF THE CONGO
BURUNDI
Bujumbura
Kananga
Mbuji-Mayi

Lake Victoria
Dodoma
TANZANIA
Dar es Salaam

SEYCHELLES
Victoria

Ascension (U.K.)

ATLANTIC OCEAN

Luanda
Kolwezi
Lubumbashi

ANGOLA

ZAMBIA
Lusaka

MALAWI
Lilongwe

COMOROS
Moroni
Mayotte (France)

INDIAN OCEAN

St. Helena (U.K.)

Okavango R.
Zambezi R.
Harare
ZIMBABWE

MOZAMBIQUE

Mozambique Channel

MADAGASCAR
Antananarivo

MAURITIUS
Port Louis
Réunion (France)

20°S

TROPIC OF CAPRICORN

NAMIBIA
Windhoek

BOTSWANA
Gaborone
Tshwane (Pretoria)
Johannesburg
Bloemfontein

Maputo
Mbabane
SWAZILAND
Maseru
LESOTHO
Durban

SOUTH AFRICA
Cape Town
Port Elizabeth

0   1,000 miles
0   1,000 kilometers
Lambert Azimuthal Equal-Area Conic projection

40°W   20°W   0°   20°E   40°E   60°E

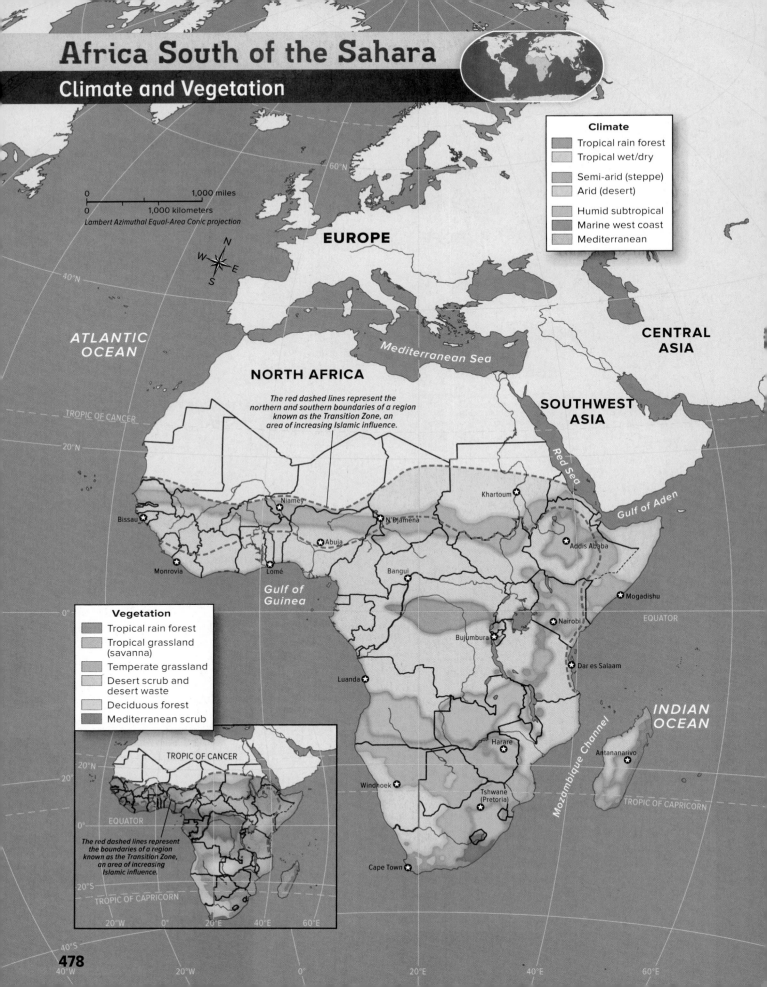

Africa South of the Sahara
Climate and Vegetation

Climate
- Tropical rain forest
- Tropical wet/dry
- Semi-arid (steppe)
- Arid (desert)
- Humid subtropical
- Marine west coast
- Mediterranean

Vegetation
- Tropical rain forest
- Tropical grassland (savanna)
- Temperate grassland
- Desert scrub and desert waste
- Deciduous forest
- Mediterranean scrub

0 — 1,000 miles
0 — 1,000 kilometers
Lambert Azimuthal Equal-Area Conic projection

EUROPE

ATLANTIC OCEAN

Mediterranean Sea

NORTH AFRICA

The red dashed lines represent the northern and southern boundaries of a region known as the Transition Zone, an area of increasing Islamic influence.

CENTRAL ASIA

SOUTHWEST ASIA

TROPIC OF CANCER

Red Sea

Gulf of Aden

Khartoum

Bissau
Niamey
N'Djamena
Abuja
Addis Ababa
Monrovia
Lomé
Bangui
Mogadishu

Gulf of Guinea

EQUATOR

Nairobi
Bujumbura

Luanda
Dar es Salaam

INDIAN OCEAN

Harare
Antananarivo

Windhoek

Mozambique Channel

Tshwane (Pretoria)

TROPIC OF CAPRICORN

Cape Town

TROPIC OF CANCER

20°N

EQUATOR

The red dashed lines represent the boundaries of a region known as the Transition Zone, an area of increasing Islamic influence.

20°S

TROPIC OF CAPRICORN

20°W   0°   20°E   40°E   60°E

40°S
40°W   20°W   0°   20°E   40°E   60°E

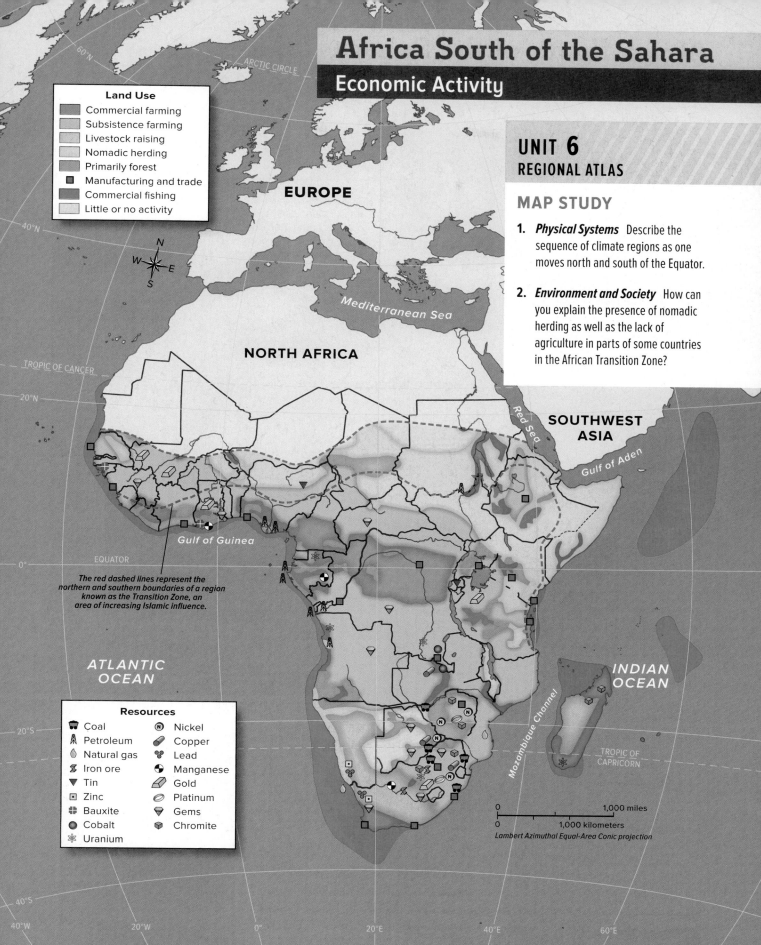

Africa South of the Sahara
Economic Activity

UNIT 6
REGIONAL ATLAS

Land Use

- Commercial farming
- Subsistence farming
- Livestock raising
- Nomadic herding
- Primarily forest
- Manufacturing and trade
- Commercial fishing
- Little or no activity

MAP STUDY

1. *Physical Systems* Describe the sequence of climate regions as one moves north and south of the Equator.

2. *Environment and Society* How can you explain the presence of nomadic herding as well as the lack of agriculture in parts of some countries in the African Transition Zone?

EUROPE

Mediterranean Sea

NORTH AFRICA

TROPIC OF CANCER

SOUTHWEST ASIA

Red Sea

Gulf of Aden

Gulf of Guinea

EQUATOR

The red dashed lines represent the northern and southern boundaries of a region known as the Transition Zone, an area of increasing Islamic influence.

ATLANTIC OCEAN

INDIAN OCEAN

Mozambique Channel

TROPIC OF CAPRICORN

Resources

- Coal
- Petroleum
- Natural gas
- Iron ore
- Tin
- Zinc
- Bauxite
- Cobalt
- Uranium
- Nickel
- Copper
- Lead
- Manganese
- Gold
- Platinum
- Gems
- Chromite

0   1,000 miles
0   1,000 kilometers
Lambert Azimuthal Equal-Area Conic projection

Africa South of the Sahara
Population Density

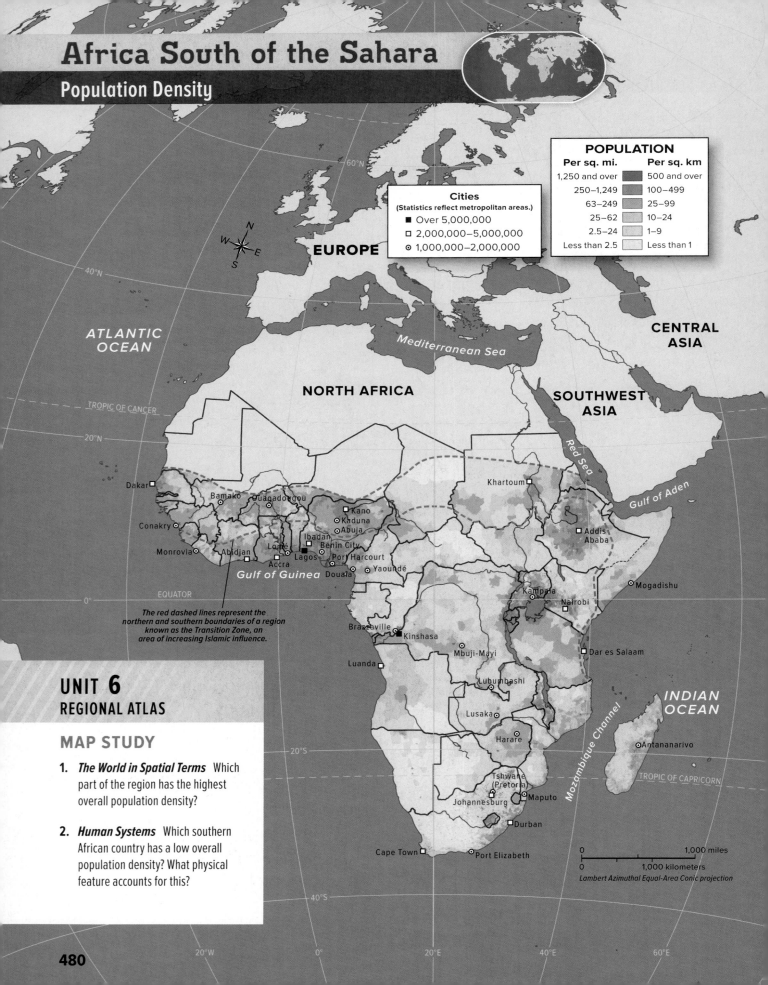

EUROPE

ATLANTIC OCEAN

Mediterranean Sea

NORTH AFRICA

CENTRAL ASIA

SOUTHWEST ASIA

Red Sea

Gulf of Aden

Cities
(Statistics reflect metropolitan areas.)

- ■ Over 5,000,000
- □ 2,000,000–5,000,000
- ⊙ 1,000,000–2,000,000

POPULATION

Per sq. mi.		Per sq. km
1,250 and over		500 and over
250–1,249		100–499
63–249		25–99
25–62		10–24
2.5–24		1–9
Less than 2.5		Less than 1

TROPIC OF CANCER

60°N

40°N

20°N

Dakar
Bamako
Ouagadougou
Conakry
Kano
Kaduna
Abuja
Monrovia
Abidjan
Lomé
Ibadan
Benin City
Accra
Lagos
Port Harcourt
Douala
Yaoundé
Khartoum
Addis Ababa
Mogadishu
Kampala
Nairobi
Brazzaville
Kinshasa
Mbuji-Mayi
Dar es Salaam
Luanda
Lubumbashi
Lusaka
Harare
Antananarivo
Tshwane (Pretoria)
Maputo
Johannesburg
Durban
Cape Town
Port Elizabeth

Gulf of Guinea

EQUATOR

0°

The red dashed lines represent the northern and southern boundaries of a region known as the Transition Zone, an area of increasing Islamic influence.

INDIAN OCEAN

Mozambique Channel

TROPIC OF CAPRICORN

20°S

40°S

UNIT 6
REGIONAL ATLAS

MAP STUDY

1. *The World in Spatial Terms* Which part of the region has the highest overall population density?

2. *Human Systems* Which southern African country has a low overall population density? What physical feature accounts for this?

0   1,000 miles

0   1,000 kilometers

Lambert Azimuthal Equal-Area Conic projection

480

20°W   0°   20°E   40°E   60°E

The Transition Zone

ESSENTIAL QUESTION • *How do physical systems and human systems shape a place?*

Why Geography Matters
Diffusion: Muslim and non-Muslim Cultures

Lesson 1
Physical Geography of the Transition Zone

Lesson 2
Human Geography of the Transition Zone

Lesson 3
People and Their Environment: The Transition Zone

Geography Matters...

The Transition Zone is a challenging land because of transitions in both the human and physical geography. It is a mix of Muslim, Christian, and animist cultures. It is also a transition from the Sahara to the Tropics. Climate and geography, as well as the legacy of colonialism, have led to turmoil in the subregion.

In recent decades, the Transition Zone has experienced extended periods of drought. Rapid population growth has contributed to the problem. People are trying to find new ways to cope with or to reverse the damage to the land and face the environmental challenge.

◄ A Muslim woman in Lomé, Togo

Godong/age fotostock

481

diffusion: Muslim and non-Muslim cultures

Africa is a continent with great diversity in its physical, cultural, and economic geography. The Transition Zone is a region that lies south of the Sahara and has been increasingly influenced by Islam. At the same time, Christian and traditional religions are also practiced. Some countries in this zone are predominantly Muslim, while others are more religiously mixed.

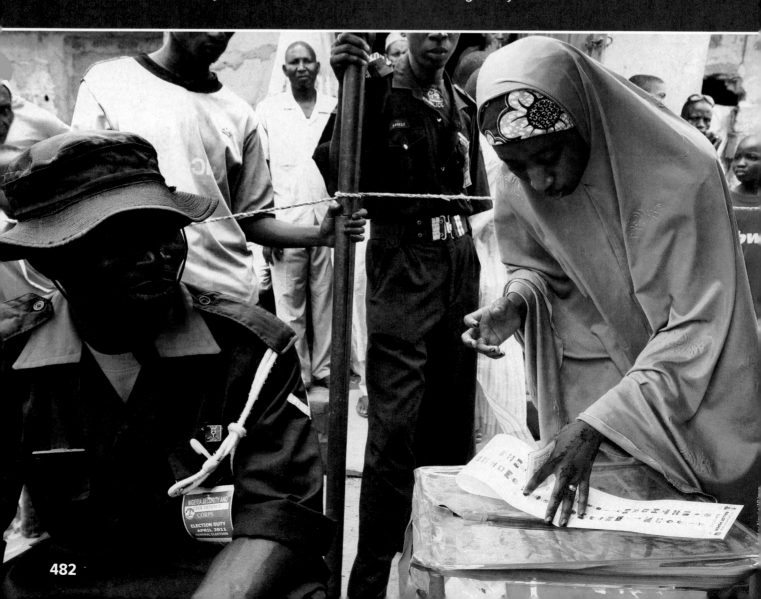

Why do Muslim and non-Muslim cultures exist in Africa?

The Sahel, which means "shore" in Arabic, serves as a cultural and ecological gateway. To the north are primarily Muslim countries and more arid land. To the south are mostly non-Muslim countries with less arid to tropical environments. This cultural gateway exists because of the continent's long history of interaction with other cultures.

When trade flourished in the West African kingdoms and East African kingdoms between 800 and the 1500s, the region encountered new cultures. Arab and Berber traders exchanged trade goods with the people of these kingdoms. They also shared ideas, including their Islamic faith. Christianity also spread in different parts of the region through trade with merchants from the Mediterranean. In Axum (Ethiopia), an East African trading state, Christianity became the official religion in the fourth century.

1. **The World in Spatial Terms** How did the location of Africa's kingdoms and trading states influence its cultural and religious diversity?

What happens when these distinct cultures encounter each other?

The slave trade, European colonialism, and independence also contributed to cultural diffusion in Africa. All of the following countries make up the Transition Zone: Somalia, Djibouti, Guinea, Guinea-Bissau, Gambia, and Senegal, as well as parts of Eritrea, Ethiopia, Sudan, Chad, Cameroon, Niger, Nigeria, Benin, Togo, Ghana, Burkina Faso, Côte d'Ivoire, Liberia, Sierra Leone, Mali, and Mauritania. Ethnic conflicts occurred in post-colonial Africa because Europeans drew these political borders with no regard for ethnic borders.

As a result, the Transition Zone has seen cultural and ethnic conflicts in many countries. In Nigeria, a civil war rooted in ethnic tensions erupted not long after independence. The "Biafran War" resulted in the deaths of about a million Ibo people, mainly from starvation. In Senegal, a rebellion spurred by religious and cultural differences began in 1984. The area's residents were mainly Christian or followers of traditional religions—unlike the majority of Senegalese, who were Muslim. The conflict ended in 1998.

2. **Human Systems** Has cultural diffusion in Africa through interaction with other cultures been positive or negative for the Transition Zone? Explain.

How are distinct cultures encountering one another today?

The Transition Zone is still a hotbed for conflict. Sudan, for example, has experienced ongoing conflicts since its independence in 1956. The government has oppressed its Christian population, which mainly lives in the south. Even though South Sudan became independent from Sudan in 2011, conflict continues.

In the Darfur region of Sudan, fighting broke out in 2003 between Muslim villagers, who identified themselves as African, and government-backed militias. The militias were made up of Muslim nomads who identified themselves as Arab. Since then, nearly 3 million people have been displaced and hundreds of thousands killed. Many of the Darfur refugees have fled to the neighboring country of Chad, but refugee camps there have swelled beyond capacity. Peace deals have been signed and a peacekeeping force tries to maintain peace. Critics predict, however, that change will be slow.

3. **Environment and Society** Write a paragraph explaining how the environment of Darfur might be contributing to this conflict.

THERE'S MORE ONLINE

VIEW an interactive image of Lake Chad • **EXPLORE** a map of religions in the Transition Zone

networks

There's More Online!

☑ **IMAGE** Lake Chad in 1973

☑ **IMAGE** Lake Chad in 2001

☑ **MAP** The African Transition Zone

☑ **INTERACTIVE SELF-CHECK QUIZ**

☑ **VIDEO** Physical Geography of the Transition Zone

Reading **HELP**DESK

Academic Vocabulary

- area
- benefit

Content Vocabulary

- transition zone
- Sahel
- delta
- harmattan

TAKING NOTES: *Key Ideas and Details*

IDENTIFYING As you read about the geography of the African Transition Zone, use a graphic organizer like the one below to note the characteristics of the land, water, and climate of the area.

The African Transition Zone		
Land	Water	Climate

LESSON 1
Physical Geography of the Transition Zone

ESSENTIAL QUESTION · *How do physical systems and human systems shape a place?*

IT MATTERS BECAUSE
The African Transition Zone lies south of the Sahara, almost entirely within tropical latitudes, and marks the shift from desert to tropical forests, as well as from Muslim to Christian and animist cultures. The subregion is known for its remarkable wildlife and extraordinary physical geography. It also has the world's fastest-growing population and hundreds of ethnic groups.

Landforms

GUIDING QUESTION *What is the Transition Zone?*

Africa immediately south of the Sahara is an **area** of land that transitions between the desert climate of North Africa and the tropical savanna of Equatorial Africa. This subregion, with a total area of 1,178,800 square miles (3,053,200 sq. km), in the widest part of Africa is known as the African Transition Zone. A **transition zone** is a physical area in which the land undergoes a radical change such as from arid to tropical.

The African Transition Zone also represents a transition between the Islamic cultures of North Africa and the Christian and animist cultures to its south. Extending east to west from Senegal to Somalia, this region is home to a very diverse population. Hundreds of ethnic groups coexist in an area influenced by the African cultures that originated in the region and by the Arab and European cultures that came later.

Part of the Transition Zone is a geographical area called the **Sahel**, which means "shore" or "coast" in Arabic. The word describes the appearance of vegetation in the Sahel as a "coastline" marking the border of the sands of the Sahara. The topography is mainly flat with a series of plateaus that range in elevation from about 650 feet (200 m) to about 1,300 feet (400 m). This steppe region runs in a band across Africa from the Atlantic Ocean in the west to the Red Sea in the east. It spans more than 3,000 miles (4,800 km) and ranges in width from about 125 miles (200 km) to 250 miles (400 km). East of the Sahel is a region known as the Horn of Africa for its shape.

The Sahel receives little rainfall. On average, 4 to 8 inches (10 to 20 cm) of rain falls annually, mostly in June, July, and August when the sun is high in the sky and the temperatues are hot. The rest of the year is a little cooler but very dry, so only low-growing grasses, shrubs, and acacia trees can grow there. Most people in the Transition Zone have traditionally herded livestock. The short grasses of the Sahel provide one of the largest pastoral and herding zones in the world. Overgrazing, however, has stripped the land bare in some places.

✓ READING PROGRESS CHECK

Identifying Describe the human and geographic features that characterize the African Transition Zone.

Water Systems

GUIDING QUESTION *What rivers flow through the Transition Zone?*

The African Transition Zone experiences dryness, infrequent rainfall, and often drought. A drought in the 1970s caused a famine that killed thousands of people. Over the past century, annual precipitation has decreased. This, in turn, is shrinking some of the subregion's natural lakes. However, the Sahel still has a number of lakes, rivers, and wetlands. These water resources, in addition to rain, are important to the livelihood of many people.

At the southern edge of the Sahara, Lake Chad is bordered by Nigeria, Niger, Chad, and Cameroon. It was once the second-largest wetland in Africa and supported a rich diversity of animal and plant life. It has been important to

area a geographical region; the amount of space that the surface of a place covers

transition zone an area in which the properties of the land undergo a radical change

Sahel steppe region extending from Senegal to Somalia that receives little rainfall

GEOGRAPHY CONNECTION

The African Transition Zone is both a physical transition between desert and savanna and a cultural transition from Muslim cultures to Christian and animist cultures in the south.

1. **THE WORLD IN SPATIAL TERMS** What three bodies of water does the Transition Zone meet in the east?

2. **HUMAN SYSTEMS** What countries in the Transition Zone may be more influenced by Muslim culture than by Christian culture? Explain.

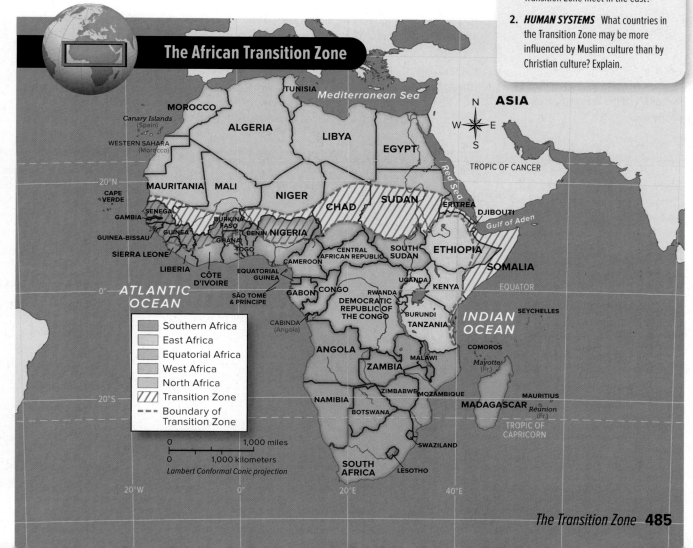

The African Transition Zone

Legend:
- Southern Africa
- East Africa
- Equatorial Africa
- West Africa
- North Africa
- Transition Zone
- Boundary of Transition Zone

0   1,000 miles
0   1,000 kilometers
Lambert Conformal Conic projection

the livelihoods of more than 20 million people but is threatened with extinction. Although fed by three rivers—the Chari, the Logone, and the Yobe—the lake is shrinking. The drought of the 1970s completely dried up the northern portion of the lake. Even during years of normal rainfall, the water level remains low. Because of the dry climate, much of the lake's water evaporates or becomes increasingly saline. The land is left dry and unable to support life. As Lake Chad shrinks, the desert expands.

Lake Volta in Ghana is one of the largest human-made lakes in the world. It is about 250 miles (400 km) long and covers 3,275 square miles (8,482 sq. km). The lake itself lies entirely in Ghana. However, six countries—Benin, Burkina Faso, Côte d'Ivoire, Ghana, Mali, and Togo—share the Volta River system. Lake Volta was created in the 1960s by damming the Volta River south of Ajena in Ghana. The new lake flooded more than 700 villages, forcing more than 70,000 people to find new homes. It was originally created as a reservoir to store water for generating hydroelectric power. People **benefit** from the lake because it supplies irrigation for farming and has several fisheries. It also serves as a transportation route linking different parts of the country. The hydroelectric plant also generates electricity used throughout Ghana.

Unfortunately, there are negative consequences of this reservoir. These include reduced agricultural productivity because annual floods no longer occur and bring new silt, and because the land higher up on the current lake shore is not as fertile as the land now on the lake bottom. Also, the ecology of the river is completely different because running water and annual floods were replaced by standing water. This has resulted in an increase in disease.

The Niger (NY•juhr) River is known by many names along its course. All of them have roughly the same meaning—"great river." The Niger is the main artery in western Africa, extending about 2,600 miles (4,183 km) in length. Originating in the highlands of Guinea, the river forms an arc flowing northeast, then curving southeast to its mouth on the Nigerian coast. In addition to being important to agriculture, the Niger River is a major means of transportation.

This great river does not flow as one well-defined stream all the way along its course to the Atlantic Ocean. In central Mali, the river spreads out across a

benefit to gain

delta an often triangular-shaped section of land formed as the waters of a river slow down and split into many channels as they deposit sand and silt that has been carried downriver

Once nearly the size of Lake Erie, the surface area of Lake Chad has declined dramatically in the last 50 years. The photo on the left was taken in 1973 and the one on the right in 2001.

▶ **CRITICAL THINKING**

1. Assessing By about what percentage does the surface area of Lake Chad appear to have shrunk?

2. Making Connections How do you suppose the lake's shrinkage has affected the surrounding land?

USGS/Science Source

broad plateau and inland **delta**, an area where a river slows down and spreads out into many smaller channels. These small river channels create a large wetland that supports wildlife. This inland delta also provides local farmers with water. Further downstream, at Aboh in southern Nigeria, the Niger splits into a delta that stretches 150 miles (241 km) north to south and extends to a width of about 200 miles (322 km) where the river empties into the Gulf of Guinea.

The Senegal River is 1,015 miles (1,633 km) long in West Africa and forms the border between Senegal and Mauritania. The sources of the river are in Guinea and the wetter, southwestern part of Mali. Its drainage basin encompasses some 174,000 square miles (450,000 sq. km). Roughly 3.5 million people live near the river. In 1972 Mali, Mauritania, and Senegal founded an organization to manage the river basin. Guinea joined in 2005. Two dams have helped the inhabitants of the area make better use of the river. The Manantali Dam, built in 1986, helps prevent flooding during the rainy season and provides freshwater during the dry season. In 2002 the dam began generating hydroelectric power. The Diama Dam was built near the mouth of the river in 1988. It helps prevent the intrusion of salt water during the dry season. The dam also creates reserves for irrigation.

Two major tributaries of the Nile River pass through the Transition Zone. These are the Blue Nile and the White Nile. The Blue Nile, the main source of the fertile soil along the banks of the Nile River, stretches 850 miles (1,368 km) and originates in the mountains of Ethiopia. In 2012 Ethiopia began construction of a 6000-megawatt hydroelectric dam on the river. Still under construction, the Grand Ethiopian Renaissance Dam will be the largest hydroelectric power plant in Africa. It will also create Ethiopia's largest artificial lake.

The White Nile River, which is longer than the Blue Nile River, originates in the mountains to the west of Lake Victoria in Burundi. The Blue Nile and the White Nile meet at the city of Khartoum in Sudan. From there, the famous Nile River flows northward into Egypt on a winding course across the desert to the Mediterranean Sea.

☑ **READING PROGRESS CHECK**

Summarizing How is Lake Volta used by the residents of the region?

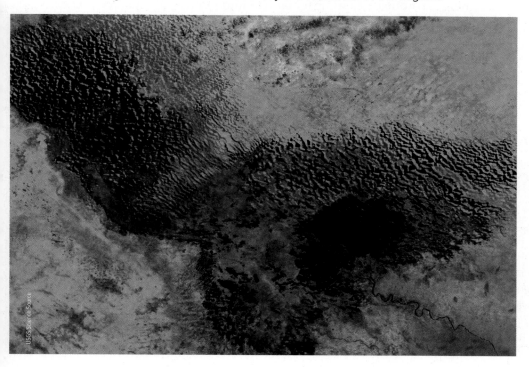

USGS/Science Source

Geometry

The Sahel is a deceptively huge area—1,178,800 square miles (3,053,200 sq. km). Consider the area of Texas, Tennessee, and the contiguous United States in relation to the area of the Transition Zone.

Texas:
268,800 square miles (696,200 sq. km)

Tennessee:
42,180 square miles (109,200 sq. km)

United States:
3,718,710 square miles (9,631,420 sq. km)

Compare Tennessee and Texas. How many times larger is Texas than Tennessee? Now answer the questions below.

ANALYZING How many times larger is the Sahel than Texas? How many times smaller is the Sahel than the United States?

Climates, Biomes, and Resources

GUIDING QUESTION *How does the climate affect life and resources in the Transition Zone?*

There are two hot seasons in the African Transition Zone. One occurs between February and April and the other between September and October. The rainy season occurs between May and August, with most of the rain falling in the southern part of the subregion. Some years, this results in a fair amount of rain. However, the rain may be isolated to one area or be so intense that it damages crops. Most of the rain occurs during the summer months. During December and January, a hot, dry wind known as the **harmattan** blows from the north. It often carries dust and sand from the Sahara. In western Africa, the harmattan is sometimes referred to as "the doctor" because its dryness—in contrast to the typical humid air of the subregion—is considered invigorating.

harmattan a hot, dry wind that blows from the northeast or east in the western Sahara

In most countries of the Transition Zone, water is such a precious resource that rain and life are one and the same. Rain helps determine climate, and in turn vegetation, in most parts of the subregion. In many areas, rainfall is the only water source. Over the past century, annual precipitation has decreased. For example, parts of interior Eritrea get almost no rainfall. Sadly, a vicious cycle of soil erosion, insufficient water, deforestation, and drought has plagued the area for decades. Long periods of drought and overuse of the land destroyed vegetation, making the land unusable. Droughts have caused crops to fail, killed livestock, and led to famines.

This vast biome is also an important home to a diverse population of flora and fauna. For much of the year, the African Transition Zone is a vast expanse of dry soil. During the rainy season, however, the Sahel comes alive with plant life. In addition to grasslands, the Sahel is home to the baobab tree and the jujube, whose fruit is used to feed herds. Although the summer months provide abundant food for animal life, at other times of the year wildlife must remain constantly on the move, foraging for water and vegetation. The diverse population of animals includes migratory birds, which use the wetlands as a rest and feeding area. Wild dogs, cheetahs, lions, elephants, giraffes, warthogs, and gerbils are some of the many different animals that roam the African Transition Zone.

Farming and nomadic herding remain the important traditional economies of the subregion. Unfortunately, deforestation and drought are making these lifestyles more difficult. However, new discoveries and better use of oil, natural gas, and coal, as well as the mining of uranium, gold, and iron deposits may help the people of the Transition Zone develop new economies.

✓ **READING PROGRESS CHECK**

Determining Importance Why is it important that the people of the Transition Zone manage their resources in the most efficient way?

LESSON 1 REVIEW

Reviewing Vocabulary

1. *Making Connections* Why is the Sahel called a transition zone?

Using Your Notes

2. *Summarizing* Use your graphic organizer on the geography of the African Transition Zone to write a paragraph summarizing how the land supports the people, flora, and fauna of the subregion.

Answering the Guiding Questions

3. *Finding the Main Idea* What is the Transition Zone?

4. *Classifying* What rivers flow through the Transition Zone?

5. *Identifying Central Issues* How does the climate affect life and resources in the Transition Zone?

Writing Activity

6. *Informative/Explanatory* In a paragraph, discuss how the climate affects the people and the flora and fauna of the Transition Zone.

networks

There's More Online!

☑ **IMAGE** Dogon Mask

☑ **IMAGE** Mine Workers

☑ **MAP** Kingdoms and Empires of the Transition Zone

☑ **INTERACTIVE MAP** Crisis in Darfur

☑ **TIME LINE** Droughts and Hunger

☑ **INTERACTIVE SELF-CHECK QUIZ**

☑ **VIDEO** Human Geography of the Transition Zone

LESSON 2
Human Geography of the Transition Zone

ESSENTIAL QUESTION • *How do physical systems and human systems shape a place?*

Reading HELPDESK

Academic Vocabulary

- **enhance**
- **potential**

Content Vocabulary

- **domesticate**
- **animist**
- **sanitation**
- **patriarchal**
- **clan**
- **nuclear family**
- **oral tradition**
- **subsistence farming**

TAKING NOTES: *Key Ideas and Details*

IDENTIFYING As you read about the human geography of the African Transition Zone, use a graphic organizer like the one below to identify examples of how history, culture, and economics have worked together to create the Transition Zone of today.

IT MATTERS BECAUSE

The Transition Zone has been home to many peoples, empires, and kingdoms for millennia. European colonization in the late 1800s and early 1900s had lasting effects on the subregion. The peoples of the Sahel maintain their distinct cultures and traditions but face many similar challenges.

History and Government

GUIDING QUESTION *What factors influenced the formation of the countries of the Transition Zone?*

Between about 9000 B.C. and 4000 B.C., the northern half of Africa received greater amounts of rain than today. At that time, what is now the Sahara was a savanna of extensive grasslands, lakes, and rivers. Nomadic people who once hunted for food settled in places, adopted agriculture, and **domesticated**, or tamed, animals. Around 4000 B.C., however, a climate shift began to occur. It became hotter and drier. Some scientists think that this shift happened quite abruptly, perhaps in as few as 400 years beginning in about 2000 B.C. As a result of the changes, the people in the farming communities migrated south.

Although much of the area became desert, the Nile Valley remained well watered, giving rise to Egypt and its civilization. Between about 2000 B.C. and 1000 B.C., the Egyptians pushed south, bringing various peoples along the Nile under their control. When Egypt's authority weakened, these peoples rose to power.

One of the civilizations Egyptians encountered as they pushed south was Kush. Around 2000 B.C., the Kush river civilization became a powerful kingdom in what is now Sudan. From about 2000 B.C. to about 1500 B.C., the Kushites controlled the Egyptian territory to the north. After retreating from Egypt, the Kushites pushed south and built a civilization around a new capital, Meroë. Kush flourished until about A.D. 300, when Meroë was attacked by Axum, a trading empire in Ethiopia. Axum was a great trading power from about A.D. 100 to A.D. 700.

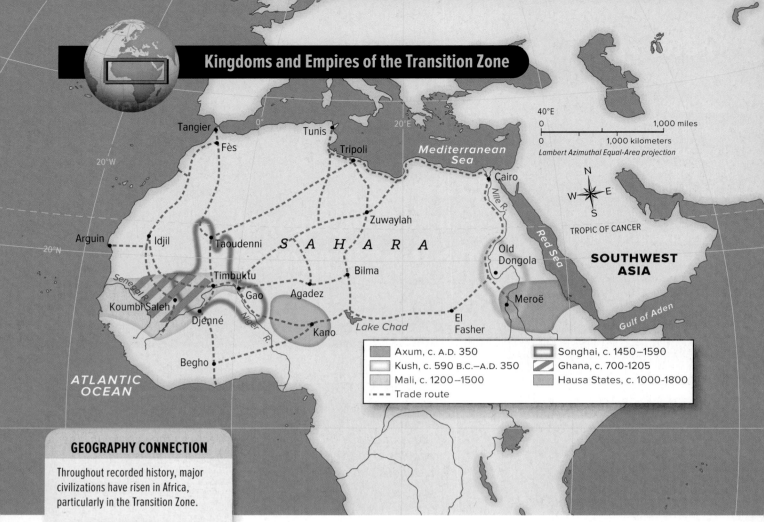

Kingdoms and Empires of the Transition Zone

Tangier
Fès
Tunis
Tripoli
Mediterranean Sea
Cairo

40°E
0 ——— 1,000 miles
0 ——— 1,000 kilometers
Lambert Azimuthal Equal-Area projection

TROPIC OF CANCER

Arguin
Idjil
Taoudenni
S A H A R A
Zuwaylah
Old Dongola
Red Sea
SOUTHWEST ASIA

Timbuktu
Bilma
Koumbi Saleh
Gao
Agadez
Meroë
Djenné
Niger R.
Kano
Lake Chad
El Fasher
Gulf of Aden

Begho

Senegal R.

ATLANTIC OCEAN

Legend	
Axum, c. A.D. 350	Songhai, c. 1450–1590
Kush, c. 590 B.C.–A.D. 350	Ghana, c. 700-1205
Mali, c. 1200–1500	Hausa States, c. 1000-1800
Trade route	

GEOGRAPHY CONNECTION

Throughout recorded history, major civilizations have risen in Africa, particularly in the Transition Zone.

1. **ENVIRONMENT AND SOCIETY** How did location affect the rise of these early kingdoms?

2. **HUMAN SYSTEMS** Which kingdom overtook the Kushites?

domesticate to adapt plants and animals from the wild for human use

enhance to improve or increase

potential possible or likely

Empires of the Transition Zone

In addition to the kingdoms of Kush and Axum in the eastern part of the Transition Zone, new trading empires gained strength in the western part. The present-day country of Mali is named after one of them, the Mali Empire. One of Mali's early kings, Sundiata Keita, helped Mali flourish by expanding trade routes for gold and salt. He also conquered many of the surrounding territories. The most famous ruler of the Mali Empire was Mansa Musa. A Muslim, Musa **enhanced** the prestige and power of Mali through a famous pilgrimage to Makkah (Mecca). Accompanied by a huge entourage, Musa apparently dispensed so much gold on his journey to Makkah that the price of gold fell drastically in places he visited.

The Mali Empire became legendary in the Islamic world and Europe, and helped Islam spread. Mali, which extended west to the Atlantic, had as its center the wealthy city of Timbuktu. Another empire, Songhai (SAWNG•hy), grew rich from the gold-for-salt trade started in the western empire of Ghana. Songhai eventually took over Mali and expanded east. It prospered until about A.D. 1600, when the Moroccans overran it.

Word of the wealth of Africa's kingdoms reached Europe. As early as the 1200s, Timbuktu became an important center of trade in gold and salt between Africa and Europe. By the 1600s and 1700s, Europeans were trading extensively with Africa. They traded for African gold and other goods and for enslaved people.

Colonization and Independence

In the 1800s, European powers regarded the region as a source of raw materials and as a **potential** market for finished goods. European countries also laid claim to African territory. A conference, known as the Berlin Conference, was held between 1884 and 1885 to regulate European colonization. All of the Sahel was under European control by 1914.

European rulers knew little about Africa's political and social systems, and no Africans participated at the Berlin Conference. As a result, Europeans created colonial boundaries that often cut across cultural, religious, or traditional boundaries, merging **animist** cultures with Muslim societies. Animist cultures believe all elements of nature, such as animals and mountains, have spirits. The introduction of Christianity added to the tension. Religious friction set African peoples against one another and strengthened European rule.

In the mid-1900s, Africans began to demand a share in government. Educated Africans launched independence movements. In the second half of the century, the colonies became independent of European rule.

These new countries faced difficult challenges, often the result of their colonial legacy. European powers, for example, set up colonial economies that met European, rather than African, needs. In addition, colonial governments did not involve Africans much in government, nor did they give Africans models for democracy. At independence, many of the new countries kept the political boundaries set by the colonial powers. Within the new countries, rival ethnic groups struggled for power. Civil wars erupted.

Conflicts also arose between countries in the Sahel. Border disputes have lasted for years between Somalia and Ethiopia. The collapse of governments as a result of warring factions, drought, and famine further weakened the newly independent countries. While some countries in the African Transition Zone continue to struggle, new opportunities from the discoveries of oil, gas, and uranium deposits may help to bring stability if governments can manage these resources properly.

Conflict in Sudan

The countries in the Transition Zone are teeming with cultural differences. At times, these differences have led to conflicts such as between Sudan's north and south. Arabic-speaking Muslims live mostly in the northern cities and favor Islamic-oriented governments. People in the south live mostly in rural areas, are focused on a subsistence economy, and prefer a secular government. These differences led to a conflict in which nearly 300,000 people died and an estimated 2.7 million people were displaced between 1983 and 2005.

A peace agreement was finally signed in 2005, which provided considerable independence for Sudan's southern provinces. Although the agreement ended the conflict between the north and the south, it did not address the conflict in the western Darfur region of the country. The civil war in Darfur occurred because non-Arab Sudanese accused the government of favoring Sudanese Arabs. Finally, in 2011, the southern provinces of Sudan held a referendum and voted for independence. South Sudan is now an independent country, with Juba as its capital. However, violence between government soldiers and civilians continues.

More recently, in January 2011, South Sudan shut down all of its oil fields after a disagreement about the fees Sudan demanded to transport the oil. In May 2011, Sudan seized control of Abyei, a disputed oil-rich border region, after three days of clashes with South Sudanese forces. On September 27, 2011, the presidents of Sudan and South Sudan signed agreements of cooperation. The status of Abyei, however, was not addressed. The future of the region is still uncertain.

☑ **READING PROGRESS CHECK**

Identifying Cause and Effect In what ways has colonialism affected the countries of the Transition Zone?

Analyzing
PRIMARY SOURCES
The Conflict in Sudan

"We left because of war. For the last one and a half years we have been bombed by planes every day. We lived in the forest; there was no chance for school for the children, no healthcare or medicine. We got food from the ground, but not corn. We would collect water in the early mornings. This has happened all seasons."

—a 36-year-old mother of nine at the Jamam refugee camp in South Sudan, from Médecins Sans Frontières, January 18, 2013

DBQ *IDENTIFYING CAUSE AND EFFECT* In what ways did the people of Sudan suffer from the conflict between the north and the south?

animist pertaining to traditional religious beliefs in which nature and objects, such as animals and mountains, are thought to have spirits

Population Patterns

GUIDING QUESTION *What are the population patterns of the Transition Zone?*

The African Transition Zone is more than just a geographical transition—it is also a cultural transition between Muslim North Africa and the animist and Christian south. Many different cultural groups inhabit the changing environment of the Transition Zone.

There are not only differences in religious beliefs, but also differences in ways of life. Herders, farmers, nomads, and city dwellers make up the people of the Transition Zone. Despite rapid population growth, population density currently remains low. The average population density of the subregion is about 58 people per square mile (23 people per sq. km). Only a small portion of the total land area, however, is suitable for agriculture. Thus, population throughout the subregion is not evenly distributed.

According to some studies, the population of the countries in the Transition Zone will reach 100 million by 2020 and 200 million by 2050. More than half of these people are expected to live in Burkina Faso, Mali, and Niger. It is difficult to imagine how the environment can support such a large population.

While there are relatively few cities, there are several important ones. The ancient city of Timbuktu, now known as Tombouctou, is still standing. Nouakchott, the capital of Mauritania, has a population of about 1 million people. Sudan has two important cities as well. Khartoum, its capital, has a population of over 5 million. The second, Port Sudan, handles the bulk of Sudan's external trade.

Cities in the Sahel will continue to grow in the years to come. Increasing drought and environmental damage have spurred people who once lived in rural areas to migrate to cities such as Niamey in Niger and Bamako in Mali.

☑ **READING PROGRESS CHECK**

Gathering Information Which religions can be found in the Transition Zone?

©Stapleton Collection/Corbis

TIME LINE ⌄

DROUGHTS
and HUNGER ➔

For centuries, the peoples of the Sahel have experienced famine brought about by droughts. As the droughts have become more severe in recent years, climate scientists have collected evidence to determine if they are, in part, caused or intensified by human activities.

▶ **CRITICAL THINKING**

1. Identifying Cause and Effect What have been some of the effects of the droughts in the Sahel?

2. Analyzing What efforts have been made to deal with the effects of drought?

1600 ➔

European travelers observe a major drought in the Sahel and create the first modern record.

1640

1740s–1750s The Great Famine in northern Nigeria, Niger, and Mali leads to massive migration and disruption of trade.

1800 ➔

1820s–1830s Major droughts lead to famine from Senegal to Chad and contribute to decline of Bornu Empire

Society and Culture Today

GUIDING QUESTION *What cultural conflicts affect life in the Transition Zone?*

As you have seen, the Transition Zone includes more than 20 countries. There are more than 200 distinct ethnic groups in Chad alone. These groups coexist in an area influenced by Arab, European, and indigenous African cultures. Some of the major ethnic groups include the Mandé peoples of Senegal and Mali, the Wolof of Senegal, and the Hausa of Niger. The Fulani and the Berber peoples both live throughout the Sahel. The many ethnic groups of the Transition Zone speak languages from several African language groups: Afro-Asiatic, Nilo-Saharan, and Congo-Kordofanian. Because of European colonial rule, French is also widely spoken throughout the subregion.

The dominant religion among the peoples of the Transition Zone is Islam. Christianity, however, is also practiced in varying degrees in Chad, Sudan, Niger, and Senegal. Additionally, many ethnic groups in the region have maintained many of their indigenous religious practices. These usually involve belief in the existence of a supreme being and a hierarchy of lesser beings.

Many issues essential for development and for meeting basic needs revolve around education, health, and urbanization. Rural-to-urban migration has increased in the Transition Zone as drought and overworked land force farmers to look for work in cities. Sometimes men leave their families behind. Money sent home by these migrant workers is often the most important source of income in rural households. But many families choose to move to the cities as well.

Poverty is a key factor in access to health care in the Transition Zone. Some of the major health concerns are high mortality rates and infectious diseases. Lack of adequate health care during pregnancy and childbirth also results in high female and infant mortality rates. Only a small number of rural Africans have access to clean water, and only one-fourth live where there is adequate **sanitation**, or disposal of waste products.

sanitation the disposal of waste products

1920s–1950s
Sustained wet period leads to population growth and expansion of agriculture

Rainy season ends 15-year drought in Sahel; aid programs promote long-term solutions

1985
Scientists predict increasing drought in Sahel in the future

About 10 million people in eastern Sahel face hunger due to severe drought.

2007

2010

2000 →

Fifteen-year drought begins, causing famine and deaths of more than a million people

1970

1973
UN Sahelian Office addresses long-term problems related to drought in the subregion.

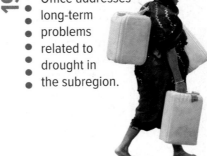

Humanitarian aid organizations provide relief as Sahel faces major food crisis

2012

patriarchal a family that is headed by a male family member

clan a large group of people descended from the same ancestor

nuclear family a family unit made up of a husband, wife, and children

oral tradition the practice of passing down stories from generation to generation by word of mouth

Dogon masks represent certain animals that symbolize their ancestors' spirits. This Dogon mask is worn at a rite that celebrates the passage of the dead into the spirit world.

▼ CRITICAL THINKING

1. *Analyzing Visuals* Describe the dominant features of the Dogon mask.

2. *Making Connections* How does this mask compare with modern abstract artistic representations of humans and animals?

School enrollment and literacy rates in the Transition Zone are generally low. In countries such as Niger and Mali, only a small percentage of children go to school. When children do go to school, they still face unemployment upon graduation. The education system has not yet adapted to the economic conditions and development that lead to jobs. However, governments are working to increase school enrollment.

Family and the Status of Women

No matter how different their ways of life, most Africans in the Transition Zone value strong family ties. Most people still live in extended families in rural areas. Women are very involved in supporting the family, often doing much of the farmwork. However, most families are **patriarchal**, or headed by a male family member, and descent is traced through the male line. Families are organized into **clans**, large groups of people descended from an early common ancestor. People often marry only within their clan. In the cities, however, extended families are more difficult to maintain. As a result, the **nuclear family**—made up of husband, wife, and children—is rapidly replacing the extended family.

The Arts

African art comes in many forms, from ritual masks to rhythmic drum music to folktales. These various art forms often express traditional religious beliefs. Visual arts include the ceremonial masks and wooden figures of the Dogon people of Mali. Music is also an important part of Dogon culture. Traditionally, the Dogon use music at ceremonies honoring their loved ones who have died. Rich musical traditions of the region include percussion and the five-string guitar. The talking drum is also popular, so called because it can reproduce the tone changes that are part of the Dogon language. By combining different tones and rhythms, the drumming creates messages that can be understood like the language.

Literature has also become an important art in the Transition Zone. Notable writers include Nafissatou Niang Diallo, whose 1975 autobiography was one of the first literary works to be published by a Senegalese woman. Chinua Achebe, from Nigeria, was a celebrated novelist and poet. He is most well known for his 1958 novel *Things Fall Apart* and later *Anthills of the Savannah* (1987). In his writings, Achebe explores colonialism and traditional life versus modernity.

African cultures of the Transition Zone also have a strong **oral tradition**. This is the practice of passing down stories and history from generation to generation by word of mouth. It is evident in folktales, myths, and proverbs. Oral literature is chanted, sung, or recited.

☑ READING PROGRESS CHECK

Identifying Central Issues What challenges do the people of the Transition Zone face today?

Economic Activities

GUIDING QUESTION *What natural resources are available in the Transition Zone for economic development?*

The empires that once occupied this region built their kingdoms with trade. Salt and gold added to the wealth of empires such as Mali. Today, the subregion also has rich deposits of oil and gas, iron, phosphates, copper, tin, and uranium to explore and develop. Oil has been particularly important in attracting outside investment.

Daphne Ouwersloot/Alamy

◀ CRITICAL THINKING

1. *Analyzing Visuals* How would you characterize the level of technology in this gold mine?

2. *Making Connections* What can you infer about the mine, given the level of technology used?

Immense oil reserves make Nigeria the region's only member of the Organization of Petroleum Exporting Countries (OPEC). In 2003 Chad became one of the subregion's oil-producing countries. Another country in the Sahel that has attracted investors is Niger. It has some of the world's largest deposits of uranium. Niger also has oil, gold, coal, and other mineral deposits, while Eritrea has gold and copper. In Senegal, the mineral industry—primarily petroleum products and phosphates—makes up about 20 percent of the country's exports.

Nonetheless, most people make their living in the traditional economies of the Transition Zone, primarily seminomadic herding and **subsistence farming**. In subsistence farming, the farmer and his or her family consume most of what is produced, leaving little to sell at market. Usually, subsistence farms are small and farming technology is labor-intensive. Typically, yields are low. Subsistence farmers decide what to plant based on what the family will need rather than on market prices. Much of what is grown in the Transition Zone is millet and sorghum. However, for countries such as Eritrea, subsistence farming often does not yield enough food even for their own populations.

subsistence farming
farming that provides the basic needs of a family with little surplus

☑ READING PROGRESS CHECK

Categorizing What economic activities can be found in the Transition Zone?

LESSON 2 REVIEW

Reviewing Vocabulary

1. *Making Connections* Write a paragraph using the terms *patriarchal*, *clan*, and *nuclear family* to describe families in the Transition Zone.

Using Your Notes

2. *Describing* Use your graphic organizer about the human geography of the Transition Zone to write a paragraph about the complexity of the subregion and the difficulties it faces.

Answering the Guiding Questions

3. *Drawing Conclusions* What factors influenced the formation of the countries of the Transition Zone?

4. *Classifying* What are the population patterns of the Transition Zone?

5. *Exploring Issues* What cultural conflicts affect life in the Transition Zone?

6. *Evaluating* What natural resources are available in the Transition Zone for economic development?

Writing Activity

7. *Informative/Explanatory* In a paragraph, discuss how the people of the Transition Zone, both past and present, have shaped the subregion.

Eric Feferberg/AFP/Getty Images

networks

There's More Online!

☑ **IMAGE** A Town in Darfur

☑ **IMAGE** Combating Desertification

☑ **INFOGRAPHIC** Great Green Wall

☑ **MAP** The Sahel's Vulnerable Zone

☑ **INTERACTIVE SELF-CHECK QUIZ**

☑ **VIDEO** People and Their Environment: The Transition Zone

LESSON 3
People and Their Environment: The Transition Zone

ESSENTIAL QUESTION • *How do physical systems and human systems shape a place?*

Reading HELPDESK

Academic Vocabulary

- **circumstance**
- **stress**

Content Vocabulary

- **desertification**

TAKING NOTES: *Key Ideas and Details*

IDENTIFYING As you read about people and their environment in the Transition Zone, use a graphic organizer like the one below to identify environmental problems faced by the people of the subregion and different ways in which people are trying to address these problems.

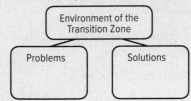

It Matters Because

The Transition Zone is faced with multiple challenges. These include drought, overpopulation, and the loss of arable land. These conditions can lead to poverty, hunger, conflict, and war. Finding solutions to these problems may help to reverse the loss of land and end both health and political crises that afflict the countries of the Transition Zone.

Managing Resources

GUIDING QUESTION *What water challenges do people of the Sahel face?*

Recall that the most vital resources of the Transition Zone, water and arable land, are at risk. While precipitation has decreased and become erratic, the number of people depending on it has increased. In the last several decades, the Sahel has experienced a rapid growth in population. Increasing at about 3 percent a year, the population will double about every 20 years. This rapid population growth makes the problem of inadequate rainfall that much worse.

One problem with chaotic shifts in rainfall is that extremely dry ground cannot absorb a heavy rain fast enough. Water that is not absorbed and does not drain pools up on the surface, causing crops to rot at the stem. By far the biggest problem, however, is a water shortage, which leads to **desertification**. Desertification is the destruction of land in arid and semi-arid areas, often caused by variations in climate. Human activity adds to the destruction. Desertification often causes poverty, food insecurity, and further water shortages. In these **circumstances**, the best hope for the subregion is to adopt new ways to manage existing water supplies. Furthermore, the adoption of agricultural methods that put less **stress** on the land could produce positive results.

☑ **READING PROGRESS CHECK**

Understanding Relationships What are the effects of the chaotic shifts in rainfall?

Human Impact

GUIDING QUESTION *What causes desertification and water shortages?*

Desertification puts pressure on the people that live in the Transition Zone. Agriculture and livestock suffer, as does the natural biodiversity of the area. In addition to infrequent rainfall, various human activities also cause desertification. One example of harmful human management is overgrazing.

Overgrazing is caused by an excessive number of livestock feeding too long in one area. Overgrazing kills plant roots. Too many animals in one area also compact the soil, thus reducing its capacity to hold water. Other detrimental human activities include poor agricultural practices and deforestation, which include the stripping of trees for firewood and clearing of land for farming. The main trigger of these harmful human activities is overpopulation. When more people live on the land, more livestock and crops are needed for food and more trees are needed for use as fuel. Just as drought and poor land use practices led to the Dust Bowl in the Great Plains of the United States, these factors have hastened desertification in the Sahel. A growing population dependent on the land in an already arid region puts further stress on that land. This stress triggers even more harm to the environment, in turn leading to more desertification.

Human activity in the Transition Zone has exacerbated desertification and created water management problems. For example, the availability of water in the Lake Chad Basin has decreased not only because of climate change but also because of over-demand. Planners have recognized that because of very high evaporation rates, it will not be enough to efficiently manage the water supply. The quality of the water that remains is also a concern. Commercial cotton and rice farmers are using agricultural chemicals that are polluting the water in Lake Chad. In other areas, poorly planned irrigation projects have pumped out too much groundwater, which causes wells to go dry. After the drought of the late

desertification the destruction of land in arid and semi-arid climates

circumstance an event or fact that accompanies or determines another

stress pressure or strain

GEOGRAPHY CONNECTION

The risk for desertification in the Sahel varies but overall is increasing.

1. *THE WORLD IN SPATIAL TERMS* Which biomes border the fragile zone?

2. *ENVIRONMENT AND SOCIETY* Why would it be important for people and governments to know which areas were in the high risk fragile zone?

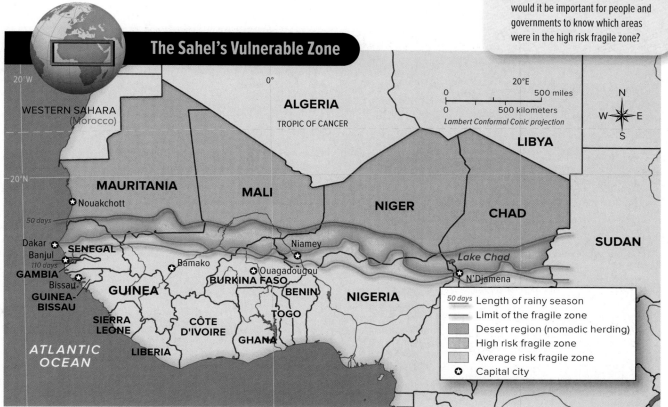

The Sahel's Vulnerable Zone

WESTERN SAHARA (Morocco)
ALGERIA
TROPIC OF CANCER
LIBYA
MAURITANIA
Nouakchott
MALI
NIGER
CHAD
SUDAN
Dakar
SENEGAL
Niamey
Lake Chad
Banjul
GAMBIA
Bamako
Ouagadougou
N'Djamena
Bissau
GUINEA
BURKINA FASO
BENIN
NIGERIA
GUINEA-BISSAU
SIERRA LEONE
CÔTE D'IVOIRE
TOGO
GHANA
ATLANTIC OCEAN
LIBERIA

50 days Length of rainy season
Limit of the fragile zone
Desert region (nomadic herding)
High risk fragile zone
Average risk fragile zone
Capital city

0   500 miles
0   500 kilometers
Lambert Conformal Conic projection

1960s—which lasted until the 1980s—humanitarian aid poured into the region in an effort to create water resources for the people. Though well-intentioned, the results were not beneficial. The irrigation systems that were built attracted disease-carrying insects and could not withstand the frequent droughts. The systems also interrupted the natural flow patterns of water. Boreholes and wells that were dug also added to the problem. Many boreholes and wells were dug in areas unsuitable for livestock and agriculture. This led to larger herds occupying smaller areas and resulted in further overgrazing and desertification.

☑ READING PROGRESS CHECK

Summarizing How has human activity affected the Transition Zone?

Addressing the Issues

GUIDING QUESTION *How are environmental problems in the Sahel being addressed?*

Once an ecosystem is damaged, it is difficult to reverse the effects. However, by improving water management and land use practices, some remediation may be possible. Water management improvements include better crop varieties and new water conservation technologies. Many of the efforts to stave off desertification are headed by international groups such as the International Atomic Energy Association (IAEA). In June 2012, the IAEA began a project with 13 countries in the region. The project aims to enhance knowledge and understanding of the five large aquifers in the Sahel, the source of most freshwater in the subregion.

The United Nations Convention to Combat Desertification (UNCCD) also works to stop desertification of the Sahel. The UNCCD helps to reduce the effects of drought and desertification and to restore land productivity in dry regions. Within the UNCCD, groups such as the Committee on Science and Technology and a Roster of Experts create action plans or programs to combat specific aspects of desertification.

This aerial view of a town in the Darfur region of Sudan shows the barren landscape produced by desertification.

▼ CRITICAL THINKING

1. Analyzing Visuals As shown in this photograph, describe the landscape of areas affected by desertification.

2. Drawing Conclusions Why is it important to reverse desertification in the Transition Zone?

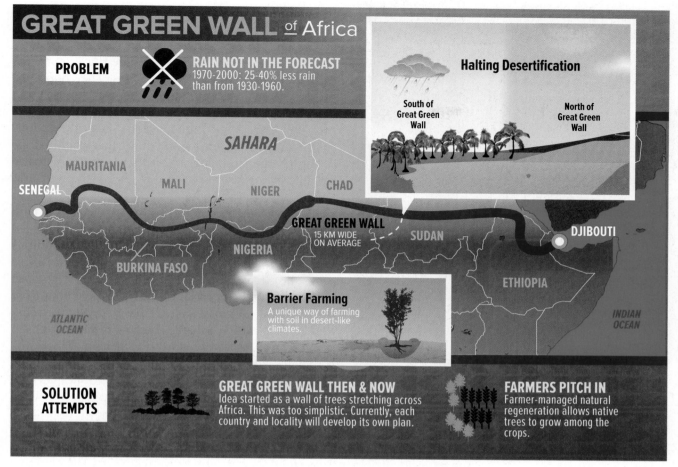

GREAT GREEN WALL of Africa

PROBLEM

RAIN NOT IN THE FORECAST
1970-2000: 25-40% less rain than from 1930-1960.

Halting Desertification

South of Great Green Wall

North of Great Green Wall

SAHARA

MAURITANIA

MALI

NIGER

CHAD

SENEGAL

GREAT GREEN WALL
15 KM WIDE ON AVERAGE

SUDAN

DJIBOUTI

NIGERIA

BURKINA FASO

ETHIOPIA

ATLANTIC OCEAN

INDIAN OCEAN

Barrier Farming
A unique way of farming with soil in desert-like climates.

SOLUTION ATTEMPTS

GREAT GREEN WALL THEN & NOW
Idea started as a wall of trees stretching across Africa. This was too simplistic. Currently, each country and locality will develop its own plan.

FARMERS PITCH IN
Farmer-managed natural regeneration allows native trees to grow among the crops.

The Great Green Wall is an ambitious project on the part of 11 countries to combat desertification. It will stretch across the continent. Expectations are high. One way the wall is expected to help is by blocking desert winds, which dry the soil.

▲ **CRITICAL THINKING**

1. Analyzing Visuals How will the Great Green Wall combat desertification?

2. Drawing Conclusions Why is it important for farmers to support the project?

Another group is the Permanent Interstate Committee for Drought Control in the Sahel (CILSS). The CILSS was established in 1973 following the severe drought that struck the region. Current member states include Gambia, Guinea-Bissau, Mauritania, Senegal, Burkina Faso, Mali, Niger, Chad, and Cape Verde. The CILSS works to improve food security and living conditions, water and land management, and natural resources management in the Sahel.

Small-scale efforts that local farmers can employ have been the most successful changes to date. Exchange of information on best farming and water management practices can be particularly helpful. Simple water-saving irrigation techniques, such as drip irrigation, have also proved useful. However, the challenge to adapt new technologies to local circumstances remains.

On a small scale, local people are also discovering ways to improve their land. After the drought in the 1980s, farmers in Burkina Faso started to experiment with traditional planting pits to reclaim severely dry land. These farmers increased the depth and width of the pits and began adding manure at the bottom before planting. These changes improved soil fertility and production. The techniques allowed farmers to increase their yield, prevent further soil erosion, and improve the soil quality. This technique has spread throughout the surrounding countries and has helped to rehabilitate between 500,000 and 750,000 acres (200,000 and 300,000 ha).

Various projects are underway to combat desertification. This man in Tombouctou, Mali, is irrigating a field.

▲ CRITICAL THINKING

1. *Analyzing Visuals* What purpose do you suppose the shallow trenches serve?

2. *Classifying* How would you describe the environment shown in the photograph?

There are differing techniques and views on how to combat desertification, and the problem is not confined to the Sahel. In some areas, such as Australia, scientists are monitoring when and where desertification occurs to better understand why it happens. They hope this information will aid in finding an appropriate solution that can be applied no matter where desertification is a problem. Others are thinking about radical solutions such as adding bacteria to sand dunes that would calcify, or harden, the sand. The hope is that the hardened sand would create a wall between the deserts and semi-arid land on the other side.

Others think that applying better water and land use management can reverse the problem. This has led to a unique initiative by many of the African countries within the Sahel to stop the spread of the Sahara. It is called the Great Green Wall, and it is a $2 billion project backed by the United Nations. The name comes from the plan that calls for a wall of trees to divide the semi-arid Sahel from the Sahara. The goal is for the trees to stop the southward spread of the Sahara. When it is completed, the Great Green Wall will stretch 4,970 miles (8,000 km) from Senegal to Djibouti in a band roughly 9 miles (15 km) wide. Planners expect that the barrier will be wide enough to block desert winds and will help hold moisture in the soil and also help to reduce soil erosion.

Senegal was one of the first countries to support the Great Green Wall project, and it has already benefited from the initiative. Senegal hopes that the project will help increase food security and prevent poverty. The planners hope to support local community efforts to sustainably manage the trees that are planted. To help build local support, seven varieties of acacia tree—including one that produces a fruit that can be used as animal food—were planted. People in local communities have harvested vegetables from gardens planted near the trees and fenced to keep out livestock. People have improved their nutrition, and they been able to sustain themselves throughout the year without having to leave for work in towns.

☑ READING PROGRESS CHECK

Making Connections How is the Permanent Interstate Committee for Drought Control in the Sahel working to combat desertification in the Transition Zone?

frans lemmens/Alamy

LESSON 3 REVIEW

Reviewing Vocabulary

1. *Determining Importance* Write a paragraph discussing some of the causes of desertification.

Using Your Notes

2. *Applying* Use your graphic organizer on the environment in the Transition Zone to write a paragraph evaluating the problems and some possible solutions.

Answering the Guiding Questions

3. *Identifying Central Issues* What water challenges do people of the Sahel face?

4. *Identifying Cause and Effect* What causes desertification and water shortages?

5. *Finding the Main Idea* How are environmental problems in the Sahel being addressed?

Writing Activity

6. *Argument* In a paragraph, explain what may happen if the environmental problems of the Transition Zone are not resolved.

Directions: On a separate sheet of paper, answer the questions below. Make sure you read carefully and answer all parts of the questions.

Lesson Review

Lesson 1

1 *Defining* Why is the region immediately south of the Sahara called the Transition Zone?

2 *Identifying Cause and Effect* How has the climate of the Sahel affected Lake Chad?

3 *Problem Solving* What are some ways that countries in the Transition Zone manage water resources?

Lesson 2

4 *Describing* Describe how culture changes as one moves from north to south in the Transition Zone.

5 *Making Predictions* Why are urban populations in the Sahel likely to grow in the years to come?

6 *Identifying Cause and Effect* What are some ways that poverty affects people in the Transition Zone?

Lesson 3

7 *Explaining* How has human activity in the Transition Zone contributed to desertification?

8 *Evaluating* After the drought that lasted from the late 1960s to the 1980s, irrigation systems were built to create water resources for people. Why were these systems ineffective?

9 *Analyzing* What steps have international organizations taken to reduce desertification in the Sahel?

Critical Thinking

10 *Explaining* In the mid-1900s, colonies in the Sahel gained independence from European powers. How did the legacy of colonialism create problems for the new countries?

11 *Comparing and Contrasting* Compare the Great Green Wall project to the proposal to create a wall of hardened sand. Which project do you think is more likely to stop the Sahara from moving south?

21st Century Skills

Use the chart to answer the questions that follow.

Country	Imports	Exports
Burkina Faso	Capital goods, foodstuffs, petroleum	Gold, cotton, livestock
Chad	Machinery & transport equipment, industrial goods, foodstuffs, textiles	Oil, cattle, cotton
Senegal	Food & beverages, capital goods, fuels	Fish, groundnuts, petroleum products, phosphates, cotton
Sudan	Foodstuffs, manufactured goods, refinery & transport equipment, medicines & chemicals, textiles, wheat	Gold, oil & petroleum products, cotton, sesame, livestock, groundnuts, sugar

Source: CIA World Factbook

12 *Economics* What general statement can you make about the types of commodities these four countries export?

13 *Compare and Contrast* How do Chad's imports differ from Senegal's imports? How are Senegal's imports similar to those of Burkina Faso?

Need Extra Help?

If You've Missed Question	**1**	**2**	**3**	**4**	**5**	**6**	**7**	**8**	**9**	**10**	**11**	**12**	**13**
Go to page	484	486	486	491	492	493	497	498	498	491	500	501	501

CHAPTER 20 Assessment

Directions: On a separate sheet of paper, answer the questions below. Make sure you read carefully and answer all parts of the questions.

College and Career Readiness

14 ***Problem Solving*** As a researcher for the United Nations Convention to Combat Desertification, you have been asked to report on how farmers in Burkina Faso used traditional planting pits to reclaim severely dry land. Write a two-page report describing their techniques. Do you recommend that these techniques be used more widely? Why or why not?

DBQ Analyzing Primary Sources

Use the document to answer the following questions.

Long regarded as a failed state, Somalia has had persistent problems dealing with hunger.

PRIMARY SOURCE

"Food is power in Somalia. Militia groups have routinely descended on the arable lands of central Somalia during harvest and claimed the crops for themselves. Pirates on the Indian Ocean have waylaid dozens of foreign vessels bearing food aid. Food prices were high here even before last year's worldwide spike, thanks to drought, militia roadblocks, and a devalued currency. The result is that millions now depend on food aid. The fresh fighting is pushing the country toward an unprecedented humanitarian crisis."

—Robert Draper, "Shattered Somalia," *National Geographic Magazine,* September 2009

15 ***Interpreting*** Why does the author say that in Somalia "food is power"?

16 ***Speculating*** How might geography have affected the availability of food in Somalia?

17 ***Evaluating*** Is food aid an effective way to address the problem of hunger? Why or why not?

Applying Map Skills

Use the Unit 6 Atlas to answer the following questions.

18 ***Physical Systems*** Use your mental map of Africa to describe the location of two or more major rivers in the Transition Zone. What countries do they connect?

19 ***Environment and Society*** What regions in the Transition Zone are most likely to be affected by desertification in the coming years?

20 ***Human Systems*** How do bodies of water affect population density in the Transition Zone?

Exploring the Essential Question

21 ***Making Connections*** Recall what you have learned about the problem of water shortage in the Transition Zone. Create a poster showing how drought, human activity, and desertification interact as part of a cycle.

Research and Presentation

22 ***Research Skills*** Use Internet and other resources to gather information and write a report about recent developments in the conflict in Sudan. What has happened since the presidents of Sudan and South Sudan signed agreements of cooperation? Has the conflict in Darfur been resolved?

Writing About Geography

23 ***Informative/Explanatory*** Use standard grammar, spelling, sentence structure, and punctuation to write an essay describing how drought has affected population patterns in the Transition Zone. How do these patterns reflect changes in traditional rural life and in the extended family?

Need Extra Help?

If You've Missed Question	14	15	16	17	18	19	20	21	22	23
Go to page	499	502	502	502	476	476	480	497	491	492

East Africa

ESSENTIAL QUESTION • *How do physical systems and human systems shape a place?*

Why Geography Matters
Export Crops and East Africa

Lesson 1
Physical Geography of East Africa

Lesson 2
Human Geography of East Africa

Lesson 3
People and Their Environment: East Africa

Geography Matters...

East Africa is a land of magnificent landscapes, from the volcanic peak of Kilimanjaro to the lakes of the Great Rift Valley. It is also, however, a land of desert and drought.

East Africa has attracted many countries over the centuries. Its prime location on the Red Sea and Indian Ocean drew the interest of people looking for trade outposts and entrances into the continent. Outsiders colonized almost all the land. Today, East Africa is working toward improving its economies, environments, and standards of living in the face of challenges such as population growth, corruption, and environmental degradation.

◀ Many young East Africans enjoy both tradition and modern ways of life.

Roy Toft/National Geographic Stock

export crops
and East Africa

About 80 percent of East Africa's population relies on subsistence farming. However, the subregion's per capita agricultural income and production is one of the lowest in the world. Still, agriculture is of vital importance and makes up about 40 percent of the subregion's gross domestic product (GDP). Even when local farmers try to produce agricultural commodities, the global market is highly competitive, which makes it difficult for them to access.

Why do farmers in East Africa grow certain crops?

Many of the crops that are grown in East Africa are the legacy of colonialism. Colonial powers made the economic decisions about what crops would be grown for consumption and for export. They based these decisions on profit for themselves, not for the benefit of the African people. Many colonial governments set up plantations that each produced a single crop for export because export crops brought in the most income for the colonial government.

In East Africa, export crops included tea, coffee, tobacco, cotton, cashews, copra, and sisal. After decolonization, it was difficult or even impossible for the newly independent governments to modify this system. The result is the current situation, in which these countries are dependent on agricultural exports and there is little support for small-scale subsistence farmers.

1. **Human Systems** How has East Africa's colonial legacy influenced what crops are produced?

How has growing single crops for export affected the people in East Africa?

Reliance on the success of a single export crop can be risky. For example, if climate conditions are extreme—too hot or cold, or too dry or wet—a crop might be wiped out or limited for a season. Sometimes the single crop grown for export requires more labor or resources such as water and fertilizer to cultivate and harvest. Essentially, the crop requires a large investment in terms of human and capital resources. These factors can affect the return on investment.

Another factor is the fluctuation in market demand and price. If demand for a product is low, then the price at which a farmer can sell will be low as well. The decision to produce one single crop over another may result in the loss of resources that could have been used elsewhere. If the government decides that the best land should be devoted to growing one export crop, that land might be unavailable for growing food. Because many farmers in East Africa rely on subsistence farming for their livelihood, they are most affected by agricultural decisions.

2. **The World in Spatial Terms** How have agricultural decisions about export crops affected subsistence farmers in East Africa?

What steps have been taken to help farmers?

To achieve better crop yields in East Africa, some organizations are working on solutions. For example, the goal of the Smallholder Cash and Export Crops Development Project is to help small growers of coffee in Rwanda by setting up cooperatives. The hope is that by working together, they can produce high-quality coffee that will fetch a high price. In another project, public land is being made available to small growers of tea.

Also, East African governments have encouraged the use of modern farm technology, such as genetically engineered seeds, soil conservation, and fertilizers. However, these actions have created new marketing problems. Some countries in Europe have boycotted certain crops from East Africa to protest the use of some of these techniques.

3. **Environment and Society** Write a paragraph evaluating the actions taken to improve crop yields in East Africa. Propose an idea that governments or other organizations can initiate to help farmers in the subregion.

THERE'S MORE ONLINE

READ about cash crops in East Africa • *EXPLORE* a map of East African natural resources

networks

There's More Online!

☑ **INFOGRAPHIC** The Great Rift Valley

☑ **IMAGE** Lake Turkana

☑ **IMAGE** Subtropical Climate

☑ **IMAGE** Arid Climate

☑ **INTERACTIVE SELF-CHECK QUIZ**

☑ **VIDEO** Physical Geography of East Africa

LESSON 1
Physical Geography of East Africa

ESSENTIAL QUESTION • *How do physical systems and human systems shape a place?*

Reading HELPDESK

Academic Vocabulary

- **constitute**
- **survive**

Content Vocabulary

- **rift valley**
- **fault**
- **escarpment**
- **cataract**

TAKING NOTES: *Key Ideas and Details*

IDENTIFYING As you read about the physical geography of East Africa, use a graphic organizer like the one below to identify examples of its key physical features.

East Africa's Physical Systems		
Landforms	Water Systems	Climates, Biomes, and Resources

IT MATTERS BECAUSE

With a variety of landscapes stretching from the Red Sea to south of the Equator and from the heart of the African continent to the Indian Ocean, East Africa can claim many natural wonders. These natural features range from mountains to deep rift valleys and from deserts to great lakes. It is an area of rich resources and, at the same time, a land of dusty, barren desert.

Landforms

GUIDING QUESTION *What physical features are part of the diverse landscape of East Africa?*

East Africa has a considerable range of landforms that include plains of differing types, volcanic mountains, and plateaus cut by a tremendous valley, marked by rivers and lakes. The subregion is home to the Serengeti Plain, as well as Kilimanjaro, Mount Kenya, and the Great Rift Valley. This subregion, which includes Burundi, Ethiopia, Kenya, Rwanda, Tanzania, and Uganda, has a strikingly diverse landscape.

A significant part of the landscape of East Africa is the Great Rift Valley. This long geologic feature begins well north of the subregion in Syria in Southwest Asia, and it extends south of the subregion to Mozambique (moh•zahm•BEEK) in the southeastern part of Africa. This natural wonder cuts through much of East Africa and is a defining feature of the landscape. A **rift valley** is a crack in Earth's surface formed by shifting and separating tectonic plates. The formation of the Great Rift Valley began millions of years ago when **faults**, or breaks in Earth's crust, were formed by the movement of plates below Earth's surface.

In East Africa, the Great Rift Valley is made up of two branches, each of which lies within a set of fault lines. The Western Rift Valley cuts through Tanzania, Burundi, Rwanda, and Uganda. Running the length of the rift, through the center of the Western Rift Valley, is Lake Tanganyika, one of the deepest and longest freshwater lakes in the world. The Eastern Rift Valley cuts through Tanzania and Kenya and north into Ethiopia. It is flanked in the east by two major volcanic mountains, Kilimanjaro and Mount Kenya.

East Africa has long coastlines along both the Red Sea and Indian Ocean. This coastal access has been of great economic importance. For example, Kenya and Tanzania have major ports on the Indian Ocean. The major rivers of Kenya and Tanzania drain into the Indian Ocean, linking the interior with the coast.

East Africa is made up of plateaus and cliffs. Plateaus, ranging in elevation from 500 feet (152 m) in the west to 8,000 feet (2,438 m) or more in the east, **constitute** much of the landscape. They are edged in **escarpments**, or steep cliffs. Rivers crossing the plateaus plunge down the escarpments in **cataracts**, or large waterfalls. The Ethiopian Plateau includes gorges, river channels, and *ambas,* or steep-sided, flat-topped land. East Africa also contains many grassy plains. Most notable is the Serengeti Plain, much of which is now a nature preserve in Tanzania and Kenya.

A mountainous region of East Africa is known as the Eastern Highlands. These highland areas include the Ethiopian Highlands, as well as the volcanic Kilimanjaro and Mount Kenya. These volcanoes are no longer active. East Africa is also home to the Ruwenzori Mountains, which divide Uganda and the Democratic Republic of the Congo. Covered with snow and cloaked in clouds, they are also known as the "Mountains of the Moon."

rift valley a crack in Earth's surface created by the shifting of tectonic plates

fault a crack or break in Earth's crust

constitute to compose or form

escarpment a steep cliff or slope between a higher and lower land surface

cataract a large waterfall

✔ **READING PROGRESS CHECK**

Identifying What are some unique features of the plateaus in East Africa?

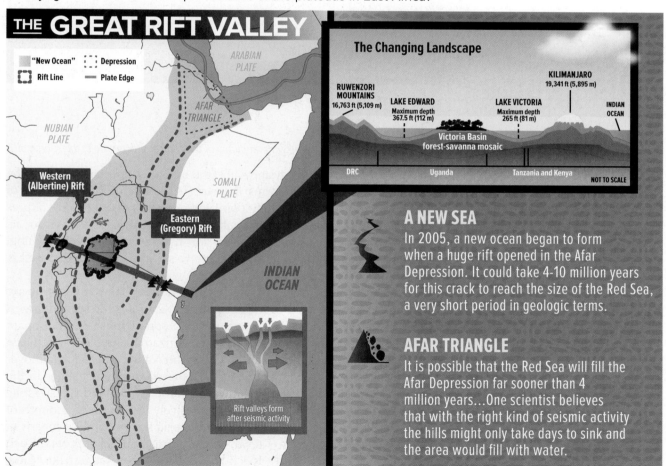

The Great Rift Valley formed millions of years ago and is a defining feature of the East African landscape.

▲ **CRITICAL THINKING**

1. *Analyzing Visuals* What happened beneath Earth's surface during the formation of the Great Rift Valley?

2. *Identifying Cause and Effect* How did the formation of the Great Rift Valley change the face of East Africa?

South Island in Lake Turkana is the largest island in the lake. Located at the southern and deepest part of the lake, it is part of a ridge of extinct volcanoes and hills poking above the surface.

▶ CRITICAL THINKING

1. *Making Inferences* How might the geography around Lake Turkana make it a good site for generating hydroelectric power?

2. *Identifying Cause and Effect* What caused Lake Turkana to separate from Lake Baringo?

Water Systems

GUIDING QUESTION *How have the water features of East Africa affected life in the subregion?*

Throughout East Africa, water systems abound and are important to life in the subregion. The Indian Ocean forms the eastern border of Tanzania and of Kenya. The ocean provides an entry point for products, people, and cultures through these countries and into the subregion. East Africa is known for its many lakes and rivers, which are used for hydroelectric power, fishing, transportation, supporting agriculture, and even as tourist attractions, as well as supporting natural ecosystems.

The two arms of the Great Rift Valley contain many lakes, and some are considered great lakes because of their size. Lake Victoria is the second-largest freshwater lake in the world, with an area of 26,828 square miles (69,485 square km). It is located along the borders of Uganda, Tanzania, and Kenya, between the eastern and western branches of the Great Rift Valley. It was formed over 100,000 years ago when the flow of water was shifted through changes in the western rift. Lake Victoria lies at the headwaters of the White Nile River. Despite its large surface area, Lake Victoria is comparatively shallow with a depth of only 270 feet (82 m). Since the 1900s, Lake Victoria has provided a means of transportation between Uganda, Kenya, and Tanzania with ferry ports in each country. Lake Victoria is nearly twice the size of Lake Tanganyika. Lake Tanganyika, in the Western Rift Valley, is one of the deepest freshwater lakes in the world. It is located between Tanzania and the Democratic Republic of the Congo and is within the drainage basin of the Congo River. The rich soil around Lake Victoria and the abundant fishing on Lake Tanganyika make the land around these lakes among the most heavily populated areas in Africa.

Lake Turkana, in Kenya, with its northern border in Ethiopia, is a source of growing wind-power and hydroelectric industries in Kenya. Lake Turkana has rocky shores in the east and south due to volcanic outcrops. It lies in a part of the Eastern Rift Valley that gets so little rain it is a hot desert. The lake had once been a part of a larger lake, along with Lake Baringo, but dry conditions caused the lakes to shrink and become two distinct bodies. In the west and north, it is marked by sand dunes, sandpits, and mudflats and has no outlet for drainage.

☑ **READING PROGRESS CHECK**

Determining Importance Describe key features of Lake Victoria and Lake Tanganyika.

Climates, Biomes, and Resources

GUIDING QUESTION *How does the climate affect life in East Africa?*

East Africa covers a wide range of latitudes and elevations, resulting in a diverse set of climates and biomes. Tropical heat characterizes much of the subregion, with savanna vegetation predominant. North of the savanna is the semi-arid steppe. North of the semi-arid steppe is the arid desert. Finally, the high mountains of the subregion have warm, humid subtropical climates.

Hot, dry weather prevails in the desert areas of East Africa. The northeastern area of the subregion, known as the Horn of Africa, is largely desert. Located in the Horn of Africa are parts of Eritrea, Ethiopia, Djibouti, Somalia, Kenya, and Uganda. Some areas receive only limited summer rain. Parts of Kenya may receive almost no rain. Thus the vegetation and wildlife must be able to **survive** on limited water.

Separating the deserts from the tropical savanna is a semi-arid, steppe transition zone. The northern steppe is called the Sahel—which means "shore" or "edge" in Arabic. The Sahel has pastures of low-growing grasses, shrubs, and acacia trees.

survive to manage to stay alive

These images demonstrate different climate zones in East Africa. East Africa's climate is as diverse as its landscape.

▼ **CRITICAL THINKING**

1. ***Exploring Issues*** How does climate impact life in East Africa?

2. ***Contrasting*** How are these climate zones different?

What is in a name? Names of places have meaning. Sometimes places are named for famous people. Sometimes names describe something about a place or its location. In 1858 Lake Victoria was named for Britain's Queen Victoria by British explorer John Speke. Lake Turkana used to be known as Lake Rudolf. This European name came from an Austrian explorer who named it for an Austrian prince. After Kenya gained independence, it changed the name to Lake Turkana, after the people who lived there. The lake has also been known as the Jade Sea, due to its color, and Basso Narok, which means "black lake." Anam Ka'alakol, meaning "sea of many fish," was the name given to the lake by the Turkana people.

SUMMARIZING Why has Lake Turkana had so many names?

On average, 4 to 8 inches (10 to 20 cm) of rain falls annually, mostly in June, July, and August. The rest of the year is very dry. This band of dry land extends across the whole continent, from Senegal to Sudan, and into Eritrea and Ethiopia. The Sahel has been plagued with soil erosion, deforestation, desertification, and drought. Rapid population growth in the area has made these problems worse.

Savanna, or tropical grassland with scattered trees, lies south of the steppe and extends east to west across the subregion. Rainfall is seasonal in this tropical climate zone, with alternating wet and dry seasons. The wettest areas are closest to the Equator. There, six months of almost daily rain is followed by a six-month dry season. Average annual rainfall is about 35 to 45 inches (90 to 115 cm). These tropical areas include Kenya, Uganda, Tanzania, Rwanda, and Burundi.

Vegetation varies in the savanna depending on the amount of rainfall and the length of the dry season. Trees are the main feature of the landscape in parts of the savanna with greater rain and a mild dry season, while tall grasses cover areas that are drier. In general, the soils are not very fertile. On the Serengeti Plain, one of the world's largest savanna plains, there are three types of grasses: short, medium, and tall. These grasslands make this area a suitable home for millions of animals such as zebras, gazelles, hyenas, lions, giraffes, and cheetahs.

Although less extensive than the tropical and dry climate regions, mid-latitude climates are also found in East Africa. The midlatitude climate zone is found in the highlands of Ethiopia and western Tanzania. The highland areas enjoy a moderate climate with comfortable temperatures and adequate rainfall for farming. Temperatures are somewhat lower, snow is not uncommon at high elevations, and there is plenty of vegetation. The highland areas, with green farmlands and protected forests, can seem almost lush.

The diversity of landforms and climates of East Africa contributes to its diversity of resources. Tanzania contains significant deposits of gold and natural gas. Diamonds and tanzanite, used as a gemstone, are found in Tanzania. Ethiopia and Burundi have stores of gold. Copper is found in Uganda. Kenya is also rich in minerals, such as soda ash, which is used in glassmaking. The biggest agricultural exports of the subregion are coffee and cotton. In Kenya, agriculture is a major part of the economy due to the fertile soil in locations like the Lake Victoria basin. Burundi has rich pastureland as well as farmland, and is well-known for its coffee. Rwanda's major resources are agricultural, but natural gas has also been discovered there.

☑ **READING PROGRESS CHECK**

Summarizing Where are the wettest areas of East Africa?

LESSON 1 REVIEW

Reviewing Vocabulary
1. ***Explaining*** Write a paragraph explaining how faults are connected to the formation of a rift valley.

Using Your Notes
2. ***Describing*** Use your graphic organizer on East Africa's physical systems to write a paragraph identifying the major aspects of the subregion's physical geography.

Answering the Guiding Questions
3. ***Classifying*** What physical features are part of the diverse landscape of East Africa?

4. ***Identifying Cause and Effect*** How have the water features of East Africa affected life in the subregion?

5. ***Differentiating*** How does the climate affect life in East Africa?

Writing Activity
6. ***Informative/Explanatory*** In a paragraph, discuss the impact of the Great Rift Valley and its features on life in East Africa.

netw⊙rks

There's More Online!

☑ **TIME LINE** A Cultural Crossroads

☑ **IMAGE** Traditional Villagers

☑ **IMAGE** Colorful City Bus

☑ **IMAGE** Mining

☑ **INTERACTIVE SELF-CHECK QUIZ**

☑ **VIDEO** Human Geography of East Africa

LESSON 2
Human Geography of East Africa

ESSENTIAL QUESTION • *How do physical systems and human systems shape a place?*

Reading **HELP**DESK

Academic Vocabulary

- **approximately**
- **incorporate**
- **export**

Content Vocabulary

- **indigenous**
- **lingua franca**
- **overfarming**

TAKING NOTES: *Key Ideas and Details*

SUMMARIZING As you read about the human geography of East Africa, use a graphic organizer like the one below to identify key events in the history of the subregion.

Key Events in the
History of East Africa

IT MATTERS BECAUSE

The population of East Africa dates far back in the archaeological record to the first humans. Much later, traders introduced cultures from other parts of the world, and eventually Europeans colonized the subregion. The current countries of East Africa gained independence beginning in 1962.

History and Government

GUIDING QUESTION *What cultures have influenced life in East Africa?*

East Africa contains fossils of prehumans that date back millions of years. Archaeologists have uncovered such fossils that are more than 3.2 million years old in the Awash River Valley in Ethiopia. In Kenya, in the area of Lake Turkana, they have found prehuman fossils that are 2.6 million years old. Other groundbreaking discoveries have been made in East Africa, including footprints of prehumans who walked the land approximately 1.5 million years ago.

In more recent times, East Africa's location, with borders on the Red Sea and the Indian Ocean, has made trade important. Living along the Red Sea coast near the Arabian Peninsula, its peoples have long had trade relationships with Arabian, Asian, and Mediterranean peoples. This contact resulted in a culturally and genetically diverse population. People from southern Arabian kingdoms invaded and absorbed Eritrea in 1000 B.C., establishing the Kingdom of Axum. This kingdom stretched west into the Ethiopian Highlands. By A.D. 600 the power center shifted to Ethiopia, which then controlled much of the Red Sea coast, linking Axum to Mediterranean cultures.

Traders from the Arabian Peninsula established colonies along the coast of East Africa in the A.D. 700s, bringing with them the Arabic language and the Islamic religion. They also established the slave trade. Persians, too, began settling along coastal areas from about the A.D. 700s. Even Chinese explorers made contact with East Africa in the early 1400s. Then, in the late 1400s, the Portuguese claimed control of the area.

networks

There's More Online!

☑ **IMAGE** Refugees in Rwanda

☑ **IMAGE** Mountain Gorilla

☑ **IMAGE** White Rhinoceros

☑ **IMAGE** Grevy's Zebra

☑ **IMAGE** African Elephant

☑ **INTERACTIVE SELF-CHECK QUIZ**

☑ **VIDEO** People and Their Environment: East Africa

LESSON 3
People and Their Environment: East Africa

ESSENTIAL QUESTION • *How do physical systems and human systems shape a place?*

Reading HELPDESK

Academic Vocabulary

- **intervene**
- **prohibit**

Content Vocabulary

- **habitat**
- **carrying capacity**
- **poaching**

TAKING NOTES: *Integration of Knowledge and Ideas*

SUMMARIZING As you read about the people of East Africa and their environment, use a graphic organizer like the one below to summarize what you learn about how humans have affected and changed this subregion.

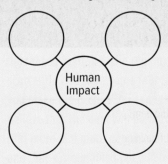

Human Impact

IT MATTERS BECAUSE

Overcoming adversity in East Africa continues to remain a significant challenge for its people. Factors such as conflict, corruption, and mismanagement compound the issues of illiteracy, health crises, and climatic problems. The will of the people to overcome adversity remains a powerful tool in meeting these challenges.

Managing Resources

GUIDING QUESTION *What makes effective resource management in East Africa especially challenging?*

Poverty, hunger, drought, corruption, and strife all create a precarious outlook for the people of East Africa. Rising above such challenges has tested the will of these countries for decades. Survival tends to take precedence over environmental concerns. However, the failure to deal with environmental issues contributes to poverty and hunger. The Human Development Index (HDI) is a measure created by the United Nations that ranks countries based on their health, education, and living standards. East African countries rate among the lowest on this scale. Of 187 countries measured, Burundi ranked 184, while Kenya ranked the highest in the subregion at only 145. This means life expectancy, literacy, and standard of living in East Africa are among the lowest in the world.

Developing sustainable agriculture is a challenge in this subregion. Poor cultivation practices and frequent drought have only exacerbated the problem. With the exception of Ethiopia, all East African countries were at one time European colonies. The end of the colonial period had an impact on the economy and environment in East Africa. In Tanzania, for example, under German control, farming coffee, rubber, and cotton for export flourished. After World War I and a British blockade of this German colony, the economy was left in shambles. When the Germans withdrew from their colony, the people returned to subsistence farming, rather than farming for export, and that is not sufficient to maintain a national economy.

After Eritrea gained its independence from Ethiopia in 1993, farmers in both countries worked to improve the land. In the Ethiopian province of Tigray, farmers terraced more than 250,000 acres (101,172 ha) of land. Then they planted 42 million young trees to hold the soil in place. They also built earthen dams to store precious rainwater. Grain crops thrived in their fields. In Eritrea, crops were so abundant that the government was able to reduce its request for relief from other countries by 50 percent. This all changed when Ethiopia and Eritrea went to war over their shared border. At the end of that war, Eritrea faced many challenges. The government controlled the economy, and the military was slow in allowing agriculturalists to help get farming back on track. In addition to a devastating drought, subsistence farmers were not able to produce enough food to feed themselves. The country is still working to undo this damage.

War continues to be a major cause of hunger and malnutrition in East Africa. Since 1990, conflicts in countries such as Rwanda have halted economic development, caused widespread starvation, and cost the lives of countless Africans. Huge refugee populations fleeing war-torn areas and crossing borders into neighboring countries strain already limited food resources. The civil war in Rwanda in the 1990s between the Hutu and Tutsi resulted in the genocide of over 800,000 people, most of whom were Tutsi. Over 2 million people became refugees. Since then, Rwanda has **intervened** in neighboring countries to limit the ethnic extremism that led to genocide within its borders. This involvement, however, has taken a toll on the country by preventing attention from being devoted to managing resources and getting its economy moving. While Rwanda is a heavily agricultural society, its exports focus on coffee and tea, which are processed there.

intervene to take action to change what is happening

East Africa has also seen government corruption. In Burundi, government corruption has kept much-needed aid from flowing into the cash-strapped country. Even in Kenya, the regional trade and finance center for East Africa, corruption has been rampant. The International Monetary Fund (IMF) stopped lending money to Kenya because it failed to put anticorruption measures in place. There was a change in government in 2002. Just when it looked as if the corruption tide was turning, Kenya went through another major corruption scandal. Despite the government's failure to prevent corruption in its ranks, lending has resumed, and the economy is slowly recovering.

☑ **READING PROGRESS CHECK**

Making Connections How has conflict impacted farming in East Africa?

The war in Rwanda, as with many conflicts in the subregion, created a refugee crisis.

◀ **CRITICAL THINKING**

1. ***Analyzing*** What has been the human and economic impact of war in East Africa?

2. ***Drawing Conclusions*** How do refugees destabilize surrounding countries?

Human Impact

GUIDING QUESTION *What environmental challenges does East Africa face today?*

In 1990, tropical forests covered almost 1.5 billion acres (608 million ha) in the region. By 2000, 126 million acres (51 million ha) had disappeared, mostly due to loggers and farmers clearing the land. On the continent as a whole, about half the original tropical forests are gone. The most valuable woods have been logged for export. Also, the lack of infrastructure in many countries has resulted in the need for wood to make charcoal for cooking and heating. Lack of electricity in villages in Tanzania, for example, means people have to depend on wood for light and heat. One goal of the effort to expand access to electricity in the country is to cut back on deforestation. Various countries have also created forest reserves to protect tropical forests. Logging companies are also becoming involved, using scientific tree farming and replanting projects to protect and renew forests.

Another consequence of deforestation is the destruction of animal **habitats**, or living areas, causing many species to be in danger of extinction, or disappearance from Earth. As the subregion's population grows, farmers have moved into and cleared some forested areas. Some savannas—home to huge herds of animals, such as elephants, giraffes, antelopes, and lions—are also being plowed for farming. Hunting also threatens wildlife. During the colonial period, European hunters reduced animal populations significantly. In recent years, hunters have continued to pursue African game for sport, profit, and food.

A combination of environmental factors and human factors, such as severe drought and poor farming practices, has been a major cause of desertification in areas of East Africa. Outdated agricultural techniques have stripped the soil of nutrients, while destruction of forests and wetlands has caused erosion. Rapid population growth has made these problems worse. As a result, in this subregion **carrying capacity**—the number of people an area of land can support on a sustained basis—has already been greatly exceeded. Despite this fact, the population growth rate remains high, further straining resources. Kenya was the first country in the subregion to recognize the problem and to put a population policy in place. Family size has been somewhat reduced since, but is still larger than in the West.

habitat area with conditions suitable for certain plants or animals to live

carrying capacity the population that an area will support without undergoing deterioration

Wildlife like the mountain gorilla, the white rhinoceros, Grevy's zebra, and the African elephant face loss of habitat and possible extinction as a result of human activities like hunting and poaching.

▼ CRITICAL THINKING

1. Drawing Conclusions How has the face of East African wildlife changed due to the impact of humans?

2. Hypothesizing How can human beings change the fate of these species?

Because traditional pastoralists have been pushed out of the area, the soil that their animals used to fertilize while grazing has become nutrient-deprived. The lack of sustainable pasture management plans has led to the overgrazing of livestock, which has taken a toll on soil quality and prevented vegetation growth. The semi-arid areas of East Africa are most at risk for desertification.

In the early 1970s, 2 million elephants roamed the subregion. Today fewer than 600,000 remain, largely because of **poaching**, or illegal hunting. Despite being **prohibited**, the poaching of elephants for their ivory tusks causes as many as 80,000 elephants to be killed each year. Such drastic reduction in numbers put the African elephant on the endangered species list starting in 1989. Other animals at risk include the mountain zebra, the mountain gorilla, and the rhinoceros. Rhinoceroses face extinction because of the demand for their horns, which are used in traditional Asian medicine. The price of rhinoceros horn far exceeds the price of gold, which makes them a prime target for poaching.

Pollution of East Africa's shorelines and waterways is creating new challenges for the subregion. The East African coast is well traveled by oil tankers, and this is causing pollution in the reef-fishing zone. Without adequate environmental safeguards in place, increasing urbanization and industrialization are also causing damage to East Africa's waterways. Sewage treatment is inadequate. Agricultural chemicals such as fertilizers and pesticides pollute runoff water, which flows unchecked into waterways.

As the population increases in the fertile areas around East Africa's lakes, so does pollution. Lake Victoria, which provides fishing and freshwater for Uganda, Tanzania, and Kenya, has become polluted. Uganda's supply of drinking water has been extremely threatened by this pollution. Authorities have warned that part of the lake suffers from severe pollution, which will be difficult to combat. "As more algal blooms, phosphates, nitrates, heavy metals, and fecal matter all pile into the lake, it's going to be harder and harder to clean the water," warned Gerald Sawula, deputy executive director of Uganda's state-run National Environmental Management Authority (NEMA).

✅ **READING PROGRESS CHECK**

Understanding Relationships Describe one way in which humans have affected the environment in East Africa.

Analyzing
PRIMARY SOURCES

Contrasting Views of Tanzania's Landscape

"We walked for miles over burnt out country. . . . Then I saw the green trees of the river, walked two miles more and found myself in paradise."

—Stewart Edward White, American hunter describing the Serengeti in Tanzania, 1913

"Since it takes wood to produce charcoal, Tanzania suffers an annual loss of 400,000 hectares of forests with the main culprit behind this depletion of its major natural resources being the domestic fuel demand."

—IPP Media, describing Tanzania, September 28, 2012

DBQ **CONTRASTING** How has deforestation changed Tanzania over time?

Ecotourism is a growing industry in East African countries.

Addressing the Issues

GUIDING QUESTION *What steps are being taken to combat these environmental challenges?*

East African countries are taking major steps to preserve the environment. For example, the protection of tropical forests has become a priority. In 1999, leaders from six central African countries signed an agreement to preserve the forests. Rwanda has put effort into reforestation, planting fast-growing eucalyptus trees in previously deforested areas to help combat erosion.

▲ **CRITICAL THINKING**

1. ***Drawing Inferences*** What draws people to ecotourism in East Africa?

2. ***Identifying Central Issues*** How do the profits from ecotourism help to drive conservation?

poaching illegal hunting

prohibit to forbid from doing through a law or rule

Much of the land in Uganda once devastated by deforestation has been turned into national parks. One of the ten national parks there is the Bwindi Impenetrable Forest, which is home to about half the world's population of endangered mountain gorillas. This park, along with others in Uganda, was designated a UNESCO world heritage site in 1994. This designation has been applied in other areas of East Africa, bringing more attention to the subregion.

Wildlife reserves—which include Tanzania's Serengeti National Park and Kenya's Maasai Mara—have helped some animals make a comeback. Rhinoceroses and elephants stand a chance at survival because wildlife reserves are providing protection. The battle against poaching, however, will continue as long as the demand for tusks and horns exists.

The parks also attract millions of tourists each year. Ecotourism has become a big business in parts of the subregion, bringing millions of dollars into regional economies. The benefits of income and job opportunities provided by the parks are strong motivations for the local people to continue measures aimed at protecting wildlife. Ecotourism also raises international interest in the ecological concerns of the region and brings international aid and support for conservation.

✓ **READING PROGRESS CHECK**

Identifying How are people working to solve environmental issues in East Africa?

LESSON 3 REVIEW

Reviewing Vocabulary

1. ***Summarizing*** Write a paragraph describing why East Africa is exceeding its carrying capacity.

Using Your Notes

2. ***Summarizing*** Use your graphic organizer on human impact in East Africa to write a paragraph summarizing the effects people have had on the environment of the subregion over the years.

Answering the Guiding Questions

3. ***Interpreting*** What makes effective resource management in East Africa especially challenging?

4. ***Identifying Central Issues*** What environmental challenges does East Africa face today?

5. ***Drawing Conclusions*** What steps are being taken to combat these environmental challenges?

Writing Activity

6. ***Argument*** In a paragraph, explain how human interaction with the environment has exacerbated East Africa's environmental problems.

Directions: On a separate sheet of paper, answer the questions below. Make sure that you read carefully and answer all parts of the questions.

Lesson Review

Lesson 1

1 *Discussing* Discuss the relationship between escarpments and cataracts.

2 *Comparing and Contrasting* Create a Venn diagram to compare and contrast Lake Victoria and Lake Tanganyika.

3 *Describing* Describe the proximity of the wettest areas of the savanna in East Africa to the Equator. Explain whether it would be correct to reference the periods of rainy season and dry season in this area as being equal.

Lesson 2

4 *Organizing* Create a time line that depicts invaders and power centers in East Africa from 1000 B.C. into the 1400s.

5 *Summarizing* Summarize the conflict between the Tutsi and the Hutu, and explain the significance of the conflict to the history of East Africa.

6 *Summarizing* Explain whether it would be accurate to state that literacy rates are similar across East Africa.

Lesson 3

7 *Making Connections* What is the Human Development Index (HDI)? Where does it rank the countries of East Africa?

8 *Describing* Describe the impact of deforestation on the tropical forests in East Africa.

9 *Evaluating* Why does poaching still exist even though wildlife preserves are working to halt this practice?

Critical Thinking

10 *Exploring Issues* Write a one-page essay to explain how the current-day issues in East Africa are tied to East Africa's physical geography and early history.

21st Century Skills

Use the diagrams below to answer the questions that follow.

The Great Rift Valley—A

Lake Tanganyika

The Great Rift Valley—B

11 *Using Primary Sources* Describe the tectonic action depicted in these images. Explain which of the images depicts events that occurred earlier—Image A or Image B—and how you are able to tell.

12 *Geography Skills* Does Image A depict the Western Rift Valley or the Eastern Rift Valley? How is it possible to make this determination based on the information in the image?

Need Extra Help?

If You've Missed Question	1	2	3	4	5	6	7	8	9	10	11	12
Go to page	507	508	510	511	512	515	520	522	523	506	506	506

Directions: On a separate sheet of paper, answer the questions below. Make sure that you read carefully and answer all parts of the questions.

DBQ Analyzing Primary Sources

In 1998 President Clinton arrived in Kigali, Rwanda, and spoke to survivors of the Rwandan genocide.

PRIMARY SOURCE

"The government-led effort to exterminate Rwanda's Tutsi and moderate Hutus, as you know better than me, took at least a million lives. Scholars of these sorts of events say that the killers, armed mostly with machetes and clubs, nonetheless did their work five times as fast as the mechanized gas chambers used by the Nazis.

It is important that the world know that these killings were not spontaneous or accidental. It is important that the world hear what your president just said; they were most certainly not the result of ancient tribal struggles. Indeed, these people had lived together for centuries before the events the president described began to unfold."

—President Bill Clinton, Associated Press, March 1998

13 *Interpreting* Why do you think President Clinton insists that the killings "were not the result of ancient tribal struggles"?

14 *Comparing* What does the comparison President Clinton makes to the Nazis suggest about the Rwandan genocide?

Applying Map Skills

Refer to the Unit 6 Atlas to answer the following questions.

15 *The World in Spatial Terms* Use your mental map of East Africa to describe the locations of the two branches of the Great Rift Valley. Include the following in your description: Tanzania, Burundi, Rwanda, Uganda, Lake Tanganyika, Kenya, Ethiopia, Kilimanjaro, and Mount Kenya.

16 *Places and Regions* What is the highest mountain in East Africa?

17 *The World in Spatial Terms* Estimate the distance in miles across the widest part of Tanzania.

College and Career Readiness

18 *Decision Making* You've been hired by an ecotourism company. Your boss explains that she wants to develop trips to East Africa. She believes that ecotourism to this location would be worthwhile, from both an economic perspective and an environmental perspective, and she has asked you to research the potential destinations. You have been asked to develop a memo identifying at least five potential stops and the activities ecotourists would engage in at each one. The memo should also include your opinion on whether these trips would be worthwhile based on economic and environmental considerations. Prepare the memo for your boss.

Research and Presentation

19 *Gathering Information* Read this statement from the chapter: "In Uganda, a dictator ruled for eight years in the 1970s, causing social disintegration, human rights violations, and economic decline." Conduct research to identify this dictator and explain how and why he came to power; a summary of the action this dictator took while in office; events that led to the end of the dictatorship; a summary of the impact of the dictator's rule; and an explanation as to whether a dictator with similar goals and objectives is likely to come to rule Uganda in the near future. With a partner, create a multimedia presentation to explain all the points above. The presentation should include maps and photos, as well as video and/or audio.

Writing About Geography

20 *Narrative* Use standard grammar, spelling, sentence structure, and punctuation to write a narrative paragraph that describes the area around you as you travel to the top of the Ruwenzori Mountains. Describe the border created by these mountains, and explain why you think they are referred to as the "Mountains of the Moon."

Need Extra Help?

If You've Missed Question	**13**	**14**	**15**	**16**	**17**	**18**	**19**	**20**
Go to page	513	513	476	476	476	524	526	507

West Africa

ESSENTIAL QUESTION · *How do physical systems and human systems shape a place?*

Why Geography Matters
Empowering Women in West Africa

Lesson 1
Physical Geography of West Africa

Lesson 2
Human Geography of West Africa

Lesson 3
People and Their Environment: West Africa

Geography Matters...

West Africa is a rich land with many natural resources and a diverse cultural heritage. Thousands of years of history have influenced the many cultures throughout the subregion and continue to do so today.

West Africa, however, faces many challenges. Increasing populations, drought, deforestation, reduction in the quality of farmland, and diminishing water resources place burdens on the land and the people.

West African countries have made strides in addressing these issues, but problems persist. With the help of individuals and national and international organizations, progress is being made.

◀ A Nigerian woman in Lagos dances in an annual festival held the day after Easter.

Pius Utomi Ekpei/AFP/Getty Images

527

empowering women
in West Africa

According to the United States Agency for Gender Development, many West African countries rank high in measures of gender inequality. Other international agencies have reported lower statistics, but the fact remains that many women in this subregion are economically, politically, and socially disadvantaged. At the same time, some West African countries and organizations are working to provide a more positive environment for women. The goal of these efforts is to empower women to boost economic development in West Africa.

What challenges do women in West Africa face?

Women in West Africa are equal to men under the law. In reality, however, they face many hardships. Many are forced into marrying much older men at a young age. When women marry early, they have few opportunities for education. Some of these women are in polygamous marriages. Polygamy is the practice of having more than one spouse at a time. While illegal in some parts of the world, polygamy is a cultural and religious tradition in many others. It is legal in parts of West Africa. Polygamy is particularly common in agricultural societies where women do most of the farmwork. Multiple wives and their children provide the labor necessary for a man to expand his agricultural holdings. Polygamous marriages offer little financial security to women.

Studies show that when women are able to earn income or receive benefit payments, they spend more money on education and health care for their children. This promotes economic growth and development.

1. Human Systems How does tradition in West Africa affect women's rights?

Why do these conditions exist?

Despite efforts to improve the status of women, discrimination and other challenges persist. Many gender inequalities are deeply rooted in cultural and religious beliefs and practices across West Africa. These traditions support limiting opportunities for women and girls in education, the economy, and political life. Rural areas of West Africa are more likely to hold onto traditional ways of life. Additionally, it is more difficult to monitor and enforce laws in rural areas. Fewer opportunities for education and health care are available in the countryside, and transportation to cities is often difficult and expensive. Isolated, the women of rural West Africa have few options.

These deeply held traditions and beliefs have prevented some countries from signing the Maputo Protocol (also known as the Protocol to the African Charter on Human and Peoples' Rights on the Rights of Women in Africa), which took effect in 2005. The goal of the protocol is to end discrimination and violence against women while also promoting women's rights in Africa. However, some countries have refused to sign or ratify the protocol, citing issues with its stances on polygamy, contraception, and abortion.

2. Human Systems Why have some West Africans been resistant to supporting the equality of women?

What steps have been taken to address these issues?

Organizations such as the Office of the United Nations High Commissioner for Refugees are working to change the nationality laws for states that do not provide equal rights to women. Gender inequality in nationality laws can prevent women from passing on their citizenship to their children. Children without citizenship face many problems, including little or no access to education and health care. Sierra Leone is one West African country that has made partial reform in this area.

In 2012 the New Partnership for Africa's Development (NEPAD) and the Economic Community for West African States (ECOWAS) signed an agreement for the empowerment of West African women who live in rural areas. This grant of one million euros will offer funding for an organization that supports African female entrepreneurs in agriculture and related endeavors.

3. Places and Regions How have other regions around the world worked to promote change and boost economic development for West African women?

THERE'S MORE ONLINE

VIEW a West African gender population map • READ about the changing lives of West African women

netw⦿rks

There's More Online!

- ☑ **IMAGE** The Akosombo Dam
- ☑ **IMAGE** The Senegal River
- ☑ **IMAGE** Oases
- ☑ **INFOGRAPHIC** Elevation Profile of West Africa
- ☑ **INTERACTIVE SELF-CHECK QUIZ**
- ☑ **VIDEO** Physical Geography of West Africa

LESSON 1
Physical Geography of West Africa

ESSENTIAL QUESTION • *How do physical systems and human systems shape a place?*

Academic Vocabulary

- **parallel**
- **environment**

Content Vocabulary

- **reservoir**
- **river plain**
- **conflict diamonds**

TAKING NOTES: *Key Ideas and Details*

IDENTIFYING As you read about the physical geography of West Africa, use a graphic organizer like the one below to note the characteristics of the land, water, and biomes of the subregion.

Physical Geography of West Africa		
Land	Water	Biomes

It Matters Because

West Africa has a rich and diverse landscape, which includes shoreline, lakes, rivers, low plains, and highlands. These landscapes and the subregion's resources are important to the people for many reasons. West Africa is rich in resources and minerals such as oil, natural gas, uranium, gold, and diamonds.

Landforms

GUIDING QUESTION *What landforms dominate the West African landscape?*

West Africa includes the southern half of the bulge of the African continent that extends west into the Atlantic Ocean. West African countries include Benin, Cape Verde, Côte d'Ivoire, Gambia, Ghana, Guinea, Guinea-Bissau, Liberia, Nigeria, Senegal, Sierra Leone, and Togo. The African Transition Zone cuts through the northern part of this subregion. The land in the African Transition Zone changes from deserts to a tropical savanna. All of the countries of West Africa, except Cape Verde, have part of their territory in the Transition Zone. The northern border of West Africa is the Sahara.

West Africa is a large area. Its size contributes to its considerable diversity, as does its location in the Transition Zone. West Africa is largely a tropical region with a landscape that includes desert, shoreline, low plains, highlands, and rain forests.

West Africa has a long coastline. The coastal region is made up of sandy beaches, thick mangrove swamps, lagoons, and broad coastal plains. Lagoons are shallow bodies of water that are separated from the ocean by islands, which lie **parallel** to the shoreline. The Ebrié Lagoon is the largest lagoon in West Africa with a surface area of 218 square miles (566 sq. km). Rain forests once covered much of the West African coast. However, rain forests have been drastically reduced as land was cleared for agriculture and logging of prized hardwood trees such as African mahogany and iroko, also known as African teak.

Mountains rise up behind the broad coastal plains of the subregion. In the south-central part of West Africa are the Guinea Highlands. This mountainous plateau rises several thousand feet above the coastal plains in southeastern Guinea, northern Sierra Leone, Liberia, and northwestern Côte d'Ivoire. The plateau contains the Nimba Range, the Loma Mountains, and the Tingi Mountains. These highlands are covered with savanna and rain forest.

In the southeast are the Cameroon Highlands, which lie between Nigeria and Cameroon. Forest and grasslands with a rich biodiversity cover these highlands. They form part of a chain of former volcanoes that stretch inland from the sea. The largest and only active one of these volcanoes is Mount Cameroon, which rises to a height of 13,353 feet (4,070 m).

Low plains, sandy soil, and grasslands cover much of inland West Africa. The Sahara, the largest desert in Africa, is in the north. It is a vast and nearly uninhabited region. The landforms here are shaped mostly by winds and infrequent rainstorms. They include sand dunes, salt flats, gravel plains, stone plateaus, and dry valleys.

parallel lying in the same direction with an equal distance between

☑ **READING PROGRESS CHECK**

Identifying What types of landscapes are found in coastal regions?

Water Systems

GUIDING QUESTION *How are saltwater and freshwater resources important to the people of West Africa?*

The water systems of West Africa are not only important to wildlife but to people as well. The lagoons and mangrove swamps are important parts of the coastal ecosystem of the subregion. They provide food and shelter for fish, shellfish, mollusks, wildfowl, and marine mammals. Flounder and bluefish use the

Mount Cameroon is the highest point in West Africa and an active volcano. It is much higher than the peaks in the Guinea Highlands.

▲ **CRITICAL THINKING**

1. Classifying What is the second-highest peak shown in the diagram?

2. Analyzing Visuals What appears to coat the surface of the mountain on which the hikers are standing?

marshes as nurseries and winter habitats. Ducks, geese, and other wild birds stop over during migration. Unfortunately, due to environmental disasters such as oil spills, global warming, and clearing of the mangrove for the farming of shrimp, these swamps are slowly disappearing.

The mangrove swamps are also very important to coastal people. They provide a source of food such as crabs, clams, oysters, and fish. Even the mangrove fruits themselves can be eaten. The mangrove trees are also useful. They are often collected for firewood and can be used for construction.

The Atlantic Ocean provides an important resource for the people of West Africa—fish. Fishing not only provides food for the people of the subregion, but is also a key source of revenue. Commercial fishing is particularly important in the Gulf of Guinea, which is part of the eastern tropical Atlantic Ocean system off the coast of western Africa. In 2006 a fishery committee was established to manage fishing among the countries of Liberia, Côte d'Ivoire, Ghana, Togo, Benin, and Nigeria. All these countries share fish stocks in the Gulf of Guinea.

Lake Chad, in west-central Africa, is bordered by Nigeria, Niger, Chad, and Cameroon. Lake Chad was once the second-largest wetland in Africa, supporting a great diversity of animal and plant life. Although Lake Chad has since shrunk by about 90 percent, it is still vital to the region for irrigation of farmland and fishing.

Lake Volta is a human-made lake and is the fourth-largest **reservoir** in the world. A reservoir is an artificial or natural lake where water is stored and used to supply farms, homes, and businesses in the area with freshwater. The lake was made by damming the Volta River, and its creation flooded more than 700 villages, forcing over 70,000 people to find new homes. The trade-off is that it has provided a consistent source of freshwater and electricity. Recently a project has begun to try to harvest the trees that were submerged when the lake was created. This project will provide additional revenue and work for the people of the region. Another new industry in the region is tourism. The town of Akosombo, which grew around the dam, is the starting point for many fishing excursions and for water sports. Fishing for Volta perch, African tiger fish, Nile tilapia, and several varieties of catfish is a favorite of charter expeditions.

reservoir a natural or artificial lake used as a source of water

Lake Volta, an artificial lake, was formed by the Akosombo Dam. Today the dam provides electricity for most of Ghana's needs.

▼ **CRITICAL THINKING**

1. Analyzing Visuals How do you suppose the landscape shown in the photo contrasts with its appearance before the construction of the Akosombo Dam?

2. Explaining Why is hydroelectricity an important resource in Ghana?

Bob Burch/Photoshot

The number of fish in the Senegal River has increased since the Senegal River Basin Multi-Purpose Water Resources Development project began managing irrigation and water resources in the area.

◄ **CRITICAL THINKING**

1. *Analyzing* How can the health of waterways affect surrounding communities?

2. *Making Inferences* In what other ways can water resource management benefit surrounding communities?

While the Atlantic Ocean, Lake Chad, and Lake Volta are important to the economy and people of West Africa, the subregion also has a number of significant rivers. The Volta River, the main river system in Ghana, runs 1,000 miles (1,600 km) long. The northern four-fifths of the Volta River valley are now covered by Lake Volta, which was formed by the Akosombo Dam. The Volta River Basin covers an estimated area of 154,440 square miles (400,000 sq. km).

The Senegal River flows north from the Guinea Highlands through four countries: Guinea, Mali, Mauritania, and Senegal. The best agricultural land along the river is the **river plain** between the towns of Bakel and Dagana in Senegal. A river plain, also known as an alluvial plain, is a plain formed by the deposit of sediment over a long period of time by one or more rivers. After the yearly floods have retreated, crops such as millet, rice, and vegetables are sown. While agriculture is the largest economic activity along the river, fishing is the second largest. Additionally, two dams along the river, the Manantali and the Diama, provide electricity to the subregion.

The Niger (NY•juhr) River, which runs about 2,600 miles (4,183 km), is the third-longest river in Africa. The Niger is the main river of western Africa. Commercial shipping takes place on about 80 percent of the river. Fishing is also an important industry to the people who live along the Niger. The Benue River, the longest tributary of the Niger River, is about 673 miles (1,083 km) long. A considerable amount of trade moves along the Benue River, including petroleum, cotton, and peanuts.

The Niger River has two deltas, a vast inland delta and a delta where the river empties into the Atlantic Ocean. The inland delta is in Mali, where the river spreads out across the plains, creating a large wetland and agricultural area. The delta at the river's mouth is in Nigeria. At one time, fishing and the production of palm oil were strong industries in the region. However, the discovery of oil in the 1950s in the Niger Delta region has changed that. Oil production and spills have polluted the land and water. This pollution has destroyed the livelihoods of farmers and fishers of the Niger Delta.

river plain a plain formed by the deposit of sediment over a long period of time by one or more rivers

☑ **READING PROGRESS CHECK**

Understanding Relationships How do the water systems of West Africa benefit the people of the subregion?

Climates, Biomes, and Resources

GUIDING QUESTION *How do the tropical climates of West Africa support different biomes?*

environment natural surroundings

The **environment** of West Africa changes dramatically along a north-to-south latitudinal climate pattern. In the north it begins with the desert, which transitions into the semi-arid steppe, savanna grassland, and finally tropical forest. These environments have different climates, biomes, and resources.

In the north, where some countries of West Africa border the Sahara, the climate is hot and dry with very little rainfall. Parts of the African Transition Zone are steppe, with low-growing grasses, shrubs, and acacia trees. This area receives about 4 to 8 inches (10 to 20 cm) of rainfall a year.

As one moves farther south in West Africa, the land slowly becomes savanna. During the rainy season, the average rainfall here is about 15 to 25 inches (38 cm to 64 cm) a month. The rainy season begins in May and ends in November as warm, moist air is drawn from the Gulf of Guinea. At other times of the year, when the wind reverses and blows from the Sahara, rainfall averages about 4 inches (10 cm) a month. The land is covered with grasses and trees, such as the acacia and baobab. Many types of mammals and birds—such as giraffe, gerbils, foxes, elephants, and pygmy hippos—live in the savanna.

Still farther south, West Africa has tropical forests. Its tropical wet and dry forests, also known as tropical seasonal forests, receive rainfall during the rainy season followed by the dry and hot months. These tropical dry forests have less densely growing trees than the tropical wet forests farther south. The tropical wet forests receive plentiful rain and are abundant with semideciduous trees, evergreens, or semievergreens. They are also rich in spices, nuts, and legumes. These forests are at risk for deforestation as the human population increases.

conflict diamonds

diamonds that are mined in war-torn areas and are used to finance wars

Oil, natural gas, coal, gold, and uranium deposits are just some of the natural resources found in West Africa. Another major resource is diamonds. In recent years, diamonds from Africa have been referred to as **conflict diamonds**, or blood diamonds. Conflict diamonds are diamonds that are mined in a war zone. Often the money from the sale of diamonds in these areas is used to finance war. For example, during the conflict in Sierra Leone in the 1990s, rebels used forced labor to mine diamonds. These diamonds were then sold in neighboring countries and then in European markets. The money earned from these sales was used to finance the civil war in Sierra Leone. As people around the world became aware of conflict diamonds, efforts were made to prevent their purchase.

✅ **READING PROGRESS CHECK**

Summarizing What are some of the natural resources in West Africa?

LESSON 1 REVIEW

Reviewing Vocabulary
1. ***Describing*** Write a paragraph describing how reservoirs are used in West Africa.

Using Your Notes
2. ***Summarizing*** Use your graphic organizer on the physical geography of West Africa to write a paragraph summarizing the characteristics of the land, water, and biomes of the subregion.

Answering the Guiding Questions
3. ***Categorizing*** What landforms dominate the West African landscape?

4. ***Identifying Central Issues*** How are saltwater and freshwater resources important to the people of West Africa?

5. ***Analyzing*** How do the tropical climates of West Africa support different biomes?

Writing Activity
6. ***Informative/Explanatory*** In a paragraph, discuss how the land of West Africa affects life in the subregion.

networks

There's More Online!

☑ **IMAGE** Lagos, Nigeria

☑ **INFOGRAPHIC** Today's Nigerian Family

☑ **TIME LINE** Struggle for Power

☑ **INTERACTIVE SELF-CHECK QUIZ**

☑ **VIDEO** Human Geography of West Africa

LESSON 2
Human Geography of West Africa

ESSENTIAL QUESTION • *How do physical systems and human systems shape a place?*

Reading HELPDESK

Academic Vocabulary

• **convert**
• **scope**

Content Vocabulary

• **infrastructure**
• **griot**
• **e-commerce**

TAKING NOTES: *Key Ideas and Details*

IDENTIFYING As you read about the human geography of West Africa, use a graphic organizer like the one below to identify examples of how history, a growing population, and economic activities affect life today.

IT MATTERS BECAUSE

West Africa has seen many peoples, empires, and kingdoms flourish. The peoples of West Africa are incredibly diverse. Colonial rule brought changes and challenges to West Africa, and the cultures of the subregion modified. Some aspects of traditional cultures were destroyed, while others remained intact, often incorporating elements of European culture, such as a language.

History and Government

GUIDING QUESTION *How does the legacy of colonialism affect governments in West Africa today?*

Thousands of years ago, the Sahara was a much wetter area that supported considerable plant and animal life and, in turn, human populations. As the climate of the Sahara changed, however, people adapted by moving south to greener and more habitable regions. These seminomadic peoples herded animals and developed a system of agriculture. As populations grew, settlements grew into cities and empires.

Empires of West Africa

The first empire to emerge in West Africa was the Ghana Empire. It became one of the richest trading civilizations of West Africa. Ghana profited from its location midway between the salt mines in the Sahara and the gold mines farther south. Archaeologists believe the empire began around A.D. 300. It lasted until about the thirteenth century. During its time as a trading empire, the kingdom prospered by imposing taxes on trade goods. Muslim traders from North Africa sent caravans of goods and salt across the Sahara to Ghana. Gold from Ghana was traded for salt. Muslim traders also brought Islamic beliefs and customs to the kingdom. Eventually, many Ghanaians **converted** to Islam.

The Mali Empire developed after the small state of Kangaba broke away from Ghana. Sundiata Keita, one of Mali's early kings, helped Mali flourish. He took over Timbuktu and made it an important center of trade and scholarship. By the fourteenth century, the power of the Mali Empire began to wane and another empire, Songhai (SAWNG•hy), arose.

Songhai broke from Mali after the death of Mali's most well-known king, Mansa Musa. Sunni Ali Ber conquered the cities of Timbuktu and Djenné, expanding his empire to include most of the West African savanna. The empire prospered until about A.D. 1600, when it fell to the Moroccans.

The Hausa city-states were located between the Niger River in what is now northern Nigeria and ended at Lake Chad in the east. These independent states, the first of which emerged around A.D. 1000, formed loose alliances. The Hausa city-states remained independent for the most part, until they were conquered in the early 1800s by the Fulani. Each of the states served a different role within the alliance. Some provided goods, others soldiers or access to trade. As a result, the states became important to international trade.

The kingdom of Benin, which occupied the area of present-day Nigeria, developed into an important power from the thirteenth to the nineteenth century. Benin became an extremely organized state by the late 1400s. Between the fifteenth and eighteenth centuries, the kingdom grew in wealth and power through trade with the Portuguese and Dutch. By the nineteenth century, however, infighting weakened the kingdom. At the same time, suppression of the slave trade, from which Benin had greatly profited, led to the kingdom's decline.

The Colonization of Africa

As contact with Europe increased, word of the riches to be found in Africa spread. Europeans soon began to view Africa as a source of resources and opportunities. As European countries laid claim to territories in Africa, disputes over territory arose. At the Berlin Conference between 1884 and 1885, 14 European countries met in an effort to sort out how territory would be divided.

With no African input at the Berlin Conference, colonial boundaries were drawn with little regard for African ethnic boundaries. One example of this is in Nigeria, where boundaries were drawn that merged Muslim societies with animist cultures. As Christianity was introduced, additional tensions emerged. These and other issues set African peoples against one another and strengthened European rule.

©Patrick Robert/Corbis

TIME LINE ⌄

Struggle for POWER ➔

The oldest republic in Africa, Liberia has experienced a lengthy period of warfare as warlords vie for power.

▶ **CRITICAL THINKING**

1. Explaining How was Liberia influenced by the United States?

2. Analyzing How would you characterize the political situation in Liberia since 1980?

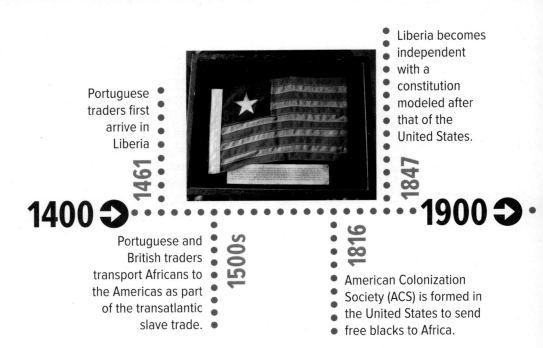

Portuguese traders first arrive in Liberia — **1461**

Liberia becomes independent with a constitution modeled after that of the United States. — **1847**

1400 ➔

1900 ➔

Portuguese and British traders transport Africans to the Americas as part of the transatlantic slave trade. — **1500s**

American Colonization Society (ACS) is formed in the United States to send free blacks to Africa. — **1816**

The one country Europeans never colonized is Liberia. African Americans freed from slavery established it on land purchased for them by the American Colonization Society. The African Americans who relocated to Liberia had a different culture from the indigenous people who already lived in the area. Over time, the differences widened, causing conflicts between the people of Liberia. In 1847 Liberia declared independence. Soon after, a constitution was written.

Decolonization and Difficulties

Europeans maintained their colonies in West Africa until the mid-1900s. But even after attaining independence, these new countries faced daunting challenges. Years of strife afflicted many of the countries. In Nigeria, for example, the ill-conceived boundaries drawn by the colonial powers led to increasing tensions between Muslims and Christians. In 2009 the Nigerian government began fighting against a militant Islamist group that sought to establish an Islamist state in Nigeria.

Another problem is corruption. In Nigeria, for example, the discovery of oil and gas made the country particularly vulnerable. Nigerian leaders used their offices to steal billions of dollars paid by foreign companies for oil extraction rights. To give an idea of the **scope**, or extent, of the problem, some estimate that close to $400 billion was stashed in foreign bank accounts between 1960 and 1999. In the early 2000s, Nigerians and new leaders waged a war against corruption.

scope the extent of an activity or influence

Changes in infrastructure and culture produced by European colonization persisted even after independence. The countries' economies were still dependent on providing raw materials to European countries. They were largely unable to revive their traditional cultural knowledge and practices that had been lost during colonization. As a result, they had to adopt the market economy and consumer culture of the European powers. These lasting changes made it impossible for the new countries to chart an independent course that would be more beneficial to their own interests.

✓ READING PROGRESS CHECK

Evaluating In what ways has colonialism affected the countries of West Africa?

Samuel Doe leads military t, toppling the government and seizing control

1980

Charles Taylor is elected president of Liberia.

1997

Charles Taylor found guilty of war crimes and sentenced to 50 years in jail

2012

2000 →

1903

After years of conflict, Great Britain and Liberia settle the border between Liberia and British territories in West Africa.

1989

Charles Taylor overthrows Doe-led government; civil war erupts

2003

Charles Taylor charged with crimes against humanity in connection with the civil war

Population Patterns

GUIDING QUESTION *How is a booming population affecting life in West Africa?*

Africa's population is increasing rapidly. Most people in West Africa live along the coast and river plains. About half of the people in West Africa live in crowded urban locations. West Africa is one of the most populous regions of Africa, and the country of Nigeria has the largest population of any country in Africa. Lagos, the commercial center of Nigeria, is Africa's largest city with an estimated population of more than 21 million. In 2015, over 180 million people lived in Nigeria. By 2025, the country's population is projected to exceed 220 million.

Demands created by climate change, deforestation, increasing population, and decreasing food supplies have prompted many Nigerians to move to urban areas. This pattern is also occurring in other parts of West Africa. People are moving to urban areas in hopes of finding better job opportunities, health care, public services, and education.

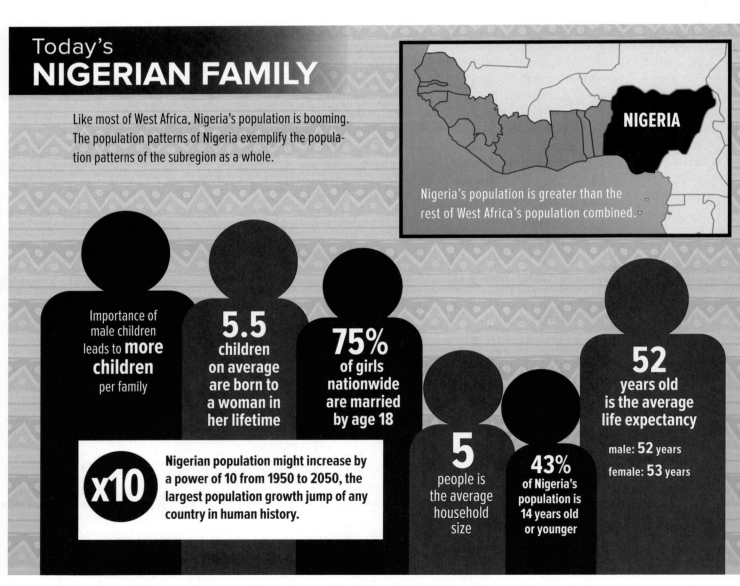

Today's
NIGERIAN FAMILY

Like most of West Africa, Nigeria's population is booming. The population patterns of Nigeria exemplify the population patterns of the subregion as a whole.

NIGERIA

Nigeria's population is greater than the rest of West Africa's population combined.

Importance of male children leads to **more children** per family

5.5 children on average are born to a woman in her lifetime

75% of girls nationwide are married by age 18

x10 Nigerian population might increase by a power of 10 from 1950 to 2050, the largest population growth jump of any country in human history.

5 people is the average household size

43% of Nigeria's population is 14 years old or younger

52 years old is the average life expectancy

male: **52** years

female: **53** years

Nigeria is Africa's most populous nation and continues to grow rapidly. The pace of population growth threatens to undermine efforts to improve the standard of living for Nigerians.

▲ **CRITICAL THINKING**

1. **Analyzing** What are the consequences of having a very young population?

2. **Problem Solving** What steps could Nigeria's government take to slow population growth?

The Nigerian city of Lagos is the most populous city in Africa and one of the fastest growing.

◀ **CRITICAL THINKING**

1. Analyzing Visuals How does the photograph illustrate the effects of rapid urbanization?

2. Drawing Inferences How does rapid urbanization affect the daily lives of a city's residents?

West Africa's rate of urbanization, or movement of people from rural areas to cities, is among the world's fastest. New population growth has caused cities to spread out into the countryside. This expansion of population in urban areas in West African countries has placed a strain on existing **infrastructure**—which includes electricity, water, roads, and information and communications technology. Limitations on infrastructure reduce productivity and limit economic development. However, with the help of international investors, West African countries are attempting to improve infrastructure to meet the demands of their ever-increasing populations.

infrastructure the set of systems that affect how well a place or organization operates, such as telephone or transportation systems, within a country

☑ **READING PROGRESS CHECK**

Evaluating How has West Africa been affected by population growth?

Society and Culture Today

GUIDING QUESTION *How is life in West Africa a mix of the ancient and the modern?*

West Africa is home to a very large number of ethnic groups—about 500 in Nigeria alone. This immense diversity includes peoples that have lived in West Africa for centuries. It also includes groups that have settled there more recently. Some ethnic groups that live in West Africa have been divided into different countries by political boundaries originally established by European colonial powers.

Of the ethnic groups in West Africa, the five largest are the Yoruba, Hausa, Fulani, Ibo, and Akan. Estimated at more than 30 million, the Yoruba live in areas of Nigeria, Benin, Togo, and Ghana. Some 30 million Hausa live in the West African countries of Nigeria, Ghana, and Côte d'Ivoire as well as in Chad, Niger, and Cameroon.

Hundreds of languages are spoken in West Africa. Arabic is common in the northern areas of the subregion. Yoruba and its many dialects, which are part of the Congo-Kordofanian language group, are widely spoken. In the southern region of West Africa, Yoruba is taught in primary and secondary schools, at universities, and in television and radio broadcasting schools. It is also commonly used in printed materials such as books, newspapers, and pamphlets. Part of the legacy of colonial rule was the introduction of European languages. Thus, English and French are widely spoken. In Nigeria, English is the official language. Using English facilitates communication in a country where more than 500 languages are spoken.

Sociology is the study of the development, structure, and function of humans in society. As more West Africans move from rural to urban areas in search of better health care, sanitation, and work, social norms are changing. West Africans have had to adapt to new ways of life in an urban setting. One example of this is the shift from living with extended families to living in nuclear families. This often leads to cultural changes and lost traditions.

ANALYZING What examples of change in social norms do you see in your own society?

griots traditional oral historians, storytellers, singers, and musicians of West Africa

Regardless of ethnic group or language, the diverse people of West Africa value religion and strong family ties. The dominant religions practiced in the subregion are Islam, Christianity, and the traditional animist religions of Africa. Although followers of different religions often coexist peacefully, conflict sometimes occurs.

Education in West Africa is inconsistent. Literacy rates in West Africa are one of the lowest in the world. As of 2015, for example, the literacy rate in Sierra Leone was only 48 percent. In Ghana, however, education has been a priority, and government spending on education has steadily increased since the 1960s. As of 2015, the literacy rate in Ghana was 77 percent, one of the highest in the subregion.

Poverty is a key factor in determining access to health care in West Africa. Health care is uneven and limited. As a result, infant mortality rates are high throughout the subregion. In Sierra Leone, for example, 92 children out of every 1,000 die before they reach the age of five. High death rates are caused by poor sanitation, health conditions, and inadequate nutrition.

Family and the Status of Women

Many people in West Africa still live in extended families made up of parents, children, grandparents, and sometimes aunts, uncles, and cousins. Generally in these traditional societies, men and women have different roles. Traditionally, women look after children and the home while men earn a living to support the family. In the cities, however, the nuclear family—made up of husband, wife, and children—is rapidly replacing the extended family. The role of women is also beginning to change.

Women in West Africa play a vital role in the family. In recent years, men have begun to move away from home to find jobs, leaving women and children behind. Yet, in traditional societies, there are few jobs available for women that would provide income to care for a family. Things are beginning to change, however. Women are beginning to enter into professions and establish small businesses like selling vegetables in local markets or opening a local beauty salon. These small businesses give women control over their own lives and pave the way for participation in other aspects of their society such as politics. For example, the Women Peace Security Network Africa (WIPSEN-Africa) was established in 2006 in Ghana to promote women's strategic participation and leadership in peace and security governance in Africa.

The Arts

Much of the art in West African culture expresses religious beliefs. Music and dance are a part of everyday life, and entire communities participate. Dancers wear masks honoring specific deities, the spirits of their ancestors, or to honor special occasions, such as a birth. West African music has become popular around the world and has influenced contemporary Western music. Kanye West and Sting are only two of the popular Western musicians who have borrowed from African music. In fact, the entire blues and jazz tradition of the United States has its roots in the music enslaved Africans brought with them. **Griots** are an ancient tradition and an important part of the history of art in West Africa. They are oral historians, storytellers, singers, and musicians. Their songs, music, and stories help the people remember their cultural heritage.

Another art form is weaving. The Ashanti of Ghana are expert weavers, known for kente cloth. Kente is a brightly colored cloth consisting of bands of fabric sewn together. The Ashanti king and royal family have worn this cloth for centuries. Today people around the world also wear kente. Kente has become a symbol of Africa for many African Americans.

☑ **READING PROGRESS CHECK**

Evaluating What ancient tradition can be found in West Africa today?

Economic Activities

GUIDING QUESTION *What natural resources are available in West Africa for economic development?*

People in West Africa earn their living in many different ways. Some West Africans run small businesses selling locally made products such as baskets, art, and jewelry. Using **e-commerce**, or buying and selling on the Internet, people sell their products to customers around the world. The Internet broadens the market for locally made products. Agriculture, however, is the main economic activity of more than half of the people in West Africa.

Many of the people in West Africa live by subsistence farming in which people focus on growing food to feed themselves and their families. These people are largely self-sufficient, growing their own food and building their own houses. They are generally not involved in a wage-earning economy.

A small percentage of the population works at commercial farming. Commercial farmers produce crops on a large scale. These cash crops are grown and sold for profit. One important cash crop in West Africa is cacao, which is used to make cocoa powder and chocolate. Côte d'Ivoire is the world's largest exporter of cacao, as well as a major exporter of palm oil and coffee beans. Cacao trees are native to Central and South America, and the crop was introduced to West Africa during the colonial period. Small family-run farms have the greatest success in growing cacao. A very small percentage of people along the coast also work in commercial fishing.

Although difficult and risky, mining is also an important economic activity in the region. Coal, gold, uranium, and natural gas are some of the natural resources found in West Africa. Oil has also been discovered in some countries. Immense oil reserves make Nigeria the region's only member of the Organization of Petroleum Exporting Countries (OPEC). Nigeria's economy has been heavily dependent on this single resource, but Nigerian leaders have tried to diversify the economy in recent years. Diamonds, another important resource for West Africa, have a sordid history. Conflict diamonds have financed wars in Sierra Leone, Liberia, and Côte d'Ivoire.

Trade groups have helped increase trade within the region. Various countries have formed regional trade associations, such as the Economic Community of West African States (ECOWAS). This group of 15 countries was founded in 1975 to promote economic integration across the region. ECOWAS is expanding trade within the region and outside of Africa. In 2012 ECOWAS signed an agreement with China for cooperation in infrastructure development, trade, and investment.

e-commerce buying and selling on the Internet

✓ **READING PROGRESS CHECK**

Evaluating What economic activities can be found in West Africa?

LESSON 2 REVIEW

Reviewing Vocabulary
1. *Describing* Write a paragraph describing how griots contribute to the culture of West Africa.

Using Your Notes
2. *Analyzing* Use your graphic organizer about the human geography of West Africa to write a paragraph about the complexity and difficulties the subregion faces.

Answering the Guiding Questions
3. *Drawing Conclusions* How does the legacy of colonialism affect governments in West Africa today?

4. *Evaluating* How is a booming population affecting life in West Africa?

5. *Making Connections* How is life in West Africa a mix of the ancient and the modern?

6. *Listing* What natural resources are available in West Africa for economic development?

Writing Activity
7. *Informative/Explanatory* In a paragraph, discuss how the people of West Africa, both past and present, have shaped the subregion.

Conflict
DIAMONDS

Diamonds are one of Africa's most important natural resources. Today, about $8.5 billion annually worth of rough diamonds are from Africa. Groups have found ways to exploit this natural resource for their own gain. Conflict diamonds, also known as blood diamonds, are ones that are mined in rebel-controlled areas. The miners, and those who lived near the mines, have often been threatened, harmed, or even killed.

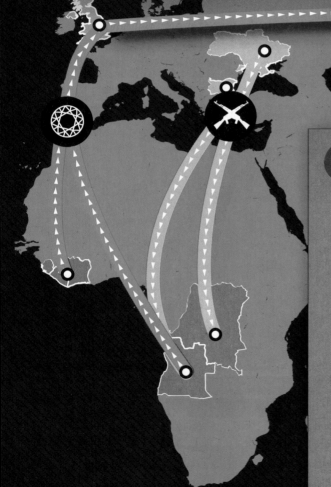

1. Funding War with Diamonds

Workers use shovels, picks, and their own hands to dig diamonds out of tunnels, open pits, and mines. In Angola, Democratic Republic of the Congo, Côte d'Ivoire, Liberia, Sierra Leone, and other African countries, rebel groups use force to steal these diamonds or the mines themselves. These same groups then sell these rough, uncut diamonds and use the profits to purchase weapons from Eastern European states like Ukraine and Bulgaria, or to fund civil wars and government takeovers.

2. Selling Conflict Diamonds

Some conflict diamonds are smuggled into neighboring African countries. These diamonds are then mixed with diamonds hat have been mined legally, making the illegal gems undetectable. Buyers in London and other large diamond markets then purchase the diamonds unaware of their source.

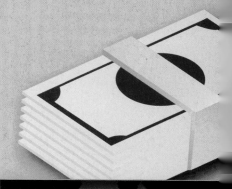

3. Cutting Diamonds

Diamonds are then cut and polished and sold to consumers throughout the world. Some of these diamonds are used in jewelry while others are used for industrial purposes.

Establishing the Kimberley Process

The United Nations established the Kimberley Process to reduce the amount of conflict diamonds. In this process a diamond must be certified that it is "conflict-free" before it can be sold.

Making Connections

1. **Places and Regions** How does the demand for diamonds in other parts of the world fuel civil war and conflict in Africa?

2. **Human Systems** Why has it been difficult to stop the flow of conflict diamonds?

3. **The Uses of Geography** Using what you know about the trade in conflict diamonds, write a one-page letter to the United Nations suggesting additional steps that could be taken to address the issue of conflict diamonds.

*Interact with **Global Connections** Online*

Reading **HELP**DESK

Academic Vocabulary

- **goal**
- **demonstrate**

Content Vocabulary

- **carrying capacity**
- **erosion**
- **fishery**

TAKING NOTES: *Key Ideas and Details*

IDENTIFYING As you read about the people and their environment in West Africa, use a graphic organizer like the one below to identify problems and the different ways in which groups are trying to address these problems.

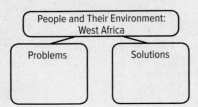

LESSON 3
People and Their Environment: West Africa

ESSENTIAL QUESTION • *How do physical systems and human systems shape a place?*

IT MATTERS BECAUSE

West Africa is faced with multiple challenges. These include drought, increasing population, deforestation, and the degradation of water resources. These challenges have contributed to poverty, hunger, conflict, and war. Finding solutions to these problems may help to reverse the losses and end the crises that afflict the countries of West Africa.

Managing Resources

GUIDING QUESTION *What challenges resource management in West Africa?*

The poor health and wellness conditions of many West Africans makes survival a chief **goal**, or aim—well ahead of environmental concerns. And yet, many of the environmental issues contribute to poor health and wellness. The ongoing water and sanitation crisis kills large numbers of people and limits economic development. The poor, especially women and children in rural areas, are most at risk. Water and sanitation problems also threaten the growing informal settlements near cities where the poor have migrated seeking work.

Sanitation problems include the lack of toilets and polluted water supplies. These problems result in illness, premature death, and loss of productivity while people are sick or looking for health care. Some of the illnesses that result from poor sanitation are infant diarrhea, cholera, malaria, and respiratory infections. Poor sanitation also costs countries money through such things as health care and lost hours of work. For example, poor sanitation costs Benin $104 million per year. Ghana loses $209 million per year, and the cost to Nigeria is $3 billion per year. Tackling these problems by providing better health care, education, and sanitation would help improve the quality of life for everyone in the subregion.

One obstacle to change and economic efficiency in some countries of West Africa is corruption. Corruption has a negative impact on social and economic development as well as on the environment and resource management. Corruption in West African countries has often led to violent political conflict. These conflicts are financed by the unlawful sale of weapons or the illegal extraction and sale of natural resources such as diamonds, gold, and timber. The use of money and natural resources to fund these conflicts hinders economic development. Fortunately, national and international groups are making efforts in West African countries to address corruption, health, sanitation, and access to education and health care.

These are not the only problems West Africans face, however. Difficulty in meeting the needs of the people also leads to the mismanagement of natural resources. Such mismanagement is part of a legacy of colonialism, during which massive numbers of people were forced to migrate from mainly rural areas to other rural areas or even urban areas. This resulted in the abandonment of long-held practices learned over generations that were no longer relevant in the new areas and which ultimately caused environmental harm in some places.

Many of the countries in West Africa depend on natural resources for economic development. However, the failure to properly manage soils, forests, lakes, rivers, and wildlife will only lead to further problems. There is great concern that the mismanagement of resources will lead to food insecurity for large numbers of West Africans.

☑ READING PROGRESS CHECK

Explaining How does poor sanitation affect the people and countries of West Africa?

Human Impact

GUIDING QUESTION *What environmental challenges does West Africa face today?*

The population of countries south of the Sahara is increasing at a rate of about 3 percent a year. While that may not sound high, it means—if the rate stays the same—that the population will double in about 22 years. This rate of population

Ghana has the highest rate of deforestation in Africa.

▼ CRITICAL THINKING

1. Analyzing Why is it important to stop deforestation?

2. Identifying Cause and Effect How has the issue of management of resources affected deforestation?

"There are constant arguments over territory between fishermen. . . . It's difficult to determine boundaries on water, yet the gendarmes [police from Cameroon and Chad] always come after us and seize our fishing nets and traps and we have to pay heavily to get them back."

—Muhammad Sanusi, a fisher in Dogon Fili, "Lake Chad fishermen pack up their nets," BBC News, January 15, 2007

DBQ *IDENTIFYING CAUSE AND EFFECT* How has drought and desertification affected fishers in some areas?

carrying capacity the maximum population of any given species that an environment can sustain

erosion the wearing away of soil

fishery a place for catching fish; the fishing industry

increase is one of the highest in the world. Such rapid growth has placed tremendous pressure on the **carrying capacity** of the land. Carrying capacity is the maximum population of a species the environment can sustain without depleting or degrading the resources available. The growth in population has already led to an environmental crisis in many parts of Africa. One hopeful idea is that humans can increase the carrying capacity of the land by employing more sustainable technologies. Examples include solar cooking technology that reduces the need to burn wood and small-scale irrigation pumps that work like bicycle pedals and allow subsistence farmers to irrigate their fields.

In addition to the heavy load on the environment caused by the increase in population, the breakdown of traditional systems of agriculture is also causing environmental deterioration on a massive scale. This deterioration, in turn, contributes to widespread poverty, malnutrition, and famine. In some countries, the breakdown in farming is contributing to political instability and civil war.

Overgrazed and overworked soils make farming difficult and can damage the long-term health of the soil. The use of heavy farm machinery, frequent tilling, and the clearing of forests have caused **erosion** and desertification. Together with the mismanagement of the land, drought and infrequent rainfall in West Africa also cause desertification. Once desertification begins, it puts pressure on the people that live in the area. Agriculture and livestock suffer, as well as the area's natural biodiversity. All these problems result in food production that has fallen short of the needs of the West African population.

Another problem facing West Africa is deforestation. Logging creates serious consequences, but the lumber industry maintains a relatively small output. Coastal countries with rain forests, such as Ghana and Côte d'Ivoire, export significant amounts of valuable hardwoods, such as teak, ebony, African walnut, and rosewood. Forests are also cut down to make room for farming, mining, and settlement. Unfortunately, most of this logging is not done sustainably, which threatens the natural biodiversity of the area and also contributes to desertification and erosion.

Fishing is an important industry in West Africa. **Fisheries** in the West African Marine Ecoregion (WAMER) generate about $400 million a year, making them a key source of revenue. In Senegal alone, over 600,000 people depend directly on fishing and related industries. Here, individual fishers catch over 80 percent of the fish while industrial fishing fleets catch about 20 percent. However, in Senegal and other West African countries, unsustainable fishing practices are leading to dwindling stocks in the ocean fisheries. More and more boats are searching for fewer fish. The competition is leading to even more destructive methods of fishing. These methods include the use of chemicals, dynamite, and bottom trawling. Pollution has also negatively affected fishing in countries such as Nigeria. Oil spills in the Niger Delta, where the local people are fishers and farmers, has led to dwindling fish stocks and reduced soil quality.

Lake Chad is another resource that has been affected by human activity. Drought, desertification, and climate change have caused Lake Chad to shrink. But there are other causes for the changes to the size of the lake as well. The need for water to irrigate farmland as well as for consumption has led to the damming of the rivers that supply Lake Chad. This has also contributed to the reduction in the size of the lake. The shrinking lake has led to smaller fish stocks and, in turn, to arguments over territory between fishers. Conflicts also arise between farmers and fishers over the use of the water from Lake Chad for irrigation. At times, these arguments have led to violence.

☑ **READING PROGRESS CHECK**

Summarizing How does human activity impact the carrying capacity of the land?

Addressing the Issues

GUIDING QUESTION *What steps are being taken to combat these environmental challenges?*

West African countries face many environmental problems. These include the degradation of land, desertification, drought, pollution, and the reduction of fish stocks and water resources that reduce the availability of food for both humans and animals. These problems increase the likelihood for conflicts to arise over competition for natural resources. Many individuals, cities, countries, and national and international organizations are working to combat the problems the subregion is facing. Some farmers, for example, have started to practice conservation farming, a land-management technique that helps protect farmland. By planting different crops where they will grow best, farmers protect farmland. Better fertilizers, seeds, and irrigation practices have increased crop yields and production.

Some farmers have begun to revive traditional land-management practices to reverse desertification and have **demonstrated** that these practices can be effective. Such practices include using planting pits to reclaim severely dry land. Additionally, help has come from Oxfam International, a nongovernmental organization (NGO). Oxfam International is a confederation of 17 organizations working together in more than 90 countries around the world to end poverty. Together with Oxfam, farmers in the Yatenga Province of Burkina Faso are building stone contour bunds, or embankments, to control rainwater runoff. This technique has rehabilitated about 740,000 acres (300,000 ha) of land. The additional food produced on the land helps feed about 500,000 people. Other farmers in the region have started to plant native plants in and around crop fields. These food-producing, drought-resistant trees and shrubs help improve soil fertility, retain moisture, and reduce erosion.

Lack of sanitation and access to clean water are major problems for some areas in West Africa.

▲ **CRITICAL THINKING**

1. Analyzing How can poor sanitation affect a community?

2. Making Connections How would people benefit from better management of water resources?

demonstrate to clearly show

In Nigeria, President Goodluck Jonathan is attempting to address water shortages in the state of Benue through the Greater Makurdi Water Works project. The water treatment plant is designed to bring clean water daily to the people of Benue. However, corruption is hampering this and other projects within the region. For example, in the case of the Makurdi Water Works project, the government council agreed to a water treatment plant with a capacity of 26 million gallons (98 million l). However, the governor of Benue state signed a contract for a treatment plant of half the capacity at the same price of full capacity, raising questions about corruption. Additionally, the plant and the state of Benue do not have the pipe structure necessary to supply the water. The people of Benue will continue to suffer with insufficient clean drinking water until more contracts are signed and the government spends additional money.

On a larger scale, the Global Water Initiative (GWI) works in various regions around the world to improve access to clean water and sanitation. In West Africa, GWI works with Burkina Faso, Ghana, and Senegal as well as Mali and Niger. One of the main objectives of the GWI in West Africa is to improve access to clean water and sanitation for vulnerable populations in the region. The group has six projects in West Africa that focus on these objectives in the countries surrounding the Gambia, Volta, and Niger River basins. In villages in Ghana where the GWI initiated projects, the rate of access to safe water has increased by an estimated 50–80 percent.

The United Nations (UN) is working on the use and availability of water in the subregion through the Integrated Water Resources Management (IWRM) project. The IWRM recognizes that water is key to economic and social development in West Africa and elsewhere. The project promotes the coordinated development and management of water, land, and related resources in order to maximize the economic and social welfare of people in the region without harming the environment.

Other projects around the region are working to improve access to clean water, thereby improving health and sanitation. Efforts by farmers and larger organizations to improve soil quality will help increase crops, thus reducing hunger. These improvements will encourage governments to make greater efforts to reduce corruption and to increase access to health care and education. These changes and others will help bring stability and security to the people of the region and reduce the incidences of conflict.

☑ **READING PROGRESS CHECK**

Evaluating In what ways are West Africans trying to combat their environmental problems?

LESSON 3 REVIEW

Reviewing Vocabulary
1. *Describing* In what ways has the human population put pressure on the land?

Using Your Notes
2. *Summarizing* Use your graphic organizer on the people and their environment in West Africa to write a summary of the problems the subregion faces and to propose possible solutions.

Answering the Guiding Questions
3. *Synthesizing* What challenges resource management in West Africa?

4. *Evaluating* What environmental challenges does West Africa face today?

5. *Classifying* What steps are being taken to combat these environmental challenges?

Writing Activity
6. *Informative/Explanatory* In a paragraph, write about the cause-and-effect relationship between people and the environment in West Africa.

Directions: On a separate sheet of paper, answer the questions below. Make sure you read carefully and answer all parts of the questions.

Lesson Review

Lesson 1

1 *Explaining* Why are mangrove swamps an important part of the ecosystem of West Africa?

2 *Evaluating* How successful was the damming of the Volta River at providing benefits for people in the area?

3 *Comparing and Contrasting* Describe how biomes and climates change as one moves from north to south in West Africa.

Lesson 2

4 *Identifying Cause and Effect* How did the colonial boundaries drawn by European countries at the Berlin Conference affect people in the region?

5 *Assessing* Why has the discovery of oil and gas reserves in Nigeria made the country more vulnerable?

6 *Summarizing* West Africa has one of the fastest rates of urbanization in the world. What are some of the reasons for this population trend?

Lesson 3

7 *Describing* Describe how the ongoing sanitation and water crisis affects the economy and health of people in the subregion.

8 *Making Connections* How does the clearing of forests contribute to desertification in West Africa? What are some of the economic and social consequences of this process?

9 *Problem Solving* What steps are some West African farmers taking to manage resources more effectively?

Critical Thinking

10 *Hypothesizing* How might diamonds purchased in Europe fuel political and social instability in West Africa?

11 *Drawing Conclusions* Why do you think e-commerce has become a popular way for West Africans to market locally produced goods?

21st Century Skills

Use the chart to answer the questions that follow.

People Vulnerable to Food Insecurity, March 2012	
Country	**Number of People Food Insecure**
Burkina Faso	2,852,280 food insecure
Chad	3,622,200 food insecure, of which 1,180,300 severe
Gambia	713,433 in areas at risk
Mauritania	700,000 food insecure, of which 290,000 severe
Niger	6,112,089 food insecure, of which 1,916,855 severe
Senegal	850,000 food insecure

Source: www.oxfam.org

12 *Geography Skills* Which country in West Africa has the highest number of people vulnerable to severe food insecurity? What is the number?

13 *Creating and Using Graphs, Charts, Diagrams, and Tables* How does food insecurity in Senegal compare to Burkina Faso? Why does the table describe Gambia as at risk?

College and Career Readiness

14 *Problem Solving* Imagine that you have been hired by Oxfam to help identify potential food shortages in West Africa. Choose a country from the region. Use the Internet to research food production and population growth in that country. How fast is the population growing? Does the production of food meet the needs of the growing population? Write a brief response assessing whether or not the country faces a food shortage.

Need Extra Help?

If You've Missed Question	**1**	**2**	**3**	**4**	**5**	**6**	**7**	**8**	**9**	**10**	**11**	**12**	**13**	**14**
Go to page	532	532	534	536	537	538	544	546	547	534	541	549	549	547

Directions: On a separate sheet of paper, answer the questions below. Make sure you read carefully and answer all parts of the questions.

DBQ Analyzing Primary Sources

Use the document to answer the following questions.

The ongoing food crisis in West Africa has many causes and consequences. Below are the observations of a journalist who has written about Oxfam's work in West Africa.

PRIMARY SOURCE

"Some of my most vivid images of the Sahel food crisis are of hands and feet.

When I traveled to Senegal recently to document Oxfam's work on the crisis, I met with women farmers who lost their last harvest to erratic rains. Several times I noticed an injury to a woman's foot or finger—usually something simple that without medical care had become so serious that it was disabling or worse: from the look of it, some would require amputation. But none of the women had any food stocks left and, with prices on the rise, none could afford to purchase enough for her family to eat, so a visit to the doctor was out of the question."

—Elizabeth Stevens, "Sahel Food Crisis: The Cost of Climate Change," October 2012

15 *Interpreting* How does the author's observation of women farmers in Senegal help you understand the title of the article?

16 *Making Connections* Describe how climate change, inadequate medical care, and food shortages combine to create problems in productivity.

17 *Analyzing Primary Sources* Why do you think the author begins her description of the food crisis with the image of hands and feet?

Applying Map Skills

Use the Unit 6 Atlas to answer the following questions.

18 *Places and Regions* Use your mental map of West Africa to describe the length and location of the Niger River and the Niger Delta.

19 *Human Systems* How has rapid population growth caused cities to spread out into the countryside in West Africa? Give examples.

20 *Environment and Society* What are some of the freshwater resources in West Africa? How important are they to the economies of the subregion?

Exploring the Essential Question

21 *Analyzing* Use the Internet and library resources to research the colonial history of a country in West Africa. Who colonized the region? Did colonial boundaries cut across ethnic lines? What happened when colonial rule ended? Create a time line showing how colonial rule affected people in the country you chose.

Research and Presentation

22 *Research Skills* The Niger Delta is one of the world's most important ecosystems. However, the discovery of oil in the 1950s has created problems. Use the Internet to research the environmental and social impact oil has had on the Niger Delta. How severe is the problem of pollution? How have oil companies and the Nigerian government responded? Write your conclusions in paragraph form and share them with the class.

Writing About Geography

23 *Informative/Explanatory* Use the Internet to research challenges facing a city in West Africa. Using standard grammar, spelling, sentence structure, and punctuation, write a two-page essay describing the ways rapid urbanization creates pressure on infrastructure.

Need Extra Help?

If You've Missed Question	**15**	**16**	**17**	**18**	**19**	**20**	**21**	**22**	**23**
Go to page	550	550	550	476	476	476	537	533	538

Equatorial Africa

ESSENTIAL QUESTION · *How do physical systems and human systems shape a place?*

A chief from Ndian and Koupe-Manengouba in Cameroon

Thomas Imo/Photothek/Getty Images

Why Geography Matters
South Sudan: Independence and Conflict

Lesson 1
Physical Geography of Equatorial Africa

Lesson 2
Human Geography of Equatorial Africa

Lesson 3
People and Their Environment: Equatorial Africa

Geography Matters...

Equatorial Africa lies in the heart of Africa. Great expanses of rain forest teem with wildlife. Through the forest flows the mighty Congo River—one of the longest in the world—and its many tributaries. These waterways form a vast transportation network that the people of Equatorial Africa have used for centuries.

Along these streams are many villages. Most of the people of these villages make their living as farmers. Others work in the region's many mines, extracting gold, diamonds, and other minerals. Still others live and work in huge port cities, where the rivers of the heart of Africa flow into the Atlantic Ocean.

South Sudan: independence *and* conflict

South Sudan declared its independence from Sudan in 2011, after centuries of ethnic and religious conflict. The celebration of newly won freedom did not last long, however. This break only fueled tensions, as both countries continue to dispute borders, which affects in part the economic survival of South Sudan.

Sudan and South Sudan Conflicts

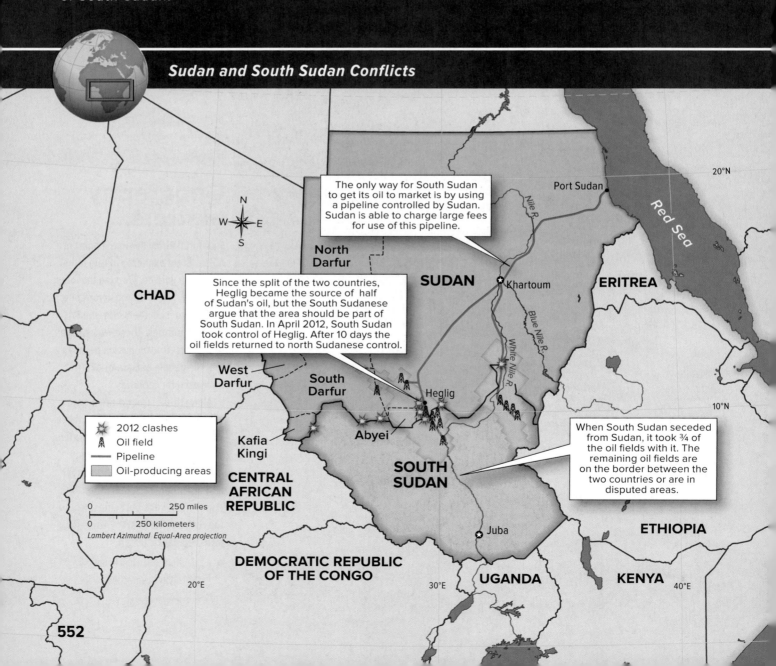

The only way for South Sudan to get its oil to market is by using a pipeline controlled by Sudan. Sudan is able to charge large fees for use of this pipeline.

Since the split of the two countries, Heglig became the source of half of Sudan's oil, but the South Sudanese argue that the area should be part of South Sudan. In April 2012, South Sudan took control of Heglig. After 10 days the oil fields returned to north Sudanese control.

When South Sudan seceded from Sudan, it took ¾ of the oil fields with it. The remaining oil fields are on the border between the two countries or are in disputed areas.

Legend:
- 2012 clashes
- Oil field
- Pipeline
- Oil-producing areas

0 — 250 miles
0 — 250 kilometers
Lambert Azimuthal Equal-Area projection

CHAD · North Darfur · West Darfur · South Darfur · Kafia Kingi · CENTRAL AFRICAN REPUBLIC · Abyei · Heglig · SUDAN · Khartoum · ERITREA · Red Sea · Port Sudan · Nile R. · Blue Nile R. · White Nile R. · SOUTH SUDAN · Juba · DEMOCRATIC REPUBLIC OF THE CONGO · UGANDA · KENYA · ETHIOPIA

20°N · 10°N · 20°E · 30°E · 40°E

What is the history of Sudan?

The modern country of Sudan emerged from a history that dates back thousands of years to the Kush civilization in the fourth century B.C. The kingdoms of Kush, Nubia, Axum, and Egypt vied for power in the region until the A.D. 600s, when the Arab Muslim conquest of Egypt and what is now northern Sudan began. By the mid-1300s, Arab Muslims had dethroned the last Christian king of Nubia, replacing him with a Muslim ruler. The last Christian kingdom in Sudan fell in the early 1500s.

In the 1800s Egyptian forces—and later, British and Egyptian forces—conquered Sudan. In 1899 the two countries consolidated their rule of Sudan. In 1916 they annexed Darfur to Sudan. Even though Britain and Egypt shared control of Sudan, the British sought to limit Egyptian power. Sudan was essentially a British colony. After decades of protests and uprisings, Britain recognized Sudan as an independent, sovereign country in 1956.

1. **Human Systems** How did other civilizations and countries influence Sudan's history?

What conditions led to the country splitting in two?

Civil war broke out in 1983 when the Sudanese government tried to enforce Islamic law throughout the country. The seeds of this conflict were planted centuries ago. The northern part of the country was dominated by Arabic-speaking Muslims. In the southern part of Sudan, however, the majority of the people were Christians or animists and spoke a variety of African languages. Differences had long brewed between the northern and southern regions of Sudan. Under British colonial rule, this tension increased. Sudanese Arabs in the north racially discriminated against the black Sudanese farmers in the south. In response to the attempt to impose Islamic law on non-Islamic southern Sudan, leaders in the south organized the Sudan People's Liberation Movement. Civil war and strife continued for decades until a peace agreement was signed in 2005.

Finally, after nearly 4 million people had been displaced and nearly half that number had died, the vast majority of registered voters in southern Sudan voted for secession from Sudan in January 2011. The Republic of South Sudan was born, but independence did not solve all the region's problems.

2. **Places and Regions** In what ways did northern Sudan differ from southern Sudan?

What challenges does the new country of South Sudan face?

Refugees who fled Sudan during the civil war have returned to their homeland, but the country faces many challenges. One hurdle is the country's devastated economy, which is linked to its landlocked status. After South Sudan seceded, economic and political conflict erupted over rights to its oil fields, because the pipeline to export oil runs through Sudan. South Sudan struggled to pay the fees to use the pipeline. In 2012 leaders from both countries met to sign a series of agreements. One part of the agreement would enable South Sudan to export oil through the pipeline in Sudan.

Other challenges have slowed efforts to rebuild. Armed rebels and militias and warring ethnic groups continue to challenge efforts to promote economic and political stability. Poverty is widespread but could be reduced or controlled if both countries cooperate and make economic decisions that will benefit the region.

3. **Places and Regions** How can Sudan and South Sudan work together to achieve economic growth and strength for the region?

THERE'S MORE ONLINE

CONNECT with the lives of refugees in Sudan • *EXAMINE* a map of the Darfur crisis

networks

There's More Online!

☑ **IMAGE** Fishing at Boyoma Falls

☑ **INFOGRAPHIC** Anatomy of the Rain Forest

☑ **MAP** Physical Geography: The Congo Basin

☑ **INTERACTIVE SELF-CHECK QUIZ**

☑ **VIDEO** Physical Geography of Equatorial Africa

LESSON 1
Physical Geography of Equatorial Africa

Reading HELPDESK

Academic Vocabulary

- navigable
- abundant

Content Vocabulary

- basin
- montane
- canopy
- understory

TAKING NOTES: *Key Ideas and Details*

IDENTIFYING Use a graphic organizer like the one below to take notes about landforms and water systems in Equatorial Africa.

Landforms	Water Systems

ESSENTIAL QUESTION • *How do physical systems and human systems shape a place?*

IT MATTERS BECAUSE

Equatorial Africa is a huge area that displays some of the most stunning landscapes on the planet. One of the world's great rivers courses through the heart of this region of plains, hills, mountains, and dense rain forests. It is a land of abundant natural resources and exotic wildlife. Everything from active volcanoes to lagoons can be found in this hot and humid tropical subregion.

Landforms

GUIDING QUESTION *What huge landform covers most of Equatorial Africa?*

Equatorial Africa is another of the five subregions that make up the larger region of Africa south of the Sahara. It is located on and near the Equator in central Africa. This tropical subregion is also called Central Africa or the Heart of Africa. Much of the land is covered by thick rain forest and is home to some of Africa's most famous and colorful wildlife.

The subregion covers a huge area. It stretches from about ten degrees north to ten degrees south of the Equator and covers about 2.6 million square miles (6.7 million sq. km)—nearly the size of Australia. It includes the countries of the Democratic Republic of the Congo, Cameroon, Central African Republic, Gabon, Equatorial Guinea, Republic of the Congo (or Congo), South Sudan, and the southern part of Chad. The island country of São Tomé and Príncipe, located in the Gulf of Guinea, is also part of the subregion. The Democratic Republic of the Congo, at about 1.5 million square miles (3.9 million sq. km), is by far the largest country in Equatorial Africa.

This large subregion also has a diverse landscape. The Atlantic coastal region of the Republic of the Congo and Gabon is mostly a low plain with lagoons and beaches, while the coast of Cameroon is mountainous. The two islands that comprise São Tomé and Príncipe are both extinct volcanoes. Each island is steep in the west and gradually slopes to the northeast.

Equatorial Africa is dominated by the Congo Basin. A **basin** is an area of land that is drained by a river and its tributaries. Water from rainfall in the Congo Basin flows downhill and collects into streams. These streams meet with larger streams, which flow into rivers that empty into the Congo River. The Congo Basin is the second-largest river basin on Earth. Most of the basin is flatland covered by rain forest.

The Congo Basin is surrounded by higher country. To the southeast lie the Congo highlands. These highlands include plateaus and mountains. South of the Congo Basin is a high, flat area called the Katanga Plateau. Southwest of the Congo Basin lie the highlands of Gabon and the Republic of the Congo. This is a region of low hills, plateaus, and mountain ranges, including the Chaillu Mountains.

North of the Congo Basin lies another plateau. It makes up most of the Central African Republic. This plateau has an average elevation of about 2,000 feet (600 m). To its east, on the northeast side of the Congo Basin, highlands separate the Congo Basin from a plain of another drainage basin, that of the Nile River. The Ruwenzori Mountain range is found here. Covered with snow and cloaked in clouds, they are also called the "Mountains of the Moon." Moist air from the Indian Ocean creates the clouds that wrap around the Ruwenzori Mountains. The highest mountain in the region is Mount Stanley, named after the famous British explorer, Sir Henry Morton Stanley. The top of Mount Stanley, Margherita Peak, soars 16,763 feet (5,109 m) into the sky.

To the northwest of the Congo Basin lie the Gotel Mountains of western Cameroon. At the southern end of this range, Mount Cameroon, at 13,353 feet (4,070 m), is Cameroon's highest point. Mount Cameroon is an active volcano that lies just a few miles inland from the Atlantic coast.

basin an area drained by a river and its tributaries

✓ **READING PROGRESS CHECK**

Describing What is the Congo Basin?

GEOGRAPHY CONNECTION

The Congo Basin is covered by lush tropical rain forests, swamps, and plains.

1. ***PLACES AND REGIONS*** What are the major tributaries of the Congo River?

2. ***HUMAN SYSTEMS*** What might encourage human settlement in the Congo Basin? What might discourage it?

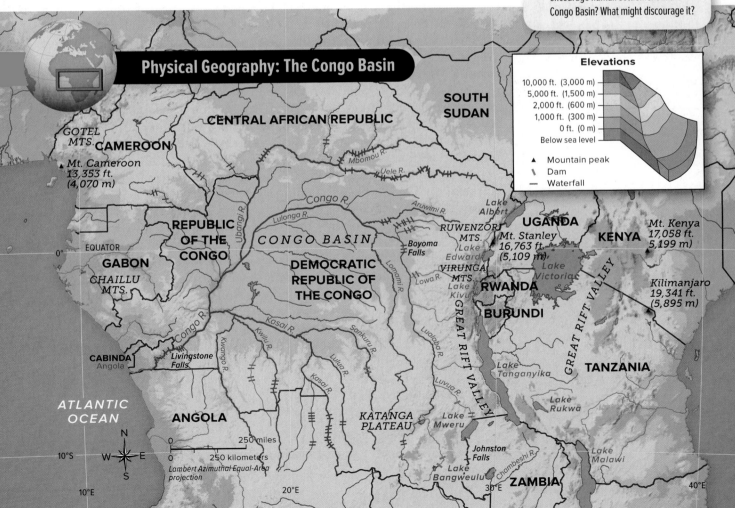

Physical Geography: The Congo Basin

Elevations
- 10,000 ft. (3,000 m)
- 5,000 ft. (1,500 m)
- 2,000 ft. (600 m)
- 1,000 ft. (300 m)
- 0 ft. (0 m)
- Below sea level
- ▲ Mountain peak
- \ Dam
- — Waterfall

Local people have been trapping fish at Boyoma Falls on the Lualaba River for centuries.

▲ CRITICAL THINKING

1. **Analyzing** Why are the fish traps so large?

2. **Inferring** What can you infer by the gender of the people fishing?

navigable able to be traveled by boat

Water Systems

GUIDING QUESTION *How is the Congo River central to life in Equatorial Africa?*

The part of the Atlantic Ocean that borders Equatorial Africa is called the Gulf of Guinea. It is the part of the ocean south of where Africa bulges to the west. A gulf is a part of the ocean that is partially enclosed by land. The island country of São Tomé and Príncipe lies in the heart of the gulf. European colonizers once used the islands to prevent the Africans they enslaved from escaping. Today, many people on the islands work on fishing boats.

On the mainland, five countries of Equatorial Africa have coastlines on the Gulf of Guinea. Fishing is important in all of them. Major ports include Libreville in Gabon and Pointe-Noire in the Republic of the Congo. At the southern edge of Equatorial Africa's coast, where the Democratic Republic of the Congo borders Angola, the freshwater of the world's mightiest river, the Congo, flows into the salt water of the Atlantic Ocean.

The Congo River serves as much of the border between the Republic of the Congo and the Democratic Republic of the Congo as it journeys to its outlet in the Atlantic Ocean. The Congo reaches the Atlantic through an estuary (EHS•chuh•WEHR•ee), or passage where freshwater meets seawater. The Congo's estuary is 7 miles (11 km) wide, and oceangoing ships can navigate its deep waters. The Congo River is nearly 3,000 miles (4,800 km) long, which makes it the second-longest river in Africa and the fifth-longest river in the world. More water flows in the Congo than any other river except for the vast Amazon River in South America.

The Congo River begins where the Lualaba and the Luvua Rivers meet. Downstream, the water flows through Boyoma Falls, near Kisangani. Boyoma Falls is a series of seven cataracts that stretch for 60 miles (97 km) down the river. For generations, local people have used the falls to catch fish in giant triangular traps.

Downstream from Kisangani, the Congo River flows in a great arc. Near Kinshasa, the Congo widens, creating a lake called Stanley Pool. Like Mount Stanley, the pool is named for the British explorer, Sir Henry Morton Stanley, who traveled the length of the Congo River in 1877.

The Congo River and its many tributaries form a large network of "natural highways" that are **navigable** by smaller boats. Some parts, however, have rapids and waterfalls that present serious obstacles to boat traffic and have limited the flow of goods and people along the full length of the river. The river plunges almost 900 feet (274 m) in numerous cataracts not far from where it meets the Atlantic Ocean.

The eastern border of Equatorial Africa is marked, in part, by the second-largest and second-deepest lake in the world. Lake Tanganyika is 420 miles (680 km) long and is nearly a mile deep in some places. Fish from the lake serve as a major source of food. Like the Congo, the huge Lake Tanganyika also serves as a means of transportation, as ferry boats cross its blue waters.

☑ READING PROGRESS CHECK

Explaining How is the Congo River an important transportation system?

SuperStock

Climates, Biomes, and Resources

GUIDING QUESTION *How is the rain forest structured and how does this affect plant and animal life?*

Equatorial Africa is called "equatorial" because it is located on and near the Equator. It is a region of high temperatures and tropical climates. Most of the subregion has a tropical rain forest climate. There is a smaller area that has a tropical wet/dry (savanna) climate. The highland areas that surround most of the Congo Basin experience highland or **montane** climates. The word *montane* comes from the Latin word for "mountain," and it simply means "of mountainous country."

montane referring to a mountainous area

Warm temperatures prevail in the tropical wet climate zone of Equatorial Africa. More than 60 inches (150 cm) of rainfall per year soak the land. Rainfall varies seasonally, but there is not a truly dry season. Daily, rain falls on an amazing number and variety of life-forms. The tropical rain forests of Equatorial Africa are among the biggest—covering about half of the subregion—and densest in the world.

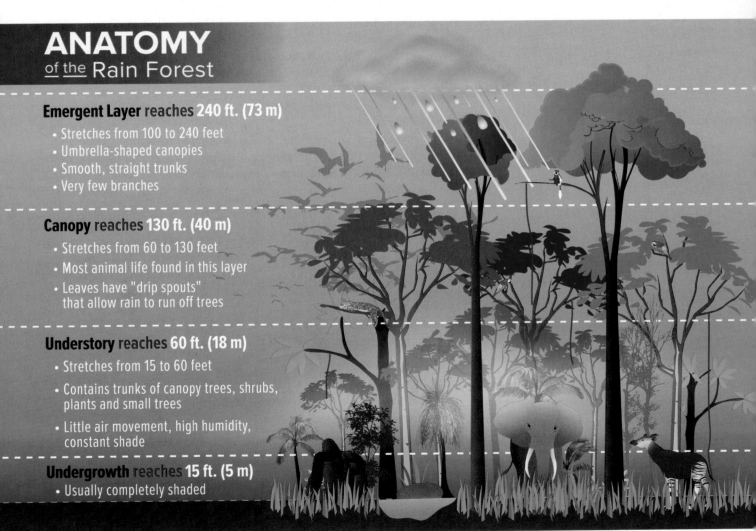

ANATOMY
of the Rain Forest

Emergent Layer reaches **240 ft. (73 m)**
- Stretches from 100 to 240 feet
- Umbrella-shaped canopies
- Smooth, straight trunks
- Very few branches

Canopy reaches **130 ft. (40 m)**
- Stretches from 60 to 130 feet
- Most animal life found in this layer
- Leaves have "drip spouts" that allow rain to run off trees

Understory reaches **60 ft. (18 m)**
- Stretches from 15 to 60 feet
- Contains trunks of canopy trees, shrubs, plants and small trees
- Little air movement, high humidity, constant shade

Undergrowth reaches **15 ft. (5 m)**
- Usually completely shaded

The rain forest is not a simple forest, but a complex web of life. Distinct forms of vegetation grow at different layers.

▲ **CRITICAL THINKING**

1. Hypothesizing Why might there be little growth on the rain forest floor?

2. Differentiating What is the difference between the understory and the canopy?

Rain forests are complex biomes. They are home to a wide variety of animal life, including birds, bats, rodents, monkeys, baboons, chimpanzees, elephants, wild boar, and okapi. A huge number and variety of insects also live there. Some of these, such as the tsetse fly and mosquito, carry diseases.

Rain forests have four basic layers. The topmost layer of the rain forest is the emergent layer, where trees stick out of, or emerge from, the canopy—the next highest layer. The **canopy** is a layer of trees and leaves with a maximum height of 130 feet (39.6 m). Orchids, ferns, and mosses grow among the branches of the canopy, which provide most of the plant food for rain forest animals. Below the canopy is the layer called the **understory**. Shrubs, ferns, and mosses grow together at this level of the rain forest, which rises 15 to 60 feet (4.6 to 18 m). The lowest layer is the undergrowth, where little sunlight penetrates and consequently few shrubs and herbs can grow. The soils in the tropical rain forest biome are typically not very fertile because the heavy rains leach, or dissolve and carry away, nutrients from the soil. Here animals feed mostly on insects.

Tropical grassland with scattered trees—known as savanna—covers much of southern Equatorial Africa. Rainfall is seasonal in this climate zone, with alternating wet and dry seasons. In the wettest areas, which are closest to the Equator, six months of almost daily rain is followed by a six-month dry season. Average annual rainfall is about 35 to 45 inches (90 to 115 cm).

Mineral resources are **abundant** through much of Equatorial Africa. In the Democratic Republic of the Congo alone, the following are mined: cadmium, cobalt, copper, gold, manganese, petroleum, silver, tin, and zinc. The value of these mineral resources is estimated to be about $25 trillion. The Congo Basin also holds major diamond deposits. Gabon, the Republic of the Congo, Cameroon, and Equatorial Guinea have oil and natural gas reserves.

Water is a major natural resource in the subregion. Equatorial Africa receives abundant rainfall. Yet controlling water for practical uses, such as irrigation and hydroelectric power, is difficult because the amounts of rainfall are irregular and unpredictable. These challenges, combined with a lack of financial support, result in untapped hydroelectric power potential in parts of the subregion.

Solar power is another renewable energy source that has been harnessed in Equatorial Africa. Rural electrification programs involve installing small-scale solar power systems. These continue to expand in some parts of the subregion.

The rain and the sun together give rise to the massive tropical rain forests, which may be Equatorial Africa's most important natural resource. Tropical rain forests are a rich source of food, medicine, fibers, oils, rubber, and, wood. Deforestation, however, is threatening many of the forests in the subregion.

☑ READING PROGRESS CHECK

Differentiating What is the difference between the tropical rain forest and tropical wet/dry areas of Equatorial Africa?

canopy top layer of a rain forest, where the tops of tall trees form a continuous layer of leaves

understory a lower layer of the rain forest

abundant present in large amounts; plentiful

LESSON 1 REVIEW

Reviewing Vocabulary
1. *Describing* Describe the layers of a rain forest using the words *canopy* and *understory*.

Using Your Notes
2. *Identifying* Use your graphic organizer to write a paragraph describing the major land and water features of Equatorial Africa.

Answering the Guiding Questions
3. *Identifying* What huge landform covers most of Equatorial Africa?

4. *Describing* How is the Congo River central to life in Equatorial Africa?

5. *Explaining* How is the rain forest structured and how does this affect plant and animal life?

Writing Activity
6. *Informative/Explanatory* Think about the physical features of Equatorial Africa. Write a paragraph describing how specific landforms affect the course of the subregion's rivers.

networks

There's More Online!

☑ **GRAPH** Democratic Republic of the Congo 2014 GDP

☑ **IMAGE** Cell Phone Use

☑ **IMAGE** Harvest in Central Africa

☑ **MAP** The Atlantic Slave Trade

☑ **MAP** Independent Africa

☑ **TIME LINE** Conflict in the Congo

☑ **INTERACTIVE SELF-CHECK QUIZ**

☑ **VIDEO** Human Geography of Equatorial Africa

LESSON 2
Human Geography of Equatorial Africa

ESSENTIAL QUESTION • *How do physical systems and human systems shape a place?*

Reading HELPDESK

Academic Vocabulary

- **intermittent**
- **hub**

Content Vocabulary

- **cash crop**
- **plantation**

TAKING NOTES: *Key Ideas and Details*

IDENTIFYING Use a graphic organizer like the one below to take notes about the human geography of Equatorial Africa.

Human Geography of Equatorial Africa

History and Government ⟶ ☐

Population Patterns ⟶ ☐

Society Today ⟶ ☐

IT MATTERS BECAUSE

Equatorial Africa reflects a wide variety of cultures. The arrival of Europeans fundamentally affected life for the Africans and led to a long struggle for freedom. The legacy of European colonization is still felt in the subregion. Today, Equatorial Africa is a rich mix of African and European influences, ancient and modern lifestyles, and challenges and hope.

History and Government

GUIDING QUESTION *What challenges does democracy face in Equatorial Africa?*

Thousands of years ago, people lived nomadic lifestyles across Africa. They eventually settled to grow food and livestock. By the time Europeans began to arrive in the 1400s, peoples of the region had established enduring cultures.

The first inhabitants of the Congo Basin are thought to be the Mbuti, once known as Pygmies, whose descendants still live in the region today. They were already living there when the great migration of Bantu-speaking peoples arrived in the region. The Bantu-speaking peoples, who were farmers and herders, had established settlements by A.D. 800. Although the origins of the Bantu and their migration routes are debated, historians believe that they spread across one-third of the continent. The Bantu founded the kingdoms of Kongo (Congo), Luba, and Lunda.

Slavery and European Colonization

When European explorers landed along Equatorial Africa's coasts in the late 1400s, they were interested in trade. They sought gold, ivory, textiles, and enslaved Africans. European ships carried away enslaved people from areas that are now the Republic of the Congo, Gabon, and Cameroon. Gabon later became a center of the slave trade.

Millions of people from the African interior were sold into slavery. Nzinga Mbemba, the king of Kongo, deplored the actions of some African rulers. He also complained to the king of Portugal about Portuguese slave merchants:

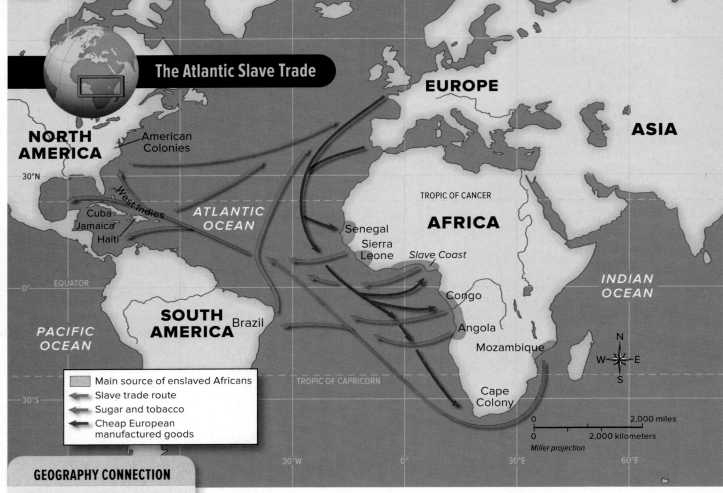

The Atlantic Slave Trade

NORTH AMERICA

American Colonies

30°N

West Indies

Cuba
Jamaica
Haiti

ATLANTIC OCEAN

EQUATOR 0°

PACIFIC OCEAN

SOUTH AMERICA

Brazil

30°S

EUROPE

ASIA

TROPIC OF CANCER

AFRICA

Senegal
Sierra Leone

Slave Coast

Congo

Angola

Mozambique

INDIAN OCEAN

Cape Colony

TROPIC OF CAPRICORN

30°W · 0° · 30°E · 60°E

N W E S

0 — 2,000 miles
0 — 2,000 kilometers
Miller projection

Main source of enslaved Africans
Slave trade route
Sugar and tobacco
Cheap European manufactured goods

GEOGRAPHY CONNECTION

The Atlantic slave trade began in the 1500s and continued into the 1800s.

1. **HUMAN SYSTEMS** What were the major destination points for enslaved people leaving Africa?

2. **THE USES OF GEOGRAPHY** How can maps help historians understand the past?

cash crop a farm crop grown to be sold or traded rather than used by the farm family

❝[They] seize upon our subjects, sons of the land and sons of the noblemen, and cause them to be sold; and so great, Sir, is their corruption and licentiousness that our country is being utterly depopulated.❞
—Nzinga Mbemba, the king of Kongo, quoted in *East Along the Equator*, 1987

Once captured and sold, enslaved Africans faced a terrible trip across the Atlantic Ocean as human cargo in a ship's hold. The passage from Africa claimed millions of lives. The loss of so many young people to the slave trade did great harm to the societies they left behind.

Large areas of Equatorial Africa were not colonized until the 1800s. Various obstacles had previously prevented colonization, including malaria and the Congo River cataracts. By the 1900s, however, European powers held colonies throughout the subregion. Colonizers were Spain, Germany, France, Belgium, Britain, and Portugal.

France gained control of what is now the Republic of the Congo in the late 1800s by engineering treaties with local rulers who wanted protection against Belgium, France's main rival in the region. The French changed the local economy into one based on income from resource extraction and growing **cash crops** for export. In 1878 King Leopold II of Belgium began establishing trade posts along the Congo River. He soon convinced other European powers to grant him control of what was then termed the Congo Free State and what is today called the Democratic Republic of the Congo. This territory and its people were considered the king's personal property. He treated the people terribly.

European rulers, traders, and missionaries promoted European culture, weakened traditional African cultures, and often treated Africans harshly. African village life was also disrupted by the replacing of locally centered agriculture with large commercial farms, called **plantations,** that grew cash crops. This harsh treatment of native peoples by European colonists led to the many struggles for independence across Africa during the twentieth century.

Independence

Resistance to colonial rule in Africa was growing by the mid-1900s. In one year alone, 1960, all of the countries of Equatorial Africa achieved independence. Thus, 1960 is remembered as "the year of independence."

People in most of the new countries experienced periods of ethnic strife, harsh rule, and human rights abuses after independence. In the Democratic Republic of the Congo, serious instability led to the reign of the dictator Mobutu Sese Seko from the late 1960s until the 1990s. Although the countries of the subregion were technically democracies, this period saw human rights abuses, one-party rule, and **intermittent** civil war.

Today the governments of Equatorial Africa have varying degrees of stability. Gabon is a stable republic. It has been developing its economy through the creation of a national park system (to promote ecotourism), developing industry, and educating its residents to work in financial services, education, and health care. In contrast, the Central African Republic has experienced military coups and periods of instability. Although today it is nominally a democracy, rebel groups, not the government, control land and people in the countryside. Abundant natural resources in the subregion have helped some countries achieve relative stability, but problems persist. Oil revenues helped Equatorial Guinea achieve greater stability, but most residents have a low standard of living, and the government has a history of human rights abuses.

✅ **READING PROGRESS CHECK**

Theorizing Why might the countries of Equatorial Africa have suffered harsh rule after independence?

Population Patterns

GUIDING QUESTION *How does geography influence where the people of Equatorial Africa live?*

Life for people in Equatorial Africa is influenced by the subregion's tropical climates, thick tropical forests, and tropical grasslands. Equatorial Africa is mostly rural, despite having prominent urban centers. The dense growth of natural vegetation in this equatorial region makes large-scale intensive plantation agriculture difficult. Most people exist by subsistence agriculture or raising cattle. In some places, coffee, rubber, and cacao are grown for export.

As a whole, Equatorial Africa is one of the least densely populated regions on the continent. The northern part of the Democratic Republic of the Congo, for example, is tropical forest. Large tracts of government forest reserves discourage cities, although there are scattered villages along the Congo and its tributaries. The southern part of the country, however, has millions of people. Kinshasa, the capital city, is the political, cultural, and economic **hub** of the region. It is home to nearly 10 million people. Just across the Congo River is Brazzaville, the capital of the Republic of the Congo, which is home to 1.3 million. In contrast, the largest city in São Tomé and Príncipe is the capital, São Tomé, with only 60,000 residents. Overall, Gabon is the most urbanized country in Equatorial Africa, with 86 percent of its population living in urban areas. At the other extreme is the newly independent South Sudan, where only 17 percent of the population lives in urban areas.

✅ **READING PROGRESS CHECK**

Explaining Why is much of Equatorial Africa thinly populated?

plantation a large commercial farm growing crops for export

intermittent occurring at irregular intervals; occasional

hub the center of an activity or region

Society and Culture Today

GUIDING QUESTION *How does conflict shape life in Equatorial Africa?*

Equatorial Africa contains hundreds of ethnic groups. The Democratic Republic of the Congo alone has more than 200 groups. The largest ethnic groups in that country are Bantu, peoples who speak one of the hundreds of Bantu languages. One indigenous group, the Mbuti, lives in the rain forest of eastern Democratic Republic of the Congo and is relatively unchanged by outside influences. In the northwestern part of the country, Fang and Bantu ethnic groups are the dominant ethnic groups.

The approximately 200,000 people of the islands of São Tomé and Príncipe are from many different ethnic groups. Many of the people are descendants of Portuguese settlers and people freed from slavery in Africa. These Portuguese Africans are known as *forros* and compose the country's elite. Much of the population is descended from groups that migrated to the islands in the late 1400s.

Along with hundreds of ethnic groups come hundreds of languages. About 700 local languages are spoken in the Democratic Republic of the Congo alone. French is widely spoken throughout most of Equatorial Africa, reflecting its colonial heritage. It is the official language of the Democratic Republic of the Congo and the Republic of the Congo. People from different sublanguage groups often communicate in pidgin, a simplified speech used among people who speak different languages.

Christianity, Islam, and indigenous religions are practiced in Equatorial Africa. Traditional indigenous religions are numerous and diverse, but they have many common elements. For example, most traditional religions profess a belief in the existence of a supreme being and a ranked order of deities. Animism, or the belief that natural or inanimate objects possess spirits, is also widespread.

European colonialism greatly influenced Equatorial Africa's religious practices. In many areas today, a majority of people observe different forms of Christianity. Roman Catholicism, brought by Belgian and French colonists and

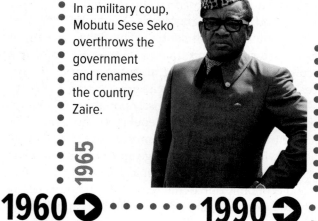

©Jacques Pavlovsky/Sygma/Corbis

TIME LINE ⌄

Conflict in the CONGO →

Since independence, the history of the Democratic Republic of the Congo has been one of corruption and civil war.

CRITICAL THINKING ▶

1. *Describing* How would you describe the political climate of the Democratic Republic of the Congo from 1960 to 2006?

2. *Analyzing* How has the history of the Democratic Republic of the Congo been affected by nearby countries?

1960 → .... **1990 →**

1965 — In a military coup, Mobutu Sese Seko overthrows the government and renames the country Zaire.

1960 — Republic of the Congo becomes an independent country.

1996 — Mobutu leaves the country for medical treatment; Tutsi rebels capture eastern Zaire.

1997 — Tutsi rebels oust Mobutu; Zaire becomes Democratic Republic of Congo, and Laurent Kabila becomes president.

missionaries, is common. About half of the population of the Democratic Republic of the Congo is Roman Catholic. The subregion is home to an increasing number of Muslims, as well.

Religion and family life are interwoven in Equatorial Africa. Some ethnic groups in the Democratic Republic of the Congo, for example, live according to a mix of indigenous and Christian beliefs. Many observe strict divisions of labor between men and women, depending on whether they farm or raise livestock. Several generations—and even people from different families— might share the same dwelling.

During the colonial era, religion dominated education systems in the subregion. After countries gained their independence, efforts were made to modernize education, but many rural areas still lacked sufficient support. Today, literacy ranges from about 37 percent in the Central African Republic to about 95 percent in Equatorial Guinea. Literacy among women is commonly lower than that of men.

Under Spanish rule, Equatorial Guinea achieved a high literacy rate as well as a decent health-care system. In general, however, Equatorial Africa's countries lack the resources to halt preventable diseases that Western countries have been able to stop. The lack of safe drinking water, the shortage of vaccines for curable diseases, and the rising rate of HIV infection and number of AIDS victims remain primary health concerns.

Ethnic conflicts continue to plague parts of Equatorial Africa. In the Democratic Republic of the Congo (once known as Zaire) the long dictatorship of Mobutu Sese Seko was followed by ethnic strife and one of the bloodiest conflicts since World War II. Ethnic conflict spiked after 1994, when the country received a heavy influx of refugees who were fleeing the fighting in the neighboring countries of Rwanda and Burundi.

✓ READING PROGRESS CHECK

Hypothesizing How might Equatorial Africa's ethnic and linguistic diversity be an obstacle to political stability?

(c)Themba Hadebe/AP Images, (b)Brennan Linsley/AP Images, (br)Lionel Healing/AFP/Getty Images

Analyzing PRIMARY SOURCES

Crime in Kinshasa

"Most reported criminal incidents in Kinshasa involve crimes of opportunity, which include pick-pocketing and petty theft. The majority of the crimes are committed by 'sheggehs,' who are generally homeless street children. Travel in certain areas of Kinshasa, Kisangani, Lubumbashi and most other major cities is generally safe during daylight hours, but travelers are urged to be vigilant against criminal activity that targets non-Congolese."

—U.S. Department of State, International Travel for U.S. Citizens

DBQ **ANALYZING PRIMARY SOURCES** Who are "sheggehs"?

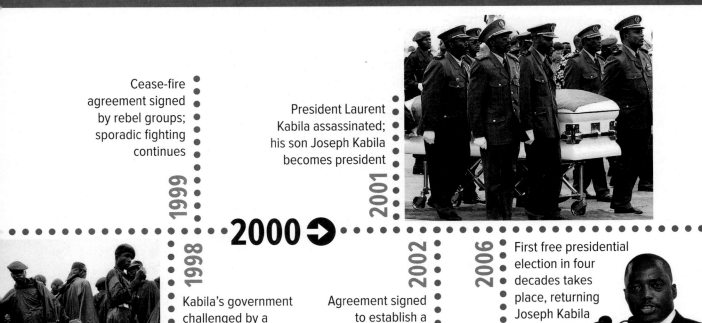

Cease-fire agreement signed by rebel groups; sporadic fighting continues

1999

President Laurent Kabila assassinated; his son Joseph Kabila becomes president

2001

2000 ➜

1998

Kabila's government challenged by a rebellion backed by Rwanda and Uganda

2002

Agreement signed to establish a government of national unity

2006

First free presidential election in four decades takes place, returning Joseph Kabila to office.

DEMOCRATIC REPUBLIC OF THE CONGO 2014 GDP

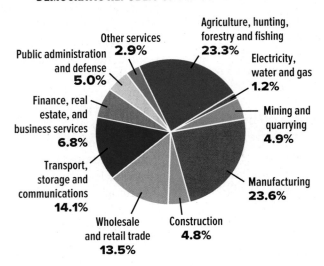

Other services **2.9%**

Public administration and defense **5.0%**

Finance, real estate, and business services **6.8%**

Transport, storage and communications **14.1%**

Wholesale and retail trade **13.5%**

Construction **4.8%**

Manufacturing **23.6%**

Mining and quarrying **4.9%**

Electricity, water and gas **1.2%**

Agriculture, hunting, forestry and fishing **23.3%**

Source: African Economic Outlook, www.africaneconomicoutlook.org

Together, agriculture and manufacturing are the most important economic activities in the Democratic Republic of the Congo, accounting for nearly half of the country's GDP.

▲ CRITICAL THINKING

1. *Analyzing Visuals* What is the country's smallest economic sector?

2. *Drawing Conclusions* How does the economy of the Democratic Republic of the Congo exemplify Equatorial Africa's colonial legacy while showing recent steps to modernize?

Economic Activities

GUIDING QUESTION *Why is the economic activity of Equatorial Africa so varied?*

The lives of people throughout Equatorial Africa are changing. The region is becoming more closely tied to the global economy. Like the countries in which they live, individuals face challenges, yet changing economic activities also offer new opportunities.

Farming is the main economic activity in Equatorial Africa. Some countries still depend on single-crop economies; others produce a variety of agricultural goods. Most farmers are subsistence farmers, providing primarily for the needs of their family or village. After they meet their own needs, farmers often sell or trade any extra harvest or livestock at a local market. Common crops are bananas, cacao, cassava (a root), corn, sweet potatoes, and yams.

A smaller percentage of the population works in commercial farming, in which farms produce crops on a large scale. These cash crops are grown and exported to global markets. Most commercial farms are large, foreign-owned plantations. They produce palm oil, peanuts, cacao, and sisal (a vegetable fiber used in rope, drywall and car interiors).

The colonial economic systems played an important role in the growth of commercial crops in Equatorial Africa. Today these same crops are the subregion's main agricultural exports. Most cash crops leave Africa to be processed elsewhere, just as during the colonial period. In the Democratic Republic of the Congo, crops and other goods are exported from Matadi, a port on the Congo River.

Logging has serious consequences, but the lumber industry maintains a relatively small output. Coastal countries with rain forests, such as Gabon, export significant amounts of ebony, mahogany, and natural rubber. Wood from the Okoume tree—found mainly in Gabon, the Republic of the Congo, and Equatorial Guinea—is famous as a good raw material for plywood.

Fishing also represents a portion of the subregion's economic activity. The Congo and its tributaries are also teeming with fish. At Boyoma Falls, the local fishers use conical baskets to trap fish as they pass through the rapids. The shallow ocean area near the coast are excellent fishing waters. Along the southwestern coast, commercial fishing vessels catch herring, sardines, and tuna for export. Few countries, however, build and support large commercial fishing fleets.

The subregion is rich in mineral resources. Equatorial Guinea, the Republic of the Congo, and the Democratic Republic of the Congo are all major oil producers. Mining of copper and other minerals has been important in the Democratic Republic of the Congo since the colonial period. In spite of rich mineral resources, however, many people in these countries do not benefit directly. Governments have often mismanaged the income from mineral wealth. In the Democratic Republic of the Congo, for example, Mobutu Sese Seko appropriated a huge fortune from the state-owned mining company, Gécamines, for his own private fortune. In other cases, foreign mine owners send their profits abroad.

Most of Equatorial Africa never developed manufacturing industries and lacks the infrastructure to process natural resources. Today many countries are receiving foreign loans to industrialize. Progress, however, is slow and unemployment is a major problem. In 2012, for example, the unemployment rate in the Republic of the Congo was a staggering 53 percent.

Demand for manufactured goods has increased, and locally produced goods have replaced some imported items. Today the subregion's industrial workers process food or produce textiles, paper goods, leather products, and cement. Some assemble electric motors, tractors, electronics, and automobiles.

Equatorial Africa faces many obstacles to industrialization. Educational systems are still developing, and more people must gain new skills. Hydroelectric resources are plentiful but untapped, and electrical outages occur. Political conflicts interrupt economic planning and divert resources from development projects.

The use of cell phones and smart phones has grown dramatically in Equatorial Africa.

Well-developed transportation and communications systems are essential to industry and trade. However, creating and maintaining such systems is difficult. Roads and railways must cross vast distances and varied terrain in the region. Many countries lack modern roads and transport systems except in urban areas.

Some Africans run small businesses selling local products such as baskets, art, and jewelry. Using e-commerce, or selling and buying on the Internet, people sell their products. Thus, the Internet broadens the market for local products.

The subregion has long relied on radio, with state-run stations providing broadcast programming. Television reaches fewer people because the land-relay systems for transmitting TV signals are costly and less available outside urban areas. Satellite technology is helping to expand the reach of television.

Land-line telephone service is also limited, especially in rural areas. However, cell phone usage is increasing. According to a United Nations report, Africa has the highest growth rate of cell phone subscriptions. Satellite and wireless technology have broadened access to phone service and the Internet. Today, more than half of the people in large cities in the region have access to the Internet. Because it is less expensive to set up cell phone towers than to connect telephone wire over vast distances, some areas now have cell phone service where they never before had access to a telephone. Increasingly, these cell phones are smart phones with Internet access which opens new opportunities.

▲ **CRITICAL THINKING**

1. *Considering Advantages and Disadvantages* What advantages do you think cell phones offer over land lines in this region? Might there be any disadvantages to their use?

2. *Making Predictions* Would you expect smart phone use to increase or decrease in the future? Why?

☑ **READING PROGRESS CHECK**

Identifying Cause and Effect Why has it been difficult to develop transportation and communications networks in Equatorial Africa?

LESSON 2 REVIEW

Reviewing Vocabulary
1. *Explaining* What effect did the introduction of colonial cash crops have in Equatorial Africa?

Using Your Notes
2. *Summarizing* For each section of the graphic organizer, write one sentence that summarizes your notes.

Answering the Guiding Questions
3. *Expressing* What challenges does democracy face in Equatorial Africa?

4. *Identifying Cause and Effect* How does geography influence where the people of Equatorial Africa live?

5. *Identifying Central Issues* How does conflict shape life in Equatorial Africa?

6. *Assessing* Why is the economic activity of Equatorial Africa so varied?

Writing Activity
7. *Informative/Explanatory* Write a paragraph describing how technology is changing how people in Equatorial Africa communicate.

Giovanni Mereghetti/age fotostock

networks

There's More Online!

☑ **IMAGE** Cooking With Biofuels

☑ **IMAGE** Refugee Camp

☑ **INFOGRAPHIC** Poaching of the African Elephant

☑ **INTERACTIVE SELF-CHECK QUIZ**

☑ **VIDEO** People and Their Environment: Equatorial Africa

LESSON 3

People and Their Environment: Equatorial Africa

Reading HELPDESK

Academic Vocabulary

- priority
- hamper

Content Vocabulary

- **internally displaced person**
- **biofuel**

TAKING NOTES: *Key Ideas and Details*

IDENTIFYING Use a graphic organizer like the one below to take notes about threats to natural resources in Equatorial Africa.

Equatorial Africa: Managing Resources and Human Impact	
Natural Feature	Threat

ESSENTIAL QUESTION • *How do physical systems and human systems shape a place?*

IT MATTERS BECAUSE

The dense rain forest of Equatorial Africa is an area of abundant resources. Banana, orange, and mango trees grow wild. The rain forest is alive with the sounds of wildlife, and a vast treasure in mineral wealth lies beneath the ground. Such resources can provide a bright future to this often-troubled region, but managing them in a sustainable way poses challenges.

Managing Resources

GUIDING QUESTION *What makes managing resources in Equatorial Africa especially challenging?*

The people of Equatorial Africa face tremendous difficulties in achieving a better life. Many environmental challenges threaten the subregion's supply of food, the health of its people, and its plant and animal life. Yet its people, like their neighbors around the globe, look to the future with hope.

Today many of the people in Equatorial Africa must focus on survival. While there is wealth in the subregion, it tends to be concentrated in the hands of the few. Poverty and hunger are bitter enemies for most of the population. When the focus is on survival, management of natural resources can be less of a **priority**.

And yet, Equatorial Africa has a great number of natural resources. The Democratic Republic of the Congo, for example, has tremendous amounts of fertile land. Rapidly moving streams are resources that could be harnessed to generate electricity. The country also has many mineral resources, including large reserves of copper, cobalt, and industrial diamonds. Although mining is extensive, the full economic benefits that could be gained from mining have not been realized.

Extensive conflict in the subregion is a major cause of misery. Fighting and looting severely **hamper** food distribution and normal economic activity. In the Democratic Republic of the Congo, rival factions fight for control of regions and wealth. In South Sudan, ethnic groups bent on eliminating other groups terrorize the countryside. In the Central African Republic, pockets of the countryside are lawless.

The Office of the United Nations High Commissioner for Refugees estimates the "population of concern" in Equatorial Africa is about four million people. These are refugees and **internally displaced persons**, individuals who have been forced from their homes but remain in their own country. Humanitarian organizations such as Doctors Without Borders (Médecins Sans Frontières) and the International Red Cross have helped by sending medical teams and relief workers, but the persistence of the conflicts limits their effectiveness.

Along with conflict, poverty takes an enormous toll on the people of the subregion. Africa south of the Sahara is the poorest region in the world, and some of the countries in Equatorial Africa are the poorest of the poor. With a per capita gross domestic product (GDP) of $800, the Democratic Republic of the Congo is arguably one of the poorest countries in the world. For comparison, the United States has a per capita GDP of about $56,300. The reality of the region's deep and widespread poverty demands that attention be focused on meeting immediate human needs rather than on managing natural resources.

A further challenge to the management of natural resources is widespread corruption in the governments of the region. Government officials who might work to manage natural resources are subject to bribery by businesses who want to maximize profits. Enforcement of environmental regulations is also hampered by corruption.

✓ READING PROGRESS CHECK

Identifying Cause and Effect Why has effective resource management been given a low priority in Equatorial Africa?

Human Impact

GUIDING QUESTION *How have human activities impacted the environment in Equatorial Africa?*

Equatorial Africa is a beautiful region of forests, rivers, mountains, plains, and wildlife. The natural resources of this land are vast. Unfortunately, human activities have placed much of this land in jeopardy. Water pollution, for example, is a major issue in much of the region. This is partly due to the dumping of raw sewage. In the Central African Republic, Equatorial Guinea, and the Republic of the Congo, tap water is not potable, or suitable for drinking.

Soil erosion has become a major ecological problem in Equatorial Africa. In the Democratic Republic of the Congo, for example, soil erosion affects agricultural productivity and thus causes food shortages. Soil erosion also leads to landslides and mud flows. These can have devastating consequences when they happen near rapidly growing cities, such as Bamenda City in Cameroon. Mudslides can destroy buildings that were hastily and poorly constructed.

These children live in a refugee camp in the Democratic Republic of the Congo. The country has endured decades of civil war and ethnic strife.

▲ CRITICAL THINKING

1. *Analyzing Visuals* Based on the photograph, what hardships are experienced by refugees from ethnic conflicts? What other basic needs might these children lack?

2. *Making Connections* How does ethnic strife contribute to the region's poverty?

priority something given or meriting attention before competing alternatives

hamper to make difficult; to impede

internally displaced person a refugee within his or her own country

The Congo Basin is one of the great forest regions of the world. Forests cover close to half of the basin's 2 million square miles (5.2 million sq. km). These forests, however, are under threat. Bananas, pineapples, cocoa, tea, coffee, and cotton are grown as cash crops on large commercial plantations. As farmers clear more land to grow these crops, the rain forest is seriously denuded, or stripped of trees. Commercial loggers also diminish the rain forest, as do local people who clear the land for small farming plots. Millions of refugees and internally displaced persons are forced to live off the land. They take wood for heating and cooking fires. About 3,500 square miles (9,100 sq. km) of the forest are lost each year.

Deforestation does not mean simply the loss of trees. It means less wood for the lumber industry, a loss of biodiversity, and accelerating climate change. It also leads to desertification, which in this case refers not to the advancing of a desert, but to the degradation of a formerly forested area. Many areas of Equatorial Africa that used to be forestland are now considered part of the savanna. Deforestation also contributes to soil erosion.

Air pollution, both inside and outdoors, is also a growing issue. In many parts of Equatorial Africa, household cooking is still done by burning **biofuels**—fuels such as charcoal or wood that are made from living matter. These pose a health threat as the smoke they produce pollutes household air. Household air pollution is much higher in countries that burn biofuels than in industrialized countries. Women, because they do most of the cooking, are especially at risk for the respiratory illnesses that pollution from burning biofuels can cause. Outdoor air pollution is also a problem. Vehicle emissions in growing, crowded cities, such as Kinshasa, pose a health threat to residents.

The Ivory Trade

African elephants once roamed in great numbers across the continent. Biologists estimate that in 1930, Africa was home to between 5 and 10 million elephants. During the last century, however, elephants were slaughtered by the tens of thousands for meat, for sport, and especially for their ivory tusks. Both male and female African elephants grow tusks—the world's main source of ivory. Ivory is a form of dentin that is excellent for carving and valued for its beauty. The trade in ivory dates back thousands of years. In the 1800s and 1900s, however, the demand for ivory from India, Europe, and the United States grew dramatically. By the mid-1900s, elephant populations had declined sharply. When the price of ivory soared in the 1970s, elephant tusks increased in value. Poachers began illegally killing elephants for their tusks. As many as 80,000 elephants a year were shot.

The ivory trade has had devastating consequences for the African elephant. During the 1980s, approximately half of Africa's elephant population was wiped out by poachers and only about 600,000 elephants remained. In 1989 African elephants were placed on the endangered species list by the

biofuel fuel created from living matter, such as trees

Millions of people in Equatorial Africa, such as these Mbororo women in Cameroon, rely on wood to make fires for cooking and heating. This accelerates deforestation and causes indoor air pollution.

▼ CRITICAL THINKING

1. *Analyzing Visuals* Based on the photo, what dangers are posed by indoor cooking with biofuels?

2. *Problem Solving* What technologies might be used to combat this problem?

Heiner Heine/age fotostock

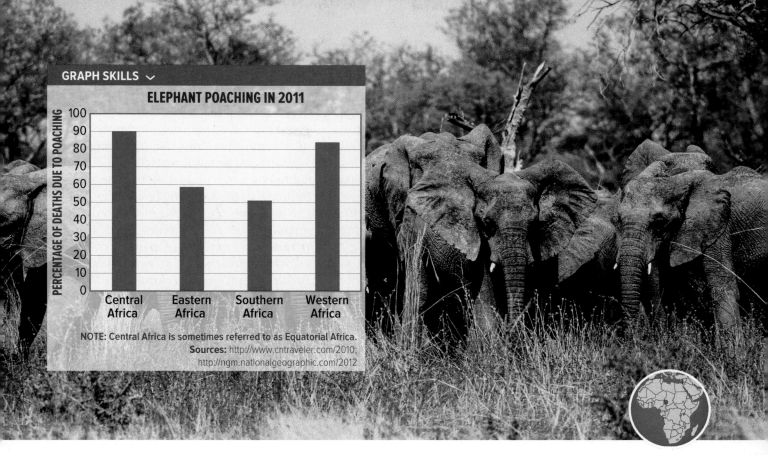

GRAPH SKILLS ⌄

ELEPHANT POACHING IN 2011

PERCENTAGE OF DEATHS DUE TO POACHING

Central Africa • Eastern Africa • Southern Africa • Western Africa

NOTE: Central Africa is sometimes referred to as Equatorial Africa.

Sources: http://www.cntraveler.com/2010; http://ngm.nationalgeographic.com/2012

Convention on International Trade in Endangered Species (CITES). Trade in ivory was banned worldwide. Organizations were established to coordinate the ban on the ivory trade and help African countries stop the poaching of elephants. The ban has only been partly successful, however. In recent years, the demand for ivory has surged. This is due in part to the huge demand for ivory ornaments in China.

Most elephants in Equatorial Africa are found in Cameroon. Armed poachers travel to Cameroon on horseback from Sudan and Chad across the Central African Republic, where almost all the elephants have already been wiped out. Even in protected areas, such as Cameroon's Bouba N'Djida National Park, more than half of the park's 400 elephants have been killed by poachers.

Hunting

In Equatorial Africa, local people do not hunt wild animals for sport as much as they hunt them for food. Warfare in the subregion has led to a widespread availability of firearms. It has also displaced millions of people who are extremely poor. This has proved a deadly combination for the subregion's wildlife. People hunt to feed themselves, and a growing number of people hunt to sell the meat to others.

The term *bushmeat* refers to the meat of animals hunted in "the bush," or forest. Among these animals are endangered or threatened species. Hunted animals include monkeys, antelopes, gorillas, and bonobos. In the Democratic Republic of the Congo, it is estimated that more than one million pounds of bushmeat are consumed each year.

✓ **READING PROGRESS CHECK**

Identifying Cause and Effect How has conflict in Equatorial Africa put pressure on the subregion's resources?

An area's elephant population will decline if more than 50 percent of the deaths are due to poaching. In 2011 all subregions in Africa south of the Sahara exceeded 50 percent. Some experts believe this rise in poaching could lead to the extinction of elephants by 2025. The loss of elephants could lead to ecosystem collapse in some areas.

▲ **CRITICAL THINKING**

1. *Exploring Issues* Do you think that the poachers or the people who buy ivory from poachers should be targeted by officials trying to stop the ivory trade? Explain.

2. *Hypothesizing* In what region are the most elephants killed by poachers? Why do you suppose more elephants are killed there than in other parts of Africa?

Addressing the Issues

GUIDING QUESTION *What steps are being taken to combat these environmental challenges?*

Increasingly, the protection of tropical rain forests is a priority in the subregion. Various countries have created forest reserves to protect tropical forests. Logging companies are also getting involved, using scientific tree farming and replanting projects to protect and renew forests. In one program, more than 10 million trees were planted on a tree plantation outside Virunga National Park in the Democratic Republic of the Congo. The Central African Forest Commission (COMIFAC) is an international organization coordinating the conservation and sustainable management of the forest ecosystem in the region. COMIFAC works in conjunction with a network of member countries, international institutions, and nongovernmental organizations (NGOs).

Some NGOs and other charitable organizations, along with Western governments and entrepreneurs, are working with local African communities to solve the indoor air pollution problem. New technologies may offer a solution to this serious health issue. Some organizations are inventing ways to make biofuel cookstoves cleaner and more efficient. Others have developed solar-powered cookstoves. Some are developing clean technologies to provide lighting. One of the most recent innovations is the development of gravity-powered lighting. This type of light uses gravity to power a small light for a short period of time. It has the potential to offer cheap lighting that is safe for both people and the environment.

To counter the ivory trade, governments in Equatorial Africa have deployed troops and special police forces to combat poachers. Unfortunately, they are often too few in number and poorly trained. Private groups like the World Wildlife Fund support local people's efforts to oppose poachers and pressure governments in the region to devote resources to antipoaching efforts. Nongovernmental organizations and conservationists have, however, come under criticism for allegedly disregarding the poverty and human suffering that exist in the region. The widespread economic hardship in Equatorial Africa creates a situation in which poaching is a tempting way of making a living. This makes the antipoaching campaign even more of a challenge.

Similarly, decreasing the trade in bushmeat is a difficult task given the ongoing political and economic struggles in the region. Humanitarian and development efforts aim to increase people's standard of living in hopes that this will in turn decrease the demand for hunted wildlife.

✓ READING PROGRESS CHECK

Identifying What actions are being taken to protect Africa's tropical forests?

LESSON 3 REVIEW

Reviewing Vocabulary
1. *Expressing* What problem does the use of biofuels create?

Using Your Notes
2. *Expressing* Use the notes in your graphic organizer to write a paragraph about environmental issues in Equatorial Africa.

Answering the Guiding Questions
3. *Identifying Central Issues* What makes managing resources in Equatorial Africa especially challenging?

4. *Identifying Cause and Effect* How have human activities impacted the environment in Equatorial Africa?

5. *Listing* What steps are being taken to combat these environmental challenges?

Writing Activity
6. *Argument* Write an e-mail to a United Nations official arguing for or against the creation of more protected wildlife areas in Equatorial Africa.

Directions: On a separate sheet of paper, answer the questions below. Make sure you read carefully and answer all parts of the questions.

Lesson Review

Lesson 1

1 *Explaining* What physical feature dominates Equatorial Africa? What functions does it serve?

2 *Making Connections* What is the economic value of the subregion's rain forests?

3 *Identifying Cause and Effect* Why do Equatorial Africa's water resources remain underdeveloped, despite abundant rainfall?

Lesson 2

4 *Sequencing* Create a graphic organizer showing the rise and fall of colonial powers in what is now the Republic of the Congo.

5 *Explaining* What role did colonialism play in the development of commercial agriculture?

6 *Identifying Central Issues* Describe the quantity of mineral resources in Equatorial Africa. Explain why the people of the area see few benefits from these resources.

Lesson 3

7 *Summarizing* Why is Africa south of the Sahara the poorest region in the world, despite its many resources?

8 *Identifying Cause and Effect* Develop a diagram with labels to show the causes and effects of soil erosion in Equatorial Africa. Write a paragraph to explain the information in your diagram.

9 *Making Connections* Discuss the connection between deforestation and desertification in Equatorial Africa.

Critical Thinking

10 *Organizing* After studying Equatorial Africa, you have decided you would like to write a book about the people, land, and climate of this subregion. Write a one-page outline for your book to show the major topics you will include.

21st Century Skills

Use the cartoon below to answer the questions that follow.

" 'Born in conservation,' if you don't mind. 'Captivity' has negative connotations."

11 *Evaluating Primary Sources* Explain why you think elephants were chosen as the topic of this cartoon. What conservation efforts are taking place in Equatorial Africa?

12 *Understanding Relationships Among Events* What relationship does poverty have to conservation in Equatorial Africa?

Applying Map Skills

Use the Unit 6 Atlas to answer the following questions.

13 *The World in Spatial Terms* Use your mental map of Equatorial Africa to describe the spatial relationship among the Congo Basin, the Katanga Plateau, and the Republic of the Congo.

14 *Places and Regions* What is the largest country in Equatorial Africa? Provide the approximate size of this country in square miles.

15 *Environment and Society* Discuss the connections between the population of Kinshasa, Democratic Republic of the Congo, and the pollution levels from vehicle emissions.

Need Extra Help?

If You've Missed Question	1	2	3	4	5	6	7	8	9	10	11	12	13	14	15
Go to page	554	558	558	560	564	564	567	567	568	551	571	571	476	476	476

Directions: On a separate sheet of paper, answer the questions below. Make sure you read carefully and answer all parts of the questions.

DBQ Analyzing Primary Sources

Conflict is a major contributor to poverty in Equatorial Africa. Read the news report below to see how conflict in the Central African Republic affects access to and production of food.

PRIMARY SOURCE

"*The U.N. World Food Program [WFP] expressed concern about a potential food crisis in CAR [the Central African Republic]. It said conflict that erupted during a rebellion . . . had sparked the situation.*

The WFP said trade has been interrupted in parts of the country controlled by the rebel group. Basic food prices for areas under government control, meanwhile, have increased 40 percent since December.

WFP called on the rebel group to let humanitarian workers into its parts of CAR. An estimated 800,000 people live in areas under rebel control.

'We are very concerned about prospect for the 2013 growing season, which is due to start in just a few weeks,' Rockaya Fall, the Food and Agriculture Organization's country representative, said in a statement. 'Land preparation, which should have begun, is behind schedule in many places, due to insecurity.'"

— "Food Crisis Plagues War-Torn CAR," United Press International, February 18, 2013

16 *Identifying Cause and Effect* In what ways is conflict contributing to a food crisis in the Central African Republic?

17 *Differentiating* Distinguish between the roles played by the rebels and the government in the food crisis.

Exploring the Essential Question

18 *Synthesizing* Create a short guide in the form of a brochure to provide an overview of the ways physical systems and human systems shape Equatorial Africa. Within your brochure, include the following headings: *History, Politics, Population, Society, Culture,* and *Economics.*

College and Career Readiness

19 *Clear Communication* Imagine that you are applying for a position as an intern with the Office of the United Nations High Commissioner for Refugees (UNHCR). The focus of this internship is on helping refugees and internally displaced persons in Equatorial Africa. Write a one-page essay to explain why you should be chosen to serve in this position. Within your essay, provide the following: your knowledge of the relevant problems in the subregion, including the conditions that have led to the problems and the relevance of the work of Doctors Without Borders (Médicins Sans Frontières) and the International Red Cross in the subregion.

Research and Presentation

20 *Gathering Information* With a partner, conduct research to learn more about conflict diamonds, also known as blood diamonds, mined from Africa. Develop a multimedia presentation to answer the following questions: What are conflict diamonds? Why are they known as conflict diamonds? What are some issues related to conflict diamonds? What steps are being taken to address these issues? Include the following in your multimedia presentation: audio and video of relevant news reports, maps, diagrams, and excerpts from printed news reports.

Writing About Geography

21 *Narrative* Use standard grammar, spelling, sentence structure, and punctuation to write a paragraph that describes a visit to the equatorial rain forest. In your paragraph, include details that relate to all five senses and to the wildlife, plant life, and resources of the rain forest.

Need Extra Help?

If You've Missed Question	16	17	18	19	20	21
Go to page	572	572	551	567	566	558

Southern Africa

ESSENTIAL QUESTION • *How do physical systems and human systems shape a place?*

Why Geography Matters
Southern Africa and HIV/AIDS

Lesson 1
Physical Geography of Southern Africa

Lesson 2
Human Geography of Southern Africa

Lesson 3
People and Their Environment: Southern Africa

Geography Matters...

Southern Africa is a difficult land to categorize. Most of it is located on a continent, but it also consists of a large island and several small islands. The continental part is surrounded on three sides by ocean, but most of its landforms are plateaus and highlands.

The people of Southern Africa have a history that goes back tens of thousands of years. Yet their culture has been heavily influenced by Europeans who arrived just a few centuries ago. Some people in Southern Africa live in densely packed cities. Others live in the desert much the same way as their ancestors did. Christianity is widespread, but many traditional religions are still practiced as well.

◄ Jeffreys Bay in South Africa is one of the world's top destinations for surfers.

Kelly Cestari/ASP/Getty Images

Southern Africa
and HIV/AIDS

In Africa, HIV/AIDS has reached epidemic proportions. Southern Africa has the highest number of people infected with HIV/AIDS on the continent. Nearly one in five people in the region is suffering from the disease. New incidences of HIV infections have declined somewhat, but the disease is still devastating the economic and social health of Southern Africa.

World Distribution of HIV 2014

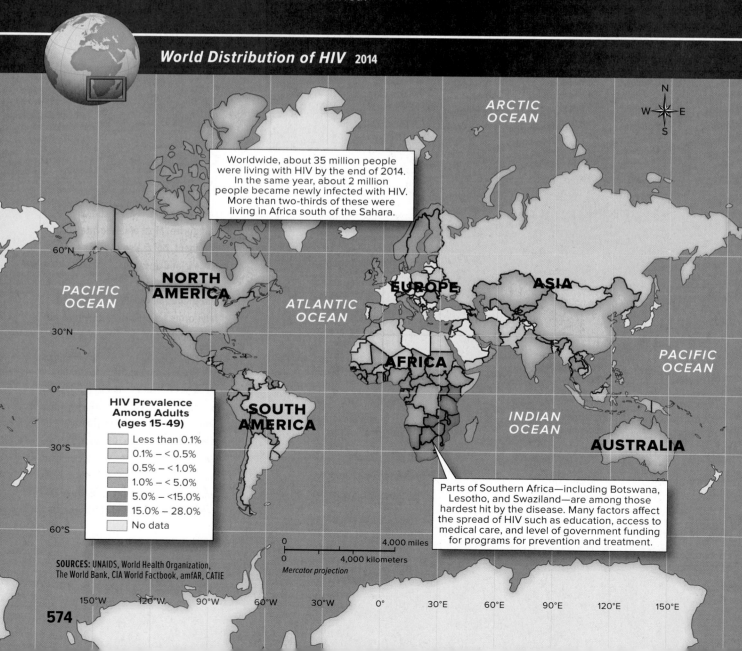

Worldwide, about 35 million people were living with HIV by the end of 2014. In the same year, about 2 million people became newly infected with HIV. More than two-thirds of these were living in Africa south of the Sahara.

HIV Prevalence Among Adults (ages 15-49)

- Less than 0.1%
- 0.1% – < 0.5%
- 0.5% – < 1.0%
- 1.0% – < 5.0%
- 5.0% – <15.0%
- 15.0% – 28.0%
- No data

Parts of Southern Africa—including Botswana, Lesotho, and Swaziland—are among those hardest hit by the disease. Many factors affect the spread of HIV such as education, access to medical care, and level of government funding for programs for prevention and treatment.

SOURCES: UNAIDS, World Health Organization, The World Bank, CIA World Factbook, amfAR, CATIE

0   4,000 miles
0   4,000 kilometers
Mercator projection

ARCTIC OCEAN

NORTH AMERICA

PACIFIC OCEAN

EUROPE   ASIA

ATLANTIC OCEAN

AFRICA

PACIFIC OCEAN

SOUTH AMERICA

INDIAN OCEAN

AUSTRALIA

574

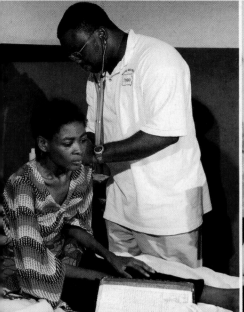

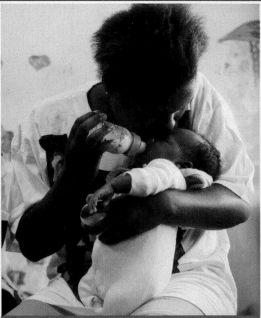

What are HIV and AIDS?

Why have the people of Southern Africa been so badly affected by HIV/AIDS?

What is being done to combat HIV/AIDS in Southern Africa?

HIV (human immunodeficiency virus) is an infection that can be spread from one person to another. Over time, the disease attacks a person's immune system and can advance to AIDS (acquired immunodeficiency syndrome). The disease can be spread through blood or other bodily fluids, as well as from a pregnant mother to her unborn child. In a few instances, the disease has spread through organ transplants. Symptoms of HIV/AIDS range from flulike symptoms to none at all. In some cases, people who are infected with HIV show no symptoms for years, so without annual medical exams it can be difficult to diagnose in its early stages.

Southern African countries have the highest number of people living with HIV/AIDS on the continent. It is believed to have emerged in the Democratic Republic of the Congo in the 1970s. However, the disease did not become an epidemic in Africa until the following decade. It first spread east and then south within the continent. By the late 1980s, the virus had become a regional epidemic.

1. **Human Systems**   Why is HIV/AIDS so dangerous to humankind?

Much of the population of Southern Africa lives in poverty. Literacy is low. Medical care is limited or too expensive for the majority of the population. These factors weigh heavily into the high rates of HIV/AIDS in the subregion. In some countries of Southern Africa, about 40 percent of the adult population is living with HIV/AIDS. Percentages are over 40 percent in some urban areas. More than a million Southern Africans die each year because of HIV/AIDS, resulting in lower life expectancy and a rise in the number of orphans. Some scientists believe that differences between HIV subtypes have an effect on transmission rates. There is also some evidence that genetic factors and parasitic worm infections, which are common in Africa south of the Sahara, may contribute to the epidemic in the region.

There are other factors that may contribute to the high rate of HIV/AIDS cases in Southern Africa. These include social instability, rapid urbanization, labor migration, gender inequality, and sexual violence. Additionally, ineffective political leadership failed to address the crisis in the 1990s when the disease became an epidemic in the region.

2. **Human Systems**   Why are the people of Southern Africa so vulnerable to HIV/AIDS?

Some governments in Southern Africa, as well as organizations such as the World Health Organization (WHO), have been working to educate people on how the disease is contracted and passed to others. This includes counseling programs, controlling incidences of other sexually transmitted diseases, which can cause the spread of HIV, and caring for people living with HIV/AIDS. Efforts have also been made to improve the safety of blood supplies.

Early testing for HIV has helped identify cases that can then be treated with antiretroviral drugs (ARVs). These drugs delay the progression of HIV to AIDS. WHO began a program in 2003 to treat 5 million people in the region with ARVs. Even though WHO failed to reach its goal, it set a precedent for countries in Southern Africa such as Botswana, Namibia, and South Africa to make ARVs accessible to those living with HIV.

3. **Human Systems**   How are people working to combat HIV/AIDS, and how successful are their efforts?

THERE'S MORE ONLINE

SEE a map of HIV infection rates in Africa   •   *READ* a quote about efforts to fight HIV/AIDS in Southern Africa

netw⊙rks

There's More Online!

☑ **CHART** Wildlife Reserves in Southern Africa

☑ **IMAGE** Baobab Tree

☑ **IMAGE** The Long Wall of Namib

☑ **IMAGE** Moremi Game Reserve

☑ **INTERACTIVE SELF-CHECK QUIZ**

☑ **VIDEO** Physical Geography of Southern Africa

Reading **HELP**DESK

Academic Vocabulary

- **altitude**
- **source**

Content Vocabulary

- **escarpment**
- **delta**
- **Mediterranean climate**
- **rain shadow**

TAKING NOTES: *Key Ideas and Details*

IDENTIFYING As you read about the physical geography of Southern Africa, use a graphic organizer like the one below to keep track of its physical features.

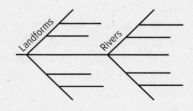

LESSON 1
Physical Geography of Southern Africa

ESSENTIAL QUESTION · *How do physical systems and human systems shape a place?*

It Matters Because

The terrain of Southern Africa produces a wide variety of biomes—tropical, desert, and temperate regions. Life and culture are very different in each of these areas. In addition, Southern Africa's rivers are essential for both the natural environment and human activity in the subregion.

Landforms

GUIDING QUESTION *What is the most dominant physical feature of Southern Africa?*

Southern Africa consists of 14 countries, 10 of which are on the mainland. The other four are island countries. Imagine a line drawn roughly along the northern borders of Angola, Zambia, Malawi, and Mozambique. Everything south of that line is part of Southern Africa. Extend the line out into the Indian Ocean, and it runs just above Comoros and Madagascar. Mauritius is east of Madagascar, while Seychelles is northeast of Madagascar.

If Angola, Zambia, Malawi, and Mozambique form the "top row" of Southern Africa, then Namibia, Botswana, Zimbabwe, and the southern half of Mozambique form the "second row." At the southern tip of the continent is South Africa. The country of Lesotho (luh•SOO•too) lies entirely within South Africa. Swaziland is on the South Africa-Mozambique border.

The mainland of Southern Africa is surrounded on three sides by ocean. Most of the land sits at a high **altitude**, or height above sea level, of over 2,000 feet (610 m). Along the coasts of Angola, Namibia, and South Africa lies the coastal plain, a narrow strip of land that varies from a few miles wide to a few dozen miles wide. Looming up behind this coastal plain are high steep cliffs known as the Great **Escarpment**. These cliffs form an almost unbroken U-shape. They run from the west coast south to the Cape of Good Hope and then curve northeast to the South Africa-Mozambique border. The Drakensberg Range is part of the Great Escarpment. The range rises to more than 11,000 feet (3,353 m) and runs along the southern edge of South Africa. As the escarpment reaches Mozambique, it turns north to follow along Mozambique's western border and on through Malawi. Because most of Mozambique lies outside of the Great Escarpment, its landscape is mainly made up of soft, rolling hills. Only in the northwestern interior of the country does the land rise above 600 feet (183 m).

The land inside the Great Escarpment is mostly hills and plateaus. A plateau, or "tableland," is a stretch of flat land that is higher than its surroundings. The most striking part of the Southern African landscape is a plateau that is over a mile high. The Highveld, as it is called, sits more than 50 miles (80 km) from the coast. However, given its 6,000 foot (1,829 m) altitude, it is visible from a long distance.

The highest parts of Namibia and Angola are near the coast, along the Great Escarpment. Zimbabwe's highest terrain is in the center of the country. Zambia's high points are in the northeast, where they merge with Mozambique's highlands along the Great Escarpment. In contrast, most of Botswana sits in a great basin. While still much higher than the coastal lands, it is lower than the surrounding landforms.

The landforms surrounding Botswana on the west into Namibia or northwest into Angola are once again hills and plateaus. This undulating terrain continues westward until it rises up over the Great Escarpment. It then drops down onto the narrow Atlantic coastal plain.

In a sense, the terrain of Madagascar is a smaller version of that on the mainland. The middle of the island, like the mainland, is a series of plateaus surrounded by an escarpment. Unlike the mainland, groups and masses of volcanoes are strewn among the Madagascar plateaus. Tsiafajavona, one of these volcanic peaks, is 8,671 feet (2,643 m) high.

In the north, the plateaus give way to volcanoes that slope down to the sea. The southern edges of the plateaus tower above the Indian Ocean. To the east and west are escarpments. The escarpments, some of which are impassable, separate the highlands from the lower areas on either side. The eastern part of the island is a narrow coastal strip, rather like the Atlantic coast of Southern Africa. The western side is more varied, with low plateaus and rolling hills.

✔ **READING PROGRESS CHECK**

Describing Where is the highest terrain in Southern Africa found?

Water Systems

GUIDING QUESTION *How do people harness the power of the subregion's rivers?*

The Okavango River runs southeast from central Angola to northern Botswana. It starts on a plateau, and for a while it flows along the border between Angola and Namibia. Then it turns south and heads into Botswana. Most rivers empty into a lake or ocean, but the Okavango ends inland. The river just spreads wider and wider until it forms an inland **delta** and swamps.

The Orange River flows nearly from one side of Southern Africa to the other. It starts in Lesotho, on the Highveld, just over 100 miles (161 km) from the coast. It flows west across South Africa, then forms part of the border between South Africa and Namibia before it empties into the Atlantic Ocean.

altitude height above sea level

escarpment a long cliff that separates lands at two different altitudes

delta an often triangular-shaped section of land formed as the waters of a river slow down and split into many channels as they deposit silt and sediment

The Long Wall of Namib is a section of the Namib Desert that runs along the coast of southern Namibia.

▼ **CRITICAL THINKING**
1. ***Making Connections*** Less than 100 miles (161 km) inland from this spot is fertile land. Why does that fertile land not extend to the ocean?
2. ***Analyzing Visuals*** Why is the water in this picture not being used to irrigate the arid land?

Wildlife abounds in the Moremi Game Reserve in Botswana.

▲ CRITICAL THINKING

1. Constructing a Thesis How can an area that is as wet as a swamp be part of a desert?

2. Identifying Cause and Effect Why do the waters of the Okavango River collect in northern Botswana and not flow into some other area?

The Zambezi River comes into contact with several countries along its course to the sea. Starting in eastern Angola, it flows south through western Zambia. It forms the border between Zambia and the countries of Namibia, Botswana, and Zimbabwe. Along the southern stretch of the border with Zimbabwe, the river suddenly drops straight down, forming Victoria Falls. Once it has cleared the northern tip of Zimbabwe, the Zambezi heads east through Mozambique, where it flows near the southern tip of Malawi before emptying into the Indian Ocean.

The Limpopo River begins in northern South Africa and flows north for a distance before it turns eastward and forms the border between South Africa and Botswana and the border between South Africa and Zimbabwe. Then it flows across Mozambique to the ocean.

None of these rivers are navigable, except for short stretches across the coastal plain. They originate in highlands and as they flow through the steep, terrain, rapids and waterfalls mark their courses. These conditions are not suitable for transportation, but they are useful for generating electricity. Two of the largest hydroelectric dams in Africa—the Kariba Dam and the Cahora Bassa Dam—are located on the Zambezi River.

Aquatic wildlife such as hippopotamuses and crocodiles thrive in and along these rivers. The northeastern part of the Okavango Delta is set aside as the Moremi Game Reserve. Hundreds of species including lions, cheetahs, hippopotamuses, buffalo, wild dogs, and many types of birds and fish live there. Two wildlife areas on either side of the Limpopo, in South Africa and in Mozambique, recently joined with each other and with several sanctuaries in Zimbabwe to form the Great Limpopo Transfrontier Park. This extended preserve provides a safe haven for lions, leopards, hippopotamuses, elephants, giraffes, and many other species.

☑ READING PROGRESS CHECK

Explaining What makes a river suitable for generating electricity?

Climates, Biomes, and Resources

GUIDING QUESTION *How do the climates of Southern Africa affect its biomes?*

Southern Africa is large with many different landforms. The result is a great variety of climates. These different climates and landforms create a number of different biomes with distinct characteristics.

Climate Regions and Biomes

The coastal areas of Southern Africa have marine climates, which means they are greatly affected by weather conditions and systems that blow in from the open ocean. The Cape of Good Hope and the area to its immediate northeast have a **Mediterranean climate** similar to that of Greece and Italy. It is not typical of Southern Africa, however. Moving up the eastern coast to Mozambique, the climate becomes tropical wet/dry. The winters are warm and the summer rainy season stretches from November through March. The western coast up through Namibia and Angola has an arid, or desert, climate.

The interior of the subregion is generally hot, although temperatures can dip below freezing in the higher elevations. The eastern areas experience a fair amount of precipitation, but toward the west the rainfall drops off. The deserts of Botswana and Namibia receive very little rain.

The climate of Madagascar is determined by its central plateau and the warm, wet winds off the open ocean. The eastern coast of the island has a tropical wet climate, while the interior plateau has a highland climate. The western side is in a **rain shadow**, and much of it is desert.

Given the number of different climates in the subregion, Southern Africa has many different biomes. Much of the coastal plain along the Atlantic coast is a desert biome. On the Indian Ocean side, grassland biomes give way to forest as one moves north. In the interior of the continent, on the highlands inside the Great Escarpment, the vast majority of the land falls into one of two types, savanna or desert.

The north and the east are covered with savanna—vast grasslands dotted with small stands of trees. Many of the most recognized African mammals, such as giraffes, zebra, and jackals, live on the savanna. This biome is also home to animals known as the Big Five: lions, leopards, elephants, Cape buffalo, and rhinoceroses. These animals became known as the Big Five because they were so dangerous and difficult to hunt. However, they are now the five species most tourists want to see while on safari.

The south and the west of the inland area are mostly desert biome. The Kalahari Desert occupies much of the Botswana basin. It stretches southwest to where Botswana meets Namibia and South Africa. Here, it blends into the Namib Desert. The Namib continues down to the coast, then runs north between the ocean and the Great Escarpment through all of Namibia and into southern Angola.

Most of the Namib Desert is quite arid. The southern parts are covered with seemingly endless sand dunes—brick red inland and yellow along the coast— some of which can be 800 feet (244 m) tall and 20 miles (32 km) long. Inland, bushes and tall grasses have adapted to grow in the sand dunes. Antelope and ostriches live here as well. Farther north in the interior desert, rivers can be found, and with them elephants, rhinoceroses, hyenas, and more.

The arm of the Namib Desert that runs along the coast is quite different. It is almost completely arid, with little or no plant life. Some reptiles and insects have adapted to this biome, but no larger animals live here. As in the interior portion, though, the most northern reaches have more water. Succulents grow here, and it is home to various marine birds, such as pelicans, flamingos, and even penguins.

In the southern part of the Kalahari, rain is scarce. Drought-tolerant grasses and scrub are all that can take root. Herds of antelope such as wildebeest and springbok roam the area. The central part of the Kalahari gets some rain, and shrubs can grow there. Acacia trees provide homes for birds, rodents, and insects. The northern Kalahari is hardly desertlike at all because of the rivers that flow through it. Many smaller animals, such as wild dogs, foxes, anteaters, and porcupines, live here. Plants such as pond lilies and reeds thrive here as well.

Connecting Geography to MATH

Output of the Kariba Hydroelectric Dam

The Kariba Dam, located on the Zambezi River between the countries of Zambia and Zimbabwe, produces 6.7×10^8 kilowatt-hours (kWh) of electricity per year. Some of that electricity goes to Zambia and some to Zimbabwe. Average per capita electricity use in Zimbabwe is approximately 850 kWh.

APPLYING If half the output of Kariba Dam were sent to Zimbabwe, how many people's electricity needs could be met? Round your answer to the nearest thousand.

Mediterranean climate
a climate marked by warm, dry summers and cool, rainy winters

rain shadow a condition created when winds blow in primarily one direction over mountains or other elevated terrain: the altitude change causes clouds to drop their precipitation on the near side of the mountain, leaving the land on the far side, where winds descend, with little rain or snow

The fruit of the baobab tree is eaten or used to make a soft drink. The bark is used to make rope, fabric, and strings for instruments. The leaves are eaten or used to make medicines.

▲ **CRITICAL THINKING**

1. **Constructing a Thesis** Why might Africans like to use the baobab tree as a symbol for the entire continent?

2. **Drawing Conclusions** Why might someone living in or near the desert like to have a baobab tree growing nearby?

source a point of origin

The most striking tree found in the northern Kalahari is the baobab. The baobab tree is often used as a symbol for all of Africa. Its thick trunk stretches high before splitting into a tangle of skinny branches, making the tree look as if it is upside-down with its roots in the air. Baobabs can live for hundreds of years and reach a diameter of 30 feet (9 m). The trunks can be hollowed out and used to collect rainwater or even as a shelter.

Natural Resources

Southern Africa's most important resource is its vast mineral wealth. Gold and copper are mined today, just as they were by ancient peoples. Coal, nickel, iron, cobalt, manganese, and uranium are found in abundance. Deposits of gemstones, especially diamonds, have drawn miners for several centuries. Additionally, Southern Africa's mineral wealth made it attractive to countries for colonization.

Unfortunately, Southern Africa's valuable resources are a **source** of controversy. In most cases, foreign companies own the mines that extract minerals and gemstones. They hire local workers and pay them very little money to work in dangerous conditions. The profits the foreign companies make are taken out of Southern Africa and do little to benefit the subregion.

One major exception is found in Botswana. In 1978 the government formed a partnership with a multinational company called De Beers. The partnership, known as Debswana, mines and sells Botswana's diamonds. The profits are split between the country and the company. This has given Botswana one of the healthiest, fastest-growing economies in all of Southern Africa.

The history of diamond mining can be seen in Kimberley, South Africa, at a site called the Big Hole. It was a hill until diamonds were discovered. From 1871 to 1914, 22.5 million tons (20.4 million t) of dirt and rock were removed, with picks and shovels. Today, the 700-foot- (213-m-) deep hole is a tourist destination.

☑ **READING PROGRESS CHECK**

Evaluating What types of biomes are found in Southern Africa?

2009 PhotoStock-Israel/Getty Images

LESSON 1 REVIEW

Reviewing Vocabulary

1. **Applying** Write a paragraph describing the altitudes of the various areas in Southern Africa. Be sure to mention the Great Escarpment.

Using Your Notes

2. **Synthesizing** Use your graphic organizer on the features of Southern Africa's physical geography to write a paragraph describing the four major rivers of the subregion and their courses.

Answering the Guiding Questions

3. **Identifying** What is the most dominant physical feature of Southern Africa?

4. **Explaining** How do people harness the power of the subregion's rivers?

5. **Applying** How do the climates of Southern Africa affect its biomes?

Writing Activity

6. **Informative/Explanatory** In a paragraph, describe the major biomes of Southern Africa.

LESSON 2
Human Geography of Southern Africa

ESSENTIAL QUESTION • *How do physical systems and human systems shape a place?*

Reading HELPDESK

Academic Vocabulary

- **policy**
- **prohibit**

Content Vocabulary

- **urbanization**
- **sanitation**
- **subsistence farming**
- **commercial farming**

TAKING NOTES: *Integration of Knowledge and Ideas*

IDENTIFYING As you read about the human geography of Southern Africa, use a graphic organizer like the one below to identify examples of how history, population, culture, and economics have worked together to create the subregion as it is today.

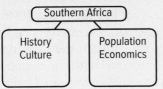

Southern Africa

| History Culture | Population Economics |

IT MATTERS BECAUSE

Southern Africa's human geography blends components from the very earliest humans, traditional African cultures, and European influences. Southern Africa today is full of contrasts—white and black, traditional and modern, wealth and poverty, urban and rural.

History and Government

GUIDING QUESTION *What cultures have influenced life in Southern Africa?*

Fossils that predate those of modern humans can be found in Southern Africa. Fossils of the first true humans are found north of the subregion in the valleys of Tanzania and Kenya. The first people to live in Southern Africa were the San, who arrived more than 20,000 years ago. Today, their descendants live in Botswana, Namibia, and Angola.

The Bantu peoples originated in central Africa, but began spreading across the continent some 3,500 years ago. The term *Bantu* refers to a group of about 500 related languages and to the various peoples who speak them. By about A.D. 300, Bantu peoples had migrated to Southern Africa.

One of the Bantu peoples, the Shona, established a city called Great Zimbabwe. By A.D. 1000, Great Zimbabwe had a population of between 12,000 and 20,000 people. The inhabitants farmed, raised cattle, and mined and traded gold. For 400 years, Great Zimbabwe was the center of a huge trading empire. Sometime in the 1400s, however, the city was abandoned. No one knows why. All that remains are ruins in southeastern Zimbabwe.

Madagascar's population is also the result of migration. Around A.D. 800, a small group of people sailed in outrigger canoes from islands in Southeast Asia to Madagascar. These people were the Malagasy, and their descendants populated the entire island, later mixing with migrants from the African continent. DNA tests confirm the ancestry of Madagascar is evenly split between Indonesia and East Africa. Despite sharing ancestry, as well as the many political and economic ties to Africa, many Malagasy do not consider themselves to be African.

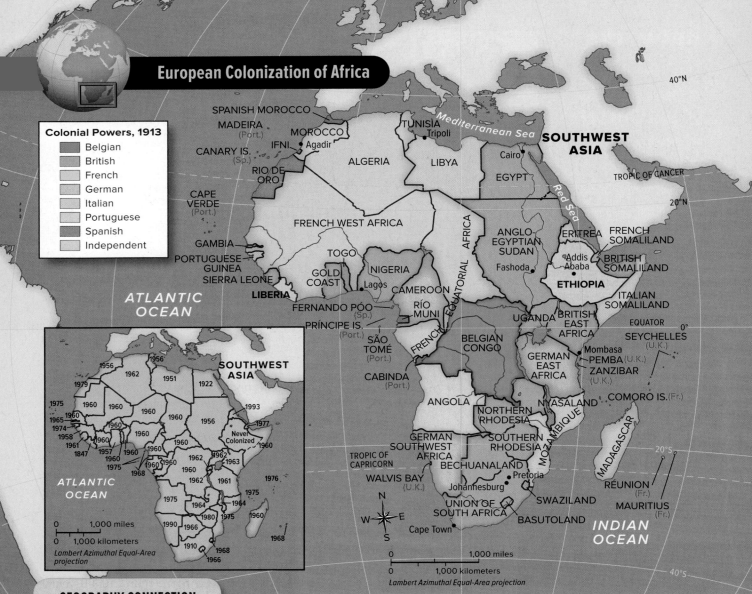

European Colonization of Africa

Colonial Powers, 1913
- Belgian
- British
- French
- German
- Italian
- Portuguese
- Spanish
- Independent

SPANISH MOROCCO
MADEIRA (Port.)
MOROCCO
IFNI • Agadir
CANARY IS. (Sp.)
RIO DE ORO
CAPE VERDE (Port.)
GAMBIA
PORTUGUESE GUINEA
SIERRA LEONE
LIBERIA
TOGO
GOLD COAST
NIGERIA
FERNANDO PÓO (Sp.)
PRÍNCIPE IS. (Port.)
SÃO TOMÉ (Port.)
CABINDA (Port.)
ANGOLA
GERMAN SOUTHWEST AFRICA
WALVIS BAY (U.K.)
TROPIC OF CAPRICORN
BECHUANALAND
Pretoria
Johannesburg
UNION OF SOUTH AFRICA
BASUTOLAND
Cape Town
SWAZILAND

TUNISIA
Tripoli
Mediterranean Sea
Cairo
ALGERIA
LIBYA
EGYPT
TROPIC OF CANCER
SOUTHWEST ASIA
Red Sea
FRENCH WEST AFRICA
ANGLO-EGYPTIAN SUDAN
Fashoda
ERITREA
FRENCH SOMALILAND
Addis Ababa
BRITISH SOMALILAND
ETHIOPIA
ITALIAN SOMALILAND
CAMEROON
RÍO MUNI
FRENCH EQUATORIAL AFRICA
UGANDA
BRITISH EAST AFRICA
EQUATOR
SEYCHELLES (U.K.)
Mombasa
PEMBA (U.K.)
ZANZIBAR (U.K.)
BELGIAN CONGO
GERMAN EAST AFRICA
COMORO IS. (Fr.)
NYASALAND
NORTHERN RHODESIA
SOUTHERN RHODESIA
MOZAMBIQUE
MADAGASCAR
RÉUNION (Fr.)
MAURITIUS (Fr.)

ATLANTIC OCEAN
INDIAN OCEAN

40°N
20°N
0°
20°S
40°S

1,000 miles
1,000 kilometers
Lambert Azimuthal Equal-Area projection

Inset map (independence dates):

SOUTHWEST ASIA
ATLANTIC OCEAN
1956
1956
1962
1951
1922
1979
1975
1960
1960
1960
1993
1965
1960
1960
1956
1974
1960
1977
1958
1960
Never Colonized
1961
1957
1960
1960
1847
1960
1960
1962
1960
1962
1963
1975
1968
1960
1962
1961
1976
1975
1975
1964
1964
1960
1975
1980
1975
1990
1966
1910
1968
1966
1968

0   1,000 miles
0   1,000 kilometers
Lambert Azimuthal Equal-Area projection

GEOGRAPHY CONNECTION

European control varied from colony to colony. For example, Cape Colony (South Africa) had a strong central government. In Angola, however, the Portuguese had little control over the people who lived in the inland areas.

1. *HUMAN SYSTEMS* Which European country governed Cape Colony?

2. *HUMAN SYSTEMS* Which European country controlled most of inland Southern Africa?

European Influences

Great Zimbabwe had faded away by the time the first Europeans arrived in Southern Africa. In the 1480s, Portuguese explorers, priests, and traders sailed into the Kingdom of Kongo in what is now the northern tip of Angola. At first, relations between the Portuguese and Kongo were peaceful, but it did not last.

By the mid-1500s, an active slave trade was running throughout Southern Africa. Coastal African kings sent raiding parties inland to capture people. The captives were then traded to the Portuguese for firearms and other manufactured goods.

The Portuguese established slave and trading posts on both the east and west coasts of Southern Africa. They either traded or warred with the various kingdoms they encountered, and they shipped much gold, silver, and ivory back to Europe. In the 1600s, other European powers expanded into Africa, causing Portuguese power to wane. By the mid-1700s, Dutch, British, and local African forces had confined the Portuguese to Angola and Mozambique. Only small settlements near the coast remained subject to Portuguese control.

In 1652 the Dutch East India Company established a settlement on the Cape of Good Hope. Dutch settlers took more and more land from the local African inhabitants, expanding well beyond the influence of the Dutch East India Company.

The Dutch were aided in their expansion by disease. As in the Americas, the local population had no immunity to European diseases and many died. After a few generations, the Dutch settlers referred to themselves as Afrikaners, which means "Africans." They were also called Boers, the Dutch word for "farmers."

Most of the Boers used slave labor on their farms. At first, enslaved people were brought in from areas farther north, but later the Boers enslaved local people as well. In order to communicate, the Afrikaners and the Africans developed a pidgin dialect that eventually evolved into Afrikaans, a new language.

At the start of the 1800s, Cape Colony was home to 22,000 whites, 25,000 enslaved blacks, and tens of thousands of free blacks. In 1806 Great Britain seized control of Cape Colony. A year later, Britain outlawed the slave trade in all its colonies. This had a positive effect in that the British and Afrikaners were now prohibited from capturing and enslaving Africans. It did not benefit people already enslaved, who were not freed by the ending of the trade.

European contact with Madagascar was sporadic and varied. The Portuguese, British, and French tried to gain influence on the island. The French established a settlement in 1642, but it did not last. Trade continued, and both the French and the British allied themselves with various local groups and leaders. In 1896 Madagascar became a French colony.

In 1884 Germany's new colonial goals led to the establishment of two colonies on the mainland of Africa. One, German East Africa, was farther north on the continent but did include some of what is today Mozambique. The other, German Southwest Africa, later became the country of Namibia. After losing World War I, Germany was stripped of all its African colonies. German East Africa was divided between Great Britain and Belgium. German Southwest Africa was renamed Southwest Africa, and it was placed under the control of South Africa.

Shaka

One of the most important Africans in the history of Southern Africa was Shaka. He was the son of a Zulu chief and a Langeni princess. Shaka was raised by his mother among the Langeni and the Mthethwa. Both Shaka and his mother were treated cruelly and resented by both groups.

Shaka's father died in 1816, when Shaka was about 30 years old. He took over the Zulu clan his father had led. He reorganized, rearmed, and retrained the Zulu army. He also instituted a draft system.

The new weapons and tactics were devastatingly effective against the other clans. Shaka's army killed hundreds of thousands of other African people. Refugees fled inland and those who remained were incorporated into the Zulu empire.

After his assassination in 1828, the Zulu empire lived on, ruled by various relatives of Shaka. Over time, the Boers and the British began encroaching on Zulu territory. The Boers captured much of southern Zululand, but were forced to give it back. Then, in 1879, the British declared war. The Zulu army of 50,000 soldiers won the first battle in an enormous victory, but it prompted the British to send even more soldiers and supplies. The war ended with the division of the Zulu empire into 13 smaller territories. Even so, Zulu resistance to British rule continued until the first decade of the 1900s.

Cecil Rhodes

One of the most important Europeans in the history of Southern Africa was Cecil Rhodes. He moved from England to Cape Colony (South Africa) in 1870, at the age of 17. He bought up gold and diamond mines throughout the area. By 1891, the company he started, De Beers, produced 90 percent of the world's diamonds.

Shaka created a modern African empire and built an army that fought colonial powers for more than 50 years.

▼ CRITICAL THINKING

1. *Identifying Perspectives* What do you think Shaka thought of the Europeans in Africa?

2. *Assessing* Was Shaka a successful leader? Why or why not?

Directions: On a separate sheet of paper, answer the questions below. Make sure you read carefully and answer all parts of the questions.

College and Career Readiness

15 *Problem Solving* You are an environmentalist working for the Nature Conservancy in Madagascar. Use the Internet to research the causes and consequences of severe deforestation. Write a brief report exploring ways to address the problem.

DBQ Analyzing Primary Sources

Use the document to answer the following questions.

Before becoming South Africa's first black president, Nelson Mandela struggled for decades to end apartheid in his country. In 2005 he delivered a speech in London comparing apartheid to poverty today.

PRIMARY SOURCE

"But in this new century, millions of people in the world's poorest countries remain imprisoned, enslaved, and in chains.

They are trapped in the prison of poverty. It is time to set them free.

Like Slavery and Apartheid, poverty is not natural. It is man-made and it can be overcome and eradicated by the actions of human beings.

And overcoming poverty is not a task of charity, it is an act of justice. It is the protection of a fundamental human right, the right to dignity and a decent life."

—Nelson Mandela, February 3, 2005

16 *Identifying Perspectives* Why does Mandela insist that poverty is man-made rather than natural?

17 *Analyzing Arguments* What metaphor does Mandela use to describe the effects of poverty on people? Why does he use this metaphor?

18 *Identifying Perspectives* Why do you think Mandela makes a distinction between charity and justice?

Applying Map Skills

Use the Unit 6 Atlas to answer the following questions.

19 *Places and Regions* Use your mental map of Southern Africa to name the northernmost countries in the subregion.

20 *Environment and Society* Identify areas in Southern Africa where people are most likely to rely on wells to access groundwater. Why might people in these areas have health problems?

21 *Human Systems* Describe how the location of mineral resources has affected the development of cities in Southern Africa. Give examples.

Exploring the Essential Question

22 *Making Connections* Choose one of the countries in Southern Africa. Use what you have learned about its history, politics, art, and economics to create a poster illustrating how human systems have shaped it as a place. Posters should be visual and can include photos, graphs, charts, and maps.

Research and Presentation

23 *Research Skills* Research the legacy of colonial rule in Zimbabwe. What factors led President Robert Mugabe to institute a forced land reform program? How effective was the program? What challenges does the country face today? Write a multi-paragraph summary of your findings and present it to the class.

Writing About Geography

24 *Argument* Using Botswana as an example, write an essay suggesting steps other African countries can take to create a healthy economy. What measures have the people of Botswana taken to strengthen the country's economy? How might other countries benefit from this example? Remember to use standard grammar, spelling, sentence structure, and punctuation in your response.

Need Extra Help?

If You've Missed Question	**15**	**16**	**17**	**18**	**19**	**20**	**21**	**22**	**23**	**24**
Go to page	590	594	594	594	476	476	476	576	584	580

South Asia

UNIT 7

Chapter 25
India

Chapter 26
Pakistan and
Bangladesh

Chapter 27
Bhutan, Maldives,
Nepal, and Sri Lanka

1 Religion These children have been painted as part of a religious festival. Hinduism and Buddhism began in the region, which also has three of the four most populous Muslim countries in the world.

EXPLORE the REGION

Though seemingly isolated from the rest of Asia by mountains, SOUTH ASIA has long-practiced religious and cultural traditions that have influenced—and been shaped by—other parts of Asia. With roots in ancient civilizations, the region today is home to vibrant software and film industries, rising economic powers, and countries that play an important role in contemporary international issues.

THERE'S MORE ONLINE

3 Rivers Major rivers that drain the region—such as the Indus, Ganges, and Brahmaputra—flow from their source high in the Himalaya in the north to the floodplains in the south, carrying fertile soil used to grow food for the region's large population.

4 Climate Strong winds traveling across warm seas bring the wet season to South Asia, providing rain vitally needed by the region's farms but also causing flooding in low-lying areas.

2 Mountains As the Indian subcontinent moves slowly and steadily into Asia, it pushes up the highest mountains in the world, the 1,500-mile- (2,414-km-) long Himalaya, which have individual peaks that reach more than 5 miles (8 km) high.

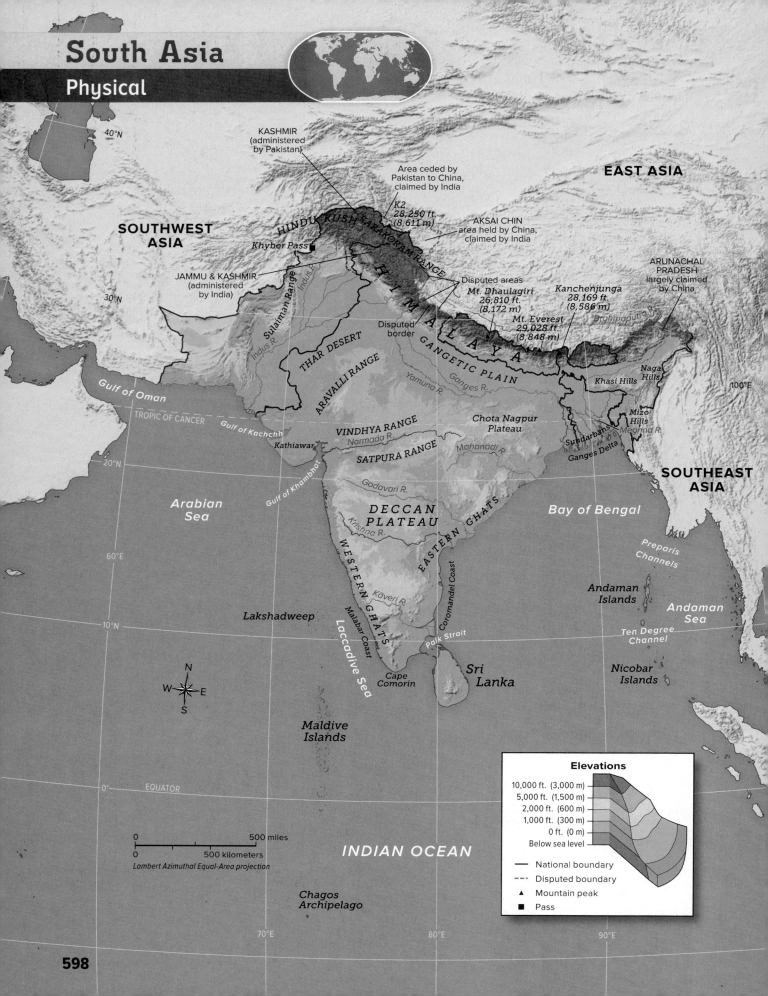

South Asia
Physical

EAST ASIA

40°N

KASHMIR
(administered
by Pakistan)

Area ceded by
Pakistan to China,
claimed by India

K2
28,250 ft.
(8,611 m)

AKSAI CHIN
area held by China,
claimed by India

**SOUTHWEST
ASIA**

HINDU KUSH KARAKORAM RANGE

Khyber Pass

JAMMU & KASHMIR
(administered
by India)

ARUNACHAL
PRADESH
largely claimed
by China

30°N

Sulaiman Range

Indus R.

Disputed areas

Mt. Dhaulagiri
26,810 ft.
(8,172 m)

Kanchenjunga
28,169 ft.
(8,586 m)

H I M A L A Y A

Disputed
border

Brahmaputra R.

Mt. Everest
29,028 ft.
(8,848 m)

GANGETIC PLAIN

THAR DESERT

ARAVALLI RANGE

Yamuna R.

Ganges R.

Khasi Hills

Nagal
Hills

100°E

Gulf of Oman

TROPIC OF CANCER

Gulf of Kachchh

VINDHYA RANGE

Narmada R.

Chota Nagpur
Plateau

Mizo
Hills

Meghna R.

20°N

Kathiawar

SATPURA RANGE

Mahanadi R.

Sundarbans

Ganges Delta

**SOUTHEAST
ASIA**

Gulf of Khambhat

Godavari R.

*Arabian
Sea*

**DECCAN
PLATEAU**

Krishna R.

EASTERN GHATS

Bay of Bengal

60°E

WESTERN GHATS

Preparis
Channels

Lakshadweep

Malabar Coast

Kaveri R.

Coromandel Coast

*Andaman
Islands*

*Andaman
Sea*

10°N

*Laccadive
Sea*

Ten Degree
Channel

Polk Strait

Cape
Comorin

**Sri
Lanka**

*Nicobar
Islands*

N
W   E
S

*Maldive
Islands*

0°   EQUATOR

Elevations		
10,000 ft. (3,000 m)		
5,000 ft. (1,500 m)		
2,000 ft. (600 m)		
1,000 ft. (300 m)		
0 ft. (0 m)		
Below sea level		

── National boundary
- - - Disputed boundary
▲ Mountain peak
■ Pass

0   500 miles
0   500 kilometers
Lambert Azimuthal Equal-Area projection

INDIAN OCEAN

*Chagos
Archipelago*

70°E   80°E   90°E

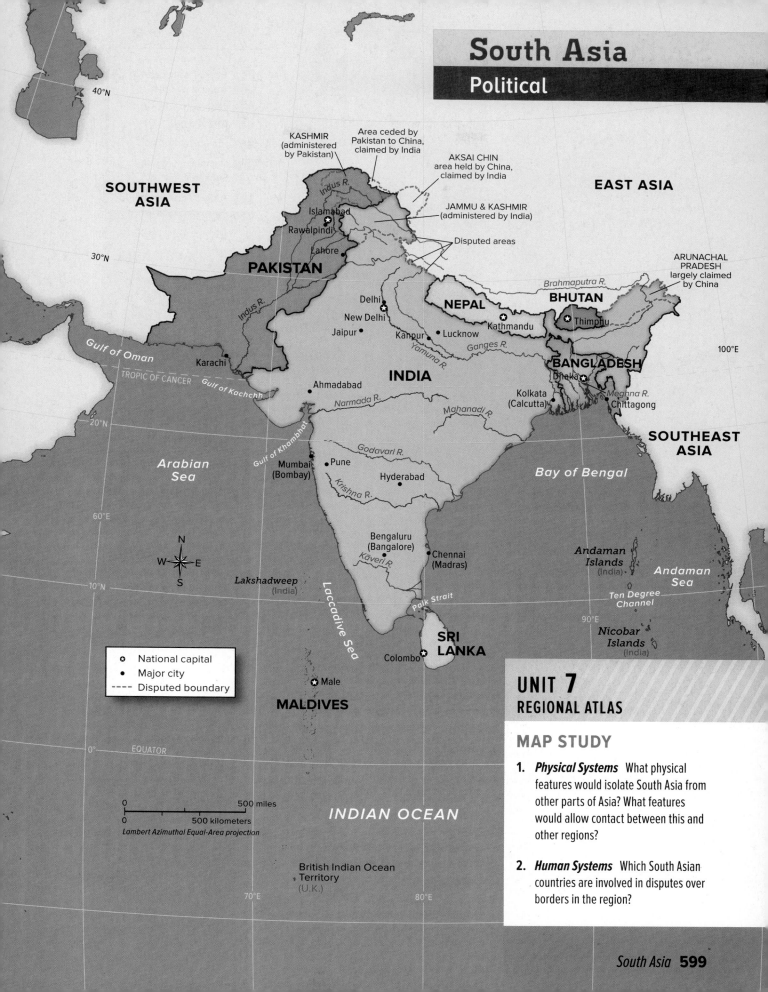

South Asia
Political

KASHMIR
(administered
by Pakistan)

Area ceded by
Pakistan to China,
claimed by India

AKSAI CHIN
area held by China,
claimed by India

JAMMU & KASHMIR
(administered by India)

Disputed areas

SOUTHWEST
ASIA

EAST ASIA

Indus R.

Islamabad
Rawalpindi

Lahore

PAKISTAN

ARUNACHAL
PRADESH
largely claimed
by China

Brahmaputra R.

Delhi
New Delhi

NEPAL

BHUTAN

Kathmandu

Thimphu

Jaipur

Kanpur
Lucknow

Yamuna R.

Ganges R.

BANGLADESH

Dhaka

Indus R.

Karachi

TROPIC OF CANCER

Gulf of Oman

Gulf of Kachchh

INDIA

Ahmadabad

Narmada R.

Kolkata
(Calcutta)

Meghna R.

Chittagong

SOUTHEAST
ASIA

Gulf of Khambhat

Mumbai
(Bombay)

Pune

Hyderabad

Godavari R.

Mahanadi R.

Krishna R.

*Arabian
Sea*

Bay of Bengal

Bengaluru
(Bangalore)

Chennai
(Madras)

Kaveri R.

*Andaman
Islands*
(India)

*Andaman
Sea*

Lakshadweep
(India)

*Laccadive
Sea*

Ten Degree
Channel

Palk Strait

SRI
LANKA

*Nicobar
Islands*
(India)

Colombo

⊙ National capital
• Major city
--- Disputed boundary

⊙ Male

MALDIVES

MAP STUDY

1. *Physical Systems* What physical
features would isolate South Asia from
other parts of Asia? What features
would allow contact between this and
other regions?

2. *Human Systems* Which South Asian
countries are involved in disputes over
borders in the region?

EQUATOR

0   500 miles
0   500 kilometers
Lambert Azimuthal Equal-Area projection

INDIAN OCEAN

British Indian Ocean
Territory
(U.K.)

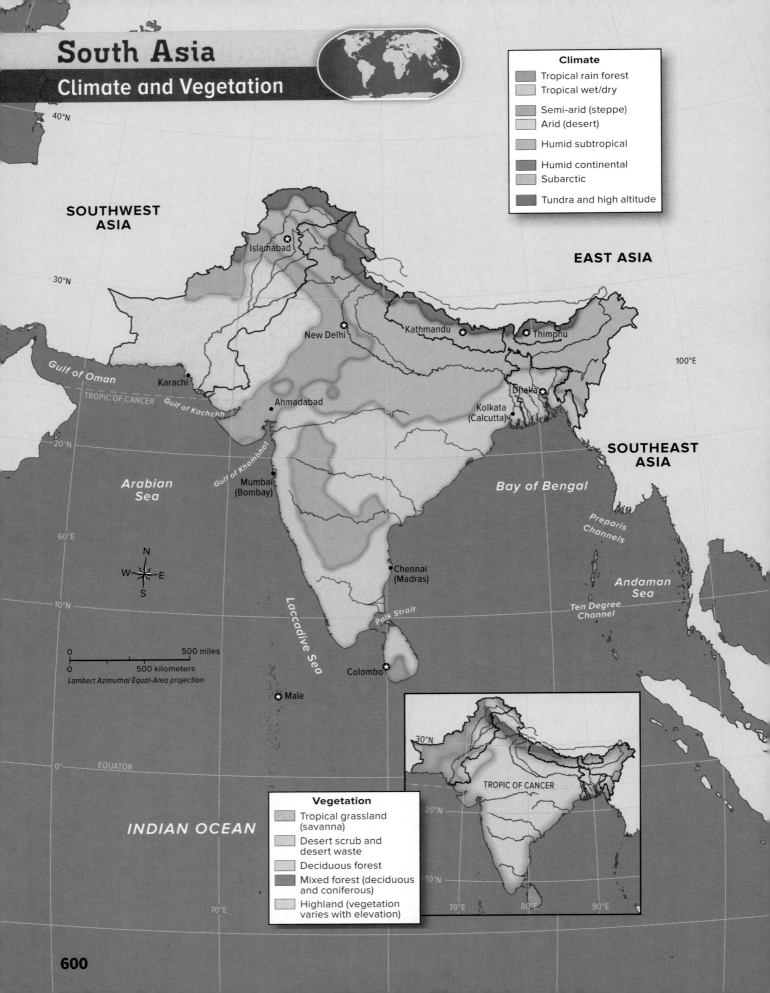

South Asia
Climate and Vegetation

SOUTHWEST ASIA

EAST ASIA

SOUTHEAST ASIA

40°N

30°N

20°N

TROPIC OF CANCER

60°E

20°N

10°N

0° EQUATOR

70°E

100°E

Islamabad

New Delhi

Kathmandu

Thimphu

Dhaka

Kolkata
(Calcutta)

Karachi

Ahmadabad

Mumbai
(Bombay)

Chennai
(Madras)

Colombo

Male

Gulf of Oman

Gulf of Kachchh

Gulf of Khambhat

Arabian Sea

Laccadive Sea

Palk Strait

Bay of Bengal

Preparis Channels

Andaman Sea

Ten Degree Channel

INDIAN OCEAN

N
W E
S

0   500 miles
0   500 kilometers
Lambert Azimuthal Equal-Area projection

Climate
- Tropical rain forest
- Tropical wet/dry
- Semi-arid (steppe)
- Arid (desert)
- Humid subtropical
- Humid continental
- Subarctic
- Tundra and high altitude

Vegetation
- Tropical grassland (savanna)
- Desert scrub and desert waste
- Deciduous forest
- Mixed forest (deciduous and coniferous)
- Highland (vegetation varies with elevation)

30°N

TROPIC OF CANCER

20°N

10°N

70°E   80°E   90°E

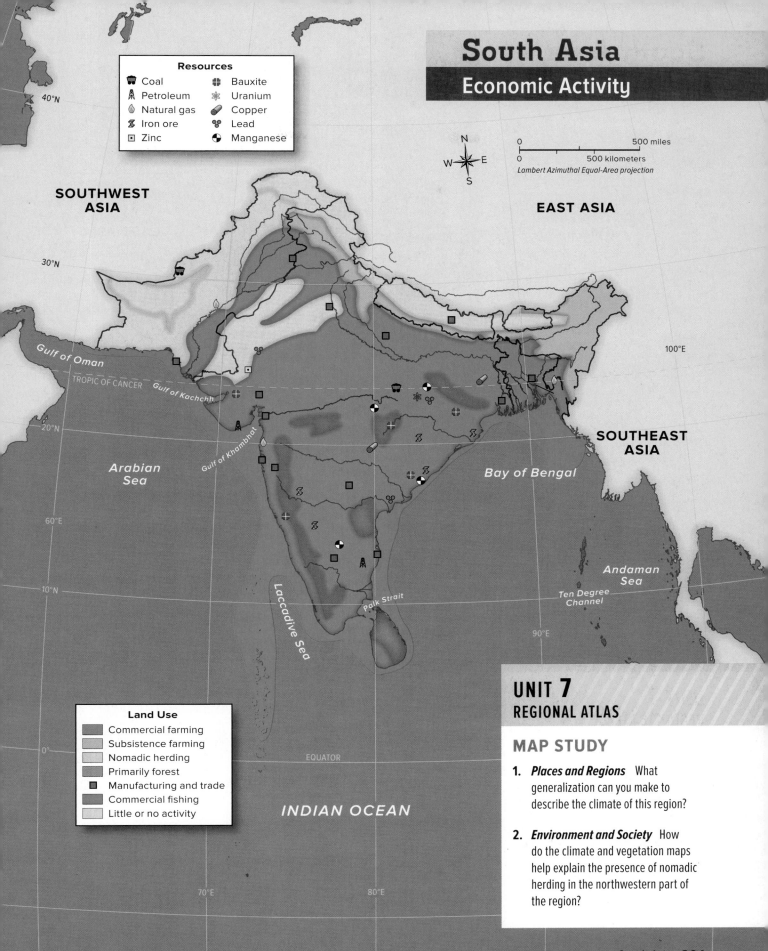

Resources

Coal	Bauxite
Petroleum	Uranium
Natural gas	Copper
Iron ore	Lead
Zinc	Manganese

SOUTHWEST
ASIA

EAST ASIA

N
W · E
S

| 0 | 500 miles |
| 0 | 500 kilometers |

Lambert Azimuthal Equal-Area projection

40°N

30°N

Gulf of Oman

TROPIC OF CANCER   Gulf of Kachchh

Gulf of Khambhat

20°N

Arabian
Sea

Bay of Bengal

SOUTHEAST
ASIA

100°E

60°E

Andaman
Sea

Ten Degree
Channel

10°N

Laccadive Sea

Palk Strait

90°E

Land Use

- Commercial farming
- Subsistence farming
- Nomadic herding
- Primarily forest
- Manufacturing and trade
- Commercial fishing
- Little or no activity

0°

EQUATOR

INDIAN OCEAN

70°E   80°E

UNIT 7
REGIONAL ATLAS

MAP STUDY

1. **Places and Regions** What generalization can you make to describe the climate of this region?

2. **Environment and Society** How do the climate and vegetation maps help explain the presence of nomadic herding in the northwestern part of the region?

South Asia
Population Density

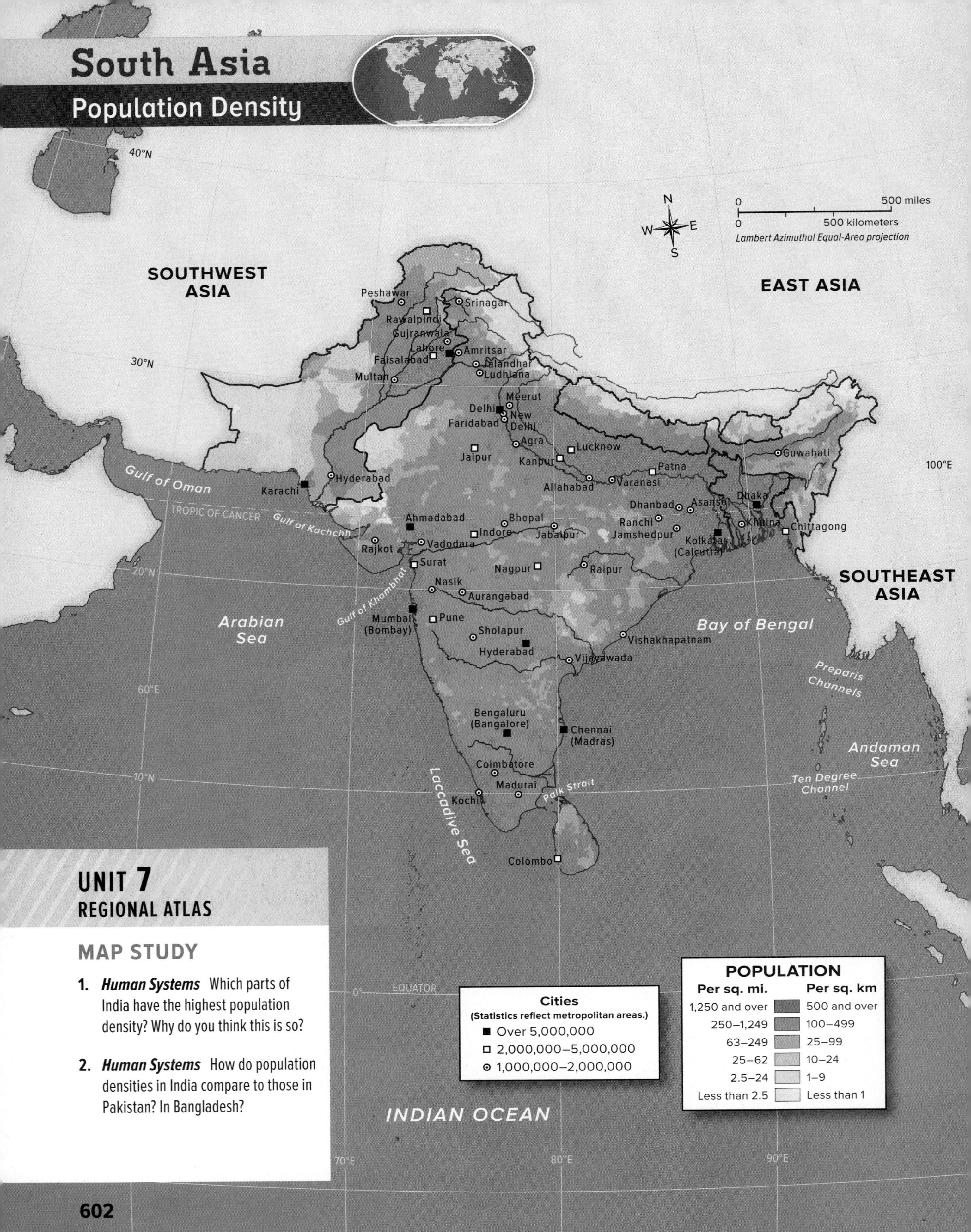

40°N

SOUTHWEST ASIA

EAST ASIA

N W E S

| 0 | 500 miles |
| 0 | 500 kilometers |

Lambert Azimuthal Equal-Area projection

30°N

Peshawar
Srinagar
Rawalpindi
Gujranwala
Lahore
Amritsar
Faisalabad
Jalandhar
Multan
Ludhiana
Meerut
Delhi
New Delhi
Faridabad
Agra
Jaipur
Lucknow
Kanpur
Guwahati
100°E
Gulf of Oman
Karachi
Hyderabad
Allahabad
Varanasi
Patna
TROPIC OF CANCER
Gulf of Kachchh
Ahmadabad
Bhopal
Dhanbad
Asansol
Dhaka
Ranchi
Khulna
Chittagong
Indore
Jabalpur
Jamshedpur
Rajkot
Vadodara
Kolkata (Calcutta)
Surat
Nagpur
Raipur
20°N
SOUTHEAST ASIA
Nasik
Gulf of Khambhat
Aurangabad
Arabian Sea
Mumbai (Bombay)
Pune
Sholapur
Bay of Bengal
60°E
Hyderabad
Vishakhapatnam
Preparis Channels
Vijayawada
Bengaluru (Bangalore)
Chennai (Madras)
Andaman Sea
10°N
Coimbatore
Ten Degree Channel
Madurai
Kochi
Palk Strait
Laccadive Sea
Colombo
0° EQUATOR

70°E 80°E 90°E

INDIAN OCEAN

UNIT 7
REGIONAL ATLAS

MAP STUDY

1. *Human Systems* Which parts of India have the highest population density? Why do you think this is so?

2. *Human Systems* How do population densities in India compare to those in Pakistan? In Bangladesh?

Cities
(Statistics reflect metropolitan areas.)

■ Over 5,000,000
□ 2,000,000–5,000,000
⊙ 1,000,000–2,000,000

POPULATION

Per sq. mi.		Per sq. km
1,250 and over		500 and over
250–1,249		100–499
63–249		25–99
25–62		10–24
2.5–24		1–9
Less than 2.5		Less than 1

India

ESSENTIAL QUESTION • *How do physical systems and human systems shape a place?*

Why Geography Matters
India's Population Structure

Lesson 1
Physical Geography of India

Lesson 2
Human Geography of India

Lesson 3
People and Their Environment: India

Geography Matters...

India is an ethnically diverse country whose population practices a wide range of religions and speaks more than a dozen major languages. Laws have helped to lessen the injustice formerly faced by members of "untouchable" castes, ethnic minorities, and women. But many problems and tensions remain. Despite the challenges of great economic inequality, India's economic power is growing. It has one of the world's largest pools of trained scientists and engineers. India's films, literature, and music have attracted a global following. While some 70 percent of its people live in rural villages, India has three of the world's largest cities.

◄ This woman in Pushkar, India, is wearing traditional dress and jewelry.

©Jens Kalaene/dpa/Corbis

603

India's population structure

In the third decade of this millennium, India may become the world's most populous country. China is in the lead, but its policies to limit population growth are rigorous, and the population growth rate has declined. India's population growth rate also has fallen. But the population is still large and, unlike China, young. Together, these two countries contain nearly 40 percent of the world's population.

What are the main characteristics of India's population?

India's population growth rate is higher than the world average. In 2015 the number of male children age 4 or under was about 62 million. This age group alone grew by about 9 percent between 1991 and 2015. India's youngest age groups dominate. Males and females ages 10–24 make up as much as one-third of the country's population.

Over the past decade, India's growth rate has declined. But in 2015 the growth rate, 1.2 percent, led to population increase. This rate still exceeds the world's average annual growth rate of 1.08 percent in 2015. Fertility rates also have dropped. Based on the data, the U.S. Census Bureau predicts that by 2025 India will outrank China as the world's most populous country.

---

1. **Human Systems** Why does India have such a large young population?

What are the consequences of this population structure?

The structure and size of India's population produces multiple effects. The population will continue to grow as the large number of young people start families of their own. Additionally, a large population in a poor country leads to poverty. About one-third of the total population lives in poverty. Social services, health care, and education are difficult to provide to a population of this size. With such a young population, India's workforce has expanded, but so too has its need to provide jobs.

At the same time, India faces what some experts think may be a positive effect of its large and growing population. They think a young and growing workforce could fuel India's economy. Much of the young population yearns for educational opportunities and training. However, the country faces a monumental task in providing education and social services as an investment for the future.

---

2. **Places and Regions** How might its young, growing population benefit India's future?

How is India responding to its population situation?

To reduce population growth, India's national government has launched an initiative to distribute contraceptives. India's National Family Planning Program has trained service providers to introduce other birth control methods. However, action at the national level may not be enough.

Some Indian states are responding to India's population challenges. They have issued incentives to control the birthrate and offered payments for people who delay having children. In some states, people who have many children are barred from running for political office. However, not all states are implementing programs to slow population growth. This uneven response could thwart efforts to achieve national goals.

---

3. **Human Systems** How are governments in India important in controlling population growth?

THERE'S MORE ONLINE

EXPLORE a graph of population structure in India's largest cities • *COMPARE* India and U.S. population structure

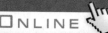

1 person is born per second (approx.)

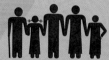

Average household size is 5 PEOPLE

Age Structure

MALE FEMALE

0-14 years: 28.1%

Median Age: 27.3 years

15-24 years: 18.1%

25-54 years: 40.7%

55-64 years: 7.2%

65 years and over: 6.0%

India total population: 1,314,100,000

Second Highest Population In The World (China Is Higher, U.S. Is Third)

Population Growth 1.2%

98TH IN THE WORLD

Urban Population 32%

Largest Cities

Delhi 25.7 Million

Mumbai 21.0 Million

Kolkata 11.8 Million

Labor Force 502.1 million

49% agriculture

31% services

20% industry

2nd largest in the world

Unemployment Rate 7.1%

Population below the poverty line 29.8%

networks

There's More Online!

- ☑ **IMAGE** Forest in Goa
- ☑ **IMAGE** Pilgrims at the Ganges
- ☑ **MAP** South Asia: Monsoons
- ☑ **INTERACTIVE SELF-CHECK QUIZ**
- ☑ **VIDEO** Physical Geography of India

LESSON 1
Physical Geography of India

ESSENTIAL QUESTION • *How do physical systems and human systems shape a place?*

Reading HELPDESK

Academic Vocabulary

- **fluctuate**
- **annual**

Content Vocabulary

- **subcontinent**
- **alluvial plain**
- **monsoon**
- **cyclone**
- **tsunami**

TAKING NOTES: *Key Ideas and Details*

LISTING As you read about the physical geography of India, use a graphic organizer like the one below to list factors that influence the climate of India.

Factors	Influence on Climate

IT MATTERS BECAUSE

Rugged mountains have isolated parts of India, while its rivers and monsoon winds have brought both benefits and catastrophic floods. Its huge population depends on the land, water, and other resources of India to meet its growing needs.

Landforms

GUIDING QUESTION *Why is the Gangetic Plain important?*

India's shape and position is the result of the movement of the Earth's tectonic plates. About 160 million years ago, geologists think that a large piece of land broke away from the landmass that is now Africa. This landmass collided with the southern edge of Asia about 50 million years ago. Over time, the force of this collision thrust up the world's highest mountains, the Himalaya. This mountain range stretches about 1,500 miles (2,414 km) from west to east and includes Mount Everest, Kanchenjunga, and over 100 other peaks that are more than 24,000 feet (7,300 m) above sea level. The Himalaya meet the Karakoram Range in northwestern South Asia, which in turn connects to the Hindu Kush farther west. These mountains separate South Asia from the rest of Asia, forming a **subcontinent**. India occupies most of the subcontinent.

At the foot of the Himalaya lies the Gangetic Plain. This plain is the world's longest **alluvial plain**, an area of fertile soil deposited by river floodwaters. It is also India's most densely populated area. West of the alluvial plain is the Thar Desert, bordered by the mineral-rich Aravalli Range. In eastern India, the Chota Nagpur Plateau is another mineral-rich region.

The Vindhya and Satpura Ranges divide India into northern and southern regions. To the south lies the Deccan Plateau, which has a relatively flat surface but rises to hundreds of feet high. It is covered with rich, black soil. Bordering the Deccan Plateau on the east are the Eastern Ghats, a low mountain range. Another range of low mountains, the Western Ghats, borders the western edge of the Deccan Plateau. The Nilgiri Hills, a fertile region for growing tea and coffee, are part of the Western Ghats.

India's long coastline includes the Malabar Coast on the west and the Ganges Delta and Coromandel Coast in the east. India also includes Lakshadweep, a group of about three dozen islands in the Arabian Sea, as well as the Andaman and Nicobar Islands at the southeastern edge of the Bay of Bengal. The southernmost tip of the subcontinent, Cape Comorin, marks the division between the Arabian Sea and the Bay of Bengal.

✓ READING PROGRESS CHECK

Explaining Why is South Asia called a subcontinent?

Water Systems

GUIDING QUESTION *How does water affect life in India?*

From sources high in the Himalaya, three major river systems—the Ganges, the Brahmaputra, and the Indus—fan out over the northern part of the subcontinent. These rivers carry fertile soil from the mountain slopes of the Himalaya and Karakoram onto the floodplains as the rivers swell with seasonal rains.

The Ganges River draws waters from a basin covering about 400,000 square miles (about 1 million sq. km). The Ganges flows throughout the year, even during the dry season from December to June. During the summer **monsoon**, heavy rains can cause devastating floods along the Ganges. Named for the Hindu deity Ganga, the Ganges is sacred to Hindus. A number of major cities, including Kanpur, Varanasi, and Kolkata (Calcutta), are located on its shores. The Brahmaputra River forms a broad delta as it joins the Ganges in Bangladesh and empties into the Bay of Bengal. The Indus flows mainly through Pakistan and empties into the Arabian Sea. The Indus is an important source of irrigation and a major transportation route.

Many other rivers begin in the Western Ghats. These rivers' many rapids and gorges make navigation impossible but make them suitable for hydropower. Many of India's major cities, such as Mumbai (Bombay) on the Arabian Sea and Chennai (Madras) on the Bay of Bengal, are port cities. People fish all along India's coastline and on nearly all of its rivers. However, it is getting harder for small fishing families to compete with larger commercial enterprises. Aquaculture, of both fish and shrimp, is increasing.

✓ READING PROGRESS CHECK

Identifying Which rivers come together to form a delta before emptying into the Bay of Bengal?

Analyzing
PRIMARY SOURCES

Hindu Pilgrims

"Lines of Indian pilgrims walk barefoot through the alleys, drawn by occasional glimpses of the holy river. At last, the alleys fall away, and the sluggish green river appears, smooth as a sheet of glass. From here the view extends to the distant eastern bank, hazed with brown dust. This year, the monsoon rains have been below average, and the Ganges lies low and tame between the banks.

Tens of narrow steps shine wetly. The pilgrims sigh, picking their way down the steps to the water's edge. It's sunrise, the most fortunate hour, and they're here to take a dip in the Ganges."

—Anika Gupta, "The Holy City of Varanasi," *Smithsonian.com* August 20, 2009

DBQ *IDENTIFYING CENTRAL ISSUES* Why do the pilgrims sigh as they walk to the river's edge?

subcontinent a large landmass that is part of a continent but still distinct from it

alluvial plain a floodplain on which flooding rivers have deposited silt

Pilgrims gather at the Ganges.

◄ CRITICAL THINKING
1. *Interpreting* What draws pilgrims to the Ganges River?
2. *Identifying* The Ganges is sacred to people of which religion?

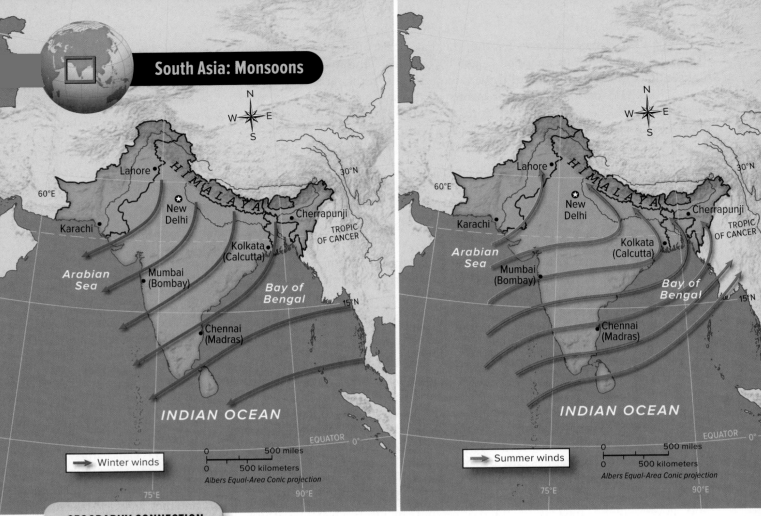

South Asia: Monsoons

Winter winds

0 — 500 miles
0 — 500 kilometers
Albers Equal-Area Conic projection

Summer winds

0 — 500 miles
0 — 500 kilometers
Albers Equal-Area Conic projection

GEOGRAPHY CONNECTION

Seasonal wind patterns called monsoons influence the subregion's climate.

1. **PHYSICAL SYSTEMS** What generalization can you make to describe the climate of South Asia?

2. **PHYSICAL SYSTEMS** Which monsoon winds are likely to bring wet weather? Which monsoons bring drier weather? Why?

monsoon in Asia, a seasonal wind that brings warm, moist air from the oceans in summer and reverses direction in the winter, bringing cold, dry air from Asia's interior

fluctuate to ebb and flow in waves

Climates, Biomes, and Resources

GUIDING QUESTION *How do monsoons affect India's weather and climate?*

India's climate is influenced by the shape of the subcontinent, its proximity to the Equator, and the Indian Ocean. The difference in temperatures of the air over the ocean and the air over the landmass creates the monsoon winds. These winds cause rainfall and temperature to **fluctuate**, creating wet and dry seasons.

From about mid-June to the end of September, the winds bring hot, wet weather. This occurs during the summer period because heated air rises over the land, which pulls in the moist ocean air from the south and southwest. The monsoon winds cause heavy rains and flooding across the subcontinent during these months. Then, from early October to February, the winds change direction. In these winter months, air from the Asian interior in the north is drawn across the subcontinent toward the ocean. The air is dry and cool compared to the wet season. The weather turns hot and humid with little rainfall from March to mid-June and the winds turn calm. The weather becomes hot with little rainfall but high humidity in the south and dry conditions in the north.

Landforms and location affect the amount of rain that monsoon winds bring to different areas. Different amounts of rain create different biomes. When rain sweeps over the Ganges-Brahmaputra Delta, the Himalaya block rain clouds from moving north. Instead, the rain moves west to the Gangetic Plain. Farmers in this densely populated plain depend on these rains.

When summer monsoons rise over the Western Ghats, the air cools and releases heavy rains that create rain forests in this part of the country. As the winds go over the mountains, they lose most of their moisture. This rain shadow makes a dry area of scrub and deciduous forests. The center of the Deccan Plateau, between the Eastern and Western Ghats, is semi-arid steppe. The driest area of India is the Thar Desert, east of the Indus River. The vegetation here is desert scrub of low, thorny trees, and grasses. The rest of the area around the Thar Desert is semi-arid grassland. Few trees grow in this steppe. The northeastern coast of India is quite different, however. It receives considerable rainfall and has a forested biome, including the world's largest protected mangrove forest.

Natural Disasters

Temperature and rainfall affect agriculture in the subregion. High temperatures and ample water allow farmers to produce plentiful crops, especially on the alluvial plains. These crops include the rice that many people in India depend on year-round. The extreme heat, however, can dry the fields, and without rainfall there is drought.

The summer monsoons bring problems as well as benefits. Rainfall waters crops, but areas outside the path of the monsoon may suffer from drought. Large amounts of rain from the monsoons cause flooding in low-lying land. The floods deposit rich silt on the floodplains which renews soil fertility. But the floods can also kill people and livestock, ruin crops, and leave thousands homeless.

Cyclones are a natural hazard in South Asia. They begin in the Bay of Bengal or the Arabian Sea. Much like hurricanes that originate in the Atlantic Ocean, they can be dangerous and destructive. Tropical cyclones bring torrential rains and winds that can reach speeds of more than 100 miles (161 km) per hour. Cyclones often cause storm tides, or surges of high water along the coasts.

Coastal regions also face the threat of **tsunamis**. Tsunamis are triggered by underwater earthquakes. The huge waves of the 2004 Indian Ocean tsunami destroyed entire villages and killed hundreds of thousands of people.

cyclone a storm with heavy rains and high winds that blow in a circular pattern around an area of low atmospheric pressure

tsunami originally a Japanese term, is a huge sea wave caused by an undersea earthquake

This forest is located in the Indian state of Goa.

▼ **CRITICAL THINKING**

1. Analyzing Visuals What is the dominant type of tree that grows in this forest? What does that tell you about its climate?

2. Making Connections Why is it difficult for India to protect its forests?

Climates and Biomes

The climates of different areas of India make diverse biomes of vegetation and animal life. About one-fourth of India's land is forest, but the types of trees and plants vary with rainfall. The heartland of northern India is a humid subtropical zone. Areas that receive more than 80 inches (203 cm) of rainfall on an **annual** basis have generally tropical evergreen and mixed evergreen-deciduous forests. Areas with 60 to 80 inches (152 to 203 cm) of annual rainfall have mostly tropical deciduous trees, while areas that receive less than 60 inches (152 cm) are dry deciduous forests. Tropical palms and bamboo are found throughout the country. Much of the land along the coast is tidal marsh or wetland, with mangrove forests in some areas. Wetlands are used for growing rice and are important wildlife habitats.

Northwestern India receives less than 20 inches (51 cm) of rain in some desert areas. Regions in the northwest and the central southern subcontinent that receive 20 to 40 inches (51 to 102 cm) of rain are steppe, covered largely with thorny scrub, acacia, and palm trees.

India has a wide diversity of animals. Among the large mammals are the Indian elephant, the Indian rhinoceros, and several species of tiger. The numbers of some of these mammals have decreased to such low levels that the animals are in danger of extinction. There are many species of monkeys and other primates. India is home to more than 1,200 species of birds, including cranes, herons, flamingos, and peacocks, the national bird. Many species of lizards and snakes—especially cobras—are widespread. Crocodiles and turtles are found in the country's rivers, swamps, and coastal regions.

annual occurring once a year

Natural Resources

India's mineral resources are numerous and widespread. Iron ore is abundant. Other important mineral resources include copper, bauxite (used to make aluminum), zinc, lead, gold, and silver. Petroleum is one of India's most valuable resources. Oil reserves are located along India's northeast coast, near the Ganges Delta, and in the Arabian Sea. India produces a slight surplus of coal from about 500 mines in many different areas of the country. India also produces a small amount of uranium.

The country has large areas of rich soil that support agriculture and timber. Timber resources include India's sandalwood, sal, and teak woods. Nearly any type of wood is used for fuel, often illegally. More than half of India's total land area is used for growing crops. Given the population of India, however, the country does not have a large amount of high-quality cropland.

✓ **READING PROGRESS CHECK**

Applying What kind of weather would you expect to find in the Gangetic Plain in July?

LESSON 1 REVIEW

Reviewing Vocabulary

1. *Summarizing* Write a paragraph summarizing weather patterns created by monsoon winds.

Using Your Notes

2. *Listing* Use your graphic organizer on India's climate to write a paragraph about the factors that influence rainfall in India.

Answering the Guiding Questions

3. *Drawing Conclusions* Why is the Gangetic Plain important?

4. *Speculating* How does water affect life in India?

5. *Finding the Main Idea* How do monsoons affect India's weather and climate?

Writing Activity

6. *Informative/Explanatory* In a paragraph, explain how the Himalaya affect the climate of India.

610

Reading **HELP**DESK

Academic Vocabulary

- **dominate**
- **neutral**

Content Vocabulary

- *jati*
- **mercantilism**
- **imperialism**
- **reincarnation**
- **karma**
- *panchayat*
- **green revolution**
- **cottage industry**

TAKING NOTES: *Key Ideas and Details*

ORGANIZING As you read about the human geography of India, use a graphic organizer like the one below to organize information about India's culture and economy.

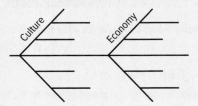

Culture   Economy

LESSON 2
Human Geography of India

ESSENTIAL QUESTION • *How do physical systems and human systems shape a place?*

IT MATTERS BECAUSE

As the world's most populous democracy, India is rich in human resources and has large amounts of natural resources. Meeting the basic needs of many of its people is a challenge even though it is one of the largest economies in the world.

History and Government

GUIDING QUESTION *How did modern India gain its independence?*

India's history dates back more than 4,500 years to the Indus Valley civilization, located in what is now modern-day Pakistan. Many historians think that a group of hunters and herders who spoke Indo-Aryan languages came from the northwest and settled in India in the 2000s B.C. Over time, a new Indian civilization emerged in the region with a rigid social structure based on caste. A caste, or *jati*, is the social position into which a person is born. The sacred writings of this new civilization, the Vedas, formed the basis of what became the Hindu religion. India's religions are key to understanding its history and culture.

Around A.D. 320, the Gupta Empire united much of India and built one of the world's most advanced civilizations. In the 700s Muslim invaders began arriving. A Muslim dynasty of mixed Mongol and Turkish heritage, the Moguls, ruled most of India from 1526 to the 1800s. During this period of Muslim rule, many Indians converted to Islam.

The final group of invaders were Europeans. Beginning in the 1490s, Europeans came to trade. Eventually, the British East India Company tightened control over India. They employed a policy of **mercantilism**. Under this economic system, colonies supplied raw materials to the colonizing country, which then sold finished goods back to to the colony. The **imperialist** policy of the British government led it to take direct control of India. The British called their Indian empire the British raj, the Hindi word for "empire." The British **dominated** India. They introduced the English language, restructured the educational system, built railroads, and developed a civil service and judiciary. Indians, however, were not treated as equal citizens and were forced to pay the costs of British domination.

jati in India, a group that defines one's occupation and social position

mercantilism the theory or practice of merchant or trading pursuits

imperialism the actions by which one country is able to extend power to control another country

dominate to have a commanding position

neutral not favoring either side in a quarrel, contest, or war

By the late nineteenth century, the Indian National Congress was calling for independence. After British troops fired on unarmed protesters at Amritsar in 1919, Mohandas K. Gandhi led Indians to seek freedom using nonviolent methods of civil disobedience. India won its independence in 1947. Britain divided the region into Hindu India and Muslim Pakistan. Conflicts between Pakistan and India over the territorial division of the state of Kashmir continue today.

Following independence, India launched a series of five-year plans to guide economic development under its first prime minister, Jawaharlal Nehru. Throughout the Cold War era, Nehru tried to keep India nonaligned, or **neutral**. He wanted to maintain friendly relations with both the United States and the Soviet Union. Two years after Nehru's death in 1964, his daughter, Indira Gandhi, became prime minister and continued this policy. She held office for a total of 15 years until her assassination in 1984. Her son Rajiv then became prime minister. He pursued a closer relationship with the United States until his assassination in 1991.

Today India is the world's most populous democracy. Ruled by a coalition, the government has been trying to improve life for farmers and rural Indians. The government approved the country's twelfth Five-Year Plan in October 2012.

☑ **READING PROGRESS CHECK**

Interpreting What was the attitude of Jawaharlal Nehru and Indira Gandhi toward the United States?

ANALYZING PRIMARY SOURCES

TIME
From the pages of TIME

The Last Straw: Salt Satyagraha, 1930

Mohandas Gandhi's Salt March is considered one of the most influential protests in history. Gandhi demonstrated how civil disobedience could be used to fight social and political injustices. His peaceful methods were a strong influence on U.S. civil rights leader Martin Luther King, Jr.

❝Britain's centuries-long rule over India was, in many ways, first and foremost a regime of monopolies over commodities like tea, textiles and even salt. Under colonial law, Indians were forbidden to extract and sell their own salt and instead were forced to pay the far higher price of salt processed in and imported from the U.K. In March 1930, Mohandas Gandhi, the charismatic and enigmatic independence leader, embarked on a 24-day march from the city of Ahmedabad to the small seaside town of Dandi, attracting followers along the way. The assembled throngs watched as he and dozens of others dipped into the sea to obtain salt. That act—for which more than 80,000 Indians would be arrested in the coming months—sparked years of mass civil disobedience that came to define both the Indian independence struggle and Gandhi himself. Known as the salt *satyagraha*—a Sanskrit term loosely meaning 'truth-force'—it carried the emotional and moral weight to break an empire.❞

—Ishaan Tharoor, "Top 10 Most Influential Protests," *Time*, June 28, 2011

TIME and the TIME logo are registered trademarks of Time Inc. used under license.

Mohandas Gandhi (center) leads fellow Indians in the 1930 Salt Satyagraha.

DBQ ▲ **CRITICAL THINKING**

1. *Identifying Central Issues* In what act of civil disobedience did Gandhi and his followers engage? Why do you think they chose this action?

2. *Identifying Cause and Effect* What actions by the British government led to the protests? What do you think motivated British policy?

Population Patterns

GUIDING QUESTION *Why are population patterns so important in India?*

One of the most significant characteristics of India's population is its size. With over 1.3 billion people, it is second only to China in population. It is estimated that India will surpass China to become the world's most populous country in the next 10 years.

Most Indians belong to one of two ethnic groups. About 25 percent, most of whom live in the south of India, speak Dravidian languages. About 72 percent speak Indo-Aryan languages, a subgroup of the Indo-European language family. Most of the Indo-Aryan-speaking people live in the north. Indians traditionally identify themselves by their religion as Hindus, Muslims, Buddhists, Sikhs, Jains, or Christians.

India has a relatively high average population density of about 1,145 people per square mile (442 per sq. km), but it is much higher in some places. The distribution of population varies from place to place due to factors such as climate, vegetation, and physical features that affect the number of people the land can support. The population density on the Gangetic Plain can be more than 2,000 people per square mile (800 people per sq. km). On the other hand, the Thar Desert and some mountain regions are very sparsely populated.

India's cities are among the world's largest and most densely populated. More than 50 urban areas in India have populations of more than a million. There are several megalopolises—super-sized cities that may include a chain of closely linked metropolitan areas—that are far larger. India's largest megalopolises are Delhi (DEH•lee), Mumbai, and Kolkata, each with populations of between 15 and 22 million people.

Although the cities are large, most of India's population is rural—about 68 percent of the people live in villages. They farm and work to grow enough food for their families. Part of their crops often goes to the owners of the fields they farm. In recent years, growing numbers of Indians have been migrating to urban areas, drawn by the hope of better jobs and higher wages. As urban populations grow, they strain public resources and facilities.

✓ READING PROGRESS CHECK

Identifying Central Issues How can India's cities be so large when most people live in rural areas?

Passengers travel on an overcrowded train in the Indian city of Lucknow on the occasion of World Population Day, an annual event designed to draw attention to global population issues.

▲ CRITICAL THINKING

1. Considering Advantages and Disadvantages As people move from villages to large cities, what are some advantages and disadvantages for them?

2. Making Predictions How will India's population growth affect transportation systems?

Society and Culture Today

GUIDING QUESTION *What is life in Indian villages like?*

The people of India speak hundreds of languages and dialects belonging to four different language families—Indo-European, Dravidian, Austro-Asiatic, and Sino-Tibetan. Hindi is the most widely spoken, while English is the common language for national, political, and business communication.

reincarnation rebirth in new bodies or forms of life

karma in Hindu belief, the sum of good and bad actions in one's present and past lives

panchayat village council

Indians identify with a variety of religions. Some 80 percent of people in India are Hindus, and many Hindus also identify themselves by a *jati,* a group that defines one's occupation and social position by birth. Hindus believe that after death people undergo **reincarnation**, or rebirth. By moving through multiple reincarnations, Hindus strive to overcome personal weaknesses. The law of **karma** states that good deeds—actions in accord with dharma, or rules of conduct—lead one to break the cycle of birth and death and attain salvation.

A little over 13 percent of Indians practice Islam, and much smaller numbers are Christian or Sikh. Only about 1 percent of Indians are Buddhist, even though Buddhism began in India. Most Sikhs live in northwestern India; many want an independent Sikh state there. Sikhism is a distinct and independent religion that arose out of the teachings of Guru Nanak. Sikhs believe in one God who is formless, all-powerful, all-loving, and without fear or hate towards anyone. One can achieve unity with God through service to humanity, meditation, and honest labor.

Village Life and City Life

In rural India, Hindus of the higher *jatis* live in the center of the village, while Muslims and those of lower *jatis* generally live in areas surrounding the center. Village streets are narrow, but people gather in open spaces next to temples or mosques, at a water well, or in front of the home of a wealthy or powerful villager. These open spaces could be public areas such as the **panchayat** hall, shops, a tea stall, a public radio hooked up to a loudspeaker, or a post office. The school is usually on the edge of the village so that children have room to play outside.

Most village houses are small with just one or two rooms. People and animals live in the same shelter, and few people have electricity or running water. The design of roofs and the materials used for building vary depending on an area's rainfall. Houses in rainy areas have sloped roofs and are likely to be made of bamboo and covered with metal or plastic; they may be built on stilts to protect from flooding. In drier regions, homes may be made of mud and have flat roofs. Brahmans and wealthier people also tend to live in the core areas of cities. As in

akg-images/British Library/Newscom

TIME LINE ⌄

WOMEN'S RIGHTS
in India ➜

Despite considerable progress in recent generations, women in India continue to struggle for equality in society.

▶ **CRITICAL THINKING**

1. Analyzing How can the increase in the literacy rate for women affect their quality of life in India?

2. Hypothesizing What might explain the difference between the legal status of women and the reality of their daily lives?

500s
Sati, the practice of widows killing themselves by fire, begins in parts of India.

500s ➜

1829
East India Company prohibits *sati* in British-ruled India, but enforcement proves difficult.

1917
First women's delegation meets with India's secretary of state to demand women's political rights.

1929
Child Marriage Restraint Act prohibits the marriage of girls younger than 12.

villages, streets are narrow and winding, with gathering places near businesses and public buildings. Markets in the city specialize in certain goods. For example, one street may include stores that sell books and stationery, while another market may sell cloth. Store owners and their employees may live in the store building.

Housing is a problem in India's largest cities. Real estate in Delhi, Kolkata, and Mumbai is very expensive. Many people are homeless and others live in slums. According to a 2011 report by the government of India, more than 7.6 million children under the age of 6 live in slums.

Artistic expression is important to Indians whether they live in cities or villages. The arts are as much a part of Indian life as religious practice. Two great epic poems—the *Mahābhārata* (muh•hah•BAH•ruh•tuh) and the *Rāmāyana* (rah•MAH•yah•nuh)—combine Hindu social and religious beliefs. India has numerous classical dance styles, most of which are based on themes from Hindu mythology. India's film industry is the world's largest, producing more full-length feature films each year than any other country. It is centered in Mumbai and is nicknamed "Bollywood," a combination of Bombay (Mumbai) and Hollywood.

Family Life

The family unit is very important to Indians of all religions. Many people live in extended families in which people of several generations make up a household. The oldest man in the household—whether father, grandfather, or uncle—is usually the head of the family, and his wife assigns tasks to the women of the household. Nearly all Indians marry and have children. Women are still expected to obey their husbands in all things. Divorce is rare. Parents or older relatives arrange most marriages, but more people in cities are choosing their own spouses in "love marriages." Hindus who belong to the same *jati* treat one another as relatives, even when there is no kinship connection. People are also expected to marry within their *jati*.

☑ READING PROGRESS CHECK

Comparing How are India's villages and cities similar?

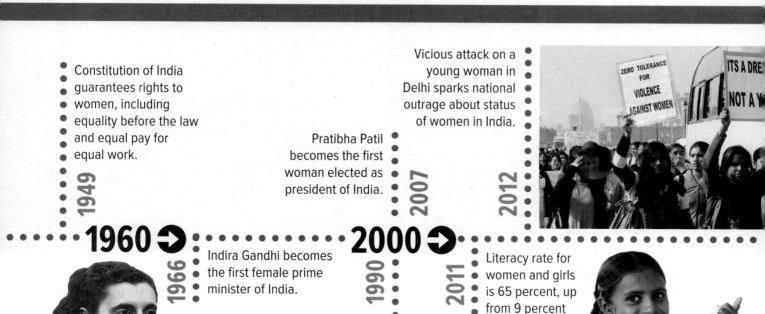

Constitution of India guarantees rights to women, including equality before the law and equal pay for equal work.

1949

1960 ➜

Indira Gandhi becomes the first female prime minister of India.

1966

Pratibha Patil becomes the first woman elected as president of India.

2007

Vicious attack on a young woman in Delhi sparks national outrage about status of women in India.

2012

2000 ➜

1990

Violent crimes against women increase by nearly 74 percent since 1980.

2011

Literacy rate for women and girls is 65 percent, up from 9 percent in 1951.

WHAT IS THE
FUTURE OF
KASHMIR?

Pakistanis, Indians, and Kashmiris today all have widely different views about who should govern Kashmir, a region in northern India. The dispute began in 1947 when India and Pakistan became independent from Britain. When Britain withdrew from the region, the Princely states ruled by maharajas were given the choice of joining either India or Pakistan. Hari Singh, the maharaja of Kashmir, wanted Kashmir to be an independent country. Many Pakistanis, however, believed that Kashmir should become part of Pakistan because it included many Muslims. As a result, Pakistanis invaded Kashmir on October 22, 1947. In response, the maharaja signed an agreement to make Kashmir part of India. In return, India agreed to provide military help to force out the Pakistanis. However, India and Kashmir were not successful. Conflict continued until a cease-fire agreement went into effect on January 1, 1949. The United Nations urged Pakistani forces to withdraw until a plebiscite, or vote, could be held to allow Kashmiris to choose between India and Pakistan. However, the Pakistanis did not withdraw and a plebiscite was not held. On July 27, 1949, Pakistan and India signed the Karachi Agreement, which established a line of control (LOC) that divided the territory. Neither Pakistan nor India, however, recognizes the LOC as an international border. Fighting between the two countries intensified when China took parts of northern Kashmir in the 1960s. In 1965 protests in Kashmir against the Indian government stirred unrest in Pakistani cities. Pakistan launched a covert operation, which quickly failed, against Indian forces in Kashmir. The United Nations Security Council became involved and India and Kashmir signed a cease-fire agreement on January 10, 1966. Today, tensions near the LOC continue as both Pakistan and India maintain a large military presence there.

Not only are Pakistan and India divided on who should govern Kashmir, but so are many Kashmiris. Some want to stay part of India. Others are tired of the conflicts and military presence in the area and want to form their own country as the maharaja wanted. Still others support the Pakistani position.

618

Map legend:

- Claimed by India, controlled by China
- Claimed by India, controlled by Pakistan
- Claimed by Pakistan, controlled by India
- - - - Disputed boundary
- —— State or province boundary
- Historic boundary of Jammu and Kashmir

0   250 miles
0   250 kilometers
Albers Equal-Area projection

A New Government is Needed

PRIMARY SOURCE

" We know that there is armed struggle going on in Kashmir since 1988, and we Kashmiris can sit down to analyse what progress we have made. India and Pakistan had armed conflicts over Kashmir, they were not only indecisive but they created more problems to all the parties to the dispute. In other words by use of force alone we cannot be successful in finding lasting solution. We need to sit down with open mind and see how we can bring peace, stability and prosperity to the region. Peace and stability can only return to Indian Sub-Continent if Kashmir dispute is resolved according to the wishes of the Kashmiri people. "

—Shabir Choudhry, Director of the Institute of Kashmir Affairs

Support for the Indian Government

PRIMARY SOURCE

" Government of India has expressed its willingness to accommodate the legitimate political demands of the people of the state of J&K [Jammu and Kashmir]. However, Pakistan sponsored terrorists have terrorised the population and hindered political dialogue by intimidating or silencing voices of moderation that wish to engage in dialogue. The human rights of the people of J&K have been systematically violated by such terror tactics and the kidnappings and killings of innocent people by terrorists.

Jammu & Kashmir is an integral part of India. There can be no compromise on India's unity and integrity. "

—Ministry of External Affairs, India

What do you think?   DBQ

1. **Drawing Conclusions** Why might Kashmiris want both Pakistan and India out of Kashmir?

2. **Identifying Central Issues** On what basis does India defend its claims to Kashmir?

3. **Hypothesizing** Why might it be dangerous for Kashmir to become an independent country?

netwrks

There's More Online!

☑ **GRAPH** Water and Sanitation in India

☑ **IMAGE** Bengal Tiger

☑ **IMAGE** Busy Street in Kolkata

☑ **IMAGE** Pollution in the Yamuna River

☑ **INTERACTIVE SELF-CHECK QUIZ**

☑ **VIDEO** People and Their Environment: India

LESSON 3
People and Their Environment: India

Reading HELPDESK

Academic Vocabulary

- **environment**
- **prohibit**

Content Vocabulary

- **pesticide**
- **ecotourism**

TAKING NOTES: *Key Ideas and Details*

FORMULATING QUESTIONS As you read about India's people and environment, use a graphic organizer like the one below to list details about India's environment. Then formulate a question about each section.

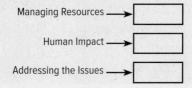

ESSENTIAL QUESTION • *How do physical systems and human systems shape a place?*

IT MATTERS BECAUSE

India is among the most biologically diverse countries in the world. Although it has less than 2.5 percent of the Earth's land area, it has nearly 17 percent of its human population. It also has the world's largest population of wild tigers and a number of other rare and endangered species. India's ability to manage its resources while raising the standard of living for its entire population will affect hundreds of millions of people, as well as many species of animals and plants.

Managing Resources

GUIDING QUESTION *Why are many resources in India not managed closely?*

In recent years, India's economy has grown rapidly, and the country has made significant advances in reducing poverty and improving the education and health of its citizens. Despite these advances, India's people still face high levels of poverty and inequality. The government's latest Five-Year Plan (2012–2017) calls for sustainable growth. In the past, India has often favored economic development over resource management and sustainability.

Development policies have put many of India's resources at risk. Forests provide a good example. India's forests are an important natural resource, providing a livelihood for some 250 million people. It is estimated, however, that 40 percent of India's forests are degraded. Deforestation began in the colonial era. The construction of the country's railways, which facilitated moving crops to market, encouraged the development of cash crop agriculture and the clearing of land.

Removal of forests also affects farmland by increasing erosion and flooding. Erosion degrades farmland by washing away the nutrients that make farmland productive. This in turn leads to further degradation because farmers overuse the soil in an effort to increase their harvests. Overuse further depletes nutrients. Erosion caused by wind is also a problem in drier areas of India.

No resources are more basic to human life than air and water. Yet these important resources are also threatened in India. Pollution of groundwater is a growing health threat. Much of India's water has high levels of chemicals that make it unsafe to drink because as water supplies are depleted, people draw water from deeper and deeper in the ground, where it is polluted by arsenic, fluoride, and heavy metals. Excessive fertilizer runoff and pollution from human and industrial wastes also contaminate water supplies.

Air pollution is another serious problem. A number of India's cities, including its capital, New Delhi, are among the world's most polluted. Overcrowding in urban areas leads to air pollution as a result of increased use of fossil fuels for transportation and industry.

☑ READING PROGRESS CHECK

Summarizing What resources in India are threatened?

Human Impact

GUIDING QUESTION *How has pollution affected the quality of life in India?*

The size and density of India's population creates a significant human impact on the **environment**. Ten percent of the world's population (about half of India's population) lives in the Ganges River valley. Although the Indian economy is growing rapidly, the country also holds a large concentration of people living in poverty, especially in the states of Chhattisgarh, Jharkhand, and Manipur. Large numbers of people cut down trees illegally because they need fuel for cooking. Removal of forests creates many other problems. In addition to soil erosion and flooding, loss of trees can cause temperatures to rise. Loss of forests destroys habitats for wildlife, including birds and other creatures that eat insects, snakes, and rodents.

India's already large cities continue to grow rapidly. Their growing populations have water and energy needs. To supply the residents of northern India with water and power, the Tehri Dam was built in the Himalaya. Located

environment the complex of physical, chemical, and biotic factors (as climate, soil, and living things) that act upon an organism or an ecological community and ultimately determine its survival

∨ GRAPH SKILLS

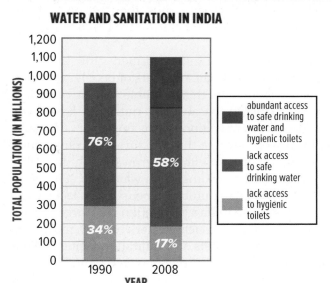

WATER AND SANITATION IN INDIA

Legend:
- abundant access to safe drinking water and hygienic toilets
- lack access to safe drinking water
- lack access to hygienic toilets

India provided more of its people with clean water and better hygiene between 1990 and 2008.

◀ CRITICAL THINKING

1. *Analyzing Visuals* Which improvement—access to safe drinking water or access to hygienic toilets—affected the greatest number of people?

2. *Synthesizing* Why is access to clean water still a problem even though over 80 percent of Indians have it?

Source: *New York Times*, March 21, 2012 "Liter by liter, Indians get cleaner water," Amy Yee

A man looks for recyclable items in the polluted waters of the Yamuna River in New Delhi.

▲ **CRITICAL THINKING**

1. Analyzing Visuals What does this photograph indicate about sanitation systems in India's major cities?

2. Identifying Cause and Effect How does India's growing population affect its environment?

on the Bhagirathi River, the Tehri Dam is India's highest dam. The dam was designed to increase the flow of water during the dry season. The poor in India's cities often also lack access to electricity; thus, they rely on biofuels. As cities grow, they increase the demand for timber to be used in construction. This further adds to deforestation.

India is rich in farmland. About half of India's farmers grow their own food, and the other half grow food to sell. The numbers of people who must use the farmland put great pressure on this resource. Much of India's farmland has become less productive because its nutrients have been depleted or because of excessive irrigation with improper drainage. Furthermore, too much plowing can change the structure and composition of the soil. Sometimes it is difficult or impossible to grow food in the soil.

India has made great progress in providing clean water and sanitation. For example, between 1990 and 2008, the number of people who lacked safe drinking water fell by half. But because India's population is so large, tens of millions of people still do not have access to clean water. By 2008, more than half of India's people still did not have toilets and sanitation facilities. Poor sewage and waste disposal threatens water supplies. Overcrowding in urban areas along with lack of sanitation leads to pollution of air and water resources as well as to the spread of disease.

The large population contributes to air pollution as well. Fires used for cooking and heating create smoke. In densely populated cities, construction dust and auto emissions create smog that can cause respiratory and heart problems, especially in children and older people. New Delhi has more than 7 million registered vehicles. Each day, another 1,400 cars and trucks get on the city's roads.

One of the worst environmental disasters in history occurred in December 1984 in Bhopal, a city in central India. A dangerous chemical leaked from a factory that made insecticides. It drifted over densely populated areas around the plant and killed between 15,000 and 20,000 people. Hundreds of thousands of others were blinded or suffered respiratory problems or other injuries. By the early twenty-first century, testing found that the soil and water around the factory site was still contaminated. This soil pollution has been blamed for high rates of birth defects and chronic health problems among people living near the factory.

☑ READING PROGRESS CHECK

Making Generalizations How did the 1984 Bhopal accident continue to affect people near the site in the early 2000s?

Addressing the Issues

GUIDING QUESTION *What steps has the Indian national government taken to combat pollution?*

Although India's environmental problems are severe, the national government has been involved in environmental protection. In 1976 India's constitution was amended to require the government to protect the environment. India's Supreme Court has made many decisions designed to protect India's air, water, ecosystems, and wildlife. It has considered a proposal to charge an extra tax on diesel cars because of the pollution they cause. In 2011 the Indian Supreme Court **prohibited** the production, distribution, and use of a **pesticide** called endosulfan because of its toxicity and its persistence in the environment.

prohibit to prevent from doing something

pesticide a chemical used to kill insects, rodents, and other pests

Almost 5 percent of India's land area falls within protected areas to ensure that the habitat of Bengal tigers is preserved.

▼ CRITICAL THINKING

1. Speculating How has the growth of India's human population caused the Bengal tiger to become an endangered species?

2. Defending Do you think it is important for India's government to protect tigers, elephants, and snow leopards? Defend your position.

Mike Ledwith/Moment/Getty Images

In 1972 the government of India passed a law to protect wildlife from poaching and smuggling. In 2003 this law was amended to increase punishment and penalties. The law was also expanded to protect plants and ecosystems as well as endangered animals. India has the world's largest population of wild tigers and many other endangered species such as the Asian elephant, the one-horned rhinoceros, and snow leopards. While the African elephant is endangered primarily as a result of poaching of the animals for their ivory tusks, the Asian elephant is at risk due to loss of its natural forest habitat. The clearing of forests for timber and agricultural purposes continues to threaten the viability of this species. To protect its unique biological heritage, India has created a network of 668 protected areas. These include national parks, wildlife sanctuaries, conservation reserves, and community reserves.

ecotourism the practice and business of recreational travel based on a concern for the environment

In 2012 the court restricted **ecotourism** in India's tiger reserves. Although ecotourism is designed to protect wildlife and ecosystems by increasing appreciation, environmentalists argued that tourists were threatening tiger habitat. After a temporary ban on any tourism, the court allowed tourism in the reserves but with far more restrictions.

In 2006 India's government passed the Forest Rights Act. This law gives authority to people living in the forest areas to prohibit forest clearance by large timber and mining companies. This law can trace its origins to a movement in the 1970s among Himalayan women. In what is considered one of India's first environmental movements, the women protected trees that were marked for logging. The women were determined to protect the trees that provided them with firewood and prevented soil erosion.

India continues its efforts to improve the quality of life for its people. The government, however, recognizes that sustainable development and economic growth are related. These principles guided India's Planning Commission as it prepared the near final draft of its twelfth Five-Year Plan. This latest plan states that India must improve living standards "through a growth process which is faster than in the past, more inclusive, and also more environmentally sustainable." The draft went on to say that India cannot "neglect the environmental consequences of economic activity, or allow unsustainable depletion and deterioration of natural resources."

☑ **READING PROGRESS CHECK**

Analyzing What perspective does India's twelfth Five-Year Plan take toward economic growth?

LESSON 3 REVIEW

Reviewing Vocabulary

1. *Hypothesizing* Write a paragraph to form a hypothesis about why India's Supreme Court put a temporary ban on the pesticide endosulfan.

Using Your Notes

2. *Formulating Questions* Choose one of the questions you formulated as you completed your graphic organizer. List three additional questions you might ask to begin researching the answer to the first one.

Answering the Guiding Questions

3. *Identifying Cause and Effect* Why are many resources in India not managed closely?

4. *Finding the Main Idea* How has pollution affected the quality of life in India?

5. *Classifying* What steps has the Indian national government taken to combat pollution?

Writing Activity

6. *Informative/Explanatory* Based on the information covered in this lesson, what environmental challenges are posed by India's rapid population growth?

Directions: On a separate sheet of paper, answer the questions below. Make sure that you read carefully and answer all parts of the questions.

Lesson Review

Lesson 1

❶ Diagramming Create a diagram to show the following movements that affected India's shape and position: a major tectonic event about 160 million years ago, a major tectonic event about 50 million years ago, and the thrusting up of the Himalaya.

❷ Assessing What is a monsoon? Discuss the positive and negative effects of monsoons in India.

❸ Identifying Cause and Effect Identify three types of timber resources in India. Explain the conditions that affect the supply of these resources in India.

Lesson 2

❹ Analyzing Cause and Effect Discuss the effects of British imperialist policy in India. Explain how the eventual push by Indians for self-rule was in response to this policy.

❺ Summarizing Provide an overview of the concept of *jati*.

❻ Identifying Central Issues What is the green revolution? Why has it given rise to controversy?

Lesson 3

❼ Identifying Cause and Effect What is the approximate percentage of farmers in India that grow their own food or grow food to sell? Create a graphic organizer to show the cause and effect of this pressure on India's farmland.

❽ Summarizing Summarize efforts taken by the Indian government to address environmental protection since 1976. Explain why these efforts have not achieved all of the desired results.

❾ Evaluating How did an act passed by India's government in 2006 connect to action among Himalayan women during the 1970s? What does this tell you about the determination of these Himalayan women? Within your response, include the name of the referenced act.

21st Century Skills

Use the document to answer the following questions.

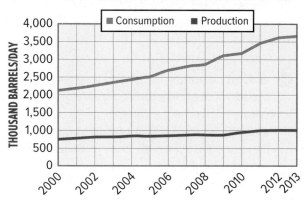

INDIA: OIL PRODUCTION AND CONSUMPTION

Source: U.S. Energy Information Administration

❿ Compare & Contrast Compare and contrast India's oil production and consumption in 2000 and in 2013.

⓫ Using Graphs Explain whether it would be logical to conclude, based on the data in the graph, that India was reliant on oil imports from 2000 to 2013.

Critical Thinking

⓬ Exploring Issues Suppose you are a geologist who is writing a full-page blog to identify and describe the three major river systems in India and their importance to life in India. Within your blog, include an explanation of at least four types of pollution that could threaten these river systems—and action that might be taken to address the threats.

Need Extra Help?

If You've Missed Question	❶	❷	❸	❹	❺	❻	❼	❽	❾	❿	⓫	⓬
Go to page	606	608	610	611	611	617	622	623	624	625	625	607

Directions: On a separate sheet of paper, answer the questions below. Make sure that you read carefully and answer all parts of the questions.

Applying Map Skills

Use the Unit 7 Atlas to answer the following questions.

13 **The World in Spatial Terms** Use your mental map of India to describe the spatial relationship among the following: Deccan Plateau, Eastern Ghats, Western Ghats, and the Vindhya and Satpura Ranges.

14 **Places and Regions** Name the location of highest elevation in India.

15 **Human Systems** Identify the plain that is the most densely populated area of India.

Exploring the Essential Question

16 **Making Connections** Use what you have learned about human systems—history, politics, population, society, culture, and economics—to write a one-page essay explaining how the 1984 Bhopal disaster and its aftermath are representative of ways the physical systems and human systems of India have shaped the country.

College and Career Readiness

17 **Reaching Conclusions** Imagine that you work for an organization that is interested in advancing environmental protection efforts in India. Your boss has asked you to research recent legislation and grassroots efforts related to environmental concerns in India. Provide a one-page memo to identify and explain the focus of the legislation and grassroots local efforts that you find the most effective. Your memo should explain why you find these to be the most effective.

Writing About Geography

18 **Informative/Explanatory** Do research and then use standard grammar, spelling, sentence structure, and punctuation to write an article explaining why the Gupta Empire was one of the world's most advanced civilizations of its time—and why it ceased to exist.

DBQ Analyzing Primary Sources

Use the document to answer the following questions.

PRIMARY SOURCE

"The new constitution gives no scope for retrenchment and therefore no scope for measures of social reform except by fresh taxation, the heavy burden of which on the poor will outweigh all the advantages of any reforms.

The reformed constitution keeps all the fundamental liberties of person, property, press, and association completely under bureaucratic control. All those laws which give to the irresponsible officers of the Executive Government of India absolute powers to override the popular will, are still unrepealed."

—C. Rajagopalachar, Introduction to *Freedom's Battle: Being a Comprehensive Collection of Writings and Speeches on the Present Situation,* 1922

19 **Understanding Historical Interpretation** What criticisms does the author make of the new constitution?

20 **Interpreting Significance** Explain whether the author's views align with Gandhi's general political and social views regarding the people of India.

Research and Presentation

21 **Gathering Information** With a partner, conduct research to learn more about the slum problem in India. Create a multimedia presentation to explain why and how these slums developed, the living conditions within the slums, and the complex problems that have resulted in them. Propose at least two actions that might ultimately lead to the decline of slums and a better quality of life for those who currently live in slums. Within your presentation, include video and audio clips, maps, photographs, and diagrams or graphs.

Need Extra Help?

If You've Missed Question	**13**	**14**	**15**	**16**	**17**	**18**	**19**	**20**	**21**
Go to page	598	598	598	623	624	611	626	626	615

Pakistan and Bangladesh

ESSENTIAL QUESTION • *How do physical systems and human systems shape a place?*

networks

There's More Online about Pakistan's and Bangladesh's geography.

CHAPTER 26

Why Geography Matters
Flood-Prone Pakistan and Bangladesh

Lesson 1
Physical Geography of Pakistan and Bangladesh

Lesson 2
Human Geography of Pakistan and Bangladesh

Lesson 3
People and Their Environment: Pakistan and Bangladesh

Geography Matters...

Pakistan and Bangladesh are lands of contrast. Northern Pakistan is home to K2, the world's second-tallest mountain, which is nearly 5.4 miles (8.7 km) high. Coastal Bangladesh is at sea level. Bangladesh has enough water to grow rice. In Pakistan, farmers must use dry-farming methods to grow crops.

Culturally, Pakistan and Bangladesh have similarities. They are both highly populous countries with strong Muslim cultures. Both share some of their history with neighboring India. Both countries are also struggling to develop their economies to overcome extreme poverty.

◄ Boy in Cox's Bazar, Bangladesh

flood-prone Pakistan
and Bangladesh

Floods are a common and major natural disaster in these two countries. In 2010 more than 18 million Pakistanis experienced terrible floods. In 1984 about 30 million Bangladeshis suffered through floods. Other major floods hit in 1992, 1993, 1995, 1998, 2004, 2005, 2007, and 2012. Why are floods so terrible here? What can Pakistan and Bangladesh do to mitigate them?

THERE'S MORE ONLINE

VIEW a chart about flooding in Pakistan and Bangladesh • EXPLORE a map of flood zones in Bangladesh

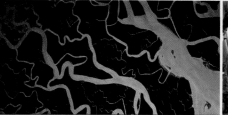

Why do floods threaten Pakistan and Bangladesh?

What do these countries do to reduce flood damage?

How has flooding harmed these two countries?

Pakistan and Bangladesh experience at least two floods a year, many of which are devastating. In Pakistan, heavy rains cause the Indus River to overflow its banks. Poor water management adds to the problem. Since population density is high along the river, these floods affect millions of people.

Many factors combine to cause floods in Bangladesh. The country has a very low elevation. Four major rivers—the Ganges, the Brahmaputra, the Jamuna, and the Meghna—flow through Bangladesh's plains, raising the probability of flooding. These floods strike when snow in the Himalaya melts, filling the rivers. The problem is made worse during the wet monsoon season or when cyclones hit.

Flooding causes a host of problems in Pakistan and Bangladesh. Hundreds of people can die in a flood. In 1988 floods in Bangladesh resulted in the deaths of more than 2,000 people. In 2010 record flooding in Pakistan killed some 1,700 people and destroyed almost 2,000 homes. The economic cost can rise to billions and billions of dollars.

The danger comes from more than the flood waters alone. The water washes away crops, reducing harvests, and stored food becomes wet and spoils more quickly. With reduced food supplies, people suffer from hunger. Ironically, too much water means too little drinkable water. River water is muddy and contaminated by chemicals washed off farmland and by the decomposition of the bodies of dead animals and other waste. This kind of contamination also raises the likelihood of deadly diseases.

Other countries faced with frequent flooding have taken steps to prevent it. They have built dikes and levees to hold water back, and put limits on building on floodplains or other flood-prone areas. They also have evacuation plans.

In 2005 world leaders met to discuss ways to prevent and recover from disasters like the floods in Pakistan and Bangladesh. They adopted a set of recommendations called the Hyogo Framework for Action. Both Pakistan and Bangladesh have tried to implement some of the recommendations. However, their efforts are hampered by the lack of funding.

1. **Physical Systems** In which country do you think flooding occurs more often? Why?

2. **Environment and Society** How are these problems affected by high population density in these countries?

3. **Human Systems** What conditions prevent Pakistan and Bangladesh from taking more steps to address flooding?

LESSON 1
Physical Geography of Pakistan and Bangladesh

ESSENTIAL QUESTION · *How do physical systems and human systems shape a place?*

Reading **HELP**DESK

Academic Vocabulary

- **generate**
- **maintain**

Content Vocabulary

- **delta**
- **floodplain**

TAKING NOTES: *Key Ideas and Details*

IDENTIFYING CAUSE AND EFFECT
As you read about the physical geography of Pakistan and Bangladesh, use a graphic organizer like the one below to identify examples of how physical systems shape human systems.

Cause	Effect

It Matters Because

Pakistan and Bangladesh are two of the ten most populous countries in the world. The landforms and climates of this subregion pose major challenges to the people living there.

Landforms

GUIDING QUESTION *How do landforms in Pakistan and Bangladesh affect human activities?*

Pakistan lies toward the northwestern edge of the Indian subcontinent, and Bangladesh lies toward the northeastern edge. Pakistan borders Afghanistan in Central Asia to the west and China to the north. Bangladesh is almost surrounded by India, but also shares a small border with Myanmar (Burma).

Physically, the two countries have few similarities and many differences. Bangladesh is almost entirely low and flat. Pakistan has large areas of low, flat plains as well as high mountains.

According to the theory of continental drift, the Indian subcontinent is slowly and steadily pushing into the southern edge of Asia. This collision thrust up the Himalaya mountain ranges. The Himalaya cover part of northern Pakistan, as do, farther north, the Karakoram Mountains. Farther west, the Hindu Kush range frames the northwest edge of Pakistan. In Pakistan, these mountain systems range from about 13,000 feet (3,962 m) to more than 19,500 feet (5,944 m) high. A few peaks soar more than 25,000 feet (7,620 m). One peak in the Karakoram, named K2, towers 28,250 feet (8,611 m). It is the second-highest mountain in the world.

These mountains are generally impassable, but the Hindu Kush have several wide, very high passes. The most important is the Khyber Pass in northwest Pakistan along the Afghanistan border. This pass has allowed armies to enter the Indian subcontinent. A highway through the Khunjerab Pass in the Karakoram is being modernized by China as a link to Pakistan.

Three lower mountain ranges run in a north-to-south direction from the western half of Pakistan and toward the Arabian Sea. These are, from north to south, the Salt Range, the Sulaiman Range, and the Kirthar Range.

West of the Kirthar Range is a highland region, the Baluchistan Plateau. The mountainous northern and western regions of Pakistan are prone to earthquakes. A major earthquake in northwestern Pakistan in 2005 killed more than 80,000 people.

To the east is the Indus River valley, which has rich alluvial soil. The Indus and its tributaries form two alluvial plains, the Punjab in the north and the Sind in the south. The Sind includes part of the sandy Thar Desert.

Bangladesh is very flat and low, reaching only about 30 feet (9 m) above sea level. The land is cut deeply by the Ganges and the Brahmaputra. In Bangladesh, these rivers are called the Padma and Jamuna, respectively. After joining, the rivers flow through central Bangladesh to empty into the Bay of Bengal in a **delta** system that is the largest in the world. Islands are scattered along the Bay of Bengal. The land is low, so the rivers often change course, altering the landscape.

Bangladesh has few hilly areas. The Chittagong Hills are the largest and the highest. Mount Keokradong, the highest point in the country, is located here, rising about 4,035 feet (1,230 m) high. Bangladesh also has lowland depressions. The Haor Basin, which covers 9,459 square miles (24,500 sq. km), has hundreds of depressions that form wetlands in the rainy months.

Along the southwest coast of Bangladesh is a massive wetland called the Sundarbans. This region is crossed by many waterways that are part of the Ganges Delta. It is thick with mangrove trees that live in the mixture of salt water and freshwater found in this coastal marshland.

delta an alluvial deposit at a river's mouth that is shaped roughly like the Greek letter *delta* (Δ)

✅ **READING PROGRESS CHECK**

Comparing How are the landforms of Pakistan and Bangladesh similar?

GEOGRAPHY CONNECTION

Pakistan has some of the highest and most rugged mountains in the world.

1. ***PLACES AND REGIONS*** Which pass allows travel into Pakistan through the Karakoram Range?

2. ***ENVIRONMENT AND SOCIETY*** Which pass was most likely used by armies invading South Asia from Central Asia?

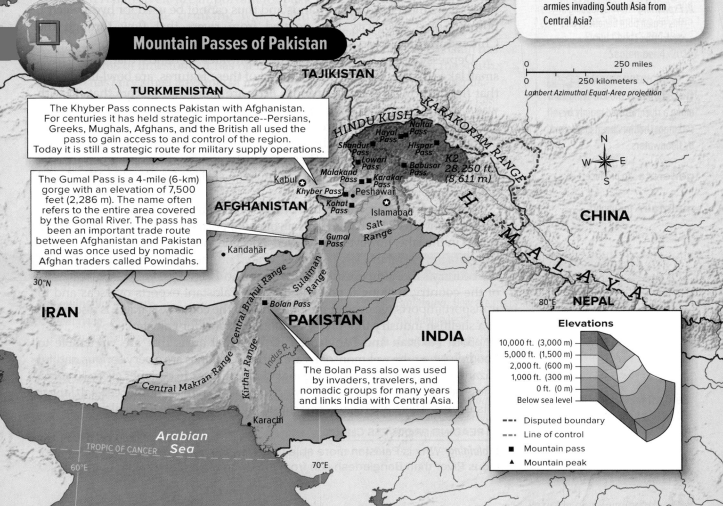

Mountain Passes of Pakistan

The Khyber Pass connects Pakistan with Afghanistan. For centuries it has held strategic importance--Persians, Greeks, Mughals, Afghans, and the British all used the pass to gain access to and control of the region. Today it is still a strategic route for military supply operations.

The Gumal Pass is a 4-mile (6-km) gorge with an elevation of 7,500 feet (2,286 m). The name often refers to the entire area covered by the Gomal River. The pass has been an important trade route between Afghanistan and Pakistan and was once used by nomadic Afghan traders called Powindahs.

The Bolan Pass also was used by invaders, travelers, and nomadic groups for many years and links India with Central Asia.

TAJIKISTAN
TURKMENISTAN
HINDU KUSH
KARAKORAM RANGE
Nahai Pass
Hayal Pass
Shandur Pass
Hispar Pass
Lowari Pass
Babusar Pass
K2 28,250 ft. (8,611 m)
Malakand Pass
Karakar Pass
Kabul
Khyber Pass
Peshawar
AFGHANISTAN
Kohat Pass
Islamabad
Salt Range
Gumal Pass
HIMALAYA
Kandahār
CHINA
30°N
Central Brahui Range
Sulaiman Range
NEPAL
80°E
Bolan Pass
IRAN
PAKISTAN
INDIA
Kirthar Range
Indus R.
Central Makran Range
Karachi
Arabian Sea
TROPIC OF CANCER
60°E
70°E

0 — 250 miles
0 — 250 kilometers
Lambert Azimuthal Equal-Area projection

Elevations
10,000 ft. (3,000 m)
5,000 ft. (1,500 m)
2,000 ft. (600 m)
1,000 ft. (300 m)
0 ft. (0 m)
Below sea level

--- Disputed boundary
--- Line of control
■ Mountain pass
▲ Mountain peak

LESSON 2
Human Geography of Pakistan and Bangladesh

ESSENTIAL QUESTION • *How do physical systems and human systems shape a place?*

Reading HELPDESK

Academic Vocabulary

- **infrastructure**
- **available**

Content Vocabulary

- **total fertility rate**
- **jute**

TAKING NOTES: *Key Ideas and Details*

COMPARING AND CONTRASTING As you read about the human geography of Pakistan and Bangladesh, use a graphic organizer like the one below to identify similarities and differences in the countries' histories.

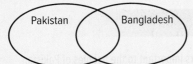

Pakistan   Bangladesh

IT MATTERS BECAUSE

Pakistan and Bangladesh share a similar history of Muslim influence, British colonialism, and the pursuit of independence. Both have struggled since independence to build stable governments and overcome poverty.

History and Government

GUIDING QUESTION *How did contemporary Pakistan and Bangladesh form?*

The area we know today as Pakistan has been home to civilizations for thousands of years. One of the earliest settlements here was at Mehrgarh, near the Bolan Pass. Evidence dating back to 7000 B.C. shows that agriculture was being practiced. Around 2500 B.C. a great civilization arose in the Indus River valley. This culture developed a writing system, a strong central government, and a thriving overseas trade. Ruins of the walled city of Mohenjo Daro include evidence of plumbing and other advanced technology.

Beginning in the A.D. 600s, a Buddhist kingdom ruled over what is today Pakistan and Bangladesh for several centuries. Hinduism from India was also practiced. However, the diversity of the two countries suggests that both were crossroads for other groups seeking trade and territorial control. Over the centuries, many peoples crossed or settled in the regions that later became Pakistan and Bangladesh.

Coming of Islam

Muslim invaders and traders brought Islam to southeast Pakistan in the A.D. 700s. About a century later, Muslim traders brought Islam to Bangladesh. Over time, Islam became the majority religion in both areas.

From the 1500s to the 1800s, Pakistan and Bangladesh were part of the Mogul Empire that also ruled much of India. Early in this period, Sikhism, which blends elements of Hinduism and Islam, arose in the Punjab in northwestern India. In the early 1800s, a Sikh named Ranjit Singh established a kingdom in northern Pakistan, but it fell apart after his death.

In the middle 1800s, the British included both Pakistan and Bangladesh within their colony of India. Together they formed a single colonial administrative unit called India. The British introduced English, restructured the educational system, built railroads, and developed a civil service. During their rule, tension between Muslims and Hindus in South Asia grew. While Hindus formed a group to campaign for independence, Muslims, who feared domination by the majority Hindus, created a similar organization of their own.

Pakistan Since Independence

By the 1930s, the idea of a Muslim state separate from India had taken hold among South Asia's Muslims. When Hindu and Muslim leaders could not agree on a constitution for a single state, in 1947 the British granted independence to two states based upon the dominant religion. India was formed as a predominantly Hindu state and Pakistan as a predominantly Muslim state. The latter consisted of two sections, known at the time as East Pakistan and West Pakistan. They were separated by about 1,000 miles (1,609 km) of Indian territory.

Independence from Britain was marked by a massive movement of Hindus and Muslims between India and Pakistan. During this time, there were outbreaks of violence between followers of the two religions. Adding to the tension, both countries claimed the area of Kashmir. It was part of India by tradition, but had a majority Muslim population. The two countries fought wars for control of Kashmir in 1948, 1965, and 1999. Tensions over Kashmir and other issues became more worrisome in the 1990s after India and Pakistan both developed nuclear weapons. Relations between the two countries have improved in recent years, but the situation remains uneasy.

Pakistan is a parliamentary republic, but instability and military rule have prevailed since 1971. In the early 2000s, General Pervez Musharraf allied Pakistan with the United States in its war on terror. His rule became increasingly unpopular, however, and he was forced to step down in 2008. Asif Ali Zardari, the widower of former prime minister Benazir Bhutto, was elected president in 2008. He introduced reforms to try to establish civilian government, but poverty and rebel activity in the west plague the country. In 2013 Nawaz Sharif took office as Pakistan's prime minister after a sweeping victory of his Pakistan Muslim League party.

Bangladesh: Independence and After

The people of East Pakistan were culturally different from the people in West Pakistan long before independence from the British. In Bangladesh they are ethnic Bengali and speak Bangla. After independence, West Pakistan wanted to impose a national language, Urdu, on all of Pakistan. Bengali leaders believed that their ethnic majority was treated unfairly by the government, which was dominated by leaders from West Pakistan. They formed a protest movement.

After wins in the 1970–1971 elections, Bengali nationalists pushed for self-rule. Pakistan sent its army to suppress the nationalists, which prompted them to declare independence for Bangladesh—which means "Bengal country." India entered the war on behalf of independent Bangladesh, and Pakistan surrendered.

Bangladesh is also a parliamentary republic. Political and ethnic rivalries have made stable rule difficult. Discontent has continued in recent years.

☑ **READING PROGRESS CHECK**

Making Connections How was the movement that led to the independence of Bangladesh similar to the movement that led to the formation of Pakistan?

Connecting Geography to **HISTORY**

Natural Disaster

When does a storm create a political movement? Since Pakistan's independence in 1947, many people in East Pakistan were unhappy with domination by those in West Pakistan. In the 1960s, some groups began calling for an independent state. Then a devastating cyclone smashed into East Pakistan in 1970, flooding large areas and killing as many as 500,000 people. The West Pakistan-dominated government was slow to provide desperately needed relief to the suffering Bengalis. The smoldering anger over this slow response contributed to the Bengali declaration of independence the following year and the establishment of Bangladesh.

EVALUATING What effect did the 1970 cyclone have on the history of Bangladesh?

Population Patterns

GUIDING QUESTION *How does high population density affect life in Pakistan and Bangladesh?*

Pakistan, with its population numbering nearly 200 million, is the sixth most populous country in the world. Bangladesh, with a population of 160 million, ranks seventh. In Pakistan, more than one-third of the people are under the age of 15. In Bangladesh, just under one-third are under age 15.

Physical geography shapes settlement in Pakistan, where most of the population lives in the Indus River valley. Although Pakistan is one of South Asia's most urbanized countries, only about 38 percent of the population lives in urban areas. Many people are migrating to cities, however, including Islamabad, Lahore, and Karachi. This growing urban population is straining resources. Migrants to the cities are forced to live in makeshift structures pulled together from scrap material. The **infrastructure** that makes urban life more livable, such as public water and sewage removal, is undeveloped in these poor areas.

In Pakistan the **total fertility rate**, or the average number of children a woman has in her lifetime, is 3.8. This is 50 percent higher than the world average. As a result, its population is growing rapidly, at 1.5 percent a year.

Bangladesh is the most densely populated country in South Asia, with 3,362 people per square mile (1,298 people per sq. km). The highest population densities occur in cities, such as Dhaka. However, density is generally high throughout the country. Only the Sundarbans in the southwest and the Chittagong Hills in the southeast have lower population densities.

To encourage Bengali women to have fewer children, private and government programs give women small loans to start their own businesses. The programs have achieved some success. While the total fertility rate was 4.4 in 1991, by 2015 it had declined to 2.2, just below the world average. Fertility rates have decreased as women become more educated and have more economic opportunities.

infrastructure system of public services such as power, water and sewage, transportation and communication networks, and schools and health care facilities

total fertility rate the average number of children a woman has in her lifetime

✓ **READING PROGRESS CHECK**

Explaining How did the government of Bangladesh address its high fertility rate?

Bert Brandt/AFP/Getty Images

TIME LINE ⌄

UNITY
and Division ➜

The state of Pakistan was born out of a desire of Muslims in India for national unity, but geographic and demographic challenges caused further disunity, leading to the creation of Bangladesh.

▶ **CRITICAL THINKING**

1. Explaining Why did the Muslim League call for a separate state to be created for Muslims?

2. Comparing In what ways are the histories of Pakistan and Bangladesh similar?

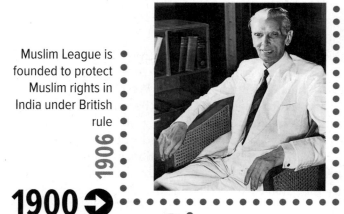

Muslim League is founded to protect Muslim rights in India under British rule

1906

1900 ➜

1940
Muslim League demands that India be partitioned to create a separate Muslim state called Pakistan

Pakistan is established, but West and East Pakistan are about a thousand miles apart. First governor-general is Mohammed Ali Jinnah.

1947

1948
Pakistan goes to war against India over disputed territory of Kashmir

Society and Culture Today

GUIDING QUESTION *How is life in Pakistan similar to and different from life in Bangladesh?*

Pakistan and Bangladesh became separate countries because of their diverse ethnic heritages. Pakistan is diverse in large part because it experienced invasions and migrations over many centuries. Pakistan has six main ethnic groups: Punjabis, Pashtuns, Sindhis, Sariakis, Muhajirs, and Balochis. Punjabis make up about 45 percent of the country's people. No other group has more than 16 percent. Urdu is the official language of Pakistan, but only 8 percent of the population speaks it. More Pakistanis speak Punjabi than any other language. English, also an official language, is typically spoken by members of the elite, including government officials.

In Bangladesh the majority of the people are Bengali, a term describing both an ethnic and a language group. It is something they share with some of their Hindu neighbors in the Indian state of Bengal. Non-Bengalis, mostly smaller indigenous groups, make up only a small percentage of Bangladesh's population.

Islam is the main religion in both countries. The two countries also have some Hindus and small Christian populations. Most Muslims are Sunnis, but about one-fifth of Pakistan's Muslims are Shias.

Literacy rates in Pakistan and Bangladesh are very low, around 60 percent. In Bangladesh, the literacy rate for men is about 6 percentage points higher than for women. In Pakistan, the gap between male and female literacy is much greater. Schooling in Bangladesh is free and required for all children to age ten, but only about half of children actually attend. School is not required in Pakistan, where the educational system is a mix of government-run, Islamic, and private schools.

Lack of health care is a major problem in both countries, mainly because of their large populations and high poverty rates. Spending on health care ranks toward the bottom of priorities, and there are relatively fewer doctors or hospital beds than are needed. Several diseases, including tuberculosis and malaria, have long been problems in both countries. Bangladesh has succeeded in reducing malaria but still faces many public health issues.

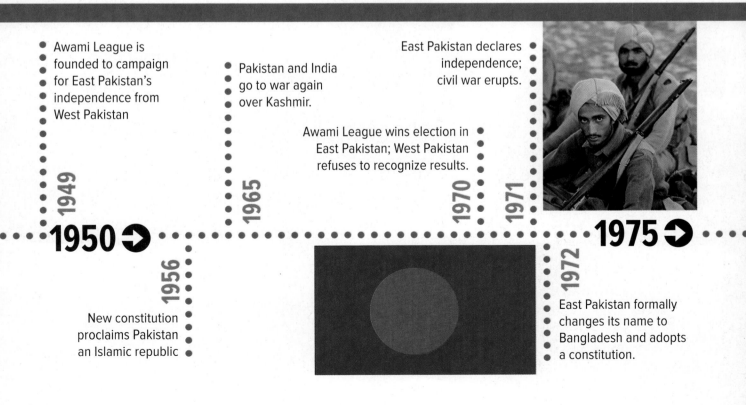

1949 — Awami League is founded to campaign for East Pakistan's independence from West Pakistan

1950 ➜

1956 — New constitution proclaims Pakistan an Islamic republic

1965 — Pakistan and India go to war again over Kashmir.

1970 — Awami League wins election in East Pakistan; West Pakistan refuses to recognize results.

1971 — East Pakistan declares independence; civil war erupts.

1972 — East Pakistan formally changes its name to Bangladesh and adopts a constitution.

1975 ➜

Family and the Status of Women

Family is the social basis in both countries. Extended families live close, often in the same home. Many marriages are arranged, though increasing numbers of educated people choose their own partners. After marriage, the wife typically lives with the husband's family.

In traditional families in Pakistan, women stay in separate parts of the house when nonfamily males visit. This is more common among urban and middle-class Pakistanis than rural or poor people, where women must work to help the family survive. Some Muslim women also wear the burka, a loose garment that covers the face and body, when they are in public.

Both Pakistan and Bangladesh have had women leaders. Benazir Bhutto was twice prime minister of Pakistan. Sheikh Hasina served as prime minister of Bangladesh from 1996 to 2001 and took that office again in 2009.

The Arts

Literature and dance are very important in Bangladesh. In 1913 Bengali Rabindranath Tagore became the first non-European writer to win the Nobel Prize in Literature. The poetry and plays of Kazi Nazrul Islam, "the voice of Bengali nationalism," have inspired poor farmers with themes about the oppression of Muslims. Bangladesh also has developed original and creative traditional dances in classical as well as folk styles.

Music and literature are the richest of all Pakistani art forms. *Qawwali,* a form of devotional singing, is popular. People recite poetry at public *musha'irahs* that are organized like music concerts. The classical music tradition can be traced to the thirteenth-century poet and musician Amir Khosrow, who composed the traditional rhythmic form known as the raga.

✓ READING PROGRESS CHECK

Contrasting How are the people of Pakistan more diverse than those of Bangladesh?

These men at a shipyard in Bangladesh, on the Bay of Bengal, work in the ship-breaking industry. They face dangerous conditions, including accidents and exposure to toxic materials.

▼ CRITICAL THINKING

1. Drawing Conclusions Why do people take such dangerous jobs?

2. Making Inferences Why do you think Bangladesh has a ship-breaking industry, but not a ship-building industry?

Economic Activities

GUIDING QUESTION *What are the dominant economic activities in Pakistan and Bangladesh?*

Pakistan and Bangladesh have traditionally relied on agriculture. Industrial activity is increasing in the twenty-first century. Poverty is widespread. The economic situation has worsened in recent years due mainly to inflation.

About 25 percent of Pakistan's gross domestic product (GDP) comes from agriculture, and about 40 percent of the workforce is in agriculture. Cash crops, including rice, cotton, and sugarcane, bring much-needed income. Pakistan also has a fishing industry, exporting shrimp, lobster, and fish. Despite reforms in the last 50 years, land distribution remains highly unequal. Most farmers continue to use draft animals on small farms. Since the 1980s, service industries have grown. Today, more than half the country's GDP is from the service sector.

Industry constitutes about one-fourth of Pakistan's GDP. The most important are cotton textiles and clothing for export. Other exports include rice, leather, sporting goods, chemicals, and carpets. Most exports are shipped out of the Port of Karachi. Pakistan has become a major trading partner of the United States. Small-scale production, or cottage industries, has played an important role in Pakistan's industrialization. They employ many craftspeople and provide at-home employment for women.

As in India, railways are the principal mode of transportation. The state-owned Pakistan Railways moves both people and cargo throughout Pakistan. Highways are increasing in importance. The Makran Coastal Highway runs along the Arabian Sea and integrates economic activities in the area.

Rural areas in both countries often lack electricity and other modern services. The spread of cell phones has made communications more widely **available** even in rural areas. That is more the case in Pakistan than Bangladesh, however.

available able to be obtained

In Bangladesh most people are sharecroppers. Rice is the major crop. In some areas, farmers can grow three crops a year, alternating rice, wheat, and other crops. **Jute**, a fiber used to make string, rope, and cloth, is also a major cash crop.

The garment industry has expanded and clothing is now Bangladesh's top export. As in Pakistan, much of Bangladesh's textile manufacturing relies on cottage industries. Dhaka is a center for weaving muslin, a lightweight cotton cloth. Nearly 2 million women work in Bangladesh's garment industry. Other craft goods include jute products, such as upholstery, and leather goods.

jute a fiber used to make string, rope, and cloth

Bangladesh is one of the world's largest aquaculture-producing countries. It has many inland fisheries that cultivate fish and shrimp. These proteins are part of a staple diet that also relies heavily on rice and lentils.

☑ **READING PROGRESS CHECK**

Explaining How are cottage industries important in Pakistan and Bangladesh?

LESSON 2 REVIEW

Reviewing Vocabulary

1. *Identifying* Write a paragraph explaining what jute is and why it could be called the "golden crop" of Bangladesh.

Using Your Notes

2. *Comparing and Contrasting* Use your graphic organizer on the human geography of Pakistan and Bangladesh to write a paragraph explaining how their histories are similar and different.

Answering the Guiding Questions

3. *Explaining* How did contemporary Pakistan and Bangladesh form?

4. *Assessing* How does high population density affect life in Pakistan and Bangladesh?

5. *Comparing and Contrasting* How is life in Pakistan similar to and different from life in Bangladesh?

6. *Identifying* What are the dominant economic activities in Pakistan and Bangladesh?

Writing Activity

7. *Argument* In a few paragraphs, discuss challenges Pakistan and Bangladesh face and how they might be tackled.

South ASIA on the BRINK

The nuclear weapons programs of India and Pakistan began in the late 1960s and early 1970s and produced nuclear weapons for both countries. India and Pakistan have over 100 missiles between them. India's Agni-V missile is capable of striking a target more than 3,100 miles (4,989 km) away. Although the two countries' nuclear arsenals are small relative to those of the United States and Russia, they nonetheless pose a threat to the rest of the world.

REGIONAL NUCLEAR STOCKPILES

PAKISTAN

INDIA

1,550 mi (2,500 km)

CHINA
200

PAKISTAN
60

INDIA
50

THREATS

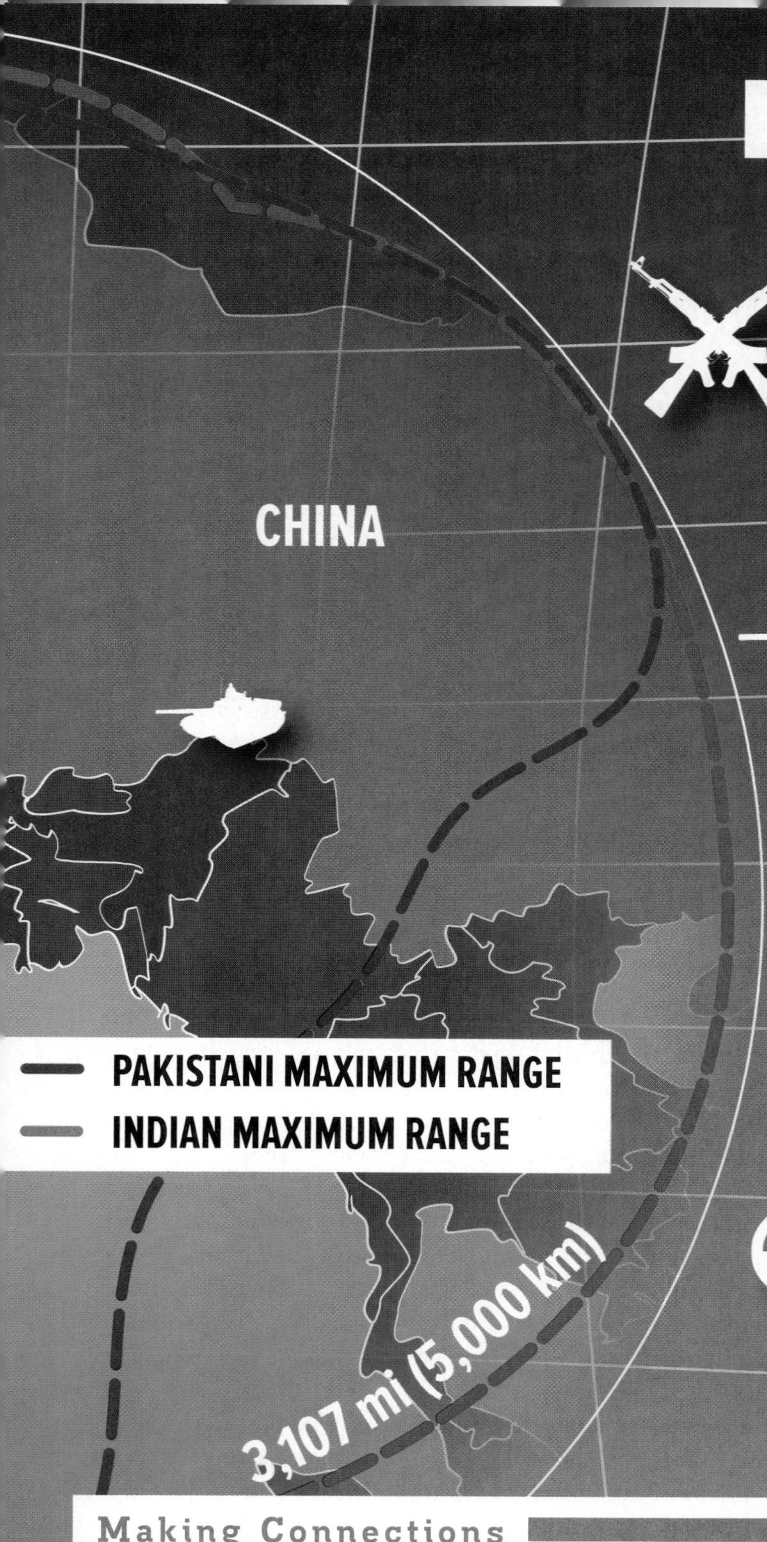

KASHMIR DISPUTE

India and Pakistan have both claimed ownership of the northern Kashmir border region. War has erupted over the area twice in the past, and there are worries that if there was ever a third it would go nuclear.

CHINA

CHINA DISPUTE

China and India too have border disputes. Both countries having nuclear arms raises fears of nuclear war.

TERRORISM

Pakistan's political instability causes some concern. Should its government fall, nuclear materials could fall into the hands of terrorist cells in neighboring countries and within Pakistan itself.

— PAKISTANI MAXIMUM RANGE
— INDIAN MAXIMUM RANGE

NUCLEAR WINTER

Should both India and Pakistan detonate their arsenals, it could result in a 20% depletion of global ozone levels. The resulting increase in UV radiation could seriously impact human health across the globe.

3,107 mi (5,000 km)

Making Connections

1. **Places and Regions** What ongoing regional conflicts increase the risk of nuclear war?

2. **The World in Spatial Terms** According to the map, about how many countries are in range of Pakistan's nuclear weapons?

3. **Environment and Society** What effects could a nuclear conflict have on the environment and the people of the region?

*Interact with **Global Connections** Online*

netw⊙rks

There's More Online!

- ☑ **GRAPH** World's Worst Cities for Air Pollution
- ☑ **IMAGE** Compressed Natural Gas Vehicle
- ☑ **IMAGE** The Sundarbans
- ☑ **IMAGE** Water Pollution in Bangladesh
- ☑ **INTERACTIVE SELF-CHECK QUIZ**
- ☑ **VIDEO** People and Their Environment: Pakistan and Bangladesh

Reading HELPDESK

Academic Vocabulary

- principle
- accumulate

Content Vocabulary

- sustainable development
- hydroelectric power

TAKING NOTES: *Key Ideas and Details*

IDENTIFYING As you read about the environments of Pakistan and Bangladesh, use a graphic organizer like the one below to identify threats to the environment and responses to those threats.

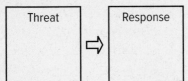

Threat ⇨ Response

LESSON 3

People and Their Environment: Pakistan and Bangladesh

ESSENTIAL QUESTION · *How do physical systems and human systems shape a place?*

It Matters Because

Pakistan and Bangladesh are two of the most populous countries in the world. They are also among the world's poorest countries. Their governments need to find solutions to environmental problems to provide their people with health and opportunity. Finding the resources for these solutions is difficult, however.

Managing Resources

GUIDING QUESTION *Why has sustainable development been given a low priority in Pakistan and Bangladesh?*

The economies of Pakistan and Bangladesh are heavily agricultural. The water from their mighty rivers and their fertile soils are very important resources. Agriculture does not generate a great deal of wealth, however. As a result, these countries have huge numbers of people living in poverty and little national income to use to build stronger economies.

In the 1970s and 1980s, people around the world began to notice the environmental damage caused by modern economic activities. Factory smokestacks and vehicle exhaust fouled the air. Chemical and industrial waste poisoned the water. Pumping oil from the ground and mining coal and other resources scarred the landscape. In response, economists and environmentalists began to advocate, or promote, the idea of sustainable development.

When first introduced, **sustainable development** was defined as economic growth that meets the needs of present populations without hampering the ability of people in the future to meet their own needs. Over time, a third goal was added. This was to promote progress in less developed countries and to reduce the gap between those who are wealthy and those who are poor. Sustainable development, then, rests on these **principles**:

- promoting economic development
- protecting the environment
- promoting social fairness

While many people agree on the goals of sustainable development, achieving them is not an easy task. Poorer countries—which need economic development to give their people a better standard of living—have few financial resources to invest in building a more modern economy. For instance, such countries lack the technology or the knowledge needed to create and use more efficient or cleaner production methods.

Less developed countries face another obstacle in achieving sustainable development. For survival, people living in these low-income countries often make short-term choices that can have long-term negative consequences. For example, to gain money by selling valuable wood, they may cut down trees too rapidly to replace them. To expand agricultural output, they may use fertilizers and pesticides that, over time, hurt the soil or leach into the water, damaging drinking supplies.

Both Pakistan and Bangladesh need to develop their economies. One-fourth of Pakistan's people and one-third of Bangladesh's population live in poverty. The rankings of these countries in education and health statistics also reflect their low standards of living. At the same time, both countries suffer from obstacles to development. These include an overreliance on agriculture, a lack of capital to invest, corruption, and the unwise use of resources.

Perhaps the most threatened environmental region in these two countries is the Sundarbans. This wetland area in southwestern Bangladesh and the northeastern coast of India has the world's largest mangrove forest. This forest is threatened due to rising sea levels and the constant floods of salt water from the Bay of Bengal. Mangroves can grow in somewhat salty water, but if the concentration of salt increases too much, they will not thrive. The forest is threatened by action upriver as well. Because of deforestation in the Himalaya, the rivers that flow into Bangladesh now carry more silt and soil than before. Some of this soil reaches the Sundarbans, where it is building up and reducing the easy flow of water on which the trees depend.

sustainable development economic growth that meets the needs of present populations without hampering the ability of people in the future to meet their own needs and that benefits people and societies

principle a fundamental characteristic

The most threatened area in Bangladesh—the Sundarbans— is also the site of the largest mangrove forest in the world. These boats are hauling harvested mangrove for export.

▼ CRITICAL THINKING

1. *Making Connections* What problems threaten this vast forest?

2. *Identifying Cause and Effect* What might be some effects of the death of the mangrove forest?

Tim Laman/National Geographic/Getty Images

Local deforestation is a problem for the mangrove forest as well. About half the trees have been cut down over the years as people seek wood for fuel or clear the land for other uses. The forest faces yet another threat. Wastewater from shrimp farms located just north of the forest is raising concerns about damage it might do to the water quality of the Sundarbans.

Another resource at risk in Pakistan and Bangladesh is water. Most of the population of Pakistan does not have regular access to potable, or drinkable, water. A lack of clean water leads to a host of health problems.

Bangladesh could benefit more than it does from its abundant supply of natural gas by exporting it. Most of it, however, is used locally. Bangladeshis burn natural gas as a fuel, and much of the rest is used in fertilizer factories.

Hydroelectric power, an important source of energy in Pakistan, is also at risk. Pakistan's two hydroelectric power dams in the north are affected by physical geography. The Indus River carries a heavy load of soil as it flows south. While this fertile soil benefits farmers, it **accumulates**, or builds up, behind the dam. This clogs the Tarbela Dam, making it unable to generate power.

✓ **READING PROGRESS CHECK**

Explaining What are the three principles on which sustainable development rests?

Human Impact

GUIDING QUESTION *What kinds of human activity have affected the environment in Pakistan and Bangladesh?*

The threatened Sundarbans is just one example of the way the people of Pakistan and Bangladesh affect the physical environment of these countries. The struggle to survive has created several environmental issues. For example, Pakistan suffers from severe water pollution, which has several different causes. Sewage treatment facilities are scarce, and raw sewage enters the country's rivers and water supply. Industrial pollution is another contributor, as is agricultural runoff of pesticides and fertilizers from farmland. Water pollution is a problem that plagues Bangladesh as well.

With its fairly dry climate, Pakistan has to manage its water. Farmers use irrigation to bring water to areas that receive little rainfall but have fertile soil. However, because of the relatively high temperatures during Pakistan's growing season, some of this water evaporates, leaving minerals and salt in the soil, which makes the land less fertile.

Along with salinity, or saltiness, of the soil, these countries also have problems with soil erosion. This is particularly a problem in Pakistan, where erosion occurs in highlands that are losing their tree cover.

---

hydroelectric power
electrical energy generated by falling water

accumulate to build up

Agricultural runoff and industrial pollutants in both countries lead to water pollution. People, like this boy in Dhaka, Bangladesh, still rely on fish from the dying rivers for food.

▼ **CRITICAL THINKING**

1. *Explaining* What are other causes of water pollution in these countries?

2. *Identifying Cause and Effect* How do you think this kind of pollution affects humans?

©mhasan/Demotix/Corbis

The crowded cities of Pakistan have serious air pollution problems. Car ownership has increased at a rapid rate in the past few decades. Many vehicles burn diesel fuel or leaded gas, which are heavy pollutants.

Pakistan's people cannot escape this problem by staying indoors. Air within homes is polluted, too, because people burn wood and animal dung for fuel. Both of these release smoke and dust that harm the respiratory system. A World Health Organization (WHO) report relates heavy indoor air pollution to the high rates of death and pneumonia that afflict many of Pakistan's children.

Deforestation is not only a problem in Bangladesh's Sundarbans, but also in Pakistan. Several decades of tree cutting in the north has left mountain slopes bare. In one area, about 15 percent of the forest was lost in just 10 years. This deforestation has had an impact on Pakistan's people. The loss of trees means that heavy rains are more likely to cause devastating floods.

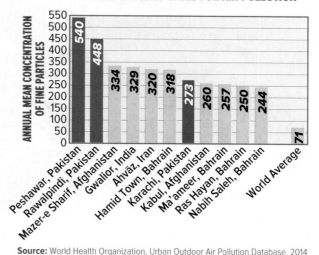

WORLD'S WORST CITIES FOR AIR POLLUTION

ANNUAL MEAN CONCENTRATION OF FINE PARTICLES

- Peshawar, Pakistan — 540
- Rawalpindi, Pakistan — 448
- Mazer-e Sharif, Afghanistan — 334
- Gwalior, India — 329
- Ahvāz, Iran — 320
- Hamid Town, Bahrain — 318
- Karachi, Pakistan — 273
- Kabul, Afghanistan — 260
- Ma'ameer, Bahrain — 257
- Ras Hayan, Bahrain — 250
- Nabih Saleh, Bahrain — 244
- World Average — 71

Source: World Health Organization, Urban Outdoor Air Pollution Database, 2014

✓ **READING PROGRESS CHECK**

Evaluating What human actions have hurt the environments of these countries?

Addressing the Issues

GUIDING QUESTION *What steps have the governments of Pakistan and Bangladesh taken to improve the environment?*

Pakistan and Bangladesh have not been able to promote major efforts of sustainable development. They have not completely ignored environmental issues, however. Both countries have taken several steps to address their environmental issues, but the efforts have been inconsistent and the success marginal.

Pakistan began making efforts to improve the environment in the 1990s, when it issued the National Conservation Strategy Report. This report set goals that included protecting the soil, increasing the efficiency of irrigation to prevent water loss, improving forest management, and a host of other steps. The country also issued an action plan in 2001 and a new environmental policy in 2005. However, ongoing political turmoil in Pakistan and the continuing emphasis on economic growth have not allowed these ambitious plans to be fully implemented.

Pakistan's government, using funds from the World Bank, tried to solve the problem of soil salinity in the Indus River valley by digging a long channel parallel to the river. The goal was to carry salt-heavy water away from fields to empty into the Arabian Sea. Unfortunately, the engineering at the outlet was poorly executed. Instead of carrying salty water away, even more saline seawater was allowed to enter the channel. So much water came in that crops and fisheries in the southeastern corner of Pakistan were destroyed. The people of the region asked the government to shut down the channel. As late as 2012, the polluted water still threatened southern areas.

To combat air pollution, Pakistan has promoted the use of compressed natural gas (CNG) vehicles, which pollute less than other vehicles. The country is among the world's leaders in CNG vehicles. However, it lacks enough fueling stations to encourage consumers to buy more of these vehicles.

Many scientists have argued that the world's climate is undergoing significant change. This change, they say, may result in warmer overall temperatures, changes in patterns of rainfall, as well as more frequent and more powerful storms.

Air pollution is high in some of Pakistan's cities.

▲ **CRITICAL THINKING**

1. ***Analyzing Visuals*** Which cities in Pakistan rank among the world's worst cities for particulate air pollution?

2. ***Explaining*** Why is this air pollution a problem for Pakistan's people?

Pakistan is a leading country in the use of CNG vehicles. However, more filling stations are needed to encourage people to buy more of the vehicles.

▲ **CRITICAL THINKING**

1. *Identifying* What are the benefits of CNG vehicles?

2. *Drawing Conclusions* Why do you think that Pakistan has not yet installed enough filling stations for CNG vehicles?

While all scientists do not agree on these conclusions, climate change could be a problem for Bangladesh. The low-lying country could experience worse damage and a higher loss of life if floods increase or cyclones grow stronger. Rising sea levels caused by climate change could also be catastrophic.

Bangladesh moved to adopt an environmental policy in the 1990s and adopted a climate-change strategy and plan in 2009. Bangladesh's prime minister, Sheikh Hasina, has urged countries around the world to take more active steps to fight climate change. Her government points out that Bangladesh produces relatively low amounts of the greenhouses gases said to cause this climate change.

Bangladesh has also taken steps at home. Rather than allowing animal waste to release climate-changing methane gas into the atmosphere, the people of Bangladesh are composting the waste to make organic fertilizer. A by-product of that process is a gas, which is used for cooking, that pollutes less when burned. The country has also made progress in reducing other gases that damage the air and water. The growing ship-breaking industry, however, is a major source of air pollution. Bangladesh recently issued regulations to try to reduce that effect, but they have not yet been enforced.

Bangladesh has also moved to protect the Sundarbans. The government has set aside several areas of the vast forest as reserves. These are aimed, not only at protecting the trees of the Sundarbans, but also the remaining population of Bengal tigers that live and hunt there. Bangladesh has targeted 5 percent of its land and sea area to be set aside as nature reserves. The country has made progress toward this goal, but has not reached a full 5 percent. This gap is attributed to a lack of resources to attain the goal.

✔ **READING PROGRESS CHECK**

Explaining Why have these governments not acted to implement all their plans to improve the environment?

LESSON 3 REVIEW

Reviewing Vocabulary

1. *Identifying* Write a paragraph explaining what sustainable development is and why it is especially challenging to countries with limited resources.

Using Your Notes

2. *Summarizing* Use your graphic organizer on the environment of Pakistan and Bangladesh to write a paragraph describing one effective response and one ineffective response to an environmental threat one of them faces.

Answering the Guiding Questions

3. *Identifying* Why has sustainable development been given a low priority in Pakistan and Bangladesh?

4. *Summarizing* What kinds of human activity have affected the environment in Pakistan and Bangladesh?

5. *Evaluating* What steps have the governments of Pakistan and Bangladesh taken to improve the environment?

Writing Activity

6. *Informative/Explanatory* Write a letter to the prime minister of either Pakistan or Bangladesh, suggesting what can be done to address environmental issues even when their resources are limited.

Directions: On a separate sheet of paper, answer the questions below. Make sure you read carefully and answer all parts of the questions.

Lesson Review

Lesson 1

1 ***Explaining*** How does the theory of continental drift explain the formation of the Himalaya mountain ranges?

2 ***Identifying Cause and Effect*** Where do the three major river systems in Pakistan and Bangladesh originate? What causes them to flood seasonally?

3 ***Summarizing*** What are some ways that South Asian rivers help people meet their basic and economic needs?

Lesson 2

4 ***Explaining*** Why did Bengali nationalists push for independence from Pakistan?

5 ***Comparing and Contrasting*** How are population patterns in Pakistan similar to and different from those in Bangladesh?

6 ***Hypothesizing*** Why do you think health care in Bangladesh and Pakistan is a major problem?

Lesson 3

7 ***Defining*** What are the characteristics and goals of sustainable development?

8 ***Identifying Cause and Effect*** How does deforestation in the Himalaya affect the mangrove forest of the Sundarbans?

9 ***Evaluating*** What are some steps Pakistan has taken to improve the environment? How effective have they been?

Critical Thinking

10 ***Analyzing*** Why is poverty a major obstacle to sustainable development in Pakistan and Bangladesh?

11 ***Assessing*** Which one of the environmental challenges facing Pakistan is most urgent? Explain your answer.

21st Century Skills

Review the graph and answer the questions that follow.

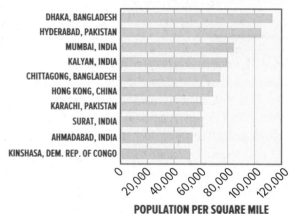

MOST DENSE WORLD URBAN AREAS OVER 2.5 MILLION POPULATION: 2015

POPULATION PER SQUARE MILE

Source: Demographia World Urban Areas; www.demographia.com/

12 ***Compare and Contrast*** How does population density in Dhaka compare to that of Karachi? About how many people live per square mile in each city?

13 ***Create and Analyze Arguments and Draw Conclusions*** What does the bar graph suggest about overall population patterns in Bangladesh?

14 ***Creating and Using Graphs, Charts, Diagrams, and Tables*** Why do you think the bar graph focuses on cities with populations of more than 2.5 million?

College and Career Readiness

15 ***Change and Continuity in Economics*** As an economist for the United Nations, you have been asked to report on the role of cottage industries in Pakistan. What benefits do cottage industries provide to local populations? How do they affect the environment and economy? Use the Internet to conduct research, and write a two-page report describing your findings.

Need Extra Help?

If You've Missed Question	1	2	3	4	5	6	7	8	9	10	11	12	13	14	15
Go to page	630	632	633	635	636	637	642	643	645	643	644	636	636	636	639

Directions: On a separate sheet of paper, answer the questions below. Make sure you read carefully and answer all parts of the questions.

DBQ Analyzing Primary Sources

Use the document to answer the following questions.

Population density is already a major concern in Bangladesh, the most densely populated country in South Asia.

PRIMARY SOURCE

"*So imagine Bangladesh in the year 2050, when its population will likely have zoomed to 220 million, and a good chunk of its current landmass could be permanently underwater. That scenario is based on two converging projections: population growth that, despite a sharp decline in fertility, will continue to produce millions more Bangladeshis in the coming decades, and a possible multifoot rise in sea level by 2100 as a result of climate change.*

'*Globally, we're talking about the largest mass migration in human history,' says Maj. Gen. Muniruzzaman. . . . 'By 2050 millions of displaced people will overwhelm not just our limited land and resources but our government, our institutions, and our borders.'*"

—Don Belt, "The Coming Storm," *National Geographic*, May 2011

16 *Making Connections* Based on the article, how is climate change likely to affect population density in Bangladesh in the coming decades?

17 *Interpreting* What does Maj. Gen. Muniruzzaman mean when he says that Bangladesh will be overwhelmed by "the largest mass migration in human history"?

18 *Making Predictions* How might the mass migration of Bangladeshis affect the resources and infrastructure of neighboring countries?

Applying Map Skills

Use the Unit 7 Atlas to answer the following questions.

19 *Environment and Society* Describe how the physical geography of Bangladesh affects population density. Give examples.

20 *Places and Regions* Use your mental map of South Asia to describe the location of the major mountain ranges in Pakistan: the Himalaya, the Hindu Kush, and the Karakoram.

21 *Environment and Society* How does agriculture in Pakistan compare with that in Bangladesh? Which country relies more heavily on agriculture?

Exploring the Essential Question

22 *Exploring Issues* Pakistan and Bangladesh face many obstacles to sustainable development. Use what you have learned about the human and physical systems in these countries to create a poster showing the challenges they face. Posters should be visual and can include photos, graphs, charts, and maps.

Research and Presentation

23 *Research Skills* Use the Internet to research relations between Bangladesh and Pakistan. What ethnic, linguistic, and political factors led Bengali nationalists to declare independence? What was the role of India? What are relations between Pakistan and Bangladesh like today? Explain your findings in a short research report.

Writing About Geography

24 *Informative/Explanatory* Use standard grammar, spelling, sentence structure, and punctuation to write a multi-paragraph essay describing the problem of water pollution in Pakistan. What factors contribute to the problem? How does water pollution affect people? How does it affect the environment and economy? What steps has the government taken to solve the problem? How effective have they been?

Need Extra Help?

If You've Missed Question	**16**	**17**	**18**	**19**	**20**	**21**	**22**	**23**	**24**
Go to page	636	636	636	598	598	598	643	635	644

Bhutan, Maldives, Nepal & Sri Lanka

ESSENTIAL QUESTION · *How do physical systems and human systems shape a place?*

networks

There's More Online about the geography of Bhutan, Maldives, Nepal, and Sri Lanka.

CHAPTER 27

Why Geography Matters
Nepal's Role as a Buffer State

Lesson 1
Physical Geography of Bhutan, Maldives, Nepal & Sri Lanka

Lesson 2
Human Geography of Bhutan, Maldives, Nepal & Sri Lanka

Lesson 3
People and Their Environment: Bhutan, Maldives, Nepal & Sri Lanka

Geography Matters...

The four countries of South Asia profiled in this chapter offer a great variety of natural beauty and human culture and more. They include the world's highest mountain, tropical forests, and beautiful coral reefs. Mountainous Nepal and Bhutan and tropical Sri Lanka and Maldives possess cultures that are centuries old as well as natural wonders that have existed for millions of years.

◀ This Hindu holy man in Nepal wears traditional facial painting.

©Lee Frost/Robert Harding World Imagery/Corbis

649

Nepal's role as a buffer state

Nepal is located high in the rugged Himalaya. Nepal's geographic isolation makes trade and travel difficult. In order to survive, Nepal has forged important political and economic alliances with its neighbors. At times, maintaining friendly relationships with surrounding countries has proved challenging.

Buffer States and Border Disputes in South Asia

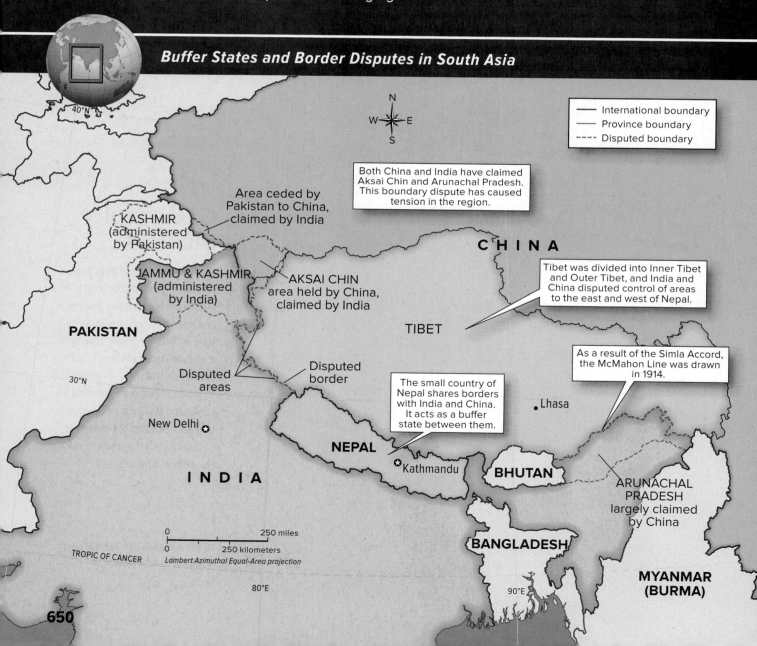

International boundary
Province boundary
Disputed boundary

Both China and India have claimed Aksai Chin and Arunachal Pradesh. This boundary dispute has caused tension in the region.

Area ceded by Pakistan to China, claimed by India

KASHMIR (administered by Pakistan)

JAMMU & KASHMIR (administered by India)

AKSAI CHIN area held by China, claimed by India

CHINA

Tibet was divided into Inner Tibet and Outer Tibet, and India and China disputed control of areas to the east and west of Nepal.

PAKISTAN

TIBET

Disputed areas

Disputed border

As a result of the Simla Accord, the McMahon Line was drawn in 1914.

30°N

The small country of Nepal shares borders with India and China. It acts as a buffer state between them.

New Delhi

. Lhasa

NEPAL

INDIA

Kathmandu

BHUTAN

ARUNACHAL PRADESH largely claimed by China

250 miles
250 kilometers

TROPIC OF CANCER

Lambert Azimuthal Equal-Area projection

BANGLADESH

80°E

90°E

MYANMAR (BURMA)

40°N

650

What is a buffer state?

A *buffer* is something that acts as a protective barrier. A *buffer state* is a smaller, independent country sandwiched between larger, more powerful rivals. In the 1800s, Thailand—then called Siam—was a buffer between the British empire in India and the French empire in Southeast Asia. A buffer state serves to prevent conflict between its neighbors. To function as a buffer state, a country must remain neutral and resist pressure from neighboring powers. A buffer state gains trading partners and other advantages from its alliances. It also literally keeps the peace in a region.

A buffer state is vulnerable to instability. It can be caught in the middle if a simmering conflict between its neighbors heats up. Being heavily dependent on other countries, a buffer state also loses the freedom to act as it chooses. A buffer state that is more influenced by one country can lose its independence and become a satellite state.

1. **The World in Spatial Terms** How does the location of a country make it a buffer state, and what purpose does a buffer state serve?

Why is Nepal considered a buffer state?

Small, landlocked Nepal stands between two much larger countries, China and India. In 1775 a Nepalese king described his country's location as "a yam between two boulders." Nepal is considered a buffer state because these two "boulders"—which combined have more than one-third of the world's population—are rivals.

Their hostility largely began with the 1959 Tibetan uprising against China, when the Dalai Lama and other Tibetans took refuge in India. In 1914 British colonial officials had drawn the McMahon Line, which divided Tibet and gave more territory to British-ruled India. India accepted the boundary as legal, but China did not. Since then, the two countries have disagreed about control of areas to the east and west of Nepal and have had an unclear border. In October 1962, Chinese soldiers launched a surprise invasion, crossing the Himalaya and seizing India's territory. Although this Sino-Indian War ended quickly, these quarreling countries have maintained an uneasy peace ever since, putting Nepal's interests at risk.

2. **Places and Regions** What physical and political characteristics of a buffer state does Nepal exhibit?

What does Nepal's status as a buffer state mean for its future?

For years, Nepal benefited from close relations with both China and India. With each country, Nepal shares historical, economic, cultural, and religious bonds. Yet being in the middle means that Nepal must constantly perform a delicate balancing act. Political and economic changes in China and India affect Nepal and its future. Although trading partners, China and India still view one another with suspicion. India's awareness that China is capable of launching a surprise attack led to militarization of the region. Having Nepal as a buffer reduces tensions.

After years of depending heavily on India for trade, Nepal is strengthening its economic ties with China. In 2007, for example, China began constructing a railway extension to link Tibet and Nepal and improve trade. As China and India themselves develop stronger economic ties, there is hope that one day Nepal will become a bridge between them.

3. **Environment and Society** In what way does being a buffer state affect Nepal's future?

THERE'S MORE ONLINE

READ a time line of Nepal's history • **SEE** a map of disputed territory claimed by China, India, and Nepal

networks

There's More Online!

☑ **IMAGE** Kathmandu, Nepal

☑ **IMAGE** Mount Everest

☑ **Infographic** Ecotourism

☑ **INTERACTIVE SELF-CHECK QUIZ**

☑ **VIDEO** Physical Geography of Bhutan, Maldives, Nepal & Sri Lanka

LESSON 1
Physical Geography of Bhutan, Maldives, Nepal & Sri Lanka

Reading HELPDESK

Academic Vocabulary

- **potential**
- **exploit**

Content Vocabulary

- **aquifer**
- **cowrie shell**

TAKING NOTES: *Key Ideas and Details*

IDENTIFYING As you read about the physical geography of Bhutan, Maldives, Nepal, and Sri Lanka, use a graphic organizer like the one below to give examples of how a landform, water system, or resource affects life in each country.

How Physical Geography Affects Life

Country	Physical Characteristics	Impact on Life

ESSENTIAL QUESTION • *How do physical systems and human systems shape a place?*

IT MATTERS BECAUSE
The northern and southern fringes of South Asia are among the highest and lowest countries in the world. Mountainous Nepal and Bhutan contain the sources of some of South Asia's mighty rivers. Sri Lanka is an island country. Maldives, a collection of low islands, faces the danger of rising sea levels.

Landforms

GUIDING QUESTION *What landforms help shape life in Bhutan, Maldives, Nepal, and Sri Lanka?*

While these four countries are unique, their locations on the fringe of South Asia have meant that their physical and human geographies have been influenced by the South Asian core. Nepal and Bhutan sit on the northern fringe of South Asia, amid the lofty Himalaya. They are bordered by India to the south and China to the north. Far to the south, in the Indian Ocean, sit the island countries of Maldives and Sri Lanka. Sri Lanka is off the southeastern coast of India. Maldives is a collection of islands southwest of India. Its islands stretch over 500 miles (805 km) from north to south and 80 miles (129 km) from east to west. The country's nearly 1,200 islands have a total area of about 115 square miles (298 sq. km).

The Indian subcontinent is a large landmass that is slowly and steadily colliding with the southern edge of Asia. This collision thrust up several mountain ranges, including the Himalaya, which dominate much of Nepal and Bhutan. These ranges spread about 1,500 miles (2,414 km) across the northern edge of South Asia and are hundreds of miles from north to south. The Himalaya are divided into three ranges: the Greater Himalaya to the north, the Lesser Himalaya in the center, and the Outer Himalaya to the south. The Greater Himalaya soar the highest, averaging over 20,000 feet (6,096 m) in elevation. In Nepal, the Greater Himalaya range has eight of the highest mountains in the world. Among them is Mount Everest, the world's highest mountain, which rises to 29,028 feet (8,848 m) above sea level along the border between Nepal and China.

Nepal has four geographic zones: the Tarai Plain on its southern border, the Churia and Mahabharat Ranges north of it, the Lesser Himalaya north of them, and the Greater Himalaya on its northern rim. The Tarai was originally forested marshland, but much of the plain has been drained for farming. The low Churia Range includes flat valleys that have been cleared of trees for farming. Between the Mahabharat Range and the Lesser Himalaya are more valleys where cities such as Kathmandu are located. Few people live in the Greater Himalaya.

Bhutan has four geographic zones. A narrow lowland called the Duars Plain sits on the southern edge of the country. This area receives heavy rainfall and is thick with vegetation and wild animals. On its northern edge are several passes into the mountains called *dwars*, meaning "doors," which gave their name to the plain. The southern part of the Duars Plain is flat land used for growing rice. A dramatic change in altitude occurs as the Lesser Himalaya and then the Greater Himalaya rise steeply above the plain. Many of Bhutan's people live in the several wide, level valleys that cut through the Lesser Himalaya. The valleys of the Greater Himalaya, found from 12,000 to 18,000 feet (3,658 to 5,486 m) above sea level, are sparsely populated. The Black Mountain range runs north and south from the Greater to the Lesser Himalaya in west-central Bhutan. Few people live in these rugged mountains.

Sri Lanka has two regions. The first is a triangle-shaped mountain range in the south-central part of the island called the Central Highlands. The second is a plain that covers most of the island. The Central Highlands has a few peaks that reach about 7,000 feet (2,134 m), including the island's highest point, Mount Pidurutalagala, with an elevation of 8,281 feet (2,524 m). The plain ranges from sea level at the coast to about 1,000 feet (305 m) high. Many inlets called lagoons lie along the coast, as well as Mannar Island. To the north of the Central Highlands, low ridges cross the plain. To the west, there are higher ridges and deeper valleys. Highland areas are used for growing tea, and the wetter southwest is home to plantations growing rubber plants. Rice is grown in the wet southwest, but irrigation projects have increased rice cultivation in other areas.

Maldives is a scattered collection of about 1,200 islands grouped into 13 atolls with barrier reefs. These atolls are coral reefs formed at the edge of volcanoes that have since sunk beneath the ocean. Colonies of tiny polyps secrete calcium carbonate to form the coral that builds the reefs. Over time, rubble from the reefs and sediment build up low islands. These islands have a very low elevation, mainly reaching no more than 6 feet (1.8 m) above the surrounding water.

Kathmandu, Nepal's capital and most populous city, sits in a valley more than 4,300 feet (1,311 m) above sea level.

◄ CRITICAL THINKING

1. *Identifying Cause and Effect* How do the mountains of Nepal and Bhutan influence where people live in those countries live?

2. *Hypothesizing* Do you think these two countries have much farmland? Why or why not?

"From Dingboche, Ed Viesturs, Jangbu, and I went ahead of the team to check on the condition of the trail [up Mount Everest]. Reports had reached us that fresh snow had fallen between the settlement of Lobuche and Base Camp, making the final leg of the route impossible for yaks. The yak drivers were refusing to lead their animals through the deep drifts. Hazard pay can tempt the Sherpas [people of Nepal who assist mountain climbers], but they won't risk their yaks for any price. We had nearly one hundred of the animals, but the loads would have to be carried by porters."

—Jamling Tenzing Norgay,
Touching My Father's Soul, 2002

DBQ *DRAWING INFERENCES*
Why were the Sherpas unwilling to use their yaks?

potential having a capacity that could be developed

aquifer an underground water-bearing layer of porous rock, sand, or gravel

The highest point on the islands is less than 8 feet (2.4 m) above sea level. Barrier reefs off many of these islands protect the sandy beaches from high seas. Many of the atolls are tiny; no island is larger than 5 square miles (13 sq. km). Only about 200 atolls are inhabited; most have fewer than 1,000 people. With little land for growing crops, fishing is a major activity.

☑ **READING PROGRESS CHECK**

Comparing How are the landforms of Bhutan, Maldives, Nepal, and Sri Lanka similar?

Water Systems

GUIDING QUESTION *How are rivers and the sea vital to human life in Bhutan, Maldives, Nepal, and Sri Lanka?*

Three major river systems—the Ganges, the Brahmaputra, and the Indus—fan out across the northern part of the Indian subcontinent. These rivers begin high in the Himalaya and carry fertile soil from mountain slopes onto their floodplains as the rivers swell with seasonal rains. The rivers of Nepal and Bhutan help feed the Ganges and the Brahmaputra.

Several rivers cut through the mountains of Nepal and Bhutan, generally flowing from the north to the south. The chief rivers in Nepal are the Baghmati, the Kosi, the Gandak, and the Karnali. These rivers eventually feed into the Ganges. Some of Nepal's rivers cut deep gorges running north and south through the mountains. Many rivers have fast-flowing streams through rocky channels that make them ideal for white-water rafting. This activity attracts many tourists who are eager for adventure. Nepal's fast rivers have great **potential** for hydroelectric power, but that resource has not been fully developed. The main rivers of Bhutan are the Torsa, the Raidak, the Sankosh, and the Manas. All join the Brahmaputra.

Sri Lanka has a dozen major rivers, most of which flow from the Central Highlands directly to the sea. The exception is the longest river, the Mahaweli, which begins flowing west of the Central Highlands but then follows a northward and eastward path to empty on the northeast coast. The Mahaweli, which swells in the rainy season, is tapped for water to irrigate fields. Rivers that drop from high elevations in the Central Highlands to lower-lying areas give Sri Lanka many spectacular waterfalls. Nearly 20 of them fall more than 348 feet (100 m).

Tiny Maldives has no rivers, but it does have underground aquifers that provide it with freshwater. **Aquifers** are underground water-bearing layers of porous rock, sand, or gravel.

The Indian Ocean surrounds both Maldives and Sri Lanka. The ocean provides rich resources to the people of these islands and affects the climate of all four countries, even far-off Nepal and Bhutan. Ocean waters also represent a threat to low-lying areas of Sri Lanka and to all of Maldives. Powerful cyclones—storms like hurricanes—can bring not only heavy rains but also high seas that can flood Sri Lanka's coastal lowlands and Maldives' low-lying islands.

Another danger comes from earthquakes that occur on the floor of the Indian Ocean. These events can unleash towering waves called tsunamis that can travel and flood far inland. A 2004 tsunami killed more than 31,000 people in Sri Lanka.

The ocean may cause problems for these countries in the future as well. Some climate scientists are forecasting that sea levels may rise as much as several feet during this century. If that does happen, the people of Maldives face a grim future, as the waters could submerge these low islands.

☑ **READING PROGRESS CHECK**

Drawing Inferences Why are aquifers important in Maldives?

Climates, Biomes, and Resources

GUIDING QUESTION *How does topography affect the climate in Nepal and Bhutan?*

As with the rest of South Asia, the monsoons strongly affect the climate of these four countries. However, the locations and landforms of these countries give them different climate zones. Elevation is also an important factor in climate.

Nepal and Bhutan have four climate zones. The Tarai in Nepal and lowlands in Bhutan have a humid subtropical climate. The lower hills and mountains have a warm temperate climate with summer rain and dry winters. The higher mountains have a highland climate with cool temperatures. Areas above 16,000 feet (4,877 m) have an arctic climate with year-round ice. Winters in the Himalaya are long and harsh. Bhutan is directly in the path of wet monsoon winds.

Vegetation varies by elevation and is related to the climate zones. Tropical forests cover the lowlands. Deciduous trees are found at the next elevation level, and conifers appear higher up. Grasses cover the next level, up to about 15,000 feet (4,572 m). Above that is the snow line, where nothing grows. Elephants, rhinoceroses, deer, and tigers roam the lowland areas.

ECOTOURISM ɪɴ NEPAL

LOW IMPACT
Ecotourism is responsible travel to areas of natural significance that attempts to conserve the environment and improve the well-being of local people.

INTERNATIONAL
Ecotourism is a growing industry, increasing by 5% each year. It represents 6% of the world's gross domestic product and 11.4% of all consumer spending worldwide.

PRE-TOURIST
Prior to the 1920s, the area was inhabited by the Sherpa people and largely undisturbed by Westerners. Climbing began in the 1920s, and the first successful attempt was in 1953.

TOURIST
Since 1990, climbing has become a popular tourist activity. Unfortunately, these tourists have made a habit of littering the mountain with gear and equipment after use.

ECOTOURIST
Today climbers are encouraged to be more responsible with their waste and unused gear. Since 2008, yearly "eco-climbs" occur to remove trash and raise awareness.

Climbing tourism generates **$500 million** in revenue for Nepal annually. Yet climbers also leave behind **120 tons of trash** each year. **Ecotourists hope to find a balance.**

Ecotourism based on the climate and natural beauty of the Himalaya is a growing industry. Mount Everest, the tallest mountain in the world, is an especially popular destination.

▲ CRITICAL THINKING

1. *Considering Advantages and Disadvantages* How does Nepal benefit from tourism? What are the negative effects of tourism?

2. *Hypothesizing* Why might someone decide to take part in an "eco-climb"?

Maldives has a humid tropical climate with a hot monsoon season and a warm dry season. Coconut palms and breadfruit trees grow on the islands. Tropical fish dart among the coral reefs, and tuna and sea turtles swim in the deeper waters.

Sri Lanka has a similar climate in lowland areas. Temperatures are warm or hot throughout the year. The Central Highlands are much cooler. The country has a tropical wet climate in the west which supports rain forests, but becomes tropical dry in the east with evergreen and deciduous forests and upland grasslands. Drier areas have low bushes and scrub plants. Sri Lanka's forests are teeming with wildlife, including elephants, sloth bears, buffalo, and a variety of birds.

Nepal's rivers have great potential for hydroelectric power, but only a tiny fraction of the country's energy is provided in this way. It does have some power-generating dams, and plans for more are underway. If the country can develop this resource, it could export surplus electricity to India. Mineral resources include deposits of coal, iron ore, copper, and limestone. Nepal's forests contain a variety of trees, but overcutting has resulted in massive soil erosion, prompting the government to implement conservation and reforestation projects.

exploit to make use of a resource

Bhutan has been much more successful in **exploiting** its hydroelectric potential. Electricity is its most valuable export. The Chhukha hydroelectric project began operating in the 1980s. Proceeds from selling power to India paid for the project. Another project completed in 2007 has also been successful, and the country has signed agreements with India to build four more dams. Bhutan has mineral resources, but transportation difficulties have limited its ability to use them.

Bhutan has worked to preserve its forests and has been more successful than Nepal in protecting this resource. While logging is important, it has not resulted in deforestation. Bhutan has taken cautious steps toward promoting tourism. While the government has encouraged tourists, it has also limited their number. The government fears damage to the land or to its culture if too many tourists visit.

Sri Lanka is a major producer of graphite, the material used for the "lead" in pencils. Other mineral resources include precious and semiprecious stones, which are important exports. Among Sri Lanka's most valuable resources, though, are its climate and rich variety of plant and animal life. These resources can attract tourists. Much of the forests have been cut down for farming or for timber, however.

cowrie shell the protective outer covering of a small sea creature, once used as money in Asia and Africa

Most of Maldives's resources derive from its ocean location. Tourism and fishing are the top two economic activities. For many centuries, the people of Maldives took another resource from the sea: **cowrie shells**. These shells were once sold to traders who used them for money in India, China, and Africa.

☑ **READING PROGRESS CHECK**

Explaining How are the climate zones of Nepal and Bhutan related to their landforms?

LESSON 1 REVIEW

Reviewing Vocabulary
1. *Identifying* Write a paragraph defining the term *aquifer* and explaining why aquifers are important to Maldives.

Using Your Notes
2. *Summarizing* Use your graphic organizer on these countries' physical systems to write a paragraph explaining how life in Nepal and Bhutan differs from that in Sri Lanka and Maldives.

Answering the Guiding Questions
3. *Drawing Conclusions* What landforms help shape life in Bhutan, Maldives, Nepal, and Sri Lanka?

4. *Identifying* How are rivers and the sea vital to human life in Bhutan, Maldives, Nepal, and Sri Lanka?

5. *Drawing Conclusions* How does topography affect the climate in Nepal and Bhutan?

Writing Activity
6. *Informative/Explanatory* In a paragraph, explain how the mountains of Nepal or Bhutan or the oceans surrounding Sri Lanka or Maldives both benefit and cause problems for people of those countries.

netw⊙rks

There's More Online!

☑ **IMAGE** Prayer Flags

☑ **CHART** South Asian Human Development Index

☑ **TIME LINE** Ethnic Strife

☑ **INTERACTIVE SELF-CHECK QUIZ**

☑ **VIDEO** Human Geography of Bhutan, Maldives, Nepal & Sri Lanka

LESSON 2
Human Geography of Bhutan, Maldives, Nepal & Sri Lanka

Reading **HELP**DESK

Academic Vocabulary

- portion
- recover

Content Vocabulary

- lama
- mantra
- stupa
- dzong

TAKING NOTES: *Key Ideas and Details*

IDENTIFYING CENTRAL ISSUES As you read about the human geography of the subregion, use a graphic organizer like the one below to identify the main points about the history, government, people, and economy of one country.

I. Bhutan
    A.
    B.
    C.
II. Maldives
    A.
    B.
    C.
    D.
III. Nepal
    A.
    B.
    C.
    D.
IV. Sri Lanka
    A.
    B.
    C.
    D.

ESSENTIAL QUESTION · *How do physical systems and human systems shape a place?*

IT MATTERS BECAUSE
The cultural geography of Bhutan, Maldives, Nepal, and Sri Lanka is widely varied. Each country's unique history and geographic location have determined the current political situation and religious practices.

History and Government

GUIDING QUESTION *What factors influenced the present-day governments of Bhutan, Maldives, Nepal, and Sri Lanka?*

There is little recorded Bhutan history before the introduction of Buddhism by monks fleeing Tibet in the A.D. 800s. These Tibetan monks gained control over the region. In the 1600s, a Tibetan high-level monk, or **lama**, named Ngawang Lopsang Gyatso merged religious and political power and created a system of law. After his death, however, chaos enveloped the country until 1885, when Ugyen Wangchuck began to establish order and unify the country. In 1907 he was proclaimed king. He adopted ideas from Britain regarding foreign affairs. Bhutan remained isolated. After Indian independence in 1947, India gained influence over Bhutan.

Bhutan began to emerge from its isolation in the 1950s. In 1972 Jigme Singye Wangchuck, great-grandson of the first king, took the throne and began to modernize the country. However, his policies favoring Buddhist culture angered thousands of Nepalese Hindus. There were violent protests in the 1990s; some Hindus fled to Nepal. The king instituted more democratic reforms and abdicated the throne in 2006. His son Jigme Khesar Namgyal Wangchuck became king with less power. In 2008 Bhutan became a constitutional monarchy and held its first general elections.

For thousands of years, small and medium kingdoms have existed in Nepal. Modern Nepal was founded in 1769, when Prithvi Narayan Shah conquered rival lands and declared a dynasty. While his heirs continued to rule as kings, the family isolated the country from the world. Nepal's skilled, disciplined soldiers, called Gurkhas, were recruited to serve in the British and Indian armies.

In the 1950s, the Narayan descendants were overthrown. The king then ruled directly. For the next 40 years, Nepal's government wavered between representative government and direct rule, which provoked democratic protests. In 1990 the king yielded and allowed election of a parliament, but party rivalries prevented stability. This gave Communist Mao rebels the chance to increase their activities. In 1996 a conflict between Maoist rebels and government troops lasting more than a decade left thousands dead. In 2008 Nepal abolished the monarchy, but political parties have been unable to agree on a new constitution.

Buddhists from South Asia first settled the Maldive Islands, followed by Muslims in the twelfth century. The islands came under the control of the Portuguese in the late 1500s and later the Dutch. In 1887 the islands became a British protectorate. After gaining independence in 1965, Maldives became a republic, but had the same person and political party in power for four decades. In 2008 the country adopted a democratic government, but rival groups struggle for power.

The Sinhalese probably arrived in Sri Lanka from India during the 400s B.C. Subsequently, the Tamil people settled in Sri Lanka from southern India. For nearly a thousand years, these two groups have been rivals for control of the island.

From the 1500s, European powers fought for control of Sri Lanka and its strategic location along trade routes. Portuguese merchants were followed by Dutch traders. The British had control of the entire island by 1815. They developed an economy based on growing coffee, tea, rubber, and coconuts. They brought in more Tamils to work on plantations. In 1948 Britain gave independence to Sri Lanka, at that time called Ceylon. The country adopted a parliamentary form of government. In 1972 it changed its name to the traditional form of Sri Lanka.

Sri Lanka was torn for decades by a civil war. The Tamil minority demanded a separate Tamil state, and many died in the fighting. The government declared an end to hostilities in 2009. Since then, Sri Lanka has had peace, but the underlying issues are unresolved and tension remains.

☑ **READING PROGRESS CHECK**

Analyzing Why have these countries had difficulties forming democratic governments?

National Maritime Museum, London/The Image Works

TIME LINE ⌄

ETHNIC
Strife ➔

Historically, the Sinhalese and the Tamils competed for power and influence in Sri Lanka, but after independence, their relationship grew increasingly hostile.

▶ **CRITICAL THINKING**

1. Evaluating What has been the relationship between the Sinhalese and the Tamils in Sri Lanka?

2. Explaining How did the end of the colonial era change the relationship between the Sinhalese and the Tamils?

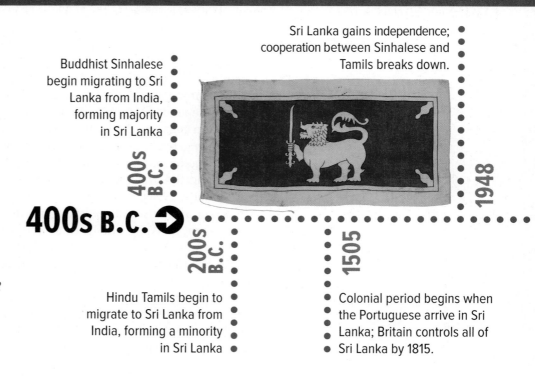

Buddhist Sinhalese begin migrating to Sri Lanka from India, forming majority in Sri Lanka

400s B.C.

400s B.C. ➔

200s B.C.

Hindu Tamils begin to migrate to Sri Lanka from India, forming a minority in Sri Lanka

1505

Colonial period begins when the Portuguese arrive in Sri Lanka; Britain controls all of Sri Lanka by 1815.

Sri Lanka gains independence; cooperation between Sinhalese and Tamils breaks down.

1948

Population Patterns

GUIDING QUESTION *How do population density maps of Nepal and Bhutan reflect the rugged landscape of the two countries?*

Of the four countries in the subregion, Nepal is the most populous country with 28 million people. As population density maps of the country show, the population distribution reflects Nepal's physical geography. Nearly half of all Nepalese live in the Tarai Plain in the southeast, and just over 40 percent live in the midlevel elevation hills and valleys, such as the valley of Kathmandu.

Bhutan, with about 800,000 people, has a rugged terrain with people living in isolated pockets. Some parts of the Duars Plain are less populated even though they are in lowlands because they are swampy and harbor malaria-carrying mosquitoes. About 35 percent of Bhutan's people are of Nepalese ancestry. They live mainly in the south of the country, speak Nepali, and practice Hinduism. In 1988 Bhutan's government declared that many of these people were living in Bhutan illegally. Protests and demands for equal rights arose, and thousands of people fled. Today about 85,000 of these former residents of Bhutan live in Nepal in refugee camps.

Sri Lanka has a population of 21 million people. Its highest population density is along the southwest coast near the capital, Colombo. Other areas of high density are around the Central Highlands city of Kandy, an important religious and commercial center, and on the northern part of the island. A chain of smaller cities runs along the west coast from Negombo to the north. Another cluster of cities is in the southeast from Batticoloa to Akkaraipattu.

About 200 of the islands in Maldives are inhabited. Another 100 islands have been reserved as tourist resorts. The most populous settlement is Male, the capital, with more than a third of the country's population. Maldives has 300,000 people, with another 100,000 living and working in foreign countries.

✔ **READING PROGRESS CHECK**

Identifying Cause and Effect What is the connection between the landforms and population density patterns in Nepal and Bhutan?

1956 Prime Minister Solomon Bandaranaike declares Sinhalese the official language. Tamils rise in violent protest.

1970 Sirimavo Bandaranaike, widow of Solomon, comes to power and reinstates pro-Sinhalese policies.

2002 Cease-fire agreement signed. Ban against Tamil Tigers ends; Tamils drop demand for a separate state.

➡ **1950**

➡ **2000**

1965 Tamil-dominated United National Party wins election.

1983 Thirteen Sinhalese soldiers killed in ambush by Tamil Tigers; civil war erupts

2006 New clashes erupt between government forces and Tamil Tigers.

Genetics

Social scientists research where the people living in a place come from—and how peoples from different areas mix. Now biochemists have a new tool to help them: detailed comparison of the genetic makeup of different peoples. Previous studies of the Nepalese showed that they have genetic connections to ancient populations from both South Asia and East Asia. Still unknown was whether the East Asian peoples came directly or went to India first, with their descendants moving into Nepal. In a study published in 2012, a team of scientists concluded, based on genetic similarities between Nepalese and Tibetans, that the migrants came directly from Tibet.

SPECULATING What other tools can geographers use to study the connections between peoples?

mantra sacred words or phrases that are repeated in prayers or chants

stupa a dome-shaped structure that serves as a Buddhist shrine

dzong a fortified monastery that also served as an administrative and commercial center

portion a share of a whole

Society and Culture Today

GUIDING QUESTION *How do different cultures and religions coexist in Bhutan, Maldives, Nepal, and Sri Lanka?*

The populations of these four countries reveal ethnic and religious diversity as the result of centuries of population movement and cultural influence from surrounding regions. Ethnic diversity has sometimes led to clashes between peoples. It has also contributed to unique and fascinating cultures.

The two main ethnic groups in Nepal are the Indo-Nepalese, whose ancestors migrated from India, and the Tibeto-Nepalese, whose ancestors came from Tibet. Sherpa are a Tibeto-Nepalese people known for their mountaineering skills. About half the people speak Nepali, but other languages are also spoken. Most of Nepal's people are Hindu. About 10 percent are Buddhists. Some aspects of Nepal's culture blend both Buddhism and Hinduism. Some Hindus and Buddhists recite **mantras**, sacred words or phrases that are repeated in prayers or chants. About four out of every five people live and work in rural areas, where they farm crops they sell in villages, herd animals, or produce traditional handicrafts.

The Bhote, who are also called Bhutia and Ngalops, make up about half the population of Bhutan and are descendants of Tibetan peoples. They practice Tibetan Buddhism, as do the Sharchops of eastern Bhutan. The official language of Bhutan, called Dzongka, is of Tibetan origin. Most people, however, speak other languages. Buddhism has left a strong mark on the country, which is home to about 4,500 monks, and thousands of **stupas**, or Buddhist shrines. In addition, the countryside is dotted by ***dzongs***, fortified monasteries that once served as administrative and commercial centers. A little over one-third of Bhutan's people live in cities, the largest of which is the capital of Thimphu, with only about 40,000 people. Rural people farm and raise livestock. Illiteracy has long been a problem in Bhutan. The government has recently devoted a major **portion**, or share, of its national budget to education. As a result, school enrollment and literacy rates are on the increase.

Sri Lanka's two main ethnic groups are deeply divided along religious as well as ethnic lines. The Sinhalese majority, found mainly in the southwest, are Buddhist and speak Sinhalese, the official language of Sri Lanka. The Tamils live chiefly in the northern region of the island. They are Hindu and speak Tamil, which is related to the Dravidian languages of southern India. The move for an independent Tamil state arose among members of this group. Most of Tamils are descended from Indians brought to the island in the 1800s to work on plantations.

The earliest-known settlers in the Maldive Islands were probably from southern India, followed by the Sinhalese from Sri Lanka. Later, people from East Africa and Arab traders settled in the Maldives. Today people of Maldives speak the official language of Dhivehi, which is based on Sinhalese, as well as English.

Family and the Status of Women

The chiefly rural populations of these countries followed traditional lifestyles and family patterns for many centuries. In Sri Lanka, parents typically arranged marriages and brides were generally quite young. Married children lived in the same household as their parents, although the wives in the household tended to cook separately for their own smaller family unit. These patterns have changed in recent decades. About 1.6 million Sri Lankans live and work in other countries, and about 100,000 leave their homes to work in plantations on the island. About half the overseas workers are women. Sri Lanka's government is concerned that these women have few protections in the countries where they work and can be exploited and harassed. The number of migrant workers combined with other economic changes has altered family patterns in Sri Lanka. People tend to marry later than in the past, have fewer children, and live in nuclear families rather than extended families. People are living longer than in the past as well, which creates the need for support from family members and for government pensions and health care.

In traditional cultures, women often are second-class in status. Society in Nepal has long had a caste system similar to that of India, in which social status is set at birth. Males belonging to the highest caste—a tiny minority of the country's people—dominated society. Women, members of the lower castes, Muslims, and others had few rights. The country has tried to change this by making women's equality before the law a fundamental part of the constitution. Women have yet to make much progress. Women are pushing for laws that would give them the rights to own and inherit property.

In Bhutan, women traditionally enjoyed more rights. In fact, women in the family inherited property to ensure their economic security and make sure that they could care for both children and elderly family members. Women were as likely as men to be the head of the household and their work was valued equally with that of men. However, Bhutanese women play a very small role in government and politics—only one-fourth of the seats in the national legislature are held by women, compared to one-third of those in Nepal. In both Sri Lanka and Maldives, less than 10 percent of lawmakers are women.

Women's rights activists say the situation for women is worsening in Maldives. The rise of religious fundamentalism there threatens women's rights. Violence against women is a major problem, and the government is slow to enact a law against it.

The Arts

The artistic spirit is part of daily life in each of these countries. The Buddhist stupas of Nepal and Sri Lanka and the dzongs of Bhutan are works of art and architecture. They also reveal the importance of religion to these cultures. Dance is a popular cultural expression in Bhutan and Nepal and often tells religious or mythical stories. The Newar people of Nepal have a rich architectural and artistic tradition and built many religious shrines. In literature, Michael Ondaatje, who was born in Sri Lanka, won England's prestigious Booker Prize for his novel *The English Patient*.

✔ READING PROGRESS CHECK

Making Connections Why is it thought that the Tamils of Sri Lanka originated in southern India?

In Bhutan and Nepal, people fly colorful prayer flags, and prayer wheels twirl on many corners, sending out invocations. Monks chant mantras, or repetitive prayers.

◄ **CRITICAL THINKING**

1. *Drawing Conclusions* Why is religion a strong force in these people's lives? Explain your answer.

2. *Comparing* What do other religions use that is similar to prayer flags? Why do you think Buddhists use prayer flags?

South Asian Human Development Report	
Country	**Rank**
United States (comparison)	8
Sri Lanka	73
Maldives	104
India	130
Bhutan	132
Bangladesh	142
Nepal	145
Pakistan	147

Source: United Nations Development Programme, 2014

The Human Development Index is based on three dimensions: health (life expectancy at birth); education (mean years of schooling and expected years of schooling); and standard of living (gross national income per capita).

▲ **CRITICAL THINKING**

1. *Comparing and Contrasting* Why do you think the four countries of this subregion rank where they do?

2. *Analyzing* Are there any other factors that you think should be included in a measure of a country's development? Why or why not?

recover to return to a previous level of performance after a decline

Economic Activities

GUIDING QUESTION *How important are farming and fishing to the people of Bhutan, Maldives, Nepal, and Sri Lanka?*

The countries of South Asia have relatively few resources and are somewhat remote. As a result, they are not very economically well developed. Nevertheless, all these countries are connected to the world economy as tourist destinations. Some also export goods or services, and all receive development aid from more developed countries. All joined with Bangladesh, India, and Pakistan in 1985 to form the South Asian Association for Regional Cooperation (SAARC).

In the highlands of Nepal, farmers practice terracing, making use of arable land on the steep slopes. They also cultivate the highland valleys and the low-lying Tarai Plain. Almost three-fourths of the workforce is engaged in agriculture and herding. Chief crops are rice, wheat, and corn. Bhutan, like Nepal, also practices terracing to increase its arable area. Farmers in Bhutan also grow wheat and rice, as well as barley, vegetables, and fruit.

Sri Lanka exports tea, rubber, and coconuts, as well as shrimp, lobsters, and fish. Rice is grown for domestic consumption. In 2004 a tsunami in the Indian Ocean caused widespread destruction in Sri Lanka and Maldives, including the devastation of the fishing industry. Fish catches of tuna and bonitos have declined because large-scale fishing operations from other countries are taking more fish and depleting the supply. The people of Maldives export fruits, vegetables, and coconut for food and oil.

Few major industries exist in these countries due to their lack of technical development. Sri Lanka mines and exports graphite. Precious and semiprecious stones are also mined. Food, tea, and rubber processing plants sell products globally. Bhutan has developed hydroelectric projects and sells electricity to India.

Hiking and white-water rafting are popular activities for tourists in Nepal, which also attracts visitors to its religious sites. Bhutan has moved in recent years to expand tourism. To protect the environment and its traditional culture, however, Bhutan restricts the number of tourists who may visit. In Sri Lanka, the tourism industry has **recovered** since relative peace has been restored. Tourism is a mainstay of the economy of Maldives, which attracts visitors with beautiful beaches and colorful coral reefs. The tourism industry was set back by the 2004 tsunami, which damaged some resorts, but it has since rebounded.

✔ **READING PROGRESS CHECK**

Summarizing How important is agriculture, fishing, and industry to the people of these countries?

LESSON 2 REVIEW

Reviewing Vocabulary

1. *Identifying* Write a paragraph explaining how *dzongs*, lamas, mantras, and stupas are related to one another.

Using Your Notes

2. *Considering Advantages and Disadvantages* Use your graphic organizer on one of these countries to write a paragraph explaining the country's greatest advantages and its biggest challenges.

Answering the Guiding Questions

3. *Explaining* What factors influenced the present-day governments of Bhutan, Maldives, Nepal, and Sri Lanka?

4. *Comparing and Contrasting* How do population density maps of Nepal and Bhutan reflect the rugged landscape of the two countries?

5. *Explaining* How do different cultures and religions coexist in Bhutan, Maldives, Nepal, and Sri Lanka?

6. *Comparing and Contrasting* How important are farming and fishing to the people of Bhutan, Maldives, Nepal, and Sri Lanka?

Writing Activity

7. *Argument* In a paragraph, explain what you think can be done to settle conflicts between groups in Sri Lanka or Bhutan and why you think it would work.

networks

There's More Online!

☑ **GRAPH** Deforestation

☑ **IMAGE** Deforestation in Sri Lanka

☑ **IMAGE** Underwater Cabinet Meeting

☑ **INTERACTIVE SELF-CHECK QUIZ**

☑ **VIDEO** People and Their Environment: Bhutan, Maldives, Nepal & Sri Lanka

LESSON 3

People and Their Environment: Bhutan, Maldives, Nepal & Sri Lanka

ESSENTIAL QUESTION • *How do physical systems and human systems shape a place?*

Reading **HELP**DESK

Academic Vocabulary

- **extract**
- **generate**

Content Vocabulary

- **clear-cutting**
- **organic farming**

TAKING NOTES: *Key Ideas and Details*

EVALUATING As you read about environmental issues in Bhutan, Maldives, Nepal, and Sri Lanka, use a graphic organizer like the one below to list key environmental issues facing each country.

Key Environmental Issues

Country	Issues
Bhutan	
Maldives	
Nepal	
Sri Lanka	

It Matters Because

As countries with developing economies, Bhutan, Maldives, Nepal, and Sri Lanka want to promote economic development so their people have better lives. At the same time, they have unique and fragile natural environments. They work to balance economic development and preservation of those environments.

Managing Resources

GUIDING QUESTION *Why has sustainable development been given a low priority in the subregion?*

The countries of Bhutan, Maldives, Nepal, and Sri Lanka have few resources and are not well developed economically. With limited resources, there are few opportunities to pursue traditional methods of economic development such as industrialization. However, these countries do have the ability to protect their existing natural resources by managing them wisely.

In recent years, environmentalists and some economists have pushed the idea of sustainable development. This concept emphasizes seeking economic development in concert with other goals. These other goals include a more equal distribution of the benefits of economic development and reduced damage to the environment.

There are obstacles to implementing sustainable development in the subregion. Countries are generally poor and lack advanced technology. They do not have the capital to invest in new technologies, and the population lacks the education needed to make use of the technology. People in the subregion live in scattered communities with strong local traditions that are sometimes at odds with the goals of sustainable development. Their isolation also makes communication difficult. In some places, government turmoil and conflicts have delayed development. These factors make it difficult to unite people in common approaches to the goals of sustainable development.

Nevertheless, these countries also have some advantages. The Buddhist and other traditions place value on all forms of life and teach respect for nature—ideas that are consistent with protecting the environment. The relative lack of modern technology means that they are not tied to more developed countries' production methods—methods that have often led to heavy pollution.

The idea of sustainable development has taken hold in Bhutan. In 1972 King Jigme Singye Wangchuck coined the phrase "Gross National Happiness." This phrase describes his vision for Bhutan's future, a future that focuses not only on economic growth but also on broader social, cultural, and environmental concerns. Although King Jigme Singye Wangchuck stepped down from the throne in 2006, the country continues to focus on these concerns.

Although the countries of this subregion do not have abundant resources, they do have some. Forests cover much of Bhutan, and about one-third of both Nepal and Sri Lanka's land is forested. Bhutan has been more successful in protecting its forests than Nepal and Sri Lanka, where deforestation is a major problem. Both Nepal and Sri Lanka have tried to step up their efforts to protect the environment in recent years after years of political conflict absorbed government attention. Their efforts are still too recent to yield strong results, however. Maldives has moved to ban mining of coral and dredging of sand. In 2012 it announced an ambitious plan to turn the entire country into a biological reserve that would allow only sustainable fishing and other resource use. The government hopes to achieve this goal in just a few years. All of the countries have areas of natural beauty that attract visitors. Because it is an island, Sri Lanka is home to many unique plant and animal species that face extinction unless protected. These assets will only attract tourists if their natural beauty is protected and maintained.

☑ **READING PROGRESS CHECK**

Summarizing What are the obstacles to sustainable development in this subregion?

Human Impact

GUIDING QUESTION *What environmental threats do these countries face?*

Human activity has had a devastating impact on the fragile environments of the countries in this subregion. Major environmental problems include deforestation and soil and water issues. Beaches and marine life have also been damaged.

Centuries ago, much of South Asia was covered with forests. Today the region is in an environmental crisis, as deforestation has accelerated in recent years. According to some estimates, Sri Lanka lost as much as 20 percent of its forest cover between 1990 and 2010. Damage in Nepal was even worse—a decline of 25 percent over the same period. Commercial timber companies have used **clear-cutting**, or the removal of all trees in a stand of timber, to harvest logs. Areas of forest are also cleared to make way for agriculture and expanding settlements, as well as to provide fuel for woodstoves. Burning wood indoors for cooking and heating leads to indoor air pollution and respiratory problems.

Whatever the reasons behind deforestation, the results are devastating. Environmentalists believe the destruction of mangrove forests in Sri Lanka made the coast more vulnerable to the devastation caused by the December 2004 tsunami. Losing tropical rain forests, as in Sri Lanka, has many damaging effects. Rain forests usually grow in poor soil where the complex root systems of trees efficiently absorb water and hold the topsoil in place. Additionally, as rainfall slowly filters through layers of leafy branches, the surrounding air is cooled. When rain forests disappear, soil erodes, rains produce floods, and temperatures rise.

clear-cutting the removal of all trees in a stand of timber

Soil erosion is also a problem when trees are cut in mountainous Nepal and Bhutan, where tree roots are also needed to hold soil in place to sustain agriculture.

Deforestation also affects South Asia's wildlife, which depend on the forests and other ecosystems for food and habitat. The region is home to an astonishing variety of wildlife. Elephants, water buffalo, and monkeys flourish in the rain forests of Sri Lanka. The Himalaya are home to the endangered snow leopard, musk deer, Himalayan black bear, red panda (firefox), and others. Bhutan's national animal, the takin, which is most closely related to the musk ox, is threatened by overhunting and habitat loss. All these animals are threatened by the spread of settlement that further reduces and fragments their habitat. Poaching, or illegal hunting, is also a threat.

Bhutan, Nepal, and Sri Lanka suffer from soil erosion, which means lost farmland, the possibility of mudslides, and the pollution of freshwater supplies. Industrial waste, agricultural runoff, and untreated human and animal waste threaten water supplies in Nepal and Sri Lanka. Groundwater in Nepal's Kathmandu and Pokhara valleys is so polluted that it is unhealthy to drink. The growing demand for water threatens the supply of freshwater in Maldives's aquifers.

Maldives is also seriously threatened by other problems. Coral is being **extracted**, or removed, from the coral reefs just offshore in order to make jewelry and to use as a building material. By depleting the coral reef, however, people have made the low-lying islands more vulnerable to high seas when major storms hit.

The government of Maldives has also grown increasingly concerned about the possibility of climate change. The rising sea levels would be a serious threat to a country that is generally only about 6 feet (1.8 m) above sea level. Warming also threatens the coral that protect Maldives. When ocean waters become too warm, the coral organisms that make up the reefs expel the algae that live inside them. The loss of algae causes the coral to lose its color, a process known as coral bleaching. Coral feed off the nutrients that the algae produce, and their loss often leads to death because of the reduced food supply. Since the coral is the foundation of the reef, other plants and animals are also threatened. A temperature spike in 1998 wiped out almost all the coral around Maldives. While the reefs recovered over time, high temperatures in 2010 and 2015 caused more loss.

☑ **READING PROGRESS CHECK**

Analyzing What are the most pressing environmental issues? Why?

Addressing the Issues

GUIDING QUESTION *What steps have the governments of Bhutan, Maldives, Nepal and Sri Lanka taken to improve the environment?*

The countries of this subregion have taken some steps to address environmental problems. Bhutan and Maldives have also become leading voices urging action toward sustainable development.

Bhutan's constitution requires that the country keep a minimum of 60 percent of its land forested. The government has stated its goals of relying on organic farming and having no emissions of carbon dioxide. **Organic farming** is the use

Deforestation is a major environmental problem in Sri Lanka where timber companies have used clear-cutting.

▲ **CRITICAL THINKING**

1. Analyzing Visuals How does deforestation change habitats?

2. Making Connections What other problems does deforestation cause?

extract to withdraw or remove

organic farming the use of natural substances rather than chemical fertilizers and pesticides to enrich the soil and grow crops

In 2009, to emphasize the threat posed by rising seas that might result from climate change, the government of Maldives held an underwater cabinet meeting.

▲ **CRITICAL THINKING**

1. Explaining Why would climate change be a particular threat to Maldives?

2. Analyzing What impact do you think an action like this will have on other governments? Why?

generate to bring into existence

of natural substances rather than chemical fertilizers and pesticides to enrich the soil and grow crops. Carbon dioxide emissions are thought to add to climate change. The government has also urged other countries to move away from an emphasis on economic growth while ignoring its social and environmental costs. In 2012 Bhutan led a conference promoting these ideas. Bhutan has also worked with donor countries and outside experts, including the European Union, to promote its environmental policies.

Nepal's government has also put in place policies aimed at protecting the environment and preserving resources. Critics say it has not gone far enough in actually implementing these plans. According to a recent report, the government has not followed through on its commitments. Nepal relies on foreign aid to fund more than half of its development efforts. Many of these donors have pushed for environmentally-friendly projects to bolster Nepal's efforts in this area.

Sri Lanka began to take steps to care for its environment in 1981, when it created the Central Environmental Authority. Twenty years later, it formed the Ministry of Environment and Natural Resources, raising environmental action to the cabinet level of government. Along with protecting some areas, the government has launched efforts to reduce household and plastic waste and promote environmental education among students. Recently the government began an initiative aimed at promoting sustainable development in ten areas. It points to some progress, including the fact that renewable resources **generate** more than half of its electricity, primarily hydroelectricity. It has also increased efforts to fight river water pollution.

Maldives, worried that climate change threatens its very existence, has launched programs to try to limit the use of fossil fuels. Its goal is to reduce carbon dioxide emissions by 10 percent by 2030. To achieve that goal, it is pushing for increased use of solar energy, and it is importing more electric cars. Other countries are helping by contributing to a Climate Change Trust Fund set up by the World Bank. In addition, the leaders of Maldives are very active in all international discussions of the climate change issue, urging other countries—especially developed countries—to take steps to reduce their own carbon emissions.

☑ **READING PROGRESS CHECK**

Analyzing Which country do you think has made the most progress in addressing environmental issues? Why?

LESSON 3 REVIEW

Reviewing Vocabulary

1. Identifying Write a paragraph explaining why organic farming is thought to help the environment.

Using Your Notes

2. Comparing and Contrasting Use your graphic organizer on the key environmental issues of these countries to explain which country you think faces the most serious environmental issues.

Answering the Guiding Questions

3. Summarizing Why has sustainable development been given a low priority in the subregion?

4. Explaining What environmental threats do these countries face?

5. Summarizing What steps have the governments of Bhutan, Maldives, Nepal, and Sri Lanka taken to improve the environment?

Writing Activity

6. Argument Write a speech that the leader of Bhutan or Maldives might give urging developed countries to take steps to adopt sustainable development or to address climate change.

666

Directions: On a separate sheet of paper, answer the questions below. Make sure that you read carefully and answer all parts of the questions.

Lesson Review

Lesson 1

1 *Explaining* Explain whether people living in Maldives are more likely to make a living through farming or fishing.

2 *Assessing* How have Nepal's rivers with fast-flowing streams been effectively used as a resource? What potential resource of fast rivers has not been fully developed?

3 *Drawing Conclusions* Is Bhutan's attitude toward tourism contradictory? Why or why not?

Lesson 2

4 *Identifying Cause and Effect* What was the main cause of European powers fighting for control of Sri Lanka?

5 *Making Connections* Identify three of the principal crops in Nepal. Explain whether it would be correct to state that the percentage of the Nepalese workforce engaging in agricultural work is not proportionate to the amount of farmland in Nepal.

6 *Explaining* Explain why industry is limited in Bhutan, Maldives, Nepal, and Sri Lanka.

Lesson 3

7 *Describing* What are the causes of the deforestation in Sri Lanka?

8 *Identifying* What are two major reasons the government of Maldives is extremely concerned about the possibility of climate change?

9 *Summarizing* Summarize Sri Lanka's environmental efforts from 1981 to the present day.

21st Century Skills

Use the image below to answer the questions that follow.

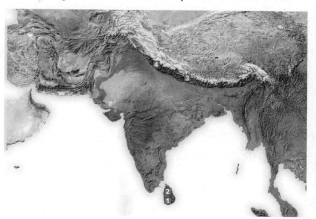

10 *Geography Skills* Identify the mountain range indicated by the number 3.

11 *Creating Diagrams* Explain the tectonic activity that resulted in this mountain range and create a diagram to show this tectonic activity.

Critical Thinking

12 *Exploring Issues* Explain how rising sea levels threaten Maldives. Discuss ways in which countries can cooperate to address this issue.

Applying Map Skills

Use the Unit 7 Atlas to answer the following questions.

13 *Human Systems* Of Bhutan, Maldives, Nepal, and Sri Lanka, identify the country with the greatest population density.

14 *The World in Spatial Terms* Use your mental map of South Asia to describe the following: the spatial relationship between the Himalaya and South Asia; the spatial relationship among the three major ranges of the Himalaya.

15 *Places and Regions* Identify the longest river in Sri Lanka.

Need Extra Help?

If You've Missed Question	**1**	**2**	**3**	**4**	**5**	**6**	**7**	**8**	**9**	**10**	**11**	**12**	**13**	**14**	**15**
Go to page	653	654	656	658	662	662	664	665	666	667	667	665	598	598	598

Directions: On a separate sheet of paper, answer the questions below. Make sure you read carefully and answer all parts of the questions.

Exploring the Essential Question

16 *Identifying Central Issues* Create a poster divided into four sections, one for each of the following: Bhutan, Maldives, Nepal, and Sri Lanka. For each country, sketch a map and include labels to identify major features such as the capital, waterways, mountains, and locations of natural resources. Provide a brief statement for each country about a major challenge faced today.

College and Career Readiness

17 *Clear Communication* You are studying anthropology and history in college, and you are applying for a position as a student assistant on a summer expedition to Nepal. The purpose of the expedition is to study the culture of Nepal and report on the current interaction between the Nepalese and their physical environment. Write a one-page application essay to your professor to explain why you should be selected for the expedition. In your essay, detail your knowledge of the culture of Nepal and current environmental benefits and challenges in the country, explaining how you will be respectful of the culture as you conduct observation and research.

Research and Presentation

18 *Gathering Information* With a partner, conduct research to learn about the roots, development, and current activity of the South Asian Association for Regional Cooperation (SAARC). Provide a multimedia presentation to explain the following: when and why the organization was formed, initial work of the organization, growth of the organization, members of the organization, current activity, and planned future activity. Within your presentation, include maps, photographs, and diagrams or graphs.

DBQ Analyzing Primary Sources

Use the document to answer the following questions.

Mount Everest has attracted climbers who hope to scale its famously steep summit for decades. It is a difficult mountain to get to, and it is even more difficult to climb. However, despite the danger, many continue to attempt to ascend Mount Everest.

PRIMARY SOURCE

"Off in the distance, Everest's three-thousand-foot summit pyramid looms, nearly invisible in the night but unmistakably there. Its steep ribs of stone and ice gullies are not so much seen as conjured up in a brain starved for oxygen weary from driving a dying body up into the upper reaches of the ionosphere. Every time I see this peak—or even get close to it—the reaction never varies: excitement, intimidation, dread."

—Tom Whittaker, *Higher Purpose: The Heroic Story of the First Disabled Man to Conquer Everest,* 2001

19 *Synthesizing* How does the quote help you understand why so many people travel to Nepal to try to climb Mount Everest?

20 *Analyzing* How is Whittaker's effort to climb Mount Everest symbolic of the struggle of the people in Nepal to try to overcome obstacles?

Writing About Geography

21 *Argument* Use standard grammar, spelling, sentence structure, and punctuation to write an editorial of five to seven paragraphs about whether or not to support increased tourism in Bhutan. Be sure to address opposing arguments in your editorial.

Need Extra Help?

If You've Missed Question	16	17	18	19	20	21
Go to page	649	660	662	668	668	656

East Asia

Chapter 28
China and
Mongolia

Chapter 29
Japan

Chapter 30
North Korea and
South Korea

UNIT **8**

1 Cities East Asia has several cities with many millions of people, where traditional culture is combined with modern skyscrapers and technology.

EXPLORE the REGION

A dynamic region of booming economies and bustling cities, **EAST ASIA** is also home to one of the world's oldest civilizations and to centuries-old traditions. Several countries in the region are major trading partners with the United States, and by virtue of their size and economic output, some countries in East Asia are among the most powerful and influential in the world.

THERE'S MORE ONLINE

3 Economy Manufacturing and trade are major economic activities in East Asia, a region that includes four of the most productive economies in the world.

4 Culture China's culture, thousands of years old, has influenced other countries in the region. But those countries have also maintained their own traditions, like this Korean dance.

2 Farming The people of East Asia have relied on its rivers for thousands of years to bring water and fertile soil to their fields. Rice, long cultivated here, is a major staple of the diet, but China is also a major producer of wheat.

East Asia
Physical

CENTRAL ASIA

RUSSIA

Khüiten Peak
14,350 ft.
(4,374 m)

Junggar
Basin

ALTAY SHAN

Hentiyn Nuruu

Kerulen R.

Amur R.

Da Hinggan Ling

Xiao Hinggan Ling

Songhua R.

Sea of
Okhotsk

TIAN SHAN

Turpan
Depression
-426 ft.
(-130 m)

G O B I

Manchurian
Plain

Liao R.

Changbai
Shan

Hokkaidō

PAMIRS

Tarim Basin

TAKLIMAKAN
DESERT

K2
28,250 ft.
(8,611 m)

Karakoram
Range

Altun Shan

KUNLUN SHAN

Muztag
25,338 ft.
(7,723 m)

Qilian Shan

Huang He
(Yellow R.)

Mu Us
Desert

Bo
Hai

Korea
Bay

Shandong
Peninsula

Korea
Peninsula

Sea of Japan
(East Sea)

Honshū

Mt. Fuji
12,388 ft.
(3,776 m)

PLATEAU OF TIBET

Chang Jiang

Mekong R.

Salween R.

Wei R.

Qin Ling

Yellow
Sea

Korea Strait

Shikoku

Kyūshū

HIMALAYA

Mt. Everest
29,028 ft.
(8,848 m)

North China Plain

Grand Canal

East
China
Sea

140°E

Brahmaputra R.

SOUTH
ASIA

80°E

Sichuan
Basin

Yunnan
Plateau

Chang Jiang (Yangtze R.)

Xiang R.

Gan R.

Xi R.

Ryukyu Islands

TROPIC OF CANCER

Taiwan Strait

Taiwan

Yü Shan
13,113 ft.
(3,997 m)

20°N

UNIT 8
REFERENCE ATLAS

Salween R.

Red R.

Leizhou
Peninsula

Gulf of
Tonkin

Hainan

Luzon Strait

South
China
Sea

Philippine
Sea

PACIFIC
OCEAN

MAP STUDY

1. **Physical Systems** What physical feature separates Mongolia from China in the southeast?

2. **Human Systems** Which Chinese cities are located along the Chang Jiang?

Mekong R.

SOUTHEAST
ASIA

EQUATOR

0°

Elevations

10,000 ft. (3,000 m)	
5,000 ft. (1,500 m)	
2,000 ft. (600 m)	
1,000 ft. (300 m)	
0 ft. (0 m)	
Below sea level	

—— National boundary

⊣⊢ Canal

▲ Mountain peak

▼ Lowest point

INDIAN
OCEAN

0 ——— 1,000 miles

0 ——— 1,000 kilometers

Lambert Azimuthal Equal-Area projection

100°E

120°E

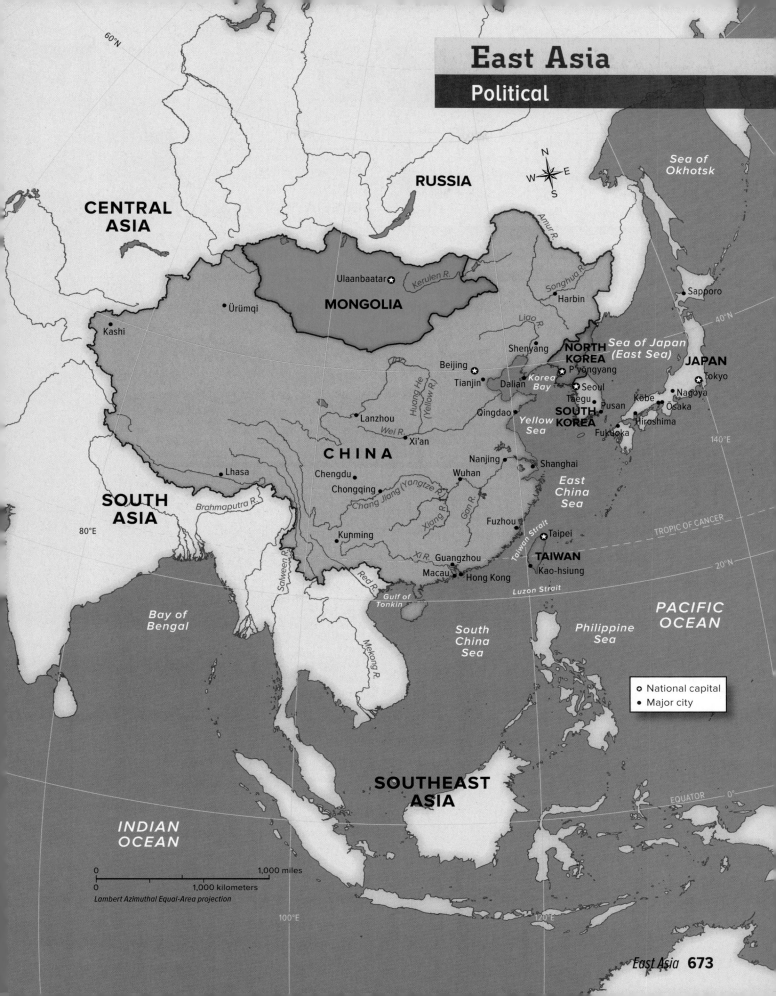

East Asia
Political

CENTRAL ASIA

RUSSIA

MONGOLIA

• Ürümqi

• Ulaanbaatar ✪

Kerulen R.

Sea of Okhotsk

• Kashi

Amur R.

Songhua R.

• Harbin

Liao R.

• Sapporo

• Shenyang

NORTH KOREA

Sea of Japan (East Sea)

JAPAN

• Beijing ✪

Huang He (Yellow R.)

• Tianjin

• Dalian

Korea Bay

P'yŏngyang ✪

Seoul ✪

SOUTH KOREA

Tōkyō ✪

• Nagoya

Kōbe

Ōsaka

• Lanzhou

Wei R.

Qingdao •

Taegu •

Pusan •

• Xi'an

Yellow Sea

Hiroshima •

Fukuoka •

CHINA

• Nanjing

• Shanghai

East China Sea

SOUTH ASIA

• Lhasa

• Chengdu

Brahmaputra R.

Wuhan •

• Chongqing

Chang Jiang (Yangtze R.)

Gan R.

Xiang R.

Salween R.

• Kunming

Xi R.

• Guangzhou

Fuzhou •

Taiwan Strait

• Taipei ✪

TAIWAN

Kao-hsiung •

Macau •

Hong Kong •

Luzon Strait

TROPIC OF CANCER

Red R.

Gulf of Tonkin

Bay of Bengal

South China Sea

Philippine Sea

PACIFIC OCEAN

Mekong R.

SOUTHEAST ASIA

INDIAN OCEAN

EQUATOR

60°N

40°N

140°E

80°E

20°N

100°E

120°E

0°

0   1,000 miles

0   1,000 kilometers

Lambert Azimuthal Equal-Area projection

✪ National capital
• Major city

East Asia **673**

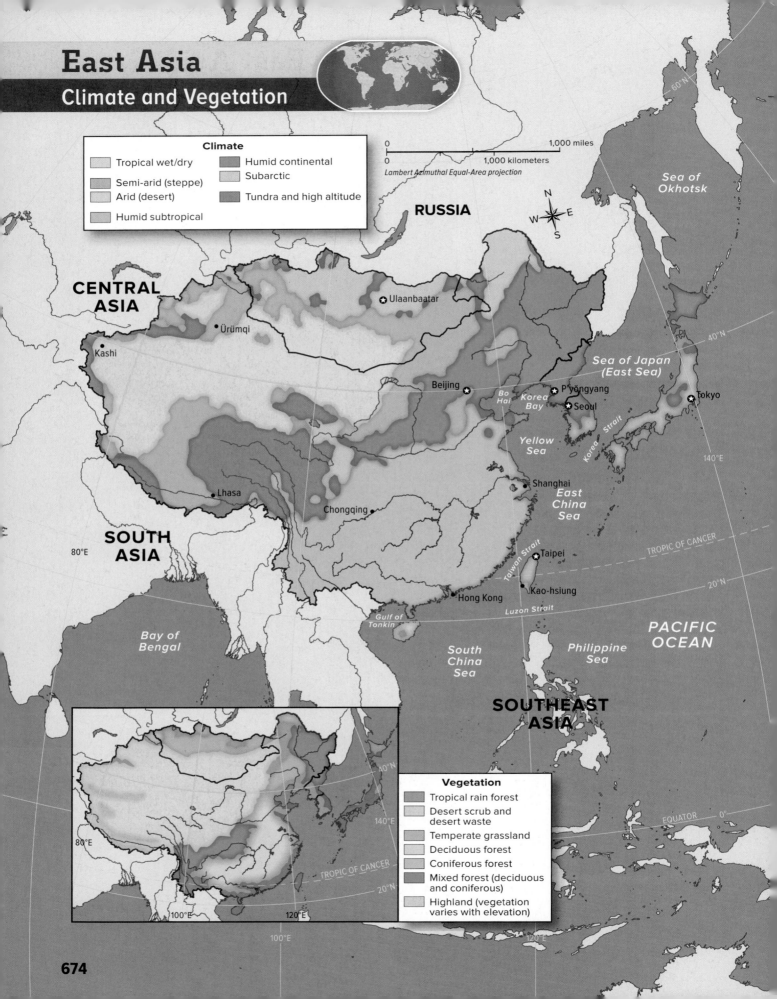

East Asia
Climate and Vegetation

Climate

- Tropical wet/dry
- Semi-arid (steppe)
- Arid (desert)
- Humid subtropical
- Humid continental
- Subarctic
- Tundra and high altitude

0   1,000 miles
0   1,000 kilometers
Lambert Azimuthal Equal-Area projection

RUSSIA

N
W   E
S

Sea of Okhotsk

CENTRAL ASIA

60°N

40°N

•Ürümqi

☆ Ulaanbaatar

Kashi•

Beijing ☆

Bo Hai
Korea Bay

☆ P'yŏngyang

☆ Seoul

Sea of Japan (East Sea)

☆Tokyo

140°E

SOUTH ASIA

•Lhasa

Chongqing•

Yellow Sea

•Shanghai

East China Sea

Korea Strait

80°E

TROPIC OF CANCER

Taipei ☆

Taiwan Strait

Hong Kong•

•Kao-hsiung

20°N

Bay of Bengal

Gulf of Tonkin

South China Sea

Luzon Strait

Philippine Sea

PACIFIC OCEAN

SOUTHEAST ASIA

EQUATOR   0°

Vegetation

- Tropical rain forest
- Desert scrub and desert waste
- Temperate grassland
- Deciduous forest
- Coniferous forest
- Mixed forest (deciduous and coniferous)
- Highland (vegetation varies with elevation)

80°E

140°E

40°N

TROPIC OF CANCER

20°N

100°E

120°E

100°E

120°E

674

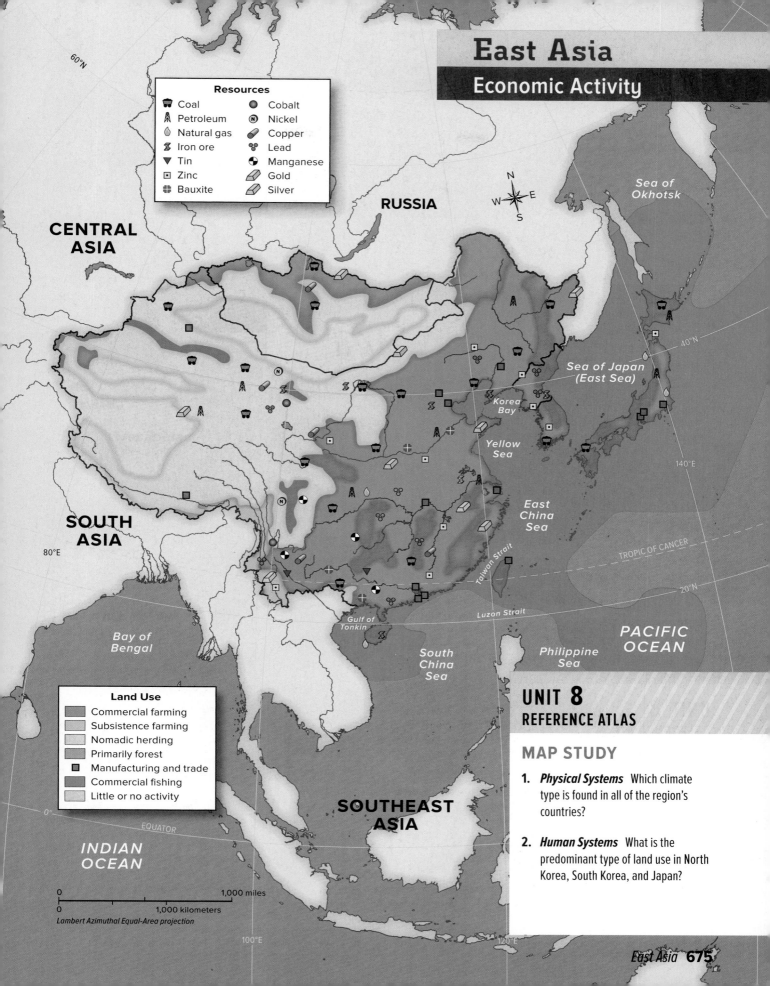

East Asia
Economic Activity

Resources

- 🛒 Coal
- ⚒ Petroleum
- 🜄 Natural gas
- ⚡ Iron ore
- ▼ Tin
- ⊡ Zinc
- ✚ Bauxite
- ⬤ Cobalt
- Ⓝ Nickel
- ⬭ Copper
- ❦ Lead
- ◐ Manganese
- ▱ Gold
- ▭ Silver

Land Use

- Commercial farming
- Subsistence farming
- Nomadic herding
- Primarily forest
- ■ Manufacturing and trade
- Commercial fishing
- Little or no activity

CENTRAL ASIA

RUSSIA

Sea of Okhotsk

SOUTH ASIA

Sea of Japan (East Sea)

Korea Bay

Yellow Sea

East China Sea

TROPIC OF CANCER

Taiwan Strait

Luzon Strait

Gulf of Tonkin

South China Sea

Philippine Sea

PACIFIC OCEAN

SOUTHEAST ASIA

Bay of Bengal

INDIAN OCEAN

EQUATOR

0   1,000 miles
0   1,000 kilometers
Lambert Azimuthal Equal-Area projection

60°N

80°E

40°N

140°E

20°N

100°E

120°E

UNIT 8
REFERENCE ATLAS

MAP STUDY

1. **Physical Systems** Which climate type is found in all of the region's countries?

2. **Human Systems** What is the predominant type of land use in North Korea, South Korea, and Japan?

East Asia
Population Density

CENTRAL ASIA

RUSSIA

SOUTH ASIA

SOUTHEAST ASIA

Sea of Okhotsk

Sea of Japan (East Sea)

Yellow Sea

East China Sea

South China Sea

Philippine Sea

PACIFIC OCEAN

Bay of Bengal

Gulf of Tonkin

Luzon Strait

TROPIC OF CANCER

EQUATOR

Cities
(Statistics reflect metropolitan areas.)

■ Over 5,000,000
□ 2,000,000–5,000,000
⊙ 1,000,000–2,000,000

0   1,000 miles
0   1,000 kilometers
Lambert Azimuthal Equal-Area projection

POPULATION

Per sq. mi.	Per sq. km
1,250 and over	500 and over
250–1,249	100–499
63–249	25–99
25–62	10–24
2.5–24	1–9
Less than 2.5	Less than 1

Cities and places labeled: Ürümqi, Qiqihar, Daqing, Jiamusi, Harbin, Changchun, Jilin, Chifeng, Shenyang, Huludao, Anshan, Hohhot, Beijing, Baotou, Tianjin, P'yongyang, Dalian, Yantai, Seoul, Incheon, Daejon, Pusan, Nagoya, Kyoto, Gifu, Osaka, Hiroshima, Shijiazhuang, Taiyuan, Jinan, Zibo, Qingdao, Gwangju, Fukuoka, Handan, Zaozhuang, Linyi, Lanzhou, Zhengzhou, Xuzhou, Suzhou, Huaiyin, Tianshui, Xi'an, Nanyang, Huainan, Nanjing, Mianyang, Hefei, Wuxi, Shanghai, Chengdu, Nanchong, Wuhan, Ningbo, Neijiang, Chongqing, Nanchang, Hangzhou, Wenzhou, Lupanshui, Guiyang, Changsha, Hengyang, Fuzhou, Kunming, Quanzhou, Taipei, Nanning, Xiamen, Guangzhou, Shantou, Kao-hsiung, Yulin, Macau, Shenzhen, Hong Kong, Zhanjiang, Sapporo, Sendai, Tokyo-Yokohama, Ürümqi, Xining

---

UNIT 8
REFERENCE ATLAS

MAP STUDY

1. *Human Systems* Is it fair to say that East Asia is a region of huge cities? Why or why not?

2. *Places and Regions* How does the population density in western China compare to the population density in eastern China?

676

China and Mongolia

ESSENTIAL QUESTION · *How do physical systems and human systems shape a place?*

networks

There's More Online about the geography of China and Mongolia.

CHAPTER 28

Why Geography Matters
China's Growing Energy Demands

Lesson 1
Physical Geography of China and Mongolia

Lesson 2
Human Geography of China and Mongolia

Lesson 3
People and Their Environment: China and Mongolia

Geography Matters...

China is one of the largest and most influential countries in the world. With a civilization thousands of years old, this country of geographical contrasts has more than a billion people.

China's mammoth economy exports goods globally. However, environmental problems resulting from the country's rapid growth have given rise to domestic and international challenges.

The center of a vast empire, Mongolia was later ruled by China. A land of windswept plains and mountains, today independent Mongolia faces challenges of urbanization and development.

◀ This woman stands in the Forbidden City located in the center of Beijing, China.

Jason Hosking/Taxi/Getty Images

677

China's growing energy demands

Since 1978 China has been shifting from an agriculture-based economy to an industrial economy. Today, Chinese workers manufacture clothes, shoes, cars, bicycles, ships, planes, washing machines, televisions, computers, cement, steel, iron, toys, and many other products for export. China's surging industrial growth has resulted in an increasing demand for energy, which is needed to power thriving cities and bustling factories from Guangzhou to Shenyang.

Why is China consuming more energy resources?

Since the 1970s, China's consumption of oil, natural gas, coal, and other sources of energy has climbed steadily. One reason for this increase in demand is China's population. With more than 1.3 billion people, China has the largest population in the world. An even more important factor is that China has developed its economy through industrialization, which causes energy demand to increase. Following this path of development, China will continue to require more energy to generate electricity for homes, schools, and businesses and to fuel cars, trucks, and buses. As of 2016, China had the second-largest economy in the world, behind the United States.

1. Human Systems What has caused increased consumption of energy in China?

How can China meet the growing demand for energy?

Currently, China uses more of the world's energy than any other country. In 2013, for example, China was the largest consumer of coal in the world. To meet its demand for energy that will keep its economy growing, China must supply enough affordable oil, coal, and other types of energy to its people and industries. To supplement its own reserves of oil, coal, and natural gas, China must also import these fuels. In addition, the Chinese are investing in products and technologies—such as solar power, LED lighting, fuel-efficient cars, and advanced coal technology—that either use or provide energy more efficiently. By improving energy efficiency, China can shrink its energy consumption and make sure it has adequate supplies of energy for the future.

2. Places and Regions How can China ensure continued economic growth?

What challenges does China face in supplying its future energy needs?

China uses coal to generate nearly 80 percent of its electricity. In 2010 its demand for coal caused a 75-mile traffic jam involving 10,000 trucks carrying coal from Inner Mongolia. The government's twelfth Five-Year Plan lays out guidelines for reducing consumption, increasing efficiency, and developing alternative sources such as solar, nuclear, water, and wind. China—one of the world's largest producers of electricity from wind power—can harness the wind energy created by its long coastline and large landmass. With many rivers and mountains, China has become the world's leading producer of hydroelectric power, or energy from flowing water.

3. Environment and Society How could developing new energy sources help China meet its energy demands?

YESTERDAY ca. 1980s

As China industrialized, it relied heavily on fossil fuels and non-renewable resources such as oil, natural gas, and coal. Emissions from vehicles remained low, as many relied on animal transportation in rural areas or bicycles in urban areas.

* 85%
- COAL
- OIL
- HYDROELECTRIC
14%
1%

TODAY

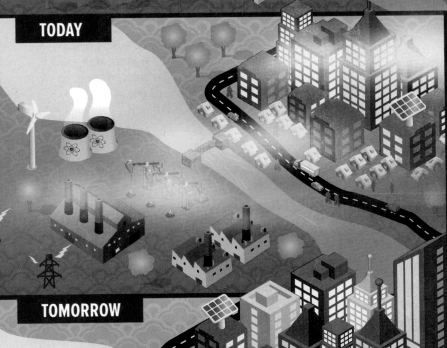

While China is making strides in the implementation of green technology, it continues to have issues with pollution due to a heavy reliance on fossil fuels. Greater access to cars and trucks has also increased emissions.

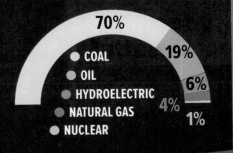

70%
- COAL
- OIL
- HYDROELECTRIC
- NATURAL GAS
- NUCLEAR
19%
6%
4%
1%

TOMORROW

In the future, China has planned to increase its usage of solar, nuclear, and wind energy sources. However, as its population continues to grow, China's ability to reduce overall emissions remains in question.

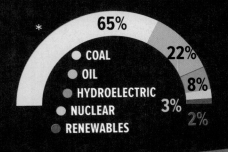

* 65%
- COAL
- OIL
- HYDROELECTRIC
- NUCLEAR
- RENEWABLES
22%
8%
3%
2%

*Data for Yesterday and Tomorrow are approximations.

networks

There's More Online!

☑ **IMAGE** Giant Panda Eating Bamboo

☑ **IMAGE** Terraced Rice Fields

☑ **MAP** China's River: Huang He

☑ **MAP** East Asian Monsoons

☑ **MAP** Physical Geography of China and Mongolia

☑ **INTERACTIVE SELF-CHECK QUIZ**

☑ **VIDEO** Physical Geography of China and Mongolia

Reading **HELP**DESK

Academic Vocabulary

- **symbol**

Content Vocabulary

- **range**
- **loess**
- **monsoon**
- **typhoon**

TAKING NOTES: *Key Ideas and Details*

IDENTIFYING China's landforms and bodies of water are regularly affected by weather systems. As you read, use a graphic organizer like the one below to note the main landforms, water systems, and climate regions.

China's Physical Geography		
Landforms	Water Systems	Climate Regions

LESSON 1

Physical Geography of China and Mongolia

ESSENTIAL QUESTION • *How do physical systems and human systems shape a place?*

IT MATTERS BECAUSE

China is a large land of many contrasts: high mountains, grassy steppes, broad plains, tropical rain forests, and cold, dry deserts. Great rivers nourish the soil to support agriculture that feeds millions, but monsoons can cause dangerous floods. Mongolia has large areas of upland plateau that are deserts, as well as mountainous regions and basins. Most of its land consists of extensive grasslands used for livestock herding.

Landforms

GUIDING QUESTION *What physical features dominate China and Mongolia, and how have they affected human geography?*

East Asia spans the snowy peaks of the Himalaya in the west all the way to the Japanese islands in the east. China makes up the majority of the landmass in this subregion. Rivers, hills, and plains occupy the east. The southwest contains high plateaus, and the northwest stretches wide into dry deserts. Mongolia is mostly an upland plateau with mountains in the north and south interrupted by basins.

The landforms of China can be divided into two parts: the mountains and plateaus of the west and the plains and hills of the east. In far western China, a mountain **range** called the Pamirs forms a node from which several other ranges radiate. This remote interior region includes the Kunlun Shan and Tian Shan ranges. Between them stretches the Tarim Basin, characterized by deserts and salt marshes. A desert, the Taklimakan, covers most of the Tarim Basin. Another massive desert, the Gobi, is found in north-central China and southern Mongolia. It is the source of frequent dust storms that plague these areas. Rain seldom falls in the Gobi—less than 3 inches (7.5 cm) a year.

The Altay Shan range, farther north, lies along part of the border between China and Mongolia. China is separated from South Asia by the Himalaya, the world's highest mountains and the location of the world's tallest peak, Mount Everest, at 29,028 feet (8,848 m) on the border of China and Nepal. The country's lowest point, the Turpan Depression, is found in western China near Mongolia and lies 426 feet (130 m) below sea level.

The highest plateau in East Asia is the Plateau of Tibet (or the Plateau of Xizang), which forms a large part of China's southwest. It has an average elevation of 16,000 feet (4,875 m). North and east of the Plateau of Tibet lie other rugged highlands, although with lower elevations. The mountains, plateaus, and great deserts in China's west and south have formed natural barriers that helped protect its inhabitants from land invasion for thousands of years, but also isolated China from peaceful contact with foreigners. For centuries China's isolation served to protect the country, allowing it to develop its unique culture.

The eastern part of China is characterized by vast lowland plains and some hills. It contains most of China's population because the land is conducive to farming and settlement. The Manchurian Plain (or Northeast Plain) and the North China Plain are two of the most important of the numerous plains areas.

In addition to the southern upland plateau and deserts, Mongolia has areas of mountains in the north and the south with basins in between. Some contain remnants of extinct volcanoes, also found on the far eastern plateau.

☑ READING PROGRESS CHECK

Identifying Which mountain ranges originate in the Pamirs?

Water Systems

GUIDING QUESTION *How do China's major rivers influence human systems?*

China's major rivers originate on the Plateau of Tibet and flow eastward, down from the plateau and across the lowland plains to the Pacific Ocean. In northern

range a group of mountains

GEOGRAPHY CONNECTION

China's immense size covers many types of landforms and physical features.

1. *THE WORLD IN SPATIAL TERMS* Approximately how wide is the Plateau of Tibet?

2. *PLACES AND REGIONS* Which desert spans both Mongolia and China? What is the elevation of this desert?

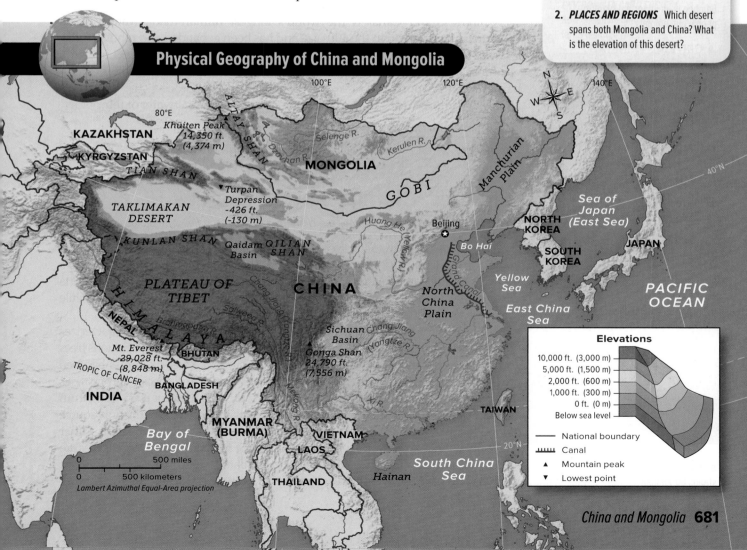

Physical Geography of China and Mongolia

Elevations

10,000 ft. (3,000 m)
5,000 ft. (1,500 m)
2,000 ft. (600 m)
1,000 ft. (300 m)
0 ft. (0 m)
Below sea level

— National boundary
� Canal
▲ Mountain peak
▼ Lowest point

There are many areas of rich farmland in China. In the photo, rice fields have been planted on terraced slopes in Guangxi.

▲ CRITICAL THINKING

1. Analyzing What is the benefit of terracing?

2. Synthesizing What geographic conditions make China such a successful crop producer?

typhoon a violent tropical storm that forms in the western Pacific Ocean, usually in late summer

annual rainfall occurs in heavy downpours from April through October. Farmers depend on these summer monsoons to water their crops. If monsoons come late or bring less rainfall, crops fail. When too much rain comes, flooding can result. From November to March, the winter monsoon brings cold, dry arctic air across China as it blows from the land onto the ocean.

Large, violent storms with high winds, known as **typhoons**, result when warm, humid air over the Pacific Ocean moves onto land. Typhoons are also called tropical cyclones and are similar to hurricanes. These intense spiral storms with their storm surges can raise coastal waters up to 20 feet (6 m) above normal, creating serious flooding along the coasts. As with hurricanes in the Atlantic and Caribbean, typhoon season peaks between late August and October.

Far from the ocean, Mongolia has a continental climate with long, cold winters and short, cool-to-hot summers. Mostly arid to semi-arid, annual rainfall in desert areas is less than 4 inches (10 cm), with 14 inches (35 cm) in the northern forested mountains.

China has a wealth of mineral resources, including iron ore, tin, tungsten, and gold. The South China Sea and the Taklimakan contain large petroleum deposits. Northeastern China has abundant coal deposits. Other natural resources include natural gas, mercury, aluminum, lead, zinc, and uranium. Mongolia has a number of minerals, including coal, gold, and copper, as well as oil and uranium deposits.

China is the world's leading producer of rice. The "rice bowl" in southern China yields two harvests of rice per year. Agricultural income as a percentage of gross domestic product (GDP) was about 9 percent in 2015. In fact, China is the world leader in the gross value of its farm output for rice, wheat, potatoes, corn, peanuts, tea, millet, barley, apples, cotton, oilseed, pork, and fish. East Asia, including China, boasts the world's biggest deep-sea fishing industries.

✓ READING PROGRESS CHECK

Making Connections How can monsoons affect China's agricultural output?

LESSON 1 REVIEW

Reviewing Vocabulary

1. Using Context Clues Write a sentence explaining how loess affects agriculture in China.

Using Your Notes

2. Comparing and Contrasting How do the water systems in China affect its climates compared to the effects of weather systems?

Answering the Guiding Questions

3. Identifying Central Issues What physical features dominate China and Mongolia, and how have they affected human geography?

4. Analyzing How do China's major rivers influence human systems?

5. Synthesizing What defines the climates, natural resources, and biomes of China?

Writing Activity

6. Informative/Explanatory In a paragraph, explain the possible effects of typhoons and monsoons on the Chinese people and their economy.

networks

There's More Online!

☑ **GRAPHS** Grain Production in China, 1950–1970

☑ **GRAPHS** U.S. Exports to China and Imports from China

☑ **IMAGE** Chinese Schoolgirl

☑ **IMAGE** Mongolian Flag

☑ **TIME LINE** Modern China

☑ **INTERACTIVE SELF-CHECK QUIZ**

☑ **VIDEO** Human Geography of China and Mongolia

Reading **HELP**DESK

Academic Vocabulary

- **impact**
- **institute**
- **dominate**

Content Vocabulary

- **dynasty**
- **aborigine**
- **ideogram**
- **atheist**
- **commune**
- **merchant marine**
- **dissident**
- **economic sanctions**
- **Special Economic Zone (SEZ)**

TAKING NOTES: *Key Ideas and Details*

DESCRIBING As you read about the human geography of China, use a graphic organizer like the one below to identify how history, population, culture, and economics have created the China of today.

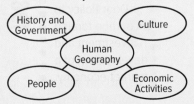

LESSON 2
Human Geography of China and Mongolia

ESSENTIAL QUESTION • *How do physical systems and human systems shape a place?*

IT MATTERS BECAUSE

China's long history and geography have shaped its culture. The ideas on which its government is based have had long-lasting effects. The country's economic life reveals how it fulfills the daily needs of its citizens and interacts with the larger world. Mongolia, once the world's largest land empire, today faces modern challenges.

History and Government

GUIDING QUESTION *What influence did ancient Chinese history have on government, culture, and daily life in modern times?*

Many powerful transformations have occurred during China's history. Under leaders ranging from nomadic warriors to long-ruling **dynasties**, the country has endured profound political and cultural changes.

China's culture spans more than 5,000 years. Archaeological evidence indicates it began in the Wei River valley. Historical records were begun when invaders established a dynasty, or ruling family, around 1766 B.C. This Shang dynasty arose on the North China Plain. They faced attacks by nomads from Central Asia, rebellions by local nobles, and natural disasters. In China, dynasties were believed to rule under the "mandate of heaven," or the approval of the gods and goddesses. When the people suffered, it was assumed that the dynasty had lost its mandate. The Shang dynasty ruled for over 700 years, coming to an end in 1046 B.C.

The Zhou (JOH) dynasty then took control of the region, and continued for the next 800 years. Under its rule, trade grew, Chinese culture spread, and the making of iron tools began. Crossbows, ox-drawn plows, and horseback riding were introduced, as were widespread irrigation and other efforts to control water. This helped to increase crop yields.

China's most famous teacher and philosopher, Confucius (or Kongfuzi), lived during the Zhou dynasty. He founded a system of thought called Confucianism. It is based on discipline and proper moral conduct, and continues to have an **impact** in China and other Asian civilizations to the present day.

ARE LIMITS ON GROWTH BEST FOR CHINA?

The high population growth rate has long been a concern in China. When the People's Republic of China came to power in 1949, its leaders wanted to slow population growth in order to reduce poverty and to stabilize the country. People were exposed to new family planning methods, and birth control was encouraged. However, once the population neared 1 billion in the 1970s, leaders wanted people to have no more than two children. By 1980, a one-child policy was put in place.

Most people were required to follow the one-child policy and were often rewarded for doing so. However, there were exceptions. Some minority groups, people whose first child had disabilities, and people in rural areas were permitted to have more than one child. Also, people were sometimes fined for each additional child. However, those who knowingly went against the policy were required to pay large fines. Some reported that if the fine was not paid, families lost their homes or jobs.

For years, demographers and others argued that China should end the one-child policy. These critics warned that if it continued, China would have a shortage of workers and fewer people available to take care of an aging population.

China's National Population and Family Planning Commission managed the one-child policy. Their goal was to make sure that the country could sustain development. To meet this goal, they monitored the effects of population changes on the economy, society, natural resources, and the environment. After 35 years, the Family Planning Commission suspended the one-child policy in 2015 and replaced it with a "one-couple, two children" policy. They were concerned about the aging population, having the resources necessary to care for the aging population, and the lack of working-age people to help the economy continue to grow.

Benefits of Population Limits

PRIMARY SOURCE

❝ Henan has much to teach the world in family planning, but it is a hard lesson to learn. Officials from Africa and India come to study what we are doing in China, but I'm not sure that they can apply it the same way. That's because they don't have a Communist party so it is difficult for them to take such strong steps. ❞

—*Liu Shaojie, Vice Director of the Population Commission in Henan province, China, quoted in The Guardian, October 25, 2011*

End Population Limits

PRIMARY SOURCE

❝ Everyone has blindly accepted the fact that population control has helped China's economy, but it has never been proven. . . . In fact, by 2030, no matter what policy China adopts the population will start to shrink. And I have never seen a country with a shrinking population and sound economic development. Western countries may have shrinking populations, but they have immigrants to make up their labour force. China does not have large immigration, so aiming for a zero or negative birth rate is very risky. ❞

—*Liang Zhongtang, a former committee member of the National Family Planning Commission, quoted in The Telegraph, September 25, 2010*

❝ We have been discussing the one-child policy since 2000. . . . It is just a matter of finding the right solution. Making the jump to two children is only a matter of time now. . . . If China sticks to the one-child policy, we are looking at a situation as bad as the one in southern Europe. Old people will make up a third of the population by 2050. ❞

—*Li Jiamin, a specialist in population studies at Nankai University, quoted in The Telegraph, October 31, 2012*

What do you think? DBQ

1. **Drawing Conclusions** Why might Chinese people have hesitated to speak out against China's one-child policy?

2. **Identifying Central Issues** According to critics of the one-child policy, what were the risks of keeping the policy as it is?

3. **Hypothesizing** Why might Chinese people in urban areas and Chinese people in rural areas have had different opinions on the one-child policy?

There's More Online!

- ☑ **GRAPH** Daily Oil Production and Consumption in China
- ☑ **IMAGE** Chinese Timber Plant
- ☑ **IMAGE** Smog and Traffic in Beijing
- ☑ **IMAGE** Electronics Factory
- ☑ **INFOGRAPHIC** Three Gorges Dam: Harnessing the Yangtze
- ☑ **INTERACTIVE SELF-CHECK QUIZ**
- ☑ **VIDEO** People and Their Environment: China and Mongolia

LESSON 3
People and Their Environment: China and Mongolia

ESSENTIAL QUESTION · *How do physical systems and human systems shape a place?*

Reading HELPDESK

Academic Vocabulary

- **utilize**
- **consequence**

Content Vocabulary

- **nuclear**

TAKING NOTES: *Integration of Knowledge and Ideas*

MAKING PREDICTIONS
As you read about the people and environment of China, use a graphic organizer like the one below to record the country's current environmental issues and predict what might happen if steps are not taken to address them.

China's Environmental Issues	Present	Future
Land		
Water		
Air		

IT MATTERS BECAUSE
China's economic success has generated many critical issues that must be addressed so that people can live healthy and safe lives. Environmental issues in both China and Mongolia affect not only people, but also animals and the natural environment.

Managing Resources

GUIDING QUESTION *How do China and Mongolia use fossil fuels as energy sources?*

China's growing economic prosperity has fueled a demand for more electric power. The country is now the number one consumer of electricity worldwide. Increasing living standards mean that more people can afford appliances and electronic devices. Meeting this increasing demand has become an issue for both the government and the public. A massive dam project known as the Three Gorges Dam was built on the Chang Jiang, or Yangtze River, to supply hydroelectric power to China's interior with its expanding population.

While hydroelectric plants provide some power to China, the country's main source of power comes from burning fossil fuels. China and Mongolia both **utilize** large local reserves of coal. Rising oil prices make it difficult to end this dependency on cheaper coal. However, burning coal has serious **consequences** for the environment, including acid rain, air pollution, and climate change.

Power from **nuclear** plants is an option for many countries, but China has been slow to embrace it. Nuclear power provides only 1 percent of China's electricity. Future plans, however, involve the construction of at least 100 more nuclear power plants.

Mongolia's valuable mineral deposits are putting its environment at risk. Mining often has long-term detrimental effects such as water sources contaminated with toxins and dead zones around open-pit mines. Yet economic development is sorely needed in the country to benefit its people.

☑ **READING PROGRESS CHECK**

Summarizing Why is China so dependent on coal?

Human Impact

GUIDING QUESTION *What special challenges come with globalization, and how will they affect China's future?*

In a world increasingly connected by the Internet and business and cultural exchanges, globalization exerts a stronger influence on Chinese life than ever before. In most periods of change in Chinese history, the country has tended to look within rather than seek out the cultures and trends of the outside world. This attitude was largely due to strong national pride, suspicion of outsiders, and more recently, rigid Communist rule.

When the Communists triumphed in 1949, the only area of the country able to maintain international connections was Hong Kong, ruled by the British. During the 1970s and early 1980s, Communist rulers instituted new policies, making an effort to open their economy to foreigners. Foreign influence began to be felt first on the coasts. Nightclubs, karaoke bars, fast-food restaurants, and theme parks began to appear. The Internet also began to have an impact on Chinese society.

The loosening of government control has resulted in the rise of regional identities, which had been suppressed before the late 1970s. This is especially apparent in southeastern China, where most people do not speak the dominant Mandarin language, making their culture distinctive. The Guangdong region of China, including Hong Kong—which was turned over to China by the British in 1997—is the main gateway for the entrance of foreign influences into the country.

Since the end of the Cold War, the balance of power in East Asia has shifted, with China rising to the top. It has the largest army in the world, a nuclear arsenal, and complex missile technology. All of this has combined to worry China's neighbors.

China's physical geography could potentially cause major problems in the future. Two of China's great rivers, the Huang He and Chang Jiang, have often produced disastrous floods. Flood control of the rivers has been attempted by constructing drainage channels and irrigation canals to carry away or redirect the water. Levees, dikes, and dams have also been built. Still, terrible flooding

STR/AFP/Getty Images

utilize to make use of

consequence the result of an action

nuclear of, relating to, or using the atomic nucleus, atomic energy, the atomic bomb, or atomic power

The increasing demand for technology has offered new challenges and benefits to China and its people.

◀ **CRITICAL THINKING**

1. *Identifying Perspectives* How has the world's demand for technology affected China's citizens?

2. *Synthesizing* How do you think the Internet might contribute to changes in Chinese society?

continues to occur. Thousands of dams built rapidly during the 1950s and 1960s are now damaged and at risk of failure.

To deal with these problems, China has invested more in flood-control projects. The Three Gorges Dam created a huge reservoir almost 400 miles (644 km) long. The project has been controversial due to the relocation of over a million people. Countless farms, villages, and historic temples were submerged. Natural ecosystems have been damaged with some animal and plant species now endangered or extinct. Changing the flow of the river may interfere with migratory paths of fish species. Pollutants in the soil and chemicals in abandoned factories may leach into the soil as the water rises in the reservoir. Supporters of the project point to the energy created by the dam and its commercial benefits. The dam will help to ease China's future power shortages that could potentially be crippling. The reservoir behind the dam holds an enormous amount of water for farming.

☑ READING PROGRESS CHECK

Understanding Relationships How does China's physical geography influence flooding?

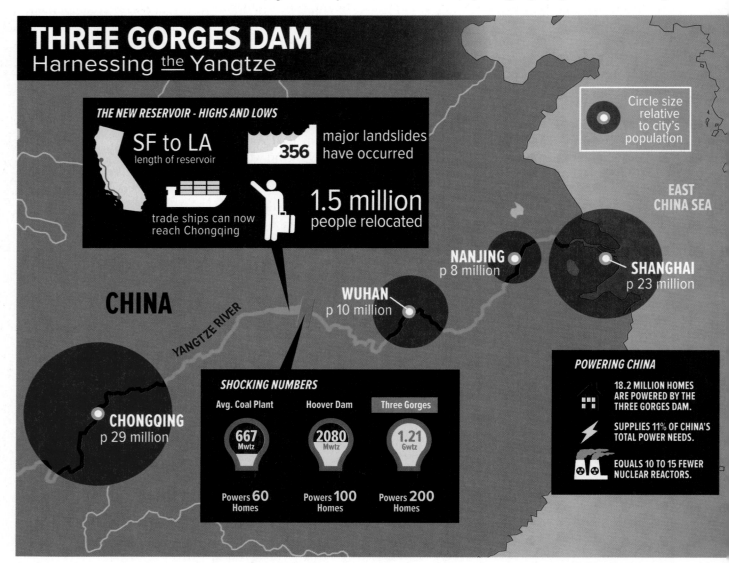

The Three Gorges Dam was a major engineering project. The dam was designed to serve several purposes: to prevent flooding along the Yangtze; to generate electricity; to increase the water supply for agriculture, industry, and homes; and to improve navigation along the river.

▲ CRITICAL THINKING

1. *Analyzing* What are the benefits of the Three Gorges Dam for China?

2. *Formulating Questions* Write three questions that you would ask the engineers of the Three Gorges Dam.

Addressing the Issues

GUIDING QUESTION *How are the Chinese people and government addressing the effects of industrial and economic growth on the environment?*

Industrial and economic growth has created very real benefits for the Chinese people. This growth has also created a host of environmental problems. The impact of China's economic boom on the environment and human health has been enormous. Urban areas are plagued by serious air pollution created by old technology in transportation and industry. One major reason for such severe pollution is China's heavy use of the country's coal supply. In addition to its abundance, coal is relatively cheap, making it attractive. But coal produces high amounts of several types of pollution. Blowing coal dust in the northern industrial areas worsens air pollution and causes many people to suffer from lung diseases. These same problems affect the inhabitants of Mongolia's few cities, where coal is burned for heating and cooking in homes and by factories.

Burning coal creates not only air pollution, but also acid rain. It has become a major problem in China and has also affected other areas of Asia. Mercury from the burning of coal has been found in the Pacific Ocean and as far away as the western United States. Ashes that remain in the coal furnaces after burning occurs, called bottom ash, must be removed and stored in large ponds that can leak or spill.

Like other rapidly urbanizing countries, China has trouble disposing of wastes. About one-third of China's population lacks access to clean water. Many tons of sewage are discharged annually into the Chang Jiang—30 billion tons in 2006. Seventy percent of the country's lakes and rivers are polluted. Two-thirds of China's cities lack clean water, forcing millions of people to boil their drinking water to make it safe to consume.

Pollution has made cancer the leading cause of death in China. Industrial waste is mostly to blame. One metal company in Shenyang spewed clouds of sulfur dioxide and other hazardous chemicals into the atmosphere for many years. City residents affected by health problems from the plant's toxic emissions prompted the government, for the first time, to shut down a state-run factory. In early 2013, a state of emergency was declared in Beijing due to terrible air quality. Mongolia's capital, Ulaanbaatar, is also among cities with the worst air pollution in the world.

Efforts to regulate polluters have been slow for many reasons. In China, one reason is the industries' reluctance to support stricter laws. More importantly, the overwhelming focus on economic growth encourages governments at both the national and local levels to ignore pollution regulations. Record-high pollution levels in Beijing in 2015 caused temporary emergency measures. The Beijing government shut down factories, reduced car use, and took heavy vehicles off the road. Despite such measures, air quality remains extremely hazardous.

China's large population and thriving industry depend on huge quantities of lumber. Each year China cuts down thousands of acres of forests, but also plants many trees to replace this resource. However, the replacement forests

Increasingly, air pollution is a major health concern in today's large Chinese cities.

▲ **CRITICAL THINKING**

1. *Identifying Cause and Effect*
   Why is coal creating a health problem in China?

2. *Analyzing* How effectively has China dealt with air pollution?

narvikk/Getty Images

Each year, China meets its demand for lumber by cutting down thousands of acres of forest.

▲ **CRITICAL THINKING**

1. Identifying Cause and Effect What are the effects of the huge demand for lumber in China?

2. Exploring Issues Why does reforestation not completely solve the problems of deforestation?

are much less diverse than the original forests. In addition, reforestation efforts have not kept up with demand for timber. China has become a major importer of lumber, pulp, and paper. Clear-cutting timber has led to soil erosion, which in turn, leads to deforestation. Trees and other types of living ground cover help to slow runoff from rain. When they are destroyed, large-scale soil erosion and flooding occur.

A succession of unusually heavy rains in the late 1990s caused flooding of the Chang Jiang and the Huang He. The consequences of such flooding included the deaths of thousands, widespread destruction of property, and damage to the landscape.

Disasters such as these prompted China to plant trees on millions of acres of deforested riverbanks. A major dam was constructed along the Huang He to control flooding. Nature and wetland reserves and wildlife protection zones have also been created.

Western China is experiencing desertification. Much of the vegetation has been depleted by over-grazing and soil erosion, leaving the land bare and dry. Desertification has worsened sandstorms and dust storms that travel from the Gobi through southern Mongolia, northern and western China, and to the Pacific Ocean.

Traditional Chinese medicines often utilize products derived from rare or exotic animals. Deer antlers, rhinoceros horns, and bear gallbladders contain desirable substances that are often worth incredible sums of money. As wealth grows in the country, the demand for these products has increased. Because China has little wildlife remaining, many animal products are imported. The consequences of this increased demand have affected exotic and endangered animal populations worldwide.

Some areas in China are protected, however. These include forests at high altitudes and bamboo groves in the western Sichuan province, where pandas need the bamboo to survive. A surviving population of Siberian tigers was found in the far northeast in 1997. Wildlands there and in northwestern Yunnan have been protected as well.

☑ READING PROGRESS CHECK

Identifying How has industrialization affected the health of Chinese and Mongolian people living in cities?

LESSON 3 REVIEW

Reviewing Vocabulary

1. *Analyzing Text Structure* How does nuclear power differ from hydroelectric or fossil fuel power sources?

Using Your Notes

2. *Organizing* Using your graphic organizer, list three environmental issues in China and efforts to address them.

Answering the Guiding Questions

3. *Analyzing* How do China and Mongolia use fossil fuels as energy sources?

4. *Hypothesizing* What special challenges come with globalization, and how will they affect China's future?

5. *Exploring Issues* How are the Chinese people and government addressing the effects of industrial and economic growth on the environment?

Writing Activity

6. *Narrative* In two or three paragraphs, discuss the possible effects of environmental problems on a fictional Chinese person. Be sure to include relevant details from the lesson.

Directions: On a separate sheet of paper, answer the questions below. Make sure you read carefully and answer all parts of the questions.

Lesson Review

Lesson 1

1 **Describing** What are some ways in which the Chang Jiang influences human systems in China?

2 **Explaining** Why do monsoons form in China? How do they change during the summer and winter months?

3 **Evaluating** China has a wealth of natural resources. How important is the production of rice to the Chinese economy?

Lesson 2

4 **Summarizing** Describe some of the challenges facing farms and cities in China as a result of urbanization.

5 **Considering Advantages and Disadvantages** What are some of the advantages and disadvantages of using ideograms for written language?

6 **Hypothesizing** Why do you think the creation of farming communes during the Great Leap Forward led to starvation?

Lesson 3

7 **Making Connections** Why has China's growing economic prosperity created more demand for electric power?

8 **Making Generalizations** How has the balance of power in East Asia shifted since the end of the Cold War?

9 **Explaining** How has burning coal affected the health of urban populations in China and Mongolia? Give examples.

Critical Thinking

10 **Analyzing** Why has globalization led to the rise of regional identities in China?

11 **Hypothesizing** China currently has a population with more boys than girls. What problems might this imbalance cause in coming years?

21st Century Skills

Use the cartoon below to answer the following questions.

PRIMARY SOURCE

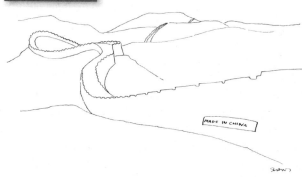

MADE IN CHINA

12 **Identifying Perspectives** What do you think the "Made in China" sign means to the cartoonist?

13 **Explaining Continuity and Change** What does the cartoon suggest about how China's booming economy will affect its tradition of isolation?

College and Career Readiness

14 **Decision Making** Imagine that you have been asked to advise the Chinese government on the construction of a new dam. Research the controversy surrounding the Three Gorges Dam. Write a two-page report summarizing the controversy and offering your recommendations. Should China go forward with the new dam project? Explain.

Need Extra Help?

If You've Missed Question	**1**	**2**	**3**	**4**	**5**	**6**	**7**	**8**	**9**	**10**	**11**	**12**	**13**	**14**
Go to page	682	683	684	688	689	690	694	695	697	695	690	690	695	696

Directions: On a separate sheet of paper, answer the questions below. Make sure you read carefully and answer all parts of the questions.

DBQ Analyzing Primary Sources

Use the document to answer the following questions.

China has taken significant measures to deal with threats to its environment. But these measures haven't kept pace with the rapid growth of its economy.

PRIMARY SOURCE

"*Given what scientists now predict about the timing of climate change, the greening of China will probably come too late to prevent more dramatic warming, and with it the melting of Himalayan glaciers, the rise of the seas, and the other horrors Chinese climatologists have long feared.*

It's a dark picture. Altering it in any real way will require change beyond China—most important, some kind of international agreement that transforms the economics of carbon. At the moment China is taking green strides that make sense for its economy. 'Why would they want to waste energy?' Deborah Seligsohn of the World Resources Institute asked, adding that 'if the U.S. changed the game in a fundamental way—if it really committed to dramatic reductions—then China would look beyond its domestic interests and perhaps go much further.' Perhaps it would embrace more expensive and speedier change. In the meantime China's growth will blast onward, a roaring fire that throws off green sparks but burns with ominous heat."

—Bill McKibben, "Can China Go Green?" *National Geographic,* June 2011

15 *Interpreting* Why does the author say that the greening of China is a "dark picture"?

16 *Problem Solving* What does Deborah Seligsohn think the United States should do to help China go green?

17 *Analyzing* How does the description of China's economy as a "roaring fire that throws off green sparks" help you understand the problem?

Applying Map Skills

Use the Unit 8 Atlas to answer the following questions:

18 *Environment and Society* What features of Mongolia's geography help explain why some areas have lower density populations than others?

19 *Human Systems* Locate the area that was submerged by the construction of the Three Gorges Dam. To what neighboring towns and cities did people most likely relocate?

20 *Physical Systems* Use your mental map of China to describe the location of the Gobi and Taklimakan deserts.

Exploring the Essential Question

21 *Making Connections* Write a paragraph describing how China's three great rivers have shaped population patterns. How have rivers influenced the location of urban centers?

Research and Presentation

22 *Understanding Relationships* Use the Internet and other resources to research the debate over human rights in China. What are the major issues? How effective have economic sanctions been at encouraging positive political reforms? Use your findings to create a poster that includes photos and quotes and present it to the class.

Writing About Geography

23 *Informative/Explanatory* Use standard grammar, spelling, sentence structure, and punctuation to write three paragraphs describing the problems of erosion and deforestation in China. How has economic growth fueled these problems?

Need Extra Help?

If You've Missed Question	**15**	**16**	**17**	**18**	**19**	**20**	**21**	**22**	**23**
Go to page	696	696	694	672	672	672	688	691	697

Japan

ESSENTIAL QUESTION • *How do physical systems and human systems shape a place?*

◄ Nobu Matsuhisa is the chef and owner of one of Tokyo's most famous restaurants.

Jeremy Sutton-Hibbert/Alamy

Why Geography Matters
Japan's Aging Population

Lesson 1
Physical Geography of Japan

Lesson 2
Human Geography of Japan

Lesson 3
People and Their Environment: Japan

Geography Matters...

The rugged island country of Japan, punctuated by volcanic mountains formed over millions of years, is home to a population that is ethnically homogeneous compared to most industrialized countries. With its ancient culture and complex traditions, Japan also has a leading world economy that is highly advanced technologically. One of Asia's most successful democracies, Japan has become a strong ally of the United States and a force of stability in the region.

Japan's aging population

After a post–World War II baby boom, Japan experienced declining birthrates and slower overall population growth. The current population is declining and aging. Aging populations mean that a smaller percentage is working to support those in retirement. This could pose a threat to Japan's economic future.

How is the population of Japan changing?

In Japan, the fertility rate is declining as women marry later, delay having children, or never marry at all. (Most Japanese babies are born to married women.) At the same time, life spans are lengthening. The Japanese have the longest life spans in the world. In 1950, on average, Japanese women lived to age 63 and Japanese men to 60. Today Japanese women are expected to live to age 88 and Japanese men to 81. In 2015 more than 60,000 Japanese were centenarians, or people who are more than 100 years old. That number is projected to be at least six times as great by 2030. The combination of lower fertility rates and lengthening life spans has caused the population of Japan to decline and age.

1. **Human Systems** What effect is Japan's current fertility rate having on its population?

What are the long-term effects of an aging population?

As Japan's population decreases and ages, the workforce shrinks. Statistics indicate that the number of Japanese workers peaked in 1999 and has shrunk 2.5 percent since then. Therefore, there is a possibility of a shortage of workers in the future, as well as an increase in the dependency ratio. The dependency ratio refers to the ratio of economically dependent persons to economically active persons. As the dependency ratio increases, more retired people are drawing on government services, such as pensions and health care, and fewer people are paying taxes to fund those services. Interestingly, the lengthening life span of the Japanese does not necessarily mean increased sickness and debility. Japan's men appear to be healthier longer than men in other countries, and its women are generally vigorous for much of old age.

2. **Human Systems** Write a paragraph describing how the increased dependency ratio may change the way the Japanese people live in the future.

How are people in Japan and across the world finding solutions?

Many Japanese men—almost one-third of the workforce—continue to work beyond the usual retirement age. If more Japanese women also worked longer, the dependency ratio and the burden on the younger workforce would be reduced. Internationally, some governments are taking steps to make it easier for people over 65 to continue working. Others are considering raising the retirement age from 65 to 67—or possibly higher—and even linking the retirement age to life expectancy. Other options include increasing taxes to pay for government pensions and social security systems or reducing payouts in these systems.

3. **Human Systems** Why is raising the retirement age in Japan and other countries an important response to aging populations?

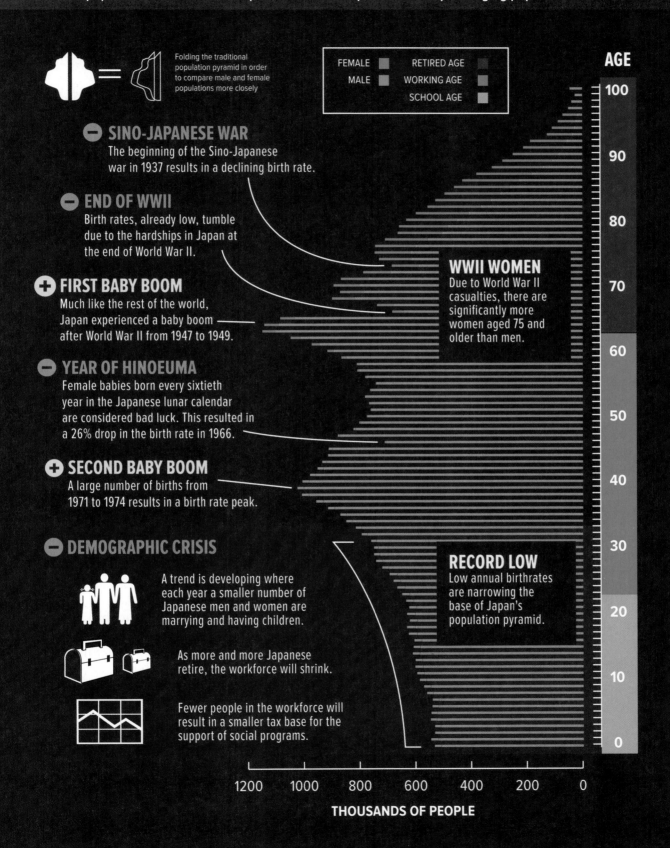

Folding the traditional population pyramid in order to compare male and female populations more closely

FEMALE ■   RETIRED AGE ■
MALE ■   WORKING AGE ■
      SCHOOL AGE ■

⊖ SINO-JAPANESE WAR
The beginning of the Sino-Japanese war in 1937 results in a declining birth rate.

⊖ END OF WWII
Birth rates, already low, tumble due to the hardships in Japan at the end of World War II.

⊕ FIRST BABY BOOM
Much like the rest of the world, Japan experienced a baby boom after World War II from 1947 to 1949.

⊖ YEAR OF HINOEUMA
Female babies born every sixtieth year in the Japanese lunar calendar are considered bad luck. This resulted in a 26% drop in the birth rate in 1966.

⊕ SECOND BABY BOOM
A large number of births from 1971 to 1974 results in a birth rate peak.

⊖ DEMOGRAPHIC CRISIS

A trend is developing where each year a smaller number of Japanese men and women are marrying and having children.

As more and more Japanese retire, the workforce will shrink.

Fewer people in the workforce will result in a smaller tax base for the support of social programs.

WWII WOMEN
Due to World War II casualties, there are significantly more women aged 75 and older than men.

RECORD LOW
Low annual birthrates are narrowing the base of Japan's population pyramid.

AGE

100

90

80

70

60

50

40

30

20

10

0

1200   1000   800   600   400   200   0

THOUSANDS OF PEOPLE

networks

There's More Online!

- ☑ **IMAGE** Fuji Fishing Harbor
- ☑ **IMAGE** Japanese Macaques
- ☑ **MAP** Physical Map of Japan
- ☑ **INTERACTIVE SELF-CHECK QUIZ**
- ☑ **VIDEO** Physical Geography of Japan

LESSON 1
Physical Geography of Japan

ESSENTIAL QUESTION • *How do physical systems and human systems shape a place?*

Reading HELPDESK

Academic Vocabulary

- **consist**
- **affect**
- **predominant**

Content Vocabulary

- **archipelago**
- **tsunami**

TAKING NOTES: *Key Ideas and Details*

SUMMARIZING As you read about the physical geography of Japan, use a graphic organizer like the one below to take notes on the climate.

(Climate of Japan) _____
              _____

IT MATTERS BECAUSE

The islands of Japan form an arc of 1,500 miles (2,400 km) across the North Pacific Ocean, rocky evidence of the Ring of Fire and its powerful volcanic action. Japan's waterways have been essential to transportation and agriculture. Although much of Japan is mountainous, its people have benefited from proximity to the ocean as a rich natural resource.

Landforms

GUIDING QUESTION *How has volcanic activity changed the land?*

Although there are more than 6,800 islands in its **archipelago**, most of Japan's land area **consists** of four main islands: Hokkaidō, Honshū, Shikoku, and Kyūshū. The islands of Japan formed during the last 15 to 20 million years. They were thrust upward and layered by the force of volcanic action from the Pacific Ring of Fire, the volcanic zone that runs along the edges of the Pacific. The resulting landscape of mostly rugged mountains **affects** where people live. The majority of the population dwells in the Japanese coastal lowlands.

Japan's location in the Ring of Fire causes some 50 Japanese volcanoes to remain active and frequent earthquakes. Japan's tallest peak, Mount Fuji, is a volcano not currently active. In a typical year more than 1,000 small earthquakes occur, with major quakes occurring less frequently but with more destruction and often causing deaths. When a powerful undersea earthquake occurs nearby, a major **tsunami**, or huge wave, can form, bringing terrible consequences. Tsunamis can travel more than 250 miles per hour (400 km per hour) and be more than 30 feet (10 m) high. In 2011 an earthquake off the coast of Honshū rocked the island and launched a series of tsunamis that flooded the coastline in massive waves. More than 19,000 people died, property damage was enormous, and a serious accident occurred at a nuclear power station. The cleanup and reconstruction effort was predicted to take ten years and cost upwards of $150 billion.

☑ **READING PROGRESS CHECK**

Summarizing How does the Ring of Fire affect Japan?

Waterways

GUIDING QUESTION *What makes the waterways in Japan unique?*

A distinctive feature of Japan is its short, swift rivers that flow from the mountains, often plunging over cliffs as stunning waterfalls. Typically, Japanese rivers rise in forested mountains, flow through steep valleys, and then cross alluvial plains to empty into the sea. The lower regions of these rivers are often used for farming rice. In the coastal lowlands, where rivers slow and pass through cities, some waterways are used for transportation, including carrying passenger water buses. These uses often involve damming the rivers or altering their original courses. Increasing demand for water—for agriculture, industry, and drinking—remains a challenge. The lack of natural or easily constructed reservoirs makes it difficult to contain rapid runoff from rainfall.

Japan's longest rivers, the Shinano and the Tone, flow from mountain heights on Honshū and can be very destructive when they flood. The Shinano is 225 miles (360 km) long. The Tone is 200 miles (322 km) long and supplies Tokyo with drinking water. On Hokkaidō, the Ishikari and the Teshio flow into the Sea of Japan (East Sea). The Yoshino River on Shikoku flows through a deep gorge and is known for its rapids.

Japan's largest lake, Lake Biwa, is in the central part of Honshū. It was created from a depression along a fault. Many of Japan's coastal lakes are simply drowned river valleys that reached the sea and were dammed over time by accumulating silt and sandbars.

✔ **READING PROGRESS CHECK**

Identifying Name three ways in which Japan's waterways have changed over time and identify whether the processes were natural or caused by humans.

---

archipelago a group or chain of islands

consist to be composed of

affect to have an effect on

tsunami a huge wave resulting from undersea earthquake or volcanic activity that gets higher and higher as it approaches the coast

---

GEOGRAPHY CONNECTION

More than four-fifths of Japan's surface area consists of mountains.

1. *PHYSICAL SYSTEMS* Why does Japan experience frequent earthquakes and occasional tsunamis?

2. *PLACES AND REGIONS* Name the four major bodies of water that border the islands of Japan.

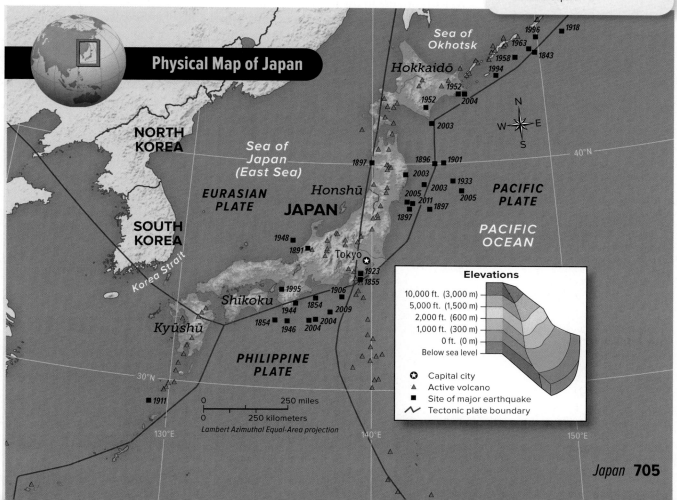

The Japanese tradition of viewing cherry tree (*sakura*) blossoms in springtime dates back a thousand years. The beautiful tree and its flowers symbolize the natural cycle of human life and became revered as a symbol of good luck. Today cherry blossom festivals are held throughout Japan, timed to match the local blooms. People picnic under the blossoming trees and hold tea ceremonies, welcoming spring together with food, drink, and song.

SPECULATING What effect might *hanami* celebrations have on Japanese society as a whole?

predominant present as the strongest or main characteristic

The Japanese macaque lives farther north than any other wild primate.

▶ **CRITICAL THINKING**

1. Analyzing Visuals How is the Japanese macaque well adapted to its natural environment?

2. Contrasting How does the appearance of the Japanese macaque differ from most other monkeys?

Climates, Biomes, and Resources

GUIDING QUESTION *How does the climate in Japan change from north to south?*

Japan generally has a mild climate and abundant rainfall. Nevertheless, there are startling contrasts in climate and biomes. The northern latitudes of Hokkaidō are much colder and experience deep winter snows and howling winds, while far to the south, Okinawa has a subtropical climate.

Climate Regions and Biomes

The **predominant** influences on Japan's climate are latitude, the proximity of the great Asian landmass to the west, the mountainous terrain, and ocean currents. The combination of these factors makes Japan monsoonal, or influenced by seasonal winds that bring either precipitation or dry air. The cold winter monsoon drops rain or snow on the western slopes of Japan's mountains, leaving the eastern region drier. The warmer summer monsoon reverses this action, bringing warm rains to the eastern slopes and dry air to the west. The warm waters of the Japan Current, which is similar to the flow of the Atlantic Gulf Stream, provide moisture for the summer monsoon. The cold Kuril Current is responsible for dense fogs off Hokkaidō. Tropical storms called typhoons, similar to hurricanes, generally occur in late summer, forming over warm tropical waters. They bring torrential rains, high winds, and frequently great destruction to Japanese islands.

Northern Japan has a humid continental climate with warm summers and cold, snowy winters. Southern Japan has a humid subtropical climate with hot summers and heavy rain. Most of Japan is forested, or was once. Semitropical rain forests in the south consist of broad-leaved evergreen trees. Transitioning north, broad-leaved evergreens mix with deciduous trees, which eventually give way to conifers in the far north. Typical Japanese trees include beeches, maples, oaks, and spruces. A variety of cherry trees, with blossoms that are a symbol of Japan, are planted throughout the country. Much of the original vegetation of Japan is gone, however, supplanted by agriculture and foreign flora.

Despite great concentrations of people in Japan, large land mammals and birds still thrive in isolated mountain regions. Bears, wild boars, deer, and antelope are common in remote mountains. The Japanese macaque, a kind of wild monkey, also dwells as far north as Honshū. Ocean animals along the coasts

Design Pics/Natural Selection Anita Weiner

include sea turtles, sea snakes, abundant water birds, fish, whales, dolphins, and porpoises. The rich sea life is made possible, in part, by the meeting of the warm and cold currents.

Natural Resources

Despite extensive mountains and poor soil, Japanese agriculture makes the most of reliable rainfall and irrigation from rivers to grow most of the rice the country needs. The main crop is rice, cultivated in flooded paddies and principally on small farms. Other crops include a variety of grains, vegetables, fruits, potatoes, and tea. Dairy cattle are raised on Hokkaidō. Beef cattle are raised in feed lots, where they are carefully nurtured to provide high-fat meat.

This busy fishing harbor lies at the base of Mount Fuji, Japan's highest mountain at 12,388 feet (3,776 m). There are about 200,000 vessels in Japan's entire fishing fleet.

▲ **CRITICAL THINKING**

1. *Analyzing Visuals* What details in the photo reflect how Japan's physical geography shapes its human geography?

2. *Hypothesizing* How may Japan's economy be affected if fisheries continue to decline?

Because Japan is an island country, the resources of the sea have always been an important source of food for the Japanese. Modern sources come from both deep sea fishing and human-made seafood farms. Japan is a world leader in harvesting and importing fish. The country consumes some 7.5 billion tons of fish annually, which amounts to nearly 15 percent of the world's catch, and reaps about $14 billion a year from the fishing industry. Ocean pollution and overfishing, however, have jeopardized the fisheries on which Japan depends. Such overfishing has seriously depleted marine life.

The mining industry is small and declining, with the exception of gold. Limestone is still widely quarried. Coal was historically the most important mineral mined in Japan. However, various factors—such as foreign competition, high production costs, and increasing reliance on oil—have lessened the importance of the industry. The last coal mine in Japan closed in 2002. Most of Japan's limited oil and natural gas resources come from Niigata Prefecture several hundred miles north of Tokyo. Since the beginning of this century, copper and iron mines have virtually ceased production, and these minerals are imported.

✓ **READING PROGRESS CHECK**

Determining Importance Which of Japan's natural resources are most closely related to Japan's being an island country? What is threatening these resources?

LESSON 1 REVIEW

Reviewing Vocabulary

1. *Making Connections* How is a tsunami different from an ordinary ocean wave?

Using Your Notes

2. *Listing* Use your graphic organizer from this lesson to list the characteristics of Japan's climate.

Answering the Guiding Questions

3. *Examining* How has volcanic activity changed the land?

4. *Describing* What makes the waterways in Japan unique?

5. *Summarizing* How does the climate in Japan change from north to south?

Writing Activity

6. *Informative/Explanatory* Write a paragraph describing how latitude affects the climate and biomes of Japan.

THE TOHOKU Earthquake AND Tsunami

The Tohoku earthquake of 2011 was the most powerful recorded earthquake ever to hit Japan. The earthquake and resulting tsunami caused incredible damage in Japan, including a meltdown at the Fukushima Daiichi Nuclear Power plant. Japan and the world will feel the effects of these disasters for years to come.

EARTHQUAKE SITE
9.0 Magnitude

NUCLEAR PLANT FAILURE

Fukushima Daiichi
200 Rads into area

EARTHQUAKE TYPE

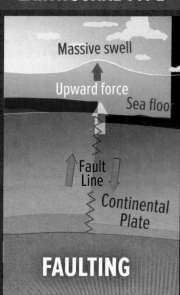

Massive swell

Upward force

Sea floor

Fault Line

Continental Plate

FAULTING

EXTENT OF THE TSUNAMI

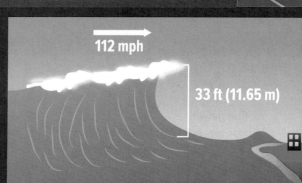

112 mph

33 ft (11.65 m)

RELIEF EFFORTS

In total, 116 countries and 28 international organizations offered assistance to Japan after the earthquake and tsunami.

As of early 2012, over 5 billion dollars for relief and recovery had been donated from people in Japan and across the globe.

700 million

Private donations from the United States exceeded 700 million dollars, the third highest U.S. donations total for any overseas disaster.

WORLD IMPACT

 **4–10 inches off axis**

At a magnitude of 9.0, the earthquake was so strong it shifted Earth on its axis by estimates of 4 to 10 inches (10–25 cm).

High tsunami waters from the Tohoku earthquake hit several other places, such as Hawaii, Peru, Chile, Oregon and California.

Millions of tons of debris washed into the ocean, some even reaching the US and Canada. A 66-ft.- (20-m-) long dock washed up on the Oregon coast in 2012.

Making Connections

1. **The World in Spatial Terms** How did the location of the Tohoku earthquake contribute to the destructive power of the tsunami?

2. **Physical Systems** Outside of Japan, the tsunami impacted areas in the southern and eastern Pacific Ocean the most. Explain why areas in the western Pacific were largely spared.

3. **Environment and Society** Numerous images and video of the Tohoku earthquake and subsequent tsunami were posted on the Internet almost immediately following the disaster. How do you think these visuals contributed to international relief efforts?

Interact with **Global Connections** *Online*

networks
There's More Online!

☑ **GRAPH** Japanese Balance of Trade with the United States

☑ **IMAGE** Commodore Perry in Japan

☑ **IMAGE** Japan and the West

☑ **IMAGE** Port of Tokyo

☑ **TIME LINE** Shifting Power

☑ **INTERACTIVE SELF-CHECK QUIZ**

☑ **VIDEO** Human Geography of Japan

Reading HELPDESK

Academic Vocabulary

- **approximate**
- **generation**
- **unique**
- **overseas**

Content Vocabulary

- **clan**
- **samurai**
- **acculturation**
- **trade surplus**

TAKING NOTES: *Key Ideas and Details*

IDENTIFYING As you read about the human geography of Japan, use a graphic organizer like the one below to identify major turning points in Japan's history.

Turning Points in Japan's History

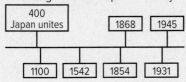

| 400 Japan unites | | 1868 | 1945 |

| 1100 | 1542 | 1854 | 1931 |

LESSON 2
Human Geography of Japan

ESSENTIAL QUESTION • *How do physical systems and human systems shape a place?*

IT MATTERS BECAUSE
Japan's early human geography reflects influences from ancient China and Korea. The Japanese civilization that developed from these two countries then isolated itself well into the 1800s. Since then, the forces of modernization and globalization have influenced Japanese society. Yet the traditional and the modern exist comfortably together.

History and Government

GUIDING QUESTION *How do ancient and modern traditions influence life in Japan?*

Japan's history combines tradition with transformation. Japan has maintained its cultural traditions while leaping toward modernization over the last century.

Japan Through the Ages

Because they are so close, China and Korea had a significant impact on Japan. Once ruled by many **clans,** in the A.D. 400s Japan united under the Yamato dynasty. Yamato rulers adopted China's philosophy, writing system, art, sciences, and governmental structure. Korean scholars also influenced the Japanese.

Japan was ruled by dynasties for centuries. Japanese society evolved, developing a government and establishing an emperor as absolute monarch. In the 1100s, warring armies brought about the shogunate, a feudal society under the control of a shogun, or military ruler. Professional warriors known as **samurai** supported the shogun's rule, although the emperor was the official ruler of Japan.

The first documented contact with the West occurred in 1542. A Portuguese ship sailing to China was blown off course and landed in Japan. Roman Catholic missionaries and European traders followed, raising concerns that military conquest might follow. To prevent this, shoguns restricted foreigners. They expelled all but a few European merchants who traded at Nagasaki.

During the 1800s, the United States worked to open Japan to trade. In 1854, following a show of strength, Commodore Matthew C. Perry of the U.S. Navy negotiated opening Japanese ports to U.S. ships. A trade agreement followed that benefited the United States. This action eventually sparked a samurai rebellion that returned full authority to the emperor.

A Changing Government

The return of authority to the emperor was a turning point in Japanese history. Called the Meiji Restoration, this period—which lasted from 1868 to 1912—involved the rapid modernization of Japanese society, including its government, economy, military, education, and legal systems. Japan became a modern country set on building an empire. In a war with China, Japan conquered the island of Formosa (Taiwan). Japan also fought Russia for control of Korea and gained rights to Manchuria and the Russian island of Sakhalin. By the time Japan took control of Korea in 1910, it had become the most powerful empire in Asia.

During World War I, Japan sided with the Allies, profiting from exports to Allied countries and expanding its influence in Asia. Following the war, Japan continued to prosper and extend its reach. Though Japan steered toward a democracy, military leaders gained influence over the government. In 1931 they invaded northeast China; in 1937 they invaded northern China. By 1939, Japan had joined with Germany and Italy as the Axis Powers. Japan's 1941 attack on Pearl Harbor drew the United States into World War II, in which some 3 million Japanese lost their lives. The war ended when Japan surrendered after the United States dropped atomic bombs on Hiroshima and Nagasaki in 1945.

The American occupation of Japan that followed stripped the empire of its territories and military. It also set the country back on the road to democracy. Out of the ruins, a vibrant economy emerged. Several decades later, Japan

clan a family group

samurai a professional warrior of preindustrial Japan

Commodore Perry's steamships, anchored in Edo Bay (now Tokyo Bay) and bristling with huge guns, convinced the Japanese government to sign a trade treaty with the United States and open two ports and a trade agreement.

▼ **CRITICAL THINKING**

1. *Identifying Central Issues* What caused the Japanese to end years of isolation abruptly?

2. *Defending* Provide a main reason to justify Commodore Perry's pressure on Japan to open trade with the United States. Then provide a main reason to justify Japanese resistance to opening their country to foreign influences.

became a global economic power. Today, Japan has the third-largest economy in the world, surpassed only by China and the United States. As a constitutional monarchy, Japan has an emperor who is the head of state.

✅ **READING PROGRESS CHECK**

Sequencing Information How has the role of the Japanese emperor changed over time?

Population Patterns

GUIDING QUESTION *How does population density influence life in parts of Japan?*

The majority of Japanese—99 percent—belong to the same ethnic group. They are descendants of migrating people from Asia who pushed out most of the indigenous people of Japan, the Ainu (EYE•noo). Some Ainu still live on Hokkaidō today.

Since most of Japan is mountainous, there is limited land area suitable for habitation. Most people are concentrated in the lowlands along the seacoasts, or in valleys and plains. The urban concentrations are great, and Tokyo-Yokohama is one of the largest megacities in the world, with more than 37 million people. The population density in this urban area is 11,300 people per square mile (4,300 people per sq. km). The average population density in Japan is about 902 people per square mile (348 per sq. km). The lowest density is on Hokkaidō.

Approximately 93 percent of Japanese live in urban areas, and a majority of these people live near the Tokaido megalopolis. This corridor is a 750-mile- (1,200-km-) long urbanization zone and is concentrated along rail lines. Extreme urbanization affects many aspects of Japanese life. This includes not only the lifestyles of the Japanese, but architecture, transportation, housing, and daily activities. In Tokyo, crowded streets with immense skyscrapers and numerous commercial areas are located within easy walking distance of subway stations. Bicycles are widely used by Japanese of all ages to ride to

approximate close to, but not exact

©The Print Collector/Corbis

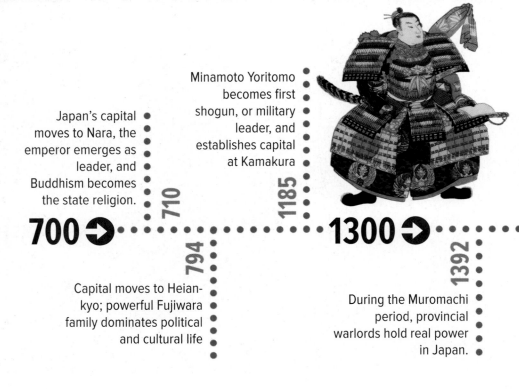

TIME LINE ⌄

Shifting
POWER →

Although Japan's monarchy has endured longer than that of any other country, the real power in Japan has shifted from emperors to shoguns to the people. The emperor plays an important symbolic role in Japan, but wields little power.

▶ **CRITICAL THINKING**

1. *Describing* In Japan's history, how has power passed from one leader to another?

2. *Analyzing* What role has the military played in Japan's government in the past?

Japan's capital moves to Nara, the emperor emerges as leader, and Buddhism becomes the state religion.

710

700 →

794

Capital moves to Heian-kyo; powerful Fujiwara family dominates political and cultural life

Minamoto Yoritomo becomes first shogun, or military leader, and establishes capital at Kamakura

1185

1300 →

1392

During the Muromachi period, provincial warlords hold real power in Japan.

school or to commute to work or subway stations. These stations link to a system of railroads. This system includes a network of long-distance, high-speed electric bullet trains. In 2012 a Japanese railroad company unveiled plans for a magnetic levitation train. It would float above the tracks, levitated by the force of powerful magnets. This train is expected to travel at 311 miles per hour (500 km per hour). Due to its emphasis on convenient public transportation, Tokyo has developed very differently from cities that are heavily reliant on privately owned vehicles. It has no slums and is considered environmentally friendly.

☑ **READING PROGRESS CHECK**

Summarizing What is the largest urban area in the world?

Society and Culture Today

GUIDING QUESTION *How do traditions influence family life and art?*

The written form of Japanese began around A.D. 400, when Chinese writing was first introduced and adapted to Japanese. The Japanese borrowed words from Chinese and have more recently borrowed thousands of English words for concepts that do not exist in Japanese.

Japan's indigenous religion, Shinto, is often practiced in a mix with Buddhism and even Christianity. Shinto emphasizes reverence for nature and is polytheistic. While it is unusual for Japanese children to receive formal religious instruction, many households have both a Buddhist altar and a Shinto altar.

Modern Japanese society has a high regard for education. By law, children must attend school until age 15. Many begin at a very young age to focus on getting into the best schools and eventually the best universities. Adults often continue to seek instruction long after formal schooling is completed, studying

1600

Tokugawa period begins, initiating a prolonged period of peace.

1900 →

1868

In response to pressure from Western nations, Japan begins rapidly modernizing during Meiji Restoration.

1912

Yoshihito becomes emperor; Japan becomes more democratic as more men gain the vote.

1926

Hirohito becomes emperor; militarism increases, eventually leading to World War II.

1989

Akihito becomes emperor; period of sluggish economic performance and political turmoil

2006

Prince Hisahito is born and becomes third in line for the Japanese throne.

The History of Origami

"Composed of the Japanese words *oru* (to fold) and *kami* (paper), origami has a rich and complex history that spans culture, class and geography.

Paper was first invented in China around 105 A.D., and was brought to Japan by monks in the sixth century. Handmade paper was a luxury item only available to a few, and paper folding in ancient Japan was strictly for ceremonial purposes, often religious in nature.

By the Edo period (1603–1868), paper folding came to be regarded as a new form of art that was enabled by the advent of paper both mass-produced and more affordable."

—"Between the Folds," *PBS Independent Lens*, November 30, 2009

DBQ *ANALYZING PRIMARY SOURCES* Why did the advent of mass-produced paper change origami from a ceremonial practice to a new form of art?

generation a group of individuals born and living at the same time

unique unlike anything else

acculturation cultural modification of an individual, group, or people by adapting to or borrowing traits from another culture

anything from foreign languages to technology. The Japanese are also avid travelers and tourists.

In addition to a high standard of living, the Japanese enjoy corresponding general good health and the longest life expectancy in the world. National health insurance covers all citizens. However, high life expectancy combined with a low birthrate means the Japanese population is aging disproportionately, which strains health and social services. A smaller workforce is a future issue for Japan.

Family and the Status of Women

Family in Japan remains a source of stability but is undergoing changes. Families are smaller than in the past. Many children are still taught the importance of being part of a group, as opposed to being an individual. They are also taught to revere their ancestors. But as new **generations** are more likely to be raised in urban settings, rather than in villages, they tend to have more modern views and changing values. In a technologically advanced and consumer-driven society, less social conformity and increasing individual choice is apparent in ordinary Japanese people. Young Japanese are among the world's major users of technology.

Some of the changes in Japanese families are due to the changing role of women in society. There is growing dissatisfaction with traditional roles in which women are expected to bear major responsibility for the household tasks. Women make up about two-fifths of Japanese workers. As they enter the workforce in larger numbers, they often delay marriage and childbearing. This has contributed to the trend of smaller families and Japan's declining population. Female role models, such as Japan's first female foreign minister, have helped to change perceptions. Many women work only part-time jobs, however, and those in careers often find it difficult to win promotion to high-level positions.

The Arts

The Japanese have developed their own art forms. Traditional Japanese art rejects what is showy and emphasizes delicacy, exquisite forms, and simplicity. Poets, writers, and artisans throughout the centuries have portrayed the beauty of the seasons, the changing environment, and the rugged mountains through poetry, theater, painting, pottery, embroidery, and silk screen.

Among the enduring and **unique** art forms of Japan is origami. Origami uses folded paper to depict the shapes of objects of nature, such as animals and birds. The Japanese tea ceremony is also a time-honored tradition. Still practiced as a hobby today, the tea ceremony combines literary and artistic traditions into a ritual of great artistry. Another tradition that remains popular is the classical Kabuki theater. It blends music, dancing, costumes, elaborate makeup, and well-timed special effects to bring characters to life.

A taste for literature among educated Japanese has a long history. Lady Murasaki Shikibu, a Japanese noblewoman, wrote one of the world's first novels, *The Tale of Genji*, around A.D. 1010. It tells the story of an imperial prince in the emperor's court. Another well-known form of Japanese poetry, the haiku, is famous for its concise and exact use of words and nature imagery that appeals to the senses.

Japan's rich artistic history does not hinder the Japanese from enjoying and excelling at art and cultural forms that developed in other places. Such **acculturation** is apparent in many ways. Western classical music is popular, and major Japanese cities may have more than one symphony orchestra. Western-style painting, sculpture, architecture, and cinema are also very popular.

✔ **READING PROGRESS CHECK**

Identifying List five of Japan's unique art forms and describe two of them.

Economic Activities

GUIDING QUESTION *How have industry and trade transformed the country?*

Japan's industrial economy is based on a mixed market system. Such a system emphasizes private ownership of the means of production and distribution. The government plays a key role, however. It carefully coordinates with industry and regulates all sectors of the economy, including imports.

Industry and Manufacturing

Japan has seen extraordinary economic growth since World War II. High demand for trade goods around the world has made it a global economic power since the 1980s. Lacking ample natural resources, such as fossil fuels and minerals, Japan has focused on trade to sustain its industries. It has highly skilled and educated workers and newly advanced technology. This has helped Japan to become a leading producer of consumer goods such as cars, electronic devices, and computers.

Telecommunications industries, particularly mobile communications leaders NTT and KDDI, rank among the largest Japanese corporations. Other huge conglomerates include Canon, which makes office and consumer products, and Honda, which builds automobiles and motorcycles. Nippon Steel is Japan's leading steel producer. Mitsubishi is a major multinational company with diverse business interests. These and many other important companies earn billions in annual profits.

Apart from their economic role, Japanese corporations play an important societal role. Although this practice is beginning to change, most employees expect to stay with their company until retirement. This culture of lifetime employment creates very strong bonds between company and workers. Politeness, sensitivity, and good manners are basic elements in the workplace. Executives and their families are ranked socially by the importance of the company for which they work.

Trade

To sustain its economy, Japan imports raw materials such as fossil fuels and minerals from other countries. Japanese industries turn these resources into finished products that they sell in huge quantities **overseas**. For example, China, the United States, South Korea, Taiwan, Singapore, Thailand, and Germany all buy Japanese cars and high-tech devices.

overseas relating to a foreign country, especially one across the ocean

Japanese Balance of Trade with the United States

⌄ GRAPH SKILLS

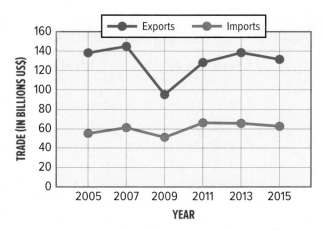

Japan relies heavily on exports to drive economic growth.

◄ **CRITICAL THINKING**

1. *Analyzing Visuals* Describe the difference between Japan's exports to and imports from the United States.

2. *Constructing a Thesis* Give a logical reason for why Japan's imports and exports both dipped in 2009. Explain your reasoning.

Source: http://tse.export.gov/TSE/MapDisplay.aspx

This freighter is docked at the Port of Tokyo. Japan's economy depends on imports of raw materials and exports of finished products.

▲ CRITICAL THINKING

1. *Finding the Main Idea* Why is Japan's economy so reliant on trade?

2. *Evaluating* Why are Japanese finished goods in high demand globally?

trade surplus earning more money from export sales than spending for imports

Although the global demand for Japanese goods is high, the Japanese government protects local industries from foreign competition. It places tariffs, or taxes, on the manufactured goods imported from foreign countries (but not on raw materials). This restricts foreign sales in Japan and helps tip the balance of trade in Japan's favor. Balance of trade is the difference between a country's imports and exports. Japan has a **trade surplus** in its balance of trade because the value of its exports is greater than the value of its imports. This makes Japan wealthy. However, it lowers profits for trading partners, such as the United States, which import from Japan more than they export. Efforts to persuade Japan to drop tariff protections on finished products have yet to be completely successful.

In the 1990s, an economic downturn occurred in Japan. This was brought about, in part, by poor banking practices. As banks failed, production slowed, exports dropped, and unemployment rose. The Japanese government took steps to reform unsafe economic practices. By 2003, the situation had stabilized and was improving. The recession that began in 2008, however, reduced global demand for Japanese goods and slowed the economy. Things began to improve again in 2013.

☑ READING PROGRESS CHECK

Inferring What might a Japanese official say to defend the practice of placing tariffs on finished goods imported from overseas countries?

LESSON 2 REVIEW

Reviewing Vocabulary

1. *Identifying* Write several sentences indicating the role of the clan and the samurai in Japan's early history.

Using Your Notes

2. *Sequencing* Use your graphic organizer to write a paragraph describing the major turning points in Japan's history.

Answering the Guiding Questions

3. *Drawing Conclusions* How do ancient and modern traditions influence life in Japan?

4. *Evaluating* How does population density influence life in parts of Japan?

5. *Making Connections* How do traditions influence family life and art?

6. *Synthesizing* How have industry and trade transformed the country?

Writing Activity

7. *Informative/Explanatory* In a paragraph, discuss the advantages and disadvantages of living in Tokyo-Yokohama, the largest urban area in the world.

Toshifumi Kitamura/AFP/Getty Images

netw⊙rks

There's More Online!

☑ **GRAPH** Total Whale Catch

☑ **IMAGE** Aquaculture

☑ **IMAGE** Bluefin Tuna

☑ **MAP** Japan's Nuclear Power Plants

☑ **INTERACTIVE SELF-CHECK QUIZ**

☑ **VIDEO** People and Their Environment: Japan

LESSON 3
People and Their Environment: Japan

Reading HELPDESK

Academic Vocabulary

- **sustain**
- **widespread**
- **issue**
- **strategy**

Content Vocabulary

- **aquaculture**
- **supertrawler**
- **chlorofluorocarbon (CFC)**

TAKING NOTES: *Key Ideas and Details*

ORGANIZING Use a graphic organizer like the one below to take notes on Japan's environment as you read the lesson.

Japan: Managing Resources

Human Activity	Impact on Environment	Addressing the Issues
Build nuclear reactor		

ESSENTIAL QUESTION • *How do physical systems and human systems shape a place?*

IT MATTERS BECAUSE
Because of its economic strength, Japan has an impact on the world disproportionate to its size. The choices the Japanese make about energy, ocean harvesting, and consumption resonate far beyond their island home.

Managing Resources

GUIDING QUESTION *What resources are at risk in Japan?*

Because of its limited supply of fossil fuels, Japan has long depended on foreign oil imports. To reduce this dependence, the Japanese people turned to nuclear technology to provide electrical energy. Today Japan has more than 40 nuclear reactors with the capacity to supply the country with some 14 percent of its energy. A series of nuclear accidents at Japanese reactors, however, exposed people to radiation. The first accident occurred in 1978. A more recent one happened in 2011.

Following the earthquake and resulting tsunami in 2011, a damaged nuclear reactor had a meltdown. The perils of operating nuclear reactors in a region prone to earthquakes and tsunamis were made clear. In response, Japan temporarily shut down most of its reactors. This prompted calls to reverse Japan's reactor-building program and focus instead on renewable energy sources. Japanese businesses and industries resisted such a change, however. They pointed out that rising fossil fuel prices could cripple the Japanese economy and add substantially to pollution. Japan continues searching for answers and alternatives that will allow it to **sustain** its energy needs safely. Investment in solar power, wind, biomass, and small hydroelectric and geothermal projects has expanded. Increasingly, Japanese companies are building wind turbines and developing solar cell technology. There is some discussion of building floating wind turbines to be anchored in the sea because Japan has limited land area for wind projects.

☑ **READING PROGRESS CHECK**

Identifying What are the pros and cons of using nuclear power in Japan?

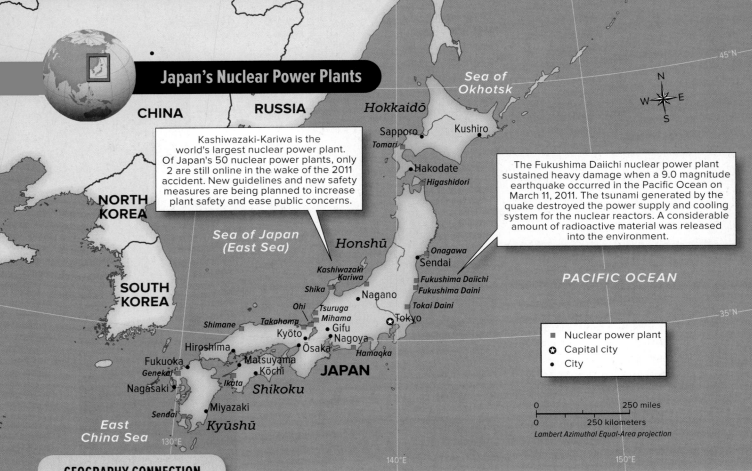

Japan's Nuclear Power Plants

Kashiwazaki-Kariwa is the world's largest nuclear power plant. Of Japan's 50 nuclear power plants, only 2 are still online in the wake of the 2011 accident. New guidelines and new safety measures are being planned to increase plant safety and ease public concerns.

The Fukushima Daiichi nuclear power plant sustained heavy damage when a 9.0 magnitude earthquake occurred in the Pacific Ocean on March 11, 2011. The tsunami generated by the quake destroyed the power supply and cooling system for the nuclear reactors. A considerable amount of radioactive material was released into the environment.

■ Nuclear power plant
✪ Capital city
• City

250 miles
250 kilometers
Lambert Azimuthal Equal-Area projection

GEOGRAPHY CONNECTION

Nuclear energy has been a priority in Japan since 1973. Following the earthquake and tsunami that caused a nuclear accident in 2011, however, this policy came under review.

1. ***THE WORLD IN SPATIAL TERMS*** Why do you think Japan's nuclear power plants are located along the coasts?

2. ***ENVIRONMENT AND SOCIETY*** What are located within a few hundred miles of the largest clusters of nuclear power plants?

sustain to continue without interruption

widespread found over a large area

Human Impact

GUIDING QUESTION *What human activities have affected the physical environment of Japan?*

Japan's rapid industrialization brought with it severe and **widespread** environmental problems. Economic growth took priority over health and safety. The result was increasing air and water pollution and declining populations of aquatic animals.

Following World War II and into the 1960s, Japan's major industries—iron and steel, cement, paper and pulp—expanded without regard to pollution. Power plants vented noxious pollutants into the air. Industry and agriculture tainted rivers, lakes, and surrounding ocean with chemicals and raw sewage. Air and water quality suffered, creating health risks. In the 1970s, smog alerts were frequent. Acid rain became a serious problem. It damaged lakes, crops, and trees; threatened animal species; and corroded buildings and bridges. By the mid-1990s, Japan was the world's fourth-largest emitter of carbon dioxide.

Starting in the 1970s, Japan began adopting stricter rules to protect air and water quality. Today, it has some of the strictest regulations in the world. Yet despite improvements in the levels of soot and smoke emissions from factory smoke stacks, motor vehicle emissions remain a serious problem. Diesel engines, in particular, foul the air of Japan's major cities. Regulations to further tighten control of emissions from diesel engines and buses have been passed. Japanese industries have been at work on fuel-cell engines, which produce no dangerous emissions.

The problem of acid rain is particularly troubling. Even as Japan has tightened regulations on emissions, the acidity of the rain has remained largely unchanged. Researchers believe that massive sulfur dioxide emissions in China contribute to acid rain in Japan. China and Japan signed a preliminary treaty to address acid

rain in 1994. Evidence suggests, however, that acid rain that originates in China will reach disastrous levels in Japan by 2020.

For centuries, the Japanese have depended on the abundance of the sea for food. Fish are an integral part of the Japanese diet, and the Japanese consume large amounts of fish and other seafood. Japan's most internationally famous food specialty, sushi, usually includes thin slices of raw fish called sashimi. Overfishing—resulting from catching too many fish of particular species—has seriously depleted fish stocks around the Japanese islands and around the world. Water pollution has also contributed to declining fish stocks.

Some 40,000 people use Japan's Inland Sea—a narrow body of salt water surrounded by Honshū, Shikoku, and Kyūshū—for fishing and **aquaculture**, which is the farming of seafood. Wild fish stocks such as red sea bream, anchovies, and mackerel have declined dramatically in this area in recent decades. The number of jellyfish—which negatively affect other species—is increasing in these waters because the plankton on which they feed thrive in polluted water.

Globally, the huge Japanese demand for seafood has contributed to declining fish stocks in fisheries in the Atlantic and the Pacific. Many countries, including Japan, rely on powerful deep-sea **supertrawlers**. These huge ships are equipped to serve as modern factories at sea, harvesting distant fisheries. Harvesting huge catches using supertrawlers leads to overfishing. It also causes the killing and waste of tons of unwanted sea life. To protect fisheries from foreign commercial fishing, many countries claim an economic zone extending to 200 nautical miles off their shores. Public opposition to supertrawlers is gaining strength globally, particularly as numerous fish stocks are rapidly declining.

Despite a 1986 international ban on commercial whaling, Japan remains the largest consumer of whale meat in the world. Centuries of overhunting by many countries caused serious declines in whale populations globally, severely endangering many species. By the 1960s, only Japan, Russia, and a few other countries continued to hunt whales on a large scale. Following the 1986 ban, Japan, Iceland, and Norway used a loophole for scientific research to continue to capture whales. Hundreds are still killed and eaten each year. Japan and the other whale-hunting countries are widely criticized for their whaling practices.

☑ READING PROGRESS CHECK

Understanding Relationships Explain why fish stocks in Japan's Inland Sea have declined in recent years.

Mature bluefin tuna reach 6.5 feet (2 m) and live up to 15 years in the wild. These endangered and valuable fish are prized for food in Japan even as their stocks have plummeted worldwide. A premium fish sold for $1.76 million in 2013.

▲ CRITICAL THINKING

1. ***Making Connections*** Why did fish become a staple in the Japanese diet?

2. ***Analyzing Ethical Issues*** Is it ethical to consume endangered species? Explain your answer.

aquaculture the cultivation of seafood

supertrawler an ocean-going factory ship with facilities for processing and freezing fish

---

Total Whale Catch

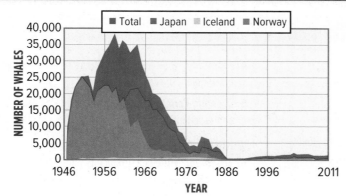

■ Total ■ Japan ■ Iceland ■ Norway

Y-axis: NUMBER OF WHALES — 40,000; 35,000; 30,000; 25,000; 20,000; 15,000; 10,000; 5,000; 0

X-axis: YEAR — 1946, 1956, 1966, 1976, 1986, 1996, 2011

Sources: Animal Welfare Institute, Whale and Dolphin Conservation

✓ GRAPH SKILLS

Whales have been hunted and eaten by the Japanese, Icelanders, and Norwegians for centuries.

◀ CRITICAL THINKING

1. ***Analyzing Visuals*** What was the largest total number of whales captured by Japan, Iceland, and Norway after 1946?

2. ***Constructing a Thesis*** Give a logical reason for why all three countries' whale hunting dipped sharply in 1986.

Sue Flood/The Image Bank/Getty Images

This fish farm is located in Ago Bay, along the south coast of Honshū. Aquaculture provides the Japanese with about 22 percent of their seafood.

▲ CRITICAL THINKING

1. **Assessing** Why would it be advantageous to create a fish farm in a bay?

2. **Making Predictions** Do you think the number of fish farms will increase in the future? Why or why not?

issue a topic for debate and discussion

chlorofluorocarbon (CFC) a chemical substance, found mainly in liquid coolants, that damages the Earth's protective ozone layer

strategy a plan of action to achieve an aim

Addressing the Issues

GUIDING QUESTION *How are environmental issues being addressed in Japan?*

Japan ignored the environmental problems caused by rapid economic growth for many years. In the 1970s, however, the Japanese government began to encourage industries to address environmental **issues** more responsibly.

Today Japan is a leader in addressing environmental issues and supporting industries that profit from sustainable growth. For example, the Japanese have taken steps to reduce emissions of **chlorofluorocarbons (CFCs)**. CFCs were widely used in industry in the liquid coolants for refrigerators and air conditioners during the mid-1900s. Released into the atmosphere, they deplete Earth's stratospheric ozone layer that protects the planet from harmful ultraviolet solar radiation. Japan is working toward a complete phase-out of CFCs.

Japan is also working to mitigate climate change. In 1997 Japan hosted an international meeting under the United Nations Framework Convention on Climate Change. The treaty that resulted is known as the Kyoto Protocol. It defined ways to reduce the carbon emissions that contribute to global warming. It was ratified by 37 industrial countries, including Japan (but not the United States). The treaty set binding targets to reduce greenhouse gas emissions, including carbon dioxide. Nevertheless, by 2014 carbon dioxide emissions in Japan had increased by more than 16 percent from 1990 levels.

One **strategy** Japan has used to offset declining fish stocks and other sea products has been to increase aquaculture. Fish farmers raise fish by floating fish cages in the sea or anchoring them to the sea floor. Floats and ropes are used to cultivate shellfish such as oysters, scallops, pearl oysters, and abalone, as well as kelp. Although aquaculture production has declined somewhat since 1990, nearly 1.1 million tons of fish and seaweed were farmed in Japan in 2012.

Another strategy for maintaining fish populations has been to prevent overfishing by establishing quotas that limit fish harvests. Like the ban on whaling, however, quotas have limited effectiveness.

☑ **READING PROGRESS CHECK**

Inferring Has Japan been generally successful in addressing environmental problems? Explain your answer.

LESSON 3 REVIEW

Reviewing Vocabulary

1. **Making Connections** Write a paragraph describing the relationship of supertrawlers to Japanese aquaculture.

Using Your Notes

2. **Transferring Information** Use your graphic organizer to identify a human activity that has been destructive to Japan's environment and write a paragraph about ways in which the issue is being addressed.

Answering the Guiding Questions

3. **Differentiating** What resources are at risk in Japan?

4. **Assessing** What human activities have affected the physical environment of Japan?

5. **Drawing Conclusions** How are environmental issues being addressed in Japan?

Writing Activity

6. **Argument** Write an introductory paragraph for a personal essay that will argue whether you think Japan is doing enough to address environmental issues and why.

Directions: On a separate sheet of paper, answer the questions below. Make sure you read carefully and answer all parts of the questions.

Lesson Review

Lesson 1

1 *Identifying Cause and Effect* How does the Pacific Ring of Fire affect human settlement in Japan?

2 *Explaining* What combination of factors makes Japan monsoonal?

3 *Assessing* Generally, where does Japan rank in oil production and oil exporting in the world? Why?

Lesson 2

4 *Describing* How did the United States "open up" Japan in 1854?

5 *Explaining* Explain whether it would be accurate to state that the emperor is the government official in Japan who makes or approves the major governmental decisions for the country.

6 *Identifying Central Issues* Where does Japan rank in life expectancy among all countries of the world? Explain the reasons that have greatly contributed to this ranking and discuss problems that are arising in relation to it.

Lesson 3

7 *Explaining* What events in 2011 caused Japan to reconsider its reliance on nuclear energy? Why?

8 *Interpreting Significance* Explain the significance of economic zones to fisheries.

9 *Evaluating* Explain whether it would be accurate to state that the level of carbon dioxide emissions in Japan in 2007 poses an irony to its role relative to the Kyoto Protocol.

21st Century Skills

Use the cartoon below to answer the questions that follow.

PRIMARY SOURCE

"Before you do something you may regret, I think you should know that I contain six parts of mercury per million."

10 *Using Primary Sources* How does the cartoon provide a whimsical look at a serious issue plaguing Japan and other countries around the world? Why might this cartoon be more effective than long lists of statistics in attracting the attention of some readers?

11 *Drawing Conclusions* What effect do you think the artist hoped to gain through this cartoon? Explain whether you think this effect is evident in Japan.

Critical Thinking

12 *Exploring Issues* You are preparing a Web site about the current economy of Japan—and the connection of the current economy to the country's history, physical geography (including natural resources), human geography (including population patterns), and environmental concerns. Lay out the main page of the Web site, including at least 10 links to information for the content detailed above. For each link, provide a short paragraph to summarize the main content that will be accessed through the link.

Need Extra Help?

If You've Missed Question	**1**	**2**	**3**	**4**	**5**	**6**	**7**	**8**	**9**	**10**	**11**	**12**
Go to page	704	706	707	710	712	714	717	719	720	721	721	701

Directions: On a separate sheet of paper, answer the questions below. Make sure you read carefully and answer all parts of the questions.

DBQ Analyzing Primary Sources

Use the excerpts to answer the questions that follow.

PRIMARY SOURCE

"*Based on a detailed investigation of all the facts, and supported by the testimony of the surviving Japanese leaders involved, it is the Survey's opinion that certainly prior to 31 December 1945, and in all probability prior to 1 November 1945, Japan would have surrendered even if the atomic bombs had not been dropped, even if Russia had not entered the war, and even if no invasion had been planned or contemplated.*"

—United States Bombing Survey, July 1, 1946,
President's Secretary's File, Truman Papers

"*. . . There are voices which assert that the bomb should never have been used at all. I cannot associate myself with such ideas. Six years of total war have convinced most people that had the Germans or Japanese discovered this new weapon, they would have used it upon us to our complete destruction with the utmost alacrity. . . . Future generations will judge these dire decisions, and I believe that if they find themselves dwelling in a happier world from which war has been banished, and where freedom reigns, they will not condemn those who struggled for their benefit amid the horrors and miseries of this gruesome and ferocious epoch.*"

—Winston Churchill, "Why Should We Fear for Our Future?"
House of Commons, August 16, 1945

13. **Understanding Historical Interpretation** Explain whether these quotes provide statements that are essentially in agreement with one another.

14. **Interpreting Significance** Explain whether you think Churchill's anticipation of potential future events related to weapons of mass destruction were valid.

Exploring the Essential Question

15. **Sequencing** Conduct research on Japan's history and the periods in which the island country was influenced by or isolated from other countries. Use your research to create a time line of at least five entries explaining major milestones.

Applying Map Skills

Refer to the Unit 8 Atlas to answer the following questions.

16. **Places and Regions** Identify the major body of water nearest Ōsaka.

17. **The World in Spatial Terms** Use your mental map of Japan to describe the spatial relationship among the islands of Hokkaidō, Honshū, Kyūshū, and Shikoku.

18. **Human Systems** Which two cities in Japan are the most densely populated?

College and Career Readiness

19. **Examining Information** You are located in Washington, D.C., and wish to host a traditional Japanese cherry blossom festival. Conduct research to determine the connection between Japan and the cherry blossom trees in your location, as well as the optimal time of the year to have a festival. Write a report explaining why you have chosen this time of the year and the activities that will be included in the experience.

Research and Presentation

20. **Gathering Information** With a partner, conduct research to learn more about the Shinkansen and the development of the bullet train and the magnetic levitation train in Japan. Create a multimedia presentation to detail development of the bullet train and the magnetic levitation train—and the importance of rapid mass transit to Japan. Include audio, video, and maps in your presentation.

Writing About Geography

21. **Argument** Use standard grammar, spelling, sentence structure, and punctuation to write an editorial to support the pro or con side of this statement: "Supertrawlers should be banned." Focus your argument on the effects of such a ban. Support your statements with information from the chapter and address opposing arguments in your editorial.

Need Extra Help?

If You've Missed Question	13	14	15	16	17	18	19	20	21
Go to page	722	722	710	672	672	676	706	713	719

North Korea and South Korea

ESSENTIAL QUESTION · *How do physical systems and human systems shape a place?*

This young Korean woman wears traditional dress.

©Topic Photo Agency/Corbis

networks

There's More Online about the geography of North Korea and South Korea.

CHAPTER 30

Why Geography Matters
Complementarity: Two Koreas

Lesson 1
Physical Geography of North Korea and South Korea

Lesson 2
Human Geography of North Korea and South Korea

Lesson 3
People and Their Environment: North Korea and South Korea

Geography Matters...

The Korean Peninsula is composed of North Korea and South Korea. The people of the peninsula share a language, a history, and a culture. War and politics, however, divided a single country into two. Political philosophies and decisions about resources have resulted in two very different economies and standards of living. Prosperity and scarcity threaten the environment in different ways. Urbanization, the increased use of technology and consumer goods, and the loss of rural culture threaten South Korea. Famine, erosion, and lack of freedom threaten North Korea.

723

complementarity: *two* Koreas

North Korea and South Korea share a peninsula, a history, and a culture, but they are enemies. The two countries are in a situation known as a complementarity in which two places are economically interdependent. North Korea has more natural resources and raw materials, while South Korea is more developed industrially and economically. Economic cooperation could occur if the countries were not bitterly divided.

How is life in both countries similar on the peninsula?

North Korea occupies the northern half of the Korean Peninsula and South Korea occupies the southern half. North Korea's rugged, mountainous terrain makes farming difficult, but yields coal and other valuable mineral resources such as gold, iron, copper, and zinc. Although North Korea has invested in mining, as well as other industries, it is one of the world's poorest countries. In contrast, South Korea's economy is one of the strongest in Asia. Like North Korea, South Korea is mountainous. However, fertile plains in the west and south are suitable for growing rice, barley, soybeans, corn, and other crops. Since the 1940s, South Korea's economy has been shifting from agricultural to industrial. Today, South Korea is the largest shipbuilder in the world. Its modern factories also manufacture cars, electronics, chemicals, and steel.

1. Places and Regions How are North Korea and South Korea similar and different?

How has a shared history affected both countries?

North Korea and South Korea have common bonds and the ability to satisfy each other's economic needs. However, these two countries have political differences that make cooperation difficult. Communist North Korea is a dictatorship. It has ties to China and Russia. South Korea is a republic with ties to the United States, Western Europe, and Japan. Since the Korean War ended in 1953, relations between the two countries have remained tense. North Korea has threatened South Korea with military strikes and South Korea has vowed to strike back. In 2010 North Korea fired on South Korean forces stationed on Yeonpyeong Island. Despite hostile relations, the two countries have been able to find some common ground. For instance, North Korea has been plagued by serious food shortages brought on by drought and flooding. To help solve this problem, North Korean leaders have accepted emergency food aid from South Korea. With few of its own resources, South Korea supports its growing industries by buying raw materials from North Korea.

2. Human Systems What prevents North Korea and South Korea from working together?

In a global economy, what do the two countries have in common?

To help one another, North Korea and South Korea can continue to work toward sharing resources. In the late 1990s, South Korea pursued the "Sunshine Policy" of economic cooperation, sharing aid and resources with North Korea. Later, the leaders of North Korea and South Korea attended inter-Korean summits to address political and economic issues. As a result, the two countries were able to launch a joint economic venture in 2006: the Kaesong Industrial Complex in North Korea. For this project, South Korea supplied money and technology. North Korea supplied land and labor. More than 100 South Korean-owned companies employed about 47,000 North Koreans. Other inter-Korean economic projects have been proposed, such as a natural gas pipeline from Russia to South Korea via North Korea. There is also a proposal to extend the Trans-Siberian Railway through North Korea to South Korea.

3. The World in Spatial Terms What are the possible benefits of economic cooperation between North Korea and South Korea?

THERE'S MORE ONLINE

EXPLORE a map of the Korean War • COMPARE the populations of North Korea and South Korea

networks

There's More Online!

☑ **CHART** Average Year-Round Temperatures in Select Korean Cities

☑ **IMAGE** Rural Town in South Korea

☑ **MAP** Waterways of the Korean Peninsula

☑ **INTERACTIVE SELF-CHECK QUIZ**

☑ **VIDEO** Physical Geography of North Korea and South Korea

LESSON 1

Physical Geography of North Korea and South Korea

ESSENTIAL QUESTION • *How do physical systems and human systems shape a place?*

Reading HELPDESK

Academic Vocabulary

- **vary**
- **cooperative**

Content Vocabulary

- **islet**
- **aquaculture**

TAKING NOTES: *Key Ideas and Details*

COMPARING As you read about the physical geography of the Korean Peninsula, use a graphic organizer like the one below to compare the landforms, climates, and natural resources of North Korea and South Korea.

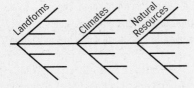

Physical Geography

Landforms   Climates   Natural Resources

IT MATTERS BECAUSE

The Korean Peninsula lies between China and Japan, at the center of one of the world's most economically and strategically important regions—East Asia. The peninsula's rugged terrain was a barrier to invaders for many centuries, and the population of the peninsula is quite homogeneous. However, the peninsula's long coastline allowed contact with other civilizations. The two Koreas have great economic and geopolitical importance today.

Landforms

GUIDING QUESTION *How does the land on the Korean Peninsula change from north to south?*

The Korean Peninsula is about half the size of California. It is bordered by Russia and China to the north and surrounded by water on the other three sides. To the east is the Sea of Japan, which the Koreans call the East Sea. To the south, the Korea Strait lies between South Korea and Japan. To the southwest lies the East China Sea, which joins the Yellow Sea on the western coast. The Koreans call this the West Sea. After World War II, the peninsula was split into the Democratic People's Republic of Korea (North Korea) and the Republic of Korea (South Korea). The two countries have been divided by the Korean Demilitarized Zone (DMZ) since 1953. The DMZ is a strip of unoccupied land about 6 miles (9.7 km) wide.

The peninsula is very mountainous. The T'aebaek Mountains begin in North Korea and stretch along the Sea of Japan following the east coast into South Korea. Both North Korea and South Korea are mountainous, but North Korea is more mountainous with highlands and plateaus across its northern and eastern areas. North Korea has some of the highest peaks. Mount Paektu, the highest peak on the peninsula, is on the border between North Korea and China. Most of the rivers on the peninsula originate in the T'aebaek range and flow westward. This creates large river valleys on the western side of the peninsula. Most people live in these lowlands or along the coastline.

(tl)Topic Photo Agency/age fotostock, (tc)©Topic Photo Agency/Corbis, (tr)©Christophe Boisvieux/Corbis

The coastline has many small inlets and bays, especially in the south. There are more than 3,000 islands off the coast of South Korea. The two largest, Cheju and Ulleungdo, were formed by volcanic lava. Cheju, in the East China Sea south of the Korean Peninsula, is beloved by honeymooners, and is known as the "Hawaii of Korea." Ulleungdo, located to the north in the Sea of Japan, has fine fishing and forests. South Korea also claims a group of rocky **islets** that Koreans call the Tok Islands. Japan also claims them.

islet a very small island

✅ READING PROGRESS CHECK

Specifying Which country of the Korean Peninsula is more mountainous?

Waterways

GUIDING QUESTION *Which major waterways drain the land in the peninsula?*

Most of the peninsula's major rivers are in South Korea. The Han, the Kŭm, and the Naktong all begin in the T'aebaek Mountains. The northernmost of these rivers, the Han, flows through South Korea's capital, Seoul. Both the Han and the Kŭm flow west to reach the Yellow Sea. South Korea's longest river, the Naktong, flows south for 325 miles (523 km) to the Korea Strait.

North Korea's longest river is the Yalu, which forms most of the border between North Korea and China. The source of the Yalu is on the southern slope of Mount Paektu. The Yalu flows some 500 miles (800 km) to Korea Bay, an inlet of the Yellow Sea, on the western coast of North Korea. Several other rivers in North Korea—the Ch'ŏngch'ŏn, Taedong, Chaeryŏng, and Yesŏng—also drain into Korea Bay. These large river valleys are important farming areas in this mountainous land.

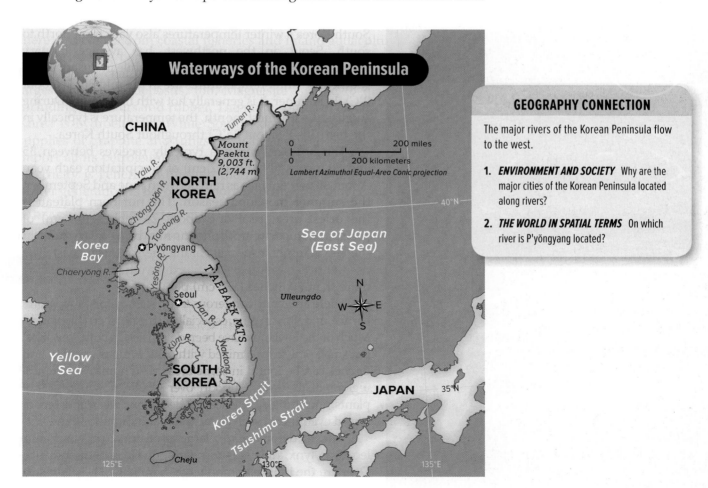

Waterways of the Korean Peninsula

CHINA
Tumen R.
Mount Paektu 9,003 ft. (2,744 m)
0   200 miles
0   200 kilometers
Lambert Azimuthal Equal-Area Conic projection
Yalu R.
Ch'ŏngch'ŏn R.
NORTH KOREA
Taedong R.
40°N
Sea of Japan (East Sea)
Korea Bay
P'yŏngyang
Chaeryŏng R.
Yesŏng R.
Seoul
Ulleungdo
N W E S
T'AEBAEK MTS.
Han R.
Kŭm R.
Naktong R.
Yellow Sea
SOUTH KOREA
JAPAN
35°N
Korea Strait
Tsushima Strait
Cheju
125°E   130°E   135°E

GEOGRAPHY CONNECTION

The major rivers of the Korean Peninsula flow to the west.

1. ***ENVIRONMENT AND SOCIETY*** Why are the major cities of the Korean Peninsula located along rivers?

2. ***THE WORLD IN SPATIAL TERMS*** On which river is P'yŏngyang located?

networks

There's More Online!

☑ **CHART** North Korean Conflicts

☑ **IMAGE** North Korean Soldiers Celebrate

☑ **IMAGE** South Korean Aircraft Factory

☑ **IMAGE** Traditional Korean Fan Dance

☑ **TIME LINE** Focus of Rival Interests

☑ **INTERACTIVE SELF-CHECK QUIZ**

☑ **VIDEO** Human Geography of North Korea and South Korea

Reading HELPDESK

Academic Vocabulary

- **principle**
- **isolation**
- **regime**
- **authority**
- **ethnicity**

Content Vocabulary

- **coup**
- **cultural divergence**

TAKING NOTES: *Key Ideas and Details*

LISTING As you read about the human geography of North Korea and South Korea, use a graphic organizer like the one below to list details about the history, population, culture, and economic activities of the two countries.

	North Korea	South Korea
History		
Population		
Culture		
Economic Activities		

LESSON 2
Human Geography of North Korea and South Korea

ESSENTIAL QUESTION • *How do physical systems and human systems shape a place?*

IT MATTERS BECAUSE

Korea's ancient culture has influenced and been influenced by the cultures of China and Japan. The tense relationship between North Korea and South Korea and their different political systems have gained the two countries different major allies—China and the United States, respectively.

History and Government

GUIDING QUESTION *How has the decision to divide the Korean Peninsula into North Korea and South Korea affected the development of each country?*

Before China expanded into Korea in 108 B.C. during the Han dynasty, the Korean people mostly belonged to village communities that were not united. Eventually three united kingdoms emerged that shared similar cultures, but were also rivals. These kingdoms traded with China and were greatly influenced by Chinese culture.

Buddhism was introduced to the peninsula from China in the fourth century A.D. It was widely adopted by ruling kingdoms. From Korea, Buddhism was introduced to Japan in the sixth century A.D. Confucianism came to the peninsula from China about a thousand years ago. At first, Korean rulers promoted Buddhism as the religion for personal enlightenment. Confucianism was adopted as a standard for political **principles**. Later, Buddhism was suppressed, and Confucianism provided the basis for a complex governmental bureaucracy.

The Silla dynasty drove out the Chinese and united Korea in A.D. 668. Other Korean dynasties came to power, which were largely supported by the Chinese and influenced by Chinese culture. The government of the Koryo dynasty, for example, used China's government as a model.

After resisting Mongol invaders, Korea became known as the Hermit Kingdom because of its **isolation**. In 1871 Korea declared an official policy of isolation, but Japanese warships arrived in 1876. This "gunboat diplomacy" by Japan forced Korea to open its ports for trade. After the Russo-Japanese War ended in 1905, Korea became a Japanese protectorate.

Japan annexed Korea in 1910, making it a colony. Japanese occupation deprived the Korean people of many rights and freedoms. The Japanese attempted to assimilate Koreans into Japanese life and culture. They used Korea's resources to grow their own economy. Anti-Japanese feelings increased the desire to gain independence.

Korea Divided

After Japan's defeat in World War II, the Korean Peninsula was jointly occupied by the Soviet Union and the United States. North of the 38th parallel, or line of latitude, the Soviets set up a Communist government. The new country that emerged was the Democratic People's Republic of Korea. The south was administered by the United States. The Republic of Korea was established there, with Seoul as its capital.

In June 1950, North Korean forces invaded South Korea. The conflict that followed became known as the Korean War. United Nations forces, mainly troops from the United States, came to the aid of the Republic of Korea (South Korea), while the Soviet Union and China supported the Democratic People's Republic of Korea (North Korea) with troops and equipment. Fighting concluded in 1953 with an armistice, or cease-fire, and the establishment of the Korean Demilitarized Zone (DMZ) separating the two countries.

United States military support of South Korea has continued since 1950. Soviet support for North Korea ceased with the fall of the Soviet Union and the end of the Cold War in 1991. Today, North Korea remains isolated from the global community. Its main supporter is China.

Kim Il Sung became the first premier under the North Korean Communist **regime** in 1948. His son succeeded him, and his grandson became premier in 2011. It is a family dynasty that rules the north. South Korea has become a democratic government. The first president of South Korea, Syngman Rhee, was elected in 1948. Strict governmental control, a military **coup**, and meeting economic over social needs continued until 1993, when civilian **authority** was restored. In 2012 South Korea elected a woman as its president for the first time.

Korean Governments Today

The Korean Workers' Party controls North Korea's elections and provides lists of approved candidates—usually only one for each position. This legislative body meets for only one or two weeks a year. Most decisions are made by the 15-member Presidium of the Supreme People's Assembly (SPA).

principle a rule or code of conduct

isolation state of being set or kept apart from others

regime a form of government

coup an overthrow of the government

authority the power to influence or command thought, opinion, or behavior

North Korean soldiers celebrate their country's successful nuclear bomb test on February 13, 2013. Experts believe that North Korea's nuclear operations are being carefully planned to avoid detection.

▼ **CRITICAL THINKING**

1. *Interpreting* What may underground nuclear testing reveal about North Korea's intentions?

2. *Making Inferences* Why might North Korea want to hide its nuclear development?

선군혁명총진군

제3차 지하핵시험성공을 열렬히 축하한다!

While North Korea and South Korea share similar cultural traits, there are also major differences between them. Government-provided medical care is free in North Korea. There is at least one clinic in every village. However, there are shortages of physicians and equipment. Most people in South Korea have medical insurance. There, the basic health care needs of the citizens are generally met. The life expectancy at birth for North Korean males was 66 years in 2015; for females it was 74 years. In South Korea, life expectancy at birth for males in 2015 was 77 years; for females it was 83 years.

Both countries value education. In North Korea, students receive a free and compulsory education for 11 years, starting with a year of preschool. Some older students are required to work while they are in school. The emphasis of all education is on science and technology. Institutions of higher education offer an additional two to six years of education. The most important institution of higher education is Kim Il-Sung University in P'yŏngyang.

Nine years of primary and middle school are compulsory for South Koreans. Nearly all who graduate from middle school go to high school or technical school. About 80 percent of high school graduates continue their education at a college or university. High school students undergo strenuous preparation for the highly competitive entrance exams to top universities. Most of the prestigious universities are in Seoul. In the twenty-first century, more and more South Koreans have pursued higher education abroad, with many studying in the United States.

Family and the Status of Women

After World War II, the occupying Soviets destroyed the lineage records of North Koreans. Since awareness of ancestry is an important component of Confucian practice, the fabric of Korean tradition was destroyed in that country. People are encouraged to spend their leisure time in activities that support the communist state. An extensive internal security apparatus monitors people's activities. The Korean Central News Agency censors newspapers. Government controls radio and television broadcasts. Internet is restricted to a few people, although illegal cell phone and Internet connections are more common in recent years.

Confucian traditions still affect family life in South Korea, and ancestral worship is central. People regard the most recent generation of ancestors as very much part of their family celebrations and rituals. South Koreans celebrate the first 100 days of a baby's life, marriages, and 61st birthdays with special family gatherings to honor their ancestors.

The roles of women in South Korea have changed as the extended nuclear family began to decline during rapid urbanization in the late twentieth century. Increasingly, women have become active in South Korea's economy. This has led to a decrease in family size and a lower birthrate. Women have gained equal rights in property ownership. In 2012 the first female president was elected in South Korea. Women in North Korea do not enjoy the same kinds of equality as those in South Korea. Many have entered the workforce out of economic necessity. Increased participation in the economy has given women more of a voice in decision making inside and outside the family.

The Arts

Cultural distinctions are noticeable in the arts. North Korea's Communist government heavily influences the country's art. In South Korea, Western culture has had a strong influence. Writers, artists, and dancers in North Korea are assigned to work for state-run theaters, orchestras, and other institutions. Their role is to promote traditional arts, commemorate the revolution, and express the superiority of Korean culture. The government maintains museums and archaeological sites to support its communist goals.

These North Korean women are performing a traditional fan dance.

◀ CRITICAL THINKING

1. *Analyzing* Why do you suppose the North Korean government supports traditional dance?

2. *Comparing and Contrasting* How are the goals of art similar and different in the two Koreas?

In South Korea, dancers, singers, and musicians preserve traditions. Dances with masked dancers and folk songs are still performed. Symphony orchestras, modern theater, rock concerts, and art galleries are also enjoyed by South Koreans.

☑ READING PROGRESS CHECK

Evaluating What are advantages of North Korea's health care system?

Economic Activities

GUIDING QUESTION *What industries have grown in importance in North Korea and South Korea?*

North Korea has a command economy in which the state plans and controls all economic production. Today, the economy's main industries produce metal products, machinery, military equipment, and chemicals. In its initial years, the focus was on development of heavy industry. Later it worked to improve technology and infrastructure. It was not until the 1970s and 1980s that North Korea's government began to pay attention to the production of agricultural products and consumer goods.

In the 1990s, a series of natural disasters occurred. These included floods and drought, which contributed to the problems of widespread starvation and malnutrition originally caused by government mismanagement. Subsequently, the North Korean government's highest priority has been to solve what it calls its "food problem."

Since the 1960s, South Korea has transformed itself from one of the world's poorest states to a highly industrialized society. Government and business

Mark Ralston/AFP/Getty Images

These workers at a South Korean aircraft factory are building a helicopter. The same factory hopes to start constructing aircraft for the U.S. Air Force.

▲ CRITICAL THINKING

1. *Making Connections* Why are exports so important to South Korea's economy?

2. *Drawing Conclusions* Why did North Korea's goal of economic self-reliance fail?

leaders developed a strategy focused on exports. They targeted specific industries for development, beginning with textiles and light manufacturing, and then moving to heavy industries such as iron, steel, chemicals, ship building, and automobiles. Manufacturing occurs in the Seoul region. Heavy industry is located in the south with access to seaports. High-tech industries, such as aerospace, electronics, and information technology, have thrived. Industries with global consumers, including smart phones, computers, and tablets, are a specialty.

Since the 1950s, the government of North Korea has stressed self-reliance as a goal in its economic planning. Nonetheless, it had trade relations with the Soviet Union and its Eastern European satellites, as well as with China. The severe economic hardships of the 1990s forced North Korea to accept foreign aid, seek foreign investment, and expand its range of trade partners. Now North Korea exports live animals and agricultural products, textiles and apparel, machinery, and mineral fuels and lubricants. It imports beverages, food and other agricultural products, mineral fuels, machinery, and textiles. Its major trading partners include China, South Korea, Russia, Japan, and Thailand.

Exports have been important to South Korea's economic growth. Many industries produce goods such as machinery, electronics, textiles, automobiles, and footwear for export around the world. South Korea's trade partners include the United States, Japan, the European Union, and the countries of Southeast Asia. South Korea imports raw materials, such as textile fibers, metal ores, and mineral fuels. The raw materials are converted into high-value exports known for their high quality.

✓ READING PROGRESS CHECK

Identifying Which country is a major trading partner with both countries?

LESSON 2 REVIEW

Reviewing Vocabulary

1. *Understanding Relationships* Write a paragraph discussing the reasons for the cultural divergence of North Korea and South Korea.

Using Your Notes

2. *Listing* Use your graphic organizer on the human geography of North Korea and South Korea to describe how physical and human systems have shaped these two countries.

Answering the Guiding Questions

3. *Making Connections* How has the decision to divide the Korean Peninsula into North Korea and South Korea affected the development of each country?

4. *Contrasting* How do settlement patterns and density differ in the two countries?

5. *Comparing and Contrasting* What are the major influences on the culture and art in each country?

6. *Identifying* What industries have grown in importance in North Korea and South Korea?

Writing Activity

7. *Informative/Explanatory* In a paragraph, explain how the Kaesong facility might have strengthened the relationship between the two Koreas had it continued to operate.

networks

There's More Online!

☑ **CHART** South Korean Exports

☑ **IMAGE** Drift-Net Fishing

☑ **INFOGRAPHIC** Facing Empty Oceans

☑ **INTERACTIVE SELF-CHECK QUIZ**

☑ **VIDEO** People and Their Environment: North Korea and South Korea

LESSON 3
People and Their Environment: North Korea and South Korea

ESSENTIAL QUESTION • *How do physical systems and human systems shape a place?*

Reading HELPDESK

Academic Vocabulary
- **impact**
- **affect**

Content Vocabulary
- **deforestation**
- **drift-net fishing**
- **fertilizer**

TAKING NOTES: *Key Ideas and Details*

IDENTIFYING CAUSE AND EFFECT
As you read about the human geography of the Korean Peninsula, use a graphic organizer like the one below to identify causes of environmental problems.

Managing Resources	
Feature	Damage
Air	
Water	
Forests	
Fish	

IT MATTERS BECAUSE
The environments of North Korea and South Korea are affected by the air quality of other countries. The actions of North Koreans and South Koreans, in turn, affect their own air and water, as well the air and water of other countries.

Managing Resources

GUIDING QUESTION *What resources are at risk on the Korean Peninsula?*

Air pollution is a serious problem, especially in South Korea's cities. Because 80 percent of the population lives in cities, this problem has an **impact** on millions of people. Dust and pollutants that are carried by the winds from China and Mongolia are a big source of the problem. Pollutants from automobiles and industries also contribute to the poor air quality. A 2012 study found that South Korea had more pollutants in its air than either Japan or the United States.

South Korea's large cities also have problems with sewage and sanitation. Sanitation systems built after the Korean War are failing because of the impact of the country's rapid urbanization. Sanitation companies are unable to keep up with and properly dispose of the increasing amount of waste materials. Instead, they pump untreated sewage directly into the Pacific Ocean. Sewer systems also overflow and pollute local rivers and streams. Pollution of the seas and other bodies of water has **affected** fish supplies.

Population density has also had a negative effect on garbage disposal and sanitation systems in Seoul and other South Korean cities. These cities have many rules for garbage disposal and recycling, but they are not enforced or followed. On some streets, uncollected garbage is left in stacks.

North Korea also continues to experience its own problems with water and air pollution. A 2004 report found that almost 30,000 cubic meters, or about 7,925 gallons, of industrial wastewater was being poured every day into the Taedong River. This river flows through the middle of

impact a significant or major effect

affect to produce an effect upon

North Korea's most densely populated city and its capital, P'yŏngyang. In North Korea, too, coal is used to heat most homes and provides power for most industries. Coal burning is a major cause of air pollution and acid rain.

☑ **READING PROGRESS CHECK**

Identifying What are two effects of pollution on natural resources in the Koreas?

Human Impact

GUIDING QUESTION *What human activities have affected the physical environment of North Korea and South Korea?*

deforestation the loss or destruction of forests, mainly from logging or farming

Deforestation has affected the entire Korean Peninsula. Forests once covered about two-thirds of South Korea and most of the western lowlands of North Korea. These forests are now nearly gone due to deforestation. Trees are cut for fuel for heating and cooking. Additionally, much land is also cleared

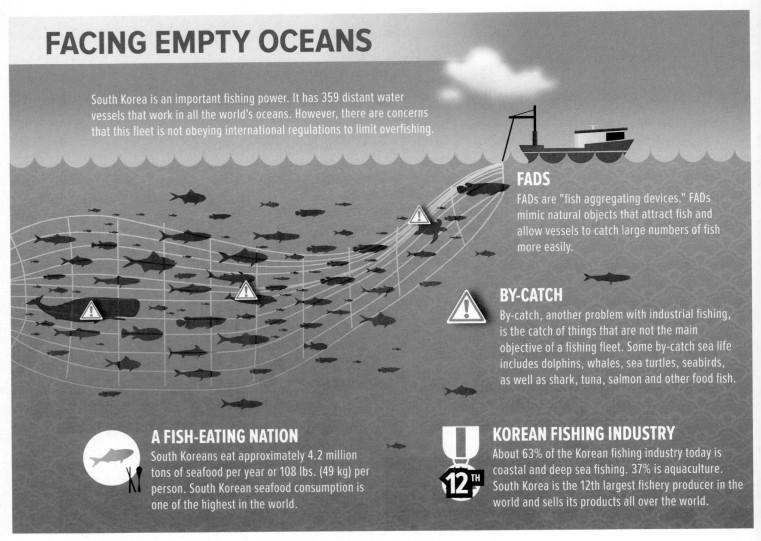

FACING EMPTY OCEANS

South Korea is an important fishing power. It has 359 distant water vessels that work in all the world's oceans. However, there are concerns that this fleet is not obeying international regulations to limit overfishing.

FADS
FADs are "fish aggregating devices." FADs mimic natural objects that attract fish and allow vessels to catch large numbers of fish more easily.

BY-CATCH
By-catch, another problem with industrial fishing, is the catch of things that are not the main objective of a fishing fleet. Some by-catch sea life includes dolphins, whales, sea turtles, seabirds, as well as shark, tuna, salmon and other food fish.

A FISH-EATING NATION
South Koreans eat approximately 4.2 million tons of seafood per year or 108 lbs. (49 kg) per person. South Korean seafood consumption is one of the highest in the world.

KOREAN FISHING INDUSTRY
About 63% of the Korean fishing industry today is coastal and deep sea fishing. 37% is aquaculture. South Korea is the 12th largest fishery producer in the world and sells its products all over the world.

Techniques commonly used by the South Korean fishing industry threaten the sustainability of fish populations in the region.

▲ **CRITICAL THINKING**

1. *Making Predictions* What could happen to South Korea's fishing industry if fishing practices are not changed?

2. *Constructing a Thesis* Do you think fish aggregating devices and drift-net fishing should be banned? Why or why not?

for cultivation and settlement as the population density increases in forested areas. The resulting loss of habitat has caused a decrease in the populations of many animals that once lived on the peninsula.

Deforestation has several terrible effects. For example, deforestation has led to soil erosion in North Korea. Trees and bushes hold soil in place during the heavy seasonal rains. Without them, soil washes down the steep hillsides of the mountainous terrain. This depletes nutrients that keep the soil healthy and make the land fertile for raising crops. Soil loss makes growing food more of a challenge for a country with ongoing food shortages.

Deforestation also leads to flooding. Aside from the direct loss of lives and property that floods can cause, flooding can also pollute drinking water. This occurs when a buildup of mud and debris is washed into the water supply. In 2012 severe floods polluted drinking water supplies in parts of North Korea. As a result, an estimated 50,000 families were left without potable, or drinkable, water.

The isolation and secrecy of the North Korean government make it difficult to monitor the environmental problems facing the country. A 2004 report by the United Nations Environment Programme makes it clear that the country faces multiple environmental problems. Data from 2012 indicate that the average temperature in North Korea had increased by almost 2°C as a result of deforestation. During this period, the world average temperature increase was only about 0.7°C.

While Japan has the world's largest fleet of drift-net fishing vessels, South Korea and Taiwan also have sizable drift-net operations. Environmentalists, the United States Coast Guard, and the Canadian Department of Fisheries and Wildlife are concerned about **drift-net fishing**, or fishing with huge nets. Drift-net fishing is condemned by many countries because it is so destructive. Sometimes called "walls of death," the huge nets are suspended vertically in the ocean and allowed to drift with the currents. Some of these nets are as much as 37 miles (60 km) long.

They are intended to trap squid and fish. However, sea mammals, such as dolphins, porpoises and whales, also die in these nets. So do turtles and birds. Animals like these die in such numbers that the nets are considered a threat to the marine ecosystem. It is estimated that about 80 percent of the animals caught in the nets are not even their intended targets.

In 1992 the United Nations created a resolution to end drift-net fishing. At the same time, the Convention for the Conservation of Anadromous Stocks (fish that migrate from salt water to freshwater) in the North Pacific Ocean was formed. The governments of the United States, Canada, Russia, and Japan signed this convention. South Korea agreed to it in 2003. However, there are few resources to monitor compliance with this agreement, and it is not clear that such fishing has declined.

✔ READING PROGRESS CHECK

Hypothesizing What might be the long-term effect of drift-net fishing on the South Korean fishing industry if it is not controlled?

Although unintended, drift-net fishing kills sea life. Several countries continue to practice drift-net fishing even though it violates a United Nations resolution.

▲ CRITICAL THINKING

1. *Drawing Conclusions* Do you think South Korea will really stop drift-net fishing while other countries still practice it? Why or why not?

2. *Identifying Cause and Effect* Why is drift-net fishing so threatening to the marine ecosystem?

drift-net fishing the use of fishing nets of great length and depth

Addressing the Issues

GUIDING QUESTION *How are environmental issues being addressed in North Korea and South Korea?*

In 2012 North Korea invited scientists from eight countries to a conference in P'yŏngyang to review the condition of the country's environment. Scientists found barren landscapes on the brink of collapse, a result of decades of environmental degradation. They developed plans for restoring the environment and increasing food security in North Korea.

Deforestation has reached alarming levels in North Korea. According to one study, North Korea was the third most deforested country in the world, after Nigeria and Indonesia. However, North Korea has also started a series of reforestation programs. Such programs are planned for Kandong province, located outside P'yŏngyang. Other programs intend to reforest the western lower hillsides and eastern highlands.

South Korea, which has succeeded in restoring forests to nearly two-thirds of the country, is also attempting to promote reforestation in North Korea. South Korea's former president, Lee Myung-bak, pledged to send seedlings to North Korea. Many private South Korean citizens have donated trees to North Korea as well. Increasing forest coverage is a good step, but these young forests are very different from the old-growth forests that were cut down. They are much less diverse and offer minimal habitat for animals.

The North Korean government has also encouraged the use of **fertilizer** to restore soil nutrients to help the trees grow back healthy. The government has provided fertilizer to farmers. However, there is evidence that, rather than use the fertilizer, farmers are selling it to cover their costs.

South Korea is also working on increasing its energy efficiency. In the twenty-first century, the government announced a plan to change every light-bulb in every public building to energy-efficient LED lights.

An ambitious stimulus package launched in 2009 allocated money to research low-carbon technologies, build high-speed railways, and expand bus lanes. The effort also supports building one million "green" homes and improving the efficiency of one million existing homes. While South Korea has yet to solve all of its environmental problems, the country is making substantial progress.

fertilizer a substance, such as manure or a chemical mixture, used to make soil more fertile

☑ **READING PROGRESS CHECK**

Predicting How might an effective reforestation program improve life for North Koreans?

LESSON 3 REVIEW

Reviewing Vocabulary

1. ***Explaining*** Write a paragraph explaining the impact of deforestation on North Korea's agriculture.

Using Your Notes

2. ***Identifying Cause and Effect*** Use your graphic organizer on the effects of the Korean Peninsula's environmental problems to write a paragraph about the cause and effect of one type of pollution on the peninsula.

Answering the Guiding Questions

3. ***Evaluating*** What resources are at risk on the Korean Peninsula?

4. ***Analyzing*** What human activities have affected the physical environment of North Korea and South Korea?

5. ***Summarizing*** How are environmental issues being addressed in North Korea and South Korea?

Writing Activity

6. ***Informative/Explanatory*** Use reliable Internet sources to locate an article about environmental issues in North Korea or South Korea over the past year. Write a paragraph summarizing the main points of the article.

Directions: On a separate sheet of paper, answer the questions below. Make sure you read carefully and answer all parts of the questions.

Lesson Review

Lesson 1

❶ *Comparing* Why do you think the capitals of North Korea and South Korea began as river ports?

❷ *Explaining* How did the creation of the Demilitarized Zone affect wildlife in the area?

❸ *Drawing Conclusions* Why does mountainous terrain make livestock production and fishing particularly important on the Korean Peninsula?

Lesson 2

❹ *Summarizing* How has North Korea's population changed since the middle of the twentieth century?

❺ *Contrasting* Describe how life expectancy in South Korea differs from life expectancy in North Korea.

❻ *Describing* What are some methods the North Korean government uses to control free speech and limit individual freedoms?

Lesson 3

❼ *Explaining* How does deforestation contribute to food shortages in North Korea?

❽ *Identifying Cause and Effect* What are some problems caused by population density in South Korea's cities?

❾ *Analyzing* Why do you think South Korea has taken steps to address the problem of deforestation in North Korea?

Critical Thinking

❿ *Making Predictions* How might reunification on the Korean Peninsula affect the economy of South Korea?

⓫ *Hypothesizing* South Korea has taken steps to improve energy efficiency. How might these steps help improve air quality in urban areas?

21st Century Skills

Use the graph to answer the following questions.

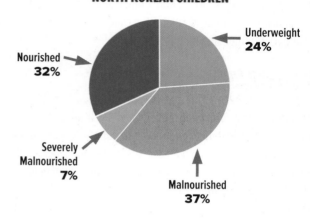

NORTH KOREAN CHILDREN

Underweight **24%**

Nourished **32%**

Severely Malnourished **7%**

Malnourished **37%**

Source: http://han-schneider.org

⓬ *Creating and Using Graphs, Charts, Diagrams, and Tables* According to the circle graph, what percentage of children in North Korea faces problems with nourishment?

⓭ *Creating and Using Graphs, Charts, Diagrams, and Tables* Why do you think the graph includes the percentage of children who are underweight?

⓮ *Identifying Perspectives and Differing Interpretations* Based on the graph, how severe do you think the problem of malnutrition is among North Korean children?

College and Career Readiness

⓯ *Problem Solving* Imagine you have been hired by the United Nations World Food Program to find solutions to the problem of food shortages in North Korea. Use the Internet to research the causes and consequences of the problem. How have natural disasters affected the availability of food in that country? What role has the government played? Write a paragraph outlining steps the North Korean government can take to solve the problem.

Need Extra Help?

If You've Missed Question	❶	❷	❸	❹	❺	❻	❼	❽	❾	❿	⓫	⓬	⓭	⓮	⓯
Go to page	728	729	729	733	734	734	739	737	740	735	740	734	734	734	734

Directions: On a separate sheet of paper, answer the questions below. Make sure you read carefully and answer all parts of the questions.

DBQ Analyzing Primary Sources

Use the document to answer the following questions.

People who have fled North Korea report that the country's economy relies on forced labor. Even children are forced to work, as a former student describes below.

PRIMARY SOURCE

"When I was between 11 and 15 years old I had to work on the government farm almost every day. . . . We finished class at 1 P.M. and had to rush back home to eat lunch because the school didn't provide food for the students. The school would announce that we'd have to meet back at the school field and bring our own farm tools. . . . They forced everyone, even the small children, to work. I felt bad because this didn't benefit our family and I had many responsibilities to do for my family but the government forced me to work for them. I was always very exhausted as a child."

—Human Rights Watch, "North Korea: Economic System Built on Forced Labor," June 13, 2012

16 *Drawing Conclusions* What does the use of children as forced labor imply about the North Korean economy?

17 *Interpreting* The students were told to bring their own tools from home. How does this detail help you understand the problem?

18 *Analyzing* What features of the farm in the description show that it is a cooperative?

Applying Map Skills

Use the Unit 8 Atlas to answer the following questions.

19 *Human Systems* Compare population density in North and South Korea. Which country is more urbanized?

20 *Places and Regions* Use your mental map of East Asia to identify the major rivers in North Korea and South Korea. Which river forms the border between North Korea and China?

21 *Environment and Society* How has the geography of the Korean Peninsula affected agricultural activity?

Exploring the Essential Question

22 *Analyzing* How has the division of Korea into two countries affected human systems on the Korean Peninsula? Write a paragraph describing the changes, including population density, use of resources, and culture.

Research and Presentation

23 *Research Skills* Using the Internet, research the development of the North Korean nuclear program. How have other countries responded to these nuclear tests? Create a time line showing your findings and share it with the class.

Writing About Geography

24 *Narrative* Use the Internet to research and write a two-page essay describing the Demilitarized Zone. How does the DMZ create a buffer between North and South Korea? What features of the DMZ make it an important nature preserve?

Need Extra Help?

If You've Missed Question	16	17	18	19	20	21	22	23	24
Go to page	733	733	729	676	672	672	730	731	729

Southeast Asia and the Pacific World

Alex Linghorn/Getty Images

① **Diversity** Southeast Asia and the Pacific World is a diverse region. Some ethnic groups have maintained their traditions while other cultures have blended together.

EXPLORE the REGION

Stretching from the eastern border of India east nearly to South America, and from China in the north to the Southern Ocean, **SOUTHEAST ASIA** and the **PACIFIC WORLD** is a vast region dominated by the sea. With populous countries like Indonesia, countries with large land areas like Australia, and small island countries with few people, this region shows great variety of landforms, people, and cultures.

THERE'S MORE ⌐ONLINE⌐

travelstock44/age fotostock

③ Environment The region has dense rain forests and unique plants and animals. Various countries are taking steps to protect these natural wonders from the threats they face.

④ Water Almost all of the countries in the region have coastlines, and some countries rely on rivers. Fishing and shipping are important activities in the region.

② Landforms The region has many different landforms, from soaring, glacier-covered mountains to low islands that rise just a few feet above the sea.

(t)John White Photos/Getty Images,
(bl)Nora Carol Photography/wwGetty Images,
(br)Jonathan Saruk/Getty Images News/Getty Images

Southeast Asia and the Pacific World
Physical

N
W · E
S

Hkakabo Razi **EAST**
19,295 ft. **ASIA**
(5,881 m)

TROPIC OF CANCER

20°N

Bay of
Bengal

Gulf of
Tonkin

**South
China
Sea**

Luzon Strait

Luzon

Wake
Island

Mariana
Islands

Guam

MICRONESIA

Marshall
Islands

For eastern
Oceania,
see inset
below

Andaman
Sea

Gulf of
Thailand

Mt. Pinatubo
5,770 ft.
(1,759 m)

Philippine
Sea

Sulu
Sea

Mindanao

Palau

Caroline
Islands

Celebes
Sea

EQUATOR

Sumatra

Borneo

Sulawesi

Moluccas

Bismarck
Archipelago

New
Guinea

Nauru

Gilbert
Islands

MELANESIA

Tuvalu

Sunda Strait

Greater Sunda Islands

Java Sea

Krakatau

Bali

Flores
Sea

Banda
Sea

Puncak Jaya
16,535 ft.
(5,040 m)

Mt. Wilhelm
14,762 ft.
(4,500 m)

Solomon
Islands

Wallis and
Futuna

Java

Lesser Sunda Islands

Timor

Timor
Sea

Arafura
Sea

Cape York
Peninsula

Coral Sea

Vanuatu

Fiji

Gulf of
Carpentaria

New
Caledonia

TROPIC OF CAPRICORN

Tonga

Elevations

10,000 ft. (3,000 m)
5,000 ft. (1,500 m)
2,000 ft. (600 m)
1,000 ft. (300 m)
0 ft. (0 m)
Below sea level

Great Sandy
Desert

WESTERN
PLATEAU

Macdonnell
Ranges

Central

GREAT DIVIDING RANGE

Great Barrier Reef

—— National boundary
▲ Mountain peak
▼ Lowest point

Gibson Desert

GREAT
ARTESIAN
BASIN

Great Victoria
Desert

Lake Eyre
(dry) ▼
-52 ft.
(-16 m)

Darling R.

Lowlands

PACIFIC OCEAN

Nullarbor Plain

Great
Australian
Bight

Murray R.

0   1,000 miles
0   1,000 kilometers

Mercator projection

100°E

Mt. Kosciuszko
7,310 ft.
(2,228 m)

Bass Strait

Tasman
Sea

Cook
Strait

North
Island

South
Island

Southern Alps

Chatham
Islands

40°S

Tasmania

Mt. Cook
12,349 ft.
(3,764 m)

120°E   140°E   160°E   180°

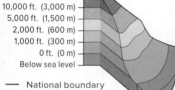

MAP STUDY

1. *Physical Systems* What large islands make up the country of Indonesia?

2. *Environment and Society* Where are most major cities in Australia located? What about the physical geography of the country explains these settlement patterns?

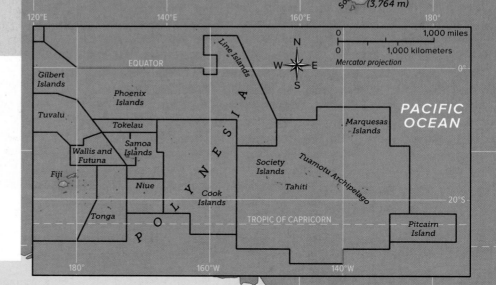

EQUATOR

Line Islands

0   1,000 miles
0   1,000 kilometers
Mercator projection
0°

Gilbert
Islands

Phoenix
Islands

N
W · E
S

Tuvalu

Tokelau

Samoa
Islands

Wallis and
Futuna

Fiji

Niue

POLYNESIA

Cook
Islands

Society
Islands

Tahiti

Marquesas
Islands

Tuamotu Archipelago

PACIFIC
OCEAN

20°S

Tonga

TROPIC OF CAPRICORN

Pitcairn
Island

180°   160°W   140°W

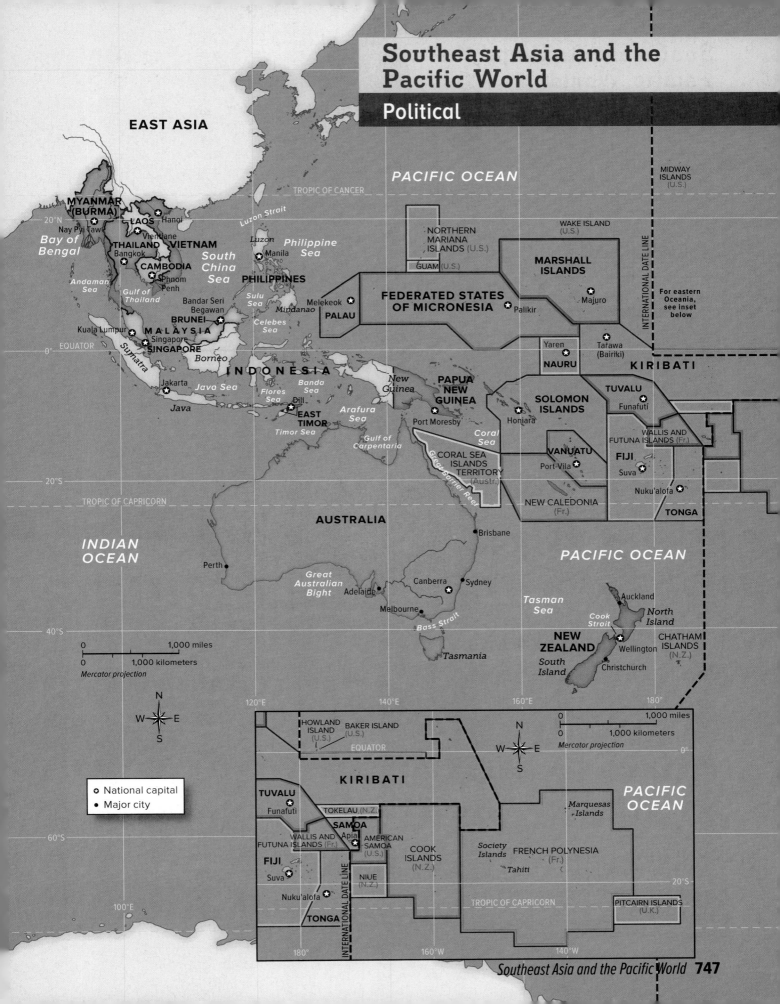

Southeast Asia and the Pacific World

Political

EAST ASIA

PACIFIC OCEAN

TROPIC OF CANCER

MIDWAY ISLANDS (U.S.)

WAKE ISLAND (U.S.)

INTERNATIONAL DATE LINE

For eastern Oceania, see inset below

MYANMAR (BURMA)
Nay Pyi Taw

LAOS
Hanoi

Vientiane

THAILAND
Bangkok

VIETNAM

Bay of Bengal

20°N

Luzon Strait

Luzon

Philippine Sea

Manila

South China Sea

CAMBODIA
Phnom Penh

PHILIPPINES

Andaman Sea

Gulf of Thailand

Bandar Seri Begawan

Mindanao

Sulu Sea

Celebes Sea

Melekeok

PALAU

NORTHERN MARIANA ISLANDS (U.S.)

GUAM (U.S.)

FEDERATED STATES OF MICRONESIA

Palikir

MARSHALL ISLANDS

Majuro

BRUNEI

Kuala Lumpur

Singapore

MALAYSIA

SINGAPORE

Sumatra

EQUATOR 0°

Borneo

Yaren

NAURU

Tarawa (Bairiki)

KIRIBATI

Jakarta

INDONESIA

Java Sea

Banda Sea

New Guinea

PAPUA NEW GUINEA

SOLOMON ISLANDS

Honiara

TUVALU

Funafuti

Java

Flores Sea

Dili

EAST TIMOR

Arafura Sea

Port Moresby

Coral Sea

WALLIS AND FUTUNA ISLANDS (Fr.)

Timor Sea

Gulf of Carpentaria

Great Barrier Reef

CORAL SEA ISLANDS TERRITORY (Austr.)

VANUATU
Port-Vila

FIJI
Suva

20°S

TROPIC OF CAPRICORN

AUSTRALIA

NEW CALEDONIA (Fr.)

Nuku'alofa

TONGA

Brisbane

INDIAN OCEAN

PACIFIC OCEAN

Perth

Great Australian Bight

Adelaide

Canberra

Sydney

Tasman Sea

Auckland

North Island

Melbourne

Bass Strait

Cook Strait

Wellington

CHATHAM ISLANDS (N.Z.)

40°S

Tasmania

NEW ZEALAND

South Island

Christchurch

0 — 1,000 miles

0 — 1,000 kilometers

Mercator projection

N W E S

● National capital
● Major city

120°E   140°E   160°E   180°

HOWLAND ISLAND (U.S.)

BAKER ISLAND (U.S.)

EQUATOR

N W E S

0 — 1,000 miles

0 — 1,000 kilometers

Mercator projection

0°

PACIFIC OCEAN

KIRIBATI

TUVALU
Funafuti

TOKELAU (N.Z.)

SAMOA
Apia

Marquesas Islands

WALLIS AND FUTUNA ISLANDS (Fr.)

AMERICAN SAMOA (U.S.)

COOK ISLANDS (N.Z.)

Society Islands

FRENCH POLYNESIA (Fr.)

FIJI
Suva

NIUE (N.Z.)

Tahiti

20°S

Nuku'alofa

TONGA

TROPIC OF CAPRICORN

PITCAIRN ISLANDS (U.K.)

60°S

100°E

180°   160°W   140°W

INTERNATIONAL DATE LINE

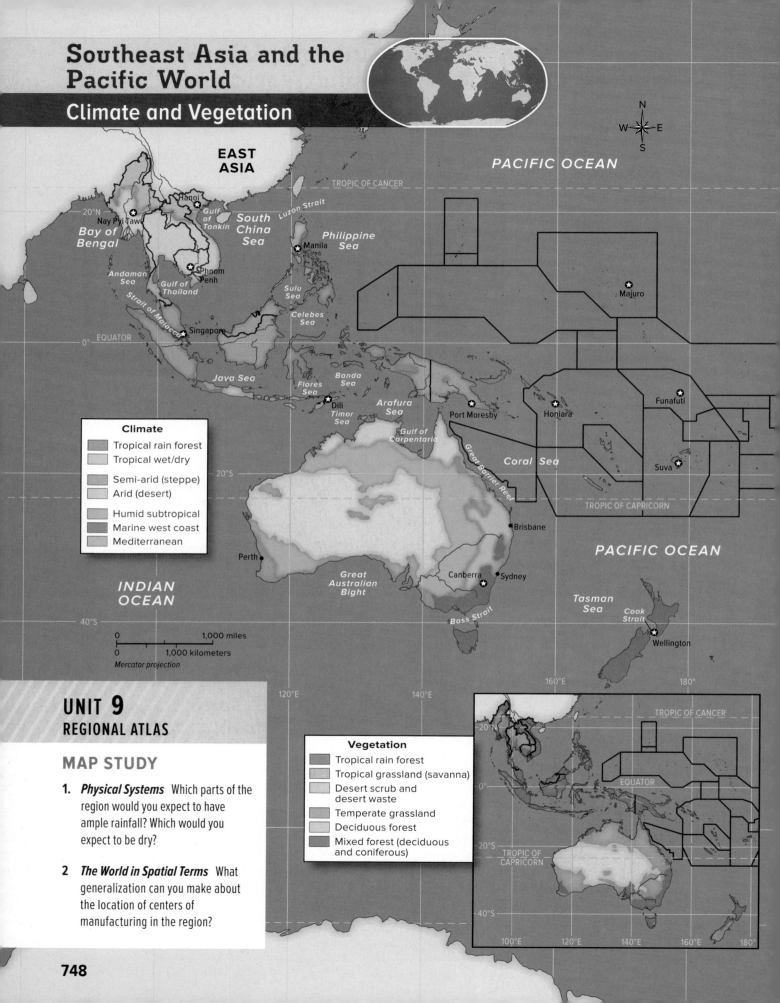

Southeast Asia and the Pacific World
Climate and Vegetation

EAST ASIA

PACIFIC OCEAN

TROPIC OF CANCER

20°N

Hanoi

Nay Pyi Taw

Gulf of Tonkin

South China Sea

Luzon Strait

Bay of Bengal

Andaman Sea

Gulf of Thailand

Phnom Penh

Manila

Philippine Sea

Majuro

Strait of Malacca

Sulu Sea

Singapore

EQUATOR

0°

Celebes Sea

Java Sea

Flores Sea

Banda Sea

Dili

Timor Sea

Arafura Sea

Port Moresby

Honiara

Funafuti

Gulf of Carpentaria

Coral Sea

Suva

Great Barrier Reef

Climate
- Tropical rain forest
- Tropical wet/dry
- Semi-arid (steppe)
- Arid (desert)
- Humid subtropical
- Marine west coast
- Mediterranean

20°S

TROPIC OF CAPRICORN

Brisbane

PACIFIC OCEAN

Perth

Great Australian Bight

Canberra   Sydney

Tasman Sea

Cook Strait

INDIAN OCEAN

40°S

Bass Strait

Wellington

0   1,000 miles

0   1,000 kilometers

Mercator projection

120°E

140°E

160°E

180°

MAP STUDY

1. **Physical Systems** Which parts of the region would you expect to have ample rainfall? Which would you expect to be dry?

2. **The World in Spatial Terms** What generalization can you make about the location of centers of manufacturing in the region?

Vegetation
- Tropical rain forest
- Tropical grassland (savanna)
- Desert scrub and desert waste
- Temperate grassland
- Deciduous forest
- Mixed forest (deciduous and coniferous)

TROPIC OF CANCER

20°N

EQUATOR

0°

20°S

TROPIC OF CAPRICORN

40°S

100°E   120°E   140°E   160°E   180°

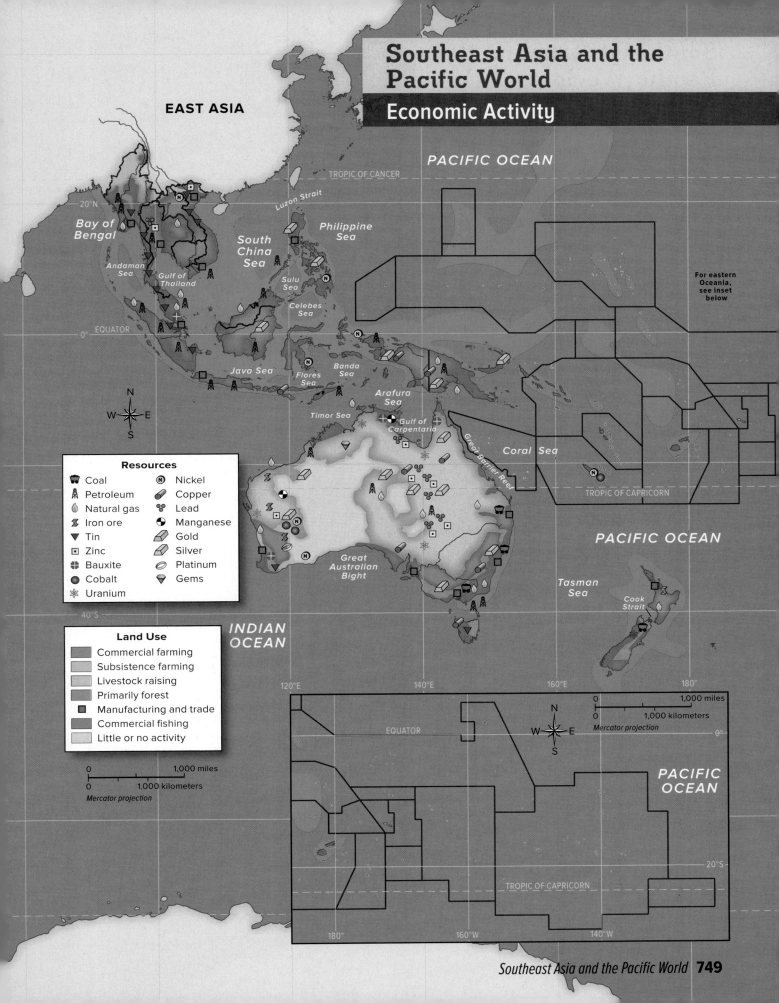

Southeast Asia and the Pacific World

Economic Activity

EAST ASIA

PACIFIC OCEAN

TROPIC OF CANCER

Luzon Strait

20°N

Bay of Bengal

South China Sea

Philippine Sea

Andaman Sea

Gulf of Thailand

Sulu Sea

Celebes Sea

EQUATOR

0°

Java Sea

Flores Sea

Banda Sea

Timor Sea

Arafura Sea

Gulf of Carpentaria

For eastern Oceania, see inset below

Coral Sea

TROPIC OF CAPRICORN

Great Barrier Reef

PACIFIC OCEAN

Great Australian Bight

Tasman Sea

Cook Strait

INDIAN OCEAN

40°S

Resources

- Coal
- Ⓝ Nickel
- Petroleum
- Copper
- Natural gas
- Lead
- Iron ore
- Manganese
- Tin
- Gold
- Zinc
- Silver
- Bauxite
- Platinum
- Cobalt
- Gems
- Uranium

Land Use

- Commercial farming
- Subsistence farming
- Livestock raising
- Primarily forest
- Manufacturing and trade
- Commercial fishing
- Little or no activity

0   1,000 miles
0   1,000 kilometers
Mercator projection

120°E   140°E   160°E   180°

0   1,000 miles
0   1,000 kilometers
Mercator projection

EQUATOR

PACIFIC OCEAN

20°S

TROPIC OF CAPRICORN

180°   160°W   140°W

Southeast Asia and the Pacific World

Population Density

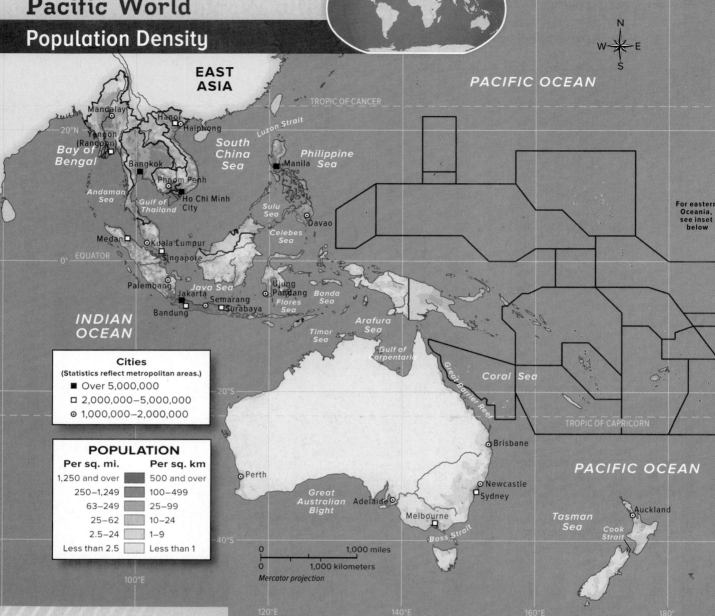

EAST ASIA

PACIFIC OCEAN

TROPIC OF CANCER

20°N

Mandalay

Hanoi
Haiphong

Yangon
(Rangoon)

Bay of
Bengal

Luzon Strait

South
China
Sea

Philippine
Sea

Manila

Bangkok

Phnom Penh

Andaman
Sea

Gulf of
Thailand

Ho Chi Minh
City

Sulu
Sea

Davao

Medan

Celebes
Sea

EQUATOR

0°

Kuala Lumpur

Singapore

Palembang

Java Sea

Jakarta

Semarang

Surabaya

Bandung

Ujung
Pandang

Flores
Sea

Banda
Sea

INDIAN
OCEAN

Timor
Sea

Arafura
Sea

Gulf of
Carpentaria

For eastern
Oceania,
see inset
below

Great Barrier Reef

Coral Sea

TROPIC OF CAPRICORN

Brisbane

PACIFIC OCEAN

Perth

20°S

Great
Australian
Bight

Adelaide

Newcastle
Sydney

Melbourne

Bass Strait

Tasman
Sea

Auckland

Cook
Strait

Cities
(Statistics reflect metropolitan areas.)

- ■ Over 5,000,000
- □ 2,000,000–5,000,000
- ◉ 1,000,000–2,000,000

POPULATION

Per sq. mi.		Per sq. km
1,250 and over		500 and over
250–1,249		100–499
63–249		25–99
25–62		10–24
2.5–24		1–9
Less than 2.5		Less than 1

0   1,000 miles
0   1,000 kilometers
Mercator projection

100°E

120°E   140°E   160°E   180°

40°S

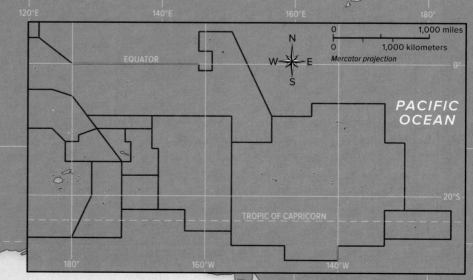

0   1,000 miles
0   1,000 kilometers
Mercator projection

EQUATOR

PACIFIC
OCEAN

20°S

TROPIC OF CAPRICORN

180°   160°W   140°W

UNIT 9
REGIONAL ATLAS

MAP STUDY

1. *Human Systems* Which countries in Southeast Asia are the most densely populated?

2. *Human Systems* Why do you think the heaviest concentration of population in Australia is along the eastern coast?

Southeast Asia

ESSENTIAL QUESTION · *How do physical systems and human systems shape a place?*

Why Geography Matters
Emerging Markets in Southeast Asia

Lesson 1
Physical Geography of Southeast Asia

Lesson 2
Human Geography of Southeast Asia

Lesson 3
People and Their Environment: Southeast Asia

Geography Matters...

Southeast Asia is one of the most diverse regions on Earth. All the world's religions, political systems, and economic systems can be found here—and so can hundreds of ethnic groups speaking hundreds of different languages. The physical geography is just as diverse, with towering volcanoes, broad river deltas, and a wide variety of tropical biomes and climates. The region includes a mainland area and two large island archipelagoes.

This young woman in Bali, Indonesia, wears an intricate gold headdress.

©Dallas and John Heaton/Free Agents Limited/Corbis

emerging markets *in* Southeast Asia

A country that is moving toward a more complex economy as the result of rapid industrialization, economic growth, and integration into the global economic system is an emerging market. In Southeast Asia, emerging markets are booming. Indonesia, Malaysia, and Vietnam—three of Southeast Asia's roaring "tigers"—are among the fastest-growing economies in the region.

Why are businesses investing in Indonesia, Malaysia, and Vietnam?

During the late twentieth century, Southeast Asia underwent an economic transformation. From the 1970s to the 1990s, traditionally agricultural countries began to industrialize. Today, Indonesia, Malaysia, and Vietnam manufacture shoes, clothing, electronics, and many other goods, most of which are exported around the world. Industrialization has led to economic growth in Southeast Asia. Between 1967 and 1997, Indonesia's gross domestic product (GDP) grew 6 percent per year. Since adopting a constitution in 1992 that encouraged the development of a market economy, Vietnam has experienced impressive economic growth. Vietnam's economy is growing about 7 percent each year. Both Vietnam and Malaysia have some of the lowest unemployment rates in the world. Thriving industries have boosted international trade and attracted European, Chinese, and American businesses to the region.

1. **Human Systems** What economic changes have taken place in these countries?

How did these three countries become classified as emerging markets?

Indonesia, Malaysia, and Vietnam share characteristics that helped them develop strong economies. First, each country has abundant natural resources. For example, Indonesia has oil, tin and other minerals, timber, coal, and natural gas. Second, these countries have plenty of highly skilled workers and low labor costs. Although several Southeast Asian countries raised the minimum wage in 2013, workers here still earn much less per hour compared to workers in Europe, North America, and many other parts of the world. Third, the location of these countries spurs trade. Southeast Asia is strategically located between the South China Sea and the Indian Ocean. The Strait of Malacca—a choke point that connects the two bodies of water—is one of the most important shipping routes in the world. Finally, the governments of Indonesia, Malaysia, and Vietnam adopted policies that foster trade and business. These policies include building new roads, airports, and ports; improving education; and streamlining processes to make it easier for people to start businesses.

2. **Places and Regions** What four factors led to Indonesia, Malaysia, and Vietnam becoming classified as emerging markets?

What are some risks of rapid economic growth?

Indonesia, Malaysia, and Vietnam have impressively expanded their economies, yet the widening gap between the rich and the poor in these countries often creates tensions. At times, economic inequality has led to labor unrest, corruption, violent clashes, and political and social instability. In 2011, for instance, copper miners at the Grasberg Mine in Indonesia went on strike for higher wages and large protests demanding political reforms erupted in Malaysia. Rapid growth can also involve uneven economic development. Indonesia, for example, has relied heavily on its oil and natural gas resources for economic growth, but its manufacturing sector has fallen behind. In addition, these emerging market countries have developed greater vulnerability to the ripple effects of the global recession. Even the most vibrant emerging markets can be negatively affected by a struggling worldwide economy.

3. **Human Systems** What are the economic challenges these countries face?

THERE'S MORE ONLINE

READ a chart of Southeast Asia's major exports • **VIEW** a graph showing GDP of Southeast Asian countries

netw⊙rks

There's More Online!

☑ **IMAGE** Mekong River Delta

☑ **IMAGE** Population Density in Southeast Asia

☑ **MAP** Volcanoes in Southeast Asia

☑ **INTERACTIVE SELF-CHECK QUIZ**

☑ **VIDEO** Physical Geography of Southeast Asia

Reading HELPDESK

Academic Vocabulary

- **access**
- **differentiate**

Content Vocabulary

- **tsunami**
- **cyclone**
- **typhoon**
- **biodiversity**

TAKING NOTES: *Key Ideas and Details*

IDENTIFYING As you read about the physical geography of Southeast Asia, use a diagram like the one below to note similarities and differences between the features of the mainland and those of the islands and archipelagoes.

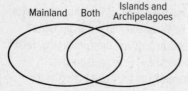

Mainland   Both   Islands and Archipelagoes

LESSON 1
Physical Geography of Southeast Asia

ESSENTIAL QUESTION • *How do physical systems and human systems shape a place?*

IT MATTERS BECAUSE
Few regions have such diverse physical features as Southeast Asia. The Pacific Ocean dominates this subregion where the combined landmass of island countries is greater than the combined landmass of countries on the mainland. Southeast Asia's physical geography also shows the effects of the Pacific Ring of Fire. Most Southeast Asian countries are mountainous, and many have volcanoes and are prone to earthquakes.

Landforms

GUIDING QUESTION *What are the major physical characteristics of Southeast Asia?*

The 11 countries of Southeast Asia can be divided into two areas: the countries of the mainland in the west, and island archipelagoes in the east. Cambodia, Vietnam, Thailand, Myanmar (Burma), Laos, and the city-state of Singapore make up most of the subregion's mainland. Malaysia is partly on the Malay Peninsula and partly on the island of Borneo. In the east, Indonesia, East Timor, Brunei, and the Philippines form island countries that stretch across the western half of the Pacific Ocean.

It is important to realize how large Southeast Asia's island countries are. Their landmass is actually larger than that of the subregion's mainland countries. Indonesia consists of more than 13,000 islands that span a distance of 1,000 miles (1,609 km) from north to south and 3,000 miles (4,828 km) from west to east—an area wider than the continental United States. The Philippines is a close second to Indonesia, with more than 7,000 islands and an area close in size to the state of Arizona. Countries on the mainland are also large. Myanmar is larger than the country of France, while Laos is almost as large as the entire United Kingdom.

Southeast Asia also includes some of the world's tiniest countries. On the Malay Peninsula, the tiny city-state of Singapore, at 433 square miles (697 sq. km), is approximately a third of the size of Rhode Island. Brunei, which borders Malaysia on the island of Borneo, is a little smaller than Delaware.

Southeast Asia's position along the Pacific Ocean places it within the Ring of Fire, a belt of volcanoes and tectonic plate boundaries that surrounds the Pacific. About 75 percent of the world's volcanoes are located along the Ring of Fire. Most of the world's strongest earthquakes also occur here because of the moving and colliding of Earth's tectonic plates. As tectonic plates move, they produce earthquakes and volcanoes. In much of Southeast Asia, earthquakes and volcanic eruptions are common. Volcanic eruptions are more typical in the subregion's island countries. One exception, however, is the large island of Borneo. Borneo is often called a mini-continent because it is so stable. This stability makes Borneo's physical geography an exception to the pattern of the rest of the subregion. Although it has mountains, Borneo has no active volcanoes and rarely experiences earthquakes.

The rest of Indonesia, in contrast, has the largest concentration of active volcanoes in the world. In 2010 Mount Merapi erupted, sending hot gases high into the air and dropping hot ash and debris on nearby villages. As Merapi's explosions grew more violent, Indonesia's government evacuated nearby villages and cut off **access** to them. It was two weeks before villagers were able to return.

Eruptions in Indonesia have changed world history. In 1815 Mount Tambora sent so much ash into the sky that the next year was dubbed "the year without a summer" because ash blocked so much solar energy. Crops failed and millions went hungry in places as far away as the United States, Europe, and northern Africa. In 1883 Krakatau exploded, killing some 36,000 people. Indonesia's most destructive volcanic eruption is thought to have occurred about 73,000 years ago, when Mount Toba exploded. That eruption changed weather patterns worldwide for the next 20 years.

Some of Southeast Asia's earthquakes occur under water. When this happens, they can produce **tsunamis**, dangerous huge waves that can flood inland areas with little warning. This is what happened in 2004, when an underwater

access a way to approach or enter a place

tsunami a dangerous large ocean wave caused by an underwater earthquake, a volcanic eruption, or a sudden displacement of the ocean floor

---

GEOGRAPHY CONNECTION

Southeast Asia is located along the Ring of Fire, an area of frequent volcanic eruptions and earthquakes.

1. *THE WORLD IN SPATIAL TERMS*
   What countries that are not part of Southeast Asia are also affected by the Ring of Fire around the Pacific Ocean?

2. *ENVIRONMENT AND SOCIETY*
   In what ways are people affected by volcanic eruptions?

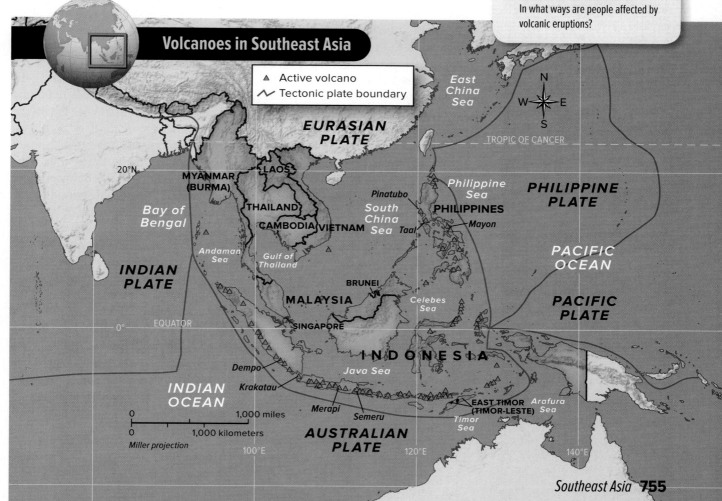

Volcanoes in Southeast Asia

Large earthquakes under the sea (those greater than 7.0 on the Richter Scale) may cause the ocean floor to move vertically, displacing water and changing the waves at the surface. At sea, waves may only be a few inches taller than usual. When these waves reach a coast, however, they may be many feet taller than usual, producing a tsunami. In 1883 earthquakes preceding Krakatau's eruption produced a 98-foot- (30-m-) high tsunami.

IDENTIFYING CAUSE AND EFFECT
What is the connection between undersea earthquakes and tsunamis?

differentiate to distinguish or recognize as different

earthquake in Indonesia triggered a devastating tsunami in the Indian Ocean. The tsunami generated a series of waves, killing more than 300,000 people.

The same tectonic forces that produce volcanoes also produced the mountains for which this subregion is famous. Not all Southeast Asian countries are as vulnerable to volcanic eruptions as Indonesia. All, however, are known for their mountains and hills. These mountains have folded together into a fanlike shape as the Indian subcontinental plate has collided with the Eurasian plate over the past 50 million years. Mountain ranges on the mainland run from north to south and form a natural backbone for most mainland countries in this region. Mountains also form natural borders. For example, the border between Thailand and Myanmar is marked by a mountain range. So is Thailand's border with China. Another major mountain range divides Laos and Vietnam. The island archipelagoes and the Malay Peninsula are also mountainous. Many elevations in the region exceed 10,000 feet (3,000 m).

☑ **READING PROGRESS CHECK**

Explaining How does the Ring of Fire affect the landforms of Southeast Asia?

Waterways

GUIDING QUESTION *What are the major waterways of Southeast Asia?*

In much of Southeast Asia, everyday life revolves around water. Rivers provide water for crops and serve as navigable arteries for trade. Much of the subregion's population lives along the rivers. This is especially the case on the mainland, where the population has become concentrated in certain areas. The Pacific Ocean is also ideal for transportation, which has made it the basis of international trade networks and much of the subregion's tourism. Southeast Asia's waterways have made it one of the most accessible regions in the world. This accessibility has shaped the region's economy and its politics. Not surprisingly, waterways have historically been important to Southeast Asian cultures as well.

Southeast Asia's most dominant waterway, the Pacific Ocean, holds the archipelagoes of the eastern half of the region. But Southeast Asia's mainland countries also depend on the ocean for trade, fishing, and tourism. Rivers on the islands are mostly short, steep, and fast-flowing. The vast majority of Southeast Asia's population is concentrated on the islands of Indonesia and the Philippines.

The mainland has five major rivers: the Mekong (sometimes called the Danube of Southeast Asia), the Red River, the Irrawaddy, the Salween, and the Chao Phraya. Three of these rivers—the Irrawaddy, the Salween, and the Mekong—originate in the mountains of Tibet in China. Southeast Asia's river systems **differentiate** the farming economies of the lowland valley and delta areas from the hill and mountain communities that separate the river valleys.

The Mekong River is the longest river in Southeast Asia and the seventh longest in Asia. It flows south for some 2,600 miles (4,200 km) across the mainland to the South China Sea. As it progresses, the Mekong passes through or along the borders of five Southeast Asian countries: Myanmar, Laos, Thailand, Cambodia, and Vietnam. It forms Laos's border with Myanmar and part of its border with Thailand. Access to the Mekong and its tributaries is the basis of several international disputes. The countries most affected by the Mekong—Laos, Thailand, Cambodia, and Vietnam—have formed a committee to study water projects involving the river.

By contrast, the mainland's other four major rivers are shorter and each is mostly (though not entirely) contained within a single country. The Irrawaddy and the Salween form part of the drainage system for Myanmar's western mountains. The Chao Phraya is the shortest of the four rivers and drains

Thailand's western mountains. Thailand's capital, Bangkok, is located on the Chao Phraya's delta. The Red River, though it has a smaller drainage area than the other rivers, provides Vietnam with a second river delta (in addition to that of the Mekong). Both of these deltas provide rich soil for farming.

☑ **READING PROGRESS CHECK**

Explaining How do the major waterways of Southeast Asia affect population distribution?

Climates, Biomes, and Resources

GUIDING QUESTION *How does climate affect human activities in Southeast Asia?*

Southeast Asia's climate is tropical and subtropical. Its seasonal changes are based more on rainfall than on temperature. Most of Southeast Asia, like many tropical oceanic regions, divides the year into just two seasons: the dry season and the rainy monsoon season. These seasons are created by the changing wind patterns over the Indian and Pacific Oceans. For most of the mainland, the rainy season is from May to September when the monsoon winds blow from the Indian Ocean. Between November and March, the winds blow in a reverse direction and make the season drier. Some parts of Malaysia, Indonesia, and the Philippines have a rainy season that lasts all year, however. They are surrounded by the Pacific and Indian Oceans so that the winds bring rains no matter which way they are blowing. During the rainy season, rainfall is quite heavy. Most of Southeast Asia gets more than 60 inches (150 cm) of rain each year. Some areas get two or three times that amount. For example, Ranong, on the coast of Thailand, receives some 160 inches (400 cm) of rain each year.

During the rainy season, some weather systems become severe tropical storms. These storms are called **cyclones** in the Indian Ocean and **typhoons** in the western Pacific Ocean. They are the same kind of storm as Atlantic hurricanes. In late 2013, Typhoon Haiyan struck in what meteorologists refer to as the "typhoon belt" of the Philippines. Haiyan was among the strongest storms ever recorded. Besides devastating winds, it produced torrential rains that caused flash floods which destroyed villages. Thousands died and many more were left homeless.

Climate Regions and Biomes

Overall, the average annual temperature in most Southeast Asian countries is around 80°F (27°C). But temperatures in Southeast Asia, like temperatures around the world, tend to vary by elevation. Mountainous regions are differentiated from river lowland deltas not only by their different landforms, but also by their cooler temperatures. Air temperatures typically drop by about 0.9°F (0.5°C) for every 328 feet (100 m) climbed.

Some tourists to Southeast Asia head for the cooler highlands. Other tourists go to the coastal areas to experience the beaches and coral reefs. They find that the heat of these areas is somewhat mitigated by the sea breezes.

This satellite image helps show population density in Southeast Asia.

▲ **CRITICAL THINKING**

1. Comparing In what areas of Southeast Asia is the population most dense? What geographic features are common in areas of high population density?

2. Identifying Do you see a connection between Southeast Asia's waterways and the borders of countries? What countries seem to have been affected by the locations of rivers?

cyclone a spiral-shaped tropical storm with winds of at least 74 miles (119 km) per hour and with heavy rain; occurs in the western South Pacific or Indian Oceans

typhoon a cyclone that occurs in the western North Pacific Ocean (west of the International Date Line)

River deltas, such as the Mekong River Delta, are important to economies in Southeast Asia.

▲ **CRITICAL THINKING**

1. *Analyzing Visuals* What purposes do river deltas serve in Southeast Asia?

2. *Explaining* Describe the ways in which everyday life revolves around water in Southeast Asia.

biodiversity the diverse life forms in a habitat or ecosystem

Southeast Asia's warm, wet tropical climate is conducive to plant growth. Plants thrive even in areas where the soil is poor. Some soils are poor, but areas where volcanoes have enriched the soil with minerals have richer soils. In mainland areas that have a dry season, the vegetation is tropical-deciduous, or monsoon, forest. In areas that are wet and rainy all the time, the vegetation is tropical rain forest. In the monsoon forests, trees shed their leaves during the dry season and grow new ones during the rainy season.

Southeast Asia is one of the few regions in the world in which equatorial rain forests can still be found. These tropical rain forests get more than 70 to 100 inches (180 to 250 cm) of rainfall every year. Rain forests around the world are known for their exceptional **biodiversity**. Southeast Asia's rain forests are especially biodiverse. Scientists think that as much as 10 percent of Earth's plant and animal species can be found in Southeast Asia's rain forests, which are mainly found in Myanmar, Thailand, Malaysia, Indonesia's island of Sumatra, the Indonesian half of New Guinea, and the Philippines. The island of Borneo (which is part of Malaysia, Indonesia, and Brunei) is also covered with rain forest.

Natural Resources

Southeast Asia's river deltas, its areas of volcanic soil, and its warm, wet climate are good for farming. Agriculture is an important part of the economy for most countries in the subregion, although living near volcanoes also means living at risk of an eruption or an earthquake—and the risk of a tsunami if near the coast. Nevertheless, farming remains a critical part of Southeast Asia's economy.

Southeast Asia is also rich in minerals, including offshore oil. Indonesia, Malaysia, Brunei, and East Timor are some of the subregion's largest producers of oil and natural gas. Cambodia also has recently discovered offshore oil deposits, and Myanmar has a developing natural gas industry. Copper, gold, iron, and gems are also mined in various parts of Southeast Asia, particularly in Cambodia and Laos. In addition, Cambodia has deposits of bauxite, a mineral that is used in the manufacture of aluminum.

☑ **READING PROGRESS CHECK**

Evaluating How does Southeast Asia's elevation affect its climate?

LESSON 1 REVIEW

Reviewing Vocabulary

1. *Describing* What is the difference between a cyclone and a typhoon?

Using Your Notes

2. *Summarizing* Use your graphic organizer on Southeast Asia's physical geography to write a paragraph summarizing the main physical differences between Southeast Asia's island archipelagoes and its mainland countries.

Answering the Guiding Questions

3. *Identifying* What are the major physical characteristics of Southeast Asia?

4. *Identifying* What are the major waterways of Southeast Asia?

5. *Making Connections* How does climate affect human activities in Southeast Asia?

Writing Activity

6. *Informative/Explanatory* In a paragraph, discuss how Southeast Asia's physical geography affects everyday life in the subregion.

Reading **HELP**DESK

Academic Vocabulary

- **dynamic**
- **interdependence**

Content Vocabulary

- **shatter belt**
- **buffer zone**
- **emerging market**
- **free port**

TAKING NOTES: *Key Ideas and Details*

IDENTIFYING As you read about the human geography of Southeast Asia, use a graphic organizer like the one below to identify examples of how history, population, culture, and economics have worked together to create the Southeast Asia of today.

Southeast Asia			
History	Population	Culture	Economics

LESSON 2
Human Geography of Southeast Asia

ESSENTIAL QUESTION • *How do physical systems and human systems shape a place?*

IT MATTERS BECAUSE

Southeast Asia's waterways, a natural transportation network, have brought waves of immigrants. The subregion's dominant religions—Islam, Buddhism, and Christianity—were brought by immigrants from other parts of the world. Some of these waves of immigration are the result of Southeast Asia's position as a buffer zone between more powerful states. In addition, all of the countries of Southeast Asia, except Thailand, experienced European colonization.

History and Government

GUIDING QUESTION *How have conflicts between powerful neighbors affected the history of Southeast Asia?*

The mainland countries of Southeast Asia are positioned near two countries that have been historical centers of power: India and China. This position alone made the region a **shatter belt**, as larger and more powerful countries invaded. When Europe began to colonize the area, Southeast Asia was caught between powerful European powers as well. A shatter belt, though, can also serve as a **buffer zone**, a neutral area separating powerful nations from each other, and thereby reduce conflict.

Early Cultures and European Conquest

At the time of European contact in the 1500s, Southeast Asia was a patchwork of small kingdoms, principalities, and sultanates. The Dutch colonized Indonesia. Portugal colonized part of eastern Timor. The British, expanding from their empire in India, colonized Burma (now Myanmar), Malaysia, and several islands in the South China Sea. The French colonized what was once known as French Indochina and divided it into regions based on the boundaries of the cultural groups already living there. Today these regions are the countries of Vietnam, Cambodia, and Laos. Spain colonized the Philippines, which became a territory of the United States after the Spanish-American War in 1898. In some cases, European powers

Population Patterns

GUIDING QUESTION *How has Southeast Asia's physical geography influenced its population patterns?*

Southeast Asia's mainland countries are not densely populated. The soil on the volcanic islands tends to be more productive than soil in the region's mainland. River valleys and flood plains of major rivers, such as the Mekong, are the exception with rich alluvial soil. Islands and river plains are where the population density is greatest. Southeast Asia's largest population center is located on Java, an island in Indonesia. Nearly 60 percent of the country's people live on this one island in this country that is entirely made up of islands. More than half of all Indonesians live on Java, and more than half of the people on Java work as farmers.

Despite Indonesia's agrarian economy, it is the fourth most populous country in the world. After Indonesia, the next most populous country in Southeast Asia is the Philippines. Southeast Asia is becoming more urban as it industrializes. Each country has at least one large city: Kuala Lumpur in Malaysia; Bangkok in Thailand; Manila in the Philippines; Yangon(Rangoon) in Myanmar; Hanoi and Ho Chi Minh City in Vietnam; and the city-state of Singapore. All of these cities have become a thriving part of the global economy.

✅ READING PROGRESS CHECK

Explaining What parts of Southeast Asia have the highest population density?

Society and Culture Today

GUIDING QUESTION *Why is Southeast Asia one of the world's most culturally diverse areas?*

Southeast Asia encompasses hundreds of different ethnic groups. Most ethnic groups have their own language. The greatest ethnic diversity is found in the island countries. For example, Indonesia's ethnic groups include Javanese, Madurese, Sundanese, and Balinese. In addition, while Malays predominate in Malaysia, they have also emigrated to nearby areas. As a result, there are enclaves of Malay minorities throughout Indonesia, and Indonesia's official language is a modified form of Malay.

The borders of Southeast Asia's mainland countries match the locations of each country's majority ethnic group. In Myanmar, which is also called Burma, Burmans form the majority. The Thai dominate Thailand. Cambodia's population is mostly Khmer, and Vietnam's population is primarily Vietnamese. Each of these countries has other, smaller ethnic groups that live in the mountains. In addition to indigenous ethnic groups, Southeast Asian countries are home to minority ethnic groups that originally emigrated from other countries. For example, Malaysia and Singapore have large Indian communities. However, the largest minority ethnic group in the region is Chinese. About 32 million Chinese live in Southeast Asia, primarily in urban areas. Nearly one-fourth of Malaysia's population is Chinese. The population of Singapore is about 77 percent Chinese.

Angkor Wat is a Cambodian temple complex built in the twelfth century. It was built as a Hindu temple, but was later turned into a Buddhist temple.

▼ CRITICAL THINKING

1. Drawing Conclusions Why do you think that the Angkor Wat temple continued to be important to Cambodians even after local religious traditions had changed?

2. Hypothesizing Why do you think an image of Angkor Wat appears on the Cambodian national flag?

Andrew Gunners/Digital Vision/Punchstock

Southeast Asia's religions reflect the waves of immigration that have reached this subregion. The most common religion in Southeast Asia is Islam. Muslim traders who traveled to the region in the twelfth and thirteenth centuries introduced the religion to Southeast Asia. Islam is the dominant religion in Southeast Asia's largest country, Indonesia. About 90 percent of Indonesian people are Muslim. So are most Malaysians. Malaysia's South Asian minority, though, tend to be Hindu, while Chinese Malaysians are mostly Buddhist. Buddhism predominates in the following mainland countries as well: Myanmar, Thailand, and Cambodia. Vietnam historically served as a religious crossroads for Buddhism, Catholicism, Daoism, Confucianism, and ancestor worship. In the last census, however, more than 80 percent of the people of Vietnam, a communist country, claimed no religion. In the Philippines, where Spanish missionaries accompanied colonists, most people are Catholic.

Family and the Status of Women

Women in Southeast Asia have traditionally been responsible for raising families but are also a major part of the workforce. This produces a double burden for women who must both work outside the home and still keep up with the daily household responsibilities. In rural areas, women often work as farmers. In cities, women frequently work in factories, especially in industries such as textiles, food processing, and electronics.

Increasingly, Southeast Asian women are rising to positions of leadership as activists for change. In Myanmar, activist Aung San Suu Kyi leads a democratic opposition party that often wins national elections. But Aung San Suu Kyi has not become president of Myanmar. Instead, she has spent much of her life under house arrest because of her outspoken opposition to the military government. She was awarded the Nobel Peace Prize in 1991 for her work to bring democracy to Myanmar. After the government liberalized policies, she was elected to the lower house of the Myanmar parliament in 2012.

In 1991 a young Indonesian woman and law student, Dita Sari, began leading factory workers to strike for higher wages and better working conditions. She was jailed and beaten, but was later released as Indonesia became more democratic. Today Sari is still a union leader, but instead of leading strikes, she lobbies Indonesia's parliament for labor law reforms.

The Arts

The arts in Southeast Asia have been heavily influenced by religion, which can be seen through the architectural style of Buddhist and Hindu temples and monuments. Chinese and Indian styles are prevalent in Southeast Asian ceramics and bronze, as well as architecture. Elaborate Chinese-style pagodas and Indian-style *wats,* or temples, dot the landscape. Traditional crafts such as weaving and other textile techniques such as batik are still practiced.

✓ **READING PROGRESS CHECK**

Identifying What are the dominant religions of Southeast Asia?

This woman is selling produce in a market in Bukittinggi on the Indonesian island of Sumatra. Originally the site of a Dutch fort and now an important commercial center, the town reflects a multiethnic heritage.

▲ **CRITICAL THINKING**

1. *Drawing Conclusions* How do you think that people of many different ethnic groups, speaking many different languages, are able to coexist peacefully and conduct transactions with each other in urban markets?

2. *Considering Advantages and Disadvantages* What are some of the advantages and disadvantages of Indonesia's ethnic diversity?

Tibor Bognar/age fotostock

Workers are drilling in preparation for the construction of a railroad that will run from China into several Southeast Asian mainland countries.

▲ **CRITICAL THINKING**

1. *Identifying Cause and Effect* Why do you think Southeast Asian countries are cooperating with China to build a railroad? How do you think the new railroad will affect trade networks in the region?

2. *Making Connections* What characteristics of Southeast Asia's geography make the construction of railroads a challenge?

emerging market an economy that may not have been very strong in the recent past, but that is in transition to becoming a stronger market

free port a place where goods can be unloaded, stored, or reshipped free of import duties

Economic Activities

GUIDING QUESTION *How have Southeast Asia's location and natural resources contributed to its economic development?*

Southeast Asia is as diverse economically as it is politically. It holds one of the world's most prosperous cities, Singapore, and a wealthy sultanate, Brunei. The subregion also includes some countries that are among the world's poorest, such as Myanmar. During the 1980s, the industrializing countries of Southeast Asia enjoyed an economic boom. This boom was based on advantages of location, natural resources, inexpensive labor, and increased foreign investment. Some countries that did not have strong economies in the past—such as Indonesia, Malaysia, and Vietnam—are now considered to be **emerging markets**, ripe for foreign investment.

Today agriculture remains Southeast Asia's leading economic activity. Most people in the subregion make their living as farmers. More than half of the subregion's arable land is used to grow rice. Thailand and Vietnam are among the world's top exporters of rice. Indonesia and Myanmar are also two of the region's major rice producers. Farmers in Southeast Asia also grow cassava, yams, corn, bananas, sugarcane, coffee, coconuts, and spices. Thailand, Indonesia, and Malaysia—the world's "rubber belt"—have many rubber plantations. Palm oil, a product of oil palm trees, is an important cash crop in Malaysia and Indonesia. Forestry and logging have also become important in Malaysia, the Philippines, Indonesia, and Thailand. Forests include teak and ebony trees, which are very valuable. The high value results in illegal logging and export to other countries. Along the coasts and rivers, many people make a living by fishing.

Resources, Power, and Industry

Southeast Asia is rich in mineral resources. Thailand, Malaysia, and Indonesia are three of the world's leading producers of tin. Iron ore is mined in Malaysia and the Philippines. Brunei, Malaysia, and Indonesia also produce oil. Indonesia once exported much of its oil. Now that it is industrializing, however, it has begun to import more oil than it exports. Malaysia uses some of its resources for industries such as manufacturing electronics, cement, chemicals, and processed foods. Indonesia's industries focus on textile and garment manufacturing.

Economic Integration

Southeast Asia has long been the crossroads of major ocean trade routes. Today, most shipping between Europe and East Asia passes through the Strait of Malacca, near Singapore. The city-state of Singapore was originally founded as a British trading colony. Not only does Singapore have a strategic location at the crossing of trade routes, but it has a large, deep natural harbor. Historically, this harbor was used by British warships to displace the economic dominance of the Dutch in the region. Today, its location enables Singapore to prosper as a **free port**, a place where goods can be unloaded, stored, and reshipped free of import duties. Its port is the largest container port in the world. These large metal containers may be stacked on ships for crossing oceans, and then carried by rail or truck after they reach ports. Singapore has also attracted foreign investors, especially in technology, consumer electronics, and pharmaceuticals. Singapore's efforts to build its economy have paid off. Its population has the highest standard of living in the subregion, and its per-capita income levels are comparable to those of the United States and Switzerland.

In contrast, the countries of Indonesia and the Philippines have much larger populations, but less than one-tenth of Singapore's per-capita income. Singapore is both a city and a country, which makes it a city-state. Of all the countries in the subregion, Singapore has the most **dynamic** economy. Because it is small, Singapore also tries to promote peaceful **interdependence** and international cooperation in the region. Other regional ports include Haiphong in Vietnam, Bangkok in Thailand, Jakarta in Indonesia, and Manila in the Philippines.

In 2010 several Southeast Asian countries signed a free trade agreement with China. They hope to cooperate to improve trade networks. Now work is underway to build railroads from the city of Kunming in China's Yunnan Province into Laos. From there, railroads will extend into Cambodia, Vietnam, Thailand, Malaysia, and Singapore. From Singapore goods can be shipped to world markets by sea.

In recent decades, Southeast Asian countries have become more interdependent. Based in the Philippines, the Asian Development Bank (ADB) was founded in 1966 to promote regional economic development. The Association of Southeast Asian Nations (ASEAN) was formed in 1967 to promote regional stability.

The ADB provides international loans to aid the economies of Asian member countries. ADB's loans support agricultural, transportation, and industrial development projects. For example, in Indonesia, ADB funds are being used to improve infrastructure such as transportation networks. ADB is also currently working to develop financing and loans that are compliant with Islamic law. ADB is attempting to be more culturally sensitive to countries that have a high percentage of Muslims, such as Indonesia and Malaysia.

Indonesia, Malaysia, the Philippines, Singapore, and Thailand are ASEAN's founding members. ASEAN's mission is to promote regional economic growth. Brunei joined in 1984. In 1992 ASEAN's member nations agreed to form the ASEAN Free Trade Area (AFTA). This meant that ASEAN members agreed to cooperate economically by opening trade between member countries and by reducing tariffs on nonagricultural products. By the late 1990s, Vietnam, Cambodia, Laos, and Myanmar had all become members. Growth is increasing in these countries, but it is slowing in Indonesia because of political instability. In 2004 ASEAN's members signed a trade agreement with China. Now ASEAN's members hope to develop a regional trading market that could operate as one interdependent unit, much like the European Union.

☑ **READING PROGRESS CHECK**

Drawing Conclusions How has Southeast Asia's location along trade routes affected the histories and economies of countries in the subregion?

Analyzing
PRIMARY SOURCES
Regional Cooperation

"To collaborate more effectively for the greater utilization of their agriculture and industries, the expansion of their trade . . . the improvement of their transportation and communication facilities and the raising of the living standards of their peoples."

— Association of Southeast Asian Nations (ASEAN), Bangkok Declaration, 1967

DBQ *MAKING CONNECTIONS*
Why are transportation and communication important areas in which to collaborate? How do you think the transportation and communication networks in one country might affect the people of another country?

dynamic energetic, characterized by constant change and progress

interdependence a condition in which people or groups rely on each other, rather than only relying on themselves

---

LESSON 2 REVIEW

Reviewing Vocabulary
1. *Identifying* Write a paragraph explaining how Southeast Asia has historically functioned both as a shatter belt and a buffer zone.

Using Your Notes
2. *Summarizing* Use your graphic organizer on Southeast Asia's human geography to write a paragraph explaining why Southeast Asia is one of the world's most ethnically diverse regions.

Answering the Guiding Questions
3. *Describing* How have conflicts between powerful neighbors affected the history of Southeast Asia?

4. *Identifying* How has Southeast Asia's physical geography influenced its population patterns?

5. *Summarizing* Why is Southeast Asia one of the world's most culturally diverse areas?

6. *Explaining* How have Southeast Asia's location and natural resources contributed to its economic development?

Writing Activity
7. *Informative/Explanatory* In a paragraph, discuss how life in small towns might be affected if transportation and communication networks in Southeast Asia are expanded to reach more rural areas.

networks

There's More Online!

☑ **IMAGE** Air Pollution

☑ **IMAGE** Reforestation Efforts
in Thailand

☑ **INFOGRAPHIC** An Oil That's
Everywhere

☑ **INTERACTIVE SELF-CHECK QUIZ**

☑ **VIDEO** People and Their
Environment: Southeast Asia

LESSON 3
People and Their Environment: Southeast Asia

Economic Development = Environmental Challenges	
Cause	Effect

ESSENTIAL QUESTION • *How do physical systems and human systems shape a place?*

IT MATTERS BECAUSE

Southeast Asia contains some of the most biodiverse areas on Earth. The subregion's equatorial rain forests provide a habitat for hundreds of unique plant and animal species, some of which have never been documented or studied by scientists. Many of the rare plant and animal species of Southeast Asia's rain forests are now facing extinction. In addition, the people of Southeast Asia depend on their environment for clean air to breathe and clean water to drink.

Managing Resources

GUIDING QUESTION *How has the management of natural resources impacted the environment in Southeast Asia?*

Minerals, metals, and rain forest timber are among the natural resources most valued by the countries of Southeast Asia. They provide important sources of income. However, mining and harvesting these resources also involve shifting some parts of Southeast Asia from a rural economy to an urban one. These changes cause deforestation and pollution. This environmental damage threatens current and future **generations**. Southeast Asia still has some of the world's last remaining rain forests. However, large sections of the rain forests are being cut down. Countries such as Laos, Thailand, Malaysia, and Myanmar rely on teak and other timber as sources of income. Economies have benefited, but the widespread cutting of trees has diminished the region's forests. Scientists predict that many unique environments will be lost to **deforestation**, the cutting down or clearing of trees, within a few years. This threatens plant and animal species that do not exist anywhere else on Earth. On the Indonesian islands of Sumatra and Borneo, logging has already destroyed much of the forests bordering national parks.

Agricultural activities pose another threat to Southeast Asia's forests. In some countries, forests are cleared to make room for rubber and palm tree plantations. Together, Malaysia and Indonesia produce 85 percent of the world's palm oil. Palm oil is a product that is used as a lubricant by industries. It is also an additive in more than half of foods sold in grocery stores. According to the United Nations, however, palm oil plantations now form the single largest threat to rain forests in these two countries. Southeast Asian palm oil plantations cover an area equivalent to the size of the country of Austria. Every day, about 30 square miles (48 sq. km) of rain forest are cleared from Indonesia's Sumatra, an island the size of Spain, and from Borneo, an island the size of Turkey that is shared by Malaysia, Indonesia, and Brunei.

The loss of forested lands is putting many unique plant and animal species in danger. Sumatra's rain forest is estimated to have 465 bird species, 194 mammal species, 217 reptile species, and 272 species of freshwater fish. It also is home to more than 10,000 plant species. Borneo is believed to have 420 bird species, 210 mammal species, 254 reptile species, and 368 freshwater fish species. Some 15,000 species of plants grow in Borneo's forests. Many unique animal species are in danger of becoming extinct. This may be avoided if governments **intervene** to protect at least some sections of the rain forests. Endangered species include

generation a group of individuals born and living at the same time

deforestation the loss or destruction of forests, mainly for logging or farming

intervene to come between so as to prevent or alter a course of events

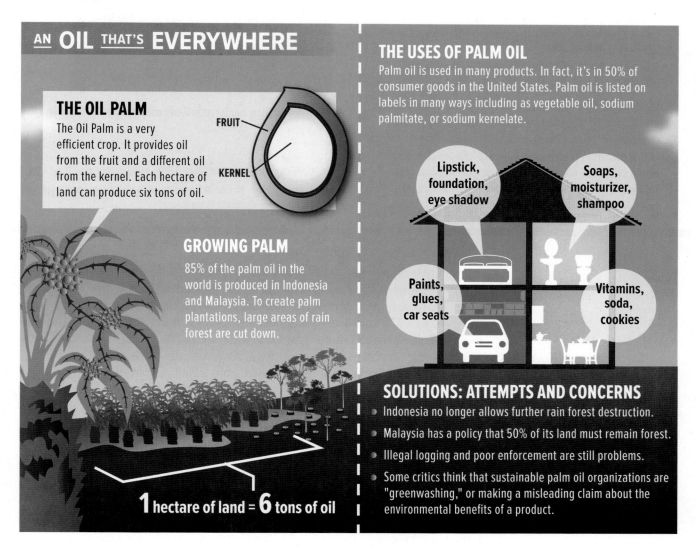

AN OIL THAT'S EVERYWHERE

THE OIL PALM

The Oil Palm is a very efficient crop. It provides oil from the fruit and a different oil from the kernel. Each hectare of land can produce six tons of oil.

FRUIT

KERNEL

GROWING PALM

85% of the palm oil in the world is produced in Indonesia and Malaysia. To create palm plantations, large areas of rain forest are cut down.

1 hectare of land = **6** tons of oil

THE USES OF PALM OIL

Palm oil is used in many products. In fact, it's in 50% of consumer goods in the United States. Palm oil is listed on labels in many ways including as vegetable oil, sodium palmitate, or sodium kernelate.

Lipstick, foundation, eye shadow

Soaps, moisturizer, shampoo

Paints, glues, car seats

Vitamins, soda, cookies

SOLUTIONS: ATTEMPTS AND CONCERNS

- Indonesia no longer allows further rain forest destruction.
- Malaysia has a policy that 50% of its land must remain forest.
- Illegal logging and poor enforcement are still problems.
- Some critics think that sustainable palm oil organizations are "greenwashing," or making a misleading claim about the environmental benefits of a product.

About 85 percent of the world's palm oil is produced on plantations in Malaysia and Indonesia. The global demand for palm oil rises by 6 to 10 percent every year.

▲ **CRITICAL THINKING**

1. *Identifying Cause and Effect* How are Southeast Asian rain forests affected by the demand for products containing palm oil?

2. *Hypothesizing* Why do you think the global demand for palm oil is rising?

Mekong Giant Catfish

The critically endangered Mekong giant catfish is the world's largest scaleless freshwater fish. Overfishing, dam building, and habitat destruction threaten the species.

"Once plentiful throughout the Mekong basin, population numbers have dropped by some 95 percent over the past century, and this critically endangered behemoth now teeters on the brink of extinction. Overfishing is the primary culprit in the giant catfish's decline, but damming of Mekong tributaries, destruction of spawning and breeding grounds, and siltation have taken a huge toll. Some experts think there may only be a few hundred adults left.

Mekong giant catfish have very low-set eyes and are silvery to dark gray on top and whitish to yellow on the bottom. They are toothless herbivores who live off the plants and algae in the river. Juveniles wear the characteristic catfish 'whiskers,' called barbels, but these features shrink as they age.

Highly migratory creatures, giant catfish require large stretches of river for their seasonal journeys and specific environmental conditions in their spawning and breeding areas. They are thought to rear primarily in Cambodia's Tonle Sap lake and they migrate hundreds of miles north of Cambodia to spawning grounds in Thailand. Dams and human encroachment, however, have severely disrupted their lifecycle.

International efforts are under way to save the species. It is now illegal in Thailand, Laos, and Cambodia to harvest giant catfish. And recently in Thailand, a group of fishers pledged to stop catching giant catfish to honor the king's 60th year on the throne. However, enforcement of fishing restrictions in many isolated villages along the Mekong is nearly impossible, and illicit and bycatch takings continue."

—"Mekong Giant Catfish,"
nationalgeographic.com

Thai fishers display a giant catfish they caught in the Mekong River.

DBQ ▲ **CRITICAL THINKING**

1. *Finding the Main Idea* What threats do the giant catfish face?
2. *Speculating* Why might the giant catfish be a prized catch for people living along the banks of the Mekong?

---

orangutans, pygmy elephants, Sumatran rhinoceroses, and Sumatran tigers. Of all these animals, biologists may be most concerned about the orangutan. Orangutans are intelligent apes. They can make tools and use them to build things like rain hats and leakproof nest roofs. Orangutans are also said to be able to distinguish between more than 1,000 different plants.

In some Southeast Asian countries, such as Laos, threats to forests come not only from big plantations, but also from small subsistence farmers who practice **shifting cultivation**. These farmers clear forests in order to plant fields. After cultivating the land for a few years, they then abandon it to move to a new spot. Farmers practice shifting cultivation deliberately. It allows the land time to become fertile again after a period of intensive farming. Farmers do not always realize that rain forest soil is frequently lacking in nutrients. Soil in these areas has often been depleted by the already lush growth in the rain forest.

Deforestation also leads to other environmental problems. Without the trees' root systems, topsoil is easily eroded by heavy rains and washed into streams. The soil clogs rivers and reduces the amount of water available for irrigation. Deforestation can also cause flooding. Without forests to absorb downpours, flash floods occur.

shifting cultivation a form of agriculture in which an area of ground is cleared of vegetation, cultivated for a few years, and then abandoned in order to move to a new area

✓ READING PROGRESS CHECK

Explaining What economic activities are increasing the rate of deforestation in Southeast Asia?

Human Impact

GUIDING QUESTION *Why do urban growth and industrialization create environmental problems in Southeast Asia?*

Industrialization and economic development in Southeast Asia have resulted in the pollution and the destruction of natural environments. Southeast Asians—like people everywhere—affect their environment. Increased manufacturing and industrialization create jobs and raise standards of living. They also produce industrial waste, however. Growing populations and crowded conditions in cities such as Bangkok, Manila, and Jakarta raise concerns about adequate housing, water supplies, sanitation, and traffic control. Bangkok, for example, has become overheated as a result of increased industrial development. In recent years, Bangkok's heat, humidity, and pollution levels have increased at levels higher than the global average. Its levels are also higher than the surrounding rural areas, making it what is known as an urban heat island.

Urbanization and industrialization also put a strain on shared local resources. Such resources include rivers and water supplies. When different communities share a river, the communities upstream have an advantage because the water flows there first. Currently, Southeast Asian communities face a dispute over how to best manage the shared Mekong River. The Mekong originates in China, and the Chinese are building a series of hydroelectric dams across it. Downstream farmers worry, however, about what will happen if the water level in the Mekong drops. Cambodians are concerned about the Tonle Sap lake, which gets its water from the Mekong. Vietnamese farmers fear that Mekong Delta rice paddies could be invaded by salt water from the sea if the Mekong's water levels drop.

The dumping of toxic waste from newly developed industries has become a widespread problem. In some cases, pollution extends into rural areas. This includes the subregion's national parks. In one of Thailand's national parks, for example, most of the freshwater wells have been contaminated by poor waste disposal. Water pollution is also severe in Thailand's coastal areas. Activities related to agriculture, as well as offshore oil and natural gas exploration, have resulted in the loss of over half of Thailand's mangrove forests.

One industry that produces substantial water pollution is mining. At Indonesia's largest gold mine, waste is dumped into the Ajkwa River that flows through part of Papua province, contaminating the surrounding water systems. Even the fishing industry causes some water pollution. Southeast Asian fishers use poisons and explosives to capture certain kinds of fish to supply Asian restaurants and the world's aquarium industry. Many of these fish come from the coral reefs. Like rain forests, coral reefs are a rich source of biodiversity. Also like rain forests, the reefs are being threatened by economic development. The live reef-fish trade generates huge profits, but current fishing methods are destroying the reefs. Once the reefs are gone, local communities will lose a primary source of food.

Deforestation has become a source of air pollution. Land is often cleared by fire, especially in Indonesia, where rain forests are being cleared to make room for palm oil plantations. Forest fires cause air pollution in rural areas, and this pollution often makes its way to cities

This river in Vietnam has been polluted by drainage from mines. Water pollution has become a serious problem in parts of Southeast Asia.

▼ CRITICAL THINKING

1. *Explaining* How does water pollution affect human populations?

2. *Problem Solving* What steps might governments in Southeast Asia take to reduce water pollution?

These children in Thailand are planting tree seedlings. Several Southeast Asian countries have introduced reforestation programs to conserve forests.

▲ CRITICAL THINKING

1. *Drawing Conclusions* Do you think that reforestation programs will be enough to save Southeast Asia's forests? Why or why not?

2. *Drawing Inferences* What might be the purpose of teaching children to plant trees?

sustainable development technological and economic growth that does not deplete the human and natural resources of a given area

reforestation the planting and cultivating of new trees in an effort to restore a forest that has been reduced by fire or by cutting

as well. Since the late 1990s, Indonesian forest fires have created pollution and respiratory problems for people as far away as mainland Malaysia and Singapore. The Malaysian government has been forced on occasion to declare a state of emergency. Singapore's government often advises residents who have heart or respiratory conditions to avoid going outside.

✔ READING PROGRESS CHECK

Identifying What environmental challenges are the results of human economic activities in Southeast Asia?

Addressing the Issues

GUIDING QUESTION *How are people and governments addressing environmental issues in Southeast Asia?*

Many Southeast Asian countries are shifting their economic policies to focus on **sustainable development**. This strategy means encouraging technological and economic growth that relies on renewable resources. For example, to prevent further loss of rain forests, Thailand, Indonesia, the Philippines, and Malaysia have limited certain timber exports. They have also introduced **reforestation** programs. In Laos, the government has encouraged some highland farmers to resettle on more fertile ground and marked some of its forests as protected areas.

Southeast Asian governments are also starting to develop ideas to reduce the impact of urban growth on the surrounding environment. Scientists are trying to come up with solutions to urban heat islands. One proposal is to create "green zones," areas within a city that are granted special environmental protection. Another idea is to ban the construction of tall buildings near the sea so that winds would be able to blow farther into the city and disperse more air pollution.

Southeast Asia's environmental issues have attracted widespread interest. The Association of Southeast Asian Nations (ASEAN) has tried to address some of these issues. In 2002 ASEAN countries formed an agreement on Transboundary Haze Pollution. Indonesia refused to ratify it. However, countries around the world are still pressuring Indonesia to conserve its rain forests. In 2012 Indonesia announced a two-year ban on forest clearing.

✔ READING PROGRESS CHECK

Identifying How are Southeast Asian governments working to limit deforestation?

LESSON 3 REVIEW

Reviewing Vocabulary

1. *Identifying* Write a paragraph explaining what sustainable development is and why it is important.

Using Your Notes

2. *Summarizing* Use your graphic organizer on the causes and effects of environmental challenges in Southeast Asia to write a paragraph explaining how economic development is connected to environmental challenges.

Answering the Guiding Questions

3. *Identifying Cause and Effect* How has the management of natural resources impacted the environment in Southeast Asia?

4. *Drawing Conclusions* Why do urban growth and industrialization create environmental problems in Southeast Asia?

5. *Identifying* How are people and governments addressing environmental issues in Southeast Asia?

Writing Activity

6. *Informative/Explanatory* In a paragraph, explain how Southeast Asia's urban and rural areas differ from each other in terms of their effects on the environment.

Directions: On a separate sheet of paper, answer the questions below. Make sure that you read carefully and answer all parts of the questions.

Lesson Review

Lesson 1

1 **Specifying** Discuss three examples of volcanic eruptions in Indonesia and how they have changed world history.

2 **Assessing** In most of Southeast Asia, the soil is poor. What causes the soil in some areas to be richer and more productive? With mostly poor soil, why does plant growth remain strong throughout most of Southeast Asia?

3 **Drawing Conclusions** Explain whether it would be accurate to state that Southeast Asia's rain forests are an example of biodiversity.

Lesson 2

4 **Identifying Cause and Effect** Discuss an example of a successful independence movement in which a Southeast Asian country gained independence from a colonial power.

5 **Summarizing** Write a summary about the religions practiced in Southeast Asia. Specify the most common religion and briefly explain how it came to be practiced in the region.

6 **Drawing Conclusions** How do Singapore and Myanmar represent the opposite extremes of economic development in Southeast Asia?

Lesson 3

7 **Analyzing** What is the greatest threat to rain forests in Indonesia and Malaysia? Why does the possibility of eliminating this threat pose a challenge to these countries?

8 **Considering Advantages and Disadvantages** Explain the advantages and disadvantages of manufacturing and industrialization in Southeast Asia.

9 **Describing** Describe two activities that Southeast Asian governments have undertaken to reduce the impact of urban growth on the surrounding environment.

21st Century Skills

Use the graph below to answer the questions that follow.

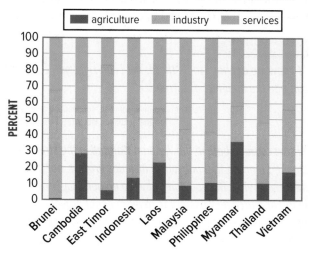

CONTRIBUTION TO GDP BY ECONOMIC SECTOR

■ agriculture ■ industry ■ services

Source: The CIA World Factbook.

10 **Using Charts** Within which countries are the contributions to GDP by industry the greatest? What is the percentage for each? Why is industry more successful there?

11 **Creating Diagrams** Explain whether Brunei's agricultural contribution to GDP seems logical when considering the size of the country.

Critical Thinking

12 **Making Connections** Write a one-page essay explaining the role of agriculture in Southeast Asia's economy. Identify specific crops and other economic activities. Explain where agriculture ranks overall economically in relation to the other activities.

College and Career Readiness

13 **Clear Communication** Suppose you want to publish biographies of Aung San Suu Kyi and Dita Sari. Write a one-page proposal about each of these women and her connection to Southeast Asia. In your proposal, explain each woman's significance in the region's recent history.

Need Extra Help?

If You've Missed Question	1	2	3	4	5	6	7	8	9	10	11	12	13
Go to page	755	758	758	760	763	764	767	769	770	771	771	764	763

Directions: On a separate sheet of paper, answer the questions below. Make sure that you read carefully and answer all parts of the questions.

DBQ Analyzing Primary Sources

Read the excerpts and answer the following questions.

There was great division of opinion regarding the way the United States entered and participated in the war in Vietnam.

PRIMARY SOURCE

"I know we oughtn't to be there, but I can't get out. . . . I just can't be the architect of surrender."

—President Lyndon B. Johnson, commenting on Vietnam to Senator Eugene McCarthy, February 1966

"I saw courage both in the Vietnam War and in the struggle to stop it. I learned that patriotism includes protest, not just military service. But you don't have to go half way around the world or march on Washington to learn about bravery or love of country. Again and again, in the causes that define our nation, we have seen the uncommon courage that is common to the American people."

—John Kerry, announcement speech, Patriots Point, South Carolina, September 2, 2003 (as prepared for delivery)

14 ***Understanding Historical Interpretation*** Explain how the quote by President Johnson indicates his attitude toward withdrawal of U.S. troops from Vietnam. How does this quote help you to assess why the United States remained involved in combat in Vietnam for such a long period of time?

15 ***Interpreting Significance*** Explain whether Kerry's statement indicates that he is aligned with Johnson's attitudes and action regarding Vietnam. How does Kerry's statement indicate a widespread attitude regarding U.S. involvement in Vietnam?

Writing About Geography

16 ***Informative/Explanatory*** Use standard grammar, spelling, sentence structure, and punctuation to write a one-page essay discussing family and the current status of women in Southeast Asia.

Applying Map Skills

Refer to the Unit 9 Atlas to answer the following questions.

17 ***Places and Regions*** Identify the five major rivers of Southeast Asia's mainland countries. Of these rivers, which is the shortest? Which is the longest?

18 ***The World in Spatial Terms*** Use your mental map of Southeast Asia to describe the spatial relationship between Thailand and Myanmar. Explain whether travel on foot would be easy or difficult between these two countries.

19 ***Human Systems*** Write a general statement to compare and contrast the populations of the Philippines and Indonesia with the populations of other countries in Southeast Asia.

Exploring the Essential Question

20 ***Organizing*** Select one of the countries from the chapter. Then suppose you are developing a website to provide information explaining how physical systems and human systems have shaped that country. Create the home page, including relevant links. For each link, write a two-to-three paragraph summary of the information to be included.

Research and Presentation

21 ***Gathering Information*** With a partner, conduct research to learn more about the Association of Southeast Asian Nations (ASEAN). Create a multimedia presentation to explain the following: members, mission, and growth of the organization, as well as its past and present activities. Explain whether you think the organization is effective currently—and whether you think the organization will be effective in the future. Within your presentation, include video, maps, photographs, and diagrams or graphs.

Need Extra Help?

If You've Missed Question	**14**	**15**	**16**	**17**	**18**	**19**	**20**	**21**
Go to page	772	772	763	746	746	746	765	765

Australia and New Zealand

ESSENTIAL QUESTION · *How do physical systems and human systems shape a place?*

netw⦿rks

There's More Online about the geography of Australia and New Zealand.

CHAPTER 32

Why Geography Matters
Australia and New Zealand: Indigenous Peoples

Lesson 1
Physical Geography of Australia and New Zealand

Lesson 2
Human Geography of Australia and New Zealand

Lesson 3
People and Their Environment: Australia and New Zealand

Geography Matters...

Australia and New Zealand have rich cultural heritages influenced by indigenous peoples, English settlers, and more recent immigrants.

The subregion faces many challenges. Environmental threats include climate change, deforestation, and the introduction of invasive plant and animal species which have no natural predators. The countries of Australia and New Zealand are making strides in addressing both cultural and environmental issues, but more work is needed. With the help of individuals, companies, organizations, and governments, change is moving forward.

◀ This young man has the Australian flag painted on his face.

©Marianna Massey/Corbis

Australia *and* New Zealand: indigenous peoples

Australia and New Zealand are home to indigenous peoples, or the people who lived in a place before it was conquered by colonial societies. Australia's Aborigines (A•buh•RIH•juh•neez) and New Zealand's Maori (MAWR•ee) have their own languages, customs, and traditions. This made it more difficult for them to live among European settlers. In the twentieth century, Aborigines and Maori protested unfair treatment. They pressed for their political, economic, social, cultural, and civil rights.

Andrew Holt/Photographer's Choice/Getty Images

How do the indigenous peoples in Australia and New Zealand differ?

Why have both populations been denied basic rights?

What steps are being taken to restore the rights of these indigenous peoples?

Both Australian Aborigines and the Maori migrated from Southeast Asia. The Aboriginal people arrived in Australia about 40,000 to 60,000 years ago. Over time, these nomadic hunters and gatherers dispersed across the entire continent. They divided into different groups and developed distinct cultures with as many as 700 languages.

Unlike the Aborigines, the Maori arrived in New Zealand sometime in the A.D. 1200s. The Maori remained in close contact with one another in a relatively small geographic area and were isolated from other societies. Therefore, Maori culture is less diffuse than Australia's Aboriginal culture. For instance, Maori today speak dialects of the same ancestral language and share similar beliefs. Although Aboriginal and Maori societies developed differently, British colonization greatly affected both peoples.

1. **The Uses of Geography**   How did geography affect the development of Aboriginal and Maori cultures?

European explorers visited Australia and New Zealand in the 1600s. The British government established a colony for convicts in Australia in 1788. Free settlers soon followed. In the 1820s, Europeans settled in New Zealand. Colonization, however, caused friction. Indigenous peoples and European colonists competed for control of valuable resources such as land and water. Social, political, and cultural differences between colonists and indigenous peoples also led to conflicts. Aborigines and Maori frequently endured discrimination. The Australian government believed the Aborigines would not survive colonization and placed many in reserve stations. It provided them with food and clothing but restricted their movements. The government also forcibly removed about 100,000 Aboriginal children from their families in order to absorb them into white society. These children are referred to as the "Stolen Generations." In New Zealand in 1840, Maori chiefs and British representatives signed the Treaty of Waitangi, but many of the Maori rights (such as land ownership) granted by the treaty were ignored, resulting in many years of war and conflict.

2. **Human Systems**   What caused Aborigines and Maori to be denied their rights?

In the twentieth century, Australia's Aborigines and New Zealand's Maori began to pressure their governments to restore their rights. Activists fought to preserve their land, language, and way of life through political protests. Eventually, they regained land, fishing, and other basic rights. Children belonging to the "Stolen Generations" received a national apology in 2008. The Australian government later adopted a formal policy aimed at preserving Aboriginal languages. In the 1990s, New Zealand's government began to honor the Treaty of Waitangi. The government transferred land and control of important fisheries and paid millions of dollars in claim settlements to the Maori. New Zealand made Maori an official language and established Maori-language schools and a Maori-language television station. These indigenous peoples, however, still suffer disproportionately from poor health and lower life expectancies due to poverty, unemployment, and other factors.

3. **Environment and Society**   What role has the environment played in helping Aborigines and Maori regain their rights?

THERE'S MORE ONLINE

WATCH a video about the lives of Aborigines • **SEE** an image of protests by Aborigines and Maori

networks
There's More Online!

☑ **IMAGE** Australian Outback

☑ **IMAGE** Hopetoun Falls, Australia

☑ **IMAGE** Kings Canyon, Australia

☑ **INFOGRAPHIC** Life in the Great Barrier Reef

☑ **INTERACTIVE SELF-CHECK QUIZ**

☑ **VIDEO** Physical Geography of Australia and New Zealand

Reading HELPDESK

Academic Vocabulary

- **resource**
- **dominate**
- **benefit**

Content Vocabulary

- **bush**
- **sunken mountains**
- **atoll**
- **caldera**
- **lagoon**
- **artesian well**
- **coral**

TAKING NOTES: *Key Ideas and Details*

IDENTIFYING As you read about the geography of Australia and New Zealand, use a graphic organizer like the one below to note the characteristics of the land, water, and biomes of the subregion.

Australia and New Zealand		
Land	Water	Biomes

LESSON 1
Physical Geography of Australia and New Zealand

ESSENTIAL QUESTION • *How do physical systems and human systems shape a place?*

IT MATTERS BECAUSE

Australia and New Zealand are part of the diverse South Pacific region, which includes sandy beaches, snow-capped mountains, volcanoes, stunning reefs, grasslands, and deserts. Among these varied physical features are unique species of wildlife that attract tourists and scientists alike.

Landforms

GUIDING QUESTION *How do the landforms of Australia and New Zealand influence the economies of these countries?*

Like an island, Australia is surrounded by water. However, geographers classify Australia as a continent because of its massive size. Australia lies between the Pacific Ocean and Indian Ocean in the Southern Hemisphere, and it has a largely flat terrain.

A chain of hills and mountains known as the Great Dividing Range interrupts the level landscape. The peaks of the range stretch along the eastern coast from the Cape York Peninsula to the island of Tasmania. The Western Plateau, a low area of flat land in central and western Australia, covers almost two-thirds of the continent. Few people live in the Western Plateau. It is made up of three deserts: the Great Sandy, the Gibson, and the Great Victoria Deserts. These areas are collectively called the Outback. Many Australians also call these sparsely inhabited areas the **bush**.

South of Great Victoria Desert is the Nullarbor Plain. The name is from the Latin *nullus arbor,* meaning "no tree." On the southern edge of this dry, treeless landscape, giant cliffs tower above the Great Australian Bight, a part of the Indian Ocean. The Western Plateau and the Great Dividing Range are separated by the Central Lowlands. This area of arid grassland and desert stretches across the east-central part of Australia. The large expanses of dry regions make most of Australia unsuitable for agriculture. As a result, it relies upon the extraction of mineral **resources** for much of its industry and exports.

New Zealand is about 1,000 miles (1,600 km) southeast of Australia. It is primarily two islands: North Island and South Island. Both islands are dotted with sandy beaches, emerald hillsides, and snow-tipped mountains. North Island's northern region includes golden beaches, ancient forests, and rich soils. A central plateau of volcanic stone features hot springs and several active volcanoes. Mount Ruapehu (roo•uh•PAY•hoo), North Island's highest point, is an active volcano.

The snowy peaks of the Southern Alps run along South Island's western edge. New Zealand's highest peak, Mount Cook, rises to 12,349 feet (3,764 m). Glaciers are responsible for carving out many of New Zealand's lakes and rivers. New Zealand also contains **sunken mountains** like those in the Marlborough Sounds and Fiordland that have sunk into the sea. Rugged cliffs, deep fjords, and coastal caves dot the western coast. The Canterbury Plains, New Zealand's flattest and most fertile land, lie on the eastern coast where livestock and agriculture flourish.

New Zealand has **atolls**, or ring-shaped islands formed by the buildup of coral reefs on the rim of submerged volcanoes. In the New Zealand territory of Tokelau, the Atafu atoll has 19 islets that rise to 15 feet (5 m) above sea level and enclose a lagoon. There is also a **caldera**, or large volcanic crater. Reporoa, a caldera on New Zealand's North Island in the Taupo volcanic zone, formed about 230,000 years ago. The last activity recorded was in 2005. **Lagoons** are shallow lakes that are at times connected to a river, another lake, or the sea. New Zealand has mainly coastal lagoons. These inland bodies of water are important to the survival of certain species of flora and fauna.

✔ READING PROGRESS CHECK

Identifying What three deserts make up Australia's Outback?

Water Systems

GUIDING QUESTION *What is the effect of water on life in Australia and New Zealand?*

Although surrounded by water, Australia is the driest inhabited continent on Earth. Freshwater is unevenly distributed, unreliable, and seasonal. Seventy percent of Australia is described as arid or semi-arid with limited precipitation.

bush a wild or sparsely inhabited region

resource a source or supply from which benefit can be gained

sunken mountains a high mountain range that has become submerged

atoll a ring-shaped island formed by the buildup of a coral reef on the rim of a submerged volcano

caldera a large volcanic crater

lagoon a shallow lake that is intermittently connected to a river, another lake, or the sea

Kings Canyon (left) and Hopetoun Falls (right) are examples of the diversity of Australia's landscape.

▼ CRITICAL THINKING

1. Analyzing Visuals Describe the climate of Kings Canyon.

2. Analyzing Visuals Describe the climate of Hopetoun Falls.

The Outback

The Outback, or *the bush*, is an important symbol of the national identity of Australians. It features prominently in Australian literature, painting, popular music, film, and food. What is so special about the bush? It was unique to Australia and very different from the European landscapes known to the early settlers. The Aborigines' skill at surviving in the bush became legendary. The idea of the Outback as integral to Australian national identity was reinforced in 1958 with the publication of *The Australian Legend*, which looked at what defined the typical Australian.

ANALYZING Why might living in regions such as the Outback shape the national identity of Australians?

coral the limestone skeleton of a tiny sea animal

benefit to promote well-being; to be useful

Its name suggests a single reef, but the Great Barrier Reef is actually a string of about 2,900 small reefs. Formed from **coral**, the limestone skeletons of a tiny sea animal, it stretches for 1,250 miles (2,000 km).

Much of New Zealand has a marine west coast climate. Ocean winds warm the land in winter and cool it in summer. Temperatures range from 65°F to 85°F (18°C to 29°C) in summer and 35°F to 55°F (2°C to 13°C) in winter.

Differences in geography also cause variations in climate. Mountainous areas exposed to western winds generally receive more rainfall than other areas. While the country as a whole averages 25 to 60 inches (64 to 152 cm) of rain annually, the Southern Alps on South Island have an average annual rainfall of 315 inches (800 cm).

New Zealand's isolation gives rise to unique plant life. Almost 90 percent of the country's indigenous plants are only found there. Manuka, a small shrub, carpets land where prehistoric volcanic eruptions destroyed ancient forests. To address erosion in deforested areas, several tree species have been imported.

The natural resources of both Australia and New Zealand enrich the economies of both countries. Water resources provide fishing and a means of transportation and trade. Only 6 percent of Australia's land is arable, but farmers grow wheat, barley, fruit, and sugarcane. In the more arid regions, ranchers raise cattle, sheep, and chickens. Aboriginal land rights have been an ongoing debate in the country.

New Zealand's fertile volcanic soil greatly **benefits** its economy. Agriculture accounts for 12 percent of the economy. About half of the land supports crops and livestock. New Zealand's sheep and wool products dominate exports, and its forests yield valuable timber. The country's rivers and dams produce abundant hydroelectric power. New Zealand also uses geothermal energy, created by water heated underground in volcanic fields, to generate power.

Australia and New Zealand have abundant mineral resources. Deposits of gold and silver as well as other minerals can be found in both countries. Mining is an important industry. One-fourth of the world's bauxite, used in aluminum production, is in Australia. The country also has some of the world's highest quality opals as well as deposits of coal, iron ore, lead, zinc, and gold. Nickel and petroleum are found mainly in western offshore sites.

New Zealand's main minerals include coal, gold, silver, iron, limestone, clay, dolomite, pumice, salt, serpentinite, zeolite, and bentonite. However, one of its most prized minerals is *pounamu*. *Pounamu* is a type of jade or greenstone that is treasured for its beauty and its spiritual significance by the Maori.

☑ **READING PROGRESS CHECK**

Describing What are some natural resources in Australia and New Zealand?

LESSON 1 REVIEW

Reviewing Vocabulary

1. *Describing* Write a paragraph describing the water resources of Australia and New Zealand.

Using Your Notes

2. *Summarizing* Use your graphic organizer on the geography of Australia and New Zealand to write a paragraph summarizing the characteristics of the land, water, and biomes of the subregion.

Answering the Guiding Questions

3. *Explaining* How do the landforms of Australia and New Zealand influence the economies of these countries?

4. *Analyzing* What is the effect of water on life in Australia and New Zealand?

5. *Identifying* How do the climates of Australia and New Zealand affect the biomes and resources of each?

Writing Activity

6. *Informative/Explanatory* In a paragraph, discuss the land of Australia and New Zealand and how it affects daily life in the countries.

networks

There's More Online!

☑ **IMAGE** The Lost Generations

☑ **MAP** Australia and New Zealand

☑ **MAP** Patterns of European Settlement

☑ **TIME LINE** Migration and Settlement

☑ **INTERACTIVE SELF-CHECK QUIZ**

☑ **VIDEO** Human Geography of Australia and New Zealand

LESSON 2
Human Geography of Australia and New Zealand

ESSENTIAL QUESTION • *How do physical systems and human systems shape a place?*

Reading **HELP**DESK

Academic Vocabulary

- **accompany**
- **impact**

Content Vocabulary

- **Aborigine**
- **clan**
- **boomerang**
- **Maori**
- **dominion**
- **national identity**
- **Strine**
- **pidgin**

TAKING NOTES: *Key Ideas and Details*

IDENTIFYING As you read about the human geography of Australia and New Zealand, use a graphic organizer like the one below to identify the factors that influenced the subregion's culture.

Indigenous Peoples	Migration	European Colonization	Power Struggles

It Matters Because

Long isolated from most of the rest of the world, Australia and New Zealand gave rise to unique indigenous cultures and wildlife. The Aborigines were the first human inhabitants of the subregion. Later, British colonists brought their culture and traditions, introducing European influences.

History and Government

GUIDING QUESTION *How did migration and geography influence the history and governments of Australia and New Zealand?*

The indigenous peoples and the settlers who arrived as part of the colonization of Australia and New Zealand have made them culturally diverse countries. These distinct groups of inhabitants have contributed to the rich cultural fabric of Australia and New Zealand and have shaped the modern countries of today.

The early inhabitants of Australia, the **Aborigines**, have the world's oldest surviving culture. The first of these nomadic hunters and gatherers is believed to have arrived in Australia about 40,000 to 60,000 years ago from Southeast Asia. Scientists believe that they migrated to the area over land bridges during the Ice Age, when ocean levels that were much lower than today had exposed land on the continental shelves. These early Aborigines led a nomadic life and used well-traveled routes to reach water and seasonal food sources. Family groups called **clans** traveled together within their territories, carrying what they needed to survive, such as baskets, bowls, spears, and sticks for digging. To hunt, Aboriginal men used a heavy throwing stick, called a **boomerang**, which soars and curves in flight. Women and children gathered plants and seeds.

Migration was **accompanied** by increased trade among the islands of the South Pacific. Research suggests that sometime in the A.D. 1200s the **Maori** left eastern Polynesia and settled the islands of New Zealand. Maori farmers lived in villages and grew traditional root crops such as taro and yams, which they had brought from their Polynesian homeland.

Aborigine the indigenous people of Australia

clan a group of close-knit, interrelated families

boomerang an Australian throwing stick that soars and curves in flight and returns near the thrower

accompany to go with

Maori the indigenous people of New Zealand

impact an effect

European Exploration and Settlement

Europeans began exploring the South Pacific in the 1500s. Captain James Cook, a British sailor, arrived in Tahiti in 1769. He then sailed southwest and, in October 1769, reached New Zealand. Over the next six months, Cook charted all of New Zealand's coasts. He decided to take a different route home. He crossed the Tasman Sea westward instead of going east around Cape Horn. In 1770 Cook came upon the southeast coast of Australia. He claimed the lands of New Zealand and Australia for Great Britain.

In 1788 Great Britain began to use Australia as a penal colony for convicts from overcrowded British prisons. By the early 1850s, free British settlers were establishing settlements along the east coast. Sheep were introduced and settlers profited from exporting wool to Britain. Another source of wealth was gold, discovered in the 1850s. Escaped convicts known as bushrangers, however, made life for the free settlers difficult. Bushrangers supported themselves by stealing from free settlers. To combat the problem, strict laws were put in effect.

During the same time, the British and other Europeans established settlements in New Zealand, which offered fishing and rich soil for farming. By the end of the 1800s, raising livestock, primarily sheep, had become a major part of the economy.

The arrival of Europeans in Australia had a disastrous **impact** on indigenous peoples. Many were forcibly removed from their land and denied basic rights. They resisted the Europeans, and conflicts were common. Violence and European diseases steadily reduced the Aboriginal population. In the mid-1800s, authorities placed many Aborigines on reserves, or separate areas.

In New Zealand, British settlement brought hardship to the Maori as well. In 1840 Britain and Maori groups of North Island signed the Treaty of Waitangi. The treaty purported to protect the rights of the Maori, but was used as a basis for British annexation of North Island. The Maori mounted an armed resistance against British rule that lasted for 15 years, but ended up losing most of their land.

Independence

dominion a largely self-governing country within the British Empire

In 1901 the Australian colonies decided to form a federation known as the Commonwealth of Australia. The new country was a **dominion**, a largely self-governing country within the British Empire. Its government blended a U.S.-style federal system with a British-style parliamentary democracy.

New Zealand became self-governing as a colony in 1853. The colony's Parliament permitted voting to men who owned property. This requirement effectively disqualified most Maori from voting since they did not own property. In 1867 New Zealand passed the Maori Representation Act, which ensured that the Maori were represented in Parliament. In 1893 New Zealand became the first country in the world to legally recognize women's right to vote, including Maori women. In 1907 New Zealand became a self-governing dominion using a British parliamentary system. Beginning in 1975 the voting laws changed, permitting Maori and people of European descent to vote in the same elections.

Both Australia and New Zealand are constitutional monarchies with parliamentary systems. They have written constitutions. The head of state, a largely ceremonial role, is the British monarch. The people elect members of a parliament. The leader of the political party with the majority of votes for parliament is the prime minister, or head of the government. A movement in Australia known as republicanism hopes to establish a republican form of government in which the British monarch would not be head of state. Instead, officials of the state would be citizens who are directly or indirectly elected or appointed.

☑ READING PROGRESS CHECK

Sequencing When did the first inhabitants of Australia and New Zealand arrive?

Population Patterns

GUIDING QUESTION *How have migration and an aging population affected Australia's and New Zealand's population patterns?*

Both New Zealand and Australia are multicultural countries. Immigrants from around the world move to both countries to make a new life. Over 7 million people have migrated to Australia since 1945. Of today's population, 44 percent were either born or have a parent who was born in another country. As a reult, many languages are spoken in the country, although English is the dominant language. In New Zealand, most early immigrants were British with some Dutch immigrants arriving in the 1950s. By the mid-1970s policy changes began to make New Zealand a more welcoming place for other immigrants. From the 1980s to today, Pacific Islanders, Asians, Africans, and immigrants from the Middle East have immigrated and now call New Zealand home.

Numerous groups make up the indigenous Aborigines of Australia. One such group is the Arrente, who have lived in central Australia for about 20,000 years. Another group is the Palawa, who have lived on the island of Tasmania for about 32,000 years. After years of harsh treatment and isolation in the Outback and other isolated areas, the Aborigines are now demanding more opportunities. In 1967 the Australian government finally recognized the Aborigines as citizens. Today, about two percent of the population is Aborigine. Growing numbers of Aborigines, particularly youth, are moving to cities.

Approximately 7 percent of New Zealand's population is Maori. The majority of the country's population is descended from British settlers. Asians and Pacific Islanders, attracted by the growing economy, have increased the diversity of New Zealand's society.

GEOGRAPHY CONNECTION

Europeans settled parts of Australia at different times.

1. **HUMAN SYSTEMS** Which areas of Australia were the last to be settled by Europeans?

2. **THE WORLD IN SPATIAL TERMS** Where were most areas of Aboriginal resistance located?

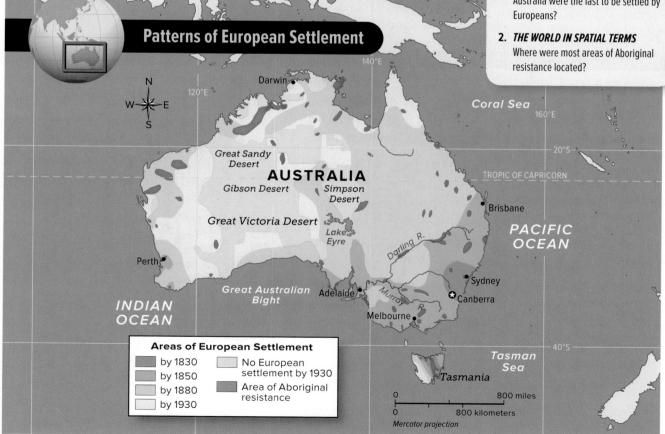

Patterns of European Settlement

Areas of European Settlement
- by 1830
- by 1850
- by 1880
- by 1930
- No European settlement by 1930
- Area of Aboriginal resistance

0   800 miles
0   800 kilometers
Mercator projection

Australia's physical geography affects the distribution of its people. Few people live in the dry central plateaus and deserts of the interior. Most people in Australia live along the southeastern, eastern, and southwestern coasts. These areas have mild climates, fertile soils, and access to the sea. The largest Australian cities are Sydney and Melbourne. Both have more than 4 million residents and both are major commercial ports.

About 85 percent of the inhabitants of New Zealand live in urban areas, mostly located along the coast. New Zealand's ports of Auckland, Christchurch, and Wellington are the country's largest cities. Both Wellington, the capital, and Auckland are located on North Island, where about 75 percent of all New Zealanders live.

The aging population of Australia and New Zealand is of growing concern for the governments of both countries. In Australia, the government expects record increases in the number of people ages 65 and over between now and the year 2021. By 2051, models indicate that between 24 and 26 percent of the population is expected to be over the age of 65. Even though New Zealand's population is aging, its birthrates have increased. As a result, the total number of young people is increasing.

An aging population means that there will be more and more people in need of long-term health care and other assistance. Without an increase in the birthrate, the ratio of persons in the labor force will decline in relation to the number of retired persons. Thus, the government and people of Australia may have to make economic choices in the future. Spending on retirement support, education, or infrastructure are among those choices. Young immigrants from countries such as Great Britain, Taiwan, China, and South Korea are also affecting the population dynamics of Australia.

☑ READING PROGRESS CHECK

Analyzing What are the concerns about the aging population in Australia and New Zealand?

TIME LINE ⌄

MIGRATION
and Settlement ➜

The history of Australia and New Zealand has been deeply affected by the relations between the indigenous peoples—the Aborigines in Australia and the Maori in New Zealand—and the Europeans who settled there.

▶ CRITICAL THINKING

1. *Describing* What are the origins of the major groups of people in Australia and New Zealand?

2. *Explaining* What are some reasons why British people immigrated to Australia?

First Aborigines arrive in Australia from Southeast Asia.

c. 40,000 B.C.

Aborigines spread throughout Australia and Tasmania.

c. 20,000 B.C.

40,000 B.C. ➜

A.D. 1200s

1642 Dutch explorer Abel Tasman is the first European known to reach New Zealand.

Maori arrive in New Zealand from different parts of eastern Polynesia.

Society and Culture Today

GUIDING QUESTION *How have immigration and migration shaped society and culture in Australia and New Zealand?*

The people of Australia and New Zealand feel a strong **national identity** toward their countries. Australia and New Zealand blend both European and indigenous elements in their cultures. In recent years, Asian influences have also increased in the region. While daily life in much of Australia and New Zealand may resemble that in Western countries, there is diversity in lifestyles across the subregion.

English is the major language spoken in both Australia and New Zealand. Australian English, called **Strine**, has a unique vocabulary made up of Aboriginal words, terms used by early settlers, and slang created by modern Australians.

Because of the rather large population of Maori in New Zealand, Maori is also spoken in certain places. Two percent of Australians—the Aborigines—speak Aboriginal languages. Aborigines also speak **pidgin** English.

Most Australians and New Zealanders, especially those in cities, have access to quality medical care and other social services. The rugged terrain and isolation of some parts of Australia can make access to health care difficult. However, modern technology allows doctors to reach and treat patients through the use of two-way radios, mobile clinics, computer displays, and by air ambulance.

Aborigines and Maori have historically received lower levels of health care, education, and government benefits. Many Aborigines are poor and suffer from malnutrition and unemployment. In recent years, the governments and private organizations of both countries have been working to make up for past injustices. The courts have recognized the claims of the Aborigine and Maori peoples to government assistance, land, and natural resources.

Both Australia and New Zealand provide free compulsory education. Literacy rates are over 99 percent, and many students attend universities. Students in Australia's remote Outback receive and submit work using the Internet, mail, and two-way radios.

national identity the sense of a being part of the whole of a country including its culture, traditions, language, and politics

Strine Australian English that has a unique vocabulary made up of Aboriginal words, terms used by early settlers, and slang created by modern Australians

pidgin a grammatically simplified form of a language

1788 Great Britain begins to send convicts to penal colony in New South Wales, Australia.

1769 British captain James Cook explores New Zealand coastline.

➜ **1800**

1815 First British missionaries arrive in New Zealand with goal of converting Maori to Christianity.

Increasing Chinese immigration leads to anti-Chinese legislation in Australian colony of Victoria.

1850s Gold rush in Victoria draws thousands to Australia, increasing the population by 300 percent.

1855

➜ **1900**

1888 Severe restrictions on Chinese immigration enacted throughout Australia

Homes Are Sought For These Children

A GROUP OF TINY HALF-CASTE AND QUADROON CHILDREN at the Darwin half-caste home. The Minister for the Interior (Mr Perkins) recently appealed to charitable organisations in Melbourne and Sydney to find homes for the children and rescue them from becoming outcasts.

I like the little girl in centre of group, but if taken by anyone else, any of the others would do, as long as they are strong.

Children of the Stolen Generations were forcibly removed from their families, who had no right of appeal.

▲ **CRITICAL THINKING**

1. *Analyzing Visuals* What effect do you think removal had on the children of the Stolen Generations?

2. *Speculating* Why do you think the Australian government removed Aboriginal children from their families?

Today Aborigines are also known as the First Australians, though they lost their land and were marginalized by European settlement. The Australian government tried to suppress the Aborigines' culture to make them more English by removing Aboriginal children from their homes and placing them with foster parents or in boarding schools. The children were not allowed contact with their families. These children are known as the Stolen Generations. About 100,000 Australian Aboriginal and Torres Strait Islander children were removed from their families between 1910 and 1971.

There have been recent improvements in relations between Aborigines and Australians of European descent. In 2008 Australian prime minister Kevin Rudd apologized to the Aborigines for policies and laws of previous governments that were responsible for the Stolen Generations. In the same year, a documentary presented the history of the Aborigines from 1778 through 1993. It showed the experiences of Aborigines during the British settlement of Australia.

In New Zealand, relations between the Maori and the Europeans were on a more equal basis. However, the Maori were still disadvantaged. In the 1970s and 1980s, the Maori become more politically active. They sought the return of land and compensation from the British government for the loss of access to natural resources. These were items promised to the Maori with the Treaty of Waitangi in 1840. By 1998, the Maori had achieved substantial compensation from the government of New Zealand. Eventually, the government also apologized for the injustices inflicted on the Maori.

For many years, both the Aborigines and the Maori suffered racism from British settlers. With increased integration into the national identity of each country and a better understanding of the cultures that make up these national identities, some think that racism will become a thing of the past.

Australians and New Zealanders have strong family ties. Most people in Australia live in a nuclear family of parents and children, but some live as extended families. In New Zealand, families are also made up of mainly parents and children. The Maori tradition has strong ties with extended families. The word for extended family in the Maori language is *whānau*.

Australian women are active in both working at home and in jobs outside the home. Women work both full and part time while they have young children. Many women have balanced full-time work with caring for a family. Between 1996 and 2005, working mothers with a child under five increased from 46 percent to 52 percent. This suggests more women are in full- and part-time work.

In New Zealand, the role of women has shifted to greater opportunities. After high school, some women find jobs in offices, shops, the health services, and as teachers. Others go on to study at a university. Some women may choose to stay home as wives and mothers.

Australians and New Zealanders traditionally used the arts—art, music, dance, and storytelling—to pass on knowledge from generation to generation. Aborigines recorded their past in rock paintings and used songs to share information about watering holes and landmarks. Maori artisans developed skills in canoe making, basketry, and woodcarving.

Movies have been filmed in New Zealand, including *The Lord of the Rings* trilogy. New Zealand has an innovative and world-renowned film industry. It is also the birthplace of several famous directors and actors including Peter Jackson, Anna Paquin, and Russell Crowe. Australian actors include Cate Blanchett, Hugh Jackman, and Nicole Kidman.

✔ **READING PROGRESS CHECK**

Analyzing How might racism become a thing of the past in Australia and New Zealand?

Economic Activities

GUIDING QUESTION *What are the characteristics of the economies of Australia and New Zealand?*

Australia and New Zealand have a close economic relationship. They have signed trade agreements such as the Australia–New Zealand Closer Economic Relations Trade Agreement (ANZCERTA) of 1983. The objective of these agreements is to improve their economic relationship by eliminating trade barriers.

Australia has a diverse economy. This includes agriculture, mining, steel, industrial and transportation equipment, food processing, and chemicals. The country also has natural resources such as gold, iron ore, copper, coal, and uranium that invite foreign investments and are a significant portion of the country's exports. Australia also exports energy and food. Agricultural products include wheat, barley, sugarcane, fruit, cattle, sheep, and poultry.

New Zealand's main industries are food processing, wood and paper products, textiles, machinery, transportation equipment, banking and insurance, tourism, and mining. Agriculture is a significant part of New Zealand's economy. Agricultural products include wool, dairy products, lamb and mutton, beef, fish, wheat, barley, potatoes, pulses, fruit, and vegetables. New Zealand also has many natural resources. These include natural gas, iron ore, sand, coal, timber, hydropower, gold, and limestone. New Zealand is a significant exporter of dairy products, meat, wool, wood products, fish, and machinery.

Both countries have large service sectors. About 75 percent of people in Australia and 74 percent of the people in New Zealand work in services ranging from government agencies to banking and tourism. Service industries are major contributors to the economies of both countries and increase their gross domestic product (GDP). Australia is ranked nineteenth in the world by GDP and New Zealand is ranked seventy-first.

✔ **READING PROGRESS CHECK**

Evaluating What economic activities can be found in Australia and New Zealand?

LESSON 2 REVIEW

Reviewing Vocabulary

1. ***Identifying*** Name four elements of the cultures of the peoples of Australia and New Zealand and describe each in a sentence.

Using Your Notes

2. ***Analyzing*** Use your graphic organizer about the human geography of Australia and New Zealand to write a paragraph about how physical systems affect both countries.

Answering the Guiding Questions

3. ***Drawing Conclusions*** How did migration and geography influence the history and governments of Australia and New Zealand?

4. ***Evaluating*** How have migration and an aging population affected Australia's and New Zealand's population patterns?

5. ***Making Connections*** How have immigration and migration shaped society and culture in Australia and New Zealand?

6. ***Explaining*** What are the characteristics of the economies of Australia and New Zealand?

Writing Activity

7. ***Informative/Explanatory*** In a paragraph, discuss how the people of Australia and New Zealand, both past and present, have shaped the subregion.

NON-NATIVE SPECIES:

RABBITS
IN
AUSTRALIA

In 1859 a man named Thomas Austin freed 24 rabbits in Australia. By 1920 there were 10 billion rabbits spread across the continent. By 1990, due to successful eradication attempts, the population was down to 600 million.

RABBIT POPULATION

- ● ABUNDANT
- ○ COMMON
- ● RARE
- ▬ RABBIT FENCE

BARWIN PARK, VICTORIA
Original site of Thomas
Austin's rabbit release.

"The introduction of a few rabbits could do little harm and might provide a touch of home, in addition to a spot of hunting."

—Thomas Austin

WHY DID THEY BECOME A PROBLEM?

- Rabbits reproduce at an extremely rapid rate.

- They can live in a variety of environments.

- They have few natural predators in Australia.

- Diseases and parasites that control the populations in Europe do not exist in Australia.

DAMAGING EFFECTS ⚠️

Rabbits are particularly damaging to the natural environment of Australia. They graze on plants, eating them down to the ground. Rabbits also eat young plants and seedlings.

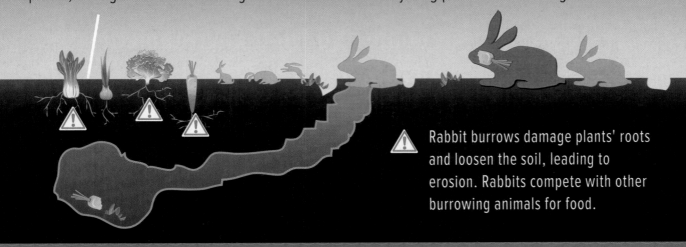

⚠️ Rabbit burrows damage plants' roots and loosen the soil, leading to erosion. Rabbits compete with other burrowing animals for food.

ATTEMPTED SOLUTIONS

DISEASE

A disease (Myxomatosis) common in South American rabbits was introduced in 1950 and killed most of the rabbits. However, the disease is carried by mosquitoes so this solution didn't work in arid areas where mosquitoes don't thrive.

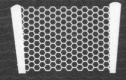

RABBIT-PROOF FENCES

About 2,023 miles (3,236 km) of rabbit-proof fences were built in Australia. They failed because parts were poorly built and rabbits climbed over or dug under the fences.

INTEGRATED RABBIT CONTROL

The best solution seems to be using a variety of tools at once. Combined techniques include poisoning, fencing, and burrow destruction and fumigation.

Making Connections

1. **Places and Regions** Why do you think rabbit species found in Australia are most commonly found in the south?

2. **Physical Systems** Few natural rabbit predators exist in Australia. Why would the introduction of a non-native predator species only add further problems?

3. **Human Systems** The threat of an accidental introduction of non-native species is still real, especially in seaports and airports. Write a one-page proposal recommending ways Australia can protect itself against such accidental intrusions.
   *Interact with **Global Connections** Online*

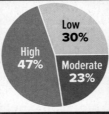

netw⊚rks

There's More Online!

- ☑ **GRAPHS** Levels of Coral Bleaching in the Great Barrier Reef
- ☑ **IMAGE** European Rabbit
- ☑ **IMAGE** Kaingaroa Forest
- ☑ **IMAGE** Whale Rescue
- ☑ **MAP** How Invasive Species Get to Australia
- ☑ **INTERACTIVE SELF-CHECK QUIZ**
- ☑ **VIDEO** People and Their Environment: Australia and New Zealand

Reading HELPDESK

Academic Vocabulary

- **adapt**
- **compound**

Content Vocabulary

- **station**
- **copra**
- **grazier**

TAKING NOTES: *Key Ideas and Details*

IDENTIFYING As you read about the people and the environment of Australia and New Zealand, use a graphic organizer like the one below to identify problems faced by the people and the different ways in which groups and individuals are trying to address these problems.

> Australia and New Zealand
>
> Environmental Issues
>
> Addressing the Issues

LESSON 3

People and Their Environment: Australia and New Zealand

ESSENTIAL QUESTION • *How do physical systems and human systems shape a place?*

IT MATTERS BECAUSE

Australia and New Zealand have strikingly beautiful landscapes. Both countries are also rich in natural resources. However, overuse and misuse have created some threats to the natural habitats in this subregion, which both countries are trying to correct. Government, national and international organizations, and individuals are working toward new ways to protect and preserve their environment.

Managing Resources

GUIDING QUESTION *What environmental impact did European practices of land management have in Australia and New Zealand?*

The remote geographic locations and challenging environments of Australia and New Zealand influence how people earn their living. Agriculture is by far the most significant economic activity. As you have read, for example, Australia and New Zealand export large quantities of farm products. Australia is the world's leading producer of wool. However, the lack of rainfall in Australia makes agriculture challenging.

Although less than 5 percent of Australians work in agriculture, much of the country's vast land area is devoted to raising livestock. Because of the generally dry climate, animals need large areas to find enough vegetation to eat. As a result, some Australian ranches, called **stations**, are huge. They can be as large as 6,000 square miles (15,540 sq. km), roughly the size of Connecticut. The impact of long-term livestock grazing in Australia can be the degradation of many of the natural ecosystems in the country.

Due to Australia's dry climate, less than 10 percent of its land is arable, or suitable for growing crops. Irrigation, fertilizers, and modern technology help Australian farmers make more productive use of the land.

In New Zealand, more than half of the land is used for agriculture. In addition to grazing, animals are fed **copra**, or dried coconut meat, mainly imported from countries such as Fiji and Papua New Guinea. Ranchers, known as **graziers** in New Zealand, raise sheep, cattle, and red deer. The country has nearly 20 times more livestock than people.

Unlike Australia, New Zealand has some of the most fertile soil in the region. This allows farmers to grow wheat, barley, potatoes, and fruits.

Although new plants and animals have helped both countries develop strong agricultural industries, the impact to the region has been detrimental in some cases. In Australia, native plants and animals were well **adapted**, or fitted, to life on an isolated continent. Since European settlement, these native plants and animals have had to compete for habitat, food, and shelter with exotic plants and animals that were introduced into the environment from other places. For example, 24 European rabbits were introduced to Australia in 1859 for hunting purposes. But rabbits had few natural predators in Australia, so by 1926, there were about 10 billion rabbits in the country. These animals destroyed native plants, exposing the soil to wind and water. The resulting soil erosion caused so much damage that some farms were abandoned. Feral foxes are also a problem. They have caused the decline of several species of native animals, and they prey on newborn lambs.

The introduction of exotic plants in Australia has also created problems. While some of these exotic plants were introduced intentionally, most were introduced accidentally. Many of them have become weeds. For example, the prickly pear was introduced in the 1900s to establish a cochineal dye industry. The prickly pear soon became an invasive species, turning 15,000 square miles (40,000 sq. km) of farmland into a sea of weeds. The weed was eventually eradicated.

Introduced species have also harmed native species in New Zealand. Feral cats and dogs have made New Zealand's native kiwi, a flightless bird, an endangered species. Weasels and ferrets introduced to control rabbits have killed native birds.

☑ READING PROGRESS CHECK

Explaining How do people in Australia and New Zealand use the land?

Human Impact

GUIDING QUESTION *How has land use in Australia and New Zealand affected their respective environments?*

Another issue of concern is the protection of forest, soil, and freshwater resources. In Australia, many forests have been cleared for farms and grazing lands. In New Zealand, forests are also converted to pastureland for livestock. Between 2000 and 2010, some 74,000 acres (30,000 ha) of forest were converted to pasture. This leaves little protection against soil erosion. Deforestation also causes soil salinity as well as increased risk of fire and floods. However, pasture continues to dominate New Zealand land use.

In Australia, soil erosion is **compounded** by overgrazing in arid areas and by the country's worst drought in over a century. Soil conservation in the region is closely linked to reducing deforestation. Australia and New Zealand are aware of the problems and are developing plans to use forest resources while reducing damage to the environment.

Land use has significantly modified the physical and chemical nature of rivers in both countries. This has caused unforeseen consequences. For example, drought, salt, irrigation, and agricultural runoff threaten Australia's freshwater sources.

The introduction of the European rabbit to Australia and New Zealand caused serious damage to the environment.

▲ CRITICAL THINKING

1. *Making Connections* How could a small mammal cause damage to the environment?

2. *Problem Solving* What steps could the governments of Australia and New Zealand take to deal with the rabbit problem?

station an Australian ranch

copra dried coconut meat fed to animal herds

grazier a New Zealand rancher

adapt to make fit to changing circumstances

compound to add to

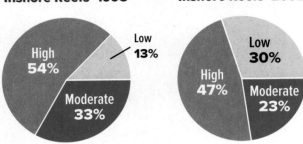

Inshore Reefs–1998

High 54%
Low 13%
Moderate 33%

Inshore Reefs–2002

Low 30%
High 47%
Moderate 23%

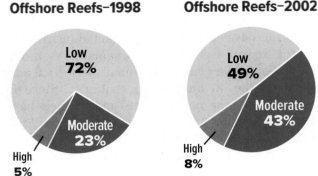

Offshore Reefs–1998

Low 72%
Moderate 23%
High 5%

Offshore Reefs–2002

Low 49%
Moderate 43%
High 8%

Note: Inshore = less than 6.2 miles (10 km) from the coast
Source: Australian Institute of Marine Science.

Warmer water temperatures can result in coral bleaching. In 1998 and 2002 massive bleaching events occurred in the Great Barrier Reef.

▲ **CRITICAL THINKING**

1. **Differentiating** In which year did inshore reefs suffer higher levels of coral bleaching?

2. **Analyzing** Which type of reef has suffered higher levels of coral bleaching?

The fertile Murray-Darling River Basin is one of the world's largest drainage basins. However, the use of water for irrigation and the increasing needs of growing city populations have dramatically reduced the rivers' flows. Large areas within the basin are also at risk from increasing soil salinity. One of the major causes of increasing salinity has been the replacement of native vegetation—which has deep root structures—with pastures and non-native shallow-rooted crops. Deep root structures prevent rainwater from reaching deep underground and causing salt to rise to the surface soil and water. In increased soil salinity, plants are unable to grow. Salinity causes building foundations, fences, and roads to crumble. Water salinity causes water to be too salty for consumption by humans and wildlife.

Agricultural runoff, chemical fertilizers, and organic waste also threaten the oceans that surround the subregion. Toxic waste, in particular, endangers Australia's Great Barrier Reef and other coral reefs. Tourists, boaters, divers, oil shale mining, and rising water temperatures increasingly stress these coral environments.

Some scientists argue that increases in Earth's temperatures could be devastating. If polar ice caps melt, ocean levels will rise. This would cause extensive flooding. Rising ocean temperatures also affect certain types of plankton and algae that grow in warm waters, causing overgrowth and the choking out of other life forms. The breakdown of the relationship between coral and the algae that provide it with nutrients causes coral bleaching. Scientists in the region are studying global warming and are hoping to discover causes, predict consequences, and provide solutions.

Pollution also affects marine life. This is especially true for the tiny organisms that make up coral reefs. Algae—on which these organisms thrive—and plankton are key parts of the ocean's food web. A food web refers to the interlinking chains of predators and their food sources in an ecosystem. As these tiny living things are destroyed, larger plants and animals that rely on them for food also die off.

☑ **READING PROGRESS CHECK**

Explaining How does human activity impact the land and waters of Australia and New Zealand?

Addressing the Issues

GUIDING QUESTION *How are groups, governments, and other organizations addressing environmental issues in Australia and New Zealand?*

Australia and New Zealand are facing tough environmental issues. Both governmental and nongovernmental organizations are working to reduce these problems. Some efforts are intended to reverse some of the environmental damage in both countries.

The Australian government has reacted to these concerns by enacting environmental laws and initiatives. It has also created organizations and education programs to combat the problems. Australia's environmental minister is responsible for managing these issues. The government created the Natural Heritage Trust in 1997. Now known as Caring for our Country, its mission is to help restore and conserve Australia's environmental and natural resources. It provides funding for environmental activities at community, regional, state, and national levels.

Caring for our Country funds various projects. Some help to protect and restore the habitat of threatened animal species. Other projects aim to reverse the decline of Australia's native vegetation. Another important area is preventing or controlling the introduction and spread of feral animals, aquatic pests, weeds, and other biological threats to biodiversity. The restoration and protection of freshwater, marine, and river ecosystems are also important to the organization.

The Australian Government Envirofund delivers funding to projects at the local level. This fund was formed to assist individuals and community groups undertaking small projects aimed at sustainable resource use and protecting biodiversity. At the regional level, governmental programs distribute money to national resource management regions (NRMs). There are 56 NRMs in Australia that support communities, farmers, and land managers to protect Australia's natural environment and increase sustainability of the country's ecosystems. At the national and state level, projects cover national priorities that cross local and regional boundaries.

Rescuers treat an injured whale before releasing it again from Hamelin Bay, south of the Australian city of Perth.

◀ **CRITICAL THINKING**

1. *Analyzing* What are the threats to the health of Australia's marine ecosystems?

2. *Problem Solving* How has the Australian government worked to protect marine life?

Tony Ashby/AFP/Getty Images

Young trees are recently planted in Kaingaroa Forest, which is the largest forest in New Zealand.

▲ CRITICAL THINKING

1. *Analyzing* Why is reforestation important to the environment and people of New Zealand?

2. *Describing* In what ways are New Zealanders trying to protect their environment?

Nongovernmental organizations also work to improve the environment in Australia. For example, the World Wildlife Fund (WWF) works to conserve the country's biodiversity by providing practical solutions to the continent's greatest environmental threats. They have been successful in improving the protection of the Great Barrier Reef by persuading the government of Australia to increase its commitment to protect this vast ecosystem. WWF has also successfully campaigned against land clearing in Queensland, Australia. In 2007 the Australian government revoked all permits for broad-scale clearing in the state. WWF continues to work with local communities and in partnership with government and industry to encourage effective conservation policies.

The WWF is also working in New Zealand to fund projects to help the environment. For example, the WWF partnered with the Tindall Foundation on the Habitat Protection Fund in 2000. Since its launch, it has partnered with volunteer groups on over 400 projects to protect areas of high conservation significance in their communities. The WWF and the Tindall Foundation are also working with the Environmental Education Action Fund. This fund supports and promotes environmental education through actions for schools and communities across New Zealand. The Conservation Innovation Fund finances activities that demonstrate innovation and conservation of the environment.

New Zealand's government is also making efforts to solve environmental problems. In 1991 it passed the Resource Management Act. Its aim is to promote the sustainable management of natural resources. One solution is through reforestation to reduce soil erosion. Efforts to reforest the exotic forests of Kaingaroa between Rotorua and Taupo began in the late 1920s. Over the last 30 years, further planting has been encouraged in order to stabilize eroding farmland. Increased log prices and improved harvesting techniques have also encouraged reforestation.

The government of New Zealand also works with the National Institute of Water and Atmospheric Research. Its goal is to improve the economic value and sustainable management of its aquatic resources and environments. The Institute works to increase understanding of the climate and the atmosphere. It hopes that additional understanding will increase resilience to weather and climate hazards. This will improve the safety of New Zealanders.

☑ READING PROGRESS CHECK

Evaluating How are Australians and New Zealanders addressing environmental problems?

LESSON 3 REVIEW

Reviewing Vocabulary

1. *Describing* Describe ranching in Australia and New Zealand.

Using Your Notes

2. *Summarizing* Use your graphic organizer on the environmental problems of Australia and New Zealand to write a summary about how the problems are being addressed.

Answering the Guiding Questions

3. *Synthesizing* What environmental impact did European practices of land management have in Australia and New Zealand?

4. *Evaluating* How has land use in Australia and New Zealand affected their respective environments?

5. *Classifying* How are groups, governments, and other organizations addressing environmental issues in Australia and New Zealand?

Writing Activity

6. *Informative/Explanatory* In a paragraph, write about the cause-and-effect relationship between people and nature in Australia and New Zealand.

Directions: On a separate sheet of paper, answer the questions below. Make sure you read carefully and answer all parts of the questions.

Lesson Review

Lesson 1

1 **Describing** What three deserts make up the Australian Outback? Why do many Australians also call this area the bush?

2 **Contrasting** How do water resources in New Zealand differ from those in Australia?

3 **Explaining** What physical features of Australia and New Zealand make tourism an important industry in both countries?

Lesson 2

4 **Evaluating** How did the arrival of Europeans in Australia and New Zealand affect indigenous peoples?

5 **Analyzing** Why has modern technology become vital to people living in remote areas of the Australian Outback?

6 **Comparing** What industries do Australia and New Zealand have in common?

Lesson 3

7 **Making Connections** Why are some Australian ranches thousands of square miles in size?

8 **Explaining** How does global warming lead to higher levels of coral bleaching?

9 **Identifying** What are some steps the government of New Zealand has taken to address the problem of soil erosion?

Critical Thinking

10 **Predicting** What problems are New Zealand and Australia likely to face if their populations continue to age?

11 **Hypothesizing** Why are native plants and animals in New Zealand and Australia especially vulnerable to introduced species?

21st Century Skills

Use the graph to answer the following questions.

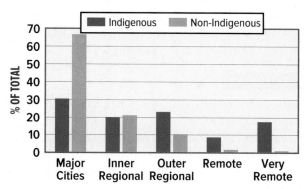

POPULATION PATTERNS IN AUSTRALIA

Source: http://www.environment.gov.au/

12 **Identifying Continuity and Change** How does the percentage of non-indigenous Australians change as areas become more remote?

13 **Explaining Continuity and Change** Why do you think most people living in very remote areas are indigenous?

14 **Identifying Continuity and Change** What does the chart suggest about changes in population patterns among Aborigines?

Need Extra Help?

If You've Missed Question	1	2	3	4	5	6	7	8	9	10	11	12	13	14
Go to page	776	778	778	782	785	787	790	792	794	784	791	795	795	795

Directions: On a separate sheet of paper, answer the questions below. Make sure you read carefully and answer all parts of the questions.

College and Career Readiness

15 *Clear Communication* Imagine that you are a journalist assigned to write a news article about the kiwi, a flightless bird native to New Zealand. Why has the kiwi become an endangered species? What measures are being taken to protect the kiwi from extinction? How successful have they been? Respond to these questions in your article.

DBQ Analyzing Primary Sources

Use the document to answer the following questions.

From 1910 to 1971, thousands of Australian Aboriginal and Torres Strait Islander children were removed from their homes and families and taken to state boarding schools.

PRIMARY SOURCE

"And anyway, these state school teachers, it was very, very different being taught by them. . . . We were separated, me and my mates, to the fair kids to the black kids. Us black ones sat in the dunce corner, cause they said we couldn't do nothing much for the black people. And for the fair ones they said they can do so much for them because they can be taught to—in school in Perth.

They used to ridicule us all the time. They even had us scrubbing— trying to scrub the black skin off our face, because they never took notice of the kids with dark skin. They concentrated on the fair skin ones you see."

—Glenys Ward, stolengenerationstestimonies.com

16 *Interpreting* Why did the state school teachers separate students by skin color?

17 *Analyzing* What do you think the teachers wanted to teach Glenys by forcing her to "scrub the black off" her face?

18 *Theorizing* Glenys says that the teachers believed they "couldn't do nothing much for the black people." Why do you think she was taken to the school?

Applying Map Skills

Use the Unit 9 Atlas to answer the following questions.

19 *Environment and Society* What percentage of land in New Zealand is arable? What percentage is arable in Australia?

20 *Human Systems* How does the physical geography of Australia help explain the uneven distribution of its population?

21 *Physical Systems* Use your mental map of Australia to describe the location and extent of the Great Barrier Reef.

Exploring the Essential Question

22 *Diagramming* Design a series of diagrams showing the causes and consequences of increased salinity in soil. Remember to show how the replacement of native plants with shallow-rooted crops allows salt to rise to the surface, making soil unsuitable for plants. Label the diagrams to describe each stage in the process.

Research and Presentation

23 *Research Skills* Using the Internet and other resources, research the history and culture of New Zealand's Maori people. Describe how the Maori were affected by British settlement. What steps did the Maori take to achieve political equality? How do the Maori live today? Create a multimedia presentation to share your findings.

Writing About Geography

24 *Argument* Use standard grammar, spelling, sentence structure, and punctuation to write a three-paragraph essay arguing why Australia's Great Barrier Reef should continue to be listed as a World Heritage Site. Discuss the features that make the reef unique, its importance to human and natural systems, and the threats it faces.

Need Extra Help?

If You've Missed Question	**15**	**16**	**17**	**18**	**19**	**20**	**21**	**22**	**23**	**24**
Go to page	791	796	796	796	746	746	746	792	782	779

Oceania

ESSENTIAL QUESTION • *How do physical systems and human systems shape a place?*

networks

There's More Online about Oceania's geography.

CHAPTER 33

Why Geography Matters
Samoa Hops the International Date Line

Lesson 1
Physical Geography of Oceania

Lesson 2
Human Geography of Oceania

Lesson 3
People and Their Environment: Oceania

Geography Matters...

Oceania is one of the most fascinating places on Earth. The physical geography is varied and ranges from volcanic mountains to blue lagoons. Oceania also includes many tiny islands and atolls. Some of these are so small that they appear on maps as tiny dots or are not on maps at all.

Outside influences on indigenous cultures have shaped Oceania's societies. Today many people blend elements of their traditional culture with those of Western cultures. Migration of people among the islands of Oceania has also shaped life on the islands today.

◀ This Yap Island woman in Micronesia wears a traditional dance costume.

©Michael DeFreitas/Robert Harding World Imagery/Corbis

797

Samoa hops *the* International Date Line

The country of Samoa is a group of islands located in the Pacific Ocean about halfway between Hawaii in the United States and New Zealand. On December 29, 2011, Samoa (along with the neighboring New Zealand territory of Tokelau) took an important step. At midnight, Samoa jumped the International Date Line, moving 24 hours ahead and losing a day.

Samoa and the International Date Line

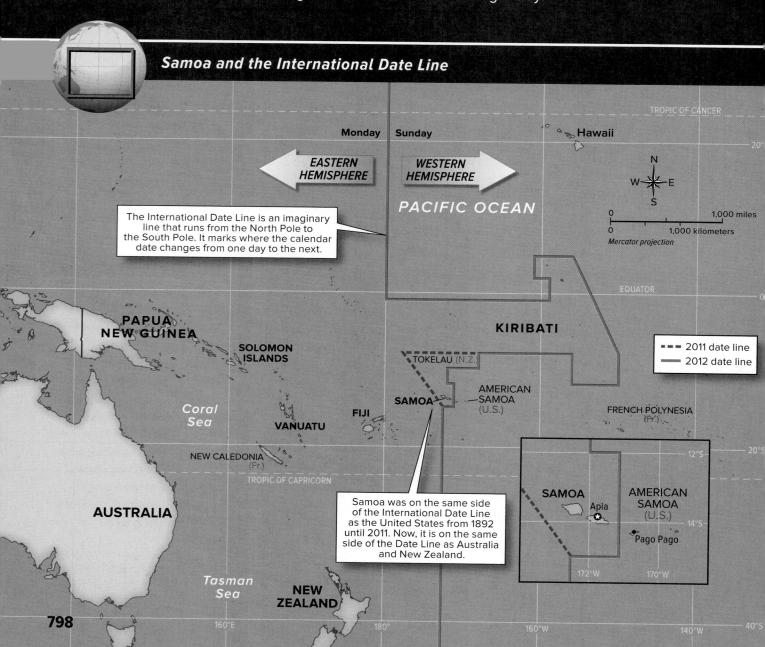

Monday

Sunday

Hawaii

EASTERN HEMISPHERE

WESTERN HEMISPHERE

PACIFIC OCEAN

The International Date Line is an imaginary line that runs from the North Pole to the South Pole. It marks where the calendar date changes from one day to the next.

TROPIC OF CANCER

N
W — E
S

0 _____ 1,000 miles
0 _____ 1,000 kilometers
Mercator projection

EQUATOR

KIRIBATI

PAPUA NEW GUINEA

SOLOMON ISLANDS

TOKELAU (N.Z.)

- - - 2011 date line
——— 2012 date line

AMERICAN SAMOA (U.S.)

Coral Sea

FIJI

SAMOA

FRENCH POLYNESIA (Fr.)

VANUATU

NEW CALEDONIA (Fr.)

TROPIC OF CAPRICORN

AUSTRALIA

Samoa was on the same side of the International Date Line as the United States from 1892 until 2011. Now, it is on the same side of the Date Line as Australia and New Zealand.

SAMOA Apia

AMERICAN SAMOA (U.S.)

Pago Pago

172°W   170°W

12°S

14°S

Tasman Sea

NEW ZEALAND

160°E   180°   160°W   140°W   40°S

What is the International Date Line?

Why did Samoa make the switch?

How was the switch accomplished and what were the results?

The International Date Line is an imaginary line that marks where the calendar date changes from one day to the next. It extends from the North Pole to the South Pole, roughly following the 180° line of longitude, in the middle of the Pacific Ocean. The line's position was initially established at the 1884 International Meridian Conference held in Washington, D.C. Delegates voted to fix the Prime Meridian, or Earth's zero longitude, at Greenwich, England. They also adopted a 24-hour clock for the world, agreeing that each day would officially begin at midnight in Greenwich. Setting the Prime Meridian helped divide the world into standard time zones—12 to the east and 12 to the west. Each time zone is one hour earlier than the time zone directly to its east. The International Date Line marked the point where a new calendar day would begin.

1. The World in Spatial Terms
What purpose does the International Date Line serve?

Until 1892, Samoa lay west of the International Date Line. Then the king of Samoa moved the line so that his country was on the eastern side. He believed that changing time zones would help promote trade with California. In June 2011, the Samoan government voted to switch back to the western side. At midnight on Thursday, December 29, 2011, the official date in Samoa jumped ahead 24 hours, skipping December 30 entirely. Samoa's decision to switch sides of the International Date Line was spurred by the desire to improve the country's economic and cultural ties with Australia and New Zealand. The date change makes it easier for Samoans to trade with these countries. According to Samoan prime minister Tuilaepa Sailele Malielegaoi, "In doing business with New Zealand and Australia, we're losing out on two working days a week. While it's Friday here, it's Saturday in New Zealand, and when we're at church on Sunday, they're already conducting business in Sydney and Brisbane." Thus, by switching to the western side of the International Date Line, Samoans have the same workweek as residents of Australia and New Zealand.

2. The Uses of Geography What prompted Samoa to switch sides of the International Date Line?

No international body governs the International Date Line. Any country can decide which side it wants to be on and move the line accordingly. After the Samoan government decided to switch sides, officials simply changed maps, charts, and atlases to show Samoa's new orientation. The date change caused some temporary problems. Travelers missed flights because they arrived at the airport on the wrong day or at the wrong time. Investors lost one day of interest from banks. Yet the switch eventually made everyday life easier for Samoans. Samoa is now three hours ahead of eastern Australia rather than 21 hours behind. Also, Samoans living in Australia or New Zealand celebrate holidays, anniversaries, and other important occasions with their families on the same day. Samoa had always been the last place on Earth to watch the sun set, but now it is the first place to see the sun rise.

3. Human Systems Imagine you were living in Samoa in 2011. How would you feel about switching sides of the International Date Line? Explain.

THERE'S MORE ONLINE

SEE the effect of time zone change on Oceania and nearby countries • *VIEW* a chart of Samoa's trading partners

networks

There's More Online!

- ☑ **IMAGE** Bora Bora in the South Pacific
- ☑ **IMAGE** High Island
- ☑ **IMAGE** Low Island
- ☑ **Infographic** From Volcano to Atoll
- ☑ **INTERACTIVE SELF-CHECK QUIZ**
- ☑ **VIDEO** Physical Geography of Oceania

LESSON 1
Physical Geography of Oceania

ESSENTIAL QUESTION • *How do physical systems and human systems shape a place?*

Reading **HELP**DESK

Academic Vocabulary

- **converse**
- **offset**

Content Vocabulary

- **high island**
- **low island**
- **coral reef**

TAKING NOTES: *Key Ideas and Details*

IDENTIFYING As you read about the physical geography of Oceania, use a graphic organizer like the one below to compare high islands and low islands.

Physical Geography of Oceania

	High Islands	Low Islands
Landforms		
Soils		
Resources		

IT MATTERS BECAUSE

Oceania, as part of the diverse South Pacific region, includes stunning volcanic mountains, low atolls, and cool blue lagoons. Among these varied physical features are unique species of wildlife that attract tourists and scientists alike.

Landforms

GUIDING QUESTION *How do the islands of Oceania affect settlement?*

Thousands of islands, differing in size and extending across millions of square miles of the Pacific Ocean, form the region of Oceania (OH•shee•A•nee•uh). Some islands were created millions of years ago by colliding tectonic plates. Other islands were formed by volcanic hot spots.

Oceania consists of three island groups: Melanesia, Micronesia, and Polynesia. These groupings are based on location, how they formed, and culture. Melanesia, meaning "black islands," lies north and east of Australia. Melanesia includes island countries such as Papua New Guinea, Fiji, the Solomon Islands, and Vanuatu, as well as New Caledonia, a self-governing French territory. The largest country is Papua New Guinea with an area of 178,850 square miles (461,693 sq. km). Micronesia, meaning "little islands," lies north of Melanesia in the western Pacific. It includes the countries of Palau, the Federated States of Micronesia, Nauru, and Kiribati. The area also includes the U.S. territories of Guam and the Mariana Islands. In the central Pacific, Polynesia, meaning "many islands," spans an area larger than either Melanesia or Micronesia. The independent countries Samoa, Tonga, and Tuvalu are found in Polynesia. Other island groups, known as French Polynesia, are French territories and include Tahiti, Polynesia's largest island. Some of the islands of Polynesia, such as those that make up the state of Hawaii, are clustered relatively close together. Other islands are extremely remote. Easter Island, also known as Rapa Nui, lies more than 2,000 miles (3,219 km) west of Chile and some 1,200 miles (1,931 km) from its nearest island neighbor.

Earthquakes and volcanic eruptions still occur on many **high islands**, one of the island types in Oceania. The landscapes of high islands feature mountain ranges split by valleys that fan out into coastal plains. The mountains can create areas that are nearly inaccessible. On many high islands, however, mountain areas have greater population density. The interior of Papua New Guinea, for example, is densely populated. This pattern is much different from Fiji, where the majority of the population centers are on the coast.

The settlement patterns of Oceania were limited by the physical geography of the region. Isolated valleys, such as those on the larger islands of Melanesia, created an environment of cultural differences between people living on the coast and those living inland. Even today the islands of Melanesia are characterized by a diversity of languages and customs.

Volcanoes shaped another type of islands, **low islands**. Low islands, such as many of the Marshall Islands, are ring-shaped islands known as atolls. They are formed by the buildup of **coral reefs** on the rim of submerged volcanoes. Atolls encircle lagoons—or shallow pools of clear water—and usually rise only a few feet above sea level. Low islands have poor soil and few natural resources. The landscape of low islands increased interaction between people since lack of physical barriers allowed for more uniform languages and cultures to develop and spread.

☑ **READING PROGRESS CHECK**

Describing What is the relationship between atolls and lagoons?

high island an island with mountain ranges and volcanic soils

low island an island formed by the buildup of a coral reef on the rim of a submerged volcano; sometimes known as an atoll

coral reef a reef made up of fragments of corals, coral sands, algal and other organic deposits, and the solid limestone resulting from their consolidation

from VOLCANO to ATOLL

Phase 1
VOLCANIC ISLAND
As magma erupts through the Earth's crust and is cooled by ocean waters, a volcanic island is slowly formed.

Phase 2
SINKING ISLAND
Once the volcano becomes dormant, the island begins to sink. As it sinks, a coral island forms around it. A small, shallow lagoon forms between the coral island and the mountain.

Phase 3
BARRIER REEF
As the island sinks further, the coral fringes form a barrier reef. The lagoon between the reef and the shore gets larger and deeper.

Phase 4
ATOLL
Once the island disappears below the waves, it becomes a seamount. The barrier reef becomes an atoll, circling an open lagoon.

SEAMOUNT

LAGOON

RISING MAGMA

CORAL ISLAND

MANTLE PLUME

OCEAN CRUST

HOT SPOT

A volcanic island forms when magma erupts from deep within the Earth. Over millions of years, a volcanic island will become an atoll.

▲ **CRITICAL THINKING**

1. Identifying What tectonic forces are required for volcanic islands to form?

2. Identifying Cause and Effect Describe how atolls are formed.

The landscapes of low islands (left) and high islands (right) are strikingly different.

▲ CRITICAL THINKING

1. Speculating Why are low islands more vulnerable to rises in ocean temperature?

2. Identifying What landforms are typically found on high islands?

converse reversed in order, relation, or action; the other way around

Water Systems

GUIDING QUESTION *How do the bodies of water surrounding the islands of Oceania affect ways of life and settlement?*

Oceania extends across millions of square miles in the Pacific. Occupying about a third of the Earth's surface, the Pacific is the largest body of water on the planet. It also has the greatest biodiversity of all the oceans. The Equator marks the division between the North Pacific and the South Pacific.

The Pacific Ocean displays diverse landforms, both above and below its surface. Some can be seen as they rise above the ocean, while others are submerged. The underwater mountain ranges vary from isolated ranges to complex systems that run for thousands of miles. Many of the ranges have exposed volcanic summits. The summits are visible as islands or chains of islands that dot the ocean. One of the longest mountain ranges forms the Hawaiian Islands of Polynesia. Its highest peak, Mauna Kea, rises nearly 13,796 feet (4,205 m) above sea level.

The world's greatest ocean depth has been measured in the Mariana Trench. Located near Guam, it reaches a maximum depth of 36,198 feet (11,033 m). Like the Mariana Trench, most trenches in the western Pacific are next to island chains. Other deep trenches in Oceania reach depths in excess of 32,000 feet (9,754 m).

Asian migrants settled Oceania in family groups along island coasts. They survived on fish, turtles, and shrimp, as well as breadfruit and coconuts. Over time, they cultivated root crops including taro and yams. They also raised livestock such as chickens and pigs. Well-built canoes made lengthy voyages possible. Thus trade gradually developed between islands. To make trading easier, people on some islands used long strings of shell pieces as money. On a few islands, shell money is still exchanged today for canned goods or vegetables at markets.

Important trade routes cross through the ocean surrounding the subregion. The majority of the exports moving from west to east and from north to south are manufactured goods on their way to markets. **Conversely**, most of the exports that are shipped from east to west and from south to north are raw materials. The small islands of Oceania depend heavily on trade—especially for basic necessities, such as foodstuffs and fuels. There are exceptions, however. Papua New Guinea generally exports more than it imports.

The bodies of freshwater in the subregion vary depending on the island type. High islands have sources of freshwater that support agriculture. The low islands typically have no freshwater sources other than places that catch rainwater.

☑ READING PROGRESS CHECK

Describing What landforms are found in the Pacific Ocean?

Climates, Biomes, and Resources

GUIDING QUESTION *How do the various climates on the islands of Oceania affect their biomes?*

Most of Oceania has a tropical wet climate and is warm year-round. The dry season features cloudless skies, but the wet season brings constant rain and high humidity. High islands are high enough to force warm, moist air to rise. It then cools and condenses. Low islands do not have this effect, so there is less rainfall.

The great expanses of open water in the Pacific Ocean influence wind and pressure patterns. These are reflected in the climatic conditions. A generally windless area called the doldrums occupies a narrow band near the Equator where the direct rays of the sun cause air to rise vertically instead of blowing horizontally. As the air rises, air from the north and south is drawn in by the low surface pressure, creating the trade winds. The flow also causes typhoons to form.

The amount of rainfall on the islands creates both arid and wet climate regions, depending on the island type and location. Only shrubs and grasses grow on dry, low islands. These islands have only a small proportion of arable land. Palms and other trees appear on islands with more rainfall.

Hot, steamy rain forests thrive where heavy rains drench island interiors. The summit of Mount Waialeale on Hawaii's island of Kauai is one of the wettest places on Earth. Rainfall averages about 450 inches (1,143 cm) of rain each year. The rich soils of high islands support a diverse group of plants and animals.

The isolation of many of the islands in Oceania makes for the presence of endemic species. Species that are endemic to a particular location are found only in that place. The intentional and accidental introduction of new species to the islands of Oceania has resulted in negative changes to the ecosystem.

Island type and location also influence resources. Low islands have poor soils and few resources beyond the sea. These mainly consist of coconut oil, dried coconut (copra), bananas, and seabed minerals. The volcanic materials of the islands of American Samoa include an abundance of pumice and pumicite. Fiji, also volcanic, has timber, gold, and a potential for offshore oil. Fiji also generates hydropower.

High islands have rich soils that support agriculture. Products grown for export include sugar, coffee, and cocoa. The high islands also have rain forests with diverse flora and fauna. The islands with forests, such as Samoa, harvest hardwoods. Islands without forests have to import lumber. In addition to forests, the Solomon Islands have a variety of other natural resources. These resources include fish and other marine animals, gold, bauxite, phosphates, lead, zinc, and nickel. Tourism helps to **offset** the lack of natural resources on many of the islands, especially the mineral-poor low islands.

offset to compensate for or serve to counterbalance

☑ **READING PROGRESS CHECK**

Naming What are some of the natural resources of Oceania?

LESSON 1 REVIEW

Reviewing Vocabulary

1. ***Describing*** Describe the significance of coral reefs to atolls.

Using Your Notes

2. ***Contrasting*** Using your graphic organizer from the lesson, contrast the soils of the low islands with those of the high islands.

Answering the Guiding Questions

3. ***Explaining*** How do the islands of Oceania affect settlement?

4. ***Identifying Cause and Effect*** How do the bodies of water surrounding the islands of Oceania affect ways of life and settlement?

5. ***Synthesizing*** How do the various climates on the islands of Oceania affect their biomes?

Writing Activity

6. ***Narrative*** Write a paragraph describing the type of activities you might participate in during a vacation to one of the islands in Oceania. Include a discussion of how the landforms and bodies of water relate to the type of activities available.

networks

There's More Online!

☑ **IMAGE** Hotel in Fiji

☑ **IMAGE** Oceanic Art

☑ **IMAGE** Subsistence Farming in Oceania

☑ **MAP** Oceania in World War II

☑ **TIME LINE** Colonization and Independence

☑ **INTERACTIVE SELF-CHECK QUIZ**

☑ **VIDEO** Human Geography of Oceania

Reading **HELP**DESK

Academic Vocabulary

- **nevertheless**
- **bulk**

Content Vocabulary

- **kinship group**
- **trust territory**
- **pidgin**

TAKING NOTES: *Key Ideas and Details*

IDENTIFYING As you read about the human geography of Oceania, use a graphic organizer like the one below to record how the different island groups were settled.

Human Geography of Oceania

Settlement

Melanesia →

Micronesia →

Polynesia →

LESSON 2
Human Geography of Oceania

ESSENTIAL QUESTION • *How do physical systems and human systems shape a place?*

IT MATTERS BECAUSE

Migrations of people among hundreds of islands over many generations shaped societies in Oceania. European colonization had a profound impact as well. Today, the islands in Oceania are reshaping themselves as independent countries.

History and Government

GUIDING QUESTION *How has life on the islands of Oceania changed over time?*

Hundreds of indigenous people lived on the many islands in the Pacific Ocean when European explorers began arriving in the 1500s. The islands had been their homes for thousands of years. Many of these cultures practiced religious beliefs that connected them to the land and sea. In many of the societies, the chief was the person of highest status. Complex systems of ranked lineages and **kinship groups** had developed. Many societies, such as that of the Marshall Islands, were matrilineal, meaning that family history was traced through the mother's family. Powerful chiefs were recognized for their ability to communicate with the gods and secure food and other valuables for the people.

Europeans brought far-reaching changes to the peoples of Oceania. When Europeans settled in Oceania in the 1800s, they developed plantations for growing sugarcane, pineapples, and other tropical products that were sold around the world. Europeans also brought new diseases that resulted in epidemics among the indigenous populations. In order to maintain a workforce, plantation owners recruited workers from East Asia and South Asia. In just a few years many of the islands became culturally diverse and dependent on the global demands for sugar, the major crop. Meanwhile, Europeans further altered traditional beliefs and customs by establishing schools, sponsoring missions, and converting people to Christianity.

During the late 1800s and early 1900s, Britain, France, Germany, Spain, and the United States struggled for control of various Pacific islands. These countries wanted to acquire or expand their influences in the region and gain new sources of raw materials. Colonial powers did not have a strong physical presence in many of the islands. In some places, islanders ignored the laws of European countries that they felt did not apply to them.

The two world wars dramatically changed Oceania's political geography. After World War I, most of Germany's territories came under Japanese colonial rule. During World War II, those same islands were the sites of fierce battles. The United States and its allies were fighting to drive the Japanese from these islands. After Japan's defeat in the war, its Pacific territories were turned over to the United States as **trust territories**. These are designated by the United Nations to be temporarily governed by another country.

Beginning in the 1960s, islands in Oceania, also called the Pacific Islands, began gaining independence. In 1962 Samoa, formerly Western Samoa, became the first Pacific Island to win freedom, after rule by Germany and New Zealand. Since the 1970s most of these islands—including Palau, the Marshall Islands, and the Federated States of Micronesia—have become independent countries. Most Pacific Islands achieved independence by the end of the 1900s.

Many islands in Oceania have a dual government system that includes the government imposed upon them by colonial powers and their traditional, indigenous systems of government. In some countries, the traditional systems have been integrated into the constitutional structure. However, in other countries, such as Papua New Guinea, indigenous governance systems have been largely sidelined in the formal government structures.

The Solomon Islands represents one example of cultural rebirth occurring in the subregion. Traditional beliefs are starting to resurface and guide decision making. These countries are looking within their borders for input on economic and social issues. Social structure is a high priority. People in the Solomon Islands are once again valuing equality for all people. They are also building a strong relationship to the land by practicing traditional farming and marketing locally grown foods rather than relying on imported, processed foods.

☑ **READING PROGRESS CHECK**

Explaining Describe how the arrival of Europeans affected the population of Oceania.

Population Patterns

GUIDING QUESTION *How and why did people spread across Oceania?*

The islands of Oceania were probably first settled by peoples from Asia around 40,000 years ago. Waves of migrants from Asia continued to arrive over many centuries. People already living there moved from island to island and settled into three major island groups: Melanesia, Micronesia, and Polynesia.

Melanesia includes Papua New Guinea, Fiji, the Solomon Islands, as well as New Caledonia. Melanesian cultures differ greatly, even among groups living on the same island. Some Melanesians are more closely related to Polynesians than to other Melanesians. The earliest people of Melanesia were the Papuans. They were hunters and gatherers that adapted to the rain forests on New Guinea. Some estimates date their arrival on the island to 40,000 years ago. Eventually the early Melanesians domesticated root crops and sugarcane. They may have also kept domesticated pigs as long as 9,000 years ago. In areas of the highlands of New Guinea, water and irrigation systems were in place 5,000 years ago.

Arriving by sea from Southeast Asia, the Austronesian peoples settled in Melanesia about 4,000 years ago. Pottery, tools, and shell ornaments date their arrival in the islands. The languages they spoke were similar to languages used in the Solomon Islands, Vanuatu, and New Caledonia. Language was used to test theories of migration among people in the region. Distances were great across ocean waters. Migration between the islands meant the people were accomplished mariners. The migration theory explained both when and how the islands were first inhabited. It also explained the cultural differences and similarities in this region.

kinship group a group of people related by blood or marriage

trust territory a dependent area that the United Nations placed under the temporary control of another country

Micronesia includes the Federated States of Micronesia, Nauru, Kiribati, Guam, and the Mariana Islands. The first people to settle in Micronesia are thought to have arrived by ocean canoes some 2,000 to 3,500 years ago from Southeast Asia and Melanesia. The islands in this region are more densely populated. A larger population suggested that more migrants arrived and that they had skills necessary to develop farming and trading economies. Language has again been used as the key to dating the arrival of the first people to these islands.

Polynesia includes Samoa, Tonga, Tuvalu, and French Polynesia. The ancestors of present-day Polynesians may have come to the islands 6,000 to 8,000 years ago. The western islands—Wallis and Futuna, Samoa, and Tonga—were settled first. These first Polynesians grew traditional crops and domesticated animals. Women gathered plants and were weavers. Many Polynesians share a similar language and culture. Today the largest population of Polynesians lives in the Samoan Islands.

Asian communities also exist in the subregion. Chinese, Japanese, and South Asian traders and laborers settled parts of Oceania during the 1800s. Today their descendants live in such places as French Polynesia and Fiji.

The population of Oceania's islands varies considerably, and many of the islands are not large enough to support people. Population is not evenly distributed among the countries of the subregion. Papua New Guinea has the largest population, with about 7 million people. Nauru, the world's smallest republic, has a population of about 10,000 people. The total land area of Oceania's 25,000 islands is 551,059 square miles (1,427,246 sq. km) spread across a vast Pacific Ocean expanse of over 20 million square miles (52 million sq. km). Urban population varies greatly. For example, the capital city of Papua New Guinea, Port Moresby, has a population of nearly 300,000. In contrast, Apia, the capital city of Samoa, has a population of approximately 36,000.

☑ READING PROGRESS CHECK

Identifying Who first settled Melanesia and from where did they come?

TIME LINE ⌄

COLONIZATION
and Independence ➔

Beginning in the 1800s, European countries and the United States colonized the islands of Oceania, but those islands became independent states in the twentieth century.

▶ CRITICAL THINKING

1. *Explaining* How did the islands of Oceania change after contact with Europeans was made?

2. *Describing* When did the pace of independence quicken in Oceania?

1600s–1700s
European navigators first contact the peoples of Oceania.

1870s
Population of the Ellice Islands (now Tuvalu) declines dramatically.

1600 ➔

1830s
European missionaries arrive in Fiji, Samoa, and Tonga to convert the population to Christianity.

1880s
Germany takes control over part of New Guinea, Marshall Islands, and Nauru.

Society and Culture Today

GUIDING QUESTION *What is life like on the islands of Oceania today?*

Today, societies in Oceania have been shaped by a variety of cultures. The subregion's countries display a blend of European, Asian, and indigenous traditions. The traditional lifestyle of indigenous people is difficult to find in the twenty-first-century world.

Before the era of modern transportation and advanced communications, vast distances of open ocean separated the peoples of the subregion from the rest of the world. As a result, isolated groups developed their own languages and belief systems without outside interference. Of the world's some 6,000 languages, 1,200 are spoken in Oceania. Some are spoken by only a few hundred people.

European colonization brought European languages to the subregion. Today French is widely spoken in Oceania. In many areas, varieties of **pidgin** English are spoken as well. This is a blend of English and indigenous words that do not follow rules of grammar, but do allow better communication among different groups.

The peoples of the subregion practice several different religions. These include Christianity, Hinduism, Islam, and Buddhism. Sometimes the religions are combined with traditional religious beliefs. Christianity is the most widely practiced religion in Oceania and was introduced by Europeans. Hinduism was introduced to Fiji when thousands of Indians immigrated to work on plantations. Over 30 percent of the population of Fiji is Hindu.

The quality of education varies throughout Oceania. In the Solomon Islands, missionary schools provided primary education until the mid-1970s. Today secondary schools and universities are common in the Solomon Islands, Fiji, and Papua New Guinea. Differences in education throughout the subregion can be seen in the varying rates of literacy. For example, over 93 percent of Fiji's population is literate, but only 57 percent of the inhabitants of Papua New Guinea are literate.

pidgin a blend of English and indigenous words to form a new language

---

1900 →

1899s
Germany buys Palau from Spain. Germany and the United States divide Samoa between them.

1941–1945
Japanese forces occupy Kiribati, Marshall Islands, Micronesia, Nauru, Palau, Papua New Guinea, and Tuvalu.

1946

United States begins nuclear weapons testing in Marshall Islands; islanders are forced to evacuate.

1962
Western Samoa becomes first colony in Oceania to become independent.

1970s
Fiji, Tonga, Papua New Guinea, Tuvalu, Federated States of Micronesia, Kiribati, and Marshall Islands become independent.

2000 →

1994
Republic of Palau becomes independent state.

Many Pacific Islanders suffer from their countries' poor economies and low standards of living. Health care varies from country to country. On remote islands, electricity, schools, and hospitals are difficult services. As independent countries, they qualify for international assistance and are making improvements.

Family and the Status of Women

The traditional family structure has consisted of many generations living in the same village and often in the same house. This structure includes large families with children being cared for by many family members. Contemporary life today has changed throughout Oceania due to modernization and loss of traditional values. Many young people migrate to Australia, the United States, and France.

The status of women is of great concern in the subregion. The status of women varies from strong matrilineal leadership to victims of violence long accepted as a part of a culture. In Oceania, there are cultural practices that make it difficult to protect women against violence. For example, in Melanesia, there is the practice of early arranged marriages and forced marriages. The practice of burning and scarring brides also occurs among some ethnic groups in the region even though it is against the law.

The Arts

The ocean, fish, and many types of plants and animals are the subjects for the arts of Oceania. Carving wood is an art mastered by many people. Skilled crafts, such as making canoes and building houses, are a tradition. Art and architecture are closely related to each other. Buildings are carved with elaborate designs. Detailed designs are also important in the symbolic art of body tattooing in Oceania. Creating decorative patterns on the skin by scarification, the intentional scarring of the skin by incision or burning, and tattooing are practiced by some culture groups.

Skilled crafts are a tradition in Oceania with artwork on buildings, canoes, and on bodies in the form of tattoos.

▼ CRITICAL THINKING

1. Identifying What are some of the common themes found in traditional Oceanic art?

2. Explaining What type of artwork is part of ritual practices in Oceania?

☑ READING PROGRESS CHECK

Explaining How did Hinduism become one of the religions practiced in Oceania?

Economic Activities

GUIDING QUESTION *What economic opportunities are available to the people of Oceania?*

The remote geographic locations and challenging environments of Oceania influence how people there earn a living. Agriculture continues to be an important economic activity in the subregion, but new industries are contributing to national economies. Tourism has gained in importance as people looking for scuba diving in pristine waters select the islands of the Pacific for their environments. Niche markets, such as call centers and web development, have increased in Oceania.

Although many islands in Oceania lack abundant natural resources, some have excellent resources for tourism. The tourism industry is still a relatively new industry in many parts of Oceania. French Polynesia, Guam, the Northern Mariana Islands, and Fiji have all experienced recent growth in this industry, and it has become vital to their economies.

Much of Oceania lacks arable land, which in turn limits agriculture. Farmers on smaller islands use traditional

©Michele Westmorland/Corbis

farming and fishing. On some islands, the village or island government owns the land and rents or leases it to farmers.

Some islands have rich volcanic soil and ample rainfall. On many of the islands, Europeans had once claimed the most productive land. Land settlements in recent years have returned the best land to indigenous people.

Most of the islands in Oceania have only small amounts of mineral resources and lack large manufacturing centers. Papua New Guinea, Fiji, and New Caledonia have mineral deposits. Papua New Guinea's rich deposits of gold and copper have only recently been exploited. Nauru's deposits of phosphates are depleted due to excessive mining. Manufacturing in Oceania is mostly limited to small-scale enterprises like clothing production. Many of the larger islands with forests have sawmills that produce lumber for export or domestic use. Some of the larger islands have facilities that process agricultural products, such as coconut oil and cane sugar.

Trade between Oceania and other parts of the world has increased because of improvements in transportation and communications, as well as the creation of new trade agreements. Some island countries are too small, too poor, or too rugged to have well-developed road or rail systems. **Nevertheless**, some governments are improving these systems. Cargo ships and planes move imports and exports to and from the subregion. Principal trading partners of Oceania include the United States, Japan, and Australia. Commercial airlines and cruise ships bring travelers. Air transport and interisland shipping are the vital lines for transportation. Many of the island groups in Oceania have international airports. Extensive road networks are limited to the larger islands.

Cellular, digital, satellite communications, and the Internet are becoming common in the subregion. In recent decades, improved transportation and communications links have increased trade between Oceania and other parts of the world. The subregion's agricultural and mining products earn the **bulk** of its export income. Countries also export handicrafts—such as baskets, floor mats, ceremonial masks, and pottery—made by island craftspeople.

Farming in Oceania frequently relies on animal power and human labor.

▲ **CRITICAL THINKING**

1. Speculating What are some of the crops that might be grown by traditional farming practices in the Solomon Islands and other parts of Oceania?

2. Contrasting How were the plantations established by the Europeans different from traditional farming?

nevertheless despite; in spite of

bulk the greater part of something

☑ **READING PROGRESS CHECK**

Making Connections What factors contribute to the lack of adequate networks of roads on many of the islands in Oceania?

Holger Leue/LOOK/Getty Images

LESSON 2 REVIEW

Reviewing Vocabulary

1. *Summarizing* Define *trust territory* and explain how it relates to Oceania.

Using Your Notes

2. *Describing* Use your notes from the graphic organizer to describe the order of settlement for Melanesia, Micronesia, and Polynesia.

Answering the Guiding Questions

3. *Identifying Cause and Effect* How has life on the islands of Oceania changed over time?

4. *Explaining* How and why did people spread across Oceania?

5. *Describing* What is life like on the islands of Oceania today?

6. *Examining* What economic opportunities are available to the people of Oceania?

Writing Activity

7. *Informative/Explanatory* Write a paragraph describing the negative impact Europeans had on the population and cultures of Oceania.

WHO OWNS THE HIGH SEAS?

In 1994 the Law of the Sea Treaty came into force. This treaty defines which countries own parts of the oceans. According to the treaty, each country's territory extends about 12 nautical miles from its coasts. Beyond this boundary a country may establish an exclusive economic zone (EEZ) that extends 200 nautical miles from shore. When two countries' ocean territories overlap, a border is set. The areas beyond these descriptions and the EEZs, however, belong to no one. They are the high seas. The high seas can be accessed by anyone with the means to do so. These areas are rich in fish and other natural resources, such as oil and minerals. People travel through the high seas, and countries mine resources and fish there.

People have different ideas about how the high seas and the natural resources within them should be managed and protected. More than 150 countries have signed the Law of the Sea Treaty, which gives the International Seabed Authority (ISA) jurisdiction over the high seas. However, some countries refuse to sign the treaty. In these countries, some fear that the ISA will impose new tax laws on travel in the oceans and reduce free movement on the waters. Another concern is that fewer economic opportunities will exist if the ISA regulates the high seas. They also worry that future projects, such as mining the sea floor, will be restricted. Other critics of the treaty think the high seas should be privately owned or even feel that no one should own and regulate these areas. Still others oppose the treaty because they want more regulation of the high seas. They do not think that the ISA and the United Nations have the authority to protect the oceans. They feel the ISA has not done enough to protect the waters from the existing threats of overfishing, pollution, pirating, and drug trafficking.

Problems of the Law of the Sea Treaty

PRIMARY SOURCE

" One of the primary missions of the United States Navy for over two centuries has been to maintain freedom of the seas for all. As a Navy veteran, I am offended to think that the Senate and the Chief of Naval Operations would even consider ceding any part of that mission to the United Nations. . . .

There is no guarantee that the treaty will remain what it is at the time of ratification. Under its terms, its content can later be changed by an amendment process that does not require the approval of the United States government. This undermines U.S. sovereignty and, to put it bluntly, is unconstitutional. "

—Colin Hanna, president of Let Freedom Ring,
quoted in "Kill the Law of the Sea Treaty,"
U.S. News, May 10, 2012

Support More Regulation

PRIMARY SOURCE

" We have the Law of the Sea [Treaty] but it does not have enough teeth to actually do anything. We also have the regional fisheries management organisation but this is not a global treaty and they are regional treaties. We see the high seas as the last frontier to be tackled. It is nobody's territory but everybody's territory to exploit. . . .

So we need to bridge this gap somehow and there is no framework yet or a treaty for this purpose. Eventually all the fish caught has to land somewhere and it becomes somebody's business and therefore this dilemma needs to be solved. "

—Gustavo Fonseca, head of natural resources
quoted in "Oceans Special: GEF Rolls Out
Investment Project to Address Issues for Areas
in the High Seas,"
June 16, 2012

What do you think?   DBQ

1. ***Drawing Conclusions*** Why might people want to make a plan for regulation and management of the high seas?

2. ***Identifying Central Issues*** According to critics of the treaty, what are the risks of regulation of the high seas?

3. ***Hypothesizing*** What might be some benefits and drawbacks of private ownership of the high seas?

networks

There's More Online!

- ☑ **CHART** United States Nuclear Tests July 1945 through September 1992
- ☑ **IMAGE** Easter Island
- ☑ **MAP** Global Sea Level Trends
- ☑ **INTERACTIVE SELF-CHECK QUIZ**
- ☑ **VIDEO** People and Their Environment: Oceania

Reading HELPDESK

Academic Vocabulary

- **nuclear**
- **grant**

Content Vocabulary

- **climate change**
- **ocean warming**

TAKING NOTES: *Key Ideas and Details*

IDENTIFYING Use a graphic organizer like the one below to take notes as you read about the issues that relate to people and their environment in Oceania.

Human Impacts on the Environment in Oceania

Deforestation	
Rising Sea Levels	
Rising Ocean Temperatures	

LESSON 3
People and Their Environment: Oceania

ESSENTIAL QUESTION • *How do physical systems and human systems shape a place?*

IT MATTERS BECAUSE

The islands of Oceania account for the emission of less than 1 percent of greenhouse gases that contribute to climate change. They are, nonetheless, among the most vulnerable to the adverse effects of greenhouse gases.

Managing Resources

GUIDING QUESTION *How are the people of Oceania using their islands' natural resources?*

The protection of forest, soil, and freshwater resources is a major concern throughout Oceania. The management of resources includes reforestation plans and implementation of "tabu," or no-take zones to control overfishing. Countries with valuable timber resources, such as Papua New Guinea, are developing plans to use forest resources without damaging the environment.

Methods used to improve agriculture production have a long history in Oceania and continue to be implemented to increase output. The labor-intensive practice of terracing is used to grow crops on the sides of hills. Terraces built into a slope increase the amount of available land for cultivating crops. Terracing helps reduce irrigation runoff and erosion. Irrigation practices such as simple flooding are used to slow down the flow of water to help trap sediment and control erosion. Crop diversification is another innovative agricultural practice used in Oceania. Planting a diversity of crops helps preserve the soil and increase productivity.

Overseas companies also have a stake in the management of Oceania's natural resources. Foreign companies are actively mining and logging on the islands and deep seabeds of the subregion. The investment in tourism by foreign companies is also a factor in managing the natural resources of the subregion. As new resorts are built, the environment is affected. Fiji has established a joint effort with hotel and resort owners to create a tourist sector that builds energy-efficient hotels, cuts carbon emissions, and reduces the impact on the environment.

Oceania's environmental challenges include managing its freshwater resources. Many coral atolls and volcanic islands hold only limited supplies of freshwater. The lack of clean drinking water keeps the standard of living low. It also poses barriers to economic growth in some countries. Improvement will come with better management of runoff and the construction of additional sanitation facilities.

The low islands of Oceania are in particular danger. **Climate change** can cause sea levels to rise and produce extreme weather. Managing the resources of these islands will help reduce their vulnerability. Programs include the protection of mangroves and seagrass habitats on these islands. In some cases, infrastructures have been relocated farther inland.

Many of the countries in Oceania have developing economies. This means that implementing programs can be difficult. Programs continue to be initiated, however, through grassroots organizations, government efforts, and international organizations. Environmental education is also an important factor in managing natural resources. Understanding issues such as climate change will help empower people to make changes and welcome new methods of balancing their needs with those of the ecosystems.

climate change any significant change in the measures of climate lasting for an extended period of time

☑ READING PROGRESS CHECK

Summarizing What issues are the most important in regard to managing the natural resources of Oceania?

Human Impact

GUIDING QUESTION *How have humans—both in Oceania and elsewhere—affected the environments of the islands of Oceania?*

The natural resources of Oceania, as in other regions, have not always been well managed. Today the subregion faces many environmental problems. Conservation efforts, however, are gaining recognition.

GEOGRAPHY CONNECTION

In recent decades, global sea levels have been rising.

1. *THE WORLD IN SPATIAL TERMS* In what areas of the globe have sea levels risen the most?

2. *THE WORLD IN SPATIAL TERMS* List three areas in which sea levels have fallen.

Global Sea Level Trends

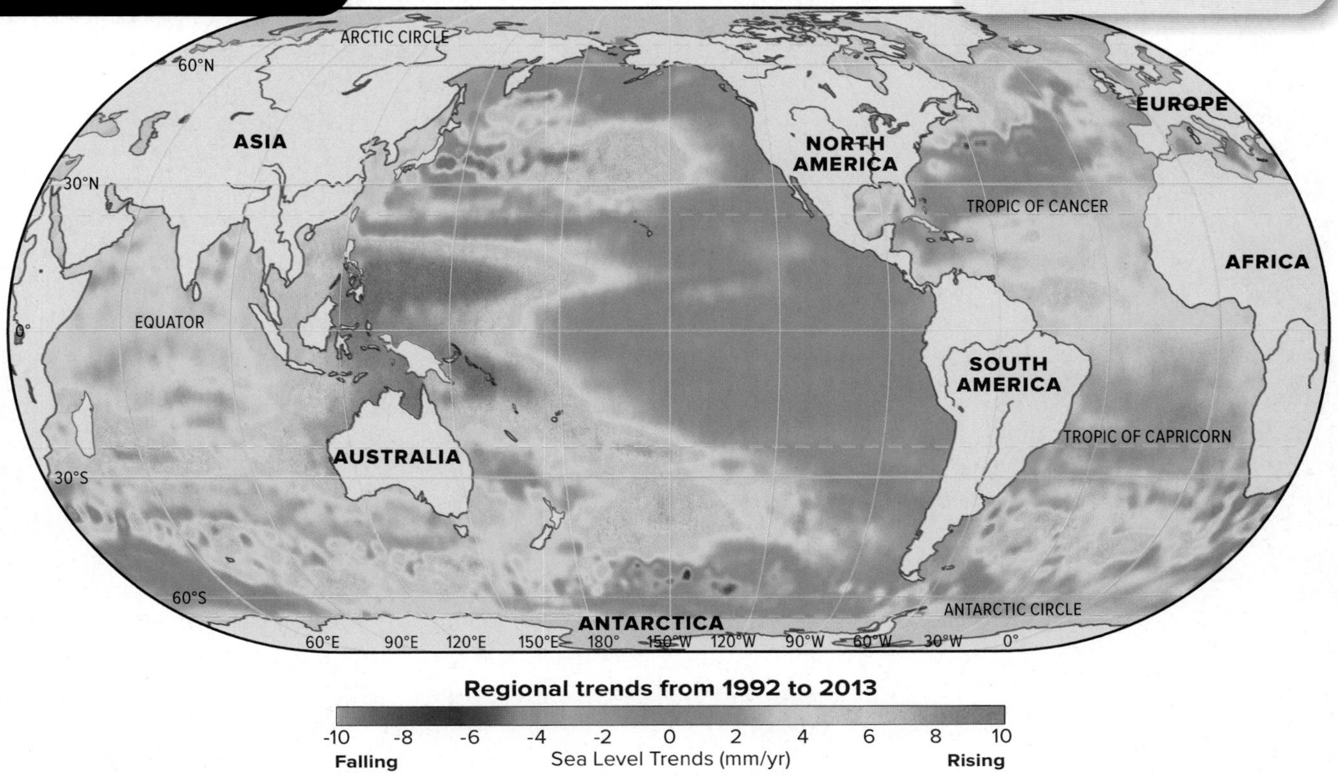

Regional trends from 1992 to 2013

-10 -8 -6 -4 -2 0 2 4 6 8 10
Falling   Sea Level Trends (mm/yr)   **Rising**

Famous for its giant carved statues,
Easter Island no longer has the
natural vegetation that once covered
the land.

▼ CRITICAL THINKING

1. Contrasting Describe how the
vegetation of Easter Island has
changed since the arrival of
Europeans.

2. Describing What are commonly
held theories for causes of
deforestation on Easter Island?

The testing of **nuclear** weapons has had major effects on parts of the
subregion's environment. In the late 1940s and 1950s, the United States and
other countries with nuclear capability carried out aboveground testing of
nuclear weapons in the South Pacific. The dangers of such testing were
gravely underestimated at the time. In 1954 the United States exploded a
nuclear device on Bikini Atoll in the Marshall Islands. The people of Bikini
Atoll had been moved to safety. However, those living downwind of the
explosion on Rongelap Atoll were exposed to massive doses of radiation.
This exposure resulted in deaths, illnesses, and genetic abnormalities.

Like other world regions, Oceania is threatened by climate change. One
resulting impact is an apparent increase in the frequency and severity of the
El Niño weather pattern. Climate and weather in Oceania are highly
sensitive to seasonal El Niño weather patterns that can cause both droughts
and powerful storms. Another impact of climate change that many scientists
consider imminent is that continued increases in Earth's temperatures could
cause a devastating rise in sea level. If ice caps and glaciers on land experience
increased melting, the resulting rise in the level of ocean waters would cause
flooding on many of Oceania's islands.

Rising sea levels not only cause loss of land but also contaminate the
limited freshwater sources of the low islands with salt water. This reduces
the amount of freshwater available for agricultural purposes as well as
human consumption. In addition, **ocean warming** affects certain types of
plankton and algae that grow in tropical waters, causing overgrowth and
the choking out of other life forms. Coral reefs are also destroyed by ocean
warming. Scientists in the region are studying global warming. They hope
to discover its causes, predict its consequences, and provide solutions.

Pollution also poses a threat to the waters of Oceania and is damaging to
the coral reefs. Agricultural runoff and inadequate sanitation cause pollution
that further threatens freshwater supplies. Agricultural runoff, chemical
fertilizers, and organic waste also threaten the subregion's oceans.

Deforestation and the loss of native species are not new to the islands of
Oceania. Easter Island, also called Rapa Nui, had already lost most of its native

plant species by the late 1700s. When European explorers first arrived on Easter Island in the late 1700s, deforestation had already taken place. They found a grassy island devoid of plants greater than 10 feet (3 m) in height. The island at one time had been covered with a subtropical forest dominated by large palms, an ecosystem entirely different from what the Europeans observed.

The exact cause of Easter Island's deforestation is not certain. However, a few theories prevail. Some scientists believe that uncontrolled populations of rodents were the cause. Others blame overexploitation of the available resources or the island's natural fragility for the deforestation. The loss of forest caused soil erosion that limited the inhabitants' ability to grow crops.

Present-day causes of deforestation include mining and logging. Although mining is beneficial to Papua New Guinea's economy it has had negative impacts on the rain forests there. Mining operations cause destruction of forest habitats and pollute waterways and oceans. The process of removing the nontarget material from the ore and the discarded waste rock can cause damage to the environment. The process of separating gold in alluvial mining releases mercury into streams and soils, which creates another source of contamination. Pollutants can cause localized fish kills and loss of vegetation in areas. Pollution in rivers and streams is harmful to agriculture as well as to animals and humans.

✅ **READING PROGRESS CHECK**

Describing What are the pros and cons of mining operations in Papua New Guinea?

Addressing the Issues

GUIDING QUESTION *How are the people of Oceania reacting to the changes in their environments?*

The U.S. nuclear testing in Oceania was stopped. The effects of radiation exposure and environmental damage, however, have continued. Today the atolls affected by the testing remain off-limits to human settlement. Recent studies, however, offer hope for eventual environmental recovery. In the 1990s, the United States government provided $90 million to help decontaminate Bikini Atoll. It also set up a $45 million trust fund to provide **grants** for blast survivors from Rongelap Atoll and their offspring. The nuclear legacy also has had political effects. Antinuclear activism is a major factor in regional politics. French plans to conduct nuclear tests on an atoll in French Polynesia caused antinuclear demonstrations. The international outcry led to an early halt to the tests.

Countries within the subregion are also addressing other environmental concerns. The Nauru Agreement Concerning Cooperation in the Management of Fisheries of Common Interest (also known as the Parties to the Nauru Agreement, or PNA) was created to manage tuna populations sustainably. The agreement requires catch reporting and maintenance of logbooks. It also mandates the installation of electronic position and data transfer devices on the fishing vessels. The members include eight Pacific Island countries that together control 25 percent of the world's supply of tuna. The PNA members are Federated States of Micronesia, Kiribati, Marshall Islands, Nauru, Palau, Papua New Guinea, Solomon Islands, and Tuvalu.

Coral reefs require clean water and sunlight to survive. The U.S. Environmental Protection Agency (EPA) has established programs to protect, restore, and maintain water quality around the coral reefs. Hawaii's Department of Land and Natural Resources (DLNR) reported that nearly 25 percent of all living coral was lost between 1994 and 2006 on Maui. Grants offered by the EPA aim at controlling pollution, managing watersheds, research, monitoring, and education.

Between 1945 and 1992, the United States carried out more than 1,000 nuclear tests. The vast majority were conducted underground, but more than 100 were done in the Pacific region.

▶ CRITICAL THINKING

1. *Hypothesizing* Why do you think the United States chose Pacific islands and atolls as locations for nuclear tests?

2. *Interpreting Significance* What do you suppose were the long-term effects of these nuclear tests on the region?

United States Nuclear Tests July 1945 through September 1992	
Location	Number of tests
Bikini	23
Christmas Island	24
Enewetak	43
Johnston Island	12
Pacific	4
Total Pacific	106

Source: United States Nuclear Tests July 1945 through September 1992, U.S. Department of Defense, December 2000

Another joint effort is in the area of sugar production. For decades sugar has been a leading export for Fiji's economy. Sugar production has, however, also damaged the ecosystems of the islands of Fiji. The Sustainable Sugar program was created by the World Wildlife Fund (WWF). The program is working with local people, companies, and the government to reduce the negative effects of sugarcane production on the coral reefs. WWF also promotes an approach to conservation and development in the region that recognizes the rights of local people to manage and benefit from use of natural resources. The approach is called Community Based Natural Resource Management (CBNRM). CBNRM supports community-based initiatives and empowerment of communities. The program seeks legislation and partnerships with public and private sectors in the use of natural resources. The goal is to create sustainable development and allow communities to benefit economically. Emphasis is also placed on maintaining traditional customs and values that have protected the natural resources for centuries.

Climate change is a major concern for all the islands of Oceania. Many of the low islands are already planning strategies to cope with rising sea levels. The plans include moving human populations to the remaining areas after loss of land due to rising sea levels. Another plan is for emigration to nearby countries. Climate change will have severe effects on the economies of the subregion as populations relocate. It will also impact the subregion's culture as people move to countries such as Australia, New Zealand, and the United States. In those countries, emigrants would not own land and might lose their social structures and traditional ways of life.

☑ READING PROGRESS CHECK

Describing What are some measures that have been taken to address environmental concerns?

LESSON 3 REVIEW

Reviewing Vocabulary

1. *Making Connections* Define climate change and ocean warming. Explain how they relate to each other.

Using Your Notes

2. *Summarizing* Use your graphic organizer to write a description of the major causes of deforestation in Oceania.

Answering the Guiding Questions

3. *Exploring Issues* How are the people of Oceania using their islands' natural resources?

4. *Evaluating* How have humans—both in Oceania and elsewhere—affected the environments of the islands of Oceania?

5. *Discussing* How are the people of Oceania reacting to the changes in their environments?

Writing Activity

6. *Argument* Suppose you are a citizen of Oceania. Write a short letter to your local government regarding environmental issues on your island and actions you would like the government to take.

Directions: On a separate sheet of paper, answer the questions below. Make sure that you read carefully and answer all parts of the questions.

Lesson Review

Lesson 1

1 **Classifying** Explain the criteria for classifying the three groups of Oceania's islands. Within your explanation, include the names of these three groups.

2 **Making Connections** What is the relationship between atolls and the development of uniform languages and culture?

3 **Identifying Cause and Effect** How do climate conditions in the doldrums contribute to typhoons?

Lesson 2

4 **Explaining** What conditions gave rise to the number of languages spoken in Oceania? Would it be accurate to state that nearly half the languages spoken in the world are in Oceania?

5 **Problem Solving** Why is health care not uniform across all the islands of Oceania? Briefly discuss the importance of health care, and suggest an idea that might help to make it accessible across the islands.

6 **Describing** What is the most important economic activity in Oceania? Describe conditions that support as well as pose threats to this activity.

Lesson 3

7 **Considering Advantages and Disadvantages** How have countries outside of Oceania helped boost the economies of some countries of Oceania but also caused damage? Discuss the efforts of Fiji to balance economic and environmental interests.

8 **Drawing Conclusions** Discuss the impact of nuclear testing on the Oceania subregion. Can the destruction caused by nuclear testing be reversed? Explain.

9 **Identifying Central Issues** Describe the environmental issues tied to gold mining and production in Papua New Guinea. Explain why these issues have not been resolved.

21st Century Skills

Use the cartoon below to answer the questions that follow.

PRIMARY SOURCE

"Gentlemen, it's time we gave some serious thought to the effects of global warming."

10 **Using Primary Sources** Explain how this cartoon illustrates a key issue in understanding why the management of resources is important to the low islands of Oceania.

11 **Analyzing** Based on the above issue, discuss the potential impacts to freshwater resources and agriculture.

Critical Thinking

12 **Making Connections** What is the relationship of "tabu" to overfishing? How does implementation of "tabu" show the rise of the countries of Oceania to overcome challenges from the past under colonialism? Write a one-page essay to explain your responses.

Exploring the Essential Question

13 **Organizing** Create a Venn diagram to compare and contrast three countries in Oceania. Include the following information: physical geography; climate and resources; population patterns; society and culture today; economic activities; managing resources; and addressing issues related to human impact. After you complete your Venn diagram, write a summary to explain it.

Need Extra Help?

If You've Missed Question	**1**	**2**	**3**	**4**	**5**	**6**	**7**	**8**	**9**	**10**	**11**	**12**	**13**
Go to page	800	801	803	807	808	808	812	814	815	817	817	812	797

Directions: On a separate sheet of paper, answer the questions below. Make sure that you read carefully and answer all parts of the questions.

Applying Map Skills

Refer to the Unit 9 Atlas to answer the following questions.

14 *Places and Regions* Identify the largest island in Melanesia.

15 *The World in Spatial Terms* Use your mental map of Oceania to describe the spatial relationship among the Marshall Islands, Palau, Vanuatu, Fiji, and Tonga.

16 *Human Systems* Of all the islands of Oceania, which has the largest population? What is the approximate population? What are the likely causes that led to this island having the greatest population?

College and Career Readiness

17 *Decision Making* Suppose you are a college student and your professor is researching subsistence farming. She has asked you to research subsistence farming in the Solomon Islands and explain how it is being practiced currently and whether it seems to have met the goals and needs of the people living there. Additionally, your professor has asked you to express and support an opinion about whether subsistence farming has been a worthwhile activity in the Solomon Islands—or whether another type of farming should replace it. Conduct research, and then prepare a summary of your findings to submit to your professor.

Writing About Geography

18 *Narrative* Suppose you are traveling at low altitude through airspace above the high islands of Oceania. Describe the scenery below using standard grammar, spelling, sentence structure, and punctuation.

DBQ Analyzing Primary Sources

Use the document to answer the following questions.

Dr. Elizabeth Lindsey, an explorer, author, filmmaker, and speaker, is the first female to become a Fellow of the National Geographic Society. She received the Visionary Award from the United Nations in 2010 for contributions in intercultural engagement and understanding.

PRIMARY SOURCE

" 'As a child, I was cared for by three old Hawaiian women while my parents worked. These elders were revered in our community for their mastery in ancient traditions. They told me that I would travel far to keep the voices of the ancestors alive and that it would take the wisdom of these elders to return the world to balance.' "

—Dr. Elizabeth Lindsey, www.elizabethlindsey.com

19 *Interpreting Significance* How does the quote help you understand the attitude of Hawaiians toward elders in their communities? Does this attitude accurately reflect the general attitude of most Americans toward senior citizens today?

20 *Drawing Inferences* What types of actions do you think Dr. Lindsey and others are taking—or should take—to return the world to balance?

Research and Presentation

21 *Gathering Information* With a partner, conduct research to learn additional information about the Mariana Trench. Create a multimedia presentation to explain the following: location of the trench, tectonic activity that formed the trench, relevance of the trench to Oceania, likelihood that the trench will give rise to future earthquakes, recent newsworthy events relevant to the trench. Within your presentation, include maps, photographs, and graphs. Additionally, develop sketched or computer-generated diagrams.

Need Extra Help?

If You've Missed Question	14	15	16	17	18	19	20	21
Go to page	746	746	750	808	800	818	818	802

Antarctica:
the land of **ICE**

Earth's fifth-largest continent, the frigid land of Antarctica is found at the bottom of the world. Since its discovery in 1820 it has fascinated explorers and scientists.

THERE'S MORE ONLINE

LAND, CLIMATE & ENVIRONMENT

Antarctica is located at the southernmost point on Earth. It is surrounded by the Antarctic Ocean, also known as the Southern Ocean. At 5.5 million square miles (14.2 million sq. km) Antarctica is the Earth's fifth-largest continent. Almost completely covered by ice, it has no indigenous inhabitants and very limited plant and animal life. The only humans to live here are researchers and scientists, and even they live here only temporarily.

SOUTHERN OCEAN

Weddell Sea

Ronne Ice Shelf

ANTARCTIC PENINSULA

WIND AND ICE

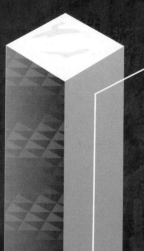

3.1 mi
(4776 m)
Thickest known piece of ice: (at Terre Adélie)

An ice shelf is a thick, floating sheet of ice that forms in coastal areas where a glacier meets a body of water.

WEST ANTARCTICA

ELLSWORTH LAND

-126° F
(-88° C)
Lowest recorded temperature

Amundsen Sea

MARIE BYRD LAND

203 mph
(327 kmph)
Highest recorded wind speed

NASA Goddard Space Flight Center Image by Reto Stöckli (land surface, shallow water, clouds). Enhancements by Robert Simmon (ocean color, compositing, 3D globes, animation). Data and technical support: MODIS Land Group; MODIS Science Data Support Team, Dave Pape; (bkgrd) rolfo/Getty Images

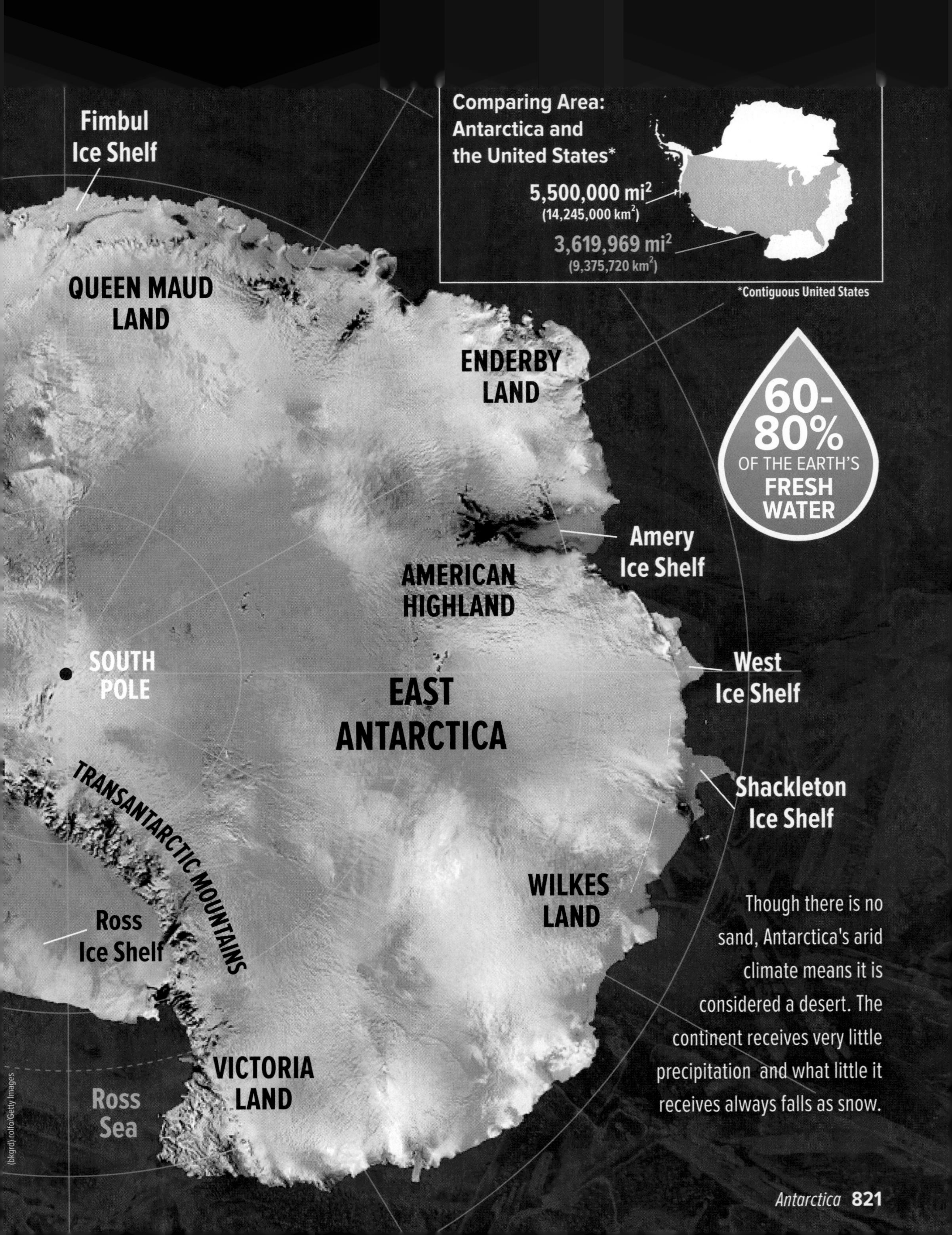
Fimbul Ice Shelf
QUEEN MAUD LAND
ENDERBY LAND
Comparing Area: Antarctica and the United States*
5,500,000 mi²
(14,245,000 km²)
3,619,969 mi²
(9,375,720 km²)
*Contiguous United States
60-80% OF THE EARTH'S FRESH WATER
Amery Ice Shelf
AMERICAN HIGHLAND
SOUTH POLE
EAST ANTARCTICA
West Ice Shelf
TRANSANTARCTIC MOUNTAINS
Shackleton Ice Shelf
WILKES LAND
Ross Ice Shelf
VICTORIA LAND
Ross Sea
(bkgrd) rolfo/Getty Images

Though there is no sand, Antarctica's arid climate means it is considered a desert. The continent receives very little precipitation and what little it receives always falls as snow.

ANTARCTIC DISCOVERIES

The frigid temperatures and isolation of Antarctica make sustaining plant and animal life difficult. The majority of animal life is found in the Southern Ocean. All warm-blooded animals in the Antarctic (seals, whales, penguins, etc.) rely on a thick layer of blubber for insulation. Most plant life is concentrated in the milder climate of West Antarctica, particularly the Antarctic Peninsula. There are no trees. Most of the vegetation consists of algae, lichens and mosses.

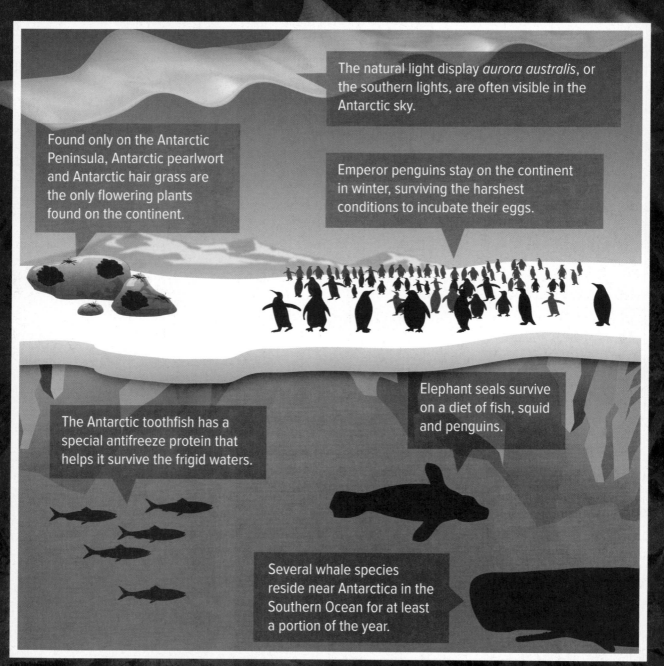

The natural light display *aurora australis*, or the southern lights, are often visible in the Antarctic sky.

Found only on the Antarctic Peninsula, Antarctic pearlwort and Antarctic hair grass are the only flowering plants found on the continent.

Emperor penguins stay on the continent in winter, surviving the harshest conditions to incubate their eggs.

Elephant seals survive on a diet of fish, squid and penguins.

The Antarctic toothfish has a special antifreeze protein that helps it survive the frigid waters.

Several whale species reside near Antarctica in the Southern Ocean for at least a portion of the year.

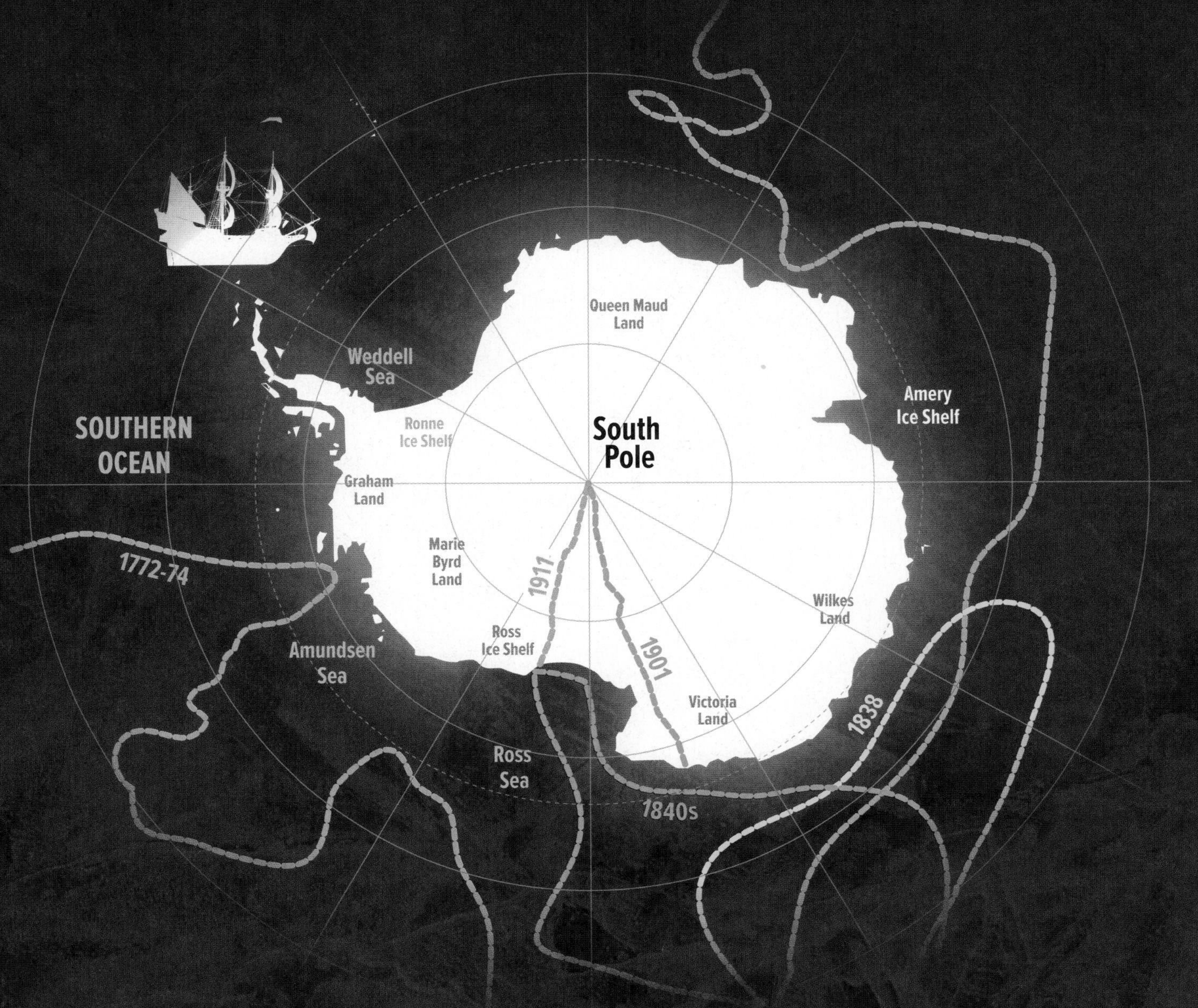

Queen Maud
Land

Weddell
Sea

Amery
Ice Shelf

Ronne
Ice Shelf

SOUTHERN
OCEAN

**South
Pole**

Graham
Land

1772-74

Marie
Byrd
Land

1911

Wilkes
Land

Amundsen
Sea

Ross
Ice Shelf

1901

1838

Victoria
Land

Ross
Sea

1840s

TIMELINE OF ANTARCTIC DISCOVERY

James Cook
1772-74

First to cross the
Antarctic Circle.

Charles Wilkes
1838

Explored the eastern
coast of Antarctica.

James Ross
1840s

Charted much of
Antarctica's coastline.

Robert F. Scott
1901

Begins first inland
exploration.

Roald Amundsen
1911

First to reach the
South Pole.

In the 1770s British captain James Cook became the first person to cross the Antarctic Circle. Cook never saw land, but discoveries were soon made in the early 1800s by scientists, explorers, and whalers. As interest in the new continent grew, several nations and individuals sponsored scientific explorations.

SCIENTIFIC RESEARCH

As interests grew in Antarctica, countries such as Argentina, Australia, Great Britain, Chile, France, New Zealand and Norway began to lay claim to sections of the continent. In 1959, 12 countries negotiated the Antarctic Treaty to preserve Antarctica for peaceful scientific research and to put all territorial claims on hold. Since then, other countries have established research programs in Antarctica and 26 new countries have signed on. Today, there are around 69 research stations operated by 30 countries that serve as bases for scientists to study physical geography, climate and wildlife. The population of these stations ranges from 4,000 people in the summer to 1,000 people in the winter.

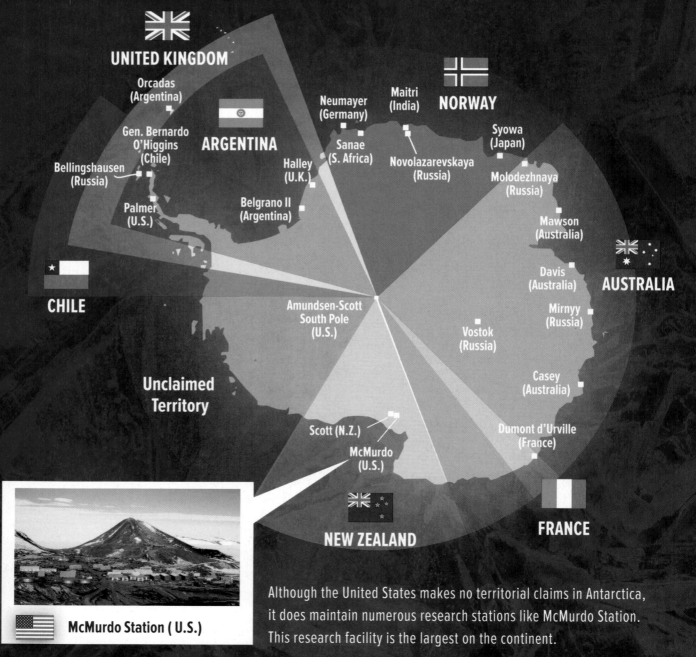

UNITED KINGDOM

Orcadas (Argentina)

Gen. Bernardo O'Higgins (Chile)

Bellingshausen (Russia)

ARGENTINA

Palmer (U.S.)

Halley (U.K.)

Belgrano II (Argentina)

Neumayer (Germany)

Sanae (S. Africa)

Maitri (India)

NORWAY

Novolazarevskaya (Russia)

Syowa (Japan)

Molodezhnaya (Russia)

Mawson (Australia)

Davis (Australia)

AUSTRALIA

Mirnyy (Russia)

CHILE

Amundsen-Scott South Pole (U.S.)

Vostok (Russia)

Casey (Australia)

Unclaimed Territory

Scott (N.Z.)

McMurdo (U.S.)

Dumont d'Urville (France)

NEW ZEALAND

FRANCE

McMurdo Station (U.S.)

Although the United States makes no territorial claims in Antarctica, it does maintain numerous research stations like McMurdo Station. This research facility is the largest on the continent.

New Frontiers

Because Antarctica's environment remains largely untouched by humans, it is a valuable location for scientific research. Compared to other places on Earth, for example, Antarctica's atmosphere, soil, and water contain little to no trace of pollution. As a result, scientists are able to use samples from Antarctica as a base for analyzing other parts of the world.

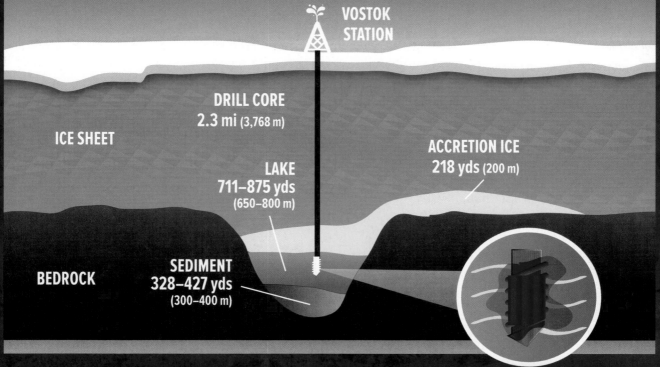

VOSTOK STATION

DRILL CORE
2.3 mi (3,768 m)

ICE SHEET

ACCRETION ICE
218 yds (200 m)

LAKE
711–875 yds
(650–800 m)

BEDROCK

SEDIMENT
328–427 yds
(300–400 m)

(bkgrd) rolfo/Getty Images

Lake Vostok

Lake Vostok, a subglacial lake, was discovered in the early 1970s. In 2012 scientists began probing the lake for study using deep drilling techniques. From this study scientists have found bacteria not known to exist anywhere else on Earth.

Impacts

Some scientists have raised concerns that antifreeze used in Lake Vostok drilling could harm the life forms discovered. Scientists must take great care to preserve Antarctica's pristine environment.

Thinking Geographically

1. **Analyzing Information** What characteristics of Antarctica's land and climate make sustaining life difficult?

2. **Identifying** What are some common characteristics shared by animal life and plant life found in and around Antarctica?

3. **Drawing Conclusions** What potential problems could result from nations' making territorial claims in Antarctica?

4. **Exploring Issues** Even though natural resource deposits have been found in Antarctica, many countries have agreed not to mine them. What do you think are the pros and cons of mining in the Antarctic?

5. **Expository** Research the type of work being done by scientists at one of the Antarctic stations. How could the work of these scientists impact your life? Write an essay detailing your findings.

World Religions Handbook

TERMS

animism—belief that spirits inhabit natural objects and forces of nature

atheism—disbelief in the existence of any god

monotheism—belief in one God

polytheism—belief in more than one god

secularism—belief that life's questions can be answered apart from religious belief

sect—a subdivision within a religion that has its own distinctive beliefs and/or practices

tenet—a belief, doctrine, or principle believed to be true and held in common by members of a group

A *religion* is a set of beliefs in an ultimate reality and a set of practices used to express those beliefs. Religion is a key component of culture.

Each religion is defined and set apart from other religions by its own special celebrations and worship styles. Most religions also have their own sacred texts, sacred symbols, and sacred sites. All of these aspects of religion help to unite followers of that faith regardless of where in the world they live.

The religions examined in this World Religions Handbook all have these sacred elements, celebrations, and worship styles. Examining these characteristics provides insight into each of these religions.

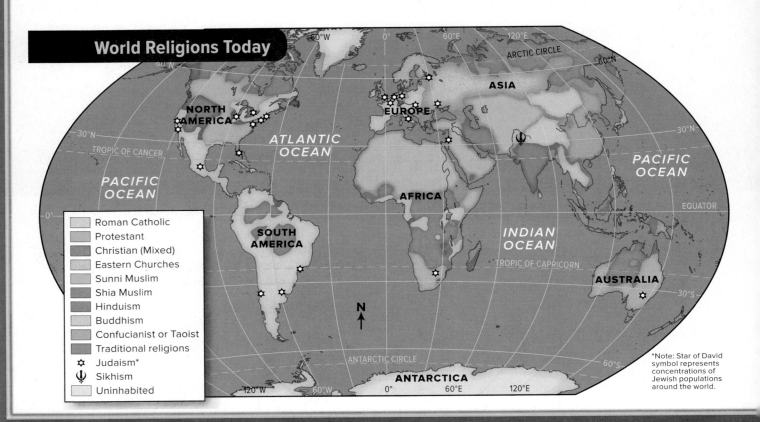

World Religions Today

- Roman Catholic
- Protestant
- Christian (Mixed)
- Eastern Churches
- Sunni Muslim
- Shia Muslim
- Hinduism
- Buddhism
- Confucianist or Taoist
- Traditional religions
- ✡ Judaism*
- ☬ Sikhism
- Uninhabited

*Note: Star of David symbol represents concentrations of Jewish populations around the world.

We study religion because it is an important component of culture, shaping how people interact with one another, dress, and eat. Religion is at the core of the belief system of a religion's culture.

The diffusion of religion throughout the world has been caused by a variety of factors including migration, missionary work, trade, and war. Buddhism, Christianity, and Islam are the three major religions that spread their religion through missionary activities. Religions such as Hinduism, Sikhism, and Judaism are associated with a particular culture group. Followers are usually born into these religions. Sometimes close contract and differences in beliefs have resulted in conflict between religious groups.

Percentage of World Population

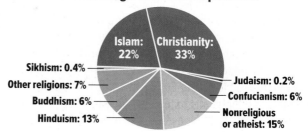

Islam: 22%
Christianity: 33%
Sikhism: 0.4%
Other religions: 7%
Buddhism: 6%
Hinduism: 13%
Judaism: 0.2%
Confucianism: 6%
Nonreligious or atheist: 15%

Note: Total exceeds 100% because numbers were rounded.
Sources: www.cia.gov, The World Factbook 2006; www.adherents.com.

Early Diffusion of Major World Religions

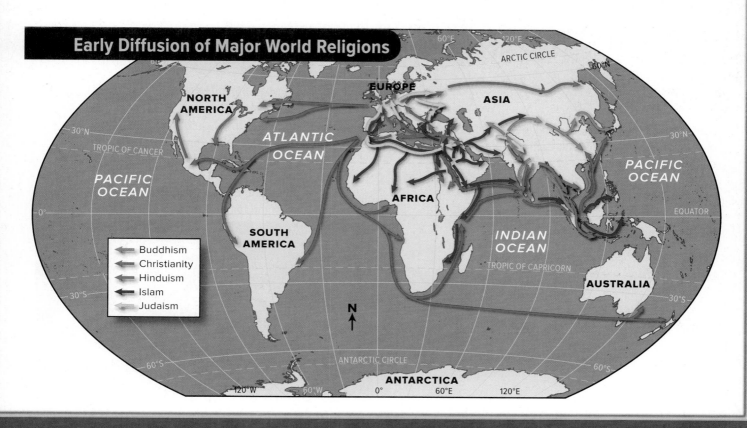

NORTH AMERICA
EUROPE
ASIA
ATLANTIC OCEAN
AFRICA
PACIFIC OCEAN
SOUTH AMERICA
INDIAN OCEAN
PACIFIC OCEAN
AUSTRALIA
ANTARCTICA

ARCTIC CIRCLE
TROPIC OF CANCER
EQUATOR
TROPIC OF CAPRICORN
ANTARCTIC CIRCLE

N

Buddhism
Christianity
Hinduism
Islam
Judaism

Buddhism

Siddhartha Gautama, known as the Buddha ("the Awakened") after his enlightenment at the age of 35, was born some 2,500 years ago in what is now Nepal. The Buddha's followers adhere to his teachings (dharma, meaning "divine law"), which aim to end suffering in the world. Buddhists call this goal Nirvana; and they believe that it can be achieved only by understanding the Four Noble Truths and by following the 4th Truth, which says that freedom from suffering is possible by practicing the Eightfold Path. Through the Buddha's teachings, his followers come to know the impermanence of all things and reach the end of ignorance and unhappiness.

Over time, as Buddhism spread throughout Asia, several branches emerged. The largest of these are Theravada Buddhism, the monk-centered Buddhism which is dominant in Sri Lanka, Burma, Thailand, Laos, and Cambodia; and Mahayana, a complex, more liberal variety of Buddhism that has traditionally been dominant in Tibet, Central Asia, Korea, China, and Japan.

Statue of the Buddha, Siddhartha Gautama

Sacred Text ▾

For centuries the Buddha's teachings were transmitted orally. For Theravada Buddhists, the authoritative collection of Buddhist texts is the Tripitaka ("three baskets"). These texts were first written on palm leaves in a language called Pali. This excerpt from the *Dhammapada,* a famous text within the Tripitaka, urges responding to hatred with love:

> ❝ *For hatred does not cease by hatred at any time: hatred ceases by love, this is an old rule.*
>
> *The world does not know that we must all come to an end here—but those who know it, their quarrels cease at once.* ❞
> —*Dhammapada 1.5–6*

Sacred Symbol ▾

The *dharmachakra* ("wheel of the law") is a major Buddhist symbol. Among other things, it signifies the overcoming of obstacles. The eight spokes represent the Eightfold Path—right view, right intention, right speech, right action, right livelihood, right effort, right mindfulness, right concentration—that is central for all Buddhists.

Sacred Site ▲

Buddhists believe that Siddhartha Gautama
achieved enlightenment beneath the
Bodhi Tree in Bodh Gayā, India. Today,
Buddhists from around the world flock to
Bodh Gayā in search of their own spiritual
awakening.

Worship and Celebration ▶

The ultimate goal of Buddhists is to achieve
Nirvana, the enlightened state in which
individuals are free from ignorance, greed,
and suffering. Theravada Buddhists believe
that monks are most likely to reach Nirvana
because of their lifestyle of renunciation,
moral virtue, study, and meditation.

Christianity

Christianity claims more members than any of the other world religions. It dates its beginning to the death of Jesus in A.D. 33 in what is now Israel. It is based on the belief in one God and on the life and teachings of Jesus. Christians believe that Jesus, who was born a Jew, is the son of God and is fully divine and human. Christians regard Jesus as the Messiah (Christ), or savior, who died for humanity's sins. Christians feel that people are saved and achieve eternal life by faith in Jesus.

The major forms of Christianity are Roman Catholicism, Eastern Orthodoxy, and Protestantism. In 1054, disputes over doctrine and the leadership of the Christian Church caused the church to divide into the Roman Catholic Church, headed by the Bishop of Rome, also known as the Pope, and the Eastern Orthodox Churches, led by patriarchs. Protestant churches emerged in the 1500s in an era known as the Reformation. Protestants disagreed with some Catholic doctrines and were critical of the Pope's authority. Despite their different theologies, all three forms are united in their belief in Jesus as savior.

Stained glass window depicting Jesus

Sacred Text

The Christian Bible is the spiritual text for all Christians and is considered to be inspired by God. This excerpt, from Matthew 5:3-12, is from Jesus' Sermon on the Mount.

Sacred Symbol

Christians believe that Jesus died for their sins. His death redeemed those who follow his teachings. The statue *Christ the Redeemer,* located in Rio de Janeiro, Brazil, symbolizes this fundamental belief.

> *Blessed are the poor in spirit, for theirs is the kingdom of heaven.*
> *Blessed are those who mourn, for they shall be comforted.*
> *Blessed are the meek, for they shall inherit the earth.*
> *Blessed are those who hunger and thirst for righteousness, for they shall be satisfied.*
> *Blessed are the merciful, for they shall obtain mercy.*
> *Blessed are the pure in heart, for they shall see God.*
> *Blessed are the peacemakers, for they shall be called sons of God.*
> *Blessed are those who are persecuted for righteousness' sake, for theirs is the kingdom of heaven.*
> *Blessed are you when men revile you and persecute you and utter all kinds of evil against you falsely on my account.*
> *Rejoice and be glad, for your reward is great in heaven, for so men persecuted the prophets who were before you.*

PHOTOS: (t) ©Tom Grill/Corbis; (b) Antonello/Moment/Getty Images; TEXT: Scripture quotations are from the ESV® Bible (The Holy Bible, English Standard Version®), copyright © 2001 by Crossway, a

Sacred Site ▶

The Gospels affirm that Bethlehem was the birthplace of Jesus. Consequently, it holds great importance to Christians. The Church of the Nativity is located in the heart of Bethlehem. It houses the spot where Christians believe Jesus was born.

Worship and Celebration ▼

Christians celebrate many events commemorating the life and death of Jesus. Among the most widely known and observed are Christmas, Good Friday, and Easter. Christmas is often commemorated by attending church services to celebrate the birth of Jesus. As part of the celebration, followers often light candles.

Confucianism

Confucianism began more than 2,500 years ago in China. Although considered a religion, it is actually a philosophy. It is based upon the teachings of Confucius, which are grounded in ethical behavior and good government.

The teachings of Confucius focused on three areas: social philosophy, political philosophy, and education. Confucius taught that relationships are based on rank. Persons of higher rank are responsible for caring for those of lower rank. Those of lower rank should respect and obey those of higher rank. Eventually his teachings spread from China to other East Asian societies.

Students study Confucianism, Chunghak-dong, South Korea

Sacred Symbol ▼

Yin-yang symbolizes the harmony offered by Confucianism. The light half represents *yang,* the creative, firm, strong elements in all things. The dark half represents *yin,* the receptive, yielding, weak elements. The two act together to balance one another.

Sacred Text ▼

Confucius was famous for his sayings and proverbs. These teachings were gathered into a book called the *Analects* (see image above) after Confucius's death. Below is an example of Confucius's teachings:

Confucius said:

> ❝ *To learn and to practice what is learned time and again is pleasure, is it not? To have friends come from afar is happiness, is it not? To be unperturbed when not appreciated by others is gentlemanly, is it not?* ❞

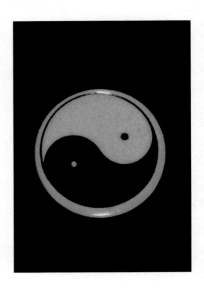

Sacred Site ▲

The temple at Qufu is a group of buildings dedicated to Confucius. It is located on Confucius's ancestral land. It is one of the largest ancient architectural complexes in China. Every year followers gather at Qufu to celebrate the birthday of Confucius.

Worship and Celebration ▶

Confucianism does not have a god or clergy, but there are temples dedicated to Confucius, the spiritual leader. Those who follow his teachings see Confucianism as a way of life and a guide to ethical behavior and good government.

Hinduism

induism is the oldest of the world's major living religions. It developed among the cultures in India as they spread out over the plains and forests of the subcontinent. It has no single founder or founding date. Hinduism is complex; it has numerous sects and many different practices among its followers. Most Hindus believe in one god, whose qualities are represented by various divinities. Among the more famous Hindu divinities are Brahma, Vishnu, and Shiva, who represent respectively the creative, sustaining, and destructive forces in the universe. Major Hindu beliefs are reincarnation, karma, and dharma.

Hindus believe the universe contains several heavens and hells. According to the concept of rebirth or reincarnation, which is central to their beliefs, souls are continually reborn. In what form one is reborn is determined by the good and evil actions performed in his or her past lives. Those acts are karma. A soul continues in the cycle of rebirth until release is achieved.

Sacred Text ▾

The Vedas consist of hymns, prayers, and speculations composed in ancient Sanskrit. They are the oldest religious texts in an Indo-European language. The Rig Veda, Sama Veda, Yajur Veda, and Atharva Veda are the four great Vedic collections. Together, they make up one of the most significant and authoritative Hindu religious texts.

Statue of Vishnu

> *Now, whether they perform a cremation for such a person or not,*
> *people like him pass into the flame,*
> *from the flame into the day,*
> *from the day into the fortnight of the waxing moon*
> *from the fortnight of the waxing moon into the six months when the sun moves north,*
> *from these months into the year, from the year into the sun,*
> *from the sun into the moon, and from the moon into the lightning.*
> *Then a person who is not human—he leads them to Brahman.*
> *This is the path to the gods, the path to Brahman.*
> *Those who proceed along this path do not return to this human condition.*
> —The Chandogya Upanishad 4:15.5

Sacred Symbol ▾

One important symbol of Hinduism is actually a symbol for a sound. "Om" is a sound that Hindus often chant during prayer, mantras, and rituals.

Sacred Site ▸

Hindus believe that when a person dies his or her soul is reborn. This is known as reincarnation. Many Hindus bathe in the Ganges and other sacred rivers to purify their soul and to be released from rebirth.

Worship and Celebration ▾

Holi is a significant North Indian Hindu festival celebrating the triumph of good over evil. As part of the celebration, men, women, and children splash colored powders and water on each other. In addition to its religious significance, Holi also celebrates the beginning of spring.

Islam

Followers of Islam, known as Muslims, believe in one God, whom they call Allah. The word *Allah* is Arabic for "god." The founder of Islam, Muhammad, began his teachings in Makkah (Mecca) in A.D. 610. Eventually the religion spread throughout much of Asia, including parts of India to the borders of China, and substantial portion of Africa. According to Muslims, the Quran, their holy book, contains the direct word of God, revealed to their prophet Muhammad sometime between A.D. 610 and A.D. 632. Muslims believe that God created nature and without his intervention, there would be nothingness. God serves four functions: creation, sustenance, guidance, and judgment.

Central to Islamic beliefs are the Five Pillars. These are affirmation of the belief in Allah and Muhammad as his prophet; group prayer; tithing, or the giving of money to charity; fasting during Ramadan; and a pilgrimage to Makkah once in a lifetime if physically and financially able. Within Islam, there are two main branches, the Sunni and the Shia. The differences between the two are based on the history of the Muslim state. The Shia believed that the rulers should descend from Muhammad. The Sunni believed that the rulers need only be followers of Muhammad. Most Muslims are Sunni.

The Dome of the Rock, Jerusalem

The Quran

Sacred Text

The sacred text of Islam is the Quran. Preferably, it is written and read only in Arabic, but translations have been made into many languages. The excerpt below is a verse repeated by all Muslims during their five daily prayers.

> *In the name of the merciful and compassionate God.*
>
> *Praise belongs to God, the Lord of the worlds, the merciful, the compassionate, the ruler of the day of judgment! Thee we serve and Thee we ask for aid. Guide us in the right path, the path of those Thou art gracious to 1; not of those Thou art wroth with; nor of those who err.*
>
> —The Quran (The Opening Chapter)

Sacred Symbol

Islam is often symbolized by the crescent moon. It is an important part of Muslim rituals, which are based on the lunar calendar.

Sacred Site ▶

Makkah is a sacred site for all Muslims. One of the Five Pillars of Islam states that all who are physically and financially able must make a hajj, or pilgrimage, to the holy city once in their life. Practicing Muslims are also required to pray facing Makkah five times a day.

Worship and Celebration ▼

Ramadan is a month-long celebration commemorating the time during which Muhammad received the Quran from Allah. It is customary for Muslims to fast from dawn until sunset all month long. Muslims believe that fasting helps followers focus on spiritual rather than bodily matters and creates empathy for one's fellow men and women. Ramadan ends with a feast known as Eid-al-Fitr, or Feast of the Fast.

Judaism

Judaism is a monotheistic religion. In fact, Judaism was the first major religion to believe in one God. Jews trace their national and religious origins back to God's call to Abraham and the revelation of Torah to Moses at Mount Sinai. Jews believe that they have a covenant with God, who expects them to pursue justice and live ethical lives and will one day usher in an era of universal peace.

Over time Judaism has separated into branches, including Orthodox, Reform, Conservative, and Reconstructionist. Orthodox Jews are the most traditional of all the branches.

El Ghriba Synagogue, Jerba, Tunisia

The Magen David

The Torah scroll

Sacred Symbols ▾

One of the oldest symbols of the Jewish people is the menorah, used in the celebration of Hanukkah, commemorating the re-dedication of the Temple of Jerusalem following the Maccabees' victory over the Syrian Greeks. Another important Jewish symbol is the Star of David, also known as the Magen David, or Shield of David. It has been popular since the 17th century.

Sacred Text ▾

The Torah is the five books of Moses, which tell the story of the origins of the Jews and explain Jewish laws. The remainder of the Hebrew Bible contains the writings of the prophets, Psalms, and ethical and historical works.

> *❝ I am the Lord your God, who brought you out of the land of Egypt, out of the house of slavery; you shall have no other gods before me. ❞*
>
> —Exodus 20:2–3

PHOTOS: (t) Juergen Ritterbach/F1online/Getty Images; (cl) ©Tetra Images/age fotostock; (cr) ©Purestock/SuperStock; (b) ©JUPITERIMAGES/Comstock Images/Alamy; TEXT: Exodus 20:2-3. From The Hebrew Bible in

Sacred Site ◄

The Western Wall is what remains of the structure surrounding the Second Jerusalem Temple, built after the Jews' return from Babylonian captivity. It is considered the most sacred spot in Jewish religious tradition, in the holiest city for Judaism. Worldwide Jews pray facing Jerusalem morning, afternoon, and evening, and within Jerusalem, Jews face toward the Western Wall.

Worship and Celebration ▼

The day-long Yom Kippur service ends with the blowing of the ram's horn (shofar). Yom Kippur is the holiest day in the Jewish calendar. During Yom Kippur, Jews do not eat or drink for 25 hours. The purpose is to reflect on the past year, repent for one's sins, and gain forgiveness from God. It falls in September or October, ten days after Rosh Hashanah, the Jewish New Year.

Sikhism

Sikhism emerged in the mid-1400s in the Punjab, in northwest India. Sikhism is a distinct and independent religion that arose out of the teachings of Guru Nanak. Sikh tradition says that Guru Nanak's teachings were revealed directly to him by God.

Sikhs believe in one God who is formless, all-powerful, all-loving, and without fear or hate towards anyone. One can achieve unity with God through service to humanity, meditation, and honest labor. While about 76 percent of the world's 27 million Sikhs live in the Punjab, Sikhism has spread widely as many Sikhs have migrated to new homes around the world.

Sikh men often wear long beards and cover their heads with turbans.

Guru Nanak.

Sacred Symbol ▲

One of the sacred symbols of the Sikhs is the *khanda*. It is composed of four traditional Sikh weapons.

Sacred Text ▾

The primary scripture for Sikhs is the Guru Granth Sahib. Compiled from the mid-1500s through the 1600s, it includes contribution from Sikh Gurus and from some persons also claimed as saints by Hindus and Muslims, such as Namdev, Ravidas, and Kabir.

> ❝ *Enshrine the Lord's Name within your heart. The Word of the Guru's Bani prevails throughout the world, through this Bani, the Lord's name is obtained.* ❞
> —Guru Amar Das, page 1066

Central Symbol ▲

Ek Onkar is one of the central Sikh symbols. It represents the belief that there is one God for all people, regardless of religion, gender, race, or culture.

PHOTOS: (l) Aloysius Patrimonio/Alamy Stock Photo; (tr) ©Photosindia/Corbis; (br) ©Chirag Pithadiya/Corbis; TEXT: From Sikhism and Indian Civilization, by R. K. Pruthi. Published by Discovery Publishing House, 2004

Sacred Site ▸

Darbar Sahib (also known as the Golden Temple) is located in Amritsar, Punjab. It is one of the most popular Sikh houses of worship because of its historical significance.

Worship and Celebration ▾

Vaisakhi is a significant Punjabi and Sikh festival in April. Sikhs celebrate Vaisakhi as the day Guru Gobind Singh, the 10th Guru, established the Khalsa, the community of people who have been initiated into the Sikh religion. In Punjab, Vaisakhi is celebrated as the New Year and the beginning of the harvest season.

Indigenous Religions

There are many varieties of religious belief that are limited to particular ethnic groups. These local religions are found in Africa as well as isolated parts of Japan, Australia, and the Americas.

Most local religions reflect a close relationship with the environment. Some groups teach that people are a part of nature, not separate from it. Animism is characteristic of many indigenous religions. Natural features are sacred, and stories about how nature came to be are an important part of religious heritage. Although many of these stories have been written down in modern times, they were originally transmitted orally.

Africa The continent of Africa is home to a variety of local religions. Despite their differences, most African religions recognize the existence of one creator in addition to spirits that inhabit all aspects of life. Religious ceremonies are often celebrated with music and dance.

These Turkana men from Kenya are performing a traditional jumping dance.

Rituals are an important part of African religions. These Masai boys are wearing ceremonial dress as part of a ritual.

Masks are a component of ritual and ceremony in many African religions.

Japan Shinto, founded in Japan, is the largest indigenous religion. It dates back to prehistoric times and has no formal doctrine. The gods are known as kami. Ancestors are also revered and worshiped. Its four million followers often practice Buddhism in addition to practicing Shinto.

Shinto shrines, like this one, are usually built in places of great natural beauty to emphasize the relationship between people and nature.

This Shinto priest is presiding over a ritual at a Japanese temple. These priests often live on shrine grounds.

Australia The Australian Aboriginal religion has no deities. It is based upon a belief known as the Dreaming, or Dreamtime. Followers believe that ancestors sprang from the Earth and created all people, plant, and animal life. They also believe that these ancestors continue to control the natural world.

These Aborigine women are blessing a newborn with smoke during a traditional ritual intended to ensure the child's health and good fortune.

Aborigines, like these young girls, often paint their faces with the symbols of their clan or family group.

Indigenous Religions

Native Americans The beliefs of most Native Americans center on the spirit world; however, the rituals and practices of individual groups vary. Most Native Americans believe in a Great Spirit who, along with other spirits, influences all aspects of life. These spirits make their presence known primarily through acts of nature.

The rituals, prayers, and ceremonies of Native Americans are often centered on health and good harvest and hunting. Rituals used to mark the passage through stages of life, including birth, adulthood, and death are passed down as tribal traditions. Religious ceremonies often focus on important points in the agricultural and hunting seasons. Prayers, which are offered in song and dance, also concentrate on agriculture and hunting themes as well as health and well-being.

Rituals are passed down from generation to generation. These Native Americans are performing a ritual dance.

There are many different Native American groups throughout the United States and Canada. This Pawnee is wearing traditional dress during a celebration in Oklahoma.

Totem poles, like this one in Alaska, were popular among the Native American peoples of the Northwest Coast. They were often decorated with mythical beings, family crests, or other figures. They were placed outside homes.

Assessment

Reviewing Vocabulary

Match the following terms with their definitions.

1. sect
2. monotheism
3. polytheism
4. animism
5. atheism

  a. belief that spirits inhabit natural objects and forces of nature

  b. belief in one God

  c. a subdivision within a religion that has its own distinctive belief and/or practices

  d. belief in more than one god

  e. disbelief in the existence of any god

Reviewing the Main Ideas

World Religions

6. Which religion has the most followers worldwide?

7. On a separate sheet of paper, make a table of the major world religions. Use the chart below to get you started.

Name	Founder	Geographic distribution	Sacred sites
Buddhism			
Christianity			
Confucianism			
Hinduism			
Islam			
Judaism			
Sikhism			
Indigenous			

Buddhism

8. According to Buddhism, how can the end of suffering in the world be achieved?

9. What is Nirvana? According to Buddhists, who is most likely to achieve Nirvana and why?

Christianity

10. In what religion was Jesus raised?

11. Why do Christians regard Jesus as their savior?

Confucianism

12. What is Confucianism based on?

13. What does yin-yang symbolize?

Hinduism

14. Where did Hinduism develop?

15. What role do Hindus believe karma plays in reincarnation?

Islam

16. What are the two branches of Islam? What is the main difference between the two groups?

17. What role does Makkah play in the Islamic faith?

Judaism

18. What is the Torah?

19. What is the purpose of Yom Kippur?

Sikhism

20. Where do most Sikhs live? Why?

21. What is Vaisakhi?

Indigenous Religions

22. Why would local religions feature sacred stories about the creation of people, animals, and plant life?

23. Which of the indigenous religions has the largest membership?

Problem-Solving Activity

24. **Research Project** Use library and Internet sources to research the role of food and food customs in one of the world's major religions. Create a presentation to report your findings to the class.

Gazetteer

A Gazetteer (GA•zuh•TIHR) is a geographic index or dictionary. It shows latitude and longitude for cities and certain other places. This Gazetteer lists most of the world's largest independent countries, their capitals, and several important geographic features. The page numbers tell where each entry can be found on a map in this book. As an aid to pronunciation, many entries are spelled phonetically.

A

Abidjan (AH•bee•JAHN) Capital and port city of Côte d'Ivoire, Africa. 5°N 4°W (p. 477)

Abu Dhabi (AH•boo DAH•bee) Capital of the United Arab Emirates, on the Persian Gulf. 24°N 54°E (p. 361)

Abuja (ah•BOO•jah) Capital of Nigeria. 8°N 9°E (p. 477)

Accra (AH•kruh) Capital and port city of Ghana. 6°N 0° longitude (p. 477)

Aconcagua (AH•kohn•KAH•gwah) Highest peak of the Andes and of the Western Hemisphere, in western Argentina near the Chilean border. 32°S 76°W (p. 168)

Addis Ababa (AHD•dihs AH•bah•BAH) Capital of Ethiopia. 9°N 39°E (p. 477)

Adriatic (AY•dree•A•tihk) **Sea** Arm of the Mediterranean Sea between the Balkan Peninsula and Italy. (p. 244)

Aegean (ee•JEE•uhn) **Sea** Arm of the Mediterranean Sea between Greece and Turkey. (p. 242)

Afghanistan Country in Central Asia, west of Pakistan. (p. 361)

Ahaggar Mountains Highest plateau region in the central Sahara. (p. 360)

Albania Country on the east coast of the Adriatic Sea, south of Serbia and Montenegro. (p. 243)

Algeria Country in North Africa. (p. 361)

Algiers (al•JIHRZ) Capital of Algeria. 37°N 3°E (p. 361)

Alps Mountain system extending through central Europe. (p. 242)

Altay Shan Mountain system between western Mongolia and China and between Kazakhstan and southern Russia. (p. 672)

Amazon River River flowing through Peru and Brazil in South America and into the Atlantic Ocean. (p. 168)

Amman Capital of Jordan. 32°N 36°E (p. 361)

Amsterdam Capital of the Netherlands. 52°N 5°E (p. 243)

Amu Dar'ya River in Turkmenistan in central and western Asia. (p. 360)

Amur River River in northeast Asia. (p. 672)

Andes Mountain system along western South America. (p. 168)

Andorra (an•DAWR•uh) Country in southern Europe, between France and Spain. (p. 243)

Angola (ang•GOH•luh) Country in Africa, south of the Democratic Republic of the Congo. (p. 477)

Ankara (AHN•kuh•ruh) Capital of Turkey. 40°N 33°E (p. 361)

Antananarivo (AHN•tah•NAH•nah•REE•voh) Capital of Madagascar. 19°S 48°E (p. 477)

Antigua Island in the West Indies, part of independent Antigua and Barbuda. 18°N 61°W (p. 169)

Apennines (A•puh•NYNZ) Mountain range in central Italy. (p. 242)

Appalachian Mountains Mountain system in eastern North America. (p. 112)

Arabian Sea Part of the Indian Ocean between India and the Arabian Peninsula. (p. 361)

Aral Sea Inland sea between Kazakhstan and Uzbekistan. (p. 360)

Argentina Country in South America, east of Chile. (p. 169)

Arkansas River River in south-central United States, emptying into the Mississippi River. (p. 112)

Armenia (ahr•MEE•nee•uh) Southeastern European country between the Black and Caspian Seas. (p. 361)

Ashkhabad (ASH•kuh•BAD) Capital of Turkmenistan. 40°N 58°E (p. 361)

Asmara (az•MAHR•uh) Capital of Eritrea. 16°N 39°E (p. 477)

Astana Capital of Kazakhstan. 52°N 72°E (p. 361)

Asunción (ah•SOON•SYOHN) Capital of Paraguay. 25°S 58°W (p. 169)

Athens Capital of Greece. 38°N 24°E (p. 243)

Atlas Mountains Mountain range on the northern edge of the Sahara. (p. 360)

Australia Country and continent southeast of Asia. (p. 747)

Austria Country in central Europe, east of Switzerland. (p. 243)

Azerbaijan (A•zuhr•by•JAHN) European-Asian country on the Caspian Sea. (p. 361)

B

Baghdad Capital of Iraq. 33°N 44°E (p. 361)

Bahamas Independent state comprising a chain of islands, cays, and reefs southeast of Florida and north of Cuba. 24°N 76°W (p. 169)

Bahrain (bah•RAYN) Independent state in the western Persian Gulf. (p. 361)

Baku Capital of Azerbaijan. 40°N 50°E (p. 361)

Balkan Mountains Mountain range extending across central Bulgaria to the Black Sea. (p. 242)

Balkan Peninsula Peninsula in southeastern Europe bordered on the west by the Adriatic Sea. (p. 242)

Baltic Sea Arm of the Atlantic Ocean in northern Europe that connects with the North Sea. (p. 242)

Bamako (BAH•mah•KOH) Capital of Mali. 13°N 8°W (p. 477)

Bangkok Capital of Thailand. 14°N 100°E (p. 5)

Bangladesh (BAHNG·gluh·DESH) Country in South Asia, bordered by India and Myanmar. (p. 599)

Bangui (bahng·GEE) Capital of the Central African Republic. 4°N 19°E (p. 477)

Banjul Capital of Gambia. 13°N 17°W (p. 477)

Barbados Island country between the Atlantic Ocean and the Caribbean Sea. 14°N 59°W (p. 169)

Barbuda Island in the West Indies, part of independent Antigua and Barbuda. 18°N 62°W (p. 169)

Barents Sea Part of the Arctic Ocean, north of Norway and Russia. (p. 242)

Bay of Bengal Part of the Indian Ocean between eastern India and Southeast Asia. (p. 598)

Beijing Capital of China. 40°N 116°E (p. 5)

Beirut (bay·ROOT) Capital of Lebanon. 34°N 36°E (p. 361)

Belarus (BEE·luh·ROOS) Eastern European country west of Russia. (p. 243)

Belgium (BEHL·juhm) Country in northwestern Europe, south of the Netherlands. (p. 243)

Belgrade Capital of Serbia. 45°N 21°E (p. 243)

Belize (buh·LEEZ) Country in Central America. (p. 169)

Belmopan (BEHl·moh·PAHN) Capital of Belize. 17°N 89°W (p. 169)

Benin (buh·NEEN) Country in western Africa. (p. 477)

Ben Nevis Peak in the highlands region of the Grampian Mountains in Scotland. 54°N 5°W (p. 242)

Bering Sea Part of the north Pacific Ocean, extending between the United States and Russia. (p. 112)

Berlin Capital of Germany. 53°N 13°E (p. 243)

Bern Capital of Switzerland. 47°N 7°E (p. 243)

Bhutan (boo·TAHN) Country in the eastern Himalaya, northeast of India. 27°N 91°E (p. 599)

Bishkek Capital and largest city of Kyrgyzstan. 43°N 75°E (p. 361)

Bissau (bih·SOW) Capital of Guinea-Bissau. 12°N 16°W (p. 477)

Black Sea Sea between Europe and Asia. (p. 242)

Bloemfontein (BLOOM·FAHN·TAYN) Judicial capital of the Republic of South Africa. 29°S 26°E (p. 477)

Bogotá (BOH·goh·TAH) Capital of Colombia. 5°N 74°W (p. 169)

Bolivia Republic in west-central South America. (p. 169)

Bosnia-Herzegovina (BAHZ·nee·uh HERT·suh·goh·VEE·nuh) Southeastern European country between Serbia and Croatia. (p. 243)

Bosporus Strait between European and Asian Turkey, connecting the Sea of Marmara with the Black Sea. (p. 413)

Botswana (baht·SWAH·nuh) Country in Africa, north of the Republic of South Africa. (p. 477)

Brahmaputra River River that begins in Tibet, flows through northeast India and Bangladesh, and empties into the Bay of Bengal. (p. 598)

Brasília (bruh·ZIHL·yuh) Capital of Brazil. 16°S 48°W (p. 169)

Bratislava (BRAH·tuh·SLAH·vuh) Capital and largest city of Slovakia. 48°N 17°E (p. 243)

Brazil Largest country in South America, in east-central South America. (p. 169)

Brazzaville (BRA·zuh·VIHL) Capital of Congo. 4°S 15°E (p. 477)

Brunei (bru·NY) Country on the northern coast of the island of Borneo. (p. 747)

Brussels Capital of Belgium. 51°N 4°E (p. 243)

Bucharest (BOO·kuh·REHST) Capital of Romania. 44°N 26°E (p. 243)

Budapest Capital of Hungary. 48°N 19°E (p. 243)

Buenos Aires (BWAY·nuhs AR·eez) Capital of Argentina. 34°S 58°W (p. 169)

Bujumbura (BOO·juhm·BUR·uh) Capital of Burundi. 3°S 29°E (p. 477)

Bulgaria (BUHL·GAR·ee·uh) Country in southeastern Europe, south of Romania. (p. 243)

Burkina Faso (bur·KEE·nuh FAH·soh) Country in western Africa, south of Mali. (p. 477)

Burundi (bu·ROON·dee) Country in central Africa at the northern end of Lake Tanganyika. (p. 477)

C

Cairo (KY·roh) Capital of Egypt. 31°N 32°E (p. 5)

Cambodia (kam·BOH·dee·uh) Country in Southeast Asia, south of Thailand. (p. 747)

Cameroon (KA·muh·ROON) Country in West Africa, on the northeast shore of the Gulf of Guinea. (p. 477)

Canada Country in northern North America. (p. 113)

Canberra Capital of Australia. 35°S 149°E (p. 747)

Cape Town Legislative capital of the Republic of South Africa. 34°S 18°E (p. 5)

Cape Verde Republic consisting of a group of volcanic islands in the Atlantic Ocean. 15°N 26°W (p. 477)

Caracas (kah·RAH·kahs) Capital of Venezuela. 11°N 67°W (p. 169)

Caribbean (KAR·uh·BEE·uhn) **Sea** Part of the Atlantic Ocean, bounded by the West Indies, South America, and Central America. (p. 170)

Carpathian Mountains Mountain range in eastern Europe in Slovakia and Romania. (p. 242)

Caspian (KAS·pee·uhn) **Sea** Salt lake between Europe and Asia. (p. 242)

Caucasus (KAW·kuh·suhs) **Mountains** Mountain range in southwestern Russia. (p. 360)

Central African Republic Country in central Africa, south of Chad. (p. 477)

Central Siberian Plateau Tableland area in Siberia. (p. 242)

Chad Country in north-central Africa. (p. 477)

Chang Jiang (CHAHNG JYAHNG) River in north-central and eastern China, also known as the Yangtze River. (p. 672)

Chao Phraya (chow PRY·uh) River in Thailand, flowing south into the Gulf of Thailand. (p. 746)

Chile (CHIH•lee) Western South American country, along the Pacific Ocean. (p. 169)

China (People's Republic of China) Country in eastern and central Asia. (p. 673)

Chişinău (KEE•shee•NOW) Capital and largest city of Moldova. 47°N 29°E (p. 243)

Colombia Republic in northern South America. (p. 169)

Colombo Capital of Sri Lanka. 7°N 80°E (p. 599)

Colorado Plateau Highlands region in the western United States. (p. 112)

Colorado River River in the western United States that flows through the Grand Canyon. (p. 112)

Columbia Plateau Flat plains area primarily in western Washington State in the United States. (p. 112)

Comoros (KAH•muh•ROHZ) **Islands** Island country in the Indian Ocean between the island of Madagascar and Africa. 13°S 43°E (p. 477)

Conakry (KAH•nuh•kree) Capital of Guinea. 10°N 14°W (p. 477)

Congo, Democratic Republic of the African country on the Equator, north of Zambia and Angola. (p. 477)

Congo, Republic of the Country in Equatorial Africa. (p. 477)

Congo River River that runs through the Democratic Republic of the Congo. (p. 476)

Copenhagen (KOH•puhn•HAY•guhn) Capital of Denmark. 56°N 12°E (p. 243)

Costa Rica (KAWS•tah REE•kuh) Central American country, south of Nicaragua. (p. 169)

Côte d'Ivoire (KOHT dee•VWAHR) West African country, south of Mali. (p. 477)

Croatia (kroh•AY•shuh) Southeastern European country on the Adriatic Sea. (p. 243)

Cuba Island country southeast of Florida. 21°N 80°W (p. 169)

Cyprus Island country in the eastern Mediterranean Sea, south of Turkey. 35°N 31°E (p. 243)

Czech (CHEHK) **Republic** Central European country south of Germany and Poland. (p. 243)

D

Dakar Capital of Senegal. 15°N 17°W (p. 6)

Damascus Capital of Syria. 34°N 36°E (p. 361)

Danube River River in Europe that begins in Germany and flows into the Black Sea. (p. 242)

Dardanelles Strait between European and Asian Turkey, connecting the Sea of Marmara with the Aegean Sea. (p. 413)

Dar es Salaam (DAHR EHS suh•LAHM) Commercial capital of Tanzania. 7°S 39°E (p. 477)

Darling River River in southeast Australia. (p. 746)

Dead Sea A landlocked body of salt water located between Israel and Jordan. (p. 360)

Deccan Plateau The peninsula of India south of the Narmada River. (p. 598)

Denmark Country in northwestern Europe, between the Baltic and North Seas. (p. 243)

Dhaka Capital of Bangladesh. 24°N 90°E (p. 599)

Djibouti (juh•BOO•tee) Country in East Africa, on the Gulf of Aden. (p. 477)

Dnieper (NEE•puhr) **River** River that begins in Russia, flows through Belarus and Ukraine, and then drains into the Black Sea. (p. 242)

Dniester (NEE•stuhr) **River** River in south-central Europe that begins in Ukraine and flows southeast to the Black Sea. (p. 242)

Dodoma (doh•DOH•MAH) Political capital of Tanzania. 7°S 36°E (p. 477)

Doha (DOH•hah) Capital of Qatar. 25°N 51°E (p. 361)

Dominica Island republic in the West Indies, lying in the center of the Lesser Antilles. 15°N 61°W (p. 169)

Dominican Republic Republic occupying the eastern two-thirds of Hispaniola Island in the West Indies. 19°N 70°W (p. 169)

Don River River in southwestern Russia. (p. 242)

Drakensberg (DRAH•kuhnz•BUHRG) Range Mountain range in South Africa. (p. 476)

Dublin Capital of Ireland. 53°N 6°W (p. 243)

Dushanbe (doo•SHAM•buh) Capital and largest city of Tajikistan. 39°N 69°E (p. 361)

E

East Timor (Timor-Leste) Island country in the Indonesian archipelago, northwest of Australia. (p. 747)

Eastern Ghats Mountain range in India. (p. 598)

Ecuador (EH•kwuh•DAWR) Country in South America, south of Colombia. (p. 169)

Egypt Country in northern Africa on the Mediterranean Sea. (p. 361)

Elbe River River in central Europe. (p. 243)

Elburz Mountains Mountain range in northern Iran parallel to the shore of the Caspian Sea. (p. 360)

El Salvador (ehl SAL•vuh•DAWR) Country in Central America, southwest of Honduras. (p. 169)

Equatorial Guinea (EE•kwuh•TOHR•ee•uhl GIH•nee) Country in western Africa, south of Cameroon. (p. 477)

Eritrea (EHR•uh•TREE•uh) Country in northeast Africa, north of Ethiopia. (p. 477)

Estonia (eh•STOH•nee•uh) Northern European country on the Baltic Sea. (p. 243)

Ethiopia (EE•thee•OH•pee•uh) Country in eastern Africa, north of Somalia and Kenya. (p. 477)

Euphrates (yu•FRAY•teez) **River** River in southwestern Asia that flows through Syria and Iraq and joins the Tigris River. (p. 360)

F

Fiji (FEE•jee) Country comprising an island group in the southwest Pacific Ocean. 19°S 175°E (p. 747)

Finland Country in northern Europe, east of Sweden. (p. 243)

France Country in western Europe. (p. 243)

Freetown Capital and port city of Sierra Leone, in western Africa. 9°N 13°W (p. 477)

French Guiana Overseas department of France on the northeast coast of South America. (p. 169)

G

Gabon (ga·BOHN) Country in western Africa, on the Atlantic Ocean. (p. 477)

Gaborone (GAH·boh·ROH·nay) Capital of Botswana, in southern Africa. 24°S 26°E (p. 477)

Gambia Country in western Africa. (p. 477)

Ganges (GAN·JEEZ) **River** River in northern India and Bangladesh that flows into the Bay of Bengal. (p. 598)

Gangetic (gan·JEH·tic) **Plain** A fertile plains region in northern India traversed by the Ganges River. (p. 598)

Georgetown Capital of Guyana. 8°N 58°W (p. 169)

Georgia Asian/European country bordering the Black Sea, south of Russia. (p. 361)

Germany (Federal Republic of Germany) Country in north-central Europe. (p. 243)

Ghana (GAH·nuh) Country in western Africa, on the Gulf of Guinea. (p. 477)

Gobi Desert in Central Asia. (p. 672)

Godavari River River in central India. (p. 598)

Gran Chaco Region in south-central South America located in Paraguay, Bolivia, and Argentina. (p. 168)

Great Britain Kingdom in western Europe comprising England, Scotland, and Wales. (p. 242)

Great Dividing Range Chain of hills and mountains, on Australia's eastern coast. (p. 746)

Great Plains Rolling treeless area of central North America. (p. 112)

Great Salt Lake Large saltwater lake in Utah in the United States that has no outlet. (p. 112)

Great Slave Lake A lake in the south-central mainland of the Northwest Territories in Canada. (p. 112)

Greece Country in southern Europe, on the Balkan Peninsula. (p. 243)

Greenland Island in the northwestern Atlantic Ocean. 74°N 40°W (p. 112)

Grenada Island in the self-governing West Indies. 17°N 61°W (p. 169)

Guam Island in the western Pacific. It is an unincorporated United States territory. 13°N 144°E (p. 747)

Guatemala (GWAH·tuh·MAH·luh) Country in Central America, south of Mexico. (p. 169)

Guatemala Capital of Guatemala and the largest city in Central America. 15°N 91°W (p. 169)

Guinea (GIH·nee) West African country on the Atlantic coast. 11°N 12°W (p. 477)

Guinea-Bissau (GIH·nee bih·SOW) West African country on the Atlantic coast. 12°N 20°W (p. 477)

Gulf of Aden Arm of the Indian Ocean between the Arabian Peninsula and Africa. (p. 360)

Gulf of Mexico Gulf on the southern coast of North America. (p. 112)

Gulf of Thailand Inlet of the South China Sea. (p. 746)

Guyana Republic in northern South America. (p. 169)

H

Hainan (HY·NAHN) Island province of China in the South China Sea. 19°N 109°E (p. 672)

Haiti (HAY·tee) Republic occupying the western third of Hispaniola Island in the West Indies. 19°N 72.25°W (p. 169)

Hanoi Capital of Vietnam. 21°N 106°E (p. 747)

Harare (huh·RAH·ray) Capital of Zimbabwe. 18°S 23°E (p. 477)

Havana Capital of Cuba. 23°N 82°W (p. 169)

Helsinki Capital of Finland. 60°N 24°E (p. 243)

Himalaya (HIH·muh·LAY·uh) Mountain system in South Asia, bordering the Indian subcontinent on the north. (p. 598)

Hindu Kush Mountain range in Central Asia. (p. 360)

Honduras (hahn·DUR·uhs) Central American republic. (p. 169)

Hong Kong Administrative district and port in southern China. 22°N 115°E (p. 7)

Huang He (HWAHNG HUH) Major river in central China, also known as the Yellow River. (p. 672)

Hudson Bay Inland sea in east-central Canada. (p. 112)

Hungary Central European country, south of Slovakia. (p. 243)

I

Iberian (eye·BIHR·ee·uhn) **Peninsula** Peninsula in southwestern Europe. (p. 242)

Iceland Island country between the north Atlantic and Arctic Oceans. 65°N 20°W (p. 243)

India South Asian country south of China. (p. 599)

Indochina Southeast peninsula of Asia. (p. 3)

Indonesia (IHN·duh·NEE·zhuh) Group of islands that forms the Southeast Asian country of the Republic of Indonesia. 5°S 119°E (p. 747)

Indus River River in Asia that rises in Tibet and flows through Pakistan to the Arabian Sea. (p. 598)

Iran (ih·RAHN) Southwest Asian country, formerly called Persia. (p. 361)

Iraq (ih·RAHK) Southwest Asian country, south of Turkey. (p. 361)

Ireland Island west of England, occupied by the Republic of Ireland and by Northern Ireland. 54°N 8°W (p. 243)

Irrawaddy River River in central Myanmar formed by the confluence of the Mali and Nmai Rivers. (p. 746)

Irtysh River River in northeast Kazakhstan and the western part of Russia, in Asia. (p. 242)

Gazetteer

Islamabad (ihs•LAH•muh•BAHD) Capital of Pakistan. 34°N 73°E (p. 599)

Israel (IHZ•ree•uhl) Country in Southwest Asia, south of Lebanon. (p. 361)

Isthmus of Panama Narrow strip of land that forms the link in Central America between North America and South America. (p. 168)

Italy Southern European country, south of Switzerland and east of France. (p. 243)

J

Jakarta Capital of Indonesia. 6°S 107°E (p. 5)

Jamaica (juh•MAY•kuh) Island country in the West Indies. 18°N 78°W (p. 169)

Japan Country in East Asia, consisting of four main islands of Hokkaidō, Honshū, Shikoku, and Kyūshū, plus thousands of small islands. 37°N 134°E (p. 673)

Jerusalem (juh•ROO•suh•luhm) Capital of Israel and a holy city for Jews, Christians, and Muslims. 32°N 35°E (p. 361)

Jordan Country in Southwest Asia. (p. 361)

Juba Capital of South Sudan. 5°N 31°E (p. 477)

Jutland Peninsula extending north from Germany. (p. 242)

K

K2 (Godwin Austen) Himalayan mountain in Jammu and Kashmir. 35°N 76°E (p. 598)

Kabul Capital of Afghanistan. 35°N 69°E (p. 361)

Kalahari Desert Plateau and part desert located in the southern part of Africa. (p. 476)

Kamchatka Peninsula Peninsula in northeast Russia, in Asia. (p. 242)

Kampala (kahm•PAH•lah) Capital of Uganda. 0° latitude 32°E (p. 477)

Kara Sea Arm of the Arctic Ocean north of Russia. (p. 242)

Kathmandu (KAT•MAN•DOO) Capital of Nepal. 28°N 85°E (p. 599)

Kazakhstan (KA•zak•STAN) Large Asian country south of Russia, bordering the Caspian Sea. (p. 361)

Kenya (KEH•nyuh) Country in eastern Africa, south of Ethiopia. (p. 477)

Khartoum Capital of Sudan. 16°N 33°E (p. 477)

Khyber Pass Mountain pass between Afghanistan and Pakistan. 34°N 71°E (p. 360)

Kiev See KYIV (KIEV)

Kigali (kee•GAH•lee) Capital of Rwanda, in central Africa. 2°S 30°E (p. 477)

Kilimanjaro Highest mountain in Africa, located in Tanzania. 3°S 37°E (p. 476)

Kingston Capital of Jamaica. 18°N 77°W (p. 169)

Kinshasa (kihn•SHAH•suh) Capital of the Democratic Republic of the Congo. 4°S 15°E (p. 5)

Kiribati (KIHR•uh•BAS) One of the two Federated States of Micronesia. 5°S 170°W (p. 747)

Korean Peninsula Peninsula on which both North and South Korea are located. (p. 672)

Kosovo (KAW•saw•VOH) Country in southeastern Europe, between Serbia and Montenegro. (p. 243)

Krishna River River of the Deccan Plateau in south India. (p. 598)

Kuala Lumpur (kwah•luh LUM•PUR) Capital of Malaysia. 3°N 102°E (p. 747)

Kunlun Shan Mountain ranges in western China on the north edge of the Plateau of Tibet. (p. 672)

Kuwait (ku•WAYT) Country between Saudi Arabia and Iraq, on the Persian Gulf. (p. 361)

Kyiv (Kiev) (KEE•EHF) Capital of Ukraine. 50°N 31°E (p. 243)

Kyrgyzstan (KIHR•gih•STAN) Small Central Asian country on China's western border. (p. 361)

L

Labrador Sea Part of the Atlantic Ocean south of Baffin Bay off the coast of Newfoundland. (p. 112)

Lagos Port city of Nigeria. 6°N 3°E (p. 477)

Lake Baikal Lake in southern Siberia, Russia. It is the largest freshwater lake in Eurasia. (p. 242)

Lake Chad Reservoir located in Chad. (p. 476)

Lake Erie One of the Great Lakes of the United States and Canada. (p. 112)

Lake Huron One of the Great Lakes of the United States and Canada. (p. 112)

Lake Malawi Lake in southeast Africa. (p. 476)

Lake Michigan One of the Great Lakes of the United States and Canada. (p. 112)

Lake Ontario The easternmost and smallest of the Great Lakes of the United States and Canada. (p. 112)

Lake Superior One of the Great Lakes of the United States and Canada. (p. 112)

Lake Tanganyika Lake in east-central Africa. (p. 476)

Lake Titicaca Lake on the border between Peru and Bolivia. Highest navigable lake in the world. (p. 168)

Lake Victoria Freshwater lake in Tanzania and Uganda. (p. 476)

Lake Volta Reservoir located in Ghana. (p. 476)

Lake Winnipeg Lake in south-central Manitoba, Canada. (p. 112)

Laos (LOWS) Southeast Asian country, south of China and west of Vietnam. (p. 747)

La Paz (lah PAHZ) Administrative capital of Bolivia, and the highest capital in the world. 17°S 68°W (p. 169)

Latvia (LAT•vee•uh) Northeastern European country on the Baltic Sea, west of Russia. (p. 243)

Lebanon (LEH•buh•nuhn) Country on the Mediterranean Sea, south of Syria. (p. 361)

Lena River River in east-central Russia. (p. 242)

Lesotho (luh•SOH•toh) Country in southern Africa. (p. 477)

Liberia (ly•BIHR•ee•uh) West African country, south of Guinea. 7°N 10°W (p. 477)

Libreville (LEE·bruh·VIHL) Capital and port city of Gabon. 1°N 9°E (p. 477)

Libya (LIH·bee·uh) North African country on the Mediterranean Sea, west of Egypt. (p. 361)

Liechtenstein (LIHK·tuhn·STYN) Small country in central Europe. (p. 243)

Lilongwe (lih·LAWNG·gway) Capital of Malawi. 14°S 34°E (p. 477)

Lima (LEE·muh) Capital of Peru. 12°S 77°W (p. 169)

Lisbon Capital of Portugal. 39°N 9°W (p. 243)

Lithuania (LIH·thuh·WAY·nee·uh) European country on the Baltic Sea, west of Belarus. (p. 243)

Ljubljana (lee·oo·blee·AH·nuh) Capital of Slovenia. 46°N 14°E (p. 243)

Llanos Vast plains in northern South America. (p. 168)

Loire River River in Europe that rises in southeastern France and empties into the Bay of Biscay. (p. 242)

Lomé (loh·MAY) Capital and port city of Togo in Africa. 6°N 1°E (p. 477)

London Capital of the United Kingdom, on the Thames River. 52°N 0° longitude (p. 243)

Luanda Capital of Angola. 9°S 13°E (p. 477)

Lusaka Capital of Zambia. 15°S 28°E (p. 477)

Luxembourg (LUHK·suhm·BUHRG) European country between France, Germany, and Belgium. (p. 243)

M

Macau (muh·KOW) Administrative district and port in southern China. (p. 673)

Macedonia (MA·suh·DOH·nee·uh) Republic in southeastern Europe, north of Greece. Macedonia also refers to a geographic region in the Balkan Peninsula. (p. 243)

Mackenzie River River in the western portion of the Northwest Territories in Canada. (p. 112)

Madagascar (MA·duh·GAS·kuhr) Island in the Indian Ocean, southeast of Africa. (p. 477)

Madrid Capital of Spain. 40°N 4°W (p. 243)

Malabo (mah·LAH·boh) Capital of Equatorial Guinea. 4°N 9°E (p. 477)

Malawi (muh·LAH·wee) Southeastern African country, south of Tanzania and east of Zambia. (p. 477)

Malaysia (muh·LAY·zhuh) Federation of states in Southeast Asia on the Malay Peninsula and the island of Borneo. (p. 747)

Maldives (MAWL·DEEVZ) Island country in the Indian Ocean near South Asia. 5°N 42°E (p. 599)

Mali Country in western Africa, south of Algeria. (p. 361)

Malta An independent state consisting of three islands in the Mediterranean Sea. 36°N 15°E (p. 243)

Managua (mah·NAH·gwah) Capital of Nicaragua. 12°N 86°W (p. 169)

Manila (muh·NIH·luh) Capital and port city of the Republic of the Philippines. 15°N 121°E (p. 747)

Marshall Islands Independent group of atolls and reefs in the western Pacific Ocean. 11°N 108°E (p. 747)

Maseru (MA·suh·ROO) Capital of Lesotho, in southern Africa. 29°S 27°E (p. 477)

Masqat Capital of Oman. 23°N 59°E (p. 361)

Mato Grosso Plateau Highlands area in southwest Brazil. (p. 168)

Mauritania (MAWR·uh·TAY·nee·uh) West African country, north of Senegal. (p. 361)

Mauritius (maw·RIH·shuhs) Island country in the Indian Ocean east of Madagascar. 21°S 58°E (p. 477)

Mbabane (EHM·bah·BAH·nay) Capital of Swaziland, in southeastern Africa. 26°S 31°E (p. 477)

Mediterranean Sea Inland sea enclosed by Europe, Asia, and Africa. (p. 244)

Mekong River River in Southeast Asia that flows south through Laos, Cambodia, and Vietnam. (p. 746)

Melekeok New capital of Palau. 7°N 134°E (p. 747)

Meseta The plains of central Spain. (p. 242)

Mexico Country in North America, south of the United States. (p. 169)

Mexico City Capital and most populous city of Mexico. 19°N 99°W (p. 169)

Minsk Capital of Belarus. 54°N 28°E (p. 243)

Mississippi River River in the central United States that rises in Minnesota and flows southeast into the Gulf of Mexico. (p. 112)

Missouri River River in the central United States that joins the Mississippi River. (p. 112)

Mogadishu (MAH·guh·DIH·shoo) Capital and major seaport of Somalia, in eastern Africa. 2°N 45°E (p. 477)

Moldova (mahl·DOH·vuh) European country between Ukraine and Romania. (p. 243)

Monaco (MAH·nuh·KOH) Independent principality in southern Europe, on the Mediterranean. (p. 243)

Monaco Capital of Monaco. 44°N 8°E (p. 243)

Mongolia (mahn·GOHL·yuh) Country in Asia between Russia and China. (p. 673)

Monrovia (muhn·ROH·vee·uh) Capital and major seaport of Liberia, in western Africa. 6°N 11°W (p. 477)

Mont Blanc The highest mountain of the Alps, in southeastern France. 46°N 7°E (p. 242)

Montenegro (mahn·tuh·NEH·groh) European country between the Adriatic Sea and Serbia. (p. 243)

Montevideo (MAHN·tuh·vuh·DAY·oh) Capital of Uruguay. 35°S 56°W (p. 169)

Morocco (muh·RAH·koh) Country in northwestern Africa on the Mediterranean Sea and the Atlantic Ocean. (p. 361)

Moscow Capital of Russia. 56°N 38°E (p. 243)

Mount Ararat Mountain in eastern Turkey. 39°N 44°3 (p. 360)

Mount Elbrus Highest point in the Caucasus Mountains. 43°N 42°E (p. 242)

Mount Everest (EHV·ruhst) Highest mountain in the world, in the Greater Himalaya mountain range between Nepal and Tibet. 28°N 87°E (p. 598)

Mount Fuji Peak in south-central Honshū, Japan. It is the highest peak in Japan. 35°N 138°E (p. 672)

Mount Logan Peak in northwest Canada. 60°N 140°W (p. 112)

Mount McKinley Highest peak in North America, located in Denali National Park in Alaska. 63°N 151°W (p. 112)

Mount Pinatubo Active volcanic mountain in the Philippines. 15°N 170°E (p. 746)

Mount Whitney Peak in the Sierra Nevada range in central California. 36°N 118°W (p. 112)

Mozambique (MOH•zuhm•BEEK) Country in south-eastern Africa, south of Tanzania. (p. 477)

Murray River River in Australia. (p. 746)

Myanmar (MYAHN•MAHR) Country in Southeast Asia, south of China, also called Burma. (p. 747)

N

Nairobi Capital of Kenya. 1°S 37°E (p. 477)

Namib Desert Arid region along the coast of Namibia in southwestern Africa. (p. 476)

Namibia (nuh•MIH•bee•uh) Country in southwestern Africa, on the Atlantic Ocean. (p. 477)

Narmada River River in central India that flows into the Gulf of Khambat in the Arabian Sea. (p. 598)

Nassau (NA•SAW) Capital of the Bahamas. 25°N 77°W (p. 169)

Nauru (nah•OO•roo) One of the two Federated States of Micronesia. 32°S 166°E (p. 747)

Nay Pyi Taw New capital of Myanmar. 20°N 96°E (p. 747)

N'Djamena (uhn•jah•MAY•nah) Capital of Chad. 12°N 15°E (p. 477)

Nepal (nuh•PAWL) Mountain country between India and China. (p. 599)

Netherlands Western European country on the North Sea. (p. 243)

New Delhi Capital of India. 29°N 77°E (p. 599)

New Zealand Major island country in the south Pacific, southeast of Australia. 42°S 175°E (p. 747)

Niamey (nee•AH•may) Capital and commercial center of Niger, in western Africa. 14°N 2°E (p. 477)

Nicaragua (NIH•kuh•RAH•gwuh) Republic in Central America. (p. 169)

Nicosia (NIH•kuh•SEE•uh) Capital of Cyprus. 35°N 33°E (p. 243)

Niger (NY•juhr) Landlocked country in western Africa, north of Nigeria. (p. 361)

Nigeria (ny•JIHR•ee•uh) Country in western Africa, south of Niger. (p. 477)

Niger River River in western Africa. (p. 476)

Nile River Longest river in the world, flowing north and east through eastern Africa. (p. 476)

Northern European Plain Plain that sweeps across western and central Europe into Russia and includes most of European Russia. (p. 242)

North Korea Asian country in the northernmost part of the Korean Peninsula. (p. 673)

North Sea Arm of the Atlantic Ocean extending between the European continent on the south and east and Great Britain on the west. (p. 242)

Norway Country on the Scandinavian Peninsula. (p. 243)

Nouakchott (nu•AHK•SHAHT) Capital of Mauritania. 18°N 16°W (p. 361)

Nullarbor Plain Dry, treeless area that lies south of the Great Victorian Desert in Australia. (p. 746)

O

Ob' River River in western Russia. (p. 242)

Ohio River Major river in the midwestern United States, emptying into the Mississippi River. (p. 112)

Oman (oh•MAHN) Country on the Arabian Sea and the Gulf of Oman. (p. 361)

Orinoco River River in Venezuela. (p. 168)

Oslo Capital of Norway. 60°N 11°E (p. 243)

Ottawa Capital of Canada. 45°N 76°W (p. 114)

Ouagadougou (WAH•gah•DOO•goo) Capital of Burkina Faso, in western Africa. 12°N 2°W (p. 477)

P

Pakistan South Asian country on the Arabian Sea, northwest of India. (p. 599)

Palau (puh•LOW) Island country in the western Pacific Ocean. 7°N 135°E (p. 747)

Pamirs Mountainous region of Central Asia. (p. 360)

Pampas Plains area of South America. (p. 168)

Panama (PA•nuh•MAH) Republic in south Central America, on the Isthmus of Panama. (p. 169)

Panama Capital of Panama. 9°N 79°W (p. 169)

Papua New Guinea (PA•pyuh•wuh noo GIH•nee) Independent island country in the south Pacific Ocean. 7°S 142°E (p. 747)

Paraguay (PAR•uh•GWY) Country in South America, north of Argentina. (p. 169)

Paraguay River River in south-central South America. (p. 168)

Paramaribo (PAR•uh•MAR•uh•BOH) Capital and port city of Suriname. 6°N 55°W (p. 169)

Paraná River River in southeast central South America. (p. 168)

Paris Capital and river port of France. 49°N 2°E (p. 243)

Patagonia Plateau region of South America primarily in Argentina. (p. 168)

Peace River River in western Alberta, Canada. (p. 112)

Persian Gulf Arm of the Arabian Sea between Iran and Saudi Arabia. (p. 360)

Peru Country in South America, south of Ecuador and Colombia. (p. 169)

Philippines (FIH•luh•PEENZ) Country in the Pacific Ocean, southeast of China. (p. 747)

Phnom Penh (NAHM PEHN) Capital of Cambodia. 12°N 106°E (p. 747)

Podgorica (PAWD•GAWR•eet•sah) Capital of Montenegro. 42°N 19°E (p. 243)

Poland Country on the Baltic Sea in eastern Europe. 52°N 18°E (p. 243)

Po River River in northern Italy that flows to the Adriatic Sea. (p. 243)

Port-au-Prince (POHRT•oh•PRIHNTS) Capital of Haiti. 19°N 72°W (p. 169)

Port Moresby (MOHRZ•bee) Capital of Papua New Guinea. 10°S 147°E (p. 747)

Porto-Novo (POHR•toh•NOH•voh) Capital and port city of Benin, in western Africa. 7°N 3°E (p. 477)

Portugal (POHR•chih•guhl) Country on the Iberian Peninsula, south and west of Spain. (p. 243)

Prague (PRAHG) Capital of the Czech Republic. 50°N 15°E (p. 243)

Pretoria See **Tshwane (Pretoria)**

Puerto Rico Island in the West Indies. It is a self-governing commonwealth in union with the United States. 18°N 66°W (p. 169)

P'yŏngyang (PYAWNG•YAHNG) Capital of North Korea. 39°N 126°E (p. 673)

Pyrenees Mountain range extending along the border of France and Spain. (p. 242)

Q

Qatar (KAH•tuhr) Country on the southwestern shore of the Persian Gulf. (p. 361)

Qin Ling Mountain range in northern China. (p. 672)

Quito (KEE•toh) Capital of Ecuador. 0° latitude 79°W (p. 169)

R

Rabat Capital of Morocco. 34°N 7°W (p. 361)

Red River River in the south-central United States, emptying into the Mississippi River. (p. 112)

Red River River in Vietnam that empties into the South China Sea. (p. 746)

Red Sea Inland sea between the Arabian Peninsula and northeast Africa. (p. 360)

Reykjavík (RAY•kyuh•VIHK) Capital of Iceland. 64°N 22°W (p. 243)

Rhine River River in western Europe that flows to the North Sea. (p. 242)

Rhône River River in Switzerland and France. (p. 242)

Riga Capital of Latvia. 57°N 24°E (p. 243)

Río de la Plata Estuary of the Paraná and Uruguay Rivers between Uruguay and Argentina. (p. 168)

Rio Grande/Río Bravo del Norte River forming part of the boundary between the United States and Mexico. (p. 112)

Riyadh (ree•YAHD) Capital of Saudi Arabia. 25°N 47°E (p. 361)

Rocky Mountains An extensive mountain system in western North America. (p. 112)

Romania (ru•MAY•nee•uh) Country in eastern Europe, south of Ukraine. (p. 243)

Rome Capital of Italy. 42°N 13°E (p. 243)

Rub' al-Khali Desert region in the southern Arabian Peninsula, also called the Empty Quarter. (p. 360)

Russia Largest country in the world, covering parts of Europe and Asia. (p. 243)

Rwanda (roo•AHN•dah) Country in Africa, south of Uganda. (p. 477)

S

Sahara Vast region of deserts and oases in North Africa. (p. 360)

St. Lawrence River River in southern Quebec and southeast Ontario, Canada. (p. 112)

St. Lucia Independent island state in the Caribbean Sea. 13°N 60°W (p. 169)

St. Vincent Principal island of St. Vincent and the Grenadines, south of St. Lucia. 13°N 61°W (p. 169)

Samoa Group of independent islands in the southwest Pacific Ocean. 13°S 172°W (p. 747)

Sanaa (sa•NAH) Capital of Yemen. 15°N 44°E (p. 361)

San José Capital of Costa Rica. 10°N 84°W (p. 169)

San Marino (SAN muh•REE•noh) Small European country, located on the Italian Peninsula. (p. 243)

San Salvador (san SAL•vuh•DAWR) Capital of El Salvador. 14°N 89°W (p. 169)

Santiago Capital of Chile. 33°S 71°W (p. 169)

Santo Domingo (SAN•tuh duh•MIHNG•goh) Capital of the Dominican Republic. 19°N 70°W (p. 169)

São Francisco River River in eastern Brazil flowing into the Atlantic Ocean. (p. 168)

São Tomé and Príncipe (SOWN tuh•MAY and PRIHN•sih•pee) Small island country in the Gulf of Guinea off the coast of central Africa. 1°N 7°E (p. 477)

Sarajevo (SAR•uh•YAY•voh) Capital of Bosnia and Herzegovina. 43°N 18°E (p. 243)

Saskatchewan River River in south-central Canada that flows into Lake Winnipeg. (p. 112)

Saudi Arabia (SOW•dee uh•RAY•bee•uh) Country on the Arabian Peninsula. (p. 361)

Scandinavia A peninsula in northern Europe. (p. 242)

Sea of Japan (East Sea) Branch of the Pacific Ocean between Japan and the Korean Peninsula. (p. 672)

Sea of Marmara Sea in northwest Turkey. (p. 413)

Sea of Okhotsk An inlet of the Pacific Ocean on the eastern coast of Russia. (p. 672)

Seine (SAYN) **River** French river that flows through Paris and into the English Channel. (p. 243)

Senegal (seh•nih•GAWL) Country on the coast of western Africa, on the Atlantic Ocean. (p. 477)

Seoul (SOHL) Capital of South Korea. 38°N 127°E (p. 5)

Serbia (SUHR•bee•uh) European country between Macedonia and Hungary. (p. 243)

Gazetteer

Seychelles (say•SHEHLZ) Small island country in the Indian Ocean near East Africa. 6°S 56°E (p. 477)

Siberia An area in the region of north-central Asia, primarily in Russia. (p. 242)

Sierra Leone (see•EHR•UH lee•OHN) Country in western Africa, south of Guinea. (p. 477)

Sierra Madre del Sur Mountain range along the coast of southern Mexico. (p. 168)

Sierra Madre Occidental Mountain range running parallel to the Pacific Ocean coast in Mexico. (p. 168)

Sierra Madre Oriental Mountain range running parallel to the Gulf of Mexico coast in Mexico. (p. 168)

Sierra Nevada Mountain range in eastern California in the United States. (p. 112)

Sinai Peninsula Peninsula in northeast Egypt between the Gulf of Suez and the Gulf of Aqaba. (p. 360)

Singapore Multi-island country in Southeast Asia near the tip of the Malay Peninsula. 2°N 104°E (p. 747)

Skopje (SKAW•pyeh) Capital of the Republic of Macedonia. 42°N 21°E (p. 243)

Slovakia (sloh•VAH•kee•uh) Central European country south of Poland. (p. 243)

Slovenia (sloh•VEE•nee•uh) Small central European country on the Adriatic Sea, south of Austria. (p. 243)

Sofia Capital of Bulgaria. 43°N 23°E (p. 243)

Solomon Islands Independent island group in the west Pacific Ocean. 8°S 159°E (p. 747)

Somalia (soh•MAH•lee•uh) Country in East Africa, on the Gulf of Aden and the Indian Ocean. (p. 477)

South Africa Country at the southern tip of Africa. (p. 477)

South China Sea Part of the Pacific Ocean extending from Japan to the tip of the Malay Peninsula. (p. 672)

South Korea Country in Asia on the Korean Peninsula between the Yellow Sea and the Sea of Japan. (p. 673)

South Sudan Country in eastern Africa, west of Ethiopia. (p. 477)

Spain Country on the Iberian Peninsula. (p. 243)

Sri Lanka (sree LAHNG•kuh) Island country in the Indian Ocean south of India. 9°N 83°E (p. 599)

Stockholm Capital of Sweden. 59°N 18°E (p. 243)

Strait of Gibraltar Passage connecting Mediterranean Sea to the Atlantic Ocean. (p. 360)

Strait of Hormuz Strait between the northern tip of Oman, the southeastern Arabian Peninsula, and the southern coast of Iran. (p. 360)

Strait of Malacca Ocean trade route running between Indonesia and Malaysia, near Singapore. (p. 746)

Sucre (SOO•kray) Constitutional capital of Bolivia. 19°S 65°W (p. 169)

Sudan Northeast African country on the Red Sea. (p. 477)

Suriname Republic in South America. (p. 169)

Suva Capital of Fiji. 18°S 177°E (p. 747)

Swaziland (SWAH•zee•LAND) South African country west of Mozambique. (p. 477)

Sweden Northern European country on the eastern side of the Scandinavian Peninsula. (p. 243)

Switzerland (SWIHT•suhr•luhnd) European country in the Alps, south of Germany. (p. 243)

Syr Dar'ya River in west-central Asia in Kyrgyzstan, Uzbekistan, and Kazakhstan. (p. 360)

Syria (SIHR•ee•uh) Country in Asia on the eastern side of the Mediterranean Sea. (p. 361)

T

Taipei (TY•PAY) Capital of Taiwan. 25°N 122°E (p. 673)

Taiwan (TY•WAHN) Island country off the southeast coast of China, claimed by China. 24°N 122°E (p. 673)

Tajikistan (tah•JIH•kih•STAN) Central Asian country north of Afghanistan. (p. 361)

Taklimakan Desert in western China. (p. 3)

Tallinn (TA•luhn) Capital and largest city of Estonia. 59°N 25°E (p. 243)

Tanzania (TAN•zuh•NEE•uh) East African country on the coast of the Indian Ocean. (p. 477)

Tashkent Capital of Uzbekistan. 41°N 69°E (p. 5)

Tasman Sea Part of the south Pacific Ocean between Australia and New Zealand. (p. 746)

Taurus Mountains Mountain range in southern Turkey. (p. 360)

Tbilisi (tuh•BEE•luh•see) Capital of the Republic of Georgia. 42°N 45°E (p. 361)

Tegucigalpa (tuh•GOO•suh•GAL•puh) Capital of Honduras. 14°N 87°W (p. 169)

Tehran (TAY•RAN) Capital of Iran. 36°N 52°E (p. 5)

Thailand (TY•LAND) Southeast Asian country south of Myanmar. (p. 747)

Thames (TEHMZ) **River** River in southern England that flows into the North Sea. (p. 243)

Thar Desert Region of sandy desert in northwest India and southeast Pakistan. (p. 598)

Thimphu (thihm•POO) Capital of Bhutan. 28°N 90°E (p. 599)

Tian Shan Mountain range in western China. (p. 672)

Tierra del Fuego Archipelago off southern South America. 54°N 68°W (p. 168)

Tiranë (tih•RAH•nuh) Capital of Albania. 42°N 20°E (p. 243)

Togo West African country between Benin and Ghana, on the Gulf of Guinea. (p. 477)

Tokyo Capital of Japan. 36°N 140°E (p. 5)

Tonga South Pacific island country. 20°S 175°W (p. 747)

Trinidad and Tobago (TRIH•nih•DAD tuh•BAY•goh) Independent republic comprising the islands of Trinidad and Tobago, located in the Atlantic Ocean off the northeast coast of Venezuela. 11°N 61°W (p. 169)

Tripoli Capital of Libya. 33°N 13°E (p. 361)

Tshwane (Pretoria) Administrative capital of the Republic of South Africa. 26°S 28°E (p. 477)

Tunis Capital of Tunisia. 37°N 10°E (p. 361)

Tunisia (too•NEE•zhuh) North African country on the Mediterranean Sea between Libya and Algeria. (p. 361)

Turkey Country in southeastern Europe and western Asia. (p. 361)

Turkmenistan (tuhrk•MEH•nuh•STAN) Central Asian country on the Caspian Sea. (p. 361)

Tuvalu Independent island group in the western Pacific Ocean. 8°S 178°E (p. 747)

U

Uganda (oo•GAHN•duh) East African country south of Sudan. (p. 477)

Ukraine (yoo•KRAYN) Large eastern European country west of Russia, on the Black Sea. (p. 243)

Ulaanbaatar (OO•LAHN•BAH•TAWR) Capital of Mongolia. 48°N 107°E (p. 673)

United Arab Emirates Country of seven states on the eastern side of the Arabian Peninsula. (p. 361)

United Kingdom Country in western Europe made up of England, Scotland, Wales, and Northern Ireland. (p. 243)

United States Country in North America located between Canada and Mexico. (p. 113)

Ural Mountains Mountain range in Russia that marks the traditional boundary between European Russia and Asian Russia. (p. 242)

Ural River River in eastern Europe and western Asia, originating in the Ural Mountains. (p. 242)

Uruguay (UR•uh•GWY) South American country, south of Brazil on the Atlantic Ocean. (p. 169)

Uzbekistan (uz•BEH•kih•STAN) Central Asian country south of Kazakhstan. (p. 361)

V

Vanuatu (VAN•WAH•TOO) Country made up of islands in the Pacific Ocean, east of Australia. 17°S 170°W (p. 747)

Vatican (VA•tih•kuhn) **City** Headquarters of the Roman Catholic Church, located in the city of Rome, Italy. 42°N 13°E (p. 243)

Venezuela Republic in northern South America. (p. 169)

Verkhoyanski Mountains Mountain range in northeastern Russia, just east of the Lena River. (p. 242)

Vienna Capital of Austria. 48°N 16°E (p. 243)

Vientiane (vyehn•TYAHN) Capital of Laos. 18°N 103°E (p. 747)

Vietnam Southeast Asian country, east of Laos and Cambodia. (p. 747)

Vindhya Range Mountain range in central India. (p. 598)

Vistula River River in southwestern Poland that flows north into the Baltic Sea. (p. 243)

Volga River River in western Russia that flows south into the Caspian Sea. (p. 242)

W

Warsaw Capital of Poland. 52°N 21°E (p. 243)

Washington, D.C. Capital of the United States, near the Atlantic coast. 39°N 77°W (p. 114)

Wellington Capital of New Zealand. 41°S 175°E (p. 747)

Western Ghats Mountain range in southern India. (p. 598)

Western Sahara Territory in northwest Africa. (p. 361)

West Siberian Plain Area of flat land that stretches from the Arctic Ocean to the grasslands of Central Asia. (p. 242)

Windhoek (VIHNT•HUK) Capital of Namibia, in southwestern Africa. 22°S 17°E (p. 477)

X

Xi (SHEE) **River** River in southeast China, known in its upper course as the Hongshui. (p. 672)

Y

Yablonovyy Range Mountain range in southern Russia. (p. 242)

Yamoussoukro (YAH•muh•SOO•kroh) Second capital of Côte d'Ivoire, in western Africa. 7°N 6°W (p. 477)

Yangon (Rangoon) Former capital of Myanmar. 17°N 96°E (p. 750) See also Nay Pyi Taw

Yaoundé (yown•DAY) Capital of Cameroon, in western Africa. 4°N 12°E (p. 477)

Yellow Sea Large inlet of the Pacific Ocean between northeast China and the Korean Peninsula. (p. 672)

Yemen (YEH•muhn) Country on the Arabian Peninsula, south of Saudi Arabia. (p. 361)

Yenisey River River in western Russia that flows north into the Kara Sea. (p. 242)

Yerevan (YEHR•uh•VAHN) Former capital and largest city of Armenia. 40°N 44°E (p. 361)

Yucatán Peninsula Peninsula including parts of southeastern Mexico, Belize, and Guatemala in Central America. (p. 168)

Yukon River River in the Yukon Territory, Canada. (p. 112)

Z

Zagreb Capital and largest city of Croatia. 46°N 16°E (p. 243)

Zagros Mountains Mountain system in southern and southwestern Iran. (p. 360)

Zambezi River River in south-central Africa. (p. 476)

Zambia (ZAM•bee•uh) Country in south-central Africa, east of Angola. (p. 477)

Zimbabwe (zihm•BAH•bwee) Country in south-central Africa, southeast of Zambia. (p. 477)

Gazetteer

GLOSSARY/GLOSARIO

- Content vocabulary terms in this glossary are words that relate to geography content. They are highlighted yellow in your text.
- Words below that have an asterisk (*) are academic vocabulary terms. They help you understand your school subjects and are **boldface** in your text.

*abandon • alluvial plain

ENGLISH	A	ESPAÑOL

*abandon to give up completely (p. 137)

aborigine/Aborigine an area's original inhabitants (p. 688); the indigenous people of Australia (p. 781)

absolute location the exact position of a place on the Earth's surface (p. 17)

*abundant present in large amounts; plentiful (p. 558)

*access a way to approach or enter (pp. 295, 755)

*accompany to be together with something (p. 781)

accretion slow process in which an oceanic plate slides under a continental plate, creating debris that can cause continents to grow (p. 47)

acculturation cultural modification of an individual, group, or people by adapting to or borrowing traits from another culture (p. 714)

*accumulate to build up (p. 644)

*accurate free from error (p. 303)

*achieve to carry out successfully; to accomplish (p. 259)

acid deposition wet or dry airborne acids that fall to the ground (p. 286)

acid rain precipitation carrying large amounts of dissolved acids, which kills wildlife and damages buildings, forests, and crops (p. 135)

*acquire to gain possession (p. 341)

*adapt to make fit to changing circumstances (p. 791)

*adequate satisfactory or acceptable (p. 406)

*administer to manage; to supervise (p. 592)

*advocate to publicly recommend or support (p. 151)

*affect to have an effect on; to produce an effect upon (pp. 704, 737)

agribusiness an industry engaged in agriculture on a large scale, sometimes including the manufacture and distribution of farm supplies (p. 281)

alluvial plain floodplain, such as the Gangetic Plain in India, on which flooding rivers have deposited silt (p. 606)

*abandonar dejar por completo (pág. 137)

aborígenes habitantes originarios de una región (pág. 688); indígenas de Australia (pág. 781)

localización absoluta ubicación exacta de un lugar en la superficie terrestre (pág. 17)

*abundante presente en grandes cantidades; copioso (pág. 558)

*acceso ruta para acercarse o entrar a un lugar (págs. 295, 755)

*acompañar estar junto a alguien o algo (pág. 781)

acreción proceso lento en el cual una placa oceánica se desliza por debajo de una placa continental, de modo que se forman detritos que originan el crecimiento de los continentes (pág. 47)

aculturación cambio cultural de un individuo, grupo o pueblo que se presenta cuando este se adapta a las características de otra cultura o las adquiere (pág. 714)

*acumular amontonar (pág. 644)

*preciso sin errores (pág. 303)

*lograr realizar exitosamente; cumplir (pág. 259)

deposición ácida partículas ácidas húmedas o secas que caen al suelo (pág. 286)

lluvia ácida precipitación con grandes cantidades de ácidos disueltos, que destruye la vida silvestre y daña edificaciones, bosques y cultivos (pág. 135)

*adquirir obtener la posesión de algo (pág. 341)

*adaptarse ajustarse a las circunstancias cambiantes (pág. 791)

*adecuado satisfactorio o aceptable (pág. 406)

*administrar dirigir; supervisar (pág. 592)

*defender recomendar o apoyar públicamente (pág. 151)

*afectar tener un efecto en algo; influir en algo (págs. 704, 737)

agroindustria industria relacionada con la agricultura a gran escala, que en ocasiones incluye la manufactura y distribución de insumos agrícolas (pág. 281)

llanura aluvial planicie en la cual las inundaciones fluviales depositan limo, como la llanura del Ganges en la India (pág. 606)

Glossary/Glosario

alluvial soil rich soil made up of sand and mud deposited by running water (p. 369)

suelo aluvial suelo fértil compuesto de arena y lodo depositados por una corriente de agua (pág. 369)

*alter to change partly (p. 120)

*alterar cambiar parcialmente (pág. 120)

*alternative different from the usual or regular (p. 236)

*alternativa diferente de lo usual o común (pág. 236)

altiplano Spanish for "high plain," a region in Peru and Bolivia encircled by the Andes (p. 218)

altiplano llanura alta; región de Perú y Bolivia rodeada por los Andes (pág. 218)

*altitude height above sea level (p. 576)

*altitud altura sobre el nivel del mar (pág. 576)

animist pertaining to the religious beliefs of animism in which nature, such as animals and mountains, has spirits (p. 491)

animista relacionado con las creencias religiosas del animismo, según las cuales los elementos de la naturaleza, como los animales y las montañas, tienen espíritu (pág. 491)

*annual occurring once a year (p. 610)

*anual que ocurre una vez al año (pág. 610)

*approximate close to but not exact (p. 712)

*aproximado cercano pero no exacto (pág. 712)

*approximately about or nearly (p. 514)

*aproximadamente casi o alrededor de (pág. 514)

aquaculture the cultivation of seafood (pp. 148, 719, 729)

acuicultura cultivo de especies acuáticas (págs. 148, 719, 729)

aqueduct a channel or pipeline for carrying a large quantity of flowing water (p. 138)

acueducto canal o tubería que transporta una gran cantidad de agua corriente (pág. 138)

aquifer underground water-bearing layers of porous rock, sand, or gravel (pp. 54, 380, 654)

acuífero estrato de roca permeable, arena o grava que almacena agua subterránea (págs. 54, 380, 654)

archipelago a group or chain of islands (pp. 197, 704)

archipiélago grupo o cadena de islas (págs. 197, 704)

*area a geographical region; the amount of space that the surface of a place covers (p. 484)

*área región geográfica; espacio que ocupa la superficie de un lugar (pág. 484)

arid excessively dry (p. 434)

árido excesivamente seco (pág. 434)

artesian well a well that brings pressurized water to the surface without pumping (p. 778)

pozo artesiano pozo por el cual el agua sometida a presión asciende a la superficie, sin necesidad de bombearla (pág. 778)

*assume to gain or acquire (p. 418)

*asumir contraer o adquirir (pág. 418)

atheist a person who does not believe in God (p. 689)

ateo persona que no cree en Dios (pág. 689)

atmosphere a thin layer of gases that surrounds the Earth (p. 42)

atmósfera capa delgada de gases que rodea la Tierra (pág. 42)

atoll a ring-shaped island formed by the buildup of a coral reef on the rim of a submerged volcano (p. 777)

atolón isla en forma de anillo compuesta por la acumulación de arrecifes de coral alrededor de un volcán sumergido (pág. 777)

*attribute to explain by indicating a cause (p. 204)

*atribuir explicar indicando una causa (pág. 204)

*authority power to influence or command thought, opinion, or behavior (pp. 88, 731)

*autoridad poder de influenciar o dominar el pensamiento, la opinión o el comportamiento (págs. 88, 731)

autocracy system of government in which one person rules with unlimited power and authority (p. 88)

autocracia sistema de gobierno en el cual una persona dirige con poder y autoridad ilimitados (pág. 88)

*available able to be obtained (p. 639)

*disponible que se puede utilizar (pág. 639)

avalanche a large mass of ice, snow, and rock that slides down a mountainside (p. 273)

avalancha gran masa de hielo, nieve y roca que se desliza por una ladera (pág. 273)

Glossary/Glosario

ENGLISH	ESPAÑOL

average daily temperature the average of the daily high temperature and the overnight low; often used for comparison across climate regions (p. 69)

temperatura media diaria promedio de la temperatura máxima diurna y la temperatura mínima nocturna; por lo general se usa para hacer comparaciones entre las regiones climáticas (pág. 69)

axis an imaginary line that runs through the center of the Earth between the North and South Poles (p. 60)

eje línea imaginaria que atraviesa el centro de la Tierra del Polo Norte al Polo Sur (pág. 60)

B

Balkanization division of a region into smaller hostile regions (p. 320)

balcanización división de una región en zonas más pequeñas, que tienen relaciones hostiles (pág. 320)

basin an area drained by a river and its tributaries (p. 554)

cuenca área drenada por un río y sus tributarios (pág. 554)

bedouin member of the nomadic desert peoples of North Africa and Southwest Asia (p. 375)

beduino miembro de los pueblos nómadas de los desiertos de África del Norte y el Sudoeste Asiático (pág. 375)

*****benefit** to gain (p. 486); to promote well-being; to be useful (p. 780)

*****beneficiar** ganar (pág. 486); promover el bienestar; ser útil (pág. 780)

biodiversity biological diversity in an environment as indicated by numbers of different species of plants and animals (p. 199); the diverse life forms in a habitat or ecosystem (p. 758)

biodiversidad diversidad biológica de un medioambiente expresada mediante la cantidad de especies vegetales y animales diferentes (pág. 199); las diversas formas de vida de un hábitat o un ecosistema (pág. 758)

biome major type of ecological community defined primarily by distinctive plant and animal groups (p. 69)

bioma área extensa ocupada por una comunidad ecológica definida principalmente a partir de las especies vegetales y animales que predominan allí (pág. 69)

biofuel fuel created from living matter, such as trees (p. 568)

biocombustible combustible que se obtiene de materia orgánica, como los árboles (pág. 568)

biosphere the part of the Earth where life exists (p. 42)

biosfera parte de la Tierra donde hay vida (pág. 42)

birthrate number of births per year for every 1,000 people (p. 82)

tasa de natalidad número de nacimientos al año por cada 1,000 habitantes (pág. 82)

black market illegal trade of scarce or illegal goods, usually sold at high prices (p. 346)

mercado negro comercio ilegal de productos ilícitos o escasos, que por lo general se venden a precios altos (pág. 346)

boomerang an Australian throwing stick that soars and curves in flight and returns near the thrower (p. 781)

bumerán utensilio de madera originario de Australia que al ser lanzado hace un movimiento curvo y vuelve a quien lo lanzó (pág. 781)

brain drain the loss of highly educated and skilled workers to other countries (p. 225)

fuga de talentos migración de trabajadores altamente calificados e instruidos hacia otros países (pág. 225)

break-of-bulk act of unloading, transferring, or distributing part or all of a shipment (p. 258)

descarga acción de descargar, transferir o distribuir parte de un cargamento o su totalidad (pág. 258)

buffer zone a neutral area serving to separate powerful nations or nations that are hostile to each other (p. 759)

zona de amortiguación área neutral que separa naciones poderosas o que tienen relaciones hostiles (pág. 759)

*****bulk** the greater part of something (p. 809)

*****grueso** la mayor parte de algo (pág. 809)

bush a sparsely inhabited region (p. 776)

monte región poco poblada (pág. 776)

C

caldera a large volcanic crater (p. 777)

caldera gran cráter volcánico (pág. 777)

canopy top layer of a rain forest, where the tops of tall trees form a continuous layer of leaves (p. 558)

dosel capa superior de una selva tropical, donde las copas de los árboles más altos forman una cubierta de hojas (pág. 558)

Glossary/Glosario

ENGLISH

cap-and-trade a method for managing pollution in which a limit is placed on emissions and businesses or countries can buy and sell emissions allowances (p. 287)

capitalism a system in which factors of production are privately owned (p. 95)

carrying capacity the population that an area will support without undergoing deterioration (p. 522); the maximum population of any given species that an environment can sustain (p. 546)

cash crop farm product grown to be sold or traded rather than used by the farm family (pp. 181, 560)

cataract a large waterfall (p. 507)

central place theory geographical theory that seeks to explain the number, size, and location of human settlements in an urban system (p. 104)

cereal any grain like barley, oats, or wheat, grown for food (p. 457)

*challenge to arouse or stimulate especially by presenting with difficulties (p. 340)

chernozem (cher·nuh·ZYAWM) rich, black topsoil found in the Northern European Plain, especially in Russia and Ukraine (p. 336)

chinook a seasonal warm wind that blows down the Rockies in late winter and early spring (p. 147)

chlorofluorocarbon (CFC) a chemical substance, found mainly in liquid coolants, that damages the Earth's protective ozone layer (p. 720)

choke point a strategic, narrow waterway between two larger bodies of water (p. 443)

*circumstance an event or fact that accompanies or determines another (p. 497)

*cite to summon to appear in a court of law (p. 588)

city-state an independently governed community consisting of a city and the surrounding lands, notably present in ancient Greece (p. 299)

clan a large group of people descended from the same ancestor; a family group (pp. 494, 710); a group of close-knit, interrelated families (p. 781)

clear-cutting the removal of all trees in a stand of timber (pp. 134, 664)

climate weather patterns typical for an area over a long period of time (p. 60)

ESPAÑOL

comercio de derechos de emisión método de control de la polución según el cual se establece un límite a las emisiones, y las empresas o países pueden comprar y vender permisos de emisión (pág. 287)

capitalismo sistema en el que los medios de producción son de propiedad privada (p. 95)

capacidad de carga población que puede soportar un medioambiente sin deteriorarse (pág. 522); población máxima de una especie dada que puede soportar un medioambiente (pág. 546)

cultivo comercial producto agrícola que se cultiva para venta o intercambio, no para el consumo de las familias que lo cultivan (págs. 181, 560)

catarata cascada de gran tamaño (pág. 507)

teoría de los lugares centrales teoría geográfica que busca explicar el número, tamaño y ubicación de los asentamientos humanos en un sistema urbano (pág. 104)

cereal grano que se cultiva para el consumo, como la cebada, la avena o el trigo (pág. 457)

*desafiar retar o estimular especialmente planteando dificultades (pág. 340)

chernozem tierra negra y rica en humus que se encuentra en la llanura noreuropea, particularmente en Rusia y Ucrania (pág. 336)

chinook viento estacional cálido y seco que sopla desde las montañas Rocosas a finales de invierno y comienzos de primavera (pág. 147)

clorofluorocarbono (CFC) sustancia química empleada principalmente en los refrigerantes líquidos, que deteriora la capa de ozono que protege la Tierra (pág. 720)

punto de estrangulamiento vía fluvial reducida entre dos cuerpos de agua más grandes que tiene fines estratégicos (pág. 443)

*circunstancia suceso o hecho que se relaciona con otro o lo determina (pág. 497)

*citar convocar para comparecer ante un tribunal (pág. 588)

ciudad-estado comunidad política independiente que incluye una ciudad y el territorio circundante, propia de la antigua Grecia (pág. 299)

clan grupo numeroso de personas que descienden de un mismo antepasado; grupo familiar (págs. 494, 710); grupo de familias estrechamente relacionadas (pág. 781)

tala generalizada corte de todos los árboles de un área (págs. 134, 664)

clima patrones de tiempo atmosférico característicos de un área en un periodo largo (pág. 60)

Glossary/Glosario

ENGLISH	ESPAÑOL
climate change any significant change in the measures of climate lasting for an extended period of time (p. 813)	**cambio climático** cualquier cambio significativo del clima en un periodo prolongado (pág. 813)
***coincide** to happen in the same place and at the same time (p. 447)	***coincidir** ocurrir en el mismo lugar y al mismo tiempo (pág. 447)
Cold War the power struggle between the Soviet Union and the United States after World War II (p. 277)	**Guerra Fría** conflicto de poder entre Estados Unidos y la Unión Soviética después de la Segunda Guerra Mundial (pág. 277)
command economy a system of resource management in which decisions about production and distribution of goods and services are made by a central authority (p. 95)	**economía planificada** sistema de la administración de recursos en el cual una autoridad central toma las decisiones relacionadas con la producción y distribución de bienes y servicios (pág. 95)
commercial farming growing large quantities of crops or livestock in order to sell them for a profit (p. 587)	**agricultura comercial** producción de grandes cantidades de cultivos o ganado para obtener ganancias (pág. 587)
commune a collective farming community whose members share work and products (p. 690)	**comuna** comunidad agrícola colectiva cuyos miembros comparten el trabajo y los bienes que producen (pág. 690)
communism society based on equitable distribution of wealth, land, and public ownership of production (p. 276)	**comunismo** sociedad basada en la distribución equitativa de la riqueza, la tierra y la propiedad pública de los medios de producción (pág. 276)
***community** people with common interests living in a particular area (p. 82)	***comunidad** personas con intereses comunes que viven en un área específica (pág. 82)
complementarity relationship between two places in which one produces something the other needs, resulting in an exchange (p. 302)	**complementariedad** relación entre dos lugares en la cual uno produce algo que el otro necesita, de modo que se produce un intercambio (pág. 302)
***compound** to make something more extreme or intense by adding something to it (p. 791)	***combinar** hacer que un elemento sea más extremo o intenso agregando algo a este (pág. 791)
***comprehensive** covering completely or broadly; inclusive (p. 279)	***exhaustivo** que abarca completamente o ampliamente; inclusivo (pág. 279)
***comprise** to make up; to constitute (p. 317)	***comprender** conformar; constituir (pág. 317)
***concentrated** less dilute (p. 469)	***concentrado** menos diluido (pág. 469)
condensation the process of excess water vapor changing into liquid water when warm air cools (p. 51)	**condensación** proceso mediante el cual el exceso de vapor de agua se convierte en agua líquida cuando el aire caliente se enfría (pág. 51)
***conflict** a competition or struggle (p. 126)	***conflicto** competencia o lucha (pág. 126)
conflict diamonds diamonds that are mined in war-torn areas and are used to finance one or more parties involved (p. 534)	**diamantes de guerra** diamantes que se extraen de zonas devastadas por la guerra y con los cuales se financia una o más partes involucradas (pág. 534)
conic projection a map created by projecting an image of Earth onto a cone placed over part of an Earth model (p. 15)	**proyección cónica** mapa que se crea proyectando una imagen de la Tierra sobre un cono situado en una parte de un modelo terrestre (pág. 15)
coniferous referring to vegetation having cones and needle-shaped leaves, including many evergreens, that keep their foliage throughout the winter (p. 71)	**conífera** vegetación con conos y hojas en forma de aguja, como muchos árboles de hoja perenne, que mantiene su follaje durante el invierno (pág. 71)
connectivity the directness of routes linking pairs of places (p. 102)	**conectividad** efectividad de las rutas que conectan dos lugares (pág. 102)
conquistador Spanish for "conqueror"; Spanish soldier who participated in conquest of indigenous peoples of Latin America (p. 181)	**conquistador** soldado español que participó en la conquista de los pueblos indígenas de América Latina (pág. 181)
***consequence** the result of an action (p. 694)	***consecuencia** resultado de una acción (pág. 694)

ENGLISH	ESPAÑOL
*consist to be composed of or made up of (pp. 270, 704)	***consistir** estar compuesto o formado por (págs. 270, 704)
*constant unchanging (p. 51)	***constante** que no cambia (pág. 51)
*constitute to compose or form (p. 507)	***constituir** componer o formar (pág. 507)
continental relating to or characteristic of a continent (p. 256)	**continental** relacionado con un continente o característico de este tipo de formación (pág. 256)
continental drift the theory that the continents were once joined and then slowly drifted apart (p. 45)	**deriva continental** teoría según la cual los continentes alguna vez estuvieron unidos y luego se separaron lentamente (pág. 45)
continental shelf part of a continent that extends out underneath the ocean (p. 43)	**plataforma continental** parte de un continente que se extiende por debajo del océano (pág. 43)
continentality effect of extreme variation in temperature and very little precipitation within the interior portions of a landmass (p. 340)	**continentalidad** efecto de la variación extrema de la temperatura y la precipitación escasa en las áreas interiores de un continente (pág. 340)
*contribute to give or supply (p. 136)	***contribuir** dar o suministrar (pág. 136)
*controversy the presence of opposing views (p. 148)	***controversia** presencia de puntos de vista opuestos (pág. 148)
*converse reversed in order, relation, or action (p. 802)	***contrario** opuesto en orden, relación o acción (pág. 802)
*convert to change from one system, use, or method to another (p. 535)	***convertir** cambiar de un sistema, uso o método a otro (pág. 535)
*cooperation a common effort (p. 211)	***cooperación** esfuerzo colectivo (pág. 211)
*cooperative an organization, often a farm, whose members work together and share expenses and profits (p. 729)	***cooperativa** organización, por lo general una granja, cuyos miembros trabajan juntos y comparten gastos y ganancias (pág. 729)
copra dried coconut meat fed to animal herds (p. 790)	**copra** pulpa seca del coco que sirve de alimento para los rebaños (pág. 790)
coral the limestone skeleton of a tiny sea animal (p. 780)	**coral** esqueleto calcáreo de un animal marino diminuto (pág. 780)
coral reef a reef made up of fragments of corals, coral sands, algal, and other organic deposits, and the solid limestone resulting from their consolidation (p. 801)	**arrecife de coral** arrecife compuesto por fragmentos de corales, arena de coral, algas y otros depósitos orgánicos, y la piedra caliza que se forma cuando estos elementos se consolidan (pág. 801)
cordillera parallel chains or ranges of mountains (p. 218)	**cordillera** cadenas de montañas paralelas (pág. 218)
core innermost layer of the Earth made up of a super-hot but solid inner core and a super-hot liquid outer core (p. 44)	**núcleo** centro de la Tierra compuesto por un núcleo interno sólido y un núcleo externo líquido, ambos extremadamente calientes (pág. 44)
Coriolis effect the resulting diagonal movement, either north or south, of prevailing winds caused by the Earth's rotation (p. 66)	**efecto Coriolis** movimiento diagonal de los vientos predominantes, en dirección norte o sur, causado por la rotación de la Tierra (pág. 66)
*corporate formed into an association and endowed by law with the rights and liabilities of an individual (p. 188)	***corporativo** que se ha convertido en una asociación a la cual la ley ha otorgado los derechos y responsabilidades civiles de un individuo (pág. 188)
*corresponding showing a direct connection between two things (p. 340)	***correspondiente** que muestra una relación directa entre dos elementos (pág. 340)
cottage industry a business that employs workers in their homes (pp. 205, 617)	**industria artesanal** empresa en la cual las personas trabajan desde su hogar (págs. 205, 617)
coup an overthrow of the government (p. 731)	**golpe de Estado** derrocamiento de un gobierno (pág. 731)
cowrie shell the protective outer covering of a small sea creature, once used as money in Asia and Africa (p. 656)	**cauri** caparazón de una criatura marina pequeña que se usaba como moneda en Asia y África (pág. 656)

Glossary/Glosario

861

ENGLISH	ESPAÑOL
*create to bring into being or cause to exist (p. 45)	***crear** dar vida o hacer que exista (pág. 45)
*crucial vitally important (p. 65)	***crucial** de vital importancia (pág. 65)
crust outer layer of the Earth, a hard rocky shell forming Earth's surface (p. 44)	**corteza** capa exterior de la Tierra, rocosa y sólida, que forma la superficie terrestre (pág. 44)
cultural boundary a geographical boundary between two different cultures (p. 90)	**frontera cultural** límite geográfico entre dos culturas diferentes (pág. 90)
cultural diffusion the spread of culture traits, material and non-material, from one culture to another (p. 80)	**difusión cultural** expansión de características culturales, materiales e inmateriales, de una cultura a otra (pág. 80)
cultural divergence a separation of people or societies, with regard to beliefs, values, and customs, due to a division under different political systems (p. 733)	**divergencia cultural** separación de personas o sociedades con respecto a sus creencias, valores y costumbres debido a la división causada por sistemas políticos diferentes (pág. 733)
culture way of life of a group of people who share similar culture traits, including beliefs, customs, technology, and material items (p. 78)	**cultura** modo de vida de un grupo de personas que comparten características culturales similares, tales como creencias, costumbres, tecnología y elementos materiales (pág. 78)
*culture the customary beliefs, social forms, and material traits of a racial, religious, or social group (p. 180)	***cultura** creencias usuales, actitudes sociales y rasgos materiales de un grupo racial, religioso o social (pág. 180)
culture hearth a center where cultures developed and from which ideas and traditions spread outward (pp. 80, 416)	**centro cultural** centro donde se desarrollaron las culturas y desde los cuales se difundieron las ideas y tradiciones (págs. 80, 416)
culture region division of the Earth in which people share a similar way of life, including language, religion, economic systems, and values (p. 79)	**región cultural** división en la cual las personas comparten un modo de vida similar, tal como el idioma, la religión, los sistemas económicos y los valores (pág. 79)
cuneiform wedge-shaped symbols that were pressed into clay tablets (p. 416)	**cuneiforme** símbolos en forma de cuña inscritos en tablas de arcilla (pág. 416)
current cold or warm stream of seawater that flows in the oceans, generally in a circular pattern (p. 65)	**corriente** movimiento del agua marina fría o cálida en un patrón circular (pág. 65)
cyclone a storm with heavy rains and high winds that blow in a circular pattern around an area of low atmospheric pressure (p. 609)	**ciclón** tormenta con lluvias y vientos fuertes que circula alrededor de un área con baja presión atmosférica (pág. 609)
cyclone, typhoon a spiral-shaped tropical storm in which the winds have reached speeds of at least 74 miles per hour (119 km per hour). Storms in the Northwestern Pacific, west of the International Date Line, are called typhoons; storms in the Southwest Pacific or Indian Oceans are referred to as cyclones. (p. 757)	**tifón (ciclón)** tormenta tropical en forma de espiral cuyos vientos han alcanzado una velocidad mínima de 74 millas por hora (119 km/h). Las tormentas del Pacífico Noroeste, al oeste de la línea internacional de cambio de fecha, se llaman tifones; las del Pacífico Sur y el océano Índico se conocen como ciclones. (pág. 757)
cylindrical projection a map created by projecting Earth's image onto a cylinder (p. 15)	**proyección cilíndrica** mapa que se crea proyectando la imagen de la Tierra en un cilindro (pág. 15)
czar ruler of Russia until the 1917 revolution; originally from Latin word *Caesar*, title of Roman emperors (p. 341)	**zar** gobernador de Rusia hasta la revolución de 1917; se deriva de la palabra latina *Caesar*, título de los emperadores romanos (pág. 341)

D

death rate number of deaths per year for every 1,000 people (p. 82)	**tasa de mortalidad** número de fallecimientos en un año por cada 1,000 habitantes (pág. 82)
deciduous falling off or shed seasonally or periodically; trees such as oak and maple, which lose their leaves in autumn (p. 71)	**caducifolio** relacionado con la vegetación que muda según la estación o periódicamente; árboles que pierden sus hojas en otoño, como los robles y los arces (pág. 71)
*decline to become less in amount (p. 344)	***disminuir** reducir en número o cantidad (pág. 344)

Glossary/Glosario

ENGLISH	ESPAÑOL
deforestation the loss or destruction of forests, mainly for logging or farming (pp. 188, 738, 766)	**deforestación** pérdida o destrucción de bosques principalmente por la explotación maderera y agrícola (págs. 188, 738, 766)
delta a triangular section of land formed by sand and silt carried downriver (p. 487); a triangular-shaped area of silt and sediment deposit found at the mouth of a river (p. 578); an alluvial deposit at a river's mouth that is shaped roughly like the Greek letter delta (Δ) (p. 631)	**delta** terreno triangular que se forma donde una corriente fluvial deposita arena y limo (pág. 487); terreno triangular en la desembocadura de un río formado por un depósito de limo y sedimento (pág. 578); depósito aluvial en la desembocadura de un río cuya forma se parece a la letra griega delta (Δ) (pág. 631)
democracy system of government in which leaders rule with consent of the citizens (p. 89)	**democracia** sistema de gobierno según el cual los líderes dirigen con el consentimiento de los ciudadanos (pág. 89)
demographic transition the model that uses birthrates and death rates to show how populations in countries or regions change over time (p. 82)	**transición demográfica** modelo que se vale de las tasas de natalidad y mortalidad para mostrar la manera como cambia la población de los países o regiones con el paso del tiempo (pág. 82)
*****demonstrate** to clearly show the value of (p. 547)	*****demostrar** mostrar claramente el valor de algo (pág. 547)
*****derive** to obtain something from (pp. 66, 368)	*****derivar** obtener algo de otro elemento (págs. 66, 368)
desalination the removal of salt from seawater to make it usable for drinking and farming (pp. 53, 447)	**desalinización** proceso que elimina la sal del agua marina para que sea potable y apta para la agricultura (págs. 53, 447)
desertification process in which arable land becomes desert (p. 404); the destruction of land in arid and semiarid climates (p. 496)	**desertización** proceso en el cual la tierra cultivable se convierte en un desierto (pág. 404); destrucción del suelo en áreas de climas áridos y semiáridos (pág. 496)
devolution the granting of self-rule to local and regional authorities (p. 277)	**descentralización** proceso de devolver la autonomía a las autoridades regionales y locales (pág. 277)
dialect local form of a language used in a particular place or by a certain group (p. 203)	**dialecto** variedad local de una lengua que comparte un grupo específico o se habla en un lugar particular (pág. 203)
*****differentiate** to recognize features that make similar regions, communities, or people different from each other (p. 756)	*****diferenciar** reconocer las características que hacen que regiones, comunidades o personas similares se diferencien unas de otras (pág. 756)
dike large bank of earth and stone that holds back water (p. 272)	**dique** gran muro de tierra y piedra que retiene el agua (pág. 272)
*****diminish** to make less or cause to appear less (p. 159)	*****reducir** hacer menos o hacer parecer menos (pág. 159)
dissident a citizen who speaks out against government policies (p. 691)	**disidente** ciudadano que manifiesta su oposición con respecto a las políticas gubernamentales (pág. 691)
*****distinct** recognizably different (p. 70)	*****distinto** evidentemente diferente (pág. 70)
*****diverse** differing from one another (p. 180)	*****diverso** que se diferencia de otro (pág. 180)
divide a high point or ridge that determines the direction rivers flow (p. 121)	**divisoria** terreno alto o cresta que determina la dirección de las corrientes fluviales (pág. 121)
*****document** to write to provide information (p. 443)	*****documentar** escribir para suministrar información (pág. 443)
doldrums a frequently windless area near the Equator (p. 66)	**calmas ecuatoriales** área con vientos muy suaves cercana al ecuador (pág. 66)
domesticate to adapt plants and animals from the wild for human use (pp. 372, 489)	**domesticar** adaptar especies vegetales y animales silvestres para el beneficio humano (págs. 372, 489)
*****dominate** to have a commanding position in (p. 611); to exert influence over (pp. 689, 779)	*****dominar** tener una posición de mando (pág. 611); ejercer influencia (págs. 689, 779)
dominion a partially self-governing country with close ties to another country (p. 150); a largely self-governing country within the British Empire (p. 782)	**dominio** país con autonomía parcial que tiene vínculos estrechos con otra nación (pág. 150); país autónomo que formaba parte del Imperio británico (pág. 782)

Glossary/Glosario

863

ENGLISH

doubling time the number of years it takes for a population to double in size (p. 83)

drift-net fishing the use of fishing nets of great length and depth (p. 739)

dry farming a farming method used in dry regions in which crops are grown that rely only on the natural precipitation (p. 127)

dune a mound or ridge of sand formed by the wind (p. 435)

*dynamic energetic, characterized by constant change and progress (p. 765)

dynasty a ruling house or continuing family of rulers, especially in China (p. 685)

dzong a fortified monastery that also served as an administrative and commercial center (p. 660)

ESPAÑOL

tiempo de duplicación número de años que tarda una población en doblar su tamaño (pág. 83)

pesca con redes de deriva uso de redes de pesca de gran tamaño y profundidad (pág. 739)

agricultura de secano método agrícola que se emplea en regiones secas, en el cual se siembran cultivos que dependen solo de las precipitaciones naturales (pág. 127)

duna colina o montículo de arena que se forma por el viento (pág. 435)

*dinámico enérgico, que se caracteriza por el cambio y el progreso constantes (pág. 765)

dinastía casa real o serie de gobernantes de una misma familia, particularmente en China (pág. 685)

jong fortaleza-monasterio que también servía como centro administrativo y comercial (pág. 660)

E

e-commerce buying and selling on the Internet (p. 541)

economic sanction a restriction placed on trade (p. 691)

*economy an ordered system for the production, distribution, and consumption of goods and services (p. 317)

ecotourism the practice and business of recreational travel based on concern for the environment (pp. 205, 264, 624)

elevation the height of a land surface above the level of the sea (p. 18)

El Niño a periodic reversal of the pattern of ocean currents and water temperatures in the mid-Pacific region (p. 67)

embargo a ban on trade (p. 421)

*emerge to rise from an obscure position or condition; to become visible (p. 251)

emerging market an economy that may not have been very strong in the recent past, but that is in transition to becoming a stronger market (p. 764)

emigrate to leave one's own country to settle permanently in another (p. 153)

*emphasis importance or special consideration (p. 327)

enclave a distinct territorial or cultural area that is within a foreign territory (p. 463)

*energy usable power (p. 198)

*enhance to improve or increase (p. 490)

*enormous gigantic; exceedingly large (p. 52)

*ensure to make sure or certain (p. 288)

comercio electrónico compra y venta de productos por Internet (pág. 541)

sanción económica restricción al comercio (pág. 691)

*economía sistema ordenado para la producción, la distribución y el consumo de bienes y servicios (pág. 317)

ecoturismo práctica y empresa relacionadas con los viajes recreativos que se basan en el cuidado del medioambiente (págs. 205, 264, 624)

altura (altitud) distancia de un punto en la superficie terrestre con respecto al nivel del mar (pág. 18)

El Niño cambio periódico en los patrones de las corrientes marinas y la temperatura del agua en la región del Pacífico Medio (pág. 67)

embargo prohibición al comercio (pág. 421)

*emerger aparecer desde una posición o condición de oscuridad; hacerse visible (pág. 251)

mercado emergente economía que no fue sólida en el pasado reciente, pero que se encuentra en el proceso de convertirse en un mercado más fuerte (pág. 764)

emigrar dejar el país natal para establecerse permanentemente en otro (pág. 153)

*énfasis importancia o consideración especial (pág. 327)

enclave territorio o grupo cultural que se encuentra dentro de otro territorio (pág. 463)

*energía fuerza aprovechable (pág. 198)

*realzar mejorar o aumentar (pág. 490)

*enorme gigantesco; extremadamente grande (pág. 52)

*asegurar garantizar o hacer que algo sea seguro (pág. 288)

ENGLISH	ESPAÑOL
entrepôt commercial center where goods are received (p. 258)	**almacén aduanero** centro de comercio donde se reciben mercancías (pág. 258)
*environment natural surroundings (p. 534); the complex of physical, chemical, and biotic factors (such as climate, soil, and living things) that act upon an organism or an ecological community and ultimately determine its survival (p. 621)	***medioambiente** entorno natural (pág. 534); unión de factores físicos, químicos y bióticos (como el clima, el suelo y los seres vivos) que actúan sobre un organismo o una comunidad ecológica y básicamente determinan su supervivencia (pág. 621)
equinox one of two days (about March 21 and September 23) on which the sun is directly above the Equator, making day and night equal in length (p. 61)	**equinoccio** uno de dos días (hacia el 21 de marzo y el 23 de septiembre) en los cuales el sol se encuentra directamente sobre el ecuador, de modo que la duración del día y la noche es igual (pág. 61)
erosion the movement of weathered rock and material by wind, glaciers, and moving water (p. 49); the wearing away of soil (p. 546)	**erosión** proceso por el cual el viento, los glaciares y el agua en movimiento transportan roca y material desgastados (pág. 49); desgaste del suelo (pág. 546)
escarpment a steep cliff or slope between a higher and lower land surface (pp. 219, 507, 576)	**escarpe** acantilado o pendiente entre una superficie más alta y una más baja (págs. 219, 507, 576)
*estimate to judge approximately the size or value of something (p. 437)	***calcular** determinar el tamaño o el valor aproximado de algo (pág. 437)
*ethnic of or relating to large groups of people classed according to common traits and customs (p. 321)	***étnico** perteneciente o relativo a grupos de personas extensos con características y costumbres en común (pág. 321)
ethnic cleansing the expelling from a country or genocide of an ethnic group (p. 319)	**limpieza étnica** expulsión de un país o genocidio de un grupo étnico (pág. 319)
ethnic group group of people who share common ancestry, language, religion, customs, or place of origin (p. 79)	**grupo étnico** grupo de personas que tienen la misma ascendencia, lengua, religión, lugar de origen y costumbres (pág. 79)
*ethnicity relating to races or large groups of people classed according to common traits and customs (p. 732)	***etnicidad** relativo a etnias o extensos grupos de personas con características y costumbres en común (pág. 732)
eutrophication process by which a body of water becomes too rich in dissolved nutrients, leading to plant growth that depletes oxygen (p. 137)	**eutrofización** proceso por el cual en un cuerpo de agua se presenta un aumento excesivo de nutrientes disueltos, lo cual da lugar a un crecimiento de plantas que agota el oxígeno(pág. 137)
evaporation the process of converting liquid into vapor, or gas (p. 51)	**evaporación** proceso por el cual un líquido se convierte en vapor o gas (pág. 51)
*eventually in the end, especially after a long delay (p. 63)	***a la larga** al final, especialmente luego de una demora (pág. 63)
*exceed to be greater than (p. 412)	***exceder** ser más grande que algo (pág. 412)
exclave a territory that belongs to a particular political unit but is separated from it and surrounded by another political unit (p. 462)	**exclave** territorio que pertenece a una unidad política específica pero que está separado de esta y se encuentra rodeado por otra unidad política (pág. 462)
*exhibit to demonstrate or show publicly (p. 198)	***manifestar** exponer o mostrar públicamente (pág. 198)
*exploit to make use of a resource (p. 656)	***explotar** hacer uso de un recurso (pág. 656)
*export a commodity sent from one country to another for purposes of trade (p. 517)	***exportación** conjunto de mercancías que se envía de un país a otro con fines comerciales (pág. 517)
*expose to put someone at risk for harm or injury (p. 469)	***exponer** poner a alguien en riesgo, de manera que puede resultar lastimado o perjudicado (pág. 469)
extended family household made up of several generations of family members (p. 183)	**familia extensa** grupo familiar compuesto por varias generaciones de parientes (pág. 183)
*external arising outside of (p. 49)	***externo** que se manifiesta hacia el exterior (pág. 49)
*extract to remove (pp. 158, 665)	***extraer** remover (págs. 158, 665)

Glossary/Glosario

865

F

ENGLISH	ESPAÑOL
*factor something that actively contributes to the production of a result (p. 306)	***factor** algo que contribuye activamente a la producción de un resultado (pág. 306)
fall line a boundary in the eastern United States where the higher land of the Piedmont drops to the lower Atlantic Coastal Plain (p. 121)	**línea de descenso** frontera en el este de Estados Unidos donde la meseta de Piedmont se une con la llanura litoral atlántica (pág. 121)
fault a crack or break in Earth's crust (pp. 47, 506)	**falla** fractura o ruptura de la corteza terrestre (págs. 47, 506)
faulting process of cracking that occurs when the folded land cannot be bent any further (p. 47)	**formación de fallas** proceso de fractura que ocurre cuando no se pueden formar más pliegues en los estratos rocosos (pág. 47)
*feature to have as a characteristic or as a prominent attribute (p. 177)	***distinguirse** tener una característica o un atributo prominente (pág. 177)
federal system form of government in which powers are divided between the national government and state or provincial government (p. 87)	**sistema federal** forma de gobierno en la cual el poder se divide entre el gobierno nacional y el gobierno estatal o provincial (pág. 87)
feeder stream a tributary to a larger river (p. 424)	**afluente** tributario de un río grande (pág. 424)
fertilizer a chemical or natural substance added to soil or land to increase its fertility; a substance (as manure or a chemical mixture) used to make soil more fertile (pp. 402, 740)	**fertilizante** sustancia química o natural que se agrega al suelo para que sea más fértil; sustancia (como el estiércol o una mezcla química) que se usa para que el suelo sea más fértil (págs. 402, 740)
First Nation one of the indigenous peoples of Canada who are neither Inuit nor Métis (p. 149)	**Primera Nación** uno de los pueblos indígenas de Canadá excluyendo a los inuits y los métis (pág. 149)
fishery an area in which fish or sea animals are caught (p. 148); a place for catching fish, the fishing industry (p. 546)	**pesquería** área donde se atrapan peces y animales marinos (pág. 148); lugar donde se pesca, la industria pesquera (pág. 546)
fjord (fee·YAWRD) a long, steep-sided glacial valley now filled by seawater (p. 250)	**fiordo** valle glacial largo y empinado que está cubierto de agua marina (pág. 250)
floodplain the low-lying land along a river, formed mainly by sediment that has been deposited by floodwaters (p. 632)	**planicie de inundación** tierra baja a lo largo de un río formada por los sedimentos que depositan las inundaciones fluviales (pág. 632)
*fluctuate to ebb and flow in waves (p. 608)	***fluctuar** subir y bajar en ondas (pág. 608)
*fluctuation a shift from a previous condition (p. 33)	***fluctuación** cambio de una condición anterior (pág. 33)
*focus to concentrate attention or effort (p. 279)	***enfocar** concentrar la atención o el esfuerzo (pág. 279)
foehn (FUHRN) a dry wind that blows from the leeward sides of mountains, sometimes melting snow and causing avalanches; term used mainly in Europe (p. 273)	**foehn** viento seco que sopla a sotavento de las montañas, y que a veces derrite la nieve y causa avalanchas (pág. 273)
fold a bend in layers of rock, sometimes caused by plate movement (p. 47)	**pliegue** ondulación de las capas de roca que a veces se produce por el movimiento de las placas tectónicas (pág. 47)
foreclosure legal proceeding in which a borrower's rights to a property are relinquished due to his or her inability to make payments on the loan (p. 131)	**ejecución hipotecaria** procedimiento legal mediante el cual se retiran los derechos de una propiedad a un prestatario que no puede pagar las cuotas del préstamo adquirido (pág. 131)
formal region a region defined by a common characteristic, such as production of a product (p. 30)	**región formal** región definida por una característica común, como la manufactura de un producto (pág. 30)
fossil fuel a resource formed in the Earth by plant and animal remains (p. 124)	**combustible fósil** recurso de la Tierra formado por restos de plantas y animales (pág. 124)

ENGLISH	ESPAÑOL

free enterprise a system in which private individuals or groups have the right to own property or businesses and make a profit with limited government interference (p. 94)

libre empresa sistema según el cual individuos o grupos privados tienen derecho a aposeer propiedades o empresas y obtener ganancias de los mismos con una interferencia gubernamental mínima (pág. 94)

free port a place where goods can be unloaded, stored, or reshipped free of import duties (p. 764)

zona franca lugar en el cual se pueden descargar, almacenar y reenviar mercancías sin pagar impuestos de importación (pág. 764)

free trade zone an area of a country in which trade restrictions do not apply (p. 186)

zona de libre comercio zona de un país en la cual no se aplican las restricciones comerciales (pág. 186)

***function** a special purpose (p. 102)

***función** propósito especial (pág. 102)

functional region a central place and the surrounding territory linked to it (p. 30)

región funcional lugar central y el territorio circundante que se vincula a este (pág. 30)

***furthermore** besides; in addition (p. 446)

***además** aparte de, adicionalmente (pág. 446)

G

***generate** to produce (p. 632); to bring into existence (p. 666)

***generar** producir (pág. 632); traer a la existencia (pág. 666)

***generation** a group of individuals born and living at the same time (pp. 714, 766)

***generación** grupo de individuos que nacen y viven en un mismo periodo de tiempo (págs. 714, 766)

geographic information systems (GIS) computer programs that process and organize details about places on Earth and integrate those details with satellite images and other pieces of information (p. 23)

sistemas de información geográfica (SIG) programas informáticos que procesan y organizan detalles sobre lugares de la Tierra y los integran con imágenes satelitales y otra información (pág. 23)

geometric boundary a boundary that follows a geometric pattern (p. 90); a fixed limit or extent that typically follows straight lines (p. 373)

frontera geométrica límite o frontera que tiene un patrón geométrico (pág. 90); frontera fija o extensión territorial que generalmente sigue una línea recta (pág. 373)

geopolitics government policy as it is influenced by physical, human, and economic geography (p. 444)

geopolítica políticas gubernamentales que se ven influidas por la geografía física, humana y económica (pág. 444)

geothermal energy a form of energy conversion that captures heat energy from within Earth (p. 251)

energía geotérmica forma de conversión de energía que toma energía calorífica del interior de la Tierra (pág. 251)

geyser a spring that throws forth intermittent jets of heated water and steam (p. 251)

géiser manantial que expulsa columnas intermitentes de agua caliente y vapor (pág. 251)

glaciation a process by which glaciers form and spread (p. 250)

glaciación proceso de formación y expansión de los glaciares (pág. 250)

glacier a large body of ice that moves across the surface of the Earth (p. 49)

glaciar enorme masa de hielo que se desliza por la superficie terrestre (pág. 49)

glasnost Russian term for a new openness in areas of politics, social issues, and media; part of Gorbachev's reform plans (p. 343)

glasnost término ruso para la apertura en temas políticos, asuntos sociales y medios de comunicación; parte de los planes de reforma de Gorbachov (pág. 343)

global positioning system (GPS) a navigational system that can determine absolute location by using satellites and receivers on Earth (p. 21)

sistema de posicionamiento global (GPS) sistema de navegación que puede determinar una localización absoluta mediante satélites y receptores instalados en la Tierra (pág. 21)

***goal** the aim of an activity (p. 544)

***meta** objetivo de una actividad (pág. 544)

***grant** a sum of money given to a person or organization for a particular purpose (p. 815)

***beca** suma de dinero que se le da a una persona u organización para un propósito determinado (pág. 815)

grazier a New Zealand rancher (p. 790)

grazier ganadero de Nueva Zelanda (pág. 790)

Glossary/Glosario

ENGLISH

great circle route an imaginary line that follows the curve of the Earth and represents the shortest distance between two points (p. 14)

green revolution a program begun in the 1960s to produce higher-yielding, more productive strains of wheat, rice, and other food crops (p. 617)

greenhouse effect the capacity of certain gases in the atmosphere to trap heat, thereby warming the Earth (p. 63)

griot traditional oral historians, storytellers, singers, and musicians of West Africa (p. 540)

gross domestic product (GDP) the value of goods and services produced within a country in a year (p. 185)

groundwater water located underground within the Earth that supplies wells and springs (pp. 54, 590)

guest worker a foreign laborer living and working temporarily in another country (p. 278)

H

habitat area with conditions suitable for certain plants or animals to live (p. 522)

hajj in Islam, the yearly pilgrimage to Makkah that Muslims must make at least once in a lifetime (p. 441)

*hamper to make difficult; to impede (p. 566)

harmattan a hot, dry wind that blows from the northeast or east in the western Sahara (p. 488)

headwaters the source of a stream or river (p. 121)

hieroglyphics an ancient Egyptian writing system in which pictures and symbols represent words or sounds (p. 372)

high island an island with mountain ranges and volcanic soils (p. 800)

Holocaust the mass murder of 6 million Jews by Germany's Nazi regime during World War II (p. 277)

hot spring a spring whose water issues at a temperature higher than that of its surroundings (p. 251)

*hub the center of an activity or region (p. 561)

hurricane a large, powerful windstorm that forms over warm ocean waters (p. 122)

hydroelectric power electrical energy generated by falling water (p. 644)

hydrosphere the watery areas of the Earth, including oceans, lakes, rivers, and other bodies of water (p. 42)

ESPAÑOL

ruta del círculo máximo línea imaginaria que sigue la curvatura de la Tierra y presenta la distancia más corta entre dos puntos (pág. 14)

revolución verde programa que inició en la década de 1960 con el fin de producir variedades mejoradas y más rentables de trigo, arroz y otros cultivos (pág. 617)

efecto invernadero capacidad de algunos gases presentes en la atmósfera de retener el calor y así calentar la Tierra (pág. 63)

griot historiadores orales, narradores, cantantes y músicos tradicionales de África Occidental (pág. 540)

producto interno bruto (PIB) valor de los bienes y servicios producidos en un país durante un año (pág. 185)

agua subterránea agua ubicada debajo de la superficie terrestre que forma pozos y manantiales (págs. 54, 590)

trabajador invitado persona que vive y trabaja temporalmente en un país distinto a su país de origen (págs. 278)

hábitat área con condiciones favorables para la supervivencia de ciertas plantas y animales (pág. 522)

hach en el islam, la peregrinación a La Meca que deben realizar los musulmanes por lo menos una vez en su vida (pág. 441)

*obstaculizar dificultar; impedir (pág. 566)

harmattan viento caliente y seco que proviene del nordeste o el este y que sopla en el Sahara Occidental (pág. 488)

cabecera fuente de un arroyo o un río (pág. 121)

jeroglífico antiguo sistema de escritura egipcio en el cual las imágenes y símbolos representan palabras o sonidos (pág. 372)

isla alta isla con cordilleras y suelos volcánicos (pág. 800)

Holocausto exterminio masivo de 6 millones de judíos ejecutado por el régimen nazi de Alemania durante la Segunda Guerra Mundial (pág. 277)

fuente termal manantial cuya agua brota a una temperatura mayor que la de su entorno (pág. 251)

*núcleo centro de una actividad o región (pág. 561)

huracán tormenta con vientos torrenciales que se forma sobre aguas oceánicas cálidas (pág. 122)

energía hidroeléctrica energía eléctrica producida por agua en movimiento (pág. 644)

hidrosfera capa de agua de la Tierra que incluye océanos, lagos, ríos y otros cuerpos de agua (pág. 42)

ENGLISH

Ibadhism a conservative form of Islam distinct from Sunni and Shia sects (p. 441)

ideogram a pictorial character or symbol that represents a specific meaning or idea (p. 689)

***ignorance** lack of knowledge, education, or awareness (p. 188)

***immigrate** to change residence from a country to begin living permanently in another country (p. 127)

***impact** an effect on something; a significant or major effect (pp. 685, 737, 782)

imperialism the actions by which one country is able to extend power to control another country (p. 611)

***implement** to carry out or accomplish by concrete measures (p. 229)

***incentive** something that motivates one to act (p. 96)

***incorporate** to blend or thoroughly combine (p. 514)

indigenous native to a place (p. 513)

industrial capitalism an economic system in which business leaders use profits to expand their companies (p. 276)

Industrial Revolution beginning in the 1700s, rapid major change in the economy with the introduction of power-driven machinery (p. 276)

***inevitable** incapable of being avoided or evaded (p. 178)

infrastructure the set of systems that affect how well a place or organization operates, such as telephone or transportation systems, within a country (p. 539)

***infrastructure** public services or systems such as power, water and sewage, transportation and communication networks, and schools and health care facilities (p. 636)

***institute** to organize and establish (p. 686)

***integrate** to blend into a functioning whole (p. 256)

***interdependence** a condition in which people or groups rely on each other, rather than only relying on themselves (p. 765)

***intermittent** occurring at irregular intervals; occasional (p. 561)

***internal** existing or lying within (p. 21)

internally displaced person a refugee within his or her own country (p. 567)

***intervene** to take action to change what is happening (p. 521); to come between so as to prevent or alter a course of events (p. 767)

Inuit a member of the Arctic native peoples of North America; once known as Eskimo (p. 149)

ESPAÑOL

ibadismo forma conservadora del islamismo que se distingue de las sectas suníes y chiíes (pág. 441)

ideograma carácter o símbolo pictórico que representa un significado o idea específicos (pág. 689)

***ignorancia** falta de conocimiento, educación o conciencia (pág. 188)

***inmigrar** dejar de residir en un país para establecerse de forma permanente en otro (pág. 127)

***impacto** efecto sobre algo; efecto importante o considerable (págs. 685, 737, 782)

imperialismo acciones mediante las cuales un país extiende su poder para controlar a otro (pág. 611)

***implementar** llevar a cabo o realizar algo con medidas concretas (pág. 229)

***incentivo** lo que motiva a alguien a actuar (pág. 96)

***incorporar** mezclar o combinar por completo (pág. 514)

indígena nativo de un lugar (pág. 513)

capitalismo industrial sistema económico en el cual los líderes de negocios usan las ganancias para expandir sus compañías (pág. 276)

Revolución Industrial cambio rápido e importante en la economía que se dio con la introducción de la mecanización en el siglo XVIII (pág. 276)

***inevitable** que no puede evitarse o evadirse (pág. 178)

infraestructura conjunto de sistemas dentro de un país que influyen en el buen funcionamiento de un lugar u organización, como los sistemas de telefonía o transporte (pág. 539)

***infraestructura** servicios públicos o sistemas como la electricidad, el acueducto y el alcantarillado, las redes de comunicación y transporte, y las instalaciones de escuelas e instituciones sanitarias (pág. 636)

***instituir** organizar y establecer (pág. 686)

***integrar** incorporar algo en un conjunto de partes (pág. 256)

***interdependencia** situación en la que las personas o los grupos dependen el uno del otro, no solo de sí mismos (pág. 765)

***intermitente** que ocurre a intervalos irregulares; ocasional (pág. 561)

***interno** que está en la parte de adentro (pág. 21)

persona desplazada interna refugiado dentro de su propio país (pág. 567)

***intervenir** actuar para cambiar lo que sucede (pág. 521); interponerse con el fin de evitar o alterar el curso de los hechos (pág. 767)

inuit miembro de los pueblos árticos de América del Norte, conocido en el pasado como esquimal (pág. 149)

Glossary/Glosario

Glossary/Glosario

ENGLISH

invasive species non-indigenous or non-native species that threatens ecosystems, habitats, or other species (p. 261)

islet a very small island (p. 727)

***isolate** to place or keep by itself (p. 285)

***isolation** set or kept apart from others (p. 730)

***issue** a vital or unsettled matter (p. 300)

***issue** a topic for debate and discussion (p. 720)

isthmus a narrow strip of land connecting two larger land areas (p. 197)

J

jati a group that defines one's occupation and social position (p. 611)

jazz musical form that developed in the United States in the early 1900s, blending African rhythms and European harmonies (p. 129)

jute a fiber used to make string, rope, and cloth (p. 639)

K

karma in Hindu belief, the sum of good and bad actions in one's present and past lives (p. 614)

karst terrain dominated by limestone bedrock and characterized by rocky ground, caves, sinkholes, underground rivers, and the absence of surface streams and lakes (p. 314)

kibbutz a communal farm or settlement in Israel (p. 394)

kinship group a group of people related by blood or marriage (p. 804)

Kyoto Protocol an amendment to the international treaty on climate change designed to reduce the amount of greenhouse gases emitted by specific countries (p. 287)

L

lagoon a shallow lake that is intermittently connected to a river, another lake, or the sea (p. 777)

lama Buddhist monk (p. 657)

land bridge a strip of land that connects two larger landmasses, enabling migration of plants and animals to new areas (p. 176)

land subsidence the sinking or settling of land to a lower level in response to various natural and human-caused factors (p. 189)

language family group of related languages that have all developed from one earlier language (p. 78)

ESPAÑOL

especie introducida especie no nativa que amenaza ecosistemas, hábitats y otras especies (pág. 261)

islote isla muy pequeña (pág. 727)

***aislar** dejar algo solo y separado (pág. 285)

***aislamiento** efecto de separar de otros (pág. 730)

***conflicto** asunto vital o sin resolver (pág. 300)

***asunto** tema de debate y discusión (pág. 720)

istmo franja delgada de tierra que conecta dos territorios más grandes (pág. 197)

J

jati En la India, grupo que define la ocupación y la posición social de una persona (pág. 611)

jazz género musical que se desarrolló en Estados Unidos a principios del siglo XX, que mezcla ritmos africanos y armonías europeas (pág. 129)

yute fibra que se usa para hacer cuerdas, sogas y telas (pág. 639)

K

karma en la creencia hindú, la suma de las acciones buenas y malas de la vida presente y las vidas pasadas de una persona (pág. 614)

carst terreno dominado por un lecho de roca caliza que se caracteriza por suelo rocoso, cavernas, sumideros, ríos subterráneos y la ausencia de arroyos y lagos superficiales (pág. 314)

kibutz granja comunal o asentamiento en Israel (pág. 394)

grupo familiar grupo de personas relacionadas por lazos sanguíneos o matrimonio (pág. 804)

Protocolo de Kioto enmienda al tratado internacional sobre el cambio climático diseñada para reducir la cantidad de gases de efecto invernadero que emiten países específicos (pág. 287)

L

laguna lago superficial conectado intermitentemente a un río, otro lago o el mar (pág. 777)

lama monje budista (pág. 657)

puente intercontinental franja de tierra que conecta dos continentes más grandes, lo que permite la migración de plantas y animales a otras zonas (pág. 176)

subsidencia del terreno hundimiento o desplazamiento de la tierra a un nivel inferior como respuesta a varios factores naturales o humanos (pág. 189)

familia lingüística grupo de lenguas relacionadas que se desarrollaron a partir de una misma lengua (pág. 78)

870

ENGLISH	ESPAÑOL
latifundia in Latin America, large agricultural estates owned by families or corporations (p. 205)	latifundios en América Latina, grandes propiedades agrícolas cuyos dueños son familias o corporaciones (pág. 205)
leeward being in or facing the direction toward which the wind is blowing (p. 68)	**sotavento** estar de frente a la dirección en la que sopla el viento (pág. 68)
less developed country a country that, according to the United Nations, exhibits the lowest indicators of socioeconomic development (p. 97)	**país menos desarrollado** país que, según las Naciones Unidas, muestra los indicadores de desarrollo socioeconómico más bajos (pág. 97)
lingua franca a common language used among people with different native languages (p. 514)	**lengua franca** lenguaje común que se usa entre personas que tienen diferentes lenguas nativas (pág. 514)
*link a connecting structure (p. 425)	*vínculo estructura conectora (pág. 425)
lithosphere uppermost layer of the Earth that includes the crust, continents, and ocean basins (p. 42)	**litosfera** capa más exterior de la Tierra que incluye la corteza, los continentes y las cuencas oceánicas (pág. 42)
llanos (LAH•nohs) fertile grasslands found in inland areas of Colombia and Venezuela (p. 219)	**llanos** praderas fértiles que se encuentran en zonas del interior de Colombia y Venezuela (pág. 219)
loess (LEHS) fine, yellowish, brownish topsoil made up of particles of silt and clay, carried and deposited by the wind (pp. 270, 682)	*loess* capa superior del suelo fina, amarillenta y marrón, compuesta de partículas de limo y arcilla, llevada y depositada por el viento (págs. 270, 682)
low island an island known as an atoll (p. 801)	**isla baja** isla conocida como atolón (pág. 801)
Loyalist an American colonist who remained loyal to the British government (p. 152)	**leal** colono norteamericano que permanecía fiel al gobierno británico (pág. 152)

M

ENGLISH	ESPAÑOL
magma molten rock that is located below Earth's surface (p. 45)	**magma** roca fundida ubicada debajo de la superficie de la Tierra (pág. 45)
*maintain to support (p. 632)	*mantener sostener (pág. 632)
*major greater in importance or interest (p. 80)	*principal de mayor importancia o interés (pág. 80)
mantle thick middle layer of the Earth's interior structure consisting of hot rock that is dense but flexible (p. 44)	**manto** capa gruesa intermedia del interior de la estructura terrestre formada por roca caliente densa pero flexible (pág. 44)
mantra sacred words or phrases that are repeated in prayers or chants (p. 660)	**mantra** palabras o frases sagradas que se repiten en oraciones o cantos (pág. 660)
Manufacturing Belt a concentrated region of manufacturing industries in the northeastern and midwestern United States (p. 127)	*Manufacturing Belt* **(cinturón manufacturero)** región de industrias manufactureras concentradas en el noroeste y medio oeste de Estados Unidos (pág. 127)
Maori the indigenous people of New Zealand (p. 781)	**maorí** pueblo indígena de Nueva Zelanda (pág. 781)
map projection a mathematical formula used to represent the curved surface of the Earth on the flat surface of a map (p. 15)	**proyección cartográfica** fórmula matemática que se usa para representar la superficie curva de la Tierra en la superficie plana de un mapa (pág. 15)
maquiladora in Mexico, a manufacturing plant owned by a foreign company (p. 186)	**maquiladora** en México, planta manufacturera de propiedad de una compañía extranjera (pág. 186)
market economy an economic system based on free enterprise, in which businesses are privately owned and production and prices are determined by supply and demand (p. 94)	**economía de mercado** sistema económico que se basa en la libertad de empresa, en el cual los negocios son de propiedad privada y la producción y los precios los determinan la oferta y la demanda (pág. 94)

ENGLISH

marsh a wetland typically covered with grasses (p. 424)

massif a body of mountain ranges formed by fault-line activity (p. 294)

matriarchal family ruled by a woman such as a mother, grandmother, or aunt (p. 203)

Mediterranean climate a climate marked by warm, dry summers and cool but not cold winters (p. 579)

megacity a great city that is made up of several large and small cities (p. 182)

megalopolis a large population concentration made up of several large and many smaller cities, such as the area between Boston and Washington, D.C. (p. 129)

meltwater water formed by melting snow and ice (p. 327)

mercantilism the theory or practice of merchant or trading pursuits (p. 611)

merchant marine a country's fleet of ships that engages in trade (p. 691)

mestizo refers to people of mixed indigenous and European descent (p. 181)

metropolitan area region that includes a central city and its surrounding suburbs (p. 103)

midnight sun continuous daylight, a time when the sun is visible at midnight during the summer in either the Arctic or Antarctic (p. 62)

*migrate to move from one place to another (p. 253)

migration the movement of people from place to place (p. 82)

minifundia in Latin America, small farms that produce food chiefly for family use (p. 205)

mistral a strong northerly wind from the Alps that can bring cold air to southern France (p. 273)

mixed economy a system of resource management in which the government supports and regulates enterprise through decisions that affect the marketplace (p. 95)

mixed forest forest with both coniferous and deciduous trees (p. 71)

monarchy a form of autocracy with a hereditary king or queen exercising supreme power (p. 88)

*monitor to watch closely, evaluate (p. 428)

monoculture the cultivation or growth of a single crop over a wide area for a consecutive number of years (p. 233)

monotheism belief in one God (p. 395)

monsoon in Asia, a seasonal wind that brings warm, moist air from the oceans in summer and cooler, dry air from inland in winter (pp. 436, 607, 683)

ESPAÑOL

marisma humedal usualmente cubierto de hierba (pág. 424)

macizo cuerpo de cordilleras formado por la actividad de las líneas de falla (pág. 294)

matriarcal familia controlada por una mujer, como la madre, la abuela o la tía (pág. 203)

clima mediterráneo clima que se caracteriza por veranos cálidos y secos e inviernos frescos, mas no fríos (pág. 579)

megaciudad ciudad enorme formada por varias ciudades grandes y pequeñas (pág. 182)

megalópolis gran concentración de población formada por varias ciudades grandes y pequeñas, como el área entre Boston y Washington, D. C. (pág. 129)

agua de fusión agua que se forma cuando la nieve y el hielo se derriten (pág. 327)

mercantilismo teoría o práctica de las actividades comerciales y mercantiles (pág. 611)

marina mercante flota de barcos de un país que realiza actividades comerciales (pág. 691)

mestizo persona descendiente de la mezcla de indígenas y europeos (pág. 181)

área metropolitana región que incluye una ciudad central y los suburbios que la rodean (pág. 103)

sol de medianoche luz solar continua, periodo en el cual el sol es visible a la medianoche durante el verano en el Polo Ártico o el Polo Antártico (pág. 62)

*migrar desplazarse de un lugar a otro (pág. 253)

migración movimiento de personas de un lugar a otro (pág. 82)

minifundios en América Latina, granjas pequeñas que producen alimentos principalmente para consumo familiar (pág. 205)

mistral viento fuerte del norte que sopla desde los Alpes y lleva aire frío al sur de Francia (pág. 273)

economía mixta sistema de administración de recursos en el cual el gobierno mantiene y regula las empresas mediante decisiones que influyen en el mercado (pág. 95)

bosque mixto bosque con árboles coníferos y caducifolios (pág. 71)

monarquía forma de autocracia en donde un rey o reina herederos ejercen el poder supremo (pág. 88)

*monitorizar observar de cerca, evaluar (pág. 428)

monocultivo cultivo de un solo producto en un área extensa durante varios años consecutivos (pág. 233)

monoteísmo creencia en un solo dios (pág. 395)

monzón en Asia, viento estacional que transporta aire cálido y húmedo desde los océanos en verano, y aire fresco y seco desde el interior en invierno (págs. 436, 607, 683)

ENGLISH	ESPAÑOL
montane referring to a mountainous area (p. 557)	**montañoso** relativo a un área con montañas (pág. 557)
moraine piles of rocky debris left by melting glaciers (p. 50)	**morrena** pilas de desechos rocosos que deja un glaciar cuando se derrite (pág. 50)
more developed country a country that has a highly developed economy and advanced technological infrastructure relative to other less developed nations (p. 97)	**país más desarrollado** país con un alto grado de desarrollo económico y una infraestructura tecnológica avanzada en relación con otros países menos desarrollados (pág. 97)
moshav a cooperative settlement of small individual farms in Israel (p. 394)	**moshav** asentamiento de carácter cooperativo formado por pequeñas granjas individuales en Israel (pág. 394)
mosque in Islam, a house of worship (p. 396)	**mezquita** en el islam, casa de culto (pág. 396)
mujahideen Islamic guerrilla fighters (p. 461)	**muyahidín** combatiente de la guerrilla islámica (pág. 461)

N

ENGLISH	ESPAÑOL
national identity the sense of a being part of the whole of a country, including its culture, traditions, language, and politics (p. 785)	**identidad nacional** sentido de pertenencia a un país, incluyendo su cultura, tradiciones, lenguaje y política (pág. 785)
nationalism belief in the right of each people to be an independent nation (p. 373)	**nacionalismo** creencia en el derecho de un pueblo a ser una nación independiente (pág. 373)
natural boundary a fixed limit or extent defined along physical geographic features such as mountains and rivers (p. 90); a boundary created by a physical feature, such as a mountain, river, or strait (p. 416)	**frontera natural** límite fijo o extensión definida a lo largo de formaciones geográficas como montañas y ríos (pág. 90); frontera creada por una formación física como una montaña, un río o un estrecho (pág. 416)
natural increase the growth rate of a population; the difference between birthrate and death rate (p. 82)	**crecimiento natural** tasa de crecimiento de una población; diferencia entre la tasa de natalidad y la tasa de mortalidad (pág. 82)
natural vegetation plant life that grows in a certain area if people have not changed the natural environment (p. 69)	**vegetación natural** vida vegetal que crece en ciertas áreas donde los seres humanos no han alterado el medioambiente natural (pág. 69)
*****navigable** able to be traveled by boat (p. 556)	*****navegable** que se puede atravesar en barco (pág. 556)
*****neutral** not favoring either side in a quarrel, contest, or war (p. 612)	*****neutral** que no favorece a ningún lado en una disputa, contienda o guerra (pág. 612)
*****nevertheless** despite a situation or comment (p. 809)	*****sin embargo** a pesar de una situación o comentario (pág. 809)
newly industrialized country a country that has begun transitioning from primarily agricultural to primarily manufacturing and industrial activity (p. 95)	**país recientemente industrializado** país que ha comenzado a hacer la transición de principalmente agrícola a principalmente manufacturero e industrial (pág. 95)
nomad a member of a wandering pastoral people (p. 374)	**nómada** miembro de un pueblo pastoril errante (pág. 374)
*****nonetheless** in spite of; nevertheless (p. 435)	*****no obstante** a pesar de; sin embargo (pág. 435)
*****nuclear** of, relating to, or using the atomic nucleus, atomic energy, the atom bomb, or atomic power (p. 694)	*****nuclear** relativo al uso de los núcleos atómicos, la energía atómica o la bomba atómica (pág. 694)
*****nuclear** a weapon whose destructive power derives from an uncontrolled nuclear reaction (p. 814)	*****nuclear** arma cuyo poder de destrucción se deriva de una reacción nuclear sin control (pág. 814)
nuclear family a family unit made up of a husband, wife, and children (p. 494)	**familia nuclear** unidad familiar compuesta por el esposo, la esposa y los hijos (pág. 494)
nuclear wastes by-products of producing nuclear power and weapons (p. 352)	**residuos radiactivos** subproductos de la producción de energía y armas nucleares (pág. 352)

Glossary/Glosario

ENGLISH

O

oasis small area in a desert where water and vegetation are found (p. 70)

*obtain to gain or acquire usually by planning or effort (p. 33)

*obvious easily discovered, seen, or understood (p. 205)

*occur to come into existence, happen (p. 264)

ocean warming rise in the temperature of the ocean water (p. 814)

*offset to counterbalance or to compensate for something else (p. 803)

old-growth forest complex forest that has developed over a long period of time and is relatively untouched by human activity (p. 158)

oligarchy system of government in which a small group holds power (p. 88)

oral tradition the practice of passing down stories from generation to generation by word of mouth (p. 495)

organic farming the use of natural substances rather than chemical fertilizers and pesticides to enrich the soil and grow crops (p. 665)

*output something produced; a mineral, agricultural, or industrial production (p. 401)

overfarming situation in which land is repeatedly farmed so that soil nutrients are depleted (p. 517)

overfishing harvesting fish to the point that species are depleted and the value of the fishery reduced (p. 148)

overgrazing grazing so heavily that the vegetation is damaged and the ground erodes (p. 404)

*overlap to partly cover (p. 69)

*overseas relating to a foreign country, especially one across the ocean (p. 715)

oxisol a thick, weathered soil of the humid tropics that is largely depleted of fertility and nutrients (p. 233)

P

pampas grassy, treeless plains of southern South America (p. 219)

panchayat village council (p. 614)

*parallel two items that are side by side and having the same distance between them (p. 530)

*participate to take part in (p. 418)

*partner one associated with another especially in an action (p. 377)

ESPAÑOL

O

oasis área pequeña en un desierto donde hay agua y vegetación (pág. 70)

*obtener ganar o adquirir, usualmente mediante la planeación o el esfuerzo (pág. 33)

*obvio fácil de descubrir, ver o entender (pág. 205)

*ocurrir acontecer, suceder (pág. 264)

calentamiento oceánico aumento de la temperatura del agua del océano (pág. 814)

*compensación para contrarrestar o para compensar algo más (pág. 803)

bosque primario bosque complejo que se ha desarrollado durante un periodo largo y relativamente no ha sido influenciado por la actividad humana (pág. 158)

oligarquía sistema de gobierno en el cual un grupo pequeño ejerce el poder (pág. 88)

tradición oral transmisión de historias de generación en generación mediante la palabra hablada (pág. 495)

agricultura orgánica uso de sustancias naturales en lugar de pesticidas y fertilizantes químicos para cultivar y fertilizar el suelo (pág. 665)

*producto algo producido; producción industrial, agrícola o minera (pág. 401)

sobrexplotación agraria situación en la que la tierra se cultiva repetidamente, de modo que se agotan los nutrientes del suelo (pág. 517)

sobrepesca pescar hasta el punto de agotar las especies y reducir el valor de la pesca (pág. 148)

sobrepastoreo pastoreo intensivo que daña la vegetación y erosiona el suelo (pág. 404)

*superponer cubrir parcialmente (pág. 69)

*ultramar relativo a un país extranjero, en especial al otro lado del océano (pág. 715)

oxisol suelo grueso y meteorizado de los trópicos húmedos que no es fértil ya que los nutrientes se han agotado (pág. 233)

P

pampa llanura austral de América del Sur, cubierta de hierba pero desprovista de árboles (pág. 219)

panchayat concejo comunal (pág. 614)

*paralelos dos objetos que están lado a lado a la misma distancia entre sí (pág. 530)

*participar tomar parte en algo (pág. 418)

*socio persona asociada con otra para tomar parte en una actividad (pág. 377)

ENGLISH

pastoralism the raising of animals for food (p. 415)

patois a dialect used in everyday speech that blends elements of several languages (p. 203)

patriarchal a family that is headed by a male family member (p. 494)

peninsula a portion of land nearly surrounded by water (p. 434)

perceptual region a region defined by popular feelings and images rather than by objective data (p. 30)

perestroika (PEHR·uh·STROY·kuh) in Russian, "restructuring"; part of Gorbachev's plan for reforming Soviet economy and government (p. 343)

permafrost permanently frozen layer of soil beneath the surface of the ground (pp. 72, 339)

***persistent** continuing, existing, or acting for a long time (p. 319)

pesticide chemical used to kill crop-damaging insects, rodents, and other pests (pp. 353, 402, 623)

***phenomenon** a fact or event of scientific interest that can be scientifically explained or described (p. 221)

phosphate natural mineral containing chemical compounds often used in fertilizers (p. 371)

pidgin a grammatically simplified form of a language (p. 785); a blend of English and an indigenous language (p. 807)

planar projection a map created by projecting an image of the Earth onto a geometric plane (p. 15)

plantation a large commercial farm (p. 561)

plate tectonics the term scientists use to describe the activities of continental drift and magma flow, which create many of Earth's physical features (p. 45)

poaching illegal hunting (p. 523)

polder low-lying area from which seawater has been drained to create new land (p. 272)

***policy** an overall plan that establishes goals and determines procedures, decisions, and actions (p. 585)

pollution hot spot a location where pollution and other human activities have led to the degradation, or even death, of an ecosystem (p. 307)

population density the average number of people living on a square mile or square kilometer of land (p. 85)

population distribution the variations in population that occur across a country, a continent, or the world (p. 85)

ESPAÑOL

pastoreo cría de animales para el consumo (pág. 415)

patois dialecto usado en la expresión cotidiana, en el que se mezclan elementos de varios idiomas (pág. 203)

patriarcal familia dirigida por uno de sus integrantes de sexo masculino (pág. 494)

península porción de tierra que está prácticamente rodeada por agua (pág. 434)

región perceptiva región que se define mediante sentimientos e imágenes populares en vez de información objetiva (pág. 30)

perestroika en ruso, "reestructuración"; parte del plan de Gorbachov para reestructurar el gobierno y la economía soviéticos (pág. 343)

permafrost capa de suelo permanentemente congelada debajo de la superficie (págs. 72, 339)

***persistente** durable, que existe o actúa por un tiempo largo (pág. 319)

pesticida sustancia química que se usa para matar insectos, roedores y otras plagas que dañan las cosechas (págs. 353, 402, 623)

***fenómeno** hecho o suceso de interés científico que se puede describir o explicar científicamente (pág. 221)

fosfato mineral natural que contiene compuestos químicos y se usa con frecuencia en los fertilizantes (pág. 371)

pidgin tipo de lengua gramaticalmente simplificada (pág. 785); mezcla de inglés y una lengua indígena (pág. 807)

proyección planar mapa que se crea proyectando una imagen de la Tierra en un plano geométrico (pág. 15)

plantación granja comercial grande (pág. 561)

tectónica de placas término con el cual los científicos describen las actividades de la deriva continental y el flujo de magma, que crearon muchas de las formaciones físicas de la Tierra (pág. 45)

caza furtiva caza ilegal (pág. 523)

pólder área baja de la cual se ha drenado el agua del mar para crear tierra nueva (pág. 272)

***política** plan general que establece metas y determina procedimientos, decisiones y acciones (pág. 585)

punto de emisión de contaminantes lugar donde la contaminación y otras actividades humanas han llevado a la degradación, o incluso la muerte, de un ecosistema (pág. 307)

densidad de población número promedio de personas que viven en una milla cuadrada o un kilómetro cuadrado de territorio (pág. 85)

distribución de la población cambios en la población que ocurren en un país, un continente o el mundo (pág. 85)

ENGLISH

population pressure the sum of factors within a population that reduce the ability of an environment to support the population, therefore resulting in migration or population decline (p. 202)

population pyramid a diagram that shows the distribution of a population (p. 84)

*portion share of a whole (p. 660)

*pose to come to attention as; to set forth or offer for consideration (p. 369)

postindustrial economy that emphasizes services and technology rather than industry and manufacturing (p. 131)

*potential possible or likely (p. 490); having a capacity that could be developed (p. 654)

prairie an inland grassland area (p. 71)

precipitation moisture that falls to the Earth as rain, sleet, hail, or snow (p. 51)

*predictable occurring in a way that is expected (p. 61)

*predominant the main or strongest (pp. 221, 706)

*preliminary something done in preparation for something more important (p. 406)

prevailing wind wind in a region that blows in a fairly constant directional pattern (p. 66)

*primary of first rank, importance, or value (p. 30)

primate city a city that dominates a country's economy, culture, and government and in which population is concentrated; usually the capital (p. 182)

*principal most important, consequential, or influential (p. 374)

*principle fundamental characteristic (p. 642); a rule or code of conduct (p. 730)

*priority something given or meriting attention before competing alternatives (p. 566)

privatization a change to private ownership of state-owned companies and industries (p. 346)

*prohibit to forbid somebody from doing something through law or rule (pp. 523, 585)

*prohibit to prevent from doing something (p. 623)

*promote to help something grow or develop (p. 306)

prophet person believed to be a messenger from God (p. 396)

ESPAÑOL

presión demográfica conjunto de factores dentro de una población que reducen la capacidad de un medioambiente para sustentarla, lo que resulta en la migración o la disminución de la población (pág. 202)

pirámide demográfica diagrama que muestra la distribución de una población (pág. 84)

*porción parte de un todo (pág. 660)

*plantear presentar; exponer o poner a consideración (pág. 369)

posindustrial economía que se enfoca más en los servicios y la tecnología que en la industria y la manufactura (pág. 131)

*potencial posible o probable (pág. 490); que tiene una capacidad que se puede desarrollar (pág. 654)

pradera área de pastizal en el interior (pág. 71)

precipitación agua que cae a la Tierra en forma de lluvia, aguanieve, granizo o nieve (pág. 51)

*predecible que ocurre de la forma esperada (pág. 61)

*predominante lo prevaleciente o más fuerte (págs. 221, 706)

*preliminar algo que se hace en preparación para algo más importante (pág. 406)

viento predominante viento que sopla en un patrón direccional bastante constante en una región (pág. 66)

*primordial de primer rango, importancia o valor (pág. 30)

ciudad principal ciudad que domina la economía, la cultura y el gobierno de un país y en la cual se concentra la mayor parte de su población; por lo general es la capital (pág. 182)

*principal más importante, relevante o influyente (pág. 374)

*principio característica fundamental (pág. 642); norma o código de conducta (pág. 730)

*prioridad algo a lo que se presta más atención que a otras alternativas con las que compite (pág. 566)

privatización cambio de propiedad al sector privado de las industrias y empresas públicas (pág. 346)

*prohibir impedir que alguien haga algo mediante leyes o normas (págs. 523, 585)

*prohibir impedir la ejecución de algo (pág. 623)

*promover ayudar al crecimiento o desarrollo de algo (pág. 306)

profeta persona considerada mensajera de Dios (pág. 396)

— **Q** —

qanat an underground canal first built by the ancient Persians (p. 417)

Quebecois Quebec's French-speaking inhabitants (p. 150)

qanat canal subterráneo inventado por los antiguos persas (pág. 417)

quebequenses habitantes de habla francesa de Quebec (pág. 150)

ENGLISH	ESPAÑOL
quipu (KEE•poo) knotted cords of various lengths and colors used by the Inca to keep financial records (p. 223)	**quipu** cuerdas anudadas de varias longitudes y colores que los incas usaban para llevar sus registros financieros (pág. 223)

R

*radical fundamental or extreme; drastic (p. 352)	***radical** básico o extremo; drástico (pág. 352)
radioactive material material contaminated by residue from the generation of nuclear energy or the testing of nuclear weapons (pp. 352, 469)	**material radiactivo** material contaminado por residuos de la generación de energía nuclear o de pruebas de armas nucleares (págs. 352, 469)
rain shadow result of a process by which dry areas develop on the leeward sides of mountain ranges (pp. 68, 579)	**sombra pluviométrica** resultado de un proceso mediante el cual se desarrollan áreas secas en los lados de sotavento de las cordilleras (págs. 68, 579)
*range a series of things in a row; a series of mountains (pp. 390, 680)	***cadena** serie de eslabones entrelazados; (montañas) cordillera (págs. 390, 680)
*recover to return to a previous level of performance after a decline (p. 662)	***recuperar** retornar a un nivel previo de desempeño después de un declive (pág. 662)
reforestation the planting and cultivating of new trees in an effort to restore a forest where the trees have been cut down or destroyed (pp. 212, 326, 770)	**reforestación** plantación y cultivo de nuevos árboles para restaurar un bosque al que se le cortaron o fueron destruidos (págs. 212, 326, 770)
*regime a form of government (p. 731)	***régimen** forma de gobierno (pág. 731)
*regulate to govern or direct according to rule (p. 95)	***regular** gobernar o dirigir según las normas (pág. 95)
reincarnation rebirth in new bodies or forms of life (p. 614)	**reencarnación** acción de volver a nacer en un cuerpo o forma de vida nuevos (pág. 614)
relative location location in relation to other places (p. 17)	**localización relativa** localización con relación a otros lugares (pág. 17)
*relevant having to do with the matter at hand (p. 381)	***pertinente** que tiene que ver con el asunto que se trata (pág. 381)
relief the variation in elevation across an area of Earth's land (p. 19)	**relieve** variación en la elevación de una superficie terrestre (pág. 19)
*rely to be dependent (p. 401)	***confiar** depender de algo (pág. 401)
remote sensing any technique used to measure, observe, or monitor a subject or process without physically touching the object under observation (p. 25)	**detección remota** cualquier técnica que se usa para medir, observar o monitorizar un sujeto o proceso sin tocarlo físicamente (pág. 25)
Renaissance rebirth; the period in European civilization characterized by a surge of interest in classical learning and values (p. 300)	**Renacimiento** acción de volver a nacer; periodo de la civilización europea caracterizado por el resurgimiento del interés en el aprendizaje y los valores clásicos (pág. 300)
reservoir a natural or artificial lake used as a source of water (p. 532)	**embalse** lago natural o artificial que se usa como fuente de agua (pág. 532)
*resource a usable stock or supply; a source from which benefit can be gained (pp. 298, 776)	***recursos** existencias o provisiones que se pueden usar; fuente de la cual se puede obtener beneficio (págs. 298, 776)
*restrain to hold down or back; to limit (p. 460)	***contener** reprimir; limitar (pág. 460)
*reverse the opposite (p. 456)	***inverso** opuesto (pág. 456)
revolution in astronomy, the Earth's yearly trip around the sun, taking 365¼ days (p. 61)	**revolución** en astronomía, el giro anual de la Tierra alrededor del Sol, el cual toma 365¼ días (pág. 61)
rift valley a valley formed by the separation of tectonic plates (p. 391)	**fosa tectónica** valle formado por la separación de las placas tectónicas (pág. 391)

Glossary/Glosario

877

ENGLISH

river plain a plain formed by the deposit of sediment over a long period of time by one or more rivers (p. 533)

Russification in nineteenth- and twentieth-century Russia and the Soviet Union, a government program that required everyone in the empire to speak Russian and to become a Christian; assignment of some Russian-speaking people to non-Russian ethnic regions (p. 342)

S

Sahel the band of land extending from Senegal to Somalia; also called the African Transition Zone (p. 484)

samurai a professional warrior of preindustrial Japan (p. 710)

sanitation the disposal of waste products (pp. 493, 586)

satellite a country controlled by another country, notably Eastern European countries controlled by the Soviet Union by the end of World War II (p. 342)

*scope the extent of an activity or influence (p. 537)

sedimentation the action or process of forming or depositing sediment (p. 211)

seismic relating to or caused by an earthquake (p. 176)

separatism the breaking away of one part of a country to create a separate, independent country (p. 151)

*series a number of things of the same type following one after the other in space or time (p. 144)

shamal a northwesterly wind in the Persian Gulf (p. 436)

shari'ah Islamic law derived from the Qur'an and the teachings of Muhammad (p. 441)

shatter belt a region where political alliances are constantly splintering and fracturing based on ethnicity (p. 319); a region caught between stronger political groups, which is under constant stress and is sometimes fragmented as the result of conflicts between strong rivals (p. 759)

sheikhdom territory ruled by an Arab tribal leader (p. 438)

Shia a branch of Islam that regards Muhammad's son-in-law Ali and the imams as his rightful successors (p. 441)

*shift to change the place, position, or direction of (p. 120)

shifting cultivation a system in which farmers plant a field for several years until its resources are depleted, then abandon it and clear a new field (p. 589); a form of agriculture in which an area of ground is cleared of vegetation, cultivated for a few years, and then abandoned in order to move to a new area (p. 768)

ESPAÑOL

llanura fluvial llanura que se forma por el depósito de sedimentos de uno o más ríos durante un largo periodo (pág. 533)

rusificación durante los siglos XIX y XX en Rusia y la Unión Soviética, programa gubernamental que exigía que todos los habitantes del imperio hablaran ruso y se convirtieran al cristianismo; asignación de algunos hablantes de ruso a regiones étnicas no rusas (pág. 342)

Sáhel franja de tierra que se extiende desde Senegal hasta Somalia; también se conoce como la zona de transición africana (pág. 484)

samurái guerrero profesional del Japón preindustrial (pág. 710)

recolección de basuras eliminación de desechos (págs. 493, 586)

país satélite país controlado por otro, en particular los países de Europa del Este controlados por la Unión Soviética al final de la Segunda Guerra Mundial (pág. 342)

*alcance extensión de una actividad o influencia (pág. 537)

sedimentación acción o proceso por el cual se depositan sedimentos (pág. 211)

sísmico relacionado con o causado por un terremoto (pág. 176)

separatismo doctrina que propugna la separación de una parte de un país para crear un país independiente (pág. 151)

*serie conjunto de cosas del mismo tipo que suceden unas a otras en el tiempo o en el espacio (pág. 144)

shamal viento del noroeste que sopla sobre el golfo Pérsico (pág. 436)

sharia ley islámica derivada del Corán y las enseñanzas de Mahoma (pág. 441)

zona de conflicto región donde las alianzas políticas se rompen frecuentemente por diferencias étnicas (pág. 319); región rodeada por grupos políticos más fuertes, bajo tensión constante y que algunas veces está fragmentada como resultado de conflictos entre rivales fuertes (pág. 759)

reino de unjeque territorio gobernado por un líder de una tribu árabe (pág. 438)

chiismo rama del islam que considera al yerno de Mahoma, Alí, y a los imanes, como los sucesores legítimos de Mahoma (pág. 441)

*mover cambiar de lugar, posición o dirección (pág. 120)

agricultura itinerante sistema en el cual los agricultores cultivan un campo durante varios años hasta que agotan sus recursos, luego lo abandonan y abren claros en otro campo (pág. 589); forma de agricultura en la que se abren claros en la vegetación de un área, se cultiva durante unos años y luego se abandona para hacer lo mismo en un área nueva (pág. 768)

Glossary/Glosario

ENGLISH	ESPAÑOL
*significant important (p. 271)	***significativo** importante (pág. 271)
*similar comparable (p. 78)	***similar** comparable (pág. 78)
simoom a hot, dry, suffocating wind that blows from time to time in the Arabian Peninsula (p. 436)	**simún** viento caliente, seco y sofocante que sopla de vez en cuando en la península arábiga (pág. 436)
site the specific location of a place, including its physical setting (p. 29)	**emplazamiento** ubicación específica de un lugar que incluye su configuración física (pág. 29)
situation the geographic position of a place in relation to other places or features of a larger region (p. 29)	**situación** posición geográfica de un lugar en relación con otros lugares o formaciones de una región más grande (pág. 29)
smog haze caused by the interaction of ultraviolet solar radiation with chemical fumes from automobile exhausts and other pollution sources (p. 136)	**esmog** niebla ocasionada por la interacción de la radiación solar ultravioleta con gases químicos que salen de los tubos de escape de los automóviles y otras fuentes de contaminación (pág. 136)
solstice one of two days (about June 21 and December 22) on which the sun's rays strike directly on the Tropic of Cancer or Tropic of Capricorn, marking the beginning of summer or winter (p. 62)	**solsticio** uno de los dos días (cerca del 21 de junio y el 22 de diciembre) en los que los rayos del sol caen directamente sobre el Trópico de Cáncer o el Trópico de Capricornio, lo que marca el comienzo del verano o el invierno (pág. 62)
*source a point of origin (p. 580)	***fuente** punto de origen (pág. 580)
spatial perspective a way of looking at the human and physical patterns on Earth and their relationships to one another (p. 26)	**perspectiva espacial** forma de mirar los patrones físicos y humanos de la Tierra y las relaciones entre ellos (pág. 26)
Special Economic Zones (SEZ) relatively small districts that are fully open to global commerce (p. 691)	**zonas económicas especiales (ZEE)** distritos relativamente pequeños completamente abiertos al comercio global (pág. 691)
*sphere a globe-shaped body (p. 40)	***esfera** cuerpo en forma de globo (pág. 40)
spreading process by which magma wells up between oceanic plates and pushes the plates apart (p. 47)	**expansión** proceso mediante el cual el magma brota entre las placas oceánicas y las separa empujándolas (pág. 47)
*stable not changing or fluctuating (p. 354)	***estable** que no cambia ni fluctúa (pág. 354)
stateless nation an ethnic group without a formal state (p. 397)	**nación sin estado** grupo étnico sin un estado formal (pág. 397)
station an Australian ranch (p. 790)	**estación** rancho australiano (pág. 790)
steppe wide, grassy plains of Eurasia; also, similar semiarid grassy areas elsewhere (p. 457)	**estepa** pradera amplia de Euroasia; también, praderas semiáridas similares en otros lugares (pág. 457)
*strategy a plan or method; a plan of action to achieve an aim (pp. 443, 720)	***estrategia** plan o método; plan de acción para lograr un objetivo (págs. 443, 720)
*stress pressure or strain (p. 496)	***estrés** presión o tensión (pág. 496)
Strine Australian English that has a unique vocabulary made up of Aboriginal words, terms used by early settlers, and slang created by modern Australians (p. 785)	*strine* inglés australiano compuesto por un vocabulario único que se toma de palabras aborígenes, términos de los primeros colonos y jerga de los australianos modernos (pág. 785)
*structure something constructed or arranged in a definite pattern of organization (p. 104)	***estructura** algo que se construye o se organiza con un patrón definido (pág. 104)
stupa a dome-shaped structure that serves as a Buddhist shrine (p. 660)	**estupa** estructura en forma de domo que se usa como templo budista (pág. 660)
subcontinent a large landmass that is part of a continent but still distinct from it, such as India (p. 606)	**subcontinente** gran masa de tierra que forma parte de un continente pero se distingue de él, como la India (pág. 606)

ENGLISH

subduction process by which oceanic plates dive beneath continental plates, often causing mountains to form on land (p. 46)

*subsequent occurring or being carried out at a time after something else (p. 381)

subsistence farming farming that provides the basic needs of a family with little surplus (pp. 495, 587)

Sunbelt a mild climate region in the southern and southwestern portions of United States (p. 128)

sunken mountains high mountain range that has become submerged (p. 777)

Sunni a branch of Islam that regards the first four successors of Muhammad as his rightful successors (p. 441)

supertrawler an oceangoing factory ship with facilities for processing and freezing fish (p. 719)

*survive to manage to stay alive (p. 509)

*sustain to give support to (p. 413); to continue without interruption (p. 717)

*sustainable a method of harvesting or using a resource so that the resource is not depleted or permanently damaged (p. 212)

sustainable development technological and economic growth that does not deplete the human and natural resources of a given area (p. 188)

sustainable development economic growth that meets the needs of present populations without hampering the ability of people in the future to meet their own needs and that benefits poor people and societies (p. 642); technological and economic growth that does not deplete the human and natural resources of a given area (p. 770)

*symbol a sign or image that stands for an idea (p. 682)

syncretism a blending of beliefs and practices from different religions into one faith (p. 182)

ESPAÑOL

subducción proceso que ocurre cuando una placa oceánica se mueve debajo de una placa continental, lo que ocasiona que se formen montañas en la superficie (pág. 46)

*subsecuente que ocurre o se realiza después de otra cosa (pág. 381)

agricultura de subsistencia tipo de agricultura que proporciona lo suficiente para cubrir las necesidades básicas de una familia, con poco excedente (págs. 495, 587)

Sunbelt [**Cinturón del Sol**] región de clima templado en el sur y el suroeste de Estados Unidos (pág. 128)

montañas sumergidas cordillera de montañas altas que se han sumergido (pág. 777)

sunismo rama del islam que considera legítimos a los cuatro primeros sucesores de Mahoma (pág. 441)

súper arrastrero buque factoría de pesca de arrastre con instalaciones para procesar y congelar el pescado (pág. 719)

*sobrevivir seguir viviendo (pág. 509)

*sostener dar apoyo (pág. 413); proseguir sin interrupción (pág. 717)

*sustentable método de agricultura o uso de un recurso de manera que el recurso no se agote ni se dañe permanentemente (pág. 212)

desarrollo sustentable crecimiento económico y tecnológico que no agota los recursos humanos y naturales de un área dada (pág. 188)

desarrollo sostenible desarrollo económico que satisface las necesidades de la población actual sin obstaculizar la capacidad de las generaciones futuras de satisfacer sus propias necesidades y que beneficia a las sociedades y gentes pobres(pág. 642); crecimiento económico y tecnológico que no agota los recursos naturales y humanos de un área dada (pág. 770)

*símbolo signo o imagen que representa una idea (pág. 682)

sincretismo mezcla de creencias y prácticas de diferentes religiones en una sola fe (pág. 182)

T

Taliban from Arabic for "seeker" or "student"; name of a fundamentalist Sunni Muslim group, active in Afghanistan, which controlled the Afghan government from 1996 to 2001 (p. 461)

tar sands sand or sandstone naturally impregnated with petroleum (p. 148)

thematic map a map that emphasizes a single idea or a particular kind of information about an area (p. 20)

theocracy system of government in which the officials are regarded as divinely inspired (p. 88)

talibán del árabe que significa "estudiante" o "investigador"; nombre de un grupo fundamentalista musulmán suní activo en Afganistán, que controló el gobierno afgano de 1996 a 2001 (pág. 461)

arenas de alquitrán arena o arenisca impregnada naturalmente de petróleo (pág. 148)

mapa temático mapa que se enfoca en una sola idea o un tipo particular de información sobre un área (pág. 20)

teocracia sistema de gobierno en donde se considera que los funcionarios están inspirados por una divinidad (pág. 88)

ENGLISH

*theory a plausible general principle offered to explain (p. 41)

timberline elevation above which it is too cold for trees to grow (p. 147)

total fertility rate the average number of children a woman has in her lifetime (p. 636)

*trace to follow or study in detail or step by step (p. 149)

trade surplus earning more money from export sales than spending for imports (p. 716)

traditional economy a system in which tradition and custom control all economic activity; exists in only a few parts of the world today (p. 94)

*transform to change in form or appearance (p. 393)

transition zone an area in which the properties of the land undergo a radical change (p. 484)

*transmit to send from one person, thing, or place to another (p. 24)

*trend a general movement (p. 84)

tributary a smaller river or stream that feeds into a larger river (p. 121)

*trigger to initiate, actuate, or set off as if by pulling a trigger (p. 342)

trust territory a dependent area that the United Nations placed under the temporary control of a foreign country (p. 805)

tsunami a Japanese term used for a huge sea wave caused by an undersea earthquake (p. 609); a huge wave that gets higher and higher as it approaches the coast (p. 704); a dangerous large ocean wave caused by an undersea earthquake, a volcanic eruption, or a sudden displacement of the ocean floor (p. 755)

tungsten an extremely rare heavy-metal element essential in high-tech industry (p. 298)

typhoon a violent tropical storm that forms in the Pacific Ocean, usually in late summer (pp. 684, 757)

ESPAÑOL

*teoría principio general verosímil que se ofrece para explicar algo (pág. 41)

límite del bosque altitud por encima de la cual hace demasiado frío para que los árboles crezcan (pág. 147)

tasa de fertilidad total número promedio de hijos que una mujer tiene a lo largo de la vida (pág. 636)

*rastrear seguir o estudiar en detalle o paso por paso (pág. 149)

superávit comercial situación en la que el valor de las ganancias obtenidas por concepto de las exportaciones es mayores que el de las importaciones (pág. 716)

economía tradicional sistema en el cual la tradición y las costumbres controlan las actividades económicas; en la actualidad, existe solo en algunas partes del mundo (pág. 94)

*transformar cambiar de forma o apariencia (pág. 393)

zona de transición área en la cual las propiedades de la tierra sufren un cambio radical (pág. 484)

*transmitir enviar desde una persona, cosa o lugar a otra (pág. 24)

*tendencia movimiento general (pág. 84)

tributario río o arroyo pequeño que desemboca en uno más grande (pág. 121)

*desencadenar iniciar, impulsar o provocar (pág. 342)

territorio en fideicomiso área dependiente que las Naciones Unidas pone bajo el control temporal de un país extranjero (pág. 805)

tsunami término japonés que describe una ola gigantesca ocasionada por un terremoto submarino (pág. 609); ola gigantesca que se vuelve más grande a medida que se acerca a la costa (pág. 704); ola gigante peligrosa ocasionada por un terremoto submarino, una erupción volcánica o un desplazamiento repentino del lecho marino (pág. 755)

tungsteno elemento metálico pesado extremadamente raro y esencial en la industria de alta tecnología (pág. 298)

tifón tormenta tropical violenta que se forma en el océano Pacífico, usualmente al final del verano (págs. 684, 757)

U

*undergo to go through, experience (p. 261)

Underground Railroad a network of safe houses in the United States that helped thousands of enslaved people escape to freedom (p. 126)

understory a lower layer of the rain forest (p. 558)

uneven development condition in which some places do not benefit as much as others from social and economic advancement (p. 227)

*someterse a experimentar, padecer (pág. 261)

Tren Clandestino red de casas seguras en Estados Unidos que ayudó a escapar a miles de personas esclavizadas (pág. 126)

sotobosque capa inferior de la selva tropical (pág. 558)

desarrollo desigual condición por la que algunos lugares no se benefician de los avances sociales y económicos tanto como otros (pág. 227)

ENGLISH

*unique being the only one; without a like or an equal; distinctive; unlike anything else (pp. 87, 454, 714)

unitary system form of government in which all key powers are given to the national government (p. 87)

urban sprawl spreading of urban developments on undeveloped land near a city (p. 102)

urbanization the relocation of people from rural areas to urban areas (p. 586)

*utilize to make use of (p. 694)

*vary to exhibit or undergo change (p. 728)

vertical climate zone a climate zone that occurs as elevation increases, with its own natural vegetation and crops (p. 178)

*virtually almost; nearly (p. 460)

*voluntary of one's own choice or consent (p. 236)

wadi in the desert, a streambed that is dry except during a heavy rain (p. 369)

water cycle regular movement of Earth's water from ocean to air to ground and back to the ocean (p. 51)

weather condition of the atmosphere in one place during a short period of time (p. 60)

weathering chemical or physical processes that break down rocks into smaller pieces (p. 49)

welfare state a state that assumes primary responsibility for the social welfare of its citizens (p. 259)

*widespread covering a wide area; prevalent (pp. 228, 718)

windward being in or facing the direction from which the wind is blowing (p. 68)

world cities cities generally considered to play an important role in the global economic system (p. 105)

Z

ziggurat a large temple built by the Sumerians (p. 420)

ESPAÑOL

*único singular; sin igual; exclusivo; como ningún otro (págs. 87, 454, 714)

sistema unitario forma de gobierno en la cual todos los poderes fundamentales son dadas al gobierno nacional (pág. 87)

expansión urbana diseminación de los desarrollos urbanos hacia las áreas sin desarrollo cercanas a una ciudad (pág. 102)

urbanización traslado de las personas de las áreas rurales a las áreas urbanas (pág. 586)

*utilizar usar (pág. 694)

*variar mostrar o experimentar cambios (pág. 728)

zona climática vertical zona climática que se presenta a medida que aumenta la altitud, con su propia vegetación y cultivos naturales (pág. 178)

*prácticamente casi; por poco (pág. 460)

*voluntario por elección o consentimiento propio (pág. 236)

uadi en el desierto, un cauce seco excepto durante una temporada lluviosa (pág. 369)

ciclo del agua movimiento regular del agua de la Tierra, desde el océano hasta el aire, luego a la tierra y de nuevo al océano (pág. 51)

tiempo atmosférico condición de la atmósfera en un lugar durante un corto periodo (pág. 60)

meteorización proceso químico o físico que rompe las rocas en pedazos más pequeños (pág. 49)

estado de bienestar estado que asume la responsabilidad primaria del bienestar social de sus ciudadanos (pág. 259)

*extendido que cubre un área extensa; generalizado (págs. 228, 718)

barlovento estar en la dirección desde donde viene el viento (pág. 68)

ciudades globales ciudades que desempeñan un papel importante en el sistema económico mundial (pág. 105)

zigurat templo grande construido por los sumerios (pág. 420)

c = chart
crt = cartoon
d = diagram
g = graph
i = infographic
m = map
p = photo
ptg = painting
q = quote

——————— **A** ———————

abaya, 442
Abdullah of Saudi Arabia, 442
Aborigine people: in Australia, 774, *p774,* 775, 780, 781, 782, 783, *c784, q796,* 843
aborigines in Taiwan, 688
Abraham, 396
Abu Hanifa Mosque, *q423*
Abu Simbel, *c382*
accretion, 47, *d48*
Achebe, Chinua, 494
acid deposition, 286
acid rain, 72; in Canada, *p160;* in China, 697; in Eastern Europe, 327, *g328;* in Japan, 718; in Northern Europe, *m263,* 328; in Northwestern Europe, 286, *g328;* in United States, 135, *m136, p160;* in Western Europe, *g328*
ACS. *See* American Colonization Society.
ADB. *See* Asian Development Bank.
Addis Ababa, Ethiopia, 514
Adriatic Sea, 295, 297, 305, 317, 318
Aegean Sea, 295, 413, 414
Aeolian Islands, 295
aerial photographs, 32
Afar Depression, *d507*
Afghanistan, 454, 455; history and government of, *m452, p453,* 460, *c461,* 463; invasions of, 452, *p453,* 461; natural resources of, 458; U.S. military campaign against, *c126, c461*
Africa: cities in, 86; as continent, 43; food shortages and population growth in, 83–84; satellite view of, *p24;* socioeconomic development in, 97; tropical rain forests in, 69; tropical savannas in, 70. *See also* East Africa; Equatorial Africa; North Africa; Southern Africa; Transition Zone; West Africa.
African Americans: *m128,* 129; enslavement of, 126; settling in Liberia, *c536,* 537; sheltered in Canada, 151
African National Congress, *c584, c585*
African tectonic plate, *m177,* 368
Afrikaners, 583, *c584*
AFTA. *See* ASEAN Free Trade Area.
Age of Exploration, *c300*
Ago Bay, Japan, *p720*
agribusiness, 281
agricultural runoff: in Australia, 792; in Eastern Europe, *g328;* in Eastern Mediterranean, 406; in Northeast, *i426,* 427; in Oceania, 814; in Pakistan, 632, *p644*
agriculture: in Arabian Peninsula, 434, 443, 447, 448; in Australia, 777, 780, 782, 787, 790; in Bangladesh, 627, 629, 632, 636, 639, 642, 643; in Canada, 147, 152, 154; in Central America and Caribbean, *p195,* 199, 205, *p206,* 209, *p210,* 211; in Central Asia, 451, *p453,* 456, *p457,* 458, 460, 462, *p464, c464,* 464–65, 466, 467, 469; in China, 682, 683, *p684,* 685, 686, 688, 690; in East Africa, 504, *p504,* 505, 508, 510, 516, 517, 520–21; in Eastern Europe, *p324,* 325, *g328;* in Eastern Mediterranean, 392, 393, 394, 401, *p404,* 405, *p406;* in Equatorial Africa, 551, 560, 561, *c564,* 566, 568; greenhouse, 448; in India, 609, 610, 613, 616–17; in Indian subcontinent, 653, 658, 660, 665; in Japan, 705, 706; in less developed countries, 97, 209–10; in Mexico, 177, 178–79, 180, 181, 182, 188; in New Zealand, 777, 778, 781, 782, 787; in North Africa, 370, 371, 374, 376, 377, 380, *c382,* 383; in Northeast, 416, 420, 425, *i426,* 427; in Northern Europe, 254, 260, 262, 263, 264, *c265;* in North Korea, 725, 729, 739, *q742;* in Northwestern Europe, 281, 285; in Oceania, 802, 803, 808, 809, 812, 816; in Pakistan, 627, 629, 632, 633, 634, 639, 643, 644; in Russia, 334, 336, *i338,* 339, 346; in South America, *p217,* 219, 221, *m228, p233,* 235; in Southeast Asia, *p758,* 762, 763, 764, 765, 768; in Southern Africa, 587, 589–90; in Southern Europe, *q297,* 298, 303, 304, 305, *c309;* in South Korea, 725, *p728;* sustainable, 520, 816; in Transition Zone, 486, 488, 489, 494, 495, 496, 497, 499, 500; in United States, 126, 128, 130, 131; in West Africa, 527, 529, 530, 531, 533, 541, 546, 547
Ahaggar, 368
Ahwaz, Iran, *q423*
Ainu people, 712
air-conditioning, 436, 446
air masses, tropical and polar, 71
air pollution: in Arabian Peninsula, *p445;* cars contributing to, 189, 211; in Central America and Caribbean, 211; in Central Asia, *p469;* in China, *p697,* 718, 737; in Eastern Europe, 326, 327–28, *p329;* in Eastern Mediterranean, 403; in Equatorial Africa, *p568,* 570; forest fires causing, 770; household (indoor), *p568,* 570, 645; in India, 622–23; in Japan, 718; in Mexico, 183, 189, 190; in Mongolia, 737; in Northeast, 425–26, 427; in Northern Europe, 263; in North Korea, 738; in Northwest Europe, 263, 285, 286, *p287,* 288; in Pakistan, *g645;* relationship of tornado activity and, *q72;* risk to plant life from, 160; in South America, *g235, p236;* in Southeast Asia, 769, 770; in Southern Africa, 591; in South Korea, 737; in United States, 136, 137
air temperature of Earth: greenhouse effect on, 63; increasing, 58, 59, 72
Ajkwa River, 769
Akhmatova, Anna, 345
Akihito, *c713*
Akosombo Dam, 532–33, *p532*
Alaska: climate of, 122, 124; migration of people to, 125
Alaska Current, 147
Alaska Range, 120
al-Assad, Bashar, 398
al-Assad, Hafez, 398
Albania, 314, 326, 330
Alberta, 152
Aleppo, Syria, 391
Alexander the Great, 453, *c460*
Alexander II, 342
Alexander III, 342
Algeria: boundaries of, 90; colonial rule of, *m373, c374;* landforms in, 368; natural resources of, 371
Aliákmon River, 297
Allied Powers, 277, *c301,* 711
alluvial plain, 606
alluvial soil, 369, 393, 415, 416, 533, 606, 631
Almaty, Kazakhstan, 462
alpine zone, 253
Alps, 270, 273, 297; climate change in, 306; water sources of, *p271,* 273, 295
al-Qaddafi, Muammar, *c375,* 384
al-Qaeda, *c126,* 461
Al-Saud, Abdul Aziz, *p441*
Altay Shan range, 455, 680
altiplano, 218, 219
Alvarado, Pedro de, 201
Amazon rain forest, 69, 215, 221, 229, *i230–31,* 232, 234, 353
Amazon River, 220, 556
Amazon River Basin, 220; forestry in, 228; tropical rain forest in. *See* Amazon rain forest.
ambas, 507
Amber Route, 318
American Colonization Society, *c536,* 537
American Revolution, 126

Amnesty International, *q423,* 433

Amsterdam, Netherlands, 279

Amu Dar'ya, 456, 462, 466–67, 468

Anatolian Peninsula, 413

Anatolian Plate, 368

Anatolian Plateau, 413, 417, 418

Andalusia, 302

Andersen, Hans Christian, 259

Andes mountain system, 47, 218, *p219,* 221, 223

Andorra, 294

Angkor Wat, *p762*

Angola, 589, 591

animals: in Arabian Peninsula, *p431,* 436, 438, 447; in Australia, 773, 777, 779, 790, *p791,* 792, *p793;* in Bangladesh, 629, 646; in Canada, *p147,* 148, 158, 162; in Central America and Caribbean, 193, *p195, i198;* in Central Asia, 451, *p457,* 462, 464, 467, 470; in China, *p682,* 698; domestication of, 372, 489; in East Africa, 510, *p522, p523, p524;* in Eastern Europe, *p318,* 327; in Eastern Mediterranean, 393, 403, 404–6; endangered, 523, 592, *g719, p768,* 793; endemic versus introduced species of, 791, 803; in Equatorial Africa, 551, 557–58, 566, 568, *g569, crt571;* extinction of, *p522;* in India, 610, 614, 616, 620, *p623,* 624; in Indian subcontinent, 655, 656, 660, 662, 664, 665; isolation of, 285; in Japan, *p706,* 718, *g719;* in Mexico, 178, 179; migration of, *p22,* 23, 147, 253; native (endemic) versus introduced, 791, 803; in New Zealand, 773, 777, 790–91; in North Africa, 369, 370, 372, 383; in Northeast, 415; in Northern Europe, 253, 263; in North Korea, 728–29, 739; in Northwestern Europe, 274, 284, 285; in Oceania, 802, 803, 815; in Pakistan, 629; poaching of, 523; in Russia, 351, 353, *q356;* in South America, 221, *i230, i231,* 232; in Southeast Asia, 758, 766, 767, *p768;* in Southern Africa, *p578,* 579, 591, *crt593;* in South Korea, 728–29, 739; in Transition Zone, 485, 488, 495, 497; in United States, 122–24; in West Africa, 531, 534, 535

animist culture, 481, 484, 491, 492, 514, 536. *See also* religions.

An Nafūd, 435

Antarctica: as continent, 43; sheet glaciers in, 50; 819–25

Antarctic Circle, daylight in, 62

Anti-Lebanon Mountains, 390, 391, 392, 394

Anyang, South Korea, 733

Apalachicola River, 132, *q133*

apartheid, *c584–85,* 585, *q594*

Apennine Mountains, 295, 297

Apia, Samoa, 806

Appalachian Mountains: in Canada, 144, 145; formation of, 121; in United States, 120–21

apportionment of representatives, 13

aquaculture, 148, 607, 639, 719, 720, 729

aquatic biome, 436, 437

aqueducts, 138

aquifers: fossil water in, 446–47; freshwater in, 54, 380, 381, *p383,* 437, 446–47; *p447,* 498, 654; illegal drilling in, *p447;* seawater drawn into, 447

Arabian Peninsula, 372, 418, 431–50, 511; agriculture in, 434, 443, 447, 448; arts in, *p442, q450;* climate regions and biomes of, 436–37; economic activities of, 433, 435, 442–43; ethnic groups in, 438, *c439;* family and status of women in, 433, 438–39, 441, 442, *q450;* fishing in, 437, 443; history and government of, 438–39, *c440–41;* immigration to, *p432, p433,* 439, 442; industrialization of, 443; landforms in, 434–35; manufacturing in, 443; migrant workers in, *p432, p433,* 439; natural resources of, 431, *p437,* 444–48, *c449;* physical geography of, 434–37; population patterns of, *c439–40;* trading partners of, 438; transportation in, 443, *q446;* urbanization of, 448; vegetation of, 436, 437, 447; water systems of, 431, 434, 435–36, *p437,* 443, 446–48

Arabian Plate, 368, 413, 435

Arabian Sea, 607, 610; cyclones developing in, 609; Indus River drainage in, 632

Arabian Shield, 434–35

Arab invasions: of Central Asia, 453, *c460;* of Equatorial Africa, 553; of North Africa, 372, *m373,* 374, 375

Arab people, 374–75, 409, 418

Arab settlers, culture of, 514

Arab Spring, 293, 371, 377, *i378–79;* 398, 400, *p419*

Arab state, Palestinian, 397

Arab traders, 483

Arafat, Yasir, *p396*

Arai, Shogo, *q93*

Aral Sea: drainage into, 456; drying of, 455, 456, *i468–69;* efforts to save, 470

Aravalli Range, 606

archipelago, 197, 704, 751, 754, 756

architecture: of classical cultures, 291, 299, *p302,* 303

Arctic Circle, daylight in, 62

arctic climate: in Canada, 152; in Indian subcontinent, 655

Arctic Ocean, 251, 350; Canadian river drainage into, 145; location of, 52, *m53;* oil drilling, *i348–49;* Siberian river drainage into, 338–39

arctic tundra, 253

Argentina: cities in, 86; climate regions and biomes of, 221; economic activities of, 227, 229; ethnic groups in, *m226;* Human Development Index score of, *g107;* landforms in, 218; natural resources in, *c222,* 227, 234

arid climates: in Australia, 778; in Eastern Mediterranean, 392, 393; in Northeast, 415; in South America, 221–22; in Southern Africa, 579; in United States, 123. *See also* deserts.

Armenia, 454, 455; ethnic groups in, 411; history and government of, 459, 462–63, 465; natural resources of, 458, 469

artesian wells, 778

arts: in Arabian Peninsula, *p442, q450;* in Australia, 780, 786–87; in Bangladesh, 638; in Canada, 153–54; in Central America and Caribbean, *p204;* in Central Asia, 451, *p463,* 464; in China, *q689;* in East Africa, 516–17; in Eastern Mediterranean, 400; in Equatorial Africa, 565; in India, 516, 603, 615; in Indian subcontinent, 660, *p661;* in Japan, 714; in Mexico,

184; in New Zealand, 786–87; in North Africa, *p376;* in Northeast, 420; in Northern Europe, *p259;* in North Korea, 734, *p735;* in Northwestern Europe, 267, *p279;* in Oceania, *p808,* 809; in Pakistan, 638; in South America, *p215, p227, q238;* in Southeast Asia, 763; in Southern Europe, 293, 299, *p302,* 303, *q310;* in South Korea, 734–35; in Southwest Asia, 464; in Transition Zone, *p494;* in United States, 129–30; in West Africa, 540, 541. *See also* culture.

Aryan people, 613

ASEAN. *See* Association of Southeast Asian Nations.

ASEAN–China Free Trade Area, 99

ASEAN Free Trade Area, 765

Ashkhabad, Turkmenistan, 455

Ashqelon, desalination in, 39

Asia: cities in, 86; as continent, 43; population density of, 85; socioeconomic development in, 97; tropical rain forests in, 69; tropical savannas in, 70

Asia Minor, 413

Asian Development Bank, 765

Asian medicines, *q356,* 523, 588–89, 698

Association of Southeast Asian Nations, 765, 770

asteroids, 41

astronomers, 41

Aswān High Dam, 369, *c382,* 383

Atacama Desert, 215, *c224*

Atafu atoll, 777

Athabasca Tar Sands, 148, 154

Athens, Greece, 302

Atlantic Coastal Plain, 121

Atlantic Forest, 232

Atlantic Ocean: Canadian lowland drainage into, 145; Canadian river drainage into, *m146;* Equatorial Africa drainage into, 551; Great Lakes linked with, 122, *m146;* location of, 52, *m53;* managing fishing in, 532, 533; Mid-Atlantic Ridge in, 47; Northwestern European drainage into, 273; population centers of North Africa around, 375; slave ship passage across, 560, 561; South American drainage into, 220; Transition Zone river drainage into, 486

Atlas Mountains, 365, 368
atmosphere, *d42, d62,* 63, 64
atolls, 653–54, 777, 797, 801, 812, 815
atomic bombs, 711, *q722*
Auckland, New Zealand, 778, 784
Aung San Suu Kyi, 761, 763
Australia, 773–96; agriculture in, 777, 780, 782, 787, 790; arts in, 780, 786–87; climate regions and biomes of, 70, 71, *q371,* 779–80; communications in, 786; as continent, 43; economic activities of, 777, 787; effects of El Niño in, 67; ethnic groups in, *p774,* 775, 780–83, *c784, q796;* family and status of women in, 786; fishing in, 777, 787; history and government of, 781–82; immigration to, 773, 783, 784, *c785,* 816; independence of, 782; industrialization of, 787; landforms in, 776–77; manufacturing in, 787; mountain ranges of, 776; natural resources of, 787, 790–94; physical geography of, 776–80, 784; population patterns of, 85, 102, 783–84, *c784–85, g795;* trading partners of, 781, 787, 799, 809; transportation in, 777; tropical savannas in, 70; urbanization of, 85, 783; vegetation of, 773, 790, 791, 792, 793; water systems of, *p777,* 778, 780, 791–92
Austria, 270; natural resources of, 274
Austria-Hungary, 277
Austro-Hungarian Empire, 313, *c320*
Austronesian people, 805
authority, governmental, 88, 89
autocracy, 88, 89–90
avalanches, 273
average daily temperature, 69, 70
average global temperature, 58, 59, 72
Awami League, *c637*
Awash River Valley, 511
axis, Earth's, 50, *d61,* 62
Axis forces, *c301,* 711
Axum empire, 489
Aymara, 223

Azerbaijan, 428, 454; ethnic groups in, 411; natural resources of, 458
Azraq wetlands, 406
Aztec Empire, 180–81, 182, 184

——— **B** ———

Bab el Mandeb, 443
Bachelet, Michelle, *p225*
Baghdad, Iraq, 414, 419
Baghmati River, 654
Bahamas, 202
Bahrain, economic activities of, 443
Baja Peninsula, 178
Baku, Azerbaijan, 462
Balboa, Vasco Nuñez de, 200
Balearic Islands, 295
Balfour, Arthur James, *q408*
Balfour Declaration, *q408*
balkanization, 320, 321
Balkan Mountains, 317, 318
Balkan Peninsula, 270, 294, 295, 297, 319, 320
Baltic gold (amber), 318
Baltic Sea, 252, 254, 261, *q262,* 285, 317, 329, 341
Bamako, Mali, 492
Bamenda City, Cameroon, 567
Bandaranaike, Solomon, *p659*
Bandaranaike, Sirimavo, *c659*
Bangkok, Thailand, 756, 762, 765, 769
Bangladesh, 607, 627–48; agriculture in, 627, 629, 632, 636, 639, 642, 643; arts in, 638; climate regions and biomes of, 632; economic activities of, 639; ethnic groups in, 635, 637; family and status of women in, 636, 638; fishing in, 632, 639, 643, *p644;* history and government of, 627, 634–35; independence of, 634, 635, *c637;* industrialization of, *p638,* 639, 646; landforms in, 630–31, *q647;* manufacturing in, 639, 644; natural resources of, 642–46; physical geography of, 630–33; population patterns of, 635–36; transportation in, 632; urbanization of, 636; vegetation of, 633; water systems of, 627, 629, 632, *q633,* 642, *p643*

banking, international: in Central America and Caribbean, 205; in East Africa, 521; in Northwestern Europe, 280, *q283*
Bantu people, 514, 559, 562, 581
baors, 632
barrier reefs, 653–54. *See also* coral reefs.
basin, 555
Basque, 302
Bastidas, Rodrigo de, 200
Bay of Bengal, 607; cyclones developing in, 609; delta system draining into, 631; river transportation from, 632; seawater floods from, 643
Bay of Biscay, 294
bedouin, 375, 438
beels, 632
Beethoven, Ludwig van, 279
Beijing, China, 688, 697
Beirut, Lebanon, 399
Belarus, 327; climate regions and biomes of, 229; ethnic groups in, *c344;* independence of, 341, 343, 347; industrialization of, 346; landforms in, 337; trading partners of, 346, 347; water systems of, 337, 338
Belgium, 270; history and government of, 276; natural resources of, 285
Belize, 199, 201
Benelux countries, 285
Benin, 483, 530
Benin, kingdom of, 536
Benue River, 533
Berbers, *p365,* 374, 483, 493
Berezina River, 338
Berlin Conference, 490, 537
Ber, Sunni Ali, 536
Bhagirathi River, 622
Bhutan, 649–68; history and government of, 657; mountain ranges of, 649. *See also* Indian subcontinent.
Bhutto, Benazir, 635, 638
Biafran War, 483
Białowieza Forest, *p327*
Bikini Atoll, 815
bin Laden, Osama, *c127,* 461
biodiversity: in Australia, 793, 794; in Central America and Caribbean, *i198,* 199, 210, 211–12; in Eastern Europe, 326, 327, 330; in Eastern Mediterranean, 405, 406; in Equatorial Africa, 568;

in Mexico, 188; in North Africa, 383; in Northern Europe, *q262;* in South America, 232, 235; in Southeast Asia, 758; in West Africa, 531
biofuels. *See* renewable and alternative energy.
biology, effects on soil of, 50
biomass fuel, peat as, 262
biome, 69, *m70;* types of, 436. *See also* climate regions.
biosphere, *d42*
biotechnology. *See* high-tech industries.
birthrate, human, 82, 83; in Australia, 784; in Canada, 149; in Central America and Caribbean, 204; in Central Asia, 461; in Eastern Europe, 323; in India, *p604;* in New Zealand, 784; in Northern Europe, 257, 259, *q260;* in Northwestern Europe, 278; in Russia, *p334, i335,* 344, 345; in Southern Europe, 301
Black Forest, 273
black market, 346
Black Mountain range, 653
Black Sea, 342, 413, 414, 427; area of, 414; Eastern European drainage into, 316, 317, 329; Northwestern European drainage into, 273; Southern European drainage into, 295; Ukraine drainage into, 337
Black Sea Lowland, 338
blood diamonds. *See* conflict diamonds.
Blue Nile River, 487; dam controversy for, 518–19, *p518*
Boers, 582, 583, 585
Bogotá, Colombia, 225
Bolivia, 218, 219; climate regions and biomes of, 221; economic activities of, 229; ethnic groups in, *m226*
"Bollywood," 615
boomerang, 781
"boomers" as edge cities, 105
borders, 90
Border 2020 Program, 190
boreal forest (taiga), 147, 158, *c159,* 262, 339, *q353*
Borneo, 754; physical geography of, 755. *See also* Southeast Asia.
Bosnia-Herzegovina, 313, 314, 330; agriculture in, 324–25
Bosporus, 414, 427

Boston, Massachusetts, 129
Boswash megalopolis, 129
Botswana: history and government of, 584; natural resources of, 580
boundaries: defining, 33–34; types of, *p89. See also individual types under their specific names.*
Boyoma Falls, 556, 564
Brahmaputra (Jamuna) River, 607, 629, 631, 632
brain drain, *p217,* 225, *i335*
Brazil: agriculture in, 228; climate regions and biomes of, 71, 221; economic activities of, 99, 229; landforms in, 219; natural resources in, *m100, c222,* 234
Brazilian Highlands, *g219*
Brazzaville, Republic of the Congo, 561
break-of-bulk, 258
British East India Company, 611, *c614*
British Empire, 126, 150, 611–12
British Isles: climate regions and biomes of, 71; landforms of, 270–71
British raj, 611
Brundtland, Gro Harlem, 279
Brunei, landforms in, 754. *See also* Southeast Asia.
Brussels, Belgium, 279
Buddhism, 614, 688, 689, 713, 734, 828. *See also* Tibetan (Lamaist) Buddhism; religions.
Buenos Aires, Argentina, 86, 225
Buffalo, New York, 129
buffer state, *m650,* 651
buffer zone, 759, 760
Bug River, 338
Bukhara, Uzbekistan, 464
Bulgaria, 277, 314; agriculture in, 324–25; history and government of, *c321,* 325; natural resources of, 326, 330
burka, 638
Burkina Faso: agriculture in, 499; in Transition Zone, 483; water systems of, 547
Burmans, 762
Burundi, 561; agriculture in, 510; water systems of, 487
bush, 776, 777, 780
Bush, George W., *p126,* 127
Bwindi Impenetrable Forest, 524

Byron, Lord, 279
Byzantine Empire, 320

———— C ————

Cahora Bassa Dam, 578
Cairo, Egypt, 375
caldera, 777
California: California State Water Project in, 138; climate regions and biomes of, 124, *q371;* megalopolis in, 129; software industry in, 131
calligraphy, 442
camanchaca **fog,** 222
Cambodia, landforms in, 754. *See also* Southeast Asia.
Cameroon: in Equatorial Africa, 554; in Transition Zone, 483
Cameroon Highlands, 531, 757
campesinos, 205
Canada, 141–64; agriculture in, 147, 152, 154; area of, 141; arts in, 153–54; climate regions and biomes of, 141, 146–48; communications in, 154, 155; economic activities of, 154–55; energy resources of, 141; ethnic groups in, 152, 153; exploration by colonialists of, 149; exports from, *c163;* growth of, 149–152, 155; history and government of, 142, 149–151, *c150–51;* human geography of, 149–57; immigration to, 149, *c150,* 153; independence of, *c151;* indigenous people of, *p142;* industrialization of, 148, 151, *c154,* 155; landforms in, 144–45; languages in, 153; literacy rate in, 153; literature in, 153; manufacturing in, 155; modern challenges for, 151; as more developed country, 97; mountain ranges in, 144, 145; natural resources in, *m100,* 146, 154–55, 158–60; physical geography of, 144–48; population patterns in, 85, 152; religions in, 153; sports in, *p141;* trading partners of, 151, 186; transportation in, *m146, c150,* 155; vegetation of, 146, 147; water systems of, 145, *m146,* 160, 161, *q164*
Canadian Charter of Rights and Freedom, 153

Canadian Rockies. *See* Rocky Mountains.
Canadian Shield, 145
Canadian–U.S. border, 144, 146, 151; population concentration near, 152
Canal Zone. *See* Panama Canal Zone.
Canary Current, *m253*
canopy, 558
Canterbury Plains, 777
cap-and-trade pollution credits, 287–88
Cape Colony, 583
Cape Comorin, 607
Cape Horn, 198
Cape of Good Hope, 511, 579, 582
Cape Town, South Africa, 586
Cape Verde, 530
Cape York Peninsula, 776
capitalism, 95
carbon-14 dating, 805
Caribbean Lowlands, 196
Caribbean Plate, *m177,* 197
Caribbean region: climate regions and biomes of, 199; Human Development Index score of, *g107;* hurricanes in, 199; tropical rain forests in, 69. *See also* Central America and Caribbean.
Caribbean Sea, 198
Carnival, *p215,* 226
Carpathian Mountains, 317, 318, 319
Carrier, Willis, 436
carrying capacity, 522, 546
Carthage, Tunisia, 377
cartographers, 14
Casablanca, Morocco, 375
Cascade Range, 120, *d121;* in Canada, 144
cash crops: in Central America and Caribbean, 205; in East Africa, 512, 517; in Equatorial Africa, 560, 564; in India, 620; in Mexico, 181, 182; in Pakistan, 639; in Southeast Asia, 764; in United States, 126; in West Africa, 541
Caspian Depression, 455
Caspian Sea, 353, 428, 455; as inland, *p456;* resorts on, *p456;* runoff, *i426,* 468, 469; Russia drainage, 337, *i338;* transportation in, 414, *i426*
castes (*jati*), 603, 611, 612, 614, 615, 661

Castilla del Oro, 200
Castro, Fidel, 201, *p203*
Castro, Raul, 88, 201
Catalan people, 411
cataracts, 507
catchments, 778
Catherine the Great, 342
Caucasus Mountains, 319, 336, 454–55
Caucasus region, 454, 455, 457–58
caudillo, 181, 224
Celsius temperature, 60
Central Africa. *See* Equatorial Africa.
Central Africa Forest Commission, 570
Central African Republic: in Equatorial Africa, 554; history and government of, 561, *q572*
Central America and Caribbean, 193–214; agriculture in, *p195,* 199, 205, *p206,* 209, *p210,* 211; arts in, *p204;* climate regions and biomes of, 199; communications in, 205; economic activities of, 198, 201, 206, 207; ethnic groups in, 201, 202, *c213;* family and status of women in, 203, 206; fishing in, 198, 199; history and government of, *m194, p195,* 200–201; immigration to, 201–2; independence of, 200–201, *c202–03;* industrialization of, 201, 205, 208, 209, 211, 212; landforms in, 193, 196, *m197;* mountain ranges of, 197, 218; natural resources of, 196, 199, 201, 208–212, *q214;* physical geography of, 176, 201; population patterns of, 202; trading partners of, 195, 198, 201, 209; transportation in, 198; tropical savannas in, 70; urbanization of, 208, 210, *p211,* 212; vegetation of, 194, 195, *i198,* 199, 209, 210; water systems of, 197–98, 207, 210, *p211,* 227. *See also individual countries and regions.*
Central Asia, 451–72; agriculture in, 451, *p453,* 456, *p457,* 458, 460, 462, *p464,* 465, 466, 467, 469; arts in, 451, *p463,* 464; climate regions and biomes of, *p457,* 458, 461, 462; communications in, 463; economic activities of, 453, 458, *c464, p465,* 467; ethnic

Index

groups in, *m452, 453,* 460, 461, *m462;* family and status of women in, 463; fishing in, 456, 470; history and government of, *m452, p453,* 459–61, 462–63, *q472;* independence of, 460–61, 462–63; industrialization of, 456, 460, 469; landforms in, 451, 454–55, *p455,* 458, 461; manufacturing in, 455, 465; mountain ranges of, 451, 454, *p455,* 457, 462, 464, 466; natural resources of, 400, 453, 458, 464, 466–70; physical geography of, 454–58; population patterns of, 461–63; trading partners of, 452, 459; transportation in, 465, 467, *i468;* urbanization of, 451, 462, 463; vegetation of, 456, 457; water systems of, 454, *p456,* 466–67, *i468,* 469, 470

Central Highlands (Central America and Caribbean), 196, 197

Central Highlands (Sri Lanka), 653, 654, 656, 659

Central Lowlands (Australia), 776, 778

Central Lowlands (U.S.), 120

central place theory, 104, *d105*

Central Plateau (Mexico), 177, 178–79

Central Powers, 277

Central Uplands (Northwestern Europe), 270

cereals, 457

Ceres, 40

CFCs. *See* chlorofluorocarbons.

Chaco War, *c224*

Chad, 369; ethnic groups in, 493; natural resources of, 495; in Transition Zone, 483

Chaeryŏng River, 727

Chaillu Mountains, 555

Chang Jiang (Yangtze River), 682, 688, 695, 696, 698

Chao Phraya River, 756

chaparral, 124, 179, 298

Chari River, 486

Chattahoochee River, 132

Chávez, Hugo, *p225*

Chechnya, 343

Cheju, 727, 728

Chenab, 632

Chennai (Madras), India, 607

Chernobyl, Russia: nuclear disaster at, *m352, q353*

chernozem, 336, 337, 339

Chhukha hydroelectric project, 656

Chiang Kai-shek, 686

Chicago, Illinois, 104, 129

child labor, *q742*

Child Marriage Restraint Act, *c614*

Chile: climate regions and biomes of, 71, 221; economic activities in, 227–28, 229; ethnic groups in, *m226;* family and status of women in, 227; landforms in, 218; natural resources in, *c222*

China, 677–700; agriculture in, 682, 683, *p684,* 685, 686, 688, 690; arts in, *q689;* climate regions and biomes of, 71, 682–84; culture hearth of, *m80;* economic activities of, 91, 96, *c97,* 99, 466, 517, 677, *p678,* 685, *c687,* 690–91, 692, 697, *q700,* 712; ethnic groups in, 688; family and status of women in, 690, *p690;* fishing in, 684; history and government of, 88, 677, 685, *c687,* 688, 711, 730; industrialization of, *p678,* 697; invasion of India, 651; Japanese invasion of, 711; Kashmir conflict for, 618; landforms in, 680–81; manufacturing in, 691; mountain ranges of, 677, 680, 681, 682, 683; natural resources of, *m101,* 684, 694, *p698;* physical geography of, 680–84, *p684;* population patterns of, 678, 688, *p692–93, q693;* rice paddies in, *p30,* 684; trading partners of, 91, 99, *c163,* 281, 517, 686, 687, *g691,* 730; transportation in, 651, 682, 685, 691, 696; urbanization of, 688, 690, *p697;* vegetation of, *p682,* 698; water systems of, 681–82, 685, 688, 691, 695–96, 697, 698. *See also* Hong Kong; Taiwan; Tibet.

Chinese explorers, 511

Chinese medicines. *See* Asian medicines.

chinook, 124, 147

Chittagong Hills, 631, 632, 633, 636

chlorofluorocarbons, 720

choke points, 366, *p367,* 421, 443, 753

Ch'ŏngch'ŏn River, 727

Choson dynasty, *c733*

Choson kingdom, *p732*

Chota Nagpur Plateau, 606

Christchurch, New Zealand, 784

Christianity, 129, 226, 256, 275–76, 302, 395–96, 830. *See also* religions.

Christopolis, Liberia, 537

Churchill, Winston, *q722*

Churia range, 653

CITES. *See* Convention on International Trade in Endangered Species.

cities: in Arabian Peninsula, 431, 440; Atlantic coast, 129; in Australia, 783, 784; in Bangladesh, 636; in Canada, 145, 146, 152, 153; capacity of, strain on, 106, 207; as centers of culture and creativity, 103; in Central America and Caribbean, 202, 205, 207, 210, *p211,* 212; in Central Asia, 462; in China, 688, 691, 696, *p697;* core, 217; in East Africa, *p514;* in Eastern Europe, 317, 322–24; in Eastern Mediterranean, 387, 391, 395, 401, 403; economic base of, 102; edge (suburban), 105; in Equatorial Africa, 551, 561, 565, 567, 568; functions of, 102–3; in India, 603, 607, 613, 614–15; in Indian subcontinent, 659, 660; in Japan, 712–13; master-planned (greenfield), 105; megalopolis, 129, 613; in Mexico, 182; nature of, 102–4; in New Zealand, 778, 784; in North Africa, 370; in Northeast, 419, *q423;* in Northern Europe, 247, 252; in North Korea, 733; in Northwestern Europe, 272, *q273,* 278, 279; in Oceania, 806; Pacific coast, 129; in Pakistan, 636, *g645;* population density of, 85, 86; problems of modern, 106; rise of, 80–81, 86, 102, 104, 106; in Russia, 341, 342, 343–44; "smart growth" in, 106; in South America, *p217,* 225, 227, 228, 234; in Southeast Asia, 762, 764; in Southern Africa, 581, 585, 586; in Southern Europe, 302; in South Korea, 733, 737; structure of, 104; in Transition Zone, 492; trends in, new, 105; in United States,127, 129, 132, 137, 138, 436, 806; in West Africa, 535, 537, 544; world, 105. *See also individual cities under their specific names.*

civil disobedience, 612

civilizations, earliest: in Australia, 781; in China, 677, 685–86; in India, 611, 613; in Northeast 413, 416, *m417,* 419; in Oceania, 805–06; in Pakistan, 634

civil war: in Angola, 589; in Bangladesh, *p637;* in China, 686; in Liberia, *p537,* 538; in Mozambique, 584, 589; in Sudan, 491, 553; in United States, 126

Civitavecchia, Italy, 297

clans, 494, 710, 781, 782

Clean Water Act, 138, 161

climate, 57–74; Earth-sun relationships affecting, 57, 60–62; effects of bodies of water on, 68; effects on human activity of, 57; factors affecting, 64–68; *d71,* 72; soil building effects of, 50; weather versus, 60. *See also climate types under their specific names.*

climate change, 58–59, 72, *q74,* 814; Canadian initiatives addressing, 161–62; Chinese initiatives addressing, *q700;* deforestation accelerating, 568; drought and, 188, 305–06; European Union standards for, 328; heat waves indicating, *q332;* natural resource extraction and, 160, 210, 402, 445; Northwestern Europe initiatives addressing, *p287;* plans to cope with, 816; rainfall reduction due to, 446; temperature and storms affected by, 645–46, 814; urbanization prompted by, 539; vulnerability of low islands to, 813

climate patterns, world, 69–72

climate regions, 69–72; in Arabian Peninsula, 436–37; in Australia, 70, 71, *q371,* 779–80; in Bangladesh, 632; in Canada, 141, 146–48; in Central America and Caribbean, 199; in Central Asia, *p457,* 458, 461, 462; in China, 71, 682–84; in East Africa, *p509,* 510; in Eastern Europe, 317–18; in Eastern Mediterranean, 392, *m393;* in Equatorial Africa, 557–58; in India, 71, *p608,* 608–10; in Indian subcontinent, 655–56; in Japan,

71, 706; in Mexico, *c178, 179;* in New Zealand, 779–80; in North Africa, 370–71; in Northeast, 415; in Northern Europe, *m253, 254,* 263–64; in North Korea, 728–29; in Northwestern Europe, 273–74; in Oceania, 803; in Pakistan, 632; in Russia, 339–40, 347; in South America, 69, 70, 218–19, *d220,* 222; in Southeast Asia, 757–58; in Southern Africa, *q371,* 579–80; in Southern Europe, 297–98, 305; in South Korea, 728–29; in Transition Zone, 488, 530; in United States, 71, 122, 125, *q371;* in West Africa, 534. *See also specific climate regions and zones.*

climate shift, 489

climatology, 66, 199

Clinton, Hillary Rodham, *q86, q386*

coal burning, 327

coal mining: in Australia, 787; in Bangladesh, 633; in China, 678, 684, 697; in Eastern Europe, 318; in India, 610; in Indonesia, 753; in Japan, 707; in New Zealand, 787; in North Africa, 377; in Northeast, 415, 428; in Northwestern Europe, *p274;* in Pakistan, 633; in Russia, 340; in Siberia, 340; in Southern Europe, 298; in Transition Zone, 495; in United States, 130

coastal lowlands in Canada, 145

Coast Mountains (Canada), 144

Coast Range (U.S.), 120

Cocos Plate, *m177*

Cold War, 277, 319, 320, *c321,* 323, 343, 469, 612, 731

Colombia: climate regions and biomes of, 221; economic activities of, 229; family and status of women in, 227; natural resources in, *c222,* 234

Colombo, Sri Lanka, 659

colonial powers: in Australia, 773, 775, 778, 781–82, *m783, c785,* 786; in Bangladesh, 635, *c636;* in Canada, 149; in Central America and Caribbean, *m194, p195,* 200–01, *c202;* in Central Asia, 460; in China, 686, *c687,* 695; in East Africa, 503, 514–15, 520; in Eastern Mediterranean, 396; in Equatorial Africa, 553, 556, 559–61, 562, 563, 564;

in India, 611, *q612;* in Indian subcontinent, 651, *c658;* in Latin America, 181; Native Americans' clashes with, 125; in New Zealand, 773, 775, 778, 782, 783, *c784–85;* in North Africa, *m373,* 374; in North America, 149–50, 300; in North Korea, 730–31, *c732;* in Oceania, 804–05, *c806–07,* 808; in Pakistan, 635, *c636;* in South America, 215, 223, 226, 300; in South Asia, 300; in Southeast Asia, 759–61, *p761,* 763; of Southern Africa, 580, 582, *ptg583,* 586; of Southern Europe, 299–300; in South Korea, 730–31; in Transition Zone, 481, 483, 489, 490–91, 493; in United States, 125–26; in West Africa, 536–37, 539

Colorado Plateau, 120

Colorado River, *p49*

Colosseum (Rome), 291, *p302*

Columbian Exchange, *m194, p195*

Columbia Plateau, 120, *d121*

Columbus, Christopher, 200

comets, 41

command economy, 95; in Central Asia, 465; in East Africa, 517, 521; incentives lacking in, 96; in North Korea, 735; in Russia, 345–46; in Soviet Union, 465

Commonwealth of Australia, 782. *See also* Australia.

Comoros Islands, *p513,* 576

communes, 690

communications: in Australia, 786; in Canada, 154, 155; in Central America and Caribbean, 205; in Central Asia, 463; in China, 685; in Eastern Europe, 325; in Eastern Mediterranean, 400, 406; in Equatorial Africa, *p565;* in India, 613; in Indian subcontinent, 663; in Japan, 714, 715; in Mexico, 186; in North Africa, 377; in Northeast, 421; in Northern Europe, 255; in Northwestern Europe, 281; in Oceania, 809; in Russia, *c355;* in Southern Africa, 586; in Southern Europe, 303; in United States, 131. *See also* phone communications.

communism, 95, 96; in Central Asia, 460, *c461;* in China, 88, 677, 686, 687, 690; in Eastern Europe,

320, *c321,* 323, 324, 327, 343; founding of, 342; in Laos, *c761;* in North Korea, 725, 731, *c733, 734, p735;* in Northwestern Europe, 276–77; in Russia, 341, *c342–43,* 345, 687; in Vietnam, *c760, c761*

Communist Party, 88, 95, *q96,* 687, *c760*

Communist revolution. *See* Russian Revolution.

community-based demining, 470

community, global, 83, 84

compadre **relationship,** 227

complementarity, 302, 724

compressed natural gas vehicles, 645, *p646*

computer technology and communications, 81; in Canada, 155; in Eastern Europe, 325. *See also* high-tech industries.

concentric zone model of urban land, *c103,* 104

condensation: caused by cooling air, 51, 66–67; in water cycle, *d52*

confederation, 88

conflict diamonds, 534

Confucianism, 685, 734, 832

Confucius (Kongfuzi), 685

Congo Free State, 560

Congo River, 508, 551; natural highways of, 556; trading posts along, 560; villages along, 561

Congo River Basin, *m555,* 559, 568

congressional districts, *p13*

coniferous trees, 71; in Canada, 147; in China, 682; in Eastern Europe, 317, 318; in India, 610; in Indian subcontinent, 655; in Japan, 706; in Northern Europe, 254; in North Korea, 728; in Northwestern Europe, 273–74; in Southern Europe, 306; in United States, 124

connectivity, 102, 103

conquistadors, 181, 200, 224

constitution: of Bangladesh, *c637;* of Chechnya, 343; of India, *c615, c626;* of Liberia, *c536,* 537; of Mexico, 181, *c182, c184;* of Mongolia, 687; of Nepal, 661; of Pakistan, *c637;* of Saudi Arabia, *c441;* of South Africa, 591; of Thailand, 761; of United States, 88; of Vietnam, 753; of Yugoslavia, 313

constitutional monarchies, 256

continental climate, 317, 337, 339; of North Korea, 728; of South Korea, 728

Continental Divide, 121, *m122*

continental drift, 45, *d46,* 251, 630; continental rebound, 250

continentality, 340

continental plates, 46, 47, *d48, d55*

continental shelf, *d43,* 402

continental society, 256

continents, 43. *See also individual continents.*

Convention on International Trade in Endangered Species, 470, 569

Cook, James, 782, *c785*

cooperative farms, 729

Copán, Honduras, 204

Copenhagen, Denmark, 258

coral reefs: in Arabian Peninsula, *p437;* in Australia, 776, 778, 780, *c792;* in Central American and Caribbean, 198, 211; in Indian subcontinent, 649, 653, 662, 664, 665; in New Zealand, 777, 778; in Oceania, 813, 814, 815, 816; ocean warming as destroying, 814; in Southeast Asia, 769

cordilleras, 218, *g219*

Córdoba, Francisco Hernández de, 200

core, of Earth, 44, *d45*

Coriolis effect, 66

Coromandel Coast, 607

Corsica, 295

Côte d'Ivoire, 483, 530

Cortés, Hernán, 181

Costa Rica, 199, 201, 208, 209, *q214,* 232

cottage industries, 205, 617

countries: governments of, 87; less developed, 76, *c97;* more developed, 76, *c97,* 106; newly industrialized, 76, *c97,* 98. *See also* government.

coup, military, 731, 761

Cousteau, Jr., Philippe, *p11*

cover crops, 235

Crete, 295

Crimea, 338

Croatia, 314; agriculture in, 324–25; history and government of, 313

crop diversification, 812

crude oil. *See* petroleum.
cruise ships, environmental effects from, 307, *p308*
Crusades, 275, *c276*
crust, of Earth, 44, *d45,* 47, 49
Cuba, 203; arts in, 204; independence of, 201, *c203*
Cuban missile crisis, *c343*
cultural boundary, *p89,* 90
cultural development and change: in contemporary world, *p81, p92–p93;* in history, 80–81
cultural divergence, 733
cultural geography. *See* human (cultural) geography.
cultural heritage, protecting, 92–*p93*
Cultural Revolution, *c686,* 688, 689
culture, *p81;* in Arabian Peninsula, *c440,* 440–42; in Australia, 775, 785–87; in Bangladesh, 637–38; in Canada, 151, 152, 153–54; in Central America and Caribbean, *p193, p195,* 200; in Central Asia, *p463,* 464; in China, 681, 682, 685, 686, 688–90, 695, 730; cities as centers of, 103, 105; in East Africa, 512; in Eastern Europe, *p323,* 324; elements of, 78–79; of Equatorial Africa, 560, 562, *q563;* globalization affecting modern, *p92–p93;* in India, 613–15; in Indian subcontinent, 649, 660, *p661;* in Japan, 713–14, 715; in Mexico, 173, 180, 181, 182, 183–84; of migrants, 81; in New Zealand, 775, 785–87; in North Africa, 375–76; in Northeast, 419, *p420;* in Northern Europe, 258, *p259;* in North Korea, 723, 724, 733–35; in Northwestern Europe, 267, *p279;* in Oceania, 801, 807–8; in Pakistan, 637–38; popular, 151; in Russia, 345; shared, 78; in South America, *p215,* 223–24, *p227, q238;* in Southern Africa, 586–87; in Southern Europe, 299–*ptg300, p302,* 303, *q310;* in South Korea, 723, *p724,* 733–35; in Transition Zone, 482, 489, 492, 493, *p494;* in United States, 117, *p118,* 126, 129–30, 151; in West Africa, 527, 539–40. *See also* human geography.
culture hearths, *m80,* 416, 417
culture region, 78, 79, 80
cuneiform, 416, 417

Cuzco, 227
cyclones, 609, 629, *p632,* 635, 646, 654, 684, 757
Cyprus, 295
Czechoslovakia, *c321*
Czech Republic, 314, 329, 330; history and government of, *c321,* 325; natural resources of, 327; population patterns of, 85

──────── **D** ────────

da Gama, Vasco, 512, *ptg513*
Dalai Lama, *p651*
Dalí, Salvador, 303
Damascus, Syria, 395
dam building: in Canada, 160; in Central Asia, 456, 466–67, 470; in China, 682, *p687,* 694, 696, 769; Danube region, 317; in East Africa, 518, *p519,* 521; in Eastern Mediterranean, 391; in India, 622; in Indian subcontinent, 656; in New Zealand, 780; in North Africa, *c382,* 383; in Northeast, 424; in Pakistan, 632; in Southeast Asia, 768; in Southern Africa, 578; in Transition Zone, 486; in West Africa, 532
Danube River: agricultural runoff into, 328; development along, 316, 320, 322, 337; drainage into Black Sea of, *m316;* economic activities along, 273, 315, 318, 325; pollution standards, 328–30
Daoism, 685-86. *See also* religions.
Dardanelles, 414
Darfur, 491
Darling River, 778
Dávila, Pedro Arias (Pedrarias), 200
da Vinci, Leonardo, 303
Dead Sea: Jordan River drainage, 392; as lowest dry land on Earth, 43, 391, 392; salt water in, 54
death rate, 82, 83; in Arabian Peninsula, *q450;* in Pakistan, 645; in Russia, 334, *i335,* 344; in Transition Zone, 493; in West Africa, 540
Debswana, 580
Deccan Plateau, 606, 609
deciduous trees, 71; in Canada, 147; in Central Asia, 458; in China, 682; in Eastern Europe, 317, 318; in India, 610; in Indian subcontinent, 655; in Japan, 706;

in Northwestern Europe, 273–74; in Russia, 339; in West Africa, 530
deforestation: in Australia, 791; in Bangladesh, 643, 645; in Central America and Caribbean, *p209,* 210, 211; in Central Asia, 467, 469; in China, 697–98; in East Africa, 522, 524; in Eastern Mediterranean, 404; in Equatorial Africa, 568; in India, 620, 622; in Indian subcontinent, 664, *p665;* in Mexico, 188; in New Zealand, 780; in North Africa, 371; in Northeast, 427; in Northern Europe, 262; in North Korea, 728, 738–39; in Oceania, 814–15; in Pakistan, 645; in South America, *i230–31,* 232, 233, 234, 235; in Southeast Asia, 766, 767, 768, 769–70; in Southern Africa, 588, 590; in Southern Europe, 308; in South Korea, 728, 738–39; in Transition Zone, 488, 497; urbanization prompted by, 539; in West Africa, 538, *p545,* 546. *See also* forests; logging; reforestation.
deindustrialization, 269. *See also* suburbs.
Delhi, India, 607; megalopolis, 613
deltas: Ganges–Brahmaputra River, 607, 608, 631; Niger River, 486–87, 533; Nile River, 365, 369; in Turkey, 415; in Vietnam, 757, *p758*
Demilitarized Zone (DMZ), 726, 729
Demko, George J., *q36*
democracy, 89, 90
Democratic People's Republic of Korea. *See* North Korea.
Democratic Republic of the Congo: economic activities of, *c564,* 566–67; emergence of HIV/AIDS in, 575; in Equatorial Africa, 554, 561; history and government of, *c562–63;* natural resources of, 566; refugees in, *p567*
democratic socialism, 96
demographic transition model, 82, *g83,* 277
demography, 301
Deng Xiaoping, 687
Denmark: family and status of women in, 259; history and government of, 255, *c256–57;* natural resources of, 254, 260, 264, *q266;* as part of Scandinavia, 250; population patterns, 257
dependency ratio, 702

desalination: brine from, *q446,* 448; environmental effects of, 39, 53–54, *q446,* 447–48; places employing, 51, *m53,* 380–81, 384, *m403,* 405; promise of, *q53;* removal of salt in, 53
desertification: in Arabian Peninsula, *p447;* in Central Asia, 467; in China, 698; in East Africa, 522; in Eastern Mediterranean, *q404,* 406; in Equatorial Africa, 568; in North Africa, 383; solutions to, 500; in South America, 232, 233–34; in Southern Europe, 306; in Transition Zone, 496, *m497, p498, i499, p500;* in West Africa, 527, 546, 547
deserts: in Arabian Peninsula, 431, 435, 436; in Australia, 776, 779, 784; as biome type, 436; in Central Asia, 451, 454, *p455,* 457, 462; in China, 680, 681, 682; in East Africa, 503, 509; in Eastern Mediterranean, 390, 391, 392, 393, 405; in India, 606, 610; in North Africa, 365, 369, 370, 376, 381, 484; in Northeast, 415; in Pakistan, 633; plants and animals in, 70; rainfall in, 70, 436; in South America, 215, 218, 225; in Southern Africa, *p577,* 579, 585; in Transition Zone, 481, 482, 484, *m485,* 486, 488, 489, 500, 530; wall between semi-arid land and, 500; in West Africa, 530, 531, 534
developing countries, 205
development, human: reasons to measure, 77; regional, *m76*
devolution, 277
Dhaka, Bangladesh, 636
Diallo, Nafissatou Niang, 494
Diama Dam, 487, 533
dictatorship, 88
Dien Bien Phu, 760
diffusion, cultural, 80, 81, 194–95
dikes, 271, *d272,* 629
Dinaric Alps, 318
direct observation and measurement, 32
dissidents, 691. *See also* human rights violations.
divide, 121–*m22*
Djenné, Mali, 536
Djibouti, in Transition Zone, 483
DMZ. *See* Demilitarized Zone (DMZ).
Dnieper River, 337, 338

Dniester River, 337

Doder, Dusko, 96

Doe, Samuel, *p537*

Dogon, *p494*

Doha, Qatar, 440

doldrums, 66

domestic trade, 99

Dominican Republic, 204, 205

Dominican Republic–Central American Free Trade Agreement, 205

dominion, 782

Dominion of Canada, 150

dos Santos, José Eduardo, 591

Dostoyevsky, Fyodor, 345

doubling time, 82–84

Drakensberg Range, 576–77

Dravidians, 613

DR-CAFTA. *See* Dominican Republic–Central American Free Trade Agreement.

drought: in Arabian Peninsula, 445; in Australia, 791; conflict caused by, *p132–33;* in East Africa, 503, 510, 513, 517, 520, 522; in Eastern Mediterranean, *q404;* impact of El Niño on, *d220;* in India, 609; in Mexico, 179, 188; in North Africa, 383; in Northeast, 425; in North Korea, 735; in Oceania, 814; in Southern Europe, 305; in Transition Zone, 481, 485, 488, 491, *c492, p493,* 496, 497–98, *q502;* in United States, 137; in West Africa, 527, 544, 547

drug cartels, 181, 186, 225

dry climates. *See* arid climates; deserts; dry subtropical climate; steppes.

dry farming, 127

dry subtropical climate: of Australia, 779; of North Africa, 370; of Pakistan, 630

Duars Plain, 653, 659

Dubrovnik, Croatia, 318

dunes, 435, 531, 579

Durban, South Africa, 586

dust storms, 427, *i468*

Dutch East India Company, 582

dwarf planets, 40

dwars, 653

dynasty, Chinese, 685, 686

dzongs, 660, 661

—————— **E** ——————

Earth: atmosphere of, *d42, d62,* 63, 64; axis of, 60, *d61,* 62; climate regions and biomes of, 57–74; core of, 44, *d45;* crust of, 44, *d45,* 47, 49; diameter of, 42; mantle of, 44, *d45;* population of, 82; relationship of sun and, 57, 60–62; revolution of, *d61,* 62, 64; in solar system, 40, *d41;* structure of, 44–46; surface air temperature of, average, 58, 59, *i231;* surface (biosphere) of, 42–43, 44, 46, 47, 49, 51, 52; effects of geopolitics on, 91; tilt and rotation of, 60, *d61,* 66; water on, 51–54

earthquakes, 44; in Asia, 48; in Central America and Caribbean, 197, *p206,* 212; in Central Asia, 455, 469; forces causing, 48, *m177;* in Iceland, 248; in Indian subcontinent, 654; in Japan, 704, *m705, i708–09,* 717, 718; in Mid-Atlantic Ridge, 248; in North Africa, 368; in North America, 48, *m177;* in Northeast, 412; in Northern Europe, 251; in Pakistan, 631; in Southeast Asia, 755; in Southern Europe, 294; study of, *q56;* tsunamis triggered by, 609, 632. *See also* Ring of Fire.

Earth Summit, 288

Earthwatch, 308

East Africa, 503–26; agriculture in, *p504–5,* 508, 510, 516, 517, 520–21; arts in, 516–17; climate regions and biomes of, *p509,* 510; economic activities of, 512, *p516,* 517, 520, 521; ethnic groups in, 514, 516; family and status of women in, *p515,* 516, 522; fishing in, 508, 517, 523; history and government of, 511, *c513,* 520; immigration to, 512; independence of, 512, *c513,* 521; industrialization of, 523; landforms in, 503, 506, *p508,* 510; mountain ranges of, 506, 507, 509; natural resources of, 510, 513, 520, *p524;* physical geography of, 506–10; population patterns of, 513, *p514;* trading partners of, 511, 514, 517; transportation in, 508; urbanization of, *p514,* 516, 523; vegetation of, 509, 510, *q525;*

water systems of, 506, *p508–9,* 510, *p518–19,* 523

East African kingdoms, 483, 511

East Anatolian Fault, 412

East China Sea, 726, 727

Easter Island (Rapa Nui), 800, *p814,* 814–15

Eastern Europe, 319–32; agriculture in, *p324,* 325, *g328;* arts in, *p323,* 324; climate regions and biomes of, 317–18; communications in, 325; economic activities of, 313, 323, 324, *q325,* 326, *c331;* ethnic groups in, *p313,* 319, *c320,* 321–22, 323; family and status of women in, 323; fishing in, 317, 325, 327, 329; history and government of, 319, *c321,* 323; immigration to, *m322,* 323; independence of, 311, 321; industrialization of, 317, 322, 323, 324, 327, 328, 329; landforms in, 314, *m315;* manufacturing in, 327; mountain ranges of, 317, 318, 322; natural resources of, 317, *p318,* 325–30; physical geography of, 311, 313–18; population patterns of, 321–23; Soviet occupation of, 342; trading partners of, 317, 323, 325; transportation in, *m316,* 322–23, 325; urbanization of, 322–23; vegetation of, 318, 326–27; water systems of, 311, *m316,* 317, 322, 325, 327, *g328,* 329, 330

Eastern Hemisphere, 17

Eastern Highlands, 507

Eastern Mediterranean, 387–408; agriculture in, 392, 393, 394, 401, *p404,* 405, *p406;* arts in, 400; climate regions and biomes of, 392, *m393;* communications in, 400, 406; economic activities of, 394, 400, *g401;* ethnic groups in, *p387, m389, c398,* 399, 400, *c407,* 411; family and status of women in, *p399,* 400; fishing in, *p392, 402,* 406; history and government of, 387, 395–98, 400; immigration to, 388, 399, 403; independence of, 388, 396, 397, 398; industrialization of, 401, 405; landforms in, 390–91; manufacturing in, 400, 401; mountain ranges of, 390–91, 393; natural resources

of, 394, 401, 402–6, *p406;* physical geography of, 390–94; population patterns of, 392, 398–99; trading partners of, 401; transportation in, 401, 403, 406; urbanization of, 403, 406; vegetation of, 393; water systems of, *p391–92, m393,* 394, 401, *m403,* 405, 406

Eastern Rift Valley, 506, 509

East Jerusalem, 388

East Timor: history and government of, 761; landforms in, 754. *See also* Southeast Asia.

Ebrié Lagoon, 530

Ebro River, 296

ecologists, 27; knowledge of, 32

e-commerce, 541

economic activities: of Arabian Peninsula, 433, 435, 442–43; of Australia, 777, 787; of Bangladesh, 639; of Canada, 154–55; of Central America and Caribbean, 198, 201, 206, 207; of Central Asia, 453, 458, *c464–65,* 467; of China, 91, 96, *c97,* 99, 466, 517, 677, *p678,* 685, *c687,* 690–91, 692, 697, *q700,* 712; of East Africa, 512, *p516,* 517, 520, 521; of Eastern Europe, 313, 323, 324, *q325,* 326, *c331;* of Eastern Mediterranean, 394, 400, *g401;* of Equatorial Africa, 560, *c564,* 565, 566–67; of India, 99, 604, 611, 616–17, 620; of Indian subcontinent, 656, 662; industrialization as, 97; of Japan, 701, *p702, ptg711,* 712, 715, *p716,* 718, 720; locations chosen for, 34; maintaining level of, 84; of Mexico, *c97,* 184–86, 189; of New Zealand, 787; of North Africa, 366, 371, 374, 376–77, 384; of Northeast, 417, 420–21, *q430;* of Northern Europe, *c97, p248,* 258, 260, *c265,* 324; of North Korea, 735–36; of Northwestern Europe, 276, 277, *m280, p283,* 324, 325; of Oceania, 799, 802, 805, 808–9, 813; of Pakistan, *c97,* 639, 645; primary, 96; quaternary, 96; of Russia, 96, 99, *c342,* 344, 345–47, 350; secondary, 96; of South America, 220, 222, 227–29, 234; of Southern Africa, *c97,* 580, *g587;* of Southeast Asia, 753, 762, 764–65, 769; of Southern Europe, *293,* 297, 300, 303–4, 325; of

South Korea, c97, 724, 725, 735, p736; tertiary, 96; of Transition Zone, 488, 494–95; of United States, c97, 130–31, p132–33, 567, 712; of West Africa, 541. See also industrialization.

Economic Community for West African States (ECOWAS), 529, 541

economic development, 96, c97; of newly industrialized countries, 97; poverty reduced by, 97; uneven, p216–17

economic geography, 94–99

economic sanctions, 691

economic system, 94–96; global, 105. See also capitalism; command economy; communism; free enterprise; market economy; socialism.

economies: interconnectedness of, 108; study of, 34; world trade and, 98–99

economy: of China, 96; command, 95–96; market, 94, 95, 96; mixed, 95; of Russia, 96; traditional, 94, 95; of Vietnam, 96. See also economic activities.

ecosystems, 31, 135; of Australia, 790, 794; of Canada, 148, 158; of Central America and Caribbean, 209; of China, 696; coastal marine, m286; of Eastern Mediterranean, 402, 406; of India, 624; in Mediterranean, 298; of North Africa, 383; of Northern Europe, 262, m263; of Northwestern Europe, 285, m286; of Oceania, 803, 813; of Russian Core, 354; of Southern Europe, 306, 307; of Transition Zone, 498; of United States, q133, 134, 135; of West Africa, 531, 532

ecotourism, 205, 264, 517, p524, 561; as growing industry, i655; restrictions on, 624

ECOWAS. See Economic Community for West African States.

Ecuador, 199; climate regions and biomes of, 221; ethnic groups in, m226; natural resources in, c222; population patterns of, 225

Edmonton, Alberta, 162

education, 153, 226, 279, 303, 323, 419, 515, 540, 563, 688, 785, 807. See also culture.

EEZ. See exclusive economic zone.

Egypt, 376, g385, q386; boundaries of, 90; colonial rule of, m373; culture hearth of, m80; history and government of, 367, c374, c397, 400; population patterns of, 86; water systems of, p366–67, 380, 384, 487, 518, p519

Egyptian civilization, 372, 489, 553

El Alto, Bolivia, 234

Elburz Mountains, 413

electricity, production of, 160, 179. See also hydroelectricity production.

elevation, influence on climate of, 64–65

elevation, map, 19, 20

Elmina Castle, 561

El Niño, q67, 199; impact on animals of, 221; impacts on climates worldwide of, d220, 221; weather patterns affected by, 814

El Salvador, 198, 201, 209

embargo, oil, 421

emergent layer of rain forest, 558

emerging markets, 99, p752–53, 764

Emi Koussi, 369

enclave, 459, 460

endangered species, 523, 569

energy resources: of Canada, 141, p142; of China, p678; of Northwestern Europe, 281; of South America, c222. See also coal mining; dam building; geothermal energy; hydroelectricity production; natural gas; petroleum; renewable and alternative energy; wind energy.

English Channel, 273

Enlightenment, 276

entrepôt, 258

environment: adapting to changing, 37; in Mexico, 187, m188, 189, p190; relationship between society and, 31. See also acid rain; air pollution; climate change; drought; ecosystems; global warming; mineral resources and mining; water pollution.

environmental conservation, 138

environmental planning and management, 264

Environmental Protection Agency (EPA), 815

environmental science, 189

EPA. See Environmental Protection Agency (EPA).

epidemics. See infectious diseases.

Equator, 17; climates at, 557; equinoxes at, 61; latitude of, 64

Equatorial Africa, 551–72; agriculture in, 551, 560, 561, c564, 566, 568; arts in, 565; climate regions and biomes of, 557–58; communications in, p565; economic activities of, 560, c564, 565, 566–67; ethnic groups in, 559, 562, 567; family and status of women in, 563; fishing in, p556, 564; history and government of, p552–53, 559–61; immigration to, 559; independence of, p552–53, 561; industrialization of, 564, 565; landforms in, 554, m555; manufacturing in, 564, 565; mountain ranges of, 554, 555, 567; natural resources of, 559, 566–70; physical geography of, 554–58; population patterns of, 561; trading partners of, 568, g569, q572; transportation in, 565; urbanization of, 551, 561; vegetation of, 558, 561; water systems of, m555, p556, 560, 561, 567

Equatorial African kingdoms, 559

Equatorial Guinea, 554; history and government of, 561

equinox, 61–62

ergs, 368

Eritrea: agriculture in, 521; in East Africa, 507, 509, 510, 512; economic activities of, 517, 521; in Transition Zone, 483

erosion, 49–50. See also soil erosion.

escarpment, 219, 507, 576–77

Estonia, 314, 325, c342

Ethiopia: history and government of, 491, 520, 521; in Transition Zone, 483; water systems of, 487, 518, p519

Ethiopian Plateau, 507

ethnic cleansing, 321

ethnic groups, 79; in Arabian Peninsula, 438, c439; in Australia, p774, 775, 780, 781, 782, 783, c784, q796; in Bangladesh, 635, 637; in Canada, 152, 153; in

Central Asia, m452, p453, 460, 461, m462; in China, 688; in East Africa, 514, 516; in Eastern Europe, p313, 319, c320, 321–22, 323; in Eastern Mediterranean, p388, m389, c398, 399, 400, c407, 411; in Equatorial Africa, 559, 562, 567; in India, 603, 613; in Indian subcontinent, 660; in Japan, 701, 712; in Mexico, 180–81; in New Zealand, p774–75, 780, 781, 782, 783, c784–85; in North Africa, 365, 373, 374; in Northeast, 409, m410, p411, 418, 425; in Northern Europe, 257; in North Korea, 732; in Northwestern Europe, 275; in Oceania, 804, 805, 807; in Pakistan, 635, 637; in Russia, c344; in South America, 223, m226, 227; in Southeast Asia, 751, 760, 762, p763; in Southern Africa, 586; in Southern Europe, 300; in South Korea, 732; in Transition Zone, 482, 483, 484, 491, 493; in United States, 127, c128; in West Africa, 537, 539–40. See also indigenous people; Native Americans.

EU. See European Union (EU).

Euphrates River, 391, 412, 414, 424, 427

Eurasia, 43

Eurasian landmass, 340

Eurasian tectonic plate, 368, 413

Europe: city populations in, 102; as continent, 43; peninsulas in, 250; population patterns of, 85; postwar division of, 277. See also Northern Europe; Northwestern Europe; Southern Europe.

European Central Bank, q283

European Environmental Agency, 307

European Union (EU), q108; countries considering membership in, 347, 417, 428; creation of, 282, c321; Eastern Europe in, c321, 325, 328, 329; economic crisis of, 282, q283, 303, 304, 325; environmental standards of, 329, 330, 428; European Environmental Agency created by, 307; formation of, 277; low-emission zones created by, 329; member countries of, m280, 282, 300, c321, 325; Northern Europe in, c257, 260, m280;

Northwestern Europe in, 277, *m280;* Regional Environmental Center funding from, 330; Southern Europe in, 300, 303–4; species protection rules of, 325; as stabilizing force, 282, *q283;* trade volume of, 281; trading partners of, 281, 383; Water Framework Directive of, 288

eutrophication, 137, 261

evaporation: sun's energy causing, 51; in water cycle, *d52*

evergreen-deciduous forests, 610

evergreen forests, 71

evergreen trees. *See* coniferous trees.

exclave, 462

exclusive economic zone, 810

extended family, 183

external forces: Earth's surface changed by, 49; types of, 49–50

Eyjafjallajökull (Eyja), *p248, d249*

Ezana, 515

F

Faeroe Islands, 251, 256, 257

Fahrenheit temperature, 60

fall line, 121, *m122*

family and status of women: in Arabian Peninsula, 433, 438, 441, 442, *q450;* in Australia, 786; in Bangladesh, 636, 638; in Canada, 153; in Central America and Caribbean, 203, 206; in Central Asia, 463; in China, 690, *p690;* in East Africa, *p515,* 516, 522; in Eastern Europe, 323; in Eastern Mediterranean, *p399,* 400; in Equatorial Africa, 563; in India, *c614–15;* in Indian subcontinent, 660–61; in Japan, 702, 714; matriarchal, 203; in Mexico, 183, *c184,* 227; in New Zealand, 786; in North Africa, *p376;* in Northeast, *p420;* in Northern Europe, 259, *q260;* in North Korea, 734; in Northwestern Europe, 277, 278, 279; in Oceania, 808; in Pakistan, 636, 638, 639; patriarchal, 494; in Russia, 334, 345; in South America, 227; in Southeast Asia, *p763;* in Southern Africa, 575, 586–87; in Southern

Europe, 303; in South Korea, 734; in Transition Zone, 494; in United States, 129; in West Africa, 203, *p528–29,* 540

famine: in East Africa, 513, 516, 520; in North Korea, 723, 735, 740, *c741;* in Transition Zone 485, 488, 491, *c492, p493,* 495; in West Africa, 546, *c549, q550*

Fang people, 562

FAO. *See* Food and Agriculture Organization (FAO).

farmland. *See* agriculture.

faulting process, 47, *d48*

fault lines, 47, *d48,* 412, 506

favelas, 234

Federal Republic of Yugoslavia, 313

federal system of government, 87, 88

Federated States of Micronesia, 800, 805, 806, *c807. See also* Oceania.

feeder systems (tributaries), 54, 121, 424, 425. *See also* rivers.

Fergana Valley, 455, 462

fertilizers, 504; in water sources, 402, 621. *See also* agricultural runoff; water pollution.

Fiji, 800, 801, 805, *c807;* agriculture in, 816; economic activities of, 808; history and government of, *c806;* natural resources of, 809, 812. *See also* Oceania.

Finland: climate regions and biomes of, 253, 254; family and status of women in, 259; history and government of, 255, *c256–57;* landforms of, 251; natural resources of, 254, 260, 262, 264; as part of Scandinavia, 250; population patterns of, 257, *g258;* water systems in, 252–53

Fiordland, 777

First Nations, 149, 150, 151, 153

fish as natural resource: in Canada, 148; in Central America and Caribbean, 198, 199, 209; in Central Asia, 470; in Eastern Europe, 327; in Indian subcontinent, 656; in Northwestern Europe, 286; in Oceania, 803; in South America, 221; in Southern Africa, 578; in United States, 124; in West Africa, 531, 532–33

fishing, commercial; in Arabian Peninsula, 437, 443, 445; in Australia, 777, 787; in Bangladesh, 632, 639, 643; in Central America and Caribbean, 198, 199; in Central Asia, 456, 470; in China, 684; drift-net, *p739;* in East Africa, 508, 517, 523; in Eastern Europe, 317, 325, 329; in Eastern Mediterranean, 402, 406; in Equatorial Africa, *p556,* 564; in high seas, 810, *q811;* in India, 607, 617; in Indian subcontinent, 662; in Japan, *p707, p719, p720,* 739; in Mexico, 178, 179; in New Zealand, 775, 777, 782, 787; in North Africa, 377, 382, 383, 384; in Northeast, 414, 420; in Northern Europe, 258, 260, 261; in North Korea, 739; in Northwestern Europe, 284–85; in Oceania, 808, 815; in Pakistan, 632, 639; in Russia, 346, 351, 354; in South America, 228; in Southeast Asia, 764, *q768,* 769; in Southern Europe, *p291,* 307; in South Korea, 737, 739; in Transition Zone, 486; in United States, 124, 134–35, 178; in West Africa, 532, *p533,* 541, *q546,* 547. *See also* overfishing.

fjords: in New Zealand, 777; in Northern Europe, 250, *p251,* 260; in Patagonia, 218

flooding: in Arabian Peninsula, 445, 446; in Bangladesh, *p628–29,* 632, *q633;* in Central America and Caribbean, 210; in China, 682, 684, 698; in Eastern Mediterranean, 392; flash, 369, 392, 445, 757, 768; hydroelectric dams causing, 209; in India, 607, 608, 609, 620; in Japan, 704; logging causing, 768; in North Africa, 369, 372, *c382;* in Northeast, 428; in North Korea, 735, 739; in Oceania, 812, 814; in Pakistan, *p628,* 629, 632, 645; in Southeast Asia, 768; in Southern Europe, 297, 298; in Transition Zone, 487; in West Africa, 533. *See also* tsunamis.

floodlands: in North Africa, 369; in Russia, 337; in Siberia, 339

floodplains, 632

Florida, climate of, 122

fluctuations, population, 33. *See also* population distribution.

foehns, 273

folds, 47

Food and Agriculture Organization (FAO), 404

food scarcity. *See* agriculture; famine.

food web, 792

foreclosure crisis, U.S., *m130,* 131

forest fires, *q394,* 779

Forest Rights Act, 624

forests: in Australia, 791; in Bangladesh, 633, *p643,* 644; in Canada, 147, 155, 158; in Central Asia, 457, 458, 469; in China, 682, 697, *p698;* clear-cutting, 134, 135; climate change affecting, 72; in East Africa, 510, 522, *q523,* 524; in Eastern Europe, 317, 319, 326, *p327;* in Eastern Mediterranean, 394, 404–5, 406; in India, 608, *p609,* 610, 620; in Indian subcontinent, 649, 655, 656, 664–65; in Japan, 705, 706; in Mexico, 187–88; in New Zealand, 777, 780, *p794;* in North Africa, 371; in Northeast, 427; in Northern Europe, 253, 254, 260, 261, 262, 264; in Oceania, 803, 809, 815; in Pakistan, 645; renewing, 570, 740; in Russia, 339, 340, 346, 353, 354; in South America, 219, 228, 232; in Southeast Asia, 758, 769; in Southern Africa, 590; in Southern Europe, 305, 306; in South Korea, 740; types of, 71; in United States, 134; in West Africa, 531, 545, 546.

Forest Stewardship Council, 308

formal region, 30

fossil fuels: in Arabian Peninsula, 437, 444, 445–46; in Canada, *p142,* 160; in Central Asia, 466; in China, 694; creation of, 437; extraction of, *p142,* 160, 427; greenhouse gases from burning, 72, 445–46; in Japan, 717; in Northeast, 415, 417, *i426,* 427; in Northern Europe, 264; in United States, 124, 130, 138; unregulated burning of, 403. *See also* natural gas; petroleum.

fossils: in East Africa, 511; in Southern Africa, 581

Fox Quesada, Vicente, c183
France, 270; economic activities
of, 280; history and government
of, 89, 276, c277, q283; natural
resources of, 274, 284, 285;
water systems of, 295
Francis Ferdinand, ptg320
Fraser River, 145
Frederick VI, ptg256
free enterprise, 94–95
free market economy, 130.
See also market economy.
free trade zones, 186
French Guiana, 221
French Polynesia, 800, 806,
808. See also Oceania.
French Revolution, 224
freshwater: access to, 208, p547,
644, 697; aquifers for, 54, 380,
381, p383, 437, 446, p447, 498,
654; bodies of, locations of,
54, 145, 339, 405; caught from
rainfall, 802; dams providing,
487; depletion of, 437, 791; in
headwater streams, 778; human
uses of, 39, 514, 523, 532, 705,
814; management improvements
for, 813; as necessary to sustain
life, 51, 52; as percentage
of Earth's water supply, 54;
as renewable resource, 778;
reservoirs holding, 532; tidal
marshes of, 262; underground,
70, 380, 381, 436, 590. See also
aquifers; glaciers; lakes; rivers.
fuel-cell engines, 718
Fulani people, 493
functional region, 30, 31

— G —

Gabon: economic activities of, 561;
in Equatorial Africa, 554
Galápagos Islands, 221
Galilee Mountains, 391
Gambia: in Transition Zone, 483, 530
game reserves, 517, 524, 646,
698, 729
Gandak River, 654
Gandhi, Indira, 612, p615
Gandhi, Mohandas K., p612
Gandhi, Rajiv, 612
Ganges Delta, 607, 610, 631
Ganges (Padma) River, p607,
629, 631, 632, 654
Gangetic Plain, 606, 613, 617
GAP. See Southeast Anatolia Project.

García Lorca, Federico, q310
Gaudí, Antoni, 303
Gaza, c396, 397
Gaza Strip, 397
GDP. See gross domestic product (GDP).
gender equality, Canada's
ranking in, 153
gender inequality in West
Africa, 528–29
genetically engineered
seeds, 504
genetics, 660
Genghis Khan, ptg460, 687
geographer, 26, p32; career
opportunities for, 27–28, 32;
focus of, branches of, 30–31;
geographer's craft, 26–34;
geographer's tools, 14–25,
c27; governments analyzed by,
79; interactions studied by, 31;
research methods of, 32–33, 37;
skills for, c27; trends studied by,
75; view of world of, 11–36; work
of climatologist and, 66. See also
types of geographers under their
specific names.
geographic information:
acquiring, organizing, and
analyzing, c27
geographic information
systems (GIS), 13, 34; database for, 23–24; layers of, d23
geographic patterns, 75
geographic perspective, 26
geography: career opportunities
in, 27, 31–32; elements of,
28–32; government influenced
by, 89–90; objective of, q36;
population, 82–86; relationships
to other subjects of, 33–34; root
of term, 26; themes of, 31; uses
of, 31–32. See also individual types
by their specific names.
geography skills: examples of,
c27; tools and technologies for,
c27; uses of, 11, 26, 28, 31
geology, effects on soil of, 50
geometric boundary, p89,
90, 373
geopolitics, 91, 444, 445
Georgia: climate regions and
biomes of, 457–58; economic
activities of, 458, 460; ethnic
groups in, 411; history and
government of, 459, 460, 464;
water systems of, p132–33, 454

geospatial technologies,
21–26; limitations of, 24–25;
transmitting satellite data, 24
geothermal energy: in New
Zealand, 780; in Northern Europe,
251, 254, 260
geothermal venting, 248
German East Africa, 583
German Southwest Africa, 583
German–Soviet Nonaggres-
sion Pact, c342
Germany, 270; economic activities
of, 324; history and government
of, 276, 277, 279, q283, c301,
c320, c342–43, 417; loss of African
colonies of, 583; natural resources
in, m101, 274, 285; population
patterns of, 84; water systems
of, 295
gerrymandering, 13
geysers, p251
Ghana: natural resources of, p545; in
Transition Zone, 483, 530
Ghana Empire, 490, 535
Gibson Desert, 776
GIS. See geographic information
systems (GIS)
glaciation, 250, 251, 270
glaciers, 49–50; in China, q700;
freshwater in, 54; loess sediment
left by, 270; melting, 59, 248,
306; in New Zealand, 777; in
North America and Greenland,
m145; in Northern Europe, 250–
51, p251, p252; in South America,
218; in Southern Europe, 306
Glåma River, 252
glasnost, 343
global economy: importance
of U.S. financial system to, 131;
Mexico in, 173, 185
globalization: effects on modern
culture of, p92, p93; increased,
173; studying changes from, 105
global positioning system
(GPS), 21–22
global warming, 58; carbon
dioxide levels and, 210;
diversifying energy sources
to address, 138; natural gas
to address, q138; release of
greenhouse gases in, 353; studies
of, 814. See also climate change;
greenhouse gases.
Global Water Initiative, 548
globe, 14, crt35
Gobi, 680, m681, 683

Golan Heights, 390
Gold Rush, p150
Gondwana, m46
Goode's Interrupted Equal-
Area Projection, m16
Gorbachev, Mikhail, p343, 346
Gotel Mountains, 555
government: corruption in, 503,
520, 521, 544–45, 567, 587,
592, 690; features of, 87–89;
influence of geography on,
89–90; levels of, 87–88; national
central, 87; regional cooperation
of, 211; services of, 702; state or
provincial, 87; terrorist attacks
against, p90, 91, p126–27; types
of, 88. See also individual countries
and government types.
GPS. See global positioning system (GPS).
graffiti, 130
grains, 457
Granada, 200
Grand Canal, 682
Grand Canyon, p49, 50, 120
Grand Ethiopian Renaissance
Dam, 487, 518, p519
Grand Millennium Dam See
Grand Ethiopian Renaissance Dam
grasslands: in Australia, 776;
in Central America and Caribbean,
199; in Central Asia, 454, 464,
467; in East Africa, 510; in Eastern
Europe, 318; in Equatorial Africa,
558, 561; in India, 609; in North
Africa, 370; in Russia, 339; in
South America, 219, 221, 228; in
Southern Africa, 579; in Transition
Zone, 488; in West Africa, 531,
534. See also savannas.
grazier, 790–91
Great Artesian Basin, 778
Great Australian Bight, 776
Great Barrier Reef, 779, 780,
c792, 794
Great Bear Lake, 146
Great Britain: economic activities
of, 280; history and government
of, 276, 277, 279; landforms of,
270–71; natural resources, 285
great circle route, 14, m15
Great Dividing Range, 776
Greater and Lesser Antilles, 197
Greater Caucasus, 455
Greater Himalaya range,
652, 653
Great Escarpment, 576–77
Great European Plain, 270

Great Famine, c492
Great Green Wall of Africa, i499, 500
Great Lakes, 122, 128; acid rain in forests around, p160; Canadian population density near, 152; formation of, 146; megalopolis sites along, 129
Great Lakes–St. Lawrence Seaway System, 122, m146
Great Lakes Water Quality Agreement, 138, 161
Great Leap Forward, c686
Great Man-Made River, 381, 384
Great Mosque, 440
Great Plains: in Canada, 144–45; in United States, 120, 123, 127
Great Rift Valley: formation of, 506, d525; water systems of, 503, 508, 517
Great Salt Lake, salt water in, 54
Great Sandy Desert, 776
Great Slave Lake, 146
Great Victoria Desert, 776
Great Wall, 686, crt699
Great Zimbabwe, 581, 582
Greece: economic activities of, 282, q283, c301, 303, 325; family and status of women in, 303; history and government of, 299, c300, c301, 302; industrialization of, 304, c309; landforms of, 295; natural resources of, p291; water systems of, 297
greenhouse effect, d62, q63
greenhouse gases, 59, d62, 63; climate change involving, 72, 287, 445–46; country rankings for emissions of, 161–62; emission regulations for, g235, p236, 288; sources of emissions, 162, 211, 353; targets for emissions of, 720; voluntary emissions targets for, 236; vulnerability to emissions of, 812
Greenland: history and government of, 251, 256, 257; sheet glaciers in, 50
Greenpeace, 288, 350, 354
green revolution, 617
griot, 540
gross domestic product (GDP), 98, 99; of Arabian Peninsula, c439; of Australia, 787; of Democratic Republic of the Congo, c564, 567; of East Africa,

504; of Eastern Mediterranean, g401; effects of Iceland volcanic eruption on global, g249; of Indonesia, 753; of Mexico, 185; of New Zealand, 787; of Northwestern Europe, 280; of Norway, c265; of Pakistan, 639; of Russian Core, g346; of Southeast Asia, g771; of United States, 567
groundwater: freshwater in, 54; pollution of, 621, 665. See also freshwater; water systems.
Guam, c203, 800, 806, 808. See also Oceania.
Guangzhou, China, 682, 688, 691
Guatemala, 202; coffee crop of, p98; history and government of, 201; petroleum refinement in, 199
Guatemala City, Guatemala, c202
guest workers: in Arabian Peninsula, p432–33; in Northern Europe, 257; in Northwestern Europe, 278
Guinea: in Transition Zone, 483, 530; water systems of, 486, 487
Guinea-Bissau: in Transition Zone, 483, 530
Guinea Highlands, 531
Gulf Coastal Plain (Mexico), 177
Gulf of Aden, 369
Gulf of Aqaba, 391, 392, 401
Gulf of Bothnia, 252, 253
Gulf of California, 178
Gulf of Finland, 341
Gulf of Guinea, 487, 554, 556
Gulf of Mexico, 124, 178, 199
Gulf of Oman, 414
Gulf of St. Lawrence, m146
Gulf Stream, 147, m253, 254, 260, 273
Gupta Empire, 611
Guyana, 221; ethnic groups in, m226
GWI. See Global Water Initiative.

——————— **H** ———————

habitat, animal, 522, 610, 624
habitat destruction, 588. See also air pollution; deforestation; logging; water pollution.
habitat protection, 813
Hainan, 683
Haiphong, Vietnam, 765

Haiti, 203; arts in, 204; development in, 206, p207; earthquake in, 197, p206; independence of, 201; slave revolt in, 201
hajj, 433, c441, 490
Hakuluki Hoar, 633
halophytes, 393
hamadas, 368
Hamas, 397, c397
Han dynasty, 686, 688, 730
Hanoi, Vietnam, 762
Han River, 727
Haor Basin, 631
haors, 632
harmattan, 488
Harris, C. D., 104
Hasina, Sheikh, 638, 646
Hausa, 493
Hausa city-states, 536
Hawaiian Islands, 800, q818; climate regions and biomes of, 122; economic activities of, 808; formation of, 121; mountain ranges of, 802
HDI. See Human Development Index.
headwaters, of river, 121
health care: in Arabian Peninsula, 433, 441; in Australia, 784, 785; in Bangladesh, 637; in Canada, 152, 153; in Central America and Caribbean, 203, 205, 206; in East Africa, 516; in Eastern Europe, 323; in Equatorial Africa, 561, 562; in India, 604; in Japan, 702; in New Zealand, 785; in Northeast, 419–20; in North Korea, 733–34; in Northwestern Europe, 279; in Oceania, 808; in Pakistan, 637; in Southern Africa, p575; in Southern Europe, 302; in South Korea, 734; in Transition Zone, q491, 493; in United States, 119, 128; in West Africa, 529, 540, 544, 548, q550
Hebrew Bible, 395, 396
Heian-kyo, Japan, c712
Hellenic Period, p300
Hezbollah, 397
Hidalgo, Miguel, c182
hieroglyphics, 372, 373
high-latitude (highland) climates, 71–72; in Canada, 147; in Central Asia, 458; in China, 680, 682; in East Africa, 507; in Eastern Mediterranean, 392; in Equatorial Africa, 555, 557; in Indian subcontinent, 662; in North

Korea, 728; in Pakistan, 644; in Southern Africa, 573, 577, 578; in West Africa, 530, 531
high seas, managing and protecting, p810, q811
high-tech industries: in Central America and Caribbean, 205; in China, p695; in Eastern Europe, 325; in Eastern Mediterranean, 401; in India, 617; in Japan, 715; in Northeast, 421; in Northwestern Europe, 280; in Southeast Asia, 764; in South Korea, 736; tungsten used in, 298; in United States, 127, 131
Highveld, 577, 587
highway systems: in Canada, 155; in Eastern Europe, 318, 325; in Mexico, 185; in Pakistan, 632, 639; in Russia, 347; in South America, 228–29; in United States, 117, 131
Himalaya mountain ranges, 46, 606, 608, 629, 630, 632, 633, 652, 655, 680
Hinduism, 612, 613, 614, 634–35, 834. See also religions.
Hindu Kush, 455, 606, 630
Hiroshima, Japan, 352, 711
Hisahito, p713
Hispanics/Latinos, m118, 119, m128
Hispaniola, 200
history and government: of Arabian Peninsula, 438–39, c440–41; of Australia, 781–82; of Bangladesh, 627, 634–35; of Canada, 142, 149, c151; of Central America and Caribbean, m194, 195, 200–201; of Central Asia, m452, 453, 459–61, 462–63, q472; of China, 88, 677, 685–c87, 688, 711, 730; of East Africa, 511, c513, 520; of Eastern Europe, 319, c321, 323; of Eastern Mediterranean, 387, 395–98, 400; of Equatorial Africa, p552–53, 559–61; of India, 460, 604, 611, p612, q612, 620, 651; of Indian subcontinent, m650, 651, 655, 657, c659, 661; of Japan, 710–c13, 716, 730–31; of Mexico, 180–81, c182–83, 184; of New Zealand, 781–82, c784–85; of North Africa, 370, 372–75, c375, 376, 377, 417; of Northeast, 416, c419; of Northern Europe, 251, 255, c257, 279; of Northwestern

Europe, 275, c277; of Oceania, 804–5; of Pakistan, 90, 460, 612, 627, 634–35; of Russia, 277, 320, 341, c343, 344, 687; of South America, 217, 223–c225; of Southeast Asia 759–61; of Southern Africa, 581–85, c385; of Southern Europe, 299–p301, 302, 307; of South Korea, 723, 726, 730, c733; of Transition Zone, 489–91, 494; of United States, 89, 119, 125–27; of West Africa, 527, 535–38. *See also* constitution; independence.

Hitler, Adolf, 88, 277

HIV/AIDS: in East Africa, q515, 516; in Equatorial Africa, 563; in Russia, 334, i335; in Southern Africa, m574, p575, 586

Ho Chi Minh, p760

Ho Chi Minh City, Vietnam, 762

Hokkaidō, Japan, 704, 705, 706, 707, 712

Holocaust, 277

Honduras, 201

Hong Kong, p34, c687, 691, 695

Honolulu, Hawaii, 806

Honshū, Japan, 704, 705, 706

Hopetown Falls, p777

Horn of Africa, 509

hostage-taking, q423

hot spot: Hawaiian Islands, 121; Iceland, 251, 254; Mexico, m177; Oceania, 800; pollution, 307

hot springs, 251, 254

Huang He (Yellow River), 681–82, 688, 695, 698

Hudson Bay, 145

Hudson's Bay Company, 149

human (cultural)
  geography: of Arabian Peninsula, 438–43; of Australia, 781–87; of Bangladesh, 634–39; of Canada, 149–57; in Central America and Caribbean, 200–205; of Central Asia, 453, 454, 459–65; of China, 685–93; of East Africa, 511–17; of Eastern Europe, 319–25; of Eastern Mediterranean, 395–401; of Equatorial Africa, 559–65; focus of, 31, 34; impact of immigration on, 119; of India, 611–17; of Indian subcontinent, 657–62; of Japan, 710–16, p716; knowledge of, 28; of Mexico, 180–86; of New Zealand, 781–87; of North

Africa, 372–77; of Northeast, 414, 416–21; of Northern Europe, 255–60; of North Korea, 730–36; of Northwestern Europe, 275–81; of Oceania, 804–9; of Pakistan, 634–39; of Russia, 341–47; of South America, 223–29; of Southeast Asia, 759–65; of Southern Africa, 581–87; of Southern Europe, 299–304; of South Korea, 730–36; of Transition Zone, 481, 489–95; of United States, 125–33; of West Africa, 535–41. *See also* arts; culture.

Human Development Index (HDI), m76; of Argentina, g107; dimensions measured by, 77, 662; of East Africa, 520; of Haiti, 206; of South Asia, c662; use by governments of, 77

human immunodeficiency virus. *See* HIV/AIDS.

human rights violations, 433, q619, p687, 691

human systems, physical systems and, p30, 31

humid continental climates, 71; in Canada, 147; in Eastern Europe, 317; in Japan, 706; in Northern Europe, 254; in United States, 123

humid subtropical climates, 71; in Australia, 779; in Bangladesh, 633; in Central Asia, 457, 464; in China, 682; in East Africa, 509; in Eastern Europe, 317; in Eastern Mediterranean, 392, 393; in India, 610; in Indian subcontinent, 655; in Japan, 706; in South America, 221; in South Korea, 728; in Ukraine, 339; in United States, 123, 132

humus, 273

Hungary, 314, 329, 330; agriculture in, p324; history and government of, c321, 325; population patterns of, 84; Regional Environmental Center funding from, 330

hunger. *See* agriculture; drought; famine.

Huns, 453

hunting: for bushmeat, 570; for sport, 569

hurricanes, 122, m123, 609; in Central America and Caribbean, 199, 212; predicting, 199

Hussein, Saddam, 88, 127, 411, 417, p418, c419, 425

Hutu people, 513, 514, 521

hybrid vehicles, 137

hydroelectricity production: in Bangladesh, 633; in Canada, 160, 162; in Central America and Caribbean, 198, 209; in Central Asia, 458, 466; in China, 678, 682, 694, 696, 769; in East Africa, 508, 509, 518; in Eastern Europe, 316, 318; in Equatorial Africa, 558, 565; in India, 607; in Indian subcontinent, 654, 656, 666; in Japan, 717; in New Zealand, 780; in North Africa, c382, 383; in Northeast, 421, 424; in Northern Europe, 251, 252–53, 254, 260, 262; in North Korea, 728–9; in Northwestern Europe, 273–74; in Pakistan, 632, 633, 644; in Russia, 337, i338; in South America, 220; in Southern Africa, 578, 579; in Southern Europe, 296, 298; in South Korea, 728; in Transition Zone, 486, 487; in Ukraine, 337; in West Africa, p532, 533

hydrosphere, d42

IAEA. *See* International Atomic Energy Association.

Ibadhism, 441

Iberian Peninsula, 294, 295, 296

ice age: glaciers of last, m145, p252; Northern Europe, 250, 251, p252

ice cap climate: in Canada, 147

ice caps: freshwater in, 54; melting, 792, 814; vegetation in, 72

Iceland: climate regions and biomes of, 253; family and status of women in, 259; history and government of, c256–57; landforms of, p248, d249, 251; natural resources of, 254, 260, 264; as part of Scandinavia, 250; population patterns of, 258; volcanic eruptions in, 248-49; water systems of, 252, 254

icons, map, 18

ideograms, 689

Ile River, 456

IMF. *See* International Monetary Fund.

immigration: to Arabian Peninsula, p432–33, 439, 442; to Australia, 773, 783, 784, c785; to Canada, 149, d150, 153; to Central America and Caribbean, 201–2;

to East Africa, 512; to Eastern Europe, m322, 323; to Eastern Mediterranean, 388, 399, 403; to Equatorial Africa, 559; to New Zealand, 783; to North Africa, m373; to Northern Europe, 257; to Oceania, 802, 804, 805, 807; to South America, 226; to Southern Africa, 582; to West Africa, c536–7; to United States, 127, c128, 129; ancestry by county, m118, c237, 258; periods of, 119; reasons for, 119

imperialism, 611, 612

impressionists, 279

Inca Empire, 223–24, 227

independence: of Australia, 782; of Bangladesh, 634, 635, c637; of Canada, c151; of Central America and Caribbean, 200–01, c202–03; of Central Asia, 460–61, 462–63; of East Africa, 512, c513, 521; of Eastern Europe, 311, 321; of Eastern Mediterranean, 388, 396, 397, 398; of Equatorial Africa, p552–53, 561; of India, q612, 618, 635; of Indian subcontinent, c658; of Mexico, 181, c182, 184; of New Zealand, 782; of North Africa, m373, c375, 376; of Northeast, 417; of Northern Europe, c256–57; of North Korea, c733; of Northwestern Europe, 277; of Oceania, 805; of Pakistan, 618, 634, 635, c636–37; of Russia, 341, 343, 345, 347; of South America, 224; of Southeast Asia, 760, p761; of Southern Africa, 584–85; of Southern Europe, 300; of South Korea, 731, c733; of Transition Zone, 491; of United States, 126; of West Africa, 537

India: agriculture in, 609, 610, 613, 616–17; arts in, 516, 603, 615; assassinations in, 612; climate regions and biomes of, 71, p609, 608–10; communications in, 613; economic activities of, 99, 604, 611, 616–17, 620; ethnic groups in, 603, 613; family and status of women in, c614–15; fishing in, 607, 617; history and government of, 460, 604, 611, p612, q619, 620, 651; independence of, q612, 618, 635; industrialization of, 617, 623; invasions of, 611; Kashmir governance issue of, p618–19, 635, c636; landforms in,

606–7; manufacturing in, 617; mountain ranges of, 606, 608, 613; natural disasters in, 609; natural resources of, *m101*, 610, 620–24; partitioning of, 90, 612, 635, *c636–37*; physical geography of, 606–10; population patterns of, 604, *g605*, *p613*; trading partners of, 611; transportation in, 607, 616; urbanization of, 613, 614–15; vegetation of, 608, 609, 610, 620; water systems of, 606, *q607*, 609, 614, 616, *g621*, *p622*, 643

Indian National Congress, 612

Indian Ocean: development along, 503, 508, 511, 512; drainage of East Africa into, 507; earthquakes and tsunamis in, 756; effects on climate of, 654; location of, 52; monsoon winds created over, 608; outlet of Persian Gulf to, 414

Indian Plate, 46

Indian subcontinent, 606, *q619*, 630, 652, 662; agriculture in, 653, 658, 660, 665; arts in, 660, *p661*; climate regions and biomes of, 655–56; communications in, 663; economic activities of, 656, 662; ethnic groups in, 660; family and status of women in, 660–61; fishing in, 662; history and government of, *p650–51*, 657, *c659*, 661; independence of, *c658*; industrialization of, 662, 663; landforms in, 652–54, 655; mountain ranges of, 649, 652–3, *m667*; natural resources of, 656, 663–66; physical geography of, 652–56, 659; population patterns of, 659; trading partners of, 656; transportation in, 656; vegetation of, 655, 656; water systems of, 653, 654, 666.

indigenous people: of Australia, 773, *p774–75*, 781, 782, *g795*; of Canada, *p142*, 149, *d151*, 152; of Central America and Caribbean, *m194*, 195, *p204*, 209, *q212*; of East Africa, 513; of Equatorial Africa, 562; of Mexico, *p173*, 180–81, 182, 184, 190; of New Zealand, 773, *p774–75*, 781, 782; of North Africa, 365, 374–75; of Northern Europe, 257; of Oceania, 797, 804, 807;

rights of, *i142–43;* of South America, 215, 221, 223, *m226*, 227; of Southeast Asia, 760; of Tasmania, *c784;* of Transition Zone, 493; of United States, 118, 125, 127; of West Africa, 537. *See also* Native Americans.

indigenous religions, 203, 493, 842

Indo-Aryan people, 613

Indochina, 759, *c761*

Indo-European people, 418

Indo-Nepalese people, 660

Indonesia: economic activities of, 99, 752–53; landforms in, 751, 754, 755–56; manufacturing in, 753; migration from, *p432*, 433, 582; natural resources in, 753. *See also* Southeast Asia.

Indus River, 607, 629, 632, 633, 644

Indus River valley, 631, 634, 636, 645

industrial capitalism, 276

industrialization: in Arabian Peninsula, 443; in Australia, 787; in Bangladesh, *p638*, 639, 646; in Canada, 148, 151, *c154*, 155; in Central America and Caribbean, 201, 205, 208, 209–12; in Central Asia, 456, 460, 469; in China, *p678*, 697; in East Africa, 523; in Eastern Europe, 317, 322, 323, 324, 327–29; in Eastern Mediterranean, 401, 405; in Equatorial Africa, 564, 565; in India, 615, 617, 623; in Indian subcontinent, 662, 663; in Japan, 715, 717, 718, 720; in Mexico, 179, 185, 188; movement from agriculture to manufacturing activities in, 97; in New Zealand, 787; in North Africa, 377; in Northeast, 421, 427; in Northern Europe, 256, 260, 261, 263, *c265;* in North Korea, 725, 735; in Northwestern Europe, 271, 276, 281, 284–86; in Oceania, 808; in Pakistan, 639; in Russia, 334, 337, *i338*, *p340*, 345, 346, 353; in South America, 217; in Southeast Asia, *p753*, 762, 764, 765, 769; in Southern Europe, 303–4, *p307*, *c309;* in South Korea, 724, 725, *p736;* in Transition Zone, 495; in United States, 126, 129, 130, 436; in West Africa, 533

Industrial Revolution, 81, 106; in Northwestern Europe, 276; in United States, 127

industry: in cities, 103; cottage, 97; heavy, 104; light, 97; manufacturing, 97; service or information, 97, 787

Indus Valley civilization, 611

infant mortality rate, 463

infectious diseases: bacterial, 637; in developing countries, 206, 227; from European explorers, *p195*, 201, 224, 583, 782, 804; insects spreading, 558, 560, 637; from poor sanitation, 544; in Russia, 334, *i335*, 353; viral, 563, *m574*, *p575;* waterborne, 383, 589, 629; from worms, 575. *See also* HIV/AIDS.

information revolution, 81

infrastructure: lack of, 564; strained, 539; undeveloped, 636; urban, 103

Inland Sea, 719

Interior Lowlands of Canada, 144–45

internally displaced persons, 567, 568

International Atomic Energy Association, 498

International Date Line, 17, 798, *p799*

International Monetary Fund (IMF), *q283*, 521

international trade. *See* world trade.

Internet, 81, 155, 186, 734, 785, 809. *See also* computer technology and communications.

Inuit, *p57*, 149, 150; Nunavut as homeland for, 152

invasive species, 261, 262, 791

Ionian Sea, 295

Iran, 420; ethnic groups in, 411; history and government of, 417, *c418, c419, c422, q423;* landforms in, 412, *m413;* natural resources of, 414, 415, 420–21, 427; water systems of, 414

Iran–Iraq War, 411, 417, *c418*

Iraq: culture hearth of, *m80;* history and government of, 417, *c418–19*, 421, 444; landforms in, 391; natural resources of, 420–21, 427, *q430;* water systems of, 414

Ireland, 270; in British Isles, 270; economic downturn in, 325; landforms in, 271; natural resources of, 274. *See also* Northern Ireland.

Irish Sea, 288

Irrawaddy River, 756

Irtysh River, 338

Ishikari River, 705

ISIS. *See* Islamic State of Iraq and the Levant (ISIL)

Islam, 396, 440–41, 611, 634–35, 836. *See also* religions.

Islamabad, Pakistan, 636

Islamic Republic (Iran), 417, 420

Islamic Revolution, *c418,* 419

Islamic State of Iraq and the Levant (ISIL), 379

Islam, Kazi Nazrul, 638

islands: in Central America and Caribbean, 197, 199; in China, 683; in Equatorial Africa, 554, 556; high, 800, 801, *p802*, 803, 809; importance of, *q31;* in India, 607; in Indian subcontinent, 652, 653, 659; in Japan, 680, 701, 704, 727; low, 801, *p802*, 803, 812, *q815*, 816, *crt817;* in New Zealand, 776–77; in Northern Europe, 250–51, 252; in Northwestern Europe, 270–71; in Oceania, 797–818; in Russia, 711; in Southern Africa, 573, 576; in Southeast Asia, 754, 762; in Southern Europe, 291, 295; in South Korea, 727; Tasmania as, 776; in United States, 121, 800, 802

islets, 727

isolation policy, 730, 731

Israel: climate regions and biomes of, 393; desalination of seawater in, 39; economic activities of, 405; history and government of, *p388*, 390, 395, *c396–97;* immigration to, 323, 388; landforms of, 391; natural resources of, 394; water systems of, 394

Israeli–Palestinian conflict, 91, 389, *m388*

İstanbul, Turkey, 105, 419, 427

Isthmus of Panama, 43, 197, *c203,* 218

Italian Peninsula, 294–95, 297

Italy: agriculture in, *q297;* economic activities of, 282, 293, *c301,* 325; family and status of women in, 303; history and government of, 299, *c300, c301, c302;* industrialization of, 304, *c309;* landforms of, 295

ivory trade, 568, *g569, crt571,* 582, 588, 624. *See also* poaching.

J

Jagland, Thorbjorn, q283
Jakarta, Indonesia, 765, 769
James Bay, 145
Jamuna River, 629
Japan, 701–22; agriculture in, 705, 706; arts in, 714; climate regions and biomes of, 71, 706; communications in, 714, 715; economic activities of, 701, p702, ptg711, 712, 715–16, 718, 720; ethnic groups in, 701, 712; family and status of women in, 702, 714; fishing in, p707, p719, p720, 739; history and government of, 710, c712, 716, 730–31; industrialization of, 715, 717, 718, 720; landforms in, 680, 701, 704, m705, 727, 755; manufacturing in, 715; mountain ranges of, m705, 706, 712; natural resources of, p706–7, 715, 717, 719; physical geography of, 704–07; population patterns of, 102, p702–3; trading partners of, 710, p711, g715, 716, 809; transportation in, 705, 712–13; urbanization of, 712–13; vegetation of, 706; water systems of, 705, 706, p716, 718, 719, p720
Japan Current, 706
Japanese invasions, 731–32, c733, c761
jati, 611, 612, 614, 615
Java, 762
jazz, 129
Jerusalem, Israel, 388, 395
Jesus, 395, 396
Jews, c407; exile of, 372–73; murder in Holocaust of, 277; origins of, 395. See also Israeli–Palestinian conflict.
Jhelum, 632
Jidda, Saudi Arabia, 440
Jinnah, Mohammed Ali, p636
Johnson, Lyndon B., q772
Jonathan, Goodluck, 548
Jordan: climate regions and biomes of, 393; history and government of, 396, 397, 398; landforms in, 390, 391; natural resources of, 394, 401
Jordan Rift Valley, 390, 391
Jordan River, 391
Juárez, Benito, c183
Judaea, control of, 275

Judaism, 395, 838. See also Jews; religions.
Jundallah, q423
jungles: in Mexico, 188; in South America, 215. See also tropical rain forests.
Jupiter, 40, d41
Jutland Peninsula, 250, 251, 255

K

Kabila, Joseph, p563
Kabila, Laurent, p563
Kaesong Industrial Complex, 725, 732
kafala system, 433
Kafka, Franz, 324
Kaingaroa Forest, p794
Kalahari Desert, 577, 579–80
Kalmar Union, c256
Kandy, Sri Lanka, 659
Karachi, Pakistan, 632, 636, 639
Karakoram Mountains, 606, 630, 631
Kara-Kum, 455, 457
Karakum Canal, 456
Kariba Dam, 578, 579
karma, 614
Karnali River, 654
Kashmir, 612, 618, m619, 635, c636; Line of Control, 618
Katanga Plateau, 555
Kathmandu, Nepal, p653
Kauai, 803
Kazakhs, c461, 462
Kazakhstan, 454; agriculture in, 460; ethnic groups in, 411; industrialization of, 460, 469; natural resources of, 458; nuclear bases in, 469
Kemi River, 252–53
kente, 540
Kenya: economic activities of, 517; natural resources of, 510; water systems of, 507, 509, 510
Kerala, India, p75
Khartoum, Sudan, 487
Khayyám, Omar, 420
Khmer people, 762
Khmer Rouge, 760
Khomeini, Ayatollah Ruhollah, c418
Khosrow, Amir, 638
Khrushchev, Nikita, c343
Khyber Pass, 630
kibbutz, 394
Kievan Rus, 341

Kiev (Kyiv), Russia, 341
Kilimanjaro, 503, 506, 507, 517
Kilwa, p512
Kimberley, South Africa, 580
Kim Il Sung, 731
Kim Jong Un, 88
King, Martin Luther, Jr., 612
Kings Canyon, p777
Kinshasa, Democratic Republic of the Congo, 561, q563, 568
kinship groups, 804. See also clans.
Kiribati, 806, c807, q815. See also Oceania.
Kirthar Range, 630
Klar-Göta River, 252
Kolkata (Calcutta), 607; as megalopolis, 613
Kongo kingdom, 582
Korea, 710, 711; history and government of, 726, 730–32, c733. See also North Korea; South Korea.
Korea Bay, 727
Korean Peninsula, 723, p724, 725; climate regions and biomes of, 728; landforms in, 726; water systems of, 727–28
Korean War, 731
Korea Strait, 726
Koryo dynasty, 730
Koryo kingdom, c732
Kosi River, 654
Kosovo, 314; history and government of, 313, 321
Krakatau (Krakatoa), 755, 756
K2 (mountain), 627, 630
Kuala Lumpur, Malaysia, 762
Kūm River, 727
kums, 455. See also deserts.
Kuna culture, p204
Kunlun Shan range, 680, 681
Kurdish nationalist movement, 411
Kurdish people, 409, 419; history and government of, 411; as multinational ethnic group, 399, 411, 418; as stateless nation, m410, 411
Kurdistan, 411
Kurdistan Workers' Party, c419
Kuril Current, 706
Kush civilization, 489, 553
Kushites, 489

Kuwait: economic activities of, 443; invasion of, c419, 426, 444
Kuwait City, Kuwait, 440
Kyoto Protocol, 162, 287, 720
Kyrgyz (Kirghiz), 462
Kyrgyzstan, 454, 463, 469, q472
Kyūshū, Japan, 704
Kyzyl Kum, 455, 457

L

Labrador Current, 147
lagoons, 530, 554, 653, 777, 797, 801
Lagos, Nigeria, p539
Lahore, Pakistan, 636
Lake al-Assad, 391
Lake Baikal, 339, 350, 353, 354
Lake Balkhash, 456
Lake Baringo, 509
Lake Biwa, 705
Lake Chad, 485, p486, p487, 532, 533, 546
Lake Chad Basin, 497
Lake Chapala, 178
Lake Erie, 122
Lake Huron, 122
Lake Lanier, 132–33
Lake Lanier reservoir, 132
Lake Managua, 200
Lake Manchar, 632
Lake Maracaibo, 220–21
Lake Michigan, 122
Lake Nasser, 383
Lake Nicaragua, 198, 200
Lake Ontario, 122, m146
lakes: in Australia, 778; in Bangladesh, 632; in Belarus, 337; in Canada, 145, 161; in Central America and Caribbean, 198, 200; in Central Asia, 456; in China, 697; in East Africa, 503, 506, p508, 508–09, 517, 523; in Eastern Europe, 325, 327, 330; in Eastern Mediterranean, 391, 403, 405; in Equatorial Africa, 556; freshwater in, 54, 556; glacial, p54, 122; in Japan, 705, 718; in Mexico, 178; in New Zealand, 777, 778; in North Africa, 383; in Northeast, 413, 414; in Northern Europe, 250–51, 252, 263; in Northwestern Europe, 286; in Pakistan, 632; in Russia, 337, 339, 350, 353; in Siberia, 339, 353; in South America, 220–21, 228, 234, q235; in Southeast Asia, 768;

Index

in Transition Zone, 485, *p486, p487;* in Ukraine, 337, 338; in United States, 122; in West Africa, *p532–33,* 545. *See also individual lakes.*

Lake Superior, 122

Lake Tanganyika, 506, 508, 556

Lake Tiberias, 405. *See also* Sea of Galilee.

Lake Titicaca, 221, *q234*

Lake Turkana, *p508,* 509, 511

Lake Vänern, 252

Lake Victoria, 487, 508, 510, 523

Lake Volta, 486, *p532,* 532

Lakshadweep islands, 607

lama, 657

land area: of Canada, 141; of Democratic Republic of the Congo, 554; of Greenland, 252; of Maldives, 654; of South America, 218; of Transition Zone, 484; of United States, 120

land bridge, 176, 177, 196, 758

landfills, *p104, p405*

landforms: in Arabian Peninsula, 434–35; in Australia, 776–77; in Bangladesh, 630–31, *q647;* in Canada, 144–45; in Central America and Caribbean, 193, 196, *m197;* in Central Asia, 451, 454, *p455,* 461; in China, 680–81; on Earth, 43; in East Africa, 503, 506–08, 510; in Eastern Europe, 314, *m315;* in Eastern Mediterranean, 390–91; in Equatorial Africa, 554, *m555;* in India, 606–7; in Indian subcontinent, 652–54, 655; in Japan, 680, 701, 704, *m705,* 727, 755; in Mexico, 176–77; in New Zealand, 776–77; in North Africa, 365, 368–69; in Northeast, 412, *m413;* in Northern Europe, 248, *i249,* 250–52; in North Korea, 726–27; in Northwestern Europe, 270–71; in Oceania, 800–*p802;* in Pakistan, 630, *m631;* in Russia, 336–37; in South America, 218, *p219;* in Southeast Asia, 754–56; in Southern Africa, 576, *p577;* in Southern Europe, 294, *p295;* in South Korea, 726–27; in Transition Zone, 484–85, 530; in United States, 120; in West Africa, 530–31

landmines, 411, 469, 470

land subsidence, 189

language families, 78, *m79. See also* languages.

language in cultural development, 78

languages: in Arabian Peninsula, 440; in Australia, 774, 775, 783, 785; in Bangladesh, 635, 637; in Belarus, 343; in Canada, 153; in Central America and Caribbean, 200, 203; in Central Asia, 463; in China, 688–89, 695; dialects of, 203; in East Africa, 511, 514; in Eastern Europe, 313, 322; in Eastern Mediterranean, *c398,* 399; in Equatorial Africa, 553, 562; in India, 603, 611, 613; in Indian subcontinent, 659, 660; indigenous, 514; in Japan, 713; loss of rare, 92; in Mexico, 180; in New Zealand, 774, 775, 785; in North Africa, 374, 376; in Northeast, 411, 419; in Northern Europe, 255; in North Korea, 733; in Northwestern Europe, 279; in Oceania, 801, 805, 806, 807; in Pakistan, 635, 637; pidgin, 562, 583, 785, 807; in Russia, 344; in South America, 226; in Southeast Asia, 751, 762; in Southern Africa, 581, 586; in Southern Europe, 302; in South Korea, 733; in Transition Zone, 493, 494; in United States, 119; in West Africa, 539–40

Laos, landforms in, 754, 756. *See also* Southeast Asia.

Laozi (Lao-Tzu), 686

latifundia, 205

Latin America: cities in, 86; colonial rule in, 181; Human Development Index score of, *g107;* impact of El Niño in, *d220;* oil reserves in, *c222;* socioeconomic development in, 97; U.S. immigrants from, 129

latitude (parallels), 16, *m17;* effects on vegetation of, *d71;* influence on climate of, 64, *d71*

Latvia, 314, 318; history and government of, *c321,* 325, *c342*

Law of the Sea Treaty, 810, *q811*

Lebanon: climate regions and biomes of, 393; ethnic groups in, 411; history and government of, *c396,* 397; landforms of, 390, 391; natural resources of, *q394,*

401, 404–5, 406; water systems of, 394

Lebanon Mountains, 391

Lee Myung-bak, 740

leeward side of mountain, *d68*

legend (key), map, 18

Lempa River, 198, 209

Lena River, 338

Lenin, Vladimir, *c342*

León, Nicaragua, 200

Leopold II, 560

Lerma River, 178

less developed countries, 97, *c97*

Lesser Caucasus, 455

Lesser Himalaya range, 652, 653

Levant, 390

Liberia: history and government of, 536–37; independence of, *c536;* in Transition Zone, 483, 530

Libreville, Gabon, 556

Libya: boundaries of, 90; climate regions and biomes in, 371; colonial rule of, *m373;* history and government of, 293, 374, 384, 400; landforms in, 368, 369; natural resources of, 371, 380, 381; terrorist attack in, 91; water systems of, 380, 381, 384

Libyan Desert, 381

Liechtenstein, 270

life expectancy, 77

Limpopo River, 578

line of control (LOC). *See* Kashmir.

lingua franca, 514

Lisbon, Portugal, 302

lithosphere, *d42;* earthquakes in, 48; landforms in, 43

Lithuania, 314; history and government of, *c321,* 325, *c342*

Livingstone, David, 512

llanos, 219, 221

loans and start-up capital, 516, 586–87, 636

location: absolute, 17, 29; determining, 16–17; relative, 17, 29

loess, 270, 681–82

logging: in Canada, 158, 161; in Central America and Caribbean, 210; corporate, 188; in East Africa, 522; in Eastern Europe, 326; in Equatorial Africa, 564, 568, 570; in India, 624; in Indian subcontinent, 656; illegal, 326; in Mexico, 188; in New Zealand, 794; in Northern

Europe, 262; in Oceania, 812, 815; in Southeast Asia, 764; in Southern Africa, 588; in West Africa, 530. *See also* deforestation; forests; rain forests; reforestation; timber.

Logone River, 486

Loire River, 273

Loma Mountains, 531

London, England, 279

longitude (meridians), *m17*

Long Wall of Namib, *p577*

Louvre Museum, 267

lowlands: in Australia, 776, 778; in Bangladesh, 631, 632; in Canada, 144–45, 152; in Central America and Caribbean, 196; in Central Asia, 455, 457; in China, 681; in Indian subcontinent, 653, 654, 655, 656; in Japan, 705, 712; in Northeast, 415; in Northern Europe, 251; in North Korea, 726, 733; in Russia, 336; in South America, 221; in Southeast Asia, 757; in South Korea, 726; in Ukraine, 338; in United States, 120

Loyalists, 152

Lualaba River, 556

Lucknow, India, *p613*

lumber. *See* forests; logging; timber.

Luther, Martin, 276, *p277*

Luvua River, 556

Luxembourg, 270; natural resources of, 285

— M —

Maasai Mara, 524

Macau, China, 682

Macedonia, 313, 314, *c321,* 330; agriculture in, 324–25; economic activities of, 324

Machu Picchu, 223, 227

Mackenzie River, 145

Madagascar, 576; history and government of, 581, 583; landforms of, 577

Maghreb, 368

magma, 45, *d48,* 49

Mahābhārata, 615

Mahaweli River, 654

Main-Danube Canal, *m316*

Main River, 316

Malagasy people, 581

Malay Peninsula, 754, 756

Malay people, 761, 762

Malaysia: economic activities of, 752, *p753;* history and government of, 753; industrialization of, 97. *See also* Southeast Asia.

Maldives, 649–68; history and government of, 660. *See also* Indian subcontinent.

Male, Maldives, 659

Mali: in Transition Zone, 483; water systems of, 486, 487

Malielegaoi, Tuilaepa Sailele, 799

Mali Empire, 490, 535–36

Malta, 295

Manantali Dam, 487, 533

Manchuria, Japanese invasion of, 711

Manchurian Plain (Northeast Plain), 681

Mandé, 493

Mandela, Nelson, *c584, p585, q594*

Mangla Dam, 632

Manila, Philippines, 762, 765, 769

Manitoba, 152; wetlands in, 158–59

Mannar Island, 653

mantle, of Earth, 44, *d45*

mantras, 660, 661

manufacturing: in Arabian Peninsula, 443; in Australia, 787; in Bangladesh, 639, 644; in Canada, 155; in Central Asia, 455, 465; in China, 691; in Eastern Europe, 327; in Eastern Mediterranean, 400, 401; in Equatorial Africa, 564, 565; in India, 617; in Japan, 715; in Mexico, 185, 186, *g191;* in North Africa, 376; in Northeast, 421; in Northern Europe, 260; in North Korea, 733; in Northwestern Europe, 278; in Oceania, 802, 809; in Pakistan, 639; in Russia, 346, 350, 353; in South America, 228; in Southeast Asia, *p753,* 769; in Southern Africa, 587; in Southern Europe, 303, 307; in South Korea, *p725, p736;* in United States, 97, 126, 127, 131; in urban areas, 102

Manufacturing Belt, 127, 128

Maori people, *p774,* 775, 781, 782, 783, *c784–85,* 784–85

Mao Zedong, *p686,* 688

map, 14; parts of, 18–19; purpose of, 18; scale of, *m19;* types of, 19–20, 33; uses of, *p28,* 33, 41

map projection, 15, *m16;* conic, 15; cylindrical, 15; interrupted, 16; planar, 15

Mapuche, 223

Maputo, Mozambique, 586

Maputo Protocol. *See* Protocol to the African Charter on Human and Peoples' Rights of Women in Africa.

maquiladoras, *q186*

Margaret of Denmark, *c256*

Mariana Islands, 800, 806, 808. *See also* Oceania.

Mariana Trench, 43, 802

marine west coast climates, 71; in Canada, 147; in New Zealand, 780; in Northern Europe, 253; in Northwestern Europe, 273; in Southern Europe, 297; in United States, 124

Maritsa River, 297

market economy, 94, 95; in Canada, 154; in Central Asia, 465; in Eastern Europe, 323, 324; in Japan, 715; in Russia, 96, 346; in United States, 130

Marlborough Sounds, 777

Mars, 40, *d41*

marsh: in Bangladesh, 631, 632; in China, 681; in India, 610; in Northeast, 424, 425, *p428;* in West Africa, 531

Marshall Islands, 801, 805; history and government of, *c806, c807. See also* Oceania.

Marsh Arabs, 425, *p428*

Marx, Karl, 342

Masai people, 516

massifs, 294, *p295,* 369

mathematics, in study of maps, 18

Mato Grosso Plateau, 219

Matsuura, Koïchiro, *q93*

Mauna Kea, 802

Mauritania, 368; in Transition Zone, 483; water systems of, 487

Mauritius, 576

Maya civilization, 180, 181, 184, *q192*

Mbuti people, 559, 562

McMahon Line, 651

Mecca (Makkah), Saudi Arabia, 396, 433, 440

Medina (Madinah), Saudi Arabia, *c440*

Mediterranean climates, 71; in Australia, *q371;* in Central Asia, 457; in Eastern Europe, 318; in Eastern Mediterranean, 392, 393; in North Africa, *q371;* in Northeast, 415; in Northwestern Europe, 273; in Southern Africa, *q371,* 579; in Southern Europe, 297–98, 305–6; in United States, 124, *q371*

Mediterranean Sea, 291, 297, 391, 413; Eastern Europe channel to, 317; environmental problems in, 284, 305, 307, 402; fish populations in, 383, 402, 406; geographic crossroad for cultures around, 302, 392; North Africa drainage into, 487; Northwestern European drainage into, 273; population centers around, 375; separating Africa and Europe, 369; size of, 52; Southern European drainage into, 295; Suez Canal at, *p366,* 367

megacity, 182

megalopolis: in California, 129; in India, 613

Meghna River, 629

Meiji Restoration, 711, *c713*

Mekong River, 756, 757, *q768,* 769

Mekong River Delta, 757, *p758,* 769

Melanesia, 800, 801, 805, 808. *See also* Oceania.

Melbourne, Australia, 784

meltwater, 327, *g328*

MENA. *See* Middle East and North African countries.

mental maps, *m20–21,* 29

mercantilism, 611, 612

Mercator projection, 15, *m16*

merchant marine, 691

Mercury, 40, *d41*

Merkel, Angela, 279

Meroë, 489

Mesopotamia, 414, 416, *m425*

Mesopotamian Marsh, 424

mestizos, 181

meteoroids, 41

meteorologists, 27

Métis, 149

Metro Grand Paris Plan, *p269*

metropolitan areas, 103. *See also* cities; megalopolis; urbanization.

Mexican Plateau, 176–77

Mexican Revolution, 181

Mexican–U.S. border, 176; migration across, *m185*

Mexico, 173–92; agriculture in, 177, 178–79, 180, 181, 182, 188; arts in, 184; cities in, 177; climate regions and biomes of, *c178,* 179; coffee crop of, *q98;* communications in, 186; culture hearth of, *m80;* economic activities of, *c97,* 99, 184–86, 189; ethnic groups in, 180–81; family and status of women in, 183, *c184,* 227; fishing in, 178, 179; growth of, *p174,* 189; history and government of, 180–81, *c182–83,* 184; independence of, 181, *c182,* 184; industrialization in, 179, 185, 188; landforms in, 176–77; manufacturing in, 185, 186, *g191;* mountain ranges in, 176, 180, 188; natural resources in, 173, *p179,* 185, 187–90, *c222;* as newly industrialized country, 97; physical geography of, 176–79; population patterns in, 176–77, 181–83; trading partners of, 151, 185, 186; transportation in, 185, 190; treaty of United States and, 90, *c182;* urbanization of, 177, 182, 188–89, 189; vegetation of, 179; water systems of, 176, 178, 188, 189

Mexico City, Mexico, 86, 180, 182–83, 184, 185, 189

Michelangelo, 303

Micronesia, 800, 805–6, *c807,* 808. *See also* Federated States of Micronesia; Kiribati; Nauru; Oceania.

Mid-Atlantic Ridge, 47, 248

Middle Ages, 300

Middle East and North African countries, 384

Middle Passage, 561

midlatitude climates, 510, 778. *See also* humid continental climates; humid subtropical climates; marine west coast climates; Mediterranean climates.

midlatitude zone, 64

midnight sun, 62

migrant workers: in Arabian Peninsula, *p432,* 433, 439; in Indian subcontinent, 660–61; in Transition Zone, 493. *See also* guest workers.

migration of human populations, 81, 82, *q86;* to Australia, 782, 783, *c784–85,* 816; in Central America and Caribbean, 193, 205; from Central Asia, *i468;* in China, 688; from East Africa, 582; to East Africa, 512; in Eastern Europe, 319, *m322,* 323; to Eastern Mediterranean, 388, 399; to Equatorial Africa, 559; to Indian subcontinent, *c658;* from Mexico, *m185;* to New Zealand, 781, 782, *c784–85,* 816; from North Africa, *m292,* 293, 375; to North Africa, *m373;* in Northern Europe, 257; in Northwestern Europe, 267, 277, 278; from Oceania, *q815,* 816; in Oceania, 797, 805, 815; in Pakistan, 636; in Russia, 334, *i335,* 344; in South America, 225, *m226, c237;* from Southeast Asia, 775; in Southern Africa, 581, 582, 586; in Southern Europe, *m292,* 293, 301, 304; in Transition Zone, 489, *q491, c492,* 493; to United States, 816; in West Africa, 535, 545

military dictator, 761. *See also* caudillo.

Milošević, Slobodan, 313

Minamoto,Yoritomo, *p712*

mineral resources and mining: in Arabian Peninsula, 437; in Australia, 776, 780, 782, 787, 792; in Bangladesh, 633; in Canada, 148; in Central America and Caribbean, 199; in Central Asia, 453, 458, 465, 466, 469; in China, 684; in East Africa, 510, 517; in Eastern Europe, 317, 318, 325, 328, 330; in Eastern Mediterranean, 394; in Equatorial Africa, 551, 558, 564, 566; in high seas, 810; in India, 610, 617, 624; in Indian subcontinent, 656, 662, 664; in Indonesia, 753; in Japan, 707; in Mexico, *p179,* 181, 187; in New Zealand, 780, 787; in North Africa, 371, 377; in Northeast, 415; in Northern Europe, 254; in North Korea, 725, 729; in Northwestern Europe, *p274,* 284; in Oceania, 803, 809, 812; in Pakistan, 633; in Russia, 336, 350, 354; in South America, 222, 224, 228–29, 234; in Southeast Asia, 758, 764, 766, 769; in Southern Africa, 580, 585, *g587;* in Southern Europe,

p298; in Transition Zone, 488, 490, 491, 494–95; in Ukraine, 347; in United States, 124; in West Africa, 534, 541. *See also* coal mining.

Ming dynasty, 686

minifundia, 205

missionaries: in China, 686; in Equatorial Africa, 560; in Japan, 710; in Oceania, *c806;* in Philippines, 763; in Southern Africa, 586

mistral, 273

mixed economy, 95

mixed forests, 71

mobile communications. *See* phone communications.

Moche, 223, *p227*

Moguls (Mughals), 611, 634

Mohenjo Daro, Pakistan, 634

Moldova, 314

Monaco, 270, 273

monarchy, 88, 89, 256, 438

Mongolia: ethnic groups in, 688; independence of, *p687;* languages in, 92; population patterns in, 688

Mongol invasions, 319, 341, *c460,* 730, *c733*

monoculture, 233

monotheism, 395, 396

Monrovia, Liberia, 537

monsoon: in Arabian Peninsula, 436; in Australia, 779; in Bangladesh, 629, 632, 633; in China, 683; in India, 607, 608, 609; in Japan, 706; in Pakistan, 633; in South Asia, 655; in Southeast Asia, 757; in South Korea, 728

monsoon cycle, 683

monsoon forests, 757

montane climates, 557

Mont Blanc, 270

Montenegro, 314, 321, 330; agriculture in, 325; natural resources of, 322

Montreal: growth of, 146; as industrial and shipping center, 152

Montserrat, 212

moraines, 50

more developed countries, 97, *c97*

Moremi Game Reserve, *p578*

Morocco: climate regions and biomes in, 371; economic activities of, 377; independence of, *c374;* mountain ranges in, 368;

trading partners of, 383; water systems of, *p381,* 384

Moro people, 761

mortality rate. *See* death rate.

mortgages, negative equity, *m130,* 131

Moscow, Russia, 342, 347

moshav, 394

Moskva River, 353

mosque, 396

Mosquito Coast of Nicaragua, 201

mountain glaciers, 50

mountain ranges: in Australia, 776; in Canada, 144, 145; in Central America and Caribbean, 197, 218; in Central Asia, 451, 454, *p455,* 457, 462, 464, 466; in China, 677, 680, 681, 682, 683; in East Africa, 506, 507, 509; in Eastern Europe, 317, 318, 322; in Eastern Mediterranean, 390–91, 393; in Equatorial Africa, 554, 555, 567; in India, 606, 608, 613; in Indian subcontinent, 649, 652, 653, *m667;* in Japan, *m705,* 706, 712; in Mexico, 176, 180, 188; in New Zealand, 777, 778, 780; in North Africa, 368–69; in Northeast, 411, 412, *m413,* 414, 415; in Northern Europe, 251–53; in North Korea, 725, 726, 727, 729; in Northwestern Europe, 270, 271; in Oceania, 797, 800, 801; in Pakistan, 630, 631; rain shadow effect on, *d68;* in Russia, 336, *p337;* in South America, 47, 218, *p219,* 221, 225; in Southeast Asia, 755–56, 757; in Southern Africa, 576–77; in Southern Europe, 294, *p295,* 296, 297; in South Korea, 725, 726, 729; in United States, 120, 802, 803; in West Africa, 531

Mountains of the Moon. *See* Ruwenzori Mountains.

Mount Ararat, 413

Mount Cameroon, 531, 555

Mount Cook, 777

Mount Damāvand, 413

Mount Elbrus, 336

Mount Etna, 295

Mount Everest, *p37, q668;* climbers of, *q654;* as highest point on Earth, 43, 52, 652, 680; in Himalaya mountain range, 606

Mount Fairweather, 144

Mount Hermon, 390

Mount Kanchenjunga, 606

Mount Kenya, 506, 507

Mount Keokradong, 631

Mount McKinley, 120

Mount Merapi, 755

Mount Paektu, 726

Mount Pidurutalagala, 653

Mount Ruapehu, 777

Mount Stanley, 555, 556

Mount Tahat, 368

Mount Tambora, 755

Mount Toba, 755

Mount Waialeale, 803

Muévete en Bici, 190

Mugabe, Robert, 584

Mughal Empire. *See* Moguls (Mughals).

Muhammad, 396, 422, *c440*

mujahideen, 461

mullahs, 417

multinational organizations, headquarters of, 98

multiple nuclei model of urban land use, *c103,* 104

Mumbai (Bombay), India, *p90,* 607; as megalopolis, 613

Muonio River, 253

Murray-Darling River Basin, 792

Murray River, 778

Musa, Mansa, 490, 536

Musharraf, Pervez, 635

music: in Central America and Caribbean, *p193,* 204; in East Africa, 516–17; in Eastern Europe, 323–24; in India, 603; in Japan, 714; of Native Americans, 129; in North Africa, 376; in North Korea, 734; in Pakistan, 638; in South America, 227; in South Korea, 734, 735; in Transition Zone, 494; in West Africa, 540

Muslim League, *c636*

Muslims. *See* Islam; Shia Muslims; Sunni Muslims; *see also* religions.

Myanmar (Burma), landforms in, 754, 756. *See also* Southeast Asia.

N

NAFTA. *See* North American Free Trade Agreement (NAFTA).

Nagasaki, Japan, 710, 711

Nagorno-Karabakh, 462–63, 467

Nairobi, Kenya, 514, 561

Naktong River, 727

Namib Desert, *p577,* 579

Nara, Japan, *c712*

Nashua River, 138

Nasser, Abdel, *c374*

National Family Planning Program, *p604*

national identity, 785

nationalism, 90, 373

National Liberation Front, *c374*

National Population and Family Planning Commission, 692

nation-states, 410, 411

Native Americans: in Canada, 152; conflicts of colonial settlers and, 126; influence of location and climate on, 125; land claims of, 127; music of, 129; traditions of, *p117;* in United States, *m118,* 125, 127, 844. *See also* ethnic groups; indigenous people.

native people. *See* indigenous people.

native plants versus introduced plants, 791

NATO. *See* North Atlantic Treaty Organization.

natural boundary, *p89,* 90, 416, 417

natural disasters: in India, 609; in North Korea, 735; in Pakistan, 635

natural gas: in Bangladesh, 633, 644; in Central Asia, 400, 453, 458, 465, 466, 467; in China, 678, 684; in East Africa, 510; in Eastern Europe, 318; in Eastern Mediterranean, 394, 400, 401; in Indonesia, 753; in Japan, 707; in Mexico, 179; in New Zealand, 787; in North Africa, 304, 371, 377, 384, *g385,* 400; in Northeast, 414, 415, 420, *i426,* 427, *q430;* in Northern Europe, 254, 257; in Northwestern Europe, 274, 287; pipelines for, 465, 466, 467, 725; in Russia, 346, 347, 350, *p354;* in Southeast Asia, 758, 769; in Southern Europe, 304; in Southwest Asia, 400; studies of, *q138;* in Transition Zone, 491, 494

natural increase. *See* population growth.

natural resources: in Arabian Peninsula, 431, *p437,* 444–48, *c449;* in Australia, 787, 790–94; in Bangladesh, 642–46; in Canada, *m100,* 146, 154–55, 158–60; in

Central America and Caribbean, 196, 199, 201, 208–212, *q214;* in Central Asia, 400, 453, 458, 464, 466–70; in China, *m101,* 684, 692, 694–*p98;* distribution of, 96, 98; in East Africa, 510, 513, 520–24; in Eastern Europe, 317, *p318,* 325–30; in Eastern Mediterranean, 394, 401, 402–06; in Equatorial Africa, 559, 566–70; in high seas, 810; in India, *m101,* 610, 620–24; in Indian subcontinent, 656, 663–66; in Japan, *p706–07,* 715, 717–20; in Mexico, 173, *p179,* 185, 187–90, *c222;* in New Zealand, 787, 790–94; in North Africa, 304, 371, 376, 377, 380–84, *g385;* in Northeast, 415, 421, 424–28, *q430;* in Northern Europe, 173, *p179,* 185, 187–90, *c222;* in North Korea, 725, 729, 737–40; in Northwestern Europe, *p274,* 276, 280, 284–88; in Oceania, 803, 812–16, *q818;* in Pakistan, 642–46; patterns of, *m100–01;* in Russia, 333, 340, 346, 350–54, *q356;* in South America, *m100,* 217, 219, *c222,* 223, 228–29, 231–33, 236; in Southeast Asia, 758, 764, 766–70; in Southern Africa, *m101, p580,* 585, 588–92, *crt593;* in Southern Europe, *p298,* 304, 305–08; in South Korea, 729, 737–40; in Transition Zone, 488, 496–*p500;* in United States, *m100,* 124, 126, 129, 130, 134–38; in West Africa, 527, 534, 544–48. *See also individual countries and types of resources.*

natural vegetation of biome, 69, 70

Nature Conservancy, 308, 592

Nauru: history and government of, 805, *c806;* natural resources of, 809. *See also* Micronesia; Oceania.

Nazca Plate, 47, *m177*

Nazis, *c320*

Negev Desert, 390, 391

Nehru, Jawaharlal, 612

Neman River, 338

NEPAD. *See* New Partnership for Africa's Development.

Nepal, 649–68; as buffer state, *m650,* 651; history and government of, 651, 657–58; mountain ranges of, 649. *See also* Indian subcontinent.

Neptune, 40, *d41*

Netherlands, 270; economic activities of, 280; landforms of, 271–72; natural resources of, 274, 285

neutrality, political, 612

New Caledonia, 800, 805; natural resources of, 809. *See also* Oceania.

New Delhi, India: air pollution from vehicles in, 623; water systems of, *p622. See also* Delhi.

New Guinea, 805; history and government of, *c806. See also* Oceania; Papua New Guinea.

newly industrialized countries, 97, *c97*

New Partnership for Africa's Development, 529

New Testament, 396

"New Urbanism," 106

New Zealand, 773–96; agriculture in, 777, 778, 781, 782, 787; arts in, 786–87; climate regions and biomes of, 779–80; economic activities of, 787; ethnic groups in, *p774,* 775, 780–3, *c784–85;* family and status of women in, 786; fishing in, 775, 777, 782, 787; history and government of, 781–82, *c784–85;* immigration to, 783, 816; independence of, 782; industrialization of, 787; landforms in, 776–77; mountain ranges of, 777, 778, 780; natural resources of, 787, 790–94; physical geography of, 776–80; population patterns of, 783–85; trading partners of, 787, 799; transportation in, 777; urbanization of, 778, 784; vegetation of, 780; water systems of, 777, 778, 780, 791

NGOs. *See* nongovernmental organizations (NGOs).

Niamey, Niger, 492

Nicaragua, 200, 201, 208, 212

Nicholas II, *c342*

Niger: landforms in, 368, 369; in Transition Zone, 483

Nigeria: economic activities of, 541; history and government of, 537; natural resources of, 495, 537; population growth in, 539; in Transition Zone, 483, 530

Niger River, 486, 533

Nile Delta, 365, 369, *p370*

Nile River: development along, 369, 372, 376; drainage basin of, 555; flooding and dams of, 269, *c382,* 383; tributaries of, 487

Nilgiri Hills, 606

Nimba Range, 531

nitrate concentrations: agricultural runoff in Europe, *g328*

Nobel Peace Prize, 282

nomads: in Arabian Peninsula, 431, 438, 440; in Australia, 775, 781; in Central Asia, *p457,* 459, 460, 464; in China, 685; in East Africa, 514; in Eastern Mediterranean, 391; in North Africa, 374, 375, 376; in Transition Zone, 483

nongovernmental organizations (NGOs): in Australia, 794; in Central America and Caribbean, 206; in Central Asia, 470; in Eastern Europe, 329–30; in Eastern Mediterranean, 400, 406; in Equatorial Africa, 570; in New Zealand, 794; in Northeast, 428; in Northwestern Europe, 288; in Southern Africa, 591–92; in Southern Europe, 308; in West Africa, 547

Nordic countries, 255, 259. *See also individual Scandinavian countries.*

Nordic model, 260

Norrland, 252

North Africa, 365–86; agriculture in, 370, 371, 374, 376, 377, 380, *c382,* 383; arts in, *p376;* climate regions and biomes of, 370–71, 484; communications in, 377; economic activities of, 292–93, 366, 371, 374, 376–77, 384; ethnic groups in, 365, 373, 374; family and status of women in, *p376;* fishing in, 377, 382, 383, 384; history and government of, 370, 372–75, 376, 377, 417; immigration to, *m373;* independence of, 373, *c374–75,* 376; industrialization of, 377; invasions of, 372, *m373;* landforms in, 365, 368–69; manufacturing in, 376; mountain ranges of, 368–69; natural resources of, 304, 371, 376, 377, 380–84, *g385,* 400; physical geography of, 368–71; population patterns of, 374–75, 383; trading partners of, 365, *p366,* 367, 377; transportation in,

Index

365, 372, 377, 393; urbanization of, 376, 380; vegetation of, 369, 370, q371, 372, 383; water systems of, p366, p367, p369, p370, 372, 375, 377, 380–81, c382, 384, 487, 518–19

North America: coastal precipitation in, 67; climate of, 71; colonial exploration of, 149–50; as continent, 43; earthquakes in, 48, m177; land bridge of, 176; satellite view of, p24; urban population of, 85. See also Canada; Mexico; United States.

North American Free Trade Agreement (NAFTA), 138, i156, q157, 186

North American tectonic plate, m177

North Anatolian Fault, 412

North Atlantic Current, 273

North Atlantic Drift, m253

North Atlantic Ocean, 250, 273, 288

North Atlantic Treaty Organization, 91, c321; in Northeast, 417

North Carolina, biotechnology industry in, 131

North China Plain, 681, 682, 685

Northeast, 409–30; agriculture in, 416, 420, 425, i426, 427; arts in, 420; climate regions and biomes of, 415; communications in, 421; economic activities of, 417, 420–21, q430; ethnic groups in, 409, m410, 411, 418, 425; family and status of women in, p420; fishing in, 414, 420; history and government of, 416–19; independence of, 417; industrialization of, 421, 427; invasions of, c419; landforms in, 412, m413; manufacturing in, 421; mountain ranges of, 411, 412–413, 414, 415; natural resources of, 415, 421, 424–28, q430; physical geography of, 412–15; population patterns of, 418–19; trading partners of, 421; transportation in, 414, 415, 421, 427; urbanization of, 418, 419; vegetation of, 415; water systems of, 412–3, p414, 415–16, 421, 424–25

Northeast Plain. See Manchurian Plain.

Northern Europe, 247–66; agriculture in, 254, 260, 262, 263, 264, c265, 280; arts in, p259; climate regions and biomes of, m253–54, 263–64; communications in, 255; economic activities of, c97, p248, 258, 260, c265, 325; ethnic groups in, 257; family and status of women in, 259, q260; fishing in, 258, 260, 261; history and government of, 251, 255–57; immigration to, 257; independence of, c256–57; industrialization of, 256, 260, 261, 263, c265, 280; integrated, 256; landforms in, p248, 250–52; manufacturing in, 260, 280; mountains of, 251, p252; natural resources of, 254, 260, 261–64; physical geography of, 250–54; population patterns of, 257, g258, 259, q260; trading partners of, 256, 260; transportation in, p248, i249; urbanization of, 258, 279; vegetation of, 253, 254, 261, 262, 264; water systems of, 250, p251, 252–53, 254, 261, 263, 279

Northern European Plain, 270, 273, 336–37, 341

Northern Hemisphere, 17; seasons in, 61, 62

Northern Ireland, 277

northern lights, 247

Northern Mariana Islands, 808

Northern Plateau (Mexico), 176–77

North Island (New Zealand), 777

North Korea, 723–42; agriculture in, 725, 729, 739, q742; arts in, 734, p735; climate regions and biomes of, 728–29; economic activities of, 735–36; ethnic groups in, 732; family and status of women in, 734; fishing in, 729; history and government of, 723, 725, 726, 730–33; independence of, c733; industrialization of, 725, 735; landforms in, 726–27; manufacturing in, 733; mountain ranges of, 725, 726, 727, 729; natural resources of, 725, 729, 737–40; physical geography of, 726–29; population patterns of, 732–33; trading partners of, 736; transportation in, 728;

urbanization of, 733; vegetation of, 728; water systems of, 726–28, 737. See also Korea; Korean Peninsula.

North Pole: midnight sun at, 62, 64

North Sea, 250, 261, 272, 274, 285, 288

Northwestern Europe, 267–90; agriculture in, 281, 285; arts in, 267, p279; climate regions and biomes of, 273–74; communications in, 281; economic activities of, 276, 277, m280, 280–83, 325; ethnic groups in, 275; family and status of women in, 277, 278, 279; fishing in, 284–g285, 288; history and government of, 275–77, 417; immigration to, 267, 275, 278; independence of, 277; industrialization of, 271, 276, 281, 284, 285, 286; landforms in, 270–71; manufacturing in, 278; mountain ranges of, 270, 271; natural resources of, p274, 276, 280, 284–88; physical geography of, 277; population patterns of, m278, 279; suffrage in, 279; trading partners of, 276, 281–83, 377; transportation in, p268–69, 271, d272, 273, 276, 281, 285; urbanization of, p268–69, 276, 277, 279; vegetation of, 273–74; water systems of, 270, 271–73, 276, q288

Northwest Territories, 158–59

Norway: economic activities of, c265; family and status of women in, 259; history and government of, 255, c256–57, 279; landforms of, 251; natural resources of, 254, 260, 264; as part of Scandinavia, 250; population patterns of, 257; water systems of, 252

Nubian Sandstone Aquifer System, 384

nuclear contamination, m351, m352

nuclear energy: in China, 694; in Japan, 704, 717, 718; in Northwestern Europe, 274; in Russia, m351, 353; in South Korea, 729

nuclear safety standards, 352–53

nuclear threat: in Cold War, 343; Iranian, c419, 420, 421

nuclear wastes, 352

nuclear weapons testing, 635, 732, p807, 814, 815, c816

Nullarbor Plain, 776

Nunavut, 152

────── O ──────

oasis, 70, 370, p381, 457

Obama, Barack, p88

Ob' River, 338

ocean currents, 65, m67; effects on Canada's climates and animals of, 147; influences on weather of, 66–67; patterns of, 66; reversal by El Niño of, 67

Oceania, 288, 797–818; agriculture in, 802, 803, 808–9, 812, 816; area of (land), 806; arts in, p808, 809; climate regions and biomes of, 803; communications in, 809; economic activities of, 799, 802, 805, 808–9, 813; ethnic groups in, 804, 805, 807; family and status of women in, 808; fishing in, 802, 809, 815; history and government of, 804–5; immigration to, 802, 804, 805, 807; independence of, 805; industrialization of, 808; landforms in, 800, p802; manufacturing in, 802, 809; mountain ranges of, 797, 800, 801; natural resources of, 803, 812–16, q818; nuclear weapons testing in, p807, 814, 815, c816; physical geography of, 797, 800–03; population patterns of, 801, 805–6, 815, 816; trade, 799, 802, 806, 809; transportation in, 809; vegetation of, 803, p814, 815; water systems of, 801, 802, 812–14, 815

oceans: climate change involving, 72; humidity levels affected by, 71; rising levels of, 210; salt water in, 52; temperature levels of, d220, 221, 814. See also oceans under their specific names.

Oder River, 317

offshoring industries, 131

oil. See petroleum.

oil pollutants, 307

oil spills: in North Africa, 382–83, 384; in Northeast, 427; in Northern Europe, 263–64; in Russia, 350, 354; in West Africa, 531, 533

Okavango Delta, 578

Okavango River, 577–78, 590
Okinawa, Japan, 706
old-growth forests, 158, 159
Olid, Cristóbal de, 200–01
oligarchy, 88, 89
Oman, economic activities of, 443
Ondaatje, Michael, 661
one-child policy, 688, 690, *p692,* 693
Ontario, wetlands in, 158–59
OPEC. *See* Organization of Petroleum Exporting Countries.
oral tradition, 494, 516; griot, 540
Orange River, 578, 590
organic farming, 665
Organization of Petroleum Exporting Countries, 418, 421, 443, 495, 541
Orinoco River, 222
Oslo Accords, *c396*
Oslo, Norway, 259
Oslo Opera House, 259
Osman's Dream, 420
Ottawa, growth of, 146
Ottoman Empire, 277; in Arabian Peninsula, 438–39; in Eastern Europe, 313, *c320;* in Eastern Mediterranean, 396; ending of, 417; in North Africa, 373; in Northeast, 411, 416; scope of, 417
Outback, 776, 778, 780, 785
Outer Himalaya range, 652
overfarming, 517
overfishing: in Canada, 148, 155, 159, 161, *q164;* in Eastern Europe, 318, 325; in Eastern Mediterranean, 402; in high seas, 810; in Japan, 719, 720; in North Africa, 383, 384; in Northern Europe, 261; in Northwestern Europe, 284, *g285;* in Oceania, 812; in West Africa, 546
overgrazing, 233, 404, 427, 497, 546, 698, 791
overlap of biomes, 69, 70
overpopulation: in East Africa, 516; in Transition Zone, 481, 496; in West Africa, 527. *See also* population density.
Oxfam International, 547, *c549, q550*
oxisols, 233, 235

P

Pacific Lowlands, 196
Pacific Ranges, 120, *d121;* high altitude climate of, 124, 147; rainfall from ocean air in, 147; tectonic plate collisions forming, 144
Pacific Ocean, 754; accretion of North America into, 47; Canadian river drainage, 145; Chinese river drainage into, 681; international trade networks across, 756; landforms in, 802; location of, 52, *m53;* Ring of Fire on, 755; trade winds over, 67; underwater mountain ranges in, 802
Pacific Plate, 176, *m177*
Pakistan, 607, 627–48; agriculture in, 627, 629, 632, 633, 634, 639, 643, 644; arts in, 638; climate regions and biomes of, 632; culture hearth of, *m80;* East and West sections of, 635, *c636–37;* economic activities of, *c97,* 639, 645; ethnic groups in, 635, 637; family and status of women in, 636, 638, 639; fishing in, 632, 639; history and government of, 90, 460, 612, 627, 634–35; independence of, 618, 634, 635, *c636–37;* India partitioned to create, 90, 612, 635, *c636–37;* industrialization of, 639; Kashmir governance issue of, *p618,* 619, 635, *c636;* landforms in, 630, *m631;* manufacturing in, 639; mountain ranges of, 630, 631; natural resources of, 642–46; physical geography of, 630–33; population patterns of, 635–36; trading partners of, 632, 634, 639; transportation in, 632, 639; urbanization of, 636; vegetation of, 633; water systems of, 629, 632, 642, 644, 645
Palau: economic activities of, major, *c97;* history and government of, 805, *c807. See also* Oceania.
Palestine: division of, 91, 397; history and government of, 396–97, *q408*
Palestinian territories, *m388,* 388–389, 394, *c397*
Pallina River, 234
Pamirs, 455, 680
pampas, 219

PAN. *See Partido Acción Nacional.*
Panama, 199, 201, 232
Panama Canal, 198, *p201, c203*
Panama Canal Zone, 198, 201
Panama City, Panama, 200
Pan-American Highway, 228
panchayat, 614
Pangaea, 45, *m46*
Papua, 761
Papua New Guinea, 800, *c807;* economic activities of, 802; natural resources of, 812, 815; population patterns of, 801, 805. *See also* Oceania.
Papuan people, 805
Paraguay, 228, 234, 235
Paraguay River, 220
Paraná River, 220
Paris, France, *m19,* 267, *p268–p69,* 270, 279
Parthenon, 299, *p302*
Partido Acción Nacional, 181, *c183*
Partido Revolucionario Institucional (PRI) , 181, *c183*
Pasadena, California, 105
Pasternak, Boris, 345
pastoralism, 415, 456, 485, 509, 523
Patagonia, 218
patois, 203
Patuca River, 209
peat deposits, 254, 274
Peña Nieto, Enrique, *c183*
peninsula, 434, 435. *See also individual peninsulas.*
Pentagon, terrorist attack against, 91
People's Republic of China. *See* China.
perceptual regions, 30, 31; of United States, *m29*
Peres, Shimon, *c396*
perestroika, 343
permafrost, 72; in Canada, 148; in Russia, 339
Perón, Juan, *p225*
Perry, Matthew C., 710, *ptg711*
Persian Empire, 411, 416, 417
Persian Gulf, 414, 417, 426, 431, 435–36, *q446*
Persian Gulf War, *p419,* 426, 444
Persians (Iranians), 409, 511, 634
Peru, 199; climate regions and biomes of, 221; ethnic groups in,

m226; landforms of, 218; natural resources in, *c222,* 234
pesticide: banning of, 623; overuse of, 353, 402, 469, 617
Peter the Great, 341
petroleum: in Arabian Peninsula, 431, 434, 437, 439, 442–43, 444–45, 446; in Australia, 780; in Bangladesh, 633; in Canada, *p142,* 148; in Central America and Caribbean, 199; in Central Asia, 400, 453, 458, *p465,* 466; in China, 678, 684; in Eastern Europe, 318; in Eastern Mediterranean, 394, 396, 400, 401; in Equatorial Africa, 553, 561; in high seas, 810; imports to Eastern Europe of, 317; in India, 610, 617, *g625;* in Indonesia, 753; in Japan, 707; in Mexico, 179, 187; of Middle East and North Africa countries, 384; in North Africa, 304, *p366–67,* 371, 376, 377, 381, 382, 400; in Northeast, 414, 415, 416, 417–18, 420–21, *i426, q430;* in Northern Europe, 254, 257; in Northwestern Europe, 274; in Pakistan, 633; pipelines for, *p142,* 415, 421, *i426,* 436, 458, 465, 466, 553; production versus consumption of, *g429;* in Russia, 340, 343, 346, 347, 350; in Saudi Arabia, 100, *m101;* shipping, *p366,* 367, 436, *q446;* in South America, *c222;* in Southeast Asia, 758, 764, 769; as trade commodity, *p99;* in Transition Zone, 491, 494–95; in United States, 124, 130; in West Africa, 533, 541. *See also* fossil fuels.
Philippines: history and government of, 759–60, *c761;* landforms in, 754; migration from, *p432,* 433; and Typhoon Haiyan, 757. *See also* Southeast Asia.
Phoenician civilization, 416
phone communications: in Australia, 785; in Canada, 155; in Central America and Caribbean, 205; in Central Asia, 463; in China, 695; in Eastern Mediterranean, 400; in Equatorial Africa, *p565;* in Japan, 715; in Mexico, 186; in Northeast, 421; in Oceania, 809; in Russia, *c355;* in United States, 131; in West Africa, 541. *See also* communications.
phosphates, 271

physical geography: of Arabian Peninsula, 434–37; of Australia, 776–80, 784; of Bangladesh, 630–33; of Canada, 144–49; of Central America and Caribbean, 176, 201; of Central Asia, 454–58; of China, 680–84; of East Africa, 506–10; of Eastern Europe, 311, 313–18; of Eastern Mediterranean, 390–94; of Equatorial Africa, 554–58; focus of, 30–31; of India, 606–10; of Indian subcontinent, 652–56, 659; of Japan, 704–07; knowledge of, 27; of Mexico, 176–79; of New Zealand, 776–80; of North Africa, 368–71; of Northeast, 412–15; of Northern Europe, 250–54; of North Korea, 726–29; of Northwestern Europe, 277; of Oceania, 797, 800–803; of Pakistan, 630–33; of Russia, 333, 336–40; of South America, 218–22, 225; of Southeast Asia, 754–58; of Southern Africa, 576–80; of Southern Europe, 294–98; of South Korea, 726–29; of Transition Zone, 481, 484–88; of United States, 120–24, 125; of West Africa, 530–34

physical map, 19

physical systems, human systems and, 30–31

physical world, 37–56

Picasso, Pablo, 303

pidgin, 785, 807. *See also* languages.

Piedmont, 121

pirating, 810

plains: alluvial (river), 533, 606; in Australia, 776; in Bangladesh, 629; in Canada, 144–45; in Central America and Caribbean, 196; in Central Asia, 455, 457; in China, 680, 682, 685, 688; in East Africa, 506, 507, 510; in Eastern Europe, 317, 319, 322; in Eastern Mediterranean, 391; in Equatorial Africa, 554, 567; in India, 606, 608, 613; in Indian subcontinent, 653, 659; in Japan, 712; in Mexico, 177, 188; in New Zealand, 777; in North Africa, 369, 370; in Northeast, 412–13, *p414*; in Northern Europe, 250, 251; in Northwestern Europe, 270, 271, 273; in Oceania, 800; in Pakistan, 630; in Russia, 336, 337, 340; in Siberia, 337; in South America,

221; in Southern Africa, 576; in Southern Europe, 294, 295, 297, 298; in United States, 117, 120, 121, 123, 125, 127, 128; in West Africa, 530, 531, 538

planets of solar system, 40–*p41;* dwarf, 40; inner, 40; outer (gas giant), 40, 41; terrestrial, 41

plantations: in East Africa, 505, 512; in Equatorial Africa, 561, 564; in Indian subcontinent, 653; in New Zealand, 778; in Oceania, 804, 807; in Southeast Asia, 766–67

plant life. *See* vegetation.

plateau(s): in Arabian Peninsula, 435; in Australia, 776, 784; in Central America and Caribbean, 196; in Central Asia, 455; in China, 680; in East Africa, 505, 507; in Eastern Europe, 322; in Eastern Mediterranean, 390; in Equatorial Africa, 555; in India, 606; in Mexico, 176, 177, 178, 180; in New Zealand, 777, 778; in North Africa, 368, 369; in Northeast, 411, 413; in North Korea, 728; in Northwestern Europe, 271; in Pakistan, 631; in Russia, 336, 337; in South America, 218, 219; in Southern Africa, 573, 577, 579; in Southern Europe, 294; tableland, 577; in Transition Zone, 483, 486; in United States, 120, *d121;* in West Africa, 531

Plateau of Tibet (Plateau of Xizang), 681, 682, 683

plate tectonics, 44, 45, *q46. See also* continental drift; continental plates; tectonic plates.

Pluto, 40, *d41*

poaching: caviar, 470; ivory, 568, *g569,* 570, *crt571;* protection against, 524, 624; threatening extinction of species, 522, 523, 588–89

Podgorica, Montenegro, 322

Pointe Noire, Republic of the Congo, 556

Poland, 314; economic activities of, *q325, c331;* history and government of, *c321,* 325, *c342;* natural resources of, 322, 327

polar easterlies, 66

polders, 272, *d272*

political boundary, 90

political conflict, *p90*–91

political geography, 12, 87–90

political map, 19

politics and geography, interactions of, 12

pollution hot spot, 307

pollution, urban, 106. *See also* air pollution; oil pollutants; water pollution.

polygamy, 529

Polynesia, 800, 805, 806, 808. *See also* Oceania.

Polynesian people, 805

Pontic Mountains, 412–13

population density, 85–86; in Bangladesh, 636, *g647, q648;* in Canada, 85, 152; in Central America and Caribbean, 202; in Central Asia, 462; in China, 682; comparison of highest, *g647;* in East Africa, 514; in Eastern Europe, 322; in Eastern Mediterranean, 398; in Egypt, 86; in Equatorial Africa, 561; in Europe, 85; in India, *p604,* 613; in Indian subcontinent, 659; in Japan, 712; in Mexico, 176; in North Africa, 376; in Northeast, 418; in Northern Europe, 257, 258; in North Korea, 739; in Northwestern Europe, *m278;* in Oceania, 801, 806; in Pakistan, *g646;* in Russia, 343; in South America, 225; in Southeast Asia, *m757;* in Southern Europe, 301, 305; in South Korea, 733, 737, 739; in Transition Zone, 492; in United States, 128–29; in West Africa, 538–39

population distribution, 84, 85; in Arabian Peninsula, *c439,* 440, 445; in Australia, 784; in Central America and Caribbean, 206; in Central Asia, 461; in China, 688; in East Africa, 508, *p514;* in Eastern Europe, 322; in Eastern Mediterranean, 392, 398; in India, 603, *p604, g605,* 613; in Indian subcontinent, 663; in Japan, 702, 712; in Mexico, 176–77, 182; in New Zealand, 784; in North Africa, 369, 370, 375; in Northeast, 419; in Northern Europe, 257, *g258, q260;* in Oceania, 806, 816; in Pakistan, 636; in South America, 225, *m226;* in Southeast Asia, 756, 762; in Southern Africa, 585; in Southern Europe, 302, *q693;*

in Transition Zone, 487; in United States, 128–29

population, emigrant, 86

population geography, 82–86

population growth: affecting environment, 383, 427, 446, 496, 522, 590, *q648;* agriculture fostering, 416; challenges of, 83–84, 202, 503, 517; concentrated in India and China, 604, 613; demographic transition model, 82–83; doubling time, 83, 545–46; environment affecting, *c493;* limited by government, 688, 690, *p692,* 692–693; negative, 84, *g258,* 278, *p293,* 301, *p334, i335,* 344, *q693, p702,* 714; policies to limit, 604; social welfare policies affecting, 259; stressing natural resources, 588, 590; total fertility rate, 636; urbanization prompted by, 539

population pressure, 202

population pyramids, *c84–85*

Po River, 297

Port-au-Prince, Haiti, 197, 206

Port Elizabeth, South Africa, 586

Port Moresby, Papua New Guinea, 806

ports: in China, 682, 686, 691; in Eastern Mediterranean, 401; in Equatorial Africa, 551, 556; free, 764; in India, 607; in Japan, 711; in New Zealand, 784; in North Africa, 377; in North Korea, 728, 730; in Northwestern Europe, *d272;* in Pakistan, 632; in Russia, 341, 342; in South America, 229; in Southeast Asia, 764, 765; in Southern Europe, 294, 295, 297, 302; in South Korea, 728, 730; in Ukraine, 347

Portugal: economic activities of, 325; history and government of, 300, 302; industrialization of, *c309;* landforms in, 294; natural resources of, 298

Portuguese explorers, 511–*ptg13,* 582, 710

postindustrial economy, 131, 155

potable water. *See* freshwater; water systems.

Prairie Provinces, Canadian, 152, 154

prairies, 71

precipitation: changes in, 72; clouds releasing, 51, 66; impact of El Niño on, 67, *d220;* in water cycle, *d52. See also* drought.

Pretoria, South Africa, 586

prevailing winds, 66

PRI. *See Partido Revolucionario Institucional. (PRI).*

primary crop, 30

primate city, 182–83

Prime Meridian, 17, 799

Pripyat' River, 338

privatization of companies, 346, 347, 616

prophets, 396

Protestant Reformation, 276, *c277*

Protocol to the African Charter on Human and Peoples' Rights of Women in Africa, 529

Puch'on, South Korea, 733

Puerto Rico, *c203*

pull factors, 86

Punjab, 632, 634

Punjab plain, 631

Pusan, South Korea, 728

push factors, 86

Pushkin, Aleksandr, 345

Putin, Vladimir, 334, 343, 350

Pygmies. *See* Mbuti people.

P'yŏngyang, North Korea, 728, 733, 734, 738

pyramids of Egypt, 372, 376

Pyrenees, 294, 296; climate change in, 306

— Q —

qanats, 417

Qatar, 446, 447, 448

Qing dynasty, 686, 687

Qin Ling, *m683*

Qin Shihuangdi, 686

Quebec: growth of, 146; separatism movement of, *d151*

Quinones, Sam, *q98*

quipus, 223, 224

Quito, Ecuador, 65

Quran, 376, 396, 442

— R —

Rabin, Yitzhak, *c396*

radioactive materials, 352, 353, 469

railroads: in Canada, *c150;* in Central Asia, 465, 467; in China, 651; in East Africa, 512; in Eastern Europe, 317, 325; in India, 611, *p613, i616,* 620; in Japan, 712–13; in Mexico, 185; in Northwestern Europe, *p269,* 285; in Pakistan, 632, 639; in Russia, 342; in South America, 229; in Southeast Asia, *p764,* 765; in United States, 127

rain forests: in Africa, 69; Amazon River Basin, 69, 215, 221, 229, *i230–31,* 232, 234, 353; in Asia, 69; in Bangladesh, 633; in Central America and Caribbean, 69, 196, 197, 199, 209, 211, 232; in China, 683; in Equatorial Africa, 551, 555, 557–58, 561, 562, 564, 566, 568, 570; in India, 608; in Indian subcontinent, 664–65; in Japan, 706; layers of, 558; in Mexico, 179; in Oceania, 803, 805, 815; in South America, 69, 215, 221, 225, 229, *i230–31,* 232, 234; in Southeast Asia, *q371,* 758, 766, 767, 770; tropical, 69, 557–58, 758; in West Africa, 530, 531, 546

rain shadow effect, *d68,* 579, 608, 682

Rāmāyana, 615

ranching: in Australia, 778, 780, 790, 791; in New Zealand, 790–91; in South America, 228, *i231,* 233–34. *See also* overgrazing.

Rapa Nui. *See* Easter Island (Rapa Nui).

Ravi, 632

Realism, 279

recession, global, 421

recycling: in Northern Europe, 264; research in, 138

Red Cross, International Committee of, 384

REDD+ program. *See* Reducing Emissions from Deforestation and Forest Degradation program.

redistricting, congressional, 13

Red River, 756–57

Red Sea, 369, 401, 432, 507; as aquatic biome, *p437;* development along, 435–36, 503, 511; filling Afar Depression, *d507;* fish farming in, *p392;* Suez Canal at, *p366,* 367

Reducing Emissions from Deforestation and Forest Degradation program, 190, 212

reforestation, 211, *q212,* 264, 326, 524, *p770,* 794, 812. *See also* deforestation.

refugees: in Europe, 278, 302; in Central Asia, 463; in China, 733; in East Africa, *p521;* in Equatorial Africa, 553, *p567,* 567; in Indian subcontinent, 659; Palestinian, 397, 399, 400; in Southern Africa, 583; in Transition Zone, 483, *q491;* in United States, *q86*

regional cooperation in Southeast Asia, *q765*

regional geographer, 28

region, defining, 29–30

regs, 368

reincarnation, 613

relief, map, 19, 20

religions: in Arabian Peninsula, 433, 438, *c440–41;* in Bangladesh, 627, 634–35, 637; in Canada, 153; in Central America and Caribbean, 200, 203; in Central Asia, 459, 463; in China, 689, 690; in culture, 78, 826; in East Africa, 511, *c512,* 514–15; in Eastern Europe, 313, 323; in Eastern Mediterranean, 387, 388, 395–97, 399, 400; in Equatorial Africa, 553, 562–63; in India, 603, *q607,* 611, 613, 615; in Indian subcontinent, 657, 658, 659, 660, 661, 664; in Japan, *c713;* in Mexico, 182; in North Africa, 372–73, 375–76; in Northeast, 409, 418, 419, *p422–p23;* in Northern Europe, 256, 258; in North Korea, 730, 733, 734; in Northwestern Europe, 275–76, 279; in Oceania, 804, *c806,* 807; in Pakistan, 627, 634–35, 637; in Russia, *p333,* 342, 344; in South America, 226; in Southeast Asia, 751, 761, 762, 763; in Southern Africa, 573, 586; in Southern Europe, 293, 302; in South Korea, 730, 733, 734; traditional, in Transition Zone, 481, *p482,* 485, 491, 493; in United States, 129; in West Africa, 529, 535, 537, 540. *See also* specific religions.

remittance, 293

remote sensing, 24

Renaissance, 300, 303

renewable and alternative energy: air pollution from, 568; Canadian initiatives for, 160, 162; Chinese initiatives for,

p678; Japanese initiatives for, 717; Mexican initiatives for, 179; Northern European initiatives for, 254; Northwestern European initiatives for, 274, 281, 287; Southeast Asian initiatives for, 770; United States initiatives for, *p137,* 138, *q140. See also* geothermal energy; wind energy.

Reporoa, 777

republic, 89

Republic of China, 686. *See also* Taiwan.

Republic of Korea. *See* South Korea.

Republic of the Congo, 554

research methods of geographers, 32–33

Research Triangle region, 131

reservoir, 532

resource management: in New Zealand, 794; in United States, 135. *See also* natural resources; reforestation; soil building and rebuilding; *individual regions.*

retirement age, raising, 702

revolution of Earth, *d61,* 62, 64

Reykjavík, Iceland, 258

Rhine River, 272, 316

Rhodes, Cecil, 583, *ptg584*

Rhodesia, 584

Rhône River, 273

Ribero, Diego, *c513*

rift valley, 391, 506. *See also* Great Rift Valley.

Rila National Park, 330

Ring of Fire, 48; in Central America, 196; in Japan, 704; in Pacific Ocean, 755; in South America, 218; in Southeast Asia, 755; volcanoes in, 755; on western side of Mexico, 176

Río Bravo del Norte, 178

Rio de Janeiro, Brazil, 226

Río de la Plata, 220

Rio Grande, 138, 178

Rivera, Diego, 184

rivers: in Arabian Peninsula, 436; in Australia, 778, 791–92; in Bangladesh, *p629,* 630, 631, 632, 642, 643; in Canada, 145, *m146, c150,* 160, 161, *q164;* in Central America and Caribbean, 197–98; in Central Asia, 456, 466–67; in China, 677, 680, 682, 688, 691, 695–96, 697; in East Africa, 506,

508, 518–19; in Eastern Europe, *m316*, 317, 325, 327–28, 330; in Equatorial Africa, 551, *m555*, 561; fall line affecting flow of, 121, *m122;* freshwater in, 54; in India, *p607*, 610, 622; in Indian subcontinent, 654; in Japan, 705, 706, 718; in Mexico (Rio Bravo del Norte), 178; in New Zealand, *777*, 780, 791; in Northeast, 412, 413, *p414*, 424, 427; in North Africa, 365, 369, 375, 380–81, 383, 384; in Northern Europe, 248, 250, 251, 252, 262; in North Korea, 726–28; in Northwestern Europe, 270, 271, 272, *q273*, 279, 288; in Oceania, 815; in Pakistan, *p629*, 630, 631, 632, 642, 644; in Russia, 337–*i38*, 351, 353; in South America, 220, 225, 228, 229, 234, *q235*, 236, 556; in Southeast Asia, 756, *q768*, 769; in Southern Africa, 577–78, 579, 580, 589; in Southern Europe, 295, *i296*, 297, 298, 305; in South Korea, 726–28, 737; in Transition Zone, 485, 486–87; tributaries of, 54, 121, 424, 425, 466, 487, 551, 555, 561, 632, 682, *q768;* in United States, 121, *m122*, 132–33; in West Africa, 532, *p533*, 545

Riyadh, Saudi Arabia, 440
Robinson projection, 15, *m16*
Rocky Mountains: in Canada (Canadian Rockies), 144, 147, 176; high altitude climate of, 124; mineral resources in, 124; in South America, 218; in United States, 120, *d121*
Roman Empire, 316, 320, 459
Romania, 314, 318; history and government of, 319, *c321*, 325; natural resources of, 326
Romanovs, 342
Roman Republic, 299, *c300*
Romanticism, 279
Rome, Italy, 297, 302
Rongelap Atoll, 815
Rosslyn-Ballston Corridor, Virginia, 105
rotation of Earth, 60–*d61,* 66
Rotterdam, the Netherlands, *d272*
Royal Society for the Conservation of Nature, 405
RSCN. *See* Royal Society for the Conservation of Nature.

Rubáiyát, 420
Rub' al-Khali (Empty Quarter), 435, 436
Rudd, Kevin, 786
Russell Glacier, *p32*
Russia, 333–56; agriculture in, 334, 336, *i338,* 339, 346; arts in, *q345;* climate regions and biomes of, 339–40, 347; communications in, *c355;* economic activities of, 96, 99, 343, 344, 345–47, 350; ethnic groups in, *c344;* family and status of women in, 334, 345; fishing in, 346, 351, 354; history and government of, 277, 320, 341–343, 344, 687; immigration to, 334, *i335;* independence of, 341, 343, 345, 347; industrialization of, 334, 337, *i338, p340,* 345, 346, 353; landforms in, 336–37; manufacturing in, 346, 350, 353; mountain ranges of, 336, *p337;* natural resources of, 333, 340, 346, 350–54, *q356;* physical geography of, 333, 336–40; population patterns of, *p334, i335,* 339, 343–44; trading partners of, 341, 346, 347; transportation in, 337, *i338,* 342, *p354;* urbanization of, 102; vegetation of, 262, 353; water systems of, 333, 337–39, *p340,* 343. *See also* Soviet Union.
Russian Core, 333–56; environmental issues in, *m351;* ethnic groups in, *c344;* water systems of, 333, 337–39
Russian Empire invasions, 460
Russian Plain, 337
Russian Revolution, *c342,* 460
Russification, 342
Russo-Japanese War, 731
Ruwenzori Mountains (Mountains of the Moon), 507, 555
Rwanda, 561; agriculture in, 504, 510, 521; genocide in, 513, 563

─────── S ───────

SAARC. *See* South Asian Association for Regional Cooperation
Sahara, 365, 368; climate regions and biomes of, 370, 530, 535; landforms in, 531; size of, 370; Transition Zone denoted by, 481, 482, 484; wall of trees to prevent spread of, 500

Saharan Atlas, 368
Sahel, 370; area of, 487; border disputes in, 491; climate regions and biomes of, 485, 509–10; colonization of, 490–91; as cultural and ecological gateway, 483; drought and hunger in, *c492–93;* natural resources of, 499; urbanization of, 492; vegetation border in, 484. *See also* Transition Zone, African.
St. Lawrence River, 145, *m146*
St. Lawrence Seaway System, 122, *m146*
St. Petersburg, Russia, 341
salinization: of lakes, 456; of seas, 437, *q446,* 469; of soil, 233, 469, 644, 790, 791
salt flats, 531
Salt Range, 631
Salt Satyagraha (March), *p612*
salt water, prevalence of, 51, 52
Salween River, 756
Samarqand, Uzbekistan, 459, 464
Sami, 257
Samoa, 800, 803, 805, 806; history and government of, *c806, c807;* International Date Line jumped by, 798–99. *See also* Oceania.
Samoan Islands, 806
samurai, 710, 711
Sanaa, Yemen, *p442,* 447
San Andreas Fault, 47, *d55*
sandstorms, 445
sanitation. *See* waste management.
San Juan River, 198
San people, 581
Santa Anna, Antonio López de, *c182*
Santa Barbara, California, 129
São Paulo, Brazil, 217, 225, 234, 236
São Tomé and Príncipe, 554, 556, 562
Sardinia, 295
Sari, Dita, 763
Saskatchewan, 152
satellite images, 32
satellites: for GPS triangulation, 22; images of, *p24;* uses of, 23
sati, ptg614
Satpura range, 606

Saturn, 40, *d41*
Saud family, 438–39
Saudi Arabia: desalination in, 53; economic activities of, 433, 442–43; landforms in, 391
Saunders, Doug, *q86*
savannas, 70; in East Africa, 509, 510, 522; in Equatorial Africa, 484, 558, 568; in North Africa, 370; in Southern Africa, 579; in Transition Zone, 484, *m485,* 489; in West Africa, 531, 534
Save the Rhino Trust Namibia, 592
scale, map, 18, *m19*
Scandinavian Peninsula, 250, 252, 254, 255, *p263*
Scotland, history and government of, 277
Scythians, 453
sea levels, rising, 59, 445, 643, 646, *q648,* 654, *p666, q700,* 792, 813, *q815,* 816
Sea of Galilee, 391, 392
Sea of Japan (East Sea), 705, 726
Sea of Marmara, 414
seaports. *See* ports.
Seas At Risk, 288
seasons, changes in Earth's, *d61,* 62
seawalls, 629
sector model of urban land, *c103,* 104
sedimentation of reef system, 211
sediment load, 681
Seine River, 270, 273, 279
seismic activity: in Mid-Atlantic Ridge, 248; in Ring of Fire, 176; by tectonic plate collisions, *m177.* *See also* earthquakes; volcanic eruptions.
Seko, Mobutu Sese, 561, *p562–63,* 564
semi-arid climates: in Australia, 778; in Canada, 147; in Central Asia, 457; in East Africa, 509; in India, 608; in Northeast, 415; in United States, 123; in West Africa, 534
seminomadic people, 535
Senegal: climate regions and biomes of, 370; natural resources, 500; in Transition Zone, 483, 530
Senegal River, 487, *p533*
Seoul, South Korea, 727, 728, 733, 734, 736, 737

Index

Serbia, 314, 330; agriculture in, 324–25; history and government of, 313, 321

Serengeti National Park, 517, 524

Serengeti Plain, 510

service industries. *See* ecotourism; tourism.

Seychelles, *c513*

SEZ. *See* Special Economic Zones.

shadow effect, 123

Shah, Prithvi Narayan, 657

Shaka: as leader of Zulu, *p583,* 583

shamal, 436

Shang dynasty, 685

Shanghai, China, 688, 691

shari´ah law, 441

Shatt al Arab River, 414

shatter belt region, 319, 320, 759, 760

sheikdoms, 438, 439

Shia Muslims, 411, 419, 422–*p23, c441*

shifting cultivation, 589–90, 768

Shikibu, Murasaki, 714

Shikoku, Japan, 704, 705

Shinano River, 705

Shinto, 843. *See also* religions.

shogun, 710, *ptg712*

Shona people, 581

Siberia, 336; climate regions and biomes of, 343; history and government of, 342; landforms of, 337; meteoroid or comet in, *q41;* water systems of, 338–39, 353

Sicily, 295

Sierra Leone: history and government of, 534; in Transition Zone, 483, 530

Sierra Madre Occidental mountains, 176, 177

Sierra Madre Oriental mountains, 176, 180

Sierra Nevada Range, 120

Sihanouk, Norodom, *c761*

Sikhism, 614, 634, 840. *See* religions.

Silicon Valley, 131

Silk Road, *c459,* 464

Silla dynasty, 730

Silla kingdom, *c732*

Silver Belt, 179

simoom, 436

Sinai Peninsula, 369, 391

Sind plain, 631

Singapore, 751, 762; economic activities in, 764–65; landforms in, 754. *See also* Southeast Asia.

Singh, Hari, 618

Singh, Ranjit, 634

Sinhalese people, *c658–59,* 660

sinkholes, 189

Sino–Indian War, 651

site, 29

situation, 29

skyscape, 41

slave forts, 561

slavery: in Central America and Caribbean, 201; in Equatorial Africa, 556, 559, *q560;* kingdoms benefiting from, 536, 582; in South America, 224, 226; in Southern Africa, 582; in Transition Zone, 483; in United States, 126. *See also* slave trade.

slave trade, 483, 511, 536, 559–*q60,* 561, 582

Slavs, 319, 321–22

Slovakia, 314; economic activities of, 324, 325; history and government of, *c321*

Slovenia, 314; history and government of, 313, *c321,* 325

smog, 72, 136, 718. *See also* air pollution.

socialism, 95, 96

social problems: of modern cities, 106

Society for the Protection of Nature, 406

sociology, 540

software development, 131

soil building and rebuilding, 50, 233, 235, 383, 499, 589, 740

soil conservation, 504

soil erosion: in Australia, 791; in China, 697–98; in East Africa, 510, 524; in Equatorial Africa, 567, 568; in India, 620–21; in Indian subcontinent, 664–65; in New Zealand, 794; in North Africa, 383; in Northeast, 427; in North Korea, 739; in Oceania, 815; in Pakistan, 644; in South America 232–33, 235; in Southern Africa, 590; in Southern Europe, *m306;* in Transition Zone, 488; in West Africa, 547

soil salinity, 427, 791, 792

soil science, 124

soil temperature, effects of, 445

solar energy. *See* renewable and alternative energy.

solar radiation, *d62,* 63

solar system, 40, *d41*

Solomon Islands, 800, 805, 807. *See also* Oceania.

solstices, summer and winter, 62

Solzhenitsyn, Aleksandr, 345

Somalia: food scarcity in, *q502;* history and government of, 491; migration from, 257; in Transition Zone, 483

Songhai Empire, 490, 535–36

Songnam, South Korea, 733

South Africa: apartheid in, *c584–c85, q594;* natural resources in, 591; townships of, *q586*

South America, 215–38; agriculture in, *p217,* 219, 221, *m228, p233,* 235; arts in, *p215, p227, q238;* climate regions and biomes of, 69, 70, 218–19, *d220,* 222; coastal precipitation in, 67; as continent, 43; economic activities of, 220, 222, 227–29, 234; ethnic groups in, 223, *m226–27;* family and status of women in, 227; fishing in, 221, 228; history and government of, 217, 223, *c225;* immigration to, 226; independence of, 224; industrialization of, 217; land bridge of, 176; land disputes in, *c224;* landforms in, 218, *p219;* manufacturing in, 228; mountain ranges of, 47, 218, *p219,* 221, 225; natural resources of, *m100,* 217, 219, *c222,* 223, 228–29, 231–33, 236, 353; physical geography of, 218–22, 225; population patterns of, 225, *m226;* trading partners of, 228, 229; transportation in, 217, 225; tropical rain forests in, 69, 215, 218, 221, 225, 229, *i230–31,* 232, 234; tropical savannas in, 70; urbanization of, 85, *p217,* 225, 228, 234, *q235,* 236; vegetation of, 69, 221–22, *i230–31,* 232, 353; water systems of, *m220,* 221, 225, 228–29, 234, *q235,* 236, 556

South American tectonic plate, *m177*

South Asian Association for Regional Cooperation, 662

South Asia, natural disasters in, 609. *See also* Indian subcontinent; *individual countries by name.*

South China Sea, 684

Southeast Anatolia Project (GAP), 424, 428

Southeast Asia, 751–72; agriculture in, *p758,* 762, 763, 764, 765, 768; arts in, 763; climate regions and biomes of, 757–58; economic activities of, 753, 762, 764–65, 769; effects of El Niño in, 67; ethnic groups in, 751, 760, 762, *p763;* family and status of women in, *p763;* fishing in, 764, *q768,* 769; history and government of, 759–61; independence of, 760, *p761;* industrialization of, *p753,* 762, 764, 765, 769; landforms in, 754–56; manufacturing in, *p753,* 769; mountain ranges of, 755–56, 757; natural resources of, 758, 764, 766–70; physical geography of, 754–58; population patterns of, 762; trading partners of, 753, 756, 765; transportation in, 753, 756, *p764,* 765; urbanization of, 762, 770; vegetation of, *q371,* 757, 758, 766–67; water systems of, 753, 756–57, *p758,* 769

Southern Africa, 573–94; agriculture in, 587, 589–90; climate regions and biomes of, *q371,* 579–80; communications in, 586; economic activities of, *c97,* 580, *g587;* ethnic groups in, 586; family and status of women in, 575, 586–87; history and government of, 581–85; immigration to, 582; independence of, 584–85; landforms in, 576, *p577;* manufacturing in, 587; mountain ranges of, 576–77; natural resources of, *m101, p580,* 585, 588–92, *crt593;* physical geography of, 576–80; population patterns of, 585–86; trading partners of, 581, 582, 585; transportation in, 585; urbanization of, 575, 586, 590; vegetation of, 579, 580; water systems of, *p577,* 578, 579, 580, *p589, p590*

Southern African Development Community, 592

Southern Alps, 777, 780

Southern Europe, 291–310; agriculture in, q297, 298, 303, 304, 305, c309; arts in, 293, 299, p302, 303, q310; climate regions and biomes of, 297–98, 305; communications in, 303; economic activities of, 293, 297, 300, 303–4; ethnic groups in, 300; family and status of women in, 303; fishing in, p291, 307; history and government of, 299, p301, 302, 307; immigration to, m292, 293, 301, 304; independence of, 300; industrialization of, 303–4, p307, c309; landforms in, 294–p95; manufacturing in, 303, 307; mountain ranges of, 294–p95, 296, 297; natural resources of, p298, 304, 305–08; physical geography of, 294–98; population patterns of, p293, 301–2, q693; trading partners of, 295, 303–4; transportation in, 307, p308; urbanization of, 301, 302; vegetation of, 306; water systems of, 291, 293, 294–97, 298, 307, p308

Southern Hemisphere, 17; seasons in, 61, 62

Southern Highlands (Mexico), 177

Southern Ocean, 52, m53

South Island (New Zealand), 777

South Korea, 723–42; agriculture in, 725, p728; arts in, 734–35; climate regions and biomes of, 728–29; Communist invasion of, q731; economic activities of, c97, 724, 725, 735, p736; ethnic groups in, 732; family and status of women in, 734; fishing in, 737; history and government of, 723, 726, 730–c33; independence of, 731, c733; industrialization of, 724, 725, p736; landforms in, 726–27; manufacturing in, p725, p736; mountain ranges of, 725, 726, 729; natural resources of, 729, 737–40; physical geography of, 726–29; population patterns of, 732–33; trading partners of, c733, 736; transportation in,

728; urbanization of, 723, 733; vegetation of, 728; water systems of, 726–28, 737. See also Korea; Korean Peninsula.

South Pole, midnight sun at, 62, 64

South Sudan, 491, 567; economic activities of, 553; in Equatorial Africa, 554; independence of, p552, 553; natural resources of, 553

Southwest Asia, natural resources in, 400

Southwestern Europe, 291–310

Soviet Era, c342–c43

Soviet Union: in Cold War, 277, 319, 343, 731; economic activities of, q96, 469; fall (breakup) of, 334, q335, 341, c343, 460, c461; history and government of, c342–43, 460, c461, 687, 731; industrialization of, 351; invasions by, 453, 460; Russian Revolution initiating, c342; trading partners of, 323. See also Russia.

Spain: agriculture in, q297; economic activities of, 282, q283, 293, c301, 325; family and status of women in, 303; history and government of, 300, 302; industrialization of, 303–4, c309; landforms in, 294, 295; natural resources of, p298

Spanish Empire, 181, 184, c300

Spanish Inquisition, 372–73

spatial diffusion, m194, 195

spatial perspective, 26, 29

Special Economic Zones, 691

Speke, John, 510

sphere, 42

Split, Croatia, 318

spreading, oceanic plate, 47, d48

Sri Lanka, 649–68; history and government of, c658–59. See also Indian subcontinent.

Stalin, Joseph, p342, c343

standard of living, 77

Stanley, Henry Morton, 555, 556, 560

Stanley Pool, 556

stateless nation, 397

stations, 790, 791

statistics, analyzing, 33

steppe climates, 70; in Canada, 147; in China, 683; in East Africa, 509; in Eastern Europe, 317; in Eastern Mediterranean, 391,

392, 393; in India, 609, 610; in Mongolia, 688; in Northeast, 415; in Russia, 339; in Transition Zone, 484; in West Africa, 534

steppes, 457, 458, 464, p467

Stockholm, Sweden, 257, 258

"Stolen Generations," 775, p786

Stoltenberg, Jens, q260

Strait of Gibraltar, 294

Strait of Hormuz, 414, 421, 443

Strait of Malacca, 753

Strait of Tirān, 401

streams, 54

stupas, 660, 661

subarctic climates, 71–72; in Canada, 147; in Northern Europe, 253; in Russia, 339; in United States, 124

subcontinent. See Indian subcontinent.

subduction, 46–47, d48

subsistence farming: in East Africa, 504, 505, 520; in Equatorial Africa, 561, 564; in Oceania, 808, p809; in Southern Africa, 587, 588, 589; in Transition Zone, 495; in West Africa, 541

suburbs, 103, 105, p268, 269, 279

Sudan: climate regions and biomes of, 370; history and government of, q491, p552–53; independence of, 553; in Transition Zone, 483. See also South Sudan.

Sudan People's Liberation Movement, 553

Suez Canal, p366–p67, c374, 436

Sukkur, Bangladesh, q633

Sulaiman Range, 630

Sumer civilization, 416, 420

sun: heat of ultraviolet rays of, 72; relationships between Earth and, 57, 60–62

Sunbelt, 128

Sun City, Arizona, 105

Sundarbans, 631, 633, 636, p643, 644, 646

Sundiata Keita, 490, 535

sunken mountains, 777

Sunni Muslims, 411, 419, 422, q423, c441

"Sunshine Policy," 725

Sun Yat-sen, 686

supertrawlers, 346, 351, 719

Suriname, 221; ethnic groups in, m226

surrealist art, 303

sustainable development projects, 188, 211, 212, 642–43, 645, 663–64, 770, 816

Sustainable Model for Arctic Regional Tourism, 264

Sutlej, 632

Suwŏn, South Korea, 733

Svalbard, 251

swamps: in Central America and Caribbean, 196, 197; in Equatorial Africa, 555; in India, 610; in Northeast, 413; in Russia, 337; in Siberia, 339; in United States, 122; in West Africa, 530, 531, 532

Sweden: climate regions and biomes of, 253; economic activities of, c97, 324; family and status of women in, 259; history and government of, 255, c256–57; landforms of, 251; natural resources of, 254, 260; as part of Scandinavia, 250; population patterns of, 257, 258

Switzerland, 270; natural resources of, 274

Sydney, Australia, 784

syncretism, 182, 226

Syngman Rhee, 731

Syr Dar'ya, 456, 468, 470

Syria, 390, 392; agriculture in, q404; climate regions and biomes of, 393; ethnic groups in, 411; history and government of, 396–98, 400; natural resources of, 394

Syrian Desert, 390, 391, m393

—————— T ——————

Taarab, 516–17

T'aebaek range, 726–27

Taedong River, 727, 737–38

Tagore, Rabindranath, 638

Tagus River, 296

Tahiti, 800. See also Oceania.

taiga. See boreal forest (taiga).

Taiwan (Formosa): climate regions and biomes of, 682; immigration to, 688; Japan's government of, 711; reunification with China of, 687

Tajikistan, 454, 455, 466–67

Taklimakan, 680, m681, 682, 684

Taliban, 453, c461, 463

Tamil people, c658–59, 660

Tamil Tigers, p659

Tang dynasty, 686, 689

Index

Tanzania: natural resources of, 510, 520; Women's Bank, 516

Tarahumara, 180

Tarai Plain, 653, 655, 659, 662

Tara River, 330

Tarbela Dam, 632, 644

tar sands in Canada, 142, 148, 154, 160

Tashkent, Uzbekistan, 462

Tasman, Abel, c784

Tasmania: ethnic groups in, c784; mountain ranges of, 776

Taupo volcanic zone, 777

Taurus Mountains, 412–13

Taylor, Charles, p537, 538

Tbilisi, Georgia, 462

Tchaikovsky, Pyotr (Peter), 345

technology, 34, 137, p137, p354, 381, p383, p590, i696. See also communications; geospatial technologies; Internet.

tectonic forces, mountains formed by, 120–21, 144, 219

tectonic plates, m47; breakup of, 434; collisions of, 176, 197, 294, 295, 413, 455, 606, 652, 755, 756, 800; convergence of, mountains from, 412, 413, 455, 606, 652, m667, 756; folds and faults in, 47–48, 251, 506; in Iceland, 248, 254; meeting in deep-sea terrains, 402; by Mexico, 176, d177; North African landscape formed by four, 368, p369; oil extraction enabled by, 222; rifts along, 506, d507; separation of, rift valleys formed by, 391; world, m47. See also continental drift; earthquakes; fault lines; Ring of Fire; volcanoes.

Teghut Forest, 469

Tehran, Iran, 419, 427

Tehri Dam, 622

Tell Atlas, 368

temperate climate: in Australia, 779; in Indian subcontinent, 655; in New Zealand, 779

temperature: elevation affecting, 64–65; tilt of Earth affecting, 60

Tenochtitlán, Mexico, 180

terracing, 662, 812

terrorist attack, p90–91; in Kashmir, q619; in United States, p126, 127, 453

terrorist groups, 453

Teshio River, 705

Texas territory, c182

Thailand (Siam), 651; landforms in, 754. See also Southeast Asia.

Thai people, 761, 762

Thames River, 272, 279

Thar Desert, 606, 609, 613, 633

Thatcher, Margaret, 279

thematic map, 19, 20, 22–23

theocracy, 88, 89

theory, 42

thermal electric plants: in South Korea, 729

Thjórsá River, 252

Three Gorges Dam, 682, p687, 694, 696

Tiananmen Square, p687

Tianjin, China, 688, 691

Tian Shan, 455, 680, 681

Tiber River, 297

Tibesti Mountains, 369

Tibet: history and government of, 651, 688; rule in Bhutan, 657

Tibetan (Lamaist) Buddhism, 657, 689

Tibetan invasions, 657

Tibeto-Nepalese people, 660

tierra caliente, d178, 199, 221

tierra fria, d178, 199, 221

tierra helada, d178, 199, 221

tierra templada, d178, 199, 221

Tigris–Euphrates River Basin, m425

Tigris River, 412, 414, 424, 427

timberline, 147, 274, 655, 683

timber resources: in Canada, 147, 148; in Central America and Caribbean, 196; in China, 687–98; in Eastern Europe, 317; in India, 610, 622, 624; in Indian subcontinent, 656, p665; in Indonesia, 753; in Mexico, 179, 188; in New Zealand, 787; in Northern Europe, 264; in Northwestern Europe, 274; in Oceania, 812; in South America, 228, 229; in Southeast Asia, 766, 770; in Southern Africa, 590; in United States, 134. See also deforestation; forests; logging; natural resources.

Timbuktu, Mali, 490, 536

Tindall Foundation, 794

Tingi Mountains, 531

Tito, Josip Broz, 313

Togo: in Transition Zone, 483

Tokaido corridor, 712

Tokelau, 798

Tok Islands, 727

Tokugawa period, c713

Tokyo (Edo) Bay, ptg711

Tokyo, Japan, 712–13

Tokyo-Yokohama area, 712

Tolstoy, Lev (Leo), 345

Toluca Basin, 178

Tone River, 705

Tonga, 800; history and government of c806, c807. See also Oceania.

Tong, Anote, q815

Tonle Sap Lake, q768, 769

topography, effects on soil of, 50

Torah, 395

tornado activity: relationship of air pollution and, q72; in United States, 123

Torne River, 253

Toronto, Ontario, 152

Torres Strait Islanders, 796

tourism: in Arabian Peninsula, 436, 443; in Australia, 776, 778, 792; in Canada, 155; in East Africa, 508, 517, 524; in Eastern Europe, 316, 325, 330; in Eastern Mediterranean, 390, 394, 401, 404; environmental concerns for, 307, p308; in India, 624; in Indian subcontinent, 656, 659, 662; in New Zealand, 776, 778, 787; in North Africa, 371, 377; in Northern Europe, 248, 264; in Northwestern Europe, 271, 280, 285; in Oceania, 803, 808, 812; in Russia, 336; in Southeast Asia, 757; in Southern Africa, 579, 580, q586; in Southern Europe, 297, 304, 307–8; in West Africa, 532. See also ecotourism.

Toussaint-Louverture, François, 201, p202

trade: in Arabian Peninsula, 438; in Australia, 781, 787, 799, 809; in Canada, 155; in Central America and Caribbean, 195, 198, 201, 209; in Central Asia, 452, 459; in China, 91, 99, c163, 281, 517, 686, 687, g691, 730; in East Africa, 511, 514, 517; in Eastern Europe, 317, 323, 325; in Eastern Mediterranean, 401; in Equatorial Africa, 568, g569, g572; in India, 611; in Indian subcontinent, 656; in Japan, 710, p711, g715, 716, 809; in Mexico, 151, 185, 186; in New Zealand, 787, 799; in North Africa, 365, p366–67, 377;

in Northeast, 421; in Northern Europe, 256, 260; in North Korea, 736; in Northwestern Europe, 276, 281–83, 377; in Oceania, 799, 802, 806, 809; in Pakistan, 632, 634, 639; in Russia, 341, 346, 347; in South America, 228, 229; in Southeast Asia, 753, 756, 765; in Southern Africa, 581, 582, 585; in Southern Europe, 295, 303–4; in South Korea, c733, 736; in Transition Zone, 483, 489, 490, 494, c501; in United States, 122, 125, 129, 135, g691, g715, 716, 809; in West Africa, 533, c536, 541. See also North American Free Trade Agreement.

trade surplus, 716

trade winds, 66, 179

traditional economy, 94, 95

Trans-Amazonian Highway, 228–29

Trans-Canada Highway, 155

Transition Zone, African, 481–502; agriculture in, 486, 488, 489, 494, 495, 496, 497, 499, 500; area of, 484, m485; arts in, p494; climate regions and biomes of, 488, 530; economic activities of, 488, 494–95; ethnic groups in, 482, 483, 484, 491, 493; family and status of women in, 494; fishing in, 486; history and government of, 489–91, 494; independence of, 491; industrialization of, 495; landforms in, 484–85, 530; natural resources of, 488, 496–p500; physical geography of, 481, 484–88; population patterns of, 492; trading partners of, 483, 489, 490, 494, c501; urbanization of, 492, 493; vegetation of, 484, 485, 488; water systems of, 485–87, 496, 497, 498, 499. See also West Africa.

Transoceanic Highway, 229

transportation: in Arabian Peninsula, 443, q446; in Australia, 777; in Bangladesh, 632; in Canada, m146, 155; in Central America and Caribbean, 198; in Central Asia, 465, 467, i468; in China, 651, 682, 685, 691, 696; in East Africa, 508; in Eastern Europe, m316, 322–23, 325; in Eastern Mediterranean, 401, 403, 406; in Equatorial Africa, 565; in India,

Index

607, 616; in Indian subcontinent, 656; in Japan, 705, 712–13; in Mexico, 185, 190; in New Zealand, *777*; in North Africa, 365, 372, 377, 392; in Northeast, 414, 415, 421, 427; in Northern Europe, *p248*, *d249*; in North Korea, 728; in Northwestern Europe, *p268*, 269, 271, *d272*, 273, 276, 281, 285; in Oceania, 809; in Pakistan, 632, 639; in Russia, 337, *i338*, 342, *p354*; in South America, 217, 225; in Southeast Asia, 753, 756, *p764*, 765; in Southern Africa, 585, 589; in Southern Europe, 307, *p308*; in South Korea, 728; in United States, 126, 131, 137; in West Africa, 529, 533, 539

Trans-Siberian Railway, 725
treaties, 91
Treaty of Guadalupe Hidalgo, *c182*
Treaty of Waitangi, *p775*, 782
trend, population, 84
triangulation, 22
tribal groups, 452
tributaries, 54, 121, 424, 425, 466, 487, 551, 555, 561, 632, 682, *q768*
Tripoli, Libya, 375, 384
tropical climates: in Bangladesh, 633; in Central America and Caribbean, 199; in China, 683; in East Africa, 510; in Equatorial Africa, 557; in Indian subcontinent, 656; in Oceania, 803; in South America, 221; in Southeast Asia, 757; in United States, 122; in West Africa, 534; wet and dry, 69–70, 199, 221, 557, 579
tropical evergreen forests, 610
tropical rain forests. *See* jungles; rain forests.
tropical savanna climate, 70; of Australia, *779*; of Equatorial Africa, 484, 558; of North Africa, 370; of West Africa, 530
tropical seasonal forest, 534
Tropic of Cancer, 62, 64, 66
Tropic of Capricorn, 64, 66
Tropics, 481
trust territories in Oceania, 805
Truth and Reconciliation Commission, 585, *c585*
Tsiafajavona, 577

tsunamis: earthquakes triggering, 609, 632, 654, 704, 717, 718, 755–56, 758; fishing industry destroying, 662
tundra as biome type, 436
tundra climates, 72; in Canada, 147, 148; in Russia, 339
Tunisia: climate regions and biomes in, 371; history and government of, 374, *c375*; independence of, 374; mountain ranges of, 368; population density in, 85
Tunis, Tunisia, 375
Turan Plain, 455
Turkey: economic activities of, 99, 421; ethnic groups in, 411; history and government of, 417; landforms in, 412, *m413, p414*; natural resources of, 415, 427; as newly industrialized country, 97; water systems of, *p414*
Turkish Straits, 414
Turkmenistan, 454, 455; natural resources of, 458
Turks, *p409*, 418, 453
Turpan Depression, 680
Tutsi people, 513, 521, *c562,* 563
Tuvalu, 800, 806, *c807. See also* Oceania.
Typhoon Haiyan, 757
typhoons, 123, 683–84, 706, 728, 757
Tyrrhenian Sea, 295, 297

U

Uganda: history and government of, 513; natural resources of, 510, 524
Ukraine: agriculture in, 347; ethnic groups in, *c344*; independence of, 341, 343, 347; landforms in, 337; natural resources of, 340; trading partners of, 347; water systems of, 337, 347
ul-Haq, Magbub, 77
Ulleungdo, 727
Ullman, E. L., 104
UN. *See* United Nations (UN).
Underground Railroad, 151
understory, 558
UNESCO. *See* United Nations Educational, Scientific and Cultural Organization (UNESCO).
uneven development among countries, 227

Union of Soviet Socialist Republics. *See* Soviet Union.
unitary system of government, 87, 88
United Arab Emirates: desalination of salt water in, 53; economic activities of, 443
United Kingdom, 270; Biodiversity Action Plan of, 288; economic activities of, 324; immigration to, 375. *See also individual countries.*
United Nations (UN), 91, 384; actions in labor disputes by, 433; actions in Palestine by, 388, *c397;* dependence of refugees on, 463; ratification of Law of the Sea Treaty by, 810
United Nations Convention to Combat Desertification, 498
United Nations Development Programme, 77, 326, 406, 739
United Nations Educational, Scientific and Cultural Organization (UNESCO), 92, *q93;* World Heritage Sites of, 330, *p442,* 524, 581, 780
United Nations Environment Program, 592
United Nations Framework Convention on Climate Change, 720
United Nations Girls' Education Initiative, *q515*
United Provinces of Central America, 201, *c202*
United States, 117–40; agriculture in, 126, 128, 130, 131; ancestry by county in, largest, *m118;* area of, 120, 487; arts in, 129–30; city populations in, 102; Civil War in, 126; climate regions and biomes of, 71, 122–24, *q371;* in Cold War, 277, 731; colonial regions of, 125–26; communications in, 131; desalination of salt water in, 53; economic activities of, *c97,* 130–31, *p132–33,* 567, 712; economic downturn in, 131; effects of El Niño in, 67; ethnic groups in, 127, *c128;* family and status of women in, 129; federal (national) system of government in, 87–88;

fishing in, 124, 134–35, 178; growth of, 125–27; history and government of, 89, 119, 125–27; human geography of, 125–33; immigration to, *m118, p119,* 127, *c128,* 129, *c237,* 258, 816; imports from China to, 99; independence of, 126; industrialization of, 126, 129, 130, 436; landforms in, 120; manufacturing in, 126, 127, 131; as more developed country, 97; mountain ranges in, 120, 802, 803; natural resources in, *m100,* 124, 126, 129, 130, 134–38; perceptual regions of, *m29;* physical geography of, 125; pollution in, 135–37; population patterns in, 128–29, 132; refugees in, *q86;* Regional Environmental Center funding from, 330; terrorism in, *p126;* trading partners of, 91, 151, 281, *g691, g715,* 716, 809; transportation in, 126, 131, 137; treaty of Mexico and, 90; urbanization of, 127, 128, vegetation of, 122–23; War of Independence (Revolutionary War) of, *c150,* 152, 224; water systems of, 121–22, 134, *m135*
United States Agency for International Development, 470
unsustainable fishing. *See* overfishing.
Ural Mountains, *i337;* separating European and Siberian Russia, 336
Uranus, 40, *d41*
urban geographer, 103
urban geography, 102–6. *See also* cities.
urban growth. *See* cities; megalopolis; urbanization.
urban heat island, 769, 770
urbanization, 106; in Arabian Peninsula, 448; in Australia, 85, 783; in Bangladesh, 636; in Canada, 146, 151, 152; in Central America and Caribbean, 208, 210, *p211,* 212; in Central Asia, 451, 462, 463; in China, 688, 690, *p697;* in East Africa, *p514,* 516, 523; in Eastern Europe, 322–23; in Eastern Mediterranean, 403, 406; in Equatorial Africa, 551, 561; in India, 613, 614–15; in Japan, 712–13; in Mexico, *p174–75;* in

New Zealand, 778, 784; in North Africa, 376, 380; in Northeast, 418, 419; in Northern Europe, 258; in North Korea, 733; in Northwestern Europe, p268–p69, 276, 277, 279; in Oceania, 808; in Pakistan, 636; in Russia, 102; in South America, 85, p217, 225, 228, 234, q235, 236; of Southeast Asia, 762, 770; in Southern Africa, 575, 586, 590; in Southern Europe, 301, 302; in South Korea, 723, 733; in Transition Zone, 492, 493; in United States, 125, 806; in West Africa, 538, p539, 540

urbanization patterns,104–5

urban land use models, c103, 104

urban planning, 28, 31

urban sprawl, 102, 103, 106, 182

urban structure, 104

Uruguay, 219; agriculture in, 228, 235; desertification in, 234; pollution monitoring in, 236

Uruguay River, 220

USAID. See United States Agency for International Development.

U.S.-Colombia Free Trade Agreement, 229

U.S. Constitution, 88

USSR. See Soviet Union.

Uyghur people, 411

Uzbekistan, 454; economic activities of, 464

—— **V** ——

Valdai Hills, 337

Valley of the Kings, 377

Vancouver, British Colum-bia, as trade center, 152

Vanuatu, 805. See also Oceania.

Vardar River, 297

Vasnetsov, Viktor, 345

Vatican City, 302

Vatnajökull, 251

Vedas, 611

vegetation: in Arabian Peninsula, 436, 437, 447; in Australia, 773, 790, 791, 792, 793; in Bangladesh, 633; in Canada, 146, 147; in Central America and Caribbean, 194, 195, i198, 199, 209, 210; in Central Asia, 456, 457; in China, p682, 698; by climate type, 69–72; in East Africa, 509, 510, q525; in

Eastern Europe, 318, 326–27; in Eastern Mediterranean, 393; in Equatorial Africa, 558, 561; in India, 608, 609, 610, 620; in Indian subcontinent, 655, 656; in Japan, 706; in Mexico, 178; in New Zealand, 780; in North Africa, 369, 370, q371, 372, 383; in Northeast, 415; in Northern Europe, 253, 254, 261, 262, 264; in North Korea, 728; in Northwestern Europe, 273–74; in Oceania, 803, p814, 815; in Pakistan, 633; relationships of latitude and climate on, 64, d71, 147; in Russia, 262, 353; in South America, 69, 221–22, i230–31, 232, 353; in Southeast Asia, q371, 757, 758, 766–67; in Southern Africa, 579, 580; in Southern Europe, 306; in South Korea, 728; in Transition Zone, 484, 485, 488; in United States, 122–23; in West Africa, 547. See also agriculture; cash crops; forests; logging.

Venezuela: climate regions and biomes of, 221; natural resources in, c222

Venice, Italy, 297

Venus, 40, d41

vertical climate zones, c178, 179, 221

Victoria Falls, 578

Vietnam: economic activities of, 96, 752, p753; history and government of, c760, c761; landforms in, 754. See also Southeast Asia.

Vietnamese people, 762

Vietnam War, q772

Viking Age, 255

Vindhya range, 606

Virunga National Park, 570

Vistula River, 317, 328, 330

Vladivostok, Russia, 342

Volcanic Axis, 196

volcanic eruptions, 44, q48, 49; in Central America and Caribbean, 196, 212; hot spots for, 121, m177, 251, 800; in Iceland, p248, d249, 251; in Mexico, m177; in New Zealand, 780; in North Africa, 369; in Southeast Asia, 755, 756, 758; volcanic ash, 248, d249. See also Ring of Fire.

volcanic necks, 369

volcanic summits, 802

volcanoes: in Arabian Peninsula, 435; in Australia, 776; in East Africa, 503, 507, p508, 509; in Equatorial Africa, 554, 555; formation of, 176; in Japan, 701, 704, 755; in New Zealand, 777; in Northeast, 412; in Oceania, 801; in Ring of Fire, 755; soil enriched by, 757; in Southeast Asia, 755, 757, 758; in Southern Africa, 577; in South Korea, 727; submerged, 801; in West Africa, 531

Volga River: development along, i338, 351, 353; as draining western Russia and Iran into Caspian Sea, 337, i338, 414

Volta River Basin, 486, 532, 533

—— **W** ——

wadis, 369, 392, 436

Wales, history and government of, 277

Wallis and Futuna, 806. See also Oceania.

Wangchuck, Jigme Singye, 657, 664

Wangchuck, Ugyen, 657

War of the Pacific, c224

War on Terror, 126

Warsaw, Poland, 317

Washington, D.C., 129

waste, reducing, 137

waste management: composting for, 646; disregard for, 718, 769; sanitation adequacy for, 493, 544, p547, 548, 567, 586, 622, 737; for tourist industry, 307; toxic, 769, 792; unavailable, 590, p591, g621, 644, 697; undeveloped, 636

water: conflicts arising from lack of, p132, 133, 466–67; effects on climate of, 68; purification of contaminated, p590. See also drought; water scarcity.

water cycle: constant volume of water in, 51, 52; movement of water in, 51, d52

water erosion, 50

waterfalls: along fall line, 126; in Australia, p777; in Equatorial Africa, 556; hydroelectricity from, 251; in Indian subcontinent, 654; in Northern Europe, 252; in Southern Africa, 578

water pollution, 136–37; acid rain contributing to, 161; in

Arabian Peninsula, 437, 445; in Australia, 792; in Canada, 158, 160, 161; in Central America and Caribbean, p210; in Central Asia, i468, 469, 470; in East Africa, 523; in Eastern Europe, 327, g328; in Eastern Mediterranean, 402; in India, g621, p622; in Indian subcontinent, 666; in Japan, 707; in Mexico, 183; in North Africa, 383; in Northeast, 425, i426; in North Korea, 737; in Northwestern Europe, 284; in Oceania, 814, 815; in Pakistan, 644, 645; in Russia, 353; risk to plant life from, p160; in Southeast Asia, 769; in Southern Africa, 591; in South Korea, 737; in West Africa, 533, 544, p547

water recycling, i403, 405

water scarcity, p38–39; impact on economic activity of, 39; leading to desertification, 496; technology addressing, 39; in United States, m135, 137. See also deserts; desertification.

water stratification, 261

water systems: in Arabian Peninsula, 431, 434, 435–36, p437, 443, 446–48; in Australia, p777–78, 780, 791–92; in Bangladesh, 627, 629, 632, q633, 642, p643; in Canada, 145, m146; in Central America and Caribbean, 197–98, 207, 210, p211; in Central Asia, 454, p456, 466–67, i468, 469, 470; in China, 677, 681–84, 685, 688, 691, q693, 695–96, 697, 698; in East Africa, 506, p508–9, 510, p518, 519, 523; in Eastern Europe, 311, m316, 317, 322, 325, 327, g328, 329, 330; in Eastern Mediterranean, p391, 392, m393, 394, 401, m403, 405, 406; in Equatorial Africa, m555, p556, 560, 561, 567; in India, 606, q607, 609, 614, 616, g621, p622, 643; in Indian subcontinent, 653, 654, 666; in Japan, 705, 706, p716, 718, 719, p720; in Mexico, 176, 178, 188, 189; in New Zealand, 777, 778, 780, 791; in North Africa, p366, 367, p369, p370, 372, 375, 377, 380–81, c382, 384; in Northeast, 412, 413, p414, 415, 416, 421, 424–25; in Northern Europe, 250, p251, m253, 254, 261, 263, 279;

in North Korea, 726–28, 737; in Northwestern Europe, 270, 271–73, 276, q288; in Oceania, 801, 802, 812–14, 815; in Pakistan, 629, 632, 642, 644, 645; in Russia, 333, 337–39, p340, 343; in South America, m220, 221, 225, 228, 229, 234, q235, 236; in Southeast Asia, 753, 756–57, p758, 769; in Southern Africa, p577, 578, 579, 580, p589, p590; in Southern Europe, 291, 293, 294–97, 298, 307, p308; in South Korea, 726–28, 737; in Transition Zone, 485–p487, 496, 497, 498, 499; in United States, 121–22, 132, 134, m135; in West Africa, p532, 533, 546, p547. See also individual types and bodies.

weather, 61; climate versus, 60; increase in severe and extreme, 59, 72. See also individual weather event types.

weathering, physical and chemical, 49–50

Wei River valley, 685

welfare states: in Northern Europe, 259, 260; systems in Southern Europe, 301, 302, 304

Wellington, New Zealand, 784

West Africa, 527–50; agriculture in, 527, 529–31, 533, 541, 546, 547; arts in, 540, 541; climate regions and biomes of, 534; economic activities of, 541; elevation profile of, c531; ethnic groups in, 537, 539–40; family and status of women in, 203, p528–29, 540; fishing in, 532, p533, 541, q546, 547; history and government of, 527, 535–38; immigration to, c536–37; independence of, 537; industrialization of, 533; landforms in, 530–31; mountain ranges of, 531; natural resources of, 527, 534, 544–48; physical geography of, 530–34; population patterns of, 538–39; trading partners of, 533, c536, 541; transportation in, 529, 533, 539; urbanization of, 538, p539, 540; vegetation of, 547; water systems of, p532–33, 546, p547. See also Transition Zone, African.

West African kingdoms, 483

West Bank, 388, m389, 392, c396, 397, 399

westerlies, 66

Western Dvina River, 338

Western Ghats range, 606, 607, 608, 620

Western Hemisphere, 17

Western Plateau, 776

Western Rift Valley, 506, 508

Western Sahara, 368, 380

Western Samoa, 805, c807. See also Oceania; Samoa.

West Siberian Plain, 337

wetlands: in Australia, 779; in Bangladesh, 631, 632, 643; in Canada, 158–59; in China, 698; in East Africa, 522; in Eastern Mediterranean, 403; in India, 610, 643; in Northeast, 424–25, 428; in Northern Europe, 262, 263; in Transition Zone, 487; in United States, 122–23, 134–35, 137

Wetlands International, 428

whaling, 719

White Nile River, 487, 508

WHO. See World Health Organization (WHO).

wildlife. See animals.

Wildlife Conservation Society, q356

wildlife reserves. See game reserves.

wind: cyclone, 609; influence on climate of, 64, 306, 436; influence on weather of, 65–66, 273; tornado, 445; trade, 66, 179; uneven temperatures creating, 65. See also individual types.

wind energy, p137; in Canada, 162; in China, 678; in East Africa, 509; in Japan, 717; in Northern Europe, 260, p264, q266; in Northwestern Europe, 274, 281, 287

wind erosion, 49, 681

windmills, d272

wind patterns, m65–66; climate change involving, 72, 199; humidity levels affected by, 71; latitudes affecting, 65; monsoon, 436, 607–9, 683, 728; mountains affecting, 147, 273; seasons determined by, 633

windward side of mountain, d68

Winkel Tripel projection, 15, m16

Wolof, 493

women, empowering: in business; West Africa, 540; education, 515; rights, in India, c614–15

women's suffrage, 279, 442, 782

women, status of. See family and status of women.

World Bank, 384, 467, 470, 645

World Bank Sustainable Forestry Pilot project, 354

world city, 105

World Health Organization (WHO), 575, 645

World Heritage Monument, 561

World Population Day, 613

world trade, 98–99; of United States, 129. See also trade.

World Trade Center, attacks against, 91

World Trade Organization (WTO), 91, c687, 691

World War I: in East Africa, 520; in Eastern Mediterranean, 396; in Equatorial Africa, 553; in Japan, 711; in North Africa, 373; in Northeast, 417; in Northwestern Europe, 277; in Russia, c320, 342

World War II: in Eastern Europe, c321, 323; in Eastern Mediterranean, 388; in Japan, 711; in Northwestern Europe, 277, q283; in Oceania, 805, c807; in Russia, c342–43; in Southeast Asia, 760; in Southern Europe, c301

World Wildlife Fund (WWF), 308, 329–30, q394, 405, 570, 794, 816

WTO. See World Trade Organization (WTO).

Wuhan, China, 691

WWF. See World Wildlife Fund (WWF).

X

Xi River, 682, 688

Xyzyl language, 92

Y

Yalu River, 727

Yamato dynasty, 710

Yamuna River, p622

Yangon (Rangoon), Myanmar, 762

Yellow (West) Sea, 726, 727

Yemen: economic activities of, 443; history and government of, 439, 447, 448; water systems of, 448

Yenisey River, 338

Yerevan, Armenia, 462

Yesŏng River, 727

Yi-fu, Tuan, q31

Yoruba people, 411

Yoshihito, p713

Yoshino River, 705

Yucatán Peninsula, 180

Yugoslavia, q283, m312, p313, 320, 321, 330. See also Federal Republic of Yugoslavia.

Yukon River, 150

Z

Zagros Mountains, 413, 417

Zaire. See Democratic Republic of the Congo

Zambezi River, 578, 590

Zambia Wildlife Authority, 592

Zardari, Asif Ali, 635

Zard Kūh, 413

Zero Deforestation Law, 235

Zheng He, 686

Zhou dynasty, 685

ziggurats, 420

Zimbabwe: economic activities of, c97, 587; history and government of, 584, 591

Zionism, 388

Zulu empire, 583

Index